2008 13th International Power Electronics and Motion Control Conference

Poznan, Poland
1-3 September 2008

Pages 2071-2566

IEEE Catalog Number: CFP0834A-PRT
ISBN 13:   978-1-4244-1741-4

**Copyright © 2008 by The Institute of Electrical and Electronics Engineers, Inc.
All Rights Reserved**

Copyright and Reprint Permissions: Abstracting is permitted with credit to the source. Libraries are permitted to photocopy beyond the limit of U.S. copyright law for private use of patrons those articles in this volume that carry a code at the bottom of the first page, provided the per-copy fee indicated in the code is paid through Copyright Clearance Center, 222 Rosewood Drive, Danvers, MA 01923.

For other copying, reprint or republications permission, write to IEEE Copyrights Manager, IEEE Operations Center, 445 Hoes Lane, Piscataway, New Jersey USA 08854. All rights reserved.

IEEE Catalog Number: CFP0834A-PRT

ISBN 13:   978-1-4244-1741-4

ISSN: 2007906910

Additional Copies of This Publication Are Available from:

IEEE Service Center
445 Hoes Lane
Piscataway, NJ 08854
Phone: (800) 678-IEEE
                       (732) 981-1393
Fax: (732) 981-9667
E-mail: customer-service@ieee.org

2008 13th International Power Electronics and Motion Control Conference

Poznan, Poland
1-3 September 2008

IEEE Catalog Number: CFP0834A-POD
ISBN: 978-1-42441-741-4

Table of Contents

Electric Drive System for Automatic Guided Vehicles Using Contact-free Energy Transmission 1
Marcel Jufer

State-of-the-Art High Power Density and High Efficiency DC-DC Chopper Circuits for HEV and FCEV Applications 7
Atsuo Kawamura, Martin Pavlovsky, Yukinori Tsuruta

Current-Based Condition Monitoring of Electrical Machines in Safety Critical Applications 21
Thomas G. Habetler

The Essence of Three-Phase AC/AC Converter Systems 27
J. W. Kolar, T. Friedli, F. Krismer, S. D. Round

An Analysis on Turn-off Behaviour of 1.2kV NPT-CIGBT under Clamped Inductive Load Switching 43
S.T. Kong, L.Ngwendson, M. Sweet, E.M. Sankara Narayanan

Turn-off behaviour of high voltage NPT- and FS-IGBT 48
Hans-Guenter Eckel, Karl Fleisch

Exact Circuit Power Loss Design Method for High Power Density Converters Utilizing Si-IGBT/SiC-Diode Hybrid Pairs 54
Kazuto Takao, Hiromichi Ohashi

A Forward Converter with a Monolithic Cascode Device: Design and Experimental Investigation 61
F. Chimento, S. Musumeci, A. Raciti, L. Abbatelli, S. Buonomo, R. Scollo

Switching and conducting performance of SiC-JFET and ESBT against MOSFET and IGBT 69
André Knop, W. Toke Franke, Friedrich W. Fuchs

In-Service Life Consumption Estimation in Power Modules 76
Mahera Musallam, C Mark Johnson, Chunyan Yin, Hua Lu, Chris Bailey

Measurement Of Temperature Sensitive Parameter Characteristics Of Semiconductor Silicon And Silicon-Carbide Power Devices 84
Mietek Nowak, Jacek Rabkowski, Roman Barlik

Unsymmetrical Gate Voltage Drive for High Power 1200V IGBT4 Modules Based on Coreless Transformer Technology Driver 88
Piotr Luniewski, Uwe Jansen

A Novel RESURFed Double Gates IGBT with Superior Performance 97
Dongming Wu, Kaihang Li, Lingling Yang

An Empiric Approach to Establishing MOSFET Failure Rate Induced by Single-Event Burnout 102
Jeroen van Duivenbode, Bart Smet

Comparative Study on Paralleled vs. Scaled Dc-dc Converters in High Voltage Gain Applications 108
Pawel Klimczak, Stig Munk-Nielsen

A Low-Loss Dc-Dc Converter For A Renewable Energy Converter 114
David S. Thompson, Otu A. Eno

A Single Active Edge-Resonant Snubber Cell-assisted ZCS Half-Bridge DC-DC Converter with Constant Frequency Asymmetrical PWM Scheme 119
Tomokazu Mishima, Mutsuo Nakaoka, Eiji Hiraki

A New Approach to High Efficiency in Isolated Boost Converters for High-Power Low-Voltage Fuel Cell Applications 127
Morten Nymand, Michael A.E. Andersen

New Modulation Strategy with Low Switching Frequency and Minimum Baseband Distortion 132
N. E. Ruger, O. Schnick, W. Mathis, A. Mertens

Table of Contents

A Bit-Stream Based PWM Technique for Variable Frequency Sinewave Generation 139
N. D. Patel, U. K. Madawala

Control Strategies of the Quasi-Resonant DC-Link Inverter 144
Slawomir Mandrek, Piotr J. Chrzan

Consideration for Input Current-Ripple of Pulselink DC-AC Converter for Fuel Cells 148
Kentaro Fukushima, Tamotsu Ninomiya, Masahito Shoyama, Isami Norigoe, Yosuke Harada, Kenta Tsukakoshi

New Practical Approach to Input Current Shaping in AC-DC Power Converters 154
Kuno Janson, Viktor Bolgov, Lauri Kütt, Ants Kallaste, Heigo Mölder

LLCC-PWM Inverter for Driving High-Power Piezoelectric Actuators 159
Rongyuan Li, Norbert Fröhleke, Joachim Böcker

Modelling and Analysis of a Matrix-Reactance Frequency Converter Based on Buck-Boost Topology by DQ0 Transformation 165
Pawel Szczeniak, Zbigniew Fedyczak, Marius Klytta

A Modular AC/DC Rectifier Based on Cascaded H-bridge Rectifier 173
H. Iman-Eini, Sh. Farhangi, JL. Schanen

Low Loss Soft Switching Boost Converter 181
So-Ri Park, Sang-Hoon Park, Chung-Yuen Won, Yong-Chae Jung

Methods for Experimental Assessment of Component Losses to Validate the Converter Loss Model 187
Yi Wang, Sjoerd de Haan, Jan Abraham Ferreira

Modified multistage semiconductor-Fitch generator topology with magnetic compression 195
Stanislaw Kalisiak, Marcin Holub

Modeling and Measuring Results of a Shunt Current Source Active Power Filter with Series Capacitor 201
P. Parkatti, M. Salo, H. Tuusa

A Multi-Drive System Based on a Two-stage Matrix Converter 207
Dinesh Kumar, Patrick W Wheeler, Jon C Clare, Lee Empringham

Characteristics of the Single Active Bridge Converter with Voltage Doubler 213
Andreas Averberg, Axel Mertens

Analysis of Capacitor Dividers for Multilevel Inverter 221
Oleg Sivkov, Jiri Pavelka

Space Vector Modulation for a Capacitor Clamped Multi-level Matrix Converter 229
Xu Lie, Jon C. Clare, Patrick W. Wheeler, Lee Empringham

New Family of Matrix-Reactance Frequency Converters Based on Unipolar PWM AC Matrix-Reactance Choppers 236
Zbigniew Fedyczak, Pawel Szczesniak, Igor Korotyeyev

Consideration of Conduction Losses for the Series Resonant Converter by Means of a Simple Extension to the SPA Approach 244
Alexander Bucher, Thomas Duerbaum, Daniel Kuebrich, Markus Schmid

Validation and Comparison of different PWM Converter Small Signal Models 250
Alexander Bucher, Markus Schmid, Lukas Bendkowski, Thomas Duerbaum

Dynamic Behaviour of a Series - Connected Multilevel Converter with Interleaved Switching 256
C. Fahrni, A. Rufer

Simple Analysis of a Flying Capacitor Converter Voltage Balance Dynamics for DC Modulation 260
A. Ruderman, B. Reznikov, M. Margaliot

Simulation of Simplified Seven Level Multilevel Converter Circuit 268
Gerardo Ceglia, Víctor Guzmán, Carlos Sánchez, Fernando Ibáñez, Julio Walter, María Giménez

Table of Contents

SEPP High-Frequency Inverter Incorporating an Auxiliary Switch and Its Performance Evaluation 275
H.Ogiwara, Y.Fujita, R.Urabe, M.Itoi, T.Sugai, M. kuwata, M.Nakaoka

Multiphase coupled converter models dedicated to transient response and output voltage regulation studies 281
Nadia Bouhalli, Marc Cousineau, Emmanuel Sarraute, Thierry Meynard

A 13.56 MHz Current-output-type Inverter Utilizing An Immittance Conversion Element 288
Yosei Sakamoto, Keiji Wada, Toshihisa Shimizu

Voltage Fed Zero-Voltage Zero-Current Switching PWM DC-DC Converter 295
Jaroslav Dudrik, Vladimír Ru1scin

PWM Spectrum Evaluation and Over-Modulation Phenomena in a Three-Phase Inverters - Analytical Approach 301
Miro Milanovic

Experimental Study of a Matrix Converter Excited Doubly-Fed Induction Machine in Generation and Motoring 307
Ivan Shapoval, Jon Clare, Eduard Chekhet

Effect of Type and Interconnection of DG Units in the Fault Current Level of Distribution Networks 313
H.R. Baghaee, M. Mirsalim, M. J. Sanjari, G.B. Gharehpetian

An Isolated Full-Bridge DC/DC Converter..with Bidirectional Communication Capability 320
Lon-Kou Chang, Ru-Shiuan Yang

Efficiency and Power Losses in PM BLDC Motor with Variable Bridge/half-bridge Structure Electronic Commutator 326
K. Krykowski, A. Bodora

Analysis of a device for converting a unipolar input voltage into two symmetric bidirectional output voltages with a magnetically coupled coil 331
Felix. A. Himmelstoss, Wilhelm Kraeftner

Invariant Modulation Strategy for Two-stage Direct Power Converter 337
Radiy Bekbudov

Experimental Study of A Multicell ac/ac Converter Balancing Circuit 345
Robert Stala, Andrzej Mondzik

A Comparison and Optimum Design of Reluctance-Controlled Classical Load-Resonant Converters 350
Stefan V. Mollov, Michael P. Theodoridis

Capacitor Clamped Multilevel Matrix Converter Controlled with Venturini Method 357
Janina Rzasa

Reliability Consideration for a High Power Zero-Voltage-Switching Flyback Power Supply 365
Arash Rahnamaee, Jafar Milimonfared, Kaveh Malekian, Mohammad Abroushan

The Traction Drive Topology Using the Matrix Converter with Middle-Frequency Transformer 372
Martin Pittermann, Pavel Drábek, Marek Cédl

Analysis of Multipulse Rectifiers with Modulation in DC Circuit in Vector Space Approach 377
Andrzej KAPLON and Jaroslaw ROLEK

High Efficiency Soft Switching Boost Converter for Photovoltaic System 383
Gil-Ro Cha, Sang-Hoon Park, Chung-Yuen Won, Yong-Chae Jung, Sang-Hoon Song

A Power Converter For Fault Tolerant Machine Development In Aerospace Applications 388
Liliana de Lillo, Patrick Wheeler, Lee Empringham, Chris Gerada, XiaoyanHuang

Optimal Bus Capacitance Design for System Stability in On-Board Distributed Power Architecture 393
Seiya Abe, Masahiko Hirokawa, Masahito Shoyama, Tamotsu Ninomiya

Table of Contents

Steady State Analysis of Hysteretic Control Buck Converters 400
L.K. Wong, T.K. Man

A Novel Control Method for IGBT Current Source Rectifier 405
Longcheng Tan, Yaohua Li, Ping Wang, Congwei Liu, Zixin Li, Yonggang Chen, Wei Xu

A procedure to optimize the inductor design in boost PFC applications 409
Florent Liffran

Electric Vehicle Drive Inverters Simulation Considering Parasitic Parameters 417
Wen Huiqing, Liu Jun, Zhang Xuhui, Wen Xuhui

DC-DC Converters with FPGA Control for Photovoltaic System 422
Jan Leuchter, Pavel Bauer, Vladimir Rerucha, Petr Bojda

Control of a Converter with Superconductive Energy Storage Inductor 428
Rozanov Yurie Konstantinovich, Lepanov Michail Gennadevich, Kiselev Michail Gennadevich

FPGA-based Controllers for Switching Converters 432
Karel Jezernik

Gamesa DAC converter: the way for REE grid code certification 437
Itziar Martinez, Daniel Navarro

Flatness-Based Voltage-Oriented Control of Three-Phase PWM Rectifiers 444
J. Dannehl, F.W. Fuchs

Control of a single phase H-Bridge multilevel inverter for grid-connected PV applications 451
Elena Villanueva, Pablo Correa, Jose Rodriguez

Switching and Voltage Controls for a Flyback Switch-Mode Rectifier 456
Yuan-Chih Chang, Chang-Ming Liaw

Method Of Designing ZVS Boost Converter 463
Miroslaw Luft, Elzbieta Szychta, Leszek Szychta

A New DC-DC Converter with Multi Output: Topology and Control Strategies 468
Arash A Boora, Firuz Zare, Gerard Ledwich, Arindam Ghosh

Maximum Frequency for Hysteretic Control COT Buck Converters 475
L.K. Wong, T.K. Man

Current Control Method Based on Hysteresis Control Suitable for Single Phase Active Filter with LC Output Filter 479
Yukinori Kobayashi, Hirohito Funato

Optimal Slope Compensation for step load in peak current controlled dc-dc Buck Converter 485
Susovon Samanta, Pradipta Patra, Siddhartha Mukhopadhyay, Amit Patra"

Performances of a PLL Based Digital Filter for double-conversion UPS 490
Armando Bellini, Stefano Bifaretti

10A 12V 1 chip digitally-controlled DC/DC converter IC with high resolution and high frequency DPWM 498
Kazutoshi Nakamura*, Toshiyuki Naka*, Yuki Kamata*, Toyoki Taguchi, Takaaki Shimizu, Yoshiko Ikeda, Akio Nakagawa, Dragan Maksimovic

Modelling and Modulation of Voltage Source Converter 504
Grzegorz Radomski

Sliding Mode Control of DC/DC Multiphase Power Converters 512
Vadim Utkin

A New Digital Control Method for High Performance 400 Hz Ground Power Unit 515
Zixin Li, Ping Wang, Haibin Zhu, Yaohua Li, Longcheng Tan, Yonggang Chen, Fanqiang Gao

Table of Contents

Single-phase 50-kW 16.7-Hz Four-Quadrant Line-Side Converter for Railway Traction Application 521
C. Heising, R. Bartelt, V. Staudt, A. Steimel

Technique to Improve IGBT Converter Efficiency and Transient Response 528
Robert W. Turner, Simon Walton, Richard Duke

The control of voltage converter rectifiers 536
Krzysztof Szubert

Load Voltage Regulation and Line Loss Minimization of Loop Distribution Systems Using UPFC 542
Mahmoud A. Sayed, Takaharu Takeshita

Control of Traction Single-Phase Current-Source Active Rectifier under Distorted Power Supply Voltage 550
Jan Michalík, Jan Molnár, Zdenck Peroutka

Simulation Model Of Neural Network Based Synchronous Generator Excitation Control 556
Damir Sumina, Neven Bulic, Gorislav Erceg

Predictive Current Control of a 7-level AC-DC back-to-back Converter for Universal and Flexible Power Management System 561
Stefano Bifaretti, Pericle Zanchetta, Florin Iov, Jon C. Clare

Predictive Stator Current Control For Three-Level Voltage-Source Inverters With Output LC-Filters 569
Tomasz Laczynski, Axel Mertens

Research on Dimming Control Method of Electronic Ballast for the Automotive HID Headlight 576
P. Dong, K.W.E.Cheng, S.L.Ho

Control Method for a Three-Port Interface Converter Using an Indirect Matrix Converter with an Active Snubber Circuit 581
Koji Kato, Jun-ichi Itoh

Precise Digital Control Method with Multi-rate deadbeat control for Single Phase Utility Interactive Inverter with FPGA based Hardware Controller 589
Kenta Hayashi, Tomoki Yokoyama

A Digital Current Controller for Zero-Current Transition Bidirectional Converter 595
Nobuyuki Kasa, Takahiko Iida

Control Method for a Single Phase Arbitrary Waveform-output Inverter 600
Satoshi Taniguchi, Keiji Wada, Toshihisa Shimizu

Elimination of Harmonics in Multilevel Inverters with Non-Equal DC Sources Using PSO 606
A. K. Al-Othman, Tamer H. Abdelhamid

Improved PFC Circuit Having Ladder Type Filter with Only Passive Devices 614
Kenji Ando, Keiju Matsui, Nobuhito Takeuchi, Masaru Hasegawa

Fuel Cell Current Ripple Minimization using a bi-Buck Power Interface 621
Nicu Bizon, Marian Raducu, Mihai Oproescu

Power Control Strategy of Parallel Inverter Interfaced DG Units 629
H.R. Baghaee, M.Mirsalim, M. J. Sanjari, G.B. Gharehpetian

Implementation of Nonlinear power flow controllers to control a VSC 637
Nelson L. Díaz, Fabián H. Barbosa, Cesar L. Trujillo

Harmonic Distortion Reduction Technique for Uninterruptible Power Supplies with DC Voltage Boost Technique 643
Juei Lung Shyu

Energy-based Modulation Error Control for High-Power Drives with Output LC-Filters and Synchronous Optimal Pulse Width Modulation 649
Tomasz Laczynski, Timur Werner, Axel Mertens

Table of Contents

Voltage Harmonic Control of Z-source Inverter for UPS Applications 657
Arkadiusz Kulka, Tore Undeland

A Method of Optimal Control for Switched-Mode Power Converters 663
Anatoly Bekishev, Albert Iskhakov, Leonid Klyachko, Vladimir Pospelov, Sergey Skovpen

Experiment results with modified Hybrid PWM method for three phase induction motor 669
Daniel Lewandowski, Grzegorz Lisowski

Optimized Design of a Delay line based Analog to Digital Converter for Digital Power Management Applications 674
Mukti Barai, Sabyasachi Sengupta, Jayanta Biswas

Overmodulation Region of Multi-Phase Inverters 682
S. Halasz

Optimal Control of Induction Motor Using High Performance Frequency Converter 690
Jerkovic Vedrana, Spoljaric Zeljko, Valter Zdravko

Power Electronic Converter for the Reluctance Pump Drive 695
B. J. Szymanski, K. Kompa, N. Michalke, H. Kuß, U. Schuffenhauer

A Predictive Control Scheme for Current Source Rectifiers 699
Pablo Correa, Jose Rodriguez

Analysis and Design of New Switching Table for Direct Power Control of Three-Phase PWM Rectifier 703
Abdelouahab Bouafia, Jean-Paul Gaubert, Fateh Krim

Improvement of the performance for DC-DC Converter 710
X..She, Yun She

A Drive System With High-Speed Single-Phase Supplied Three-Phase Induction Motor 714
T. Binkowski, M. Grad, M. Latka, W. Malska, D. Sobczynski

A Pulse Width Modulation Technique for a Multilevel Converter in High Voltage High Frequency Applications 718
Jafar Adabi, Hamid Soltani, Firuz Zare

Bidirectional Positive Buck-Boost Converter 723
Arash A Boora, Firuz Zare, Gerard Ledwich, Arindam Ghosh

Control system of power electronics current modulator utilized in diode rectifier with sinusoidal source current 728
Michal Gwózdz, Michal Krystkowiak

Design and control of a half-bridge converter to drive piezoelectric actuators 731
Oriol Gomis-Bellmunt, Josep Rafecas-Sabate, Daniel Montesinos-Miracle, Josep-Maria Fernandez-Mola, Joan Bergas-Jane

Online Diagnosis of PEM Fuel Cell 734
Abdellah Narjiss, Daniel Depernet, Denis Candusso, Frederic Gustin, Daniel Hissel

Application of Kalman filters to the control of independent power electronic voltage sources 740
Ryszard Porada, Lukasz Nyczkowski

Verification of the load sharing characteristics in Autonomous Decentralized UPS system using FPGA based Hardware Controller 744
Nobuaki Doi, Tsuyoshi Saito, Tomoki Yokoyama

Fault Current Reduction in Distribution Systems with Distributed Generation Units by a New Dual Functional Series Compensator 750
H.R. Baghaee, M. Mirsalim, M. J. Sanjari, G.B. Gharehpetian

Dynamic Simulation of PM Motor Drive System based on Reluctance Network Analysis 758
Kenji Nakamura, Osamu Ichinokura

Table of Contents

Performance Improvement of Direct Torque Controlled Interior Permanent Magnet Synchronous Motor Drive by Considering Magnetic Saturation 763
Behrooz Majidi, Jafar Milimonfared, Kaveh Malekian

Condition Monitoring for Mechanical Faults in Fully Integrated Servo Drive Systems 769
Jesus Arellano-Padilla, Mark Sumner, Chris Gerada

Feed-forward Compensation of Load and Parameter Variations of Electric Drive 776
Alon Kuperman, Yoram Horen, Saad Tapuchi, Uri Suissa

Thermal Effect of Short-Circuit Current in Low Power Induction Motors 782
Leo.s Beran

Generalized Model for a Class of Switched Reluctance Motors 787
Constantin Pavlitov, Yassen Gorbounov, Radoslav Rusinov, Alexandar Alexandrov, Kliment Hadjov, Dimitar Dontchev

Neural Network based Fault Detection of PMSM Stator Winding Short under Load Fluctuation 793
J. Quiroga, D.A. Cartes, C.S. Edrington, Li Liu

Review of Electrical Machine in Downhole Applications and the Advantages 799
Anyuan Chen, Ravindra. B. Ummaneni, Robert Nilssen, Arne Nysveen

Broken Rotor Bar Impact on the Closed Loop and Sensorless Control of Induction Machine 804
Piotr Kotodziejek, Elzbieta Bogalecka

Coupled Magnetic Circuit Method and Permeance Network Method Modeling of Stator Faults in Induction Machines 810
Amin Mahyob, Mohamed Y. Ould Elmoctar, Pascal Reghem, Georges Barakat

Explosion Protected Electrical Drives - Risk Assessment and Technical Diagnostics 818
Ivica Gavranic, Drago Ban, Damirarko Zarko

The effect of subharmonics on induction machine heating 826
Piotr Gnacinski, Marcin Peplinski, Mariusz Szweda

Influence of Saturation Effects in a Transverse Flux Machine 830
M. Siatkowski, B. Orlik

A Model of Semiconductor Converter-Fed Asynchronous Machines Taking into Account Energy Losses and Thermal Processes 837
M. Pronin, O. Shonin, Y. Koskin, A. Vorontsov, P. Kalatchikov

Use of an AC Self-excited Switched Reluctance Generator as a Battery Charger 845
Abelardo Martínez, Estanislao Oyarbide, Javier Vicuña, Francisco Perez, Eduardo Laloya, Bonifacio Martín-del-Brío, Tomás Pollán, Beatriz Sánchez, Juan Lladó

Direct Thrust Controlled Linear Induction Motor Including End Effect 850
Berrin Susluoglu, Vedat M. Karsli

Analysis of Short-Circuit Forces at the Top of the Low Voltage U-Type and I-Type Winding in a Power Transformer 855
Leonardo Strac, Franjo Kelemen, Damir Zarko, Josipa Mokrovica

Anisotropy Comparison of Reluctance and PM synchronous Machines for Low Speed Position Sensorless Applications 859
H.W. de Kock, M.J. Kamper, R.M. Kennel

Analysis of VSI-DTC Fed 6-phase Synchronous Machines 867
Ibrahim Abuishmais, Waqas M. Arshad, Sami Kanerva

Optimal Rotor Flux Shape for Multi-phase Permanent Magnet Synchronous Motors 874
Roberto Zanasi, Federica Grossi

Table of Contents

Modelling of Electrical Machines Using the Modelica Bond-Graph Library 880
Mieczyslaw Ronkowski

Induction Motor Parameters Identification using Genetic Algorithms for Varying Flux Levels 887
Konstantinos Kampisios, Pericle Zanchetta, Chris Gerada, Andrew Trentin, Omar Jasim

Study of the sudden symmetrical short-circuit using the mathematical models of the synchronous machine and the numerical methods 893
Petropol Serb Gabriela, Petropol Serb Ion, Campeanu Aurel, Sonia Degeratu, Anca Petrisor

Analytical Method of Calculation of the Current and Torque of a Reluctance Stepper Motor Using Fourier Complex Series 899
Pavel Zaskalicky, Maria Zaskalicka

Bearing Damage Analysis by Calculation of Capacitive Coupling between Inner and Outer Races of a Ball Bearing 903
Jafar Adabi, Firuz Zare, Gerard Ledwich, Arindam Ghosh, Robert D.Lorenz

The Model of the Squirrel Cage AC Motor including Rotor Slot Harmonics 908
Eleonora Darie, Costin Cepisca, Emanuel Darie

Identification of mathematical model induction motor's parameters with using evolutionary algorithm and multiple criteria of quality 912
Hudy Wiktor, Jaracz Kazimierz

Simulation Study on Control of Ultrahigh Speed Drives in Waste Energy Recovery Systems 916
Péter Stumpf, Miklós G. Simon, Rafael K. Járdán, István Nagy

Adaptive Back EMF Parameter Adjustment of Simplified Vector Control for Position Sensorless Permanent Magnet Synchronous Motors 924
Kiyoshi Sakamoto, Yoshitaka Iwaji, Daigo Kaneko, Toshihiro Takeuchi, Tsunehiro Endo, Atsuo Kawamura

Identification and Control of Precision XY Stages with Active Vibration Suppression System 932
Mayumi Nitta, Seiji Hashimoto

Sensitivity of the Currents Input-Output Decoupling Vector Control of the DFIM versus Current Sensors Fault 938
Meriem Abdellatif, Maria Pietrzak-David, Ilhem Slama-Belkhodja

Extended Back EMF model for PM synchronous machines with different inductances in d- and q-axis 945
Andreas Eilenberger, Manfred Schroedl

Gait generation of a two-legged robot by using adaptive network based fuzzy logic control 949
Umit Onen, Mete Kalyoncu, Mustafa Tinkir, Fatihm. Botsali

Walking robot HEXOR® II - a versatile platform for engineering education 956
M. Sajkowski, T. Stenzel, B. Grzesik

Motion Control of Steel Sheet Shears with Rocking Knife Mechanism 961
Jan Fetyko, Frantisek Durovsky, Viliam Fedak

Intelligent Adaptive Control and Monitoring Of Band Sawing 967
Ilhan Asiltürk, Ali Ünüvar

Hierarchical adaptive network based fuzzy logic controller design for a single flexible link robot manipulator 974
Mete Kalyoncu, Mustafa Tinkir

Digital Controlled High Speed Synchronous Motor 982
Zdenk Cerovský, Jaroslav Novák, Martin Novák, Marek Cambál

Analysis of combustion engine - electric Linear generator set operation 988
Jirí Pavelka

Table of Contents

Closed Loop Control of AC Drive with LC Filter..994
Jaroslaw Guzinski

Sensorless IPMSM based drive for reciprocating compressor...1002
Anton Dianov, Kim Young-Kwan, Lee Sang-Joon, Lee Sang-Taek, Yoon Tae-Ho

Controlling system of electrodynamic drive..1009
Josef Cernohorský

Expert System for Electric Drive Design...1017
Juhan Laugis, Valery Vodovozo

Improvement of Moving Characteristics of Cableless Micro-actuator and Consideration of Reversible Motion..1020
Hiroyuki Yaguchi, Kazumi Ishikawa, Toshihiro Zamma, Koichi Funayama

Sensorless Control of AC Machines using High-Frequency Excitation...1024
Heiko Zatocil

Adaptive PF Speed Control of SRAM Drives..1033
Laszlo Szamel

A Very Simple Fuzzy Control System for Inverter Fed Synhronous Motor...................................1040
Pawel Fabijanski, Ryszard Lagoda

Distributed control system of DC servomotors for six legged walking robot................................1044
D. Belter, K. Walas, A. Kasinski

Optimization of Starting Process of the Frequency Controlled Induction Motor...........................1050
I.Ya. Braslavsky, A.V. Kostylev, D.P. Stepanyuk

3-Axes Satellite Attitude Control Based on Biased Angular Momentum......................................1054
Azam Ghaedi, Mohammad Ali Nekoui

Modelling and simulation of a signal injection self-sensored drive...1058
Alen Poljugan, Mark Sumner, Chris Gerada, Qiang Gao

Robust PI Cascade Control for a Multi-Mass System Optimized by Evolutionary Algorithms.........1064
M. Joost, K. Zielinski, B. Orlik, R. Laur

Permanent Magnet Synchronous Servo-Drive with State Position Controller................................1071
Lech M. Grzesiak, Tomasz Tarczewski, Slawomir Mandra

Closed-Loop Control of Virtual FPGA-Coded Permanent Magnet Synchronous Motor Drives using a Rapidly Prototyped Controller...1077
Christian Dufour, Vincent Lapointe, Jean Bélanger, Simon Abourida

Speed Sensorless Nonlinear Control Of Induction Motor In The Field Weakening Region..............1084
MiroslawWlas, Haithem Abu-Rub, Joachim Holtz

Comparison of Dynamic Performances of Speed Control System Containing Time - Minimal Speed Controller with Control System Containing PI Speed Controller...1090
Andrzej Andrzejewski, Marian Roch Dubowski

Optimisation of Real-Time Complex Path Generation in Constrained Intelligent Motion Applications Based on IPM Motor Drives..1097
Silverio Bolognani, Roberto Petrella, Fabio Stefanutti, Piero Stocco

PMSM Sliding Mode Observer for Speed and Position Estimation Using Modified Back EMF........1105
Ilioudis Vasilios C., Margaris Nikolaos I.

Optimal Control of Electrical Drives with Induction Motors for Variable Torques.........................1111
Corneliu Botan, Marcel Ratoi, Vasile Horga

Table of Contents

An Optimal Control for Saturated Interior Permanent Magnet Linear Synchronous Motors Incorporating Field Weakening 1117
Mohammad Abroshan, Jafar Milimonfared, Kaveh Malekian, Arash Rahnamaee

Improved Direct Torque Control for Induction Machine Drives using Fuzzy Logic and Particle Swarm Optimization 1123
Mohammad Mehdi Rezaei, Mojtaba Mirsalim, Kaveh Malekian

Design and Implementation of High Performance Full-Digital Spindle Drives 1128
Liu Yang, Zhao Jin

Semi hierarchical adaptive network based fuzzy logic controller design for a multi-straight-line path tracing flexible robot manipulator with rotating-prismatic joint 1132
Mete Kalyoncu, Mustafa Tinkir

Control System with the Set Point Observation 1140
Algirdas Baskys, Vitoldas Gobis, Valerijus Zlosnikas

Electropneumatic Servo System with Adaptive Force Controller 1144
Arunas Grigaitis, Vilius Antanas Gele~evicius

New fault tolerant DTC control for induction machine drives 1149
A.Ben Abdelghani Bennani, M. Ghodbane Cherif, I. Slama Belkhodja

Stability Analysis of the Natural Field Orientation Controlled Induction Machine Drive 1155
G. Mirzaeva, A. Rojas

Control of SR motor EV by instantaneous torque control using flux based commutation and phase torque distribution technique 1163
Ayumu Nishimiya, Hiroki Goto, Hai-Jiao Guo, Osamu Ichinokura

Simulation of IPM Motor by Nonlinear Magnetic Circuit Model for Comparing Direct Torque Control with Current Vector Control 1168
Hiroki Goto, Kensuke Kimura, Hai-Jiao Guo, Osamu Ichinokura

A Simplified Model for Induction Machines with Faults to Aid the Development of Fault Tolerant Drives 1173
O. Jasim, C. Gerada, M. Sumner, J. Arellano-Padilla

About the Experimental Results of an Electric Driving System Based on Asynchronous Motor and PWM Converter 1181
Petre-Marian Nicolae, Dan-Gabriel Stanescu, Ioana-Gabriela Sîrbu

Real-World Force Feedback Control for Mobile-Hapto 1187
Wataru Yamanouchi, Yuki Yokokura, Seiichiro Katsura, Kiyoshi Ohishi

The new numerical integration routine applied in sensorless drives 1193
Arkadiusz Gardecki, Krystyna Macek-Kaminska

Application of Fuzzy Logic Techniques To Robust Speed Control of PMSM 1198
Tomasz Pajchrowski, Krzysztof Zawirski

Optimal control of current commutation of high speed SRM drive 1204
Jan Deskur, Tomasz Pajchrowski, Krzysztof Zawirski

Comparison Between Direct Torque Control and Vector Control of a Permanent Magnet Synchronous Motor Drive 1209
Rafa Souad, Houcine Zeroug

Detection and self-tuning compensation of periodic disturbances by the control of DC motor 1215
Michael Ruderman, Frank Hoffmann, Johannes Krettek, Torsten Bertram

A Linear Switched Reluctance Motor Based Position Tracking System 1221
S. W. Zhao, N. C. Cheung, Y. Lu, W. C. Gan, Z. G. Sun

Table of Contents

Mobile Robot Navigation with Obstacle Avoidance Capability 1225
Anca Sorana Popa, Mircea Popa, Ioan Silea

Requirements for Power Electronics in Solid Oxide Fuel Cell System 1233
T. Riipinen, V. Väisänen, M. Kuisma, L. Seppä, P. Mustonen, P. Silventoinen

Power Supply for a IGBT-Driver with High Insulation Voltage based on a Printed Planar Transformers 1239
Günter Schmitt, Wolf Kusserow, Ralph Kennel

Variable Motor Operating Point by Integration of Power Electronic Device into Rotor 1243
Adrian Tulbure, Hans-Peter Beck, Mircea Risteiu

Magnetic Material Comparisons for High-Current Gapped and Gapless Foil Wound Inductors in High Frequency DC-DC Converters 1249
Marek S. Rylko, Brendan J. Lyons, Kevin J. Hartnett, John G. Hayes, Michael G. Egan

Feasibility Study of Half- and Full-Bridge Isolated DC/DC Converters in High-Voltage High-Power Applications 1257
Dmitri Vinnikov, Tanel Jalakas, Mikhail Egorov

Evaluation of Different Loss Calculation Methods for High-voltage IGBT-s Under Small Load Conditions 1263
T. Jalakas, D. Vinnikov, J. Laugis

Control of Power Supply Unit for Military Vehicles Based on Four-Leg Three-Phase VSI with Proportional-Resonant Controllers 1268
Tomál Glasberger, Zdenek Peroutka

Optimal Design of a Half Wave Cockroft-Walton Voltage Multiplier with Different Capacitances per Stage 1274
Ioannis C. Kobougias, Emmanuel C. Tatakis

Calculation of Leakage Inductance of Core-Type Transformers for Power Electronic Circuits 1280
Reinhard Doebbelin, Marcel Benecke, Andreas Lindemann

Enhanced Current Pulsation Smoothing Parallel Active Filter for Single Stage Grid-connected AC-PV Modules 1287
A.C. Kyritsis, N.P. Papanikolaou, E.C. Tatakis

Outline of the Design of a Cascaded H-bridge Medium Voltage STATCOM 1293
R.E. Betz, B.J. Cook, T.J. Summers, R. Fisher, A. Bastiani, S. Shao, P. Stepien, K. Willis

Investigation of High Frequency Effects on Layered Coils 1301
Georgios S. Dimitrakakis, Emmanuel C. Tatakis

Soft Switching PWM Inverter for Induction Heating Applied to Heating of Ferromagnetic Metal 1309
Sachio Kubota, Muneo Sato, Fumio Ito, Yoshihiro Shimaoka, Kunihiro Nishioka

Corona Treatment System with Resonant Inverter - Selected Proprieties 1316
Mucko Jan

Power supply unit for an electric discharge machine 1321
Wojciech Mysinski

High Power, High Voltage, High Frequency Transformer / Rectifier for HV Industrial Applications 1326
T. Filchev, D. Cook, P. Wheeler, A. Van den Bossche, J. Clare, V. Valchev

Small Power Laboratory Model and High Power Prototype of the Four-Level VSI 1332
Ryszard Michal Strzelecki, Pawel Szczepankowski, Andrzej Kasprowicz, Genady Stepanovic Zinoviev, Krzysztof Zymmer, Zbigniew Zakrzewski

AC Voltage Regulator Using PWM Technique and magnetic flux distribution 1337
A.M. Dabroom

Minimum Reactive Power Filter Design for High Power Converters 1345
Alex-Sander Amavel Luiz, Braz Jesus Cardoso Filho

xiii

Table of Contents

Injection of a carrier with higher than the PWM frequency for sensorless position detection in PM synchronous motors 1353
Roberto Leidhold, Peter Mutschler

Parallel Fixed Point FPGA Implementation of Sensorless Induction Motor Torque Control 1359
Jacek D. Lis, Czeslaw T. Kowalski

Design of an FPGA-Based Real-Time Simulator for Electrical System 1365
I. Bahri, M-W. Naouar, E. Monmasson, I. Slama-Belkhodja, L.Charaabi

A New, Ultra-low-cost Power Quality and Energy Measurement Technology 1371
Alex McEachern, Andreas Eberhard

Rotor Time Constant Adaptation Using Radial Basis Function Network 1375
Pavel Brandltetter, Ondfej Skuta

Application of Speed and Load Torque Observers in High Speed Train 1382
Jaroslaw Guzinski, Marc Diguet, Zbigniew Krzeminski, Arka diusz Lewicki, Haithem Abu-Rub

Position Estimator including Saturation and Iron Losses for Encoder Fault Detection of Doubly-Fed Induction Machine 1390
Kai Rothenhagen, Friedrich W. Fuchs

Wide Range Low Noise Current Sensor 1398
F. Richter, C. Sourkounis

Transducerless Speed Control with Initial Position Detection for Low Cost PMSM Drives 1402
Roman Filka, Peter Balazovic, Branislav Dobrucky

Study About the Possibility of Electrodes Motion Control in the EAF Based on Adaptive Impedance Control 1409
Manuela Panoiu, Caius Panoiu, Sorin Deaconu

Asynchronous machine stator resistance estimation using integrated PWM modulator and sampler unit as FPGA application 1416
Dag Samuelsen, Waldemar Sulkowski

Development of Monitoring System for Series HEV Bus with Touch Panel 1421
Tae-Won Chun, Quang-Vinh Tran, Uk-Don Choi, Heung-Gun Kim

A Development System for Testing Integrated Circuits Used for Power and Energy Measurements 1426
Vladimir Cuk, Aleksandar Nikolic, Aleksandar Zigic

State and parameter estimation in a hydraulic system - moving horizon approach 1432
Jerzy Baranowski, Andrzej Tutaj

Technologies of Current Sensors Suitable for Hot High Density Power Electronics 1440
Filip Grecki, Grzegorz Iwanski, Wlodzimierz Koczara, Jozef Lastowiecki

Nonlinear dynamical feedback for motion control of magnetic levitation system 1446
Jerzy Baranowski, Pawel Piatek

Speed and position estimation of SRM 1454
Konrad Urbanski, Krzysztof Zawirski

Potential of Digital Gate Units in High Power Appliations 1458
Harald Kuhn, Thies Koneke, Axel Mertens

Disturbance Currents of Inverters 1465
Petr Vrana, Jiri Javurek

Improvement of the Energy Recovery of Traction Electrical Drives using Supercapacitors 1469
Diego Iannuzzi

Table of Contents

A Multi-Core PC-based Simulator for the Hardware-In-the-Loop Testing of Modern Train and Ship Traction Systems..1475
Christian Dufour, Guillaume Dumur, Jean-Nicolas Paquin, Jean Bélanger

Energy Saving Control of Tram Motors Taking Light Signalling and City Disturbances into Account...................1481
Stanislaw Rawicki

Characterization and Improved Control of a Brushless DC Drive with In-Wheel Motor...........................1491
Manuele Bertoluzzo, Giuseppe Buja, Alessandro Pavoni

Supply of Electric Vehicles via Magnetically Coupled Air Coils...1497
Slawomir Judek, Krzysztof Karwowski

Sliding-Mode Approach to Control Design for Induction Motor Drive fed by a Three-Level Voltage-Source Inverter...1505
Sergey Ryvkin, Richard Schmidt-Obermoeller, Andreas Steimel

Analysis and configuration of supercapacitor based energy storage system on-board light rail vehicles.................1512
R. Barrero, X. Tackoen, J. Van Mierlo

Design of High Power Electronic Building Block based on Parallel of IGBTs for Electric Vehicle...........................1518
Wen Huiqing, Liu Jun, Zhang Xuhui, Wen Xuhui

Stability Analysis on the DC Power Distribution System of More Electric Aircraft................................1523
H. Zhang, C. Saudemont, B. Robyns, N. Huttin, R. Meuret

Design Considerations for Control of Traction Drive with Permanent Magnet Synchronous Machine.....................1529
Zden..k Peroutka, Karel Zeman

Control of Primary Voltage Source Active Rectifiers for Traction Converter with Medium-Frequency Transformer...1535
Vojtech Blahník, Zdenek Peroutka, Jan Molnár, Jan Michalík

Energy management strategy for Coupling Supercapacitors and Batteries with DC-DC converters for hybrid vehicle applications...1542
M.B. Camara, F. Gustin, H. Gualous, A. Berthon

Dual-Source Fed Multiphase Traction System with Standard and Non-Standard Control Regimes Based on Synchronized PWM...1548
Valentin Oleschuk, Marian P. Kazmierkowski

Analysis of a H-NPC topology for an AC Traction Front-End Converter...1555
I. Etxeberria-Otadui, A. Lopez-de-Heredia, J. San-Sebastian, H. Gaztañaga, U. Viscarret, M. Caballero

Hybrid - type system of power supply for a trolleybus with an asynchronous motor...........................1562
Zygmunt Gizinski, Marcin Gasiewski, Ireneusz Mascibrodzki, Michal Zych, Krzysztof Zymmer, Marcin Zulawnik

Control of rotor flux in AC tram drive during sudden braking operation...1568
Andrzej Debowski, Piotr Chudzik

A New Novel Power Electronic Circuit to Reduce Stray Current and Rail Potential in DC Railway.....................1575
Reza Fotouhi, Siamak Farshad

Slip Control Upgrades for Light-Rail Electric Traction Drives...1581
Madis Lehtla, Hardi Hõimoja

Practical Aspects on the Improved DC Driving System Used in Electric Urban Traction...........................1585
Petre Marian Nicolae, Ioana-Gabriela Sîrbu, Ileana-Diana Nicolae, Lucian Mandache

The study of using the traction drive topology with the middle-frequency transformer...........................1593
Martin Pittermann, Pavel Drábek, Marek Cédl, Jiŕí Foŕt

Control of a Linear Switched Reluctance Motor as a Propulsion System for Autonomous Railway Vehicles...........................1598
L. Kolomeitsev, D. Kraynov, S. Pakhomin, F. Rednov, E. Kallenbach, V. Kireev, T. Schneider, J. Böcker

Table of Contents

Motion Copying System Based on Real-World Haptics in Variable Speed..1604
Yuki Yokokura, Seiichiro Katsura, Kiyoshi Ohishi

Adaptive Fuzzy Control of magnetically suspended Rotary Table 1610
Thomas Schallschmidt, Denis Draganov, Frank Palis

Wideband Force Sensing for Haptic Energy Transmission Utilizing FPGA.....................................1614
Seiichiro Katsura, Masaki Kondo, Kiyoshi Ohishi

On the development of BLDC motor control run-up algorithms for aerospace application 1620
Vladimir Hubik, Martin Sveda, Vladislav Singule

Rotor Levitation by Active Magnetic Bearing Using Digital State Controller...............................1625
Chip Rinaldi Sabirin, Andreas Binder

Dynamical Torque-Speed-Curve Adaption To Damp Load Peaks Occuring In Drive Trains Of Shredding
Plants...1633
Constantinos Sourkounis

Traction vehicle distributed control computer system architecture with auto reconfiguration features and
extended DMA support 1638
Jiri Zdenek

Analysis and Position Control of a Linear Switched Reluctance Actuator Based on Sliding Mode Control 1646
António Espírito Santo, Maria R. A. Calado, Carlos M. P. Cabrita

Development and Control for a Reaction Wheel System Driven by Permanent Magnet Synchronous
Motor 1652
Ming-Chang Chou, Chang-Ming Liaw, Sywe-Bin Chien, Fa-Hwa Shieh, Jih-Run Tsai, Hao-Chi Chang

Nonlinear control design for magnetic bearings via automatic differentiation.............................1660
Stefan Palis, Mario Stamann, Thomas Schallschmidt

Design of Energy Harvesting Generator Base on Rapid Prototyping Parts...................................1665
Zdenek Hadas, Jan Zouhar, Vladislav Singule, Cestmir Ondrusek

Control of Bouc-Wen hysteretic systems: Application to a piezoelectric actuator 1670
Oriol Gomis-Bellmunt, Faycal Ikhouane, Daniel Montesinos-Miracle

Electric drive for carding machine draft device..1676
Martin Diblík

Two-level and Multilevel Converters for Wind Energy Systems: A Comparative Study 1682
R. Melício, V. M. F. Mendes, J. P. S. Catalão

A Stand-alone Photovoltaic Supercapacitor Battery Hybrid Energy Storage System 1688
M.E. Glavin, Paul K.W. Chan, S. Armstrong, W.G Hurley

Integrated contactless power transmission systems with high positioning flexibility 1696
Daniel Kürschner, Christian Rathge

A Transformerless Interface Converter for a Distributed Generation System................................1704
Tzung-Lin Lee, Zong-Jie Chen

A Comprehensive Analysis and Comparison Between Multilevel Space-Vector Modulation and Multilevel
Carrier-Based PWM 1710
Constantinos Sourkounis, Ahmad Al-Diab

Identification of Electrical Parameters in a Power Network Using Genetic Algorithms and Transient
Measurements 1716
Wei. Dong, Pericle Zanchetta, David W.P. Thomas

On Acoustic Noise Reduction Procedure for Inverter-Fed Induction Machines 1722
Weiss Helmut, Zaucher Peter, Xiao Jian

Table of Contents

Cascaded Doubly Fed Induction Generator for Mini and Micro Power Plants Connected to Grid 1729
Marek Adamowicz, Ryszard Strzelecki

Contactless power transmission with new secondary converter topology 1734
Matthias Dockhorn, Daniel Kürschner, Rudolf Mecke

Modeling Approach of a Generator with Non-linear Load in Embedded Electrical Network 1740
Nicolas Amelon, Mourad Ait-Ahmed, Mohamed-Fouad Benkhoris

Optimal Use of the 14 V Alternator in 42 V Automotive Supply Systems 1748
Vasile Comnac, Mihai Cernat, Adrian Mailat

New Dual Channel Quasi Resonant DC-DC Converter Topologies for Distributed Energy Utilization 1755
J. Hamar, I. Nagy, P. Stumpf, H. Ohsaki, E. Masada

Output Filtering of the Customer-end Inverter in a Low-Voltage DC Distribution Network 1763
Pasi Peltoniemi, Pasi Nuutinen, Pasi Salonen, Markku Niemelä, Juha Pyrhönen

Power Flow Control through a Multi-Level H-Bridge based Power Converter for Universal and Flexible Power Management in Future Electrical Grids 1771
Stefano Bifaretti, Pericle Zanchetta, Yue Fan, Florin Iov, Jon Clare

Energy Storage Systems The Flywheel Energy Storage 1779
Tomasz Siostrzonek, Stanislaw Piróg, Marcin Baszynski

Analysis of Wide Area Integration of Dispersed Wind Farms Using Multiple VSC-HVDC Links 1784
S. González-Hernández, E. Moreno-Goytia, O. Anaya-Lara

Generator Selection for Offshore Oscillating Water Column Wave Energy Converters 1790
D.L. O' Sullivan, A.W. Lewis

A Novel Approach To Photovoltaic Powered Water Pumping Design 1798
Michael James Case, Ernest Edward Denny

Direct Controls in Voltage-Source Converters - Generalizations and Deep Study 1803
Karoly Veszpremi, Istvan Schmidt

Multipolar double fed induction wind generator with a single phase secondary winding 1811
Leonids Ribickis, Guntis Dilevs, Nikolajs Levins, Vladislavs Pugachevs

The measurement on the solar cells in Liberec city 1815
Jiri Kubin

Rotor Turn-to-Turn Faults of doubly-fed Induction Generators in Wind Energy Plants - Modelling, Simulation and Detection 1819
Vincenz Dinkhauser, Friedrich W. Fuchs

Static and Dynamic Response of a Photovoltaic Characteristics Simulator 1827
Anastasios Ch. Nanakos, Emmanuel C. Tatakis

Modeling and Optimal Sizing of Hybrid Renewable Energy System 1834
Rachid Belfkira, Cristian Nichita, Pascal Reghem, Georges Barakat

Photovoltaic System MPPTracker Investigation and Implementation using DSP engine and Buck- Boost DC-DC converter 1840
Dimosthenis Peftitsis, Georgios Adamidis, Panagiotis Bakas, Anastasios Balouktsis

Multi Objective Distributed Generation Planning Using NSGA-II 1847
Muhammad Ahmadi, Ashkan Yousefi, Alireza Soroudi, Mehdi Ehsan

Testing of the Grid-connected Photovoltaic Systems Using FPGA-based Real-Time Model 1852
Robert Stala

xvii

Table of Contents

Output Maximization Using Direct Torque Control for Sensorless Variable Wind Generation System Employing IPMSG...1859
Yukinori Inoue, Shigeo Morimoto, Masayuki Sanada

Improving Connection and Disconnection of a Small Scale Distributed Generator Using Solid-State Controller 1866
M.M.R. Ahmed

Research control of electric systems in wind generator systems...1872
Stefan Winternheimer, Artem Kolesnikov, Evgeny Glushkin, Alexander Bukatov

Stand-alone Photovoltaic Generation System with Combined Storage using lead Battery and EDLC 1877
Hiroaki Nakayama, Eiji Hiraki, Toshihiko Tanaka, Noriaki Koda, Nobuo Takahashi, Shuji Noda

Active Filter Action of Inverter Exciting Induction Generator for Wind Power Generation 1884
Noriyuki Kimura, Tomoyuki Hamada, Katsunori Taniguchi, Toshimitsu Morizane

The Operation of Power Electronic Converters in Photovoltaic Drive Systems 1890
Marek Niechaj

Experimental results of a hybrid wind/hydro power system connected to isolated loads 1896
Mehdi Nasser, Stefan Breban, Vincent Courtecuisse, Arnaud Vergnol, Benoît Robyns, Mircea M. Radulescu

Grid Connection of Multi-Megawatt Clean Wave Energy Power Plant under Weak Grid Condition........1904
Kai Rothenhagen, Marek Jasinski, Marian P. Kazmierkowski

Improved sizing method of storage units for hybrid wind-diesel powered system 1911
A.M. Tankari, B. Dakyo, C. Nichita

A Research Platform for a Smart-Blade Wind Generation System 1918
J. Davey, Udaya K. Madawala, R. Sharma

Soft Switching Multi-Phase Boost Converter for Photovoltaic System..1924
Joo-Hyuk Lee, Jae-Hyung Kim, Chung-Yuen Won, Su-Jin Jang, Yong-Chae Jung

Soft Switching Boost Converter for Photovoltaic Power Generation System 1929
Doo-Yong Jung, Young-Hyok Ji, Jae-Hyung Kim, Chung-Yuen Won, Yong-Chae Jung

Optimisation Of Wind Power Pmsm To Grid Conversion System 1934
Ince Kayhan, Weiss Helmut

Analysis of Wind Farm and Multilevel Converter Interactions in Medium Voltage Networks Under Steady-State and Transient Conditions 1941
J. Sosa-Ruiz, E. Moreno-Goytia, O. Anaya-Lara

A Simple, Low Cost Design Using Current Feedback to Improve the Efficiency of a MPPT-PV System for Isolated Locations 1947
Herman Fernández, Abelardo Martínez, Víctor Guzmán, María Isabel Gímenez

A Single-Phase Active Power Filter Based in a Two Stages Grid-Connected PV System 1951
Kleber C.A. De Souza, Denizar C. Martins

Wide Bandwidth Power Flow Control Algorithm of the Grid Connected VSI under Unbalanced Grid Voltages...1957
Zoran Ivanovic, Marko Vekic, Stevan Grabic, Evgenije Adzic, Vladimir Katic

The use of Switched Reluctance Generator in wind energy applications 1963
Eleonora Darie, Costin Cepisca, Emanuel Darie

Active Line Shaping of a Single Phase Rectifier using the Switching Function Technique 1967
Christos Marouchos

Control of Reactive Power in Double-Fed Machine Based Wind Park 1975
Elzbieta Bogalecka, Michal Kosmecki

xviii

Table of Contents

A Novel Hybrid Modulation Method for Cascaded H-bridge Active Power Filter 1981
Yonggang Chen, Ping Wang, Yaohua Li, Zixin Li, Longcheng Tan

Apparent Power Ratio of the Shunt Active Power Filter 1987
A. Kouzou, B.S Khaldi, S. Saadi, M.O. Mahmoudi, M.S. Boucherit

Shunt Active Power Filter with Improved Dynamic Performance 1995
Krzysztof Piotr Sozanski

The Research on the Active Power Filter Based on the Cascaded H-bridge Converter 2000
Yonggang Chen, Junling Chen, Ping Wang, Yaohua Li, Longcheng Tan, Zixin Li, Wei Xu

E-laboratory in the Field of Electrical Drives 2005
H.Hõimoja, A.Rosin, T.Möller, M. Müür

Laboratory Setup for Studying Ultracapacitors in Industrial Applications 2011
I. Roasto, D. Vinnikov, T. Lehtla

Synchronous machine direct axis parameters estimation module from an iterative strategy 2015
Emile Mouni, Slim Tnani, Gérard Champenois

Determination of the Characteristic Life Time of Paper-insulated MV-Cables based on a Partial Discharge and tan(..) Diagnosis 2022
I. Mladenovic, Ch. Weindl

Elimination of Increased Excitation of Common- Mode Oscillations in Electrical Drive Systems with Active Front End and Long Motor Cables 2028
Thomas Weidinger

Internal Short Circuit in a Tooth Wound PMSM with Stranded Conductors 2037
Damien Birolleau, Christian Chillet, Laurent Albert

Implementation of a Virtual Laboratory for Low Power Electrical Drives 2043
Gh. BALUTA, V. HORGA, C. LAZAR

DQ-Transformation Approach for Modelling and Stability Analysis of AC-DC Power System with Controlled PWM Rectifier and Constant Power Loads 2049
K-N Areerak, S.V. Bozhko, G.M. Asher, D.W.P. Thomas

Genetic Identification of Parameters the Sandwich Piezoelectric Ceramic Transducers for Ultrasonic Systems 2055
Pawel Fabijanski, Ryszard Lagoda

The Impact of Higher-Order Harmonics on Tripping of Residual Current Devices 2059
Stanislaw Czapp

Estimation of the Untapped Regenerative Braking Energy in Urban Electric Transportation Network 2066
Leonards Latkovskis, Linards Grigans

Performance Evaluation of Electric Power Steering with IPM Motor and Drive System 2071
Hamidreza Akhondi, Jafar Milimonfared, Kaveh Malekian

Optimal Control: Load Frequency Control of a Large Power System 2076
Sílvio José Pinto Simões Mariano*, Luís António Fialho Marcelino Ferreira

LCL-Load Modular Converter For Induction Heating 2082
Maciej A. Dzieniakowski, Jan Fabianowski, Robert Ibach

On-line PID Controller Tuning Using Genetic Algorithm and DSP PC Board 2087
Pawel Fabijanski, Ryszard Lagoda

Regulation Properties of Pumping Station Control System In The Highest Efficiency Range 2091
Szychta Leszek

xix

Table of Contents

Inner Gas Pressure Measurement Based Life-span Estimation of Electrolytic Capacitors.........................2096
A. Riz, D. Fodor, O. Klug, Z. Karaffy

Robust Control Methodologies for Optical Micro Electro Mechanical System - New approaches and Comparison 2102
Alireza Izadbakhsh, S.M.R. Rafiei

Modeling a Buck-Based Switching Amplifier for Sinusoid Wide Band Tracking by Using a Nonlinear Time Varying Map 2108
A. El Aroudi, E. Alarcón, E. Rodriguez, R. Leyva

Single Inductor Multiple Outputs Interleaved Converters Operating in CCM...2115
Luis Benadero, Vanessa Moreno-Font, Abdelali El Aroudi, Roberto Giral

Control of a two-cell dc/dc converter in presence of saturating duty cycle 2120
Moez Feki, Abdelali El Aroudi, Bruno Gerard Michel Robert, Nabil Derbel

Bifurcations and Chaotic Dynamics in a Linear Switched Reluctance Motor 2126
M.R. De Castro, B.G.M. Robert, C. Goeldel

Modular Architecture for Decentralized Hybrid Power Systems 2134
E. Ortjohann, M. Lingemann, O.Omari, A. Schmelter, N. Hmasic, A. Mohd, W. Sinsukthavorn, D. Morton

Design of a power management system for an active PV station including various storage technologies 2142
Di Lu, Tao Zhou, Hicham Fakham, Bruno Francois

Energy Management and Power Flow of Decoupled Generation System for Power Conditioning of Renewable Energy Sources...2150
Wlodzimierz Koczara, Zdzislaw Chlodnicki, Nazar Al-Khayat, Neil L.Brown

Inversion Based Control of a Diesel Fed Low Temperature Fuel Cell System 2156
Daniela Chrenko, Marie-Cecile Pera, Daniel Hissel

Power Management in an Autonomous Adjustable Speed Large Power Diesel Gensets 2164
Grzegorz Iwanski, Wlodzimierz Koczara

Cost evaluation of Generator-set with Energy Storage for 4Q-load 2170
Freek J.F.Baalbergen, Pavol Bauer

Integrating renewable energy sources and storage into isolated diesel generator supplied electric power systems 2178
Chad Abbey, Jonathan Robinson, Géza Joós

Performance comparison of different wind generator based hybrid systems 2184
Vincent Courtecuisse, Benoit Robyns, Marc Petit, Bruno Francois, Jacques Deuse

First Approach for a Fault Tolerant Power Converter Interface for Multi-Stack PEM Fuel Cell Generator in Transportation Systems 2192
Alexandre De Bernardinis, Gérard Coquery

Development of Electrical System for Hybrid Vehicles Using the Free-swinging Piston Engine and Oscillating Rotating Generator 2200
Sigitas Kudarauskas

Power flow control in different time scales for a wind/hydrogen/super-capacitors based active hybrid power system 2205
ZHOU Tao, LU Di, FAKHAM Hicham, FRANCOIS Bruno

Neuro-Fuzzy Adaptive Control of the IM Drive with Elastic Coupling 2211
Teresa Orlowska-Kowalska, Krzysztof Szabat, Mateusz Dybkowski

Control of Flexible Drive with PMSM employing Forced Dynamics...2219
Vittek Ján, Bris Peter, Makys Pavol, Stulrajter Marek, Vavrus Vladimír

Table of Contents

The problems of high dynamic drive control under circumstances of elastic transmission 2227
Jan Deskur, Roman Muszynski

Protective Predictive Control of Electrical Drives with Elastic Transmission 2235
Mario Vasak, Nedjeljko Peric

Low-Cost High-Performance Predictive Control of Drive Systems with Elastic Coupling 2241
Marcin Cychowski, Kieran Delaney, Krzysztof Szabat

Development of an Expert System for Identification, Commissioning and Monitoring of Drives 2248
Mario Pacas, Sebastian Villwock

Control of Axial Flux Permanent Magnet Motor by the PIPCRM Method at Standstill and at Low Speed 2254
Janusz Wisniewski, Wlodzimierz Koczara

Zero Speed Position Estimation of a Matrix Converter Fed AC PM Machine using PWM Excitation 2261
Q. Gao, G. M. Asher, M. Sumner

Sensorless Direct Torque and Flux Control of an IPM Synchronous Motor at Low Speed and Standstill 2269
Gilbert Foo, S. Sayeef, M.F. Rahman

Sensorless Control of PM Synchronous Motors Using a Predictive Current Controller with Integrated INFORM and EMF Evaluation 2275
Manfred Schrödl, Christian Simetzberger

Torque Sensorless Control of Induction Motor 2283
Karel Jezernik, Miran Rodic

Application of the induction motor torque - observer to the control of turbo - machines 2289
Andrzej Debowski, Daniel Lewandowski

Observer of induction motor speed based on exact disturbance model 2294
Zbigniew Krzeminski

Experimental Performance Evaluation for Low Speed and Regenerating Operation of Sensor-less Vector Control System of Induction Motor Using Observer Gain Tuning 2300
Kazuhiro Ohyama, Greg Asher, Mark Sumner

Application of the Stator Current-based MRAS Speed Estimator in the Sensorless Induction Motor Drive 2306
Mateusz Dybkowski, Teresa Orlowska-Kowalska

State and Parameter Estimation in Induction Motors using Sliding Modes 2312
Sachit Rao, Martin Buss, Vadim Utkin

Torque Transient Alleviation in Fixed Speed Wind Generators by Indirect Torque Control with STATCOM 2318
Marta Molinas, Jon Are Suul and Tore Undeland

Flicker Study on Variable Speed Wind Turbines with Permanent Magnet Synchronous Generator 2325
Weihao Hu, Zhe Chen, Yue Wang, Zhaoan Wang

Power Output Characteristics Analysis of Wind Energy Converter Control Methods 2331
Bingchang Ni, Constantinos Sourkounis

A Cooperative Control Method for Output Power Smoothing and Hydrogen Production by Using Variable Speed Wind Generator 2337
Rion Takahashi, Hirotaka Kinoshita, Toshiaki Murata, Junji Tamura Masatoshi Sugimasa, Akiyoshi Komura, Motoo Futami, Masaya Ichinose, Kazumasa Ide

A new interconnecting method for wind turbine/generators in a wind farm and basic characteristics of the integrated system 2343
Shoji Nishikata, Fujio Tatsuta

Educational aspects of mechatronic control course design for collaborative remote laboratory 2349
Andreja Rojko, Darko Hercog, Karel Jezernik

Table of Contents

PEMCWebLab - Distance and Virtual Laboratories in Electrical Engineering: Development and Trends...............2354
Pavol Bauer, Viliam Fedák, Otto Rompelman

Integrated multimedia educational program of a DC servo system for distant learning...............2360
Gabor Sziebig, Istvan Nagy, Rafael Kalman Jardan, Peter Korondi

Electromechanical Actuators WEB-lab...............2368
Dusan Maga, Jan Sitar, Juraj Dudak, Rene Hartansky, Peter Siroky, Jan Halgos, Pavol Bauer

Power Quality and Active Filters as Web-Controlled Experiment in the frame of PEMC WebLab...............2371
Volker Staudt, Andreas Steimel, Pavol Bauer, Vítezslav Hájek

Distant learning of Pulse Width Modulation Techniques for Voltage Source Converters...............2378
Bartlomiej Kamiski, Dariusz Sobczuk

Modern design optimisation exploiting field simulation...............2383
Jan K. Sykulski

Transmission-Line Modelling of Wave Propagation Effects in Machine Windings...............2385
Herbert De Gersem, Olaf Henze, Thomas Weiland, Andreas Binder

An efficient field-circuit coupling method by a dynamic lumped parameter reduction of the FE model...............2393
F. Henrotte, E. Lange, K. Hameyer

Coupled field-circuit-mechanical model of an electromagnetic actuator operating in error actuated control system...............2400
Lech Nowak

Simulation and Investigation of Magnetorheological Fluid Brake...............2406
Wieslaw Lyskawinski, Wojciech Szelag, Cezary Jedryczka

Field and Field-Circuit Description of Electrical Machines...............2412
Andrzej Demenko, Kay Hameyer

Interaction between Thermal Impedance and Parasitics in Power Sections...............2420
Stefan Forster, Andreas Lindemann

Discussion of Internal and External High Frequency Common Mode Noise Current on a Chopper Circuit...............2428
Tetsuya Mitani, Keiji Wada, Toshihisa Shimizu, Hiromichi Ohashi

A Novel Digital Control Method for DC-DC Converter...............2434
Fujio Kurokawa, Masashi Okamatsu, Yuichi Sumida, Yasuhiro Mimura, Masahiro Sasaki

A Novel Single/Three-phase Matrix Converter For High Power Integration...............2439
Makoto Saito

An Effective Design Method for High Power Density Converters...............2445
Yusuke Hayashi, Kazuto Takao, Toshihisa Shimizu, Hiromichi Ohashi

Power Devices in Polish National Silicon Carbide Program...............2452
Mariusz Sochacki, Andrzej Kubiak, Zbigniew Lisik, Jan Szmidt

SiC Power Semiconductor Devices for new Applications in Power Electronics...............2457
Dominique Planson, Dominique Tournier, Pascal Bevilacqua, Nicolas Dheilly, Herve Morel, Christophe Raynaud, Mihai Lazar, Dominique Bergogne, Bruno Allard, Jean-Pierre Chante

Silicon carbide Schottky diodes and MOSFETs: solutions to performance problems...............2464
Owen J. Guy, Michal Lodzinski, Ambroise Castaing, P. M. Igic, Amador Perez-Tomas, Michael R. Jennings, Philip A. Mawby

Characterization of the Static and Dynamic Behavior of a SiC BJT...............2472
M.M.R. Ahmed, N-A.Parker-Allotey, P.A. Mawby, Muhammed Nawaz, Carina Zaring

An active network control method using distributed energy resources in microgrids...............2478
Takayuki Tanabe, Yoshinobu Ueda, Toshihisa Funabashi, Shigeo Numata, Kimio Morino, Eisuke Shimoda

Table of Contents

Energy Management in Solar Photovoltaic Plants based on ESS 2481
M. Lafoz, L. García-Tabarés, M. Blanco

A Method of Three-Phase Balancing in Microgrid by Photovoltaic Generation Systems 2487
Masahide Hojo, Yuta Iwase, Toshihisa Funabashi, Yoshinobu Ueda

**Development of HILS(Hardware In-Loop Simulation) System for MMS(Microgrid Management System)
by using RTDS** 2492
Jin-Hong Jeon, Jong-Yul Kim, Seul-Ki Kim, ong-Bo Ahn, JuneHo Park

Power Quality Analysis of Jeju Island Power System with Wind Farm and HVDC System 2498
Jae-Hong Kim, Eel-Hwan Kim, Se-Ho Kim, Jaeho Choi, Gil-Soo Jang, Seung-Ho Song

A New Control Method for Power Turbine Generators Using an Accurate Ship Plant System Model 2504
Nobumasa Matsui, Fujio Kurokawa, Keiichi Shiraishi

Voltage profile support in distribution networks - influence of the network R/X ratio 2510
B. Bla~ic, I. Papic

**Modeling, Simulation and Analysis of Conducted Common-Mode EMI in Matrix Converters for Wind
Turbine Generators** 2516
S. Zhang, K.J. Tseng

Design of Frequency Shift Acceleration Contol for Anti-islanding of an Inverter-based DG 2524
Seul-Ki Kim, Jin-Hong Jeon, Heung-Kwan Choi, Jonng-Bo Ahn

Integrated Power Converter for Photovoltaic and Fuel Cell Systems in Home 2530
Yasuyuki Nishida, Shinichiro Sumiyoshi, Hideki Omori

A Comparison of Position Control Structures for Ironless Linear Synchronous Motor 2538
Martin Hrasko, Pavol Makys, Marek Franko, Jozef Kuchta

A Comparison of Sliding Mode Approaches to a Nanometre Position Control Application 2543
Paul Andreas Stadler, Stephen James Dodds

Sliding Mode Control of PMSM Drives Subject to Torsion Oscillations in the Mechanical Load 2551
Stephen J. Dodds, Jan Vittek

Sliding Mode Vector Control of PMSM Drives with Minimum Energy Position Following 2559
Stephen J. Dodds

Performance Evaluation of Electric Power Steering with IPM Motor and Drive System

Hamidreza Akhondi[†]   Jafar Milimonfared   Kaveh Malekian

Amirkabir University of Technology (Tehran Polytechnic),
Center of Excellence in Power Systems, No. 424, Hafez Ave., Tehran 15914, Iran.

† Corresponding author: akhondi@aut.ac.ir

Abstract—As the development of microprocessors, power electronic converters and electric motor drives, electric power steering system which uses an electric motor came to use a few year ago. Electric power steering systems have many advantages over traditional hydraulic power steering systems in engine efficiency, space efficiency, and environmental compatibility. This paper deals with design and optimization of an interior permanent magnet motor for power steering application. Direct search method is used for optimization. After optimization and finding motor parameters, an interior permanent magnet motor and drive with mechanical parts of electric power steering system are simulated and performance evaluation of system is done.

Index Terms—Electric power steering, interior permanent magnet motor, optimization.

I. INTRODUCTION

ELECTRIC power steering (EPS) in newer vehicles is becoming an alternative to the hydraulic power steering (HPS) because of the recent advances in electrical motors, power converters , sensors and digital control systems [1]. In the EPS, an electric motor is connected to the steering rack via a gear mechanism. Some sensors measure the torque on the steer and the angular position of the hand wheel. A control system receives these signals from the sensors, together with vehicle speed and turning rate, and gives operating commands to the electric motor drive, controlling steering direction and dynamics and driver effort. The control unit determines both the amount of steering assist required, which has to be also modified according to vehicle speed to maintain good steering feel.

Electric power steering is an advanced steering system which eliminates the need for a hydraulic power steering pump, hoses, hydraulic fluid, drive belt and pulley on the engine, therefore the total system is lighter than a comparable hydraulic system through the use of compact system units. Also, since EPS is an on-demand system that operates only when the steering wheel is turned, the fuel efficiency of vehicle equipped with such system is up to 3% better than that of automotive equipped with an equivalent-output hydraulic system [2]. As a result, Electric power steering systems have many advantages over traditional hydraulic power steering

systems in engine efficiency, space efficiency, and environmental compatibility. This motivates the great increase of EPS-equipped automotive recently [2].

Electric power steering basically consists of a torque sensor and motor actuator couple. The sensor is attached to the steering column and measures the torque applied by the driver when he moves the steering wheel. This torque signal is transmitted to a control power card that sends an amplified proportional power signal to the electric motor (in this paper we use interior permanent magnet (IPM) synchronous motor), which is engaged to the steering rack bar.

An EPS system has the following two functions. First, it can reduce steering torque and present various steering feels. The steering torque (or driver torque) is defined as the one which a driver experiences (or a driver applies to the steering column) when turning the steering wheel. When an appropriate assist torque from an EPS system is applied in the same direction as the driver's steering direction, the amount of steering torque required by a driver for steering can be significantly relieved. In addition, adjustment of the characteristics of assist torque allows the driver to experience various steering feels.

Second, the EPS system can improve return to center performance of a steering wheel when it is steered. While the steering wheel is turned and then released during cornering, it returns to the center position by the so-called self-aligning torque exerted on the tires by the road. Since this torque increases with vehicle speed, at high vehicle speeds the steering wheel may exhibit excessive overshoot and subsequent oscillation. The EPS system can eliminate this phenomenon by providing active damping capability and thus enhance return ability characteristics.

This paper presents an Electric Power Steering system using IPM motor and drive systems which are widely being applied in automotive applications. Due to factors such as high power density and efficiency, maintenance, and extremely wide operating speed range, permanent magnet synchronous motors (PMSM) are the subject of development for traction drive applications [3,4].

Optimization with Direct Search [5] method is done on the IPM motor structure considering the EPS requirements and constraints to obtain the motor parameters. So the paper deals

978-1-4244-1741-4/08/$25.00 ©2008 IEEE

with the design and performance evaluation of an IPM motor for EPS system. The components of system such IPMSM electrical and mechanical parts, power electronic converter, steering mechanism and controller are integrated as entire model of EPS using Simulink environment for analyzing the system performance with interactions between each component. The block diagram of EPS with IPMSM drive system is shown in Fig. 1.

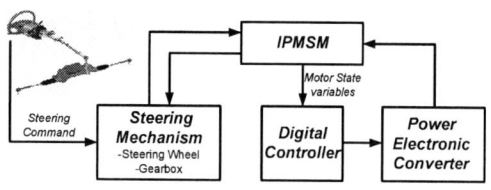

Fig. 1. Block Diagram of Overall EPS System.

II. OPTIMIZATION OF THE IPM MOTOR PARAMETERS

The IPM motor presents many advantages over the other motors. Among them, it exhibits high torque density, yielding a minimum size and weight, good power to weight ratio as demanded in automotive applications [6] and a high efficiency, even under reduced loads. In addition, its manufacture is easy, because the PMs are simply introduced in suitable holes in the rotor and the motor has capability to operate in flux weakening region. So for the IPMSM, the magnetic circuit has been fully designed using the optimization from analytic and finite-element based software.

In this design, the objectives to be reached are the reduction of the volume and total weight as well as the reduction of the size of the magnets to decrease the cost [7]. An easiness of the manufacturing process must be kept in mind for a future industrial application. Other objectives as the dynamic behavior and the torque ripple will be examined in the future with the power electronic and control interactions.

Taking into account the important number of design variables, an optimization under constraints is chosen. Variables are classified into discrete and continuous one. If discrete variables are fixed (for example number of stator slots) a non-linear mathematic algorithm can be used to optimize the machine structure with geometrical constraints. A number of optimization variables noted U are selected in order to find optimal values noted U* as shown below.

U* minimizes an objective function F and verifies the feasibility domain under constraints:

- Minimize $F(U)$
  $$U^* \in U$$
- Subject to
  $$H_i(U^*) = 0$$
  $$G_i(U^*) \geq 0$$
  $$U_{\ell i} \leq U_i \leq U_{ui}$$

U* must permit to reach the desired goal with the minimization criterion. It must also verify the equality and inequality constraints while keeping in the range of allowed values. For example, if the power to weight ratio has to be limited, it is necessary to choose:

$U \rightarrow$ motor parameter (magnet flux linkage & dimensions)
$F(U) \rightarrow$ power to weight ratio
$G_1(U) \rightarrow$ external diameter $\leq D_{max}$
...
$G_i(U) \rightarrow$ motor torque $\geq T_{min}$

The method is based on an analysis and optimization parts. The analysis part uses the parametric model with the variables U to calculate the energetic values of the machine according to design specifications. The analysis part treats three domains (Fig.2). The magnetic domain is the first and the central one because it is coupled with the two others. It allows the evaluation of inductance and back-*emf*. This brings to electromechanical performances. The thermal domain gives temperature of magnets and copper to estimate flux density and resistances. The electrical domain gives the relationship between the current reference and the real current according to the voltage limit.

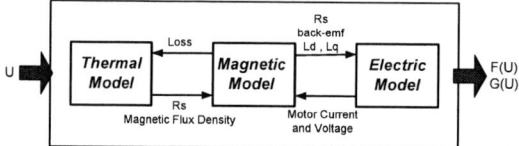

Fig. 2. Analysis module in optimization process.

The optimization part manages the variables U on the basis of F(U) and G(U) information given by the analysis part. In this paper the Direct Search Method is used in optimization part. Because the search domain is restricted with EPS requirements and constraints. The final machine parameters after optimization with this method is given in table 1.

III. MODELING OF EPS WITH IPM MOTOR AND DRIVE

Typical EPS system is shown in Fig. 3. The major components are a torque sensor, an electric motor, a reduction gear and control unit. Torque sensor is located between steering wheel and steering column and measures the applied torque by converting difference of twisted angle to electric signal. The electric motor is attached in steering column through reduction gear box. Control unit calculates motor target torque and current from the signal of torque sensor and vehicle velocity. So the calculated torque is applied to steering column by motor through reduction gear.

There are four different type of EPS, column, pinion and rack-assisted or a fully steer by wire type [8]. In this paper, a

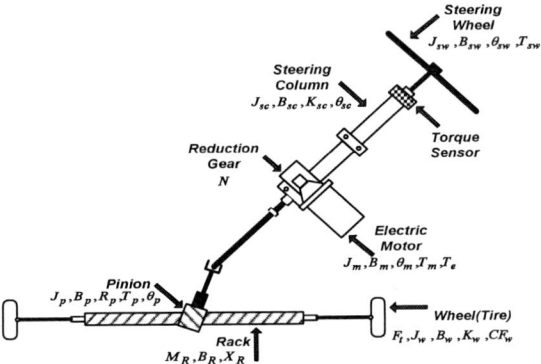

Fig. 3. Simple description of EPS Mechanism.

column-assist-type EPS shown in Fig. 3 is used for modeling and simulation. The equilibrium equations of steering wheel, pinion, rack and motor are:

$$J_{sw}\ddot{\theta}_{sw}+B_{sw}\dot{\theta}_{sw}+K_{sc}(\theta_{sw}-\theta_{sc})=T_{sw} \quad (1)$$

$$J_p\ddot{\theta}_p+B_p\dot{\theta}_p+K_{sc}(\theta_p-\theta_{sw})=N\,T_m-T_p \quad (2)$$

$$M_R\ddot{X}_R+B_R\dot{X}_R+F_t=\frac{T_p}{R_p}\quad,\quad X_R=R_p\cdot\theta_p \quad (3)$$

$$J_m\ddot{\theta}_m+B_m\dot{\theta}_m+T_m=T_e\quad,\quad \theta_m=N\cdot\theta_p \quad (4)$$

Once the Equations (3) and (4) are rearranged about θ_P and substitute it to Equation (2), we can get new equations as below:

$$(J_p+R_p^2 M_R+N^2 J_m)\ddot{\theta}_p+$$
$$(B_p+R_p^2 B_R+N^2 B_m)\dot{\theta}_p=K_{sc}(\theta_{sw}-\theta_p)+T_e-R_p F_t \quad (5)$$

If $J_{eq}=J_p+R_p^2 M_R+N^2 J_m$, $B_{eq}=B_p+R_p^2 B_R+N^2 B_m$ then we have below equation:

$$J_{eq}\ddot{\theta}_p+B_{eq}\dot{\theta}_p=K_{sc}(\theta_{sw}-\theta_p)+T_e-R_p F_t \quad (6)$$

The equation of wheel and tire loads is the same as:

$$F_t=J_w\ddot{X}_R+B_w\dot{X}_R+K_w X_R+CF_w\,sign(\dot{X}_R) \quad (7)$$

where CF_w is Coulomb friction breakout force on road wheel. Thus final equation is expressed as:

$$(J_p+R_p^2(M_R+J_w)+N^2 J_m)\ddot{\theta}_p+$$
$$(B_p+R_p^2(B_R+B_w)+N^2 B_m)\dot{\theta}_p+ \quad (8)$$
$$R_p^2 K_w\,\theta_p+R_p CF_w\,sign(\dot{\theta}_p)=K_{sc}(\theta_{sw}-\theta_p)+T_e$$

Using these equations, the mechanical part of EPS which incorporates the mechanical part of PMSM is modeled. The block diagram of EPS system with IPM motor and drive is shown in Fig. 4. In this block diagram the motor parameters is obtained from design and optimization procedure that mentioned in previous section. The voltage source inverter is constructed with IGBT for more accurate simulation. A current phase control technique with simple PI controller is used for simulation including a PWM module. SVPWM model is implemented by unified voltage modulation techniques [9] And a dead time function is implemented using delay block. The control strategy which is applied to IPM motor is maximum torque per ampere up to rated speed and maximum torque per voltage (flux) over the rated speed.

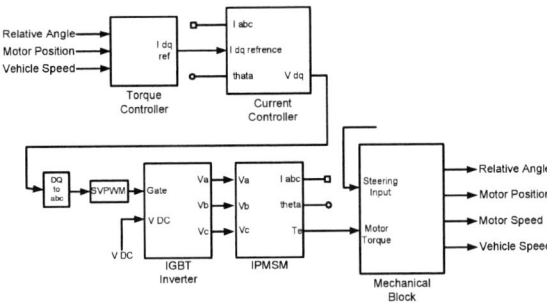

Fig. 4. A model of EPS with IPMSM Drive System.

IV. SYSTEM SIMULATION AND RESULT

Table I shows the parameters of IPMSM for EPS system simulation. These parameters are obtained by the information of motor structure, dimension and material that obtained in optimization procedure and FEM (Finite Element Method) analysis is done to obtain parameters. Stator resistance is depends on winding turns and material.

In simulation of electric power steering system with mechanical parts, the torque controller is simple PI controller which uses relative angle (difference between steering angle and rotor angle), motor position and vehicle speed to generate the reference value of q-axis current. d-axis current is generated from q-axis current and maximum torque per ampere or maximum torque per flux (voltage). The current controller uses this reference values to generate d-q reference voltages. Using these reference voltages, switching signals for IGBT inverter is constructed and system can control the IPM motor

TABLE I
MOTOR PARAMETERS OBTAINED FROM OPTIMIZATION

Motor Parameter		VALUE
Magnet Flux Linkage	(Wb)	0.105
Stator Resistance	(Ω)	0.02
d-Axis Inductance	(mH)	8.6
q-Axis Inductance	(mH)	22.5

torque in order to produce needed assisted torque. DC voltage link in inverter is 12V as in vehicle.

In the following simulation results, the goal is to control the rack displacement. Fig. 5 shows the results when rack displacement of 20mm is desired driver steering torque of 1.5 N.m is applied to steering wheel. It is obvious that motor toque is applied to rack after passing through the gear.

Fig. 6 shows the results when we want to evaluate the system return ability. In this test the rack displacement is commanded to reach 20mm and then return to its primary position. In all of the simulations in column-assist-type EPS the rack displacement and motor angular displacement have the same shape.

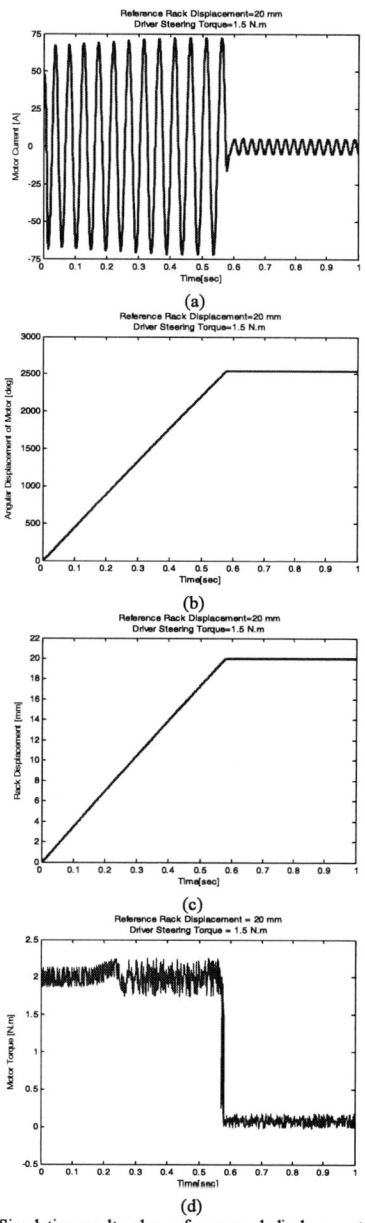

Fig. 5. Simulation results when reference rack displacement is 20 mm and driver steering torque is 1.5 N.m. (a) Motor Current. (b) Motor angular displacement (c) Rack displacement. (d) Motor torque.

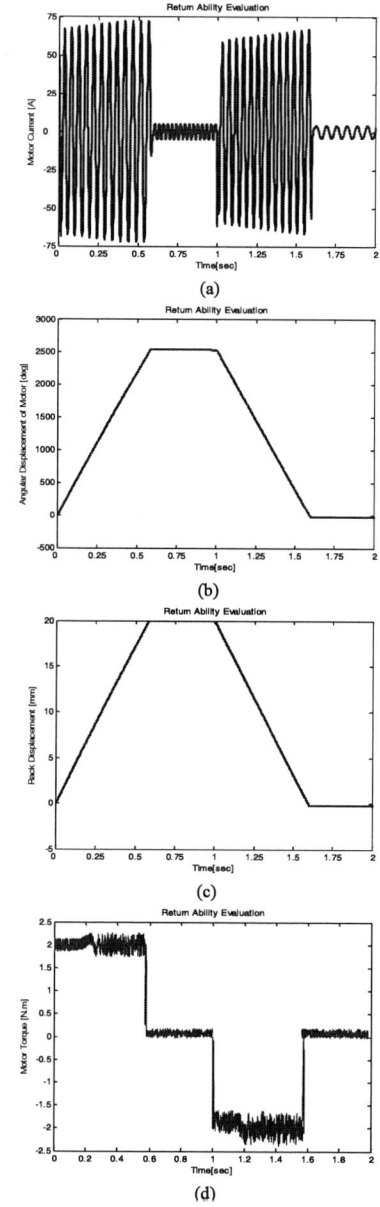

Fig. 6. Evaluation of return ability of rack in presence of motor torque (a) Motor Current. (b) Motor angular displacement (c) Rack displacement. (d) Motor torque.

Fig. 7 is the simulation results left and right movement of rack is desired. The rack moves right up to 20mm, return back, moves left up to 20mm and finally back to its first position. We can notify that this modeling of EPS system facsimiles the characteristics and behavior of system very similar to which we desire and command.

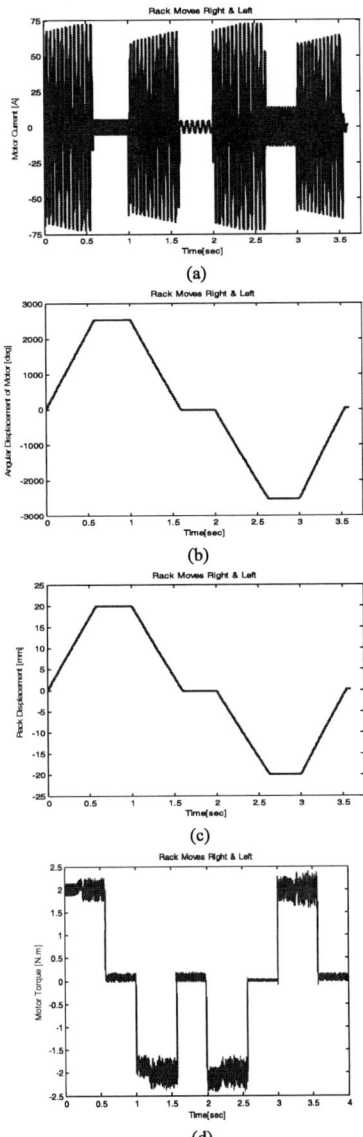

(a)

(b)

(c)

(d)

Fig. 7. Rack moves right, return back, moves left and finally return to primary position: (a) Motor Current. (b) Motor angular displacement (c) Rack displacement (d) Motor torque.

V. CONCLUSION

In this paper, design and optimization of interior permanent magnet motor for electric power steering application is studied. IPM motors advantages in automotive systems are discussed. Optimization is done with different objective function such as power to weight and magnet volume. After obtaining the motor parameters, we evaluate the performance of entire system through some different tests. Simulation results show that the system performance can improved using an IPM motor and drive because of its high torque density and capability of flux weakening operation. With the EPS logic for the reduction in steering torque, the driver can turn the steering wheel with a significantly reduced steering torque. With the EPS control of return to center performance, a quick response of steering wheel without overshoot after cornering can be obtained by proper control of assist motor.

REFERENCES

[1] S. Mir, M. Islam, and T. Sebastian, "Role of electronics and controls in steering system," in *Proc. 29th IEEE IECON*, Roanoke, VA, pp. 2859–2864, Nov. 2–6, 2003.

[2] Y. Gene Liao ,H. Isaac Du, "Modeling and analysis of electric power steering system and its effect on vehicle dynamic behavior", *Int. J. of Vehicle Automotive System (IJVAS)*, vol. 1, no. 2, pp. 153-166, 2003.

[3] Dave Wilson, "Electric power steering: one good turn deserves another", "Bush: No quick energy fix" *Associated Press article*, The Arizona Republic, April 28, 2005.

[4] T. Jahns, G. Kliman, and T. Neumann, "Interior pm synchronous motors for adjustable speed drives," *IEEE Trans. Ind. Appl.*, vol. IA-22, no. 4, pp. 738–747, Jul./Aug. 1986.

[5] D.T.Pham, and D. Karaboga, *"Intellligent Optimisation Technique"*, Springer-Verlag London Limited, 2000.

[6] L. Chédot, G. Friedrich, "Optimal control of interior permanent magnet synchronous integrated starter-generator," *Eur. Power Elec. Drives Conf. (EPE'2003)*, CD Proceedings, Toulouse (France), 2003.

[7] T. Sebastian, S. Mir, M. Islam, "Electric Motors for Automotive Applications," *EPE Journal*, vol. 14, no. 1, February 2004.

[8] Hooman Mohammadi and Reza Kazemi, "Simulation of Different Types of Electric Power Assisted Streering (EPS) to Inverstigate Applied Torque Positions' Effects", *SAE*, paper No. 2003-01-0585, 2003.

[9] N. Mohan and et al., *"Power Electronics and Variable Frequency Drives"*, IEEE Press, pp. 400-453, 1997.

Optimal Control: Load Frequency Control of a Large Power System

Sílvio José Pinto Simões Mariano[*], Luís António Fialho Marcelino Ferreira[†]

[*]University of Beira Interior/Department of Electromechanical Engineering, Covilhã, Portugal, e_mail: sm@ubi.pt
[†]Instituto Superior Técnico, Energy Section, Lisbon, Portugal, e-mail: lmf@ist.utl.pt

Abstract—This paper addresses the stabilization and performance of the load frequency regulator. The problem is solved by using the theory of the optimal control. An algorithm, based on the new technique, proposed by the authors in [1], to overcome the difficulties of specifying the weighting matrices Q and R, is presented. The algorithm here proposed considers the multi-area electric energy system. The paper presents illustrative, numerical simulation results for this model. The results indicate that the obtained controller exhibits better performance then those based on classic control.

Keywords—Generation of electrical energy, Optimal control, Control methods for electrical systems.

I. INTRODUCTION

In recent years the electric power industry has experiencing significant changes towards deregulation and competition. One of the scenarios is a competitive, fully deregulated market, with centralized system operation (pool operation). A power pool is an interconnection of the power systems of individual utilities. Each company operates independently within its own jurisdiction, but there are contractual agreements about intercompany exchange of power through the tie-lines and others agreements about operating procedures to maintain system frequency. The basic principle of pool operation is that in normal steady state, (1) scheduled interchanges of tie-line power are maintained and (2) each area absorbs its own load changes [2, 5, 6, 7].

Application of the optimal control theory to power systems has shown that an optimal load frequency controller can improve the dynamic stability of a power system [4]. However, some difficulties to apply this technique still remain, mainly due to the large system dimension, because the nature of multi-area interconnected power systems, and to the specification of matrices Q and R.

The problem of selecting the weighting matrices Q and R has shown up in optimal control design [3,4]. Assigning appropriate values to the matrices Q and R in a systematic manner is a major concern and is the main point addressed in this paper.

This paper develops a new centralized control algorithm that is based upon the authors previous results [1]. For the purpose of this paper, our interest in the optimal centralized state feedback control solution is for carrying out the specification of the weighting matrices Q

and R and for benchmark purpose. First we shall develop an effective optimal centralized state feedback controller, and then, as future work, seek controllers of the more practicable decentralized structures whose performance are such that they result in costs (as measured by the performance index) and performance nearly as small as that caused by this optimal centralized controller.

II. BACKGROUND

To design a linear optimal control, a linearized model of the power system is sought. The linear model of the system is described in state space form, as

$$\dot{x} = A\,x + B\,u \qquad x \in \Re^{n_x},\ u \in \Re^{n_u},\ y \in \Re^{n_y}, \qquad (1)$$
$$y = C\,x$$

where A and B are system matrices, x the state vector, u the control signal vector, y the output vector and n_x, n_u and n_y are the number of state, control and output variables, respectively.

Consider first that the complete state x is available for feedback. Then it is possible to allocate the set of closed-loop system poles to an arbitrary position by means of a suitable linear state feedback. Thus the control law is

$$u = -K\,x, \qquad (2)$$

where k is the feedback gain matrix.

When the number of output variables n_y is less than the order of the system n_x, then it is always possible to allocate n_y poles of the closed-loop system. Unfortunately, we can not assign the location of the remaining $(n_x - n_y)$ poles of the closed-loop system [3].

As the location may be arbitrary, the first step in pole assignment is to decide on their location. To choose the location of the poles, one should bear in mind that the required physical effort is dependent on the distance to the imaginary axis. Particularly, zeros in open loop attract poles, being very difficult to remove the poles near zeros. Consequently, for an arbitrary location, the control vector could lead to physically or economically undesired action. These considerations lead to the linear optimal regulator problem.

For the linear system described in (1), the optimal control signal u that minimizes the performance index

$$J = \int_0^\infty \left[x^T Q\,x + u^T R\,u \right] \mathrm{d}t \qquad (3)$$

978-1-4244-1741-4/08/$25.00 ©2008 IEEE

is a linear function in terms of the system state variables x as [2, 4]

$$u = -Kx = -R^{-1}B^T Px , \qquad (4)$$

where Q and R are the weighting matrices and P is the solution of the matrix Riccati equation

$$A^T P + PA - PBR^{-1}B^T P + Q = 0 . \qquad (5)$$

Riccati equation is the key to obtain the optimal control. Once the matrices Q and R are known, the matrix P can be obtained by solving the Riccati equation. From equation (4) one gets the optimal control vector u.

So far it was assumed that the complete state x is available for feedback. That assumption is unrealistic for the most of the systems. Usually, only a reduced number of state variables, or a linear combination thereof, are available — the so-called output vector y. In that case, it is necessary to recur to output feedback.

Therefore, suppose that the control vector must be directly obtained from the system output [8]:

$$u = -Fy , \qquad (6)$$

where F is the feedback gain matrix. The linear model of system (1) can be rewritten as

$$\dot{x} = [A - BFC]x . \qquad (7)$$

Let $A^* \triangleq [A - BFC]$. The F matrix, for which the control vector u is optimal, minimizes the performance index (3):

$$F = R^{-1}B^T K^* L^* C^T \left[CL^* C \right]^{-1} . \qquad (8)$$

The matrix K^* is a positive semi-definite solution of

$$K^* A^* + A^{*T} K^* + Q + C^T F^T RFC = 0 \qquad (9)$$

and L^* is a positive definite solution of

$$L^* A^{*T} + A^* L^* + I = 0 . \qquad (10)$$

In the special case that C^{-1} exists, the result is identical to that obtained with the optimal state feedback, and (8) reduces to

$$F = R^{-1}B^T K^* C^{-1} \qquad (11)$$

and (9) becomes

$$K^* A + A^T K^* + Q - K^* BR^{-1}B^T K^* = 0 . \qquad (12)$$

To obtain the optimal gain matrix F, it will be necessary to apply an iterative method. The following computer algorithm can be used [8].

Let

$$F_{n-1} = R^{-1}B^T K_{n-1} L_{n-1} C^T \left[CL_{n-1} C^T \right]^{-1} , \qquad (13)$$

where K_n is solution of

$$K_n [A - BF_{n-1}C] + [A - BF_{n-1}C]^T K_n + Q + \\ + C^T F_{n-1}^T RF_{n-1}C = 0 \qquad (14)$$

and L_{n-1} is solution of

$$L_{n-1}[A - BF_{n-1}C]^T + [A - BF_{n-1}C]L_{n-1} + I = 0 . \quad (15)$$

Start by guessing the initial value for F_0. Then, by (14) and with $n = 1$, obtain K_1. Substitute K_1 into (13). F_1 is still unknown, but it depends only on L_1. Substituting F_1 into (15) one gets a nonlinear equation in only one unknown, L_1. Equation (15) is then solved for a positive definite L_1. This value of L_1 is substituted into (13) to give F_1. This completes the first iteration. The value of F_1 initiates the next iteration [8].

The foregoing has been a brief review of optimal control theory as it applies to the optimal load frequency regulator problem. However, in order to design the load frequency regulator, a major problem still remains: how to select the matrices Q and R. Next we present a technique for specifying the matrices Q and R in the context of optimal state feedback.

III. PROPOSED APPROACH

In this paper, without loss of generality, the weighting matrix R is specified as an identity matrix. Thus, the problem then is to specify the weighting matrix Q.

The most common procedure to specify the matrix Q is search by trial and error. Different matrices Q_i result in different gain matrices K_i, which in turn result in different dynamic performances for the closed-loop system. There are many possibilities to test by trial and error, which makes this procedure burdensome.

In order to simplify and accelerate this procedure of trial end error, or even avoid it altogether, a new algorithm, based on [1], to select the weighting matrix Q is proposed.

When the location of the systems poles is assigned, and a suitable gain matrix K is computed, then it is possible to judge on the weight of each state variable upon the control vector. An initial weighting matrix Q_0 may then be selected. This matrix plays a decisive role in making the system poles move close to the desired pole locations, as done by conventional pole assignment techniques.

Algorithm

Step 1- Specify desired location for system poles.

Step 2- Compute the corresponding gain matrix K to allocate the poles Λ in the desired location.

Step 3- Let $Q_0 = diag(|K|)$ (i.e. Initialize Q).

Step 4- Vary each value of Q_0, one by one, and compute the eigenvalues of the closed-loop system:

for $i=1 : n$

$\quad Q_i = Q_0$

$\qquad$ for $j=1 : \dfrac{\max \text{ value for } Q_i(i,i)}{\max \text{ value of } Q_0}$

$\qquad\quad Q_i(i,i) = Q_i(i,i) \times j$

$\qquad\quad$ compute Λ_{ij}

$\qquad$ end

end

Step 5- Retrieve Q for the desired pole location.

Notes:

The more negative the real parts of the eigenvalues specified in step 1 are, the larger the elements values of Q_0 will be.

Step 4 imposes a maximum value for the elements of the weighting matrix Q, as expressed by max *value for* $Q_i(i,i)$. This prevents feedback gains from becoming too large.

As an example, the value of Q_i is computed ten times with an increment in j of five, for max *value for* $Q_i(i,i) = 100$.

To select the value for $Q_i(i,i)$ (step 5), it is convenient to draw the locus of the eigenvalues.

IV. TEST RESULTS

For an interconnected system with a large number of generating units, modeling the power control system of each individual generator is complex. Although, simpler models of lower dimension give more understandable results. In general, we may identify groups of generators which are closely coupled internally but relatively weakly coupled between groups. The model and its usefulness are enhanced when the boundaries of the coherent groups of generators coincide with those of an operating company or utility. In this case, under a single operating jurisdiction, the generators are frequently controlled as single unit and in such case the groups are called control areas.

To illustrate the proposed algorithm a four different interconnected areas are considered: one area with hydro power plants–area 1, two with fossil power plants, one with nonreheat steam system configuration–area 2 and the other one with tandem compound, single reheat steam system configuration–area 3 and the last area is considered to have infinite kinetic energy–area 4, as shown in Fig. 1.

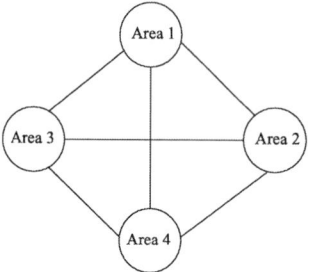

Fig.1 – Case study – four interconnected control areas

This case study can be compared to an interconnection of the power systems of individual utilities. Each company operates independently within its own jurisdiction, but there are contractual agreements about intercompany exchanges of power through the tie-lines and other agreements about operating procedures to maintain system frequency. The adjustment of the command increment, ΔP_{r_i}, is done automatically and according to the contractual agreements.

We have combined the referred *four-area system* and the model was obtained. This model is shown in Fig. 2, using a block diagram representation, and the system is defined by the symbols and parameters shown in Appendix, where $K_p = 1/D_i$ and $T_p = 2H_i/D_i$.

As a result a linear model of the system, in state space form, is obtained (the state matrix has dimension $n_x = n_y = 17$, and the control inputs and load deviation has dimension $n_u = 6$). The state vector is defined as

$$
\begin{aligned}
x = [\ & \Delta P_{G_1} \quad \Delta P_{G_2} \quad \Delta P_{G_3} \vdots \\
& \Delta P_{M_1} \quad \Delta P_{M_2} \quad \Delta P_{T_{31}} \quad \Delta P_{T_{32}} \quad \Delta P_{M_3} \vdots \\
& \Delta F_1 \quad \Delta F_1 \quad \Delta F_1 \vdots \\
& \Delta P_{\text{int},1} \quad \Delta P_{\text{int},2} \quad \Delta P_{\text{int},3} \vdots \\
& \Delta P_{r_1} \quad \Delta P_{r_2} \quad \Delta P_{r_3} \ ]
\end{aligned}
$$

Applying the proposed algorithm, for the parameters given in appendix, we can specify the weighting matrix Q, and the corresponding gain matrix, K:

$$
\begin{aligned}
Q = diag([\ & 8.5 \quad 8.5 \quad 8.5 \vdots 68 \quad 68 \quad 68 \quad 68 \quad 68 \vdots \\
& 85 \quad 85 \quad 85 \vdots 8 \quad 8 \quad 8 \vdots 8 \quad 8 \quad 8 \ ])
\end{aligned}
$$

$$K = \begin{bmatrix} 12.9 & 0.2 & 0.1 \\ 0.2 & 8.3 & -0.1 \\ 0.1 & -0.1 & 5.9 \\ 55.3 & 0.3 & -0.1 & -1.0 & -0.9 \\ 4.8 & 11.3 & -0.5 & -0.8 & 0.5 \\ 1.9 & -0.2 & 8.7 & 6.7 & 1.7 \\ 18 & 0.7 & 1.5 & 0.3 & 0.4 & -1.4 \\ 2.1 & 14.9 & -0.6 & 0.8 & 0.8 & -1.1 \\ 0.85 & -0.11 & 6.9 & 1.0 & 1.0 & 0.8 \\ 1.9 & -0.17 & -0.4 \\ 0.1 & 1.9 & -0.3 \\ 0.5 & 0.4 & 1.9 \end{bmatrix}$$

The corresponding eigenvalues for the system without and with optimal control are shown in Table I.

TABLE I.
SYSTEM EIGENVALUES FOR BOTH CONVENTIONAL AND OPTIMAL CONTROL

System eigenvalues with conventional control	System eigenvalues with optimal control applied
-24.75	-22.64
-10.80	-5.43
-5.28	-3.35
-6.19	-0.84
-3.30	-0.16
-2.72	$-8.10 \pm J5.00$
-1.68	$-7.85 \pm J2.61$
$-0.35 \pm J4.25$	$-5.10 \pm J3.53$
$-0.25 \pm J3.22$	$-2.22 \pm J0.33$
$-0.30 \pm J1.56$	$-2.21 \pm J2.64$
$-0.87 \pm J1.34$	$-1.27 \pm J1.02$
$-0.20 \pm J0.38$	

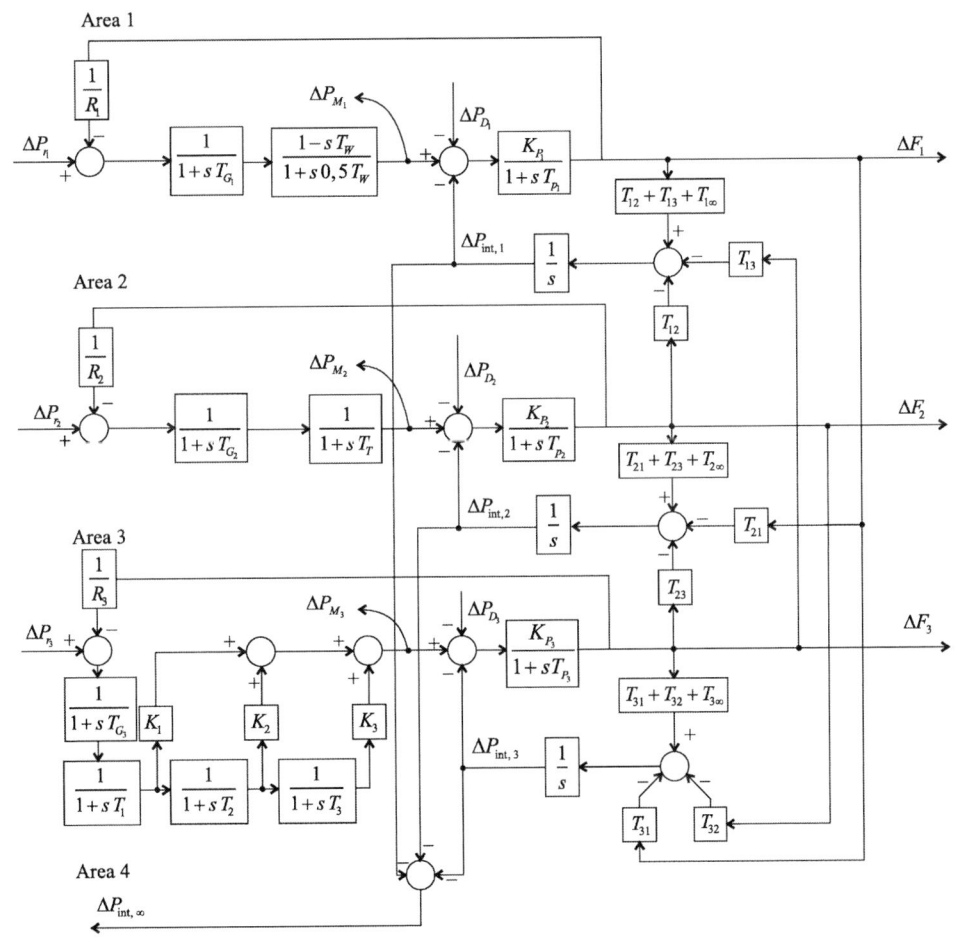

Fig.2 – Block diagram of four-area perturbation model.

A 10% step-load increase ($\Delta P_D = 0.1\,pu\,MW$) is applied to area 1 and subsequent deviations of Δf_1, Δf_2, Δf_3 and $\Delta P_{int,1}$, $\Delta P_{int,2}$, $\Delta P_{int,3}$, $\Delta P_{int,\infty}$ are studied. Fig. 3 and Fig. 4 illustrate the time evolution of the system frequency deviations, for each area, with classic control and with optimal control, respectively. Fig. 5 and Fig. 6 illustrate the time evolution of the tie-line power deviation, for each area, with classic control and with optimal control, respectively.

From these results, along with many others obtained but not discussed herein, some conclusions can be made: (1) the performance of the system with centralized optimal state feedback controller, Fig. 4 and Fig. 6, exhibits better damping than with the conventional control – optimal controllers are especially effective in damping fast the oscillatory tie-line load and frequency deviations; (2) considering the system eignevalues, was expected the increase of the stability margin of optimal solution as Fig. 4 and Fig. 6 shows.

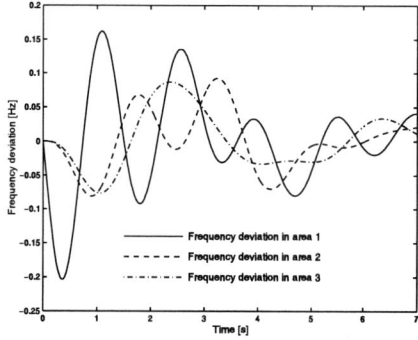

Fig.3 – Step response of system frequency deviations with classic control

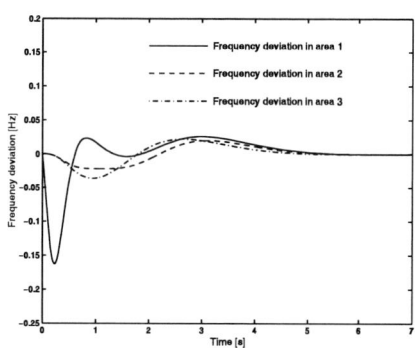

Fig.4 – Step response of system frequency deviations with optimal centralized state feedback controller

The longer term effects, as shown in Fig. 3 and Fig. 4, shows that, since the outside world is infinite (one area considered to have infinite kinetic energy), frequency is essentially back to normal and therefore the powers of each area are back to normal, by increasing the

mechanical power on area 1 – the basic principle of pool operation (scheduled interchanges of tie-line power are maintained and each area absorbs its own load changes) are satisfied. Thus, the interconnection is effective on delivering power to the control area 1, specially with optimal control where the step-load increase in area 1 is satisfied simultaneous and in equal parts, in the first second, by the other areas and, in the next seconds, by the area considered to have infinite kinetic energy.

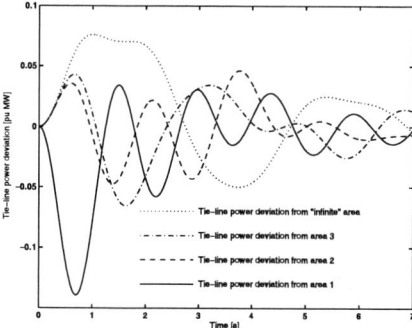

Fig.5 – Step response of system tie-line power deviation with classic control

Also, by examining the optimal gains, it is seen that an area control is much more influenced by the states from its own area, with less significant contribution of the states from other interconnected areas. This is a very promising result to seek controllers with decentralized structures.

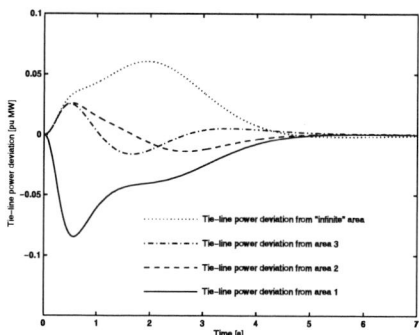

Fig.6 – Step response of system tie-line power deviation with optimal centralized state feedback controller

V. CONCLUSION

The frequency control problem of a large power system (a multi-area electric energy system) was solved through the application of a new technique that allows the specification of the weighting matrices Q and R. As a conclusion, the optimal state feedback, together with the proposed algorithm for specifying Q and R, yields results that are robust, and according to the desired performance.

REFERENCES

[1] S.J.P.S. Mariano, L.A.F.M. Ferreira, *A Procedure to Specify Q and R for an Optimal Voltage Regulator*, Proceedings of the IASTED International Conference, Las Vegas, USA, 1999

[2] L. A. F. M. Ferreira, *Teaching Power Systems Control and Operation Using General-Purpose Software Technologies*, 11th Power Systems Computation Conference, PSCC-93, Avignon, France, Sep 1993.

[3] Milan Calovic, *Linear Regulator Design for a Load and Frequency Control*, IEEE Trans. on PAS, Vol. PAS-91, pp.2271-2285, 1972.

[4] Charles E. Fosha e Olle I. Elgerd, *The Megawatt-Frequency Control Problem: A New Approach Via Optimal Control Theory*, IEEE Trans. on PAS, Vol. PAS-89, No. 4, Abril 1970.

[5] Working Group on Prime Mover and Energy Supply, *Dynamic Models for Fossil Fueled Steam Units in Power System Studies*, IEEE Trans. on Power Systems, Vol. 6, No. 2, Maio 1991.

[6] A report of the AGC Task Force of the IEEE/PES/PSE/System Control Subcommitte, *Understanding Automatic Generation Control*, Trans. on Power Systems, Vol. 7, No. 3, Agosto 1992.

[7] J.A. Rovnak, R. Corlis, *Dynamic Matrix Based Control of Fossil Power Plants*, IEEE Trans. on Energy Conversion, Vol. 6, No. 2, June 1991.

[8] William S. Levine, Michael Athans, "On the Determination of the Optimal Constant Output Feedback Gains for Linear Multivariable Systems", *IEEE Trans. on Automatic Control*, vol. AC-15, Nº 1, February 1970.

APPENDIX

System symbols and parameters

Δf_i frequency deviation in area i

ΔP_{M_i} generated real power (mechanical power) deviation in area i

$\Delta P_{T_{31}}$ and $\Delta P_{T_{31}}$ fraction generated power in tandem compound, single reheat turbine

ΔP_{D_i} incremental load demand change in area i

ΔP_{r_i} command signal to governor speed changer in area i

$\Delta P_{\text{int},i}$ net incremental real power exported from area i

$\Delta P_{\text{int},\infty}$ net incremental real power exported from area considered to have infinite kinetic energy

H_i inertia constant ($H_1 = 0.06$, $H_2 = 0.1$ and $H_3 = 0.16$ in seconds)

D_i system damping ($D_1 = D_2 = D_3 = 0.01$ in $pu\,MW/Hz$)

T_{G_i} governor time constant ($T_{G_1} = 0.08$, $T_{G_2} = 0.1$ and $T_{G_3} = 0.2$ in seconds)

R_i drop constant ($R_1 = 5$, $R_2 = 2.5$ and $R_3 = 2.6$ in $Hz/pu\,MW$)

T_{ij} transmission constant (considered equal for all interconnections, $T_{ij} = 0.5$ in $pu\,MW/Hz$)

T_w hydro turbine time constant ($T_w = 0.1$ in seconds)

T_T nonreheat turbine time constant ($T_T = 0.3$ in seconds)

T_1, T_2 and T_3 tandem compound, single reheat turbine time constants ($T_1 = 0.2$, $T_2 = 6$ and $T_3 = 0.4$ in seconds)

K_1, K_2 and K_3 tandem compound, single reheat cylinder fractions ($K_1 = 0.3$, $K_2 = 0.4$ and $K_3 = 0.3$)

LCL-Load Modular Converter
For Induction Heating

Maciej A. Dzieniakowski[*], Jan Fabianowski[†] and Robert Ibach[†]

[*] Warsaw University of Technology, Institute of Control and Industrial Electronics
Warsaw, Poland, e-mail: *mad@isep.pw.edu.pl*
[†] ABP Induction Systems GmbH, Dortmund, Germany
e-mail: *jan.fabianowski@abpinduction.com, robert.ibach@abpinduction.com*

Abstract - the paper presents novel resonant converter with LCL load, applied for induction heating. The control structure and power electronics circuit of VS inverter fed medium frequency induction furnace are described. The two resonant load topologies are investigated and applied. The control system assures fully synchronous operation in whole resonant load parameter changes. The control method and proposed converter topology results in resonant load independence on inverter parameters. It allows simple, modular converter construction. The present applications of the system cover range power to 1MW.

Keywords - Voltage Source Inverter, resonant converter, parallel operation, induction heating.

I. INTRODUCTION

The presented new converter type [1] [2] is the result of scientific and developing research done by WUT-ABP team. The principle assumption of the project was creating the modular system allowing extremely simple expanding of output power. The converter topology, the kind of resonant load circuit and the converter control method were the basic research goals defined and tested in the project first stage. These general issues consist many of consequent tasks that have been described and solved. Some of them will be discussed further on.

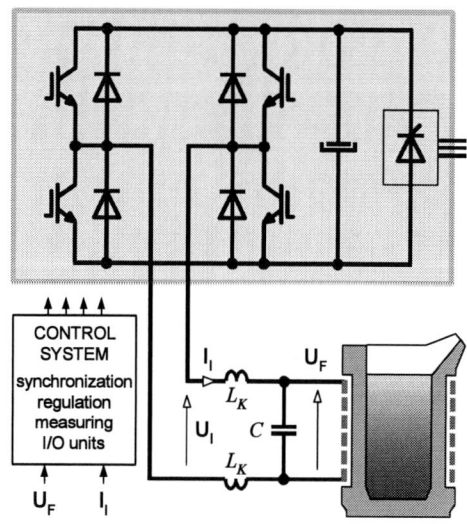

Fig. 1

Fig. 2

The Fig.1 shows the simplified diagram of the system under consideration. Its equivalent scheme is presented on Fig.2. The DC link is equipped in capacitance filter only – without serial chock. The DC link special construction guaranties minimalisation of connections residual inductance. It solves over-voltage peek problem, eliminates need of inverter snubber circuits using and limits IGBTs commutation losses. The full-bridge, one phase inverter supplies resonant load which inherent part is the induction furnace. The furnace charge physical state determines all principal parameters of the resonant circuit and regulation process. The control system follows load parameters changes assuring regulation of output power, voltage and current according to reference signals.

The LCL output load [3,4,5,6,7] creates resonant circuit (Fig.2) that having the operation frequency:

$$f_o = \frac{1}{2\pi\sqrt{(L_K L_F C)/(L_K + L_F)}} \qquad (1)$$

The expression (1) is the consequence of not only *LCL* circuit topology, but also typically applied control method e.g. frequency changing or PDM. The authors

978-1-4244-1741-4/08/$25.00 ©2008 IEEE

proposed phase control system that results in decoupling of the resonant circuit parameters from inverter parameters including L_K.

II. LOAD

The principal output circuit of the system has the *LCL* topology, however the resonant load is created by elements C-L_F-R_F only (Fig.3).

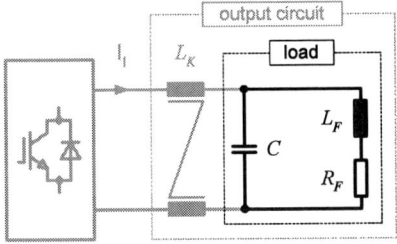

Fig. 3. Basic Load

The coupled inductors L_K act as the buffer separating VSI output and classical parallel resonant load. The system operating frequency is equal to load resonant frequency and is described by expression:

$$f_o = \frac{1}{2\pi\sqrt{L_F C}} \sqrt{\frac{L_F/C - R_F^2}{L_F/C}} \qquad (2)$$

The presented topology and the applied control method results in decupling of inverter and resonant load parameters ($\omega_o, \theta, d\theta/d\omega$ etc), which have influence on regulation process quality.

For the basic load topology the furnace voltage is limited by the DC link voltage or inverter valve parameters. In many applications there is the need of load voltage higher then DC or max. IGBT voltage. The changing of inverter scheme or changing of furnace construction are typical solutions of this problem, however they do not guarantee optimal utilization of system elements. The authors proposed and applied slight modification of the load (Fig.4) without changing the inverter main circuit and control principle.

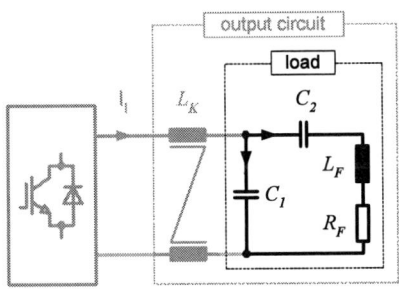

Fig. 4. Modified Load

The phasor diagram illustrating principle of voltage U_F increasing is shown on Fig.5. The appropriate choosing of C_1-C_2 value assures expected U_F/U_{C1} voltage ratio.

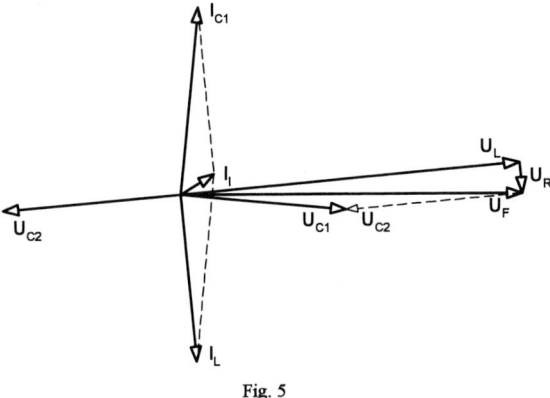

Fig. 5

Like in the case of the basic load the resonant circuit parameter are independent on power electronic system parameters. The resonant frequency is given by equation:

$$f_o = \frac{1}{2\pi} \sqrt{\frac{L_F(C_1+C_2) - R_F^2 C_1 C_2}{L_F^2 C_1 C_2}} \qquad (3)$$

The correct operation of the system with modified load can be disturbed by C_1 constant voltage component occurrence. This effect leads to drastic reduction of output power in the whole range of C_1-C_2 variation. The theoretical analysis of this phenomena resulted in regulation subsystem, which do not allow the constant voltage component existence.

III. CONTROL SYSTEM

There are many control methods guaranteeing acceptable system regulation. They are different not only in principles and resulting realization solutions but also in regulation features. In described system the phase control was proposed. The exemplary inverter current and voltages waveforms are shown on Fig.6. It illustrates system controlling way. The signals defining conducting time (t_m) of VSI diagonal transistors are generated synchronously to positive zero-crossing of voltage U_{C1}. It allows regulation of chosen physical values (power, voltage, ..) determining of the furnace charge heating process. The actual value of modulation factor $m=2t_m/T_0$ is evaluated by regulation system.

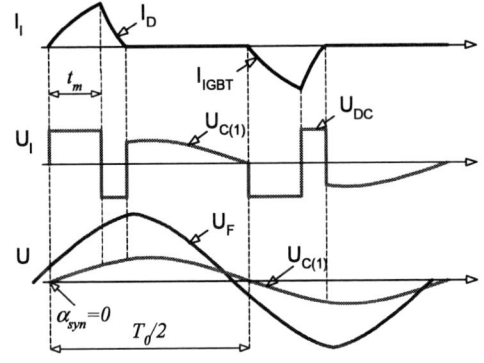

Fig. 6

2083

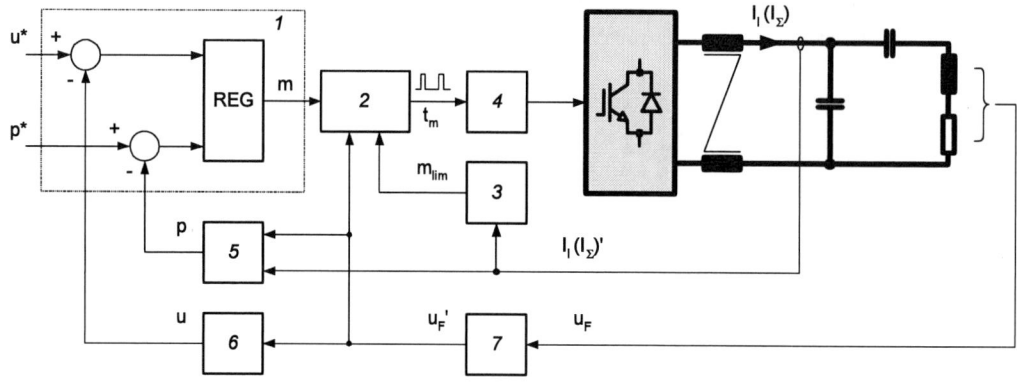

Fig. 7

The simplified control system diagram is shown on Fig.7. It consists of the complex regulators structure (*1*), synchronous pulse width modulator (*2*) – it allows to fix synchronization angle α_{syn} among other, modulation factor limiter (*3*), IGBT control signals distributor (*4*), fast real power measuring block (*5*), voltage measuring block (*6*) and voltage measuring preprocessor (*7*).

The regulation structure is not depended on applied load - basic or modified. The system guarantees output power stabilizing according to reference value (p^*) and simultaneously limits furnace voltage on setting level (u^*). In some heating applications (e.g. through motion charge) not power but the voltage must be regulated and stabilized, what changes the controllers hierarchy and goals. The sophisticated regulators structure (*1*) realizes all those tasks continuously.

Fig. 8

The control hardware (Fig.8) is based on advanced digital and analogue technology. The digital core (FPGA, µC-32b) is designed as the removable sub-board what results in easy system upgrading and checking.

The control system assures very precise regulation in steady and dynamic states and also fast communication with override and supervising systems.

IV. MODULAR CONVERTER

The converter topology and the described control method result in mutual independency of load and inverter parameters. This allows inverter output parallel connection without any changes of converter or load circuit. The Fig.9 shows applied way of converter modules connection.

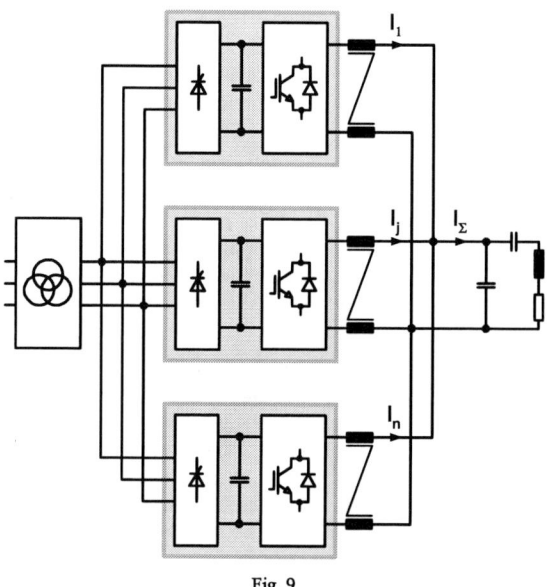

Fig. 9

The module consists of inverter, rectifier supplying DC link and DC link filter. The typical 250kW converter module is presented on Fig.10.

Fig. 10

The only one, common regulator system is needed for controlling multiple converters operating in parallel. The inverters valves control signals are distributed directly, using fiber-optic, from control system to IGBT drivers. There is no need to take into account individual switching and delay time of IGBTs and control signal transferring paths. This is thanks to proposed output circuit with the buffer inductance L_K, which eliminates inverters output current sharing problem during commutation.

V. INVESTIGATION RESULTS

The set of simulation and experimental investigations were done. Most of the experimental tests have been made on the final customers sides during the systems erection. The oscillograms bellow show exemplary waveform shapes of inverter output current and voltages for different resonant circuit topology – basic (Fig.11) and modified (Fig.12). In both cases the output power increasing is presented.

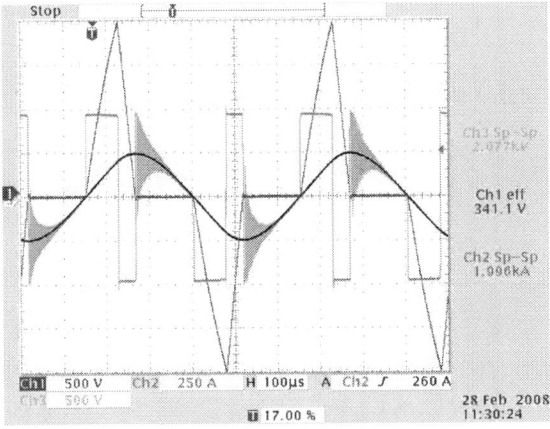

Fig. 11a. Basic Load – P=150kW

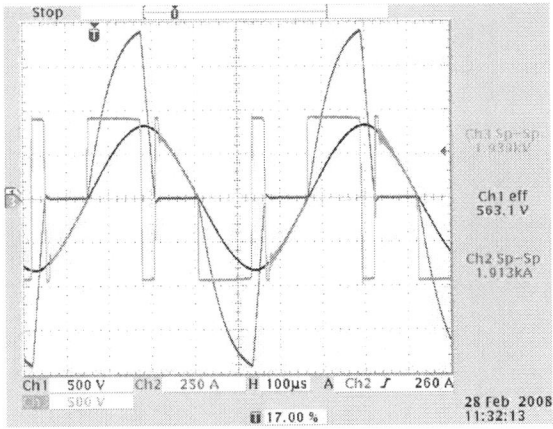

Fig. 11b. Basic Load – P=250kW

Fig. 11 legend:
Ch1: capacitor and furnace voltage $U_C=U_F$ (blue)
Ch2: inverter output current I_I (red)
Ch3: inverter output voltage U_I (green)

All the measurements were done on the melting systems with physically stabile, non liquid furnace charge. The observed, low energy voltage oscillations results from output circuit parasitic elements existence. They have no influence on the system operation, output power, inverter losses or efficiency.

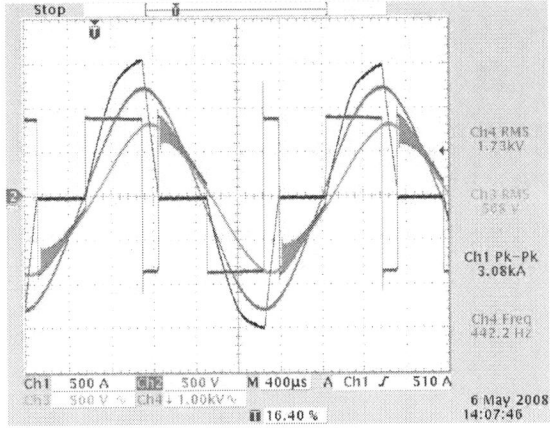

Fig. 12a. Modified Load – P=750kW

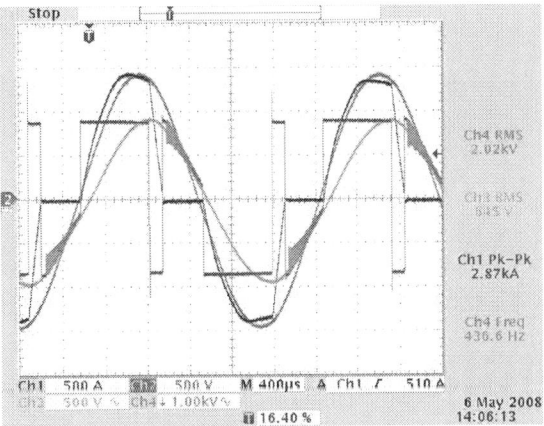

Fig. 12b. Modified Load – P=1000kW

Fig.12 legend:
Ch1: inverter output current I_I (blue)
Ch2: inverter output voltage U_I (red)
Ch3: capacitor C_1 voltage U_{C1} (green)
Ch4: furnace voltage U_F (magenta)

The modified load capacitors C_1-C_2 allow U_{C1} voltage increasing to match expected furnace voltage value. In the presented case the multiplication factor was fixed to three.

VI. CONCLUSION

The new concept of modular VSI for induction heating was presented and described. The converter topology, the load circuit topology, the control method and the regulation tasks have been defined and solved. The analytical and simulation results were verified by laboratory and industrial tests finally.

The robust construction, high operational reliability, low power losses, very good regulation behaviors and the extremely simple power extension possibility can define future trends in induction heating converters designing.

REFERENCES

[1] R. Ibach, J. Fabianowski and M. A. Dzieniakowski, „Verfahren zur speisung eines induktinsofens oder induktors", European Patent Office (EPO), EP 1 590 990 B1, 2006

[2] R. Ibach, J. Fabianowski and M. A. Dzieniakowski, "Method For Feeding An Induction Furnace or Inductor", World Intellectual Property Organization (WIPO), WO/2004/071132, 2004

[3] S.V. Mollov, M. Theodoridis and A. J. Forsyth, "High frequency voltage-fed inverter with phase-shift control for induction heating", IEE Proc.-Electr. Power Appl., Vol.151, No.1, January 2004, pp.12-18

[4] C.-S. Wang, G. A. Covic and O. H. Stielau, "Investigating an LCL Load Resonant Inverter for Inductive Power Transfer Applications", IEEE Trans. on Power Electronics, Vol.19, No.4, July 2004, pp.995-1002

[5] A. Schönknecht and R. W. De Doncker, „Novel Topology for Parallel Connection of Soft Switching, High Power, High Frequency Inverters", IEEE IAS, 36th Annual Meeting, Chicago, 2001, pp. 1477-1482

[6] S. Dieckerhoff, M. J. Ryan and R. W. De Doncker, "Design of an IGBT-based LCL-Resonant Inverter for High-Frequency Induction Heating", Proc. of the 34th Annual Meeting of the IEEE Industry Application Society (IAS), Phoenix (Arizona), 1999, pp.2039-2045

[7] K. Louati and D. Sadarnac, "Analysis of Resonant Converter with LCL-type Commutation", Proc. of EPE'93, Brighton, 1993, pp.141-146

On-line PID Controller Tuning Using Genetic Algorithm and DSP PC Board.

Paweł Fabijański[*], Ryszard Łagoda[†]

[*]Institute of Control and Industrial Electronics, Warsaw University of Technology, Warsaw, Poland
e-mail :pawel@isep.pw.edu.pl
[†]Institute of Control and Industrial Electronics, Warsaw University of Technology, Warsaw, Poland
e-mail : lagoda@isep.pw.edu.pl

Abstract—In this paper we uses a software pack, which allows to adjust PID controller automatically in real time way in the case of changes parameters of used object. PID settings are optimized in real time by using genetic algorithm and are compared with generally accepted quality indicators for control systems like rise time, % overshoot and settling time of step response or such coefficient like oscillations frequency, reduced damping and gain of second order object. A digital signal processor (DSP) is a specialized microprocessor designed specifically for digital signal processing generally in real time applications requiring high computation. The simulation results were verified by some laboratory test.

Keywords—Control of drive, genetic algorithm , DSP, PID controller

I. INTRODUCTION

Genetic algorithm is one of the efficient tools that are employed for tuning PID coefficients in real time control system.

Our automatic control system consists of:
- second-order controlled object, which could be, represented either ways as oscillatory system characterized by parameters: k, ξ, ω or as inertial object characterized by gain k and T1,T2 time constants was connected to the DSP PC board, transfer function of our controlled object is described by (4),
- PID controller settings (also on the DSP PC board) can be adjusted in real time so the control system matches quality indicators in relation to time requirements, PID transfer function is described by (6).

Control system with PID controller and second order object is shown on Fig.1.

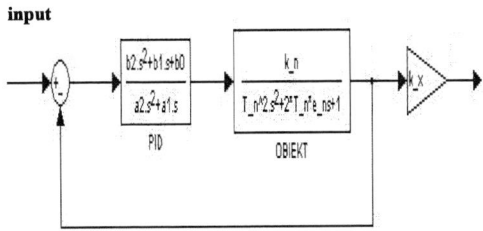

Fig 1. Control system – PID controller and second-order object

II. SHORT DESCRIPTION OF CLOSED LOOP CONTROL SYSTEM

Second-order objects are represented by second-order differential equations

$$b_2\frac{d^2y(t)}{dt}+b_1\frac{dy(t)}{dt}+b_0y(t)=a_0x(t) \qquad (1)$$

Assuming that initial conditions equal zero transfer function of the object takes form as follows:

$$G_o(s)=\frac{a_0}{b_0}\frac{1}{\frac{b_2}{b_0}s^2+\frac{b_1}{b_0}s+1} \qquad (2)$$

Denominator section can result in giving either real or imaginary roots. When real roots occur step response of the object has an aperiodic form and transfer function is:

$$G_o(s)=\frac{k}{(1+sT_1)(1+sT_2)} \qquad (3)$$

When imaginary roots occur transfer function is described as:

$$G_o(s) = \frac{k\omega^2}{s^2 + 2\xi\omega s + \omega^2} \text{ or}$$

$$G_N(s) = \frac{k_n}{T_n^2 s^2 + 2T_n\xi_n s + 1} \qquad (4)$$

where :

ω - oscillations frequency of undamped system ,

ξ - reduced damping coefficient ,

k - gain $T_n = 1/\xi$

Transfer function of PID controller:

$$G_{PID}(s) = G_P(s) + G_I(s) + G_D(s) =$$

$$= k_p + \frac{k_I}{T_I s} + \frac{T_D s}{\mathcal{E}T_D s + 1} = \qquad (5)$$

$$= \frac{(k_P T_I \mathcal{E}T_D + T_I T_D)s^2 + (k_P T_I + k_I \mathcal{E}T_D)s + k_I}{\mathcal{E}T_I T_D s^2 + T_I s}$$

could be also described as:

$$G_{PID}(s) = \frac{b_2 s^2 + b_1 s + b_0}{a_2 s^2 + a_1} \qquad (6)$$

Final transfer function for the circuit closed by loop- $G_{CL}(s)$ as illustrated below takes a form as per equation 7.

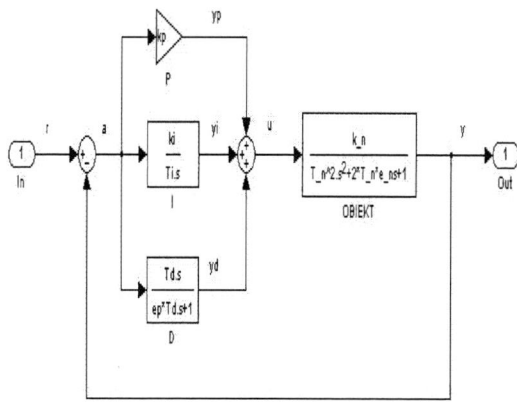

Fig.2 – detailed diagram of controller designed by use of DSP1104 board

and then closed-loop transfer function of entire circuit is:

$$G_{CL}(s) = \frac{G_N(s)G_{PID}(s)}{1 + G_N(s)G_{PID}(s)} \qquad (7)$$

III. SHORT DESCIPTION OF DS1104 SOFTWARE CONTROL PANEL IN USE.

Figure3 shows sections of the control panel as follows:

At the left-hand side from the top as follows:
-Results, PID controller parameters, Object parameters, Parameters of requested response

At the right-hand side:
-Data Receiver settings, Time settings, Additional info

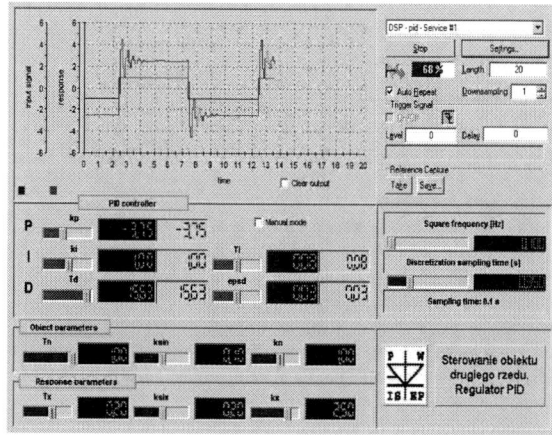

Fig.3. Control panel structure used to preset PID controller and second-order regulated object.

When our system start to work controller parameters are being adjusted automatically by calculation based on transfer function and step equation figures. When manual mode is active controller settings such us P, I and D should be adjusted manually.

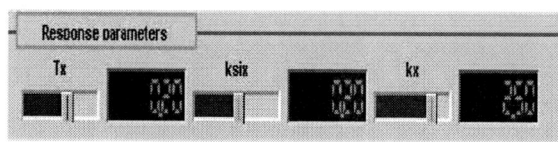

Fig.4. Response parameters section

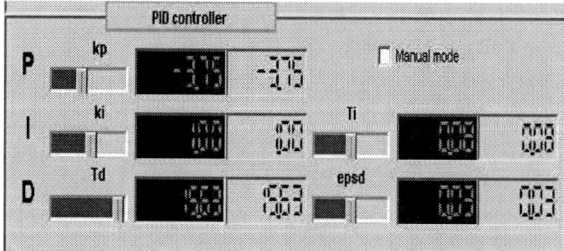

Fig.5. PID settings section

IV. SHORT DESCRIPTION OF PID CONTROLLER TUNED BY GENETIC ALGORITHM.

Genetic algorithm will be using population of n-number individuals. Every each of them contains subset of three PID controller settings as follows: Kp, Tp and Td. The number of variables is defined by variable "POPSIZE". Function "evpop" is responsible for generating population of random subsets. Fitness function associated with every subset is calculated based on overshoot of PID corresponding to step response. As it is clear, these fitness functions are proposed to achieve small value.

```
regParameters algen()
{
  regParameters retval;
  int i,j,k;
chrom popcurrent[POPSIZE];
popnext[POPSIZE];
    evpop(popcurrent);

  for(i=0;i<generationsNum;i++){
    for(j=0;j<(POPSIZE-unused);j++)
      popnext[j]=popcurrent[j];
      pickchroms(popnext);

  selection(popnext);
    crossover(popnext);
    mutation(popnext);
    for(j=0;j<(POPSIZE-unused);j++)
      popcurrent[j]=popnext[j];
    }
  }
  pickchroms(popcurrent);

  retval.Kp=x(popcurrent[0]).Kp;
  retval.Tp=x(popcurrent[0]).Tp;
  retval.TD=x(popcurrent[0]).TD;

  return retval;
```

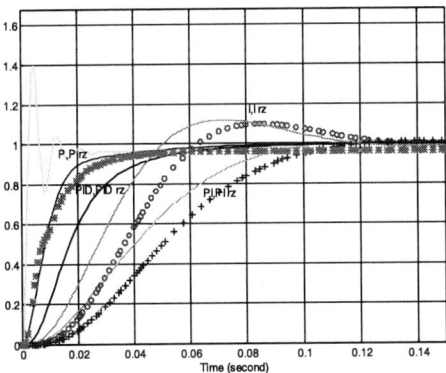

Fig. 6. Step response for some T_{period} in which genetic algorithm tuned parameters of PID controller

(o o o o o) -> step response for accepted quality indicators for control systems: rise time, % overshoot and settling time of step response

(* * * * * *) (+ + + + +) -> temporaly amount for maximum overshoot and small steady error, as much as possible. Relationship of fitness and error figures is defined as inversely proportional.

As reproduction is a next step in GA new population has to be generated. New individual are chosen based on roulette-wheel selection. In this case the fitness level is used to associate a probability of selection with each individual chromosome.

If function F(p) is the fitness of individual in the population, its probability of being selected is
p= (F(p) / ΣF(p)) * 100%. While candidate solutions with a higher fitness will be less likely to be eliminated, there is still a chance that they may be.

During the next stage of our algorithm randomly changes are made in our genotype. This operation is called a mutation and it's performed by the program function "mutation". Making slight changes in individual solutions with low probability as 1% makes sure that algorithm won't become convergent to soon. Higher probability used in the above process brings opposite effect and might destroy genotype.

Operations bellows are repeated until fixed number of populations is reached. Number of generations is defined by variable "generationsNum". Result of last iteration is a subset of optimal settings Kp, Tp and TD in terms of highest fitness value. Those settings are used with our PID controller.

It is clear that higher number of generations or size of initial population would result in giving either higher quality or even ideal settings. This could appear then as a time consuming process and more likely should be considered when it's used as a part of a real-time system where time is most important figure.

V. SIMULATION TEST

The proposed control structure under application of genetic algorithm was tested with two-order real object connected to the DSP PC board.

On Fig. 7–9 is shown step response parametres during work of genetic algorithm

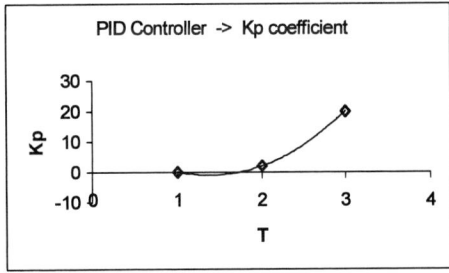

Fig. 7. Genetic algorithm tuning of PID controller coefficient -> kp

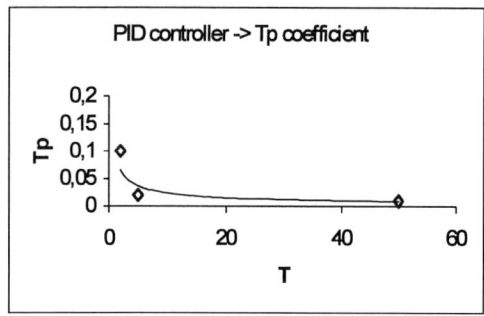

Fig. 8. Genetic algorithm tuning of PID controller coefficient -> Tp

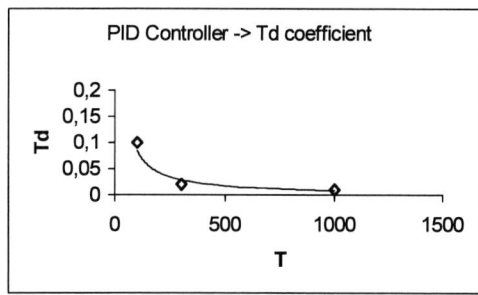

Fig. 9. Genetic algorithm tuning of PID controller coefficient -> Td

VI. CONCLUSION

The most important results of our investigation is to find a simply genetic algorithm. The simulated result were compared during the laboratory test on two order real object connected to the DSP PC board with the simulation test. The experimental results show that the proposed algorithm has the feature of simplicity.

VII. REFERENCES

[1]   Goldberg D., Genetic Algorithm, Addison Wesley, 1988

[2]   Guo Qingding, Wang Limei and Luo Ruifu . ;   " Completely digital PMSM Servo System based   on   new self-tuning pid algorithm "   Proceedings of the IEEE International Conference on industrial Technology, 1999

[3]   Porter B., Jones A.H. Genetic Tuning of Digital PID Controllers IEEE Electronics Letters Vol. 28, pp.843-844, Apr 1992

[4]   Seyedkazaemi   M.,   Akbarimajd   A.,   Rahnamaei   A., Baghbanpourasi A. A Genetically Tuned Optimal PID Controller Proceedings of the 6th WSEAS Int.Conf. on Artificial Intelligence, Knowledge Engineering and Data Bases, Corfu Island, Greece, February 17-19, 2007

Regulation Properties of Pumping Station Control System In The Highest Efficiency Range

Szychta Leszek

Technical University of Radom, Radom, Poland, e-mail: *l.szychta@pr.radom.pl*

Abstract—Three systems of control the work of water-supply pumping station were introduced in the article: the cascade, converter and highest efficiency range - worked out by the author. The motors of all pumps are supply by frequency converters in the system in the highest efficiency range. This makes possible the work of pumps in so called the recommended area in which the largest efficiencies step out. The author gives the conditions of the continuous and interrupted operation of the pump. The changes of the rotation speed were qualified stay in the function efficiency at the constant head for pumps characterized by specific speed n_q=42. The author suggests applying the re-dimensioning pumps in the relation to the rated conditions of the demand.

Keywords—Control of Drive, Efficiency, Industrial application.

I. INTRODUCTION

Outlet pressure of the pumping station conforms with its set value where capacity of pumping station Q corresponds to water demand Q_x. The capacity Q is adapted to the demand Q_x by means of various control systems of pumping station operation. These systems are based on measurements of water pressure or capacity in the pumping outlet collector. In the most cases, the pressure is measured immediately downstream of the pumping station [1, 2, 7, 10, 11].

Set values of pressure in the outlet collector can be obtained by changing a group of pumping units $H(Q)$ curve or pumping plant $H(Q)$ curve, where H is pump head. The factors most commonly affecting changes of a group of pumping units $H(Q)$ curves include:
- number of pumping units in parallel (less often in series),
- rotational speed of pump(s),
- pump capacity due to changes of air volume supplied to the suction collector.

Changes of a pumping plant $H(Q)$ curves are affected by:
- the degree to which the valve in the pumping line is opened (throttling regulation),
- the degree to which the valve in the discharge line is opened (discharge regulation).

In view of the power losses associated with changes of pumping sets $H(Q)$ curves, it is assumed, for purposes of further analysis, that output pressure of the pumping station is controlled with variation of a group of pumping units $H(Q)$ curves. Depending on curve, two principal control systems can be distinguished:
- cascade,
- at variable rotational speed of one or several pumps.

II. PROPERTIES OF PUMPING STATION CONTROL SYSTEMS

For purposes of comparative analysis, three pumping station control systems are postulated: cascade, converter, and highest efficiency range – the latter developed by the author. The system at the highest efficiency range has been named *PNS* by author. Each configuration includes one, the so-called guiding pump whose capacity adjusts to current demand by possible turn-on and turn-off (cascade control) or variation of rotational speed (converter or *PNS* control). The concept of guiding pump is functional and not permanently associated with any given pump. Beside the guiding pump, a pumping set comprises the so-called supplementary pumps that operate at constant rotational speed. The concept of guiding pump is not commonly used in cascade control. However, for clarity of analysis in the article, this principle has been abandoned [3, 4 ,8].

A. Cascade control system

During cascade control, all pumps have constant rotational speed (equal to rated) (Fig. 1). The number of pumps in operation depends on the value of demand. Pressure variation in the outlet collector is allowed within a set range $p \in (p_{min}; p_{max})$. The assumed pressure variation is obtained by turning on or off of the guiding pump. If the guiding pump is off, pressure in the water system is reduced. When pressure variation in the outlet collector reaches its minimum value p_{min}, the guiding pump is switched on. The pressure rises in the outlet collector and the pump's head H moves towards increasing values of the pump $H(Q)$ curve. When pressure in the outlet collector reaches its maximum value p_{max}, the guiding pump is switched off [4, 8].

978-1-4244-1741-4/08/$25.00 ©2008 IEEE

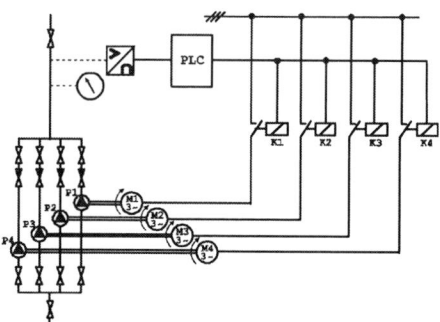

Fig.1. Structure of hydraulic and electric connections in pumping station during cascade control, P1÷P4 – pumps, M1÷M4 – motors, PLC – Programmable Logic Controller, K1÷K4 – network contactors.

B. Converter control system

During converter control, the guiding pump is supplied from a frequency converter and operates at variable rotational speed n, while supplementary pumps have a constant, rated rotational speed n_n (Fig.2). There are drive systems where a frequency converter is permanently associated with a specific motor. In the circumstances, the same pump always fulfils the function of the guiding pump. Where the structure of electric connections in the power circuit enables the frequency converter to be switched between motors (Fig.2), the pump whose drive changes the rotational speed becomes the guiding pump. The number of pumps in operations depends on the water demand. Variation of the guiding pump's rotational speed affects capacity of the pumping station Q, which is adapted to the actual demand Q_x. Smooth capacity changes result in the pumping collector variation to become close to its set value p_{set} [5, 6, 8].

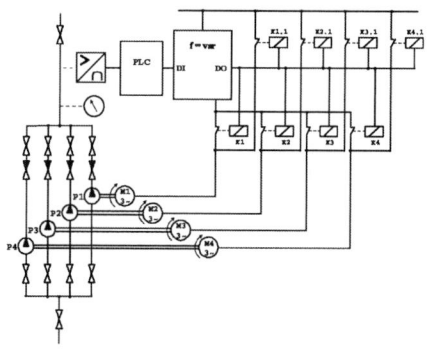

Fig.2. Structure of hydraulic and electric connections in pumping station during converter control, PLC – Programmable Logic Controller, K1÷K4 – converter contactors, K1.1÷K4.1 – network contactors.

At low capacities Q, associated with low rotational speeds n, the pump operates at low efficiency η. In the event, energy consumption e in pumping 1m³ of water increases. It becomes then beneficial to switch off the guiding pump and induce the so-called 'sleeping' mode (Fig.3). Pressure in the outlet collector reduces to minimum value p_{min}, and the guiding pump starts operating to reach its set value p_{set}.

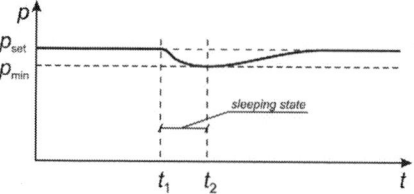

Fig.3. Output pressure of the pumping station in the "sleeping" mode.

The unit energy consumption e of the pumping unit in the process of pumping 1m³ water is obtained:

$$e = \frac{1}{367}\frac{H}{Q}\sum_{i=1}^{m}\frac{Q_i}{\eta_{zi}}\quad\left[\frac{\text{kWh}}{\text{m}^3}\right]\qquad(1)$$

where:

$\eta_z = \eta\eta_s\eta_f$ - efficiency of pumping unit   (2)

η – pump efficiency,
η_f – efficiency of frequency converter,
η_s – efficiency of induction motor.

C. Control system in the highest efficiency range

Structure of hydraulic and electric connections in pumping station during *PNS* control, highest efficiency range, is illustrated in figure 4. Two control types are possible in the *PNS* system: *PNS_H*, at constant pump heads $H = H_{set} = $const, and *PNS_N*, at optimum pump rotational speeds $n = n_{opt}$ [8, 9].

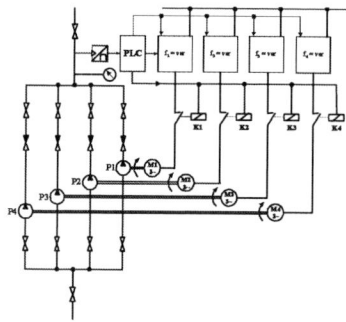

Fig.4. Structure of hydraulic and electric connections in pumping station during *PNS* control, PLC – Programmable Logic Controller, K1÷K4 – converter contactors.

During *PNS_H* control, all pumps operate at variable rotational speed within a limited range $n^* \in \left(n_{min}^*; n_{max}^*\right)$. Relative rotational speed n^* is defined:

$$n^* = \frac{n}{n_n}\qquad(3)$$

where:
n – rotational speed [rpm],
n_n – rated rotational speed [rpm],

The variation range of rotational speed depends on the assumed minimum efficiency η_{min} of the pump and determines variations of the pump's capacity

$Q \in (Q_{\min}; Q_{\max})$ within the recommended operating region (Fig. 5).

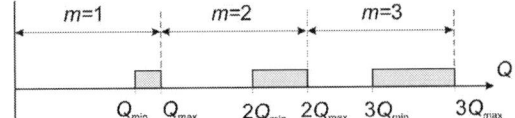

Fig.5. Recommended operating region of the pump during *PNS_H* control, m – number of pumps in operation.

According to the similarity theory, minimum $Q_{\min}$ and maximum $Q_{\max}$ capacity depend on the rotational speeds $n_{\min}^*$, $n_{\max}^*$, respectively:

$$\frac{Q}{Q_1} = n^* \qquad (4)$$

where:

Q_1 –pump capacity during rated rotational speed n_n.

There is defined relative pump efficiency η^*:

$$\eta^* = \frac{\eta}{\eta_{opt}} \qquad (5)$$

where:

η – pump efficiency,

η_{opt} – optimum pump efficiency.

Decrease of the set minimum efficiency $\eta_{\min}^*$ increases the recommended range of rotational speed $\left(n_{\min}^*, n_{\max}^*\right)$ and consequently increases the range $\left(Q_{\min}, Q_{\max}\right)$. It is therefore possible that recommended capacity ranges (operating regions) will coincide at turn-on of m pumps. Under such circumstances, a pumping set operates continuously.

During continuous operations of a pumping set, demand Q_x equals the sum total of pump capacities within the range $Q \in (Q_{\min}, Q_{\max})$. This obtains always when the condition is fulfilled:

$$2 \cdot Q_{\min} \le Q_{\max} \qquad (6)$$

Water demand Q_x in recommended operating regions meets the condition:

$$mQ_{\min} < Q_x < mQ_{\max} \quad \text{dla} \quad m=1, 2, \ldots \qquad (7)$$

The number of pumps in operation m is defined:

$$\begin{cases} m = Int\left(\dfrac{Q_x}{Q_{\max}} + 1\right) & \text{for} \quad \dfrac{Q_x}{Q_{\max}} \notin N \\[3mm] m = Int\left(\dfrac{Q_x}{Q_{\max}}\right) & \text{for} \quad \dfrac{Q_x}{Q_{\max}} \in N \end{cases} \qquad (8)$$

where: Q_x – total capacity of pumping station equal to water demand,

N – set of natural numbers.

Int function separates the integral part of the expression in brackets.

Interrupted operation of a pumping station consists in interrupted operation of the guiding pump. The guiding pump is turned off at the moment the pressure in the outlet collector corresponds to head H_{set}, and capacity of the guiding pump is $Q_{\min}$. Supplementary pumps work at maximum acceptable capacity $Q_{\max}$. An interruption in the guiding pump operation causes the total capacity of the pumping station $(m-1)Q_{\max}$ is lower than the average demand Q_x. Pressure in the water plant decreases. When pressure in the outlet collector corresponds to minimum head $H_{\min}$, the guiding pump is turned on. The pump operates in the range of head $H \in \left(H_{\min}, H_{set}\right)$ at optimum rotational speed n_{opt}^*. Momentary capacity of the pumping station Q_{opt} is then greater than demand Q_x at the discharge opening, thus, pressure in the water plant gradually rises until the set head H_{set} is reached. Continuing growth of the pumping station's output pressure is not allowed, therefore, rotational speed falls to reach $n_{\min}^*$, and then the pump is switched off. The variation range of rotational speed n^*, associated with capacity reduction from Q_{opt} to $Q_{\min}$, is assumed to be negligible and short in comparison to the overall operation cycle. Interrupted operation is thus analogous in nature to the cascade control, the difference being that the pump operates at optimum rotational speed n_{opt}^*. During interrupted operation, total demand fulfils the condition:

$$(m-1)Q_{\max} \le Q_x \le mQ_{\min} \quad \text{for} \quad m=1, 2, \ldots \qquad (9)$$

Operation of a pumping station is interrupted more often in *PNS_N* control than is the case during *PNS_H* control. This results from the fact that supplementary pumps always have optimum rotational speeds n_{opt}^*, which assures the greatest efficiency η_{opt}^*. The guiding pump may work at variable rotational speed in the range $n^* \in \left(n_{\min}^*; n_{\max}^*\right)$. Frequent changes of the pumping outlet pressure are a drawback of this control system that may impair reliability of the whole water plant. Great pressure variations increase power consumption and impair water demand conditions, increasing frequency of pumps' turn-on.

III. PUMP'S CALCULATION ROTATIONAL SPEED AT CONSTANT HEAD

With the aid of equations in [8], variation of calculation speed n_{rcalc}^* can be determined as a logarithm function of rising characteristic curve of calculation efficiency η_r^*.

$$\eta_r^* = a\ln(x) + b \qquad (10)$$

where:

a, b – constant parameters depends on the value of specific speed n_q,

x – variable depends on rotational speed n_{rcalc}^*,

$$n_q = \frac{n}{60} \frac{\sqrt{Q_{opt}}}{H_{opt}^{3/4}} \quad \text{– specific speed} \tag{11}$$

Figure 6 presents variation of n_{rcalc}^* at h^*=0,6; h^*=0,8; h^*=1 in the full range of rising characteristic curve of the pump's efficiency $\eta_r^* \in \left(\eta_{min}^*, \eta_{opt}^*\right)$. The relative pump head h^* is defined:

$$h^* = \frac{H}{H_n} \tag{12}$$

where:

H – pump head [m],

H_n – rated pump head [m].

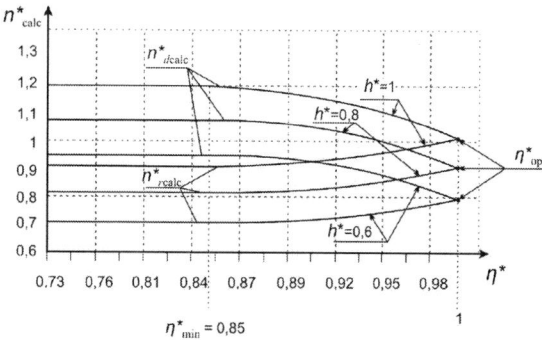

Fig.6. Calculation rotational speed n_{calc}^* as a function of η^* of the pump at specific speed n_q=42.

If the head h^* reduces, the characteristic curve of calculation rotation speed's is shifted towards lowering values (fig.6). Within the variation range of the pump's efficiency $\eta_r^* \in (0,73 \div 0,85)$, relative calculation rotation speed varies by $\Delta n^* = 0,01$, or approximately 10% of total rotational speed variation. Great dynamics of efficiency variation obtain in this region. Limited variation range of n_{rcalc}^* causes on slight variation of Q. Therefore, operations in the region of greater rotational speeds then optimum speed n_{opt}^* is recommended.

On the basis of dependencies in [9], calculation rotational speed n_{dcalc}^* can be analysed in relation to the descending characteristic curve of calculation efficiency η_d^* at h^*= 0,6; h^*= 0,8; h^*= 1:

$$\eta_d^* = ax^2 + bx + c \tag{13}$$

where:

a, b, c – constant parameters depends on the value of specific speed n_q,

x – variable depends on rotational speed n_{dcalc}^*.

In this event, maximum range of rotational speed variation is $\Delta n_{dcalc}^* \cong 0,18$ in relation to optimum rotational speed n_{opt}^*, at various head h^*. There is low

range of rotational speed n_{dcalc}^* for efficiencies η^* below $\eta_{min}^* = 0,85$ and head h^*=1. This means that, with limited range of pump's capacity regulation, a great variation range of efficiency, and the resultant high energy consumption of pumping occur.

Under dependencies (10), (13) it is possible to determine an algorithm of calculation rotational speed n_{calc}^* control as a function of efficiency η^* taking account of both the rising and descending branch. Assuming random efficiency η^* and h^*, recommended variation of the pump's calculation rotational speed n_{calc}^* can be determined (Fig. 6). At h^*=0,8 and $\eta_{min}^* = 0,85$, the variation range of rotational speed is in the range $n^* \in (0,83; 1,06)$, where great pumping efficiency can be achieved in the pump's capacity range $Q \in \left(Q_{min}; Q_{max}\right)$.

At $h^* \in (0,6; 1)$, minimum capacities Q_{min} are lower than rated value Q_n, while maximum capacity Q_{max} may be greater than Q_n. For instance, at h^*=1 and $\eta^* = \eta_{min}^*$, the pump can operate in the recommended region at maximum rotational speed n^*=1,18 and capacity $Q_{max} \cong 1,6Q_n$. Such high capacities cause significant motor overloading of even $P \cong 1,6P_n$ and lead to occurrence of cavitation.

IV. CONCLUSION

The foregoing analysis leads to the conclusion that maximum capacity and power consumption in the recommended operating region may exceed acceptable parameters of a pumping unit. The recommended operating region can be utilised when the pump works at a head that meets the condition: h^*<1. For example, when the pump operates at h^*=0,8, the variation $\Delta n^* \cong 0,25$ (Fig.6) is acceptable, that is, 25% of the full range of rotational speed. On the other hand, at h^*=1, the variation range of rotational speed is only $\Delta n^* \cong 0,1$. Therefore for $h^* = 1$ operations in the recommended region is not fully utilised.

In light of this analysis, it is reasonable to propose that pumps should be over-dimensioned in relation to rated parameters of a pumping system. This re-dimensioning involves application of pumps of rated head greater than the maximum required in a pumping system. This reduces relative head h^* and consequently shifts the optimum operations point (at efficiency η_{opt}^*) towards decreasing rotational speeds (Fig.6). As a result, the region of recommended operations is extender in the direction of decreasing capacities that also do not exceed the rated value Q_n.

In the *LCC* (Life Cycle Costs) perspective, a pump's re-dimensioning generates higher investment expenditure which can, however, be compensated with reduced operating costs in effect of lower demand for electricity. It must be emphasised that the analysis has been based on a single characteristic curve of one-degree, one-stream pump of specific speed $n_q \approx 42$. Similar conclusions in regard of other pump types of different n_q can only be drawn from analysis of appropriate characteristic curves.

REFERENCES

[1] Bonetyński K., Wiszniewska-Oraczewska I., "Strategia zapobiegania niekorzystnym skutkom uderzenia hydraulicznego w sieciach wodociągowych" (Strategy for prevention of adverse effects of watter hammer in water supply systems), Gaz, Woda i Technika Sanitarna, 8/2001.

[2] Dongwei S., Jianbo D., Yong Z., "Investigation of pressure in pipe subjected to axial-symmetric pulse loading", International Journal of Impact Engineering 25/2001.

[3] Jędral W., "Zmiana sprawności pompy wirowej przy zmianie jej prędkości obrotowej", (Change of Impeller Pump Efficiency at Variable Rotational Speed), Problemy badawcze energetyki cieplnej, Warsaw, 1-3 December 1999.

[4] Jędral W., "Pompy wirowe" (Impeller Pumps). PWN, 2001.

[5] Luft M., Łukasik Z., "Podstawy teorii sterowania" (Fundamentals of control theory), Zakład Wydawniczy Politechniki Radomskiej, 2007.

[6] Mielcarzewicz E.W., "Obliczanie systemów zaopatrzenia w wodę" (Calculations of water supply systems), Arkady Sp. z o. o., Warszawa 2000.

[7] Szychta E., "Multirezonansowe przekształtniki ZVS napięcia stałego na napięcie stałe" (Multiresonant ZVS DC-DC converters), Oficyna Wydawnicza Uniwersytetu Zielonogórskiego, 2006.

[8] Szychta L., "Zasady doboru systemu sterowania pompowni wodociągowych" (Principles of selection of control systems for water mains pumping stations), Zakład Poligraficzny Politechniki Radomskiej, 2006.

[9] Szychta L., "Analysis of pump energy consumption in the upper slope region of efficiency characteristic curve". The Archive of Mechanical Engineering IV 2003.

[10] Wood D.J., Asce S., Jones S.E., "Water-hammer charts for various types of valves", Journal of the Hydraulics Division, Proceedings of the American Society of Civil Engineers, January 1973.

[11] Wylie E.B., Streeter V.L., Suo L., "Fluid transients in systems", 1993.

Inner Gas Pressure Measurement Based Life-span Estimation of Electrolytic Capacitors

A. Riz[*], D. Fodor[*] and O. Klug[**], Z. Karaffy[**]

[*] University of Pannonia/Department of Automation, Veszprém, Hungary
[**] EPCOS Electronic Parts and Components, Hungary/Development Laboratory, Szombathely, Hungary

Abstract—As the wide scale usage of aluminum electrolytic capacitors it is very important to be familiar with them. The research and development of new and long sight better capacitors is very essential for circuit designers and costumers in general hence this is a very important issue to deal with. In the presented paper a summarize of the buildup and the useful life of the capacitors and a special procedure for the life-span analysis will be shown. The presentation is also deals with the related part of the Electrolytic Capacitor Measurement System (ECMS) implemented in the Aluminum Electrolytic Capacitor Development Laboratory of EPCOS Electronic Parts and COmponentS, Hungary. The results show, thanks to the introduction of the mentioned system, the efficiency of special measurements like the analysis of the useful life of aluminum electrolytic capacitors have been increased considerably.

Keywords—measurement automation, test automation, passive electronic components, electrolytic capacitor, life-span estimation, gas pressure measurement

I. INTRODUCTION

Capacitors are widespread all over the world. They can be found in every electronic device around us used as energy storage elements, filters, advancers and decouplers. From this reason capacitors play a very important role in our world.

Nowadays the ceramic, the foil, the tantalum and the electrolytic types are the frequently used capacitors by the industry. Coupling and radio frequencies, power supply, power storage and power electronics are the most significant application fields for capacitor technologies. Except for the first application field the aluminum electrolyte capacitors can be applied for so this type of capacitors is real prevalent.

Withal the wide application range the main advantages of the aluminum electrolyte capacitor is their high volumetric efficiency which enables the production of capacitors with up to one Farad capacitance, and the fact that an Al electrolytic capacitor provides a high ripple current capability together with a high reliability and an excellent price/performance ratio.

The paper deals with the buildup of the aluminum electrolyte capacitors and a development environment which was designed and implemented testing and measuring the parameters and properties of this type of capacitors via practical kind of life-span estimation method.

The buildup of the aluminum electrolyte capacitors and a short summary about the useful life are given in *section II*. An overview of the related applied development system including the registration and measurement software will be shown in the *section III*. The mentioned special life-span estimation method is summarized in the *section IV*. The conclusion is drown in *section V*.

II. ALUMINUM ELECTROLYTIC CAPACITOR

A. Buildup of aluminum electrolytic capacitor

As is the case with all capacitors, an Al electrolytic capacitor comprises two electrically conductive material layers that are separated by a dielectric layer. One electrode (the anode) is formed by an aluminum foil with an enlarged surface area. The oxide layer that is built up on this is used as the dielectric. In contrast to other capacitors, the counter electrode (the cathode) of Al electrolytic capacitors is a conductive liquid, the operating electrolyte. A second aluminum foil, the so-called cathode foil, serves as a large-surfaced contact area for passing current to the operating electrolyte (see Fig. 1).

The anode of an Al electrolytic capacitor is an aluminum foil of extreme purity. The effective surface area of this foil is greatly enlarged by electrochemical etching in order to achieve the maximum possible capacitance values. The type of etch pattern and the degree of etching is matched to the respective requirements by applying specific etching processes.

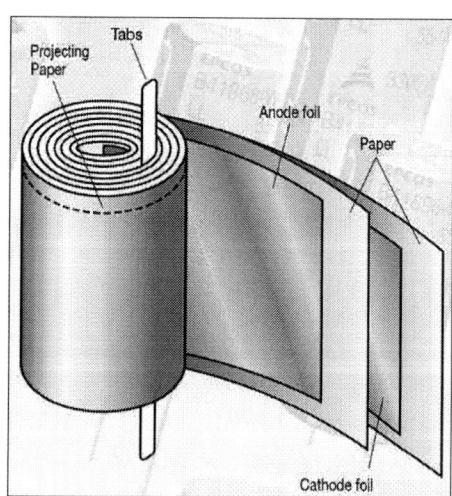

Fig. 1. Buildup of aluminum electrolytic capacitor

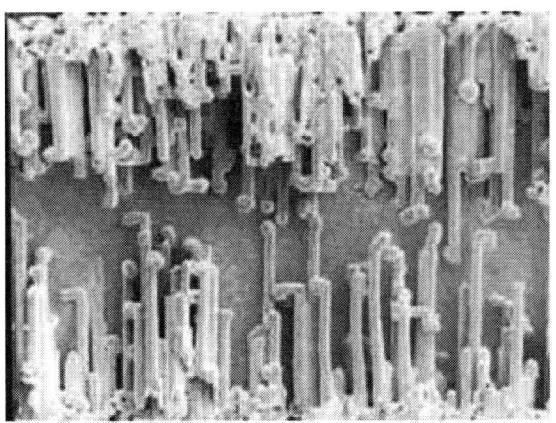

Fig. 2. Ethcing of anode foil for high-voltage capacitors

Etched foils enable very compact Al electrolytic capacitor dimensions to be achieved and are the form used almost exclusively nowadays.

The dielectric layer of an Al electrolytic capacitor is created by anodic oxidation (forming) to generate an aluminum oxide layer on the foil. The layer thickness increases in proportion to the forming.

Even for capacitors for very high voltages, layer thicknesses of less than 1 m are attained, thus enabling very small electrode spacing. This is one reason for the high volumetric efficiency achieved.

During the forming process the very fine pits of the etched foils will encrust partially in proportion to the forming voltage and thus also to the achieved layer thickness (see Fig. 2). Due to this effect, the final operating voltage range must already be taken into account when the foils are etched.

The two aluminum foils are separated by paper spacers. The paper serves various purposes, it serves as a container for the electrolyte – the electrolyte is stored in the pores of the absorbent paper – and also as a spacer to prevent electric short-circuits, as well as ensuring the required dielectric strength between the anode and cathode foils.

B. Useful life

Useful life (also termed service life or operational life) is defined as the life achieved by the capacitor without exceeding a specified failure rate. Total failure or failure due parametric variation is considered to constitute the end of the useful life.

Depending on the circuit design, device failure due to parametric variation does not necessarily imply equipment failure. This means that the actual life of a capacitor may be longer than the specified useful life. Data on useful life have been obtained from experience gained in the field and from accelerated tests.

The useful life can be prolonged by operating the capacitor at loads below the rating values (e.g. lower operating voltage, current or ambient temperature) and by appropriate cooling measures.

The useful life examined in this paper applies to Al electrolytic capacitors with natural cooling, i.e. the heat generated in the winding is dissipated through the casing and by natural convection. It is possible to increase the permissible ripple current and/or prolong the useful life by using additional cooling by heat sinks, water or forced ventilation.

III. THE CAPACITOR MEASUREMENT SYSTEM

The researching and developing process of the new capacitors is time consuming. Usually the process can be separated two distinct parts. On one side it is sure that the phase of testing, measuring, acquiring and analyzing is more important than the preparing phase. But on the counter side the accurate preparing process can be facilitate the development considerably. The material and parameter registration and the data storage can be instanced because these are fundamental parts of a research.

The base of the whole software system is a framework which was originally designed to provide a common user interface for the different measurement and registry programs. In spite of all the measurements have an individual character, they were integrated into the above mentioned framework, in order to manage the ports and instruments, and provide the parallel run of the programs.

The measurement system includes at least 27 different measurements, whose presentation can not be done here only those which related to the special life-span estimated method.

A. Project registry

In accordance with the whole development process is project orientated define a new project for the research or connect the measurements or tests to an existing one is a requisite. Thereupon during the report generation and cost keeping processes all related scheduled jobs can be grouped easily (see Fig. 3).

B. Capacitor registry

As a consequence of the huge amount of the capacitors the individual registration of them is a must. A special program guides the engineers through this process. All of the components of a capacitor are stored in the database of the development laboratory for further analysis.

The registration process can be separated into two steps. The first one is the registration of the semi finished capacitors which finally defines several groups of capacitors with different content. For testing and measuring all of the capacitors are finalized once a change in one or more component was applied satisfy the special needs.

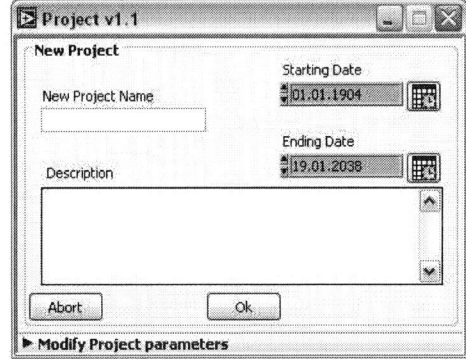

Fig. 3. Registration of projects

2097

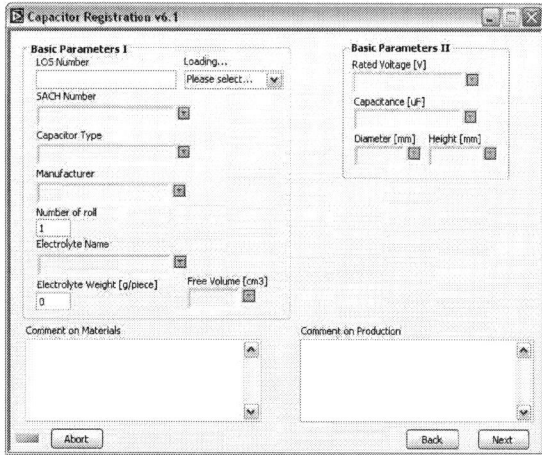

Fig. 4. Registration of capacitor components

At the end all capacitors are labeled with an individual identification number which is only used before the test phases. After this the identification number remains in hide used only by the development system to trace the walk of life of a capacitor.

C. Test initialization

During the development process after a capacitor group was registered the accurate initialization of the tests and measurements is also necessary.

The number of tests and measurements can be very high and of course one capacitor can be a part of several sequential tests. To identify the tests and all the covered capacitors the introduction of a new identification number was a must. This number is individually identifying all capacitors and the test which part of it.

As it was mentioned the parameters of the tests and measurements can also be registered through the initialization process (see Fig. 5). Firstly it is necessary to select the related one or more projects then to select the capacitors as parts of the test. The next step is the setting of the test parameters (e.g. temperature, applied voltage) then the setting of the parameters of the sequential measurements. After this initialization process the mentioned test identification numbers are created for all individual capacitor (see Fig. 6).

The structure of the database gives the possibility to register and trace the milestones in the life of a capacitor.

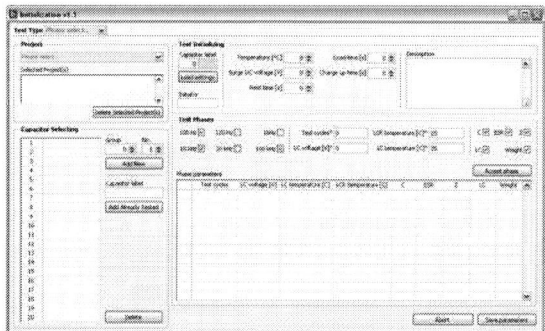

Fig. 5. Test initialization

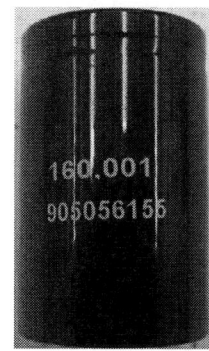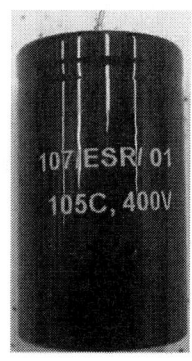

Fig. 6. Individual identification number (left) and the test identification number (right) of a capacitor

D. Test monitoring

After the necessary preparation process of the capacitors the tests can be started. As it was mentioned there are several tests running at the same time at the development laboratory most in special industrial oven.

In the ovens many tests can be running with different test parameters so it can be difficult to allocate accurately the location of the capacitors and in addition the length of these tests can be thousands of hours hence the tracing of a test is also complicated. With this object it was aimed to develop and implement a monitoring application which guides true a test and manage all of them (see Fig. 7).

The software divided into two modules such as the long term test parameters and the task manager module. The two modules of the application can start, suspend and close the tests, send notifications, e-mails if necessary and manage the database beyond the long term tests (see Fig. 8).

Fig. 7. Long term test environment

2098

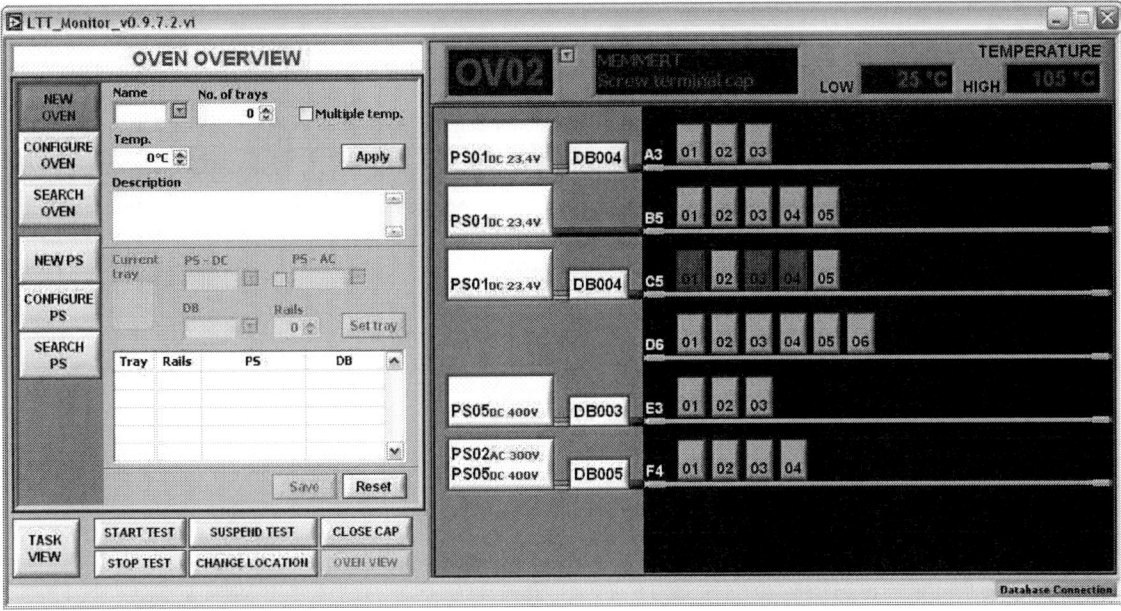

Fig. 8. Long term test environment parameters

E. Gas pressure measurement

Once all of the parameters have been set and a test has been started the next move is the data acquisition. There are several automated measurements. There are more than ten applications for measuring the parameters of a capacitor like the leakage current, the equivalent series resistance (ESR), the capacitance and the impedance, weight etc. One of the most important among these measurements is the gas pressure measurement. In this paper the gas pressure measurement and a related research will be heighten.

The gas pressure measurement is a fully automated application thanks to the initialization process since the parameters of the test had already been saved to the database. To start the measurements only the selection of the actual test identification number and placing the capacitors into the ovens are necessary (see Fig. 9).

Fig. 9. Inner gas pressure measurement during a running long term test

During the measurement the user can choose between two main tab controls: the first page contains the set parameters of the measurement, while the other tab shows the results of the measurement and the remaining time before the next phase via graphs and displays.

Firstly the program sets up the temperature. On the temperature graphs the measured temperature of the oven can be followed up. Then the program flow enters into a cycle and executes the same steps cyclically. Through measuring the gas pressure phase the measured values are stored locally and the pressure as a function of time are showed on digital displays. The remaining time before the end of the experiment is showed during the measurements. The program can be stopped whenever it is needed. Finishing the measurement the results of the experiment are showed. If the results are correct then saved into database otherwise the results will be deleted.

F. Search engine

All measurement data are migrated to a relational database system from where with the help of a searching engine can be retrieved in a pre-defined or ad-hoc way. The results below have been retrieved from the database with the aid of the searching program (see Fig. 10).

IV. LIFE-SPAN ESTIMATION BASED ON RESULTS

In the following an estimation method in the case of screw terminal capacitors under direct current load will be introduced focusing on the life cycle of a device in a real working environment.

The inner gas pressure is the most significant factor determining the life-span of a screw terminal electrolytic capacitor under direct current load. Generally at least 90% of the failures are caused by gas pressure increase.

In spite of the many components only a few of it influence the gas generation primarily. Previous tests show that the electrolyte, the paper, the anode type and the

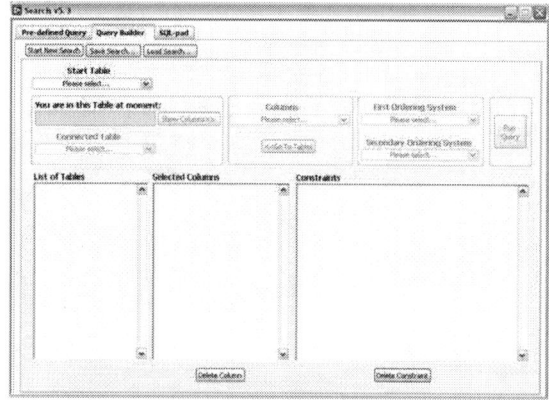

Fig. 10. Database search engine

forming voltage which influence mainly the gas generation. Changing only one component among these at a time the gas pressure as a function of this component can be determined.

According to the theory the inner pressure is defined by the continuous pressure generation and diffusion the following quantities are emphasized during the analysis:

- the diffusion;
- amount of gas arisen during using the capacitor.

A. Gas generation and diffusion

Inside in a capacitor during operation gases arise and the internal pressure will increase. The amount of gases depends on the building component of the capacitor. Usually the water content of electrolyte dissolves to hydrogen and oxygen. The oxygen fuse with the aluminum to regenerate the aluminum oxide layer and hydrogen gas develops. The inner pressure can be more than five bars. This arisen gas can diffuse thanks to the structure of the capacitor. The higher the diffusion the lower the inner pressure. Experimental studies show that the diffusion is mainly on safety vent therefore testing the safety vents is also a must in estimation of the life-span. The gas can also diffuse on the rim of the aluminum case but previous tests show that this diffusion is irrelevant

During the measurements the capacitors are brought under artificial aging and when the main physical parameters alter considerably from the initial values (when they were manufactured), than can be drown that the device can't provide further the required specifications.

Before effectuating the tests the special gas sensors and the gas tap were placed into the capacitor cap and the capacitor cases were filled with electrolyte to simulate the real mechanism. To accelerate the whole process the capacitor can had to be pressurized (2.5 bars). Because the operating temperature can be different the test were taken at several temperatures (room, 50ºC, 85 ºC, 105 ºC).

The pressure of capacitors as a function of time can be seen in Fig. 11. Different electrolytes cause different gas pressure so each line represents a capacitor with different type of electrolyte. We can see two types of safety vents and a capacitor without went on multiple temperatures. With the help of this graph the alteration of the inner pressure as a function of the pressure can be determined as we can see in Fig. 12.. There are three different vents is on the graph. With a good vent, as the pressure arises, the diffusion arises too. So with the standard vent, this is the red one, the pressure is continuously arising, because the diffusion is almost zero. The black and the blue one is much better.

B. Burst pressure

When a capacitor reaches high pressure the safety vent bursts, sometimes the capacitor blows up and the capacitor becomes unusable. Foremost to analyze the burst pressure of the safety vents (see Fig. 13) is also necessary. Consequently the mainly used safety vents are tested and the pressure which can not be tolerated by the vents is determined.

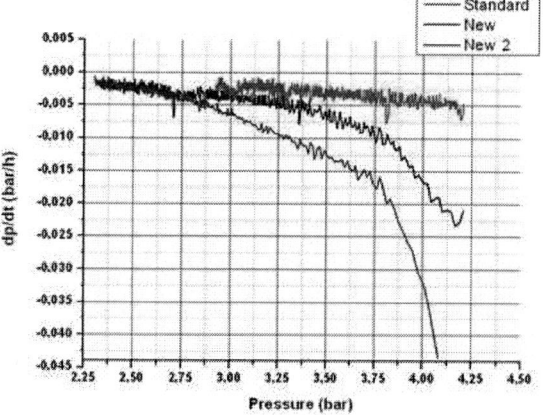

Fig. 12. Diffusion graphs

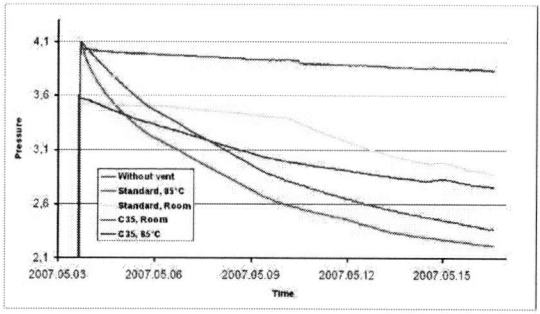

Fig. 11. Gas pressuer vs. time

Fig. 13. Cover disc and the safety vent of a capacitor

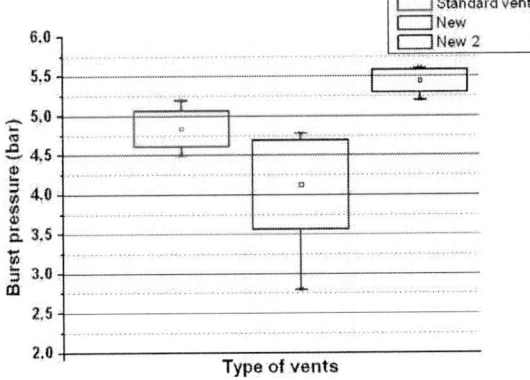

Fig. 14. Burst pressure graph

During these experiments different vent types were tested by a special measurement tool which was developed for special tests like this. With a numerous samples from the different types the burst pressure graph can be determined for each type.

In the graph below (see Fig. 14) three vent types can be seen. For example at the standard vent (red) case the maximum (5.2 bars), the minimum (4.5 bars) and the average (4.8 bars) burst pressure of the standard vent can be seen. As the height of the rectangles the standard deviation of the three different vent types can be read too. Creating a model the minimum burst pressure will be taken into account.

C. Estimation method of life-span

After the gas pressure and the burst pressure analysis it can be possible to determine a model which depends on the most significant parameters as in (1). where the temperature (T), the Boltzmann factor (k) and the free volume (V_{free}) are given, will contain a polynomial which is calculated by the approximation of the diffusion graphs and the amount of the generated gas (g_{prod}).

$$\frac{\partial p}{\partial t} = T \cdot \frac{k}{V_{free}} \cdot \left[A \cdot g_{prod} + \left(a_0 + a_1 p + a_2 p^2 + a_3 p^3 + a_4 p^4 \right) \right] \quad (1)$$

Thanks to this model the generated gas as a function of time ($P(t)$) can be calculated and with the burst pressure of the safety vent (P_B) the life-span of the capacitor (t_B) with a good approximation can be determined (see Fig. 15).

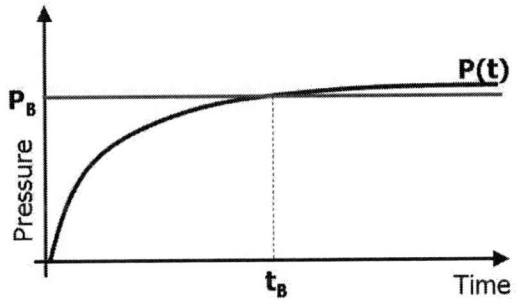

Fig. 15. Life-span estimation methode

V. CONCLUSION

The aim of this paper was to present partly the measurement automation system of an Electrolytic Capacitor Development Laboratory at EPCOS Hungary. The paper introduced only a few application of the entire system. More then 27 different electrolyte and capacitor measurements, where automated. All the measurements have been implemented in a similar manner. Firstly the user initializes the measurement, sets the measurement parameters, launch the execution and leave the program to run on its own, sending the results of the measurement to a database system, from where the data can be retrieved in a predefined or a non-predefined way.

After validation of the system, there are many advantages as making the measurements more precise and more reliable, fault tolerant (i.e.: missing of line voltage, open gate...), running multiple measurements in parallel, which all contribute to speed up the research and development of new component and devices

Now, after the initialization of a measurement it works on its own. The developers do not have to encroach. The developed measurement system controls and harmonizes the different devices and supervises their work. The user can simply check the measurement phase by glance at the screen. The programs estimate and display the running time of the experiments, allowing for the researchers working on the laboratory to manage the instrumental resources in time and to schedule in advance (for hours, weeks and months) the new measurements. Another big advantage is the database system, which stores the result of each measurement in an easy searchable way, reports and diagrams can be made automatically and the results can be reused in later research. The effectiveness of the system is presented via the inner gas pressure measurement of electrolytic capacitors to estimate the life-span of the capacitors. According to the experiments very useful results can be drown.

ACKNOWLEDGMENT

The author's whish to tanks the support of the applied research grant under the code *GVOP-3.1.1.-2004-05-0029/3.0*.

REFERENCES

[1] D. Fodor, O. Klug, I. Bálint, A. Horváth, A. Riz, "Electrolyte measurements automation for capacitor research and development", *in Proc. Of the 12th International Power Electronics and Motion Control Conference*, pp. 1277-1282, Portoroz, Slovenia, August 30 - September 1 2006.

[2] D. Fodor, O. Klug, I. Balint, A. Horváth, "Aluminium electrolytic capacitor research and development time optimization based on a measurement automation system", *submitted for presentation at the 10th International Conference on Optimization of Electrical and Electronic Equipment's, OPTIM'06*, Brasov, Romania, May 18-19 2006.

[3] K.H. Theisbürger, "Der Elektrolyt-Kondensator", FRAKO Kondensatoren- und Apparatbauen G.m.b.H Teningen (inner documentation)

[4] Per-Olof Fägerholt, "Passive components", 1999 (inner documentation)

Robust Control Methodologies for Optical Micro Electro Mechanical System - New approaches and Comparison

Alireza Izadbakhsh*, S.M.R. Rafiei, Senior Member, IEEE †

* Shahrood University of Technology, Shahrood, Iran, *izadbakhsh_alireza@hotmail.com*
† Shahrood University of Technology, Shahrood, Iran, *rafiei@ieee.org*

Abstract—The paper presents a comparison of three new robust control approaches for a micro electro mechanical system (MEMS) optical switch, considering uncertainties such as system parameter variations, external disturbance, and un-modeled dynamics. The control approaches are Sliding Mode Control, High Gain Control, and Model Free Based Control. Two of these approaches are quite novel in the field of MEMS control design. The comparative results highlights the main advantages of the proposed approaches which are simplicity, practicably, low computation burden, and low simulation time to control of electro mechanical systems.

Keywords—High gain control, model- free control, robust MEMS control, micro electro mechanical systems.

I. INTRODUCTION

The problem of robust control of micro electro mechanical system has been an interesting subject over the past decades [1-5]. To overcome uncertainties, many control schemes were proposed such as adaptive [6], neural networks [7], and variable structure control (VSC) [8]. VSC scheme is a powerful strategy against external disturbances, quickly varying parameters and un-modeled dynamics. However, a conservative design may be obtain, since the VSC scheme should be design to treat the worst situation of uncertainties [9]. Alternatively, fuzzy/neural network based control methods are also known as effective and robust approaches for uncertain systems. Neural networks provide powerful abilities such as adaptive learning, parallelism, fault tolerance, and generalization to the fuzzy controller. However, it is very difficult to guarantee the stability and robustness of neural network control systems [10]. In addition, some fuzzy/neural networks methods require predefined and fixed fuzzy rules or NN structure, which reduce the flexibility of the controller [10].

This paper is organized into six sections. Following the introduction, in section II, sliding mode control is presented. Sections III and IV are the main results of this paper. In Section III, the high gain control scheme is designed. Model-free control scheme is constructed in section IV. Illustrative purpose confirmed by a computer simulation in section V. Finally, conclusions are drawn in section VI.

II. SLIDING MODE CONTROL

Consider the dynamic equations of the micro electro mechanical system described in [11, 12] as follows

$$m\ddot{X} + d\left(X,\dot{X}\right) + k(X) = f\left(V,X\right) \quad (1)$$

Where m is the effective moving mass of the shuttle, d is a function describing losses such as damping and friction, k is the stiffness of the suspension, f is the electrostatic force acting on the model, V is voltage applied over the electrodes, and x is the shuttle position [12]. On the other hand we assume that $f(V, X)$ can be written as

$$f\left(V,X\right) = g(X).u \quad (2)$$

Where u is new control input vector according to voltage and g(x) is a nonsingular function. Based on the last assumptions and some mathematical works, the system's dynamic equations can be represented as follows

$$M\ddot{X} + D\left(X,\dot{X}\right) + K(X) = u \quad (3)$$

Where

$$M = g^{-1}(X)m \ , \ D\left(X,\dot{X}\right) = g^{-1}(x)d\left(X,\dot{X}\right)$$
$$K(x) = g^{-1}(x)k\left(X\right) \quad (4)$$

It can be easily shown that equation (3) can also be rewritten as

$$\begin{pmatrix} \dot{X}_1 \\ \dot{X}_2 \end{pmatrix} = \begin{pmatrix} X_2 \\ \Lambda(X) + \Psi(X)u \end{pmatrix} \quad (5)$$

Where $\Psi(x)$ and $\Lambda(x)$ are nonsingular and nonlinear functions of x, respectively as follow

$$\Lambda(X) = -M^{-1}\left(D\left(X_1,X_2\right) + K(X_1)\right) , \Psi(X) = M^{-1} \quad (6)$$

Let us consider bounded disturbance Δ exerted on the system with known upper limit as

$$\begin{pmatrix} \dot{X}_1 \\ \dot{X}_2 \end{pmatrix} = \begin{pmatrix} X_2 \\ \Lambda(X) + \Psi(X)u + \Delta \end{pmatrix} , \ \left|\Delta\right| \leq \rho \quad (7)$$

Where $\rho \succ 0$ is the disturbance bound. The first step in designing a variable structure controller is to choose an appropriate state vector function as

$$s = \dot{e} + \lambda e \quad (8)$$

Where λ is scalar and e and $\dot{e}$ are defined as

$$e = X_1 - X_{1d} \quad, \quad \dot{e} = X_2 - X_{2d} \qquad (9)$$

Where x_d is desired displacement of actuator. It must be noted that, equating of last function to zero takes the state vector to sliding surface. Now, switching control can be designed by application of the Lyapunov stability theorem, where the candidate Lyapunov function is

$$V = \frac{1}{2}s^2 \qquad (10)$$

In the sliding mode, it is required that

$$\dot{V} = s\dot{s} = 0 \qquad \text{for} \qquad s=0 \qquad (11)$$

Or equivalently

$$\dot{V} = s\left(\Lambda(X) + \Psi(X)u - \dot{X}_{2d} + \lambda\dot{e}\right) = 0 \qquad (12)$$

It must be noted that, for variable structure control design, we designed a two components control input:

 1.Equivalent control u_e

 2.Switching control u_s

Equivalent control can be obtained as

$$u_e = \Psi(X)^{-1}\left(-\Lambda(X) + \dot{X}_{2d} - \lambda\dot{e}\right) \qquad (13)$$

In which, it is responsible to keep the system states on the sliding surface. Furthermore, the convergence condition requires that

$$\dot{V} = s\dot{s} \prec 0 \qquad \text{for} \qquad s \neq 0 \qquad (14)$$

With the derived equivalent control, we will have

$$\dot{V} = s\ \Psi(X)\ u_s \prec 0 \qquad (15)$$

Hence, a possible choose for switching control is

$$u_s = -k\ \Psi(X)^{-1}\ sign(s) \qquad (16)$$

Where *"sign"* is the signum function and $k \succ 0$ is determined by disturbance bound. It must be note that u_s is designed for rejecting the external disturbances while the system's states are successfully driven towards the sliding surface. The same work is used to suppress effects of external disturbances in [11]. From the control strategy point of view, VSC is a powerful strategy to overcome external disturbances, quickly varying parameters and un-modeled dynamics. However, a conservative design is usually obtained, since the VSC scheme should be designed to treat the worst situation of uncertainties [9]. Moreover, as we know, an important challenge here is the parameter identification and dynamic modeling required, which limits the operation range of the model-based control techniques. This problem becomes hypersensitive in the case of micro system. Therefore, even if possible, it leads to very high expensive ways. Therefore, design of a controller that solves the above problems has been the subject of interest for many researches over the last decade. In the next section, we propose a high gain control (HGC) strategy for MEMS control in presence of external disturbances and unstructured uncertainty based on feedback linearization.

III. HIGH GAIN CONTROL OF MEMS

Consider a generic nonlinear dynamic system as equation (3) in presence of external disturbances $\delta(t)$ as below

$$M\ddot{X} + D\left(X,\dot{X}\right) + K(X) = u + \delta(t) \qquad (17)$$

The robust controller design with state feedback linearization consist of finding a control law of the form

$$u = \hat{M}v + \hat{D}\left(X,\dot{X}\right) + \hat{K}(X) \qquad (18)$$

Where v is an auxiliary input. With substituting equation (18) in (17), and a few mathematic simplification we will have

$$\mathrm{v} = \ddot{X} + \Xi \qquad (19)$$

Where

$$\Xi = \hat{M}^{-1}\left(\Delta M\ddot{X} + \Delta D\left(X,\dot{X}\right) + \Delta K(X) - \delta(t)\right)$$
$$\Delta M = M - \hat{M}$$
$$\Delta D\left(X,\dot{X}\right) = D\left(X,\dot{X}\right) - \hat{D}\left(X,\dot{X}\right) \qquad (20)$$
$$\Delta K(X) = K(X) - \hat{K}(X)$$

It must be noted that, Ξ includes the uncertainties such as external disturbances, un-modeled dynamics, and parametric uncertainties. Now, we design, an outer feedback control loop as

$$\mathrm{v} = \ddot{X}_d + K_v\dot{e} + K_pe \qquad (21)$$

Where

$$e = X_d - X \qquad (22)$$

Substituting equation (21) in (19) leads to

$$\ddot{e} + K_v\dot{e} + K_pe = \Xi \qquad (23)$$

At a glance, the last equation is not robust against time variable disturbances. To overcome this shortcoming, we change the inner control law in equation (18) to

$$u = \hat{M}\left(\mathrm{v}+\Gamma\ \Xi\right) + \hat{D}\left(X,\dot{X}\right) + \hat{K}(X) \qquad (24)$$

Where Γ is a very large positive gain. With this new controller and using equation (19), we will have

$$\ddot{e} + K_v\dot{e} + K_pe = (1+\Gamma)^{-1}\ \Xi \qquad (25)$$

Therefore, it is sufficient that set Γ have a very large positive gain. The schematic diagram of the considered control scheme is shown in Fig. 1. Although this approach has not problems arisen from variable structure control (VSC) and fuzzy/neural network based control, however, it is very difficult to find an even approximate model of micro systems. Hence, we suggest using the previous work [13] to control micro electro mechanical systems.

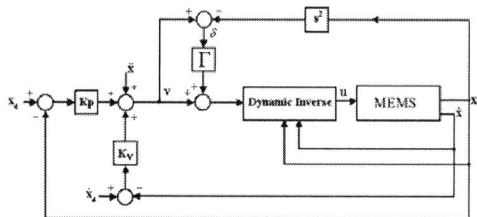

Fig. 1. High gain control scheme.

IV. MODEL-FREE CONTROL DESIGN

Suppose that equation (7) rewritten as

$$\dot{x} = Ax + Bu + \eta \tag{26}$$

Where

$$x = \begin{pmatrix} X_1 \\ X_2 \end{pmatrix} , \quad A = \begin{pmatrix} 0 & 1 \\ 0 & 0 \end{pmatrix} , \quad B = \begin{pmatrix} 0 \\ 1 \end{pmatrix}$$

$$\eta = \begin{pmatrix} 0 \\ \Delta + \Lambda(X) + (\Psi(X) - 1)u \end{pmatrix} \tag{27}$$

Now, we design, a state feedback controller as

$$u = -kx + k_0 r \tag{28}$$

This leads to a closed-loop system in the presence of uncertainties shown by state-space equations of the form

$$\dot{x} = (A - Bk)x + Bk_0 r + \eta \tag{29}$$

In addition, let us define a desired closed-loop state-space equation as below

$$\dot{x}_d = (A - Bk)x_d + Bk_0 r_d \tag{30}$$

Where r_d is the reference displacement. Subtracting (30) from (29) yields

$$\dot{\varepsilon} = (A - Bk)\varepsilon + Bk_0 \aleph + \eta \tag{31}$$

Where

$$\varepsilon = x - x_d , \quad \aleph = r - r_d \tag{32}$$

Here, we assume that the dynamic structure of η can generally be modeled by a p-order ordinary differential equation of the form [14]

$$\eta^{(p)} = \sum_{j=1}^{p} b_j \eta^{(p-j)} \tag{33}$$

Therefore, p-order derivative of (31) results in

$$\varepsilon^{(p+1)} - \sum_{j=1}^{p} b_j \varepsilon^{(p-j+1)} = (A - Bk)\left\{ \varepsilon^{(p)} - \sum_{j=1}^{p} b_j \varepsilon^{(p-j)} \right\}$$
$$+ Bk_0 \left\{ \aleph^{(p)} - \sum_{j=1}^{p} b_j \aleph^{(p-j)} \right\} + \left\{ \eta^{(p)} - \sum_{j=1}^{p} b_j \eta^{(p-j)} \right\} \tag{34}$$

Now, we suggest a control law as

$$\aleph^{(p)} - \sum_{j=1}^{p} b_j \aleph^{(p-j)} = -C \sum_{j=1}^{p} \mu_j \varepsilon^{(p-j)} - \mu_0 \left(\varepsilon^{(p)} - \sum_{j=1}^{p} b_j \varepsilon^{(p-j)} \right) \tag{35}$$

Where the c vector defines as $c = [1 \ 0]$. Substituting (35) and (33) into (34) results in the equations of motion in the error space

$$\varepsilon^{(p+1)} - \sum_{j=1}^{p} b_j \varepsilon^{(p-j+1)} = (A - Bk - Bk_0 \mu_0)\left\{ \varepsilon^{(p)} - \sum_{j=1}^{p} b_j \varepsilon^{(p-j)} \right\}$$
$$- Bk_0 C \sum_{j=1}^{p} \mu_j \varepsilon^{(p-j)} \tag{36}$$

The final step is to adjust the control input in (29) to suppress the effects of modeling errors as well as external disturbances as follows

$$r = \aleph + r_d \tag{37}$$

According to [15], good tracking accuracy can be achieved with relatively low uncertainty model error (p=1 or 2). The schematic diagram of the considered control scheme is shown in Fig. 2 The main advantages of the proposed approach are simplicity, practicably, and low computation burden for control of electro mechanical systems without any additional treatment.

V. EXAMPLES AND SIMULATION RESULTS

For illustrative purposes, a 2nd order micro electro mechanical system is considered. The dynamic model of the under study MEMS can be found in [11, 12]. Also, the desired positions are r_d=5μm and 23μm. Also, we consider a disturbance exerted on the system as Δ=D.Sin(ωt), where D=10^{-6}. Figs. 3-5 are giving the responses of the system, sliding surface and applied voltage for short displacement of actuator, respectively. The parameters of the controller in here are λ=-2, k=5×10^{-6} and ρ=10^{-6} [11]. In addition, Figs. 6-8 give the mentioned outputs for long displacement of actuator (r_d=5μm). It must be noted that available voltage for control is between 0-35V. As a test of robustness and highlight some weakness of VSC scheme, we exerted a disturbance on system as Δ=10^{-5} Sin(ωt), in which ω=π. Fig. 9 and Fig. 10 show the responses of the system for short and long displacement of actuator, respectively.

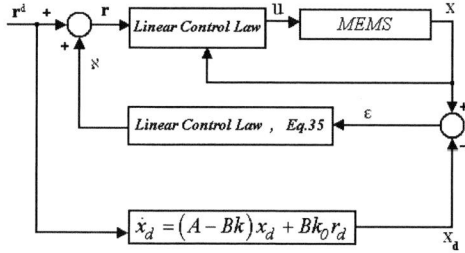

Fig. 2. Model-free control scheme.

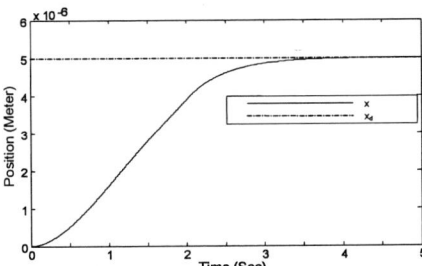

Fig.3. Responses of the system to desired r_d=5μm.

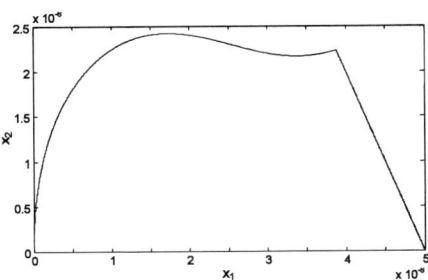

Fig. 4. Sliding Surface.

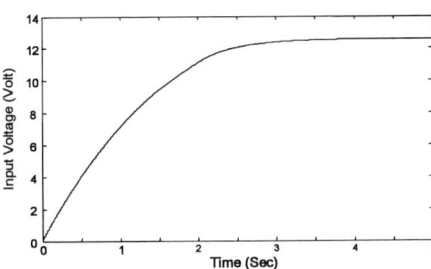

Fig. 5. Input Voltages.

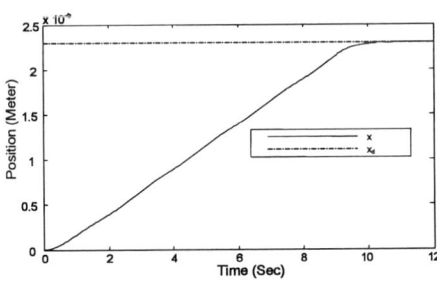

Fig. 6. Responses of the system to desired r_d=23μm.

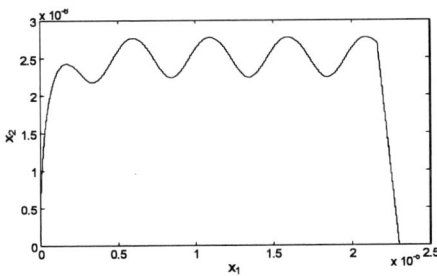

Fig. 7. Motion on Sliding Surface.

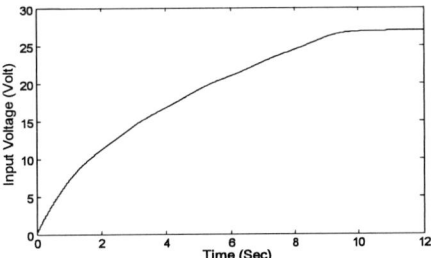

Fig. 8. Input Voltages.

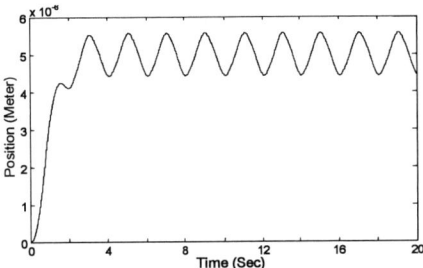

Fig. 9. Responses of the system to desired r_d=5μm.

The external disturbances are usually unknown in the most cases. This situation will fall worse, when the magnitude of the disturbances is larger than what expected. To overcome this problem, we used the high gain control approach in presence of parametric uncertainties. By following the procedure described in section III, control gains are set as k_p=25, k_v=10, and G=5×10^5.The results of simulations for short and large displacement of actuators are presented in Figs. 11-14. They indicate the position and applied voltage under 50 percent parametric uncertainties and $\Delta=10^{-3}$Sin(ωt). It must be noted that the identification process of stiffness of the suspension is very simpler to other parameters identification, therefore we set $\hat{k}$ to k. Now, we assume Δ set to be zero and there is 200 percent uncertainty in model parameters. It can be show that, we have the same results as before for both of short and large displacement states. As seen, in front of VSC approach, we have an excellent answer in presence of uncertainties. Finally, based on what described in section IV, and in presence of $\Delta=10^{-3}$ Sin(ωt), we draw the state feedback vector K and μ and b_1 as

$$K=\begin{pmatrix} 22500 & 300 \end{pmatrix}, \mu=\begin{pmatrix} 2000 & 16.3 & 0.0356 \end{pmatrix}, b_1=0 \quad (38)$$

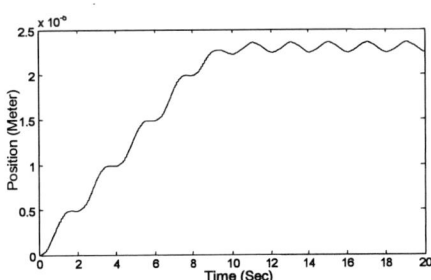

Fig. 10. Responses of the system to desired r_d= 23μm.

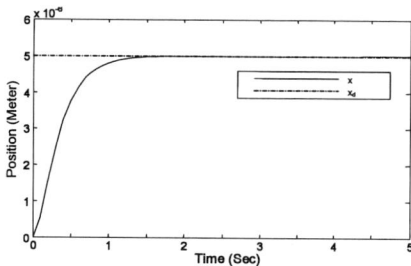

Fig. 11. Responses of the system to desired r_d=5μm.

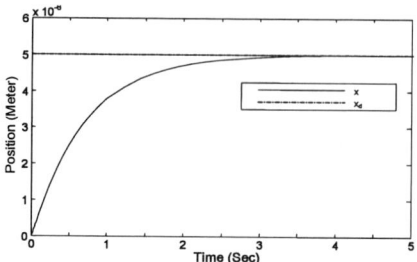

Fig. 15. Responses of the system to desired r_d=5μm.

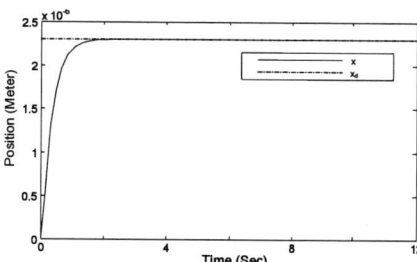

Fig. 12. Responses of the system to desired r_d= 23μm.

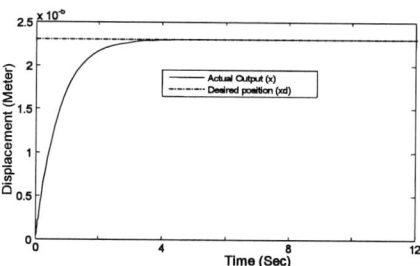

Fig. 16. Responses of the system to desired r_d= 23μm

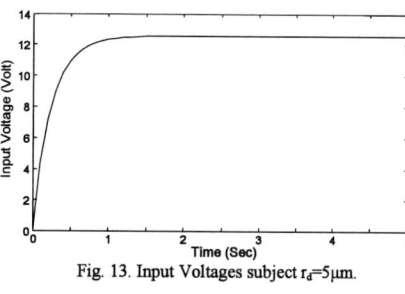

Fig. 13. Input Voltages subject r_d=5μm.

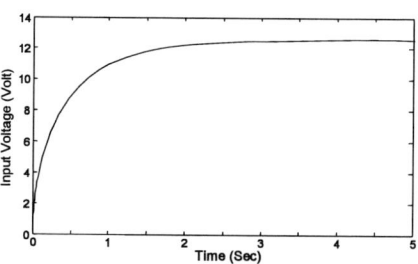

Fig. 17. Input Voltages subject to r_d=5μm.

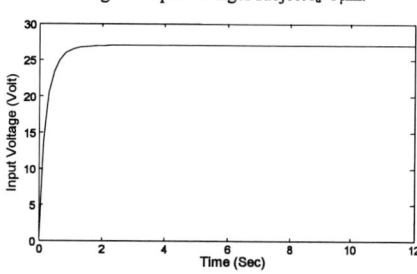

Fig. 14. Input Voltages subject to r_d=23μm.

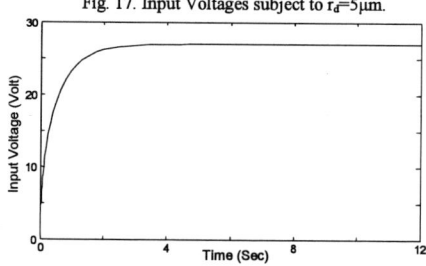

Fig. 18. Input Voltages subject to r_d=23μm.

By choosing p=1 for the uncertainty model and set the uncertainty equation to zero, we have really modeled all the uncertainty by an arbitrary constant (a step function). Figs. 15 and 16 show the displacement of the actuator with the presence of the disturbances including external disturbances and modeling error. It is observed that the output converges well to the desired set point. The actuators voltages are satisfactorily limited as shown in Figs. 17 and 18. In the next simulations, we compare the controllers in terms of design complexity, and theirs robustness against both additional external disturbances and modeling error. In the case of design complexity problem, it must be

mentioned that, sliding mode control have required the bounds of uncertainty. So that a few changes in the assumed bound, influences the performance of SMC approach. Therefore, to have a good response we are forced to design controller for new conditions. This problem is easier in the high gain control approach. So that we must only changes the Γ parameter and consequently, the robustness of this approach is also established under bad conditions of external disturbances. Finally, model free approach is simpler than aforementioned approaches. Because it is designs for an optional model of system, by modeling of uncertainties with a p-order ordinary differential equation. Based on which mentioned to them in above, Figs. 19 and 20 shows the performance of 3-control system in rejecting of disturbances under $\Delta=10^{-3}\text{Sin}(\omega t)$, and 50 percent parametric uncertainty for high gain control approach, for short and long displacement. The control gains are set as before. As seen, the response of HGC and MFC approach is stable in presence of external disturbances and modeling error, however for this magnitude of uncertainty, SMC will be un-stable. Finally, it can be shown that, two novel proposed approach are robust subject to $\Delta=0.1 \text{ Sin}(\omega t)$ in which, is a very large disturbance for micro electro mechanical system.

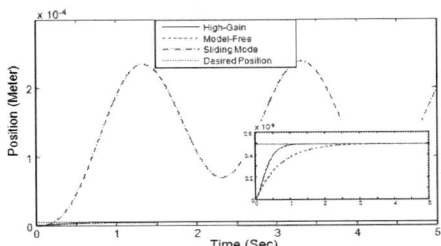

Fig. 19. Responses of 3 Control approach subject to r_d=5μm.

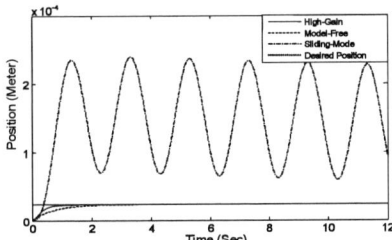

Fig. 20. Responses of 3 Control approach subject to r_d=23μm.

VI. CONCLUSIONS

Three novel robust control approaches consist of VSC control, high gain control, and model-free Control approaches have been proposed to control MEMS systems. The system stability was verified analytically and verified through simulations. It was shown that the control system is sufficiently robust against some uncertainties including modeling errors and external disturbances for almost all controllers. The main advantages of the high gain and model-free approaches are simplicity in design and/or modeling, practicably, low computation burden and low simulation time that make the aforementioned controllers quite interesting systems for control of electro mechanical systems.

REFERENCES

[1] X. Huang, R. Nagamune, and R. Horowitz "A Comparison of Multirate Robust Track-Following Control Synthesis Techniques for Dual-Stage and Multi-Sensing Servo Systems in Hard Disk Drives", Proceedings of the 2006 American Control Conference, USA, pp. 1290-1296.

[2] G. Zhu, J. Penet, and L. Saydy, "Robust Control of an Electrostatically Actuated MEMS in the Presence of Parasitics and Parametric Uncertainties", ", Proceedings of the 2006 American Control Conference, USA, pp. 1233-1238.

[3] J. Fei and C. Batur," A Novel Adaptive Sliding Mode Controller for MEMS Gyroscope", 46th IEEE Conference on Decision and Control, USA, 2007, pp. 3573-3578.

[4] G. Zhu, J. L´evine, and L. Praly, "On the Differential Flatness and Control of Electrostatically Actuated MEMS", 2005 American Control Conference, USA, pp.2493-2498.

[5] X. Huang, R. Horowitz, and Y. Li," Design and Analysis of Robust Track-Following Controllers for Dual-Stage Servo Systems with an Instrumented Suspension", 2005 American Control Conference, USA, pp. 1126-1131.

[6] N´estor O. P´erez Arancibia, Neil Chen, Steve Gibson, and Tsu-Chin Tsao, "Adaptive Control of a MEMS Steering Mirror for Suppression of Laser Beam Jitter", 2005 American Control Conference, 3586-3591.

[7] S. Jagannathan and Q. Yang,"A Robust Controller for the Manipulation of Micro-Scale Objects", 2005 American Control Conference, USA, pp. 4154-4159.

[8] N. Yazdi, H. Sane, T. D. Kudrle, and C. H. Mastrangelo, "Robust sliding mode control of electrostatic torsional micromirrorres beyond the pull-in limit", International Conference on Solid State Sensors, Actuators and Microsystems, Boston. June 8-12. 2003, pp. 1450-1453.

[9] Y. C. Chang, " Intelligent Robust Control for Uncertain Nonlinear Time-Varying Systems and Its Application to Robotic Systems" IEEE Transaction on systems, man, and cybernetics-part B: cybernetics, Vol. 35, NO. 6, pp. 1108-1119, December. 2005.

[10] M. Joo Er, and Y. Gao, "Robust adaptive control of robot manipulators using generalized fuzzy neural networks", IEEE Transactions on industrial electronics, vol. 50, NO. 3, PP. 620-628, June 2003.

[11] B. Ebrahimi, M. Bahrami, "Robust sliding-mode control of a MEMS optical switch", Journal of Physics: Conference Series 34 (2006) 728–733.

[12] B. Borovic, C. Hong, A. Q. Liu, L. Xie, and F. L. Lewis, "Control of a MEMS Optical switch," 43rd IEEE Conf. Decision and Control, 2004.

[13] A. Izadbakhsh, M. M. Fateh, "A Model-Free robust control approach for robot manipulator", International Journal of Mechanical Systems Science and Engineering, Vol. 1, No.1, pp. 32-37, 2008, ISSN 1307-7473.

[14] G.F. Franklin and A.Emami-Naeini, "A foundation of the multivariable information servomechanism problem,"Int. Rep, Stanford Univ, Informaion Science Lab., 1983.

[15] C. D. Doyle and G. Stein, "Multivariable feedback design: Concepts for a classical/modern synthesis,"IEEE Trans. Automat. Contr., vol. AC-26, no. 1, pp.4-16, Feb.1981.

Alireza Izadbakhsh was born in Garmsar, Iran, in 1980. He received the B.S degree in electronic Eng. from the Islamic Azad University, Garmsar Branch and the M.S.c degree in control Eng. from the Shahrood University of Technology, Shahrood, Iran, in 2003, 2007 respectively. His current research is on robust control of robots.

Mohammad-Reza Rafiei received the B.Sc. degree with honor from the Sistan and Baluchistan University, Zahedan, Iran, in 1991, received the M.Sc., and Ph.D. degrees from the Ferdowsi University of Mashhad, Mashhad, Iran, in 1995 and 2000, respectively all in electrical engineering. His research interests are power electronics and control systems, and computer science.

Modeling a Buck-Based Switching Amplifier for Sinusoid Wide Band Tracking by Using a Nonlinear Time Varying Map

A. El Aroudi*, E. Alarcón[†], E. Rodriguez[†], and R. Leyva*

*Departament d'Enginyeria Electrònica, Elèctrica i Automàtica (DEEEA),
Universitat Rovira i Virgili Tarragona, Spain, e-mail: [abdelali.elaroudi,ramon.leyva]@urv.net
[†]Departament d'Enginyeria Electrònica,
Universitat Politècnica de Catalunya, Barcelona, Spain, e-mail: [ealarcon,enricrv]@eel.upc.edu

Abstract—In this paper we derive a discrete time model to characterize the nonlinear dynamic behavior of a buck-based switching amplifier controlled by fixed frequency and PWM with a proportional-integral (PI) corrector. Numerical simulations show that the system can present period doubling bifurcation at the fast scale (switching frequency). As the switching frequency and the modulating frequency become closer, quasi-static approximation is not valid and the results based on it lead to erroneous conclusions about the stability of the system. We derive an exact solution discrete-time model able to predict accurately the nonlinear dynamical behavior of the system. The model is obtained without making quasi-static approximation and it can be used to obtain the useful region in the design parameter space from time domain simulations in a very fast and accurate manner.

Index Terms—Wideband Tracking, Discrete Time Model, Bifurcations, DC-AC Buck, Switching Amplifier, Instability, Large Signal, Modeling and Simulation.

I. INTRODUCTION

Switching power converters are widely used in the power management area, due to their potential of high efficiency, low cost and small size [1]. In most of the applications, these systems are used in situations where there is a need to stabilize a voltage to a desired constant value, but there also exist more advanced power management applications where it is required to add dynamics to its reference. Examples are audio amplifiers, envelope trackers in polar RF transmitters [2]-[13], line drivers for power line communications and dc-ac inverters [14].

One of the main drawbacks in switched power converters is the intrinsic nonlinear and switched nature of their control, which can result in nonlinear behavior such as bifurcations and chaos. It is well documented that accurate analytical models based on discrete-time recurrent nonlinear maps that describe the piecewise linear dynamic behavior allow to predict any kind of instability but this at the price of complex mathematical involvement [15]-[21]. Recent studies demonstrate the occurrence of different kind of instabilities and bifurcations in power electronics circuits. A large variety of complex nonlinear behaviors are shown to be possible for simple power electronics circuits like DC-DC converters. These studies allowed a deep understanding of their fundamental properties and dynamic behavior. They are based on obtaining accurate

mathematical models of these circuits allowing to predict the nonlinear behaviors that the system can undergo. Obtaining these mathematical models is a traditional challenge for the power electronics engineers. There are many efforts devoted to this research area. Nevertheless, the work in this field has hitherto been focused on regulation applications [15]-[21] and boost AC-DC power factor correction circuits with time varying input voltage [22]-[34]. In the case of a time varying input voltage, instabilities can occur at both fast (switching) and slow (varying line voltage) scale [33].

The work in this paper focuses instead in the nonlinear dynamics in the case of a time-varying reference voltage in DC-AC tracking applications. A typology of representative cases for tracking applications could include four categories, namely a) Quasi-static variation of the reference voltage b) Low-frequency sinusoidal reference variation, c) High-frequency sinusoidal reference, close to filter and switching frequency and d) Noise-like wideband signal. We will give an accurate discrete-time model that can be applied in all three first cases. The last case requires a statistical approach and it is beyond the scope of this work. The simplest case consists of a quasi-static variation of the output voltage level of the converter depending on system-level power requirements. Representative application-level examples include a slow case of Dynamic Voltage Scaling (DVS) for supplying microprocessors depending on computing workload, and quasi-static system-level control of supply voltage in radio frequency power amplifiers. Therefore its instantaneous stability will depend on the output level required in each instant. An example that is conceptually similar to case b) of a quasi-static dynamic reference in power switching converter is presented in [28]-[34], where the stability of power factor correction (PFC) is evaluated by a reference sinusoidal signal, the frequency of which is very low (50Hz). This work focuses in the stability and behavior of the non-linear switched system in the context of high frequency tracking applications, representative of audio amplifiers, envelope trackers in polar RF transmitters and line drivers for power line communications. In this case, reference has a wide-band frequency range and its amplitude covers the converter output voltage range.

978-1-4244-1741-4/08/$25.00 ©2008 IEEE

Quasi-static approximation is not valid in this case and an accurate model is required. In this paper we will first present some nonlinear phenomena that can occur in a simple DC-AC buck switching amplifier and then we will derive a nonlinear time varying map able to detect these nonlinear phenomena. The remainder of the paper is organized as follows: Section II presents the power switching amplifier based on buck converter under a proportional-integral (PI) feedback. Second III explores possible nonlinear phenomena by varying the design-space parameters especially focusing in the effect of the PI corrector gain on the reference signal tracking. Time domain simulations are given to put in evidence most of the phenomena that the system can undergo. In section IV we derive a discrete-time model able to predict the instability detected in the system from the continuous time switched model. We will proceed a careful modeling of a DC-AC half bridge buck switching amplifier being the approach similar in other switching time varying circuits like AC-DC and AC-AC converters. We will deal with the reference voltage as sinusoidal signal and no quasi-static approximation is done. In section V, the model is used to obtain the stability boundary in a suitable design parameter space. In section VI, we show that the large signal startup transient can also obtained from the same model. Finally, our conclusions are given in the last section.

II. BUCK-BASED SWITCHING AMPLIFIER SYSTEM DESCRIPTION AND CONTINUOUS TIME MODELING

A. Modeling of the power stage circuit

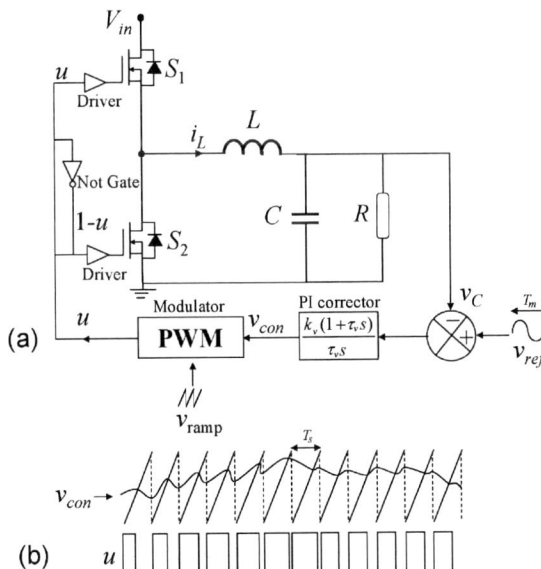

(a)

(b)

Fig. 1. Buck based switching amplifier under fixed frequency voltage mode control.

A typical buck-based switching amplifier consists of a modulator that convert a time varying signal into a

high frequency PWM signal followed by a half bridge or a full bridge power converter. The schematic diagram of half bridge DC-AC buck based switching amplifier is shown in Fig. 1. The activation of the switches S_1 and S_2 is done in a complimentary fashion in such a way that when S_1 is open S_2 is closed and vice-versa. In order to get perfect reference tracking, the output voltage is sensed and the voltage error represented by the difference voltage $v_{\text{ref}}(t) - v_C$ is processed by a PI corrector. The output $v_{con}(t)$ of this controller is connected to the non inverting input of the comparator whereas to the inverting input a sawtooth ramp generator is applied in such away that in the steady state the switch S_1 is activated each switching cycle while it is turned OFF whenever the ramp voltage $v_{\text{ramp}}(t)$ crosses the control signal $v_{con}(t)$. The circuit is designed in order to give periodic waveforms of state variables. There are two forcing periods that characterize the dynamics of the system in steady state, the switching period $T_s = 1/f_s$ and the reference voltage period $T_m = 1/f_m$. The stationary periodic attractor is settled therefore into a two-dimensional torus characterized by two periods: the switching period and the reference period. The system of state equations for the power stage during ON phase are given by:

$$\frac{dv_C}{dt} = -\frac{1}{RC}v_C + \frac{1}{C}i_L \tag{1}$$

$$\frac{di_L}{dt} = \frac{1}{L}\left(V_{\text{in}} - v_C\right) \tag{2}$$

whereas during OFF phase we have the following equations for the state variables:

$$\frac{dv_C}{dt} = -\frac{1}{RC}v_C + \frac{1}{C}i_L \tag{3}$$

$$\frac{di_L}{dt} = -\frac{1}{L}v_C \tag{4}$$

where L is the inductance of the inductor, C is the capacitance value, R is the load resistance, $v_C := x_1$ is the output capacitor voltage, $i_L := x_2$ is the inductor current and V_{in} is the input voltage.

B. Modeling of the output voltage controller

With the purpose that the T_s-averaged value of the output voltage v_C tracks perfectly a desired sinusoidal reference $v_{\text{ref}}(t) = V_0 + V_m \sin(\omega_m t)$ ($\omega_m = 2\pi f_m$), an outer voltage loop controller in the form of a PI compensator is used. The crossover frequency $f_v = 1/\tau_v$ is a design parameter that should be adjusted together with the gain k_v of the PI controller to get the desired behavior. In the frequency domain, the small signal transfer function of this controller can be written in the following form:

$$H_v(s) := \frac{\tilde{v}_{con}(s)}{\tilde{v}_{\text{error}}(s)} = k_v \frac{1 + s\tau_v}{s\tau_v} \tag{5}$$

where $v_{\text{error}} = v_{\text{ref}}(t) - v_C$ and k_v is the corrector gain. This dynamic controller adds a third state variable for the system whose dynamic behavior is given by:

$$\frac{dx_3}{dt} = V_0 + V_m \sin(\omega_m t) - v_C \tag{6}$$

C. Compact form expression of the exact switched model

The switched model of the system can be written in the following form:

$$\begin{aligned}\dot{\mathbf{x}} &= \mathbf{A}\mathbf{x} + \mathbf{B}_1(t) \quad \text{during ON phase}\\ \dot{\mathbf{x}} &= \mathbf{A}\mathbf{x} + \mathbf{B}_2(t) \quad \text{during OFF phase}\end{aligned} \tag{7}$$

where $\mathbf{x} \in \mathbb{R}^3$ is the vector of state variables whose entries are the capacitor voltage x_1, inductor current x_2, the integral of the error voltage x_3. $\mathbf{A} \in \mathbb{R}^{3 \times 3}$ and $\mathbf{B}_1 \in \mathbb{R}^3$ are the system matrices and vectors given by:

$$\mathbf{A} = \begin{pmatrix} -\frac{1}{RC} & \frac{1}{C} & 0 \\ -\frac{1}{L} & 0 & 0 \\ -1 & 0 & 0 \end{pmatrix} \tag{8}$$

$$\mathbf{B}_1(t) = \begin{pmatrix} 0 \\ \frac{1}{L}V_{\text{in}} \\ V_0 + V_m \sin(\omega_m t) \end{pmatrix} \tag{9}$$

$$\mathbf{B}_2(t) = \begin{pmatrix} 0 \\ 0 \\ V_0 + V_m \sin(\omega_m t) \end{pmatrix} \tag{10}$$

A compact form expression for the switched model of the system can be expressed by the following nonlinear time-varying model:

$$\dot{\mathbf{x}} = \mathbf{A}\mathbf{x} + (\mathbf{B}_1(t) - \mathbf{B}_2(t))u(t) + \mathbf{B}_2(t) \tag{11}$$

Taking into account the switching decision and that the binary command signal can take values in the set {0,1}, the switched model of the system can be written as:

$$\dot{\mathbf{x}} = \mathbf{A}\mathbf{x} + (\mathbf{B}_1(t) - \mathbf{B}_2(t))[u(v_{con}(t) - v_{\text{ramp}}(t))] + \mathbf{B}_2(t) \tag{12}$$

where $u(\Delta)$ stands for the Heaveside step function that describes the switching action when $v_{con}(t)$ and $v_{\text{ramp}}(t)$ intersect. Unlike the DC-DC converters, the duty cycle is time-varying from cycle to cycle even in steady state operation. The duty cycle corresponding to a switching period can be obtained by solving the equation resulting from the crossing of the ramp signal $v_{\text{ramp}}(t)$ with the control voltage $v_{con}(t)$ in this period. In this paper, we consider a ramp signal with offset value V_l, maximum value V_u and positive slope. In this case, the switching condition during one period can be written as:

$$\sigma(\mathbf{x}, t) := v_{con}(t) - V_l - (V_u - V_l)f_s t = 0 \tag{13}$$

where

$$v_{con}(t) = k_v(V_0 + V_m \sin(\omega t) - v_C) + \frac{k_v}{\tau_v}x_3 \tag{14}$$

If Eq. (13) is not feasible and $v_{con}(T_s) > V_u$ then d is set to 1 whereas it is set to 0 if $v_{con}(T_s) < V_l$. From the expression of the switched model (12), it can be deduced that the system is characterized by different scales of frequencies. Namely, the resonance frequency f_o due to the RLC output filter and it is contained in the state matrix $\mathbf{A}$, the switching frequency f_s contained in the expression of the ramp voltage and the modulating frequency f_m due

to the time varying sinusoidal reference voltage in $\mathbf{B}_1$ and $\mathbf{B}_2$. Let us define:

$$\Gamma_m = \frac{f_m}{f_0}, \ \Gamma_s = \frac{f_s}{f_0}, \ \gamma = \frac{\Gamma_s}{\Gamma_m} \tag{15}$$

In practice we have $f_m < f_0 << f_s$ in such away that $\Gamma_m < 1$, $\Gamma_s >> 1$ and $\gamma << 1$. In this case, the quasi-static approximation can be used. In wideband tracking applications we have $f_m < f_0 < f_s$ and γ is an important parameter that appear in the system model. In this case, quasi-static approximation, which not take into account this parameter, can give erroneous information about the dynamics of the system. Figure 2 shows a bode plot of the RLC output filter and example of output voltage spectrum in the frequency domain. This voltage contains spectral peaks at f_m, f_s, and $nf_s + mf_m$, where $n \in \mathbb{N}$ and $m \in \mathbb{Z}$

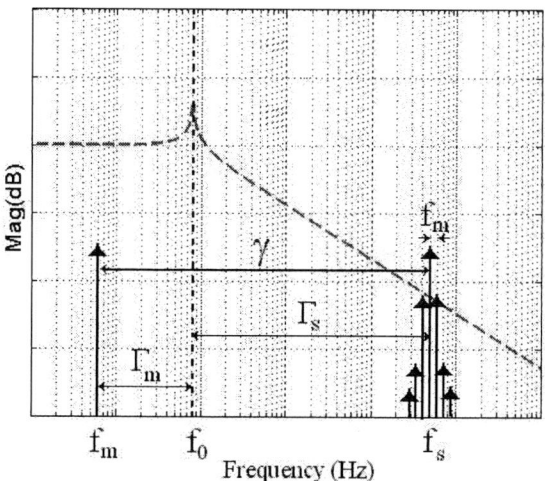

Fig. 2. Different scales of frequency in the system.

III. PERIOD DOUBLING OF THE SYSTEM FROM THE SWITCHED MODEL

Some interesting simulations will be shown in this section from the above described continuous-time switched model. The main steady-state periodic waveforms of the system is depicted in Fig. 3-a.

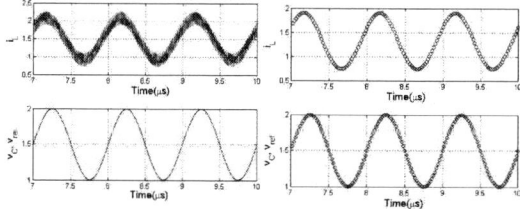

Fig. 3. Right: Waveforms of a DC-AC buck switching amplifier showing normal periodic behavior. Left: without sampling. Left: T_s- sampled time domain waveforms. $k_v = 24$. Other parameters are from Table I.

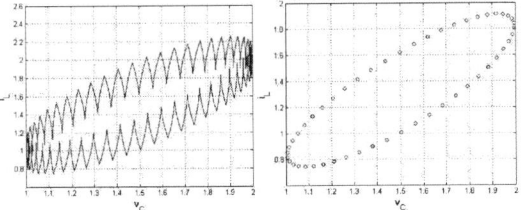 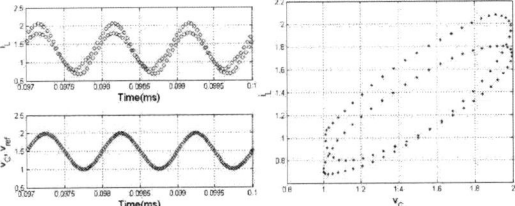

Fig. 4. Trajectory of a DC-AC buck switching amplifier showing normal periodic behavior. Left: without sampling. Right: Poincaré section at $t = nT_s$. $k_v = 24$. Other parameters are given in Table I.

Fig. 5. Left: T_s-sampled waveforms of a DC-AC buck switching amplifier showing period doubling behavior. Right: Poincaré section at $t = nT_s$ showing period doubling. $k_v = 27$.

TABLE I

PARAMETER VALUES USED IN NUMERICAL SIMULATIONS.

Parameter	Value
V_{in}	3 V
L	40 nH
C	100 nF
R	1 Ω
f_r	1 MHz
f_s	50 MHz
V_l	0
V_u	1 V
V_m	1.5 V
V_0	1 V
τ_v	1 μs
k_v	variable

The system trajectory evolves to a two-dimensional torus characterized by two periods that are the switching period and the reference sinusoidal voltage period. In order to make clear the figures, we sample the state variable at the switching period T_s. This is equivalent to taking a first order Poincaré section. The result is shown in Fig. 3-b. A detail of the inductor current and control signal waveforms is shown in Fig. 3-c where we can see clearly that the normal periodic behavior is characterized by switching periodically once every switching period as it is desired in a practical circuit. The first order Poincaré section of the steady-state trajectory of the system is an invariant closed curve indicating that the attractor in the continuous-time domain is effectively a torus (Fig. 4). The value of the circuit parameters used for Fig. 3 and Fig. 4 are shown in Table I. It is worth noting here that the dynamics range of the reference voltage $v_{\text{ref}}(t)$ must lie strictly between zero and the input voltage V_{in} in order that the system operates in normal operation (as it corresponds to a buck converter). In order to achieve output voltage tracking, the control parameters τ_v and k_v should be adjusted appropriately in terms of the output power and the dynamic margin of the reference voltage. Improperly adjusted values can give rise to different kinds of instabilities as it will be shown later. In a practical design procedure, the aim is to decrease the quasi-static error at the output and to maintain stable behavior. Note for example, from Fig. 3, that good tracking exists between the reference signal and the output voltage. In order to achieve a fast response to line and load disturbances while rejecting switching noise, the gain k_v and time constant τ_v should be taken in such a way that the crossover frequency be much smaller than the switching frequency.

Under parameter changes, the desired periodic solution could loose stability resulting in a behavior of the system with different dynamic characteristics to that of Fig. 3. Torus bifurcations may occur. A well known bifurcation phenomenon in power electronics circuits is period doubling bifurcation at the switching period. In a DC-AC switching amplifier, this bifurcation is possible if some parameters are not well adjusted. An example of waveforms in this case is shown in Fig. 5-a. The period doubling phenomena can be clearly observed in

the T_s-sampled inductor current i_L waveforms while it is not appreciated in the T_s-sampled output voltage v_C. The Poincaré section at $t = nT_s$ in Fig. 5-b and the continuous time control voltage $v_{con}(t)$ in Fig. 5-c show more clearly that a period doubling has occurred in the dynamical behavior of the system. This 2-periodic behavior can again lose stability under further parameter change and could result in another dynamic behavior. Usually, this period doubling phenomenon culminates in a chaotic behavior through a successive torus doubling [27]. However, due to the nonsmooth character of the system, the period doubling cascade is suddenly interrupted by the appearance of a chaotic behavior (Fig. 6).

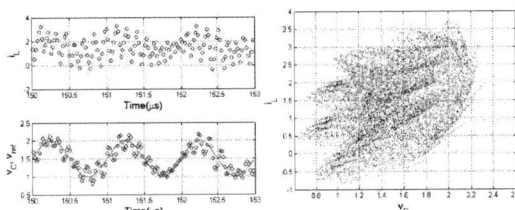

Fig. 6. T_s-sampled time domain waveforms showing chaotic behavior. Poincaré section at $t = nT_s$ showing chaotic behavior.

It should be noted that the period doubling instability at the switching period is due to the switching action and it can not be detected by using the averaged model. The detection of this instability requires the derivation of a discrete-time model. In the next section we will give a systematic procedure to obtain this model for the voltage

2111

mode controlled DC-AC switching amplifier under a PI corrector (Fig. 1). This model is shown be to able to predict most of the dynamical behaviors that can be predicted by the exact switched model. It should be noted that the ratio between the switching frequency and the reference frequency is a parameter that have an important effect on the dynamics of the system. More precisely, reference frequencies close to the switching frequency tends to stabilize the system but tracking is lost in this case. As a consequence, the quasi-static approximation which is valid in low frequency case will give erroneous conclusions about the system dynamics.

IV. DESCRIPTION OF THE SYSTEM DYNAMICS BY A DISCRETE-TIME MODEL

Accurate discrete-time models are derived from regular sampling of the state variables of the continuous-time dynamics [28], [29]. This model does not assume the approximations taken in the averaged modeling approach such as small ripple and high switching frequency and it is shown to be an accurate model in predicting the different kinds of instabilities of power electronics systems. Being characterized by two different forcing periods, two kinds of discrete-time modeling for the system can be obtained. If we are concerned with the dynamics of the system within the switching cycle, the first order discrete-time model can be defined from the current switching cycle to the next one. This is the mapping that relates the state variables from the beginning to the end of a switching cycle. If we look for the dynamics of the system during a reference cycle then the second order discrete-time model is considered [23]. Next, we will obtain the first order discrete-time model.

The system configuration during each switching sub-interval is linear and time varying as it is mentioned earlier. During ON phase the trajectory of the system, starting from the initial condition $\mathbf{x}_n$, is expressed by:

$$\mathbf{x}(t) = e^{\mathbf{A}(t-nT_s)}\mathbf{x}_n + \int_{nT_s}^{nT_s+t} e^{\mathbf{A}\tau}\mathbf{B}_1(\tau)d\tau \quad (16)$$

At time instant d_nT the system switches from ON phase to OFF phase. This instant can be determined from the equation resulting from the crossing of the ramp signal $v_{\text{ramp}}(t)$ with the control voltage $v_{con}(t)$. This equation can be written as follows:

$$\sigma(\mathbf{x}, d_n) := k_v(V_0 + V_m \sin(\omega_m t) - v_C(d_nT_s)) + \frac{k_v}{\tau_v}x_3(d_nT_s) - V_l - (V_u - V_l)d_n = 0 \quad (17)$$

At the instant $(n+1)T$ the system switches to ON phase again. The state of the system at time instant $(n+1)T_s$ is given by:

$$\mathbf{x}_{n+1} = e^{\mathbf{A}T_s}\mathbf{x}_n + e^{\mathbf{A}(1-d_n)T_s}\int_{nT_s}^{(n+d_n)T_s} e^{\mathbf{A}\tau}\mathbf{B}_1(\tau)d\tau + \int_{(n+d_n)T_s}^{(n+1)T_s} e^{\mathbf{A}\tau}\mathbf{B}_2(\tau)d\tau \quad (18)$$

It is worth noting here that (17) is a transcendental equation and that a root finding algorithm must be applied in order to obtain the duty cycle for each switching period.

In order to simplify the writing of the first order map, let us write it in the following form:

$$\mathbf{x}_{n+1} := \mathbf{P}(\mathbf{x}_n, n) = \mathbf{\Phi}\mathbf{x}_n + \mathbf{\Phi}_2\mathbf{\Psi}_1(n) + \mathbf{\Psi}_2(n) \quad (19)$$

where

$$\mathbf{\Phi} = e^{\mathbf{A}T}, \quad \mathbf{\Phi}_1 = e^{\mathbf{A}d_nT}, \quad \mathbf{\Phi}_2 = e^{\mathbf{A}((1-d_n)T)}$$
$$\mathbf{\Psi}_1(n) = \int_{nT}^{(n+d_n)T} \mathbf{\Phi}_1\mathbf{B}_1(\tau)d\tau, \quad (20)$$
$$\mathbf{\Psi}_2(n) = \int_{(n+d_n)T}^{(n+1)T} \mathbf{\Phi}_2\mathbf{B}_2(\tau)d\tau$$

Note that the vectors $\mathbf{\Psi}_k(n)$ are time-dependent, making the first order Poincaré map time-varying. Once the matrices $\mathbf{\Phi}_k$ are obtained, an expression for matrices $\mathbf{\Psi}_k(n)$ can be found from Eq. (20). The exponential matrices $\mathbf{\Phi}$, $\mathbf{\Phi}_1$ and $\mathbf{\Phi}_2$ can be obtained by using the Cayley-Hamilton theorem and the Putzer method [30]. For simplicity of integration of the system equations during each phase, let us write the vectors $\mathbf{B}_1(t)$ and $\mathbf{B}_2(t)$ as the sum of a constant term $\mathbf{B}_a$ and a T_m-periodic time-varying term $\mathbf{B}_b \sin(\omega_m t)$:

$$\mathbf{B}_1(t) = \mathbf{B}_{a,1} + \mathbf{B}_b \sin(\omega_m t) \quad (21)$$
$$\mathbf{B}_2(t) = \mathbf{B}_{a,2} + \mathbf{B}_b \sin(\omega_m t) \quad (22)$$

where

$$\mathbf{B}_{a,1} = \begin{pmatrix} 0 \\ \frac{V_{\text{in}}}{L} \\ V_0 \end{pmatrix}, \quad (23)$$

$$\mathbf{B}_{a,2} = \begin{pmatrix} 0 \\ 0 \\ V_0 \end{pmatrix}, \quad (24)$$

$$\mathbf{B}_b = \begin{pmatrix} 0 \\ 0 \\ V_m \end{pmatrix} \quad (25)$$

By computing the integral term in (16), $\mathbf{\Psi}_1(n)$ can be expressed as:

$$\mathbf{\Psi}_1(n) = (\mathbf{A}^2 + \omega_m^2\mathbf{I})^{-1}[\omega_m\mathbf{\Phi}_1\mathbf{B}_b\cos(n\omega_m T_s) + \mathbf{\Phi}_1\mathbf{A}\mathbf{B}_b\sin(n\omega_m T_s) - \omega_m\mathbf{B}_b\cos((n+d_n)\omega_m Ts) \\ \mathbf{A}\mathbf{B}_b\sin((n+d_n)\omega_m T_s)] + \int_{nT}^{(n+d_n)T} e^{\mathbf{A}\tau}d\tau\mathbf{B}_{a,1} \quad (26)$$

The vector $\mathbf{\Psi}_2(n)$ can be obtained in the same way. Note that different expressions for $\mathbf{\Psi}_k$ $(k = 1, 2)$ would be obtained if the reference voltage was considered constant (quasi-static approximation). Note also that the function $\mathbf{\Psi}_k$ corresponding to a constant reference voltage can be obtained by putting $f_m = 0$ in Eq. (26). Once we get both of $\mathbf{\Phi}_k$ and $\mathbf{\Psi}_k(n)$ and combining with Eq. (17), the expression of $\mathbf{P}$ is obtained from Eq. (19). $\mathbf{P}$ is the mapping that relates the vector of the state variables $\mathbf{x}_n$ at the beginning of the switching cycle to $\mathbf{x}_{n+1}$ that at the end of the same cycle. Note that it is a non-linear map in the state variables and periodically time-varying in the discrete-time domain. This non-linear time-varying map has a periodic orbit (not a fixed point) as a nominal operating regime. The period N of this discrete-time periodic orbit is Γ_m/Γ_s if this ratio is an integer number. If Γ_m/Γ_s is a rational number, the period will be the denominator of this number. If the periodic orbit is stable, and starting form a point $\mathbf{x}_0$ belonging to this periodic orbit, the map will take entirely N periods to return to the same point.

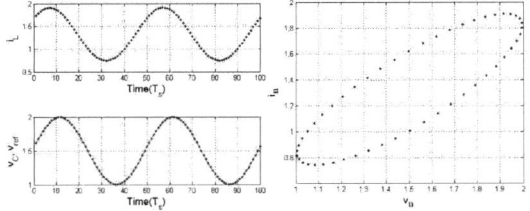

Fig. 7. Left: waveforms of a DC-AC buck switching amplifier from the discrete-time model showing normal periodic stable. Right: Corresponding Poincaré section behavior.

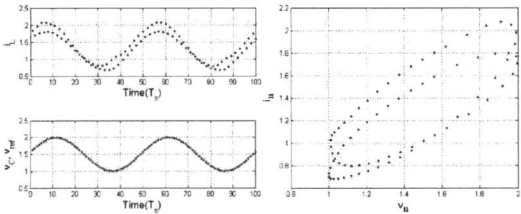

Fig. 8. Left: waveforms of a DC-AC buck switching amplifier from the discrete-time model showing period doubling at the switching period. Right: Corresponding Poincaré section.

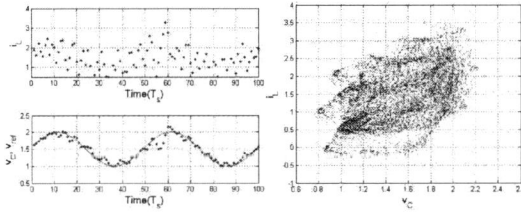

Fig. 9. Left: waveforms of a DC-AC buck switching amplifier from the discrete-time model showing chaotic oscillations. Right: Corresponding Poincaré section.

Numerical simulations are carried out using this map in a MATLAB code and an iterating procedure to generate the discrete-time waveforms. The results are obtained by using the same parameter values for their corresponding time domain simulations from the switched model. Simulations corresponding to these cases are shown in Fig. 7, Fig. 8 and Fig. 9. Figure 7 corresponds to the stable operation of the system while Fig. 8 and Fig. 9 correspond to the subharmonic oscillations at switching period resulting from a period doubling bifurcation and chaotic behavior respectively. As it can be observed there is a perfect agreement between the results obtained from the numerically integrated switched model and our derived analytical discrete-time model. However, the results obtained from our model are much obtained faster than those obtained by using the switched model as it is always more easy to iterate a recurrence equation than to numerically integrate a differential equation.

V. STABILITY BOUNDARY FROM THE DISCRETE TIME MODEL

In this section we will obtain the useful region (region of stable periodic behavior) of the DC-AC buck switching amplifier in the parameter space. As design parameters we will now choose Γ_m, Γ_s in addition to k_v the gain of the PI controller. All other parameters are considered constants as shown in Table I. We vary Γ_m and Γ_s in a suitable grid and for each point in this grid we increase the value of k_v till obtaining a bifurcation. The critical value is recorded and the process is repeated for the whole points pertaining to the selected grid in the (Γ_m, Γ_s) plane. The results are obtained both from both the switched model and the discrete time model. Figure 10 shows the stability boundary as a two-dimensional surface in the design parameter space. Note that the reference frequency is swept from low frequencies to the resonance frequency since higher frequencies will results in inadequate tracking in average. The simulations show also a higher influence of the switching frequency compared to that of the modulating frequency. It is worth noting that low frequencies could be considered a conservative limit that guarantees that in higher reference frequencies the output is stable even near the resonance frequency. Note also that the quasi-static approximation does not take Γ_m as a design parameter as it is considered extremely small.

CONCLUSIONS

We have presented a characterization of dynamic behavior of a buck-based switching converter with output voltage PI feedback in a signal tracking configuration with a sinusoidal reference. This configuration is considered to be indicative of wide band switching amplifiers as required in audio amplifiers, envelope trackers in polar RF transmitters and line drivers for power line communications. The simulation results presented throughout the paper show that subharmonic oscillations can occur at the switching period scale if the system parameters are not well selected. After exploring the different dynamic behaviors that the system can present, we observed that subharmonic oscillations in the form of period doubling require a discrete-time model to be detected. Such model was derived and it was shown to be able to accurately predict the instability phenomena. The model was derived without making traditional assumptions like the quasistatic approximation. Future works will deal with the application of the approach used in this paper to other more complex time-varying circuits such as multi-phase and multi-level power electronics switching amplifiers.

ACKNOWLEDGMENTS

This work was partially supported by the Spanish MED under grant TEC-2007-67988-C02-01 & 02. EU FENDER funds are also acknowledged.

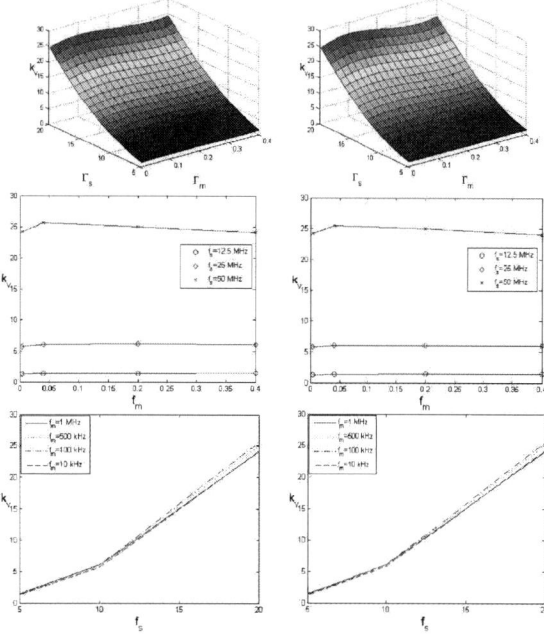

Fig. 10. Stability boundary in the design parameter space $(\Gamma_m, \Gamma_s, k_v)$, Top left: from the switched model, Top right: from the discrete time model, Middle left:, Middle right, Bottom left, Bottom right: projection in the two-dimensional parameter space (f_s, k_v)

REFERENCES

[1] R. W. Erickson and D. Maksimovic, "Fundamentals of Power Electronics", Springer, 2001.

[2] N. O.Sokal and A. D.Sokal, "Class E, a new class of high efficiency tuned single-ended switching power amplifiers," IEEE J. Solid-State Circuits, vol. SC-10, pp. 168-176, June (1975).

[3] F. H. Raab, P. Asbeck, S. Cripps, P. B. Kenington, Z. B. Popovic, N. Pothecary, J. F. Sevic, and N. O. Sokal, "Power amplifiers and transmitters for RF and microwave," IEEE Trans. Microw. Theory Tech., vol. 50, no. 3, pp. 814-826, Mar. 2002.

[4] P. Midya, K. Haddad, L. Connell, S. Bergstedt, and B. Roeckner, "Tracking power converter for supply modulation of RF power ampli- fiers," in Proc. IEEE Power Electron. Specialists Conf. (PESC'01), vol. 1, 2001, pp. 1540-1545.

[5] J. Staudinger, B. Gilsdorf, D. Newman, G. Norris, G. Sadowniczak, R. Sherman, and T. Quach, "High efficiency CDMA RF power amplifier using dynamic envelope tracking technique,"in IEEE MTT-S Int. Microw. Symp. Dig., vol. 2, Jun. 2000, pp. 873-876.

[6] S. Abedinpour,K. Deligoz, J. Desai, M. Figiel, and S. Kiaei, "Monolithic supply modulated RF power amplifier and DC-DC power converter IC," in IEEE MTT-S Int. Microw. Symp. Dig., vol. 1, Jun. 2003, pp. A89-A92.

[7] D. K. Su and W. J. McFarland, "An IC for linearizing RF power am- plifiers using envelope elimination and restoration,"IEEE J. Solid-State Circuits, vol. 33, no. 12, pp. 2252-2258, Dec. 1998.

[8] D. R. Anderson and W. H. Cantrell, "High-efficiency high-level modulator for use in dynamic envelope tracking CDMA RF power ampli- fiers,"in IEEE MTT-S Int. Microw. Symp. Dig., vol. 3, May 2001, pp. 1509-1512.

[9] P. Midya, K. Haddad, and M. Miller, "Buck or boost tracking power converter,"IEEE Power Electron. Lett., vol. 2, no. 4, pp. 131-134, Dec. 2004.

[10] A. Soto, J. A. Oliver, J. A. Cobos, J. Cezon, and F. Arevalo, "Power supply for a radio transmitter with modulated supply voltage,"in Proc. IEEE Appl. Power Electron. Conf., vol. 1, Feb. 2004, pp. 392-398.

[11] P. Midya, "Linear switcher combination with novel feedback,"in Proc. IEEE Power Electron. Specialists Conf., vol. 3, Jun. 2000, pp. 1425-1429.

[12] W. Feipeng, A. Ojo, D. Kimball, P. Asbeck, and L. Larson, "Envelope tracking power amplifier with pre-distortion linearization for WLAN 802.11g,"in IEEE MTT-S Int. Microw. Symp. Dig., vol. 3, Jun. 2004, pp. 1543-1546.

[13] A.H. Feipeng Wang Yang, D.F. Kimball, L.E. Larson and P.M. Asbeck, "Design of Wide-Bandwidth Envelope-Tracking Power Amplifiers for OFDM Applications," F. Wang, A. Yang, D. Kimball, L.E. Larson and P. M. Asbeck, IEEE Trans. Microwave Theory and Techniques, pp. 1244-1255, 53, 4, (2005).

[14] C. Meza; J. J. Negroni; F. Guinjoan and D. Biel, "Inverter Configurations Comparative for Residential PV-Grid Connected Systems", IEEE Industrial Electronics, IECON 2006 - 32nd Annual Conference on Volume , Issue , Nov. 2006 Page(s):4361 - 4366.

[15] D. C. Hamill and D. J. Jeffries, "Subharmonics and chaos in a controlled switched-mode power converter", IEEE Transactions on Circuits and Systems, vol. 35, no. 8, pp. 1059–1061, Aug. 1988.

[16] J. H. B. Deane and D. C. Hamill, "Analysis, simulation and experimental study of chaos in the buck converter", in Power Electronics Specialists Conference, 1990. PESC '90 Record., 21st Annual IEEE, San Antonio, TX, Jun. 1990, pp. 491–498.

[17] M. di Bernardo, F. Garefalo, L. Glielmo, and F. Vasca, "Switch-ings, bifurcations, and chaos in DC/DC converters", Circuits and Systems I: Fundamental Theory and Applications, IEEE Transac-tions on [see also Circuits and Systems I: Regular Papers, IEEE Transactions on], vol. 45, no. 2, pp. 133–141, feb 1998.

[18] C. K. Tse and M. di Bernardo, "Complex behavior in switching power converters", Proceedings of the IEEE, vol. 90, no. 5, pp. 768–781, May 2002.

[19] K. Chakrabarty, G. Poddar, and S. Banerjee, "Bifurcation behavior of the buck converter", IEEE Transactions on Power Electronics, vol. 11, no. 3, pp. 439–447, May 1996.

[20] H. H. C. Iu and C. K. Tse, "Bifurcation behavior in parallel-connected buck converters", Circuits and Systems I: Fundamental Theory and Applications, IEEE Transactions on [see also Circuits and Systems I: Regular Papers, IEEE Transactions on], vol. 48, no. 2, pp. 233–240, Feb. 2000.

[21] Orabi, M.; Ninomiya, T.; "Nonlinear Dynamics of Power-Factor-Correction Converter", IEEE Trans. on Industrial Electronics, Vol. 50, Issue: 6, 2003, pp. 1116 - 1125.

[22] H. H. C. Iu, Zhou and C. K. Tse, "Fast-Scale Instability in a Boost PFC Converter Under Average Current Control", International Journal of Circuits Theory and Applications, Vol. 31, pp. 611-624, 2003.

[23] S. K. Mazumder, M. Alfayoumi, A. H. Nayfeh, and D. Borojevic, "An Investigation Into Fast- and Slow-Scale Instabilities of a Single Phase Bidirectional Boost Converter", IEEE Trans. on Power Electronic, Vol. 18, No. 4, pp. 1063–1069, 2003.

[24] A. El Aroudi, M. Orabi, L. Martínez-Salamero and T. Ninomiya, "Investigating Stability and Bifurcatios of a Boost PFC Circuit Under Peak Current Control", in Proceedings IEEE International Symposium on Circuits and Systems ISCAS'05, 2005.

[25] A. El Aroudi, M. Orabi, L. Martínez-Salamero and T. Ninomya, "Bifurcations in a Boost PFC Circuit Under Different Control Strategies", NOLTA 2004, pp. 317-322, Fukuoka, Japan.

[26] Mohamed Orabi and Abdelali El Aroudi, "Different Frequency Instabilities of Averaged Current Controlled Boost PFC AC/DC Regulators" IEEE International Telecommunications Energy Con-ference INTELEC 2006, Rhode Island, USA, 2006

[27] Munehisa Sekikawa, Tetsuya Miyoshi, and Naohiko Inaba, "Suc-cessive Torus Doubling", IEEE Trans. Ciruicts and Systems I, Vol. 48, No. 1, pp. 28-34 2001.

[28] A. El Aroudi, M. Debbat, G. Olivar, L. Benadero, E. Toribio and R. Giral, "Bifurcations in DC-DC Switching Converters: Review of Methods and Applications", International Journal of Bifurcations and Chaos, Vol. 15, No 5, pp. 1549-1578, May 2005.

[29] G. C. Verghese, M. E. Elbuluk, and J. G. Kasakian, "A General Ap-proach to Sampled-Data Modeling for Power Electronic Circuits," IEEE Trans. on Power Electronics, Vol. 1, pp. 76–89, 1986.

[30] Tom. A. Apostol, "Calculus", Vol. 2, 2nd edition, John and Wiely Sons, New York, 1988.

2114

Single Inductor Multiple Outputs Interleaved Converters Operating in CCM

Luis Benadero[*], Vanessa Moreno-Font[*], Abdelali El Aroudi[†] and Roberto Giral[†]

[*] Dep. Física Aplicada, Universitat Politècnica de Catalunya (UPC), Barcelona, Spain, *luis@fa.upc.edu*
[†] Dep. d'Enginyeria Electrònica, Elèctrica i Automàtica, Universitat Rovira i Virgili (URV), Tarragona, Spain

Abstract—Feasibility of single inductor multiple outputs (SIMO) dc-dc converters, able to operate in continuous-conduction mode (CCM), is shown in this paper. The power stage combines simple structures based on boost for non-inverted outputs and buck-boost for inverted outputs. Individual switches associated to each of the outputs are current mode controlled through respective stages as called channels. The set of dynamical references for every channel is obtained by means of a matrix arrangement whose input is the set of signals provided by proportional-integral (PI) blocks applied to each of the input-output errors. Finally, phase-shifted (interleaved) compensating ramps are added to those references. Analysis of dynamics stability is provided by means of an averaged model and direct simulation.

Keywords—Switched-mode power supply, PWM control, Single Inductor Converters, Interleaving.

I. INTRODUCTION

Single inductor multiple outputs dc-dc converters is an emerging solution for switching mode power supplies (SMPS) for portable applications supporting several outputs (such as PDAs) fed by a single battery, in order to fulfill requirements of power efficiency, cost and size [1-9]. Traditional SMPS implementation using a transformer with several windings has the problem that only one of the outputs is, in general, controlled. The drawback of handling independent converters for each of the outputs is the higher number of switches and external components. Besides those options and other dual dc-dc converter configurations [10-12], single inductor dc-dc converters are, in general, convenient solutions for low power applications. The use of a single inductor shared for several outputs is somehow dual to the use of several channels for a single output in parallel converters for high current applications. Several topologies of single-inductor multiple outputs (SIMO) dc-dc converters, which use just one inductor to transfer energy from the battery to several outputs, are described in [1]. Some of them are single-polarity as those derived from buck (step down) or from boost (step up) converters. Multiple bipolar outputs are also possible for instance by combining boost and buck-boost structures, then having bipolar converters.

The following question to be considered is the problem of control and regulation of SIMO converters. Time multiplexing is a simple and safe option [2], but to avoid cross regulation between outputs and to guarantee their stability, the inductor current must be zero within some duration between specific intervals for outputs, so in fact the outputs are independently controlled, regardless sharing the same inductor, and the converter must operate in discontinuous conduction mode (DCM). Pseudo-continuous conduction mode (PCCM) [3] uses a constant value for the inductor current as virtual zero, so although being CCM, its operating rules are equivalent to DCM. A different approach by using CCM operation and digital control with separate regulation of common and differential modes for the output voltages is in [4].

An alternative CCM approach for the regulation of two bipolar outputs by means of the time adjustment of three intervals of the modulating period is in [5]. In this work, the inductor is firstly fed by the input generator, secondly the output with higher current load is connected to the inductor and finally both outputs are connected in series with the inductor. The control acts in such a way that any kind of perturbation would produce automatic variation of time intervals so as to keep the regulation of the outputs. The stability analysis through averaged models [13-14] was also given in the paper. Discrete models provide a more accurate description of bifurcations of non-autonomous piecewise dynamical systems [15-18], so approximated Poincaré maps for the single-inductor dc-dc converter with single-phase control has been obtained and further details for instabilities given in [19-20].

To extend the mode of operation to a general multiple output case, a multi-phase current mode control is proposed in this paper. From the set of input-output errors, a set of dynamical references, which are associated to each of the channels, is obtained. Each of those reference signals is added to a phase-shifted compensating ramp, so that in the steady state regime, a sequence of time intervals, once per channel, will be produced. These intervals are associated to the transfer of energy from the input generator to the corresponding output like in the equivalent basic converter. The major advantage of this proposal is that the system operates naturally in CCM, being the cross regulation integrated as a part of the solution. On the other hand, as a drawback of the proposal, the stability margin is expected to decrease if the number of outputs is increased, while a detailed analysis of stability might become progressively difficult.

The rest of the paper is organized as follows. Section II deals with the description of several topologies for SIMO dc-dc converters based on boost and buck-boost elementary converters. In Section III, the basis for control is presented. In Section IV the averaged model is given and in Section V several results obtained by numerical simulations are carried out. Some conclusions are given in the last section.

978-1-4244-1741-4/08/$25.00 ©2008 IEEE

II. TOPOLOGIES FOR SIMO DC-DC CONVERTERS BASED ON BOOST AND BUCK-BOOST

The first schematic diagram in Figure 1 shows a dc-dc converter with a single inductor for two outputs with opposite polarity. Both loads, considered here as equivalent resistances R_1 and R_2, can be fed from the source V_{IN} by generating a sequence of signals driving switches S_A and S_B. One trivial case, the boost converter itself, would correspond to S_A always ON. In this case, a positive voltage in R_1 is given by controlling S_B. Conversely, the buck boost converter would correspond to S_B always ON, so that a negative voltage in R_2 is given by means of S_A. But the goal is the regulation of both positive and negative outputs accordingly to corresponding inputs, and this can be given following two different strategies, both of them using CCM operation, as will be demonstrated in the next section.

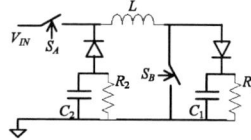

Fig. 1. Schematic diagram of a single inductor dc-dc converter with positive (1) and negative (2) outputs based on boost and buck-boost.

The extension to a four outputs converter is in Fig. 2. Two positive or non-inverted outputs (1, 3) and two negative or inverted outputs (2, 4) can be regulated by means of respective switches S_1 and S_3 combined with S_B, and besides them, S_2 and S_4 combined with S_A. Details of the control will be also given in the next section. A generic SIMO dc-dc converter with N_P positive outputs and N_N negative ones is straightforward from Fig. 2. For instance, a diagram for three outputs of single polarity based on boost is in Fig. 3.

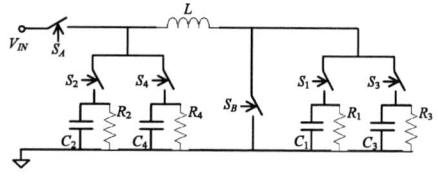

Fig. 2. SIMO dc-dc converter based on boost (positive outputs: 1, 3) and buck-boost (negative outputs: 2, 4)

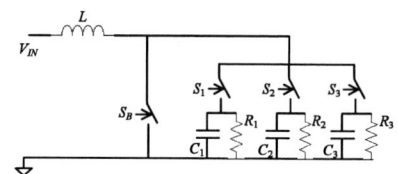

Fig. 3. SIMO dc-dc converter based on boost with three outputs

III. STRATEGIES FOR CONTROL

One control strategy for the single inductor dual output (SIDO) topology (Fig. 1a), similar to that proposed in [5], is shown in Fig. 4 (a). Based on current mode control, the general idea is the use of two comparators, one per each channel, everyone using a different modulating signal to be compared with another signal (v_I) which is proportional to the inductor current. The particular

modulating signal is formed by a single compensating ramp and some combination of the output of the PI error filtering blocks. It is demonstrated in the same paper that in order to achieve stable behavior, a certain channel should not use the output of the PI corrector of the same channel. That is switch S_A, which controls the current provided to the negative output, must be driven accordingly to the error produced only by the positive output. Likewise, switch S_B, which affects the positive output, is driven by the error of the negative output. The regulation is possible because of the crossed effect produced when considering the whole system (power stage plus control). Standard operation means period one behavior (same period than ramp) in the steady state regime. In general, three intervals (Fig. 5(a)) can be identified along this limit cycle. First one, which is triggered by the clock connected to the set input of two SR flip-flops, corresponds to S_A and S_B closed, so that the inductor is fed by the source. The second interval is triggered by one of the comparators (connected to the reset input of the flip-flop), so the related switch is open and the load will be connected to the inductor. The favored load will be that with higher current requirement. Finally, in the third interval (triggered by the other comparator), both switches are left open so that the inductor is kept connected in series with both outputs, thus meaning an equivalent current supply for them.

An alternative control strategy based on interleaving modulation is shown in Fig. 4 (b). The difference lies in the use of a multi-phase modulation, so compensating ramps with progressive delay are used for each of the channels. For that reason, the inputs of the SR flip-flops must be activated by pulses synchronized with the corresponding ramp. In Fig. 5 (b), the steady state regime is represented. The instants for transitions are given at the intersections of the modulating signals with the inductor current. This interleaving strategy can be easily extended for a generic number of outputs, for instance, the diagrams for control of the two positive and two negative outputs (Fig. 2) and the three positive outputs (Fig. 3) are represented in Fig. 6 and Fig. 7 respectively. These quite simple schemes for control, which is able to regulate each of the outputs, are possible because in boost and buck-boost dc-dc converters, the inductor is connected only to the voltage source during the ON interval.

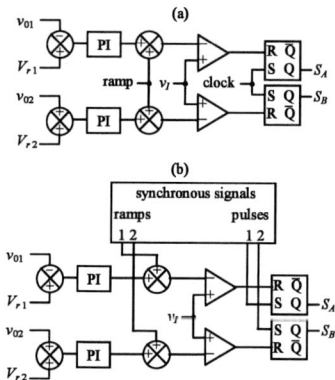

Fig. 4. Strategies of control of single inductor dual output (SIDO) in Fig.1 based on (a) single-phase or (b) multi-phase modulation. The feedback current is $v_I = r_S i_L$, being r_S the sensing resistance

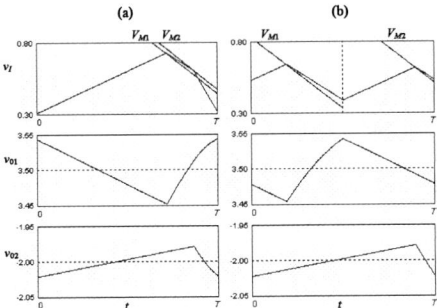

(a)   (b)

Fig. 5. Steady state behavior for SIDO converter (Fig. 1), (a) and (b) correspond to control in Fig. 4 (a) and (b) respectively, with parameters: V_{IN}=1.5V, L=10μH, R_S=0.1Ω, R_1=22Ω, R_2=33Ω, C_1=C_2=5μF, references V_{r1}=3.5V (positive), V_{r2}=2.0V (negative), ramp with amplitude V_U–V_L=1V and period T=4μs. The switching instants are determined by the intersection of modulating signals V_{M1} and V_{M2} (ramp plus dynamical level) with feedback current (v_I)

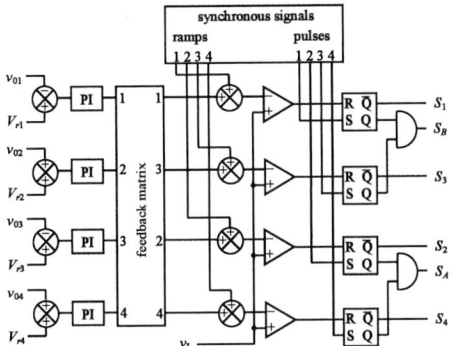

Fig. 6. Diagram for control of the SIMO converter in Fig. 2

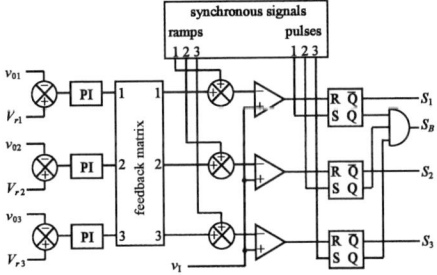

Fig. 7. Control diagram of the three outputs converter in Fig. 3

In the interleaving operation, the cycle is divided into an equivalent number of intervals to that of the outputs (two in this example). Every channel is therefore associated to a specific output and it operates during its corresponding time interval, which in turn is sub-divided into two sub-intervals. During the first (ON) sub-interval, the inductor is connected to the source (S_A and S_B in the ON state) and during the second (OFF) sub-interval, which is triggered by the corresponding comparator, the related main switch (S_A or S_B) will be OFF, and then, the inductor will be connected to its output by closing the

respective switch. The action of the control is given by means of the automatic selection of the instant of time at which the transition ON-OFF for each interval is produced, thus causing a direct effect on one specific output during the OFF sub-interval. But the key for the global stability of the system is the indirect effect on the rest of the outputs that is given through the duration of the ON subintervals, along which the inductor current is partly recovered from the source. The total time per cycle to charge the inductor is determined by the addition of all partial ON sub-intervals, thus being a combined action of the channels. Consequently, the regulation of each output is achieved through the total ON interval besides the duration of the respective OFF sub-interval.

IV. AVERAGED MODEL AND STABILITY

Under assumption of small period of the modulating signal, the dynamics can be approximated by the smooth averaged model. The averaged model gives the equations for a weighted field taking into account the duty ratios (considered as continuous variables) for each topology. A simple model can be obtained from the switched model by substituting the binary signals (ON/OFF switches) by their respective duty ratios, then the set of $2N+1$ equations, where N is the number of outputs, are obtained:

$$\frac{d\bar{v}_{0i}}{dt}=\frac{1}{C_i}\left(\bar{i}_L d_i - \frac{\bar{v}_{0i}}{R_i}\right)$$

$$\frac{d\bar{\sigma}_i}{dt}=\frac{1}{\tau_i}\left(\bar{v}_{0i}-V_{ri}\right)$$

$$\frac{d\bar{i}_L}{dt}=\frac{1}{L}\left((1-d_N)V_{IN}-\sum_{i=1..N}v_{0i}d_i - R_L\bar{i}_L\right)$$

The first subset of N equations refers to the averaged dynamics of each voltage output v_{0i} (i=1..N), which, in the case of an inverted (negative) output has been considered positive in these and following equations for the sake of simplicity. For each channel, R_i is the resistive load, C_i is the filter capacitance and the duty ratio d_i is defined here for its corresponding OFF sub-interval divided by period T. Additional subset deals with N equations for each of the integral terms σ_i, being V_{ri} the reference voltage (also inverted for the negative outputs):

$$\sigma_i = \frac{1}{\tau_i}\int(v_{0i}-V_{ri})dt$$

where τ_i is the time constant for each PI transfer function:

$$H_i(s)=g_i\left(1+\frac{1}{s\tau_i}\right)$$

The last equation deals with inductor current dynamics, being d_N the addition of the ratios for OFF sub-intervals (d_i) associated to the inverting channels. This is dur to the fact that the voltage applied to the inductor during each sub-interval is V_{IN} if this is ON and if it is OFF will be either V_{IN}-v_{0i} or –v_{0i} for non-inverting or inverting cases.

The ratio d_i for each OFF sub-interval, which is given by the action of the comparator, can be expressed as a function of the averaged state variables as follows:

$$d_i = \frac{1}{V_U-V_L}\left(\sum_{j=1..N}\alpha_{ij}g_j\left(\bar{v}_{0j}-V_{rj}+\bar{\sigma}_j\right)+r_S\bar{i}_L - V_L\right)$$

2117

being r_S the sensing coefficient used for the inductor current feedback and α_{ij} a generic coefficient of the feedback matrix i.e., output i versus input j. By means of the set of N equations for OFF duty ratios, the above set of $2N+1$ state equations (averaged field) is then in closed form and it can be used to obtained approximated information of the whole system, in particular stability features. First of all, the periodic orbit for the system corresponds to the equilibrium point of the continuous averaged system, which can be calculated by imposing the null field condition to the $2N+1$ equations of the field, giving the following results:

$$V_{0i} = V_{ri}$$

$$D_i = \frac{V_{0i}}{I_L R_i}$$

$$I_L = \frac{V_{IN}}{2R_L}\left(1 - \sqrt{1 - \frac{4R_L}{V_{IN}^2}\sum_{i=1..N}\frac{V_i V_i'}{R_i}}\right)$$

where V_i' is either V_i or $V_{IN}+V_i$ for positive (non-inverted) or negative (inverted) outputs respectively. The voltage outputs are therefore imposed by the correction of the PI block due to the inductor current self-adjustment in order to satisfy that condition.

For instance, with parameters used in Fig. 5, the averaged model gives the following values for equilibrium: V_{01}=3.5V (non-inverted), V_{02}=2V (inverted), D_1=0.30, D_2=0.114 and I_L=0.53A. These results are valid regardless of interleaving is used or not. It has to be noted that in Fig. 5 (a), the duty ratio D_1 includes the second and third sections while D_2 refers only to the third section.

The existence of an equilibrium point is not a sufficient condition for a proper behavior. In addition to this, stability has to be demonstrated for this point, what can be achieved by means of the linearization of the system (Jacobian) in the equilibrium point. After a convenient partition of the Jacobian, the following nine sub-matrices are obtained:

$$J_{i,j} = \frac{1}{C_i}\left(\beta_{ij} + \frac{\alpha_{ij} g_j I_L}{V_U - V_L}\right)$$

$$J_{i,N+j} = \frac{1}{C_i}\left(\frac{\alpha_{ij} g_j I_L}{V_U - V_L}\right)$$

$$J_{i,2N+1} = \frac{1}{C_i}\left(D_i + \frac{r_S I_L}{V_U - V_L}\right)$$

$$J_{N+i,j} = \gamma_{ij}$$

$$J_{N+i,N+j} = 0$$

$$J_{N+i,2N+1} = 0$$

$$J_{2N+1,j} = \frac{1}{L}\left(-D_j - \frac{1}{V_U - V_L}\sum_{k=1..N}\alpha_{jk} g_k V_k'\right)$$

$$J_{2N+1,N+j} = \frac{1}{L}\left(-\frac{1}{V_U - V_L}\sum_{k=1..N}\alpha_{jk} g_k V_k'\right)$$

$$J_{2N+1,2N+1} = \frac{1}{L}\left(-R_L - \frac{r_S}{V_U - V_L}\sum_{k=1..N}V_k'\right)$$

being i=1..N, j=1..N, β_{ij}=$-R_i$ for j=i and β_{ij}=0 for $j\neq i$ and γ_{ij}=$-1/\tau_i$ for j=i and γ_{ij}=0 for $j\neq i$ and V_U-V_L the amplitude of the ramp.

For the system in Fig. 5 with following control parameters: r_S=1Ω, g_i=0.1, τ_i=15μs (i=1..2), α_{ij}=0 if j=i and α_{ij}=1 if $j\neq i$, and V_U-V_L=1V, all the eigenvalues of the Jacobian have a negative real part ($-9.4\cdot10^3$ s^{-1} is the largest one), so the equilibrium is stable. This averaged model is valid for both cases, single-phase and interleaving. Complementary information with a qualitative analysis of the influence of several parameters in the stability is provided in [5].

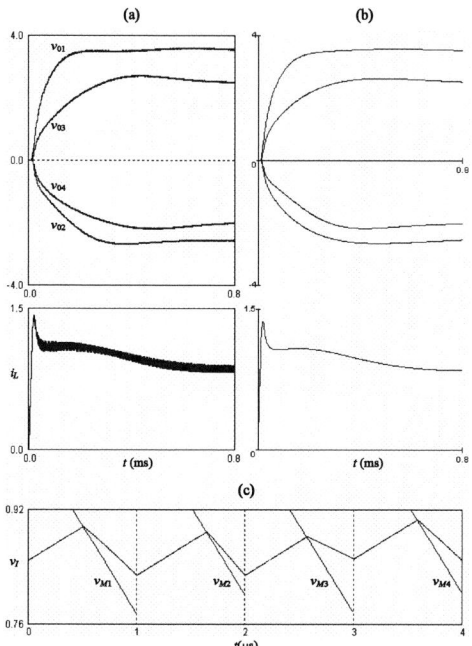

Fig. 8. Simulation results of SIMO dc-dc converter in Fig. 2 with control in Fig. 6, including transient from switched model (a) and from averaged model (b) and detail of the steady state (c). Parameters: V_{IN}=1.5V, L=15μH, R_S=0.1Ω, R_1=R_2=33Ω, R_3=27Ω, R_4=22Ω, C_i=10μF (i=1..4), V_{r1}=3.5V (+), V_{r2}=2.5V (−), V_{r3}=2.5V (+), V_{r4}=2.0V (−), g_1=0.1, g_2=g_3=0.075, g_4=0.05, τ_1=150μs, τ_2=τ_3=100μs, τ_4=75μs, α_{ij}=0 if j=i, α_{ij}=1 if $j\neq i$ (i, j=1..4) and ramp with lower and upper values V_L=0V, V_U=1V and period T=4μs. Modulating signals V_{Mi} (i=1..4) include ramp plus dynamical level

V. SIMULATION RESULTS

Stable behavior has also been found for SIMO converters with higher number of outputs and convenient parameters for control. Figures 8 and 9 deal with two illustrative cases and include the transient for voltage outputs and inductor current calculated by simulation of both the switched and the averaged model. Also the dynamics of the inductor current in the steady state is shown at the bottom, together with the modulating signals, so the transitions due to switching can be seen clearly. For simplicity, evenly distributed delay has been chosen for the ramps, but in general delays can be adapted to the specific system, for instance to minimize the ripple of the inductor current.

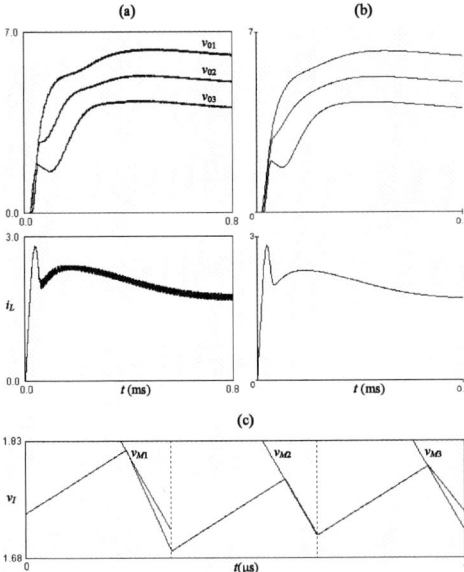

Fig. 9. Simulation results of SIMO dc-dc converter in Fig. 3 with control in Fig. 7, including transient from switched model (a) and from averaged model (b) and detail of the steady state (c). Parameters: V_{IN}=1.5V, L=15µH, R_S=0.1Ω, R_1=33Ω, R_2=39Ω, R_3=27Ω, C_i=10µF (i=1..3), V_{r1}=6V, V_{r2}=5V, V_{r3}=4V, g_1=0.18, g_2=0.15, g_3=0.12, τ_1=80µs, τ_2=60µs, τ_3=50µs, α_{ij}=0 if j=i or 1 if $j \neq i$, V_L=0V, V_U=1V, T=4µs

VI. CONCLUSIONS

Regulation of single-inductor multiple output dc-dc converters with a power stage configuration based on boost and buck-boost elementary structures is demonstrated by means of standard current mode control extended to multiple channels (one per output). Each channel uses a modulating signal with two special features: one is a reference level self-adjusted by a combination of PI correctors applied to each of the input-output errors and the other one is a selected delay to permit a sequential transference of energy through the inductor between the input generator and the outputs. Discontinuous conduction mode operation is not necessary for this proposal, so the advantages of continuous conduction mode can be accomplished.

A first attempt to analyze the stability of these multiple regulators has been made by means of a simple averaged model. According to it and also proved by direct simulations, a proper behavior is demonstrated for arbitrarily chosen examples with two, three and four outputs respectively. However, the ripple (mainly the associated to the inductor current) and saturation effects can provoke undesired instabilities (although not necessarily catastrophic), so additional research is being made in order to obtain more accurate discrete models or maps specific for single inductor multiple outputs converters with interleaving operation.

ACKNOWLEDGMENTS

This work was supported by the Spanish 'Ministerio de Educación y Ciencia' under Grant ENE2005-06934/ALT.

REFERENCES

[1] W. H. Ki, D. Ma, "Single-inductor multiple-output switching converters," *Power Electronics Specialists Conference (PESC)*, vol. 1, pp. 226-231, 2001.

[2] D. Ma, W. H. Ki, C. Y. Tsui and P. K. T. Mok., "Single-inductor multiple-output switching converters with time-multiplexing control in discontinuous conduction mode," *IEEE Journal of Solid-State Circuits*, vol. 38, no. 1, pp. 89-100, 2003.

[3] D. Ma, W. H. Ki, C. Y. Tsui, "A pseudo-CCM/DCM SIMO switching converter with freewheel switching," *IEEE Journal of Solid-State Circuits*, vol. 38, no 6, pp. 1007-1014, 2003.

[4] D. Trevisan, W. Stefanutti, P. Mattavelli, and P. Tenti, "FPGA control of SIMO dc-dc converters using load current estimation," *32nd Annual Conf. of IEEE Industrial Electronics Society (IECON)*, pp. 2243-2248, 2005.

[5] L. Benadero, R. Giral, A. El Aroudi and J. Calvente, "Stability analysis of a single inductor dual switching dc-dc converter," *Mathematics and Computers in Simulation* vol. 71, no 4-6, pp. 256-269, 2006.

[6] D. Ma, W. H. Ki, C. Y. Tsui and P. K. T. Mok, "A 1.8 V single-inductor dual-output switching converter for power reduction techniques," *Symposium on VLSI Circuits Digest of Technical Papers*, pp. 137-140, 2001.

[7] A. Sharma, Y. S. Pavan, "A single inductor multiple output converter with adaptive delta current mode control," *IEEE Int. Symposium on Circuits and Systems (ISCAS)*, pp. 5643-5646, 2006.

[8] P. Patra, S. Samanta, A. Patra, C. Amit and S. Chattopadhyay, "A novel control technique for single-inductor multiple-output dc-dc buck converters," *IEEE Int. Conf. on Industrial Technology (ICIT)*, pp. 807-811, 2006.

[9] H. P. Le, C. S. Chae, K. C. Lee, S. W. Wang and G. H. A Cho, "Single-inductor switching dc-dc converter with five outputs and ordered power-distributive control," *IEEE Journal of Solid-State Circuits*, vol. 42, no. 12, pp. 2706-2714, 2007.

[10] I. Dénes and I. Nagy, "Two models for the dynamic behaviour of a dual-channel buck and boost dc-dc converter," *Electromotion 2003*, vol. 10, no. 4, pp. 556-561, 2003.

[11] J. Hamar and I. Nagy, "Asymmetrical operation of dual channel resonant dc-dc converters," *IEEE Trans. Power Electron.*, vol. 18, no. 1, pp. 83-94, 2003.

[12] B. Buti., I. Nagy. H. Ohsaki and E. Masada, "Novel approach in stability analysis presented in controlled boost converter," *Power Conversion Conference (PCC)*, pp. 581-587, 2007

[13] R. W. Erickson and D. Maksimovic, *Fundamentals of Power Electronics*, Springer Science + Business media, LLC.

[14] P. T. Krein, J. Bentsman, R. M. Bass and B. C. Lesieutre, "On the use of averaging for the analysis of power electronic systems," *IEEE Trans. Power Electronics*, vol. 5, no. 2, pp. 182-190, 1990.

[15] M. di Bernardo, C. Budd, A. Champenys and B. Kowalczyk, *Piecewise-Smooth Dynamical Systems: Theory an Applications*, Springer, 2008.

[16] S. Banerjee, M. S. Karthik, G. H. Yuan and J. A. Yorke, "Bifurcations in one-dimensional piecewise smooth maps-theory and applications in switching circuits," *IEEE Trans. Circuits Syst. I*, vol. 47, no. 3, pp. 389-394, 2000.

[17] C. K. Tse and M. Di Bernardo, "Complex behavior in switching power converters," *Proc. IEEE*, vol. 90, no. 5, pp. 768-781, 2002.

[18] A. El Aroudi, B. Robert, L. Martinez-Salamero, "Bifurcation behavior of a three cell dc-dc buck converter," *EPE-PEMC'06*, 2006.

[19] V. Moreno, L. Benadero, "Investigating stability of a single inductor current mode controlled dual switching DC-DC converter," *EPE'05*, 2005.

[20] V. Moreno, L. Benadero, A. El Aroudi, R. Giral, J. Calvente, "Three dimensional discrete map for a single inductor current mode controlled dual switching dc-dc converter," *EPE-PEM'06*, pp. 2008-2013, 2006.

Control of a two-cell dc/dc converter in presence of saturating duty cycle

Moez Feki*, Abdelali El Aroudi†, Bruno Gérard Michel Robert‡, Nabil Derbel*

*Research Unit ICOS, Ecole Nationale d'Ingénieurs de Sfax, Tunisia,
e-mail: *moez.feki@enig.rnu.tn, n.derbel@enis.rnu.tn*
†Research group GAEI, Technical Engineering School of the Rovira i Virgili University, Spain,
e-mail: *abdelali.elaroudi@urv.cat*
‡Laboratoire CReSTIC, Université de Reims-Champagne-Ardenne, France,
e-mail: *bruno.robert@univ-reims.fr*

Abstract—In this work , we suggest a controller which is simple to construct to lead to zero current and zero voltage static error in the behavior of a two-cell dc/dc buck converter. Using nonlinear analysis we prove that zero error is achieved even in presence of duty cycle saturation.

I. INTRODUCTION

An important research area in the industry of devices, which are used in power electronics, consists in the design of semiconductor devices able to conduct high current during the ON phase and to support high voltages when they are in the OFF phase. Multi-cell power electronic converters have been developed to circumvent shortcomings in solid-state switching device ratings so that they can be applied to high-voltage electrical systems. The multi-cell converters applications include the speed variation for medium and high voltage motors, the dynamic restoration of voltage and the harmonics filtering. Because distributed power sources are expected to become increasingly prevalent in the near future, the use of multilevel converters to control the current and voltage output from renewable energy sources will provide significant advantages because of their response and autonomous control. Additionally, multi-cell converters can also control the real and reactive power flow from a utility connected renewable energy source [1].

With appropriate control strategy, it was demonstrated that multi-cells power electronic converters can have exceptional performances in transients and in steady state. The goal of the control is to balance the flying capacitor voltage to a fraction of the input and to adjust the current to a reference level [2]. This is achieved by controlling the time durations during which the switches are conducting.

It has been observed during last decade that most static converters exhibit strange behaviors apart from their normal operating mode [3], [4]. Indeed, H-bridge inverters [5], [6] as well as multi-cell dc/dc converters [7]–[9] were shown to present sub-harmonic modes and also chaotic behaviors. As a matter of fact, the chaotic behavior usually emerges from bad tuning of the proportional gain while trying to increase it seeking least static error [7]–[9].

In order to decrease the static error while avoiding the chaotic behavior, researchers tried the extended time delay feedback controller [10]–[12] as well as the proportional-integral controller [13]. The aim of this work is to design a new simple controller for a two-cell dc/dc converter that can achieve zero static error with low gain, whilst maintaining a periodic state behavior.

II. DISCRETE TIME MODEL OF A TWO-CELL CONVERTER

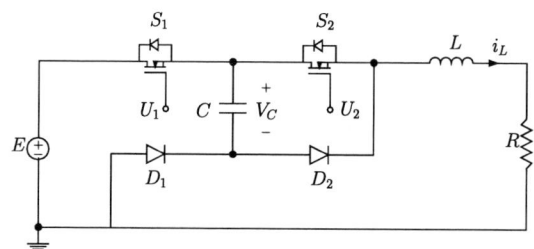

Fig. 1. A basic two-cell dc/dc buck converter.

The converter that we will deal with in this paper is depicted in Figure-1. It is based on a buck chopper modified in order to allow a higher input voltage by using two serial switches (transistors or diodes). The role of the capacitors is to balance the switch voltages. U_1 and U_2 are the outputs of a PWM modulator driven by the controller to be designed to achieve a constant voltage $V_c = \frac{E}{2}$ and a constant output current $i = I_r$. The controller is nothing but the duty cycles d_1 and d_2 of U_1 and U_2 that we will define in this work with respect to the OFF state.

In this paper, we will consider a dimensionless state space model, where the state vector $X = [i, v]^t$ is scaled with the maximum vector $[\frac{E}{R}, E]^t$ and time will be scaled with the switching period T. Therefore, the duty cycles d_1 and d_2 should vary in the unit interval $[0, 1]$. If we consider the transient time, we can define six possible operating modes each of which has a maximum of four different topologies [14]; the dynamic of each topology can be described by a linear continuous system of the form $\dot{x} = A_k x + B_k$. Let's consider for instance a scaled reference current $I_r > 0.5$, then in normal operating mode we should have $0 < d_1, d_2 < 0.5$ at steady state and the

978-1-4244-1741-4/08/$25.00 ©2008 IEEE

operating cycle is split into four intervals as shown in Figure-2.

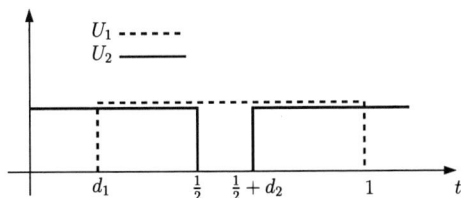

Fig. 2. Different topologies at the normal operating mode.

By integrating on each interval and stacking up solutions, we can define the state $x\big((n+1)T\big)$ as a function of $x(nT)$ and the duty cycles d_1 and d_2 in addition to the circuit parameters. It has been shown in [7] that the simplified discrete dynamic model is given by:

$$
\begin{bmatrix} x_{i,n+1} \\ x_{v,n+1} \end{bmatrix} = \begin{bmatrix} (1-\delta_L) & \Delta d\delta_L \\ -\Delta d\delta_C & 1 \end{bmatrix} \begin{bmatrix} x_{i,n} \\ x_{v,n} \end{bmatrix} + \begin{bmatrix} \delta_L(1-d_1) \\ 0 \end{bmatrix},
\tag{1}
$$

where $\delta_L = \frac{RT}{L}$, $\delta_C = \frac{T}{RC}$ and $\Delta d = d_1 - d_2$. In order to reduce the ripple current through the load and the ripple voltage across the capacitor, the circuit parameters should satisfy $\delta_L, \delta_C \ll 1$. In the sequel, simulations will be carried out with the following parameters:

$$
\delta_L = 0.1, \quad \delta_C = 0.1, \quad I_r = 0.6 \quad \text{and} \quad V_r = 0.5
$$

III. CONTROLLER DESIGN

The aim of this section is to find expressions of d_1 and d_2 that drive the state $x_{v,n}$ to V_r and $x_{i,n}$ to the reference value I_r. Hence, by defining $e_{i,n} = x_{i,n} - I_r$ and $e_{v,n} = x_{v,n} - V_r$, the aim becomes to drive the error variables to zero. To accomplish this aim, we need first to describe the error dynamics. Indeed, by simple algebra, we obtain:

$$
\begin{bmatrix} e_{i,n+1} \\ e_{v,n+1} \end{bmatrix} = \begin{bmatrix} 1-\delta_L & 0 \\ 0 & 1 \end{bmatrix} \begin{bmatrix} e_{i,n} \\ e_{v,n} \end{bmatrix} + \begin{bmatrix} \Delta d\, \delta_L x_{v,n} \\ -\Delta d\, \delta_C x_{i,n} \end{bmatrix}
$$
$$
\dots + \begin{bmatrix} -\delta_L d_1 + \delta_L(1-I_r) \\ 0 \end{bmatrix}.
\tag{2}
$$

This is a non-autonomous system due to the presence of the independent signals $x_{i,n}$ and $x_{v,n}$ and the constant term $\delta_L(1-I_r)$ that will play a key role in our design.

A. Proportional controller

Let's simply choose the proportional controller, that is the duty cycles are functions of the errors:

$$
d_1 = k_i e_{i,n} + k_v e_{v,n} \quad \text{and} \quad d_2 = k_i e_{i,n} - k_v e_{v,n}
\tag{3}
$$

As a matter of fact, the proportional controller is usually the first controller to try due to its simplicity of construction and cheapness. For instance, the realization of such controller can either be using Op-amps in case of analog realization; or in case of digital realization the PIC provides an easy way to implement the controller described in (3).

By applying the proportional controller we get the following error dynamics

$$
\begin{bmatrix} e_{i,n+1} \\ e_{v,n+1} \end{bmatrix} = \begin{bmatrix} 1-\delta_L(k_i+1) & k_v\delta_L(1+2x_{v,n}) \\ 0 & 1-2k_v\delta_C x_{i,n} \end{bmatrix} \begin{bmatrix} e_{i,n} \\ e_{v,n} \end{bmatrix}
$$
$$
\dots + \begin{bmatrix} \delta_L(1-I_r) \\ 0 \end{bmatrix}
\tag{4}
$$

This is a linear system having a unique fixed point $(e_i^*, e_v^*) = \left(\frac{1-I_r}{1+k_i}, 0\right)$. Since it is non-autonomous, we need to determine the stability conditions of the error system using the Lyaponuv theory. In fact, the error dynamics in (4) can be written in general under the cascade form:

$$
e_{v,n+1} = f_1(e_{v,n})
\tag{5}
$$
$$
e_{i,n+1} = f_2(e_{v,n}, e_{i,n})
\tag{6}
$$

It has been proved [15] that under mild conditions, the stability of system (5)-(6) can be deduced from the simultaneous stability of the following subsystems:

$$
e_{v,n+1} = f_1(e_{v,n})
\tag{7}
$$
$$
e_{i,n+1} = f_2(e_v^*, e_{i,n})
\tag{8}
$$

where e_v^* is the fixed point of (7). Now, the first subsystem is described by:

$$
e_{v,n+1} = (1 - 2k_v\delta_C x_{i,n})e_{v,n}
\tag{9}
$$

Hence, we may consider the Lyapunov function candidate $V(e_{v,n}) = e_{v,n}^2$. In discrete systems we need to show that the variation of the Lyapunov function is negative i.e. $\Delta V(e_{v,n}) = V(e_{v,n+1}) - V(e_{v,n}) < 0$.

$$
\Delta V(e_{v,n}) = \big((1-2k_v\delta_C x_{i,n})^2 - 1\big)e_{v,n}^2
\tag{10}
$$
$$
= 2k_v\delta_C x_{i,n}\big(2k_v\delta_C x_{i,n} - 2\big)e_{v,n}^2
\tag{11}
$$

Knowing that $x_{i,n}$ is always positive, then we can guarantee that ΔV is negative if the voltage gain k_v is chosen to satisfy condition (12).

$$
0 < k_v < \frac{1}{\delta_C x_{i,n}} = k_{v,cri}
\tag{12}
$$

We can also optimize the choice of k_v to obtain fast convergence of the error to zero. This is done by minimizing $\Delta_V(e_{v,n})$, therefore we need to minimize $h(k_v) = 2k_v\delta_C x_{i,n}\big(2k_v\delta_C x_{i,n} - 2\big)$ with respect to k_v and we get

$$
k_{v,opt} = \frac{1}{2\delta_C x_{i,n}}
\tag{13}
$$

Until now, we can guarantee the stability of subsystem (9) at the origin by the choice of k_v. Next, we will investigate the stability of the second subsystem:

$$
e_{i,n+1} = f_2(0, e_{i,n})
$$
$$
= \big(1 - \delta_L(1+k_i)\big)e_{i,n} + \delta_L(1-I_r)
\tag{14}
$$

2121

By choosing the Lyapunov function candidate $V(e_{i,n}) = e_{i,n}^2$, we get

$$
\begin{aligned}
\Delta V(e_{i,n}) &= e_{i,n+1}^2 - e_{i,n}^2 \\
&= \delta_L(1+k_i)\big(\delta_L(1+k_i) - 2\big)e_{i,n}^2 \\
&\quad \dots + 2\big(1 - \delta_l(1+k_i)\big)\delta_L(1-I_r)e_{i,n} \\
&\quad \dots + \delta_L^2(1-I_r)^2
\end{aligned}
$$

$\Delta V(e_{i,n})$ is a quadratic function of $e_{i,n}$ if $k_i \neq \frac{2}{\delta_L} - 1$. We can easily check that $\Delta V(e_{i,n}) = 0$ has real solutions for all $k_i > 0$. Indeed, these solutions are given by:

$$
e_{i1} = \frac{1 - I_r}{1 + k_i}, \quad \text{and} \quad e_{i2} = \frac{\delta_L(1 - I_r)}{\delta_L(1 + k_i) - 2} \quad (15)
$$

According to the value of k_i we can distinguish three cases.

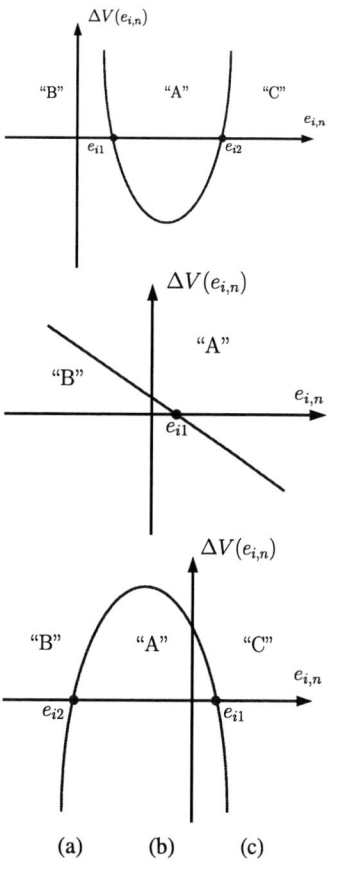

Fig. 3. Proportional control, variation of $\Delta V(e_{i,n})$ in case (a) $k_i > \frac{2}{\delta_L} - 1$ (b) $k_i = \frac{2}{\delta_L} - 1$ (c) $k_i < \frac{2}{\delta_L} - 1$

case 1: $k_i > \dfrac{2}{\delta_L} - 1$

In this case, $\Delta V(e_{i,n})$ has the curve shown in Figure 3(a), and system (14) is unstable. Indeed, when the system is initiated in regions "B" or "C", where $\Delta V > 0$, the error diverges and when the system is initiated in "A" the error decreases until it is in region "B" and then it diverges.

case 2: $k_i = \dfrac{2}{\delta_L} - 1$

In this case $\Delta V(e_{i,n}) = 0$ is a linear equation and has

only one solution which we can notice that it corresponds to e_{i1}.

$$
e_{i1} = \delta_L \frac{1 - I_r}{2}, \quad (16)
$$

Figure 3(b) depicts the variation of $\Delta V(e_{i,n})$ and we deduce that when the system is initiated in region "B", where $\Delta V > 0$, the error diverges and when the system is initiated in region "A" the error decreases until it is in region "B" and then it diverges.

case 3: $k_i < \dfrac{2}{\delta_L} - 1$

The variation of $\Delta V(e_{i,n})$ becomes as shown in Figure 3(c). The error diminishes when the system is initiated in regions "B" or "C" until it enters region "A". Therein the system tends to diverge to regions "B" or "C". Therefore, a limit set is defined which consists of two points $E = \{e_{i,n} | \Delta V(e_{i,n}) = 0\} = \{e_{i1}, e_{i2}\}$. However, we easily notice that $M = \{e_{i1}\}$ is the smallest invariant set of E because it defines the fixed point. Therefore, according to Lasalle's invariance principle we have

$$
\lim_{n \to \infty} e_{i,n} = e_{i1} = \frac{1 - I_r}{1 + k_i} \quad (17)
$$

It is clear that $\lim_{n \to \infty} e_{i,n} = 0$ only if $k_i = \infty$. In fact, we can write $e_{i1} = 1 - I_r - d_1^*$ (where we take into account that $e_v^* = 0$). Therefore, $\lim_{n \to \infty} e_{i,n} = 1 - I_r - d_1^*$. Hence, we may obtain $\lim_{n \to \infty} e_{i,n} = 0$ only if

$$
\lim_{n \to \infty} d_1(n) = d_1^* = 1 - I_r . \quad (18)
$$

Remark 1: The classic proportional controller given in (3) is applied to reach $e_{i,n} = 0$ and $e_{v,n} = 0$ which inherently leads to $d_1 = d_2 = 0$. However, we can get $e_{i,n} = 0$ only if at steady state we have $d_1 = d_2 = 1 - I_r$. This means that the proportional controller cannot reach its aim.

After having proved that $\lim_{n \to \infty} e_{i,n} = e_{i1} = \frac{1 - I_r}{1 + k_i}$, then the optimality condition on the choice of the current gain k_i can be obtained by rewriting (14) using a new variable $\epsilon_{i,n} = e_{i,n} - e_{i1}$, hence obtaining:

$$
\epsilon_{i,n+1} = \big(1 - \delta_L(1 + k_i)\big)\epsilon_{i,n} \quad (19)
$$

and the optimal value of the current gain is:

$$
k_{i,opt} = \frac{1}{\delta_L} - 1 \quad (20)
$$

Figure 4 depicts the effect of the proportional controller. We notice that a non zero static error is obtained for the current. We have shown that if we attempt to decrease the static error by further increasing the current gain, then we obtain higher periodic behavior and also an undesirable chaotic behavior. Indeed, the effect of increasing the current gain k_i is delineated on the bifurcation diagram of Fig. 5.

2122

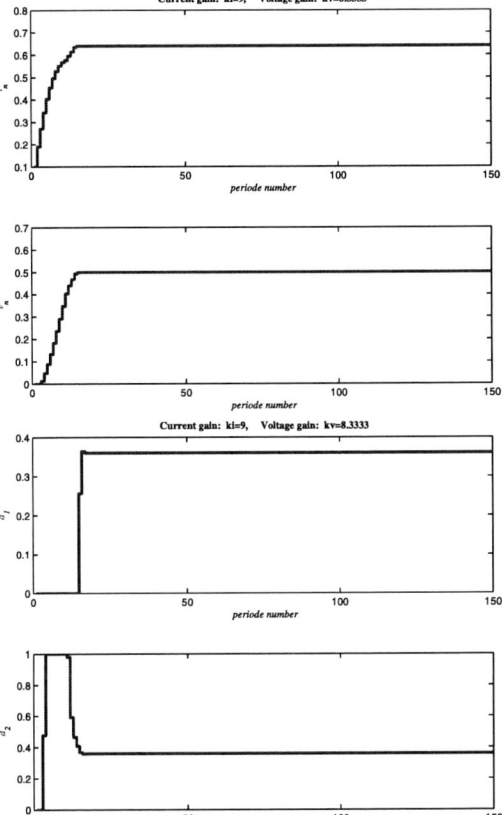

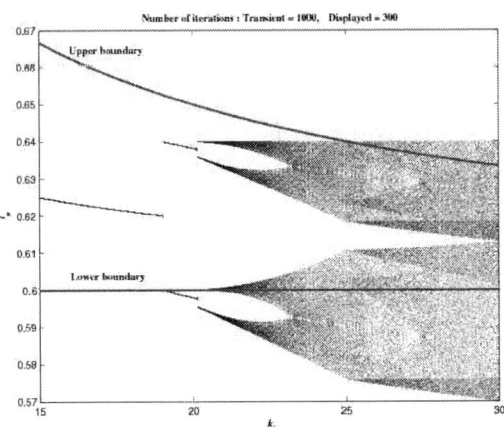

Fig. 4. Time evolution of the system under the action of the proportional controller.

Fig. 5. The bifurcation diagram with variation of k_i.

B. Zero error controller

From the previous analysis, we notice that when the error becomes nil, the duty cycles should be equal to $1 - I_r$, thus the controller we suggest here has the following form:

$$\begin{cases} d_1 & = 1 - I_r + k_i e_{i,n} + k_v e_{v,n} \ , \\ d_2 & = 1 - I_r + k_i e_{i,n} - k_v e_{v,n} \ . \end{cases} \quad (21)$$

After application of (21) the error system (2) becomes:

$$\begin{bmatrix} e_{i,n+1} \\ e_{v,n+1} \end{bmatrix} = \begin{bmatrix} 1 - \delta_L(1+k_i) & 2k_v \delta_L x_{v,n} - k_v \delta_L \\ 0 & 1 - 2k_v \delta_C x_{i,n} \end{bmatrix} \begin{bmatrix} e_{i,n} \\ e_{v,n} \end{bmatrix} . \quad (22)$$

That is the system can again be described in cascade form. Similar analysis can be carried out, and only equation (14) will be modified to get:

$$e_{i,n+1} = \big(1 - \delta_L(1+k_i)\big)e_{i,n} \quad (23)$$

which is similar to equation (19) therefore, the current gain that leads to stability is $k_i < \frac{2}{\delta_L} - 1$ and the value that leads to fast response is $k_{i,opt} = \frac{1}{\delta_L} - 1$. Figure 6 delineates the response of the system to the controller given in (21).

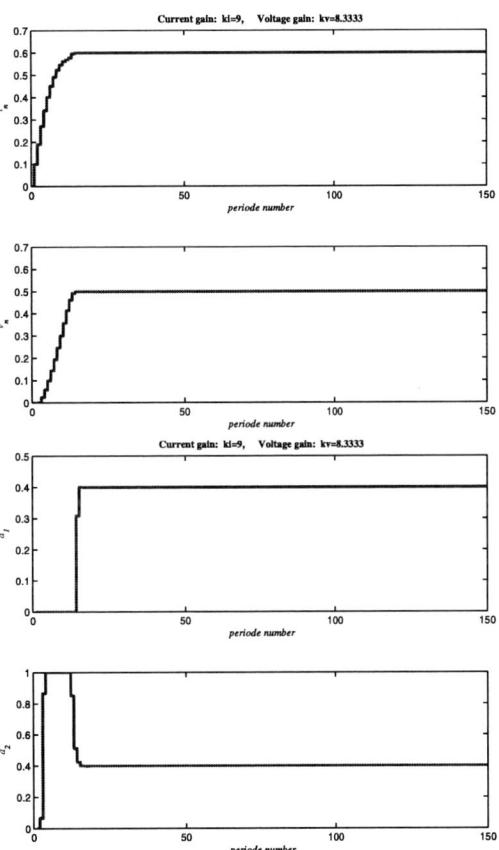

Fig. 6. Time evolution of the system under the action of zero error controller.

C. Complete stability analysis

The previous sections, have presented stability analysis of the system while neglecting the saturation of the duty cycle that appears evident in the simulation results and seems inevitable. Therefore, the previous stability analysis is not reliable. In this section, we try to present a concrete proof to show that the two-cell converter is in fact stabilized with the proposed controller even in the presence of duty cycle saturation. To our knowledge, very few works in literature have considered the saturation phenomenon.

let $\mathcal{R}$ defines the space of all possible errors:

$$
\begin{aligned}
\mathcal{R} =\{(e_{i,n}, e_{v,n})| &-I_r < e_{i,n} < 1 - I_r, \\
\ldots &-V_r < e_{v,n} < 1 - V_r\}
\end{aligned} \tag{24}
$$

As a matter of fact, the saturation of both duty cycles defines four boundaries in $\mathcal{R}$. These boundaries are in the form of lines defined by:

$$
\mathcal{L}_{11} : \ d_1 = 1 \ \Rightarrow \ -I_r + k_i e_{i,n} + k_v e_{v,n} = 0 \tag{25}
$$

$$
\mathcal{L}_{10} : \ d_1 = 0 \ \Rightarrow \ 1 - I_r + k_i e_{i,n} + k_v e_{v,n} = 0 \tag{26}
$$

$$
\mathcal{L}_{21} : \ d_2 = 1 \ \Rightarrow \ -I_r + k_i e_{i,n} - k_v e_{v,n} = 0 \tag{27}
$$

$$
\mathcal{L}_{20} : \ d_2 = 0 \ \Rightarrow \ 1 - I_r + k_i e_{i,n} - k_v e_{v,n} = 0 \tag{28}
$$

These boundary lines divide the error space into nine subspaces denoted by $\mathcal{R}_i$, such that $\bigcup_{i=1}^{i=9} \mathcal{R}_i = \mathcal{R}$ (see Fig. 7). In each subspace, the two-cell converter is described by a different model (denoted by $\mathcal{M}_i$) that can be obtained from (1) by replacing the duty cycles by their corresponding values, namely $d_i = 0$ or $d_i = 1$ or the expressions given in (21).

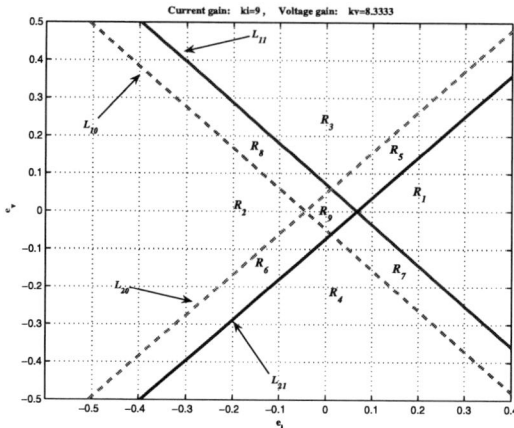

Fig. 7. The error space divided into nine subspaces.

We now consider the following Lyapunov function:

$$
V_n = e_{i,n}^2 + e_{v,n}^2 \tag{29}
$$

Then, we can verify that $\Delta V = V_{n+1} - V_n < 0$ for each model $\mathcal{M}_i$ in the subspace $\mathcal{R}_i$ except at the origin where $\Delta V = 0$ (see Fig. 8) which implies that $\mathcal{R}$ is an invariant set.

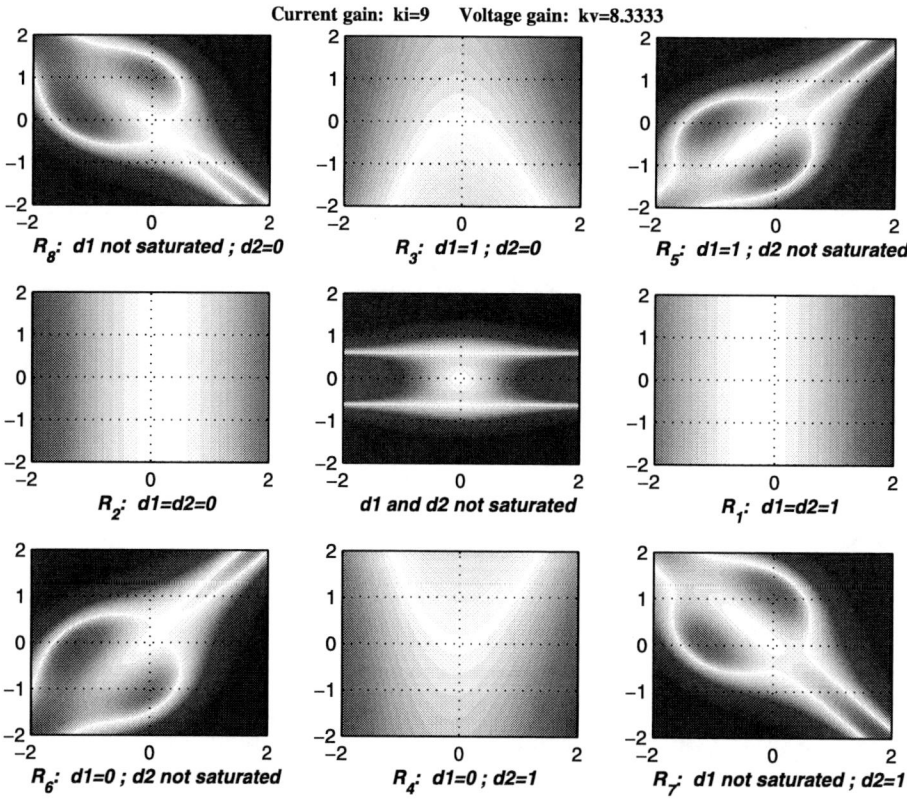

Fig. 8. The values of ΔV corresponding to each model $\mathcal{M}_i$.

Now, let $B_r(0)$ be a ball of radius r around the origin such that $B_r(0) \in \mathcal{R}_9$. Moreover, we define $\epsilon > 0$ such that

$$-\epsilon = \max_{(e_{i,n}, e_{v,n}) \in \mathcal{R} \setminus B_r(0)} \Delta V \qquad (30)$$

Thus $\Delta V \leq -\epsilon$ for all $(e_{i,n}, e_{v,n}) \in \mathcal{R} \setminus B_r(0)$, that is $V_{n+1} \leq V_n - \epsilon$ and hence $V_{n+1} \leq V_0 - m\epsilon$ for some number $m \in \mathbb{N}$. We deduce that the sequence of points $(e_{i,n}, e_{v,n})$ can only stay in $\mathcal{R} \setminus B_r(0)$ for a finite number of iterations since $V_n > 0$ for all n. Besides, since $\mathcal{R}$ is an invariant set, then the sequence $(e_{i,n}, e_{v,n})$ has to enter $B_r(0)$. Eventually, the exponential stability is deduced from the analysis presented in the previous section which in fact concerns $\mathcal{R}_9$.

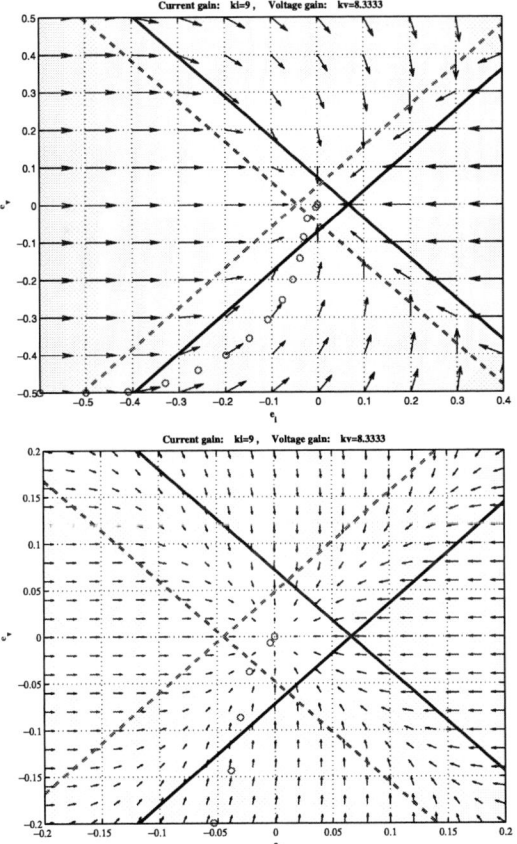

Fig. 9. The behavior of the error when $k_i = 9$ and $k_v = 50/6$; (at the bottom is a zoomed figure).

Figure 8 shows the variation of ΔV in the error space for each model $\mathcal{M}_i$. Areas in green mean that $\Delta V < 0$, areas in red mean that $\Delta V > 0$ and as the color gets lighter, ΔV approaches zero. Figure 9 shows the behavior of the system for each point in the error space. In red circles, we represent the error evolution corresponding to the simulations shown in Fig. 6.

IV. CONCLUSION

In this paper, we have proposed a simple controller to achieve zero current and voltage error for a two-cell dc/dc converter. This method has been proposed after a nonlinear analysis of the proportional controller proposed in [7] which by increasing the current gain to decrease the static error provoked a chaotic behavior. The controller proposed herein, outperforms other controllers proposed in the literature where only decrease of the static error is obtained. Moreover, we have provided a complete stability analysis where the duty cycle saturation has been taken into consideration.

ACKNOWLEDGMENT

This work is prepared within a Tunisian-Spanish cooperation framework, under grant A/6828/06.

REFERENCES

[1] L. M. Tolbert and F. Z. Peng, "Multilevel converters as a utility interface for renewable energy systems," in *Power Engineering Society Summer Meeting, 2000 IEEE*, 2000, pp. 1271–1274.

[2] T. A. Meynard, M. Fadel, and N. Aouda, "Modeling of multilevel converters," *IEEE Trans. Ind. Electronics*, vol. 44, no. 3, pp. 356–364, 1997.

[3] I. Nagy, "Nonlinear dynamics in power electronics," in *Proc. Electrical drives and power Electronics Conf.*, 2000, pp. 1–15.

[4] S. Banerjee and G. Verghese, *Nonlinear phenomena in power electronics: Attractors, bifurcations, chaos and nonlinear control.* New York: IEEE Press, 2001.

[5] B. Robert and C. Robert, "Border collision bifurcations in a one-dimensional piecewise smooth map for a pwm current-programmed h-bridge inverter," *International Journal of Control*, vol. 75, pp. 1356–1367, 2002.

[6] H. Iu and B. Robert, "Control of chaos in a pwm current-mode h-bridge inverter using time-delayed feedback," *IEEE Transactions on Circuits and Systems-I*, vol. 50, no. 8, pp. 1125–1129, 2003.

[7] B. Robert and A. E. Aroudi, "Discrete time model of a multi-cell dc/dc converter: Nonlinear approach," *Mathematics and Computers in Simulation*, vol. 71, pp. 310–319, 2006.

[8] A. El Aroudi and B. Robert, "Stability analysis of a voltage mode controlled two-cells dc-dc buck converter," in *IEEE Annual Power Electronics Specialists Conference, PESC Record*. Art. No 1581759, 2005, pp. 1057–1061.

[9] A. El Aroudi, B. Robert, and L. Martinez-Salamero, "Modelling and analysis of multicell converters using discrete time models," in *IEEE International Symposium on Circuits and Systems, IS-CAS'06*, Kos island, Greece, May 2006.

[10] B. Robert, H. Iu, and M. Feki, "Adaptive time-delayed feedback for chaos control in a pwm single phase inverter," *Journal of Circuits, Systems and Computers*, vol. 13, no. 3, pp. 519–534, 2004.

[11] B. Robert, M. Feki, and H. Iu, "Control of a pwm inverter using proportional plus extended time-delayed feedback," *Int. J. Bifurcation Chaos*, vol. 16, no. 1, pp. 113–128, 2006.

[12] M. Feki, B. Robert, and H. Iu, "A proportional plus extended time-delayed feedback controller for a pwm inverter," in *IEEE Annual Power Electronics Specialists Conference, PESC Record*, Achen, Germany, 2004, pp. 3317–3320.

[13] B. Buti, I. Nagy, and E. Masada, "Stability analysis of pwm-controlled dual channel resonant buck converter using pi controller," in *International Symposium on Power Electronics, Electrical Drives, Automation and Motion, SPEEDAM*. Art. No 1649772, 2006, pp. 208–213.

[14] A. E. Aroudi, "Modelling and dynamics of a multilevel dc-dc converter," 2006, internal report.

[15] X.-M. Bai, H.-M. Li, and X.-S. Yang, "Some results on cascade discrete-time systems," *Discrete Dynamics in Nature and Society*, vol. 2006, no. DOI 10.1155/DDNS/2006/14631, pp. 1–8, 2006.

Bifurcations and Chaotic Dynamics in a Linear Switched Reluctance Motor

M.R. De Castro, B.G.M. Robert*, C. Goeldel

University of Reims Champagne-Ardenne/CReSTIC, Reims, France, e-mail: bruno.robert@univ-reims.fr

Abstract—This paper presents first computed results of a switched linear variable reluctance motor showing non-linear phenomena as bifurcations and chaotic dynamics. Performances of two models of the motor, operating in open loop, are compared by the mean of bifurcation diagrams and Poincaré sections. The first one is a permeance model based on the flux tubes method. It is accurate but complex. The second one is simplified by using a harmonic method. The objective of this study is double. In a first time, the precise model is used to point out bifurcations and chaos. Then, qualitatively similar results obtained from the simplified model are presented to prove that non linear dynamics arise from the main non linearity of the motor and do not result from some details of motor manufacturing.

Keywords—Bifurcation, Chaos, Reluctance motor, Nonlinear model.

I. INTRODUCTION

The studied object in this work is a prototype of a linear switched reluctance motor. This kind of machine is useful in many manufacturing processes requiring high-precision linear position control as pick-and-place machines, material transfer or packaging. The low cost, robustness, due to a very simple technology, and the ability to operate in open loop for the direct-drive of precise displacements over a long traveling distance are some advantages.

The primary disadvantage of a linear switched reluctance motor (LSRM) is the higher torque ripple compared with conventional machines, which contributes to acoustic noise and mechanical vibrations. The origin of torque pulsations in an LSRM is due to the highly nonlinear and discrete nature of torque production mechanism. That nonlinear nature, in this kind of motor, can be viewed from the strongly dependence between translator position and air-gap permeance. The phenomena like saturation of the core and mutual inductances among the phases have also substantial contribution to increase the nonlinearities in the dynamical model.

Nowadays the numerical and computational methods are a very important and powered tool in the design, simulation and analysis of dynamic systems in physics and engineering. In the last decades new mathematical tools, like ergodic theory of chaos, made it easier the nonlinear systems analyze. Applying such method allows to reach a new dimension in the non linear control and open the door to a new generation of speed/position drivers and control [1][2][3][4][5]. At the present time, they address machines like variable reluctance motors, hybrid step motors in the industrial environment.

From this kind of machine, some works about step motors [6][7][8], have detected many nonlinear phenomena. This very complicated dynamics nature became this motors susceptible to different types of behaviors when is running in open loop, like usual periodic orbits and non periodic like quasi-periodic and chaotic behaviors. However, to obtain the correct numerical results it's very important to have a dynamical model that is accurate and simple.

This paper presents a dynamical nonlinear study about a prototype of a linear reluctance machine using two models for the air-gap permeance. The first model used in this work, introduced in [9], assumes that flux flows across the air gap through flux tubes shaped by straight lines and circular arcs [10][11]. The second model used is a simplified model considering the air gap permeance like a harmonic function of the position. The idea of a harmonic function for the permeance model allows proving that it's possible to obtain the same qualitative dynamical behaviors from a simple model.

This paper is organized in eight sections that are presetting in suite.

Section II presents the motor and the electromechanical parameters from which the dynamical system equations are obtained. Some considerations about the model, as saturation of iron core, mechanical frictions and the open loop command strategy are also discussed.

Section III is devoted to the model of the permeance and inductance from the flux tubes method. This method is also used to justify the mutual inductance considerations on the dynamic model.

The harmonic model, derived from the analysis of the flux tubes model, is presented in section IV.

Section V develops the dimensionless model and some simulation strategies are described. For all models the friction is neglected. In general, these parasites do not affect the qualitative results regarding the bifurcation patterns, as observed previously by [13].

Some dynamic analysis tools for the chaos detection, Feigenbaum diagrams and Poincaré sections are presented in section VI.

Section VII gives some examples of computed results and deals with comparisons of the harmonic model and the flux tube model.

Conclusions and perspectives are proposed in section VIII.

II. THE PROTOTYPE

The motor used in this work is presented in this section. The prototype is a linear reluctance variable motor with three-phases and two poles by phase. The TABLE I present the principal electrical and mechanical motor parameters used in the dynamic modeling.

978-1-4244-1741-4/08/$25.00 ©2008 IEEE

TABLE I.

Parameters	Symbol	Value	Unit
N. Phases	Nph	3	-
N.Poles/Phase	Np	2	-
Stator length	Lt	1.8	m
Rated speed	vn	0.5	m/s
Translator mass	M	10	Kg
Rated Current	Ir	8.5	A
Current Density	J	6	A/mm2
Induction	B	1.2	T
Minimum Air Gap	lg	1	mm
Mechanic Power	Pm	12.5	W

The linear machine is a kind of motor that has no need of a mechanical system to convert the rotating movement to translating movement. The constructed linear machine is equivalent to a 6/4 (stator poles/rotor tooth) rotating machine (RSRM). The Fig. 1 shows the prototype transversal cut view with the principal dimensions.

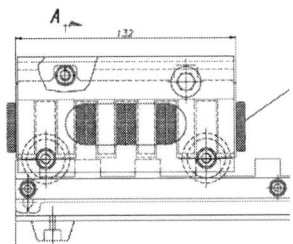

Fig. 1 - Mechanical characteristics.

The translator of the machine is the upper part. This configuration allows a more practical mechanical design. Also, it makes the attraction force in the same direction of gravitational force that is compensated by the normal force into bear bearings.

Defining $f_{ch} = 1/T_{ch}$ the switching frequency voltage applied and $f = 3 \cdot f_{ch}$, the related step frequency, for the open looping command strategy, the phase voltages can be modeled by (1).

$$V_n(t) = \frac{E}{2} \cdot sign\left[sin\left(w \cdot t - (n-1)\frac{2\pi}{3} + \frac{\pi}{6} \right) - \frac{1}{2} \right] + \frac{E}{2} \quad (1)$$

Where, E and $w = 2 \cdot \pi \cdot f_{ch}$ are the voltage amplitude and pulsation frequency. The $V_n(t)$ and $i_n(t)$, respective voltage and current by phase.

This command strategy and the motor geometric symmetry, like will be presented in the section III, allow that the mutual inductance influence can be neglected. The prototype schema is presented in the Fig. 2

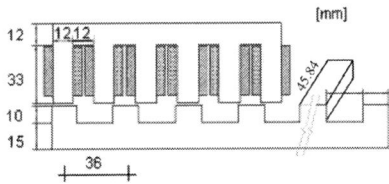

Fig. 2 - Three-phase linear reluctance variable motor model.

Typically the linear electrical machines have a non saturated magnetic circuit due to gap characteristic, resulting in a flux linkage versus current almost linear. The flux equation of one motor phase for a phase resistance r_n is given by (2).

$$\lambda_n = \int (V_n - r_n \cdot i_n) dt \quad (2)$$

From the non saturated core condition, the inductance concept can be used. So, the ratio between the flux linkage and the current defines the inductance and is given by:

$$L_n = \frac{\lambda_n}{i_n} \quad (3)$$

From the inductance concept established, the dynamic system of differential equations is obtained and presented in (4). This is a frictionless and unloaded complete five dimensional model for a three phase motor ($n = 1, 2, 3$).

$$\begin{cases} \dfrac{di_n}{dt} = \dfrac{1}{L_n(x)} \cdot \left(V_n - r \cdot i_n - \dfrac{dL_n(x)}{dx} \cdot i_n \cdot v_x \right) \\ \dfrac{dx}{dt} = v_x \\ \dfrac{dv_x}{dt} = \dfrac{1}{M} \cdot \dfrac{1}{2} \cdot \sum_{n=1}^{3} i_n^{\ 2} \cdot \dfrac{dL_n(x)}{dx} \end{cases} \quad (4)$$

III. THE FLUX TUBES METHOD

The first inductance model is obtained by flux tubes method. The flux tube shapes are position depending functions and are basically represented by arcs and lines.

The permeance of each flux tube is computed analytically using:

$$P_{tube} = \mu_o \cdot \int \frac{dA}{l} \quad (5)$$

Where, μ_0 is the vacuum permeability, l is the length of the tube and dA is the differential cross section of the tube. Evaluation of (5) for all possible cases leads to explicit position-dependent expressions for the permeance $P_{tube}(x)$ and spatial derivate of permeance $dP_{tube}(x)/dx$. For the LSRM of Fig. 1, there are six air gap regions, two identical for each pole pair, and each of these tooth regions will be covered by a number of flux tubes. A new flux tube is creating since a new minimal permeance trajectory will be possible.

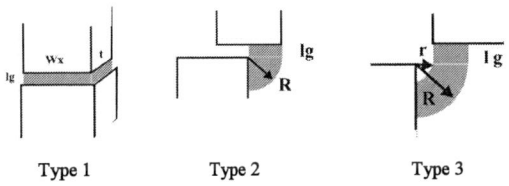

Fig. 3 - The Three types of the flux tubes used.

The respective equations for each flux tube type 1, 2 and 3 are presented below.

$$P_1 = \mu_o \cdot \frac{w_x \cdot t}{l_g} \qquad (6)$$

$$P_2 = \frac{2 \cdot \mu_o \cdot t}{\pi} \cdot \ln\left(\frac{R \cdot \pi + l_g}{2 \cdot l_g}\right) \qquad (7)$$

$$P_3 = \frac{2 \cdot \mu_o \cdot t}{\pi} \cdot \ln\left(\frac{R \cdot \pi + l_g}{r \cdot \pi + l_g}\right) \qquad (8)$$

For the LSRM were detected five configurations, or stages, to the flux tubes along the displacement. Each flux tube linked to the same pole and stage is associated in parallel with the others tubes. The Fig. 4 shows the five stages of the permeance modeling using the flux tubes method. The modeling is made for one pole and for a relative displacement of πrad. The Z axis displacement is neglected.

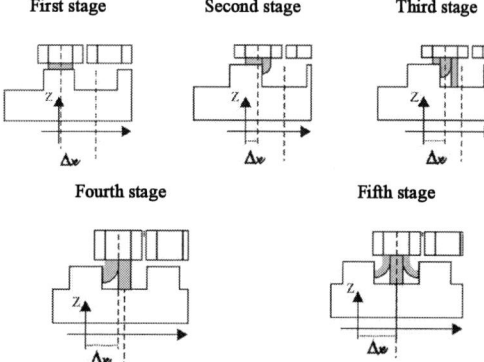

Fig. 4 - Stages to the flux tube method.

Using (6), (7) and (8) into each flux tube set and associating correctly in parallel all permeances calculated for the respective stages, the permeance values can be plotted. For example, for the second stage:

$$P_2 = P_{tube\_1} + P_{tube\_2} \qquad (9)$$

The Fig. 5 shows the waveform of the permeance flux tube model to each stage versus the displacement.

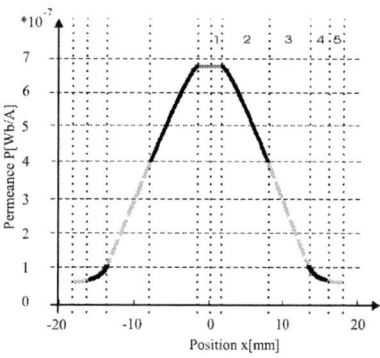

Fig. 5 - Permeance vs displacement

The Fig. 6 shows the permeance derived waveform versus displacement.

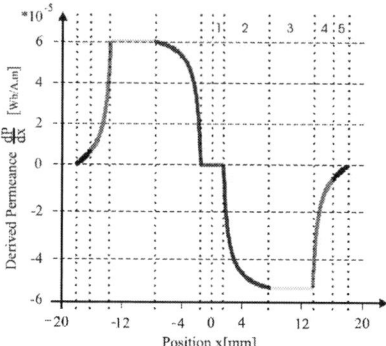

Fig. 6 - Permeance derived vs. displacement

Using the permeance model, the inductance can be obtained by (10).

$$L_n = \frac{P_n}{2} \cdot N_e^{\,2} \qquad (10)$$

The parameter N_e is the turn number in the winding by phase. P_n is the permeance into the air gap for one translator tooth.

The Fig. 7 shows the results of the computed inductance using the permeance model.

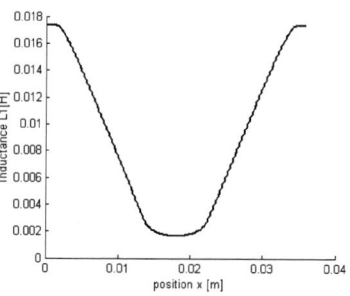

Fig. 7 - Phase1 Inductance model.

The flux tube method is also used to legitimate the neglected mutual inductance in the dynamic system equations. Combining the flux tubes analyzes and the superposition theorem, it can be shown that the mutual flux is canceled between the two poles by the same phase.

The flux produced by the pole 1 of the phase A is presented in Fig. 8.

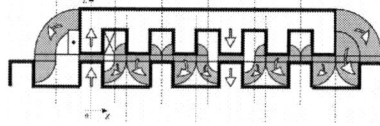

Fig. 8 - Phase 1, pole 1.

The flux produced by the pole 2 of the same phase A is presented in Fig. 9.

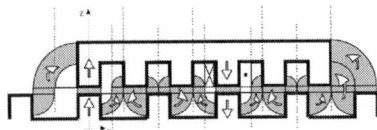

Fig. 9 - Phase 1, pole 2.

Considering infinity core permeance, the flux value in the hashed area has the same value but opposed sense in the Fig. 8 and Fig. 9. So, using the superposition theorem the result is showing by Fig. 10.

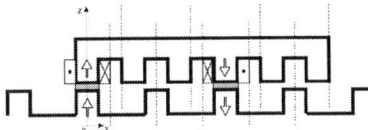

Fig. 10 - Composition by superposition to phase A.

So, in this model it is possible to assume the following hypothesis: For this motor geometric characteristics and considering a linear isotropic ferromagnetic material, the mutual inductance can be neglected.

IV. THE HARMONIC MODEL

From the modeling by flux tubes presented in the previous section, the idea of a simple model based on the harmonic function is vary attractive. The Fourier discrete transform depicted Fig. 11 shows the most important harmonic components of the inductance. These are the DC component and the first harmonic.

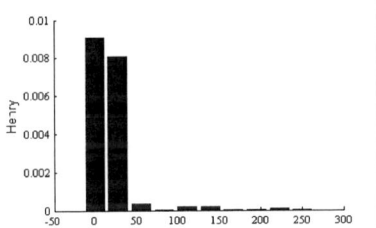

Fig. 11 - FFT to the phase Inductance.

Let be w_s the distance between two consecutive aligned positions (three steps) for the same phase.

From the FFT diagram, we deduce a harmonic model that gives the phase inductance with a good precision (11).

$$L_n(x) = L_o + L_1 \cdot \cos\left(\frac{2 \cdot \pi}{w_s} \cdot x - (n-1) \cdot \frac{2\pi}{3}\right) \quad (11)$$

By derivation of (11):

$$\frac{dL_n(x)}{dx} = -L_1 \cdot \frac{2 \cdot \pi}{w_s} \cdot \sin\left(\frac{2 \cdot \pi}{w_s} \cdot x - (n-1) \cdot \frac{2\pi}{3}\right) \quad (12)$$

V. DIMENSIONLESS MODEL

The accuracy and the stability of numerical simulations are some of the current challenges in scientific computing. The rounding errors due to the floating point arithmetic contribute to the inaccuracy of the computed results and, more important, to the instability of some numerical algorithms [12].

Considering the stable regular dynamics, this small inaccuracy is of little importance. However this problem could be more perceptible and "dangerous" when the system merges a high number of instable trajectories into a stable strange attractor. In addition the butterfly affect yields often to fake dynamic behaviors for example, the so called ghost attractor that no exists in the real experiments.

The floating-point representation with single or double precision is susceptible to the rounding errors, overflow and underflow.

To try to eliminate or to minimize these errors the first step is the scaling in order to obtain a dimensionless equations system. The second and very important action is to introduce the variables that have a recurrence behavior into the algorithmic (like the time and/or frequency) with an exact representation in the floating-point arithmetic.

The scaling results in a bounded system that allows looking for the results in the relative format becoming easier the comparison among the others simulations results with the different parameters.

For this simulation, a resistance r is placed in series with the phase to limit the maximum current.

Let be f the step switched frequency. The scaling is shown below.

$$i_n = i_n' \cdot \frac{E}{r} \quad (13)$$

$$v_x = v_x' \cdot \frac{w_s}{3} \cdot f \quad (14)$$

$$x = x' \cdot \frac{w_s}{3} \quad (15)$$

$$t = \frac{t'}{f} \quad (16)$$

$$L = L' \cdot \frac{r}{f} \quad (17)$$

$$V_n = V_n' \cdot E \quad (18)$$

Regarding the mass parameter, the scaling can be viewed like a mechanical and electrical energy relation.

$$M' = \frac{\left[\left(\frac{w_s}{3}\right) \cdot f\right]^2}{\left(\frac{E^2}{r}\right) \cdot \frac{1}{f}} \cdot M \quad (19)$$

Using the scaled variables, (15) and (17) in (11) the harmonic inductance function becomes:

$$L_n(x) = L_n'(x') \cdot \frac{r}{f} \qquad (20)$$

It yields to the relation (21) for the new inductance function after the variable changing.

$$L_n'(x') = L_o' + L_1' \cdot \cos\left(\frac{2 \cdot \pi}{3} \cdot x' - (n-1) \frac{2 \cdot \pi}{3}\right) \qquad (21)$$

And:

$$\frac{dL_n(x)}{dx} = \frac{dL_n'(x')}{dx'} \cdot \frac{3 \cdot r}{w_s \cdot f} \qquad (22)$$

The new dimensionless model is presented.

$$\begin{cases} \dfrac{di_n'}{dt'} = \dfrac{1}{L_n'(x')} \cdot \left(V_n' - i_n' - \dfrac{dL_n'(x')}{dx'} \cdot i_n' \cdot v_x'\right) \\[3mm] \dfrac{dx'}{dt'} = v_x' \\[3mm] \dfrac{dv_x'}{dt'} = \dfrac{1}{2 \cdot M'} \cdot \sum_{n=1}^{3} i_n'^2 \cdot \dfrac{dL_n'(x')}{dx'} \end{cases} \qquad (23)$$

The second action to improve precision in the computational simulation is to control the rounding errors. The objective of this work is not to present a new rounding error control technique. We perform a Runge-Kutta algorithm and reduce the rounding errors propagation by an exact floating-point representation of the initial conditions, the time step and the frequency parameter.

To have an exact representation in the floating-point mode the numbers have to be formatted as below.

$$A = \frac{B}{2^k} \text{ with } B \in N \text{ and } k \in N \qquad (24)$$

After having scaled the time as other variables ($0 \le t' \le 1$), the use of the step integration $s_i = 2^{-k}$ gives an exact representation of the time variable.

The same strategy for the stepping calculus is used for the frequency to obtain the Feigenbaum diagram.

$$s_f = \frac{(F_{max} - F_{min})}{2^k} \Rightarrow k \in N \qquad (25)$$

VI. GRAPHIC ANALYSIS TOOLS

In order to represent the dynamic behavior of the system models obtained from de previous sections, will be used the Feigenbaum diagram or bifurcation diagram and the Poincaré sections [14].

The bifurcation diagram represents an overview of all the dynamics which can be experienced by the system, including periodic, quasi-periodic and chaotic solutions.

The bifurcation diagram shows, among others, the phenomena called period doubling where one stable trajectory can change qualitatively by the bifurcation phenomena in others ones: one unstable period 1 and one stable period 2. One of the routes to chaos is by period

doubling. In this case, the period continues to double until there are no more stable periodic solutions available as seen in the Fig. 12.

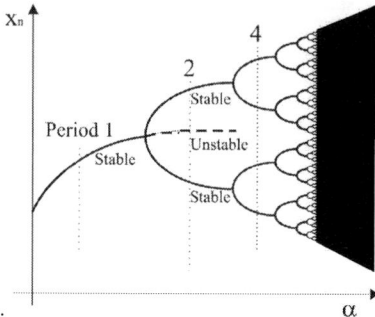

Fig. 12 The doubling period route to chaos.

A Poincaré map can be defined from a continuous trajectory n-dimensional by a stroboscopic method resulting in the (n-1)-dimensional discrete dynamical system. This reduced system of (n-1) dimensions inherits many dynamical properties from the original system. They are plotted for one parameter value and are a powerful tool to distinguish between a quasi-periodic solution and a chaotic one. This kind of analysis can not be given by the Feigenbaum diagram.

The previous tools will be used in the next section for a comparison of the harmonic and the flux tubes models.

VII. THE RESULTS AND COMPARISONS

In order to investigate all possible behaviors of the motor according to the supply frequency f, the Feigenbaum diagram and the Poincaré section are plotted. A fixed step, fifth order, Runge-Kutta algorithm is used to solve numerically the system (23) for different frequency values.

The dynamic behavior for $f = 3Hz$ is periodic. The numerical simulation for this region is showing in the Fig. 13.

This result was obtained applying the initial conditions $v_x = 1$ and $i_2 = 1$ is presented in Fig. 13.

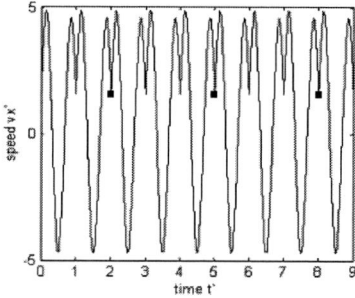

Fig. 13 speed vs. time (f=3Hz)

The Fig. 14 presents the cylindrical space built in order to obtain the Poincaré section yelling the non autonomous into an autonomous dynamic system by using the time as a new state. The respective Poincaré's section to the trajectory bellows is representing by one point in the plane speed vs. position space states and $t' = 0$.

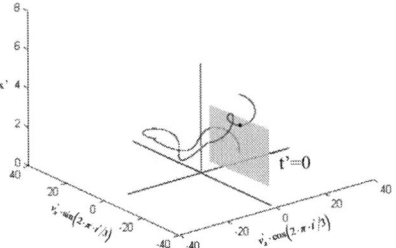

Fig. 14 Trajectory into a cylindrical space.

Fig. 15 presents the bifurcation diagram of the flux tubes model and of the harmonic model. Both diagrams present similar qualitative results. However, a precise comparison of the diagrams revel frequency shift. Periodic running modes alternate with chaotic behaviors. Inside the chaotic zones, we distinguish some periodicity windows.

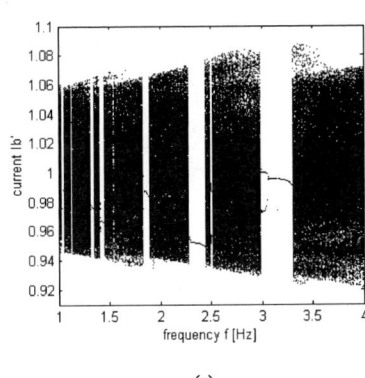

(a)

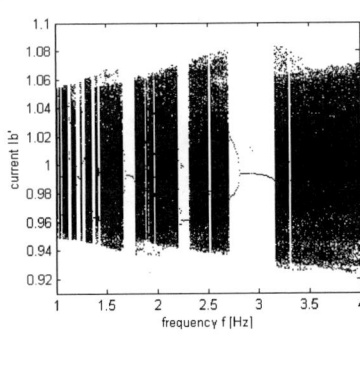

(b)

Fig. 15 – (a) Flux tube model and (b) Harmonic model. Current bifurcation diagrams.

Fig. 16 and Fig. 17 show a strange attractor that exists in both models and at the same frequency.

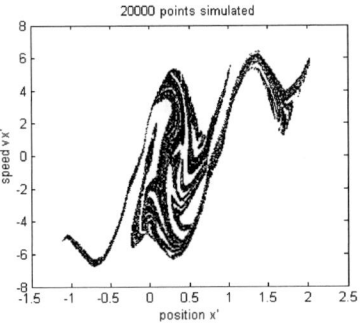

Fig. 16 Poincaré section projection to the tube the flux model in the speed vs. position plan for f = 4.203125 Hz.

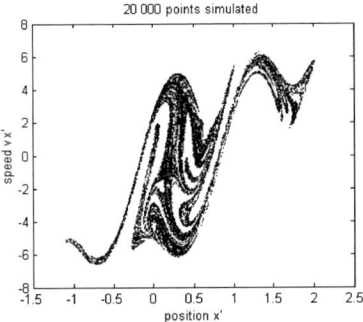

Fig. 17 - Poincaré section projection to the harmonic model in the speed vs. position plan for f = 4.203125 Hz.

The same attractor is shown in three dimensions on Fig. 18 and Fig. 19.

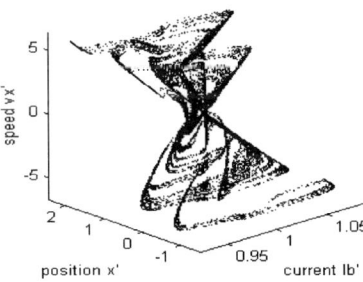

Fig. 18 - Flux tube model: Strange attractor 3D.

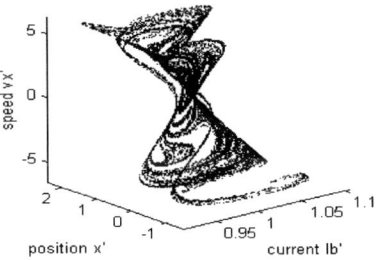

Fig. 19 Harmonic model: 3D attractor.

The others qualitative similar behavior can be observed in the two models but with a translation in the frequency axis.

An inverse period doubling is depicted on Fig. 20 for the harmonic model.

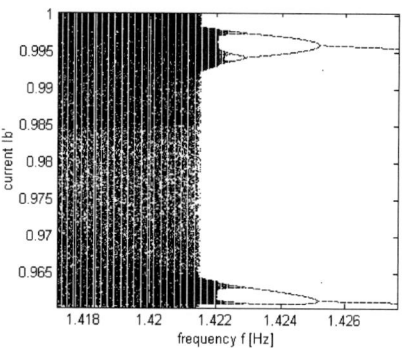

Fig. 20 - Harmonic model: Current doubling period

The Fig. 21 presents the behavior of the flux tubes model. In this figure it is difficult to say if it is a quasi-periodic or a chaotic behavior but by zooming on the attractor, Fig. 22, shows a characteristic very similar to strange attractor.

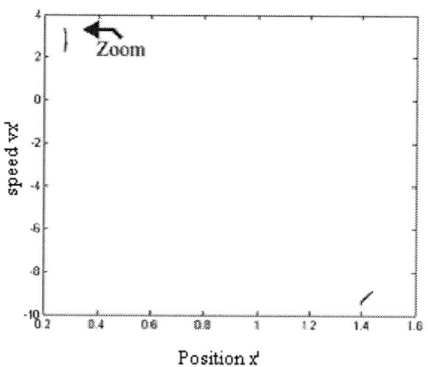

Fig. 21 - Flux tube model. Poincaré section f = 2.5Hz.

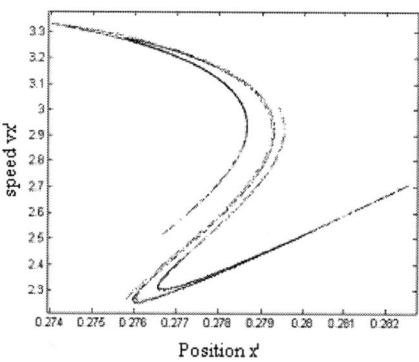

Fig. 22 - Strange attractor. Zoom from the previous figure.

The same behavior is obtained from the harmonic model but for a different frequency value. It's depicted on Fig. 23.

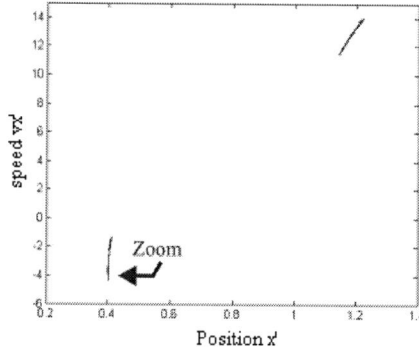

Fig. 23 - Harmonic model: Poincaré section f = 1.421875Hz.

The Fig. 24 and Fig. 25 present two successive zoom to the chaotic attractor presented in the previous figure.

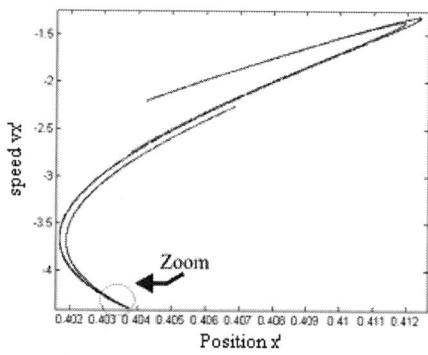

Fig. 24 - Strange attractor. First zoom from the previous figure.

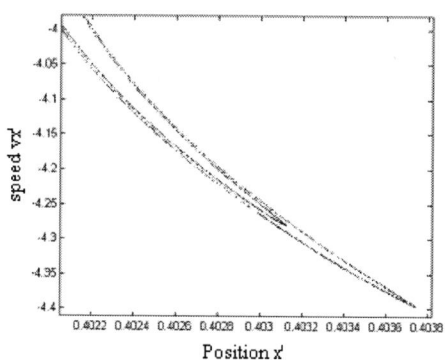

Fig. 25 – Second zoom of the strange attractor.

Fig. 26 presents the speed behavior $v_x^{'}$ to the chaotic region presented by the strange attractor in Fig. 17.

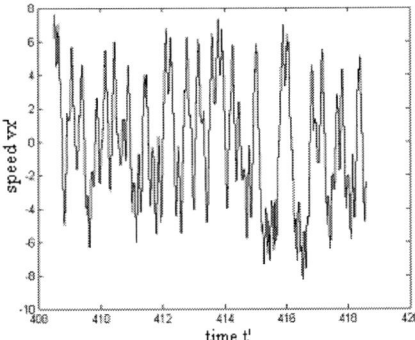

Fig. 26 - Speed waveform in the chaotic region.

VIII. Acknowlegment

This work is supported by grants from PROGRAMME ALβAN - European Union Programme of High Level Scholarships for Latin America with scholarship number: E06D104154BR.

IX. Conclusions

This work investigates the nonlinear nature of LSRM and its behaviors. It also presents a qualitative comparison between the results obtained from a tube de flux model and a simplified harmonic model in which the air gap permeance and frictions are neglected. In general this kind of parasites causes the shift in the critical points at which bifurcation occurs as observed.

The next steps of this study will be the computed estimation of the Lyapunov spectrum for some non periodic regions. After that, detailed numerical investigations will be made by changing the phase commutation instant, reference current and friction. Then simulations will be compared to the experimental time series data produced by the prototypes of the converter and the linear motor.

X. Bibliography

[1] Suto, Z.; Nagy, I., "Nonlinearity in Controlled Electric Drives: Review," Industrial Electronics, 2006 IEEE International Symposium on , vol.3, no., pp.2069-2076, July 2006

[2] Suto, Zoltan; Nagy, Istvan; Masada, Eisuke, "Nonlinear dynamics in direct torque controlled induction machines analyzed by recurrence plots," Power Electronics and Applications, 2007 European Conference on , vol., no., pp.1-10, 2-5 Sept. 2007

[3] Suto, Z.; Nagy, I., "Bifurcation phenomena of direct torque controlled induction machines due to discontinuities in the operation," Industrial Technology, 2005. ICIT 2005. IEEE International Conference on , vol., no., pp. 496-501, 14-17 Dec. 2005

[4] Alin F., Robert B., "Application de la théorie du chaos à l'approche expérimentale de la dynamique non linéaire d'un moteur pas à pas". Journées Doctoral d'Automatique JDA'01, Toulouse, France, pages. 231-236, 2001

[5] Pyragas K., "Control of chaos via extended delay feedback", Physics letters A, 1995

[6] Pera M-C., Robert B., Goeldel C., , "Quasiperiodicity and Chaos in a Step Motor". Proceedings of the 8th European Power Electronics Conference, Lausanne, Suisse. CD-ROM, 1999.

[7] Robert, B., Alin, F., Goeldel, C., "Aperiodic and Chaotic Dynamics in Hybrid Step Motor – New Experimental Results". Proceedings of the IEEE International Symposium on Industrial Electronics ISIE, Pusan, Corée. pp. 2136-2141, 2001.

[8] Reiss, J., Alin, F., Sandler, M., Robert, B.,, "A Detailed Analysis of the Nonlinear Dynamics of the Electric Step Motor". IEEE International conference on Industrial Technology ICIT'02, Bangkok, Thaïland, CD-ROM, 2002.

[9] Chai H. D., "Permeance model and reluctance force between toothed structures," Proc. 2nd Annual Symposium on Incremental Motion Control Systems and Devices, Urbana- Champaign, IL, pp. K1-K12, April 1973.

[10] Chayopitak, N.; Taylor, D.G., "Nonlinear magnetic circuit model of a linear variable reluctance motor," System Theory ". Proceedings of the Thirty-Sixth Southeastern Symposium, vol., no., pp. 170-174, 2004

[11] Jang, S.M.; Park, J.H.; You, D.J.; Choi, J.Y.; Kim, Y.H.; Sung, H.K., "Dynamic characteristics analysis of linear switched reluctance motor," Electrical Machines and Systems. ICEMS'05. Proceedings of the Eighth International Conference on , vol.1, no., pp. 529-534 Vol. 1, 27-29 Sept. 2005

[12] Alin, F., Robert, B., Goeldel, C., "On the limits of chaotic simulations by classic software – application to the step motor". IEEE International conference on Industrial Technology ICIT'02, Bangkok, Thaïland, CD-ROM, 11 au 14 décembre, 2002.

[13] Banerjee S., Chakrabarty K., "Nonlinear modeling and bifurcations in the boost converter," IEEE Trans. Power Electronics, vol. 13, no. 2, pp. 252–260, Apr. 1988.

[14] Pera M.C., Robert B., Goeldel C, "Nonlinear dynamics in electromechanical systems-application to a hybrid stepping motor". Electromotion, 7, pp 31-42, 2000.

Modular Architecture for Decentralized Hybrid Power Systems

E. Ortjohann[1], M. Lingemann[1], O. Omari[2], A. Schmelter[1], N. Hmasic[1],
A. Mohd[1], W. Sinsukthavorn[1], D. Morton[3]

[1] University of Applied Sciences South Westphalia/Division Soest,
Lübecker Ring 2, D-59494 Soest, Germany, e-mail: *Ortjohann@fh-swf.de*
[2] The Arab American University, Jenin, Palestime
[3] Bolton University, Deane Road, Bolton, UK

Abstract—Modularity offers large potentials to the rising demands of modern power supply systems. It ensures an advanced level of flexibility and adaptability required by expanding grids. In this paper an introduction to modular design strategies in combination with standardized interfaces is given. Differences in application of modular design strategies to Hybrid Power System (HPS) hardware and software are discussed. An intensified application of decentralized energy converters is expected in future. By modular design this leads to the availability of new production techniques. Wide product and various application ranges in combination with large production quantities require design strategies adjusted to this circumstances. The modular design approach support this by offering flexibility also during the entire product lifecycle. As further effect, the reliability of decentralized HPS can be improved significantly. The introduced modular design strategy was successfully applied to an exemplary PV-battery HPS. The implemented pilot system proved the practicability, efficiency and reliability of modular designed HPS for modern expanding power supply systems.

Keywords—Energy Management System, Distributed Power, Hybrid Power Integration, Renewable Energy Systems, Battery Management System.

I. INTRODUCTION

One of the projecting tasks of the near future is the adequate and stable power supply of rural areas, in remote regions of the industrialized countries, but mainly in countries of the Third World and in the developing countries [1]. Especially, in these areas the electrical power supply is one of the corner pillars of development and prosperity. This again is a necessary premise for social and political stabilization of the contingent conflict areas in those regions. Therefore, appropriate system concepts have to be developed, to build a supply structure which is adjusted according to the development status of a region. In this matter, beside the system concept, first of all the type of utilized primary energy source is very crucial. In consideration of the facts of global climate changes and the rising prices of fossil energies, the intensified utilization of renewable energy source (RES) is an accepted basic principle for the future [2]. Especially in developing countries a further significant aspect supporting renewable energy sources is the secured availability of natural energy as primary energy. No transport of fuel is needed since the natural energy is available directly at the point where it is converted without dependence on any infrastructure. In industrialized countries the main aspect is the saving of money for transport and reduction of environmental impacts not only by reduction of emission of the energy conversion but also of the primary fuel transport, which should not be underestimated [3].

Beside the use of renewable energy sources, new system concepts should be adjusted to simple, quick, reliable and inexpensive realization strategies. In the field of stand-alone power systems mainly system concepts restricted to individual demands are available [4], [5]. These systems are designed and specialized for certain supply situations and often limited to certain loads. Due to the continuous change of the power supply structures all over the world, more flexible concepts of decentralized energy conversion are needed [6]. Therefore, systems and components, that are - due to their modular structure - adaptable to the potentially existing supply structure and local circumstances, must be realized. They must be freely scalable in a certain range, regarding to the required power and the applied converters.

In connection with intensified further developments in the fields of DC/DC and DC/AC power electronic as well as control and communication systems, new concepts for DC-coupled hybrid power systems became possible. New power electronic topologies for common tasks are available at the market offering high dynamic ranges and increased efficiency. Approved and reliable technologies were improved and handling and management of the devices has been deepened but simplified [7], [8].

Due to the availability of highly developed power electronic components, new implementation strategies are possible but design criteria are still lagging behind. A basic approach that supports the mentioned future demands is the modular design strategy described in this paper. It introduces a more structure-centered view helping to improve the design, production, operation and maintenance of decentralized HPS. To prove the practicability of a modular design concept, a pilot installation of a PV/battery hybrid system using industrial components from other power electric application ranges was first validated by simulation and further realized in hardware.

To point out the potentials of modular design strategies the present limitations and future demands of decentralized power supply systems are first emphasized in the following background section. The term modularity is defined in the third section. The underlying idea if a modular design strategy and the application of standardized interfaces is described. Section four introduces the fundamental HPS structure by illustrating the concept of DC coupled HPS in hardware as well as their corresponding control and automation architectures. The idea of the introduced modular design approach applied to the DC coupled HPS are discussed in detail in section five. In this section the hardware and the software part is analyzed separately. Regarding the software modularity this paper specially concentrates on the main elements control and communication. To prove the feasibility and advantages of the introduced modular design strategy, test results of the implemented HPS are shown in section six. Concluding, effects and profits resulting from the described concepts, architectures and strategies are discussed in section seven.

II. BACKGROUND

Nowadays, most conventional DC coupled HPS are specialized for defined system structures only. The different components, as renewable energy sources, energy storages, inverters, system control and operational management, are adjusted to each other and closely linked. The systems are designed to fulfill the demands of defined energy sources in connection with certain load situations. Many manufacturers define their own standards for bus systems or special protocols for communication. This leads – mainly intended by the manufacturers – to compatibility of subcomponents to only this manufacturer but on the other hand larger and expensive design efforts and challenges in certification of the products. Fixed implementations and design strategies like this offer no flexibility and adaptability. A later change of the power sources, in example enlargement of the source structure by addition of further types of sources, or more sources is not easily possible. Re-powering (exchanging old energy sources by larger state-of-the-art products) can only be done by exchange of the major part of components or the entire system, mainly deterring the customers from this step due to huge investments needed for this step.

At the grid side, on the other hand, changes of the load structure, due to changes of the consumer behaviour or the number or location of the consumers in the grid, also may require a change of the energy storage capacities or the power electronic system structure. This is necessary to ensure a reliable power supply of the loads even in cases of poor meteorological conditions for the renewable energy sources.

Any change in the power electronic architecture and power rating of the system compulsorily leads to necessary changes in the system control and management. Additionally, different kinds of sources need different control structures to handle the systems state variables. Different variables may be needed to be measured and other control commands may be necessary. Furthermore, higher power rating also requires different security and accuracy regulations to be abided as well as an advanced energy management for the energy storages.

Latest investigations on expected changes showed that power supply structures in regions all over the world require flexibility and adaptability especially in the discussed aspects to ensure a lasting and reliable power supply in future [9]. Since decentralized HPS become a

bulk good, new concepts have to lower the barriers for the investors, therewith make the products more attractive and increase the turnover of the vendors. New future oriented designs of distributed energy supply systems therefore necessarily have to overcome the limitations of existing, mainly fixed systems. One of these future oriented design strategies is the modular architecture presented in this paper.

III. MODULARITY AND STANDARDIZATION

To achieve the desired flexibility and adaptability needed for distributed power supply systems, future HPS concepts have to be based on a structure with maximum grade of modularity. This section describes the meaning of modularity and standardization from a general point of view.

Modularity or modular design is the subdivision of a complex system into smaller units (modules) with basic functions. These modules can then be used in different systems to drive multiple functionalities [10].

The different discrete modules can be organized in a module pool containing various categories of modules regarding to their functional behaviour, characteristics, parameters or interfaces. In the design and in the manufacturing stage, suitable and available modules can be selected from a kind of catalogue. Regarding the demands of the intended final functionality, the modules can be connected by their interfaces to perform certain functionality as shown by Fig.1 [11]. To connect any modules in a selectable topology, the module interfaces have to be standardized to react on linked neighbour module actions and in reverse direction hand over information to them [12].

The main advantage of this design strategy is that the designer of the final product does not need to know all details of each module. The information needed to apply and connect the module can be simplified and abstracted in a short description and set of characteristics and parameters. In this way the main components of any complex system can be treated and designed almost as stand alone systems. By use of standardized interfaces these separated modules can be scaled independently. Modules become "black boxes". This effect leads to various benefits, some of which will be pointed out in the following.

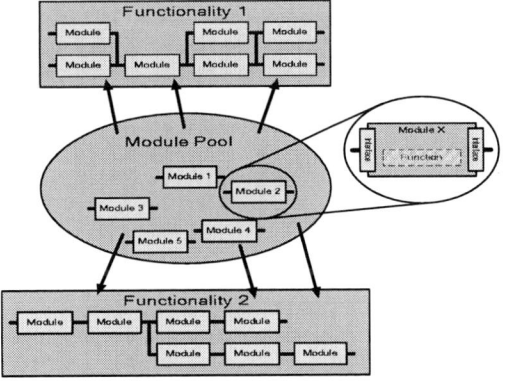

Fig. 1. Structure of the modular approach..

The production costs are reduced by completely independent manufacturing of the various modules. The discrete modules can fulfil purposes in different systems. Therefore, the number of manufactured equal modules is increased. This offers the application of mass production methods to the modules. On the other hand the wide range of applications of the modules keeps the flexibility of system design, which is comparable to the customized design strategies. Furthermore, modular design offers additional benefits such as augmentation and exclusion. An existing system can be enlarged, updated, modified or pared down in functionality by adding or excluding new sub-functional modules. The advantages of this modular design already came clear in other fields of application such as automotive and aerospace engineering [10]. Modular design is characterized by [13]: (1) Functional partitioning into discrete scalable, reusable modules consisting of isolated, self-contained functional elements; (2) Rigorous use of well-defined modular interfaces, including object-oriented descriptions of module functionality; (3) Ease of change to achieve technology transparency and make use of industry standards for key interfaces.

The basic principle of modularity can be applied to the two main fields of HPS design, the hardware and the software control design [14], [15], [16]. Although these design fields are functional closely connected to each other, they can be more or less decoupled in design by standardized interfaces or e.g. by the use of per unit (pu) values for communicated data. In the underlying project, the modular approach has been applied to the design of an exemplary PV/battery hybrid system which was first verified by simulation and later implemented in hardware as a pilot installation in the laboratory of power supply at the University of Applied Sciences South Westphalia.

IV. STRUCTURE OF THE IMPLEMENTED HYBRID POWER SYSTEM

In a first step the basic architecture of the exemplarily implemented HPS will now be introduced. The applied modular design strategy can, as described later, in the same way be applied to any other configuration of power supply systems different from the introduced test architecture.

The exemplary HPS is build of a DC coupled HPS with a photovoltaic power source as shown by Fig. 2 and a battery as energy storage since these components are easily available in the laboratory. To expand the system for further investigations, optional power and communication interfaces for integration of additional sources as wind energy converters, diesel generators, etc. or other sub-grids are provided, but not part of the investigation of this paper. In DC coupled HPS the different energy sources are connected to a common DC bus. To enable this, DC sources have to be linked via DC/DC converters and AC sources are linked by AC/DC converters.

By the application of a common DC bus or an intermediate DC-circuit, the state variables (frequency and voltage) of the grid can be completely decoupled from the sources. This means that in DC coupled HPS, the different sources can be controlled independently from the grid voltage and frequency. This allows a simplification of the power source control as well as an improvement of the

Fig. 2. PV system as marquee installation.

utilization of sources by advanced operational control. On the grid side, on the other hand, only one connection point is needed to link the DC bus to the grid. Therefore, the system has to be synchronized to the grid only at this point.

To increase the output power and to be flexible regarding later changes of the load conditions and the generated power, many inverters can be applied in parallel mode. This again increases the efficiency of the system on the grid side, by the ability to drive a defined number of inverters in their optimal point of operation. An additional significant advantage of the application of more than one inverter in parallel is the resulting redundancy if more inverters are available than needed to drive the actual needed power. Redundancy does not only secure the system stability and fault tolerance in case of failures of single inverters. It also allows the exchange of single components for upgrades, maintenance purposes etc. without interrupting the power supply system in its normal operation.

As can be seen by Fig. 3, a DC coupled hybrid power system can be created of various source combinations [17]. The functional groups are structured in the different types of sources (left), a common DC bus build by the power converters (middle) and parallel inverters on the grid side (right).

Beside the power electronic components the basic control structure is indicated. An online control unit supervises the power demand of the loads or the amount of power to be injected into the grid.

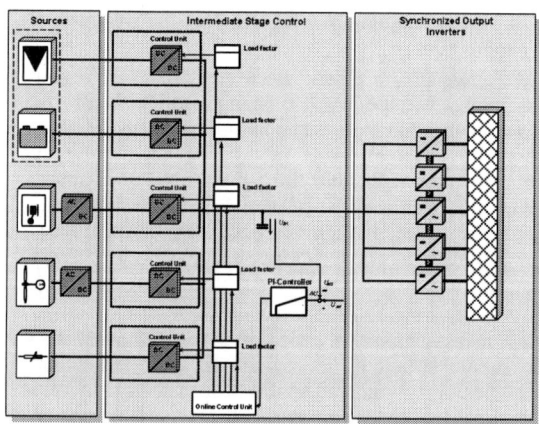

Fig. 3. DC-coupled HPS concept. [17]

It can be seen that the overall load power can be spread to the different sources by adjustment of the load factors of the DC/DC control units. In this way the different sources can be utilized regarding to their power generation potentials and the main focus can be set. The control of each energy converter itself is done local inside the converters. In addition to these control elements more superordinate control levels can be added.

They can e.g. monitor the meteorological conditions and forecast the amount of energy produced by the different sources. In this way the load factors continuously can be adjusted to the environmental condition and therewith the efficiency of the entire HPS can be improved. The architecture of the basic devices of the implemented exemplary HPS is shown in Fig. 4.

It consists of a PV generator, DC/DC converters, a battery as energy storage and DC/AC inverters as well as a control unit with separated power supply, peripheral I/Os and an industrial PC (IPC). For further enlargements, the system is also prepared to handle other sources than the PV generator in parallel to it. In comparison to other DC coupled HPS the applied configuration uses a battery directly as intermediate circuit. To control the operation and the SOC (state of charge) of the battery, the tasks of the battery converter of conventional configurations (comp. Fig. 3) are taken over by an advanced control of the PV converter and the grid inverters. By this advanced management the necessity of an additional DC/DC converter for the battery management is avoided. This decreases the size of the HPS, its installation costs and the number of critical components that could cause downtime. Further advantages and disadvantages of this topology are not part of this paper. In the design of this HPS configuration, the modular architecture approach was applied as shown in the next sections.

V. APPLICATION OF THE MODULAR APPROACH

As stated in the introduction to the modular approach, modularity can be applied to the hardware and the software. These two fields of application are now described in detail. The level and depth of the applied modular structure depends on the complexity of the product as well as the width of the product range and the potential of the final product to be reworked or upgraded. Therefore an optimal level has to be found as the case arises. In this paper only the top level of the modular approach will be discussed.

Examination of the most detailed modular structure would exceed the paper and not result in significant increase of information for the reader.

A. Hardware Modularity

The implemented hardware architecture is designed modular regarding the sources, the different power electronic devices (DC/DC converter, intermediate circuit and inverter) as well as the control hardware (I/O interfaces). All elements are separated into functional units (modules) with defined / standardized interfaces and certain functions. Regarding to the application and under consideration of the utilized energy sources and the grid connection strategy, suitable modules have been selected from the module pool. The number and class of elements has been defined by the power ratings of the load to be supplied and the potentials of the sources. The modules have been joint together to build a DC coupled HPS of the structure shown above. The modular designed hardware of the implemented HPS as introduced is shown in Fig. 5.

The hardware is mounted into a standard industrial electrical cabinet, of which various numbers can be connected next to each other. The cabinet has different slots that can be equipped with different types of DC/DC converters and DC/AC inverters. By this, the hardware can be adjusted to the required value of transmitted energy by switching converters or inverters in parallel. By application of separated inverters different numbers and types of energy sources can be handled even if the requirements of the environment change later on.

The cabinet unit keeping the selected type of DC/DC converters and active rectifiers is shown in Fig. 6. It can be seen that three components are currently installed and additional three converters can be added. Larger types of converters are available for even higher power ratings.

The inverter unit for the chosen inverter category is shown by Fig. 7 where two of the six inverters are currently installed.

The battery bank as storage of energy is located external due to its volume and weight as well as to abide special security regulations for the storage of batteries. It also can be scaled independently due to modular structure of the battery blocks.

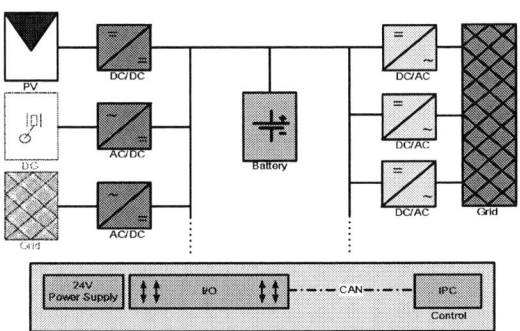

Fig. 4. Implemented HPS components. [18]

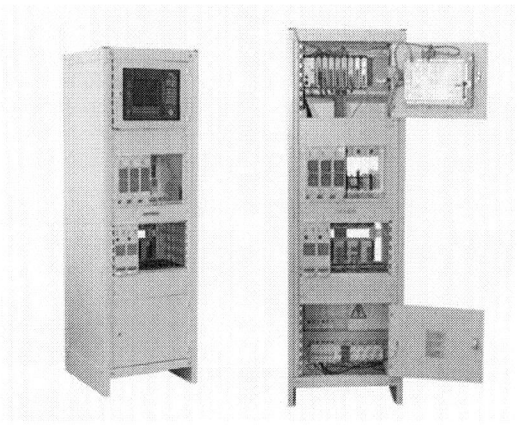

Fig. 5. Pilot unit of the modular HPS. [18]

Fig. 6. Modular standardized DC/DC converter and rectifier unit. [18]

Fig. 7. Modular standardized inverter unit. [18]

The control interfaces for the output of command values and for the input of measurement values to and by the system control is located in the top of the cabinet. It is also built of standard industrial interface modules connected by a common local bus system. These modules measure and adjust the state variables of the system depending on the type of energy source and the connection mode to the grid if applicable. In addition to the state variables measurement, further security and supervision functions as temperature monitoring, etc. could be implemented by simple addition of further measurement modules. The analog output modules set the analog desired values (mainly currents) for the converters and inverters. Digital inputs and outputs are responsible to drive relays and read back confirmation signals. In this way sources and loads can be potentially separated from the system for security or maintenance purposes. Therefore the number of desired digital I/O also depends on the HPS complexity and the number of sources and loads.

The communication of the peripheral interface module with the industrial control PC of the HPS is realized by standard CAN bus. A bus coupler module is set in front of the series connected I/O modules to manage the communication. By change of the CAN-module to modules of any other communication standard, the I/O unit can be adapted to the communication requirements of different central processing and control units. The I/O periphery interface module with the bus coupler and a local auxiliary power supply is shown in Fig. 8.

At the bottom of the cabinet a power interface unit is mounted. It includes fuses, relays and connection interfaces for the main power signals. The first task of this unit is to secure, supervise and control the power inlets and outlets of the system. In case of large HPS or expanding HPS this power interface is helpful to connect multiple cabinets. It is also possible to separate the converters and inverters into different cabinets. By parallel connection of single modules or entire cabinets large systems can be built up or small HPS can be upgraded.

Fig. 8. Modular standardized peripheral I/O units. [18]

All of these hardware modules can be easily exchanged, expanded or replaced by other module types (e.g. of different size) or components of different vendors. By standardized interfaces the linking and communication of different module types is ensured. Standardized interfaces are the basic requirement to utilize the full advantages of a modular approach.

B. *Software Modularity*

The second main part of an entire modular approach is the modularity of the software. The modular structure of the hardware as described before can only be handled in an efficient way if the software offers the same degree of modularity. The software has to be able to adopt the current configuration of the hardware. An adjustment in case of changes of the hardware has to require only minimum efforts.

In the implemented control and supervision software, all basic routines are designed to deal with scaled per unit values. By this as basic principle, the control software can be completely decoupled from the ratings of the power electronic hardware. All developed software elements are created in a clear structure of sub-functions. Regarding to the intended architecture of the control and management environment, different pre-designed functions can be selected from the software module pool and connected in a very simple way. These basic functions have been designed as "black boxes" while the internal design of each box does not have to be known by the user. The interfaces and the behaviour of each box are the only needed information for the user to apply these discrete software modules and to create the entire control architecture [19].

The software environment of the HPS can be structured into six main elements as given by Fig. 9. Some of these elements are always required in their basic structure, some have to be scaled, adjusted and chosen regarding to the hardware architecture and others are optional depending on the customers' demands.

The network interface layer (NIL) can connect the system to superordinate management systems. In this way e.g. desired values for the HPS power output can be set from the upper grid control, or the system status can be checked via the internet. The human machine interface layer (HMIL) is responsible for the visualization of the HPS status to the end user. System parameters e.g. for the power controllers or rated parameters of the hardware can be changed or checked manually. The supervision layer (SL) monitors the status of the HPS. The content of the energy storages can be supervised as well as the status of the peripheral and power electronic devices, software controllers and communication elements.

2138

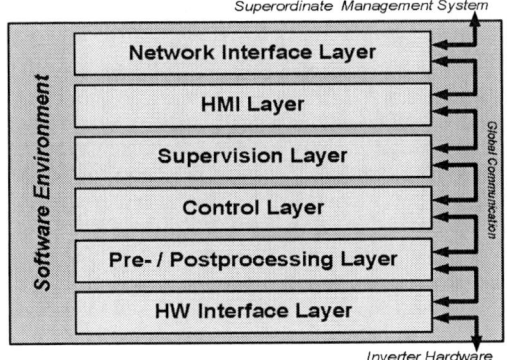

Fig. 9. Structure of the modular approach.

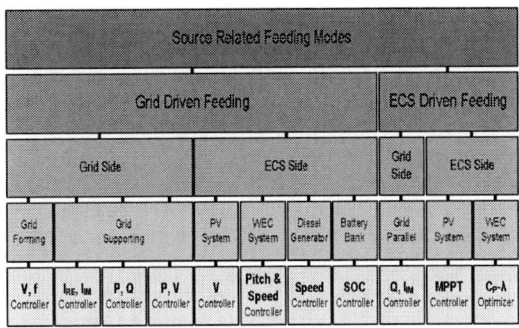

Fig. 10. Different control concepts for decentralized HPS. [17]

The control layer (CL) keeps the software controllers for the management of the power sources and storages, as well as the inverters as connection to the grid or loads. The two bottom layers are the pre- and post processing layer PPL) which is changing data type formats and value ranges, and the hardware interface layer (HIL) adjusting the direct outputs of the peripheral interfaces to the standardized per unit values. The hardware interface layer is one of the most important parts of the software to ensure a more or less hardware decoupled software implementation. The HIL transforms any measured input value to a standardized per unit range. This means that the software supervision and management functions (e.g. security limits to be abided, etc.) and the controllers can be designed independent from the (power -) ratings of the hardware. In case of enlargement of the system by addition of more devices (e.g. regarding the battery bank) the scaling factors of the HW interface can be adjusted in a central point and the supervision and control functions can deal with the same values as before. Another case is the change of measurement hardware with different standard output ranges. In this case also the parameters of the HWI can be readjusted and the system can be operated without further changes. The same aspects can be utilized concerning the digital and analog output standards and levels.

1) HPS control Layer

Special attention has to be taken on the control layer since this is the software part most closely related to the power electronic hardware as main actuator of the HPS. It is obvious that the types of power sources applied to the HPS as well as the kind of connection to the grid define the basic design of the control environment. Fig. 10 shows an extract of possible HPS configurations and the corresponding controllers of the power source side [20].

The two main operation modes are the grid driven and the ECS (Energy Conversion System) driven feeding. In grid driven mode the power transmitted to the grid is defined by the grid. In ECS driven mode the injected power is defined by the ECS. Most renewable energy sources are operated in ECS driven mode. Their amount of power that can be injected into the grid is strongly dependant on environmental conditions e.g. wind speed, irradiation, etc. To minimize the need for power storages, the power should of course be injected when available.

Therefore the ECS has to define the power rating. Conventional energy sources on the other hand are mainly operated in grid driven mode since they are completely independent from environmental conditions and can inject power to the grid whenever needed. In this way the grid can request power from the conventional sources when environmental conditions are poor. The power balance can be stabilized while the utilization of the renewable energy sources is maximized.

The second level of differentiation is the target direction of the controllers. They can be aiming to the grid or the ECS side. Aiming to the grid side, the system can control the variables of the grid and the source is treated passive. The source is reacting in a way that is necessary to influence the grid variables. In ECS related control, the controllers influence the state variables of the ECS while the grid variables are treated as control signals. Grid side and ECS side based control are possible in both - grid driven and ECS driven feeding mode.

If the control is running in grid driven feeding mode and managing the grid side, the system can be operated in grid forming or grid supporting mode. In grid forming mode the state variables of the grid (voltage and frequency) are built by the system. This control mode is needed e.g. for stand alone systems in island mode or for the main power system the grid is based on since one power source in the grid has to define and built the grid state variables. In grid supporting mode the state variables voltage and frequency are defined by the grid and the system is reacting passive on it. In this case different control modes are available regarding to the demands on the grid side.

In case of grid driven feeding mode and ECS related control, the kind of controller applied is depending on the type of energy source utilized. Different sources also require different types of controllers to act on the different state variables. The transmitted power is still defined by the grid.

In ECS driven feeding mode and control of the grid side (as shown on the right half of Fig. 10) the operational mode is called grid parallel. In this mode the reactive power and the reactive current part are controlled while the injected power is defined by the energy source.

The last category is the ECS driven feeding mode with the control related to the ECS side. This is the most typical control mode of renewable energy systems since the energy sources define the amount of injected power by control of the state variables. In this way the maximum utilization of the renewable energy source is ensured and the efficiency of the power supply system concerning the minimum use of fossil fuel and therewith the minimization of CO_2 emission can be achieved.

The shown operational modes and control aims make clear that various topologies, power sources and control strategies are available and have to be selected regarding to the local supply and grid case. By appropriate categorization and structuring of these control cases also the control environment can be designed modular.

For this, the different control architectures have to be broken down into discrete elements of controllers, limiters and supervisors. They can be arranged in a module pool regarding their category comparable like described in the hardware section. From this pool sufficient standard elements can later in the design or upgrade phase be selected, parameterized and joint together to build the control environment as needed. By Standardized interfaces and per unit values it is possible to arrange powerful control layers and entire software environments with minimum effort. Beside the control architecture, supervision and monitoring functions ca be selected in the same way.

2) Communication Layer

A further, quite often underestimated aspect here is the communication layer – mainly as interface to superordinate management systems. By standardization of data types, communication protocols and information structures any kind of the various available communication standards can be applied or even changed later on by replacement of the communication software function and hardware interface module. This aspect is very important because of two reasons. First, many different industrial communication standards exist and can be applied in the different levels of the system architecture. Second, the superordinate management system to be connected is gaining more and more influence, since small systems are growing and have to be grouped together. This leads to the demand of flexible and upgradable communication interfaces to the upper management and control level. A modular structure specially supports the possibility to freely select any communication standard or to change it later on if necessary due to new demands of the system or the entire mini-grid architecture.

It is obvious that the modular approach can be applied to all detail levels of the HPS architecture – from the main power electronic components and software functions down to discrete electronic elements and software commands [21][11]. Therefore, further description of the detailed developed structure of the modular HPS architecture would exceed this paper.

VI. TEST RESULTS

The exemplary PV/battery hybrid power system has been structured and set up based on the modular design strategy introduced in this paper. It was simulated and implemented in hardware in laboratory of power supply at the University of Applied Sciences South Westphalia, Soest Division. The system was finally tested under common meteorological conditions and with different loads. The system architecture was changed by addition of further sources and by change of the load rating. Some results of the various hardware tests will be shown in the

following to prove the practicability and feasibility of the design strategy and the implemented HPS.

The first Fig. 11 shows the boot and start phase of the HPS and the change if irradiation. At the start phase the PV system is connected and the PV voltage rises to its open circuit value (ca. 320V). After 2 seconds the MPP tracker is started and the PV voltage is adjusted to reach the maximum power point (MPP). During this phase the load is first supplied by the battery until the power of the PV system is able to take over the load.

At the point when the PV power is larger than the power of the load the power difference is used to charge the battery as can be seen by the positive battery current. In the middle of the figure the reaction of irradiation change is shown. The MPPT starts to readjust the PV voltage and reaches the new MPP after short time.

Fig. 12 shows the behaviour of the hybrid power system when load changes while irradiation is nearly constant. The load is varied in four steps, three steps up and one step down as can be seen by the bright line of the load current in the lower part of the image. The battery current directly responds on load steps and decreases. After the third step of load increase the PV system is not able to supply the load any more and the battery has to assist the system indicated by a negative battery current.

A second case of load change during nearly constant irradiation is given by Fig. 13. In this case the load has been changed with larger values. The irradiation is also lower than in the first case.

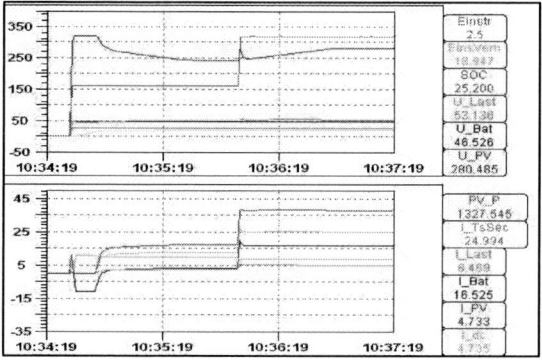

Fig. 11. Start and change of the irradiation. [18]

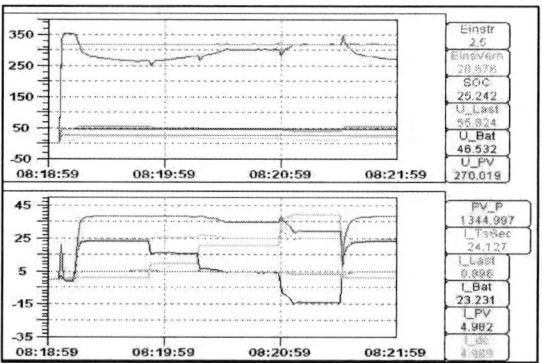

Fig. 12. Load change while irradiation is constant – Case 1. [18]

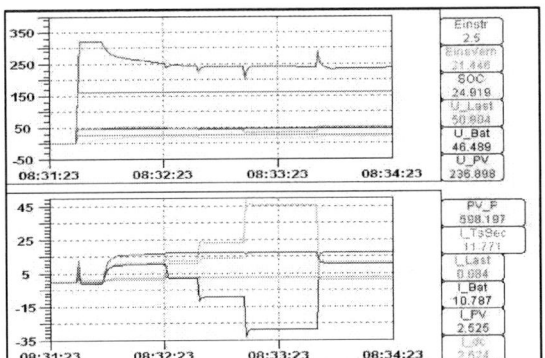

Fig. 13. Load change while irradiation is constant – Case 2. [18]

The implemented load management function also showed that the redundant modular inverter architecture can be used to increase the efficiency of the decentralized HPS by driving the desired number of inverters for each load case in such a way that each inverter is always operated in its optimal point of operation.

VII. CONCLUSION

The introduced modular approach offers many benefits to the design of Hybrid Power Systems (HPS). On the one hand various equal modules fulfilling purposes in different systems can be manufactured completely independent. This offers the application of mass production methods reducing production costs. On the other hand the wide range of application of the modules keeps the flexibility in system design comparable to customized design strategies. This means that strengths of mass production and customized production can be combined. Furthermore, modular design offers additional benefits such as augmentation and exclusion. An existing system can be enlarged, updated, modified or pared down in functionality by adding or excluding new sub-functional modules. Decentralized HPS are elements of continuously changing grids. Especially components of modern power supply systems have to be flexible and even more adaptable in future. Beside the essential flexibility and adaptability, the modular design approach also offers a high level of reliability as one of the fundamental aspects of power supply systems. This is supported the simplified ability to create redundant structures as well as to quickly replace fault elements without certain knowledge of the entire system. By the implemented pilot installation of an exemplary PV/battery HPS, the benefits of a modular design approach in comparison to conventional design strategies of power supply systems are demonstrated.

ACKNOWLEDGMENT

This research work is supported by the FHprofUnd Program of the German Federal Ministry of Education and Research. The Research and development of the application was done in cooperation with the industrial partners Ferrocontrol (GER), Tippkötter (GER), Powercorp (AUS), Energie- und Technologiebüro Westfalen (GER) and The University of Bolton (GBR).

REFERENCES

[1] World Energy Outlook, IEA (international energy agency), 2006

[2] J.K. Kaldellis: Off-Grid Solutions Based on RES and Energy Storage Configurations. International Energy Storage Conference (IRES), Gelsenkirchen - Germany, October 2006.

[3] EIA (Energy Information Administration), U.S. Government: Diesel Fuel Prices - What Consumers Should Know, April, 2007

[4] P. Thounthong, S. Rael, B. Davat: Control Algorithm of Fuel Cell and Batteries for Distributed Generation System, IEEE TOEC, Vol. 23, No1, March 2008

[5] C.-L. Chen: Optimal Wind-Thermal Generating Unit Commitment, , IEEE TOEC, Vol. 23, No1, March 2008

[6] L. Söder, L. Hofmann, A. Orths: Experience From Wind Integration in Some High Penetrated Areas, IEEE TOEC, VOl. 22, No. 1, March 2007.

[7] J. Nilsson, L. Bertling: Maintenance Management of Wind Power Systems Using Condition Monitoring Systems-Life Cycle Cost Analysis for Two Case Studies, IEEE TOEC, Vol. 22, No.1, March 2007

[8] Benning GmbH & Co. KG, www.benning.de

[9] D. Bohn, "Decentralised energy systems: state of the art and potential," International Journal Energy Technology and Policy, vol. 3, 2005.

[10] Deboe B. D: Glossary Modular Design. NESI (Net-Centric Enterprise Solutions for Interoperability), US Government, September 2007. A novel type of grid converter, EPE 2011.

[11] C&C Power, inc.: DC to AC – Modular Inverter System. http://ccpower.com/PDF/inverters_topazmod.pdf, Nov. 2006.

[12] A K W Lau, R C M Yam: A case study of product modularization on supply chain design and coordination in Hong Kong and China. Journal of Manufacturing Technology Management. Bradford: 2005. Vol. 16, Iss. 4; p. 432

[13] Glossary Modular Design. NESI (Net-Centric Enterprise Solutions for Interoperability), US Government, September 2007

[14] Guo, J.; Boroyevich, D.; Edwards, S.H: Distributed, modular, open control architecture for power conversion systems. Power Electronics Specialists Conference, 2004. PESC 04. 2004 IEEE 35th Annual, Volume 3, 20-25 June 2004, Vol.3

[15] M. Lingemann: FPGA Controlled Real-time Ethernet Interfaced Modular Inverter System. Master thesis, South Westphalia University of Applied Sciences, Soest Division, Germany and The University of Bolton, UK, February 2007.

[16] J H Mikkola: Management of Product Architecture Modularity for Mass Customization: Modeling and Theoretical Considerations. IEEE Transactions on Engineering Management. New York: Feb 2007. Vol. 54, Iss. 1; pg. 57

[17] Omari, O.; Ortjohann, E.; Saiju, R.; Morton, D, "A Simulation Model For Expandable Hybrid Power Systems," *2th European PV-Hybrid and Mini-Grid Conference, September, 2003, Ger.*

[18] Egon Ortjohann, Alaa Mohd, Andreas Schmelter, Nedzad Hamsic, Max Lingemann: Simulation and Implementation of an Expandable Hybrid Power System, 2007 IEEE International Symposium on Industrial Electronics, Vigo, Spain, June 4-7, 2007.

[19] A. Rodrigues, M.J.R. Armada: The Valuation of Modular Projects: A Real Options Approach to the Value of Splitting, GFJ, Volume 18, Issue 2, 2007, Pages 205-227, Portugal

[20] E. Ortjohann, A. Mohd, N. Hamsic, D. Morton, O. Omari: Advanced Control Strategy for Three-Phase Grid Inverters with Unbalanced Loads for PV/Hybrid Power Systems, 21th European PV Solar Energy Conference, Dresden – Germany, September 2006.

[21] Egon Ortjohann, Alaa Mohd, Nedzad Hamsic,Danny Morton, Osama Omari: Control and Representation of Three-Phase Asymmetrical Signals Used by Modular Inverters to Feed Unbalanced Loads in Hybrid Power Systems. The Great Wall World Renewable Energy Forum and Exhibition (GWREF), Beijing - China, October 2006.

[22] Tzu-Liang (Bill) Tseng and Chun-Che Huang: Design Support Systems: A Case Study of Modular Design of the Set-Top Box from Design Knowledge Externalization Perspective, Decision Support Systems (2007)

Design of a power management system for an active PV station including various storage technologies

Di LU, Tao ZHOU, Hicham FAKHAM , Bruno FRANCOIS

Laboratoire d'Electrotechnique et d'Electronique de Puissance de Lille L2EP,
Ecole Centrale de Lille, Cité Scientifique, BP 48, 59651 Villeneuve d'Ascq Cedex, France,
Phone : 33-3-20 33 54 59, Fax: 33-3-20 33 54 54
E-mail: di.lu@ec-lille.fr, tao.zhou@ec-lille.fr, hicham.fakham@ec-lille.fr, bruno.francois@ec-lille.fr

Abstract—**A hybrid power system with a PV energy conversion system is proposed with lead-acid batteries based long-term storage units and ultracapacitors based fast-dynamic storage units in a DC-coupled structure. It is necessary and important to manage the powers among those different power sources. A power management algorithm is proposed for this hybrid power system by taking into account the characteristics of each power source. It carries out the calculation and the power distribution of an adjustable power margin for the power sources. As results, the lead-acid batteries can smooth the generated fluctuating photovoltaic power in a long time range (from minutes to hours), and the ultracapacitors can smooth the photovoltaic power in a short time range (from seconds to minutes).**

Keywords—**Power Management, hierarchical control, control design, hybrid power system Photovoltaic, Lead-Acid batteries, Ultracapacitors**

I. INTRODUCTION

Faced with the challenges of energy, the demand for primary energy in the world wide is evolving, particularly in the highly developing countries like China. However, the stock of oil in our planet will soon be exhausted.

Today global warming becomes more serious due to the greenhouse effect. Some emissions of greenhouse gases come from the human activity. The production and processing of electrical energy is one of the principle sources of greenhouse gases. In December 1997, the Kyoto protocol has been established in order to reduce global emissions of greenhouse gases. In the area of power generation, this protocol promotes the renewable energy. On the other side, almost one third of the world population has no access to electricity. Electrifying those regions can be done either by extending the grids of some existing power systems or by constructing new stand-alone power systems. The extension of the existing grids is generally preferred, but it is not always feasible in remote area, due to the geographical, environmental and logistic consideration (e.g. in hilly regions, forests, deserts and islands). Power generation with renewable energy sources are proposed for remote areas where it is difficult to extend the conventional grids to those regions and/or to transport fuel and equipment needed for conventional power source, such as diesel generators [1]. By properly exploiting the renewable energy sources in local area, we can considerably reduce the operating periods of the conventional power sources,

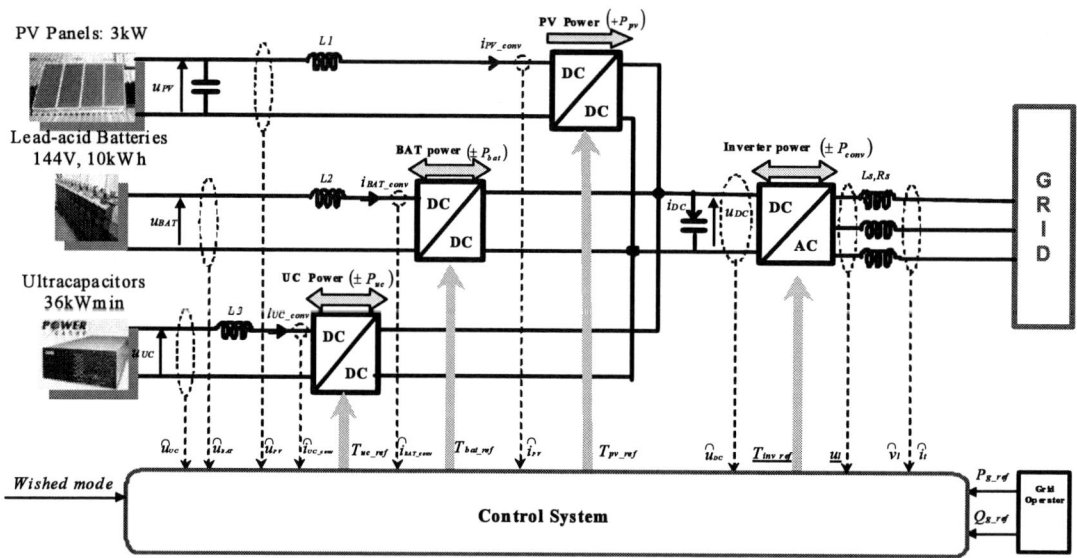

Fig. 1. Grid-connected photovoltaic power station including batteries and ultracapacitors

978-1-4244-1741-4/08/$25.00 ©2008 IEEE

thus reduce the fuel consumption and minimize the need for maintenance. As a result, the sustainability of the power generation can be improved. With the reduction of the photovoltaic panel price, the photovoltaic technologies will be widely used in the hybrid power system. However, an electrical generating system depending entirely on the renewable energy sources is not reliable because the availability of the renewable energy sources can not be constantly assured. Therefore, the hybrid power system combing the renewable energy sources and the storage units should be chosen for supplying electricity to the isolated grids. Integrating photovoltaic power sources with batteries as storage, can lead to a long-term reliable energy source [2] [3]. The ultra-capacitors have fast dynamics, thus can be added to smooth fast fluctuations of the photovoltaic power and can ensure a good supply to the grid. Thus, a power system comprising a photovoltaic power generator, batteries based long-term storage and ultra-capacitors based fast-dynamic storage is proposed in this paper (Fig. 1).

A power management system is necessary for the hybrid power system to manage the power flow among the sources in order to satisfy the load requirements throughout the whole operation period. The main purpose of the power management is to maximize the benefit of the renewable energy source and to optimize the operation of each storage unit. In this paper, such a power management algorithm is presented, by considering that grid requirements are specified as active and reactive power references.

II. HIERARCHICAL CONTROL

According to the statistics of solar illumination and the temperature during a certain period, the average generated photovoltaic power can be predicted and a power generation plan in average values can be deduced. Instantaneously, the generated power may be different from the grid power reference. Therefore, an efficient power management has to optimize the operation of both complementary storage units and PV panels of the power station.

For our application, three DC/DC power electronic converters are used to connect each power unit to the common DC bus. We have used a hierarchical control [4]

for the design of the energy management system (Fig. 2). The structure of a hierarchical control system includes 4 levels. Each one has precise control tasks depending on its hierarchical position:

-Control of Operating Modes (COM)

-Power Tracking of Sources (PTS)

-Control Algorithm and Modulation Technique (CAMT)

-Switching Control (SC)

The level COM decides the operating mode for the whole power station according to the current capacity of PV production and the states of each storage unit and the actual power demand of the grid. The level PTS calculates every power references for each source according to the measured values, the selected operating mode from the COM and transforms this power references information to the level CAMT. The level CAMT applies the control algorithm and modulation technique to each converter with current or voltage reference. The SC (switching control) gives the semiconductor signals ({-5, +15}) in order to apply wished ideal states ({0, 1}).

The multi-source power station has four power converters and thus we have also 4 SC and 4 CAMT for each converter (Fig. 2).

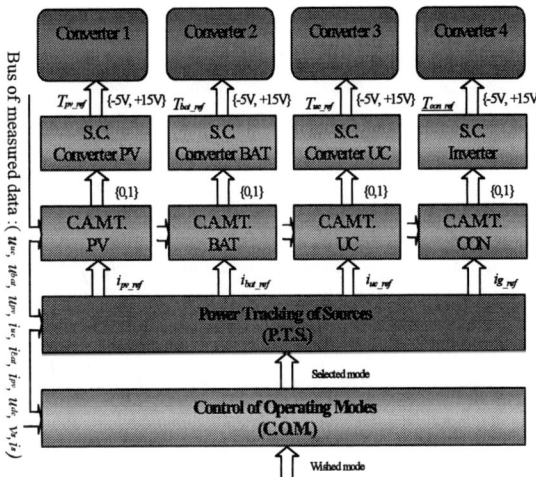

Fig. 2. Hierarchical control of the PV station

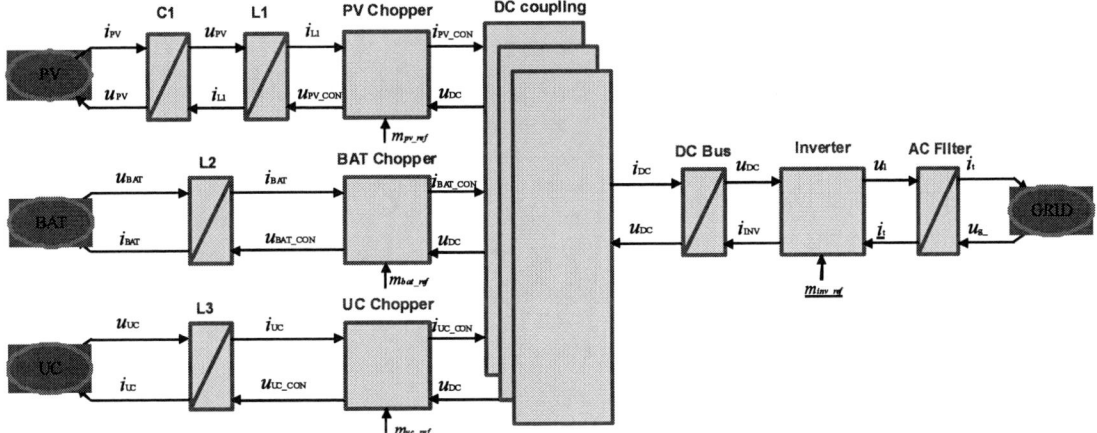

Fig.3. Energetic Macroscopic Representation of the global modeling for the multi source power station

III. MODELING OF THE POWER STATION AND DESIGN OF THE CONTROL SYSTEM

For the system modeling, we use the Energetic Macroscopic Representation (EMR) to represent the whole power station (Fig. 3). The EMR is a synthetic graphical tool based on the principle of action and reaction between connected elements [5]. It leads to a synthetic description of an overall conversion system between several sources. In such a representation, coupling devices distribute energy from upstream or downstream elements. The interest of the EMR is that we can deduce the structure of the control system, which is depicted as a Maximum Control Structure (MCS) using inversion rules. The MCS uses a maximum number of control functions and measurements. The controllers are depicted by parallelograms with an oblique bar (see Appendix). From the MCS a schema bloc representation is easily found (fig.10).

A. Photovoltaic Panels

Models of the PV panels- The photovoltaic panels are modeled as a current source (PV) [6] and are represented by a green oval in Fig. 4. It delivers a current (i_{pv}) through a filter (C1, L1).

Model of the filter - The choke is modeled as a current source (i_{L1}). This current depends on the PV capacitor voltage u_{pv} and the output voltage of the chopper u_{pv_con}:

$$\frac{di_{L1}}{dt} = \frac{1}{L_1}\left(u_{pv\_con} - u_{pv}\right)$$

L_1 is the inductance of the choke. This equation is represented by the block L1 in Fig.4.

Using a capacitor across the terminals of PV panels enables to set its terminal voltage (u_{pv}). The variation of this voltage will vary the generated PV power. This capacitor can be modeled by using the PV current i_{pv} and the current i_{L1}:

$$\begin{cases} \dfrac{du_{pv}}{dt} = \dfrac{1}{C_{buspv}}i_{cpv} \\ i_{cpv} = i_{pv} - i_{L1} - \dfrac{1}{R_{buspv}}u_{pv} \end{cases}$$

R_{buspv} and C_{buspv} are the resistance and the capacity of the DC bus of the photovoltaic panels. Both equations are represented by the bloc C1 in fig. 4.

Model of the PV choppers- The filtered current (i_{L1}) is modulated by a chopper, and then this modulated current (i_{PV_conv}) is injected into the common DC bus. The chopper is considered ideal and is modeled in mean values as:

$$\begin{cases} i_{pv\_con} = m_{pv\_ref}i_{L1} \\ u_{pv\_con} = m_{pv\_ref}u_{dc} \end{cases}$$

with the duty cycle (m_{pv_ref}).

Choppers will be represented by orange square pictograms in Fig. 4.

The EMR of the photovoltaic system is shown on the top part of Fig. 4.

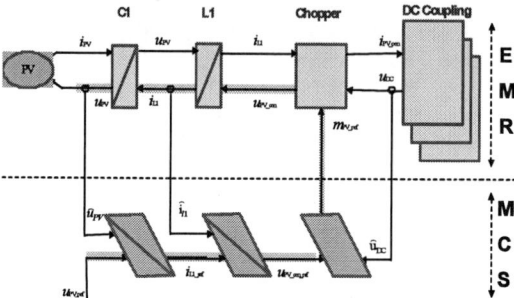

Fig. 4. EMR and MCS of the photovoltaic generation system

PV Control system: To make variable the output PV current, we have to control the PV terminal voltage (u_{pv}) by acting on the duty cycle (m_{pv_ref}). Therefore we reverse the (yellow) path between u_{pv} and m_{pv_ref} to build the control system (MCS). The control of the PV terminal voltage (u_{pv}) is achieved with:

_ Two closed loop C_{L1}, C_{C1} controls,

$$i_{L1\_ref} = C_{I1}\left(u_{pv\_ref} - \hat{u}_{pv}\right)$$

$$u_{pv\_con\_ref} = C_{C1}\left(i_{L1\_ref} - \hat{i}_{L1}\right)$$

_ A converter controller to calculate the duty cycle,

$$m_{pv\_ref} = \frac{u_{pv\_con\_ref}}{\hat{u}_{DC}}$$

where $\hat{u}_{pv}, \hat{i}_{L1}$ and $\hat{u}_{DC}$ are the measured values of u_{pv}, i_{L1} and u_{DC}

B. Battery

Model of the batteries- The batteries are modeled as a voltage source (BAT) (CIEMAT model) [7] and are represented by a green oval in Fig. 5.

Model of the inductive filter- The battery imposes a voltage (u_{bat}) to an inductive filter ($L2$). By neglecting losses, dynamic equation of $L2$ current is expressed with the battery voltage (u_{bat}) and the modulated voltage (u_{bat_con}):

$$\frac{di_{bat}}{dt} = \frac{1}{L_2}\cdot\left(u_{bat} - u_{bat\_con}\right)$$

where L_2 is the inductance of the filter.

Model of the battery chopper- The filtered current (i_{bat}) is modulated by a chopper, and then this modulated current (i_{bat_conv}) is injected into the common DC bus. The output current (i_{bat_con}) of the chopper is equal to the battery current (i_{bat}) multiplied by the duty cycle (m_{bat_ref}). In a same way, the terminal voltage of the battery (u_{bat_con}) is obtained from the DC voltage (u_{DC}) and the duty cycle (m_{bat_ref}):

$$\begin{cases} i_{bat\_con} = m_{bat\_ref}\, i_{bat} \\ u_{bat\_con} = m_{bat\_ref}\, u_{DC} \end{cases}$$

The battery system is described using the EMR (top part of Fig. 5).

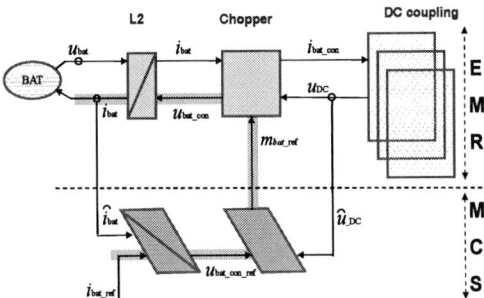

Fig. 5. EMR and MCS of the battery system

Battery control system: The objective of the battery control system is to make equal the battery current to a reference (i_{bat_ref}) by varying the duty cycle (m_{bat_ref}).The control of the battery current (i_{bat}) is achieved with:

_A closed loop control C_{L2}

$$u_{bat\_con\_ref} = C_{L2}(i_{bat\_ref} - \hat{i}_{bat}),$$

_A converter controller

$$m_{bat\_ref} = \frac{u_{bat\_con\_ref}}{\hat{u}_{DC}}.$$

C. Ultracapacitor

Model of the ultracapacitor- The ultracapacitors are also modeled as a voltage source (UC), which imposes a voltage (u_{uc}) to an inductive filter (L3) (Fig. 6). In many applications, the Zubieta and Bonert [8] model is used. Here we use a simplified ultracapacitor model, which consists in a resistance and a capacitor in series.

Model of the filter- By neglecting losses, the filter current is expressed with the following differential equations:

$$\frac{di_{uc}}{dt} = \frac{1}{L_3} \cdot (u_{uc} - u_{uc\_con})$$

Where L_3 is the inductance of the filter.

Model of the ultracapacitor chopper- The filtered current (i_{uc}) is modulated by a chopper, and then this modulated current (i_{uc_conv}) is injected into the common DC bus. The output current (i_{uc_con}) of the chopper is equal to the ultracapacitor current (i_{uc}) multiplied by the duty cycle (m_{uc_ref}). In a same way the modulated voltage is obtained with the DC voltage (u_{DC}) and the duty cycle (m_{uc_ref}):

$$\begin{cases} i_{uc\_con} = m_{uc\_ref}\, i_{uc} \\ u_{uc\_con} = m_{uc\_ref}\, u_{DC} \end{cases}$$

The EMR of the ultracapacitor system is shown on the top part of Fig. 6.

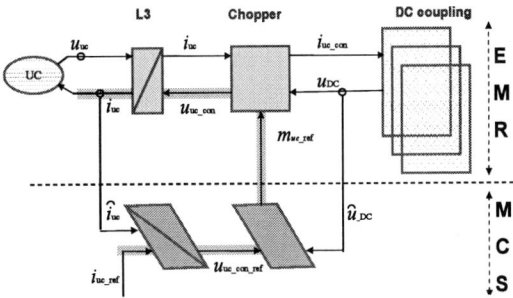

Fig. 6. EMR and MCS of the ultracapacitor system

UC control system: For regulating the ultra capacitor current, we reverse the path between this current i_{uc} and the duty cycle of the chopper m_{uc} to build the control system. The control is achieved with a closed loop control:

$$u_{uc\_con\_ref} = C_{L3}(i_{uc\_ref} - \hat{i}_{uc})$$

and a converter controller:

$$m_{uc\_ref} = \frac{u_{uc\_con\_ref}}{\hat{u}_{DC}}$$

D. Grid connection

A three phase inverter within a choke as filter is used for the grid connection. Hence, an equivalent mean modeling of the power converters is sufficient for this study. It represents fundamental voltage/current components as:

_ Dependent phase to phase voltage sources (um_{13} and um_{23}) with the DC bus voltage through modulation indexes (m_1 and m_2) and

_ A dependent current source (i_{INV}) with AC currents through identical modulation indexes (Fig. 7).

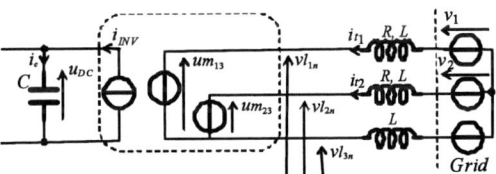

Fig. 7. Equivalent continuous electrical model of the inverter

Then mean values of modulated phase to phase voltage are expressed as:

$$um_{13} = m_1\, u_{DC}, \quad um_{23} = m_2\, u_{DC}$$
$$i_{INV} = m_1\, i_{l_1} + m_2\, i_{l_2}$$

Where m_1 and m_2 are modulation indexes. Line voltages are obtained through:

$$\begin{cases} vl_{1n} = \dfrac{2}{3}.um_{13} - \dfrac{1}{3}.um_{23} \\ vl_{2n} = -\dfrac{1}{3}.um_{13} + \dfrac{2}{3}.um_{23} \end{cases}$$

The time evolution of the DC bus voltage u_{DC} is given by the following equation:

$$\frac{du_{DC}}{dt} = \frac{1}{C} . i_{DC}$$

The filter currents are deduced from following differential equations:

$$\begin{cases} \dfrac{di_1}{dt} = \dfrac{1}{L}\left(v_{l_{1n}} - R.i_{t_1} - v_1\right) \\ \dfrac{di_2}{dt} = \dfrac{1}{L}\left(v_{l_{2n}} - R.i_{t_2} - v_2\right) \end{cases}$$

Mean values of modulated grid side voltages, filter currents and line grid voltages are respectively gathered in following vectors:

$$\underline{v_l} = \begin{bmatrix} v_{l_{1n}} \\ v_{l_{2n}} \\ v_{l_{3n}} \end{bmatrix}, \ \underline{i_t} = \begin{bmatrix} i_{t_1} \\ i_{t_2} \\ i_{t_3} \end{bmatrix}, \ \underline{v_s} = \begin{bmatrix} v_1 \\ v_2 \\ v_3 \end{bmatrix}$$

The EMR is shown on the top part of Fig.8.

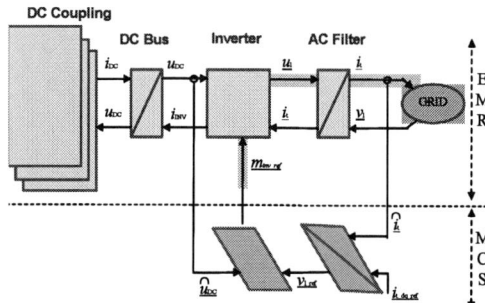

Fig .8. EMR and MCS of the ultracapacitor system

Control of grid current

Grid currents can be controlled by using a converter controller and a closed loop control of currents. To design it, a Park transform is used with synchronization with the first line voltage. Hence in this frame filter equations are written as:

$$\begin{cases} \dfrac{di_d}{dt} = \dfrac{1}{L}\left(v_{ld} - v_d - R_s\,i_{t_d} - L_s.\omega_s.i_{tq}\right) \\ \dfrac{di_q}{dt} = \dfrac{1}{L}\left(v_{lq} - v_q - R_s\,i_{t_q} + L_s.\omega_s.i_{td}\right) \end{cases}$$

The control of these currents is obtained by a compensation of grid voltages, a current decoupling and a closed loop control (Fig. 9).

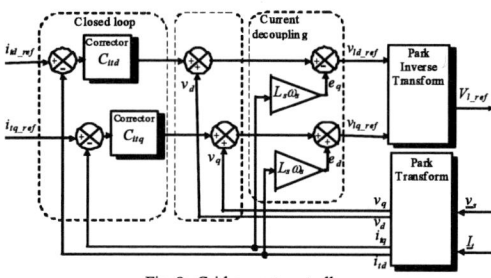

Fig. 9. Grid current controller

Phase to phase voltages are obtained with the following equations:

$$\begin{cases} um_{13\_ref} = v_{l_1\_ref} - v_{l_3\_ref} \\ um_{23\_ref} = v_{l_2\_ref} - v_{l_3\_ref} \end{cases} \quad (Ra)$$

Modulation functions are calculated by using the inverse equation:

$$m_{l1\_ref} = \frac{um_{13\_ref}}{\hat{u}_{DC}}, \ \ m_{l2\_ref} = \frac{um_{23\_ref}}{\hat{u}_{DC}} \quad (Rb)$$

The correspondent vector is defined as:

$$[m_{lmv\_ref}] = [m_{l1\_ref} \ \ m_{l2\_ref}]^T$$

Hence current references can be assumed equal to grid currents: $[i_{t\_ref}] = [i_{td\_ref} \ \ i_{tq\_ref}]^T$.

IV. POWER MANAGEMENT SYSTEM

A. Power control

A schema bloc representation of all used control algorithms is shown on fig.10. Power references for the three sources can be easily obtained by setting:

$- u_{pv\_ref} = \dfrac{P_{pv\_ref}}{\hat{i}_{pv}}$ with $\hat{i}_{pv}$ the sensed PV current and P_{pv_ref} the power reference for the PV panels,

$- i_{bat\_ref} = \dfrac{P_{bat\_ref}}{\hat{u}_{bat}}$ with $\hat{u}_{bat}$ the sensed battery current and P_{bat_ref} the power reference for the battery,

$- i_{uc\_ref} = \dfrac{P_{uc\_ref}}{\hat{u}_{uc}}$ with $\hat{u}_{uc}$ the sensed ultracapacitor current and P_{uc_ref} the power reference for the ultra capacitors.

Active and reactive grid powers can be expressed with Park components of the grid voltages and currents:

$$\begin{cases} P_g = v_{l_d}\,i_{t_d} + v_{l_q}\,i_{t_q} \\ Q_g = v_{l_d}\,i_{t_q} - v_{l_q}\,i_{t_d} \end{cases}$$

By inversion of these relations, it is possible to regulate the active and reactive grid power by setting the filter current references (fig.10) according to:

$$\begin{cases} it_{d\_ref} = \dfrac{P_{g_{ref}}.\hat{v}_{l_d} - Q_{g_{ref}}.\hat{v}_{l_q}}{\hat{v}_{l_d}^{\,2} + \hat{v}_{l_q}^{\,2}} \\ it_{q\_ref} = \dfrac{P_{g_{ref}}.\hat{v}_{l_q} + Q_{g_{ref}}.\hat{v}_{l_d}}{\hat{v}_{l_d}^{\,2} + \hat{v}_{l_q}^{\,2}} \end{cases} \quad (Rpower)$$

with $\hat{v}_{l_d}$ and $\hat{v}_{l_q}$ the Park components of the sensed grid voltages.

B. Studied Operating Mode

A hybrid power system can work in several operating modes depending on the different working conditions due to the availability of each source (photovoltaic panels, batteries, ultracapacitors) and the grid demand, such as the climatic condition (the illumination, the temperature), the storage level of each storage unit, the SOC (state of charge) for the battery and u_{uc} (ultracapacitor's voltage) for the ultracapacitors). According to the selected mode, the PTS can calculate the power references.

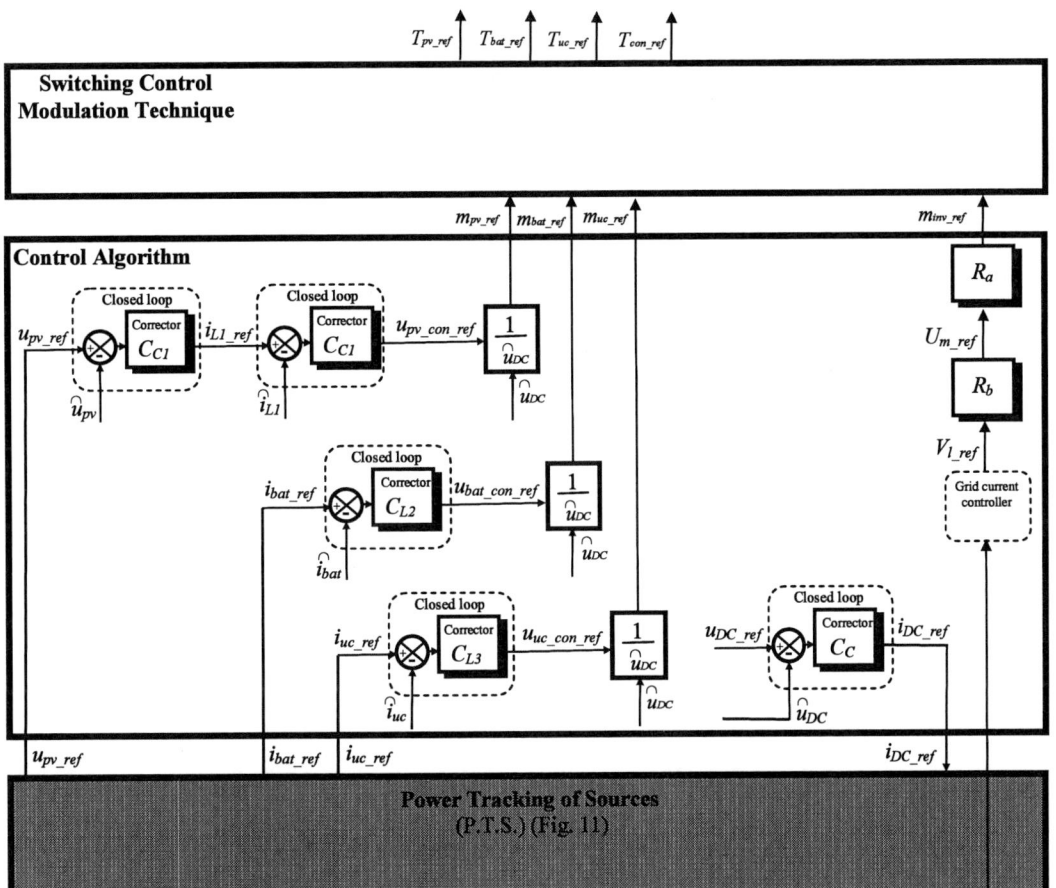

Fig. 10. Schema bloc representation of the control algorithms

In the normal mode, the photovoltaic panels produce the maximum electrical power. The inverter has to deliver prescribed power references (P_{ref}, Q_{ref}). If the produced electric power from PV panels is more or less than the power need for the grid, the batteries and the ultracapacitors are used to compensate this difference.

For this mode, the photovoltaic panels operate in MPPT (Maximum Power Point Tracking). $p_{pv\_ref} = f_{MPPT}(\widehat{u}_{pv})$ (R0). For the analysis, we assume that the batteries and the ultracapacitors are in generating mode. We define:

- P_{pv}, the produced power by the photovoltaic panels,

- P_{bat}, the exchanged power with the batteries,

- P_{uc}, the exchanged power with the ultracapacitor,

- P_{sto}, the total power exchanged with the two storage units,

- P_{conv}, the added power to the DC bus,

- p_{DC}, the exchanged power with the capacitor of the DC bus, which can be written with two terms:

$$p_{DC}(t) = p_{DC\_cha}(t) - p_{DC\_dec}(t)$$

We set $p_{DC}(t) = p_{DC\_cha}(t) > 0$ in the loading mode of the DC bus capacitor and $p_{DC}(t) = p_{DC\_dec}(t) < 0$ to decrease this capacitor DC voltage.

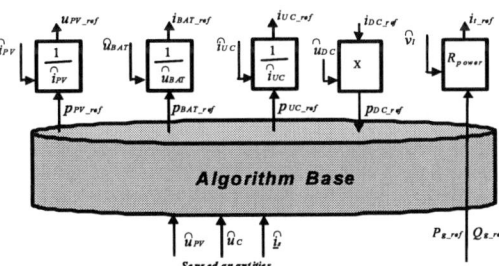

Fig. 11. Algorithm of the PTS

C. Algorithms for Power Management

The power management algorithm is based on the power balancing of the hybrid system (Fig.12). It is drawn by assuming that batteries and ultracapacitors are in a generating mode (as PV panels). Since we assume that the losses in power electronic converters are negligible, the generated power should be equal to the local absorbed power and the power provided to the grid.

2147

The exchanged power with batteries (p_{bat}) and with ultracapacitors (p_{uc}) constitute the total exchanged storage power (p_{sto_ref}, relation $R1$ in the table 1), which can be used for compensating or absorbing the produced photovoltaic power (p_{pv}). The resulting power (p_{conv}) appears on the DC bus for exchanging the power (p_{AC}) with the grid across the three-phase inverter and the filters with the losses (l_{L4}). So we can obtain the final supplied power (p_g) to the grid according to this power balancing. Corresponding modelling equations of the power flow are summarized in table 1.

For the power management algorithm, the reference of the exchanged power (p_{sto_ref}) with both storage units is calculated by using an estimation of the produced PV power ($\tilde{p}_{pv}$) (Relation R2c). A part of the power is required to regulate the DC bus (p_{DC_ref}). It is calculated by using a measurement of the DC bus voltage ($\hat{u}_{DC}$) and the current reference (i_{c_ref}) (Relation R5c). Losses in the grid filter can be estimated by $\tilde{l}_{L4} = 3Ri_s^2$. They are used to calculate the AC power reference (p_{AC_ref}) (Relation R4c).

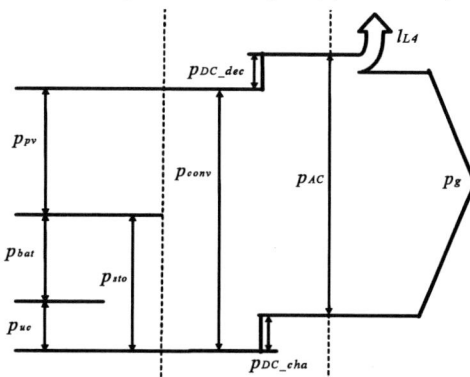

Fig. 12. Schema of the power flow in normal mode

TABLE I.
MODELING AND MANAGEMENT EQUATIONS OF POWER

Modeling equations		Power management	
$p_{sto} = p_{bat} + p_{uc}$	R1	$p_{bat\_ref} = f\left(p_{sto\_ref}\right)$ $p_{uc\_ref} = p_{sto\_ref} - p_{bat\_ref}$	R1c
$p_{conv} = p_{pv} + p_{sto}$	R2	$p_{sto\_ref} = p_{conv\_ref} - \tilde{p}_{pv}$	R2c
$p_{AC} = p_{conv} - p_{DC}$	R3	$p_{conv\_ref} = p_{AC\_ref} + p_{DC\_ref}$	R3c
$p_g = p_{AC} - l_{L4}$	R4	$p_{AC\_ref} = p_{g\_ref} + \tilde{l}_{L4}$	R4c
$p_{DC} = u_{DC} \cdot i_c$	R5	$p_{DC\_ref} = \hat{u}_{DC} \cdot i_{c\_ref}$	R5c

Power Management of Storage Units

Since the battery has low dynamic transients, we use it to supply the required power reference in a long time range $\left\{p_{bat\_ref}(t)\right\}_T$. The value of the battery power reference is calculated with the mean value of the total storage power reference:

$$\left\{p_{bat\_ref}(t)\right\}_T = \frac{1}{T}\int_{-T}^{T} p_{sto\_ref}(t)dt$$

In practice, this value is calculated with a first order filter (Fig.13).

During such a long time range, the fast power

variations exchanged with the super-capacitors are assumed zero.

The batteries can not provide a power varying as fast as the generated photovoltaic power. More over to avoid serious or earlier damage we must avoid this transient use. In our hybrid system, we use ultracapacitors for this task. Their power reference is calculated from the difference between the timing reference of the storage p_{sto_ref} and the calculated power reference of the battery:

$$p_{uc\_ref}(t) = p_{sto\_ref}(t) - \left\{p_{bat\_ref}(t)\right\}_T .$$

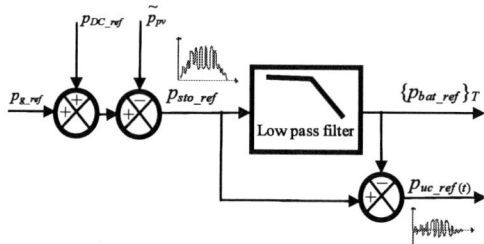

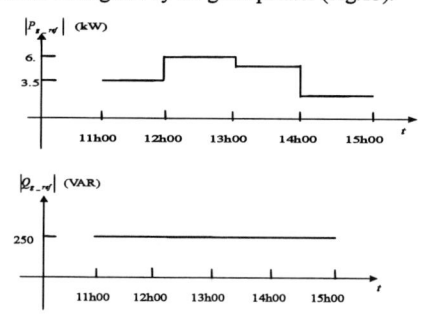

Fig. 13. Power references for storage units in the normal mode

V. SIMULATION RESULTS

We consider the following power production plan, which has been given by the grid operator (Fig.13).

Fig. 14. Power references for the power station

In the figure 14, at 12 o'clock, the power, which is required by the grid, increases very quickly to 6kW. The power, which is generated by the PV panels (about 2.6kW) (Fig. 15.1) is not sufficient. In consequence, our power management system orders the power from the storage units p_{sto_ref} (about 3.4kW). The batteries can not provide all the power because the batteries do not discharge rather quickly (Fig.15.2). Then the ultracapacitor is used to compensate the gap (Fig. 15.3). Since the total produced power (Fig.14.4) is closed to the reference power (Fig.14).

VI. CONCLUSION

In this paper, a power management system for a photovoltaic based power station including various storage technologies (with long-term storage by batteries and fast-dynamic storage by ultracapacitors) in DC-coupled structure has been introduced. Three different power management algorithms for three different operating modes of this hybrid power system have been developed. Hence the batteries can compensate the difference between the generated photovoltaic power and the grid consumption in the long time range, and the ultra-capacitors can smooth the power delivered to the grid in short time range.

VII. ACKNOWLEDGMENT:

This work has been supported by the French National Agency of the Research (ANR SuperEner Project).

VIII. REFERENCES

[1] C. Gouvello, Y. Maigne, "Decentralized rural electrification: an opportunity for mankind, techniques for the planet", Paris: Solar Systems 2002

[2] P. M. Nair, "Photovoltaic development and use in India", 17th European Photovoltaic Solar Energy Conference Proceedings, Munich, Oct 2001, CDROM

[3] H. Fakham, P. Degobert, B. François, "Control system and power management for a PV based generation unit including batteries", Electromotion'07, Bodrum,Turkey, pp.141-146, 2007

[4] B. François, J.P. Hautier, "Hierarchical control design using structural decomposition of a rectifier converter model", ELECTRIMACS 1996, p. 255-260, vol. 1, St Nazaire, FRANCE, September 17-18 1996, CDROM

[5] A. Bouscayrol, Ph. Delarue, "Weighted control of drives with series connected DC machines", *IEEE-IEMDC'03*, Madison (USA), June 2003, pp. 159-165.

[6] Xiao W., Dunford W.G., Capel A., A novel modeling method for photovoltaic cells, IEEE Power electrics specialists conference, Aachen, Allemagne, 2004, CDROM

[7] O. Gergaud, G. Robin, B. Multon, H. Benahmed, "Energy modelling of a lead-acid battery within hybrid wind/ photovoltaic systems", SATIE- Brittany Branch, EPE 2003, Toulouse, 2003, CDROM

[8] Zubieta L., Bonert R., "Characterization of double-layer capacitors for power electronics applications", IEEE Transactions on Industry Applications, Janvier-Février 2000, vol. 36, issu. 1, pp. 199-20

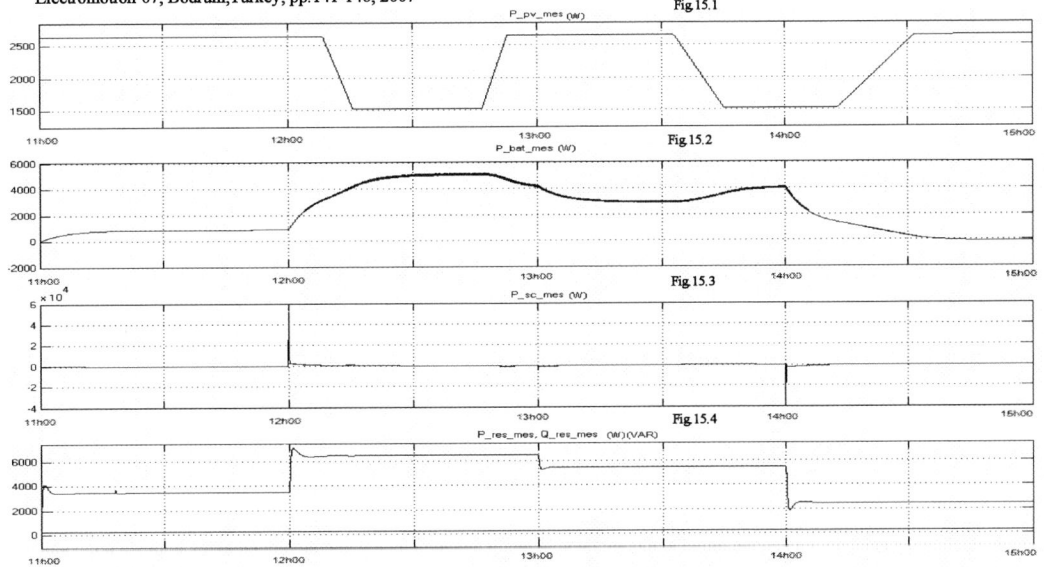

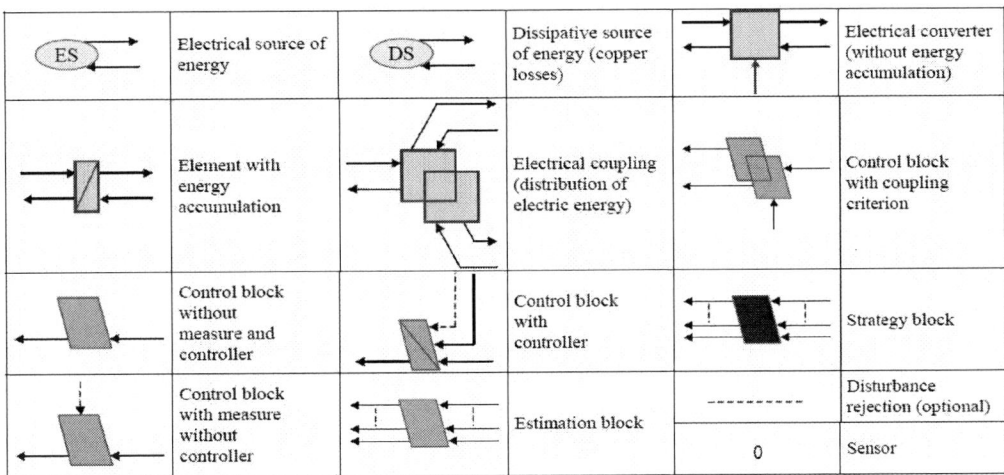

Fig. 15. Simulation results

APPENDIX: SYNOPTIC OF ENERGETIC MACROSCOPIC REPRESENTATION [5]

Energy Management and Power Flow of Decoupled Generation System for Power Conditioning of Renewable Energy Sources

Włodzimierz Koczara[*], Zdzisław Chłodnicki[†] Nazar Al-Khayat[**], Neil L.Brown[***]

[*] Warsaw University of Technology, Warsaw, Poland, e-mail: koczara@isep.pw.edu.pl
[†] Warsaw University of Technology, Warsaw, Poland, e-mail: zdzislaw.chlodnicki@cummins.com
[**] Advance Engineering, Cummins Technical Center USA, nazar-al.khayat@cummins.com
[***] Cummins Generator Technologies UK, neil.l.brown@cummins.com

Abstract— **Paper presents decoupled variable (adjustable) speed generation system and its application in renewable energy generation. The decoupled generation system consists of internal combustion engine, permanent magnet generator and AC/AC converter. Performance of single decoupled generation set are discussed supported by results of laboratory tests. The variable speed system provides high quality voltage in wide speed range when the step of load is low. However, when the step load is significant the voltage drop is not acceptable. Therefore to provide high quality produced voltage an additional energy storage, made from supercapacitor and bidirectional DC/DC converter, is applied. Such system performs very stiff voltage in any load condition. Integration of renewable wind energy system is provided via DC link of the variable speed decoupled autonomous generation system. The priority of the power draw is from the wind source. This is assured by a method of three reference signals for the DC link voltage. Speed control of the driving engine in region of low specific fuel consumption results in high efficiency conditioning of the unstable (wild) variable renewable energy source.**

Keywords— **Renewable energy variable speed power generation, power conditioning, specific fuel consumption, power electronic converters.**

I. INTRODUCTION

Renewable energy sources (RES) have non stable nature and their available power and energy are practically not predictable. However, their average potential energy is high and they will be important source of energy hungry world. Variation of the available power of the RES results in poor quality of delivered energy. Therefore to utilize the energy, produced by the RES, a power conditioners are applied. In case of RES, producing electricity the most common system, is connecting them to stiff grid or to parallel operating generating set that provides quality AC voltage. Other method of RES energy quality improvement is application of energy storage as battery or supercapacitor bank. To provide high quality energy by hybrid system the energy storage ratio of installed RES power to rated load power should be very high and energy storage capacity should be big. Moreover, in case of longer break of renewable energy the recharging the energy storage may be a problem. Therefore the parallel operated conventional power source, including synchronous generator driven by internal combustion engine, as is shown in Fig 1, is common practice. In conventional generating systems refueling is provided at any time and takes short time.

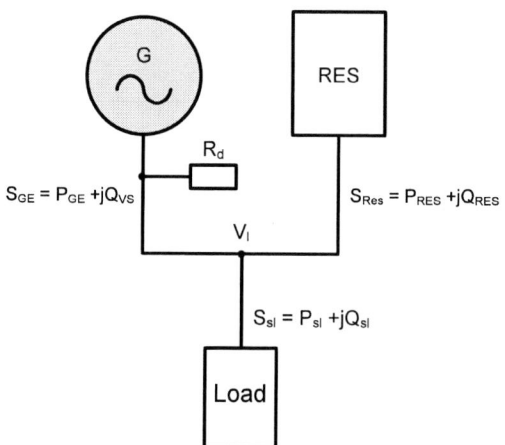

Fig. 1. Common method for energy quality of renewable energy conditioning

The generator G is sized to rated power of the load and the power produced by the RES reduces theoretically demand of the primary energy delivered by the drive to the generator. As the generating set efficiency drops significantly with loads then primary energy savings are much lower than expected. Additionally, in case of Diesel engine driving the generator, low loads results in not full fuel burning. This not burned fuel dilutes lubrication oil and shorts the engine life cycle. To extend cycle life of this not fully loaded engine an additional resistive load R_d (called commonly "dummy load") is connected to the generator. Hence a most of saved energy by the RES application is dissipated by this additional load.

The paper proposes decoupled variable speed generating system (DVSGS) application as energy saving power conditioning method which also provides high quality delivered power and assures long life cycle of the engine. Shown in Fig. 2 generating system consists of renewable power source RES and autonomous, variable speed generating system DVSGS connected parallel via AC bus.

In such system the DVSGS has to be ready to deliver full AC power and there is a need of special control arrangement to provide quality power. Another option is presented in Fig. 3, where the RES is connected to a DC bus of the autonomous decoupled variable speed generation system.

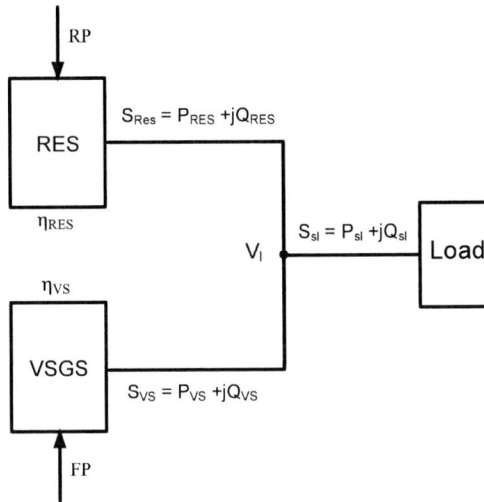

Fig. 2. Variable speed generating system (DVSGS) as power conditioner for RES – common AC bus

Therefore in the generating system (Fig. 3) the quality of the generated power depends on quality of the produced power by the decoupled variable speed generating system.

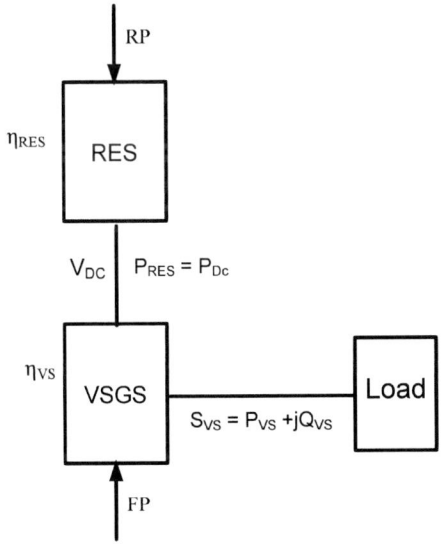

Fig. 3. Variable speed generating system (DVSGS) as power conditioner for RES – common DC bus

II. DECOUPLED VARIABLE SPEED GENERATION SYSTEM (DVSGS)

An example of a topology of the decoupled hybrid generation system DVSGS is shown in Fig. 4. The decoupled variable speed generation system is equipped in permanent magnet generator PMG, power electronic AC/AC converter and supercapacitor energy storage C_{SC}.

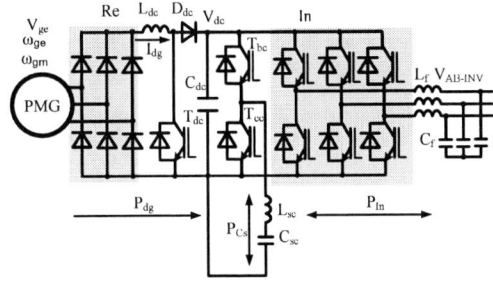

Fig. 4. Hybrid decoupled DVSGS variable speed generation system

The AC/AC converter consists of rectifier (Re), a DC/DC step-up converter (L_{dc}, T_{dc}, D_{dc}), a transistor inverter (In) and output filter (L_f, C_f). Additionally to the DC link is connected a supercapacitor energy storage CSC via controlled bidirectional DC/DC converter (C_{sc}, L_{sc}, T_{bc}, T_{cc}). The DC/DC converter, operating as step-up or as step-down chopper, provides the DC link voltage stabilization when the generator operates as variable speed voltage source. The inverter, with output filter, produces AC sinusoidal three phase voltage. According to the Fig. 4 a laboratory 5 kW decoupled variable speed hybrid generation system was built and tested.

The ability of the DVSGS to provide high quality power for low step power and without additional supercapacitor energy storage are shown in oscillograms Fig. 5, Fig. 6 and Fig. 7.

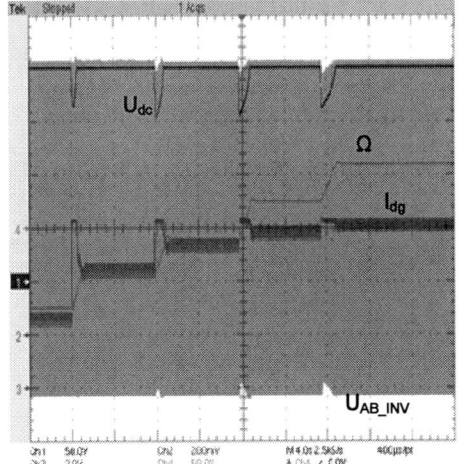

Fig. 5. Decoupled generation system speed and voltage – oscillogram of case of series low steps of power (U_{dc} – 50 V/div, I_{dg} – 20 A/div, Ω – 33 rad/sec, U_{AB_INV} – 50V/div, time – 4 sec/div)

2151

The set of small steps of load from 0 to 5 kW (Fig. 5) results increasing of the speed and the output AC voltage U_{AB_INV} is slightly effected. During any step load the generator, rectified current I_{dg} rises to its reference level limiting the generator load torque and then limits the output power which is proportional to the torque and speed. Therefore for maximum speed maximal available power is 5 kW.

However, when at low speed a full load is applied (from 0 to 5kW) then the output AC voltage U_{AB_INV} and DC link voltage U_{DC} drop significantly as is shown in Fig. 6. To recovery of the demanded DC (U_{DC}) and then the AC voltage (U_{AB_INV}) the speed Ω has to rise and it takes several seconds. Fig. 7 shows case of initial load 1.25 kW (at the same speed range settings) the speed is rising according the load 1.25 kW and when the step load from the 1.25 to 5 kW is applied then the AC voltage U_{AB_INV} is decreasing much less. However the voltage drop is still significant.

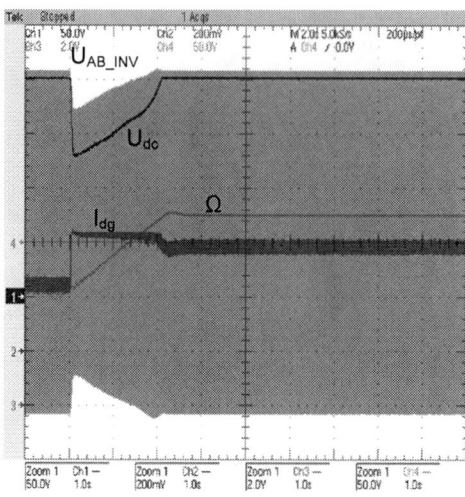

Fig. 7.   Decoupled generation system speed and voltage without energy storage – oscillogram of case of high step of load power at low speed from 1.25 to 5 kW (U_{dc} – 50 V/div, I_{dg} – 20 A/div,

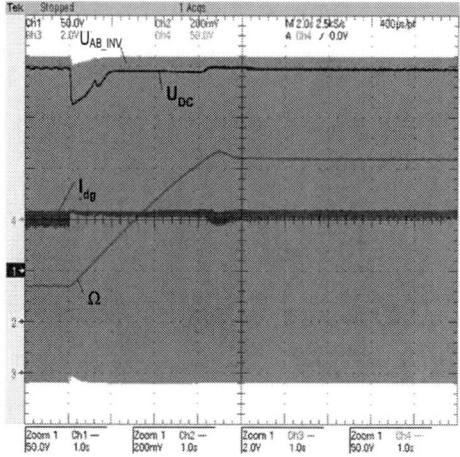

Fig. 8.   Step load from 0 to 5 kW DWSGS- hybrid - Decoupled hybrid generation system voltage and speed – oscillogram case of 0 to 5 kW step of load at low speed. U_{dc} – 50 V/div, I_{dg} – 20 A/div, Ω – 33 rad/div, time 4 sec/div

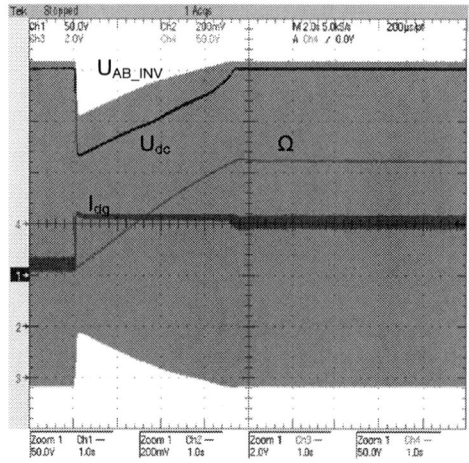

Fig. 6.   Decoupled hybrid generation system speed and voltage – oscillogram of case of high step of load power at low speed from 0 to 5kW (U_{dc} – 50 V/div, I_{dg} – 20 A/div, Ω – 33 rad/sec,

Therefore another method is applied. To reduce the voltage drop a supercapacitor energy storage, controlled by reversible DC/DC converter (Fig.4), is used. The energy storage, made with the supercapacitor C_{SC}, delivers an additional transient power to the DC link and it compensates the voltage drop caused by load step at low speed. Hence the hybrid system is proposed. Fig 8 shows the output AC voltage of the hybrid system for the same case of the load step shown in Fig. 6 i.e. step-load from 0 to 5 kW but with support of power delivered by the supercapacitor C_{SC}.

The DC link voltage U_{DC} is decreasing much less as it was in case shown in Fig. 6 and the AC voltage U_{AB_INV} drop is not significant and may be accepted. The supercapacitor current will be delivered to the DC link and its energy will be transferred to the load when the DC link voltage drops below given reference level. To illustrate the energy storage operation when the supercapacitor is discharged and recharged another test was provided.

Fig. 9 shows transient case when the same load step from 0 to 5 kW and then step-off the load from 5 to 1.25 kW is applied. In this oscillogram instead of the AC voltage is shown only the DC link voltage (only 4 channel scope was used). By action of the DC/DC converter the supercapacitor current I_{SC} is delivered instantly to the DC link but speed Ω is rising slowly according maximum torque delivered by the prime mover. When the DC link voltage and speed get their rated values then the supercapacitor current is reduced to zero. As result of step-off load from 5 to 1.25 kW the supercapacitor is recharged. As the capacitor charging load is low then the speed of the generating system decreases. And at the end of the process the load is zero and the speed going to its minimum value.

The supercapacitor delivers energy W_{SC} and for given time

$$W_{sc} = \int_{t_1}^{t_2} u_{sc}\, i_{sc}\, dt \qquad (1)$$

To provide high quality power the capacity of the energy storage has to be sized to deliver power in case of several sequences of full load steps before full recharging.

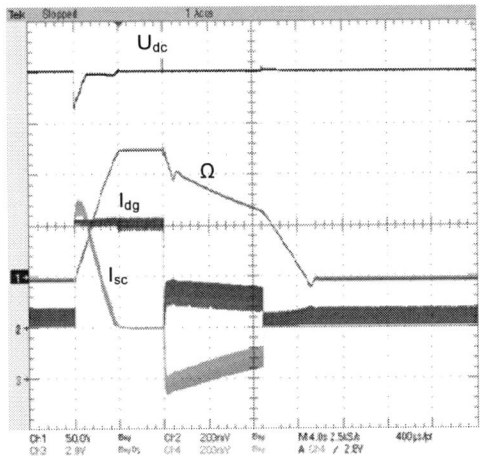

Fig. 9. Decoupled hybrid generation system voltage and speed – oscillogram case of 0 to 5kW and from 5 to 1.25 kW step of load at low speed. U_{dc} – 50 V/div, I_{dg} – 20 A/div, Ω – 33 rad/div, time 5 sec/div. I_{sc} - 20/div

III. INTEGRATION OF THE RES TO THE VARIABLE SPEED GENERATION SYSTEM

Integration of the wind turbine driven generation system RES to the variable speed generation system via DC link (Fig. 3) is shown in Fig, 10. The wind turbine generator WG supplies rectifier Rew and load torque (speed) of the turbine generator is controlled by control of the rectified current I_{dgw} by step-up chopper (transistor T_{cw}, inductor L_{cw}, diode D_{cw}) The main goal of the system is to draw as much as possible energy from the wind turbine. This may be achieved indirectly by control of the DC bus voltage V_{dc} according three reference voltages and loads (Fig.11). First the highest reference DC link voltage V_{dcr1} is related to the DC voltage produced by the step-up chopper of the wind turbine system which is limited and adjusted to wind speed, rectified current I_{dgw}. Its reference current I_{dgwr} is produced by the wind turbine controller providing wide speed range operation. The maximum power delivered by the wind turbine is ΔP_{dcr1}. Second reference DC link voltage V_{dcr2} is related to DC voltage, produced by step-up chopper of the variable speed generator controlling rectified current I_{dg}, that is set to keep its rated value. To cope with increasing load the generator speed is rising and when the speed gets its maximum value then DC link voltage is coming to $V_{dc} = V_{dc2r}$. The maximum power delivered by the generation system is ΔP_{dc2r} When $V_{dc} < V_{dc2r}$ then the DC link is

supplied by the supercapacitor energy storage which is able to deliver maximum power ΔP_{dc3r}.

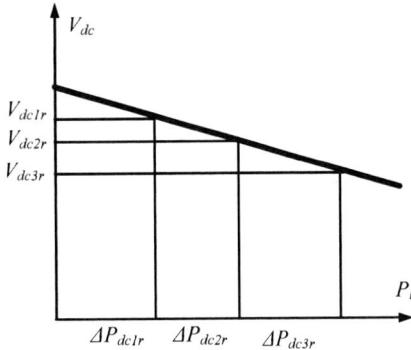

Fig. 11. DC link voltage as result of load –case of all sources rated power available

In case of low power, produced by the wind turbine $\Delta P_{dc1a} < \Delta P_{dc1r}$ and low speed of generator $\Delta P_{dc2a} < \Delta P_{dc2r}$, the DC link voltage available changes are shown in Fig. 12.

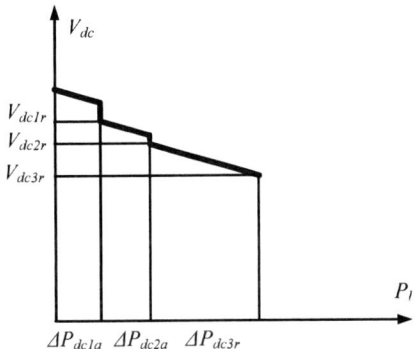

Fig.12. DC link voltage in case of low power delivered by the wind turbine and generation system for high power demand by the load.

Now, for $\Delta P_{dc} + \Delta P_{dc1a}$ the $V_{dc} < V_{dcr1}$ the Diesel engine accelerates and produces more power [13]. Hence, the DC link is maintained between reference levels $V_{dc1r} < V_{dc} < V_{dc2r}$ sufficient to produce high quality AC voltage. However, when a high power step load is produced then the engine is not able to accelerate instantly and the DC link voltage drops below third reference level V_{dcr3} which activates the supercapacitor energy storage (examples Fig. 8 and 9) that supplies the DC link until the engine accelerates to demanded speed and DC link voltage $V_{dc1r} < V_{dc} < V_{dc2r}$.

To get fuel savings the driving engine has to operate in the most efficient region. Fig. 13 presents an example of internal combustion industrial engine maximum power P_{dmax} and specific fuel consumption g_r (g/kWh) as a function of speed. The red dot line (from A to B) presents the most efficient operation and the engine controller is tuned to operates in this region. Such an arrangement results in low fuel consumption and long life cycle of the

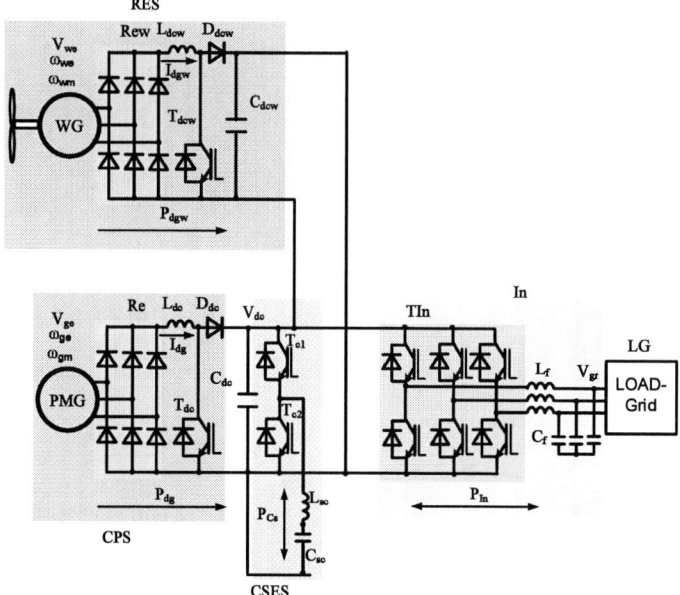

Fig. 10.   Integration of the wind turbine driven generator to the DC link op the hybrid decoupled variable speed generation systems

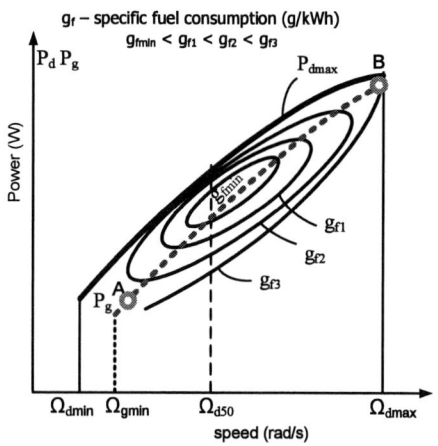

Fig. 13.   Power of the decoupled variable speed generation
system as a function of speed

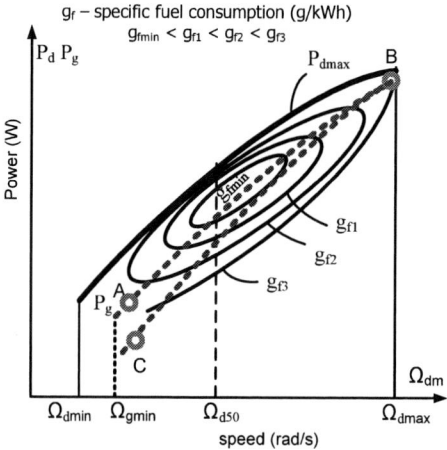

Fig. 14. Power of the decoupled variable speed generation system as
function of speed – lower power demand

engine. However, usually the difference between maximum power P_{dmax} and power responding to low specific consumption (power responding line A-B) is small. The maximum power line P_{dmax} is result of maximum driving torque T_{em} and the power P_{AB} is result of load torque T_l produced by the permanent magnet generator PMG. Both torques are function of the time and the engine speed. The operation in steady state conditions is responding by AB line being very close to P_{dmax} providing low fuel consumption but in transient states,

caused by the load steps increase, the change of speed from Ω_{ei} to Ω_e take long time from t_1 to t_2

$$\Omega_e = \Omega_{ei} + \frac{1}{J_e} \int_{t_1}^{t_2} \left[T_{em}(t, \omega_e) - T_l(t, I_g) \right] dt \qquad (2)$$

where J_e is inertia factor.

Hence, the supercapacitor energy storage has to be sized to cope with high power demand and resulting energy. When the power line is shifted from AB to CB as is shown in Fig. 14 the transient time will be shorter but the steady state operation will responds regions of higher specific fuel consumption. The Fig. 13 and Fig. 14 presents simplified curves of the specific fuel consumption. In practice each engine has these curves very nonlinear and only table function may respond to real situation. Therefore each engine load power has to be specially tuned.

The decoupled variable speed generation system operates in wide speed range what means that maximum speed ω_{dmax} (Fig.13) is higher than ω_{d50} responding to conventional synchronous generator producing 50 Hz voltage. This increase of the available engine power to P_{dmax} at the point B, permits to use the engine that's rating at the speed ω_{d50} is lower and then lower consumption of the fuel. This is a serious advantage of the variable speed generation system over the conventional synchronous generator. However, the low power engine has lower torque what results in longer transients. The longer transients must be compensated by higher energy drawn from the supercapacitor.

The power delivered to load is a sum of the power delivered by RES and DVSGS. It is advised to the DVSGS to operate continuously with minimum power no less than power marked by point A. In this point the engine torque is sufficient high to assure long life of the engine but low speed results in low specific fuel consumption.

IV. SUMMARY

The paper presents hybrid generation system made from renewable energy source and decoupled variable speed generation system. The decoupled variable speed generation system operates as conditioner, which delivers power, when the renewable energy source power is not sufficient. It is proposed to arrange common DC link for the renewable energy source and decoupled generation system. The simple decoupled variable speed generation system is perfect conditioner because it delivers power along high efficiency (i.e. along low specific fuel consumption) of the driving engine. This operation is specially advantaged when changes of the load are related not to steps of rated power. However, when high steps of load power are applied then an additional conditioner made from supercapacitor and DC/DC converter is proposed. The supercapacitor is a source of transient high power and is recharged after end of transient state with low current. Another advantage of the decoupled variable speed generation system over conventional conditioner made from synchronous generator is lower power of the driving engine. The driving engine of the conventional synchronous 4 poles 50Hz generator is sized to 1500 rpm (153,7 rad/sec) whereas the maximal speed of the decoupled variable speed generation system is much higher reaching approximately 2800 rpm. Thus the driving engine of the decoupled variable speed generation system consumes less fuel for the same amount of the produced energy for both - low load and high load operation.

REFERENCES

[1] Z. Chlodnicki, W. Koczara, Nazar Al-Khayat: "Laboratory Simulation of the Adjustable Speed Generation Systems". 12th International Power Electronics and Motion Control Conference - EPE-PEMC 2006, Portoroz, Slovenia. IEEE Conference Proceeding, pp. 2070-2076.

[2] Chandorkar, M.C.; Divan, D.M.; Adapa, R.; "Control of parallel connected inverters in stand-alone AC supply systems", Industry Applications Society Annual Meeting, 1991, 28 Sep, Page(s): 1003 –1009, Vol.1.

[3] Divan et al., "Method and Apparatus for Decentralized Signal Frequency Restoration in A Distributed UPS System", United States Patent# 5,596,492, 1997.

[4] Perreault, D.J.; Sato, K.; Selders, R.L., Jr.; Kassakian, J.G.; "Switching-ripple-based current sharing for paralleled power converters", IEEE Transactions on Circuits and Systems I: Fundamental Theory and Applications, Oct 1999, Volume: 46, Issue: 10, Page(s): 1264-1274.

[5] Perreault, D.J.; Kassakian, J.G.; "Distributed interleaving of paralleled power converters", IEEE Transactions on Circuits and Systems I: Fundamental Theory and Applications, Volume: 44, Issue: 8 , Aug 1997, Page(s): 728-734.

[6] Bolognani S., Venturo A., Zigliotto M.: Novel Control Technique for High-Performance Diesel-Driven Generator-Sets. Conference Proceedings on Power Electronics and Variable Speed Drives, 18-19 September 2000. Conference Publication No. 475 □ IEE 2000, pp. 18-523.

[7] Koczara W., Dziuba R., Leonarski J., Al-Kayat N.; "Variable speed set for embedded power generation", Proceedings of Electrical Power Quality And Utilisation, Cracow , Poland, September 2001.

[8] W. Koczara, E. Ernest, N. A. Khayat, J. A. Tayie: Smart and Decoupled Power Electronic Generation System. Proceedings of Power Electronics Specialists Conference PESC'2004 Aachen 21-15 June 2004, Germany.

[9] Marcin Moskwa, Wlodzimierz Koczara, Nazar-Al-Khayat: Load Sharing of the Parallel Operating Adjustable Speed Generation Systems without Control Signal Interconnection. Proceedings of EPE 2007 Aalborg, 2007.

[10] Koczara Wlodzimierz, Emil Ernest, Moskwa Marcin: Control System. World patent: WO200126003, 2006-11-30.

[11] Marcin Moskwa: Parallel Operations of Power Electronic Generator Sets with Alternating Output Voltage. Ph.D Thesis. Warsaw University of Technology. Warsaw, Poland 2006.

[12] Da Ponte Manuel Dos Santos L.Grzesiak, W. Koczara, P. Pospiech, A. Niedzialkowski: Hybrid generator apparatus. Panet number AT207668T. Priority numbers: WO1997EP07273 19971219

[13] Z. Chlodnicki, W. Koczara, N. Al.-Khayat: Laboratory Simulaton of Adjustable Speed Generation Systems. EPE Journal Vol 17 n° 4 Janvier 2008.

Acknowledgement. The paper is supported by Cummins Generator Technologies, Stamford, UK and Warsaw University of Technology, Poland.

Inversion Based Control of a Diesel Fed Low Temperature Fuel Cell System

Daniela Chrenko*, Marie-Cécile Péra*, Daniel Hissel*

*FCLAB, FEMTO-ST, University of Franche-Conte, CNRS, UMR 6174, Belfort, France, e-mail: *daniela.chrenko@utbm.fr*

Abstract—This document introduces the control structure development for the fuel processing unit of a pressurized, low temperature fuel cell system fed by commercial diesel. The control structure is developed using an inversion based method derived from the causal, graphic modeling approach called Energetic Macroscopic Representation. Its inversion leads to the maximum control structure. It has been applied to the fuel processing unit which transforms diesel into hydrogen rich gas. The approach contains mass flow and temperature control.

Index Terms—Fuel Cell System, Control methods for electrical systems, Modeling, Generation of electrical energy

I. INTRODUCTION

This work is accomplished as one aspect of the French national project *GAPPAC*, gathering N-GHY, Airbus and Nexter as industrial partners and LMFA, Armines, IFFI, INRETS LTN and FCLAB as research institutes. The aim of this project is to develop a fuel cell based auxiliary power unit capable of trigeneration (electricity, heat and refrigeration). Different areas of transportation applications are addressed, such as land, air, salt and fresh waters. This power generation unit with 25kWe and 30kWth power will run on commercial fuels such as diesel.

The use of fuel cell systems for transportation application is promising. However, fuel cell generators are complex multi domain systems and have to deal with a high number of different inputs. There is for example a large influence of ancillaries, and particularly the motor compressor unit, on the system performance [1]. The electrical power conditioning after the fuel cell plays a role also [2]. To be able to take a large number of those factors into consideration without being obliged to do time consuming and costly tests, system modeling is used. A good overview about system modeling approaches is given in [3]. In this case the model based control development of fuel cell systems is of special interest. Often model based control approaches are rather focused on details of the system than on the overall system [4]. Even in the work of Pukrushpan et al. [5], that shows a more complete approach, the interconnection between the fuel processing unit and the fuel cell has not yet been done. In [6] an approach of a control of a combined reformer-fuel cell system is introduced. The Energetic Macroscopic Representation (EMR) and its inversion, via the Maximum Control Structure (MCS) to the Applicable Control Structure (ACS) show the possibility not only to connect the submodules of the fuel cell system,

but also to implement strategic aspects on the overall system control. Furthermore the model can be refined and extended in following steps without changing the overall structure. An introduction of EMR and MCS in comparison with other methodologies has been given in [7]. EMR and MCS are methodologies that can be applied on graphical modeling platforms, for the moment mainly MATLAB/Simulink® has been used. The application of a low temperature fuel cell system running on hydrogen has been demonstrated [8].

Until now mainly the low temperature fuel cells which have a polymer electrolyte membrane are considered. This membrane material implies working temperatures of around 80 °C. The cooling of a system working on such low temperatures is difficult, furthermore the polymer electrolyte is sensitive to poisoning for example by carbon monoxide and hydrogen sulfide. Both elements occurring during the diesel fuel processing. During the last years more and more work has been done concerning low temperature fuel cell systems having a poly-benz-imidazole (PBI) membrane, also known as high temperature proton exchange fuel cells (HTPEM-FC). Those fuel cells work on higher temperatures (180 °C), simplifying the system cooling. Furthermore, they are less sensitive on poisoning [9]. After preliminary reliability difficulties, the new generation of HTPEM shows good potential for industrialization and are therefore considered in this work. A fuel cell system running on commercial diesel and used for trigeneration is a complex multi domain system (figure 1). One factor to run such a complex system successfully is to have a well adapted control. This will also be able to improve reliability and life time and therefore it will have a positive influence on the system costs as well. As classical control structure development is often based on experience, its development requires a high expertise.

In the following section the basic ideas of the EMR, MCS and ACS are introduced, as well as the conception of a fuel processing unit. In section III the simulation of the fuel processing unit including its ancillaries as well as the control structure development with regard to mass flows and temperature is introduced. Section IV present simulation results of the developed system. The article finishes with conclusions and perspectives.

978-1-4244-1741-4/08/$25.00 ©2008 IEEE

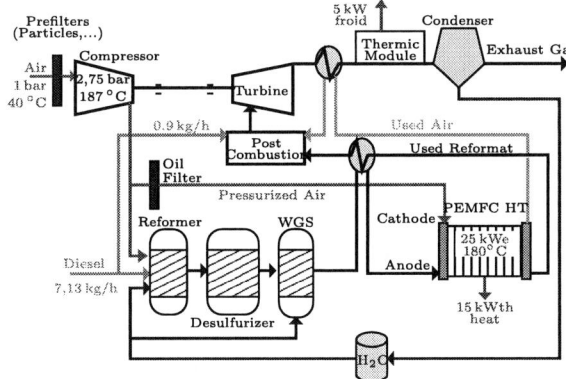

Fig. 1. Scheme of GAPPAC fuel cell system

II. METHODOLOGY

A. Diesel fuel processing unit adapted to supply HTPEM

The combination of a low temperature fuel cell with a fuel processing unit has the advantage, that no secondary fuel infrastructure has to be implemented for the use of fuel cells. At the same time the fuel processing is a complex multi stage system, that has to run in a small working window to provide hydrogen with a specified purity to the fuel cell. Furthermore it is expected that the time constant of the fuel processing will lead the time constant of the overall system. Therefore the first aspect of the system that was regarded in detail for inversion based control development was the fuel processing system.

The fuel processor unit contains the following blocks. The fuel processor is charged by diesel, water and compressed air. The diesel and water are supplied by pumps, the air is supplied by a compressor. The inflows are heated up by a heat exchanger *(Hex2)*. The preheated mixture of gas enters the reformer, where the diesel undergoes a reaction toward hydrogen (H_2), water (H_2O), carbon monoxide (CO), and carbon dioxide (CO_2) *(Reformer)*. This reaction takes place at high temperatures around 1400 °C, so that the effluent gas mixture is used as the heat source for the heat exchanger introduced above *(Hex2)*. Diesel always contains fractions of sulfur, which are poisonous for the fuel cell and has to be removed. Therefore the gas mix undergoes two purifications *(Desulfuration1)* and *(Desulfuration2)*. Furthermore carbon monoxide deactivates the catalyst of the fuel cell membrane as well and has therefore be removed. This is done by a water gas shift reaction *(Water Gas Shift)*.

B. Energetic Macroscopic Representation (EMR) and the Maximum Control Structure (MCS)

EMR is a causal, graphic modeling tool that is based on the idea that every system is describable as a combination of three basic elements, table I. The source elements indicate the interface of the system to its surrounding. Via the source, energy streams can enter and exit the system. Conversion elements calculate conversions of any kind. They can be separated in conversions in the same domain and conversions between different domains. A special case of the conversion is the coupling. With overlapping conversion elements, the coupling or decoupling of energy flows can be described. Finally, there is the accumulation element. It is the only element capable of representing time dependent behavior, for reasons of causality only integral time dependencies can be represented in EMR. The elements are connected by parameters indicating action and reaction. Those parameter pairs reflect the causality of the system. Very often the product of action and reaction is a power. A power is equal to an energy flow, therefore the methodology is named *energetic macroscopic representation*. The EMR gives only the structure of the system and the parameters used, how the calculation is put into practice inside the elements is open to the user. This permits also to have different levels of detail in the same model or to improve the representation of one element without changing the overall structure.

MCS is an inversion based control structure development based upon the EMR. First, the control chain has to be defined from a control input following the causality chain up to a system input. This control chain depends on the system but also on its specification. Then the model can be inverted block wise following the control chain. Therefore the following blocks are used: source blocks are not inverted, conversion blocks are inverted by conversion control elements, analogical is the inversion of coupling blocks. The conversion control elements (table I) contain the inverse function of the function used inside the EMR. The conversion of an accumulation cannot be realized likewise, because the inversion would lead to differential time dependence which is forbidden by definition. Therefore a controller has to be inserted. The kind of controller used depends on the choice of the developer. This MCS will contain a maximum number of control elements and measurements, furthermore it is based on the assumption that all elements are measurable. An applicable control structure can be derived from the MCS in a second step with the help of simplifications and estimations of non measurable values. Hence, the developed system control will be very close to the causality of the system. Still, to run the overall system, global aspects have to be taken into consideration. This can be done based on a breakdown of the degrees of liberty that may leave the possibility to incorporate global and strategic aspects.

C. Adaption of EMR to fuel processor unit

As introduced above, the project *GAPPAC* aims to develop a low temperature fuel cell system. For reasons of volume and efficiency it is pressurized and it will run on commercially available fuels, mainly diesel but also

2157

TABLE I
OVERVIEW OF BASIC ELEMENTS OF EMR AND MCS

	Source element		Coupling element
	Conversion element (same domain)		Accumulation element
	Conversion element (different domain)	action → ← reaction	Parameter pair
	Inversion of conversion		Inversion of coupling
	Inversion of accumulation		

biofuels are envisaged. All together those specifications formulate a challenging task. Inside the overall system one of the most interesting aspects is the fuel processor. The fuel processor unit is demanded to convert diesel into hydrogen, following the load demands of the fuel cell system. As the fuel processing contains several stages, where the flow undergoes different procedures, it has been chosen as interesting aspect for control development.

The fuel processor modeling is based on two basic laws: the mass conservation and the energy conservation.

$$0 = \sum \dot{m}_{\text{out}} - \sum \dot{m}_{\text{in}} \tag{1}$$

$$0 = \sum \left(\dot{m}_{\text{out}} \cdot h(T_{\text{out}}, p_{\text{out}}) \right)$$
$$- \sum \left(\dot{m}_{\text{in}} \cdot h(T_{\text{in}}, p_{\text{in}}) \right) + \sum \dot{Q} + \sum P \tag{2}$$

with $\dot{m}$ the mass flow in kg s^{-1} the enthalpy in J kg^{-1} depending on temperature and pressure, $\dot{Q}$ the heat crossing the system limits in W and P all remaining energy flows crossing the system limits in W. Those two laws, extended by knowledge of the reactions taking place, permit to calculate the output parameters of the respective blocks.
The individual EMR elements are connected by parameter pairs. Normally one action and one reaction parameter at each end of the EMR element are sufficient to describe an element (e.g. an electric dc system can be fully described by current i and voltage u). To describe a system of gaseous flows, three parameters are necessary to describe the system: mass or molar flow (m or n), pressure (p) and temperature (T). To keep the causality at least one of the parameters has to be an action and one other a reaction. As we are working with gas mixtures that change their rations during the reforming process, it has been decided to use vectors for the extensive values (molar flow vector $\dot{n}$) and scalars for the intensive values (temperature T). The molar flow vector has been chosen as compromise that is capable to apply the mass conservation law (using the molar masses) and the chemical reaction stoichiometry.

III. SIMULATION AND CONTROL DEVELOPMENT

A. Simulation

1) Pressure drop: There are two elements, the pressure drop element and the temperature evolution element, that appear regularly in the fuel processor model. Therefore, they are introduced here.
As gas mixtures are handled inside the fuel processor unit and it can be supposed that the gas mixture is ideal, there is a connection between the flows entering and exiting the a volume and the pressure inside this volume (figure 2).

$$p = \frac{R \cdot T}{V} \int \left(\sum \dot{n}_{\text{in}} - \sum \dot{n}_{\text{out}} \right) \mathrm{d}t \tag{3}$$

In most cases the downstream molar flow vector ($\dot{n}_{\text{out}}$) is not known. It can be calculated by the linear throttle equation using the downstream pressure, which is normally given:

$$\dot{n}_{\text{out}} = k \cdot (p_{\text{in}} - p_{\text{out}}) \tag{4}$$

with k the throttle constant in mol s^{-1} Pa^{-1}.

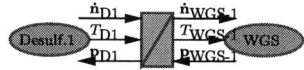

Fig. 2. Pressure drop of the Water Gas Shift reaction

2) Temperature: As the fuel processor is not isolated from the surrounding the temperature evolution is taken into consideration. Therefore the hypothesis has been made that there is a heat exchange between the gas flow and the so called box. This is the mostly metallic surrounding of the different modules. For the first approach the fuel processor has been considered to consist of several boxes. The heat exchanged between the gas flow and the box is most easily described by the entropy flow $\dot{S}$ and the temperature of the box T_{B}. The entropy flow depends on the mean temperature of the gas $\overline{T}$ and the temperature of the box as well as the heat exchange coefficient times the connection surface kA in kJ K^{-2} s^{-1}.

$$\overline{T} = \frac{T_{\text{out}} - T_{\text{in}}}{\ln \frac{T_{\text{out}}}{T_{\text{in}}}} \tag{5}$$

$$\dot{S} = kA \frac{(\overline{T} - T_{\text{B}})^2}{\overline{T} \cdot T_{\text{B}}} \tag{6}$$

$$Q = \dot{S} \cdot T_{\text{B}} \tag{7}$$

Furthermore, the boxes can be cooled by a secondary cooling system. The energy flow withdrawn by the cooling is indicated P (figure 3). The cooling is defined to be instantaneous. The time constant of the cooling system and the influence of the heat up of the cooling liquid is not taken into consideration. The heat exchange and the cooling interact with the system output temperature using the energy conservation, equation 2.

2158

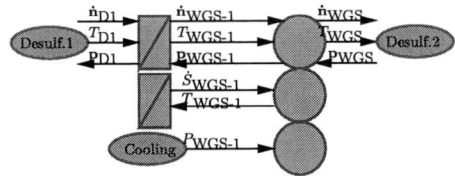

Fig. 3. Temperature development of the water gas shift block.

3) Diesel Supply: As the system shall run on diesel as a fuel, the diesel has to be supplied by to the system. The most simple diesel supply is the use of a pump, a similar approach has been introduced in [10]. To model a pump in EMR (figure 4) the following elements are used:

- **Battery:** A battery is used to supply the pump with electrical energy. A source block is used for the battery.
- **Chopper:** The mass flow provided by the pump shall be controlled. The control is realized using a chopper. A chopper is used to transform the electrical energy supplied by the battery into electrical energy needed to supply the further blocks. As the further blocks are normally voltage driven, the voltage has to be converted. Therefore the conversion coefficient m_d is used. An orange square, standing for a conversion in the same energetic domain, represents the chopper.

$$V_{ch} = m \cdot U_{Su} \qquad (8)$$

- **Electric Accumulation:** The pump is driven by an electric motor. It converts electrical energy into mechanical energy. Based on the supply voltage (V_{ch}) and the electromotive force the motor current (I_{mo}) can be calculated as:

$$(V_{ch} - e) = L\frac{dI_{mo}}{dt} + R \cdot I_{mo} \qquad (9)$$

As the electric accumulation is time dependent, a accumulation block (orange rectangle with bar) is used.

- **Electro-mechanical Conversion:** Conversion between the electric domain is governed by the electro-mechanical conversion coefficient (k). It is described by a orange circle, indicating a conversion between different domains.

$$\Gamma_{mo} = m \cdot I_{mo} \qquad (10)$$

- **Mechanic Accumulation:** An electric motor depends also on the mechanic accumulation, taking into consideration the system inertia and friction and defining the turning speed of the motor (Ω_d). It is represented by an accumulation block describing:

$$\Gamma_{mo} - \Gamma_p = J\frac{d\Omega}{dt} + f \cdot \Omega \qquad (11)$$

- **Pump:** The pump converts the mechanic energy into a volume flow and is therefore represented by a orange circle. It can be assumed that for each turn of

the motor, which is equal of one turn of the pump, a certain molar flow is transported.

$$\dot{n} = k \cdot \Omega \qquad (12)$$

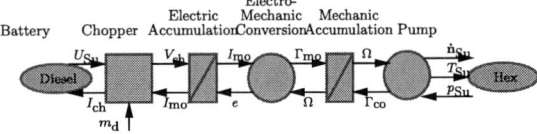

Fig. 4. EMR of a supply system.

4) Water Supply: As water is a liquid such as diesel, the water supply can be described analog to the diesel supply. Only the parameters might change due to the use of a different motor and/or pump.

5) Air supply: The air supply differs from the diesel and water supply in two aspects. First, as the air is gaseous and the fuel cell system is pressurized, it has to be supplied with the help of a compressor. As for the use of a pump an electric motor is used to run the compressor, therefore the elements: battery, chopper, electrical accumulation, electro-mechanical conversion and mechanical accumulation are equivalent, only the parameters differ [10]. Furthermore, the compressor supplies not only the air needed for the fuel processing but also the air needed to supply the fuel cell. As the fraction between the both air streams is kept constant in the first approach, a coupling element (orange overlapping rectangles) is used to split the air stream.

6) Heat Exchanger: The diesel, water and air stream enters the heat exchanger and are mixed. Afterward, the heat exchange takes place. The other partner of the heat exchange is the hot gas mixture coming from the reformer. At the end of the heat exchanger the gases leave toward the reformer and the first desulfurization unit respectively, see figure 5. The approach of the heat exchanger description can be found in [9].

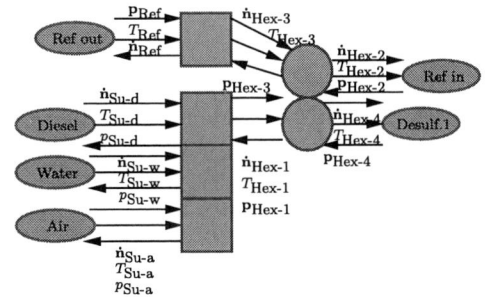

Fig. 5. EMR of heat exchanger.

7) Reformer: Inside the reformer the diesel splits into hydrogen, water, carbon monoxide and carbon dioxide. The reforming process can be split into two phases. First, during the combustion, the diesel reacts exothermically with the oxygen inside the air to form, water and carbon

2159

dioxide. This reaction does not contribute actively to the hydrogen formation, but it supplies the heat needed for the reforming reaction. The reforming reaction is endothermic and transforms the diesel into hydrogen and carbon monoxide. The temperature of the gas downstream the reformer is calculated using the energy balance introduced in equation 2. Furthermore the reformer has a pressure development block upstream of the reforming reaction and a temperature development block connected to the conversion block containing the chemical reaction (figure 6), for further information refer to [11].

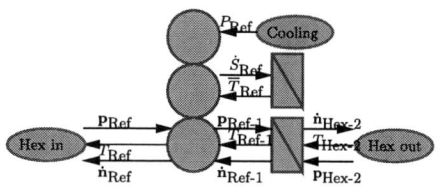

Fig. 6. EMR of reformer unit.

8) Desulfurization: As commercial diesel fuel in Europe may contain up to 10 mg kg^{-1} of sulfur compounds (DIN EN 590), but even more tolerant HTPEM fuel cells, accept only very low fraction of sulfur compounds. Hence the hydrogen rich gas has to be desulfurized. In this case two desulfurization stages are foreseen, one between the heat exchanger and the water gas shift reaction and another between the water gas shift reaction and the fuel cell stack. Both desulfurization units are prefaced by a pressure drop element and as they are incooperated in separate boxes, the temperature evolution is taken into consideration. But only the first desulfurization unit is cooled (figure 7). During the desulfurization a certain fraction of the sulfur content is filtered according to the selectivity.

$$\dot{n}_{\text{out-sulfur}} = (1 - sel)\dot{n}_{\text{in-sulfur}} \qquad (13)$$

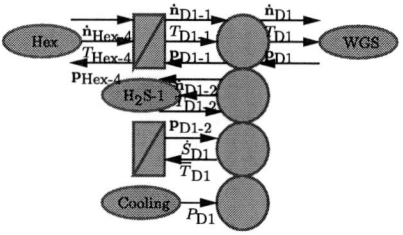

Fig. 7. EMR of desulfurization unit.

9) Water Gas Shift Reaction: Carbon monoxide degrades the fuel cell performance and has therefore to be removed. As introduced in paragraph III-A7 there is an equilibrium between the fractions of hydrogen, carbon monoxide, water and carbon dioxide, which variates with the temperature. This principle can be used to reduce the carbon monoxide content in the hydrogen rich gas, as lower temperatures favor the formation of hydrogen and

the hydrogen rich gas has to be further cooled between the reformer and the fuel cell [11], see figure 8.

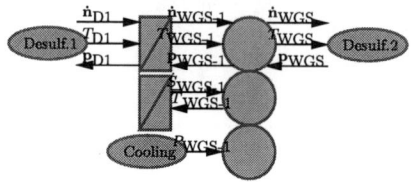

Fig. 8. EMR of water gas shift unit.

B. Maximum Control Structure Development

1) Control of Mass flow Supply: To control the mass flows needed by the fuel cell, the chain between the fuel cell entrance and the control parameter has to be inverted. The respective control parameters are the conversion coefficients of the choppers for diesel, water and air supply. As the entire system is long we first describe the inversion from the fuel cell entrance to the fuel processor unit entrance and afterward the inversion from the three supply flows.

The control chain between the fuel cell entrance and the reformer entrance consists of: the molar flow vector coming out of the second desulfurization unit $\dot{n}_{D2}$; the molar mass flow vector coming out of the pressure drop of the second desulfurization unit $\dot{n}_{D2-1}$; the molar flow vector coming out of the water gas shift unit $\dot{n}_{WGS}$; the molar flow vector coming out of the pressure drop of the water gas shift unit $\dot{n}_{WGS-1}$; the molar flow vector coming out of the first desulfurization unit $\dot{n}_{D1}$; the molar mass flow vector coming out of the pressure drop of the first desulfurization unit $\dot{n}_{D1-1}$; the molar flow vector between the heat exchanger and the first desulfurization unit $\dot{n}_{Hex-4}$; the molar flow vector between the reformer exit sided heat exchanger entrance and the heat exchanger $\dot{n}_{Hex-3}$; the molar flow vector coming out of the reformer $\dot{n}_{Ref}$; the molar flow vector coming out of the pressure drop of the reformer $\dot{n}_{Ref-1}$; the molar flow vector between the heat exchanger and the reformer $\dot{n}_{Hex-2}$; the molar flow vector coming out of the element mixing the diesel, water and air molar flow $\dot{n}_{Hex-1}$ and the three molar flow vectors entering the fuel processor unit $\dot{n}_{Su-d}$, $\dot{n}_{Su-w}$ and $\dot{n}_{Su-a}$ (figure 9). As soon as the control chain is defined the elements inside the control chain can be inverted one by one. To invert the desulfurization units the proportional connection between the final and the initial sulfur content via the known system selectivity is used. The fuel cell imposes a certain hydrogen mass flow, but from the knowledge of the gas composition at the working point of the reformer, the other mass flows can be defined. The pressure development block contains a time dependency, equation 3 and 4. Hence its inversion requests the use of a controller. As it is a first order time dependency a PI controller has been chosen. For reasons of simplicity only one controller is used to respond to the four pressure drop blocks inside the system. The water gas shift reaction can be inverted using the energy conservation law, equation 2. Therefore, the upstream temperature $T_{\text{WGS-1}}$ is measured. Inside the heat exchanger block it can be assumed that no change inside the gas composition occurs. Hence no control block is needed. To invert the reformer block, the assumption is made that the reformer has to run in a limited working window, defined by constant fractions between hydrogen output and, diesel, water and air input. Those constant fractions are used to find the mass flows of diesel, water and air corresponding to a certain, imposed hydrogen flow. Finally, the molar flow vector can be splitted inverting the mixing element, as each element of the vector can be attributed to one of the three supply flows.

The three supply mass flows (diesel, water and air) have to be controlled. They are controlled via the conversion coefficient of the chopper. As the structure of the different mass flow supplies is equivalent, the conception is introduced once. The first step is to define the control chain. The molar flow vector $\dot{n}_{\text{Su}}$ has to be controlled interacting on the conversion coefficient of the chopper. The control chain contains the speed of motor Ω, the motor torque Γ_{mo}, the motor current I_{mo} and the chopper voltage V_{ch}. Now the elements can be inverted one by one. Firstly, the pump or compressor has to be inverted using the proportional connection between the molar flow and the turning speed. Secondly, the mechanic inductance has to be inverted. As it contains a time dependent element, a controller has to be used to invert the accumulation element. A PI controller is used and parametrized. Thirdly, the electro mechanic conversion can be inverted using the proportional connection between the motor torque Γ_{mo} and the motor current. Fourthly, the electric inductance has to be inverted. As this element is time dependent, a controller has to be used to invert the accumulation element. A PI controller is used, and parametrized accordingly for a first order system. Lastly, the control parameter m can be found using the proportional connection between the chopper voltage V_{ch} and the conversion parameter m, knowing that the supply voltage is constant. This gives the mass flows at the reformer input (figure 10).

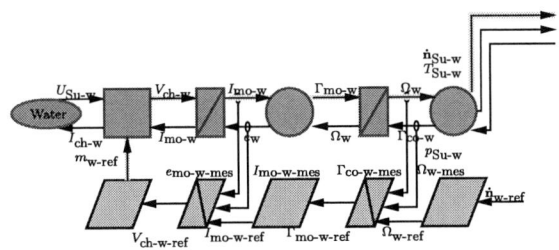

Fig. 10. EMR and MCS of the water supply system.

2) Temperature Control: As the fuel processing unit works only inside a small and well defined working window, the expected temperature after each block is well known. Hence the control chain includes only two ele-

ments, the system output temperature T and the cooling power P. They are connected via a coupling element, indicated by three or four overlapping orange circles, see figure 11. A coupling element can be inverted directly, there is no need to use a controller. Hence the temperature control is based on the energy balance, equation 2. The following values have to be measured: input molar flow vector $\dot{n}_{\text{in}}$, input temperature T_{in}, output molar flow vector $\dot{n}_{\text{out}}$ and temperature of the surrounding box T_{B}.

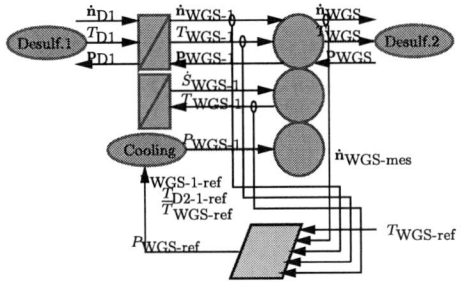

Fig. 11. EMR and temperature MCS of the water gas shift unit.

IV. RESULTS

The EMR has been applied for the fuel processor of the fuel cell system proposed by GAPPAC including temperature and molar flow evolution control (figures 10 and 11). A current profile changing in steps is applied to the system. The current profile for the study of molar flow development is given in figure 12. The current profile for the temperature evolution is equivalent with the difference to have always a 200 s period between the step changes instead of a 2 s period.

For the molar flow evolution the change of the fractions

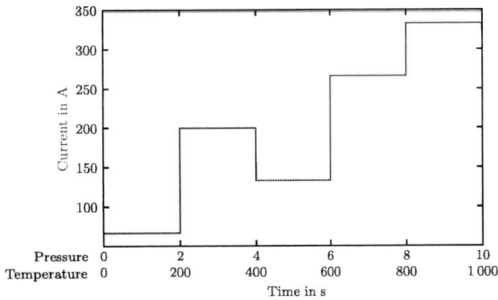

Fig. 12. Current profile submitted to the model

of hydrogen, oxygen, water and carbon monoxide is shown respectively. The molar flow evolution of hydrogen is given in figure 13. The system is supplied with diesel, which is to be transformed into hydrogen. The fraction of diesel is much smaller as the fraction of hydrogen. This can be explained by the fact that diesel is a long chain molecule containing several hydrogen atoms. The fact that no hydrogen molar flow difference can be seen between the heat exchanger and the first desulfurization unit can be seen is explained by the fact that no molar flow change

2161

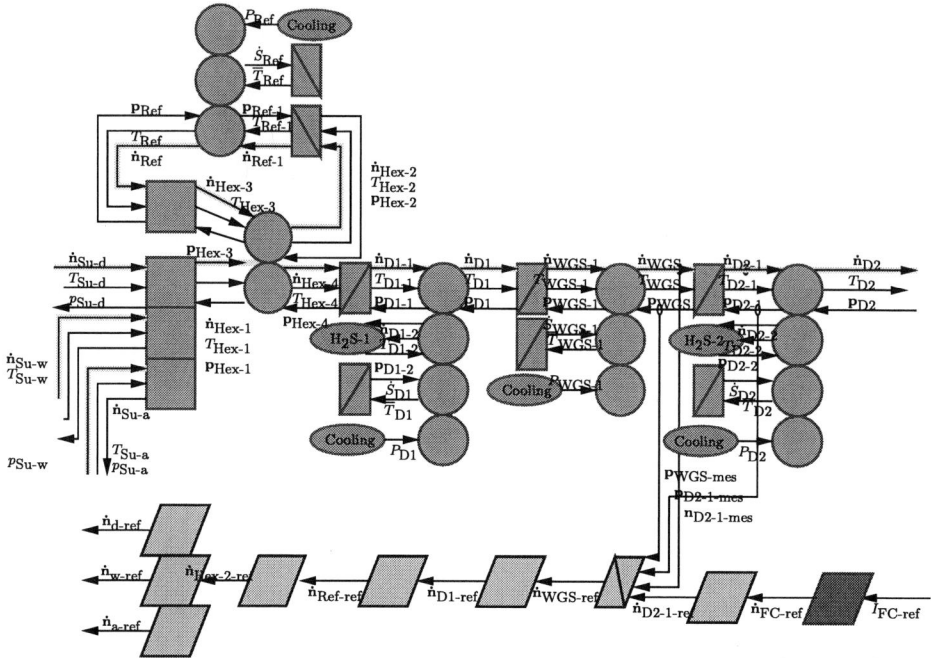

Fig. 9. EMR and MCS of the system between heat exchanger entrance and fuel cell entrance.

is imposed by the desulfurization unit. The molar flow fraction after the water gas shift and the second desulfurization unit respectively is higher than the hydrogen molar flow before the water gas shift. It can be concluded that the water gas shift increases hydrogen molar flow. The molar flows follows the changes in current. After a period of 2 s the pumps and compressors followed the load step. The parameters to describe the pumps and compressor are preliminary. As soon as final values can be derived from the real system an optimization of the control parameters has to be done. Oxygen is supplied to the fuel processor

the exit of the water gas shift unit stabilizes more rapidly than the water fractions further upstream. This can be explained by the fact that a temperature is imposed for the water gas shift unit and the downstream water fraction is defined by the equilibrium at this temperature.

Figure 15 shows that the carbon monoxide ratio is

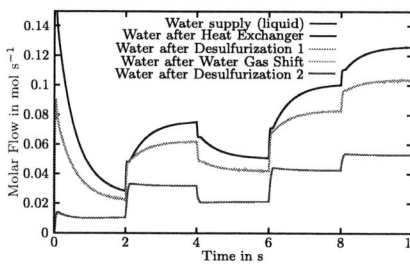

Fig. 14. Water molar flow evolution

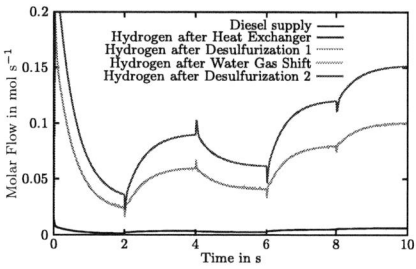

Fig. 13. Hydrogen molar flow evolution

and all consumed inside the reformer. Its supply follows the demand in a way equivalent of the hydrogen one. Water is supplied to the fuel processor in form of liquid water. It is evaporated inside the heat exchanger and it can be considered as liquid for the rest of the fuel processor. Figure 14 shows that a fraction of the water supplied is used inside the reactions. It also shows the existence of gaseous water until the fuel cell entrance. The water at

reduced largely after the water gas shift reaction.

For the temperature evolution, the temperatures of the surrounding modules have been considered to be ambient, only the combined heat exchanger reformer unit is considered to have 650 °C. The temperature limits for the gases provided for the system are: 1400 °C for the reformer outlet, 222 °C at the outlet of the water gas shift and 200 °C at the outlet of the second desulfurization unit. This model leads to the following results regarding temperature evolution (figure 16). The temperature at the reformer does not arrive at the limit value during the course of the simulation. The temperature of the first desulfurization unit is not controled. It reflects the

2162

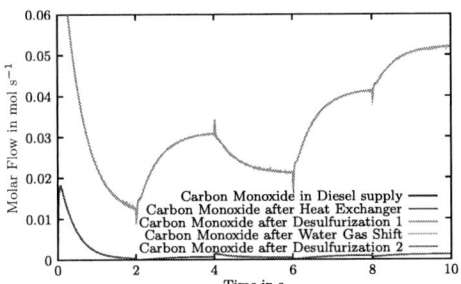

Fig. 15. Carbon monoxide molar flow development

temperature at the heat exchanger, which is cross-linked with the reformer showing complex interactions, that is why a current increase might lead to the drop in gas temperature (at 600 s). The first current is not sufficient to heat up the gas streams, as only cooling has been taken into account. A heat up strategy has to be applied. The temperatures arrive rapidly at constant values, only for the last blocks slight overshoots can be seen (Desulfurization 2). This can be explained by numerical problems. As soon as the time delay of the cooling is taken into consideration, those overshoots will disappear. Figure 17

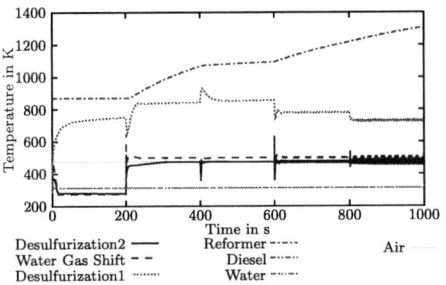

Fig. 16. Temperature evolution of the gases

shows the heat up of the boxes. The heat exchanger box shows rapid heat up (also because it has been preheated). The other boxes heat up much slower. They do not reach their final temperatures in the period of the experiment.

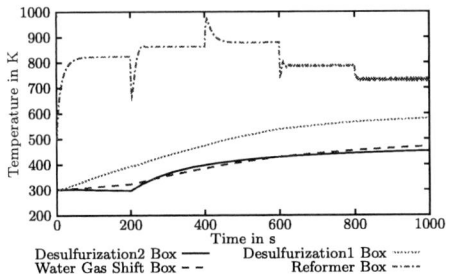

Fig. 17. Temperature evolution of the surrounding boxes.

V. CONCLUSION AND PERSPECTIVES

An inversion based control structure development methodology based on EMR and MCS has been intro-

duced. EMR has successfully been applied on the fuel processor unit taking into account the molar flow and temperature development. This is an important component of the sophisticated trigeneration fuel cell system proposed by GAPPAC. The system control with respect to molar flow and temperature development show promising results. In further steps this parameters used for the mass flow supplies have to be verified against values from the real system. Thereafter an improvement of the control parameters can be done. Furthermore, it has to be considered if the time dependency of the system cooling can be taken into account. Afterward the system has to be enlarged, first by combining the fuel processor unit with a fuel cell model, than step by step include the other auxiliaries and finally the trigeneration. EMR and MCS show the capability to be well adapted for this complex, multi domain task.

ACKNOWLEDGMENT

ANR the French National Research Agency through hydrogen program PAN-H is acknowledged for its financial support (ANR-06-PANH-012)[1].

REFERENCES

[1] M Tékin, D Hissel, M.C. Péra, and J. M. Kauffmann. Energy optimization of a fuel cell generator: Modelling and experimental results. In *EPE'03, European Power Electronics Conference*, 2003.

[2] M.C. Péra, J. Garnier, A. de Bernadinis, R Lallemand, G Coquery, and J. M. Kauffmann. High frequency power converter for pefc generator architecture based on a multi stack association for transport applications. In *EPE 2005, European Power Electronics Conference*, 2005.

[3] Denver Cheddie and Norman Munroe. Rewiew and comparison of approaches to proton exchange membran fel cell modeling. *Journal of Power Sources*, 147:72–84, 2005.

[4] Joshua Golbert and Daniel R. Lewin. Model-based control of fuel cells: Optimal efficiency. *Journal of Power Sources*, doi:10.1016/j.jpowsour.2007.04.062:12, 2007.

[5] Jay T Pukrushpan. *Control of Fuel Cell Power Systems*. Springer, 2004.

[6] Anders R. Korsgaard, Mads P. Nielsen, and Soeren K. Kaer. Part two: Control of a novel htpem-based micro combined heat and power fuel cell system. *Internaltional Journal of Hydrogen Energy*, 33:1921–1931, 2008.

[7] J. P. Hautier, A Bouscayrol, R Schoenfeld, G Dauphin-Tanguy, G.-H. Geitner, X. Guillaud, and A. and Pennamen. Different energetic descriptions for electromechanical systems. In *EPE Dresden*, 2005.

[8] D. Chrenko, M.C. Pera, A. Bouscayrol, and D. Hissel. Inversion-based control of a pem fuel cell system using energetic macroscopic representation. *ASME Journal of Fuel Cell Science and Technology*, accepted.

[9] Anders R. Korsgaard, Mads P. Nielsen, and Soeren K. Kaer. Part one: A novel model of htpem-based micro-combined heat and pozer fuel cell system. *International Journal of Heat and Mass Transfer*, 33:1909–1920, 2008.

[10] Loïc Boulon, Marie-Cécile Péra, Philippe Delarue, Alain Bouscayrol, and Daniel Hissel. Causal fuel cell system mode suitable for transportation simulation applications. In *ASME Conference of fuel cell science and technology*, 2007.

[11] Daniela Chrenko, Julien Coulié, Samuel Lecoq, Marie-Cécile Péra, and Daniel Hissel. Static and dynamic modelling of a diesel fuel processing unit for polymer electrolyte fuel cell supply. *Internaltional Journal of Hydrogen Energy*, submitted, 2008.

[1]

Power Management in an Autonomous Adjustable Speed Large Power Diesel Gensets

Grzegorz Iwanski, Wlodzimierz Koczara

Warsaw University of Technology/Institute of Control and Industrial Electronics, Warsaw, Poland,
e-mail: *iwanskig@isep.pw.edu.pl* , *koczara@isep.pw.edu.pl*

Abstract— The autonomous power generation system with DFIG dedicated for diesel driven gensets is presented. Characteristics of torque and power versus speed of internal combustion engine ICE are analyzed at an angle of application in electrical power generation. Features and advantages of variable speed operation versus fixed speed operation of internal combustion engine are described. The vector control method for generation of fixed amplitude and frequency output voltage is presented. The PLL vectorial synchronization of the actual stator voltage with the reference vector allow the rotor speed and position sensorless operation. Variable speed operation of the ICE engine reduces fuel consumption what is important in continuously operated diesel and hybrid systems with renewable sources. Results of standalone operation of 2.2kW doubly fed induction generator supplying nonlinear load are presented.

Keywords—adjustable speed generation system, doubly fed induction generator, generation of electrical energy, distributed power.

I. INTRODUCTION

The demand for diesel based power generation systems constantly grows. The main applications are the stand-by generation sets for important load in telecommunication, hospitals, military applications, etc. and island power systems, which are never connected to the grid and where the supplied load needs reliable power source. In the second case very popular solutions are combined systems with renewable energy sources (e.g. wind-diesel) in which internal combustion engine assures reliability of supply voltage, whereas the renewable source is used for reduction of engine fuel consumption.

The disadvantage of typical diesel power systems is fixed speed operation, because the internal combustion engine ICE drives the classical synchronous generator. During fixed speed operation of classical ICE genset the fuel consumption relatively weakly depends on the delivered power and for low load power the efficiency of genset is very poor. Low power of load supplied from the ICE genset requires to connect to the dump (dummy) load to provide the correct operating conditions. Operation with low torque caused by low load conditions may cause significant decrease of the engine lifetime. In general, large industrial diesels for the gensets are designed for fixed speed operation adequate e.g. to electrical generator synchronous speed and they do not like to run at less that 50% load as cylinder glazing can occur. To obtain safe operation with the low load conditions, typically in the fixed speed diesel gensets, the dump load of power up to 40% is used, which is normally disconnected but necessary in case when load is lower than critical [1].

Speed of internal combustion engine can be adjusted to the load power and thus the fuel consumption can be decreased. The variable speed operation can be obtained by the use of power electronics conversion systems which is responsible for standardization of the output voltage in standalone systems and delivered power in grid connected systems. The simplest way to obtain variable speed genset is the use of classical system equipped in the load profile tracking speed controller and additional power electronics AC/DC/AC system connected to the synchronous generator in series.

Modern mobile gensets uses ultra-light permanent magnet generators [2][3] instead of wound excited synchronous machines and power electronics converters connected in series, but such systems are designed mainly for low power gensets.

The doubly fed induction generator is excellent solution for medium and large power adjustable speed diesel engine based genset (Fig. 1) as the range of speed around the synchronous value is limited. Speed of typical diesel engine used in power genset is equal 1500 rpm. Assuming that the acceptable speed range is from 1000 to 2000 rpm, the maximum value of slip power is equal to 33%. It has to be taken into consideration, that the part of the reactive power is uncompensated by the filtering capacitances C_f and have to be provided by the power electronics converter.

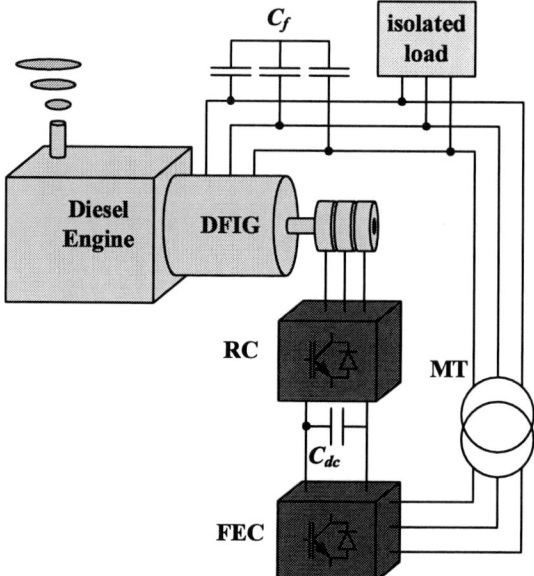

Fig. 1. Standalone power generation system with DFIG driven by variable speed diesel engine.

978-1-4244-1741-4/08/$25.00 ©2008 IEEE

II. Fixed and Adjustable Speed ICE

Fixed speed operation of industrial diesel engines are very often used in power gensets. In this case the load torque T_{load}, which varies from zero to maximum of diesel torque T_{dmax} is compensated by the diesel torque T_d to obtain fixed synchronous speed ω_{synch} (Fig. 2a). It corresponds to the frequency of the voltage generated by synchronous generator driven from the engine. The maximum power $P_{dmaxsynch}$ which can be obtained at synchronous speed ω_{synch} is significantly lower than the total maximum power P_{dmax} of the engine (Fig. 2b). The maximum power P_{dmax} is delivered at higher speed, but this feature cannot be used in the genset with classical synchronous generator.

In the characteristics of diesel engine there is a forbidden area, where for some range of speed the long-standing load torque cannot be smaller then limited, because it results in decrease of engines lifetime. It is caused by incomplete combustion of hydrocarbons, what produces some kind of lake or glaze on the diesel components like cylinders, pistons, fuel injectors, etc.

Fixed speed operation of ICE gensets require connection of additional "dummy" load like resistors in case of low load operation to increase the delivered power above the forbidden range. Small decrease of the load power below the point of the best efficiency (the lowest fuel consumption per kWh) causes rapid increasing of fuel consumption (g/kWh).

The variable speed operation of diesel engine allows to obtain a stable operation with low load conditions at low speed [2], what provide significant reduction of fuel consumption. In the steady state, the engine operates with an excess of the torque in wide speed range (Fig. 3a). The load torque T_{load} in every operating point is lowest than the maximum instantaneous diesel torque $T_{dmaxinst}$ at this point. The reserve of torque allow for the acceleration of the engine in case of the step increase of the load.

The speed range from ω_{gmin} to ω_{gmax}, in which the power can be generated stably is slightly smaller than limited values of acceptable engine speed range from ω_{dmin} to ω_{dmax}. Operation in forbidden area is not necessary and artificial "dummy" load may be eliminated from the system. Diesel engine can operate at low speed with power range from zero to several percent. Above this power the speed is increased and optimized to the load conditions (Fig. 3b).

At low speed, the diesel engine consumes significantly lower amount of fuel than at synchronous speed for the same load torque. The engine speed is adjusted to the load to obtain to the best efficiency area what causes the lower fuel consumption than in fixed speed operation [3].

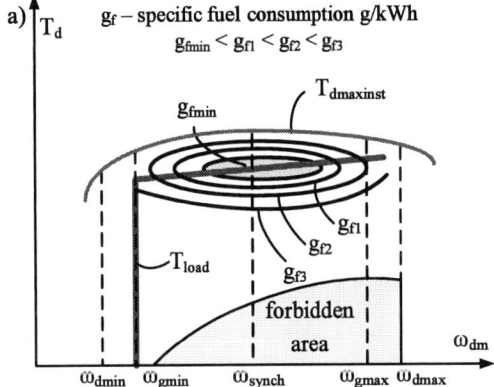

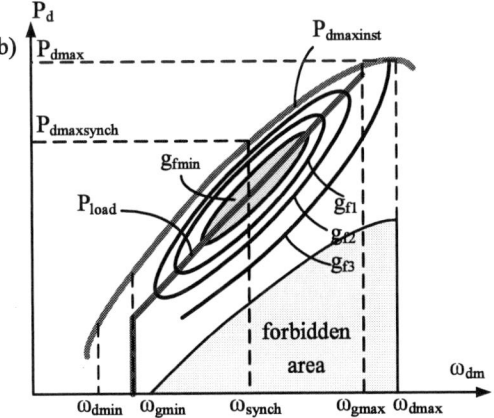

Fig. 3. Torque and power characteristics and operation range of the adjustable speed diesel engine based genset.

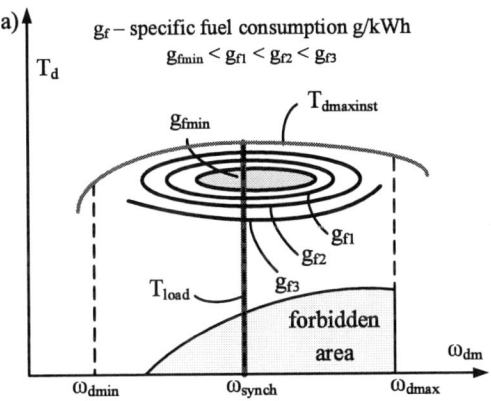

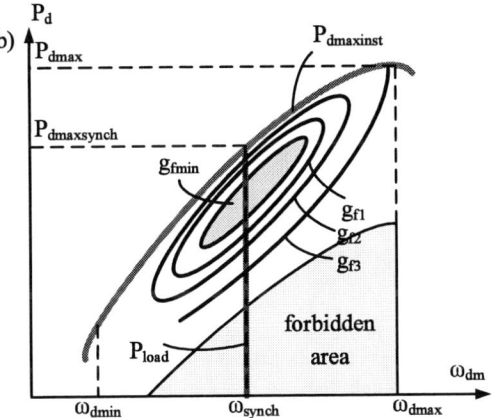

Fig. 2. Torque and power characteristics and operation range of the fixed speed diesel engine based genset.

There is one more important aspect concerning not only the fuel consumption, but also dimensions and price of the diesel engine. During variable speed operation the power which can be delivered from the engine is higher than this which is available at synchronous speed (Fig. 4). The maximum power produced by the variable speed diesel engine is equal P_{dmax} and to obtain the same power from the fixed speed engine at synchronous speed, the engine with meaningfully larger maximum power has to be used, than the variable speed operating set. Larger engine means higher costs and increase of dimensions, weight and fuel consumption for the same available electrical power range.

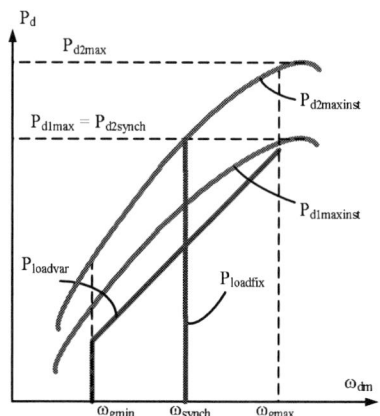

Fig. 4. Power characteristics of the fixed speed and variable speed engines providing the same power.

III. SENSORLESS STANDALONE OPERATION OF DFIG.

There are some similarities of the standalone doubly fed induction generator DFIG and synchronous generator SG [10]. Output voltage in SG is controlled by the DC current fed to the excitation winding, whereas the frequency of output voltage is controlled on the mechanical way by the regulation of mechanical speed (Fig. 5). It can be seen, that the system can operate with fixed speed to provide fixed output frequency. Mostly, the amplitude and the frequency of output voltage are calculated twice per period of the generated voltage. The step change of load produces strong disturbances in the speed of primary mover, and consequently in the output frequency, as well as in the output voltage amplitude.

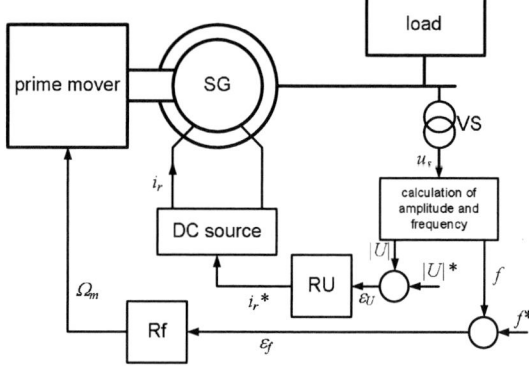

Fig. 5. Idea of the control of output voltage amplitude and frequency in fixed speed standalone synchronous generator SG.

The standalone system with DFIG has significant advantage in relation to the SG. The rotor is supplied from three phase system, therefore there can be controlled not only the amplitude of the rotor current $|i_r|$, but also its frequency. Output voltage frequency strictly equals to the sum of the frequency originated from the mechanical speed and the rotor current frequency. Thus the frequency of the stator voltage can be controlled by the change of the rotor current frequency and the speed of primary mover can be independently adjusted to the best operating efficiency. The idea of simple control of standalone DFIG is presented on Fig. 6.

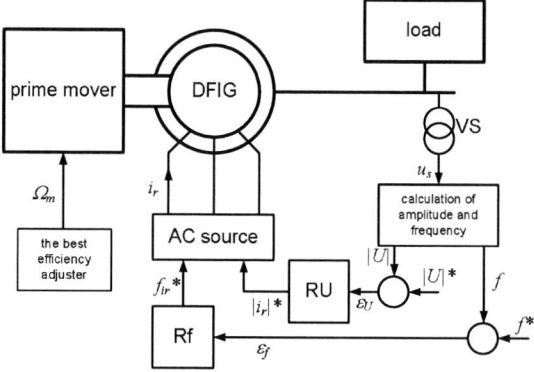

Fig. 6. Idea of the control of output voltage amplitude and frequency in adjustable speed standalone doubly fed induction generator DFIG.

Using a high performance DSP controllers, the output voltage amplitude can be calculated and controlled very fast (hundreds Hz to few kHz depending on the generator power). Moreover it can be used space vector theory to control the output voltage frequency.

The bases of standalone operation of DFIG are presented in [4][5][6]. The control system based on the vectorial PLL structure [7] allow the very fast control of output frequency in spite of the variable speed operation. Synchronization of the actual stator voltage vector with the reference vector rotating with synchronous speed is used. It makes it possible to eliminate the measured or estimated mechanical rotor speed or position. Improved system with PI vector control of the rotor current, instead of the scalar three phase rotor current control by proportional regulators [4][5], is presented in Fig. 7. Front-end converter control is neglected as it is current controlled VSI [8][9] and can be treated as additional load from the stator point of view.

Output voltage amplitude is proportional to the rotor current amplitude therefore the controller RU for this voltage parameter is quite simple. The second voltage controller $R\alpha$ synchronizes actual voltage vector with reference one and produces the reference signal of the rotor current angular speed $\Omega_{ir}{}^*$, and after integration gives rotor current position $\phi_{ir}{}^*$ related to the rotor. Advantages of PI controller can be used if the controlled variables are fixed, therefore additional transformation is applied in the control algorithm. The reference parameters of the rotor current vector (amplitude $|i_r|^*$ and position angle $\phi_{ir}{}^*$) obtained from controllers RU, $R\alpha$ need to be transformed to the xy coordinate system, in which the new rotor current vector components are fixed.

2166

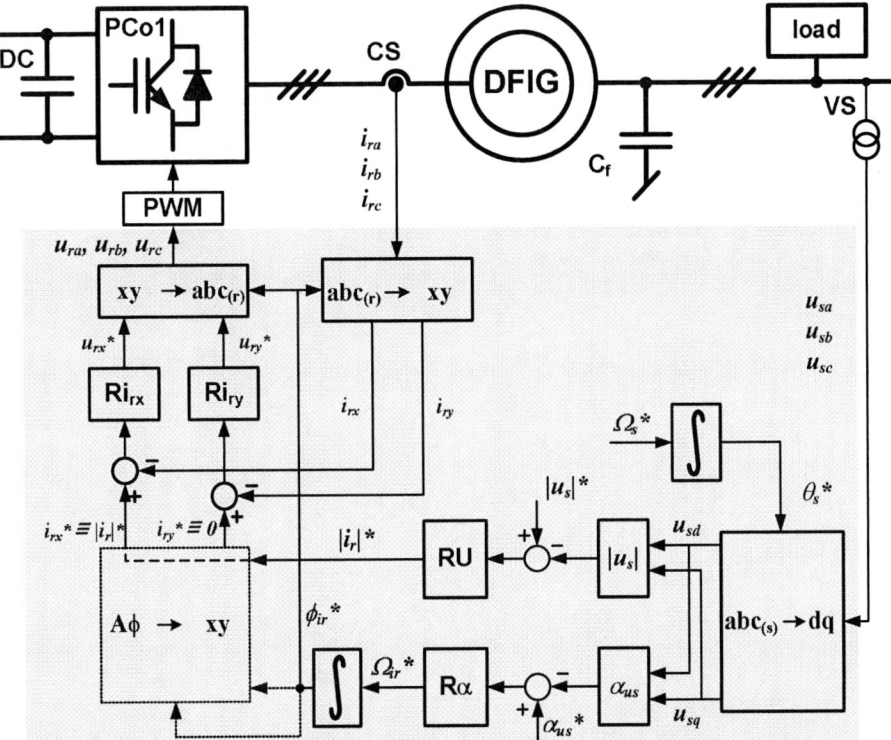

Fig. 7. Control system of standalone DFIG with PLL synchronization of the stator voltage vector and rotor current PI vector control.

The rotation of new frame xy in which the current vector components are fixed can be obtained by the use of angle ϕ_{ir}^* produced by $R\alpha$ controller.

$$\begin{bmatrix} i_{r\alpha}^* \\ i_{r\beta}^* \end{bmatrix} = \left| i_r^* \right| \begin{bmatrix} \cos \phi_{ir}^* \\ \sin \phi_{ir}^* \end{bmatrix} \quad (1)$$

$$\begin{bmatrix} i_{rx}^* \\ i_{ry}^* \end{bmatrix} = i_{r\alpha}^* \begin{bmatrix} \cos \phi_{ir}^* \\ \sin \phi_{ir}^* \end{bmatrix} + i_{r\beta}^* \begin{bmatrix} \sin \phi_{ir}^* \\ -\cos \phi_{ir}^* \end{bmatrix} = \begin{bmatrix} \left| i_r^* \right| \\ 0 \end{bmatrix}, \quad (2)$$

The characteristic feature of this selected xy frame is, that the reference i_{rx}^* component is identically equal to the reference amplitude $\left| i_r \right|^*$, whereas reference i_{ry}^* component is identically equal to zero. Thus, the calculation (1)(2) can be neglected and adequate signals used directly as the references for the rotor current controllers Ri_{rx}, Ri_{ry}. The measured signals of the rotor current are transformed to xy (3)(4) and used as feedback.

$$\begin{bmatrix} i_{r\alpha} \\ i_{r\beta} \end{bmatrix} = \frac{2}{3} \begin{bmatrix} 1 & -\dfrac{1}{2} & -\dfrac{1}{2} \\ 0 & \dfrac{\sqrt{3}}{2} & -\dfrac{\sqrt{3}}{2} \end{bmatrix} \begin{bmatrix} i_{ra} \\ i_{rb} \\ i_{rc} \end{bmatrix}, \quad (3)$$

$$\begin{bmatrix} i_{rx} \\ i_{ry} \end{bmatrix} = i_{r\alpha} \begin{bmatrix} \cos \phi_{ir}^* \\ \sin \phi_{ir}^* \end{bmatrix} + i_{r\beta} \begin{bmatrix} \sin \phi_{ir}^* \\ -\cos \phi_{ir}^* \end{bmatrix}, \quad (4)$$

In the steady state the rotor current vector i_r takes the reference position i_r^*, which causes that the stator voltage vector u_s reaches its reference position u_s^*. In the transient states caused by speed change the actual position of the rotor current i_r is instantaneously moved from reference position by angle β_{ir}. Simultaneously the rotating frame x'y' is also instantaneously displaced from the reference position xy (Fig. 8), because of the inertia of stator voltage angle controller $R\alpha$.

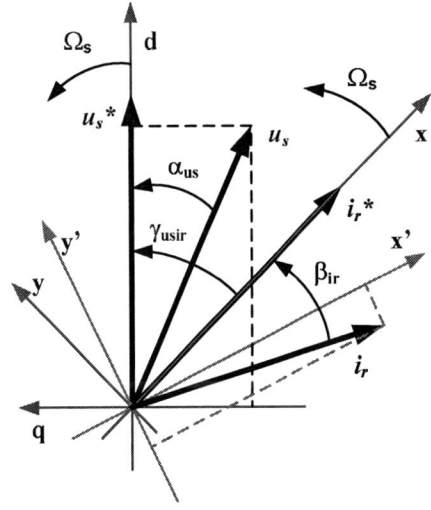

Fig. 8. Vector diagram of the transient state in autonomous DFIG

IV. SIMULATION RESULTS OD STANDALONE DFIG

Fig. 9 presents the simulation results of adjustable speed 250kW doubly fed induction generator with ideal converter connected to the rotor. The model was created in PSIM simulation software. Initial power is equal to 25% while in 0.4s additional load is connected and finally 75% of the rated load is supplied. The stator voltage has fixed frequency and amplitude in spite of the rotor speed change. The speed is adjusted to the power of supplied load.

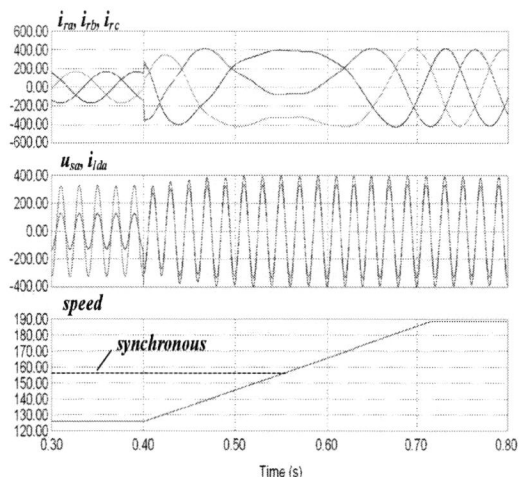

Fig. 9. Simulation results of the step loading of diesel driven autonomous DFIG.

The stator voltage and rotor current controllers were tuned using the step response method for no load operation, which is the worst case from the system stability point of view.

Fig. 10 shows the generation system response on the step change of output amplitude from 300V to 600V. The transient state is quite short and equal to about one period of the generated voltage.

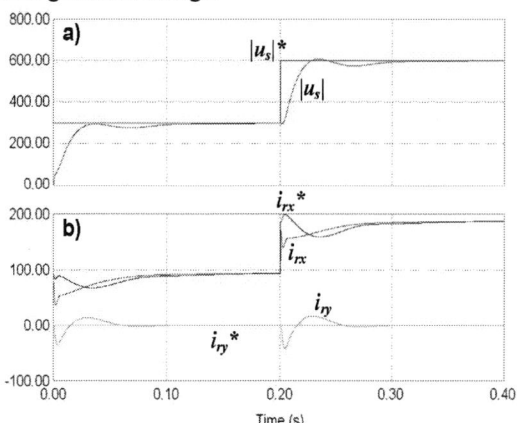

Fig. 10. Simulation results of the step response on the change of reference stator voltage amplitude $|u_s|^*$.

Fig. 11 presents the system response on the step change of the output voltage phase to the counter-phase. The transient state is than 0.1s. This shows that the voltage phase control on the electrical way in DFIG is much faster than voltage phase control on the mechanical way in SG. The phase control permits for fast synchronization of DFIG and soft connection to grid [6]. Thus the load is protected from the rapid change of the supply voltage phase.

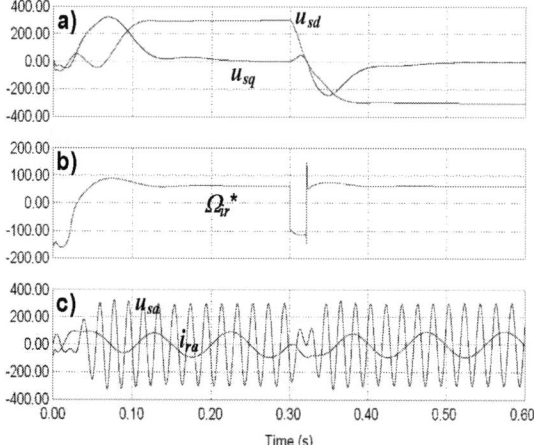

Fig. 11. Simulation results of the step response on the change of reference stator voltage phase to counter-phase.

V. LABORATORY TESTS OF STANDALONE OF DFIG.

The laboratory tests were made with 2.2kW slip ring induction machine driven by DC motor which can emulate the behavior of different variable and adjustable primary movers like wind turbines and internal combustion engines [10]. The system is controlled by DSP unit build on ADSP-21061 by Analog Devices responsible for the execution of the control algorithm and FPGA by ALTERA responsible mainly for the PWM channels, data acquisition and user interface. Oscillogram presenting results of variable speed standalone DFIG from laboratory setup is shown in Fig. 12.

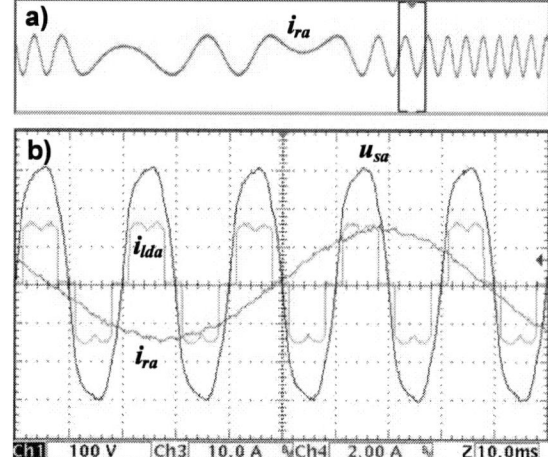

Fig. 12. Oscillogram presenting stator voltage u_{sa} (100V/div), rotor current i_{ra} (10A/div) and load current i_{lda} (2A/div) during variable speed standalone operation of DFIG.

High quality stator voltage with fixed amplitude and frequency is achieved in spite of the rotor variable speed and supply of strongly nonlinear load at power equal to 75% of the rated value. Oscillogram presenting transient of step loading of the generator from 75% of the generator rated power to zero is presented on Fig. 13, whereas oscillogram presenting step unloading from 75% to 25% is presented in Fig. 14. Transient state with voltage sag during loading and overvoltage during unloading are equal to one period of the generated voltage.

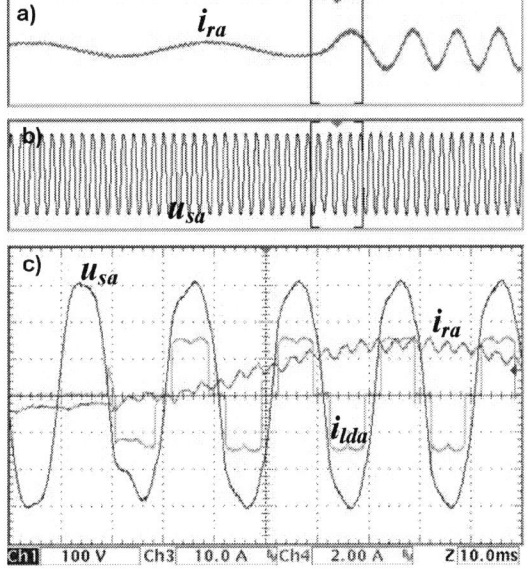

Fig. 13. Oscillogram presenting stator voltage u_{sa} (100V/div), rotor current i_{ra} (10A/div) and load current i_{lda} (2A/div) during step unloading of standalone DFIG.

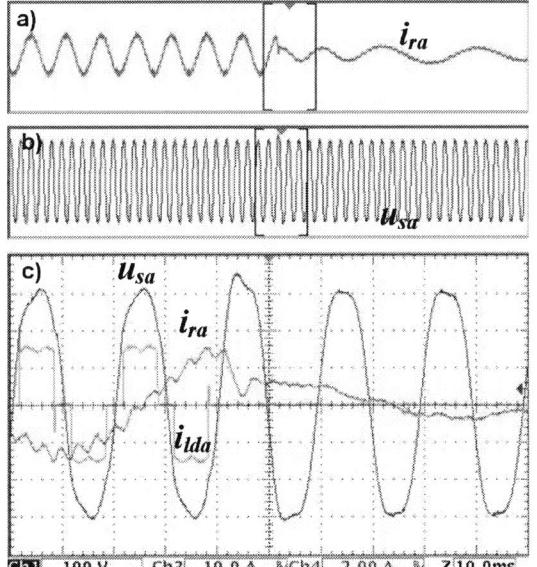

Fig. 14. Oscillogram presenting stator voltage u_{sa} (100V/div), rotor current i_{ra} (10A/div) and load current i_{lda} (2A/div) during step unloading of standalone DFIG.

VI. CONCLUSION

The standalone DFIG can replace classical fixed speed synchronous generator, what allow the variable speed operation of diesel engine and consequently significant reduction of the fuel consumption.

The fixed amplitude and frequency of output voltage can be obtained in spite of the variable speed and load and cab be obtained without any information about the mechanical speed or position.

Vectorial PLL allow to obtain controlled voltage phase, what is very important for synchronization and soft connection of loaded generator to the grid.

Adequate selection of the frames for stator voltage and rotor current vectors allow to apply the PI controllers in each control loop.

The proposed method of standalone DFIG output voltage control are proved in laboratory tests bench with 2.2kW slip ring machine.

VII. FUTURE WORKPLAN

The future research will concentrated on cooperation of standalone DFIG system with supercapacitor as energy storage applied for compensation of the mechanical power variation and improving the transients during step loading and unloading og the generation system.

REFERENCES

[1] L. M. Grzesiak, W. Koczara and M. da Ponte, "Novel hybrid load-adaptive variable-speed generating system", *IEEE International Symposium on Industrial Electronics ISIE '98*, Vol. 1, 7-10 July, 1998 pp. 271 – 276

[2] L. Grzesiak, W. Koczara, P. Pospiech and M. da Ponte, "Power quality of the Hygen autonomous load-adaptive adjustable-speed generating system", *14th Annu Applied Power Electronics Conference and Exposition - APEC '99*. Vol. 2, 14-18 March 1999, pp. 945 - 950

[3] Z. Chlodnicki, W. Koczara and N. Al-Khayat, "Laboratory Simulation of the Adjustable Speed Generation Systems", *CD Proc. of Power Electronics and Motion Control Conference – PEMC'06*, Portoroz, Slovenia, 30 Aug. – 1 Sept., 2006.

[4] G. Iwanski and W. Koczara, "Sensorless direct voltage control of the stand-alone slip-ring induction generator" *IEEE Trans. on Ind. Electron.*, Vol. 54, No. 2. Apr. 2007, pp.1237-1239.

[5] G. Iwanski and W. Koczara, "Sensorless stand alone variable speed system for distributed generation" *Proc. 35th IEEE Power Electron. Specialist Conf. – PESC*, Aachen, Germany, 2004, Vol.3, pp. 1915-1921

[6] G. Iwanski and W. Koczara: "A DFIG based Power Generation System with UPS Function for Variable Speed Applications" *IEEE Trans on Industrial Electronics.* (accepted for publication)

[7] L. Guilherme, B. Rolim, D.R. Costa and M. Aredes, "Analysis and software implementation of a robust synchronizing pll circuit based on the pq theory", *IEEE Trans. on Ind. Electron.*, vol. 53, pp. 1919-1926, Dec., 2006.

[8] M. Malinowski, M. P. Kazmierkowski and A. M. Trzynadlowski, "A comparative study of control techniques for pwm rectifiers in ac adjustable speed drives" *Proc. 27th Conf. IEEE Ind. Electron. Soc. IECON '01*, Denver, CO, USA, 2001, Vol. 2. pp. 1114-1118.

[9] T. Noguchi, H. Tomiki, S. Kondo and I. Takahashi, "Direct power control of pwm converter without power-source voltage sensors", *IEEE Trans. Ind. Appl.*, Vol. 34, issue 3, pp 473–479, 1998

[10] L. Jiao, B.T. Ooi, G. Joos and F. Zhou, "Doubly-fed induction generator (DFIG) as a hybrid of asynchronous and synchronous machines", *Electric Power Systems Research*, Vol. 76, Issues 1-3, Sept. 2005, pp. 33-37

Cost evaluation of Generator-set with Energy Storage for 4Q-load

Freek J.F.Baalbergen, Pavol Bauer

Delft University of Technology, Delft, Netherlands, e-mail: *P.Bauer@TUDelft.nl*

Abstract—Diesel generators in small electricity grids are not always used in an efficient way. The reason for this is twofold. First of all the efficiency of a diesel generator is dependent on the ratio between the average power and the peak power of the generator. The smaller this ratio, the lower the efficiency. Furthermore some loads can regenerate energy. In small grids this energy is mostly not needed elsewhere and should be dissipated. A solution solving both above mentioned problems is using an energy storage device in the system. This storage can be used for peak shaving and storing regenerated energy. This paper focuses on the costs of a generator-set with energy storage. This calculation is done for the general case and for six different power management strategies: Only Generator, Peak Power Shaving, Dynamic Solution, Max On or Off, Average Power and Only Storage. Calculation of the costs shows that adding an energy storage device lowers the cost for all methods. Calculations show that the Max On or Off method is the best choice.

Keywords—Diesel driven generators, Energy conservation, Energy storage, Power electronics, Lithium ion battery, Super capacitor, Costs.

I. INTRODUCTION

THE diesel generator in a small electricity grid is often not used in a very efficient way. The generator has to supply the peak power of the load. The average power of the load can be much smaller. When the generator is not being used at full load it gets less efficient. Diesel generators are mostly used for only 30 % of the full load [1].

Some loads uses all four quadrants of the voltage-current characteristic (4Q-load.) These loads act both as a load and as a generator. If such a load is being used in a large grid, this energy will immediately be used by one of the other loads. In small grids, however, this is not possible. These small grids often use a diesel driven generator as their main power supply.

For the first problem using a Variable Speed Constant Frequency (VSCF) generator is an option [1,2]. With such a generator, however, no regenerated energy can be recovered.

A solution solving both problems mentioned above is to incorporate an energy storage device in the system. This storage device can be used for peak shaving and storing regenerated energy. The diesel driven generator can be both: a constant speed generator or a VSCF

generator. Fig. 1 gives an overview of such a system.

A. Problem description

Fig. 1 shows an overview of the system. It is important that the loads can be controlled separately, energy is saved and the costs are lower or the same as for a generator without energy storage. The problem is to determine the costs for a typical system.

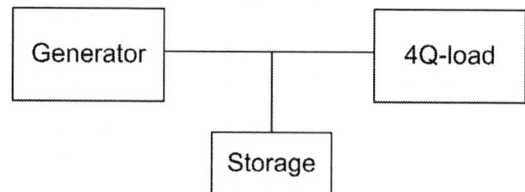

Fig. 1. Overview of generator with energy storage for 4Q-load.

This paper gives a calculation of the costs for such a system for two different storage elements: the Li-ion HP battery and the super capacitor. The calculation is done for the general case and for the six different Power Management Strategies (Only Generator, Peak Power Shaving, Dynamic Solution, Max On or Off, Average Power, and Only Storage) defined in [3, 4].

B. System topology

Because the motors should be controlled separately, only the system configurations with separate motor drives are feasible. The voltage of the Li-ion HP battery does only vary a little with the state of charge [5, 6]. Fig. 2 gives the most feasible topology for a system using a Li-ion HP battery. The Li-ion HP should be sized so that the depth of discharge stays low. The voltage of the super capacitor changes significant with the state of charge [7]. Fig. 3 gives the most feasible topology for a system using a super capacitor. The super capacitor is connected via a DC/DC-converter to the DC-bus.

II. POWER MANAGEMENT STRATEGIES

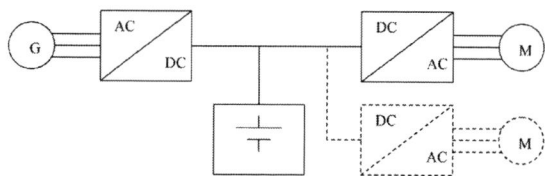

Fig. 2. System topology for generator with Li-ion HP battery.

In [3,4] different Power Management Strategies are defined. In this section a summary will be given.

In the Only Generator PMS (OG) the power is fully being supplied by a generator. This system shows no difference with the normal case and is used as a reference.

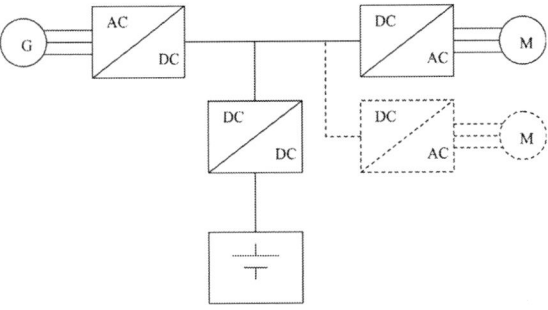

Fig. 3. System topology for generator with Super capacitor.

When the Peak Power Shaving PMS (PPS) is used, the generator supplies most of the power. The energy storage element only supplies the peak power that is needed for a short time.

The Dynamic Solution PMS (DS) is specially designed for use with a VSCF generator. Such a generator cannot react to changes in the output power instantaneously. The energy storage is used to supply/absorb the difference in needed power and power from the generator.

In the Max On or Off PMS (MOO), the generator either supplies its maximum rated power or it is turned off. The generator is dimensioned in such a way that its power rating is higher than the average power needed.

In the Average Power PMS (AV), the generator continuously supplies the average power from the power load profile for a specific duty-cycle. All variations on this average power during this load profile will be supplied or absorbed by the energy storage.

The Only Storage PMS (OS) only supplies power from an energy storage device. In this thesis it is assumed that the storage is being charged from the grid. Electricity from the grid is less expensive than electricity generated by a diesel generator. Further unbalanced cycles become an option.

III. SYSTEM FOR CASE STUDY

The Power Management Strategies will be put in practice on a case study. The system used for this case study is a crane used in the harbor. An illustration of such a crane is given in Fig. 4. The crane consists of two legs with in between a bridge. A trolley with a hoist mechanism is able to move over this bridge. With a spreader containers can be attached to the crane. The crane is able to hoist a container and to move this container over the bridge with the trolley. Furthermore

this crane can move as a whole.

The generator-set with energy storage will be designed based on the worst-case scenario for the system. The maximum power required for this profile is 1,0 pu and the average power is 0,14 pu. Furthermore it is given that the diesel generator has at least a nominal output power of 0,33 pu.

For safety reasons it is required that the system contains a breaking resistor. The system should have a lifetime of 10 years.

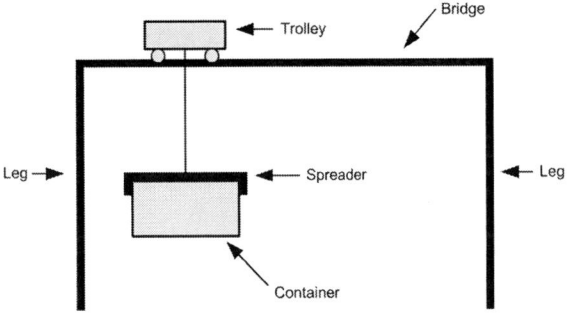

Fig. 4. Typical crane system as used for the case study.

IV. LIFETIME

The system should have a lifetime of 10 years. During this 10 years the system will perform x cycles. For each PMS, one cycle of the crane implies one important discharge cycle [3]. It is assumed that the battery will encounter during its life x discharge cycles.

A. Li-ion HP system

The lifetime of a Li-ion HP battery is depended on the depth of discharge (DOD), number of cycles and the age [6]. The dependency on the DOD is not linear, a lower DOD means a much larger cycle life [6]. This relation is given schematically in Fig. 5. The lifetime of a Li-ion battery is 10 year [6, 8].

The optimal DOD can be found by stating that cycle $= f(\text{DOD}, \text{EOL}) = x$. So the optimal DOD can be found with:

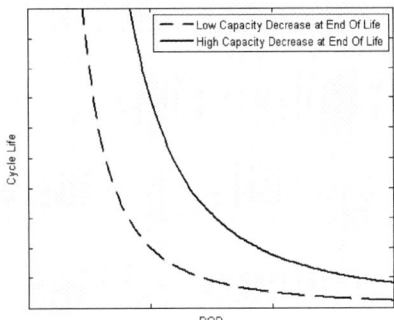

Fig. 5. Dependency cycle life on depth of discharge.

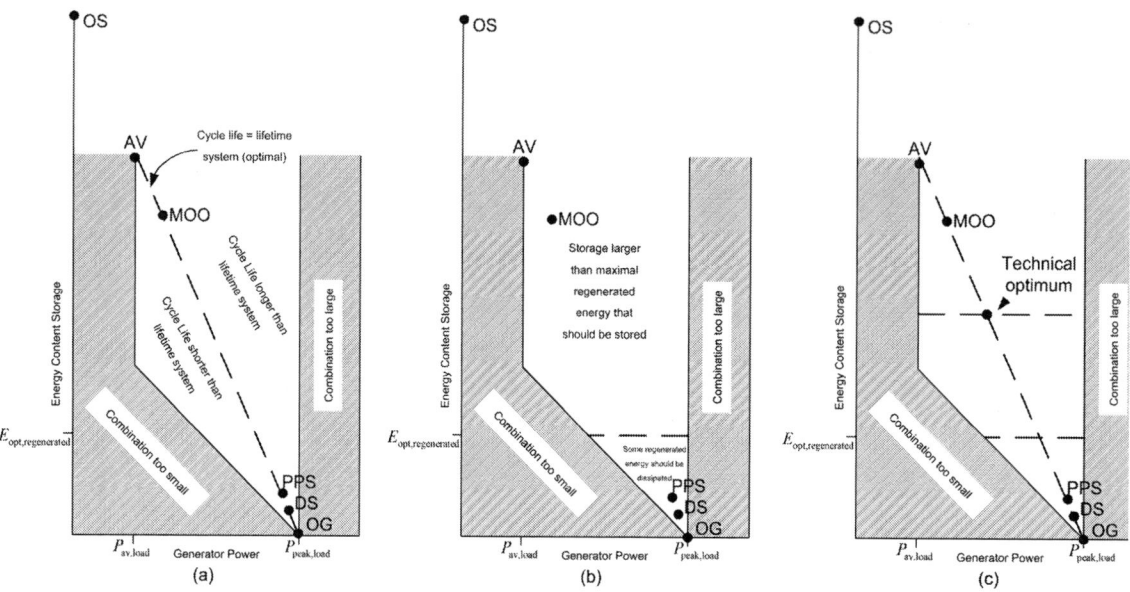

Fig. 6. Sizes of generator and storage.

$$DOD = f(x, EOL) \tag{1}$$

The oversize factor for the battery can be found with [9, 10]:

$$overSize_{Li\text{-}ion\ HP} = \frac{1}{EOL} \cdot \frac{1}{DOD} \tag{2}$$

In chapter 6 of [3] the calculation for an end of life criterion of 80 % and 60 % of the initial capacity is given. The end of life criterion (EOL) with the lowest oversize factor should be chosen. The Li-ion HP battery can, for this case study, best used with the 60 % EOL. The DOD is then 3,89 % and the battery should be oversized by a factor 42,8.

B. Super capacitor

The cycle life of the capacitor is dependent on temperature and voltage [11] and not on the DOD. It is known a super capacitor can have more than 500.000 cycles [8, 11]. It is assumed that the x cycles are no problem for the super capacitor (see [3]) The lifetime of the super capacitor is 10 years [8].

It is known from [11] that at the EOL the capacity of the super capacitor is decreased with 30 % (so EOL = 70%.) This should taken into account for the over sizing of the super capacitor. From [7] it is further known that the super capacitor should be oversized with a factor 1,56. The oversize factor for the super capacitor is thus (based on [9, 10]):

$$overSize_{super\ capacitor} = 1,56\frac{1}{EOL} = 2,23 \tag{3}$$

This is remarkably lower than the oversize factor of the Li-ion HP.

V. AREA OF INTEREST

In Fig. 6 the area of possible sizes for the generator and storages are given. This is limited by some boundaries:

1. the generator should be able to supply the average power required by the load, otherwise another source of energy should be present (as for the Only Storage PMS);
2. the potential energy of the generator and the storage element together, should be larger or the same as the total energy needed:
   $P_{nom,peak,gen} \cdot T + E_{peak,\ stor} \geq E_{need}$;
3. it makes no sense to have a generator that can supply more power than the required peak power.

Another requirement, not in Fig. 6 is: that the energy storage should be able to supply the power peaks required ($\frac{E_{need}}{t} - P_{nom,peak,gen} \leq P_{peak,storage}$).

For the lifetime for the Li-ion HP battery it was calculated in section V that the energy storage should be oversized with a factor 42,8 to get the optimal cycle life. This optimal cycle life is given by the dashed line in (Fig. 6a).

From energy saving point of view an optimum can be found. This is the line where all regenerated energy can be stored in the energy storage. This is given with the horizontal line in (Fig. 6b).

Combining the optimal lines of (Fig. 6a) and (Fig. 6b) result in (Fig. 6c). The line for the optimal energy saving should be multiplied with the factor 42,8. The intersection of the obtained line and the optimal cycle life line gives an optimum. In the figure an illustration is given for the PMSs.

The intersection found in (Fig. 6c) of the optimal cycle life line and the optimal energy line gives a theoretical technical optimum.

VI. METHOD FOR CALCULATION OF THE COSTS

To get insight in the results of the calculation of the costs, not only the costs for the power management strategies are investigated, but a continue spectrum of generator power and storage energy is investigated. Not all generator/storage combinations are possible. If $P_{nom,peak,gen}$ is the generators nominal peak power, T is the time needed for the load profile, $E_{peak,stor}$ is the storage elements energy content, E_{need} is the total energy needed (energy that is regenerated is not taken into account), $P_{peak,storage}$ is the maximum power the storage element is able to supply and $P_{av,generator}$ is the average energy needed from the generator; the calculation is bounded by the fact that:

- The energy the generator and the storage element are able to supply together, should be larger or the same as the total energy needed:
$$P_{nom,peak,gen} \cdot T + E_{peak,stor} \geq E_{need}.$$
- The energy storage should be able to supply the power peaks required:
$$\frac{E_{need}}{t} - P_{nom,peak,gen} \leq P_{peak,storage}.$$
- The generator should be able to supply the average power needed from this device:
$$P_{av,generator} \leq P_{nom,peak,gen}.$$

The costs per cycle of the hybrid generator can be calculated with:

$$Cost_{total} = Cost_{storage} + Cost_{generator} + \\ Cost_{power\ electronics} + Cost_{fuel} \qquad (3b)$$

Where $Cost_{total}$ is the total costs per cycle, $Cost_{storage}$ the costs per cycle of the storage, $Cost_{generator}$ the costs per cycle of the generator and $Cost_{power\ electronics}$ the costs per cycle of the power electronics. In the following equations x is the number of cycles in 10 year.

The costs per cycle of the storage can be calculated with:

$$Cost_{storage} = (1+r) \cdot \frac{IE_{storage}}{x} \qquad (4)$$

Where r is the number of replacements during 10 year (ideally zero) and $IE_{storage}$ is the initial expense for the storage element:

$$IE_{storage} = Price_{storage} \times overSize_{storage} \times E_{peak,stor} \qquad (5)$$

Where $Price_{storage}$ is the price per energy amount ([J] or [kWh]) for the storage, $overSize_{storage}$ is the factor with which the storage element should be oversized (calculated with (2) for the Li-ion HP battery and (3) for the super capacitor) and $E_{peak,stor}$ is the calculated energy content of the storage.

The costs per cycle of the generator can be calculated with:

$$Cost_{generator} = \frac{IE_{generator}}{x} \qquad (6)$$

Where x is the number of cycles in 10 year and $IE_{generator}$ is the initial expense for the generator. This initial expense can be calculated with the equation in Table I.

The costs per cycle of the power electronics can be calculated with:

$$Cost_{power\ electronics} = \frac{IE_{power\ electronics}}{x} \qquad (7)$$

Where $IE_{power\ electronics}$ is the initial expense for the power electronics:

$$IE_{power\ electronics} = Price_{power\ electronics} \times P_{power\ electronics} \qquad (8)$$

Where $Price_{power\ electronics}$ is the price per amount of power for the power electronics, and $P_{power\ electronics}$ is the power that the power electronics should be able to handle. For the system with Li-ion HP there is only power electronics for the generators rectifier. In this case $P_{power\ electronics} = P_{nom,peak,gen}$. In the case of the super capacitor there are power electronics at the generator and at the storage side. In this case $P_{power\ electronics} = P_{peak,load}$. Where $P_{peak,load}$ is the peak power required by the load.

The costs per cycle for the fuel can be calculated with:

$$Cost_{fuel} = fc \cdot Price_{fuel} \cdot P_{av,generator} \qquad (9)$$

Where fc is the fuel consumption, and $Price_{fuel,av}$ is the fuel price. The fuel consumption is dependent on the use of the generator. For the Average Power method, the Max On or Off method and the Dynamic Solution it is assumed that the fuel consumption is always one (optimal.) For the other methods, based on [12], the following equation is derived:

$$fc = \frac{1}{4} + \frac{3}{4}\frac{P_{av,generator}}{P_{nom,peak,gen}} \qquad (10)$$

VII. RESULTS FOR CALCULATION OF THE COSTS

A. Costs for Li-ion HP system

The above described method is used to calculate the costs for the system with a Li-ion HP battery. The battery is used untill 60 % EOL. Data used is given in Table I.

In Fig. 7 the costs in per unit are given for the different Power Management Strategies. The costs for the Only Generator method is chosen 1 pu. The battery is oversized with the optimal factor (42,8.) It is not necessary to replace the storage. For the Only Storage

TABLE I
DATA FOR COST CALCULATION

Parameter	DATA	Source
Price fuel	1,00 [€/l]	[13]
Price electricity	0,10 [€/kWh]	[14]
Price Li-ion HP	2.000 [€/kWh]	[6]
Price super capacitor	45.000 [€/kWh]	[8]
Price generator	$59 \cdot 10^3 \cdot P_{nom,gen,peak}^{pu} + 4676$ [€]	[15]
Price power electronics	75 [€/kW]	[16]
Caloric value diesel	35.700 [kJ/l]	[17]
Fuel efficiency generator	43 %	[18]

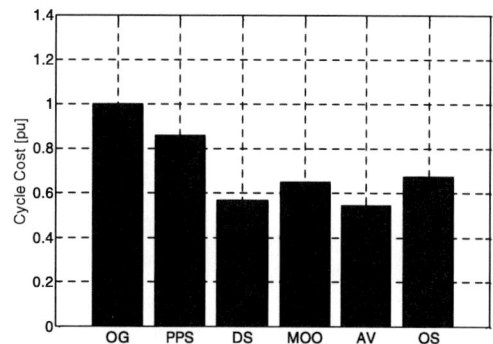

Fig. 7. Costs for the different PMSs with Li-ion (60 % EOL) as storage. OG = Only Generator, PPS = Peak Power Shaving, DS = Dynamic Solution, MOO = Max On or Off, AV = Average Power and OS = Only Storage.

method it is assumed that the power electronics is dimensioned for 1 pu. Further it is assumed for this PMS that the energy comes from the grid.

The generator is used most optimal if the relative fuel consumption is one. From the calculations it becomes clear that the Power Management Strategies using the generator most optimal (Dynamic Solution, Max On or Off, Average Power and Only Storage method), have the lowest costs. The Average Power method has the lowest costs of all.

In Fig. 8 the result for the calculation of the costs, for different generator power is given. Note that the potential energy from the generator and the storage can be larger than the required energy.

For this calculation it is assumed that the fuel consumption varies with the relative output power (equation (10).) The battery is oversized with the optimal factor (42,8 x.)

At the x-axis the nominal generator power is given (in [pu]) and at the y-axis the cycle costs (in [pu], 1 pu are the costs for an only generator system). The calculation is done for different energy contents of the storage. In the graph the average power from the load, and the minimal power from the requirements are marked. The results for the different PMSs are also marked.

When the generator becomes smaller, there is a need for more energy from the storage. This could possibly

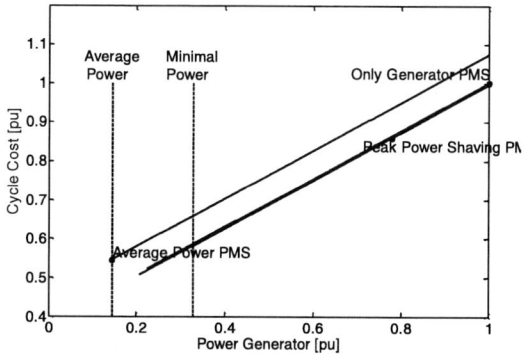

Fig. 8. Cycle costs for System with Li-ion HP at 60 % EOL after 10 years, fuel consumption generator variable. Size storages: Only generator = 0 pu; Peak Power Shaving = 42,8 x 0,021 pu; Average Power method = 42,8 x 0,145 pu.

result in the need for replacement of the storage element. During calculations it became clear that the boundaries are so strong (especially: $P_{av,generator} \leq P_{nom,peak,gen}$) that the size for the generator wherefore replacement of the storage is needed, is never reached.

It can be seen that the smaller the generator the lower the costs. Optimizing the costs with respect to the generator power is thus lowering this power.

In Fig. 8 not all Power Management Strategies are marked. This is because not all strategies have a generator that follows equation (10). For these strategies the generator has always a relative fuel consumption of one (optimal.) The results for this calculation are given in Fig. 9.

A point that becomes clear is that for small energy storages the generator should be able to supply for more than the average power.

The solid line in Fig. 10 gives the cycle costs for different sizes of the energy storage. In this graph the generator power is assumed 0,33 pu and the relative fuel consumption is one. The energy storage content is varied. The minimal energy content of the energy storage is marked. This minimal energy content of the storage is the first point where the conditions of the boundaries are met.

From Fig. 10 it becomes clear that there is an optimum for the energy content of the energy storage for a generator of 0,33 pu. This optimum is, however, for an energy content that is lower than the minimal required. But it becomes clear that the energy storage for the Max

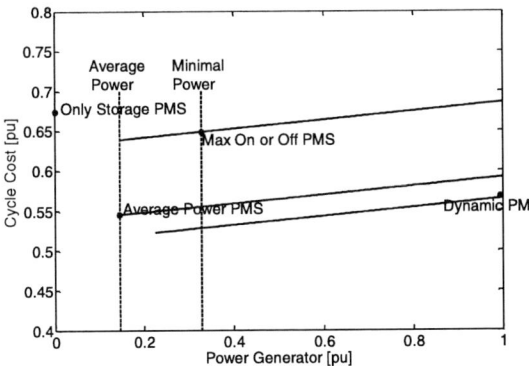

Fig. 9. Cycle costs for System with Li-ion HP at 60 % EOL after 10 years, fuel consumption one (optimal.) Size storages: Only Storage = 42,8 x 0,23 pu; Max On or Off = 42,8 x 0,21 pu; Average Power method = 42,8 x 0,145 pu; Dynamic Solution = 42,8 x 0,00092.

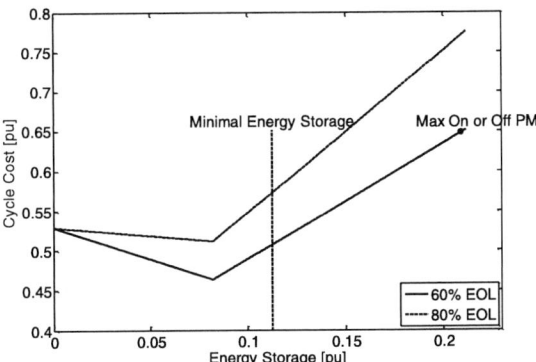

Fig. 10. Cycle costs for different content of Energy Storage for 0,33 pu Nominal Generator Power.

On or Off method can be optimized. In [3] it is shown that the the energy storage can be optimized if an energy content of 0,21 − 0,07 = 0,14 pu is taken. The costs per cycle would be 0,55 pu.

The dashed line in Fig. 10 shows the costs for a system used at until 80 % EOL. The same effect is observed as for the 60% EOL line. The costs for a Li-ion HP used until 80% EOL are higher.

B. Costs for Super Capacitor system

The above described method is used to calculate the costs for the system with a Super Capacitor. Data used is given in Table I.

In Fig. 11 the costs in per unit are given for the different Power Management Strategies. The costs for the Only Generator method is chosen 1 pu. The battery and Super Capacitor are oversized with the optimal factor (42,8 x for Li-ion HP and 2,23 x for super capacitor.) It is not necessary to replace the storage. For the Only Storage method it is assumed that the power electronics is dimensioned for 1 pu. Further it is assumed for this PMS that the energy comes from the grid.

It can be seen that the costs per cycle are higher for the super capacitor system than for the Li-ion HP system.

Varying the generator power will show again a linear dependency [3]. Further calculations will show that the super capacitor system is more expensive than the Li-ion

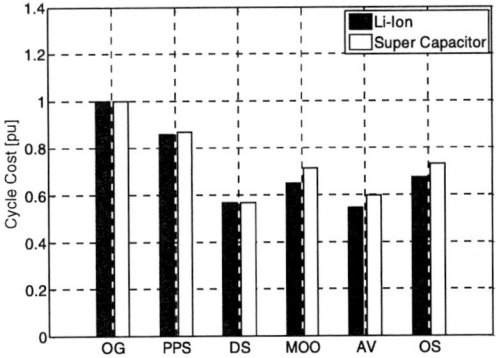

Fig. 11. Comparison for costs for the different PMSs for Li-ion system and super capacitor system. OG = Only Generator, PPS = Peak Power Shaving, DS = Dynamic Solution, MOO = Max On or Off, AV = Average Power and OS = Only Storage.

HP system although the Li-ion HP system should be oversized [3].

VIII. SENSITIVITY STUDY

In this section a sensitivity study will be carried out for the costs of a system with Li-ion HP (60% EOL) using the six defined power management strategies.

In Fig. 12 up and until Fig. 15 the costs per cycle are given for the different PMSs. In the four different graphs the (1) diesel prices (Fig. 12), (2) Li-ion prices (Fig. 13), (3) generator prices (Fig. 14) and (4) power electronic prices (Fig. 15) are varied between 0 % and 300 % of its initial value.

From these calculations it becomes clear that the fuel price is the most sensitive quantity. For the Power Management Strategies using a small energy storage (Only Generator, Peak Power Shaving and Dynamic Solution method) it is even the only element that is important from sensitivity point of view. Note that for the Only Storage method it is assumed that the energy comes from the grid.

For the systems using a larger storage (Max On or Off, Average Power and Only Storage method), the price of this storage becomes also a sensitive quantity. The larger the energy storage, the more sensitive the costs becomes for the price of the storage element.

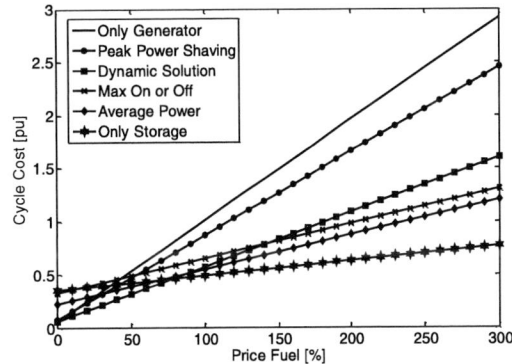

Fig. 12. Sensitivity study: Variable Fuel Prices.

2175

Based on future trends [3] the Average Power and Max On or Off methods will become more and more advantageous compared to the Dynamic Solution. After an increase of 40 % of the fuel price or a decrease of 25 % in the Li-ion HP prices, the Max On or Off method has an advantage over the Dynamic Solution.

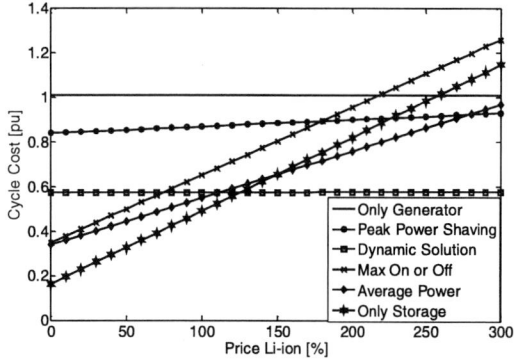

Fig. 13. Sensitivity study: Variable Li-ion Prices

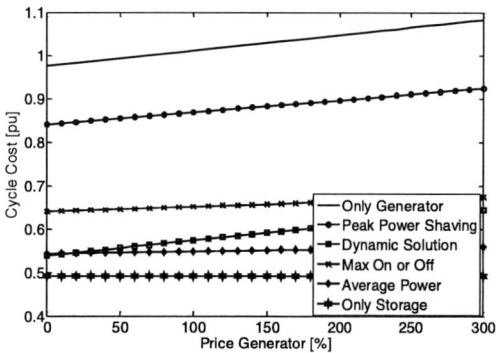

Fig. 14. Sensitivity study: Variable Generator Prices.

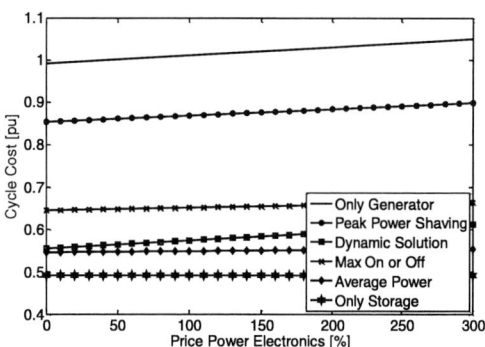

Fig. 15. Sensitivity study: Variable Power Electronics Prices.

IX. UNBALANCED CYCLES

One of the advantages of using an oversized Li-ion HP battery is the possibility to use unbalanced cycles. This means that after one cycle the battery is not necessarily

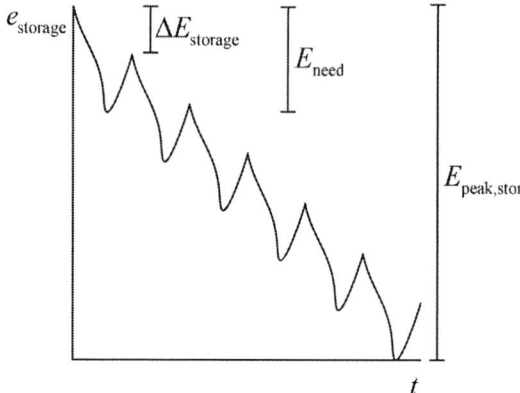

Fig. 16. Unbalanced Cycles.

charged. The storage element can be charged, for instance, after 10 cycles. This opens the possibility to use the Only Storage method and charge the battery via the grid. Electricity of the grid is much cheaper (about € 0,10 per kWh [14]) than electricity from a diesel generator (about € 0,23 per kWh [13]). In Fig. 16 the use of unbalanced cycles is illustrated.

If n is the number of unbalanced cycles, the energy content of the storage becomes:

$$E_{peak,stor} = (n-1)\Delta E_{storage} + E_{need} \qquad (11)$$

Each time the storage element should be charged, takes some time. With the current fast charging techniques (for instance Epyon uses) it is assumed that to travel to the docking station and connecting to the docking station takes much more time than the charging itself. So each charging cycle takes the same amount of time. It is assumed that a charging cycle takes three times the time needed for a hoisting cycle. The number of cycles in 10 years becomes $\frac{n}{n+3}$ times the number of cycles for the balanced cycle.

The above stated assumptions and equations (3) and (2) are used for the calculation of the cycle costs when using unbalanced cycles. In this calculation:

- x is corrected for $\frac{n}{n+3}$,
- there is no generator,
- the fuel costs are based on electricity from the grid (€ 0,10 per kWh [14]),
- it is assumed that the peak power of the power electronics is equal to 1 pu.

The result is shown in Fig. 17.

It becomes clear that the unbalanced cycles will not result in a cost effective system. The price for the larger storage element has very large effect on the system costs. Further the lower price for the fuel doesn't compensate the negative effect of less possible cycles.

X. CONCLUSION

An important observation that can be made from the calculation of the costs is that adding a storage element to

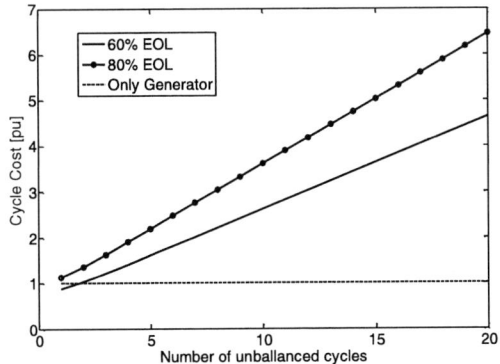

Fig. 17. Cycle cost for using unbalanced cycles.

a generator system for a typical crane system, lowers the costs.

A system based on a Li-ion HP (60% EOL) has lower costs per cycle than a system based on Super Capacitors. This battery used until 60% EOL leads to lower costs than used until 80 % EOL.

The Only Storage, Average Power, Max On or Off and Dynamic Solution has the lowest costs per cycle. The Only Storage method, however, only makes sense when used with unbalanced cycles. Using unbalanced cycles becomes too expensive.

The fuel prices and the storage prices are the most sensitive values. Based on future trends [3], the Average Power and Max On or Off method will become more and more advantageous compared to the Dynamic Solution. After an increase of 40 % of the fuel price or a decrease of 25 % in the Li-ion HP prices, the Max On or Off method has an advantage over the Dynamic Solution.

Looking at the cost issues the Max On or Off and the Average Power method are the best choice. The Average Power method, however, doesn't meet the requirement that the generator should have a minimal power of 0,33 pu.

ACKNOWLEDGMENT

The case study in the example is obtained from Epyon BV and Exendis BV, two companies working on practical energy storage systems. The authors want to thank Crijn Bouman (Epyon BV, cbouman@epyon.nl) and Evert Raaijen (Exendis BV, e.raaijen@exendis.com) for supplying this case study.

Further the authors want to thank from Epyon BV: Wouter Robers, Wouter Smit, Pelle van der Heide, Stefan Raaijmakers, Peter de Jong and Bas Molenaar for providing practical information during the research.

REFERENCES

[1] J. Leuchter, P. Bauer, O. Kurka, and V. Hajek, "Efficiency investigation of mobile power sources with VSCF technology," in *Proc. International Symposium on Power Electronics, Electrical Drives, Automation and Motion*, 2006, pp. 475–480.

[2] J. Leuchter, P. Bauer, V. Rerucha, and Z. Krupa, "Dynamic Behaviour Identification of Electrical Gen-Set," in *Proc. 12th International Power Electronics and Motion Control Conference*, Portoroz,2006, pp. 1528–1535.

[3] J.F. Baalbergen, "System Design and Power Management of a Generator-Set with Energy-Storage for a 4Q drive," M.Sc. thesis, Dept. Elect. Eng., Delft Univ. of Techn.., Delft, the Netherlands, 2007.

[4] Freek Baalbergen, P.Bauer Power Management Strategies for Generator-set with Energy Storage for 4Q-load 2008, IEEE 39th Annual Power Electronics Specialists Conference, ISBN 978-1-4244-1668-4

[5] D. Linden, T.B. Reddy, *Handbook of batteries*. 3rd ed., New York, McGraw-Hill, 2002,.

[6] A. Rufer and P. Barrade, "A supercapacitor-based energy-storage system for elevators with soft commutated interface," in *IEEE Trans. Ind. Appl.*, vol 38, no. 5, pp. 1151-1159, Sep./Oct. 2002.

[7] M. Ehsani, Y. Gao, S.E. Gay, and A. Emadi, *Modern electric, hybrid electric, and fuel cell vehicles: fundamentals, theory, and design*. 1st ed., Boca Raton, CRC Press, 2005.

[8] M. Piemontesi. (2006, May), *Alternative Energy Storage Systems for UPS*. [presentation sheets].

[9] *IEEE Recommended practice for Sizing lead-acid Batteries for Stationary Applications*, IEEE Standard 485-1997, 1997.

[10] *IEEE Recommended practice for Sizing Nickel-Cadmium Batteries for Stationary Applications*, IEEE Standard 1115-2000, 2000.

[11] Nesscap Co. (aces date: 2007, May, 23). *Nesscap Ultracapacitor Technical Guide* [Website]. Available: http://www.nesscap.com/data_nesscap/Downlad%20NessCap%20Ultracapacitor%20Technical%20Guide.pdf

[12] A. Leotta, U. Nocera, A. Raciti, "Diesel Electric Power Stations," in *Wiley encyclopedia of electrical and electronics engineering*. Vol 5, J.G. Webster, Ed. New York: Wiley, 1999, pp 368-378.

[13] Shell (aces date: 2007, June, 1). *Historisch prijzenoverzicht*. [Website]. Available: http://www.shell.com/home/Framework?siteId=nl-nl&FC2=/nl-nl/html/iwgen/shell_for_motorists/brandstofprijzen/zzz_lhn.html&FC3=/nl-nl/html/iwgen/shell_for_motorists/brandstofprijzen/historisch_prijzenoverzicht_0314.html

[14] Nuon. (aces date: 2007, September, 26). *Energietarieven elektriciteit* [Website]. Available: http://www.nuon.nl/producten-en-diensten/energietarieven/tarieven-elektriciteit.jsp

[15] Emis (aces date: 2007, June, 1). *Het Energie en Milieu Informatiesysteem voor het Vlaams Gewest*. [Website]. Available: http://www.emis.vito.be/mobiliteit/index.asp?pageChoice=Biomotor&Bc=Brandstoffen

[16] L. L. J. Mahon, *Diesel Generator Handbook*. Oxford, Butterworth-Heinemann, 1992.

Integrating renewable energy sources and storage into isolated diesel generator supplied electric power systems

Chad Abbey, Jonathan Robinson, and Géza Joós

McGill University, Department of Electrical and Computer Eng., Montreal, Canada, e-mail: *geza.joos@mcgill.ca*

Abstract— **The integration of renewable energy sources, such as wind and solar energy, into isolated power systems, supplying remote communities, offers a number of economic and environmental advantages. Energy storage systems (ESS) can be used to offset the variability of the energy supplied by renewables. This paper presents a design methodology for configuring and sizing the various components added to a diesel generator supplied electric power system. A design example is given for a remote community, including parametric studies of the economics associated with optimizing the size of battery storage, considering all the components of the system. Control options and techniques for suitable energy storage systems are presented.**

Keywords—**Renewable energy systems, Wind energy, Battery management systems, Energy storage, Energy system management.**

I. INTRODUCTION

Two important developments have occurred recently in the area of electricity generation and distribution systems. The increased deployment of renewable energy sources, mainly wind and solar energy, and more recently significant activity in the development of better and cheaper storage devices, including batteries and supercapacitors [1-3]. In electricity generation, these two developments are closely linked, since the intermittency and variability of renewable energy sources, such as wind and solar, can be made constant and dispatchable using energy storage [4,5].

Three general applications for storage that help implement cleaner energy can be considered:

1. Reducing the impact of variable generation from renewable energy sources on system stability;
2. Allowing energy generated from renewable energy sources to be dispatched to meet system demand;
3. Providing peak shaving and back-up power at the distribution substation or at a customer site.

Given the cost of electricity generated by conventional means and the complexity of implementing large scale renewable energy deployment, there are limited financial incentives at this time to include energy storage in a renewable energy system. Without these incentives, there are fewer situations where renewable energy system operators will install energy storage, unless their attempt to connect to the electricity grid has been rejected due to capacity constraints and availability of balancing power.

Off-grid or micro-grid renewable energy installations however operate under conditions where an energy storage system can provide economic benefits as well as improve the power quality of the system. Remote communities combining both diesel generator sets and renewable energy generation (also known as a hybrid power system) must generally dump a certain amount of excess energy to avoid overvoltages. This will normally occur during off-peak times when the diesel generator is operating at minimum loading and there is a high amount of renewable power being generated. Maintaining operation at a minimum threshold is required in most diesel generators to avoid excessive wear and low efficiencies. Including an energy storage system in a hybrid power system enables a larger portion of the renewable energy to be captured and the operation of the diesel generator set to be optimized for fuel efficiency, thereby reducing fuel consumption and reserve requirements [6]. This is becoming a priority due to the rising cost in fuel prices, especially in remote areas and islands not connected to the main electrical grid [7]. An example is in rural Alaska where electricity costs can exceed 40 cents per kWh [8].

Such applications offer the greatest economic potential and possibilities in the near term for energy storage, particularly in the form of batteries. As the technology evolves and improvements are made to efficiency, performance and cost, additional market opportunities will arise. The challenge at this time lies in justifying the cost of an energy storage system based on the benefits that it can bring to the utilities, operators, and consumers [9,10].

Section II of this paper presents a design methodology for configuring and sizing the various components added to a diesel generator supplied electric power system. The optimizing methodology is applied to an example remote community and economic benefits are compared for variations in system parameters. Section III reviews control techniques, power system benefits, and testing.

II. ENERGY STORAGE ECONOMICS AND SIZING

A. System Overview

A typical hybrid system will consist of a community load connected through an MV system to a diesel generator plant and a form of renewable energy generation. The system may include more than one type of renewable energy and an energy storage system.

The diesel plant often contains more than one generator so that the system is fully redundant. Unless there is a very high penetration of renewable resources then the diesel generators will normally be maintained at the minimum loading and the diesel plant will control the frequency and voltage of the system (systems with a high penetration are not considered in this paper).

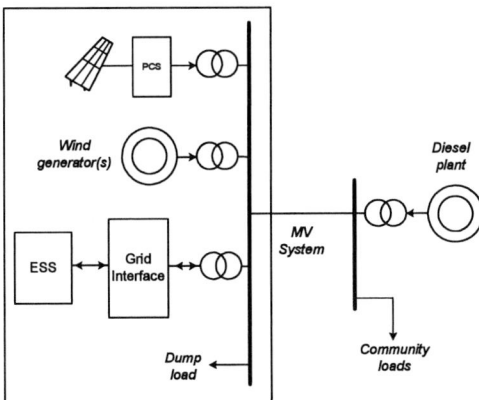

Fig. 1. Generic representation of a remote community consisting of community load, diesel plant, solar and wind energy, and dump load.

B. Energy Storage Technologies

There are various energy storage methods that vary in applicability, type, and cost. Generally storage technologies applicable to power systems are grouped depending on the time frame – Short term (capacitors, supercapacitors, flywheels), Medium Term (batteries, fuel cells, compressed air) and Long Term (pumped storage).

Most energy storage systems are connected using a power converter that controls the stored energy in the ESS and the rate flow of power into and out of the ESS.

Detailed information on the various energy storage technologies has been discussed in [4,11,1,2]

C. Energy Storage Sizing Methodology

To determine the energy storage requirements an analysis of the renewable energy resource should be carried out. The assessment will compute the return on investment of the installation based on measured data of the resource, the installation and maintenance costs, and the expected lifetime of the project.

The components of the energy storage sizing methodology outline preliminary considerations analyzing the system needs and energy storage sizing that will determine the optimal energy storage for the system.

Preliminary Considerations:

Impact Assessment – Define and quantify the impacts of variability in the renewable resource on the performance of the overall electrical system. This will determine possible problems that may arise from connecting to a weak network.

Value Streams Assessment – The system performance characteristics on which storage can add value need to be identified and benefits be quantified as precisely as possible. Areas that should be quantified include the decrease in the wasted energy, reduction in greenhouse gas emission, and mitigation of negative system impacts by increasing the system power quality and reliability. The existing storage based on rotor inertia should be considered. This analysis relates to controller design which is elaborated on in section III.

Energy Storage Sizing:

Pre-feasibility Analysis – The type of storage must be determined that will provide the most value based on the previous studies. Selection of the storage type will generally be based on the following criteria:

1. The time frame: Short term (< 10 s), Medium Term (10 – 60 min) and Long Term (1 – 24 hrs)
2. Discharging and charging rates
3. ESS Efficiency

Further information on energy storage selection has been summarized in [4,12].

Detailed Analysis – The size of storage needed should be calculated based on site data. This should take into account the various energy sources.

The sizing study will determine the optimal size of the ESS in order to minimize the costs. The cost to be minimized for a wind diesel hybrid system is based on the following parameters:

r_{wl}	Ratio of wind energy to load energy
r_{xy}	Hourly correlation of the load and wind power
π_e	Cost of energy supplied from the diesel plant (\$/kWh)
η	Efficiency of the ESS
P_{min}	Minimum load of the diesel generator (pu)
π_w	Cost of energy supplied by the wind (\$/kWh)
IC_p	Incremental cost of storage power rating (\$/kW)
IC_e	Incremental cost of storage energy rating (\$/kWh)

The incremental costs associated with the investment can be calculated using a basic calculation of annuity:

$$IC = \frac{IC_{fixed\ cost}}{365 \cdot N_a}(1+r)^{N_a} \qquad (1)$$

In this calculation the fixed cost of the energy storage is converted to the cost per kWh per day. The interest rate r and the lifetime of the project N_a will vary depending on the project.

A block diagram of the inputs into the optimization algorithm is shown in Fig. 2 below:

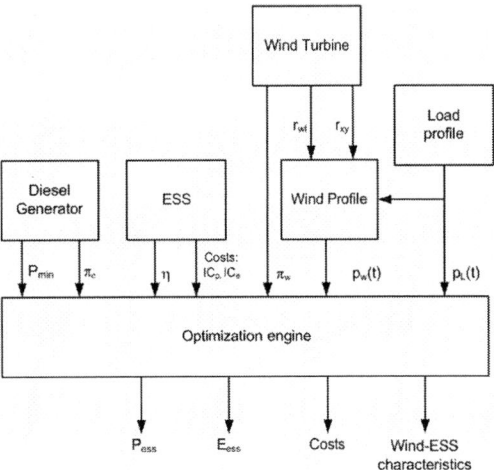

Fig. 2. Energy storage sizing methodology for integration in a wind diesel hybrid system

The ratio of wind energy to load energy, the correlation between the load and wind power, and the costs of energy for the diesel generator (including generator maintenance and diesel cost, transport, and storage) will need to be calculated or obtained from measured data at the installation.

The final step in the design of the overall system will determine the most suitable operating algorithm that will control the use of the distributed resources on the system: generation, storage, load, and the dump load. The control algorithm will use the results of the sizing study to determine the modes and time frames of operation of the different resources.

D. Optimal Sizing – Principles

This section shows the optimization algorithm required to implement the last step of the sizing methodology shown in Fig. 2 for a wind diesel hybrid system with batteries used as the energy storage system. The overall cost minimizing algorithm will be:

$$Min \left\{ Energy\ Storage\ Costs + Generation\ Costs \right\} \quad (2)$$

Stated formally, the minimization problem is:

$$\underset{\underline{x}}{Min} \left\{ IC_e E_{ESS} + IC_p P_{ESS} + \sum_{t=1}^{T} \pi_e P_{diesel,t} + \pi_w P_{w,t} \right\} \quad (3)$$

Where the variable x is a vector given by:

$$\underline{x} = [\ \underline{p}_{diesel}\ \ \underline{p}_{dump}\ \ \underline{p}_{charge}\ \ \underline{p}_{discharge}\ \ P_{ESS}\ \ E_{ESS}\ \ E_o\ ] \quad (4)$$

Where p_{charge} and $p_{discharge}$ are the ESS charging and discharging powers in the time interval and p_{dump} is the amount of power wasted. P_{ESS} and E_{ESS} are the power and energy rating of the ESS and E_o is the ESS inertial energy. Wasted energy through losses or the dump load will be reflected in increased diesel production and is not explicitly stated in the minimization function.

The optimization problem is subject to the following constraints:

Power Balance:

$$p_{diesel,t} + p_{w,t} + p_{dis,t} = p_{L,t} + p_{dump,t} + p_{ch,t} \quad (5)$$

Diesel Constraints

$$p_{diesel,t} \geq P_{min} \quad \forall t \in T \quad (6)$$

Dump Load Constraints

$$p_{dump,t} \geq 0 \quad \forall t \in T \quad (7)$$

ESS Power Constraints

$$p_{ch,t},\ p_{dis,t} \geq 0 \quad \forall t \in T \quad (8)$$

ESS Energy Rating Limits

$$0 \leq e_{ESS,t} \leq E_{ESS} \quad \forall t \in T \quad (9)$$

Energy Transition

$$e_{ESS,t} = \sum_{k=1}^{t} \left(E_0 + \eta \cdot p_{ch,k} - \frac{1}{\eta} p_{dis,k} \right) \quad \forall t \in T \quad (10)$$

And a constraint that requires that the energy at the end of the profile must be equivalent to the initial energy

$$e_o = E_T \quad (11)$$

E. Optimal Sizing - Application to a Remote Community

To demonstrate the effect of different parameters on the ESS rating the case of a remote community with increasing wind energy penetration levels ($r_{wl} = E_{wind}/E_{load}$) is considered.

The base case is simulated with the following parameters (defined in Section II part B.):

r_{xy}	-1
π_e	0.30 \$/kWh
η	0.85
P_{min}	0.3 pu
π_w	0.25 \$/kWh
IC_p	213 \$/kW
IC_e	875 \$/kWh

These represent typical ESS parameters taken from [13] and the load profile from [14].

To understand the possible economic benefits, three situations are compared with the base case:

1. *Ignoring the fixed cost of the ESS ($IC_{ESS} = 0$)* – This is equivalent to minimizing energy generation and will provide a limit case to identify conditions under which using an ESS can add economic value. It can also be used to show the benefits that can be achieved as storage costs decrease (due to lower production costs or new technology).

2. *Equal cost of diesel and wind generation ($\pi_e = \pi_w$)* – This shows how the cost of a system with an ESS compares as the cost of varies compared to diesel generation.

3. *No ESS is installed* – This will allow comparison to determine what economic advantages there are for an installation at current ESS prices.

Fig. 3 shows the resulting cost per day depending on the penetration of wind energy. It can be seen that when the cost of the ESS is not considered, economic benefits start being realized at wind penetration of about 30% and reaches the maximum at around 65% before decreasing as the wind penetration is closer to 100%.

At current ESS prices where $IC_{ESS} = IC_{base}$ there is a small economic benefit for installing an ESS at wind penetrations of 60 - 80%. The case with equal cost of wind and diesel shows that there are only economic benefits when the cost of wind generation is cheaper than diesel. In this example the cost of diesel was \$0.30/kWh however in remote areas the cost can be greater and is increasing steadily with rising fuel prices.

It should also be noted that these simulations were performed with a correlation of negative 1 between the wind and the energy usage. Simulations show that the economic benefits will decrease as the correlation increases to 1 (this result is intuitive since ESS will be most useful when renewables are generating when the load is at a minimum and therefore more energy must be dumped).

2180

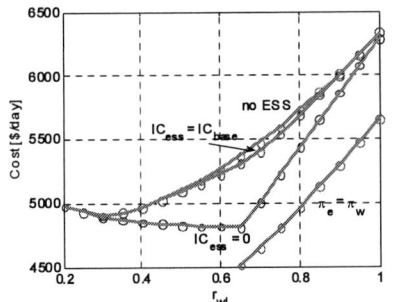

Fig. 3 . Energy Storage System sizing study results: Cost of daily energy versus ratio of wind to load energy for no ESS, $IC_{ess} = 0$, $\pi_e = \pi_w$, and for the base case

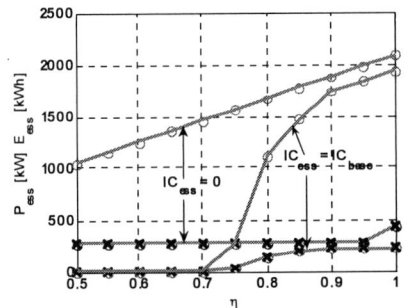

Fig. 4 Power ('x') and Energy ('o') ratings as a function of efficiency. Cases for fixed costs of storage included ($IC_{ess} = IC_{base}$) and neglected ($IC_{ess} = 0$).

Fig. 4 shows how the power and energy rating of the ESS depending on the efficiency of the ESS.

As can be seen from Fig. 4, technologies with one way conversion efficiencies of less than 0.7 are not attractive. As well, benefits of the ESS increase sharply from efficiencies of 0.7 to 0.9. This is a critical parameter to consider in an ESS sizing study.

To examine the economic benefits of an ESS in the future the cost put kWh for the diesel generator was increased to $ 0.50 per kWh while the cost for wind generation and the ESS were held constant. The resulting graph is shown in Fig. 5.

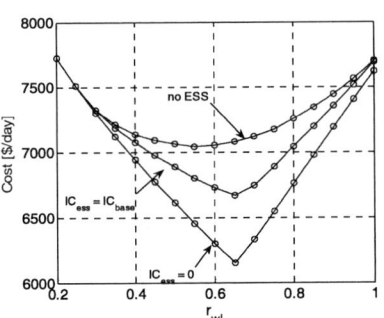

Fig. 5. Energy Storage System with a 70% increase in fuel costs (from the base case)

It is clear from Fig. 5 that as the costs of operating the diesel generator increase, the benefits for using energy storage system become more significant.

It should be noted that the study distinguishes between energy and power sizing, each with their associated incremental costs. In a wind turbine generator energy handling capacity is more critical in determining ESS sizing. This would favor technologies focusing on energy storage capacity rather than power capacity.

Overall these results demonstrate that there are potential economic benefits when installing an ESS and applicable systems will become more common in the future for any of the following reasons:

1. Rising cost of diesel fuel
2. Decrease in cost of wind generation (or other renewable generation technologies)
3. Decrease in cost and increase in efficiency of energy storage systems

III. ENERGY SYSTEM STORAGE OPERATION

A. Multi-level Storage Structure

ESS Control strategies can be designed to realize both the economic benefits of energy storage and to limit the influence of variations in power generated from the renewable energy source. In order to maximize the economic benefits, the results of the sizing study described in Section II can be used to determine the modes and time frames of operation of the different distributed resources.

Fig. 6 presents a two-level storage structure of an energy storage system controller for a system using both batteries and supercapacitors. The proposed structure can be modified to be applicable to more storage levels, in either isolated or grid-connected distribution systems.

The controller architecture presented shows two parts to the storage scheduling block. The first part involves extracting reference powers at various frequencies. The reference powers will be the frequency component of the difference between the actual turbine power and the predicted power (obtained using low pass filters).

The second part implements knowledge based techniques that will be used to determine the controller actions for various' states of the system and of the renewable resource. The controller must be designed to accommodate for inaccuracies in the prediction algorithm.

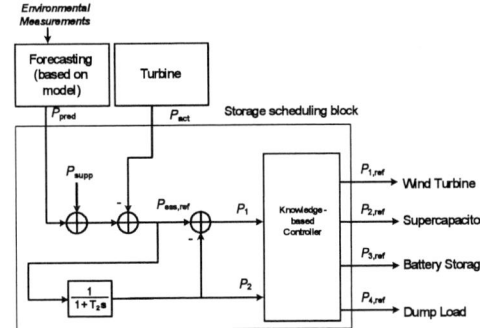

Fig. 6. Two-level storage control algorithm.

Fuzzy rule bases have been used in a number of studies of wind diesel hybrid systems and offer advantages such as the ability to adapt to disturbances and dealing with non-linearity and uncertainty [15-17]. They can also combine existing knowledge with the ability to learn using off-line simulations. This technique has also been used in a hydraulic turbine controller to dampen oscillations and maintain operation parameters in set thresholds [18].

For this application the fuzzy rule base should be balanced to reduce output fluctuations and control the energy level and regulate flow of power into and out of the storage device.

This type of structure allows allocation of the power reference between the different energy sources, based on the time frames associated with each storage device. The example in Fig. 6 shows a controller for a system with a combination of supercapacitors and batteries, but other possible combinations include flywheel and batteries and superconducting magnetic energy storage (SMES) and batteries. Furthermore, the expert system based approach (knowledge-based controller) enables contingency handling (how the reference is modified when the energy limits of one or more levels are reached). Other approaches to ESS scheduling include dynamic programming or robust control.

Fig. 7 shows the effect of a two-level energy storage system on the power fed into the electric grid by a wind plant.

B. Storage Control Strategies

Control strategies for systems with energy storage can be optimized to offset for fluctuations caused by the renewable resource generators. This may be especially important in remote communities where integrating renewable energy into a weak isolated system can lead to power and frequency fluctuations due to the unpredictability of the renewable resource. Examples of using short term energy storage design methods to minimize the effects of the variability of wind turbine generation include:

- Allowing the wind turbine speed to vary from the point of maximum power production and using the inertia of the rotor as an ESS to reduce power fluctuations [19].
- Power smoothing using a flywheel controlled by an induction machine [20].
- Supercapacitor storage in a DFIG to reinforce the DC bus during transients and to absorb energy from the generator to meet low voltage ride through requirements [21].

The control and ESS required will vary depending on the rate of change of power of the renewable energy resource. Studies have been done for wind [22], wave [23], tidal [24], and photovoltaic [25].

Other areas that should be considered when designing the ESS control system include islanding/non-islanding operation modes [26], the influence of aggregated versus distributed connection topologies [11], and low voltage ride through requirements [27].

C. Performance Evaluation - Real-Time Simulation

A powerful tool for testing and validating control algorithms is real-time simulation. The real-time simulator

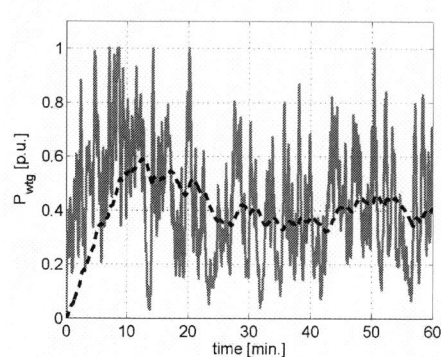

Fig. 7. Operation of a two level energy storage system for a wind turbine generator - results obtained using an RTDS. Wind power and dispatched power (dashed lines).

used in this study is a Canadian design using a PC cluster of up to 16 processors running MATLAB-Simulink and SimPowerSystems (simulation results are shown in Fig. 7). In addition to the benefits of producing simulation results faster than with offline simulations as a preliminary step to the implementation of complete systems, it allows testing of any number of individual hardware components. Examples include: batteries, controllers, and power electronic interfaces, the electric power system being represented in the real-time simulator. This would allow, at reduced cost, continuous improvement of technologies without having to wait for results from field trials.

IV. CONCLUSIONS

This paper has presented some of the potential benefits of energy storage. It has provided a systematic approach to determining the economics of energy storage. Economic studies show that energy storage can become economical in specific circumstances where the cost of electricity generated in the conventional manner is high enough to offset the capital and associated running costs. Examples include situations of high spot electricity prices in liberalized markets and the high costs associated with diesel generation in remote locations. In addition, if environmental attributes can be quantified, the benefits can be substantially increased. A control system configuration was proposed for a two-level energy storage system and various control strategies were discussed.

ACKNOWLEDGMENT

The authors acknowledge the contributions to various aspects of this work, both theoretical, simulation and real time implementations of the following graduate students: Wei Li, Carlos Martinez, John Chahwan, and Mohamed El-Chahaly, members of the McGill Energy Systems Laboratory.

REFERENCES

[1] A. Kusko and J. DeDad, "Short-term, long-term, energy storage methods for standby electric power systems," *Industry Applications Conference, 2005. Fortieth IAS Annual Meeting. Conference Record*, 2005, pp. 2672-2678 Vol. 4.

[2] R. Schainker, "Executive overview: energy storage options for a sustainable energy future," *Power Engineering Society General Meeting, 2004. IEEE*, 2004, pp. 2309-2314 Vol.2.

[3] Tongzhen Wei, Sibo Wang, and Zhiping Qi, "Design of supercapacitor based ride through system for wind turbine pitch systems," *Electrical Machines and Systems, 2007. ICEMS. International Conference on*, 2007, pp. 294-297.

[4] J. Barton and D. Infield, "Energy storage and its use with intermittent renewable energy," *Energy Conversion, IEEE Transaction on*, vol. 19, 2004, pp. 441-448.

[5] P. Lautier et al., "Off-Grid Diesel Power Plant Efficiency Optimization and Integration of Renewable Energy Sources," *IEEE Canada Electrical Power Conference*, Montreal: 2007.

[6] F. Katiraei and C. Abbey, "Diesel Plant Sizing and Performance Analysis of a Remote Wind-Diesel Microgrid," *Power Engineering Society General Meeting, 2007. IEEE*, 2007, pp. 1-8.

[7] J. Kaldellis and K. Kavadias, "Cost-benefit analysis of remote hybrid wind-diesel power stations: Case study Aegean Sea islands," *Energy Policy*, vol. 35, Mar. 2007, pp. 1525-1538.

[8] M. Devine et al., "Wind-Diesel Hybrid Options for Remote Villages in Alaska," *Proceedings of the AWEA Annual Conference*, Chicago: 2004.

[9] C. Abbey and G. Joos, "Coordination of Distributed Storage with Wind Energy in a Rural Distribution System," *Industry Applications Conference*, 2007, pp. 1087-1092.

[10] Korpaas M., Holen A.T., and Hildrum R., "Operation and sizing of energy storage for wind power plants in a market system," *International Journal of Electrical Power and Energy Systems*, vol. 25, Oct. 2003, pp. 599-606.

[11] Wei Li and G. Joos, "Comparison of Energy Storage System Technologies and Configurations in a Wind Farm," *Power Electronics Specialists Conference, 2007. PESC 2007. IEEE*, 2007, pp. 1280-1285.

[12] Chad Abbey and Geza Joos, "Energy Storage and Management in Wind Turbine Generator Systems," *Power Electronics and Motion Control Conference, 2006. EPE-PEMC 2006. 12th International*, 2006, pp. 2051-2056.

[13] L. Mears and H. Gotschall, "EPRI-DOE Handbook of Energy Storage for Transmission and Distribution Applications," 2003.

[14] C. Grigg et al., "The IEEE Reliability Test System-1996. A report prepared by the Reliability Test System Task Force of the Application of Probability Methods Subcommittee," *Power Systems, IEEE Transactions on*, vol. 14, 1999, pp. 1010-1020.

[15] R. Chedid, S. Karaki, and C. El-Chamali, "Adaptive fuzzy control for wind-diesel weak power systems," *Energy Conversion, IEEE Transaction on*, vol. 15, 2000, pp. 71-78.

[16] A. Marques, G. Taranto, and D. Falcao, "A knowledge-based system for supervision and control of regional voltage profile and security," *Power Systems, IEEE Transactions on*, vol. 20, 2005, pp. 400-407.

[17] L. Leclercq, B. Robyns, and J. Grave, "Control based on fuzzy logic of a flywheel energy storage system associated with wind and diesel generators," *Math. Comput. Simul.*, vol. 63, 2003, pp. 271-280.

[18] J. Jiang and R. Doraiswami, "Design of a real-time knowledge-based controller with applications in hydraulic turbine generator systems," *Journal of Intelligent and Robotic Systems*, vol. 2, Jun. 1989, pp. 229-244.

[19] L. Ran, J. Bumby, and P. Tavner, "Use of turbine inertia for power smoothing of wind turbines with a DFIG," *Harmonics and Quality of Power, 2004. 11th International Conference on*, 2004, pp. 106-111.

[20] R. Cardenas et al., "Control strategies for enhanced power smoothing in wind energy systems using a flywheel driven by a vector-controlled induction machine ," *Industrial Electronics, IEEE Transactions on*, vol. 48, 2001, pp. 625-635.

[21] C. Abbey and G. Joos, "Supercapacitor Energy Storage for Wind Energy Applications," *Industry Applications, IEEE Transactions on*, vol. 43, 2007, pp. 769-776.

[22] Wei Li, G. Joos, and C. Abbey, "Wind Power Impact on System Frequency Deviation and an ESS based Power Filtering Algorithm Solution," *Power Systems Conference and Exposition, 2006. PSCE '06. 2006 IEEE PES*, 2006, pp. 2077-2084.

[23] S. Muthukumar et al., "On Minimizing the Fluctuations in the Power Generated from a Wave Energy Plant," *Electric Machines and Drives, 2005 IEEE International Conference on*, 2005, pp. 178-185.

[24] J. Clarke et al., "Regulating the output characteristics of tidal current power stations to facilitate better base load matching over the lunar cycle," *Renewable Energy*, vol. 31, Feb. 2006, pp. 173-180.

[25] A. Woyte et al., "Quantifying the occurrence and duration of power fluctuations introduced by photovoltaic systems," *Power Tech Conference Proceedings, 2003 IEEE Bologna*, 2003, p. 7 pp. Vol.3.

[26] C. Abbey et al., "Impact and Control of Energy Storage Systems in Wind Power Generation," *Power Conversion Conference - Nagoya, 2007. PCC '07*, 2007, pp. 1201-1206.

[27] J. Morneau, C. Abbey, and G. Joos, "Effect of Low Voltage Ride Through Technologies on Wind Farm," *Electrical Power Conference, 2007. EPC 2007. IEEE Canada*, 2007, pp. 56-61.

Performance comparison of different wind generator based hybrid systems

Vincent Courtecuisse*(1), Benoit Robyns*(1), Marc Petit(2), Bruno Francois*(3)
and Jacques Deuse(4)

* Laboratoire d'Electrotechnique et d'Electronique de Puissance de Lille (L2EP)
(1) Ecole des Hautes Etudes d'Ingénieurs (HEI), Lille, France email :
benoit.robyns@hei.fr ;vincent.courtecuisse@hei.fr
(2) Supélec, Plateau de Moulon, Gif-sur-Yvette, France, e-mail : *marc.petit@supelec.fr*
(3) Ecole Centrale de Lille, Villeneuve d'Ascq Cedex, France, e-mail : *bruno.francois@ec-lille.fr*
(4) Suez-Tractebel, Bruxelles, Belgique, e-mail : *jacques.deuse@tractebel.be*

Abstract—In this paper a graphical modelling tool is proposed to develop a fuzzy logic based supervisor. This tool facilitates the analysis and the determination of fuzzy control algorithms adapted to complex hybrid systems. To explain this methodology, the association of wind generators, dispersed generator and storage systems are considered. The performance of this supervisor is shown with the help of simulations. A comparison of different topologies is also proposed.

Keywords—**Energy system management, Hybrid power integration, Power management, Renewable energy systems.**

I. INTRODUCTION

Hybrid renewable energy systems (HRES) are increasingly used to improve the grid integration of wind generators. Many publications have been written in this topic [1], but the majorities (almost 90%) of these papers focus on design and economic aspects. However, relatively small changes in the control strategy can significantly affect the performance of the hybrid unit.

Two methods are investigated in the literature to supervise hybrid production units:

- Power flow analysis [2],[3]; the drawback of this method is that it does not take part in the energy management. It is critical when a storage system is associated with the hybrid production.
- Fuzzy logic supervisor [4],[5],[6],[7]; this method compared with the classical method allows to consider the level of storage units. It is well adapted to represent uncertainties by fuzzification of ambiguous variable and assigning membership functions based on preferences and/or experience. Currently, no dedicated modelling tools is available.

In this paper, a graphical modelling tool is proposed in order to develop a fuzzy logic based supervisor. This modelling tool facilitates :

- the analysis of the control problem,
- the design of control algorithms,
- the experimental implementation.

To explain the methodology, association of a wind turbine, a foreseeable power source and storage systems are considered. The goal of this supervision is to track a reference power and to participate in the primary frequency control.

In this paper a comparison of different topology of Hybrid renewable energy system is proposed. A fuzzy logic based supervisor is proposed to manage the different sources. A Chart methodology developed in [8] is used to determine the control algorithm. Two objectives are considered:

- the HRES must provide the reference power imposed by the network manager with maximising the wind power,
- the HRES must participate to the primary frequency control.

II. MODELING OF THE SYSTEM UNDER STUDY

In order to generate a reference power and to ensure a power reserve, the generating system considered in this paper includes wind generators, short and long-term storage systems and foreseeable sources (like for instance gas turbines). This hybrid generating system is connected at one point of the network (Fig.1) and must be seen from the network manager as a classical source.

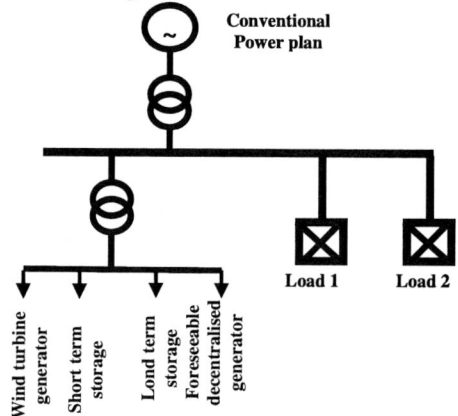

Fig. 1 System under study

978-1-4244-1741-4/08/$25.00 ©2008 IEEE   2184

A. Wind turbine generator (WTG)

The considered variable speed wind generator is based on a permanent magnet synchronous generator connected to the network via two back to back AC-DC converters. The turbine is modeled by the classical relation between wind speed and power which can be extracted:

$$P_w = \frac{1}{2} \rho \, A_r \, C_P(\lambda, \beta) \, v^3 \qquad (1)$$

where: ρ is the air density, A_r is the surface swept by the rotor, $C_p(\lambda,\beta)$ is the power coefficient, $\lambda = \dfrac{\Omega_t \, R_t}{v}$ is the speed ratio, Ω_t is the turbine speed, R_t is the turbine radius, β is the pitch angle and v is the wind speed.

The power coefficient $C_p(\lambda,\beta)$ is modeled considering an analytical expression.

B. Foreseeable decentralized generator (FDG)

As show in Fig.2, the considered FDG is modeled by a simple first order transfer function [1].

$$H(s) = \frac{1}{\tau_{DG} \, s + 1} \qquad (2)$$

P_{FDG_ref} is the refence power of the FDG, P_{FDG_max} is the maximum power of the FDG, P_{FDG_min} is the minimun power of the FDG, P_{FDG} is the Power of the FDG.

The value of the time constant τ_{DG} depends on the considered technology.

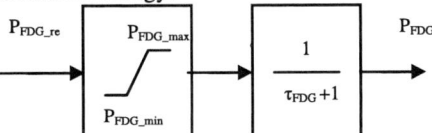

Fig. 2 FDG model

C. Storage system

In the same way, a simplified model (Fig.3) is used to modelise the short and long term storage units.

Fig.3 shows the storage system model, with $P_{refstock}$ the reference power of the storage system, W the energy stored and P_{stock} the power of the storage system. The storage system does not have a technology defined a priori, it is characterized by: P_{chmax} which is the maximum power of charge, P_{dchmax} is the maximum power of discharge, W_{max} is the maximum energy stored and η is the efficiency (= discharged energy/charged energy).

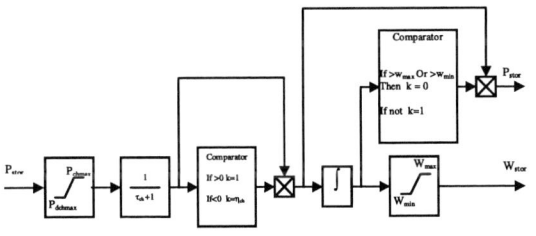

Fig. 3 Storage unit model

III. SUPERVISOR STRATEGY

To develop the supervision, five steps are considered:
(1) Determination of system work specifications: we must define objectives of the system.
(2) Supervisor design, in this steps all the inputs and outputs must be determined.
(3) Chart representation of operating modes
(4) Determination of the membership functions.
(5) Determination of the fuzzy rule.

A. Determination of system work specification

In this application, two objectives are considered; the HRES must **provide the reference power** imposed by the network manager and **maximise the wind power**. Furthermore, in order to participate in the primary frequency control [10] a primary **energy reserve must be available**.

B. Supervisor design

The supervisor of the system under study (Fig. 4) must provide the reference power (P_{ref}), which is specified by the network manager and must maintain a primary power reserve.

In Fig. 4, P_{hyb} is the power generated by the hybrid generator (which is the sum of the power generated by storage units, WTG and FDG), P_{WTG} is the power generated by the WTG, Lev_{stor_sht} is the short term storage power energy level, Lev_{stor_lgt} is the long term storage power energy level, f_{meas} is the measured frequency, f_{ref} is the reference frequency, P_{ref_FDG} is the reference power of DG, β_{ref} is the reference of pitch angle, $P_{ref\_stor\_sht}$ is the reference power of short term storage, $P_{ref\_stor\_lgt}$ is the reference power of long term storage. Due to the complexity of the system, the random behaviour of the wind speed and of the network load (which affect the frequency), the fuzzy logic is well adapted to design the supervisor [5] and [9].

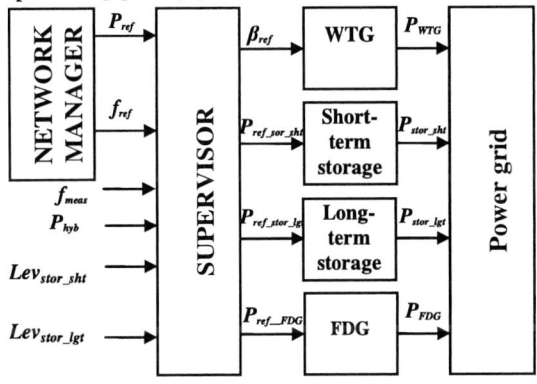

Fig. 4 System under study

The hybrid system must have a behavior similar to a traditional source with respect to the network. The storage size and the power of the FDG are fixed a priori;

their sizing will not be studied in this paper. It is considered that the primary source of the FDG is a fossil energy one, thus its use will be minimized.

The block diagram of the supervisor is shown in Fig. 5. The supervisor can be divided into two parts:
- A fuzzy logic supervisor, which manages the FDG (P_{ref_FDG}), the WTG (β_{ref}) and the storage system ($P_{ref\_stor\_sht1}, P_{ref\_stor\_lgt1}$). The input variables of this system are the powers of the hybrid generator, the reference power ($\Delta P = P_{hyb} - P_{ref}$), the level of the storage (Lev_{stor_sht}, Lev_{stor_lgt}) system, the measured and the reference frequency ($\Delta f = f_{meas} - f_{ref}$) .
- A droop characteristic, which allows to participate in primary frequency control with the short and long term storage systems ($P_{ref\_stor\_sht2}, P_{ref\_stor\_lgt2}$).

The reference power of the storage system is the sum of two terms:
$$P_{ref\_stor\_sht} = P_{ref\_stor\_sht1} + P_{ref\_stor\_sht2} ,$$
$P_{ref\_stor\_sht1}$ allows to compensate the high wind power frequency variations and $P_{ref\_stor\_sht2}$ allows to participate in the primary frequency control.

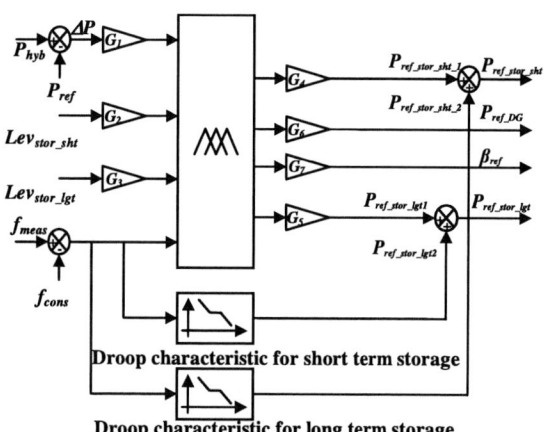

Fig. 5 Block diagram of the supervisor

The main goal of this supervisor is to follow the power reference and to guarantee an energy reserve to participate to the primary frequency control.

IV. DETERMINATION OF OPERATIONAL GRAPH

Fig. 6 shows graphically the various operating modes (represented in the rectangle with round corners) and the states of the system (represented by transitions).
In this paper, the supervision system is divided in two main parts (N1 and N2). The objective of the first part (N1) is to control the power generated by the hybrid system. To achieve this goal, we consider three possible operating modes (Fig. 6):

- N1.1: the hybrid renewable energy system (HRES) must supply the reference power with maximizes the WTG power,
- N1.2: if the storage level is high, the HRES must keep a storage capacity,
- N1.3: if the storage level is low: the HRES must keep a storage availability.

The goal of the second part (N2) is to take part in the primary frequency control.

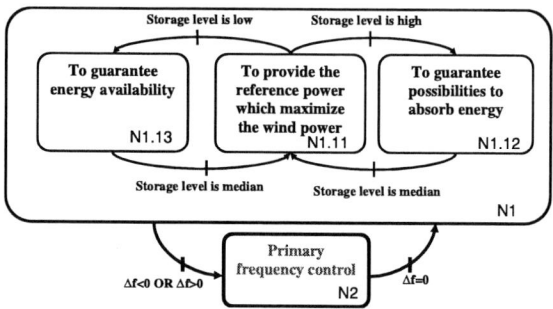

Fig. 6 Chart representation of the system

N1.11, N1.12 and N1.13 are the objectives of the supervisor; the transitions (storage level and frequency variation) represent the constraints. Finally the perturbation is represented by N2 operating mode.

Each operating modes (N1.1, N1.2, N1.3 and N2) are associated with a set of fuzzy rules. The structure is managed by the fuzzy logic, thus it is possible to be in several operating modes at the same time. In fact when several conditions are true, several fuzzy rules impact the same output. The final value of this output is the centre of gravity of a function determined by the fuzzy logic. This approach allows to find a compromise between different operating modes.

In Fig. 7 the operation of the different subsystems is detailed. The N1 operating mode is divided in two part N1.1 which allows to determine the reference power in function of the short term storage level and N1.2 which allow to determine the reference power in function of the long term storage.

N1.11 operating mode allows to control the reference power, if the short term storage level is medium then the WTG operates in maximal extraction power (MPPT) and the storage unit compensates the difference between the reference power and the WTG power.
N1.12 operation mode operates when the short term storage level is high. In this case the storage unit must be discharged and the WTG power is decreased with the help of the pitch angle. When the WTG does not operate in MPPT mode it is possible to increase this power to take part in primary frequency control. N1.13 operating mode is activated when the short term storage level is low, then the storage units must be charged and the FDG must control the reference power. When the FDG is in operation, it can take part in primary frequency control.

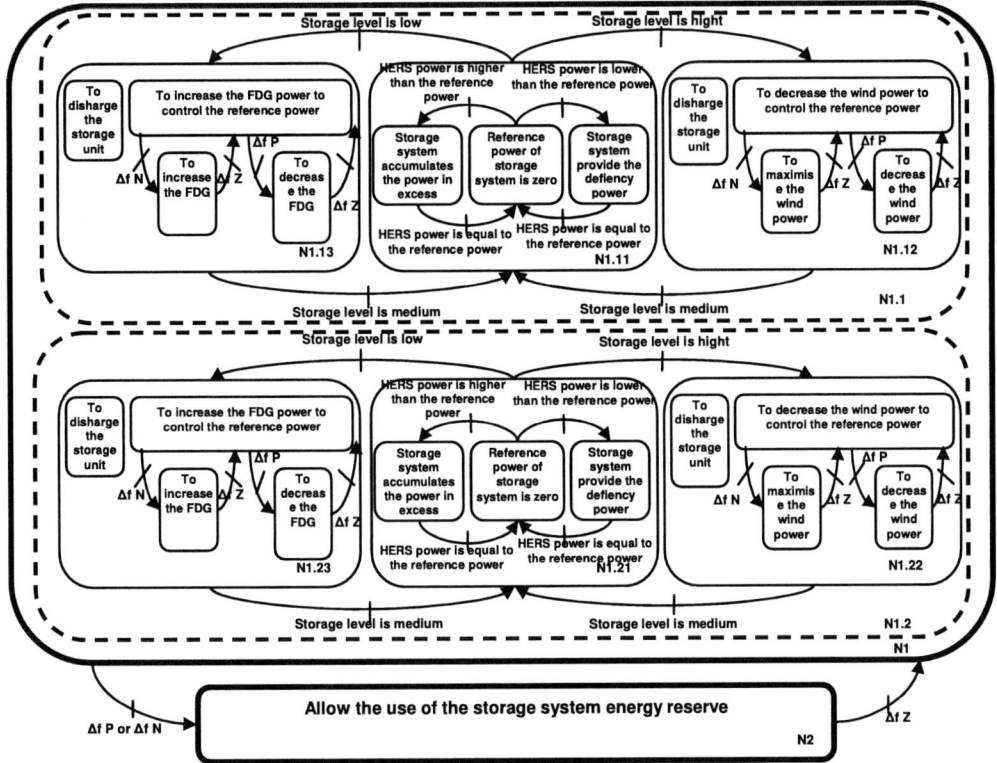

Fig. 7 Chart representation of different operating mode

A. Determination of membership functions

The membership function must divide in two parts: the membership function of the input of the system show Fig. 8, and the membership function of the output of the system (Fig. 9).

The input membership function is use as a transition between different operating mode.

Fig. 8.b and Fig. 8.d show the membership function of the storage level, three levels are considered:

- Small and high level, which represent the energy reserve required to participate in primary frequency control; in this case the short term storage level and long term storage level have a minimum of 0.05 pu in primary energy reserve.
- Medium level : in this case storage the system is used to compensate the wind power variation around the reference power (0.6 pu are dedicated to this action).

Fig. 8.c, represent the membership functions of frequency variation $\Delta f = f - f_0$. In the same way three areas are defined. When the frequency variation is negative or positive, the primary frequency control must be in operation, the trapezoidal form of the membership function zero allows the introduction of a dead band. In this case between -0.1 and 0.1 Hz the primary frequency control is not in operation.

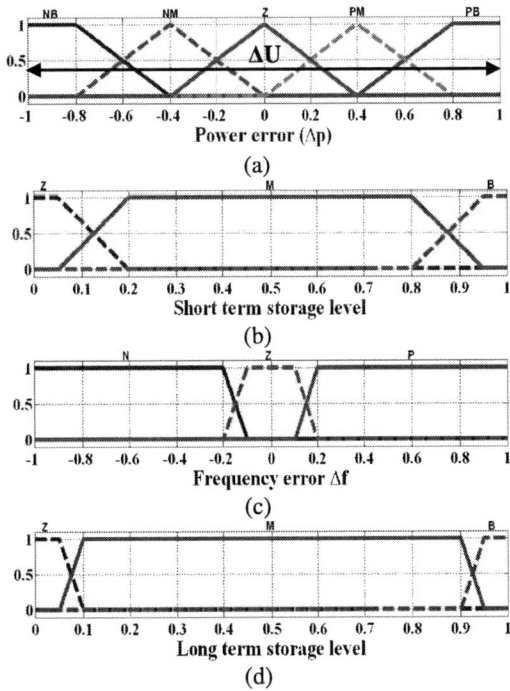

Fig. 8 Input membership function of : (a) Power error; (b) Short term storage level, (c) Frequency error and (d) Long term storage level

Fig. 9.a, Fig. 9.b, Fig. 9.c and Fig. 9.d show the output membership functions of respectively the pitch angle, the short term storage reference power, the FDG reference power and the long term storage reference power. Finally, to supervise the HRES we consider four membership functions to describe the input variables with a total of 15 membership degrees. Therefore 135 combinations are possible. Traditionally, to determine the fuzzy rules, all the cases are identified with the help of tables [11]. In this case, the table must contain 540 cases. A graphic representation enables to grasp the global performance of the system. It is also possible to decompose the system in many subsystems. This decomposition allows to determine more easily the fuzzy rules.

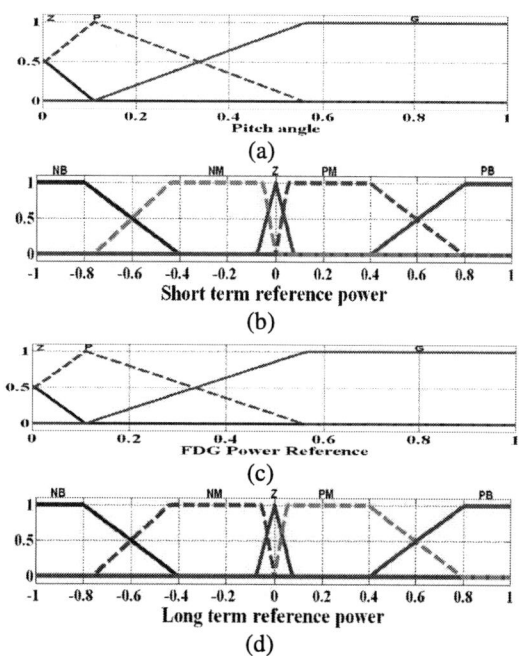

Fig. 9 Output membership function of (a) Pitch angle, (b) Short term reference power, (c) FDG reference power and (d) Long term reference power

B. Determination of operational graph

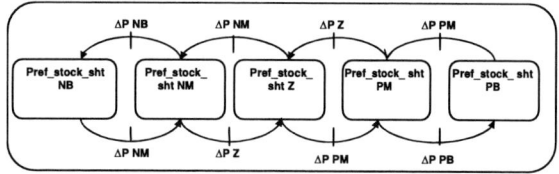

Fig. 10 Operational graph of N1.11 operating mode

To determine the fuzzy rules, each operating mode must be formulated with membership functions. To achieve this goal the functional graph must be translate with the variable of the membership function. For example, the fuzzy rules of N1.11 operating mode are deduced from the diagram shown in Fig. 10:

If Lev_{stor_sht} is M AND Δf is N AND ΔP is NB THEN $P_{ref\_stor\_sht}$ is NB
If Lev_{stor_sht} is M AND Δf is N AND ΔP is NM THEN $P_{ref\_stor\_sht}$ is NM
If Lev_{stor_sht} is M AND Δf is N AND ΔP is Z THEN $P_{ref\_stor\_sht}$ is Z
If Lev_{stor_sht} is M AND Δf is N AND ΔP is PM THEN $P_{ref\_stor\_sht}$ is PM
If Lev_{stor_sht} is M AND Δf is N AND ΔP is PB THEN $P_{ref\_stor\_sht}$ is PB

Finally, 44 fuzzy rules are necessary in place of the maximum of 540 fuzzy rules.

V. SIMULATIONS

TABLE I. PARAMETERS

Parameters	Values	Parameters	Values
P_{WTG}	750 kW	S_{conv}	3 MVA
$P_{chmax\_stor\_sht}$	300 kW	$P_{chmax\_stor\_lgt}$	230 kW
$P_{dchmax\_stor\_sht}$	-300 kW	$P_{chmax\_stor\_lgt}$	-230 kW
$\tau_{ch\_stor\_sht}$	0.5 s	$\tau_{ch\_stor\_lgt}$	5 s
$\tau_{dch\_stor\_sht}$	0.5 s	$\tau_{dch\_stor\_lgt}$	5 s
$W_{max\_stor\_sht}$	15000 kJ 4.167 kWh	$W_{max\_stor\_lgt}$	1500000kJ 416.7 kWh
P_{FDG}	750 kW	τ_{FDG}	5 s
P_{load1}	800 kW	$P_{load\,2}$	800 kW

The network considered in the simulations is shown in Fig.1. The most important parameters are given in Table I. In this table, S_{conv} is the apparent power of the conventional generator, P_{load1} is the active power of the load 1 and P_{load2} is the active power of the load 2. To show a significant contribution to the hybrid generator, without modeling large network, the total power of the network is rather small. Simulations are carried out with the help of Matlab Simulink™ software.

Simulation of Fig. 11 shows the power generated by the WTG in broken line (Fig. 11.a), the total power generated by the hybrid generator in full line (Fig. 11.a), the grid frequency (Fig. 11.b), the storage level of the short term (Fig. 11.d) and long term storage (Fig. 11.c), the FDG (Fig. 11.f), the pitch angle (Fig. 11.e). The reference power of hybrid generator is fixed at 600 kW for 0<t<1 h, 400 kW for 1<t<2 h and 800 kW for 2<t<3 h. A load of 800 kW is connected at t=20 min, t=1h20 and t= 1h40 and disconnected at t= 40 min t= 1h40 and t=2h40.

Fig. 11.a shows that the power reference is well respected in spite of the wind and load variations. When a load is connected, the energy of the hybrid generator increase rapidly, and the inverse phenomenon appears when a load is disconnected in accord with the primary frequency control. When the energy of short term storage is high, the pitch angle is controlled to reduce and smooth the wind power (Fig. 11.e). And when the energy of short term storage is low, the FDG is controlled to compensate the wind power (Fig. 11.f). In addition, the WTG and the FDG take part in frequency control when their are in operation.

VI. COMPARISON OF DIFFERENT STRUCTURE

In order to make a comparison with different topologies, the methodology develop previously has applied for different topologies.

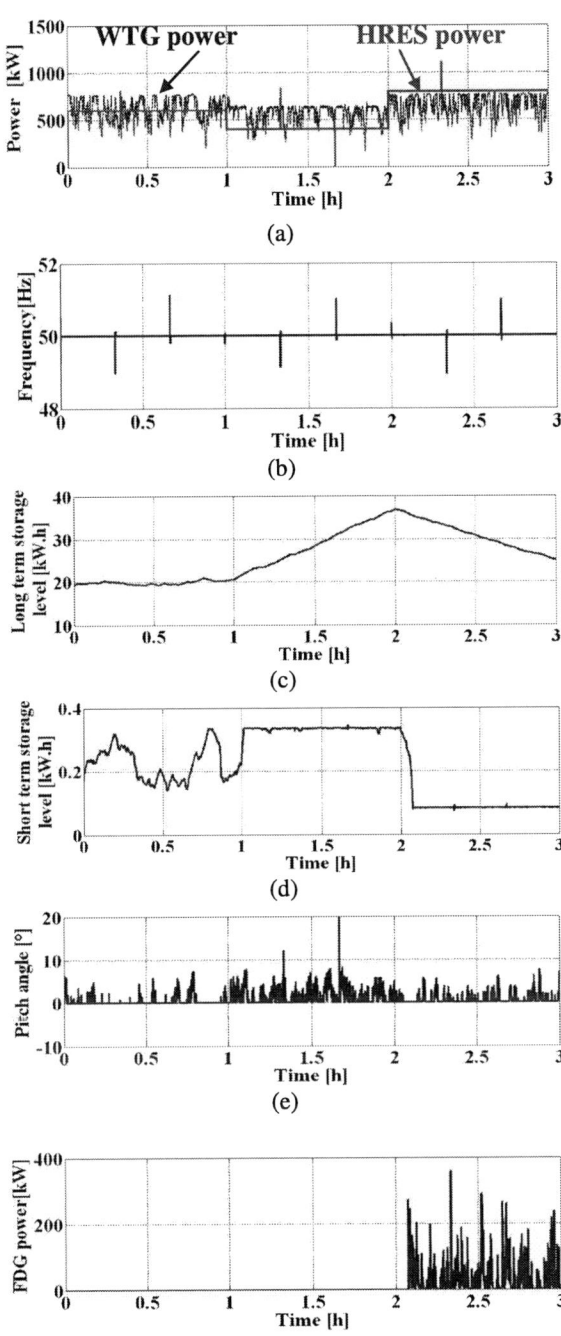

Fig. 11 Hybrid generator performance for three different reference power : (a) Wind and hybrid generator power, (b) Network frequency, (c) energy of long term storage, (d) energy of short term storage, (e) Pitch angle of wind turbine, (f) Power of the FDG units.

1) Association of WTG and FDG (Case a)

The system considered in this part includes wind generators and foreseeable source as shown Fig. 12. As the previously part the goal of the supervisor is to generate a reference power and ensure a primary power reserve.

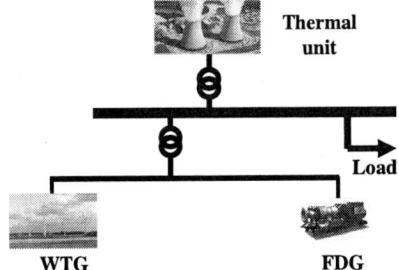

Fig. 12 Association of WTG and FDG

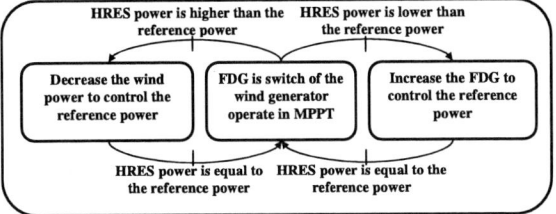

Fig. 13 Chart representation of different operating mode

Fig. 13 shows the chart representation of the system, the supervision system is divided in three part; when the HRES power is lower than the reference power the FDG must provide the difference between the reference power and the WTG power. In the same way when the HRES power is higher than the reference power the wind power must be decrease with the help of the pitch angle.

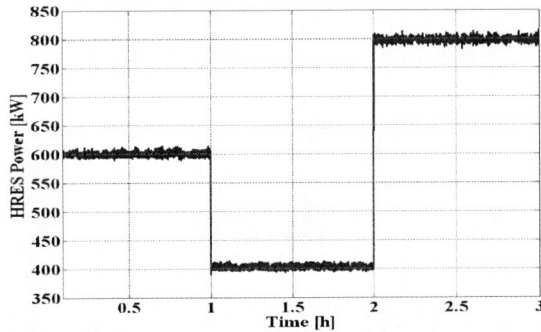

Fig. 14 Hybrid generator power compare to the hybrid reference power

In Fig. 14 curves in broken lines correspond to the reference power whereas the curve in full line corresponds to the HRES generated power. Globally, the reference power is well respected.

2) Association of WTG, short and long term storage (Case b)

The system considered in this part includes wind generators, FDG source and short term storage system as shown Fig. 15. Fig. 16 shows the chart representation of the supervisor, the supervision system can be divided in three parts when the storage level is medium, the WTG operates in maximal extraction power (MPPT) and the storage unit compensates the difference between the reference power and the WTG power.

2189

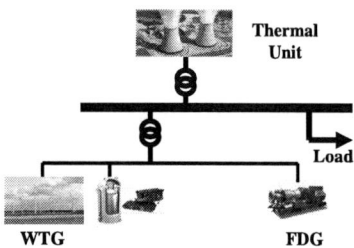

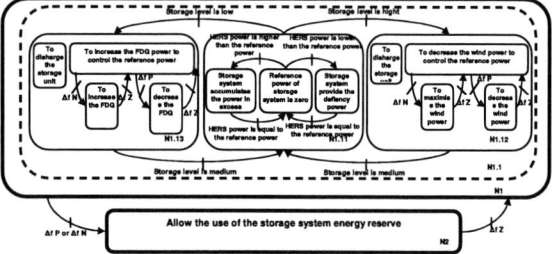

Fig. 15 Association of WTG, short and long term storage

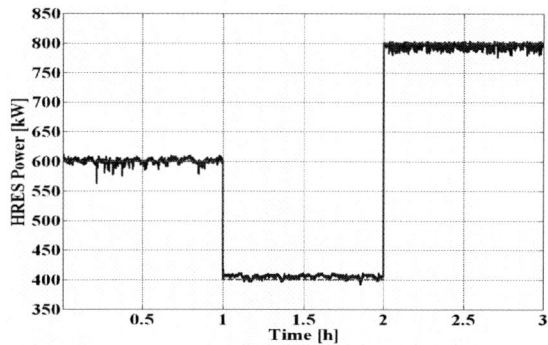

Fig. 16 Chart representation of different operating mode

When the storage is high, the storage unit must be discharged and the WTG power is decreased with the help of the pitch angle. When the storage level is low, the storage must be charged and the FDG must control the reference power. In Fig. 17 curves in broken lines correspond to the reference power whereas the curve in full line corresponds to the HRES generated power. Globally, the reference power is well respected.

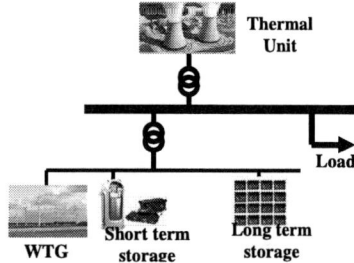

Fig. 17 Hybrid generator power compare to the hybrid reference power

3) Association of WTG, short and long term storage (case d)

Fig. 18 Association of WTG, short term storage and long term storage system

The system considered in this part includes wind generators and short and long term storage system as shown Fig. 18. Fig. 19 shows the chart representation of the supervisor, the supervision system can be divided in three parts when the storage level is medium, the WTG operates in maximal extraction power (MPPT) and the storage unit compensates the difference between the reference power and the WTG power.

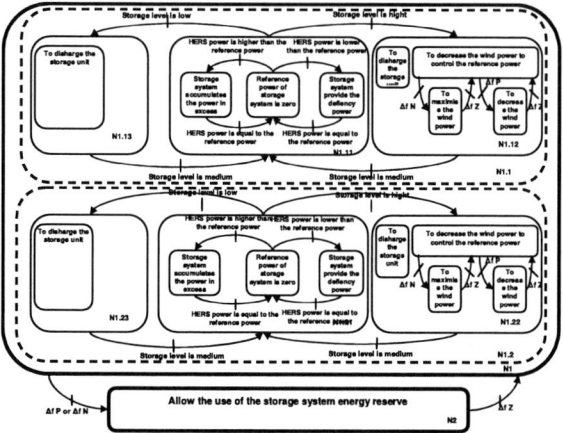

Fig. 19 Chart representation of different operating mode

When the storage is high, the storage unit must be discharged and the WTG power is decreased with the help of the pitch angle. When the storage level is low, the storage must be charged; in this case the reference power is respected in limit of storage capacity.

In Fig. 20 curves in broken lines correspond to the reference power whereas the curve in full line corresponds to the HRES generated power.

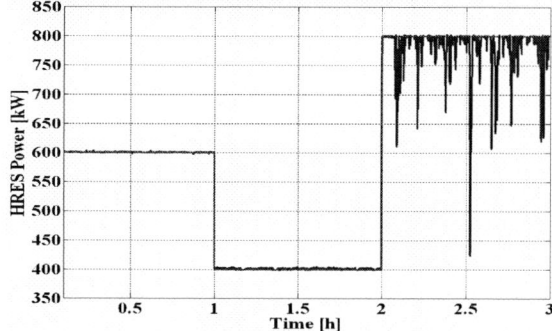

Fig. 20 Hybrid generator power compare to the hybrid reference power

4) Association of WTG and short term storage (Case e)

The system considered in this part includes wind generators and short term storage system as shown Fig. 21. Fig. 22 shows the chart representation of the supervisor, the supervision system can be divided in three parts when the storage level is medium, the short term storage control the reference power. When the storage is high, the storage unit must be discharged and the WTG power is decreased with the help of the pitch angle. When the storage level is low, the storage must be charged; in this case the reference power cannot be respected.

2190

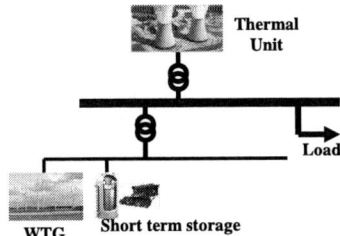

Thermal Unit

Load

WTG   Short term storage

Fig. 21 Association of WTG and short term storage

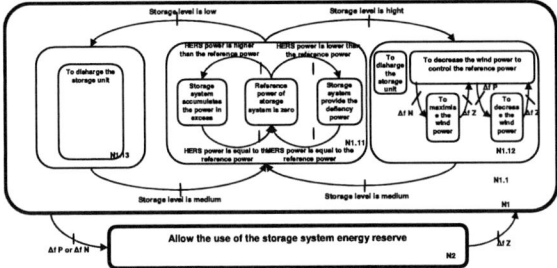

Fig. 22 Chart representation of different operating mode

In Fig. 23 curves in broken lines correspond to the reference power whereas the curve in full line corresponds to the HRES generated power. Only when the reference is smaller than the wind power, the reference power is respected. When the reference is significantly different than the wind power, the reference is respected in the limit of the storage system capacity.

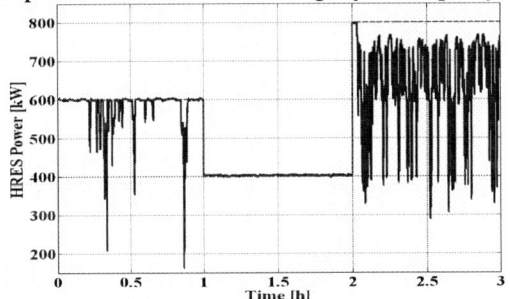

Fig. 23 Hybrid generator power compare to the hybrid reference power

TABLE II. COMPARISON BETWEEN DIFFERENT TOPOLOGIES

Case under study	Grid Energy (kW.h)	WTG Energy (kW.h)	FDG Energy (kW.h)	Average Error (kW)
(a)	1800	1563	237	0.165
(b)	1800	1636	163	0.223
(c)	1800	1823	30.92	0.0109
(d)	1783	1823	0	5.4
(e)	1635	1635	0	54.86

Five topologies are explored, WTG and FDG based HRES case (a), WTG, FDG and short term storage units based HRES case (b), WTG, FDG, short and long term storage units case (c), WTG, short term and long term storage units based HRES case (d), WTG and short term storage units based HRES case (e). Table II allows us to compare different HRES typology. When a FDG is included, similar energy is provided to the grid. In this case, the storage system enables to maximize the WTG

energy and then to minimize the FDG energy (case c). The error between the power reference and the HRES power is rather small. When the HRES does not include the FDG, the reference power can not be guaranteed when the wind power is lower than the reference power.

VII. CONCLUSION

In this paper, a methodology has been proposed to develop a fuzzy logic based supervisor. This method facilitates the analysis and the determination of fuzzy control algorithms adapted to complex hybrid systems. It allows to avoid to use of precise and intricate models of different sources and storage systems. This method allows a systematic determination of the supervisor and a minimization of the number of fuzzy rules. To explain this methodology, the association of a wind turbine based dispersed generators and storage systems are considered. The performance of this supervisor is shown with the help of simulations. A comparison of different topologies is also proposed.

ACKNOWLEDGMENT

This work was supported by a financing from the regional Council Nord-Pas de Calais, the European Regional Development Funding, SUEZ Tractebel Engineering company, Supelec and HEI.

REFERENCES

[1] M.K. Deshmukh and S.S. Deshmukh "Modeling of hybrid renewable energy systems", Renewable and Sustainable Energy Reviews, Volume 12, Issue 1, January 2008, Pages 235-249

[2] Ph. Degobert, S. Kreuawan, X. Guillaud, Micro-grid powered by photovoltaic and micro trubine, International Conference on Renewable Energy and Power Quality (ICREPQ'06), CD-ROM, Palma de Mallorca, Spain, April, 2006

[3] J. Morren, S W.H. de Haan and J.A. Ferreira, "Primary Power/Frequency control with wind turbine and fuel cells", IEEE PES General Meeting 2006, Montreal, Canada, June 2006.

[4] G. Cimuca, C. Saudemont, B. Robyns, M. Radulescu, "Control and performance evaluation of a flywheel energy storage system associated to a variable speed wind generator", IEEE Transactions on Industrial Electronics. Vol. 53, No. 4, June 2006.

[5] M. N.Eskander, T. F.El-Shatter, M. T.El-Hagry "Energy flow management of a hybrid Wind/PV/Fuel cell generation system" Energy Conversion and Management, Volume 47, Issues 9-10, June 2006, Pages 1264-1280

[6] G. Boukettaya, L. Krichen, A. Ouali, "Fuzzy logic supervisor for power control of an isolated hybrid energy production unit", International Journal of Electrical and Power Engineering".

[7] C. Abbey, G.Joos "Energy management strategies for optimization of energy storage in wind power hybrid system", Power electronics Specialists Conference (PESC), Recif, Brazil ,2005.

[8] V. Courtecuisse, M. El Mokadem, B. Robyns, B. Francois, M. Petit et J. Deuse, "Association of wind turbine based dispersed generators and storage systems to participate in primary frequency control », EPE 2007, Aalborg, 2-5 septembre, Danemark.

[9] V. Courtecuisse, B. Robyns, S. Plumel, B. Francois and J. Deuse, "Capacity of a variable speed turbine to participate in primary frequency control", Sixth international workshop on large-scale integration of wind power and transmission networks for offshore wind farms, Delft, October 26-27, 2006.

[10] Y. G.Rebours, D. S.Kirschen, M. Trotignon and S. Rossignol, „A survey of Frequency and Voltage Control Ancillary Service-Part I: Technical Feature", IEEE transaction on power system, Vol.22, N°1., February 2007

[11] P.Vas "Electrical machines and drive", Oxford science publication, ISBN0-19-859397-X, Great Britain, 1999

First Approach for a Fault Tolerant Power Converter Interface for Multi-Stack PEM Fuel Cell Generator in Transportation Systems

Alexandre De Bernardinis*, Gérard Coquery*

*INRETS – LTN, The French National Institute for Transport and Safety Research
2, Avenue du Général Malleret-Joinville 94114 Arcueil, France, e-mail: *alexandre.de-bernardinis@inrets.fr*

Abstract—This paper presents a fault tolerant architecture for connecting PEM fuel cells inside an hybrid on-board power generation system. As a first approach, the problematic of fault tolerance for multi-stack PEM fuel cell power systems dedicated to battery-hybrid transportation systems will be considered from both sides which closely interact: the fuel cell generator in case of faulty working and the power converter interface. For the first case, the use of antiparallel by-pass diodes have proved to be a solution and are tested on series assembly of 2-cell (experimental validation) and 20-cell (simulation) PEM fuel cell stacks. For the second one, the technological choice of an interleaved multi-phase boost converter interface with backup strategies and fault-tolerance management will be detailed. One of the great issues of these converter architectures is also to give new possibilities for power management and coordination with the battery inside this hybrid transportation system.

Keywords—Fault handling strategy, Fuel cell system, Power converters for FCEV, Power management, Protection device, Traction application.

I. INTRODUCTION

A fault tolerant architecture for connecting multi-stack PEM fuel cells inside a hybrid generation system is complex. It should take into account the constraints relating to the fuel cell stacks and the power electronic interface. Moreover it should be linked to the global power management of the battery-hybrid powertrain for transportation.

A. Problematic of the PEMFC Multi-Stack Association with its Power Converter Interface

In the objective of power increase for on-board traction systems (60 to 120kW) fed by PEM (Proton Exchange Membrane) fuel cells (FC), a technological solution consists in connecting several fuel cell stacks. In order to build and develop industrial power systems using fuel cell sources for transportation, reliability, availability, maintainability and safety (RAMS) have to be achieved. Availability of a minimal power is often obtained from the association of several fuel cell modules to reach a sufficient level of redundancy. This is particularly important in the transportation field to ensure a power

delivery continuation in case of a fault. However, the problematic of multi-stack association leads to critical constraints: they are issued from the fuel cell generator itself, from the multi-stack association in case of one stack failure, from the auxiliaries and also from the on-board powertrain. Thus, the design of a power converter interface which is fault-tolerant is a key issue in terms of global power system management. Fault tolerance is very important from the point of view of the system, because it implies also the management of degraded working modes.

Practically, the fuel cell generator is connected to the on-board electrical DC network through a power converter interface. Basically, "low voltage-high current" structures are necessary due to the FC electrical characteristics and the behaviour of a FC as low voltage source to be connected to a high voltage on-board DC Link. Figure 1 presents several electrical configurations for the association of fuel cells modules with their power electronic converters: indeed fuel cells can be independently connected through their static power converters (Fig. 1a and 1b), coupled in series as twin-stack fuel cell clusters (Fig. 1c) or electrically connected in series (Fig. 1d) [1].

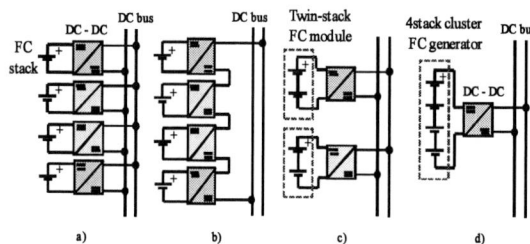

Fig. 1. Different configurations for the electrical coupling of fuel cells with power converter interface to the on-board DC Link.

Such a fuel cell power system should be fault tolerant. Fault tolerance has to be managed from the point of view of the fuel cell generator and from the power converter interface. Specific constraints relative to the fuel cells have to be considered for the design and power management of a fault-tolerant power converter structure.

978-1-4244-1741-4/08/$25.00 ©2008 IEEE

B. Specific Constraints of the PEM Fuel Cell Stack

One of the constraints is the decrease of the FC voltage in case of faulty operation of the stack. Fuel cell performances can be degraded during operation for many reasons linked to the fuel cell stack itself or to the auxiliaries. As an example, when the low temperature PEM fuel cell stack doesn't operate with pure hydrogen, a poisoning risk by pollutants, especially carbon monoxide (CO) can occur. Poisoning leads to a blocking of the active platinum catalyst areas and leads to a decrease of the global electric voltage. A same tendency can occur during the ageing process of a fuel cell stack. Figure 2 illustrates the modification of the polarisation curve due to CO poisoning or ageing phenomena.

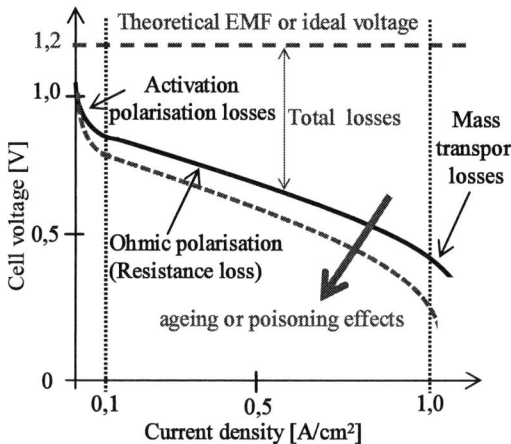

Fig. 2. Modification of the fuel cell polarisation curve owing to the influence of ageing or CO poisoning.

Moreover, in a fuel cell stack operation, some of the cells can suffer from starvation: the cells are not fed with a sufficient amount of reactants (hydrogen, air). Then the voltage of the cells will drop down to a lower level which can be critical to the stack working and lifetime. In the case of real insufficient fuel feeding, the starved cells might drop to negative values due to electrolysis current [2, 6]. Many factors like a sudden load current demand can cause starvation in a fuel cell. Among those operating problems, some severe stack degradation is leading to irreversible damage or drastic reduction of the lifetime operation.

The use of by-pass diodes, electrically connected in anti-parallel (Diode AP) to each fuel cell stack can short-circuit and isolate a stack under fault [3-5].

II. DIODE BY-PASS SOLUTION

A. Principle of the Diode By-pass for a Fault Tolerant Multi-Stack PEMFC Generator

An association of hydrogen / oxygen PEMFC modules equipped with anti-parallel diodes has already been used in the case of an air independent propulsion system for submarine application. In the Submarine Class 212 from German navy [4], eight 30-50kW PEFC modules are connected in series. If one of the FC modules fails, it is switched off, while the others continue to operate. Depending on the series connection of the FC modules, a diode connected in anti-parallel to each module is necessary to ensure the current flow while one or more FC module(s) are out of operation. This principle of FC by-pass, could also be implemented in terrestrial urban vehicles like FC-electric buses, light rail vehicles (light trams) taking into account the possible degraded working modes of the fuel cell stacks.

For the power supply of a DC on-board electrical network, a possible solution should consist in using a power generator composed of n PEFC stacks connected in series (Fig. 3). This solution leads to an increase of the total input voltage and thus the voltage constraints on the power converter are reduced. An anti-parallel diode is connected to each FC stack. The FC multi-stack power generator is connected to a DC/DC power converter, which adapts the input voltage to the DC-Link voltage amplitude. A battery imposes the DC bus potential and is also used for transient load power phases. For simplification, each FC_n stack is represented as a voltage source in series with a static resistance.

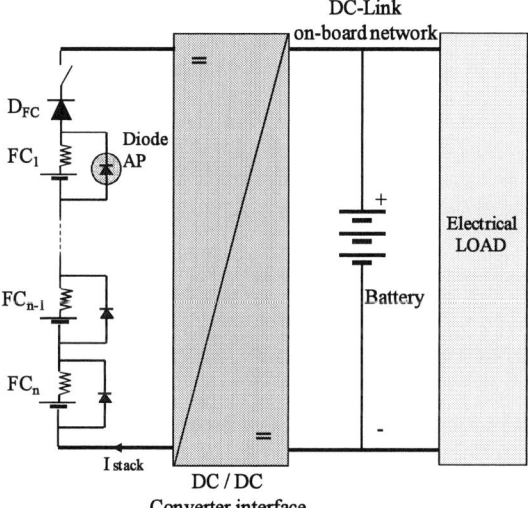

Fig. 3. Diode by-pass devices used for multi-stack PEMFC generator for fault handling management.

The fuel cell degraded working modes can have two main origins. The first cause is a possible inadequacy in the gas reactant supply to the stack or even a breakdown of the stack feeding in hydrogen and/or air. Thus, the voltages of the cells decrease dangerously and can lead to the starvation of several cells or to the starvation of the complete stack [9-10]. The second possible origin of a FC degraded working mode can be related to an electrical phenomenon such as the contactor switch-off in the external electrical circuit branch. Indeed, the shutting-down of a FC stack can occur following to one fault detection in the system. In both cases, the role of the anti-parallel diode is essential to ensure the current derivation and the electrical isolation of the faulty stack.

B. First Experimental Evaluation of a Diode By-pass on a 2-cell PEMFC Stack

In order to evaluate and characterise the electrical behaviour of the FC equipped with an anti-parallel diode, some degraded working modes have been imposed to a PEFC stack under test. Different experiments have been performed by the means of a specific set-up representing the series coupling of two FC stacks (Fig. 4). The experiments have been carried out in order to characterise and evaluate precisely the "switching on" behaviour of the by-pass diode. The experiment has been carried out at the FCLAB laboratory in Belfort, France.

The FC generator under test (first stack) is a two-cell PEMFC stack delivering a DC output voltage (UFC). A 5V constant supply source, connected in series to the stack, simulates the second FC generator in operating condition. The load current is imposed by a TDI electronic load and the current in the stack is measured by a LEM sensor. In the experimental set-up presented below, a power diode (Diode AP) is placed in anti-parallel position towards the assembly of FC stack, current sensor and on/off switch (electric contactor). In this test, the diode in series with the stack (Diode S1) is shunted. The anti-parallel diode used in the tests is an element of an IGBT power module (ref: Mitsubishi IGBT module CM350DU-5F). The stack is made of two cells assembled with commercially membranes (100 cm² active cell area), Gas Diffusion Layers (GDL) and flow distribution plates (in graphite and machined). The two cells (U1 [V] for cell n°1 and U2 [V] for cell n°2) are connected in series electrically and in parallel from the fluidic point of view.

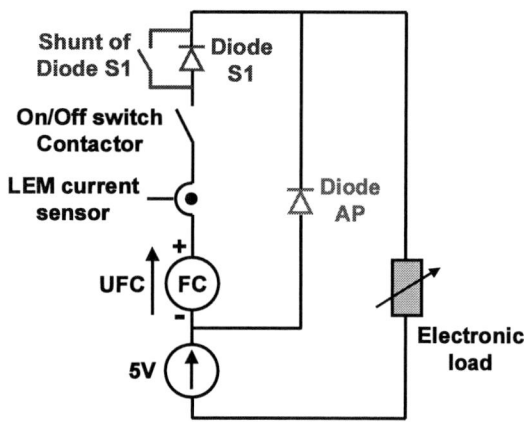

Fig. 4. Experimental set-up reproducing the series electrical coupling of two fuel cell stacks with antiparallel by-pass diode (Diode AP).

The results of the tests allow the evaluation of the by-pass diode solution in the case of specific fuel cell degraded working modes, such as breaking of the gas reactant feeding (air, hydrogen) and switching-off contactor active [6].

The experiment was carried out in the following physical conditions. The FC temperature was regulated at 30°C. This low temperature level was selected here only for basic practical reasons, especially to reach a steady-state FC operation in a very short time. The selected 30°C could correspond also to the transient temperature level of a FC start-up phase, during which one the stack can be more sensitive to some problems like cell flooding, leading generally to a load current interruption or reduction. The FC stack was fed by dry hydrogen and humidified air. The air dew point temperature was 25°C. The set of FSA / FSC (Anode and Cathode Stoichiometry Factors) was fixed to 2 / 4 and the reactive gas flow references were computed for a FC current of 5A. The FC was operated in "open mode". The air and hydrogen flows were controlled by flow regulators placed upstream of the stack. The tail gases were at atmospheric pressure (no use of the back pressure valves).

During the experiment (Fig. 5a to Fig. 5d), the following control strategy of the on/off switch was applied. When any of the cell voltages was reaching the low threshold of 380mV, the contactor was switched-off and the FC current falls to zero. At the beginning of the test, for time values less than 100s, the Open Circuit Voltages (OCV) of the two cells were approximately equal to 0.9V. At t = 100s, the amplitude of the load current imposed by the electronic load was fixed to 1A, as a precaution. A slight drop-down of the cell voltages could be observed as a consequence (Fig. 5d). The gas flows were then stopped at t = 140s by applying zero references on the flow controllers located upstream of the FC (Fig. 5b). The gas flows measured at stack outlets reached also zero values but with a delay of a few seconds due to the test bench fluidic and mechanical circuitry characteristics. From t = 140s to t = 200s, some buffer volumes of reactive gases were present in the stack. These volumes were decreasing progressively since the FC was continuing to deliver a 1A current and therefore was still consuming hydrogen and air. Two explanations can also be given concerning the decreases of the reactive gas quantities in the stack: it can be attributed on the one hand to the gas cross-over through the membranes and on the other hand to some slight internal and external leakages in the FC and along the test bench pipes as well. Anyway, the decreases of reactive gas quantities versus time were leading to a drop-down of the cell voltages. The decrease of the voltage was slightly higher for the first cell than for the second cell (Fig. 5d). This can probably be explained by a larger porosity to the reactants in the membrane of cell n°1. It should be noted that the investigated FC stack under test has already been operated during 1000 hours before this experiment and thus, some degradation has occurred in the stack. At t = 200s, cell n°1 was reaching the 380mV threshold, which induced the opening of the contactor in series with the stack in the adopted protection strategy. The FC current falls down to zero. Consequently, the voltage differential observed by the diodeAP led to its conduction. Then, the load current was totally delivered by the 5V constant voltage source. Just after the opening of the contactor, some peaks on the cell voltages could be observed during a few ten seconds (Fig. 5d). After the cell voltage peaks had occurred and until the end of the test, both cell voltages were decreasing slowly for the two reasons already given: gas cross-over through the membranes and small external leakage rate in the FC and in the test bench. At the end of the test, the voltage levels reached 0V [6].

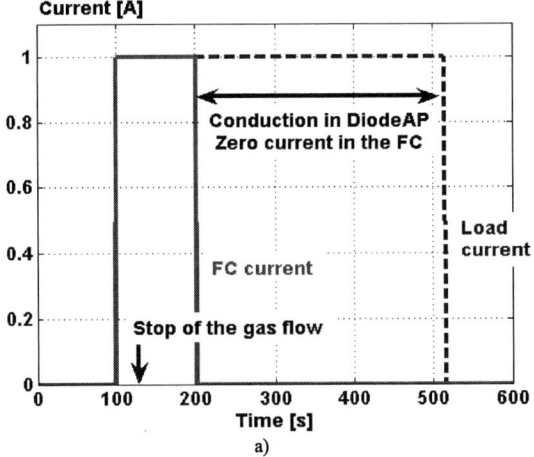

a)

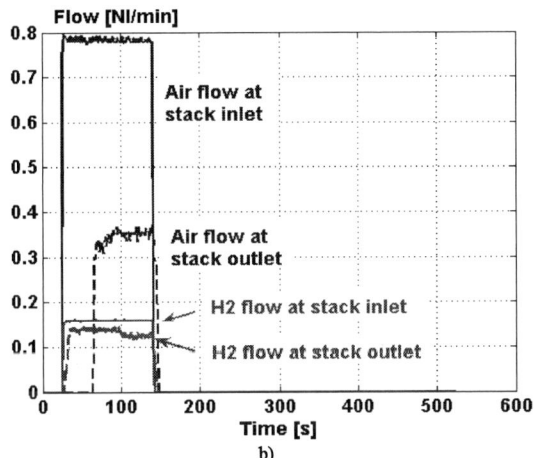

b)

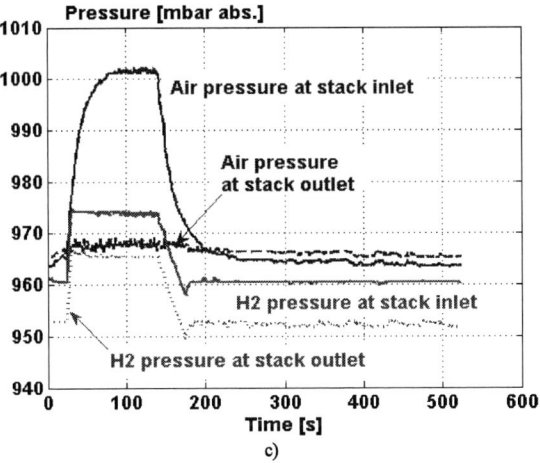

c)

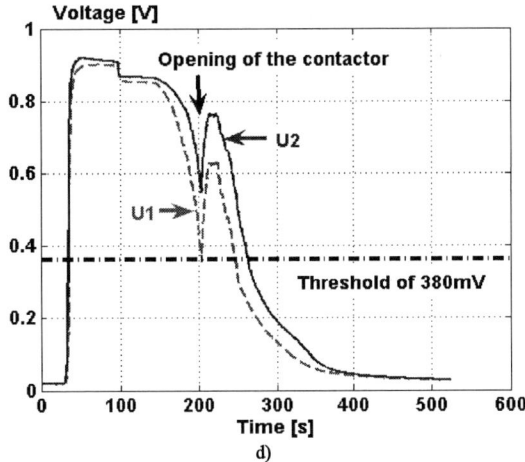

d)

Fig. 5. [6- D.Candusso *et al*] a) Evolutions of the FC and load currents during the experiment; b) Reactive gas flows; c) Air and hydrogen pressures; d) Cell voltages.

Some other experimental results performed on the 2-cell FC stack have been published in [6]. These results permit to characterize and underline the role of the DiodeAP as a protection device (for example as a by-pass in the case of an inactive switch-off of the electrical contactor).

C. Extrapolation of the First Results to a 20-cell PEMFC Stack by Numerical Simulation

The next objective of the work is to extrapolate these results to larger and more powerful stacks and to verify the behaviour of the FC stack and by-pass diode when a failure occurs. The key role of the fuel cell internal resistance and the "switching-on" capabilities of the anti-parallel diode (Diode AP) have been investigated by several numerical simulations. The topology of the electrical circuit (two PEM fuel cells in series, Diode AP and electrical load) is modelled using MATLAB/Simulink® and PowerSystemBlockset® library. The simulated fuel cell stack is a 20-cells PEMFC stack from ZSW Company. The load current is imposed using a constant current source. The model adopted for the FC is basically composed of an electrical voltage source E [V] in series with a pure resistance R_m [Ω], which represents the high frequency impedance of the FC (the resistance of the membranes). Such an (E, R_m) simplified FC model is sufficient for our study to analyse the "switching on" behaviour of the anti-parallel diode. Since the value of the membrane resistance is always lower than the total polarisation resistance, the (E, R_m) model can be considered as the "worst-case" for the triggering of the anti-parallel diode conduction [6]. The simulations are performed with the E voltage decreasing from 19V to 0V as a ramp in order to simulate a progressive break in the reactive gas feeding. The value of the global twenty cell stack resistance (R_m) is fixed to 36 mΩ. The Simulink® scheme in Fig. 7 shows the global model of the FC stacks in series adopted for the simulation. The simulations are performed for several values of the load current amplitude

(I_{Load}). The currents involved in the circuit are linked by Equ. (1) and the FC voltage is given by Eq. (2).

$$I_{Load} = I_{FC} + I_{DiodeAP} \qquad (1)$$

$$V_{FC} = E - R_m . I_{FC} \qquad (2)$$

If we neglect the internal resistance of the contactor in series with the FC stack, we get the following relation (Eq. 3) between the Diode AP voltage (V_{D_AK}) and the FC voltage (V_{FC}):

$$V_{D\_AK} = -V_{FC} = -E + R_m . I_{FC} \qquad (3)$$

When E equals zero or reaches values close to zero (during the starvation phase), the FC behaves like a pure resistance (R_m) and the current in the FC stack is dependent both on the resistance R_m and on the conducting threshold of the Diode AP (V_F [V] diode forward voltage). The current I_{FClim} through the stack is limited by Eq. (4):

$$I_{FC\,lim} = \frac{V_F}{R_m} \qquad (4)$$

Then, we come to the following relations corresponding to two different states:
$I_{Load} \leq I_{FClim}$: Diode AP is blocked; $I_{FC} = I_{Load}$ and $I_{DiodeAP} = 0$
$I_{Load} > I_{FClim}$: Diode AP is conducting; $I_{FC} = I_{FClim}$ and $I_{DiodeAP} = I_{Load} - I_{Fclim}$

The conduction of the anti-parallel diode is illustrated by the following simulation results, respectively obtained for 5A and 30A load current amplitudes. The conduction threshold of the diode AP is set to $V_F = 0.30V$ for these simulations.

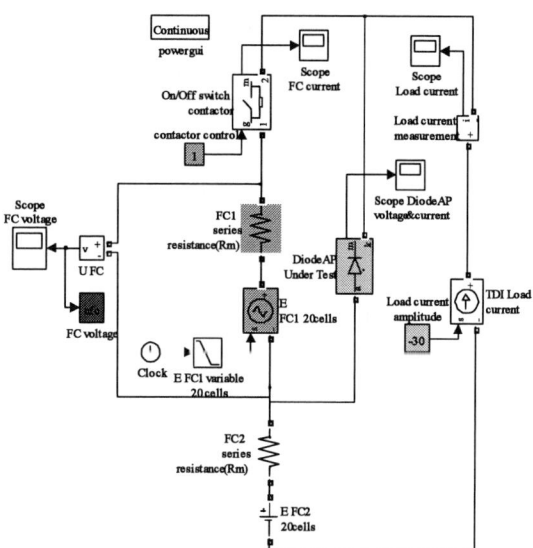

Fig. 7. Scheme of the simulation using the simplified model for the 20-cells PEMFC stack (using Simulink PowerBlockset®).

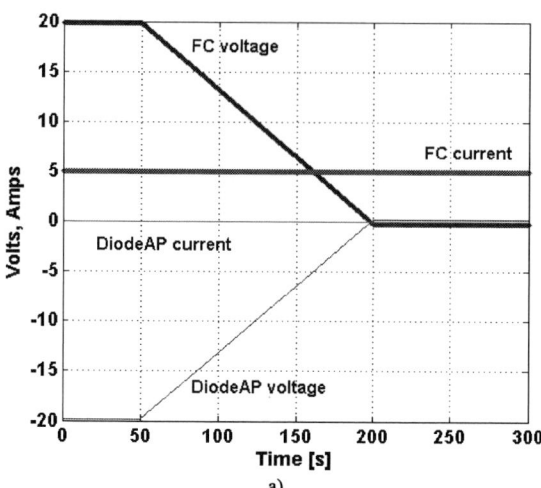

a)

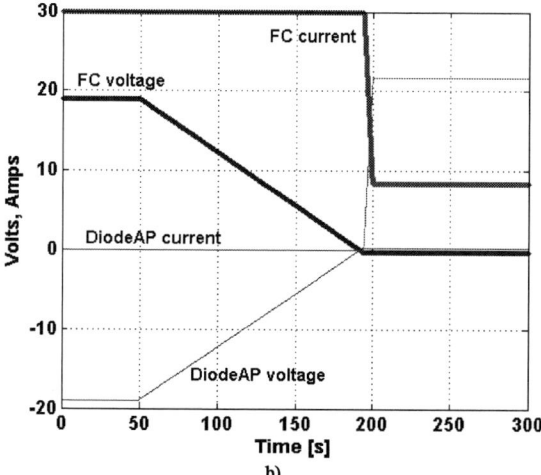

b)

Fig. 8. Simulation results: Fuel cell and DiodeAP currents, voltages; a) For a 5A load current amplitude (Diode AP not conducting); b) For a 30A load current amplitude (Diode AP in conduction state).

For a 5A load current amplitude, the Diode AP does not conduct (Fig. 8a)). The conduction threshold (V_F) of the Diode AP is not reached. For 30A load current amplitude, the Diode AP is conducting. The FC current amplitude is equal to 0.30 / 0.036 = 8.33A. The current through the Diode AP has a value of 30 - 8.33 = 21.67A (Fig. 8b)).

The presented simulations demonstrate that the results can be very different with larger stacks and higher load current amplitudes. It should be also noted that the conduction threshold of the Diode AP can be adjusted, either by choosing a suitable technology for this diode (Schottky diodes, fast recovery diodes...) depending on the current amplitude, or by associating two or more diodes in series or in parallel. All these simulation results should be obviously validated by further experiments but the results provided in this section show that the modelling tools developed can be very useful to design some electrical topologies allowing the stack protection and the power delivery in degraded working modes.

III. FAULT TOLERANT POWER CONVERTER INTERFACE

Fault tolerance has also to be considered from the point of view of the power electrical interface which interacts with the FC generator. This approach is studied by numerical simulation for a high power twin-stack PEM fuel cell system dedicated to transport applications (60 to 120 kW). Both fuel cell stacks are electrically connected in series. A lead-acid battery is connected to the on-board DC-Link which imposes the 540V_{DC} nominal voltage and delivers the transient power (Fig. 10).

Power electronics converters are devices which permit to introduce control strategies and methodologies in the energy system production and to be able to manage degraded working modes. Among a multitude of possible candidate's power converter structures, we have focused our study on an interleaved boost converter for the power converter interface between the fuel cell multi-stack generator and the on-board DC bus. The converter should meet objectives linked to the transportation environment: compactness, reliability, high efficiency, adaptability for degraded working modes. In addition, it should be fault-tolerant in case of one FC stack is out of work or a converter fault appears.

The N-phase interleaved boost converter structure has been chosen mainly because of its simplicity (it uses only a few standard components) and redundancy (if a converter phase is out of work, the others can still be used as a backup system avoiding any power delivery interruption). The choice of N=3 phases is justified by a severe specification on the FC stack current ripple (1% of the fuel cell rated current) [8]. As illustration, Figure 9 shows the normalized current ripple versus the duty cycle.

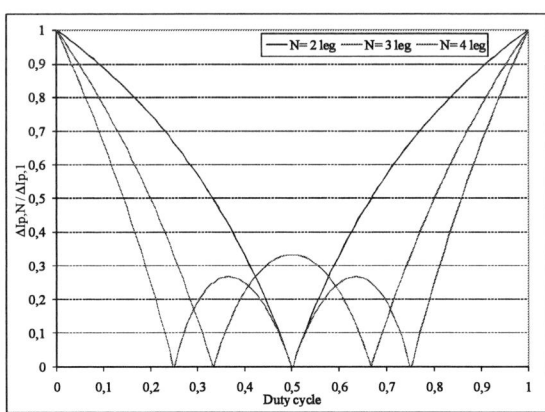

Fig. 9. Normalized ripple current cancellation for two-, three- and four- leg phases.

The following two degraded working modes have been taken into account in this study: one FC stack failure and one converter phase leg out of work. The specifications for the degraded working modes are:

- In case of one stack failure, the system must supply half of the power,

- In case of one converter phase is out of work, the converter must continue to supply the full power for a while.

The chosen converter topology is illustrated in Fig. 10. The phase-legs are identical and each phase is composed of an inductance (L_i) and a power semiconductor module (T_i+D_i). They are linked by two filtering capacitors, C_{in} and C_{out}. IGBTs gate control signals are shifted by $1/(3.f_s)$, where f_s is the switching frequency [8]. In order to fulfil the fault-tolerant requirements, bypass diodes, fuses and circuit breakers (contactors) have been added to FC stacks and converter legs, as shown in Figure 10. Switches and fuses are necessary to isolate the stacks or a converter leg in case of short-circuits. HT_p is the main coupling contactor, D_p preserves the FC stacks against feedback currents, diodes D_{AP1} and D_{AP2} act like a bypass for the out of work FC stack [3-5].

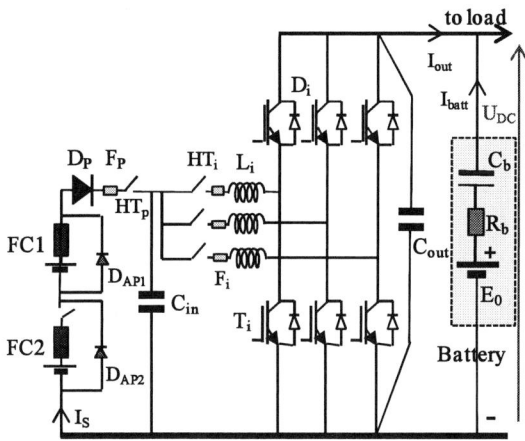

Fig. 10. Overview of the twin-stack FC generator with its power converter interface and electrical protection devices.

The battery connected to the DC-Link is modelled as a (E_0, R_b, C_b) assembly, its voltage is given by Eq. 5:

$$U_{DC} = E_0 - R_b.i_b - \frac{1}{C_b}.\int_{Tch} i_b dt \qquad (5)$$

The 1-phase current ripple is given by Eq. 6:

$$\Delta I_{S,1} = \frac{(U_{FC1} + U_{FC2})}{L_i.f_s} \alpha. \qquad (6)$$

The inductance $L_{i=1,2,3}$ is 220µH. The input and output capacitors are set to C_{in}= 500µF and C_{out} = 800µF.

The fuel cell current ripple is deducted from (6) by the following equations:

$$\left(\frac{\Delta I_{S,3}}{\Delta I_{S,1}} \right) = \begin{cases} \dfrac{1-3\alpha}{1-\alpha}, 0 < \alpha < \dfrac{1}{3} \\ \dfrac{(3\alpha-1)(2-3\alpha)}{3\alpha(1-\alpha)}, \dfrac{1}{3} < \alpha < \dfrac{2}{3} \\ \dfrac{3\alpha-2}{\alpha}, \dfrac{2}{3} < \alpha < 1 \end{cases} \qquad (7)$$

Figure 11 presents the 3 phase current ripple when the IGBT gate control signals are shifted by $1/(3f_s)$.

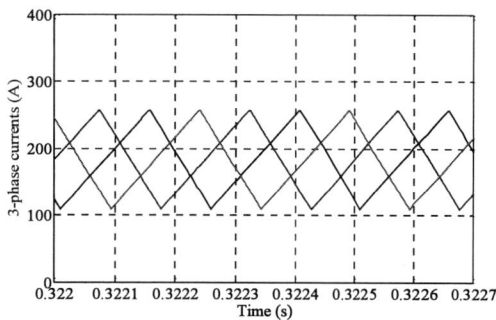

Fig. 11. Boost converter 3-phase current ripple.

IV. SIMULATION RESULTS FOR DEGRADED MODES

The simulations are performed for the degraded working modes using MATLAB/Simulink® and Power System Blockset® toolbox.

A. Case of FC Failure and Compensation by the Battery

The study simulates an accelerated breakdown in the gas reactant feeding (Fig. 12a). Each FC module is composed of 2 stacks in series and its static resistance value is fixed to 0.214 □. The DiodeAP2 forward voltage threshold is $V_F = 0.7V$. The fuel cell current is controlled to the reference value of 550 A. The FC2 voltage begins to decrease from t= 0.5s. When FC2 voltage level reaches the threshold conduction V_F, then DiodeAP2 conducts and switches with the FC2 current.

If one of the FC modules in series gets out of work, the input voltage halves and consequently the necessary deliverable power by the converter. The battery delivers the rest of the necessary power demanded by the load during the transition phase (Fig. 12b). Hereafter is shown the example of simulation in case of FC2 module fails following to a decrease of the voltage owing to a break in the gas reactant feeding. The anti-parallel diode plays its role as a bypass of the fuel cell stack under fault.

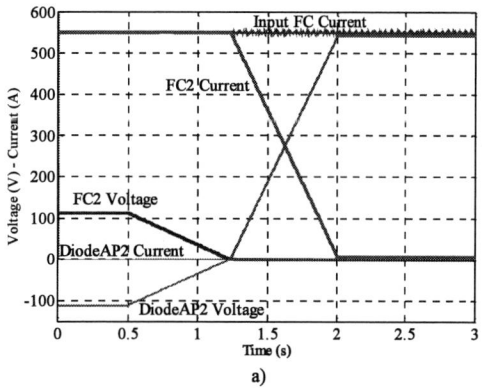

a)

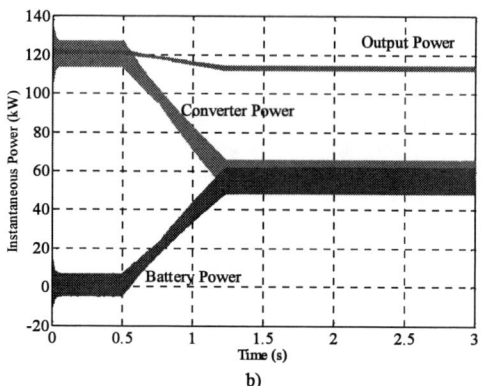

b)

Fig. 12. a)Fuel cell stack failure, b) Power compensation by the battery.

B. Case of Fuel Cell Circuit Contactor Opening

In this case the simulation concerns the impact of the opening of a contactor added in series with module FC2. At t= 1s the contactor is open simulating a sudden break in the electrical circuit. Diode AP1 is not conducting. The FC2 current falls down to zero and DiodeAP2 switches into conduction.

After the transient phase and thanks to a robust current control strategy, the fuel cell current and converter phase current remain controlled (Fig. 13b). Also in that case, the battery, acting as a power buffer, provides the complementary power to the load (Fig. 13c).

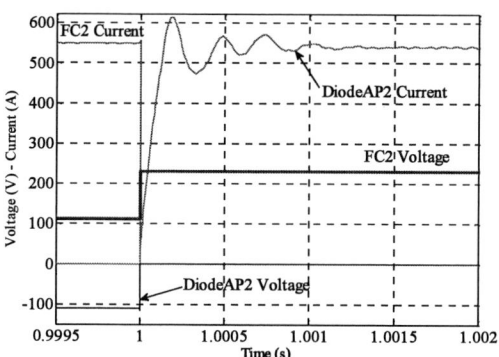

a)   Focus on the DiodeAP2 current response during switching

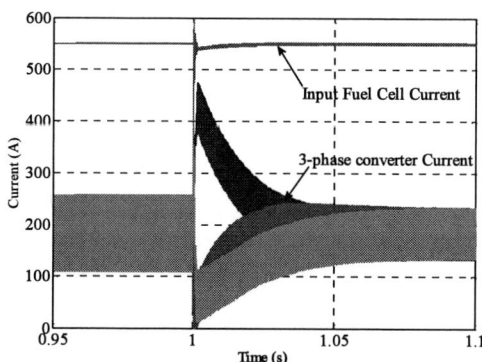

b) Three-phase boost-converter currents and FC input current in case of contactor opening

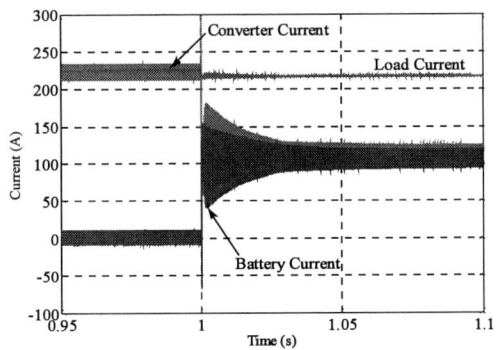

c) Battery power compensation after contactor opening

Fig. 13. Case of a contactor opening.

The power compensation is realised by the battery in case of contactor opening, the battery plays the role of a transient power buffer (Fig. 13c).

C. Boost converter phase leg out of work

The opening of the phase contactor (HT_i) or error in the gate signal which becomes zero (in case of short-circuit of the IGBT gate control signal) lead to the shutdown of one converter phase. Just after the fault detection, the 2-phase currents are controlled; however the current ripple rate has increased (Fig. 14a) (Eq. 7).

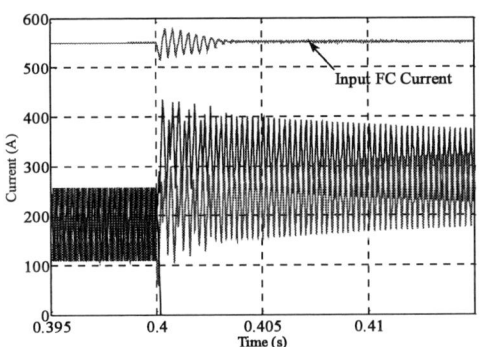

a) Two converter phases currents after the fault detection

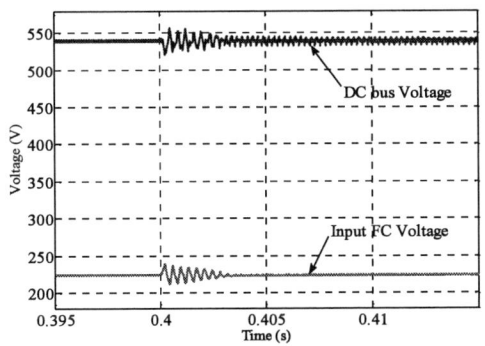

b) FC voltage and DC-bus voltage after the fault

Fig. 14. One converter phase-leg out of work.

After the localization of the fault, the corresponding phase leg is isolated and the converter continues to operate with only two active phases. This fault causes two

important changes: first, the IGBTs control shift is modified from $1/(3f_s)$ to $1/(2f_s)$, and second, the power is transferred through only two IGBTs increasing the current amplitude. Thus, the IGBTs and the switching frequency of the semi-conductors have to be adjusted from the beginning according to this constraint. A short-circuit identification and the leg isolation take time, but the battery supplies the necessary power during this transition. In case of a fault, the battery plays a key role to ensure a power compensation for the global hybrid on-board system.

V. CONCLUSION AND PERSPECTIVES

The design of a fault-tolerant electrical architecture to associate fuel cell multi-stacks inside an hybrid on-board powertrain is rather complex. However technological solutions exist to face up to the constraints relating to the multi-stacks fuel cells generator and its power converter interface. Anti-parallel diodes towards each fuel cell stacks allow to by-pass the stack in case of a fault. This solution has been evaluated for small PEMFC (2-cell) and extrapolated for more powerful stacks (20-cells). Moreover an adequate control strategy linked with power management permit to build a fault-tolerant power electronics interface. The battery plays a significant role in the global power management strategy as a transient power buffer in hybrid on-board systems. The global fuel cell- battery hybrid system should be optimized via a RAMS approach in a next step.

REFERENCES

[1] J.Garnier, A.De Bernardinis, R.Lallemand, M.C.Péra, G.Coquery, J.M.Kauffmann "High Frequency Power Converter for PEFC Generator Architecture based on a Multi-stack Association for transportation applications", EPE 2005 Conference, 11-14 September 2005, Dresden, Germany.

[2] Pierre Coddet « Etude des stratégies de sauvegarde de piles à combustible en cas d'arrêt d'urgence », Rapport de stage Master 2Recherche PROTEE, Belfort, France, mars-juin 2006.

[3] Bouvet F., Buchsbaum L., Lacarnoy A., Micoud B., and Durand-Schmutz C. "Convertisseur DC/DC non isolé pour applications à piles à combustible", Schneider Electric, Proceedings Conférence EPF 2004.

[4] Strasser K., H_2/O_2 PEM Fuel Cell module for an air independent propulsion system in a submarine, Handbook of Fuel Cells – Chapter 88, pp. 1201-1214, J.Wiley&Sons.

[5] P.Coddet, M.-C.Péra, D.Candusso, D.Hissel "Study of Proton Exchange Membrane Fuel Cell safety procedures in case of emergency shutdown" in IEEE ISIE 2007, Vigo - Spain, pp. 725-730.

[6] D.Candusso, A.De Bernardinis, M.-C.Péra, F.Harel, X.François, D.Hissel, G.Coquery, J-M.Kauffmann, Fuel cell operation under degraded working modes and study of a diode by-pass circuit dedicated to multi-stack association, Energy Conversion and Management, Elsevier, Vol. 49, Issue 4, April 2008, Pages 880-895.

[7] F. Harel, X.François, D.Candusso, D.Hissel, M.C.Péra, J-M Kauffmann, First experimental results of PEMFC durability under constant power solicitation constraint, FDFC 2004, Belfort, pp.365-370.

[8] B.Vulturescu, A.De Bernardinis, R.Lallemand, G.Coquery "Traction Power Converter for PEM Fuel Cell Multi-Stack Generator used in Urban Transportation", EPE 2007 European Conference on Power Electronics and Applications, September 2007, Aalborg, Denmark, 10p.

[9] Akira Taniguchi, Tomoki Akita, Kazuaki Yasuda, Yoshinori Miyazaki, Analysis of electrocatalyst degradation in PEMFC caused by cell reversal during fuel starvation, Journal of Power Sources, Vol. 130 (2004), pp. 42 – 49.

[10] S. D. Knights, K. M. Colbow, J. St-Pierre, D. P. Wilkinson, Ageing mechanisms and lifetime of PEFC and DMFC, Journal of Power Sources, Vol. 127 (2004), pp. 127–134.

Development of Electrical System for Hybrid Vehicles Using the Free-swinging Piston Engine and Oscillating Rotating Generator

Sigitas Kudarauskas

Klaipeda University/Department of Electrical Engineering, Klaipeda, Lithuania,
e-mail: *kudarauskas@klaipeda.omnitel.net*

Abstract—The original free-swinging piston engine with the oscillating rotating generator distinguishes by its compactness, minimum frictional losses, full-balanced mechanical system, possibility of four-stroke operation of the engine, possibility of the multifuel operation of the engine, optimal magnetic circuit of the electrical generator, etc. Because the engine can optimally operate over whole range of the output power, it enables to improve electrical system of the hybrid vehicle by usage of supercapacitors and optimised energy flows. In this way the competitive hybrid electric vehicle can be created.

Keywords—non-standard electrical machine, generation of electrical energy, hybrid electric vehicle, energy system management, supercapacitor.

I. INTRODUCTION

Persistent expansion of usage of the conventional road vehicles (that is, vehicles driven by internal combustion engines) causes some unacceptable effects, above all, the pollution of the environment. Limited efficiency of the internal combustion engine (ICE) and in consequence uneconomical consumption of the depleting fossil fuels also stimulates to look for novel technical solutions. That is why various hybrid vehicles (that is, the vehicles with the composite storage of energy) are created.

Nowadays the hybrid electric (separately marking out the fuel cell vehicles), the hybrid hydraulic vehicles are known. In spite of technical complexity of such vehicles, in some cases reduction of emitted pollutants and increased energy efficiency are achieved. It is to note that the hybrid electric vehicles on a basis of ICE are prevalent most of all. Possibilities of development of such hybrid vehicles are topical and they are analysed herein.

In the considered hybrid electric vehicles energy is stored in the fuel, the battery and/or supercapacitors. Thus, the electrical system includes electrical machines (generators and motors), devices of electrical energy storage and conversion, devices of control. The considered development of the electrical system is realised by usage of the oscillating electrical generator and storage of electrical energy by supercapacitors. Concurrent development of the ICE is analysed too, that is application of the free-swinging piston ICE.

The main goal of the discussed development is to demonstrate possibility to create a hybrid electric vehicle competitive with the conventional car.

II. ACTUAL SITUATION OF HYBRID VEHICLES

Because of prevalence of the hybrid electric vehicles with an ICE, the possible development of such vehicles is discussed below.

A. Actual Hybrid Electric Vehicles

Advantage in terms of the energy economy and the minimised pollution of the hybrid electric vehicle with ICE is achieved by the following main means:

- operation of the ICE with fixed (optimal) output power,
- start-stop regim of the ICE,
- regenerative braking of the vehicle.

The *Toyota Prius* is commonly-used hybrid electric vehicle that is treated almost as synonym of hybrid vehicles. The recent model NHW20 has an ICE (1.5 L, 57 kW), NiMH battery (25 kW, 1.31 kWh, weight 45 kg), electric motor (50 kW), curb weight 1325 kg [1].

Fig. 1 shows growth of US hybrid cars sales. We can see the prevalence of *Toyota Prius* as well as an existing of other hybrid cars in the marked. In some cases the hybrid electric car practically does not solve the problem to economise energy resources and to save the environment (e.g. the hybrid cars *Lexus* whose are about 5 time more expensive than *Prius* and with the fuel consumption in several time bigger too).

The history of short duration of the real production of the hybrid electric vehicles demonstrates some unsuccessful solutions too. For example, production of the *Honda Insight* was stopped after a few years (maybe, as overpriced two-seater), despite the fact that the fuel consumption was very small [2].

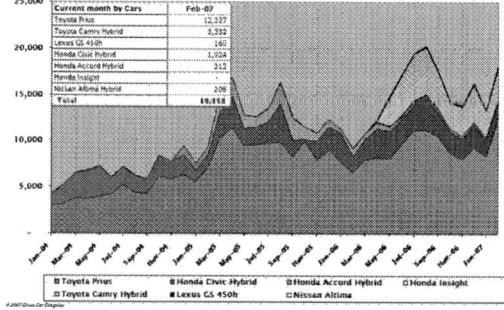

Fig. 1. US hybrid car sales 2004-2007 [1].

978-1-4244-1741-4/08/$25.00 ©2008 IEEE

Production of the *VW Golf CityStromer* was also stopped after about 500 items were sold as "too heavy, to expensive, no for the *Golf* customers segment " [3].

Thus, the complexity and costliness of the existing hybrid electric vehicles are the main their shortcoming.

B. Free-piston Engines

The applicable downsizing of ICE in the contemporary hibrid electric vehicles can not to compensate increased weight and complexity of such vehicle to compare with conventional one. Therefore it is necessary to improve all components of the hybrid vehicle, including ICE.

A long time the free-piston engines are known. In 1912, Prof. Hugo Junkers reported on the principle of a free-piston air compressor [4]. Thus, such compressors were created and manufactured in the first decades of the 20th century, and they are being used by now (Fig. 2).

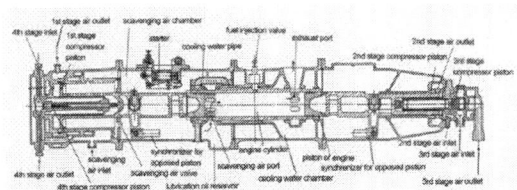

Fig. 2. Structure of hight pressure 4 stages free-piston Junkers compressor [4].

As a rule, the free-piston ICE operates in the two-stroke mode. The free-piston engine enables to avoid a transformer of mechanical movement and in this way to simplify the device with such engine. The free-piston engine enables to vary the compression ratio, therefore theoretically it can realise optimal operation in wide range of output power, and multifuel operation is possible too.

Nowadays the free-piston engine and hydraulic pump is being produced by *Innas inc.* (Fig. 3) [5]. It should be noted that such hydraulic pump is provided to use in hybrid hydraulic vehicle.

Of course, there are attempts to use the free-piston ICE with an oscillating generator for the hybrid vehicle [6], [7]. These devices are created as an unit of the free-piston engine and the oscillating linear generator. Therefore they are discussed in the next subsection.

On the other hand, there are attempts to improve ICE in other ways, for example, realising a variable compression ratio in the crankshaft ICE. The *MCE-5 DEVELOPMENT Company* (France) is proposing such solution using a complicated gear mechanism between the piston and the crankshaft (Fig. 4) [8].

Fig. 3. *Chiron Free Piston Engine* produced by *Innas Inc.* [5].

Fig. 4. Variable compression ratio engine proposed by *MCE-5 DEVELOPMENT Company* [8].

It is stated that a vehicle with the presented in Fig. 4 mechanically complex engine of mass production should have many advantages in comparison with the hybrid electric vehicle. If this statement is insufficiently reasoned, but it also accentuates shortcomings of the existing hybrid vehicles. On the other hand, a challenging remark is presented here: "Hybridization would not add anything to a thermal engine with constant efficiency over its whole range of use, except to recover a part of the vehicle's kinetic energy normally lost during braking" [8].

C. Oscillating Generators

The movable part of the operating oscillating generator (as any oscillating electric machine) is being in periodical motion, therefore it can be directly connected with the piston of the free-piston ICE (according to the direct drive principle) [9]. This situation enables to simplify the corresponding unit.

Typical sample of the oscillating linear generator driven by free-piston engine is shown in Fig. 5 [10]. Here we can see the oscillating generator with permanent magnets in the mover and having tubular structure of the magnetic circuit.

In Fig. 6 the principal layout of the analogous oscillating generator and the free-piston ICE created by consortium with leadership of *Volvo Technology Corporation* is shown [11] (the prototype was manufactured by *Innas com.*[5]).

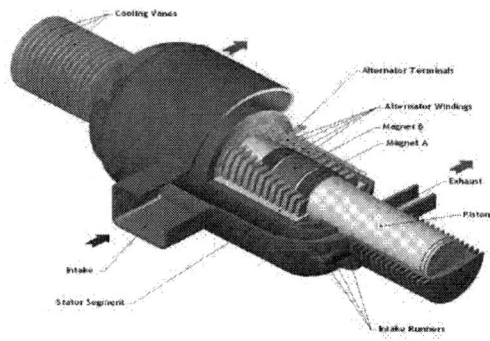

Fig. 5. So-called *Free piston linear alternator* [10].

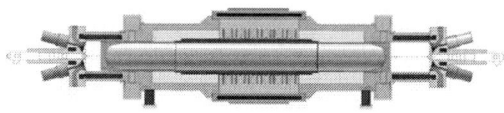

Fig. 6. Principal layout of the so-called
Free Piston Energy Converter [11].

The like structure of the oscillating linear generator is used in the aggregate of *Pempec Systems com.* (Australia) depicted in Fig. 7 [12]. Here a package of four identical units of the free-piston engine and oscillating linear generator is arranged on purpose to eliminate vibrations of the whole aggregate. Of course, the struture is complicated. In spite of such situation, the declared data of the aggregate are very interesting [12]:

Salient features of a 100kW FP3 module:

• *Size: 660x280x280mm*

• *Mass: 100kg*

• *Two-stroke operation*

• *Power Density: 1kW/kg; 2 kW/litre*

In spite of many advantages of the presented above units of the free-piston ICE and the oscillating linear generator (avoidance of movement transformer, minimised number of frictional pairs, etc.), some their shortcomings are unavoidable.

Fig. 7. *Free Piston Power Pack* of the *Pempec Systems com.* [12].

Noticeable imperfections of the mentioned above linear devices are as follows:

- problems of vibration of the frame,
- significant frictional losses in the linear guides,
- non-optimal magnetic circuit of the linear oscillating electrical machine (causing non-full usage of the permanent magnets and the windings).

But the possibility only two-stroke operation of the linear free-piston engine (with problems of the scavenging) is the main shortcoming of the devices, because optimal operation of such engine practically is possible only over limited range of the output power. The latter imperfection of the proposed devices is substantial, if the device is intended to be used as an energy source of the hybrid electric vehicle. Consequently, the linear variants of the considered devices will hardly suitable for the hybrid electric vehicles.

III. FREE-SWINGING PISTON ENGINE AND OSCILLATING ROTATING GENERATOR

In principle the free-piston ICE can be with rotating oscillation of the vane-shaped pistons (that is, the swinging pistons). An example of such device is lately proposed the *Free-swinging piston heat machine* [13] that can operate as internal or external combustion engine, and as compressor of pump.

The principle of structure of the device is shown in Fig. 8. The free swinging piston heat machine comprises a housing with a cylindrical cavity (e.g., formed by cylindrical ring 1 and covers 2, 3) in which two analogous assemblies of pistons 4, 5 are placed. The assembly of pistons consists of the cylindrical hub 6 with two or more vanes (e.g., 5', 5") having form of symmetrical star. The assemblies of pistons could independently pivot (oscillate) in opposite directions around the axis of the housing.

Minimum number of the working chambers between the assemblies of pistons hawing two vanes is four. Therefore the four-stroke operation of the free swinging piston engine can be realised, as it is depicted in Fig. 9. This is principal advantage of the free swinging piston engine to compare to the free-piston linear ICE.

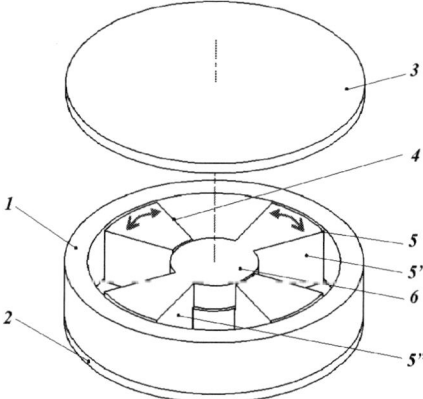

Fig. 8. Principle of structure of the free-swinging piston heat machine.

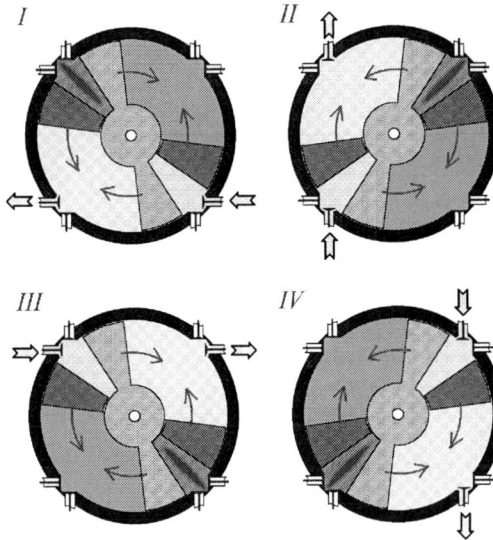

Fig. 9. Sequential stages *I – IV* of four-stroke operation of the free-swinging piston ICE.

The oscillating rotating electrical generator could be arranged concentrically with the engine (around of the engine), at it is shown in Fig. 10 (here the engine is shown without cowers). The oscillating electrical machine must have two rotors (7, 8), which are connected with the corresponding assemblies of pistons and which oscillate in opposite directions too. The stator of electrical machine 9 can be common. Thus, the non-standard electrical machine must be used in the considered device.

When the heat machine works as ICE, two rotors of the oscillating rotating electrical generator are powered by the engine. The oscillating electrical generator can pursue control of operation of the free-piston engine (the control of oscillation amplitude, position of oscillation centre, and oscillation spectrum). The same oscillating electrical machine can be used as engine starter in motor operation.

Guidance of the assemblies of pistons could be realised by rotary bearings (including roller bearings) mounted in the hubs. This structure enables to avoid friction between the swinging parts and the housing, and simplifies solution of the sealing.

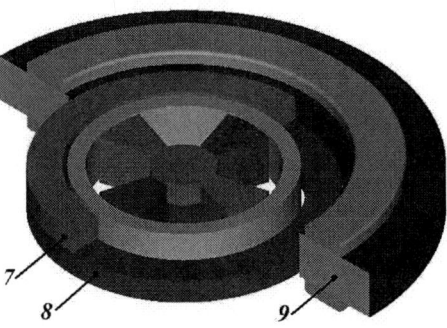

Fig. 10. Structure of unit of the free-swinging piston ICE and oscillating rotating electrical machine.

Direct connection of external devices (without any transformer of mechanical movement) to the ICE also simplifies structure of the unit and reduces mechanical losses.

The free-swinging piston ICE with the oscillating rotating generator becomes a balanced mechanical system of three bodies, because the centres of mass of two swinging in opposite directions movable parts are located in the axis of rotation. Therefore operation of such device does not provoke vibrations of the frame.

The main advantages of the considered device are as follows:

- compactness (decreased up to 50 % the weight and volume),
- minimum frictional losses in the rotary bearings,
- reduced (or absent) wear of the pistons and cylinder (housing) surfaces,
- possibilities to simplify sealing and lubrication,
- full-balanced mechanical system, without vibration of the frame,
- possibility of the four-stroke operation of the engine,
- possibility of the multifuel operation of the ICE,
- optimal magnetic circuit of the oscillating rotating electrical machine.

Fig. 11 shows the principal layout and dimensions of the estimated unit. The main data are as follows: the output power 65 kW, the frequency of oscillation 70 Hz, the weight (without auxiliaries) about 60 kg.

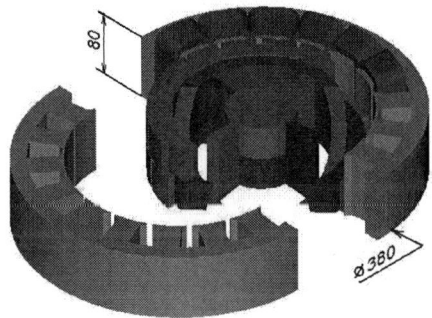

Fig. 11. Principal layout and dimensions of the estimated unit.

IV. ENERGY MANAGEMENT

Possibility of optimal operation of the considered four-stroke free-swinging piston ICE over wide range of the output power enables to arrange innovatively energy management of the hybrid electric vehicle with such device. It is possible to decrease amount of stored electrical energy, because the energy storage is necessary only for control of operation of the free-piston ICE (via the oscillating electrical machine) and for regenerative braking of the vehicle. Consequently, the supercapacitors could be used as the energy storage device instead of the expensive battery.

Fig. 12 shows the principle of energy flows in the hybrid electric vehicle with the oscillating rotating generator and the free-swinging piston ICE [7].

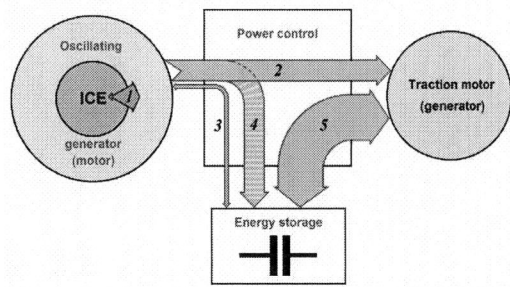

Fig. 12. Energy flows in the hybrid electric vehicle
with the considered device.

Fig 12 shows the main blocks of the energy management system of the hybrid electric vehicle: ICE & oscillating generator (as motor during start of the ICE), traction motor (operating as generator during regenerative braking), the supercapacitors as energy storage, and the power control block (including converters of the electrical energy). The corresponding arrows show energy flows: 1 – the bidirectional flow of the mechanical energy (the small reversible flow into the ICE fulfils requirements of the control, and the ICE starting), 2 – the unidirectional flow of the generator electrical energy to the traction motor (or motors), 3 – the small bidirectional energy flow between the generator and the energy storage, 4 – the unidirectional flow of the generator electrical energy to the energy storage (if necessary), 5 – the bidirectional energy flow between the traction motor and the energy storage (during operation in electric vehicle mode, during extra acceleration of the vehicle, and during the regenerative braking). Different width of the arrows relative shows various possible powers of the energy flows.

Because the considered free-swinging piston ICE in principle can optimally operate over whole range of the output power, the amount of stored energy is predetermined by the energy of regenerative braking and by the necessary energy reserve for extra-acceleration of the vehicle. This amount of energy enables to use the supercapacitors as the energy storage in the estimated hybrid vehicle.

Data of the power and stored energy level of some hybrid electric vehicles are presented in Table I. Here we can see the data of existing hybrid vehicles with a battery and the estimated vehicle with the free-swinging piston ICE and oscillating rotating generator, and with supercapacitors. In the latter case the amount of stored energy is predetermined by energy of regenerative braking depending on the vehicle speed (data of the corresponding row are presented for the initial speeds 60 and 100 km/h). Because the regenerative braking is more important for the urban cycle, the supercapacitors could be chosen according to the lower level of the initial speed.

TABLE I.
POWER AND STORED ENERGY OF THE HYBRID VEHICLES

Hybrid electric vehicle	Power		Stored el. energy	Weight (battery or supercap.)
	ICE	Battery or supercap.		
	kW	kW	kWh	kg
Toyota Prius (battery)	57	25	1.31	45
VW Golf (battery)	85	15	0.9	35
Estimated HEV (supercap., 60 km/h / 100 km/h)	65	80/210	0.08/0.18	12/30

As we can see, the estimated hybrid electric vehicle requires markedly lesser stored energy. On the other hand, increased power of the supercapacitors enables to brake vehicle more effectively in regenerative mode.

Practically some "dehybridization"of the vehicle is realised in this way, as it was mentioned above according to [8].

Consequently, the competitive hybrid electric vehicle can be created using the free-swinging piston engine and oscillating rotating generator.

REFERENCES

[1] Hybrid electric vehicle, http://en.wikipedia.org , accessed on 06/25/08.

[2] Honda Insight, http://www.hybridcars.com, accessed on 06/25/08.

[3] *Hybrid and electric vehicles.* International Energy Agency, Implementing Agreement on Hybrid and Electric Vehicle Technologies and Programmes, Annual report of the Executive Committee over 2006.

[4] M. Nakahara and H. Kohama, "Junkers High pressure Air Compressor-A Case of Technology Transfer through the Imperial Japanese Navy" in *Proc. ICBTT2002*, Kyoto, Japan, October 20-21, 2002.

[5] Chiron Free Piston Engine, http://www.innas.com/CFPE.html, accessed on 06/25/08.

[6] W.M. Arshad , T. Bäckström, P. Thelin ,Ch. Sadarangani, "Integrated Free-Piston Generators: An Overview", in *Proc. of IEEE-NORPIE 2002*, KTH, Sweden, 2002.

[7] S. Kudarauskas, "Oscillating Generators Driven by Free-piston Engine in the Hybrid Electric Vehicles: Prospects and Problems", in *Proc. of EVS 21*, Monaco, 2-6 April 2005, pap. 364.

[8] MCE-5 VCR-i: pushing back the limits of gasoline engines, http://www.mce-5.fr, accessed on 06/25/08.

[9] S. Kudarauskas, *Introduction to Oscillating Electrical Machines.* Edit. Of Klaipeda University, 2004.

[10] P. Van Blarigan, "Advanced internal combustion electrical generator", in *Proc. Of the 2001 DOE hydrogen Program Review*.

[11] E. Max, "FPEC, Free Piston Energy Converter", in *Proc. of EVS 21*, Monaco, 2-6 April 2005, pap. 159.

[12] Free Piston Power Pack (FP3), Pempec Systems, Australia, http://www.freepistonpower.com, accessed on 06/25/08.

[13] S. Kudarauskas, "Free-swinging piston heat machine", *International Application Publication Under the Patent Cooperation Treaty*, WO/2006/118437, 2006.

Power flow control in different time scales for a wind/hydrogen/super-capacitors based active hybrid power system

ZHOU Tao, LU Di, FAKHAM Hicham, FRANCOIS Bruno

L2EP, Ecole Centrale de Lille, Cité Scientifique, Villeneuve d'Ascq, France

tao.zhou@ec-lille.fr; di.lu@ec-lille.fr; hicham.fakham@ec-lille.fr; bruno.francois@ec-lille.fr

Abstract—**A hybrid power system is proposed and is based on wind energy conversion system with hydrogen based long-term storage unit and super-capacitor based fast-dynamic storage unit. A power management algorithm is organized in different time scales for its normal operating mode. Four studied cases show that, the long-term storage can compensate or absorb the difference between the generated fluctuating wind power and the grid power consumption in a long time range, and the fast-dynamic storage can smooth the power delivered to the grid in a short time range. It proves also the necessity of the three types of power sources.**

Keywords—**Renewable energy system, Hybrid power integration, Power management, Wind energy, Supercapacitors.**

I. INTRODUCTION

Due to the development of renewable energy systems and sustainable development issues (pollutant emission, rarefaction of fossil energy resources), hybrid power systems (HPSs) are one efficient solution for electrical energy generation, especially for isolated sites or for a micro-generation unit connected to weak AC grid [1]. The advancing wind power technologies have increased the use of the HPSs based on the wind power [2-5]. However, an electrical generating system depending entirely on only one renewable energy source (RES) is not reliable because

the availability of the RES can not be constantly ensured. The hybrid systems combine several energy conversion sources to overcome this problem. For remote area with difficulties in fuel and equipment transport, the association of electrochemical storages in the hybrid system allows to eliminate the diesel generator (which is commonly required in generation systems based on a single renewable energy source). Integrating wind power sources with hydrogen technologies as storage, can lead to a long-term non-polluting reliable energy source [6-9]. The super-capacitors have fast dynamics [10,11], thus can be added to smooth fast fluctuations of the wind power and to ensure a good power supply to the grid [12,13]. Thus, in this paper we propose a HPS combining a wind generator, a hydrogen based long-term storage and a super-capacitor based fast-dynamic storage [14,15] (Fig.1).

A power management system is needed for an HPS to manage the power flow among the sources in order to satisfy the load requirements throughout the whole operation period [16,17]. The main purpose of the power management is to satisfy the grid requirements qualitatively and quantitatively while maximizing the benefit of renewable energy sources and optimizing the operation of each storage unit. In this paper, such a power management algorithm is presented, by considering that grid requirements are specified as references for active

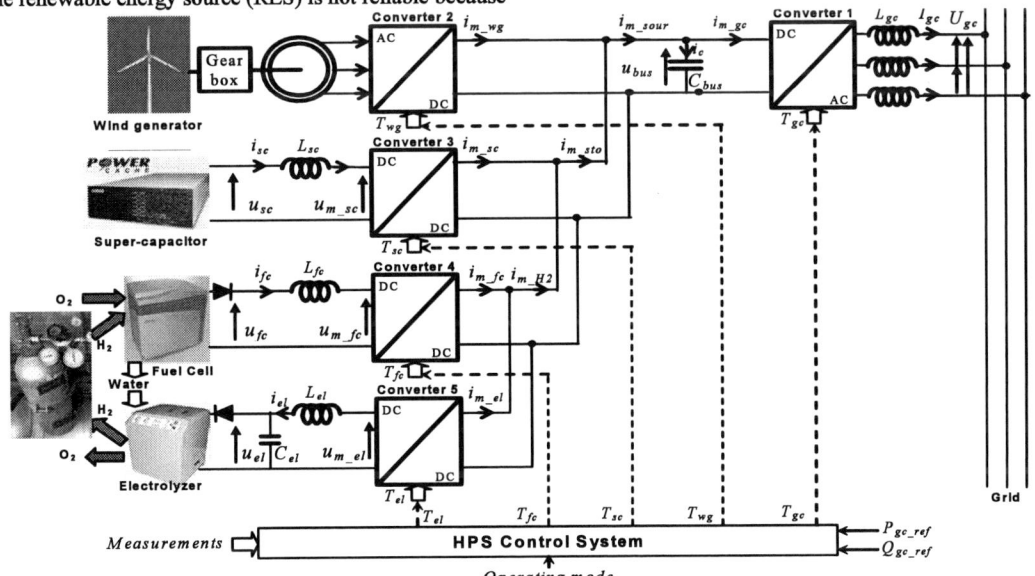

Fig.1: Structure of the multi-source hybrid power system

and reactive power. In order to prove the power management idea, some cases have been studied in simulation by using the practical wind data from a wind farm at Dunkerque in the North of France.

II. II. HYBRID POWER SYSTEM AND CONTROL SYSTEM

A. Hybrid Power System

In this HPS, all energy sources are connected to a main DC bus though each DC/DC converter and the main DC bus is connected to the grid through a main inverter [14,18] (Fig.1). This structure is flexible and expandable, since the number and the type of the energy sources may be freely chosen. Even more, the frequency can be independent from the grid through the DC-coupling, and the voltage of the main DC bus is independent to the nominal voltage of each Energy Conversion Source (ECS) through the use of each DC/DC converter.

Both the control structure and the power management are developed for this HPS consisting of a wind generator, a fuel cell, an electrolyzer, super-capacitors (Fig.1). However, the number and the type of the power sources do not alter the global control structure of the HPS and the main idea of the power management.

B. Control System

Since power electronic converters introduce control inputs for regulating powers in the HPS, we remind firstly the control structure of a single power converter (Fig.2) [19, 20]. A power electronic converter connects two dual sources (voltage source and current source) through a power axis. The control structure consists of three different levels:

- Switching Control Unit (SCU)
- Control Algorithm & Modulation Technique (CAMT)
- Power Tracking of Sources (PTS)

The SCU generates the transistor signals from the ideal states {0, 1}, via some drivers and opto-couplers. For each converter, the transistor signals are gathered into a vector T. For example, the vector gathering the transistor signals for the wind generator inverter (n°2) is called T_{wg} In the CAMT, the control algorithms determine the duty ratio through the regulation of some physical quantities and the modulation techniques determine the ideal states to send to the SCU. The main task of the PTS is to manage the power flows in the HPS for a selected operating mode.

The HPS have five power sources: the wind generator (WG), the fuel cell (FC), the electrolyzer (EL), the super-capacitors (SC) and the grid connection (GC) (Fig.1). Five power electronic converters connect each source to a common DC bus and hence enable to regulate the exchanged powers. Therefore, the control structure of a single power converter can be extended to the HPS (Fig.3), where we can find the SCUs and the CAMTs for each energy source, as well as a common PTU for the entire HPS to manage the power flows among the difference energy sources.

C. Power Management

The power management strategy of the HPS for all possible operating modes is located inside the PTS (Fig.4). With a good wind prediction, during a given

period, the grid operator may set a required grid power reference closed to the generated wind power in average value. However instantaneously, the generated wind power can be very different from the required grid power reference. Therefore, an efficient power management is necessary for the optimized use of each energy source.

The HPS may work in many operating modes depending on different working conditions due to the availability of each ECS and the grid demands, such as the climatic condition (wind speed v_{wind_mes}), the storage level of each storage unit (hydrogen pressure P_{H2} in the tank and super-capacitor's voltage u_{sc_mes})... In the normal operating mode, all power sources are available and can work normally. There exist still many other operating modes, for example, the limitation mode when one storage unit can not work normally. For example, if the H_2 tank is fully filled (in Full-H_2 mode), the electrolyzer can not work and the wind generator can not work in Maximum Power Point Tracking (MPPT) strategy all the time. Hence the generated wind power must be limited in case of strong winds (with high wind speed). For another example, if the SC is discharged (in Empty-SC mode) and cannot continue to deliver any power, we should propose a solution for the HPS to continue satisfying the grid or disconnect the HPS from the grid. The power management algorithm for the normal operating mode is presented in the next section.

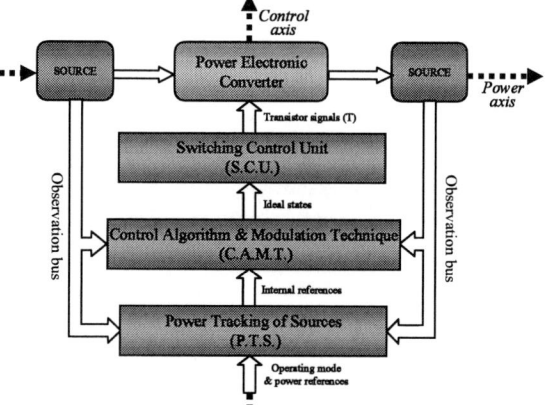

Fig.2. Control structure of a power electronic converter

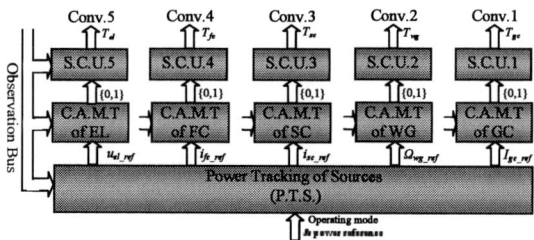

Fig.3. Control structure of the hybrid power system

III. POWER MANAGEMENT IN NORMAL OPERATING MODE

In the normal operating mode, the reference of the generated wind power P_{wg_ref} is set by a MPPT (Maximizing Power Point Tracking) control method. In order to focus on power management algorithms, the models and the controls of each power source are not detailed in this paper. We assume that each power source

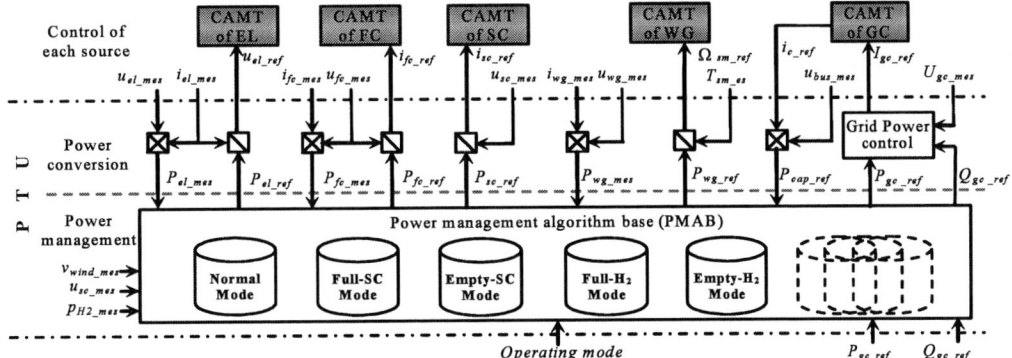

Fig.4: Control structure and power management layout

is well controlled. The power management algorithm is based on the power balancing of the hybrid system.

A. DC-bus voltage control

For this HPS structure, it is necessary to control the DC bus voltage to guarantee the system's normal operation. The voltage of the DC-bus capacitor is controlled by a voltage corrector, which sets the necessary reference for the dc bus current (i_{c_ref}). The reference of the necessary exchanged power with the DC bus (p_{cap_ref}) is found (Fig.4) by multiplying it with the measured DC-bus voltage (u_{bus_mes}).

B. Power Balancing

If we take into account the losses in the filters and in the power converters, losses estimation and compensation should be added in the PTS. For an easier presentation we assume that the losses in filters and in power converters are negligible. Thus the generated powers are equal to the local absorbed powers and the power delivered to the grid. By writing the instantaneous power balancing of the HPS (Fig.5), we find the equation describing the instantaneous power p_{gc}, which is delivered to the grid:

$$p_{gc}(t) = p_{wg}(t) + p_{fc}(t) - p_{el}(t) + \left(p_{sc}^-(t) - p_{sc}^+(t)\right) + \left(p_{cap}^-(t) - p_{cap}^+(t)\right) \quad (1)$$

with p_{wg}: the power generated by the wind generator;

p_{fc}: the power generated by the fuel cell;
p_{el}: the power consumed by the electrolyzer;
p_{sc}^-: the power extracted from the super-capacitors,
p_{sc}^+: the power stored into the super-capacitors,
p_{cap}^-:the power extracted from the DC-bus capacitor,
p_{cap}^+:the power stored into the DC-bus capacitor.

Fig. 5: Power balancing of the hybrid power system

This equation can be rewritten as:

$$p_{gc}(t) = p_{wg}(t) + p_{sto}(t) + p_{cap}(t) \quad (2)$$

where p_{sto} is the total power exchanged with all storage units:

$$p_{sto}(t) = p_{H2}(t) + p_{sc}(t) \quad (3)$$

with

$$p_{H2}(t) = p_{fc}(t) - p_{el}(t) \quad (4)$$

$$p_{sc}(t) = p_{sc}^-(t) - p_{sc}^+(t) \quad (5)$$

Hence, for a given grid power reference $p_{gc\_ref}(t)$, for a given for the DC bus regulation power reference ($p_{cap\_ref}(t)$) and for a sensed wind power $p_{wg\_mes}(t)$, we can deduce the necessary power to be used with storage units by inverting the equation (2):

$$p_{sto\_ref}(t) = p_{gc\_ref}(t) - p_{wg\_mes}(t) - p_{cap\_ref}(t) \quad (6)$$

C. Power Analysis

The wind generator provides a random and fluctuant power, which can not meet the (nearly) constant grid power reference instantaneously. An energy management strategy based on the intrinsic features (especially power dynamics) of the different sources is proposed in order to distribute the generated wind power into the different storage units and the grid in the way that each source is optimally used. So the storage units and the grid can be classified in a following way (Fig.6).

The grid can be considered as a stable power source, since the required grid power is usually stable during a period.

The supercapacitors can be used as a high dynamic power source, since they can provide fast dynamics and be used to smooth the fast fluctuations of the wind power.

The H_2 based storage unit via fuel cell and the electrolyzer has slow dynamics but, thanks to the use of a large H_2 tank it can store and provide energy in a long time range term.

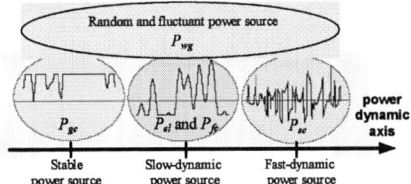

Fig.6: Waited frequency range for the operation of the different sources

D. Power Management in a Short Time Range

Power references for supercapacitors are calculated by inverting the timing power equation (3):

$$p_{sc\_ref}(t) = p_{sto\_ref}(t) - \left(p_{fc\_mes}(t) - p_{el\_mes}(t)\right) \quad (7)$$

E. Power Management in a Long Time Range

The average value of a power during a time interval (Δt) is defined as:

$$P = \{p(t)\}_{\Delta t} = \frac{1}{\Delta t}\int_{-\Delta t}^{} p(t)dt \quad (8)$$

To provide electrical power in a long time range (LTR), we use respectively the fuel cell and the electrolyzer to generate and to absorb the difference between the reference of the delivered grid average power P_{gc_ref} and the measured generated wind average power P_{wg}.

The average power, which is delivered to the grid is expressed as:

$$P_{gc} = \{p_{gc}(t)\}_{\Delta t} = P_{wg} + P_{sto} + P_{cap} \quad (9)$$

Hence by inverting this modeling equation in average value, we get:

$$\{p_{sto\_ref}(t)\}_{\Delta t} = P_{sto\_ref} = \{p_{gc\_ref}(t) - p_{wg\_mes}(t) - p_{cap\_ref}(t)\}_{\Delta t} \quad (10)$$

During a long time range $\Delta t = \Delta t_{LTR}$, we can ignore the fast power variations exchanged with the super-capacitors:

$$\{p_{sc}(t)\}_{\Delta t} = \{p_{sc}^-(t)\}_{\Delta t} - \{p_{sc}^+(t)\}_{\Delta t} \approx 0 \quad (11)$$

By inverting the exchanged average power with storage units, we can calculate the power references for H$_2$ units:

$$\{p_{H2\_ref}(t)\}_{\Delta t} \approx \{p_{sto\_ref}(t)\}_{\Delta t} \quad (12)$$

In practice the average value of the power during a Long Time Range is implemented by a low-pass filter (Fig.7). In order to get a maximum efficiency, the fuel cell and the electrolyzer must not work at the same time. A selector is inserted to assign the reference p_{H2_ref} to the fuel cell (p_{fc_ref}) or to the electrolyzer (p_{el_ref}) according to the sign of p_{H2_ref}.

$$\begin{cases} if : P_{H2\_ref} > 0, & P_{fc\_ref} = P_{H2\_ref}, \ P_{el\_ref} = 0; \\ if : P_{H2\_ref} = 0, & P_{fc\_ref} = 0, \quad\quad P_{el\_ref} = 0; \\ if : P_{H2\_ref} < 0, & P_{fc\_ref} = 0, \quad\quad P_{el\_ref} = p_{H2\_ref}. \end{cases} \quad (13)$$

Fig.7: Power management algorithms for a normal operating mode

IV. SIMULATION RESULTS AND DISCUSSION

We have studied four cases in simulation by adding in or removing some power sources in the hybrid system. In these four cases, we use the same wind data sensed in the wind farm at Dunkerque in the North of France, and the generated wind power is shown in Fig.8.

A. WG

In the 1st case, no storage unit is used, and the wind generator works in a MPPT mode. Since the DC-bus voltage is maintained, no power is exchanged in average with the DC-bus capacitor. The generated fluctuating wind power is totally delivered to the grid and does not satisfy the grid consumption qualitatively (Fig.9). It shows that the wind generator working in MPPT mode behaves like a passive generator, can not offer the required power.

B. WG + FC + EL

In the 2nd case, H$_2$ based long-term storage unit has is added via the fuel cell and the electrolyzer. The reference of the grid power is fixed (P_{gc_ref}=2000W, Q_{gc_ref}=0). We see that, when the wind generator send more than 2000W, the electrolyzer is activated to absorb the difference (Fig.10b); when the wind generator send less than 2000W, the fuel cell is activated to compensate the difference (Fig.10c). The power delivered to the grid can be regulated around 2000W, but is still very fluctuant (Fig.10a), because both the fuel cell and the electrolyzer are slow dynamic power sources, and are not able to filter fast fluctuation of the wind power.

C. WG + FC + EL + SC

Based on the 2nd case, we add the super-capacitors based fast-dynamic storage unit in the 3rd case. The fuel cell and the electrolyzer exchange the same power as in the 2nd case (Fig.10b and Fig.10c). The super-capacitors have smoothed the fast power variation (Fig.11b), and therefore, the power delivered to the grid become smooth and satisfactory (Fig.11a). This fact shows the advantage of somoothing the power to the grid by adding a fast dynamic storage by super-capacitors. Since we have the fuel cell and the electrolyzer for long-term energy storage, the voltage over the super-capacitors U_{fc} varies in a range from 75.5V to 77.3V (Fig.11c). The storage level (varying about from 49.3% to 53 %) of the super-capacitors always stays in the security zone, and the super-capacitors always work in the normal operating mode.

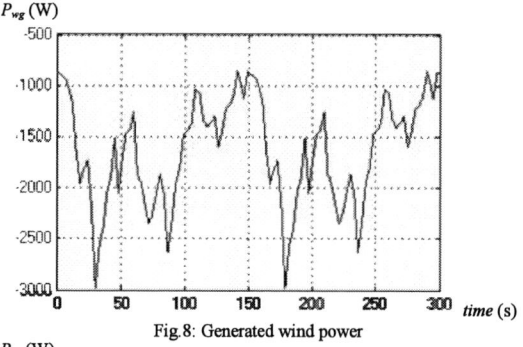

Fig.8: Generated wind power

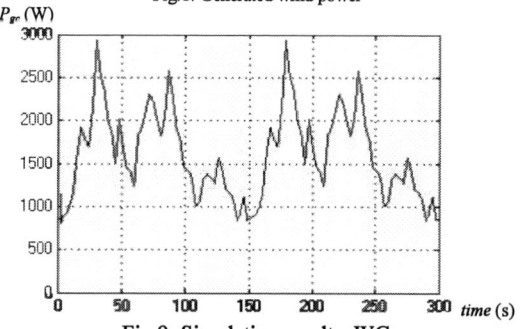

Fig.9: Simulation results: WG

D. WG + SC

Based on the 3rd case, we remove the fuel cell and the electrolyzer in the 4th case. The power delivered to the grid is still smooth and satisfactory (Fig.12a). But such a satisfaction can just be maintained for only a short-range time. The super-capacitors take in charge all the task of filtering the fluctuating wind power (Fig.12b) and its storage level varies fast (Fig.12c). If the average generated wind power during a short-rang period is different from the reference of the power delivered to the grid, the super-capacitors absorb or compensate all these power difference, and its storage level varies fast and gets out of its security range easily. In the presented simulation results, since the average power during 150s is lower than the grid reference P_{gc_ref} (2000W), the power extracted from the super-capacitors is more than that stored into it (Fig.12b). As result, after 300s' operation in normal mode, the voltage over the super-capacitors decreases from the 76.8V to 53.5V (Fig.12c), as well as the storage level from 51% to 8%. Therefore, it does not seem to be able to work in the same way during the next 150s.

E. Conclusion of the Simulation Results

RES based generation systems can not supply a prescribed power reference for the grid because of their fluctuating characteristics. Therefore, such a power generation system needs storage units to provide power for the fast power demands and to give enough energy for operating in a long time range. However, such ideal storage component does not exist in the real life if we can not oversize the storage units. A fast-dynamic storage unit is needed as a power source to smooth the fast power variations due to renewable energy; and a long-term storage unit is also needed as an energy reservoir to generate and absorb the power difference during a long period. A power management should be designed in different time scale to control the power flow with respect to their proper characteristics.

A test bench of 3kW is being realized in our laboratory in order to test this power management algorithm. The different dynamics will be taken into account to highlight the necessity of all the three kinds of sources in a hybrid power system.

V. CONCLUSION

A hybrid power generation based on wind energy conversion system is suggested for the access of electricity in remote areas, with hydrogen based long-term storage and super-capacitors based fast-dynamic storage in DC-coupled structure. A power management algorithm is proposed for this HPS in the normal operating mode, to manage the power flow among the different sources in different time scales. As results, the fuel cell and the electrolyzer can compensate or absorb the difference between the generated wind power and the grid consumption in long time rang, and the super-capacitors can smooth the power delivered to the grid in short time rang. The four cases' studies show the behaviors and the necessity of all the three type of power source: the renewable energy conversion systems; the long-term storage units and fast-dynamic storage units.

ACKNOWLEDGMENT

This work has been supported by the French National Agency of the Research (ANR SuperEner Project) and the China Scholarship Council (CSC).

REFERENCE:

[1] C.C. Chan, "The state of the art of electric and hybrid vehicles", Proc. IEEE, Feb, 2002; 90: 247–275.

[2] M.A. Elhadidy and S.M.Shaahid "Decentralized/stand-alone hybrid wind-diesel power systems to meet residential loads of hot coastal regions", Energy Convers Manage 2005; 46(15-16):2501-2513.

[3] J.S. Anagnostopoulos and D.E. Papantonis, "Pumping station design for a pumped-storage wind-hydro power plant", Energy Convers Manage 2007; 48(11):3009-3017.

[4] C.L. Chen, T.Y. Lee and R.M. Jan, "Optimal wind-thermal coordination dispatch in isolated power systems with large integration of wind capacity", Energy Convers Manage 2006; 47(18-19):3456-3472

[5] H.S. Ko, and J.Jatskevich, "Power quality control of wind-hybrid power generation system using fuzzy-LQR controller", IEEE Trans Energy Conversion 2007; 22(2): 516-527.

[6] M. Korpas and A.T Holen, "Operation planning of hydrogen storage connected to wind power operating in a power market", IEEE Trans Energy Conver 2006; 21(3): 742-749.

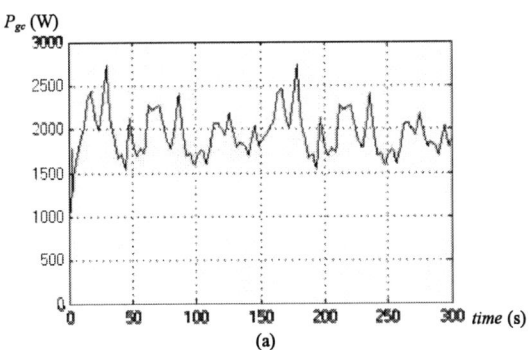

(a)

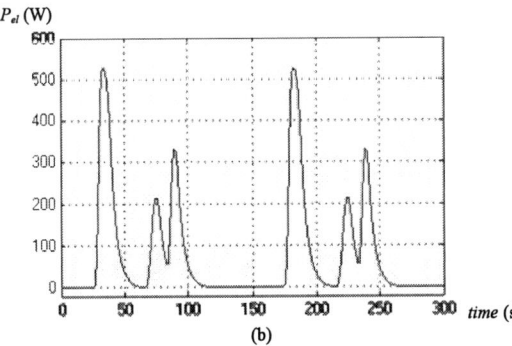

(b)

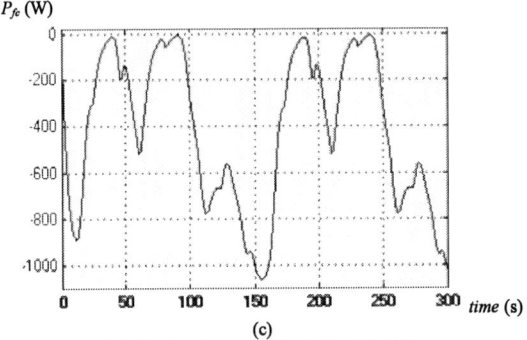

(c)

Fig.10: Simulation results: WG + FC + EL

[7] D.A. Bechrakis and P.D. Gallagher, "Simulation and operational assessment for a small autonomous wind–hydrogen energy system", Energy Convers Manage 2006; 47(1): 46-59

[8] K. Agbossou, R. Chahine, J. Hamelin, F. Laurencelle, A. Anouar, J.M. St-Arnaud, and Bose TK, "Renewable energy systems based on hydrogen for remote applications", J Power Source 2001; 96(1):168-172.

[9] A. Bilodeau and K. Agbossou, "Control analysis of renewable energy system with hydrogen storage for residential applications", J Power Source 2006; 162(2): 757-764.

[10] Y. Cheng, M.J. Van, B.P. Van, and P. Lataire, "Super capacitor based energy storage as peak power unit in the applications of hybrid electric vehicles", IET PEMD'06, Dublin; Mar 2006, p.404-408

[11] V. Paladini, T. Donateo, A. Risi, and D. Laforgia, "Super-capacitors fuel-cell hybrid electric vehicle optimization and control strategy development", Energy Convers Manage 2007; 48(11): 3001-3008.

[12] Y. Chen, V. Mierlo, V. Bossche and P. Lataire, "Using Super capacitor based energy storage toimprove power quality in distributed power generation", EPE-PEMC'06, Portoroz; Aug 2006, p.537–543.

[13] P. Degobert, S. Kreuawan and X. Guillaud, "Use of super capacitors to reduce the fast fluctuations of power of a hybrid

system composed of photovoltaic and micro turbine", IEEE-SPEEDAM'06, Taormina; May 2006. p. 1223-1227.

[14] B. François, D. Hissel and M.T. Iqbal, "Dynamic modelling of a fuel cell and wind turbine DC-linked power system", Electrimacs'05, Hammamet; Apr 2005, CD ROM

[15] T. Zhou, B. François, M. Labal and S. Lecoeuche, "Modeling and control design of hydrogen production process by using a causal ordering graph for wind energy conversion system", IEEE-ISIE'07, Vigo; Jun 2007: p. 3192-3197.

[16] T.F. El-Shatter, M.N. Eskander and M.T. El-Hagry, "Energy flow and management of a hybrid wind/PV/fuel cell generation system", Energy Convers Manage 2006; 47(9-10): 1264-1280

[17] A. Hajizadeh, and M.A. Golkar, "Intelligent power management strategy of hybrid distributed generation system", Int J Electr Power & Energy sys 2007;29(10):783-795.

[18] O. Omari, E. Ortjohann, A. Mohd and D. Morton, "An optimal control strategy for DC coupled hybrid power systems", IEEE-ISIE'07, Vigo; Jun 2007: p.2589–2594.

[19] B. François and J.P. Hautier, "Hierarchical control design using structural decomposition of a rectifier converter", ELECTRIMACS'96, Saint-Nazaire; Sep 1996: p. 255-260.

[20] D. Loriol, B. François and P. Degobert, "A DSP interface device for control algorithms implementation in the scope of power systems", IEEE-CESA'96, Lille; Jul 1996(2): p. 717-722.

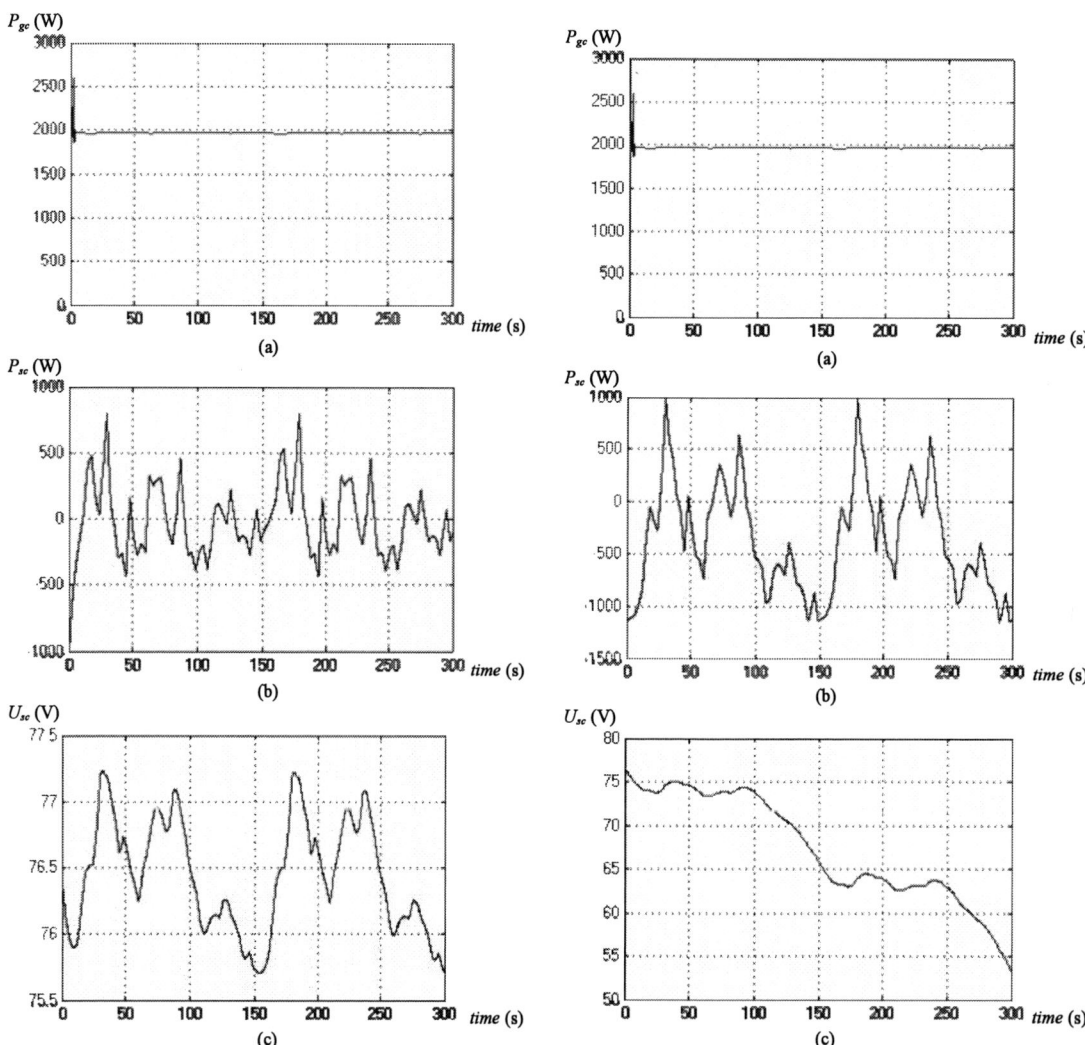

Fig.11: Simulation results: WG + FC + EL + SC

Fig.12: Simulation results: WG + SC

Neuro-Fuzzy Adaptive Control of the IM Drive with Elastic Coupling

Teresa Orlowska-Kowalska, *Senior Member IEEE*, Krzysztof Szabat, Mateusz Dybkowski

Wroclaw University of Technology, Institute of Electrical Machines, Drives and Measurements, Wroclaw, Poland

Abstract— The paper deals with two concepts of a model reference adaptive control (MRAC) of the induction motor drive with elastic joint. The adaptive speed controller uses fuzzy neural network equipped with additional option for on-line tuning its chosen parameters. In the paper PI-type and sliding-mode fuzzy logic controllers are used as the speed controllers, whose connective weights are trained on-line according to the error between the state variable of the plant and the reference model. The torsional vibrations dumping abilities of the two-mass drive system with the field oriented control of induction motor are tested and compared for both analyzed neuro-fuzzy adaptive controllers. It is shown that torsional oscillations can be successfully suppressed in the proposed control structures, using only one basic feedback from the motor speed. The simulation results are verified in experimental tests, in the wide range of motor speed and parameters changes.

I. INTRODUCTION

Modern industrial drives often require high dynamical performance, consisting in achieving the reference signal with high dynamics and well dumped transients. This is especially important in servodrives [1]-[2]. Usually, when the industrial drive is designed, the elasticity of the shaft is neglected. In the case of the standard drive such an assumption is reasonable; however, there is a large group of drives of different power, like disk drives, robot arm or space robot and manipulators, deep space antenna drives or even belt conveyers and rolling-mill drives, where characteristics of the mechanical part have to be included in the analysis [1]-[7], and the shaft elasticity must be taken into consideration. The finite stiffness of the shaft causes the torsional vibrations which affect the drive system performance significantly. The torsional vibrations decrease the product quality, system reliability; in some cases the control system can even lose stability.

The solution of the problem of mechanical oscillation dumping depends on the specific features of the system. The simplest method to avoid the system state variable oscillations consists in decreasing the dynamics of the control structures. However, this approach neglects the performance of the drive and is hardly ever utilized. The next method depends on alternative controller tuning techniques [6]. Yet, this approach can be applied successfully only in some ranges of parameters of the system masses.

If the good control system performances are required, the application of the additional feedbacks from selected state variables is necessary [7]. But in those cases the information about mechanical state variables of the drive system must be known and special state estimators, like Luenberger observers, Kalman filters or neural network-based estimators must be used [8]-[10].

When state variables of the mechanical coupling and load machine are not measured or estimated, the different control concept can be used [11]-[13]. In [12] and [13] the neuro-fuzzy controllers, operating in the Model Reference Adaptive Control (MRAC) scheme are proposed and tested in the stiff and elastic drive system with DC driven motor. Very good performances are obtained and the torsional vibration dumping is achieved. It should be checked, if in much more complicated and nonlinear drive system with induction motor (IM) these performances will be achieved also.

The main goal of this paper is to test and compare the effectiveness of two types of adaptive neuro-fuzzy speed controllers of two-mass vector-controlled induction motor drive, namely PI-type (ANFC) and sliding-mode (ASNFC). The connective weights of these controllers are trained on-line using modified gradient descent algorithm, according to the error between the measured speed of the IM and the reference model output. It is worth saying that the only measured mechanical state variable is the motor speed. Moreover, prior off-line learning of NFC is not necessary. The comparative analysis of the dynamical performance of the induction motor (IM) drive system with both types of speed controllers is performed from the point of view of their ability to torsional vibration dumping and sensitivity to rotor inertia changes in a wide range. The analysis, design, simulation and experimental implementation of the proposed control schemes are described.

II. DESCRIPTION OF THE TWO-MASS DRIVE AND CONTROL STRUCTURE

In the paper the commonly used model of the drive system with the resilient coupling is considered. The system is described by the following set of equations (in per unit system). (The way of parameter calculation from nominal to per unit values is shown in [7]).

$$T_1 \frac{d\omega_1}{dt} = m_e - m_s - m_f, \tag{1}$$

$$T_2 \frac{d\omega_2}{dt} = m_s - m_L - m_f, \tag{2}$$

$$T_c \frac{dm_s}{dt} = \omega_1 - \omega_2, \tag{3}$$

where: T_1, T_2, T_c – mechanical time constant of the motor, load and the shaft, ω_1, ω_2 – motor and load speeds, m_e, m_s, m_f – electromagnetic, torsional and non-linear friction torques.

The scheme of the considered two-mass system is presented in Fig. 1.

978-1-4244-1741-4/08/$25.00 ©2008 IEEE

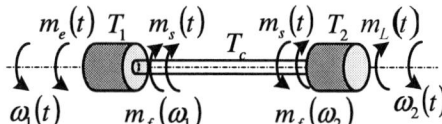

Fig. 1. The schematic diagram of the two-mass system

The classical drive control scheme consists of the two major loops. The inner loop includes of the torque controller, power converter, electromagnetic part of the motor and torque (current) sensor. The torque controller is tuned to provide sufficiently fast torque control. In the case of IM drive this could be achieved with field-oriented or DTC control methods. The outer speed control loop comprises the optimized torque control loop, mechanical part of the drive, speed sensor (usually encoder or resolver) and speed controller.

The usually applied classical PI speed controller is replaced in this study by the speed controller in the form of adaptive neuro-fuzzy network. The model reference adaptive control structure with the on-line tuned ANFC (or its sliding version) is proposed for the two-mass drive system [13]. The general diagram of the adaptation system is presented in Fig.2.

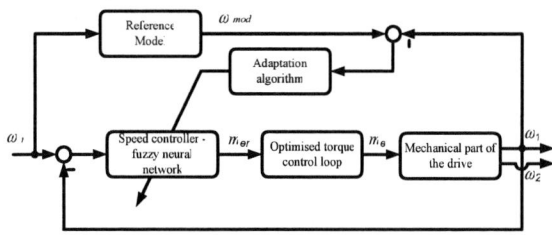

Fig. 2. Structure of the adaptive control system

The torque control of the induction motor is optimized here using the well known vector control method. In the analyzed direct field-oriented control (DFOC) structure of the IM, presented in the Fig. 3, the classical PI speed controller is replaced by ANFC of PI or sliding-mode type.

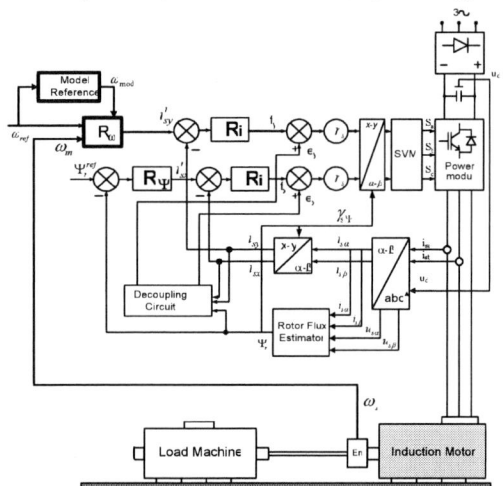

Fig. 3. Structure of adaptive speed control loop for the field oriented control of the IM drive with on-line tuned FL controller

The rotor flux vector is estimated using reduced-order Luenberger observer [14], with its eigenvalues tuned spe-

cially with the motor speed [15], to obtain the low sensitivity of the flux estimation to motor parameter uncertainties and accurate drive operation in the low speed range.

III. FUZZY NEURAL NETWORK SPEED CONTROLLER

A. Description of the fuzzy-neural network

The general diagram of the used fuzzy controller is presented in Fig. 4. It describes the relationship between the speed error e(k), its change $\Delta e(k)$ and change of the control signal $\Delta u(k)$ or u(k).

This fuzzy controller is of PI-type, if the system has the integrator in the output (marked by a dotted line in Fig. 4) [13], [16]. If this integrator is missing, the controller can be regarded a PD-type fuzzy controller. However, the control surface of the fuzzy system can be also seen as the nonlinear switching function [12]. So, this controller is treated as the fuzzy sliding-mode controller (FSC) in this study.

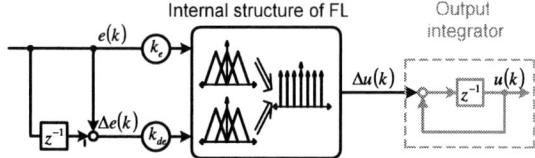

Fig. 4. A general structure of a fuzzy controller (PD-type or sliding FL – solid line; PI-type – including dotted line)

According to the literature, two main frameworks of the fuzzy sliding mode control can be distinguished. In the first one the switching surface s is calculated directly. Then, on the basis of the obtained function s, the nonlinear switching function is approximated by the fuzzy system. The second framework is presented in detail in book [17] and used in this paper. In this approach the switching surface is not directly visible. It can be calculated from the properties of the rule base, which describes the relationship between the error (e) and the change of the error (Δe). Similarly to the classical sliding mode control, the switching surface is described by:

$$s^* = \lambda^* e^* + \Delta e^* \qquad (4)$$

where λ^* represents the slope of the switching line: $\lambda^* e^* + \Delta e^* = 0$. The parameter λ^* can be calculated on the basis of the universe of discussion range for the input variable (e, Δe).

The rule base of the fuzzy controller incorporates several IF-THEN rules, in the following form:

$$R_j: \text{IF } x_1 \text{ is } A_1^j \text{ and } x_2 \text{ is } A_2^i \text{ THEN } y = w_i, \qquad (5)$$

where x_i – input variable of the system, A_i^j – input membership function, w_i – consequent function.

This controller can be realized as a general structure of neuro-fuzzy system shown in Fig. 5 [13], [16].

The functions of each layer are presented as follows:

Layer 1. Each input node in this layer corresponds to the specific input variable ($x_1 = e(k)$; $x_2 = \Delta e(k)$). These nodes only pass input signals to the second layer.

Layer 2. Each node performs a membership function A_i^j, that can be referred to as the fuzzification procedure.

Layer 3. Each node in this layer represents the precondition part of fuzzy rule and is denoted by Π which multiplies the incoming signals and sends the results out.

2212

Layer 4. This layer acts as a defuzzifier. The single node is denoted by Σ and performs the summation of all incoming signals. A defuzzification process in the neuro-fuzzy network, known as the singleton defuzzification method, is described by:

$$\Delta u = y_o = \sum_{j=1}^{M} w_j u_j \qquad (5)$$

In this paper two types of neuro-fuzzy controllers with 25 rules and triangular membership functions, shown in Fig. 4, are tested.

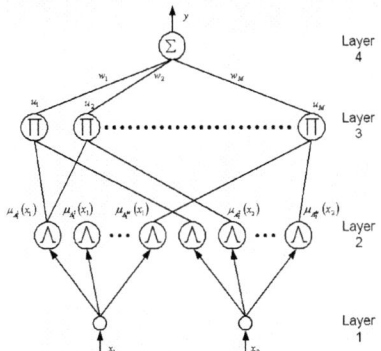

Fig. 5. General structure of a neuro-fuzzy controller

B. On-line learning algorithm

The fuzzy controller is tuned so that the actual motor speed ω_1 of the two-mass drive system can follow the output of the reference model ω_{mod} (see Scheme in Fig. 2). The tracking error is used as the tuning signal. The reference model is chosen as a standard second order term [13]:

$$G_m(s) = \frac{\omega_n^2}{s^2 + 2\zeta\omega_n s + \omega_n^2}, \qquad (6)$$

where ζ is a damping ratio and ω_n is a resonant frequency.

The supervised gradient descent algorithm is used to tune the parameters $w_1,...,w_M$ in the direction of minimizing the cost function like:

$$J(k) = \frac{1}{2}(\omega_{mod} - \omega_1)^2 = \frac{1}{2}e_m^2. \qquad (7)$$

Parameter adaptation is obtained using the following expression:

$$w_j(k+1) = w_j(k) + \Delta w_j, \qquad (8)$$

where:

$$\Delta w_j = -\gamma \frac{\partial J}{\partial w_j} = \gamma\left(-\frac{\partial J}{\partial y_o}\right)\left[\frac{\partial y_o}{\partial w_j}\right] = \gamma\delta_o u_j, \qquad (9)$$

u_j– is the normalized firing strength of *j*-th rule, γ– learning rate and

$$\delta_0 = -\frac{\partial J}{\partial y_o} = -\frac{\partial J}{\partial e_m}\frac{\partial e_m}{\partial y_o} = -\frac{\partial J}{\partial e_m}\frac{\partial e_m}{\partial \omega_1}\frac{\partial \omega_1}{\partial y_o}. \qquad (10)$$

Expression (10) involves computation of the gradient of ω_1 with respect to the output of the controller y_o, which is the reference electromagnetic torque m_{er}. The exact calculation of this gradient cannot be determined due to the uncertainty of the plant and nonlinear friction characteristic. However, it can be assumed that the change of the

drive speed with respect to the motor torque or current is a monotonic increasing process. Thus, this gradient can be approximated by some positive constant values. Owing to the nature of the gradient descent search only the sign of the gradient is critical to the iterative algorithm convergence. So the adaptation law of the controller parameters (8) can be written as:

$$w_j(k+1) = w_j(k) + \gamma\delta_o u_j \cong w_j(k) + \gamma e_m u_j \qquad (11)$$

The learning speed of the above algorithm is usually not satisfactory due to the slow convergence. To overcome this weakness, a modified algorithm based on local gradient PD control is used [13]:

$$\delta_o \cong e_m + \Delta e_m \qquad (12)$$

where Δe_m is the derivative of the e_m.

The learning rate γ can be divided into two factors k_p and k_d for e_m and Δe_m, respectively. The derivative term is used to suppress a large gradient rate. Thus, in formula (12), the similarity to the back propagation algorithm (with the learning rate and momentum factor) used for neural network training can be seen [13]. The dynamic performance of the drive system strictly depends on the value of the learning rate γ. The maximum value which guarantees the convergence of the system can be obtained by the analysis of the discrete type Lyapunov function.

IV. SIMULATION RESULTS

A. Control structure with Adaptive Neuro-Fuzzy Controller ANFC

Both adaptive controllers are tested for the two-mass system, with only one basic feedback from IM speed, where effective damping of torsional vibrations and load speed oscillations is the main problem. In the simulation study the nonlinear characteristics of the motor and load machine frictions are taken into consideration. The reference speed is set to 20% of nominal speed to avoid the electromagnetic torque limit.

The performances of the control structure strictly depend on k_d and k_p parameters of the adaptive law (12). The proper selection of those coefficients is always a compromise between the speed of the algorithm, measurement noises and parameter uncertainties [13].

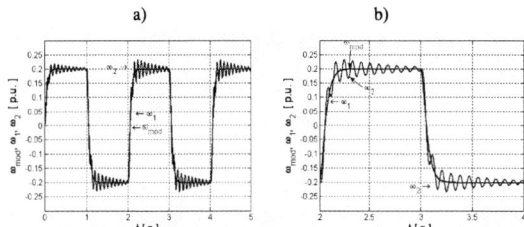

Fig. 6. Transients of the two-mass system for incorrect values of the adaptation coefficients in SNFC

The right choice of coefficients is very complicated in a two-mass system. Its big values can suppress the oscillations of the motor speed but at the same time they cause the vibrations in the load speed transient. This case is demonstrated in Fig. 6. Despite that the motor speed covers the model speed almost perfectly; the big oscillations are visible in the load speed transient. In order to select optimal values of k_d and k_p, nominal parameters of the drive are assumed. Then parameters k_d and k_p are determined using off-line optimization algorithm (genetic algorithm), to eliminate the tracking error as soon as possible.

First the adaptive control system with the neuro-fuzzy PI speed controller is investigated. In Fig. 7 transients of the system with the resonant frequency of the model set to $\omega_r = 20\text{s}^{-1}$ are presented. The main parameters of the drive are $T_1 = 100\text{ms}$, $T_2 = 100\text{ms}$, $T_c = 1.9\text{ms}$.

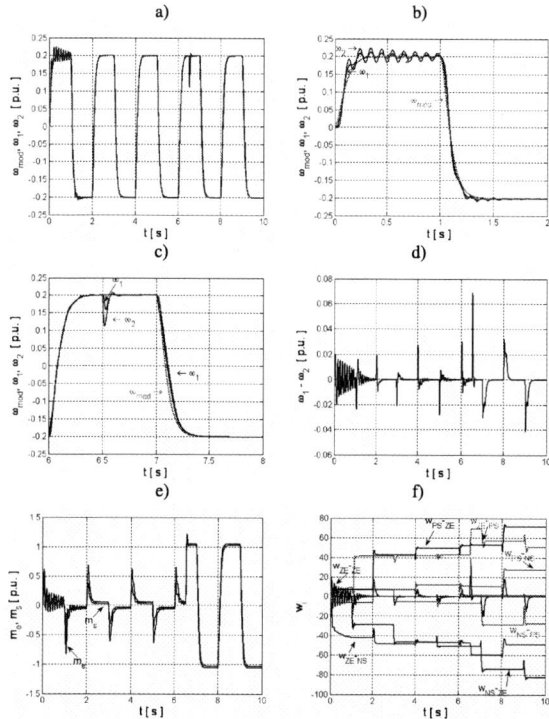

Fig. 7. Transients of the two-mass system with ANFC of PI-type for the inertia ratio $R=1$ and $T_c=1.9\text{ms}$: reference, motor and load speeds (a), fragments of the system speeds (b,c), speed difference (d), electromagnetic torque (e), selected weights (f)

The system starts with the controller weights set to zeros (Fig. 7f). It means that parameters of the drive are unknown to the controller. Despite this, the initial tracking error is small and decreases continuously (Fig. 7a). In the first operation period, the system speeds present big oscillations (Fig. 7b), which are suppressed quickly (Fig. 7a,b,d), after 2s, by the control structure. At the time $t = 6.5\text{s}$ the nominal load torque is applied to the system. It causes visible speed fall, which is eliminated fast. The system torque transients are presented in Fig. 7e.

Next the adaptive control structure is tested for different vales of the inertia ratio R. The system speeds transients for $R = 2$ and $R = 0.5$ are presented in Fig. 8. The tested system works in a stable way. In the $R = 2$ case the speeds have bigger oscillations during the first operation period. The system with smaller value of R eliminates the torsional vibrations much faster; after 1s oscillations are not noticeable in the system transients. It should be emphasized that both systems work correctly, i.e. they are

robust to the changes of the mechanical time constant of the load machine.

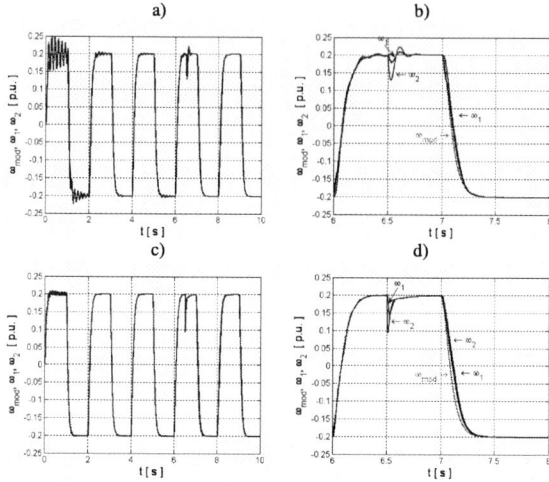

Fig. 8. Speeds of the two-mass system with the ANFC of PI-type for $T_c=1.9\text{ms}$ and the inertia ratio: $R=2$ (a,b) and $R=0.5$ (c,d)

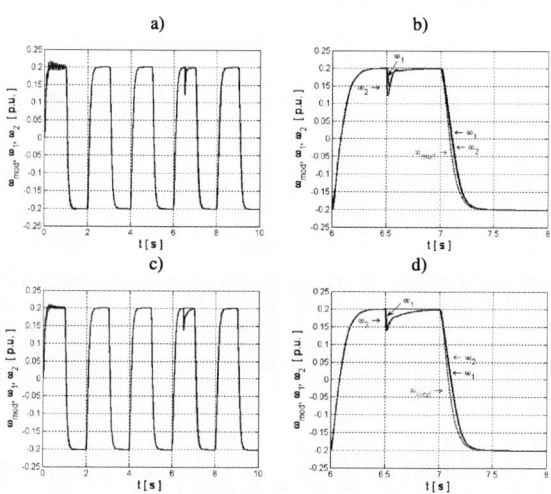

Fig. 9. Speeds of the two-mass system with the ANFC of PI-type for the inertia ratio $R=1$ and different values of the shaft time constant: $T_c=1.2\text{ms}$ (a,b) and $T_c =0.6\text{ms}$ (c,d)

Then the drive system is examined with respect to different value of the stiffness time constant T_c. The speed transients are presented in Fig. 9 for $T_c = 1.2\text{ms}$ (a,b) and $T_c = 0.6\text{ms}$ (c,d), respectively. The changes of the stiffness time constant do not influence the drive characteristics in a negative way. The smaller T_c results in smaller speed oscillations, but the frequency of the vibrations rises.

B. Control structure with the Adaptive Sliding-mode Neuro-Fuzzy Controller ASNFC

The adaptive control structure working with the sliding-mode neuro-fuzzy (ASNFC) speed controller is examined next. As in the previous case, the resonant frequency of the model is set to $\omega_r = 20\text{s}^{-1}$. In Fig. 10 the system transients are presented.

The load speed covers the reference signal almost perfectly (Fig. 10a,b,c). There is no visible difference between those two signals even in the first period of the work (Fig. 10b) despite the fact, that the fuzzy controller starts working with the weight factors set to zeros (Fig. 10f). Very small oscillations exist only during the speed reversal (Fig 10d). Also in the electromagnetic and shaft torques vibrations do not appear (Fig. 10e).

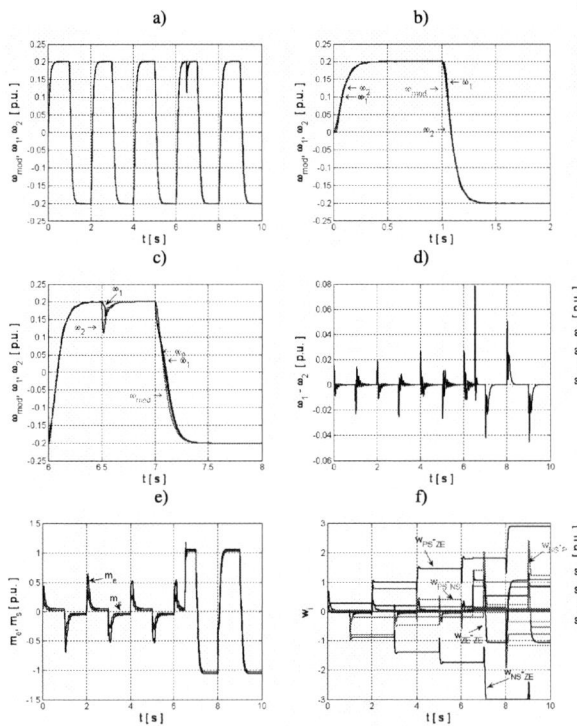

Fig. 10. Transients of the two-mass system with the ASNFC for the inertia ratio $R=1$ and $T_c=1.9$ms: reference, motor and load speeds (a), fragments of the system speeds (b,c), speed difference (d), electromagnetic torque (e), selected weights (f)

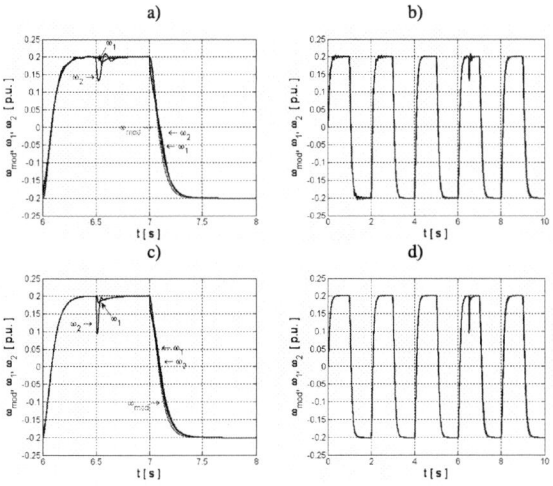

Fig. 11. Speeds of the two-mass system with the adaptive fuzzy sliding speed controller for $T_c=1.9$ms and inertia ratio $R=2$ (a,b) and $R=0.5$ (c,d)

Then the drive system is examined for different initial values of the time constant of the load machine. In Fig. 11 transients of the IM drive speeds are demonstrated, for different values of the load machine time constant. The drive system works well also for the bigger ($R=2$ – Fig. 11a,b) and smaller ($R=0.5$ – Fig. 11c,d) values of the inertia ratio.

Next the performance of the system with a changed value of the time constant of the shaft T_c is investigated. The performance of the system for the value $T_c=1.2$ms (Fig. 12a,b) and $T_c=0.6$ms (Fig. 12c,d) is checked. The changes of this parameter do not influence the drive characteristics negatively. Still the system speed covers the reference signal with high accuracy. Only the changes of the load torque disturb the speed transients. However, the speed falls are eliminated quickly and the system works correctly with the nominal torque.

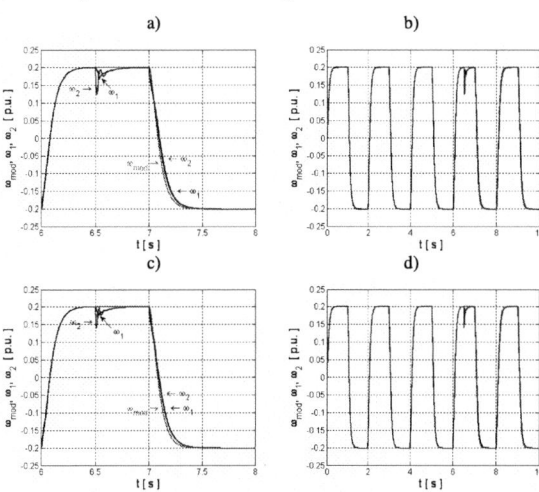

Fig. 12. Speeds of the two-mass system with the adaptive fuzzy sliding speed controller for the inertia ratio $R=1$ and different values of the $T_c=1.2$ms (a,b) and $T_c=0.6$ms (c,d)

V. CHOSEN EXPERIMENTAL RESULTS

A. Experimental set-up

The experimental set-up is composed of a IM motor fed by the SVM voltage inverter. The motor is coupled to a load machine, which is another IM supplied from an inverter (Fig.13). Both machines have the nominal power of 1.5 kW. The speed and position of the driving motor and load machine are measured by the incremental encoders with 36000 and 50000 imp./rev respectively. The control algorithm is implemented in the DS1103 card.

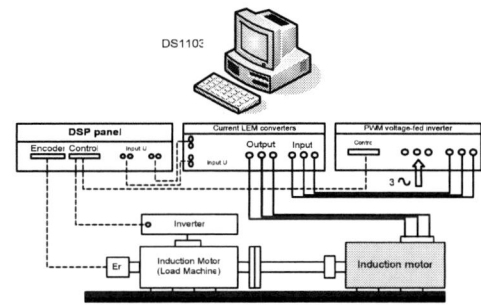

Fig. 13. Schematic diagram of the laboratory set-up

B. Control structure with the Adaptive Neuro-Fuzzy Controller ANFC

In the following section selected experimental results concerning the adaptive control structure with the fuzzy PI type speed controller are demonstrated.

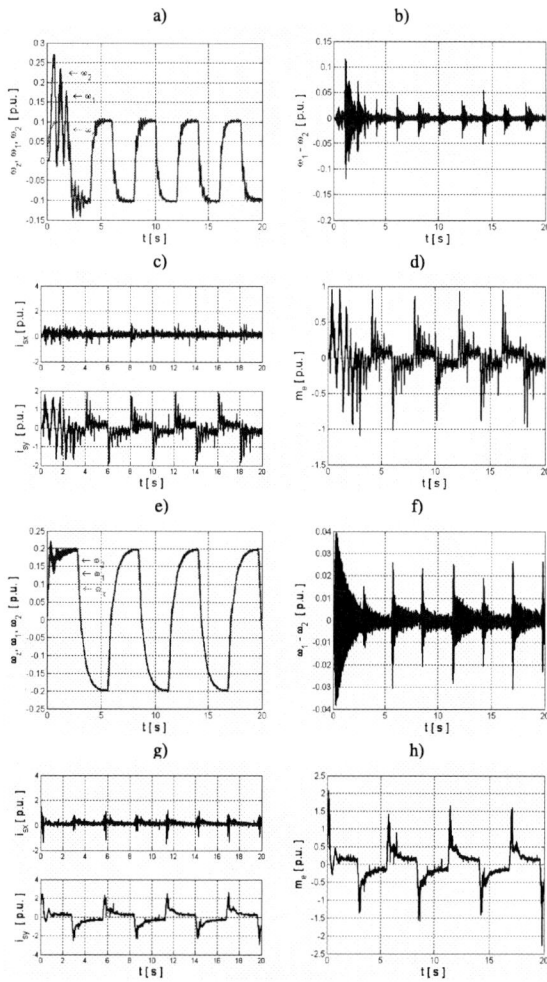

Fig. 14. Transients of the two-mass system with the ANFC of PI-type for the square-type of reference speed; for the inertia ratio R=1, T_c=1.9ms (a,b,c,d) and R=2, T_c=0.6ms (e,f,g,h): reference, motor and load speeds (a,e), speed difference (b,f), components of IM stator current (c,g), electromagnetic torque (d,h)

Firstly, the control system with the inertia ratio R=1 and the stiffness coefficient T_c=1.9 is tested. The selected system transients are shown in Fig. 14a-d. As in the simulation study, the system starts work with the weights set to zero. Therefore, during the first period of the operation in the speed transients large oscillations appear (Fig. 14a). With the course of time they are gradually suppressed. The biggest difference between the speeds exists during the first period (Fig. 14b). The sator current components of the IM driving motor are shown in Fig. 14c. This implies that the field oriented control structure of the IM works properly. The transient of the electromagnetic torque is presented in Fig. 14d, respectively.

Then the vector controlled drive system is examined for changing inertia $R=2$ and $T_c=0.6$ms. The system speeds are presented in Fig. 14e. As in the previous case, during the first operation period, performances of the system are not optimal. In speed transients the high frequency oscillation are visible (Fig. 14e,f). After one reversal period, transients of the system are much improved and the system speeds follow the reference signal with good accuracy. The components of the stator current and electromagnetic torque are demonstrated in Fig. 14 g,h.

Next the system is tested for the sinusoidal-type reference signal. In this case the reference model in the adaptive structure is replaced by the unity gain. The inertia ratio of the tested drive is $R=1$ and $T_c=0.6$ms. The selected system states are demonstrated in Fig. 15. Except of the first operation period, the system speeds cover the reference signal without a visible error (Fig. 15a). The difference between speeds is small for the whole time of the work (Fig. 15b). The presented current components and electromagnetic torque transients have required shapes.

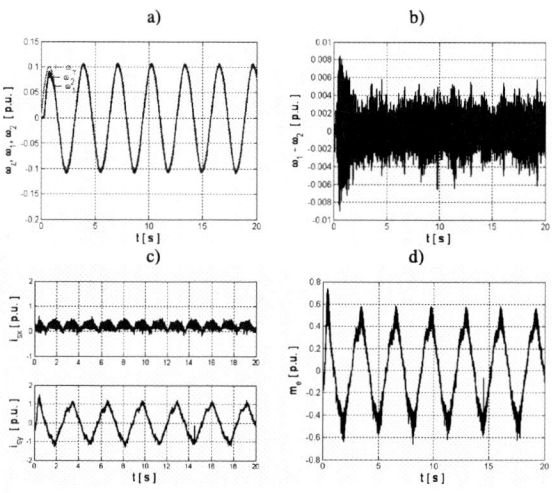

Fig. 15. Transients of the two-mass system with the ANFC of PI-type for the sinusoidal-type of reference speed; for the inertia ratio R=1 and T_c=0.6ms: reference, motor and load speeds (a), speed difference (b), components of IM currents (c), electromagnetic torque (d)

C. Control structure with the Adaptive Sliding-mode Neuro-Fuzzy Controller ASNFC

The study concerning the adaptive control structure with the on-line tuned sliding-mode fuzzy controller is presented in this section. As in the simulation study, in all presented cases the initial values of controller weights are set to zero.

First, the system with the inertia ratio R=1 and time constant of the shaft T_c=1.9ms is investigated. The selected transients of the system are presented in Fig. 16a-d. During the whole operation time the system speeds cover the reference signal. Contrary to the PI-type neuro-fuzzy adaptive controller, the system with the sliding-mode controller damps the torsional vibrations more effectively, even in the first operation period (Fig. 16a). The biggest difference between motor and load speeds exists during reversals (Fig. 16b). Transients of the stator current components are presented in Fig. 16c. Their shape proves the correct work of the field-oriented control structure. The electromagnetic torque is presented in Fig. 16d.

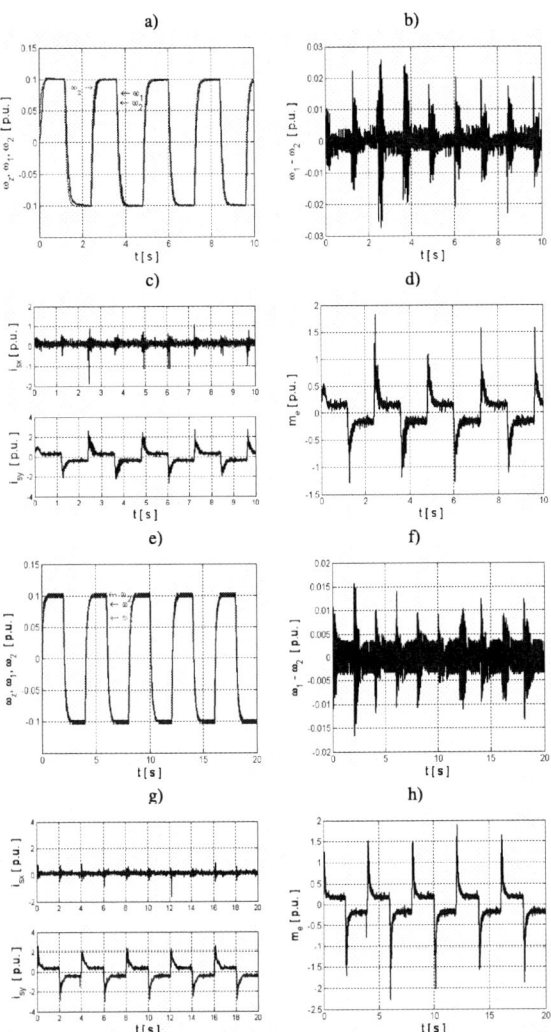

Fig. 16. Transients of the two-mass system with the ASNFC for the square-type of reference speed; for the inertia ratio $R=1$ and $T_c=1.9$ms (a,b,c,d) and $T_c=0.6$ms (e,f,g,h): reference, motor and load speeds (a,e),speed difference (b,f), components of IM currents (c,g), electromagnetic torque (d,h)

Then the system with a changed value of the inertia ratio $R = 2$ and stiffness time constant $T_c = 0.6$ms is tested. There are no visible torsional vibrations in the system speed transients (Fig.16e) and the electromagnetic torque (Fig. 16h), the difference between speeds is very small even on transients (Fig. 16f). The field-oriented structure of IM works correctly.

Next the system working with the sinusoidal-type reference signal is examined also. The selected transients of the system are presented in Fig. 17. The control structure works in a correct way. There is no visible difference between the speeds of the system and the signal of the model (Fig. 17a,b). The torsional vibrations are also not visible in transients of the other state variables of the two-mass drive.

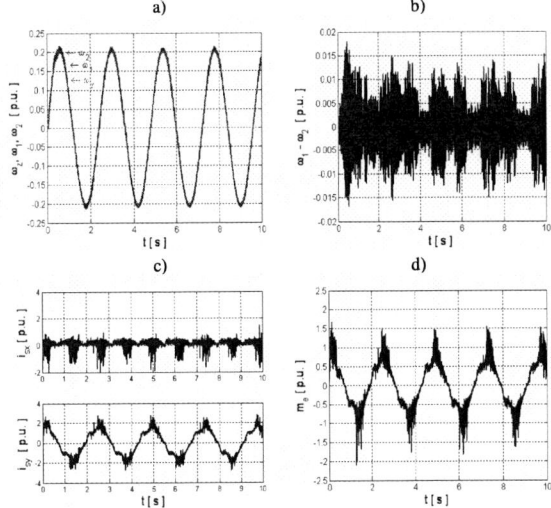

Fig. 17. Transients of the two-mass system with the ASNFC for the sinusoidal-type of reference speed; for inertia ratio $R=1$ and $T_c=0.6$ms: reference, motor and load speeds (a), speed difference (b), components of IM currents (c), electromagnetic torque (d)

VI. CONCLUSIONS

A model reference adaptive speed control using two types of on-line trained fuzzy neural networks (AFN) is presented and compared in the paper. The structure and the on-line learning algorithm of AFN speed controllers are described in detail. The performances of drive system with the proposed controllers are demonstrated through simulation and experimental results. Based on the conducted tests the following remarks can be formulated:

- The adaptive control structure working with the on-line tuned neuro-fuzzy speed controller ensures effective damping of torsional vibrations in the case of unknown mechanical parameters of the system.

- The selection of the learning rates k_p and k_d has a significant effect on the system performance. If these values are not selected properly, the learning process of the adaptive controllers does not lead to a well-defined control behavior.

- The control structure with the adaptive sliding-mode neuro-fuzzy controller (ASNFC) presents better dynamical performance than the control structure with the PI-type neuro-fuzzy controller (ANFC).

- When the initial controller weights are set to zero, the torsional vibrations are effectively damped from the first period of the drive system operation only in the control structure with the ASNFC.

- The torsional vibrations are easier suppressed in the analyzed control structure for the value of the inertia ratio $R<1$.

It should be emphasized that the torsional vibrations are suppressed effectively in both adaptive control structures with only one basic feedback from the motor speed.

ACKNOWLEDGMENT

Research work partially financed by The Ministry of Sciences and Higher Education (Poland) under Grant N510 023 32/2345 (2007-2009)

REFERENCES

[1] S. Katsura, Y. Matsumoto, K. Ohnishi, Analysis and Experimental Validation of Force Bandwidth for Force Control, *IEEE Trans. on Industrial Electronics*, vol. 53, no.3, pp. 922-928, 2006.

[2] S. N. Vukosovic, M. R. Stojic, Suppression of Torsional Oscillations in a High-Performance Speed Servo Drive, *IEEE Trans. on Industrial Electronics*, vol. 45, no.1, pp. 108-117, 1998

[3] J. Arellano-Padilla, G. M. Asher, M. Sumner, Control of an Dynamometer for Dynamic Emulation of Mechanical Loads With Stiff and Flexible Shafts, *IEEE Trans. on Industrial Electronics*, vol. 53, no.4, pp. 1250-1260, 2006

[4] K. Hace, K. Jezernik, A. Sabanovic, Improved Design of VSS Controller for a Linear Belt-Driven Servomechanism, *IEEE/ASME Transaction on Mechatronics,* vol. 10, no.4, pp. 385-390, 2005

[5] M. A. Valenzuela, J. M. Bentley, R. D. Lorenz, Evaluation of Torsional Oscillations in Paper Machine Sections, *IEEE Trans. on Industry Applications*, vol. 41, no.2 , pp. 493-501, 2005.

[6] G. Zhang, J. Furusho, Speed Control of Two-Inertia System by PI/PID Control, *IEEE Trans. on Industrial Electronics,* vol. 47, no.3, pp. 603-609, 2000.

[7] K. Szabat, T. Orlowska-Kowalska, Vibration Suppression in Two-Mass Drive System using PI Speed Controller and Additional Feedbacks – Comparative Study, *IEEE Trans. on Industrial Electronics,* vol. 54, no. 2, pp. 1193-1206, 2007.

[8] J. K. Ji, S. K. Sul, Kalman Filter and LQ Based Speed Controller for Torsional Vibration Suppression in a 2-Mass Motor Drive System, *IEEE Trans. on Ind. Electronics*, vol. 42, no.6, pp. 564-571, 1995

[9] P. Korondi, H. Hashimoto, V. Utkin, Direct Torsion Control of Flexible Shaft in an Observer-Based Discrete-Time Sliding Mode, *IEEE Trans. on Ind. Electronics*, vol. 45, no.2, pp. 291-296, 1998.

[10] T. Orlowska-Kowalska, K. Szabat, Neural Networks Application for Mechanical Variables Estimation of Two-Mass Drive System, *Trans. on Ind. Electronics*, vol. 54, no. 3, pp. 1352-1364, 2007.

[11] Y. Hori, H. Sawada, Y. Chun, Slow resonance ratio control for vibration suppression and disturbance rejection in torsional system, *IEEE Trans. on Ind. Electronics,* vol. 46, no.1, pp. 162-168, 1999.

[12] T. Orlowska-Kowalska, K. Szabat, Adaptive Fuzzy Sliding Mode Control of the Drive System With Flexible Joint, *Conf. Proc. of the 32nd Annual Confer. IECON'06,* pp. 994-999 , 2006

[13] T. Orlowska-Kowalska, K. Szabat, Control of the Drive System with Stiff and Elastic Couplings Using Adaptive Neuro-Fuzzy Approach, *IEEE Trans. on Industrial Electronics*, vol. 54, no.1, pp. 228-240, 2007.

[14] T. Orlowska-Kowalska, Induction Motor Flux Reconstruction Via New Reduced Order State Observer, *Electric Machines and Power Systems*, vol. 17, pp. 139-153, 1989.

[15] T. Orlowska-Kowalska, M. Dybkowski, Influence of the reduced-order rotor flux observer eigenvalues on dynamical properties of the sensorless induction motor drive, in *Computer Applications in Electrical Engineering*, ALWERS Poznan, pp. 144 – 153, 2006.

[16] F. J. Lin, R. F. Fung, R. J. Wai, Comparison of Sliding-Mode and Fuzzy Neural Network Control for Motor-Toggle Servomechanism, *IEEE Trans. on Mechatronics*, vol.3, no.4, pp. 302-318, 1998.

[17] R. R. Yager, D. P. Filev, Essentials of Fuzzy Modeling and Control, *Wiley*, 1994

Control of Flexible Drive with PMSM employing Forced Dynamics

Vittek Ján, Briš Peter, Makyš Pavol, Štulrajter Marek, Vavruš Vladimír

Faculty of Electrical Engineering/University of Žilina, Žilina, Slovak Republic, e-mail: *Jan.Vittek@fel.uniza.sk*

Abstract—Forced dynamics control principles are exploited to develop control algorithms for electric drive with PMSM and flexible couplings. The controller is of the cascade structure, comprising an inner speed control loop, which respects the principles of vector control and requires a load torque estimate and an outer position control loop, which is designed to control the PMSM rotor or load angle in the presence of the vibrations due to flexible coupling including an external load torques. Computation of PMSM stator voltage is based on novel algorithm exploiting forced dynamics control. Derived control laws require the estimates of the load torque and its derivatives estimated in the set of observers. Control algorithm is verified by simulations and preliminary experiments indicating good agreements with theoretical predictions, therefore further experimental verification of the designed control algorithm is recommended.

Keywords—Non-linear motion control, Permanent magnet synchronous motor, Servo-drive with flexible coupling.

I. INTRODUCTION

The combination of the elasticity and the inertia of the materials used in controlled mechanisms give to rise to vibration modes that can impair control accuracy if ignored. Some progress on active control of vibration modes in the driven mechanical loads based on the two-mass representation depicted in Fig. 1 has already been reported [1], [2]. Forced dynamics control (FDC) and observer based control has been also used for control of the rotor and load position [3], [4]. The original contribution here is the application of FDC for control of the PMSM rotor and mechanical load positions. Original cascade control structure is completed with voltage controlled inverter, including corresponding PWM, using novel algorithms for computation of the demanded stator voltages. Derived control algorithms require estimates load torque and its derivatives, what gives FDC a certain degree of robustness not only with respect to external disturbances but also to plant parameter variations since the plant parameter variations are equivalent to load torques applied to the unperturbed plant model.

The control algorithms for FDC of rotor position, θ_r and load mass angle, θ_L are developed in two steps. First an inner rotor speed control loop is formed using the FDC technique [5], which is a form of feedback linearisation [6]. The outer loop control law is designed with prescribed closed-loop dynamics for the control of the load mass angle in the presence of external load torques and the vibration mode. This approach achieves non-oscillatory control with a specified settling time of the step response. Due to page limitation impose on abstract the only substantial part of load mass angle control is included.

II. CONTROL LAW DEVELOPMENTS

The driven mechanism is a balanced mass with moment of inertia J_L, coupled to the motor shaft via a torsion spring representing the flexible coupling with spring constant, K_S, as shown in Fig. 1a together with a corresponding state variable block diagram, Fig. 1b for further mathematical analysis.

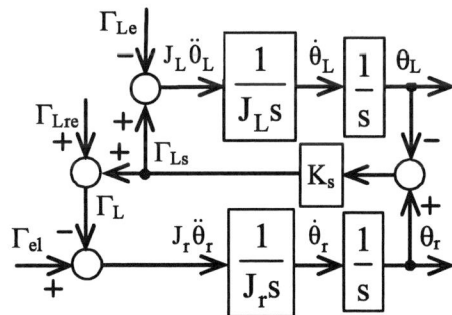

a) flexible coupling between motor and load

b) block diagram of the drive with flexible coupling

Fig. 1. Flexible coupling between motor and load

The internal electrical torque developed by the motor is Γ_{el}. Γ_{Le} and Γ_{Lre}, are load torques externally applied, respectively, to the load mass and the PMSM rotor. The shaft torsional deflection torque is Γ_{Ls} and the net load torque acting directly on the rotor is $\Gamma_L = \Gamma_{Ls} + \Gamma_{Lre}$. For generality, however, Γ_{Lre} is included to represent driven mechanisms (*such as friction torque on the motor side*) with more than one point of application of external load torques including one rigidly connected to the PMSM rotor. The differential equations corresponding to the block diagram of Fig. 1b are as follows:

978-1-4244-1741-4/08/$25.00 ©2008 IEEE

$$\dot{\theta}_r = \omega_r . \tag{1}$$

$$\ddot{\theta}_r = \frac{1}{J_r}\left[\Gamma_{el} - \Gamma_L\right], \text{ where } \Gamma_L = \Gamma_{Ls} + \Gamma_{Lre}, \tag{2}$$

$$\ddot{\theta}_L = \frac{1}{J_L}\left[\Gamma_{Ls} - \Gamma_{Le}\right], \text{ where } \Gamma_{Ls} = K_s\left(\theta_r - \theta_L\right) \tag{3}$$

The transfer function of the mechanism between the electrical torque and the rotor angle can be derived directly from Fig. 1b using Mason's formula: Here, the *encastre natural frequency*, v_n, is the frequency of the oscillations of the spring and load with the rotor held inertially fixed and with $\Gamma_{Le}=0$. The *free natural frequency*, ω_0, is the frequency of the oscillations of the combined rotor, spring and load with $\Gamma_{Le}=0$ and $\Gamma_{el}=0$.

$$F(s) = \frac{\theta_r(s)}{\Gamma_{el}(s)} = \frac{s^2 + \frac{K_s}{J_L}}{s^2 J_r\left[s^2 + \frac{K_s}{J_r} + \frac{K_s}{J_L}\right]} = \frac{1}{s^2 J_r}\frac{s^2 + v_n^2}{s^2 + \omega_0^2}, \tag{4a}$$

where $v_n = \sqrt{\dfrac{K_s}{J_L}}$, $\omega_0 = \sqrt{\dfrac{K_s}{J_r} + \dfrac{K_s}{J_L}} = v_n\sqrt{1 + \dfrac{J_L}{J_r}}$. $\tag{4b}$

A. Forced Dynamics Control of Rotor Speed

The speed control loop of the PMSM is designed using FDC while satisfying the vector control requirements [7]. The rotor speed obeys (5a) and the differential equation describing the closed loop dynamics must be of the same order. It therefore has linear first order dynamics prescribed by (5b), where T_ω is the prescribed time constant and $\omega_{r\,dem}$ is the demanded rotor speed.

$$\frac{d\omega_r}{dt} = \frac{1}{J_r}\left[c\left(\Psi_d i_q - \Psi_q i_d\right) - \Gamma_L\right], \tag{5a}$$

$$\frac{d\omega_r}{dt} = \frac{1}{T_\omega}\left(\omega_{r\,dem} - \omega_r\right) \text{ or } \ddot{\theta}_r = \frac{1}{T_\omega}\left(\dot{\theta}_{r\,dem} - \dot{\theta}_r\right). \tag{5b}$$

Here, p is the number of pole pairs and c=3p/2. Setting i_d=0 for vector control of the PMSM up to the nominal speed, [7] and equating the right hand sides of (6a) and (6b) yields the following FDC speed control law:

$$i_{d\,dem} = 0$$

$$i_{q\,dem} = \frac{1}{c\Psi_{PM}}\left[\frac{J_r}{T_\omega}\left(\dot{\theta}_{r\,dem} - \dot{\theta}_r\right) + \Gamma_L\right], \tag{6a}$$

$$\frac{d}{dt}\begin{bmatrix} i_d \\ i_q \end{bmatrix} = \begin{bmatrix} -\frac{R_s}{L_d} & p\omega_r\frac{L_q}{L_d} \\ -p\omega_r\frac{L_d}{L_q} & -\frac{R_s}{L_q} \end{bmatrix}\begin{bmatrix} i_d \\ i_q \end{bmatrix} - \frac{p\omega_r}{L_q}\begin{bmatrix} 0 \\ \Psi_{PM} \end{bmatrix} + \begin{bmatrix} \frac{1}{L_d} & 0 \\ 0 & \frac{1}{L_q} \end{bmatrix}\begin{bmatrix} u_d \\ u_q \end{bmatrix}. \tag{6b}$$

Hence $i_d=i_{d\,dem}$ and $i_q=i_{q\,dem}$ are regarded as the control variables. Given the motor governed by (6), a current controlled inverter is assumed in which the demanded stator phase voltages to, in turn, vary the stator voltage components, u_d and u_q, so that i_d and i_q follow their respective demands, $i_{d\,dem}$ and $i_{q\,dem}$, with zero dynamic

lag. This means that the stator current equations, (6b) are eliminated from the control system design.

Principles of feedback linearisation are exploited to determine PMSM stator voltage for voltage controlled inverter. If linearising functions for demanded stator currents, $i_{d\,dem}$ and $i_{q\,dem}$ are chosen as a first order time delay, (7a), then demanded stator voltages can be computed direct from stator current equations (6b) as (7b):

$$\frac{d}{dt}\begin{bmatrix} i_{d\,dem} \\ i_{q\,dem} \end{bmatrix} = \frac{1}{T_c}\begin{bmatrix} i_{d\,dem} - i_d \\ i_{q\,dem} - i_q \end{bmatrix}, \tag{7a}$$

$$u_{d\,dem} = \frac{L_d}{T_c}\left(i_{d\,dem} - i_d\right) + R_s i_d - p\omega_r L_q i_q$$

$$u_{q\,dem} = \frac{L_q}{T_c}\left(i_{q\,dem} - i_q\right) + R_s i_q + p\omega_r\left(\Psi_{PM} + L_d i_d\right) \tag{7b}$$

Derived equations (7b) are used for FDC of PMSM rotor speed exploiting voltage controlled inverter. Since the load torque appears on the right hand side of the demanded current, $i_{q\,dem}$ (6a), it is necessary to design an observer to estimate the net load torque acting on the shaft of the motor (*see corresponding section D*).

B. Forced Dynamics Control of Rotor Angle

In this case, the plant is formed by first replacing the FDC speed control loop described in the previous section by its ideal first order transfer function block and then integrating this block into Fig. 1b with its output equal to $\dot{\theta}_r$. Then $\dot{\theta}_{r\,dem}$ is treated as the control variable. Only the kinematic integrator producing the rotor angle is needed from Fig. 1, the remainder of the block diagram being irrelevant to the control law derivation. The resulting plant is shown in Fig. 2.

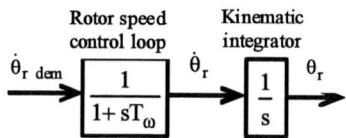

Fig. 2 Plant for the design of the rotor position control algorithm

The net load torque, Γ_L, (ref. Fig 1b), does not, of course, appear in Fig. 2 because it has already been compensated in the FDC speed control loop. The position control loop will now be formed by application of FDC to this plant.

If the control system is to have a specified settling time, $T_{s\theta}$, then the desired closed loop transfer function and the corresponding closed loop differential equation may be determined with the aid of the Dodds settling time formula [5]:

$$\frac{\theta_r(s)}{\theta_{r\,dem}(s)} = \left[\frac{1}{1 + s\frac{T_{s\theta}}{1,5(1+n)}}\right]^n_{n=2} = \frac{\frac{81}{4T_{s\theta}^2}}{s^2 + \frac{9}{T_{s\theta}}s + \frac{81}{4T_{s\theta}^2}}. \tag{8}$$

Converting (8) to the time domain yields the second order closed loop differential equation for the rotor position:

$$\ddot{\theta}_r = \frac{81}{4T_{s\theta}^2}\left(\theta_{r\,dem} - \theta_r\right) - \frac{9}{T_{s\theta}}\dot{\theta}_r \qquad (9)$$

Equating the right hand sides of (9) and (5b) and solving the equation for the control variable, $\dot{\theta}_{r\,dem}$, then yields the following FDC law:

$$\dot{\theta}_{r\,dem} = \left(1 - \frac{9T_\omega}{T_{s\theta}}\right)\dot{\theta}_r + \frac{81T_\omega}{4T_{s\theta}^2}\left(\theta_{r\,dem} - \theta_r\right). \qquad (10)$$

The resulting block diagram for the closed-loop control of the rotor angle is shown in Fig. 3.

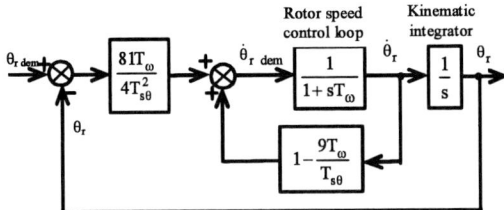

Fig. 3 Block diagram for FDC of rotor angle

C. Forced Dynamics Control of Mechanical Load Angle

FDC approach as used for control of the rotor speed is adopted to design outer position loop for control of the load mass angle. Here, the FDC speed control loop is replaced by its ideal transfer function block and integrated into Fig. 1b resulting in the plant of Fig. 4.

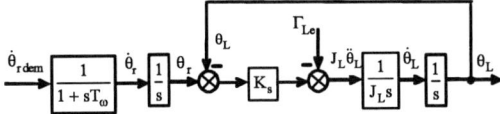

Fig. 4 Block diagram for load mass angle control system development

The load torque acting on the rotor does not appear because it has been cancelled in the rotor speed FDC loop. The plant differential equations in the time domain corresponding to Fig. 4 are as follows:

$$\ddot{\theta}_L = \frac{1}{J_L}\left[K_s\left(\theta_r - \theta_L\right) - \Gamma_{Le}\right], \qquad (11a)$$

$$\ddot{\theta}_r = \frac{1}{T_\omega}\left(\dot{\theta}_{r\,dem} - \dot{\theta}_r\right). \qquad (11b)$$

Successive differentiation of (11a) gives:

$$\dddot{\theta}_L = \frac{1}{J_L}\left[K_s\left(\dot{\theta}_r - \dot{\theta}_L\right) - \dot{\Gamma}_{Le}\right], \qquad (12a)$$

$$\theta_L^{IV} = \frac{1}{J_L}\left[K_s\left(\ddot{\theta}_r - \ddot{\theta}_L\right) - \ddot{\Gamma}_{Le}\right]. \qquad (12b)$$

Using the derived equation (12b) and substituting for $\ddot{\theta}_r$ and $\ddot{\theta}_L$ yields the fourth order output derivative equation (10) of the load mass angle for the FDC synthesis:

$$\theta_L^{IV} = \frac{K_s}{J_L T_\omega}\left(\dot{\theta}_{r\,dem} - \dot{\theta}_r\right) - \frac{K_s^2}{J_L^2}\left(\theta_r - \theta_L\right) + \frac{K_s}{J_L^2}\Gamma_{Le} - \frac{1}{J_L}\ddot{\Gamma}_{Le}. \qquad (13)$$

Using settling time formula (14a) for n=4, the following 4th order system will yield a specified settling time, $T_{s\theta L}$, of the step response without overshoot:

$$T_{s\theta L} = 1.5(1+n)\frac{1}{\omega_n} \quad\text{or}\quad \omega_n = \frac{15}{2T_{s\theta L}}, \qquad (14a)$$

$$\frac{\theta_L(s)}{\theta_{L\,dem}(s)} = \frac{\left(\frac{15}{2T_{s\theta L}}\right)^4}{s^4 + 4\left(\frac{15}{2T_{s\theta L}}\right)s^3 + 6\left(\frac{15}{2T_{s\theta L}}\right)^2 s^2 + 4\left(\frac{15}{2T_{s\theta L}}\right)^3 s + \left(\frac{15}{2T_{s\theta L}}\right)^4}. \qquad (14b)$$

Using (12a) from (11b) θ_L^{IV} can be expressed as:

$$\theta_L^{IV} = a_0\left(\theta_{L\,dem} - \theta_L\right) - a_1\dot{\theta}_L - \frac{a_2 K_s}{J_L}\left(\theta_r - \theta_L\right) + \frac{a_2}{J_L}\Gamma_{Le} - \frac{a_3 K_s}{J_L}\left(\dot{\theta}_r - \dot{\theta}_L\right) + \frac{a_3}{J_L}\dot{\Gamma}_{Le}, \qquad (15)$$

where a_i, i=0,1,2,3 are the corresponding coefficients of s^i in transfer function (14b). $\dot{\theta}_{r\,dem}$ is again treated as the control variable. If right hand sides of (13) and (15) are compared while exploiting (12a) and (12b) the following control algorithm results for control of the load mass angle (16).

Corresponding overall control system block diagram is shown in Fig. 5. As stated previously, this is in the form of a state feedback control law with the gains already determined as functions of the plant parameters and the desired closed loop system parameters. But in this case, the first and second derivatives of the external load torque also appear on the right hand side. This is a general feature of FDC when external disturbances are included in the plant model used for the synthesis of control system.

D. Load Torque Observer Design

Estimates are required of a) the net load torque, $\hat{\Gamma}_L$, acting on the PMSM rotor to feed the FDC law (6a) and b) the external load torque, $\hat{\Gamma}_{Le}$, applied to the load mass together with its first and second derivatives to feed FDC law (16). Two separate observers will be designed for this purpose and their places in the overall structure of the load mass angle control system are shown in Fig. 6.

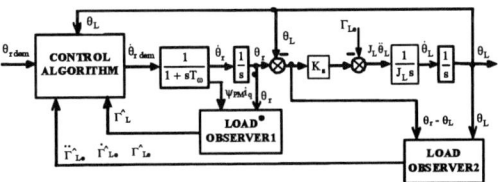

Fig. 6 Load mass angle control system showing the observers

With reference to Fig. 1b, $\Gamma_L = \Gamma_{Ls} + \Gamma_{Lre}$. The shaft torsional deflection torque, Γ_{Ls}, is state dependent and could be estimated by an observer based on a real time model of the two-mass system but Γ_{Lre} is an external load torque. Hence the observer has a real time model based on the PMSM alone and caters for the driven mechanism by estimating Γ_L directly, treating it as if it was an external load torque. This also greatly simplifies the observer '1'

2221

design. In absence of the model of the driven mechanism, and in view of the fact that the external load torque, Γ_{Lre}, directly applied to the rotor can be arbitrarily time varying, then the form of $\Gamma_L(t)$ is unknown and therefore differential equations accurately modeling the load torque and augmenting the real time model cannot be formed.

$$\dot{\theta}_{r\,dem} = \frac{1}{T_\omega}\left[\frac{J_L a_0}{K_s}\left(\theta_{L\,dem} - \theta_L\right) - \left(a_3 - \frac{1}{T_\omega}\right)\dot{\theta}_r - \left(\frac{J_L a_1}{K_s} - a_3\right)\dot{\theta}_L - \left(a_2 - \frac{K_s}{J_L}\right)\cdot\left(\theta_r - \theta_L\right) - \left(\frac{1}{J_L} - \frac{a_2}{K_s}\right)\Gamma_{Le} + \frac{a_3}{K_s}\dot{\Gamma}_{Le} + \frac{1}{K_s}\ddot{\Gamma}_{Le}\right] \quad (16)$$

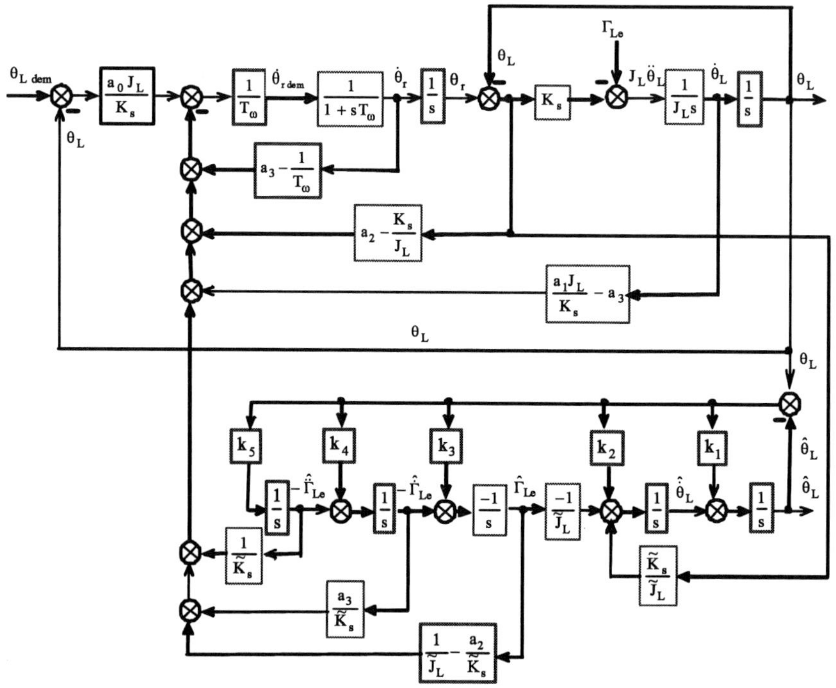

Fig. 5 Overall control system block diagram for FDC of load mass angle

Under these circumstances, Γ_L can be treated *theoretically* as if it is constant provided that its change over a period equal to the observer correction loop settling time, T_{so}, is negligible. So with sufficiently small T_{so}, the observer produces a net load torque estimate, $\hat{\Gamma}_L(t)$, that is able to track $\Gamma_L(t)$ with negligible dynamic lag. Thus, the real time plant model in the observer is based on (1) and (5a) augmented by a further state equation, $d\Gamma_L/dt = 0$. The observer correction loop is actuated by the error, $e_\theta = \theta_r - \hat{\theta}_r$, between the measured rotor position and its estimate from the observer. The observer state equations are therefore as follows:

$$\frac{d\hat{\theta}_r}{dt} = \hat{\omega}_r + k_\theta e_\theta, \quad (17a)$$

$$\frac{d\hat{\omega}_r}{dt} = \frac{1}{J_r}\left[c\left(\psi_d i_q - \psi_q i_d\right) - \hat{\Gamma}_L\right] + k_\omega e_\theta, \quad (17b)$$

$$-\frac{d\hat{\Gamma}_L}{dt} = 0 + k_\Gamma e_\theta. \quad (17c)$$

Here, k_θ, k_ω and k_Γ are the correction loop gains. Fig. 7 shows the corresponding block diagram. Again the settling time formula is used with the denominator of transfer function (8) as the characteristic polynomial. Then design equations for the observer gains may be derived to yield a correction loop settling time, T_{so}, defined as the time taken for $|e_\theta(t)|$ to fall to and stay below 5% of its peak value following a disturbance.

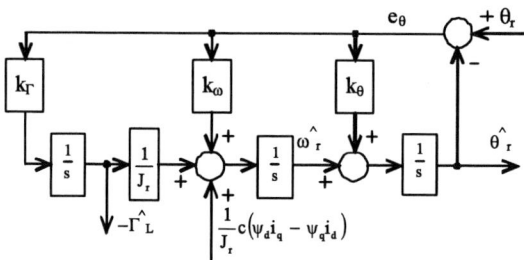

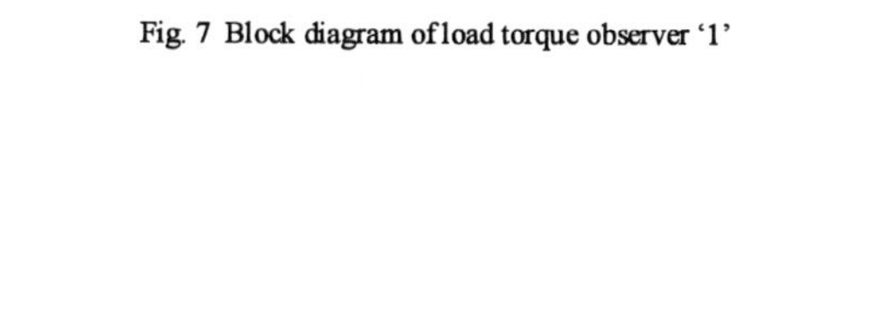

Fig. 7 Block diagram of load torque observer '1'

Equating the desired characteristic polynomial (*LHS* of (18a)) and the characteristic polynomial obtained from Fig. 9 (*RHS* of (18a)) yields the desired gains (18b) of the observer:

$$s^3 + \frac{18}{T_{so}}s^2 + \frac{108}{T_{so}^2}s + \frac{216}{T_{so}^3} = s^3 + s^2 k_\theta + s k_\omega + \frac{k_\Gamma}{J_r}, \quad (18a)$$

$$k_\theta = \frac{18}{T_{so}}, \quad k_\omega = \frac{108}{T_{so}^2}, \quad k_\Gamma = \frac{216 J_r}{T_{so}^3}. \quad (18b)$$

Although the motor load torque is assumed constant in the formulation of observer real time model, the estimate of the torque acting on the shaft of the motor, $\hat{\Gamma}_L$, will follow a time varying load torque and will do so more faithfully as T_{so} is reduced.

The second observer '2' for estimation of the external load mass torque, Γ_{Le}, and its first two derivatives is shown in Fig. 5. In this case the highest load torque derivative that needs to be estimated is treated theoretically as if it is constant provided that its change over a period equal to the observer correction loop settling time, T_{so}, is negligible. The observer correction loop is actuated by the error, $e_\theta = \theta_L - \hat{\theta}_L$, between the measured load position and its estimate from the observer. The observer state equations are therefore as follows:

$$-s\hat{\ddot{\Gamma}}_{Le} = 0 + k_5 e_\theta, \quad -s\hat{\ddot{\Gamma}}_{Le} = -\hat{\ddot{\Gamma}}_{Le} + k_4 e_\theta \quad (19a,b)$$

$$-s\hat{\dot{\Gamma}}_{Le} = -\hat{\Gamma}_{Le} + k_3 e_\theta, \quad s\hat{\theta}_L = \hat{\dot{\theta}}_L + k_1 e_\theta \quad (19c,e)$$

$$s\hat{\dot{\theta}}_L = \frac{K_s}{J_L}(\theta_r - \theta_L) - \frac{1}{J_L}\hat{\Gamma}_{Le} + k_2 e_\theta \quad (19d)$$

The observer correction loop settling time is again given by the settling time formula, $T_{so} = 1.5(1+n)/\omega_0$, where the gains, k_i, i=1,2,3,4,5 are determined by pole assignment at $s_{1-5} = -\omega_0$. Equating the desired characteristic polynomial (*RHS of (20a)*) and the characteristic polynomial obtained from Fig. 3 (*LHS of (20a)*) yields the gain equations (20b):

$$s^5 + k_1 s^4 + k_2 s^3 + (k_3/J_L)s^2 + (k_4/J_L)s + k_5/J_L =$$
$$= s^5 + 5\omega_0 s^4 + 10\omega_0^2 s^3 + 10\omega_0^3 s^2 + 5\omega_0^4 s + \omega_0^5 \quad (20a)$$

$$k_1 = 5\omega_0, \quad k_2 = 10\omega_0^2, \quad k_3 = 10 J_L \omega_0^3,$$
$$k_4 = 5 J_L \omega_0^4, \quad k_5 = J_L \omega_0^5 \quad (20b)$$

III. Verification by Simulation and Premilinary Experimental Results

The simulation and subsequent experimental results for control of the PMSM rotor angle are presented in Fig. 8 and Fig. 9, respectively. Due to problems with implementation of the second position sensor for load mass angle control algorithm this one is verified by simulation only and results are presented in Fig. 10 with a parabolic external load mass torque. The parameters of the PMSM relevant to both simulation and experiments are listed in the Appendix.

Computational step of simulations was $h_s = 1$ μs for correct simulation of PWM, while the sampling frequency for control algorithm was $1/h_c = 0.1$ ms, which corresponds to the sampling frequency achieved for digital implementation of the rotor angle control algorithms with rigid coupling. All the simulations are carried out with zero initial state variables and a step load position demand, $\theta_{L\ dem} = 10$ rad and a prescribed settling time of $T_s = 0.1$ s. The ideal position curve, $\theta_{id}(t)$ is computed from transfer function (8) as a standard of comparison and a corresponding error function is evaluated as $e_\theta = \theta_r - \theta_{id}$ (*in plots 2 times magnified*).

The external load mass torque, Γ_{Le}, is applied at t=0.25 s and simulated as a step change for rotor angle control, being zero for the time interval t<0.25 s. During experiments the load torque was developed as a starting locked torque of the induction machine rigidly coupled to the PMSM. For load angle control load torque is simulated as a parabolic function, which is limited to a maximum value of $\Gamma_{Le\ max} = 2$ Nm, to verify proper function of the observer '2' placed at a load side.

A PC equipped with a PCL812/PG PC lab card was used for control algorithm evaluation, sensing of phase currents via LEM current sensors and for control of the inverter. Supply d.c. voltage of inerter was $U_{dc} = 550$ V.

A. Forced Dynamics Control of Rotor Angle

For simulations and experiments of rotor angle control was chosen settling time of speed control loop, $T_\omega = 0.01$ s, settling time of load torque observer from the motor side was $T_{so} = 20$ ms.

For both, simulations and experiments the subplot (a) shows the response of the designed rotor angle control system together with the ideal response computed from transfer function to a step position demand of $\theta_{rdem} = 10$ rad including the 2 times magnified difference between them. Subplot (b) and subplot (c) show the rotor speed and angular acceleration as the outputs of the filtering observer '1' on the motor side taken as the inputs of control algorithm (10) for whole measure period. Estimated load torque acting on the shaft of the motor is shown in subplot (d) while this subplot for simulation shows also step change of the applied load torque. Demanded and real stator currents in rotor coupled frame, d_q are shown in subplot (e) and (f) respectively. Subplot (g) shows corresponding motor phase currents in the stator coupled frame, α_β. Applied stator voltage and resultant stator current as a function of time are shown in subplot (h).

Comparison of both, simulation and experimental results of rotor angle control shows a good agreement with the theoretical predictions made during the control law development.

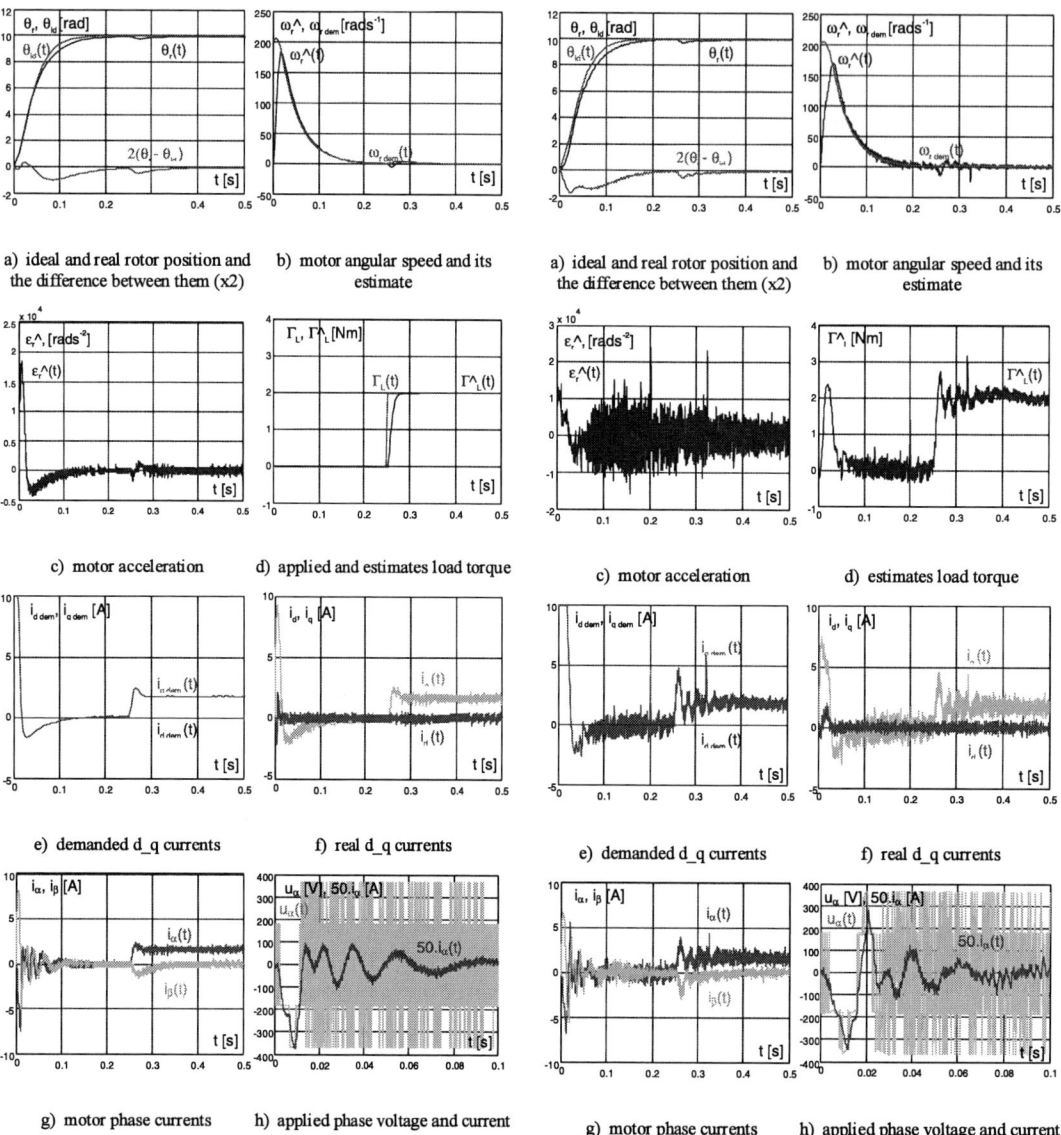

a) ideal and real rotor position and the difference between them (x2)

b) motor angular speed and its estimate

a) ideal and real rotor position and the difference between them (x2)

b) motor angular speed and its estimate

c) motor acceleration

d) applied and estimates load torque

c) motor acceleration

d) estimates load torque

e) demanded d_q currents

f) real d_q currents

e) demanded d_q currents

f) real d_q currents

g) motor phase currents

h) applied phase voltage and current

g) motor phase currents

h) applied phase voltage and current

Fig. 8 Simulation results of rotor angle control

Fig. 9 Experimental results of rotor angle control

It is also evident that in the experiments as well as in the simulations, the rotor position reaches 9,5 radians (*95% of demanded value*) at a time very close to the prescribed settling time of 0,1 s. It can be therefore concluded that position Forced Dynamics Control of the electric drive employing PMSM and respecting vector control principles was verified not only with simulations but also experimentally.

B. Forced Dynamics Control of Mechanical Load Angle

For simulations of load angle control were chosen parameters: rotor moment of inertia, J_r=0,000306 kgm^2, load moment of inertia, J_L=0,000918 kgm^2 and shaft spring constant, K_s=9 Nrad^{-1} . Following settling time constants were chosen: a) for speed control loop, T_ω=0,01 s, b) for load torque observer '1' from the motor side, T_{so}=5 ms and c) for load torque observer '2' from load side, T_{so}=15 ms.

Settling time for FDC of stator current components were set as $T_c=2$ ms. Simulation results of load mass angle control are shown in Fig. 10. All the simulations are carried out with zero initial state variables and a step load position demand, $\theta_{L\ dem}=10$ rad and a prescribed settling time of $T_s=0.1$ s. The ideal position curve, $\theta_{id}(t)$ is computed from transfer function (8) as a standard of comparison.

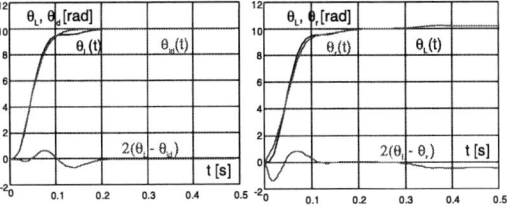

a) ideal and real load positions and   b) load and the rotor positions and
   the difference between them (x2)   the difference between them (x2)

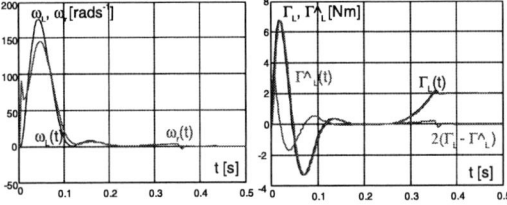

c) load mass and the rotor speed as   d) applied and estimated load torque
   a function of time   & the difference between (x2)

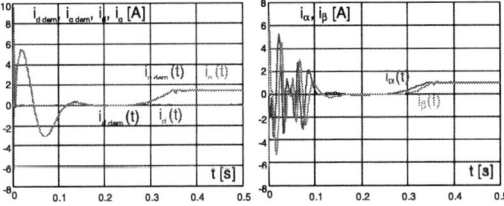

e) magnetizing and torque stator   f) stator current components in stator
   current components   fixed reference frame

Fig. 10 State variables as a function of time for load mass angle control

Subbplot (a) in Fig. 10 shows the ideal step response computed from (14b) and the simulated step response of the designed control system including the error between them which is magnified 2 times.

Subplot (b) shows the positions of the rotor and the load and the difference between them. The compensation of the external load mass torque applied at $t=0.25$ s is clearly evident from the angular displacement between the PMSM rotor and the load mass after this torque is applied. The initial differences between the rotor and load displacements are due to the acceleration and deceleration torques being applied via the torsion spring.

Subplot (c) shows the angular velocities of the rotor and the load. The applied net shaft torque and its estimate fed back to the FDC algorithm including the difference between them (*magnified twice*) are shown in subplot (d). Subplot (e) shows the demanded current $i_{d\ dem}$ and $i_{q\ dem}$ and real current components, i_d and i_q in the rotor fixed frame d_q. As can be seen i_d is kept close to zero, as required for the vector control while the initial positive and negative swings of $i_q(t)$ current producing the acceleration and deceleration torques can be seen together with the increase of $i_q(t)$ after $t=0.25$ s required to counteract the external load torque via the spring. Finally subplot (f) shows real stator current components, i_α and i_β in stator coupled frame α_β.

Applied stator phase voltage calculated from (7b) and a resulting current component in the stator coupled frame for time interval $t\in(0-0.25)$ s are shown in Fig. 11.

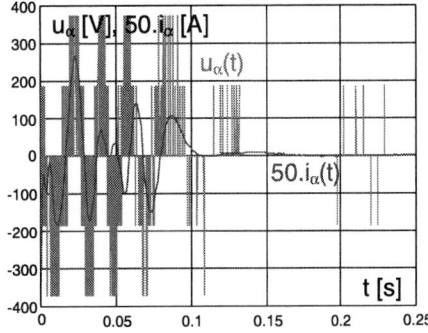

Fig. 11 Applied stator phase voltage and resulting current

Possibility to estimate net load torque and its first and second derivative as the inputs of the control algorithm shows Fig. 12 for whole computational interval.

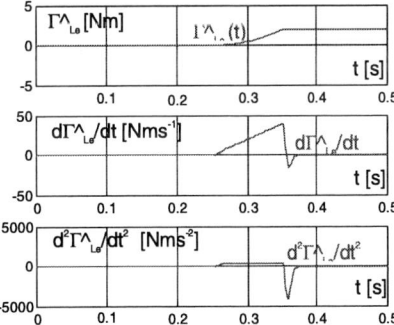

Fig. 12 Estimated load torque and its derivatives

Presented simulation results confirm possibility of Forced Dynamics Control of the load angle in spite of flexible coupling between motor and load. As can be seen from the Fig. 10, subplot (a) the load angle response follows the prescribed ideal response fairly closely. The error could be kept within the specification for particular applications by appropriate setting of T_{so}.

IV. Conclusions and recommendations

A control system for electric drives with permanent magnet synchronous motor and flexible couplings between the motor and the load based on the principles of forced dynamics control has been presented and verified by simulations and preliminary experiments.

Possibility of rotor angle control exploiting principles of Forced Dynamics Control was verified not only with simulations but also experimentally. Experimental results presented confirm that the rotor angle response follows the prescribed ideal response fairly closely.

Control algorithm for load mass angle control was verified by simulations. The results for the parabolic increase of the external load torque to its limit of 2 Nm show very effective compensation and consequential robustness of the control system to this disturbance. This implies successful operation of the observers including the estimation not only of the load torque but also its derivatives needed by the FDC algorithm.

Further investigations should be carried out with regard to dynamic motor and load parameter mismatches to exploit the robustness afforded by the load torque estimation. The robustness of the overall control system for load mass angle control could be improved further by adding an outer model reference adaptive control loop or sliding mode control loop. The experimental investigation of the proposed load mass angle control will follow as soon as possible.

Acknowledgment

The authors wish to thank Slovak Grant Agency VEGA for funding the project No.4087/07 'Servosystems with Rotational and Linear Motors without Position Sensor'.

References

[1] Korondi, P., Hashimoto, H., Utkin, V. I.: 'Direct Torsion Control of Flexible Shaft in an Observer-based Discrete Sliding Mode', IEEE Trans. on Industrial Electronics, y. 1998, vol. 45, No. 2, pp.291-296.

[2] Arellano-Padilla, J., Asher, G. M., Summer, M.: 'Robust Fuzzy-Sliding Mode Position Controller for Motor Drives Operating Variable Loads', Proc. of EPE-PEMC 2002 conf., Cavtat, Croatia, CD-Rom.

[3] Dodds, S. J. and Szabat, K, 'Forced Dynamic Control of Electric Drives with Vibration Modes in the Mechanical Load', EPE-PEMC 2006, Portoroz, Slovenia, CD-Rom.

[4] Szabat, K., Orlowska-Kowalska, T.: 'Application of Different State Estimators in Control Structure of Two-mass Drive System', Proc. of OPTIM 2004 conf., Brasov, Romania, vol. 3, pp. 43-48.

[5] Vittek, J., Dodds, S. J.: 'Forced Dynamics Control of Electric Drives', EDIS, Zilina University Publishing Centre, 2003, ISBN 80-8070-087-7, http://www.kves.uniza.sk (E-learning).

[6] Isidori, A.: 'Nonlinear Control Systems', Springer-Verlag London, 1995.

[7] Leonhard, W.: 'Control of Electrical Drives', Springer, Berlin 2001.

[8] Novotny, D. W., Lipo, T. A.: 'Vector Control and Dynamics of AC Drives', Clarendon Press Oxford, 2003.

[9] Brock, S., Deskur, J., Zawirski, K.: 'Robust Speed and Position Control of PMSM using Modified Sliding Mode Method', Proc. of EPE-PEMC 2000 conference, Kosice, Slovakia, vol. 6, pp. 6-29-6-34.

[10] Aguilar, O., Loukianov, A. G., Canedo, J. M.: 'Observer-based Sliding Mode Control of Synchronous Motor', proc. of IFAC Congress on Automatic Control, Guadalajara, Mexico 2002, CD-Rom.

[11] Ryvkin, S., Izosimov, D., Palomar-Lever E.: 'Digital Sliding Mode Based References Limitation Law for Sensorless Control of an Electromechanical Systems', Journal of Physics 2005, IOP Publ. Ltd, series 23, pp. 192-201.

[12] Comnat, V., Cernat, M., Moldoveanu, F., Ungar, R.: 'Variable Structure Control of Surface Permanent Magnet Synchronous Machine', Proceedings of PCIM'99 conference, Numberb, Germany, 1999, pp. 351-356.

[13] Dodds, S. J., 'Observer Based Robust Control', Proceedings of Advances in Computing and Technology (AC&T) 2007, University of East London, UK.

[14] Perdukova, D., Fedor, P., Timko, J., 'Modern Methods of Complex Drives Control', Acta Technica CSAV vol. 49, Czech Republic, 2004, pp. 31-45.

Appendix

Parameters of PMSM are: winding resistance, R_s=33,3 Ω, winding inductances, L_d=53,8 mH, L_q=53,8 mH, permanent magnet flux, Ψ_{PM}=0,262 Wb, number of pole pair, p=3 and moment of inertia, J_r=0,0006 kg.m^2.

The problems of high dynamic drive control under circumstances of elastic transmission

Jan Deskur, Roman Muszynski

Poznan University of Technology, Institute of Control and Information Engineering, Poznan, Poland
Jan.Deskur@put.poznan.pl   Roman.Muszynski@put.poznan.pl

Abstract— **In the paper the theoretical, design as well as experimental aspects of the control of high dynamic drive with elastic coupling are considered. It is shown that the solutions of problem of vibration damping, suitable for the drives with low natural torsional frequency, give not good result at high dynamic. In this case the low frequency can be higher than several tens or 100 Hz and the next frequency of the multi-mass model can be thousand Hz and higher. Authors show the way of suppression of oscillations, based on the adequate PID speed controller. In the solutions only one sensor i.e. position encoder is used. The paper presents experience acquired during studying the problem, designing and investigating the laboratory plant based on the PMSM.**

Keywords—**high dynamic drive, elastic coupling, control optimization, PID controller.**

I. Introduction

Torsional vibration can occur in every drive because of finite stiffness of mechanical coupling between a motor and load. Due to small inertia of modern motors the resonance frequency of drivers are relatively high. For instance the inertia of the PMSM as well as the modern induction motor is about 10 times smaller than inertia of the traditional DC machine of the same power. Optimization of mechanical construction of plants leads in many cases to decreasing also the inertia on the second side of shaft, which contribute increase of resonance frequency, as well. In many processes, with regard to their quality, there is need of consideration and damping the torsional oscillations in the range of frequency higher than up to now. In this way the problem of vibration in the drives relates to higher and higher frequency.

The high dynamic modern motors allow shortening the transient processes of the drive due to possibility of forcing the fast and great changes of torque. However, it excite the oscillation more than at low dynamics.

The natural development of the drive systems, mentioned above, has led to need of controllers that are able to dump actively the mechanical vibration in the frequency band higher than several tens and even 100 Hz. In order to reach it, the fundamental difficulties must be overcome and special solutions must be used. It is reason for which the practical realization of the control with active damping the mechanical resonance in the given band of frequency is rather scarce. Elaboration such the control with consideration only one sensor in the form of position encoder is a goal of the present work.

First, in the paper the problems are discussed on the base of drive model and reference information. Then, the disadvantages of state controller for high dynamic drive

(with resonant frequency over 100 Hz) are shown. At the end, the solution based on the suitable PID controller as well as good results of its application are given.

II. Model Of The Drive

The one-, two- or three-mass model of the drive with flexibility can be useful for designing the controller. For drive with high dynamic motor, where the inertia of the motor is sometimes smaller than coupling inertia, the three-mass model shown in Fig. 1 can be necessary. In this case the total inertia J_M on the motor side of the shaft is divided into inertia J_m of the motor and J_c of the coupling. Between these parts is elastic element with coefficient K_g of stiffness and D_g of dumping.

The block diagrams of the mechanical part for one-, two- and three-mass models are shown in Fig. 2. There is motor m_M and load m_L torque on the input and motor a_M, load a_L as well as (eventually) coupling a_c acceleration on the output of each diagram. In order to have angular velocity ω and position angle θ of the separate parts of the system, one and two integration blocks must be added to the acceleration outputs.

In Fig. 3 the great difference between course of the frequency characteristics plotted for one-, two- and three-mass model of the investigated plant is seen. For low frequency all the models become ideal with the zero phase shift and constant amplification equal to $1/(J_M+J_L)$ which was used as a 0 dB level in Fig. 3. At the frequency near by anti-resonance $ARF = 57$ Hz and resonance $NTF = 81$ Hz the two- and three-mass model characteristics differ insignificantly while at the higher frequency diverge totally. Amplification at the high frequency resonance $NTFH = 1240$ Hz for three-mass model is over thousand times higher than for one-mass and several hundreds times higher than for two-mass model. This is the reason for which two-mass model can be insufficient in some cases.

If the angular speed and position of the motor: ω_M and θ_M as well as of the load: ω_L and θ_L are chosen as the state variables, then for two-mass model the following equations can be written:

$$J_M \frac{d\omega_M}{dt} = m_M - m_{sh}, \qquad (1)$$

$$J_L \frac{d\omega_L}{dt} = m_{sh} - m_L, \qquad (2)$$

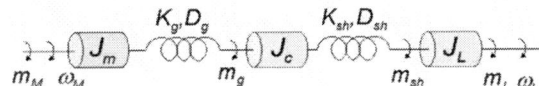

Fig.1. The three-mass model of mechanical system

978-1-4244-1741-4/08/$25.00 ©2008 IEEE   2227

$$\frac{d\theta_M}{dt} = \omega_M , \qquad (3)$$

$$\frac{d\theta_L}{dt} = \omega_L , \qquad (4)$$

$$m_{sh} = K_{sh}(\theta_M - \theta_L) + D_{sh}(\omega_M - \omega_L). \qquad (5)$$

The natural dumping is very small. For $D_{sh} = 0$ the conjugate imaginary poles and zeros of two-mass system are defined by two parameters: the anti-resonance frequency ARF and the natural torsional frequency NTF:

$$ARF = \frac{1}{2\pi}\sqrt{\frac{K_{sh}}{J_L}} , \qquad (6)$$

$$NTF = \frac{1}{2\pi}\sqrt{K_{sh}\left(\frac{1}{J_M} + \frac{1}{J_L}\right)} . \qquad (7)$$

The inertia ratio can be expressed as a function of ARF and NTF as follows:

$$R = \frac{J_L}{J_M} = \left(\frac{NTF}{ARF}\right)^2 - 1 . \qquad (8)$$

The speed control loop, shown generally in Fig. 4, has decisive significance for dynamic properties of the drive. In the diagram the continuous model **M** (it can be one of the models from Fig. 2) represents the mechanical part of the drive. Block **C** is the digital controller and **m** is the hybrid block of SISO type, representing torque control loop with discrete (digital) input m_r of reference torque and continuous output m_M of motor torque.

On the controller output during investigating a signal m_v can be added. Its step causes reaction likewise the load change, but is much simpler to do, especially for the real drive. Additionally, this "virtual load" excites mechanical oscillation more than the real load [2].

H in Fig. 4 is a model of measurement circuit. According to requirement of only one sensor (position encoder) there must be integrator (in order to count angle θ_M on the base of ω_M from the block **M**), quantizer, sampler, digital differentiation block and eventually additional digital filter in the circuit. Block **T** represents a signal delay in the main circuit.

In many cases, the blocks **H** (measurement circuit) and **m** (torque control loop) can be omitted. Instead of them, it is enough to increase suitably the delay of the block **T**. In the simplified model this total time delay in the speed control loop equals to the sum of delays of the separate elements of the real system:

$$T = T_{SH} + T_{D\theta} + T_C + T_m + T_F , \qquad (9)$$

where the separate delays are: $T_{SH} = T_s/2$ – of the sampling/holding block, $T_{D\theta} = T_s/2$ – of the differentiating block, T_C – of the proportional circuit of the controller (measured between sampling the input signal and actualising the output one), T_m – of the torque control loop, T_F – of the additional filters in the speed control loop and T_s is the sampling period.

It is suitable to use two generalised dimensionless parameters instead of four parameters (J_M, K_{sh}, J_L and T) in physical units. There are: the inertia R and the product $T \cdot NTF$ of the total delay in the speed loop and the resonant frequency. These two parameters characterize difficulties in active damping the mechanical oscillation [1]. This is the reason that the damping ability of the speed controllers is tested in the present work for the space definite by the range of R and $T \cdot NTF$.

In order to express all parameters of the drive using two generalized parameters the p.u. system must be used. One of the best choice is when the time base unit depends on the anti-resonance frequency according to formula:

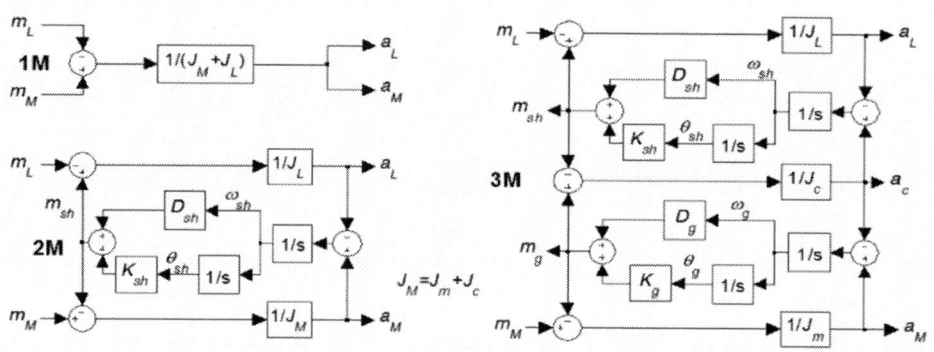

Fig..2. The diagrams of one-mass (1M), two-mass (2M) and three-mass (3M) models of mechanical system

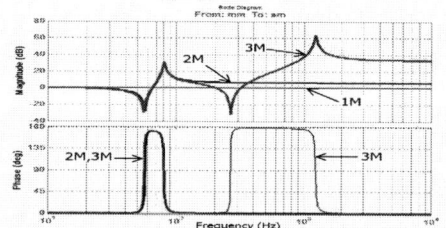

Fig.3.The frequency characteristics of one-, two- and three-mass models

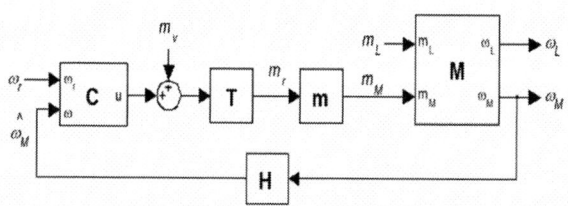

Fig. 4. Structure of the speed control loop

$$t_b = \frac{1}{ARF} = \frac{NTF}{\sqrt{1+R}} \qquad (10)$$

As a simple quality factor for estimation of the separate structures the setting time t_r can be used. Consequently to the above choice of the time base, the p.u. setting time equals to $t_r \cdot ARF$. This factor can be suitable measure for objective comparison of different control structures [1,2].

III. THE CONTROL PROBLEM OVERVIEW

The majority investigations of drives with elasticity, reported in literature, concerns the systems with low resonance frequency, below several tens Hz. In this conditions all delays introduced in the section II can be neglected as a rule. Also no account can be taken how is obtained the speed signal. Meanwhile, for the high dynamic drive, where the resonance frequency is 100 Hz or higher, the difficulties in active damping results mainly from lack of speed measurement and from delays caused by digital signal processing. If the motor is equipped with position sensor (encoder) only, then the rotational speed is not measured directly, but must be derived, which delays signal and degrades its quality.

As the control structure from Fig. 3 uses only one feedback (measured signal) and has one input ω_r as well as one output u, its controller can be treated as digital filter or polynomial controller [3]. All forms of this controller (state controller with observer, LQ controller with Kalman filter, RRC controller, PI/PID controllers with measurement signal filter) can be represented by means of two transfer functions: one for the reference path (from ω_r to u) and the second for the feedback path (from the measured quantity to u). The separate structures differ one from another only by degree of the polynomials. With increase of the polynomial degree increases ability of the controller on the one hand and radically increases difficulties in its tuning as well as decreases robustness of the controller to model errors, on the other hand. The possible advantages of the high order controllers can be not fully used in the real system also due to difficulties with introducing and co-ordinating the limitation of the inner integrals of the controller. Only the *windup-less* controller [4] with integrating the signal on output does not possess this disadvantage.

For high dynamic drives there is not recommended to install the sensors (for instance torque sensor) outside motor. It causes extra cost and decreases reliability, but first of all, can decrease stiffness of the mechanical system and introduce the additional resonance [5, 6].

Necessity of delay consideration in the drive model influences significantly utility of the controller analysis and synthesis methods. The methods based on the state space (state controller, Luenberger observer, Kalman filter) do not allow to introduce in a simply way the transportation delay. On the contrary, there is no problem with delay in the frequency methods (especially Bode diagram and Nyquist method) consisting in investigating the open-loop characteristics.

It is interesting to get acquainted with evolution of ideas of professor Hori who is engaged in control of multi-mass systems with flexibility since many years. Firstly, he prefers approach based on state space and full order observers [7]. Then, he reduces observer to the simple continuous model observer of disturbance [8] and

recently, he prefers the discrete-continuous structures designed in frequency domain [9]. In all this solutions the PI controller is the basic structure: in classical [7], two-degree-of-freedom [8] or fuzzy [9] form.

There are great measurement difficulties for the drives with resonance frequency above 100 Hz, as in this case the amplitude of oscillation is so small that can be smaller than resolution θ_{res} of the position sensor. Obtaining the speed signal by differentiating gives at the sampling period T_s the following speed resolution:

$$\Omega_{res} = \frac{\theta_{res}}{T_s} \qquad (11)$$

Formula (11) shows that for the high dynamic, where the controller must be fast, there is problem with accuracy and quantisation noise. The useful signal goes lost in the noise. As result of it the properties of all structures with explicit or hidden differentiating (observers, Kalman filter, etc.) are significantly deteriorated. This problem is reported in many works, for instance in [10]. However, the drive examined in [10] is relatively slow ($ARF = 6.9$ Hz, $NTF = 9.8$ Hz), there are difficulties with choice of elements of covariance matrix of Kalman filter. The estimates have either a great delay or high level of noise. For this reason the properties of composed structures are not better than properties of simple PI/PID controllers.

The p.u. setting time ($t_r \cdot ARF$) reached in practice for the drive with one sensor is not shorter than 1. For majority solutions presented in literature it is much greater then 1. The anti-resonance frequency can be reckoned as a boundary of dynamical ability of controllers for drives with flexibility [2]. For low dynamic drives it is possible to approach this boundary, but for drives with resonance frequency above several tens Hz it is impossible. The higher is frequency, the greater are difficulties.

In [11] ($ARF = 8$ Hz) for the system without delays at the PI/PID structures the setting time equals to about 1.5 was reached. The best results and the highest dynamic ($ARF = 65$ Hz) are in [12]. For small delay ($T \cdot NTF > 0.3$) is setting time 3.1 and for great delay ($T \cdot NTF > 0.85$) is 5.1. In [12] are also results 1.2 and 0.6 for low dynamic ($ARF = 5.7$ Hz) but they are obtained for speed equal to zero for which it is substantial damping due to friction torque.

The most exhaustive comparison of many different control structures is in [10]. Results of this work as well as many others investigations (for instance [13]), carried over in the same research team, shows that:

- even for low dynamic drive ($ARF = 6.9$ Hz) the ability of composed structures with many feedbacks are not much better than the ability of the simplest PID controller without any its improvement (as the setting time for the PID controller is about 1.7 and for structure with one and two additional feedbacs is about 1.5 and 1.2, respectively),
- it is possible to approach in this case the setting time equal to inverse of the anti-resonance frequency.

IV. THE STRUCTURE WITH STATE CONTROLLER AND OBSERVER

In order to show problems for the controller with many feedbacks the high dynamic drive with parameters $R = 2.14$, $ARF = 76.9$ Hz, $NTF = 136$ Hz and state controller was simulated. The signals of motor torque m_M (in the form of motor current i), motor ω_M and load ω_L speed as well as of elasticity torque m_{sh} were used in the controller.

First, the signals were ideal as they came directly from the model. The controller, after input step, gives very fast and good damped course of quantities shown in Fig. 5, in which (as well as in all figures in section IV) the time is in second and other quantities are related to their maximum values. In this case the setting time is very short: $t_r \cdot ARF = 0.46$.

Almost the same very good result can be reached when in the feedback paths is an ideal observer connected directly to the drive model and having the same parameters. If there is difference between drive and observer parameters or at constant controller parameters the drive ones are changed then the setting time is much longer. It can be 10 [2]. At great difference the instability is possible, as well.

Next, after the controller and to the feedback paths the filters were added with the following time constants:
- 16.7 µs on the controller output,
- 20 µs into motor and elasticity torque paths,
- 50 µs into motor and load speed paths.

As a result of it the visible deterioration of dynamical state is observed (Fig. 6) in spite of new optimum tuning the controller. If the given time constants were increased four times then it was impossible to reach stable settings of the controller. It means that the robustness of the investigated structure to delays is low.

In order to test ability of the state controller under circumstances of noise disturbance, the stochastic signal was introduced into the motor speed feedback (for the system with robustly tuned controller and ideal remaining conditions). The amplitude of positive and negative pulses of the signal with adjustable frequency is the Gaussian distribution time function with expectation equals to zero and standard deviation σ. Additionally, the random chosen pulses are amplified 10 times. Figure 7 shows influence of the disturbance on dynamical and steady state for chosen frequency and deviation. The disturbance influences mainly the motor current i and velocity ω_M. Here the coupling elasticity do not transfer speed oscillation on the load side but above $\sigma = 0.01$ the system becomes unstable [2]. The deviation should be referred to the speed signal. If to take into consideration that maximum value of speed is 1 then one can state that robustness of the controller to the noise is not great. With regard to it, the situation is much worse when the noise frequency is near by resonance [2].

V. The Universal PID Digital Controller

The above accomplished overview of the problem and quoted results of investigation as well as results obtained in [2] for other control structures (i. e. with LQ controller and Kalman filter, RRC controller, ANN controller, modal control with many feedbacks fed from the observer) allow to state as follows:
The potentially great ability of the high order controllers can be not fully used in the high dynamic drives with flexibility, where the resonance frequency is above several tens Hz.

On the other hand, the good properties of the PI/PID controllers are confirmed in [14] and [15] on condition of correct choice of their structure and suitable tuning. Ideas of these works were developed in theoretical and simulation ways. The Bode diagrams of one-, two- and

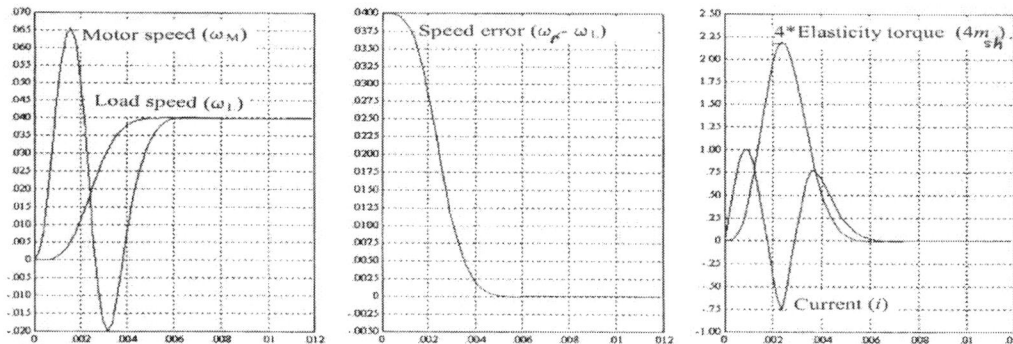

Fig. 5. The response of the system with state controller for step of reference speed at ideal conditions

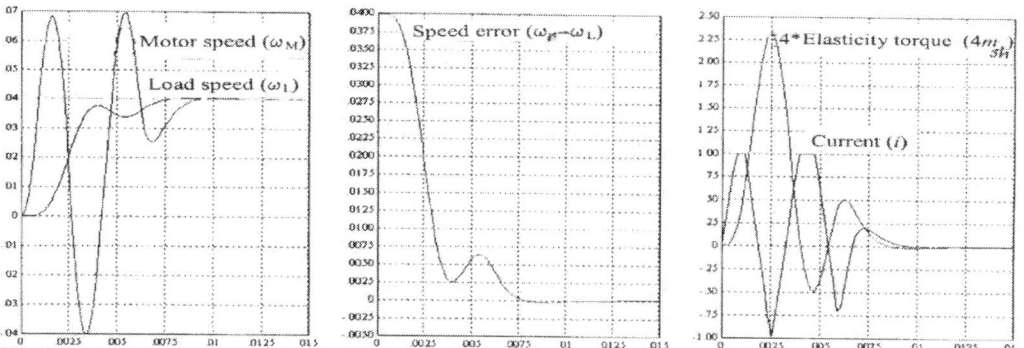

Fig. 6. The response of the system with state controller for step of reference speed when the filters are introduced

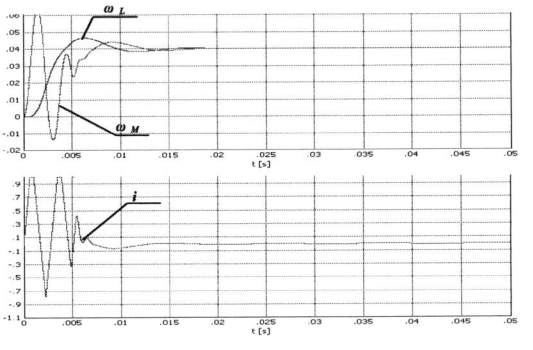
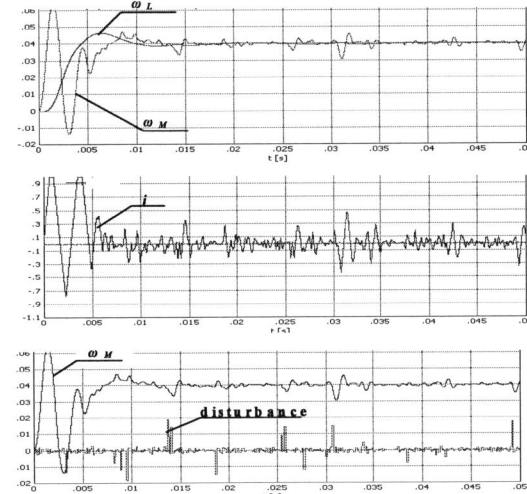

Fig. 7. The comparison of quantity course after step of reference speed in the system without disturbance (left) and under disturbance with f=3072 Hz, σ=0,001 (right)

three-mass models were basis for improving the controller. The universal PID digital controller shown in Fig. 8 is a result of it [2].

It is a digital form of PID controller with some improvements. The structure has features of 2DOF (2-Degree-Of-Freedom) controller and allows separate forming the reaction for both input and disturbance changes. Proportional gains can be different for reference (r) and feedback (y) signals. Only the integrating action is identical for both component of the error.

The integrator was placed on the output after the summing point of the controller (Fig. 8) in order to avoid the problem with co-ordination of signal limitation. Differentiating s of continuous controller is replaced with the backward difference $1 - z^{-1}$ and integrating $1/s$ with operation $1/(1 - z^{-1})$ i. e. with adding the signal to its preceding value u_{-1}. The integrator is seen in the upper part on the output of the controller in Fig. 8, where also the place of limiter location is marked. In this way the path with coefficient K_0 is the I path of the controller.

The signals in proportional (P) paths (with coefficients K_{1r} and K_1) first are differentiated, as on the output is integration, which for these paths must be compensated. On the contrary to continuous systems, the discrete differentiation introduces filtering the signals (time constant $T_s/2$). In many cases such filtering is not sufficient and for that reason the measurement path of the controller in Fig.8 is equipped with extra filter $F(z^{-1})$. The additional delay can be realised in this block. Such

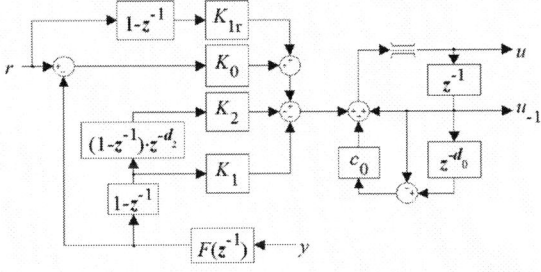

Fig. 8. The universal PID controller

delay may be useful in some difficult cases of control and was applied in present investigations (in this case the controller has dPID denotation).

Differentiating of signal y is repeated in the D path (coefficient K_2) of the controller. There is also possibility of introducing additional delay equal some sampling periods ($d_2 T_s$) into this path. The delay (additional letter d before D) helps to damp the oscillation due to decreasing the phase delay at damped frequency [2].

The additional delay can be also in the loop of positive feedback of the integrator. It is seen in the lower part on the output of the controller in Fig. 8. Due to the delay $d_0 T_s$ in this loop the lead of integrating action is $d_2 T_s$ times greater than in the integrator without it. In this way the controller PdID faster liquidates error and can operate at greater delays in the speed loop than PID one. As the PdID controller is sensitive to noise, the gain c_0 is foreseen in structure for tuning the intensity of feedback.

The presented universal PID structure allows obtaining many different controllers (for instance PI, PdI and dPID) by means of setting to zero the separate gains of the structure. In this way its properties can be good adapted to the controlled system.

For each form of the controller can be find formulas of its transfer function and parameters. They are functions of the gains (time constants) in the proportional (K_P), integrating (K_I or T_I) and differentiating (K_D or T_D) path of the continuous controller as well as of the sampling period T_s and needed delays. For the controller of PdI type the transfer function of the feedback path (from measured y to output u signal) in the discrete time domain is [2]:

$$G(z) = -\frac{K_0 + K_1\left(1 - z^{-1}\right)}{1 - \left(1 - c_0\right)z^{-1} - c_0 z^{-1-d_0}} \quad (12)$$

and the parameters are:

$$K_0 = K_I T_s\left(1 + c_0 d_0\right) = K_P \frac{T_s}{T_I}\left(1 + c_0 d_0\right), \quad (13)$$

$$K_1 = K_P\left(1 - \frac{T_s}{2T_I}\right)\left(1 + c_0 d_0\right), \quad (14)$$

2231

$$K_2 = 0. \tag{15}$$

VI. RESULTS OF THE INVESTIGATION

First the simulation examination of the separate structures of the controller was done. Figure 9 shows the

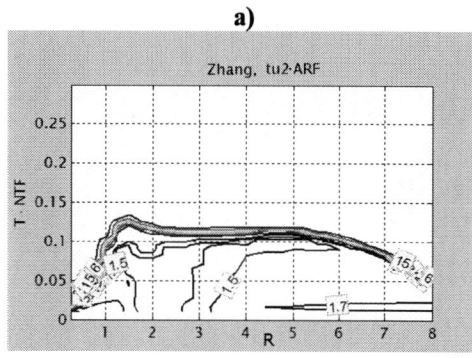

a)

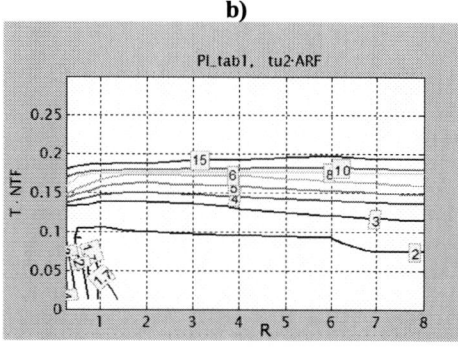

b)

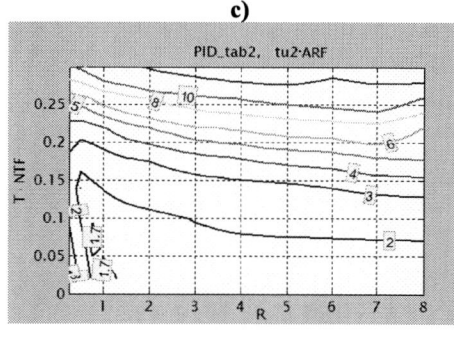

c)

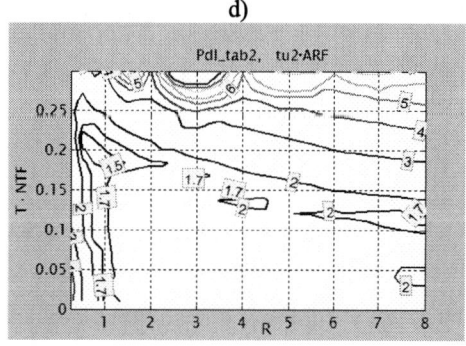

d)

Fig. 9. The lines of the same values of setting time of the speed ω_M after the step of virtual load m_v for the controller: a). PI/PID tuned according [11], b) PI, c) PID, d) PdI.

damping ability in the form of "maps", which present the lines of the same value of setting time on the plane of generalised parameters of the system: inertia ratio R and total delay time $T \cdot NTF$. It is seen in the figure that ability of elaborated unconventional controllers are better than the ability of the PI/PID one used in [11], which is one of the better solution known from the literature. At this the properties increase in sequence how in the Fig. 9: PI[11], PI, PID, PdI. However the last one is the best, its "map" is irregular, i. e. has "tops" and "valleys". It can mean difficulties during tuning the controller. One can see also that it is impossible to have better result that 1.5.

Figure 10 presents the ability of the controllers to damping the oscillation in the wide range of delays, up to

a)

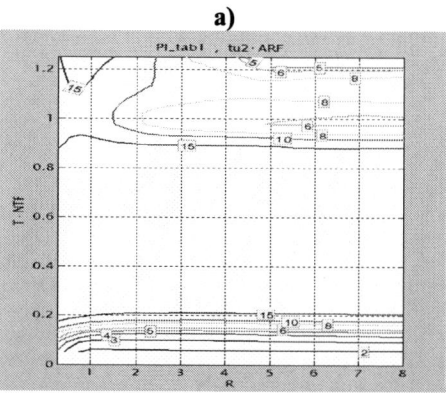

b)

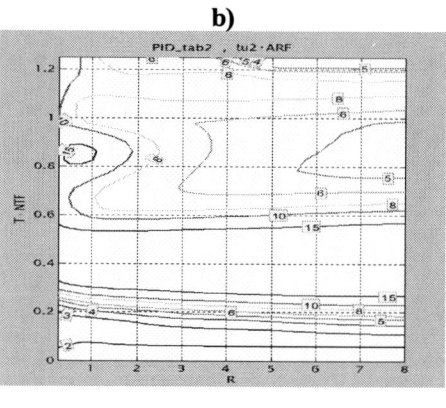

c)

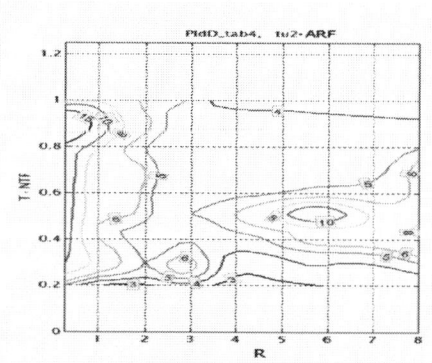

Fig. 10. The lines of the same values of setting time of the speed ω_M after the step of virtual load m_v for the wide range of delay under the controller: a). PI, b) PID, c) PIdD.

full period of the resonance and more ($T \cdot NTF$ equals 1 and more). The phase delay in the control loop at $T \cdot NTF = 1$ is 360° and stability boundary is at 540°. The setting times are above 5. They are approximately the same as in the fully stiff system (about 10). Between two stable areas: lower for small delays and upper for great delays there is a middle area in Fig. 10, in which the controllers can not operate stable. The unstable range for PI controller [11] is much wider than for the unconventional controllers. The PIdD controller (Fig. 10c) has especially good properties. However, this one can be difficult for tuning, considering its variegated "map".

Analysis of "maps" in Fig. 10 allow to formulate interesting conclusion: If the delay is so great, that the middle unstable region takes place, then introducing the next suitably chosen delay (in order to go to the upper stable region) can solve the problem. For this the discrete filter $F(z^{-1})$ in the Fig. 8 is very useful.

The above ideas were implemented in the real high dynamic drive. The drive shown in Fig. 11 bases on the PMSMs with the rated data: speed 3000 rpm, torque 3.6 N·m, rotor inertia $J_m = 2.5$ kg·cm², type: UNIMOTOR 95UMB300CASAA. The motor is equipped with the sin/cos sensor giving resolution of 4 194 304 positions /revolution. The mechanical part has four changeable shafts 0.72 m long, with the following stiffness: 6271, 737.3, 46.1 and 14.6 N·m/rad. The complete clutch of both machines has inertia $J_c = 6J_m$. There are 8 changeable discs (each with inertia $J_D = 4.2J_m$) for using. They can be easily apply on the one and/or second side of the shaft. In this way the resonance frequency reachable in the plant are between $ARF = 283.6$ Hz, $NTF = 401.1$ Hz and $ARF = 7.9$ Hz, $NTF = 11.1$ Hz.

Both the machines are fed from the converters UNIDRIVE SP. For enlarging the control ability of motor converter the universal module SM-Application was used. Apart from it, for testing the structures with LQ controller and Kalman filter, RRC controller and ANN controller, the separate laboratory converter with control system based on the DSP SHARC ADSP21061 and coprocessor ADMC201, was applied.

Some practical results obtained in the plant are given in Fig. 12, where they are compared with the simulation results for the same parameters. During the tests the step of reference speed, then the step of the virtual load m_v and at the end the step of real load m_L was applied. The figure shows three cases: with small, middle and great delay. In order to have the requested delay $T \cdot NTF$ the mechanical parameters of the drive as well as the sampling time of the controller were changed. In the case with $T \cdot NTF = 1.26$ additionally the exterior delay was used. In all cases also the high frequency resonance about $NTFH = 1300$ Hz took place and was damped. In the figure the frequencies, the structure and main parameters of the controller as well as total delay are given. The setting time can be easily read.

The excellent convergence of the simulation and experimental results for all cases in the Fig. 12 proves the high accuracy of the used models. It is seen that the higher is dynamic of the drive and the greater is the delay, the more composed controller must be used and the longer is the p.u. setting time. The obtained time is about 1.2, 2.5 and 5 for small, middle and great delay, respectively. The result is fully satisfied. In the case with great delay the controller with additional delay in the measurement path was applied in order to go from unstable to stable region. During testing good properties of the elaborated controller were confirmed.

VII. CONCLUSIONS

The generalised parameters in the form of inertia ratio R and of the product $T \cdot NTF$ of the total delay in the speed loop and the resonant frequency are suitable for investigation of the drive with elasticity.

The setting time related to the anti-resonance period ($t_r \cdot ARF$) is a simple factor for objective comparison of properties of different control structures.

The difficulties in active damping of high dynamic drive results mainly from lack of speed measurement and from delays caused by digital signal processing.

The p.u. setting time equal to about 1 is the boundary of practical reached results for systems with flexibility. At this the higher is dynamic of the drive and the greater is the delay, the more composed controller must be used and the longer is the p.u. setting time.

Due to delay and noise the possible advantages of the high order controllers based on the state space methods can be not fully used in the real system of high dynamic drive with elasticity. In this case the simply PID controllers with suitable chosen structure and parameters are good solutions of the control problem.

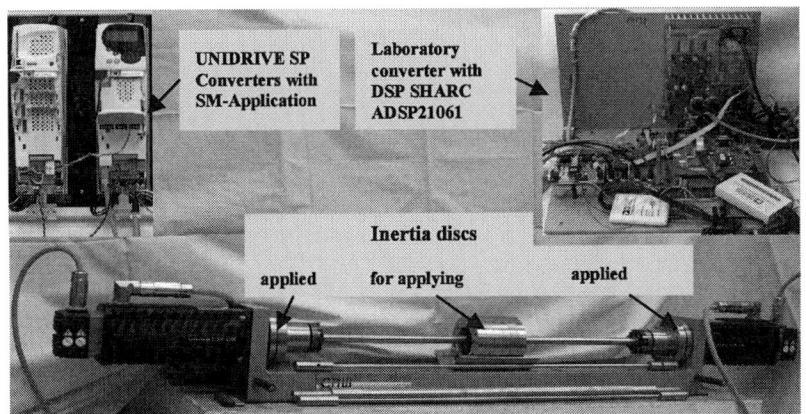

Fig. 11. The laboratory plant

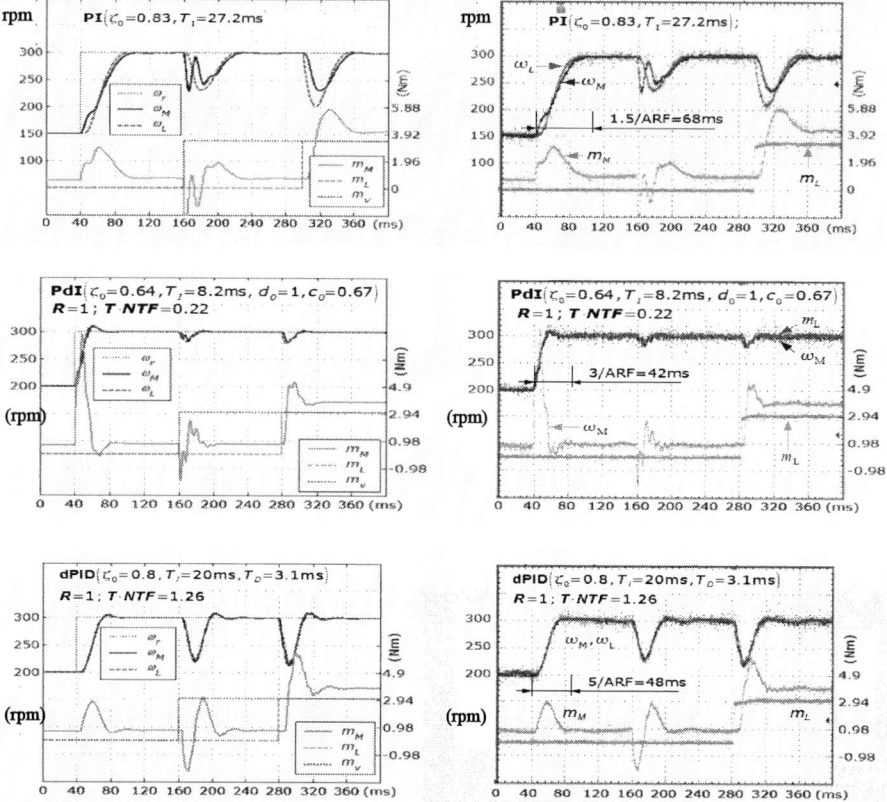

Fig. 12. The step response simulated (left) and real (right) system for: *ARF* = 20.4 Hz, *NTF* = 32.9 Hz., PI controller with T_s=250 μs (upper oscillograms); *ARF* = 69.6 Hz, *NTF* = 98.5 Hz., PdI controller with T_s=1 ms (middle oscillograms); *ARF* = 103.6 Hz, *NTF* = 146.5 Hz., dPID controller with T_s=2 ms and filter T_f=2 ms (lower oscillograms)

REFERENCES

[1] S. Brock, J. Deskur, D. Janiszewski, R. Muszyński: *Active Damping of Torsional Vibrations in Servodrives.* Power Electronics and Electrical Drives – selected problems. Oficyna Wydawnicza Politechniki Wrocławskiej, Wrocław 2007, pp. 271-290.

[2] R. Muszynski, S. Brock, J. Deskur and all: Control of the drive with flexibility, report from the realization of the project No 3T10A 02628, Poznan University of Technology, Poznan 2008 (in polish).

[3] J. Deur, N. Peric: *A Comparative Study of Servosystems with Acceleration Feedback.* Conference Record of the 2000 IEEE Industry Applications Conference vol.1, pp.1533-1540.

[4] M. Huba, M. Simunek: *Modular Approach to Teaching PID Control.* IEEE Trans. on Industrial Electronics, vol. 54, 2007, pp. 3112-3121.

[5] A. Baehr, P. Mutschler:, *Speed Acquisition Methods for High-Bandwith Servo Drives.* IAS 2005, pp.737-744.

[6] T.M.O'Sulivan, C.M. Bingham, N. Schofield: *High-Performance Control of Dual-Inertia Servo-Drive Systems Using Low-Cost Integrated SAW Torque Transducers,* IEEE Transactions on Industrial Electronics, vol.53, no.4, pp. 1226-1237, June 2006.

[7] Y. Hori; H Iseki, K. Sugiura: Basic consideration of vibration suppression and disturbance rejection control of multi-inertia system using SFLAC (state feedback and load acceleration control), *Industry Applications, IEEE Transactions on* , vol.30, no.4, pp.889-896, Jul/Aug 1994.

[8] Y. Hori, H. Sawada, Y. Chun, *Slow Resonance Ratio Control for Vibration Suppression and Disturbance Rejection in Torsional*

System, IEEE Trans. on Industrial Electronics, vol. 34, pp. 162-168, 1999.

[9] W. Li, Y. Hori: *Vibration Suppression Using Single Neuron-Based PI Fuzzy Controller and Fractional-Order Disturbance Observer.* IEEE Transactions on Industrial Electronics, vol.54, No.1, pp.117-126, 2007.

[10] K. Szabat: Control structures for electrical drive systems with elastic joint, Hability work, Scientific Papers of the Institute of Electrical Machines, Drivers and Metrology of the Wroclaw University of Technology, No 61, Monographs No 19, 2008 (in polish).

[11] G. Zhang, J. Furusho: *Speed Control of Two-Inertia System by PI/PID control.* IEEE Trans. on Industrial Electronics, vol. 47, no.3, pp. 603-609, 2000.

[12] J. Deur, N. Peric: *Analysis of speed control system for electrical drives with elastic transmission.* Proc. of the IEEE Int. Symp. ISIE'99, Bled, Slovenia, vol. 2., pp. 624-630, 1999

[13] T. Orlowska-Kowalska, K. Szabat: *Neural-Network Application for Mechanical Variables Estimation of a Two-Mass Drive System.* IEEE Transactions on Industrial Electronics, vol. 54, no.3, June 2007, pp.1352-1364.

[14] J. Deskur: Practical method for designing the speed and position controllers robust to disturbances and change of drive parameters, 5th Scientific Country Conference on Control in Power Electronics and Electrical Drive, SENE'01, Lodz 2001, pp. 105–109 (in polish).

[15] J. Deskur: The robust PID servo-drive controller, 9th Symposium on Power Electronics in Science and Didactic, Poznan 2004, pp. 45–52.

Protective Predictive Control of Electrical Drives with Elastic Transmission

Mario Vašak, Nedjeljko Perić
Faculty of Electrical Engineering and Computing
University of Zagreb
Zagreb, Croatia
Email: mario.vasak@fer.hr, nedjeljko.peric@fer.hr

Abstract—In this paper we propose a protective predictive control scheme for motor drives with elastic transmission that are subject to physical and safety constraints on their variables. Namely, we extend the classical LQR controller with a safety set obtained using invariant sets methodology. The easy implementation of this set on-line allows the correction of the LQR control signal in order to suppress the violation of constraints and thus avoid possibly dangerous situations during the drive transients. The added protective algorithm can be also used with any other drive control scheme to correct its outputted control signal.

Keywords—electrical drive with elastic transmission, LQR, constraints, the maximum controlled invariant set

I. INTRODUCTION

Quality control of electrical drives is a necessity in many manufacturing processes and many services. The reasons for that are: the outcomes quality, power savings, drive safety, prolongation of the exploitable life of the drive etc. There is a significant number of applications where the rotor of the drive and the load rotational mass are coupled with a shaft whose torsional characteristics may not be neglected during the drive control system synthesis. In such situations a satisfactory control cannot be attained using classical cascade control system with PI controllers solely due to the occurring low-damped torsional vibrations of the shaft. To suppress the vibrations the control algorithm should be more sophisticated.

Electrical drives with elastic transmission can be well controlled using polynomial controllers [1], state-space controllers [2], [3] or controllers based on neural networks and fuzzy logic [4], usually coupled with observers for hard-to-measure drive variables. However, the synthesis of such controllers does not take into account the physical and safety constraints on variables in the drive, e.g. limitations of the available motor torque and the maximal allowable torsion of the elastic shaft. Not only that disregarding those limitations in operation may outwear the drive, but may also lead to unpredicted and possibly dangerous behavior of the control system.

In order to cope with the mentioned problems in this paper the standard drive control algorithm is extended. This extension is actually a protection

algorithm which guarantees that the electrical drive system does not violate the constraints imposed on its variables. The electrical drive with elastic transmission, including the load torque effect, is modeled using a linear state-space model with the states: rotor and load speed, shaft torsional angle, the drive and the load torque. This model is extended with the reference speed of the load rotational mass. Based on the model we design a classical linear quadratic tracking controller (LQR) with a linear observer of the model states.

The existing physical and safety constraints for the drive at hand form bounded polytopic sets of the admissible model states and inputs. Guarantees on imposed constraints satisfaction are introduced by off-line computing the largest controlled invariant set [5] within the set of admissible states. In simple words, controlled invariant set is a set of system states where for each state in it there exists a feasible control input that keeps the successor state in that set. Furthermore, based on the computed controlled invariant set it is straightforward for a certain state to compute on-line the interval of feasible control inputs that allow keeping of the state in the set forever according to the model and input constraints. The mentioned computations allow implementation of a drive controller with an additional protection of the control system. Namely, the control signal obtained for a certain drive state by the drive controller (LQR or some other) is checked whether it matches the interval of feasible control inputs for that state. If the computed input lies outside the interval, it is replaced with the closest value from the interval and otherwise is left unchanged. The presented comparison between the drive control algorithm without protection (LQR) and with protection indicates that practically all violations of the constraints are suppressed for any transient due to the reference speed change. The facts that the considered protection scheme is rather easy to compute and implement, and that it can be also combined with any other drive control scheme, are also very motivating.

The paper is organized as follows. After this introductory section, Section II outlines the state-space model of the drive with an elastic shaft between its rotor and the load mass. Section III presents the drive control system design and Section IV gives a comparison of the electrical drive system behavior with and without the proposed controlled invariant set based protection. The paper contributions are summarized in the concluding Section V.

II. STATE-SPACE MODEL OF THE DRIVE WITH ELASTIC TRANSMISSION

The electrical drive with an elastic transmission between the rotor and the load mass is modeled as a two-mass elastic system [6], see Fig. 1.

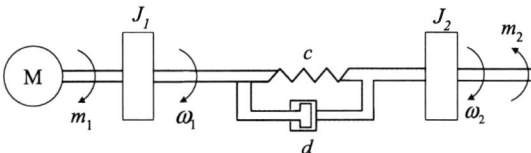

Fig. 1. Functional scheme of the considered electrical drive

It is assumed that the inner current control loop (the motor torque control loop) can be modeled as a first-order lag system, such that the following relation holds between the reference torque signal $m_{1,R}$ and the actual torque m_1:

$$\dot{m}_1 = \frac{1}{T_i}(m_{1,R} - m_1), \qquad (1)$$

where T_i is the equivalent time constant of the current control loop. The variables involved in the mechanical part of the drive, speed of the rotor ω_1, speed of the load mass ω_2, the shaft torsion angle ψ and the load torque m_2, are connected with the following relations:

$$\dot{\omega}_1 = \frac{1}{J_1}(m_1 - c\psi - d(\omega_1 - \omega_2)), \qquad (2)$$

$$\dot{\omega}_2 = \frac{1}{J_2}(c\psi + d(\omega_1 - \omega_2) - m_2), \qquad (3)$$

$$\dot{\psi} = \omega_1 - \omega_2, \qquad (4)$$

where c is the shaft stiffness, d is the damping coefficient, J_1 and J_2 are, respectively, the rotor and the load mass moments of inertia. Normalization parameters M_n, Ω_n and Ψ_n are introduced (see Table I) together with the following normalized variables: $\bar{\omega}_i = \omega_i/\Omega_n$, $\bar{m}_i = m_i/M_n$, $\bar{m}_{1,R} = m_{1,R}/M_n$, $\bar{\psi} = \psi/\Psi_n$ $(i = 1, 2)$. The normalization of variables is used due to numerical reasons in further computations with the model and due to easier presentation of the results. Parameters M_n and Ω_n are also the rated torque and speed of the drive, respectively.

The following model with normalized variables is

TABLE I
PARAMETERS FOR THE CONSIDERED ELECTRICAL DRIVE

Symbol	Value
M_n	14.8 Nm
Ω_n	210 $\frac{rad}{s}$
Ψ_n	$\frac{5\pi}{180}$ rad
$T_{M1} = \frac{M_n}{\Omega_n J_1}$	0.147 s
$T_{M2} = \frac{M_n}{\Omega_n J_2}$	0.241 s
$T_\psi = \frac{\Psi_n}{\Omega_n}$	0.42 ms
$\bar{c} = \frac{c\Psi_n}{M_n}$	900
$\bar{d} = \frac{d\Omega_n}{M_n}$	0.7

obtained:

$$\dot{\bar{\omega}}_1 = \frac{1}{T_{M1}}\left(\bar{m}_1 - \bar{c}\bar{\psi} - \bar{d}\bar{\omega}_1 + \bar{d}\bar{\omega}_2\right), \qquad (5)$$

$$\dot{\bar{\omega}}_2 = \frac{1}{T_{M2}}\left(\bar{c}\bar{\psi} + \bar{d}\bar{\omega}_1 - \bar{d}\bar{\omega}_2 - \bar{m}_2\right), \qquad (6)$$

$$\dot{\bar{\psi}} = \frac{1}{T_\psi}\left(\bar{\omega}_1 - \bar{\omega}_2\right), \qquad (7)$$

$$\dot{\bar{m}}_1 = \frac{1}{T_i}\left(\bar{m}_{1,R} - \bar{m}_1\right), \qquad (8)$$

where T_{M1}, T_{M2}, T_ψ, $\bar{c}$ and $\bar{d}$ are given in Table I. The given parameters stem from the experimental setup of the soft-coupled electrical drive described in [7].

The predictive reference tracking control law computation also requires information about the evolution of the load torque and the reference signal $\bar{\omega}_{2,R} = \omega_{2,R}/\Omega_n$. Since we do not know their behavior in the future, we simply presume they remain unchanged:

$$\dot{\bar{m}}_2 = 0, \qquad (9)$$

$$\dot{\bar{\omega}}_{2,R} = 0. \qquad (10)$$

Using the notation

$$x = \begin{bmatrix} \bar{\omega}_1 & \bar{\omega}_2 & \bar{\psi} & \bar{m}_2 & \bar{\omega}_{2,R} \end{bmatrix}^T, \quad u = \bar{m}_{1,R} \qquad (11)$$

we can simply rewrite (5)–(10) as

$$\dot{x} = Ax + Bu, \qquad (12)$$

where the structure of A and B can be deduced from (5)–(11). We denote the reference tracking error as

$$y = \bar{\omega}_{2,R} - \bar{\omega}_2 = Cx, \qquad (13)$$

where $C = [0 \ -1 \ 0 \ 0 \ 1]$. For the considered soft-coupled drive an appropriate sampling time is $T = 5$ ms [1]. Discretizing the process (12) leads to the discrete-time state-space model

$$x(k+1) = A_d x(k) + B_d u(k), \qquad (14)$$

$$y(k) = Cx(k), \qquad (15)$$

where $x(k)$, $u(k)$ and $y(k)$ are short notations for $x(kT)$, $u(kT)$ and $y(kT)$, respectively. Model (14)–(15) is the starting point for the computation of the predictive drive control strategy which is detailed in the sequel.

III. Design of the drive control system

The goal of the drive control system is to provide good load speed tracking of the reference speed. However, during its operation the control system should respect the physical and safety constraints on the electrical drive variables. Since the produced motor torque cannot be larger than its maximum available value considering the drive at hand (for example 120% of the rated torque), we pose the physical constraints

$$|\bar{m}_1| \leq 1.2, \tag{16}$$

$$|\bar{m}_{1,R}| \leq 1.2. \tag{17}$$

Constraint (17) incorporates also the physical limits of the drive inverter (maximum available inverter current rate). We introduce safety limits on the speeds of the rotor and the load mass (the boundary values are for example 110% of the rated speed):

$$|\bar{\omega}_1| \leq 1.1, \tag{18}$$

$$|\bar{\omega}_2| \leq 1.1. \tag{19}$$

Very important for long-life of the mechanical parts of the drive is to prevent excessive shaft twist from the steady-state value $\bar{\psi}_0 = \frac{\bar{m}_2 M_n}{c \Psi_n}$ that may occur during drive transients. Thus the constraint

$$|\bar{\psi} - \bar{\psi}_0| \leq 3 \tag{20}$$

is introduced, which forbids the shaft to twist more than 15° above and below the steady-state value. Finally it is assumed that the reference speed is always less or equal than rated and that the load torque is not greater than 110% of the rated torque:

$$|\bar{\omega}_{2,R}| \leq 1, \tag{21}$$

$$|\bar{m}_2| \leq 1.1. \tag{22}$$

Constraints on the drive variables are described with the following polytopes

$$\mathcal{P}^x = \{x | C^x x \leq C^1\} \subset \mathbb{R}^6, \tag{23}$$

$$\mathcal{P}^u = \{u | C^u u \leq C^2\} \subset \mathbb{R}^1, \tag{24}$$

where matrices C^x and C^1 follow from (11), (16) and (18)-(22), while C^u and C^2 follow from (11) and (17).

In the first step of the drive control system synthesis the existing constraints (16)-(22) are disregarded and and an LQR-based full-state feedback controller [8] for the motor drive presented in the previous section is designed. Besides the tracking error penalization, in the corresponding LQR cost function introduced are also the deviation of the normalized torsion angle $\bar{\psi}$ from the steady state value $\bar{\psi}_0$, as well as the deviation of the normalized reference torque $\bar{m}_{1,R}$ from $\bar{m}_2$. The cost function has the form

$$
\begin{aligned}
J &= \sum_{k=0}^{\infty} \left(1000(\bar{\omega}_{2,R}(k) - \bar{\omega}_2(k))^2 + \right. \\
&\quad + 5(\bar{\psi}(k) - \bar{\psi}_0(k))^2 + (\bar{m}_{1,R}(k) - \bar{m}_2(k))^2 \right) = \\
&= \sum_{k=0}^{\infty} \left(x^T(k)Qx(k) + u^T(k)Ru(k) + 2x^T(k)Nu(k) \right).
\end{aligned}
\tag{25}
$$

The coefficients 1000, 5 and 1 by the three penalization terms in (25) are chosen to enforce fast speed tracking with low torsional vibrations. Note that due to (9) $\bar{\psi}_0(k)$ and $\bar{m}_2(k)$ are actually constant during predictions based on the model (14), and that analog reasoning applies to $\bar{\omega}_{2,R}(k)$ due to (10):

$$
\begin{aligned}
&\bar{\psi}_0(k) = \bar{\psi}_0(0), \ \bar{m}_2(k) = \bar{m}_2(0), \ \forall \ k > 0, \\
&\bar{\omega}_{2,R}(k) = \bar{\omega}_{2,R}(0), \ \forall \ k > 0.
\end{aligned}
\tag{26}
$$

The minimization of J over $u(k) = \bar{m}_{1,R}(k)$ subject to the dynamics (14) leads to the control law

$$u(0) = K_{LQR} x(0) \tag{27}$$

with

$$K_{LQR} = [-33.91\ 15.77\ -0.56\ -0.84\ 3.33\ 18.14]. \tag{28}$$

The dominant poles of the closed-loop control system are placed at 0.96 and $0.86 \pm j0.47$ in the complex z-plane.

Since LQR control requires feedback information from all states, its implementation requires synthesis of an observer used for estimation of hard-to-measure drive states, see e.g. [2] or [9]. In our problem setup we assume that the drive states $\bar{\omega}_1$, $\bar{\omega}_2$, $\bar{\psi}$ and $\bar{m}_1$ can be measured, while the load torque $\bar{m}_2$ needs to be estimated. We employ a full-state linear current observer [10]:

$$
\begin{aligned}
&\hat{z}(k|k-1) = A_o \hat{z}(k-1|k-1) + B_o u(k-1), \\
&\hat{z}(k|k) = \hat{z}(k|k-1) + K_o(y_m(k) - C_o \hat{z}(k|k-1)),
\end{aligned}
\tag{29}
$$

where

$$
\begin{aligned}
z &= [\bar{\omega}_1\ \bar{\omega}_2\ \bar{\psi}\ \bar{m}_1\ \bar{m}_2]^T, \\
y_m &= [\bar{\omega}_1\ \bar{\omega}_2\ \bar{\psi}\ \bar{m}_1]^T,
\end{aligned}
\tag{30}
$$

$\hat{z}(k_1|k_2)$ is the estimate of z in step k_1 based on information until the moment k_2, A_o and B_o are simply formed of the corresponding rows of A_d and B_d in (14), C_o extracts values corresponding to y_m from z and K_o is selected such that the observer dominant pole is placed at 0.5 in the complex z-plane. The state feedback $x(k)$ for LQR is taken from $\hat{z}(k|k)$ and from given $\bar{\omega}_{2,R}(k)$.

The control strategy designed so far does not take into account the constraints (16)–(22) which can be very harmful for the drive. In the second design step additionally the maximum controlled invariant set $\mathcal{I}$ inside the polytope $\mathcal{P}^x$ for the system (14) subject to constraint (17) is computed off-line. The set $\mathcal{I}$ contains all the states for which there exists an admissible control input according to (17) such that the successor state is also in $\mathcal{I}$ and is the largest of all the sets in $\mathcal{P}^x$ with that property. In other words, $\mathcal{I}$ contains all the states from which there exist transients in the future that do not violate given constraints. The computational scheme for obtaining $\mathcal{I}$ is the following [5]:

1) Initialize $\mathcal{T} := \mathcal{P}^x$;
2) Compute $\mathcal{T}^+ := \{x | \ \exists u : [x\ u]^T \in \mathcal{P}_\mathcal{T}\} \cap \mathcal{P}^x$ where $\mathcal{P}_\mathcal{T} = \{[x\ u]^T | \ A_d x + B_d u \in \mathcal{T}, \ u \in \mathcal{P}^u\}$;

3) If $\mathcal{T}^+ = \mathcal{T}$ set $\mathcal{I} := \mathcal{T}$, else $\mathcal{T} := \mathcal{T}^+$, and goto step 2.

Convergence of the given procedure occurs in our problem setup, but unfortunately for a general linear model with constraints there is no guarantee that it ends in finite time. Finally, for on-line implementation the polytopic set

$$\begin{aligned}\mathcal{P}_{\mathcal{I}} &= \{[x\ u]^T |\ A_d x + B_d u \in \mathcal{I},\ u \in \mathcal{P}^u\} = \\ &= \{[x\ u]^T | H_{\mathcal{I}} x + L_{\mathcal{I}} u \le K_{\mathcal{I}}\}\end{aligned} \quad (31)$$

is needed. For any measured $x \in \mathcal{I}$ the interval $\mathcal{U}_{\mathcal{I}}(x) = \{u | L_{\mathcal{I}} u \le K_{\mathcal{I}} - H_{\mathcal{I}} x\}$ is determined. It contains all feasible control inputs u that can be applied to the drive in order for the transient that starts in x not to violate the constraints in any future time.

We also introduce some additional measures to guarantee certain level of robustness in constraints satisfaction for this control scheme. We assume that any of $\bar{\omega}_1, \bar{\omega}_2, \bar{\psi}, \bar{m}_1$ and $\bar{m}_2$ may deviate by ± 0.01 due to measurement noise or estimation error. We then account for that such that the set $\mathcal{I}$ is made smaller by implementing the so-called Minkowski difference [11]:

$$\tilde{\mathcal{I}} := \mathcal{I} \ominus \mathcal{W} = \{x \in \mathcal{I} |\ x + w \in \mathcal{I}\ \forall w \in \mathcal{W}\}, \quad (32)$$

where $\mathcal{W}$ is a lower-dimensional hypercube defined as:

$$\mathcal{W} = [-0.01, 0.01]^5 \times \{0\}. \quad (33)$$

The overall control algorithm performed at each time-step on-line is as follows:

1) Compute $u' = K_{LQR} x$;
2) Compute $\mathcal{U}_{\tilde{\mathcal{I}}}(x)$ from off-line computed matrices $H_{\tilde{\mathcal{I}}}, L_{\tilde{\mathcal{I}}}, K_{\tilde{\mathcal{I}}}$;
3) If $u' \in \mathcal{U}_{\tilde{\mathcal{I}}}(x)$ implement it, otherwise implement the edge-value of $\mathcal{U}_{\tilde{\mathcal{I}}}(x)$ that is closer to u'.

The steps 1) and 3) of the listed on-line procedure require negligible amount of processing time compared to the step 2). Implementation of the step 2) on a DSP is analyzed in the sequel.

The obtained polytope $\mathcal{P}_{\tilde{\mathcal{I}}}$ in our problem setup consists of 288 constraints and for measured x we thus need $288 \cdot 6 = 1728$ multiple-accumulate instructions to compute the feasible interval $\mathcal{U}_{\tilde{\mathcal{I}}}(x)$ defined by its boundary values. If it is required that the algorithm for computing u ends in one tenth of the sampling time, i.e. in 0.5 ms, and if we assume that a multiple-accumulate instruction takes a single instruction cycle on a DSP, the DSP clock frequency should be greater than 1728/0.5 ms = 3.46 MHz.

It should be noted that the step 1) of the on-line procedure is not bound to LQR or any other specific control strategy, i.e. the proposed protective predictive control scheme can accompany also polynomial or neuro-fuzzy drive control strategies with a mandatory state-observer.

IV. SIMULATION RESULTS

In this section we present the simulation results obtained on the considered drive model, using LQR only and LQR extended with the derived protective predictive control strategy based on invariant sets, see Fig. 2. The normalized reference signal and load torque have rectangle shapes of different frequencies, both with the amplitude equal 0.5. Without using the computed protection (from 0 to 8.5 s) it may be observed that in the transient due to reference change significant violation of the safety constraint on the shaft twist ψ may occur. However with the proposed predictive protection (from 8.5 s on) such behavior is suppressed and the constraints violation during the transients prevented. The transients in both modes, with and without protection, last comparably long which indicates that the introduced protection scheme does not deteriorate the drive control system performance regarding the reference speed tracking.

Another testing of the protective predictive control scheme is performed in Fig. 3. Here the speed reference abruptly changes from the full rated speed to the full rated reversed speed. The load torque abruptly changes between -0.4 and 0.4 asynchronously with the speed. Again the protective predictive strategy eliminates almost all constraints violations. The only non-eliminated violation happens at 19.01 s when the unpredictable load torque change induced abrupt change in the allowed torsion angle interval.

Fig. 4 gives another interesting insight into the performance of the protective algorithm since it indicates how often is the protection activated for the transients given in Fig. 3. Level 0 indicates that LQR signal is outputted by the controller, level 1 indicates activation of protection and level 2 indicates that violation occurs. It is obvious that for the presented large and abrupt reference and load torque changes the protection scheme is rather often activated, i.e. rather often limits the LQR computed control signal in order to avoid constraints violation.

V. CONCLUSION

In this work we propose a protective predictive control for motor drives with elastic transmission. We extend the classical LQR controller plus observer structure with a safety set obtained using invariant sets methodology. Implementation of this set on-line allows us to correct the LQR control signal in order to suppress the violation of safety constraints on the drive variables. The implementation of this protective control, although presented in a combination with an LQR controller, is not limited only to such type of controllers and could be realized in a DSP.

ACKNOWLEDGMENT

This work was supported by the Ministry of Science, Education and Sports of the Republic of Croatia under grant No. 036-0361621-3012.

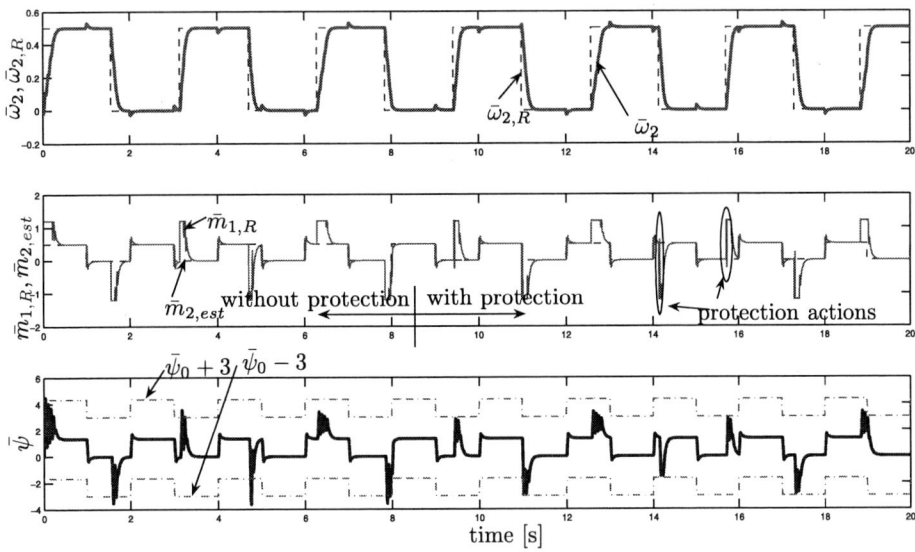

Fig. 2. Illustration of the control system operation without and with the proposed protection

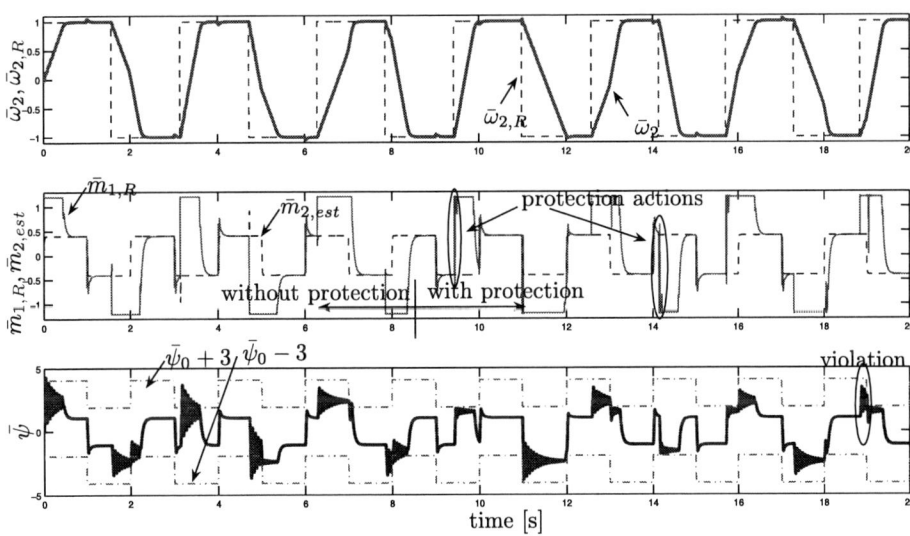

Fig. 3. Illustration of the control system operation without and with the proposed protection

REFERENCES

[1] J. Deur and N. Perić. Design of polynomial speed controller for electrical drives with elastic transmission. In *Conference Record of the 8th European Conference on Power Electronics and Applications*, pages 1–10, Lausanne, Switzerland, September 1999.

[2] J. Ji and S. Sul. Kalman filter and lq based speed controller for torsional vibration suppression in a 2-mass motor drive system. *IEEE Transactions on Industrial Electronics*, 42(6):564–571, 1995.

[3] J. Deur, T. Koledić, and N. Perić. Optimization of speed control system for electrical drives with elastic coupling. In *Proceedings of the 1998 IEEE International Conference on Control Applications*, pages 319–325, Trieste, Italy, September 1998.

[4] T. Orlowska-Kowalska and K. Szabat. Control of the drive system with stiff and elastic couplings using adaptive neuro-fuzzy approach. *IEEE Transactions on Industrial Electronics*, 54(1):228–240, 2007.

[5] F. Blanchini. Set Invarinace in Control – A Survey. *Automatica*, 35(11):1747–1767, 1999.

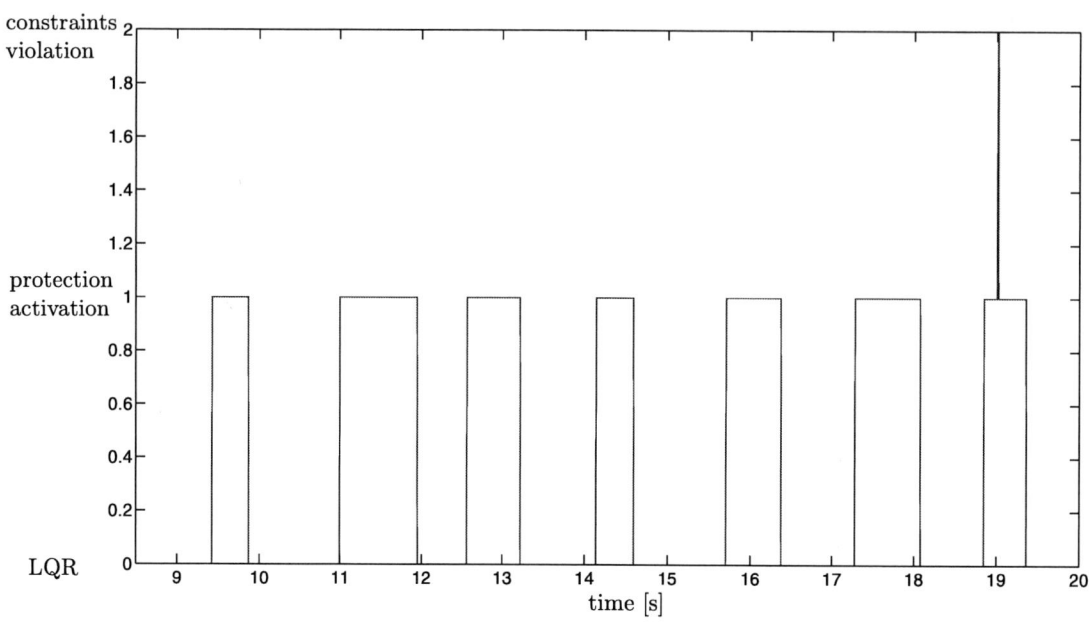

Fig. 4. Illustration of the frequency of protection activation for the transients in Fig. 3

[6] W. Leonhard. *Control of Electrical Drives.* Springer, Berlin, 2001.

[7] J. Deur, A. Božić, and N. Perić. Control of electrical drives with elastic transmission, friction and backlash – experimental system. *Automatika, Journal for Control, Measurement, Electronics, Computing and Communications*, 40(3–4):129–137, 1999.

[8] F. Lewis. *Optimal Control.* John Wiley and Sons, New York, 1986.

[9] T. Orlowska-Kowalska and K. Szabat. Neural-network ap-

plication for mechanical variables estimation of a two-mass drive system. *IEEE Transactions on Industrial Electronics*, 54(3):1352–1364, 2007.

[10] J. O'Reilly. *Observers for Linear Systems.* Academic Press, Burlington, 1983.

[11] S. V. Raković, E. C. Kerrigan, D. Q. Mayne, and J. Lygeros. Reachability Analysis of Discrete-Time Systems With Disturbances. *IEEE Transactions on Automatic Control*, 51(4):546–561, 2006.

Low-Cost High-Performance Predictive Control of Drive Systems with Elastic Coupling

Marcin Cychowski*, Kieran Delaney*, Krzysztof Szabat[†]

*Cork Institute of Technology, Cork, Ireland, e-mail: {*marcin.cychowski, kieran.delaney*}@cit.ie
[†]Wroclaw University of Technology, Wroclaw, Poland, e-mail: *krzysztof.szabat@pwr.wroc.pl*

Abstract—Effective damping of torsional oscillations while guaranteeing desired speed response characteristics and satisfaction of safety and operational constraints remains the main challenge in control of modern industrial drive systems with elastic couplings. This paper investigates the application of an explicit approach to the design of Model Predictive Control (MPC) for the drive system with elastic coupling. The resulting explicit MPC controller achieves the same level of performance as the conventional MPC, but requires only a fraction of the real-time computational machinery, thus leading to fast and reliable implementation. The main advantage of the proposed approach is that the drive physical and safety limitations can be directly incorporated into the control problem formulation and thus enforced explicitly during the operation of the controller. The results are compared to the current state of the art demonstrating the potential for notable performance improvements.

I. INTRODUCTION

Many industrial drive systems including rolling-mills, robotic arms, conveyor belts and textile or paper machine drives are characterized by excessive torsional vibrations causing significant performance degradation (decrease in reliability and product quality) or even instability if controllers are designed naively [1]-[8]. Those systems are typically composed of a motor connected to a load machine through a shaft. Shaft elasticity, when neglected, is a major contributing factor to increased torsional oscillations. Several different control approaches have been proposed for reducing the effect of torsional vibrations effectively including PI(D) control, Notch Filters, Resonance Ratio Control, structures with additional feedbacks and Model Predictive Control. The following will outline these approaches in more detail presenting the main limitations of each method and the rationale for the use of Model Predictive Control in modern industrial drive systems.

It is well known that a standard PI control algorithm is insufficient in eliminating the torsional oscillations effectively as shown in [7], [11]. For this reason, a PID control law is proposed in [7]. While the derivative part virtually minimizes the inertia of the motor increasing the damping effectiveness, it is highly sensitive to measurement noise.

The application of digital filters such as Notch-filters or Bi-filters for handling the oscillatory behavior has been demonstrated in [8], [9]. These solutions are effective and often used in industry but may undesirably affect the dynamics of the drive system.

In [10], the so-called Resonance Ratio Control structure has been presented. It is shown that the damping capability of this system increases when the ratio of the resonant to antiresonant frequency is large. The sufficient resonance ratio is achieved by using additional feedbacks from the estimated shaft torque. The main drawback of the approach is that the system time response cannot be set arbitrarily.

In [11], the analysis of control structures with different additional feedbacks has been presented. The work demonstrates that nine structures with one additional feedback can be divided into three groups with the same dynamic properties. The approach guarantees effective damping of torsional oscillations although shaping of the load speed response within a wide range of set-points requires the information from two selected state variables to be available for control purposes.

In drive systems with significant parameter variations, more sophisticated control paradigms such as non-linear or adaptive control can be adopted in order to achieve the required dynamic properties of the system. Proposals in this regard include sliding-mode control [3], neural fuzzy control [13] and non-linear adaptive control [14].

Interestingly, none of the aforementioned control techniques guarantees the satisfaction of the drive's safety and physical constraints. This problem has traditionally been dealt with by deliberately reducing the performance demands to keep the system responses away from their bounds. This may result in control actions being overly conservative unnecessarily limiting the level of achievable performance. In this paper, the application of Model Predictive Control for optimal handling of torsional oscillations and drive's operating and safety constraints is investigated.

Model Predictive Control is a control strategy that relies on solving a mathematical optimization problem at each control step or sampling time [15]. The control objectives are directly expressed in a cost function which typically penalizes any deviations of controlled variables from their set-points. Assuming a quadratic cost and linear inequality constraints, the underlying optimization problem can be posed as a quadratic programming (QP) problem. Despite the fact that QP problems can be solved efficiently using off-the-shelf solvers, the computational effort required for the implementation of the MPC algorithms on-line can be quite prohibitive for many real-time applications. This is particularly evident in systems with fast dynamics where

978-1-4244-1741-4/08/$25.00 ©2008 IEEE

sampling in the mili/micro-second scale is needed.

In [16], the authors report on the application of model predictive control to a rolling mill system. This work is interesting although very limited since the proposed MPC controller would require very complex and expensive processing hardware as well as dedicated software to enable implementation on a real system.

In this paper, low-cost MPC controllers are developed by utilizing the recent results in the field of parametric programming [17]. More specifically, the work in [17] demonstrates that the constrained MPC problem with a quadratic cost is a multi-parametric quadratic program (mp-QP), when the state vector is viewed as a parameter to the problem. It is shown that the solution (the control input) can be explicitly represented as a piecewise affine (PWA) function defined on a polyhedral partition of the state space. The resulting *explicit* MPC replaces an often demanding optimization problem with a simple look-up table solution. This makes the explicit MPC potentially useful in application areas characterized by fast sampling or low-cost processing requirements such as the drive system considered in this paper.

The reminder of this paper is organized as follows. In Section II a mathematical model of the drive system and the drive control structure are described. Section III introduces the model predictive control framework and discusses how explicit MPC solutions can be computed. A detailed design and tuning procedure for MPC is presented in Section IV accompanied by a comprehensive simulation study. Section V concludes with some remarks and plans for future work.

II. DRIVE CONTROL STRUCTURE

A. Mathematical Model of the Drive

Many mathematical modeling approaches have been considered in technical literature that can be used for the analysis of drive systems with elastic couplings. In many cases the drive system can be modeled as a two-mass system, where the first mass represents the moment of inertia of the drive and the second mass refers to the moment of inertia of the load. The mechanical coupling is treated as inertia free. Occasionally, the internal damping of the shaft is also taken into consideration. Such systems are described by the following state equation in per unit system [11] (with non-linear friction neglected)

$$
\frac{d}{dt}
\underbrace{
\begin{bmatrix} \omega_1(t) \\ \omega_2(t) \\ m_s(t) \end{bmatrix}
}
=
\underbrace{
\begin{bmatrix} 0 & 0 & \frac{-1}{T_1} \\ 0 & 0 & \frac{1}{T_2} \\ \frac{1}{T_c} & \frac{-1}{T_c} & 0 \end{bmatrix}
}_{A}
\underbrace{
\begin{bmatrix} \omega_1(t) \\ \omega_2(t) \\ m_s(t) \end{bmatrix}
}_{x_c}
$$
$$
+
\underbrace{
\begin{bmatrix} \frac{1}{T_1} \\ 0 \\ 0 \end{bmatrix}
}_{B}
m_e(t)
+
\underbrace{
\begin{bmatrix} 0 \\ \frac{-1}{T_2} \\ 0 \end{bmatrix}
}_{B_d}
m_L(t), \quad (1)
$$

where ω_1 is the motor speed, ω_2 is the load speed, m_s is the shaft torsional torque, m_e denotes the electromagnetic torque and m_L is the load torque, T_c is the stiffness time

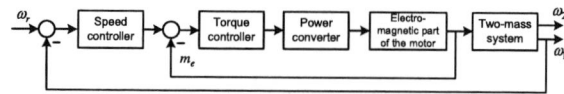

Fig. 1. Classical cascade control structure of the drive system.

constant, and T_1 and T_2 are the mechanical time constants of the motor and the load machine, respectively.

Two characteristic frequencies of the two-mass system are described by the following equations

$$
\omega_{res} = \sqrt{\frac{T_1 + T_2}{T_1 T_2 T_c}}, \quad \omega_{are} = \sqrt{\frac{1}{T_1 T_c}}. \quad (2)
$$

The value of the resonant frequency depends on the type of the drive and can vary from a few hertz in a paper machine section [5], through dozens of hertz in a rolling-mill drive [7], and can exceed hundreds hertz in a modern servo-drives [8]. The value of the antiresonant frequency can be even ten times smaller than the resonant one in a dryer [5], but usually the difference is much smaller than two [10].

The other characteristic parameter of the two-mass system is the inertia ratio defined by

$$
R = \frac{T_2}{T_1}. \quad (3)
$$

The torsional vibrations are very difficult to suppress when the value of R is small.

B. Cascade Control Structure

A typical electrical drive system is composed of a power converter-fed motor coupled to a mechanical system, microprocessor-based controllers, current, rotor speed and/or position sensors used as feedback signals. Motor speed control is typically realized using the cascade control arrangement shown in Fig. 1.

The inner (torque) control loop consists of the power converter, electromagnetic part of the motor, current sensor and respective current or torque controller. As this control loop is designed to provide sufficiently fast torque control, it can be approximated by an equivalent first order term with small time constant. For well-tuned torque controller, the drive machines could be AC or DC motors without any impact on the outer speed control loop. The outer loop consists of the mechanical part of the motor, speed sensor, speed controller, and is cascaded to the inner loop. It provides speed control according to the reference value ω_r.

The goal of this paper is to design and validate a predictive speed controller which guarantees optimal speed tracking performance and satisfaction of operating and safety limitations of the drive system. The performance of the proposed MPC solution will be verified against a variation of the commonly used PID controller known as IPD control. This concept is illustrated in Fig. 2. Since the characteristic equation of this system is of fourth order and there are only three tuning parameters (K_p, K_i and K_d), independent location of the system closed-loop poles

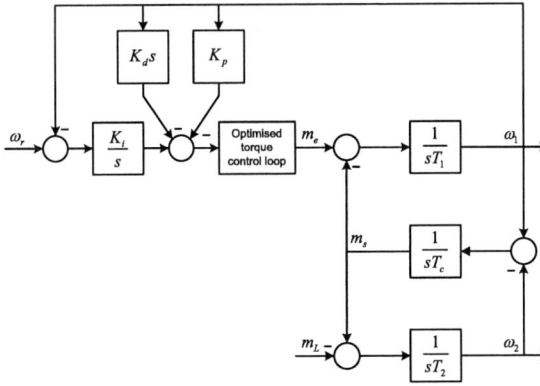

Fig. 2. Block diagram of the control structure with the IPD controller.

is impossible. This means that the performance of the drive system will never be optimal.

The IPD controller will be designed using the following rules [12]

$$K_d = \frac{T_2}{4\zeta^2} - T_1, \tag{4a}$$

$$K_p = 2\sqrt{\frac{T_1 + K_d}{T_c}}, \tag{4b}$$

$$K_i = \frac{T_1 + K_d}{T_2 T_c}, \tag{4c}$$

where ζ is the desired damping coefficient of the closed-loop system.

III. MODEL PREDICTIVE CONTROL

In this section, the model predictive control framework will be briefly introduced (see [15]-[18] for a comprehensive overview). MPC was originally developed to meet the demanding control specifications of power plants and petroleum refineries, and can now be found in many new application areas including chemicals, food processing, metallurgy, pulp and paper, automotive and aerospace [18]. The widespread acceptance of the MPC paradigm in industry is primarily due to the ease with which system constraints can be explicitly incorporated into the control problem formulation.

MPC refers to a class of computer control algorithms that utilize an explicit process model to predict the future response of a plant. In the research literature, a linear discrete-time state-space model is typically assumed

$$x(k+1) = Ax(k) + Bu(k), \tag{5}$$
$$y(k) = Cx(k),$$

where $x(k) \in \mathbb{R}^n$, $u(k) \in \mathbb{R}^m$, $y(k) \in \mathbb{R}^p$ are the state, input and output variables, respectively. Let $x_{k+i|k}$, $y_{k+i|k}$ denote the predictions of x and y at time $k+i$, made at time k. At each control interval an MPC algorithm attempts to optimize future plant behavior while respecting the plant operating constraints. This can be achieved by solving the following optimization problem [15]

$$\mathcal{J}^* = \min_U \sum_{i=k}^{k+N_p} \|y_{i|k} - r_i\|_Q^2 + \sum_{i=k}^{k+N_u} \|\Delta u_i\|_R^2, \tag{6a}$$

subject to

$$y_{k|k} = y(k), \tag{6b}$$

$$u_{min} \le u_{k+i} \le u_{max}, \quad i = 0, ..., N_u, \tag{6c}$$

$$y_{min} \le y_{k+i|k} \le y_{max}, \quad i = 1, ..., N_p, \tag{6d}$$

in which $\|v\|_P^2 = v^T P v$, Q and R denote the positive-definite weighting matrices, $U \triangleq [\Delta u_k^T, ..., \Delta u_{k+N_u}^T]^T$ is the vector of optimization variables (control actions) with $\Delta u_{k+i} \triangleq u_{k+i} - u_{k+i-1}$. N_p and N_u denote the prediction and control horizons, respectively and r_{k+i} is the reference trajectory at time $k+i$.

Let U^* be the minimizing sequence of (6). The MPC algorithm proceeds (according to the *receding horizon* principle) by sending the first element of the optimal input sequence u_k^* to the plant and the entire calculation is repeated at the next time step based on new measured (or estimated) output $y(k+1)$. This procedure is summarized in the following algorithm.

Algorithm 1 (On-line MPC Controller) *Consider the optimization problem (6):*

1) *At time k, measure (or estimate) the current system state $x(k)$*
2) *Solve (6) to obtain $U^* = [\Delta u_k^{*T}, ..., \Delta u_{k+N_u}^{*T}]^T$*
3) *Apply $u_k^* = \Delta u_k^* + u_{k-1}^*$ to the plant*
4) *Update $k \leftarrow k+1$ and return to step 1.*

Using $y_{k+i|k} = CA^i x_{k|k} + C \sum_{j=0}^{i-1} A^i B u_{k+i-1-j}$, problem (6) can be reformulated as a quadratic program

$$\mathcal{J}^* = x_{k|k}^T Y x_{k|k} + \min_U U^T H U + 2 x_{k|k}^T F U, \tag{7a}$$

subject to $\quad GU \le W + E x_{k|k}, \tag{7b}$

where H, F, Y, G, W and E can be easily obtained from Q, R, (5) and (6c)-(6d) (see [17]).

Because problem (7) depends on the initial state $x_{k|k} = x(k)$, the implementation of the MPC algorithm can be performed either by solving the quadratic program on-line for a given $x(k)$, or as shown in [17], by solving problem (7) *off-line* for all $x(k)$ within a polyhedral set of values using multi-parametric quadratic programming. As stated in Section I, solving problem (7) directly can be prohibitive and costly for real-time applications requiring fast sampling since the entire optimization routine needs to be embedded into a particular hardware platform. This would necessitate the use of advanced and often expensive digital signal processor chips in order to meet the demanding processing and storage requirements and would often involve a substantial programming effort. These drawbacks can be eliminated (to some extent) by formulating problem (7) as a multi-parametric program.

Parametric programming techniques systematically subdivide the parameter space into characteristic regions where the optimal value and the optimizer are given as *explicit* functions of the parameters. These methods result in PWA functions represented as

$$u(x) = K_r x + g_r, \quad \forall x \in \mathcal{X}_r, \ r = 1, ..., R, \quad (8)$$

where $K_r \in \mathbb{R}^{m \times n}$, $g_r \in \mathbb{R}^m$ and $\mathcal{X}_r$ are polyhedral sets defined as

$$\mathcal{X}_r = \{x | H_r x \le d_r, \ H_r \in \mathbb{R}^{N_c^r \times n}, \ d_r \in \mathbb{R}^{N_c^r}\}, \quad (9)$$

for every $r = 1, ..., R$. For the remainder of this paper, the collection of sets $\mathcal{P} = \bigcup_{r=1}^{R} \mathcal{X}_r$ will be referred to as a *partition*.

The resulting explicit MPC controller is completely characterized by $\{K_r, g_r, H_r, d_r\}_{r=1}^{R}$. The implementation of the control law (8) is simply executed according to the following algorithm.

Algorithm 2 (Explicit MPC Controller) *Given K_r, g_r, H_r, d_r, $r = 1, ..., R$:*

1) *At time k, measure (or estimate) the current system state $x(k)$*
2) *Search for the j-th region that contains x, i.e. $x \in \mathcal{X}_j$*
3) *Apply the corresponding j-th control law, i.e., $u_j(x) = K_j x + g_j$*
4) *Update $k \leftarrow k + 1$ and return to step 1.*

Algorithm 2 requires the storage of all polyhedra $\mathcal{X}_r$, i.e., $(n + 1)N_C$ real numbers and in the worst case (the state is contained in the last region of the list) it will give a solution after nN_C multiplications, $(n - 1)N_C$ sums and N_C comparisons, where $N_C \triangleq \sum_{r=1}^{R} N_c^r$ [17]. Nevertheless, the evaluation of the explicit MPC control law through the PWA mapping (8) is often several orders of magnitude more efficient than solving the quadratic program (7) directly. This gain in efficiency is crucial for demanding applications with fast dynamics or sampling times in the mili/micro second range as will be demonstrated in the next section.

IV. DESIGN AND SIMULATION RESULTS

In this section, the proposed explicit model predictive control algorithm for the drive system with elastic coupling will be evaluated through simulations. The nominal parameters of the drive system considered here are $T_1 = T_2 = 0.203\text{sec}$ and $T_c = 0.0026\text{sec}$.

A. MPC Design

In order to guarantee zero steady-state offset control in MPC, the model (1) is augmented with an additional state to take into account the effect of load disturbance m_L as follows

$$\frac{d}{dt} \begin{bmatrix} x_c(t) \\ m_L(t) \end{bmatrix} = \begin{bmatrix} A & B_d \\ 0 & 0 \end{bmatrix} \begin{bmatrix} x_c(t) \\ m_L(t) \end{bmatrix} + \begin{bmatrix} B \\ 0 \end{bmatrix} m_e(t), \quad (10)$$

in which $x_c = [\omega_1 \ \omega_2 \ m_s]^T$. Let the controlled variables be the motor speed ω_1 and shaft torque m_s. The task is to bring the motor speed to the reference value ω_r while

ensuring m_s is sufficiently damped. This can be achieved by defining the auxiliary output variables

$$y_1 = \omega_1 - \omega_r, \quad y_2 = m_s - m_L. \quad (11)$$

The choice for y_1 is rather obvious while y_2 deserves some explanation. By examining the drive model (1) and assuming no external disturbance (other than m_L) is present it is clear that when $\lim_{t \to \infty} m_e = \lim_{t \to \infty} m_s$ and $\lim_{t \to \infty} m_s = \lim_{t \to \infty} m_L$ the system is in steady-state, i.e. $\frac{d\omega_1}{dt} = 0$ and $\frac{d\omega_2}{dt} = 0$. Furthermore, $y_1 = 0$ indicates that $\omega_1 = \omega_r$ which, together with $y_2 = 0$, also implies that $\omega_2 = \omega_r$. Note that minimizing output y_2 is equivalent to the minimization of the derivative of the load speed ω_2.

For the purpose of computing an explicit MPC solution, the reference state ω_r needs to be embedded directly into the augmented system model (10). This can be achieved by defining $\frac{d\omega_r}{dt} = 0$ as an additional state yielding the overall state vector $x = [x_c^T \ m_L \ \omega_r]^T \in \mathbb{R}^5$. For the continuous-time model (10)-(11), a linear discrete-time model corresponding to a sampling interval $T_s = 0.5\text{msec}$ is obtained by applying the exact discretization using zero-order hold. The control objective is to minimize y_1 and y_2 subject to the following constraints

$$-2 \le m_e(k) \le 2, \quad -2 \le m_s(k) \le 2. \quad (12)$$

The parameters of the MPC problem (6) are given by

$$Q = \begin{bmatrix} 3 & 0 \\ 0 & 0.5 \end{bmatrix}, \ R = 10^{-5}, \ N_p = 10, \ N_u = 2, \quad (13)$$

with $m_e \to \Delta u$ and $r = 0$ in (6a). It is clear from (13) that the tracking error in motor speed is penalized more heavily than the error due to the shaft-load torque imbalance. This ensures that the speed tracking performance is prioritized over the shaft torque performance while still guaranteeing that any excessive shaft torque behavior is sufficiently damped. The general approach for selecting the prediction and control horizons is that N_p should be large enough to capture the dominant system dynamics while N_u gives the required degrees of freedom. Increasing N_u will yield more aggressive control actions. In practice, $N_u \ll N_p$ to avoid large computational burden for the on-line MPC (Algorithm 1) and large number of regions for the explicit MPC (Algorithm 2). Needless to say, keeping N_u low is crucial for obtaining low-complexity controllers. General guidelines for selection of the MPC parameters are given in [15].

In order to bound the exploration space on which $\mathcal{P}$ is defined, the condition $x_{k|k} \in \mathcal{X}_0$ is explicitly imposed during computation of the explicit MPC control laws, where

$$\mathcal{X}_0 = \{x \in \mathbb{R}^5 | -d \le x \le d\}, \quad (14)$$

with $d = [1 \ 1 \ 2 \ 1 \ 1]^T$. This guarantees that $\mathcal{P} \subseteq \mathcal{X}_0$ holds [17].

The explicit MPC controller has been computed using Multi-Parametric Toolbox (MPT) [19]. The solution consists of 55 polyhedral regions and their corresponding control laws defined on $\mathbb{R}^5$ state-space. In Fig. 3 and

2244

Fig. 4, two-dimensional cuts through the polyhedral partition corresponding to $[\omega_1, \omega_2]$ and $[\omega_1, m_s]$ are depicted for illustration. In each case, regions #1 and #2 correspond to the saturated controllers, region #6 corresponds to the unconstrained linear controller, and regions #3, #4, #5, #7, #8 and #9 are transition regions between the unconstrained and saturated controllers. This can be confirmed from Fig. 5 in which the explicit control law for the MPC controller computed in the $[\omega_1, m_s]$ subspace is shown.

B. Performance Evaluation: Nominal Case

In the nominal case it is assumed that the drive dynamics is known exactly with $T_1 = T_2 = 0.203$sec and $T_c = 0.0026$sec, and that a full state vector $x(k) \in \mathbb{R}^5$ is available at each time step k (e.g. through measurement or estimation).

Fig. 6 and Fig. 7 demonstrate the performance of the proposed explicit MPC controller in the low- and high-speed operating regions, respectively. In each case, full load ($m_L = 1$) is applied at time $t = 0.4$sec. The responses of the IPD controller computed using (4) with $\zeta = 0.8$ are also shown for comparison. It is clear from the results that the MPC controller significantly improves the dynamic properties of the drive system in comparison to the IPD controller. In the low-speed region (Fig. 6), the transient responses due to the reference speed and load disturbance are both faster for MPC and yield zero overshoot. On the other hand, the application of the IPD controller results in an overshoot of about 2% for ω_1 and ω_2. Moreover, defining

$$\bar{J} = \sum_{k=0}^{\infty}(\omega_1(k) - \omega_r(k))^2 + (\omega_2(k) - \omega_r(k))^2, \quad (15)$$

then $\bar{J}^{IPD} = 0.0056$ and $\bar{J}^{MPC} = 0.0027$. Notice that the electromagnetic and shaft torque responses stay well within their safe limits given by (12).

In the high-speed operating region (Fig. 7), the performance of the IPD controller deteriorates due to the electromagnetic torque constraint. Since the controller is not "aware" of this physical limitation, overshoots of 16% and 25% result in the motor and load speed responses. On the other hand, the MPC controller provides smooth and almost zero-overshoot speed response and very effective disturbance rejection properties. This can be attributed primarily due to the prediction and optimal constraint handling capability which the control algorithms like IPD do not possess. This point is illustrated in Fig. 7 by removing the electromagnetic torque constraint from the IPD controller output. Although the performance of the system is improved, both the electromagnetic and shaft torque responses violate the safety constraints of the drive. This is not the case for the proposed MPC controller as clearly seen in Fig. 7.

C. Performance Evaluation: Robust Case

In this section robustness of the proposed explicit MPC controller to model uncertainty will be briefly analyzed. For this purpose, it is assumed that the load time constant

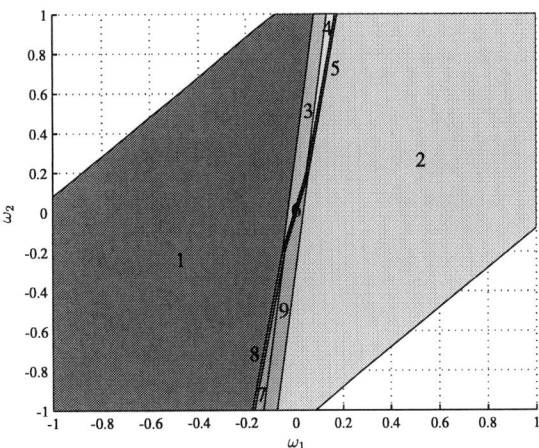

Fig. 3. State-space partition defined over $[\omega_1, \omega_2]$.

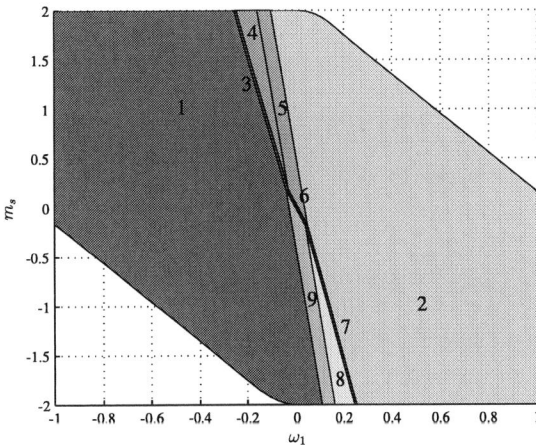

Fig. 4. State-space partition defined over $[\omega_1, m_s]$.

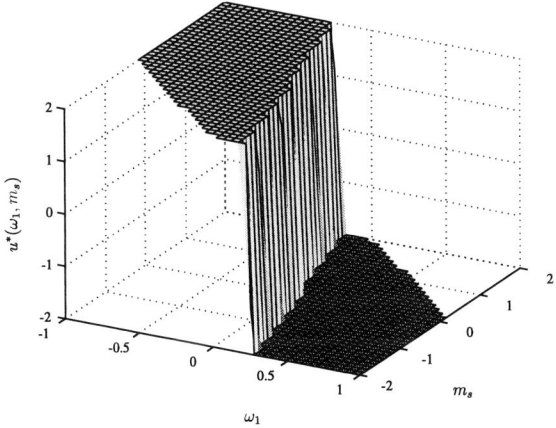

Fig. 5. Explicit control law $u^*(x)$ in $[\omega_1, m_s]$-space.

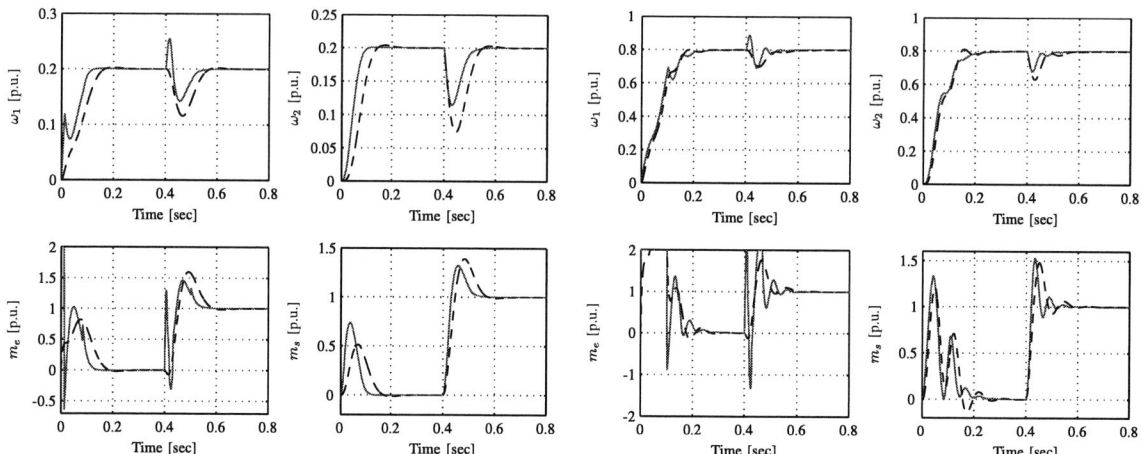

Fig. 6. Trajectories of the drive system corresponding to $\omega_r = 0.2$: explicit MPC (solid) and IPD (dashed).

Fig. 8. Trajectories of the drive system corresponding to $\omega_r = 0.8$ and $T_2 = 0.1015$sec: explicit MPC (solid) and IPD (dashed).

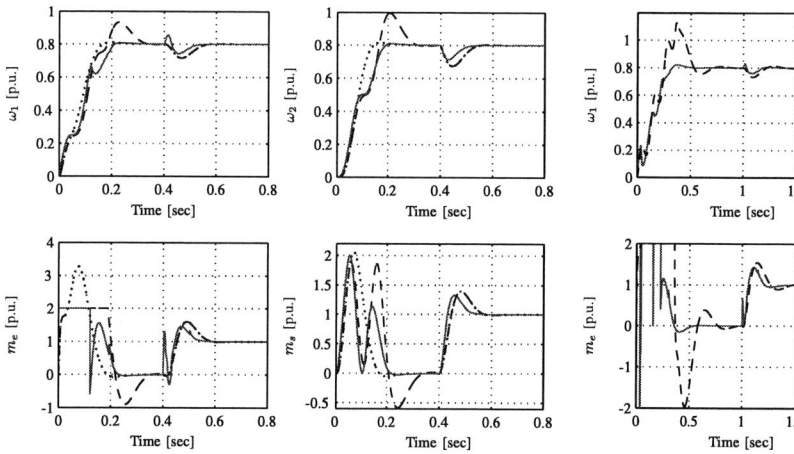

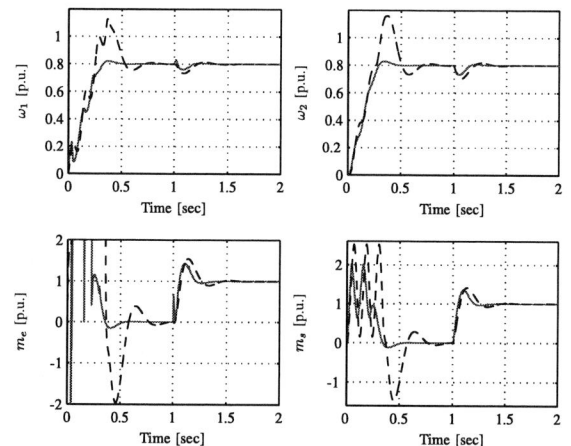

Fig. 7. Trajectories of the drive system corresponding to $\omega_r = 0.8$: explicit MPC (solid), IPD (dotted) and saturated IPD (dashed).

Fig. 9. Trajectories of the drive system corresponding to $\omega_r = 0.8$ and $T_2 = 0.406$sec: explicit MPC (solid) and IPD (dashed).

T_2 is an unknown but bounded parameter. This is a common case in industrial drive systems with variable loads [1]. The boundedness of T_2 is defined by

$$0.1015 \leq T_2 \leq 0.406, \quad (16)$$

which, according to (3), translates into $0.5 \leq R \leq 2$.

The performance of the explicit MPC controller corresponding to the reference speed $\omega_r = 0.8$ and load machine time constants $T_2 = 0.1015$sec and $T_2 = 0.406$sec are shown in Fig. 8 and Fig. 9, respectively. It is obvious from the results that a significant performance loss due to the model uncertainty can be expected from both controllers. For $T_2 = 0.1015$sec, the closed-loop dynamic properties are similar although more effective disturbance rejection properties are provided by the proposed MPC controller. This can be confirmed by evaluating the performance measure in (15) which yields $\bar{\mathcal{J}}^{IPD} = 0.064$ and $\bar{\mathcal{J}}^{MPC} = 0.055$.

When the inertia ratio is maximum (i.e. $R = 2$), the

MPC controller is faced with an infeasible solution at time 0.042sec. An infeasible solution indicates that no control action (m_e) exists that would satisfy the shaft torque constraint. In model predictive control, this problem is usually handled by using a *soft* constraint formulation (constraints that can be violated occasionally but the excess of violation is penalized in the cost function (see [15] for more details)). In this paper, a simpler strategy is adopted. When infeasibility (constraint violation) is detected, the electromagnetic torque input is set to zero. As can be seen in Fig. 8, this approach results in closed-loop responses that are almost unaffected by the uncertainty in T_2. On the other hand, the performance of the IPD controller deteriorates drastically exhibiting large overshoots in ω_1 and ω_2, oscillatory behavior and, most importantly, resulting in violation of the shaft torque safety constraint.

Calculating the performance index in (15) yields $\bar{\mathcal{J}} = 0.14$ for IPD and $\bar{\mathcal{J}} = 0.11$ for MPC.

D. Computational and Storage Complexity

The explicit MPC controller is defined on a partition of 55 regions in $\mathbb{R}^5$ state-space. The partition is characterized by $N_C = 672$ inequalities (refer to Section III for details), thus the computational complexity of evaluating the control law consists, in the worst-case, of a total of 6,720 arithmetic operations per sample (3,360 multiplications, 2,688 sums and 672 comparisons). Alternatively, binary search techniques proposed in [20] can be used to accelerate the evaluation of the explicit MPC control law. Using this method, the on-line computational burden of the proposed controller can be reduced to a total of 120 arithmetic operations per sample at the cost of a slight increase in storage requirements.

The explicit MPC controller requires the storage of all 55 polyhedral regions and the associated control laws. This translates to a total of 4,362 real numbers (4,032 for the regions and 330 for control laws). Assuming 32-bit floating point arithmetics, this requires 17.5 kB of storage to be available in a microcontroller platform. It is clear that the proposed explicit model predictive controller is a particularly appealing solution for low-cost embedded control units with as simple as 8-bit architectures.

V. CONCLUSIONS AND FUTURE WORK

The goal of this paper was to demonstrate the applicability of model predictive control for high-performance speed control and explicit constraint handling in the drive system with elastic coupling. The proposed algorithm relies on explicit parameterizations of the control law using simple affine functions defined over polyhedral regions covering the desired operating space. This enables the reduction in on-line computational complexity by several orders of magnitude since the computation is reduced to a simple function evaluation problem. The proposed MPC scheme is proven to improve the performance of the drive system with respect to IPD control and to handle system operating and safety constraints in a systematic and effective way. Current work is focusing on the implementation of the proposed explicit MPC controller on a real drive system. Topics for future work will include the development of efficient robust MPC controllers, optimal constrained state reconstruction and adaptive control.

REFERENCES

[1] J. Arellano-Padilla, G. M. Asher, M. Sumner, Control of an Dynamometer for Dynamic Emulation of Mechanical Loads With Stiff and Flexible Shafts, *IEEE Trans. on Industrial Electronics*, vol. 53, no. 4, pp. 1250-1260, 2006.

[2] S. Katsura, Y. Matsumoto, K. Ohnishi, Analysis and Experimental Validation of Force Bandwidth for Force Control, it IEEE Trans. on Industrial Electronics, vol. 53, no. 3, pp. 922-928, 2006.

[3] T. Orlowska-Kowalska, K. Szabat, Optimization of Fuzzy Logic Speed Controller for DC Drive System with Elastic Joints, *IEEE Trans. on Ind. Applications*, vol. 40, no. 4, pp. 1138-1144, 2004.

[4] A. Hace, K. Jezernik, A. Sabanovic, Improved Design of VSS Controller for a Linear Belt-Driven Servomechanism, *IEEE/ASME Trans. on Mechatronics*, vol. 10, no.4, pp 385-390, 2005.

[5] G. Ferretti, G. A. Magnoni, P. Rocco, Impedance Control for Elastic Joint Industrial Manipulators, *IEEE Trans. on Robotics and Automation*, vol. 20, pp. 488-498, 2004.

[6] M. A. Valenzuela, J. M. Bentley, R. D. Lorenz, Evaluation of Torsional Oscillations in Paper Machine Sections, *IEEE Trans. on Industry Applications*, vol. 41, no.2 , pp. 493-501, 2005.

[7] K. Sugiura, Y. Hori, Vibration Suppression in 2- and 3-Mass System Based on the Feedback of Imperfect Derivative of the Estimated Torsional Torque, *IEEE Trans. on Industrial Electronics*, vol. 43, no. 2, pp. 56-64, 1996.

[8] G. Zhang, J. Furusho, Speed Control of Two-Inertia System by PI/PID Control, *IEEE Trans. on Industrial Electronics*, vol. 47, no.3, pp. 603-609, 2000.

[9] S. N. Vukosovic, M. R. Stojic, Suppression of Torsional Oscillations in a High-Performance Speed Servo Drive, *IEEE Trans. on Industrial Electronics*, vol. 45, no.1, pp. 108-117, 1998.

[10] G. Ellis, R. D. Lorenz, Resonant Load Control Methods for Industrial Servo Drives. *Proc. of the IEEE Industry Application Society Annual Meeting*, vol. 3, pp. 1438-1445, 2000.

[11] Y. Hori, H. Sawada, Y. Chun, Slow resonance ratio control for vibration suppression and disturbance rejection in torsional system, *IEEE Trans. on Ind. Electronics*, vol. 46, no. 1, pp. 162-168, 1999.

[12] K. Szabat, T. Orlowska-Kowalska, Vibration Suppression in Two-Mass Drive System using PI Speed Controller and Additional Feedbacks - Comparative Study, *IEEE Trans. on Industrial Electronics*, vol. 54, no. 2, pp. 1193-1206, 2007.

[13] T. Orlowska-Kowalska, K. Szabat, Damping of Torsional Vibrations in Two-Mass System using Adaptive Sliding Neuro-Fuzzy Approach, *IEEE Trans. on Ind. Informatics*, vol. 4, no. 1, 2008.

[14] K. Szabat, T. Orlowska-Kowalska, Performance Improvement of the Industrial Drives with Mechanical Elasticity using Nonlinear Adaptive Kalman Filter, *IEEE Trans. on Industrial Electronics*, accepted to print, 2008.

[15] J. M. Maciejowski, *Predictive Control with Constraints*, Prentice Hall, UK, 2002.

[16] J. Wang, Y. Zhang, L. Xu, Y. Jing, S. Zhang, Torsional Vibration Suppression of Rolling Mill with Constrained Model Predictive Control, *6th World Congress on Intelligent Control and Automation*, pp. 6401-6405, China, 2006.

[17] A. Bemporad, M. Morari, V. Dua and E. N. Pistikopoulos, The explicit linear quadratic regulator for constrained systems, *Automatica*, vol. 38, no. 1, pp. 3-20, 2002.

[18] S. J. Qin, T. A. Badgwell, A Survey of Industrial Model Predictive Control Technology, *Control Engineering Practice*, vol. 11, no. 7, pp. 733-764, 2003.

[19] M. Kvasnica, P. Grieder, M. Baotic, M. Morari, Multi-Parametric Toolbox (MPT), *HSCC (Hybrid Systems: Computation and Control)*, Lecture Notes in Computer Science, vol. 2993, pp. 448-462, 2004.

[20] P. Tøndel, T. A. Johansen, A. Bemporad, Evaluation of piecewise affine control via binary search tree, *Automatica*, vol. 39, no. 5, pp. 945-950, 2003.

Development of an Expert System for Identification, Commissioning and Monitoring of Drives

Mario Pacas[*], Sebastian Villwock[†]

[*] University of Siegen - Institute of Power Electronics and Electrical Drives, Siegen, Germany, jmpacas@ieee.org
[†] Baumüller Nürnberg GmbH, Nürnberg, Germany, s.villwock@baumueller.de

Abstract— The present paper deals with identification methods for the automatic commissioning and condition monitoring of electrical drives. This works is considered as a contribution to the development of expert systems, taking into account parameter-fitting, backlash detection and the diagnosis of rolling bearing faults. The basic method is the measurement of the frequency response of the mechanical system of the drive.

Keywords— Drives, Mechatronics, Modeling, Signal Processing, System Identification, Frequency Response, Mechanical System, Bearings, Fault Diagnosis

I. INTRODUCTION

This paper intends to give an overview of our research work that aims the development of an expert system for drive commissioning, identification and condition monitoring.

Because of the importance of the matter, in the last decades many efforts were dedicated to its investigation. Some of the most important achievements are commented in the following. In [1] and [2], different methods for the identification of two-mass-systems are experimentally investigated and compared to each other. These identification strategies are confined to the detection of the resonant frequency of the plant. In these works, the determination of the other parameters of the system is not aimed. The following methods are treated in [1] and [2]: Application of Fourier analysis, identification of the resonant frequency utilizing a phase control loop, identifying the resonant frequency by using a filter with variable frequency and analytic methods.

In [3], two methods for estimating the mechanical parameters of a rotating drive system that can be treated as a resonant three-inertia-system are proposed. The first method is the least squares (LS)-estimation, which estimates the discrete transfer function by minimizing the sum of the squares of the prediction error. The second method that is proposed in [3] evaluates the fast Fourier transform (FFT) of the angular acceleration signal in a graphical approach. The system is excited by a superposition of several sinusoidal signals with different frequencies, different phases and equal amplitudes. On the basis of the estimated poles, zeros and the inertia of the whole plant, which is assumed to be a priori known, the parameters are calculated analytically.

In [4] and [5] PRBS is utilized for exciting the system. From the measurement signals, the frequency response of the mechanics is calculated. To carry out this task the authors of [4] and [5] compute the correlation functions and the power density spectra. Like [4] and [5], this paper utilizes PRBS for

stimulating the system. Yet, the identification procedure presented in this paper is much different from those mentioned above. This paper applies the Welch-method, which is a more powerful digital signal processing, to the measured signals instead of conventional methods.

In many cases the mechanical system of a drive can be modeled as two- or three-inertia-system. This method in modeling is successfully used in many industrial applications e.g. printing machines, cross cutters, steel rolling mills, robots and in many other fields of technology. In our work the same simplification is made and the basic scheme for both, identification and commissioning on the one hand and condition monitoring and diagnosis on the other hand is the measurement of the frequency response $G_{mech}(j\omega)$ of the mechanical system of the drive that is assumed to be a two- or three-inertia-system. The system is excited by pseudo random binary signals (PRBS). The measurement signals are the velocity of the motor and the phase currents. No additional sensors are necessary at all. The drive is stimulated by the electromagnetic torque and the measurement is carried out in closed loop speed control. Therefore it can be run during the normal operation of the plant. In this way, a simple control structure with only one feedback from the measured motor speed is utilized.

Usually for a subsequent optimization of the control system the parameters of the plant are required as well. For their calculation the numerical optimization method of Levenberg and Marquardt is utilized.

In general all industrial drives contain certain mechanical backlash, thus the present work addresses its identification as well. Since backlash is a nonlinearity the measurement of the frequency response will not deliver the backlash angle, which in the following is designated by 2ε. Hence, for this task a special deterministic time domain identification method is introduced.

Since backlash is a result of wear of the machine elements and can grow beyond the allowed limits during the lifespan of the machine, a self-operating identification method is desirable for diagnostics, too.

Regarding the detection of further faults in the mechanics the present paper focuses on rolling bearing damages. It is well known that they are a main factor for unplanned downtimes of machines and plants [6]. Hence, they give rise to costs for repair and maintenance. An early detection of incipient bearing failures allows therefore the reduction of these costs by initiating repair and maintenance at planned times before catastrophic failures occur.

978-1-4244-1741-4/08/$25.00 ©2008 IEEE

Many diagnosis methods utilize vibration condition monitoring of the bearings [7], [8], [9]. A considerable disadvantage of this method is the required additional sensor. An alternative method for the detection of rolling-element bearing faults in induction machines can be found in [8], [10], [11] and is based on the spectral analysis of the motor current. Of course, this method does not require an additional sensor, yet another important problem must be considered. The indicators for a bearing fault, which can be detected in the stator current, are not very pronounced and therefore not easy to distinguish.

In the present work the detection of bearing faults is obtained by frequency response analysis of the mechanical system of the drive. The obtained measured curve is compared to the frequency response of the same plant which can be obtained during the commissioning of the drive.

II. MODELING

Fig. 1a displays the CAD-drawing of the mechanical system which was utilized in order to validate the identification methods. A permanent magnet synchronous machine with a rated torque of 15 Nm drives the mechanics. The machine is fed by an inverter with field oriented control. In order to enable thorough investigations in a wide range of backlash values a special mechanical backlash element has been designed and realized. The realized module is a claw-type construction. It provides the possibility to change the backlash angle manually to certain values within a wide range. Fig. 1b shows an exploded view of the claw coupling.

Fig. 1a. Mechanical setup
Moment of inertia J_M 0,0138 kg m^2
Moment of inertia J_L 0,0966 kg m^2
Spring constant C   4141 Nm

Fig. 1b. Claw coupling, exploded view

The elementary mechanical equations of a three-mass-system are given by

$$\frac{d\omega_M}{dt} = \ddot{\varphi}_M = \frac{1}{J_M} \cdot \left(-C_1 \cdot (\varphi_M - \varphi_{L1}) - D_1 \cdot (\dot{\varphi}_M - \dot{\varphi}_{L1}) + M_M \right), \quad (1)$$

$$\frac{d\omega_{L1}}{dt} = \ddot{\varphi}_{L1} = \frac{1}{J_{L1}} \cdot (C_1 \cdot (\varphi_M - \varphi_{L1}) + D_1 \cdot (\dot{\varphi}_M - \dot{\varphi}_{L1})$$
$$- C_2 \cdot (\varphi_{L1} - \varphi_{L2}) - D_2 (\dot{\varphi}_{L1} - \dot{\varphi}_{L2})) \quad (2)$$

$$\frac{d\omega_{L2}}{dt} = \ddot{\varphi}_{L2} = \frac{1}{J_{L2}} \cdot (C_2 \cdot (\varphi_{L1} - \varphi_{L2}) + D_2 (\dot{\varphi}_{L1} - \dot{\varphi}_{L2}) - M_L), \quad (3)$$

where index L1 stands for the first load inertia and the first non-rigid shaft and L2 represents the second load inertia and the second shaft, respectively.

The transfer function of the three-mass-system is

$$G_{mech}(s) = \underbrace{\frac{1}{T_\Sigma \cdot s}}_{G_m(s)} \cdot \underbrace{\frac{a_7 \cdot s^4 + a_6 \cdot s^3 + a_5 \cdot s^2 + a_4 \cdot s + 1}{a_3 \cdot s^4 + a_2 \cdot s^3 + a_1 \cdot s^2 + a_4 \cdot s + 1}}_{G_{ms}(s)}, \quad (4)$$

with $T_\Sigma = T_M + T_{L1} + T_{L2}$ and   (5)

$$a_1 = d_1 d_2 T_{C1} T_{C2} + \frac{T_{L2}(T_M + T_{L1}) \cdot T_{C2}}{T_\Sigma} + \frac{T_M(T_{L1} + T_{L2}) \cdot T_{C1}}{T_\Sigma} \quad (6)$$

$$a_2 = \frac{T_{C1} T_{C2}}{T_\Sigma} \cdot \left(d_1 T_{L2}(T_M + T_{L1}) + d_2 T_M(T_{L1} + T_{L2}) \right) \quad (7)$$

$$a_3 = \frac{T_M T_{L1} T_{L2} T_{C1} T_{C2}}{T_\Sigma} \quad (8)$$

$$a_4 = d_1 T_{C1} + d_2 T_{C2} \quad (9)$$

$$a_5 = d_1 d_2 T_{C1} T_{C2} + T_{L2} T_{C2} + (T_{L1} + T_{L2}) \cdot T_{C1} \quad (10)$$

$$a_6 = \left((d_1 + d_2) T_{L2} + d_2 T_{L1} \right) \cdot T_{C1} T_{C2} \quad (11)$$

$$a_7 = T_{L1} T_{L2} T_{C1} T_{C2} \quad (12)$$

Fig. 2 shows the block diagram of a three-inertia-system with two non rigid shafts.

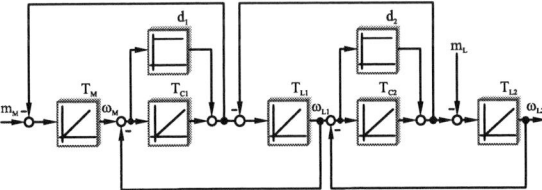

Fig. 2.   Block diagram of the three-inertia-system

III. FREQUENCY RESPONSE

The frequency response of a linear system is given by

$$G(j\omega) = \frac{S_{uy}(j\omega)}{S_{uu}(j\omega)}. \quad (13)$$

2249

The system is excited by pseudo random binary signals (PRBS). The utilization of the multifrequent PRBS has certain characteristics providing superior possibilities compared to other known test signals. In contrast to purely random signals PRBS has a periodicity T_P which is given by

$$T_P = \left(2^n - 1\right) \cdot T_t. \tag{14}$$

The PRBS is generated by using a shift register consisting of n bits. The bits of the register are shifted one step to the right after the time interval T_t. The resulting LSB is taken for the generation of the PRBS. The new MSB is the output of an XOR-gate. The PRBS is injected at the output of the PI-speed controller according to Fig. 3.

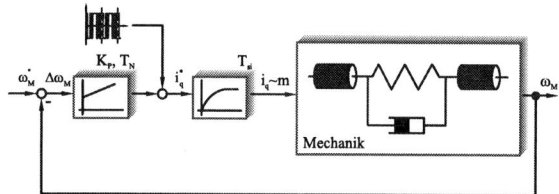

Fig. 3. Block diagram of the closed loop speed control with injection of the PRBS at the output of the PI-speed-controller

The measurement of the required signals for calculating the frequency response are the velocity of the driving machine $\Omega_M(t)$ and the torque generating component of the stator current $i_q(t)$. For doing so, the Welch-method, which is known from the field of communications engineering is utilized. Fig. 4 points out the main steps of this method, starting from the measured signals u(k), y(k), k=0,1,...,M-1.

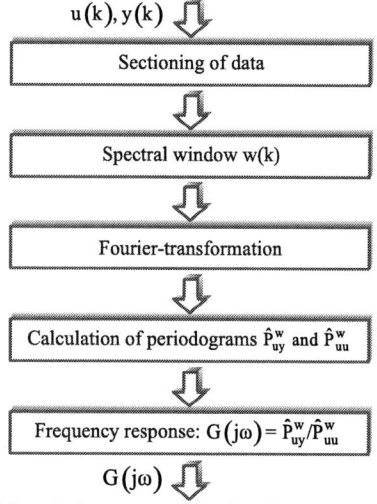

Fig. 4. Schematic diagram summarizing the main steps of the digital signal processing according to the Welch-method

A more detailed explanation of the signal processing can be found in [12], [13].

The measurement signals are displayed in Fig. 5.

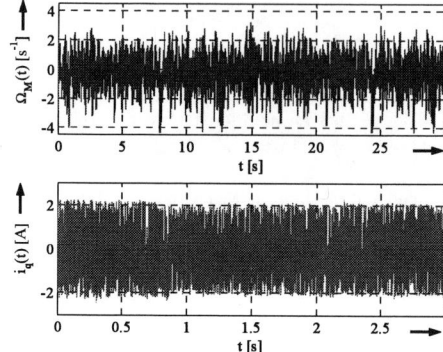

Fig. 5. Measurement signals

On basis of the measurement signals the frequency response of the plant is calculated using the Welch-method.

Fig. 6 shows the obtained frequency response of the system which is displayed in Fig. 1a. The typical characteristics of the three-inertia-system can be seen clearly: two resonant frequencies and two anti-resonant frequencies in the amplitude response. The phase response shows the corresponding behavior at these frequencies.

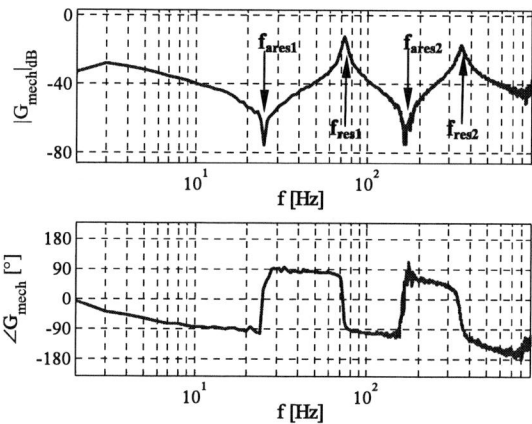

Fig. 6. Frequency response of the mechanical system of the drive

IV. PARAMETER IDENTIFICATION

Although the frequency response of the system gives a good overview of the dynamical behavior of the drive the frequency response itself is only a non-parametric model of the system. In contrast to the measurement of the frequency response the parameter-fitting requires the assumption of a certain model structure. As already explained in many cases the two-mass-model is sufficient. Then only the first resonant frequency would be considered. Further resonances would be ignored. In the present paper the algorithm is implemented for a three-inertia-system that accounts for two resonant frequencies. On basis of the frequency response the mechanical parameters of the plant can be calculated by utilizing optimization methods. For this purpose the Levenberg-Marquardt-algorithm is used [14], assuming that the run-up time of the whole system is a priori known according to

$$G_{nrs}(j\omega) = j\omega \cdot T_{\Sigma} \cdot G_{mech}(j\omega). \tag{15}$$

If T_{Σ} is not a priori known it can be obtained from the data of the frequency response in the way described in [5], in which the mechanical system of the drive is approximated as one-inertia-system for low frequencies below the anti-resonant frequency. Fig. 7 illustrates the idea.

$$T_{\Sigma} = J_{\Sigma} \cdot \Omega_N / M_N , \tag{16}$$

with Ω_N: nominal speed, M_N: nominal torque

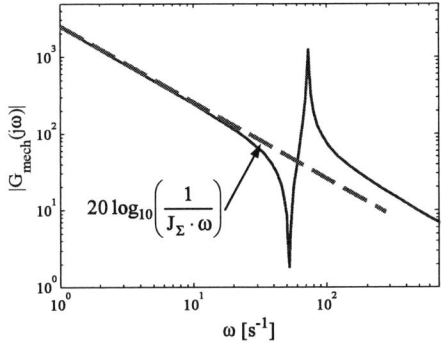

Fig. 7. Estimation of the moment of inertia of the drive being approximated as one-inertia-system

The calculation of the model function is carried out by processing N points of the frequency response data. Hence a quadratic error function χ^2 which depends on the coefficients of $G_{NRS}(s)$ follows:

$$\chi^2(\vec{a}) = \sum_{i=1}^{N} \frac{1}{\sigma_i^2} \cdot \left| G_{measure,i}(j\omega_i) - G_{nrs,i}(j\omega_i, \vec{a}) \right|^2 \tag{17}$$

$G_{measure}$ is the measured data and G_{nrs} is the model function according to (4). The minimization of the error-function proceeds iteratively.

The Levenberg-Marquardt-method can be regarded as a numerical mixture of the gradient method and the method of the inverse Hessian Matrix. This is achieved by introducing a scalar λ on the main diagonal of $\mathbf{A}$ and leads to the modified Hessian matrix $\mathbf{A}_{mod}$:

$$\mathbf{A}_{mod} = \mathbf{A} + \mathbf{A} \lambda \mathbf{I} = \begin{pmatrix} \frac{\partial^2 \chi^2(\vec{a})}{\partial a_1 \partial a_1}(1+\lambda) & \cdots & \frac{\partial^2 \chi^2(\vec{a})}{\partial a_1 \partial a_n} \\ \vdots & \ddots & \vdots \\ \frac{\partial^2 \chi^2(\vec{a})}{\partial a_n \partial a_1} & \cdots & \frac{\partial^2 \chi^2(\vec{a})}{\partial a_n \partial a_n}(1+\lambda) \end{pmatrix} \tag{18}$$

In the course of the iteration procedure λ is increased or decreased depending on whether the last iteration step has brought advancement or not. The order of matrix $\mathbf{A}_{mod}$ is equal to the number of parameters which must be calculated. The correction vector is given by

$$\delta \vec{a} = -\mathbf{A}_{mod} \, \nabla \chi^2(\vec{a}). \tag{19}$$

In Fig. 8 the obtained model function as well as the utilized measured frequency response data are displayed. It can be seen that the model function corresponds well to the measured data.

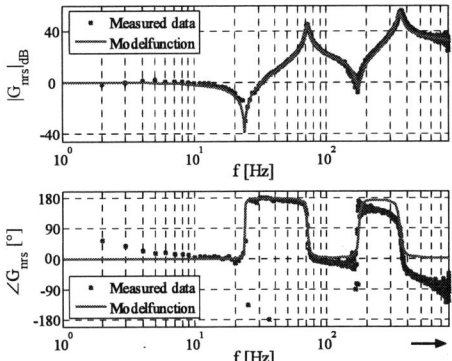

Fig. 8. Measured frequency response data and obtained model function

More details concerning the algorithm itself can be found in [15], [16].

V. ROLLING BEARING DAMAGE DETECTION

In case of mechanical defects changes of the frequency response can be expected. As an example bearing damages have been investigated. To obtain an overall surface roughness, the investigated bearings were contaminated with some corundum in the laboratory. Over a large frequency range, the measured curve of the faulty system differs significantly from the reference amplitude response so that this kind of fault can be detected as Fig. 9 points out [17].

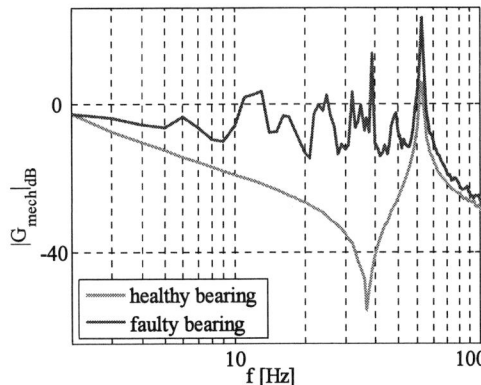

Fig. 9. Amplitude response of the mechanical system of the drive with a bearing fault

The differences between the amplitude response of the healthy system and of the faulty one, which can be seen in Fig. 9 do not occur at specific frequencies. Therefore, this type of bearing fault is called broadband damage.

In case of single point defects the differences between both amplitude responses occur at characteristic fault frequencies that can be calculated. For an outer race fault the characteristic fault frequency f_{ORF} is

$$f_{ORF} = \frac{\Omega_{ORF}}{2\pi} = \frac{z}{2} \cdot \frac{\Omega_n}{2\pi} \cdot \left(1 - \frac{d_b}{d_o} \cos(\theta)\right), \tag{20}$$

and for an inner race fault it is

$$f_{IRF} = \frac{\Omega_{IRF}}{2\pi} = \frac{z}{2} \cdot \frac{\Omega_n}{2\pi} \cdot \left(1 + \frac{d_b}{d_o} \cos(\theta)\right), \tag{21}$$

with

- d_c: pitch diameter,
- d_b: ball diameter,
- θ: ball contact angle,
- z: number of rolling elements,
- Ω_n: mechanical frequency of the drive.

Theoretically, the signals contain characteristic outer and inner race fault frequency and their ν multiples [18]:

$$f_{S,ORF}(\nu) = f_{S,ORF,\nu} = \nu \cdot f_{ORF} \tag{22}$$

$$f_{S,IRF}(\nu) = f_{S,IRF,\nu} = \nu \cdot f_{IRF}, \quad \nu = 1, 2, 3, \dots \tag{23}$$

Whether or not the multiples occur depends on the intensity of the fault.

Fig. 10 shows the bearing with an outer race fault. The damage was installed at the 6 o'clock position, which is the point of maximum radial force. The required technical data of the analyzed bearing is

Fig. 10.   Photograph of the analyzed bearing with outer race fault

$d_b = 8, 7 \, mm$, $d_o = 46 \, mm$, $z = 9$. For the investigated radial ball bearing the ball contact angle θ is zero.

In Fig. 11 the measured curve for the outer race fault of the bearing (Fig. 10) is displayed. While recording the signals the desired value of the velocity of the motor was $n_M^* = 200 \, min^{-1}$. According to (20) the resulting fault frequency is $f_{ORF} = 12 \, Hz$. For this characteristic frequency a difference between the two curves can be seen clearly. Furthermore, the multiples of f_{ORF} ($24 \, Hz$, $36 \, Hz$, $48 \, Hz$) do also indicate the fault in Fig. 10. The most significant difference can be found in the area of the anti-resonant frequency. With an inverter fed machine it is possible to choose the velocity n_M^* in order to shift the characteristic fault frequency in a way that the method indicates the damage very clearly.

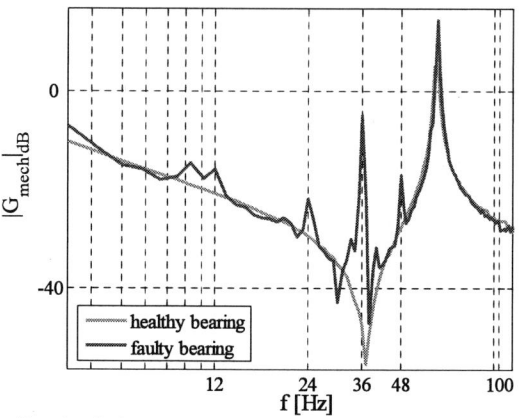

Fig. 11.   Amplitude response for bearing with outer race fault

$f_{s,ORF}(\nu = 1) = 12 \, Hz$, $f_{s,ORF}(\nu = 2) = 24 \, Hz$, $f_{s,ORF}(\nu = 3) = 36 \, Hz$,

$f_{s,ORF}(\nu = 4) = 48 \, Hz$

VI. BACKLASH IDENTIFICATION

In case of backlash larger than $2\varepsilon = 1°$ the measurement of the frequency response does not work as it is only defined for linear systems [19]. Therefore, another identification strategy is necessary in order to estimate backlash. The identification is carried out in closed speed control loop. The system is excited by triangular signals for the speed reference value, which causes an acceleration profile with either maximum or minimum value [19]. At the maximum value of the velocity of the motor $\Omega_M(t)$ the polarity of the acceleration and of the torque $i_q(t)$ changes and the shaft begins to ride through the dead zone of the backlash. From this point on, no torque is exerted on the load side and its speed can be assumed to remain constant, i.e. for $t_1 \leq t \leq t_2$ it is $\Omega_L(t)$ = constant. t_1 is the instant when the mechanics decouples and t_2 is the instant when it engages.

The identification of the backlash value is carried out by the calculation of the area between the signals $\Omega_M(t)$ and $\Omega_L(t)$ according to

$$2\varepsilon = \left| \int_{t_1}^{t_2} \left(\Omega_L(t) - \Omega_M(t) \right) dt \right| \tag{24}$$

Assuming that $\Omega_L(t) = \hat{\Omega}_M$ until the mechanics engages the calculation of the backlash value 2ε can be accomplished according to the expression

$$2\varepsilon \approx \left| T_S \cdot \sum_{k=k_1}^{k_2} \left(\hat{\Omega}_M - \Omega_M(k) \right) \right|. \tag{25}$$

T_S is the sampling time. According to (25) no load side encoder is required.

In Fig. 12 the measurement result for $2\varepsilon = 2°$ is displayed. The signals $\Omega_M(t)$ and $\Omega_L(t)$ are depicted and Fig. 13 shows a zoom, pointing out the interesting area of $\Omega_M(t)$.

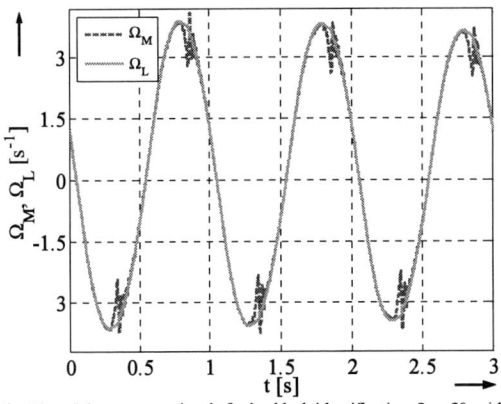

Fig. 12. Measurement signals for backlash identification $2\varepsilon = 2°$ with load side encoder

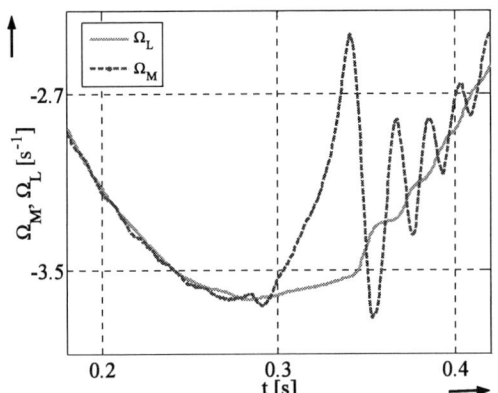

Fig. 13. Zoom of Fig: 12

A position sensor at the load side is implemented in order to analyze the behavior of the load within the dead zone. When decoupling from the motor-side the load nearly keeps the velocity of the motor at that moment until the mechanics engages. The relative error of 2ε achieved by applying this method is about 5%.

VII. CONCLUSION

The present paper deals with the development of an expert system for commissioning and monitoring of drives. The basic method is the measurement of the frequency response of the mechanical system of the drive. For doing so pseudo random sequences are used. The calculation of the plant parameters is accomplished by the Levenberg-Marquardt-algorithm. The work considers the identification of backlash and rolling bearing faults. A deterministic time domain method for identifying backlash is introduced. Various measurement results obtained on a laboratory setup point out that the proposed identification strategies are suitable for the commissioning and condition monitoring of drives.

VIII. REFERENCES

[1] J. M. Pacas, A. John, and T. Eutebach, "Automatic identification anddamping of torsional vibrations in high-dynamic-drives," in Proc. IEEE ISIE-Conf., Puebla, Mexico, 2000, pp. 201–206.

[2] T. Eutebach and J. M. Pacas, "Damping of vibrations in high-dynamicdrives," in Proc. EPE-Conf., Lausanne, Switzerland, 1999. CD-ROM.

[3] I. Müller and P. Mutschler, "Two reliable methods for estimating thevmechanical parameters of a rotating three-inertia-system," in Proc. EPEPEMC-Conf., Dubrovnik, Croatia, 2002, CD-ROM.

[4] H. Wertz, S. Beineke, N. Fröhleke, S. Bolognani, K. Unterkofler, M. Zigliotto, and M. Zordan, "Computer aided commissioning of speed and position control for electrical drives with identification of mechanical load," in Conf. Rec. IEEE IAS Annu. Meeting, 1999, pp. 2372–2379.

[5] F. Schütte, Automatisierte Reglberinbetriebnahme in elektrischen antrieben mit schwingungsfähiger Mechanik. Paderborn, Germany: Univ. Paderborn. Diss. 2002.

[6] J. R. Stack, T. G. Habetler, R. G. Harley: Fault-Signature Modeling and Detection of Inner-Race Bearing Faults, IEEE Transactions on Industry Applications, Vol. 42, No. 1, Jan./Feb. 2006

[7] J. R. Stack, R. G. Harley, T. G. Habetler: An Amplitude Modulation Detector for Fault Diagnosis in Rolling Bearings Element Bearings, IEEE Transactions on Industrial Electronics, Vol. 51, No. 5, 2004

[8] R. R. Schoen, T. G. Habetler, F. Kamran, R. G. Bartheld: Motor Bearing Damage Detection Using Stator Current Monitoring, IEEE Transactions on Industry Applications, Vol. 31, No. 6, 1995

[9] J. R. Stack, R. G. Harley, T. G. Habetler: An Amplitude Modulation Detector for Fault Diagnosis in Rolling Bearings Element Bearings, IECON-conference 2002, Sevilla, Spain

[10] J. R. Stack, Thomas G. Habetler, Ronald G. Harley: Fault Classification and Fault Signature Production for Rolling Element Bearings in Electric Machines, IEEE Transactions on Industry Applications, Vol. 40, No. 3, May/June 2005

[11] R. R. Obaid, T. G. Habetler, J. R. Stack: Stator Current Analysis for Bearing Damage Detection in Induction Motors, SDEMPED (Symposium on Diagnostics for Electrical Machines, Power Electronics and Drives), Atlanta, USA, 2003

[12] S. Villwock, M. Pacas T. Eutebach: Application of the Welch-Method for the Automatic Parameter Identification of Electrical Drives, IECON 2005, USA

[13] S. Villwock, M. Pacas: Application of the Welch-Method for the Identification of Two- and Three-Mass-Systems, IEEE Transactions on Industrial Electronics, Vol. 55, No. 1, January 2008

[14] J.M. Pacas, S. Villwock, T. Eutebach: Parameter Identification of a Two-Mass-System in the Frequency Domain, PCIM-conference, Nuremberg, Germany, 2004

[15] Press, Flannery, Tenkolsky, Vetterling: Numerical Recipes, Cambridge University Press, 1986

[16] J. Schoukens, R. Pintelon: Identification of linear systems – A practical guideline to accurate modeling, Pergamon Press, 1991

[17] S. Villwock, H. Zoubek, M. Pacas: Rolling Bearing Condition Monitoring Based on Frequency Response Analysis, The 6th IEEE International Symposium on Diagnosis for Electrical Machines Power Electronics and Drives (SDEMPED), 6. - 8. September 2007, Cracow, Poland

[18] J. R. Stack, T. G. Habetler, R. G. Harley: Fault-Signature Modeling and Detection of Inner-Race Bearing Faults, IEEE Transactions on Industry Applications, Vol. 42, No. 1, Jan./Feb. 2006

[19] S. Villwock, M. Pacas: Deterministic Method for the Identification of Backlash in the Time Domain, IEEE ISIE 2006, International Symposium on Industrial Electronics, 9. - 13. July 2006, École de Technologie Supérieure (ÉTS), Montréal, Canada

Control of Axial Flux Permanent Magnet Motor by the PIPCRM Method at Standstill and at Low Speed

Janusz Wiśniewski[*], Włodzimierz Koczara[†]

[*] Warsaw University of Technology, Warsaw, Poland, e-mail: *wisniewj@isep.pw.edu.pl*
[†] Warsaw University of Technology, Warsaw, Poland, e-mail: *koczara@isep.pw.edu.pl*

Abstract— **The paper presents novel sensorless method of rotor position detection without any mechanical sensor. Described method bases on inductance variations connected with poles position. The rotor position is obtained by applying one voltage pulse only. Efficiency of presented method has been confirmed by laboratory tests on Axial Flux Permanent Magnet Motor. Moreover described method can be used to startup the permanent magnet motor from standstill in certain direction.**

Keywords— **Sensorless control, Permanent Magnet Motor, Brushless Drive, Estimation Technique**

I. INTRODUCTION

A sensorless techniques development is observed in recent years. Main aim of sensorless methods is elimination of mechanical rotor position sensors. Such a methods are based on voltage and current measurements only. Using special algorithms rotor position can be calculated or identified. It is observed that many papers deals with subject of electrical drives without any rotor position sensor.

Methods, which are used at standstill, usually base on inductance variations connected with rotor position changes. Most known method in that group is INFORM method [1] [2] [3] [4] [5]. Flux angle is estimated by applying series of transient test voltage pulses and calculating the inductance, which depends on the rotor position.

The control strategy which is presented in that paper is based on competitive sensorless method which is called as PIPCRM. Advantages of that method is that rotor position detection is realized by using one parallel voltage pulse only. Moreover presented method has been designed particularly to operate new Axial Flux Permanent Magnet Motor. The PIPCRM method was preliminary presented by authors [6], [7], [8], [9]. However, recent experimental works permit to extend the knowledge about this new method going to be more matured and these new results will be published in the proposed paper.

II. CONCEPT OF THE PIPCRM METHOD

Using the PIPCRM method, initial rotor position can be detected without any mechanical rotor sensor. Presented method bases on saturation and inductance changes with respect to the rotor position. The presented method works through mentioned phenomena existing together. First of them has to be constrained by applying an additional stator flux. Second phenomenon can be observed directly,

because it is obtained in natural way. A natural relationship between the phase inductance and the rotor position is shown in fig. 1. Note that, natural characteristic will be different if d-axis inductance is similar to q-axis inductance. In that case difference between maximal phase inductance L_{NAT_MAX} and minimal phase inductance L_{NAT_MIN} goes down to zero.

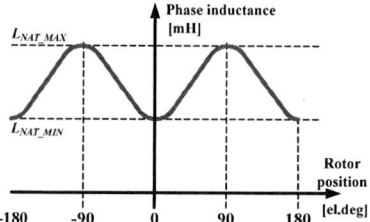

Fig. 1.   Natural changes of stator inductance with respect to the rotor position.

However, if no saliency is designed into the rotor, presented method will behave correctly, because the varying phase inductance will constrain. Granting that condition is possible by introducing stator magnetic circuits into the saturation. It causes that saturation saliency occurs. Simplified representation of the phase inductance, influenced by saturation, in function of electrical angle is shown in fig. 2. When saturation takes place, phase inductance will go down suddenly. Therefore the L_{MAX} level (fig. 2) is much lower than the L_{NAT_MIN} level (Fig. 1).

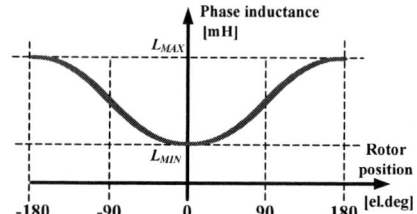

Fig. 2.   Constrained changes of the stator inductance with respect to the rotor position.

An additional stator flux is required to saturate magnetic circuits of stator windings. Applying a single voltage step U_{TEST} to the stator windings connected in parallel (fig. 3), the additional stator flux Ψ_D is achieved. A representation of parameters of stator coil magnetic circuits are magnetic set points, marked as msp_U, msp_V,

978-1-4244-1741-4/08/$25.00 ©2008 IEEE

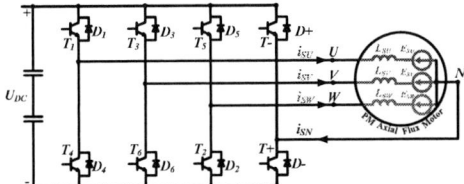

Fig. 3.   The converter topology used by the PIPCRM method

msp_W. Effect of additional stator flux applying is a transition of magnetic set points to saturation area of B-H characteristic [6] (fig. 4). Explanation of the inductance function, shown in fig. 2, is achieved by analyzing the B-H characteristic. Since magnetic circuits saturates, significant reduction of stator inductances L_{SU}, L_{SV}, L_{SW} is observed, but even than relationship between the stator inductance and the rotor position is kept. It is provoked by the feature of the magnetic set points, which maintain references between each other. Mutual position of magnetic set points is defined by the particular electrical angle φ. Therefore it allows to determine electrical position φ of the rotor.

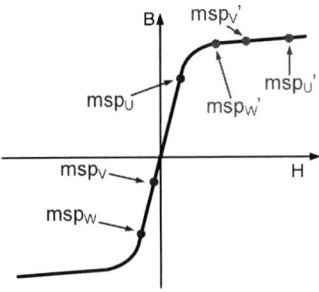

Fig. 4.   Movement of magnetic set points to saturation area caused by additional stator flux Ψ_D

As mentioned, the additional stator flux is achieved by applying the test voltage step U_{TEST}, called as PIPCRM voltage step. Such test voltage step generates current flow in stator windings. Due to parallel connection of the windings, the current response i_U, i_V, i_W is observed in each phase. Different phase inductances provoke different phase di/dt. Thus top value of each current response is not equal (Fig. 3). The PIPCRM method requires a measurement of the maximal value of current responses I_U, I_V, I_W.

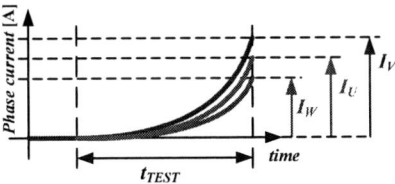

Fig. 5.   Current responses of the PICPRM method due to applying test voltage step

Analyzing measured current responses with respect to the rotor position, some regularity has been noticed. That feature allows to express few conclusions. First of them is that current responses have two components: constant component and variable component, but only one of them is function of the rotor position. Reference [7] gives description of the current responses. However an improvement of the accuracy of the PIPCRM method requires some modifications of the current response description. It has been observed that variable component depends on constant component. It means that variable component is a function of the rotor position φ and the constant component I_0 too. Thus new equations are given:

$$I_U = I_0 + I_0 \cdot m \cdot f_U(\varphi) \qquad (1)$$

$$I_V = I_0 + I_0 \cdot m \cdot f_V(\varphi) \qquad (2)$$

$$I_W = I_0 + I_0 \cdot m \cdot f_W(\varphi) \qquad (3)$$

where m is the PIPCRM position factor, constant for given topology of the motor and the converter.

Proper analysis of current response allows to identify rotor position with high accuracy [8] [9]. Reference [9] pronounces that constant component is unwanted and gives method of its elimination. Because I_0 is contained also in variable component, the method which is proposed by [9] is unprofitable. Therefore new way has been proposed [10]. It is called as proportions of current responses and is described by (1), (2), (3). Presented method allows to increase accuracy of the PIPCRM method. Theoretical waveforms of current response proportions I_{PUV}, I_{PVW}, I_{PWU} are presented in fig. 6.

$$I_{PUV} = \frac{I_U}{I_V} = \frac{I_0 \cdot (1 + m \cdot f_U(\varphi))}{I_0 \cdot (1 + m \cdot f_V(\varphi))} = \frac{f'_U(\varphi)}{f'_V(\varphi)} = f_{UV}(\varphi) \qquad (4)$$

$$I_{PVW} = \frac{I_V}{I_W} = \frac{I_0 \cdot (1 + m \cdot f_V(\varphi))}{I_0 \cdot (1 + m \cdot f_W(\varphi))} = \frac{f'_V(\varphi)}{f'_W(\varphi)} = f_{VW}(\varphi) \qquad (5)$$

$$I_{PWU} = \frac{I_W}{I_U} = \frac{I_0 \cdot (1 + m \cdot f_W(\varphi))}{I_0 \cdot (1 + m \cdot f_U(\varphi))} = \frac{f'_W(\varphi)}{f'_U(\varphi)} = f_{WU}(\varphi) \qquad (6)$$

Waveforms in fig. 5 are simplified and can be distorted in practice. Many elements influence on current response proportions. First distortion is provoked by construction of the permanent magnet motor. It means that motor can be constructed with some magnetic and mechanical asymmetry. On the other hand any asymmetry can be observed and detected from current responses. Latter distortion is back electromotive force. Since BEMF is ratable to speed, it will deform waveforms when rotor speed goes up.

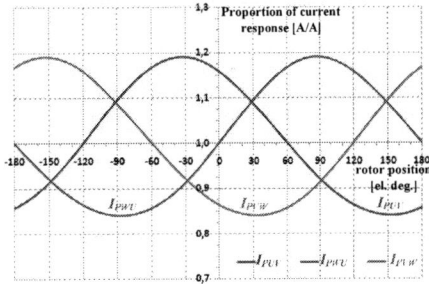

Fig. 6.   Theoretical waveforms of current response proportions.

Finally, the rotor position is calculated from one equation which describes current response proportion. Appropriate function is chosen with respect to logical dependences between (4), (5) and (6).

TABLE I.
CONDITIONS OF NOMINATION A FUNCTION TO CALCULATE ROTOR POSITION

Maximal value	Minimal value	The best function to calculate the rotor position
I_{PUV}	I_{PVW}	I_{PWU}
I_{PVW}	I_{PUV}	I_{PWU}
I_{PVW}	I_{PWU}	I_{PUV}
I_{PWU}	I_{PVW}	I_{PUV}
I_{PWU}	I_{PUV}	I_{PVW}
I_{PUV}	I_{PWU}	I_{PVW}

III. LOW SPEED CONTROL STRATEGY USING PIPCRM METHOD

Low speed control strategy bases on the "Detect and Drive" (D&D) procedure. Such algorithm decide about switching between position measurement and driving torque production. That characteristic control algorithm has three steps. First is zero-current stage. Load currents are switched off. When currents fall down to zero, next stage will occur. In that part of the D&D procedure, test voltage step is applied to the stator windings. Next, phase currents are measured and rotor position is calculated. Current responses, which are induced by test voltage, doesn't produce any torque. All energy delivered to the motor is lost. In consequence rotor is not moved. Therefore, last part of algorithm is load-current stage. Goal of that stage is to provide rotor motion. It means that torque producing currents are set with respect to the calculated electrical angle. Time sequence of Detect and Drive procedure is showed in fig. 7.

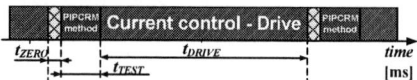

Fig. 7.   Time sequence of control based on the PIPCRM method

The current and speed regulation is realized during current control stage only. When procedure goes from drive to other stages, regulation will be stopped. In the beginning of the zero current stage, the output value of the current regulator is latched and kept until end of the position measurement. Since detecting procedure is switched on, the stator currents don't cover the reference current. Therefore the regulator error is not zero and in consequence the regulator increases its output. It is first reason to introduce such latching process. Result is regulator update with saved value at the end of the test voltage step. Next reason is an oscillations minimization of the torque producing current at beginning of the drive stage.

Using control based on the PIPCRM method, sensorless start and reverse of permanent magnet can be realized. Additionally speed control of axial flux permanent magnet motor can be applied. Simplified structure of the control

based on the PIPCRM method is shown in fig. 8. The D&D is marked as "Detect/Drive selector".

However discrete rotor position information makes some limitation. Result of discrete position information is discontinuous speed signal. Moreover position which is calculated by the PICPRM method is not ideal and it can lead some errors. Thus mentioned problems can reduce dynamic of the drive system at range of low speed.

Next problem is frequency of D&D realization. Higher frequency improves quality of position information. On the other hand, increased number of position tests reduces average torque. Thus some research has been provide to estimate the dependence between number of tests, speed and torque. All simulations have been done using PSIM 6.0 software [11]. The goal of the simulations was to confirm torque losses caused by PIPCRM measurements.

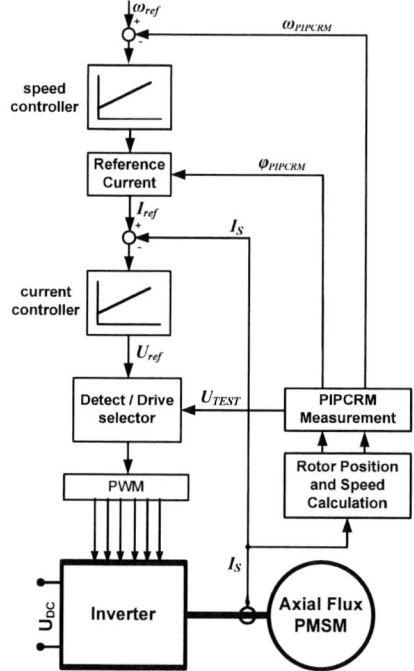

Fig. 8.   Block diagram of the sensorless control system

The PSIM built-in model of Permanent Magnet Synchronous Motor was used. Series of simulations show different number of position tests at certain speed. The load torque was set to 50 Nm. Increasing frequency of the Detect and Drive procedure, the average torque should be constant. It is possible by growing the reference current. In the other case produced torque will go down. Therefore the current has been observed and written down when steady state was achieved. The D&D frequency depends on speed Ω and number of test voltage steps per 360 electrical degrees marked as n_{IMP}. At certain speed frequency can be changed by n_{IMP} only. Total time used to position detection at set speed will be greater when higher number of tests per 360 electrical degrees are applied. It causes that losses increase themselves.

The percentage participation of the zero-current stage and measurement stage in total time of the Detect and Drive is defined as η_{PIPCRM} (7)

$$\eta_{PIPCRM} = \frac{(t_{ZERO}+t_{TEST}) \cdot n_{IMP} \cdot \Omega \cdot p}{60} \qquad (7)$$

The η_{PIPCRM} depends on the rotor speed Ω, a number of position measurements per 360 electrical degrees n_{IMP} and a number of poles p. For given DC link voltage U_{DC} and permanent magnet motor, t_{ZERO} and t_{TEST} are nearly constant.

Results of the computer simulations are presented in fig. 9, 10, 11 and 12. A comparison of carried out results is showed in the Table II.

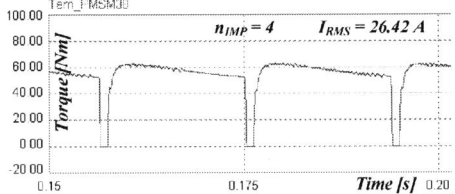

Fig. 9.   Calculated torque characteristic versus time at 100 rpm and 4 position measurements per 360 electrical degrees

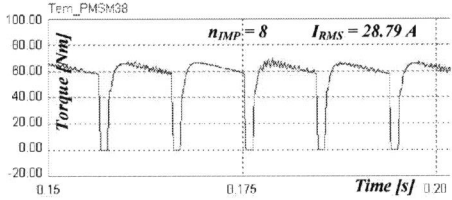

Fig. 10.   Calculated torque characteristic versus time at 100 rpm and 8 position measurements per 360 electrical degrees

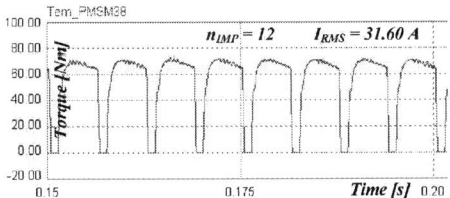

Fig. 11.   Calculated torque characteristic versus time at 100 rpm and 12 position measurements per 360 electrical degrees

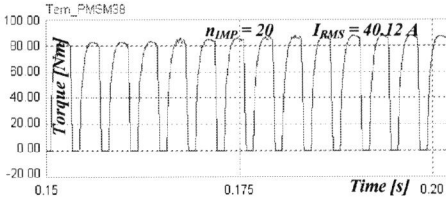

Fig. 12.   Calculated torque characteristic versus time at 100 rpm and 20 position measurements per 360 electrical degrees

TABLE II.
COMPARISON OF THE WORK CURRENT WHICH IS NEEDED TO KEEP CONSTANT SPEED (100 RPM) AT CERTAIN LOAD (50 NM) USING DIFFERENT NUMBER OF THE POSITION MEASUREMENT PER 360 ELECTRICAL DEGREES

n_{IMP}	Work current [A_{RMS}]	Current growth [%]
0	24.78	0.0
4	26.42	6.6
8	28.79	16.2
12	31.60	27.5
20	40.12	61.9
30	73.06	194.8

Presented results show that keeping reference speed with higher values of n_{IMP} requires some current growth. Figure 13 presents stator current I_S in function of n_{IMP}. Demonstrated waveforms are output value of the speed regulator at steady state for given reference speed. It should be noticed that obtaining higher speeds using above 20 number of tests per 360 electrical degrees can be problematic. Because the current can overstep its nominal value. Therefore next research has been undertaken to minimize losses which are caused by position measurements. The goal of the researches was evaluation of a minimal n_{IMP} allowing the sensorless control works fine.

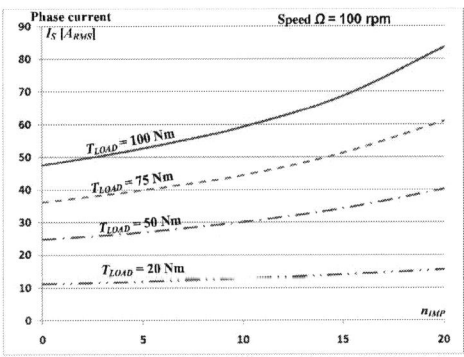

Fig. 13.   Stator current I_S which is needed to keep certain speed ($\Omega = 100$ rpm) in function of number of position tests per 360 electrical degrees n_{IMP}.

IV. LABORATORY TESTS OF THE PIPCRM METHOD

In order to confirm the presented sensorless method and sensorless control, the laboratory prototype with Axial Flux Permanent Magnet Drive System has been constructed. Laboratory inverter is built using IGBT transistors, controlled by the Analog Devices DSP Sharc 21061 processor combined with Altera Flex FPGA processor (Fig. 13). The sampling frequency is set to 16 kHz. Laboratory experiments of the PIPCRM method have been provided with Axial Flux Permanent Magnet Motor (Fig. 14). Such motor parameters are collected in Table III. The motor simulation model parameters match with the real one. The mechanical load is delivered to the drive system by a DC motor.

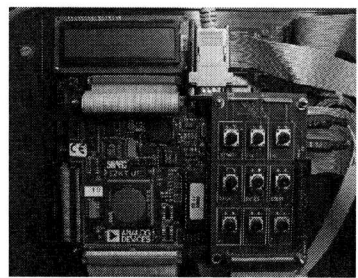

Fig. 13. Laboratory drive system: DSP/FPGA processor

TABLE III.
PARAMETERS OF THE AXIAL FLUX PERMANENT MAGNET MOTOR

rated power	P	40	kW
rated speed	Ω	3000	rpm
rated torque	T_e	127	Nm
stator resistance	R_S	0.247	Ω
stator inductance	L_S	134	μH
number of poles	p	16	

Fig. 14. Laboratory drive system: Axial Flux Permanent Magnet Motor

Current responses (Fig. 15) are result of the PIPCRM test voltage step, measured in stator coils connected parallel. Sudden current rises are observed and provoked by decreasing phase inductances. Waveforms of current responses confirm that stator coils magnetic circuits are saturated. Differences between peak values of current responses mean that phase inductances are not equal.

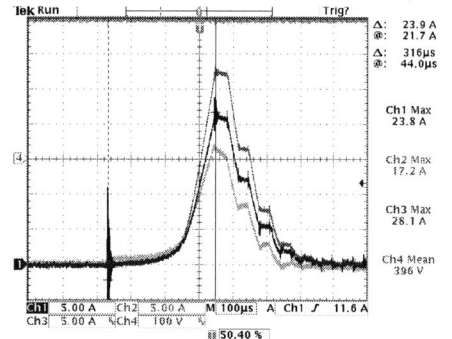

Fig. 15. Laboratory result: Measured current responses induced by the PIPCRM test voltage step

Moreover different *di/dt* is observed in each phase. That feature is used to calculate electrical position of the rotor φ.

The calculated rotor position is obtained by the special current response analysis [9][10]. In dependence of used way of rotor position calculation accuracy can be higher. Moreover using some position prediction, continuous position information is achieved. It allows to use PMSM control strategy with positive results.

Recent research permit to improve way of current response analysis. Therefore accuracy is about 1-2 electrical degrees. In case of the 16 poles motor it corresponds up to 0.25 mechanical degrees. Error of the calculated rotor position is presented in fig. 16.

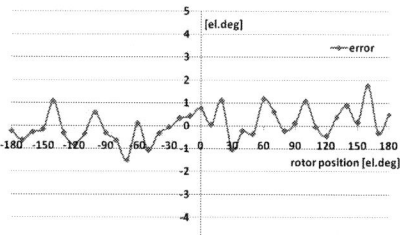

Fig. 16. Accuracy of the PIPCRM method: error of rotor position identification.

Achieved accuracy of rotor position information fulfill expectations. Nevertheless all errors of position detection influence on speed estimation. Hence some filtering of estimated speed value is required to assure stability of controlled drive system.

Few set frequencies of position information updating has been tested. As mentioned rotor position is discrete signal. So it is updated with some delay to the real position. But consensus between frequency of position measurement and current control stage is needed to achieve expected dynamic of the drive system. However delay of the sensorless rotor position information preclude achieving very high dynamic of the drive system.

The simulation results has been confirmed. A torque drop has been observed with increasing number of position tests per 360 electrical degrees. For that reason the D&D frequency should be as low as possible. Figures 17 and 18 shows different D&D frequency at same speed.

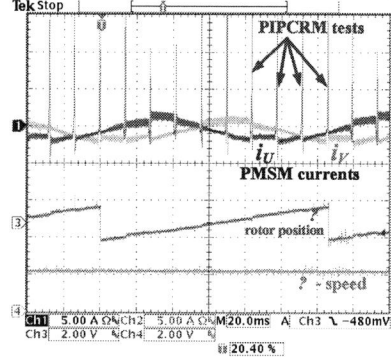

Fig. 17. Low speed (50 RPM) operation of the axial flux permanent magnet motor - 10 PIPCRM measurements per 360 electrical degrees.

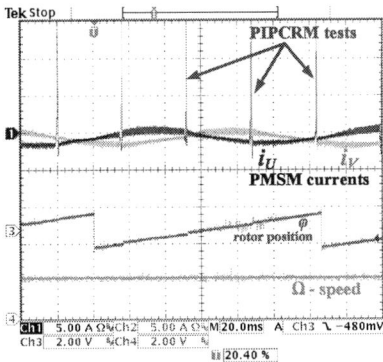

Fig. 18. Low speed (50 RPM) operation of the axial flux permanent magnet motor - 4 PIPCRM measurements per 360 electrical degrees.

Using the Detect and Drive procedure at PIPCRM speed range, startup of axial flux PMM has been provided. Following results show startup from standstill to 150 rpm (5% of nominal speed) in fig. 19 and from standstill to 240 rpm (8% of nominal speed) in fig. 20.

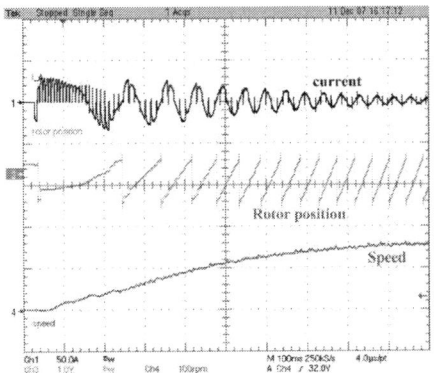

Fig. 19. Oscillogram of the axial flux permanent magnet motor startup from standstill to 150 rpm, top waveform – stator current, middle – rotor position, bottom – speed

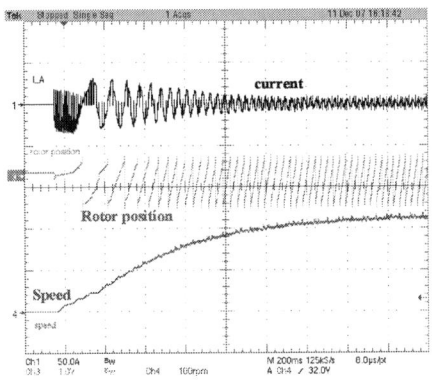

Fig. 20. Oscillogram of the axial flux permanent magnet motor startup from standstill to 240 rpm, top waveform – stator current, middle – rotor position, bottom – speed

Moreover reverse test have been introduced. There is depicted sensorless reverse of Axial Flux PMM from -150 rpm to 150 rpm in fig. 21 and load step from 0 Nm to 30 Nm in fig. 22. However carried out results show that next research should be done to achieve higher dynamic of the sensorless drive system. Presented results prove that the PIPCRM based control can be used to control the axial flux permanent magnet synchronous motor at standstill and low speed range.

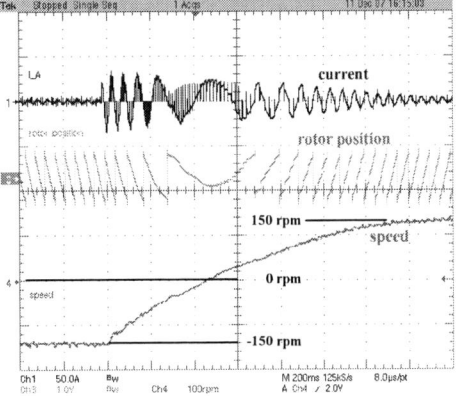

Fig. 21. Oscillogram of the axial flux permanent magnet motor reverse from 150 to -150 rpm, top waveform – stator current, middle – rotor position, bottom – speed

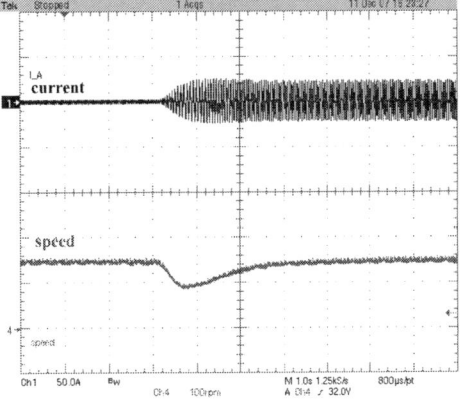

Fig. 22. Oscillogram of the axial flux permanent magnet motor load step from 0 Nm to 30 Nm, top waveform – stator current, bottom – speed

V. CONCLUSION

The paper shows that the PIPCRM method and control strategy based on that method allows to provide sensorless startup of axial flux permanent magnet synchronous motor. Also the reverse of such a motor is permissible. Presented control system is able to operate up to 250 rpm.

All theoretical considerations have been confirmed by laboratory experiments. The accuracy of rotor position is quite high, about 1-2 electrical degrees. Carried out results confirmed that presented control strategy coincides with expectations. All laboratory tests have been realized on

40kW/16 poles/3000 rpm Axial Flux Permanent Magnet Motor.

Minimization of the number of position measurements per 360 electrical allows to decrease torque losses. Moreover using lower frequency of the Detect and Drive procedure, the higher speed can be provide. Due to lower torque pulsation and longer drive stage when torque producing current is applied to the stator windings.

REFERENCES

[1] M. Schroedl, "Statistic properties of the INFORM-method in highly dynamic sensorless PM motor control applications down to standstill", *EPE Journal, Vol 13, No. 3*, August 2003

[2] E. Robeischl, M. Schroedl, K. Salutt, "Improved INFORM-measurement sequence and evaluation for sensorless permanent magnet synchronous motor drive" Vienna University of Technology A-1040 Vienna, Austria

[3] E. Robeischl, M. Schroedl: "Optimized INFORM measurement sequence for sensorless PM synchronous motor drives with respect to minimum current distortion" *IEEE Transactions on Industry Applications*, Vol. 40, No. 2, March/April 2004

[4] M. Schroedl, M. Hofer, W. Staffler: "Combining INFORM method, voltage model and mechanical observer for sensorless control of PM Synchronous Motor in the whole speed range including standstill" *International Exhibition & Conference for Power Electronics Intelligent Motion Power Quality*, PCIM 2006, May/June 2006, Nuremberg, Germany

[5] M. Schroedl, M. Hofer, W. Staffler, "Sensorless control of PM Synchronous Motor in the whole speed range including standstill using a combined INFORM/EMF model" *12th International Power Electronics and Motion Control Conference*, EPE-PEMC 2006, 30 August – 1st September , 2006, Portoroz, Slovenia,

[6] P. Jakubowski, W. Koczara, N. Al-Khayat: „PIPCRM method of sensorless control start of the permanent magnet motor", *International Conference on Power Electronics and Intelligent Control for Energy Conservation*, PELINCEC 2005, 17-19 October 2005, Warsaw, Poland

[7] P. Jakubowski, W. Koczara, N. Al-Khayat, „Method of the poles position identification for brushless axial flux permanent magnet motor drive system", *11th European Conference on Power Electronics and Applications*, EPE 2005, 11-14 September 2005, Dresden, Germany

[8] J. Wisniewski, P. Jakubowski, W. Koczara, N. Al.-Khayat, „Poles position identification of permanent magnet axial flux motor using PIPCRM sensorless method", *12th European Conference on Power Electronics and Applications*, EPE 2007, 2 - 5 September 2007, Aalborg, Denmark

[9] J. Wisniewski, W. Koczara, P. Jakubowski, "Permanent magnet synchronous motor start by the sensorless PIPCRM", *2nd International Conference Automotive Power Electronics*, APE 2007, 26-27 September 2007, Paris, France

[10] J. Wisniewski, W. Koczara, "The Sensorless Rotor Position Identification and Low Speed Operation of the Axial Flux Permanent Magnet Motor Controlled by the Novel PIPCRM Method", *IEEE 39th Power Electronics Specialists Conference*, PESC 2008, 15-19 June 2008, Rhodes, Greece

Zero Speed Position Estimation of a Matrix Converter Fed AC PM Machine using PWM Excitation

Q. Gao, G. M. Asher and M. Sumner

School of Electrical and Electronic Engineering, University of Nottingham, Nottingham, UK

Abstract— This paper describes a sensorless position estimation method of a permanent magnet motor fed from a three-phase matrix converter using only its space vector PWM (SVPWM) sequence. The scheme employs measurements of the di/dt of the motor line currents to construct orthogonal "resolver like" position signals. In principle, the SVPWM waveforms provide sufficient excitation to extract the position signal from the motor current. However, in order to improve the signal-to-noise ratio of the di/dt signals, a small modification to the SVPWM is required. Sensorless control of the permanent magnet synchronous motor at low and zero speed is demonstrated experimentally. The small input voltage to the matrix converter is also investigated to reduce the current ripple.

Keywords—Sensorless control, permanent magnet motor, matrix converter, pulse width modulation (PWM0.

I. INTRODUCTION

The saliency tracking based techniques [1,2,3,4] to estimate the rotor position of a permanent magnet synchronous motor (PMSM) have received more research efforts for their advantageous capabilities over the model-based counterparts of speed/position estimation at very low and zero speed. Among these methods, a particular one that uses the natural Space Vector PWM (SVPWM) as the excitation signals to the motor for the position estimation has been reported [5]. As the geometrical or magnetic anisotropy of an AC motor modulates the stator inductances, by observing the current change (di/dt) against the applied stator voltage it is possible to detect the position of the anisotropy. This method has been implemented on an induction machine with open slots on the rotor [5] and on a PMSM [6], both of which are driven by a standard two-level inverter. The di/dt measurement often requires a minimum pulse width (t_{min}) of the stator voltage vectors due to the high frequency oscillation in the stator current whenever a new voltage vector appears on the motor, which prohibits the prompt measurement of the di/dt, and hence the voltage vector time is extended to t_{min} whenever it is too short. This is the case when the motor operates at low and zero frequencies. The extended pulse width results in the increased current ripple.

In this paper, the feasibility of employing the Space Vector PWM waveform of a three-phase to three-phase matrix converter (MC) to the rotor position estimation of a PMSM is investigated. As a direct AC-AC converter, a MC doesn't have the bulky and expensive DC link capacitors as used in a two-level inverter [7]. It also

provides more choice of voltage vectors than a two-level, three-phase inverter, so that smaller input voltages may be employed when the position estimated is needed. In this way, the motor line ripple can be reduced.

Experimental results of position estimation at low and zero speed are given in this paper to validate the proposed principle. The effect of applying smaller input voltages on the motor line currents is also reported.

II. POSITION ESTIMATION ALGORITHM FOR A THREE-PHASE MATRIX CONVERTER

A. Introduction to the SVPWM of the Matrix Converter

The schematic of the matrix converter fed PMSM drive is shown in Fig. 1. Many modulation schemes have been developed with an aim to find the appropriate switching pattern for each of the nine power devices [7]. As one of the most general modulation schemes used in direct matrix converters, the direct SVPWM is chosen in this work. The brief introduction to its principle is given here.

While in the indirect SVPWM of the matrix converter, the modulation is split into two successive stages, i.e. the rectification stage and the inversion stage, to simulate the operation of an inverter, the direct SVPWM considers these two stages simultaneously. When maximum input

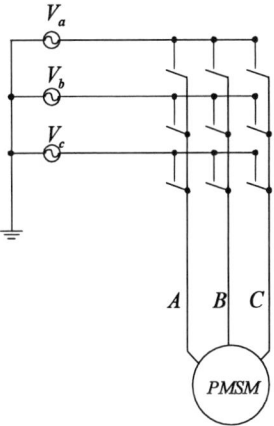

Fig. 1 Simplified PMSM drive fed by a matrix converter

978-1-4244-1741-4/08/$25.00 ©2008 IEEE

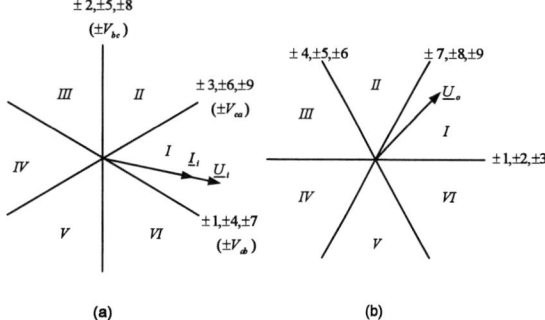

(a)

(b)

Fig. 2 Definition of the sectors for the input current vector (a) and output voltage vector (b)

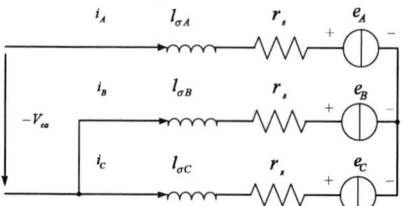

Fig. 3 The equivalent circuit of the drive applied with -3

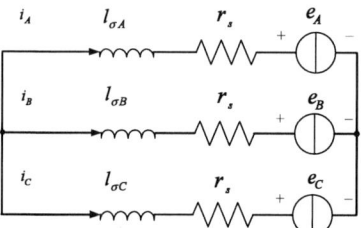

Fig. 4 The equivalent circuit of the drive applied with zero vector

line to line voltages are utilized, the plane of the input phase voltage vector $\underline{U_j}$ can be divided into six sectors as shown in Fig. 2 (a), assuming $\underline{U_j} = u_{im}e^{j\alpha}$ and the unity input power factor. The definition of the 18 stationary active vectors ($\pm 1 \sim \pm 9$) as well as three zero vectors ($O_1 \sim O_3$) can be found in [8]. The analysis is made by considering at the same time the modulation of the input current by the output currents and that of the output voltage by the input voltages, which leads to the selection of 4 active vectors corresponding to each sector combination of the input voltage vector and the output voltage vector. The duty cycles of each vector can also be determined uniquely [8] and won't be repeated here. For example, if both the input/output voltage vectors are located in sector I as in Fig. 2, the applied active vectors are +1, -3, +7 and -9. There are a variety of ways of arranging the sequence of these vectors within a PWM cycle. In order to have better harmonic performance in both the input and the output currents, usually the double-sided sequence is employed. The vector sequence in the first half PWM period for the above example is then given as O_3 (ccc) $\rightarrow$-3 (acc) $\rightarrow$ +9 (aac) $\rightarrow O_1$ (aaa) $\rightarrow$-7 (aab) $\rightarrow$+1 (abb) $\rightarrow O_2$ (bbb) , and the second half cycle is the same but the sequence is symmetrical to the first half cycle.

B. Methodology of the Position Estimation

Consider a three-phase, star-connected SMPMSM, whose stator leakage inductances are modulated by the main flux saturation and therefore can be modeled as follows.

$$l_{\sigma A} = l_0 + \Delta l \cos(2\theta_e) \tag{1}$$

$$l_{\sigma B} = l_0 + \Delta l \cos(2*(\theta_e - 2\pi/3)) \tag{2}$$

$$l_{\sigma C} = l_0 + \Delta l \cos(2*(\theta_e - 4\pi/3)) \tag{3}$$

where l_0 is the average leakage inductance, and Δl is the amplitude variation due to the modulation by the main flux with an angle of θ_e.

Again take the case of $K_v = K_i = 1$ as an example. The three-phase equivalent circuit of the drive, when vector -3

(acc) is applied, is given in Fig. 3, assuming small leakage inductances. Fig. 4 shows the equivalent circuit for $\underline{U_o} = 0$. From Fig.3 and Fig. 4, the following equations can be derived by referring to [5, 9].

$$\frac{di_A^{(-3)}}{dt} - \frac{di_A^{(0)}}{dt} = -\frac{V_{ca}}{c_1}(2 - \frac{\Delta l}{l_0}\cos 2\theta_e) \tag{4}$$

$$\frac{di_B^{(-3)}}{dt} - \frac{di_B^{(0)}}{dt} = \frac{V_{ca}}{c_1}(1 + \frac{\Delta l}{l_0}\cos(2(\theta_e - \frac{4\pi}{3}))) \tag{5}$$

$$\frac{di_c^{(-3)}}{dt} - \frac{di_c^{(0)}}{dt} = \frac{V_{ca}}{c_1}(1 + \frac{\Delta l}{l_0}\cos(2(\theta_e - \frac{2\pi}{3}))) \tag{6}$$

where

$$c_1 = 3l_0(1 - \left(\frac{\Delta l}{2l_0}\right)^2) \tag{7}$$

Similar analysis to the case of $\underline{U_o}$ = +9 leads to (8) ~ (10).

$$\frac{di_A^{(+9)}}{dt} - \frac{di_A^{(0)}}{dt} = -\frac{V_{ca}}{c_1}(1 + \frac{\Delta l}{l_0}\cos(2(\theta_e - \frac{2\pi}{3}))) \tag{8}$$

$$\frac{di_B^{(+9)}}{dt} - \frac{di_B^{(0)}}{dt} = -\frac{V_{ca}}{c_1}(1 + \frac{\Delta l}{l_0}\cos 2\theta_e) \tag{9}$$

2262

$$\frac{di_c^{(+9)}}{dt} - \frac{di_c^{(0)}}{dt} = \frac{V_{ca}}{c_1}(2 - \frac{\Delta l}{l_0}\cos(2(\theta_e - \frac{4\pi}{3}))) \quad (10)$$

It is therefore valid to define the following position scalars p_{aoff}, p_{boff} and p_{coff}, from (5), (6), (8) and (9).

$$p_{aoff} = \frac{1}{c_1}(1 + \frac{\Delta l}{l_0}\cos 2\theta_e) = -\frac{1}{V_{ca}}(\frac{di_B^{(+9)}}{dt} - \frac{di_B^{(0)}}{dt}) \quad (11)$$

$$p_{boff} = \frac{1}{c_1}(1 + \frac{\Delta l}{l_0}\cos(2(\theta_e - \frac{2\pi}{3})))$$
$$= -\frac{1}{V_{ca}}(\frac{di_a^{(+9)}}{dt} - \frac{di_a^{(0)}}{dt}) \quad (12)$$

or

$$p_{boff} = \frac{1}{c_1}(1 + \frac{\Delta l}{l_0}\cos(2(\theta_e - \frac{2\pi}{3})))$$
$$= \frac{1}{V_{ca}}(\frac{di_c^{(-3)}}{dt} - \frac{di_c^{(0)}}{dt}) \quad (13)$$

and

$$p_{coff} = \frac{1}{c_1}(1 + \frac{\Delta l}{l_0}\cos(2(\theta_e - \frac{4\pi}{3})))$$
$$= \frac{1}{V_{ca}}(\frac{di_B^{(-3)}}{dt} - \frac{di_B^{(0)}}{dt}) \quad (14)$$

Thus the estimated position vector $\underline{p}_{\alpha\beta}$ can be constructed.

$$\underline{p}_{\alpha\beta} = p_\alpha + jp_\beta$$
$$= p_{aoff} + a \cdot p_{baoff} + a^2 \cdot p_{coff} \quad (15)$$
$$= \frac{3\Delta l}{2c_1 l_0}e^{-j2\theta_e}$$

where $a = e^{j2\pi/3}$ and

$$p_\alpha = p_{aoff} - (p_{boff} + p_{coff})/2 \quad (16)$$

$$p_\beta = \sqrt{3}(p_{boff} - p_{coff})/2 \quad (17)$$

From (16) and (17) the estimated position angle can be calculated via the arctan operation. Meanwhile the input line-line voltage V_{ca} is part of the definition of p_{aoff}, p_{boff} and p_{cof}, it is viable without the measurement of V_{ca} if p_{aoff}, p_{boff} and p_{cof} are scaled by V_{ca}, and it will be cancelled out during the arctan operation finally. The inclusion of input line-line measurement, however, facilitates the usage of

di/dt measurement of other voltage vectors, as will be seen later.

The position estimation can also be accomplished if one looks at the other two active voltage vectors -7 and +1. In this case, p_{aoff}, p_{boff} and p_{cof} take the form as

$$p_{aoff} = \frac{1}{V_{ab}}(\frac{di_B^{(-7)}}{dt} - \frac{di_B^{(0)}}{dt}) \quad (18)$$

$$p_{boff} = \frac{1}{V_{ab}}(\frac{di_A^{(-7)}}{dt} - \frac{di_A^{(0)}}{dt}) \quad (19)$$

or

$$p_{boff} = -\frac{1}{V_{ab}}(\frac{di_A^{(+1)}}{dt} - \frac{di_A^{(0)}}{dt}) \quad (20)$$

and

$$p_{coff} = -\frac{1}{V_{ab}}(\frac{di_B^{(+1)}}{dt} - \frac{di_B^{(0)}}{dt}) \quad (21)$$

Therefore, (18) ~ (21) can also be used to construct $\underline{p}_{\alpha\beta}$ via (16), (17). The averaging of p_{aoff}, p_{boff} and p_{cof} from the di/dt measurement of these four vectors can be conducted. It may also be observed that the back-EMF terms do not appear in the constructed position signal, as they have been compensated by the di/dt of zero vectors that lie nearest to the active vectors. Thus the position estimation at higher speed can be achieved.

In appendix, Table 1 and Table 2 give the definition of the position scalars for each vector applied in the case of a star-connected and a delta-connected machine.

It can be found that when K_v or K_i falls into other sectors, the position vector $\underline{p}_{\alpha\beta}$ can be formed by measuring the di/dt of the available vectors selected by the SVPWM. The procedure of analysis is not repeated here. Therefore, in principle the standard SVPWM waveform provides sufficient excitation to the motor that position estimation from zero to higher speed is possible, as does the two-level inverter [5].

C. Pulse Width Extension and Compensation

In practice, the motor line currents are subject to the high frequency oscillations whenever a new voltage vector is applied due to the drive's parasitic parameters. The di/dt measurement thus can only be made after this high frequency oscillation has decayed. The time delay determines a minimum pulse width t_{min}. Under zero and low frequency operation, the durations of active voltage vectors are normally shorter than t_{min}, and thus extension of the pulse widths is necessary. In order to maintain the same output voltage, the compensation is also needed afterwards. Because of the complexity of the SVPWM waveform of the MC, compared with the two-level inverter, the scheme in [5], i.e., the edge shifting method is not directly usable here. A different approach is utilized instead to accommodate the MC's SVPWM sequence.

Still take the example of $K_v = K_i = 1$, where first half of the PWM sequence is

$O_3(ccc,t_0) \rightarrow -3(acc,t_{-3}) \rightarrow +9(aac,t_{+9}) \rightarrow O_1(aaa, t_0), \rightarrow -7(aab, t_{-7}) \rightarrow +1(abb, t_{+1}) \rightarrow O_2(bbb,t_0)$

and the di/dt measurement takes place on -3 and +9 as well as on zero vectors. If $t_{-3} < t_{min}$, it is extended to t_{min}, and the opposite vector +3 (caa) is applied to compensate for this modification. The placement of +3 (caa, t_{min}-t_{-3}) should be such that it won't introduce two commutations at the same time. Out of this consideration, the new pulse sequence in the first half period can be arranged as follows,

$O_3(ccc,t_0) \rightarrow -3(acc,t_{min}) \rightarrow +9(aac,t_{+9}) \rightarrow O_1(aaa, t_0/2) \rightarrow +3$ (caa, t_{min}-t_{-3}) $\rightarrow O_1(aaa, t_0/2) \rightarrow -7(aab, t_{-7}) \rightarrow +1(abb, t_{+1}) \rightarrow O_2(bbb,t_0)$.

If on the other hand $t_{+9} < t_{min}$, the similar extension and compensation scheme can be used and the resulting pulse sequence is given as

$O_3(ccc,t_0/2) \rightarrow -9(cca, t_{min}-t_{+9}) \rightarrow O_3(ccc,t_0/2) \rightarrow -3(acc,t_{-3}) \rightarrow +9(aac,t_{min}) \rightarrow O_1(aaa, t_0) \rightarrow O_1(aaa, t_0/2) \rightarrow -7(aab, t_{-7}) \rightarrow +1(abb, t_{+1}) \rightarrow O_2(bbb,t_0)$.

It can be seen that the compensating vectors are inserted between one of its closest zero vectors, and this way no double-switching is produced. The second half cycle of the PWM sequence remains the same. This extension and compensation scheme is also applicable for other sectors of K_v and K_i.

If both $t_{-3} < t_{min}$ and $t_{+9} < t_{min}$, the extension and compensation can be done for these two vectors at the same time. But if only low and zero speed are concerned, one can perform this scheme on one vector in one PWM period and on the other vector in another period. The same approach can be used if one measures the di/dt on -7 and/or +1.

Fig. 5 shows the output line-line voltages and line current of the matrix converter after the extension of the second active vector, where the compensation vector was applied between 7 us ~ 16 us. As can be seen, the compensation has managed to bring the current back to its target value, and therefore the measurement does not affect the current control.

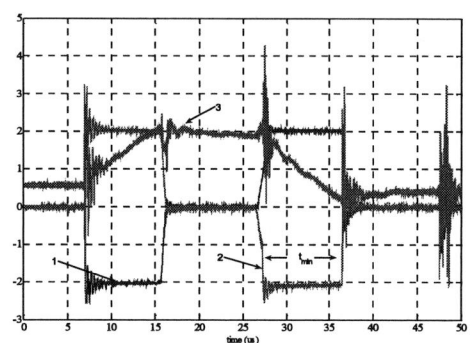

Fig. 5 Line voltages (1- V_{AB}, ×300 V, 2 – V_{BC}, ×300 V) and line current (3 – i_A, A) after the extension of the second active vector

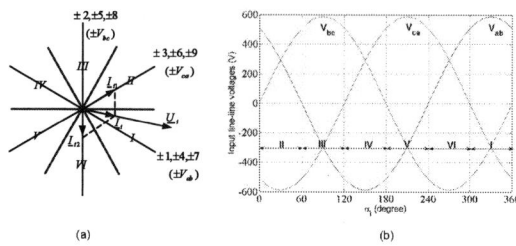

Fig. 6 New division of the input current/voltage vector plane

D. Position Estimation using Smaller Input Voltages

From (4) ~ (6), one can see that given the same t_{min}, at low and zero speeds (di(0)/dt≈0), applying reduced input voltages may contribute to smaller current ripple. The aforementioned SVPWM scheme, however, always chooses the maximum input line-line voltage, thus a new SVPWM scheme needs to be used to allow for the selection of smaller values of the input voltage. In this paper, the modulation scheme presented in [10] is utilized. This scheme relies on the new division of the input current/voltage vector plane as shown in Fig. 6, where the solid blue lines separate every two adjacent sectors. A whole PWM period of the new scheme consists of four different active vectors and three zero vectors, whose sequence is arranged in a similar way to that of the normal SVPWM in this work. For instance, when the input voltage vector falls into Sector I, the input line-line voltages V_{bc} and V_{ca} are used. If the output voltage vector also lies in Sector I, the vector sequence in the first half of a PWM cycle is set to

$O_1(aaa,t_0) \rightarrow +9(aac,t_{+9}) \rightarrow -3(acc,t_{-3}) \rightarrow O_3(ccc,t_0), \rightarrow +8(ccb, t_{+8}) \rightarrow -2(cbb, t_{-2}) \rightarrow O_2(bbb,t_0)$.

By referring to Table 1, it may be found out that if the di/dts of the vectors +9 and -3 are measured, and the position scalars formed as (22) ~ (24), the angle of the position vector $p_{\alpha\beta}$ can be calculated as in (16) ~ (17).

$$p_{aoff} = -\frac{1}{V_{ca}} \cdot \left(\frac{di_B^{(+9)}}{dt} - \frac{di_B^{(0)}}{dt}\right) \qquad (22)$$

$$p_{boff} = \frac{1}{V_{ca}} \cdot \left(\frac{di_C^{(-3)}}{dt} - \frac{di_C^{(0)}}{dt}\right) \qquad (23)$$

$$p_{coff} = \frac{1}{V_{ca}} \cdot \left(\frac{di_B^{(-3)}}{dt} - \frac{di_B^{(0)}}{dt}\right) \qquad (24)$$

One can also choose to sample the di/dts of the vector +8 and/or -2 to form the position vector. The examination of all other input and output sector combination leads to the similar selection of active vectors. Hence, no modification to the modulation sequence is needed when opting for smaller input voltages in this modulation scheme.

In the above case, it should be noted that if always choosing either +9 and -3, or +8 and -2, both the medium

2264

and smallest input voltages will be applied in the input Sector I. Therefore, in order to excite the motor with the smallest input voltages, the sampled vectors need to change from +9 and -3 if $300 < \alpha_i \leq 330$, to +8 and -2 if $300 < \alpha_i < 330$, where α_i is the position of the input voltage vector, as shown in Fig. 6 (b). On the contrary, sampling +9 and -3 if $330 < \alpha_i \leq 330$, and +8 and -2 if $330 < \alpha_i < 360$ will choose the medium input line-line voltages as the excitation to the motor. The conclusion holds true for other sectors as well. In this paper, the position estimation under minimum excitation voltage is carried out. One problem, however, may arise when always applying the smallest voltages as they may approach zero, meaning no external excitation to the motor. Also since bi-directional power switches are usually used in a matrix converter, the device drop is more influential than that of a two-level converter. To overcome these problems, a minimum excitation voltage U_{min} to the motor for the position estimation is maintained so that whenever the predicted excitation voltage drops below U_{min}, the medium voltage is applied instead. This can be realized simply by changing the sampled vectors, as mentioned above. It resembles the boost in the V/f control, so it is named the boosted excitation in this paper.

The extension and compensation scheme described in *Section C* is still applicable for this new SVPWM scheme, and will not be repeated.

III. THE EXPERIMENT SETUP AND RESULTS

A. The Experiment Setup

The schematic of the test rig is shown in Fig. 7. Both schemes using the maximum and the boosted excitation as well as their vector extension and compensation schemes have been implemented on this rig.

A 3-phase, 1.9 kW, 6-pole PMSM is used for testing under field oriented control, which is driven by a 3-phase to 3-phase MC. The DC load motor is coupled with the PMSM. A control board based on an Actel FPGA converts the measurements of the motor currents, the input voltage of the MC and the di/dt into digital signals to be used by the DSP. The timing of the di/dt measurement is realized within the FPGA. It also controls the logic of the four-step current based commutation of the MC. A TI DSP is used for the vector control for the PMSM. An improved air-cored di/dt sensor is used to measure the di/dt of the motor line currents. The PWM frequency is set to 8 kHz. The vector extension and compensation was

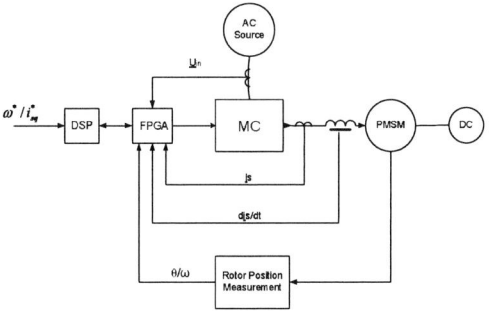

Fig. 7 The schematic of the experimental rig

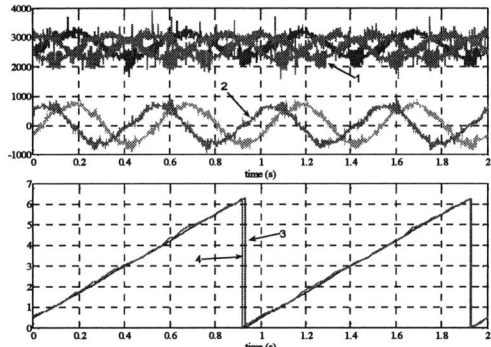

Fig. 8 Rotor position estimation at 20 rpm, 50% rated load
Top plot: 1- p_{aoff}, p_{boff} and p_{coff}, 2- filtered $p_{\alpha\beta}$
Bottom plot: 3 – measured rotor position, 4 – estimated rotor position
(rad)

performed in two consecutive PWM cycles with only one vector dealt with in each cycle.

A mechanical observer based on [11] is used to filter the high frequency noise in the position signals. Also, the angle shift between the estimated angle and the real flux angle is compensated by a lookup table [4].

B. The Experimental Results

1) Maximum Voltage Excitation

Fig. 8 shows the position estimation at 20 rpm (1.0 Hz) with 50% rated load when the motor was under sensored speed control. Three phase position signals p_{aoff}, p_{boff} and p_{coff} directly derived from the di/dt measurements along with the constructed orthogonal position vector $p_{\alpha\beta}$ after low pass filtering are given. The estimation error varies around 4 electrical degrees.

In Fig. 9, the estimated rotor position was used for the field orientation of the PMSM under torque-control. The speed of the DC drive was set to 0 rpm. Heavy load steps from rated -20% → -50% → -100% were applied consecutively. This test validates the capability of the accurate position estimation of the proposed scheme even under heavy load at standstill.

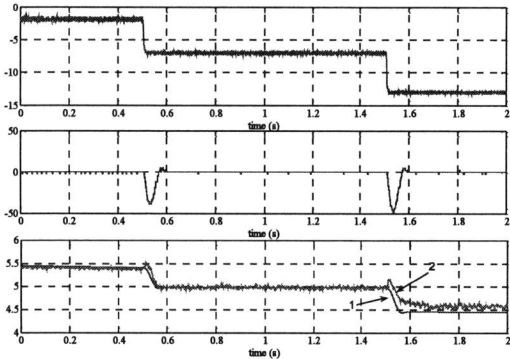

Fig. 9 Sensorless torque control with load torque changes from -20% to -50% to -100% load at zero speed
Top plot: isq (A), middle plot: measured speed (rpm),
Bottom plot: 1- measured rotor position, 2 – estimated rotor position (rad)

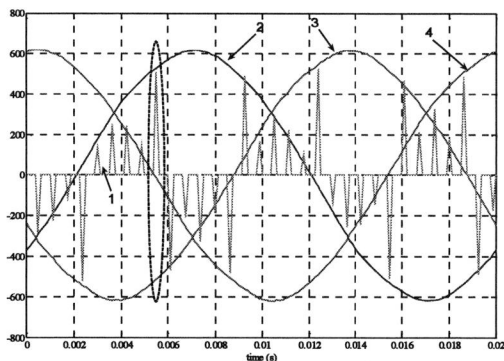

Fig. 10 The instantaneous excitation voltage to the motor and input line-line voltages (V) to the MC ~ time (s), 1 – excitation voltage, 2,3,4 – input line-line voltages

2) Boosted Excitation

In this paper the results for the boosted excitation were obtained with $U_{min} = 100$ V. Lower voltage should be allowed, but it comes with the expense of lower signal to noise ratio. The research on smaller U_{min} is still under way. Fig. 10 shows the measured excitation line voltage to the motor along with the input line voltages to the MC. It can be seen that if the minimum input line voltages are larger than U_{min}, they are enabled to be the excitation voltage. Otherwise, the medium voltages are selected. The ellipse in Fig. 7 highlights such a case. It may also be noticed that there is a delay between the instantaneous excitation voltage and the input line voltages. This is caused by the sampling delay of the excitation voltage with respect to the measurement of the input voltages in the software.

Fig. 11 shows the sensorless speed reversal at ± 5 rpm under 57% rate load. The estimation error of the position remains in the range of ± 10 electrical degrees. This is due to uncompensated position signals directly used for position/speed calculation.

3) Comparison of Current Waveforms

The current waveforms under three conditions are evaluated next, namely, without extension/compensation, with maximum excitation voltage and with the boosted

excitation voltage. The PMSM was under sensored torque control to minimize the cogging torque effect to the

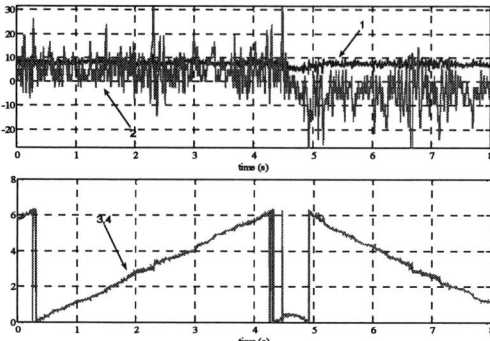

Fig. 11 Speed reversal at ± 5 rpm (0.25 Hz) with 57% load, 1 – isq (A), 2 – measured speed (rpm), 3 - measured rotor position (blue, rad), 4 – estimated rotor position (green,rad)

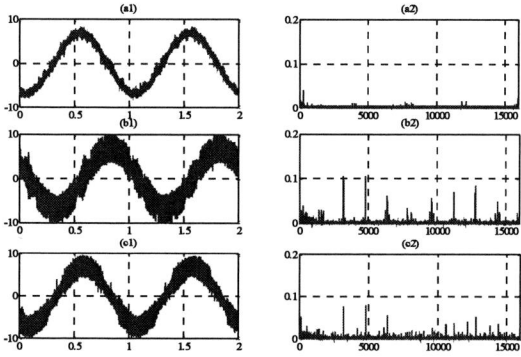

Fig. 12 Current waveforms (a1, b1, c1, A ~ s) and their zoomed FFT (a2, b2, c2, A ~ Hz) under different conditions: a1, a2 - without extension/compensation, b1, b2 – with maximum excitation, c1, c2 – with boosted excitation

current waveforms and half nominal load was applied.The DC motor was speed controlled to 20 rpm. Fig. 12 shows the corresponding motor current waveforms and their FFT. It can clearly be seen that the high frequency current ripples due to the vector extension decrease with the reduced excitation voltages. The total harmonic distortions (THD) of the three currents are 3.2%, 7.3% and 5.6% respectively.

IV. CONCLUSIONS

This paper presents a principle of the position estimation of a matrix converter fed PMSM. It has been shown that, like a two-level inverter, the SVPWM waveform of the matrix converter can also be used for the excitation source for the transient excitation approach of position estimation. To overcome the high frequency noise in the di/dt signals, a different vector extension/compensation scheme from the edge shifting technique used in a two-level inverter is also proposed. The unique topology of the matrix converter provides additional advantage over a two-level inverter in that it is capable of applying smaller input voltages to the motor when the position measurement is required. This feature may help to reduce the current ripple introduced by the pulse extension. A new SVPWM scheme is used in the work to achieve this benefit. Finally, preliminary experimental results are also given to validate the position estimation principles. Better current waveforms in terms of lower THD when using smaller input voltages are also confirmed in the experiment. Lower input voltages than 100V currently set in this paper should be possible, although its side effects, such as lower signal to noise ratio and device drop, should be addressed. This remains the further research.

ACKNOWLEDGMENT

The authors would like to thank the EPSRC for their support of the work. They would also like to thank Dr. Lee Empringham for his kind support to the FPGA development, and Prof. Jon Clare and Prof. Pat. Wheeler for their support to the matrix converter building.

REFERENCES

[1] M. Corley and R. D. Lorenz, "Rotor Position and Velocity Estimation for a Salient-Pole Permanent Magnet Synchronous Machine at Standstill and High Speeds", *IEEE Transactions on Industry Applications*, vol. 34, pp. 784-789,1998

[2] M. Schroedl, "Sensorless control of AC machines at low speed and standstill based on the INFORM method", *Proceedings of IEEE IndustrialApplication Society Annual Meeting (IAS)*, 1996, pp. 270–277.

[3] M. Linke, R. Kennel and J. Holtz, "Sensorless position control of Permanent Magnet Synchronous Machines without Limitation at

Zero Speed", *Industrial Electronics Society Annual Conference (IECON)*, 2002. on CD-ROM.

[4] C. A. Silva, G. M. Asher, M. Sumner and K. J. Bradley, "Sensorless Rotor Position Control in a Surface Mounted PM Machine Using *HF* Voltage Injection", *Proceedings of EPE – PEMC*, 2002, on CD-ROM.

[5] Gao, Q, Asher, G.M., Sumner M., and Makyš, P, "Position Estimation of AC Machines at All Frequencies Using Only Space Vector PWM Based Excitation", *Proceedings of IEE 3rd International Conference on Power Electronics, Machines and Drives (PEMD)*, 2006.

[6] Yahan, H, Asher, G.M., Sumner M. and Gao, Q., "Sensorless Control of Surface Mounted Permanent Magnet Machine using the Standard Space Vector PWM", *IAS*, 2007, on CD-ROM.

[7] P.W.Weeler, J. Rodríguez, J.C.Clare, L. Empringham and A. Weinstein, "Matrix Converter: A Technology Review", *IEEE Transactions on Industrial Electronics, Vol. 49. No. 2, April, 2002*

[8] D. Casadei, G. Serra, A. Tani and L. Zarri, "Matrix Converter Modulation Strategies: A New General Approach Based on Space-Vector Representation of the Switch State", *IEEE Transactions on Industrial Electronics, Vol, 49, No. 2, April, 2002*

[9] Q. Gao, G. M. Asher and M. Sumner, " Position Estimation of a Surface Mounted Permanent Magnet Synchronous Motor Controlled by a Three-phase Matrix Converter using the Space Vector PWM Waveform", *Proceedings of The 4th International Conference on Power Electronic , Machines and Drives (PEMD)* ,April, 2008

[10] L. Helle, K. B. Larsen, A. H. Jorgensen, S.M-Nielsen and F. Blaabjerg, "Evaluation of Modulation Schemes for Three-Phase to Three-Phae Matrix Converters", *IEEE Transactions on Industrial Electronics, Vol, 51, No. 1, Feb, 2004*

[11] P.L.Jansen and R.D.Lorenz, "Transducerless Position and Velocity Estimation in Induction and Salient AC Machines", *IEEE Transactions on Industry Applications*, Vol, 31, No.2, March/April 1995

Appendix.

	p_{aoff}	p_{boff}	p_{coff}
$\pm 1+$ O_x	$\dfrac{3}{c_1}\dfrac{\text{sgn}(\pm 1)}{V_{ab}}\cdot\left(\dfrac{di_A^{(\pm1)}}{dt}-\dfrac{di_A^{(0)}}{dt}\right)$	$-\dfrac{\text{sgn}(\pm 1)}{V_{ab}}\cdot\left(\dfrac{di_C^{(\pm1)}}{dt}-\dfrac{di_C^{(0)}}{dt}\right)$	$-\dfrac{\text{sgn}(\pm 1)}{V_{ab}}\cdot\left(\dfrac{di_B^{(\pm1)}}{dt}-\dfrac{di_B^{(0)}}{dt}\right)$
$\pm 2+$ O_x	$\dfrac{3}{c_1}-\dfrac{\text{sgn}(\pm 2)}{V_{bc}}\cdot\left(\dfrac{di_A^{(\pm2)}}{dt}-\dfrac{di_A^{(0)}}{dt}\right)$	$-\dfrac{\text{sgn}(\pm 2)}{V_{bc}}\cdot\left(\dfrac{di_C^{(\pm2)}}{dt}-\dfrac{di_C^{(0)}}{dt}\right)$	$-\dfrac{\text{sgn}(\pm 2)}{V_{bc}}\cdot\left(\dfrac{di_B^{(\pm2)}}{dt}-\dfrac{di_B^{(0)}}{dt}\right)$
$\pm 3+$ O_x	$\dfrac{3}{c_1}\dfrac{\text{sgn}(\pm 3)}{V_{ca}}\cdot\left(\dfrac{di_A^{(\pm3)}}{dt}-\dfrac{di_A^{(0)}}{dt}\right)$	$-\dfrac{\text{sgn}(\pm 3)}{V_{ca}}\cdot\left(\dfrac{di_C^{(\pm3)}}{dt}-\dfrac{di_C^{(0)}}{dt}\right)$	$-\dfrac{\text{sgn}(\pm 3)}{V_{ca}}\cdot\left(\dfrac{di_B^{(\pm3)}}{dt}-\dfrac{di_B^{(0)}}{dt}\right)$
$\pm 4+$ O_x	$-\dfrac{\text{sgn}(\pm 4)}{V_{ab}}\cdot\left(\dfrac{di_C^{(\pm4)}}{dt}-\dfrac{di_C^{(0)}}{dt}\right)$	$\dfrac{3}{c_1}\dfrac{\text{sgn}(\pm 4)}{V_{ab}}\cdot\left(\dfrac{di_B^{(\pm4)}}{dt}-\dfrac{di_B^{(0)}}{dt}\right)$	$-\dfrac{\text{sgn}(\pm 4)}{V_{ab}}\cdot\left(\dfrac{di_A^{(\pm4)}}{dt}-\dfrac{di_A^{(0)}}{dt}\right)$
$\pm 5+$ O_x	$-\dfrac{\text{sgn}(\pm 5)}{V_{bc}}\cdot\left(\dfrac{di_C^{(\pm5)}}{dt}-\dfrac{di_C^{(0)}}{dt}\right)$	$\dfrac{3}{c_1}\dfrac{\text{sgn}(\pm 5)}{V_{bc}}\cdot\left(\dfrac{di_B^{(\pm5)}}{dt}-\dfrac{di_B^{(0)}}{dt}\right)$	$-\dfrac{\text{sgn}(\pm 5)}{V_{bc}}\cdot\left(\dfrac{di_A^{(\pm5)}}{dt}-\dfrac{di_A^{(0)}}{dt}\right)$
$\pm 6+$ O_x	$-\dfrac{\text{sgn}(\pm 6)}{V_{ca}}\cdot\left(\dfrac{di_C^{\pm(6)}}{dt}-\dfrac{di_C^{(0)}}{dt}\right)$	$\dfrac{3}{c_1}\dfrac{\text{sgn}(\pm 6)}{V_{ca}}\cdot\left(\dfrac{di_B^{(\pm6)}}{dt}-\dfrac{di_B^{(0)}}{dt}\right)$	$-\dfrac{\text{sgn}(\pm 6)}{V_{ca}}\cdot\left(\dfrac{di_A^{(\pm6)}}{dt}-\dfrac{di_A^{(0)}}{dt}\right)$
$\pm 7+$ O_x	$-\dfrac{\text{sgn}(\pm 7)}{V_{ab}}\cdot\left(\dfrac{di_B^{(\pm7)}}{dt}-\dfrac{di_B^{(0)}}{dt}\right)$	$-\dfrac{\text{sgn}(\pm 7)}{V_{ab}}\cdot\left(\dfrac{di_A^{(\pm7)}}{dt}-\dfrac{di_A^{(0)}}{dt}\right)$	$\dfrac{3}{c_1}\dfrac{\text{sgn}(\pm 7)}{V_{ab}}\cdot\left(\dfrac{di_c^{(\pm7)}}{dt}-\dfrac{di_c^{(0)}}{dt}\right)$
$\pm 8+$ O_x	$-\dfrac{\text{sgn}(\pm 8)}{V_{bc}}\cdot\left(\dfrac{di_B^{(\pm8)}}{dt}-\dfrac{di_B^{(0)}}{dt}\right)$	$-\dfrac{\text{sgn}(\pm 8)}{V_{bc}}\cdot\left(\dfrac{di_A^{(\pm8)}}{dt}-\dfrac{di_A^{(0)}}{dt}\right)$	$\dfrac{3}{c_1}\dfrac{\text{sgn}(\pm 8)}{V_{bc}}\cdot\left(\dfrac{di_c^{(\pm8)}}{dt}-\dfrac{di_c^{(0)}}{dt}\right)$
$\pm 9+$ O_x	$-\dfrac{\text{sgn}(\pm 9)}{V_{ca}}\cdot\left(\dfrac{di_B^{(\pm9)}}{dt}-\dfrac{di_B^{(0)}}{dt}\right)$	$-\dfrac{\text{sgn}(\pm 9)}{V_{ca}}\cdot\left(\dfrac{di_A^{(\pm9)}}{dt}-\dfrac{di_A^{(0)}}{dt}\right)$	$\dfrac{3}{c_1}\dfrac{\text{sgn}(\pm 9)}{V_{ca}}\cdot\left(\dfrac{di_c^{(\pm9)}}{dt}-\dfrac{di_c^{(0)}}{dt}\right)$

Table 1. Position scalars definition for a star-connected motor, x = 1 ~ 3.

	p_{aoff}	p_{boff}	p_{coff}
±1+ O_x	$-\dfrac{\text{sgn}(\pm1)}{V_{ab}}\cdot\left(\dfrac{di_B^{(\pm1)}}{dt}-\dfrac{di_B^{(0)}}{dt}\right)$	$\dfrac{3}{c_2}-\dfrac{\text{sgn}(\pm1)}{V_{ab}}\cdot\left(\dfrac{di_A^{(\pm1)}}{dt}-\dfrac{di_A^{(0)}}{dt}\right)$	$-\dfrac{\text{sgn}(\pm1)}{V_{ab}}\cdot\left(\dfrac{di_C^{(\pm1)}}{dt}-\dfrac{di_C^{(0)}}{dt}\right)$
±2+ O_x	$-\dfrac{\text{sgn}(\pm2)}{V_{bc}}\cdot\left(\dfrac{di_B^{(\pm2)}}{dt}-\dfrac{di_B^{(0)}}{dt}\right)$	$\dfrac{3}{c_2}-\dfrac{\text{sgn}(\pm2)}{V_{bc}}\cdot\left(\dfrac{di_A^{(\pm2)}}{dt}-\dfrac{di_A^{(0)}}{dt}\right)$	$-\dfrac{\text{sgn}(\pm2)}{V_{bc}}\cdot\left(\dfrac{di_C^{(\pm2)}}{dt}-\dfrac{di_C^{(0)}}{dt}\right)$
±3+ O_x	$-\dfrac{\text{sgn}(\pm3)}{V_{ab}}\cdot\left(\dfrac{di_B^{(\pm3)}}{dt}-\dfrac{di_B^{(0)}}{dt}\right)$	$\dfrac{3}{c_2}-\dfrac{\text{sgn}(\pm3)}{V_{ab}}\cdot\left(\dfrac{di_A^{(\pm3)}}{dt}-\dfrac{di_A^{(0)}}{dt}\right)$	$-\dfrac{\text{sgn}(\pm3)}{V_{ab}}\cdot\left(\dfrac{di_C^{(\pm3)}}{dt}-\dfrac{di_C^{(0)}}{dt}\right)$
±4+ O_x	$\dfrac{3}{c_2}-\dfrac{\text{sgn}(\pm4)}{V_{ab}}\cdot\left(\dfrac{di_c^{(\pm4)}}{dt}-\dfrac{di_c^{(0)}}{dt}\right)$	$-\dfrac{\text{sgn}(\pm4)}{V_{ab}}\cdot\left(\dfrac{di_B^{(\pm4)}}{dt}-\dfrac{di_B^{(0)}}{dt}\right)$	$-\dfrac{\text{sgn}(\pm4)}{V_{ab}}\cdot\left(\dfrac{di_A^{(\pm4)}}{dt}-\dfrac{di_A^{(0)}}{dt}\right)$
±5+ O_x	$\dfrac{3}{c_2}-\dfrac{\text{sgn}(\pm5)}{V_{bc}}\cdot\left(\dfrac{di_c^{(\pm5)}}{dt}-\dfrac{di_c^{(0)}}{dt}\right)$	$-\dfrac{\text{sgn}(\pm5)}{V_{bc}}\cdot\left(\dfrac{di_B^{(\pm5)}}{dt}-\dfrac{di_B^{(0)}}{dt}\right)$	$-\dfrac{\text{sgn}(\pm5)}{V_{bc}}\cdot\left(\dfrac{di_A^{(\pm5)}}{dt}-\dfrac{di_A^{(0)}}{dt}\right)$
±6+ O_x	$\dfrac{3}{c_2}-\dfrac{\text{sgn}(\pm6)}{V_{ca}}\cdot\left(\dfrac{di_c^{(\pm6)}}{dt}-\dfrac{di_c^{(0)}}{dt}\right)$	$-\dfrac{\text{sgn}(\pm6)}{V_{ca}}\cdot\left(\dfrac{di_B^{(\pm6)}}{dt}-\dfrac{di_B^{(0)}}{dt}\right)$	$-\dfrac{\text{sgn}(\pm6)}{V_{ca}}\cdot\left(\dfrac{di_A^{(\pm6)}}{dt}-\dfrac{di_A^{(0)}}{dt}\right)$
±7+ O_x	$-\dfrac{\text{sgn}(\pm7)}{V_{ab}}\cdot\left(\dfrac{di_C^{(\pm7)}}{dt}-\dfrac{di_C^{(0)}}{dt}\right)$	$-\dfrac{\text{sgn}(\pm7)}{V_{ab}}\cdot\left(\dfrac{di_B^{(\pm7)}}{dt}-\dfrac{di_B^{(0)}}{dt}\right)$	$\dfrac{3}{c_2}-\dfrac{\text{sgn}(\pm7)}{V_{ab}}\cdot\left(\dfrac{di_A^{(\pm7)}}{dt}-\dfrac{di_A^{(0)}}{dt}\right)$
±8+ O_x	$-\dfrac{\text{sgn}(\pm8)}{V_{bc}}\cdot\left(\dfrac{di_C^{(\pm8)}}{dt}-\dfrac{di_C^{(0)}}{dt}\right)$	$-\dfrac{\text{sgn}(\pm8)}{V_{bc}}\cdot\left(\dfrac{di_B^{(\pm8)}}{dt}-\dfrac{di_B^{(0)}}{dt}\right)$	$\dfrac{3}{c_2}-\dfrac{\text{sgn}(\pm8)}{V_{bc}}\cdot\left(\dfrac{di_A^{(\pm8)}}{dt}-\dfrac{di_A^{(0)}}{dt}\right)$
±9+ O_x	$-\dfrac{\text{sgn}(\pm9)}{V_{ca}}\cdot\left(\dfrac{di_C^{(\pm9)}}{dt}-\dfrac{di_C^{(0)}}{dt}\right)$	$-\dfrac{\text{sgn}(\pm9)}{V_{ca}}\cdot\left(\dfrac{di_B^{(\pm9)}}{dt}-\dfrac{di_B^{(0)}}{dt}\right)$	$\dfrac{3}{c_2}-\dfrac{\text{sgn}(\pm9)}{V_{ca}}\cdot\left(\dfrac{di_A^{(\pm9)}}{dt}-\dfrac{di_A^{(0)}}{dt}\right)$

Table 2. Position scalars definition for a delta-connected motor, x = 1 ~ 3, $c_2 = l_0$.

Sensorless Direct Torque and Flux Control of an IPM Synchronous Motor at Low Speed and Standstill

Gilbert Foo *, S. Sayeef * and M.F. Rahman*
* School of Electrical Engineering & Telecommunications
University of New South Wales, Sydney, Australia
e-mail: *gilbert.foo@student.unsw.edu.au*

Abstract—This paper investigates a sensorless direct torque and flux control (DTFC) scheme for IPM synchronous motor (IPMSM) drives at low speed and standstill. Closed-loop control of both the torque and stator flux linkage are achieved using two proportional-integral (PI) controllers. A high frequency signal injection technique is utilized at low speeds to obtain the rotor position information. The hybrid current-voltage model flux observer with a smooth handover algorithm estimates the stator flux linkage over the whole speed range. The sensorless method is capable of zero-speed full load operation. Experimental results confirm the effectiveness of the proposed sensorless SVM-DTFC system.

Keywords—direct torque and flux control, low speed, sensorless, signal injection.

I. INTRODUCTION

Since the advent of direct torque control (DTC) for induction machines in the 1980's as proposed by M. Depenbrock [1] and Takahashi [2], its research has been becoming ever more prevalent in the society. In the late 1990's, the DTC was successfully implemented on interior permanent magnet synchronous motors (IPMSMs) as reported in [3-6]. The block diagram of the classical DTC is depicted in fig. 1.

When compared to conventional vector controlled drives, DTC possesses several advantages such as elimination of coordinate transformation, lesser parameter dependence and faster dynamic response [7]. As the torque and flux are regulated directly and independently, DTC features fast responses. Nevertheless, the large torque and flux ripples are its main drawbacks. Many researchers have attempted to improve the ripples as

shown in [7-10]. By using multilevel inverters [9], the resolution of the voltage vectors can be improved and hence, more smooth torque and flux responses. Nevertheless, due to the increased number of power switches, the system cost and complexity increase. Switching losses inevitably increase too and the overall system efficiency decreases. A modified DTC scheme that utilizes space vector modulation (SVM) was reported in [7]. In this method, fixed switching frequency and lower torque and flux ripples were achieved with the help of a proportional-integral (PI) controller and the SVM technique. A variable structure DTC controller was introduced in [12]. This scheme too reduces torque and ripples significantly but it is parameter dependant and requires gain scheduling.

This paper employs the direct torque and flux control (DTFC) method based on SVM for IPMSM drives. A similar controller implemented on induction motors was detailed in [10]. The torque and flux linkage are controlled independently via two PI controllers, and the reference voltage vectors are generated by the SVM unit. When compared to the classical DTC scheme, the DTFC suppresses the torque and flux ripples significantly. Furthermore, with the adoption of the SVM technique, a constant switching frequency is obtained. At the same time, all the advantages of the DTC are not jeopardized. However, akin to the conventional DTC, the low speed performance is still an issue to be resolved.

To achieve sensorless operation of an IPMSM drive at extremely low speeds and at standstill, high frequency signal injection offers a reliable solution but as the speed increases, its performance deteriorates drastically and other forms of observers have to be adopted. Numerous high frequency signal injection techniques for vector controlled IPMSM drives have been reported in recent literature [16-18]. But, similar approaches for direct torque controlled IPMSMs are still scarce.

A DTFC IPMSM drive with high frequency signal injection for low speed operation was investigated in [19]. The lowest speed sustaining full-load is 1rpm. A similar control system for PM-assisted reluctance synchronous machine was reported in [20]. It was claimed to be able of zero speed full-load operation but there was no experimental result to substantiate the claim.

This paper presents a DTFC IPMSM drive which is capable of handling full-load at zero speed. The drive uses high frequency signal injection techniques only at low speeds to extract the machine position information. Hence,

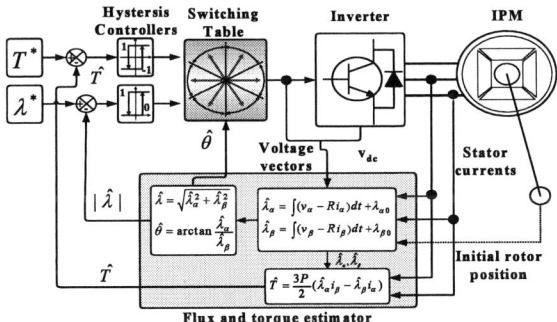

Fig. 1. Block diagram of the classical DTC.

flux and torque estimation is performed with the current model at low speeds and is handed over to the voltage model at medium and high speeds. Dead-time, forward voltage drop and other nonlinearities of the power converter is compensated via a simple look-up table to further refine the performance of the drive at low speeds.

An elaborate presentation of the proposed sensorless DTFC IPMSM drive is given, followed by ample experimental results.

II. IPM MACHINE EQUATIONS

The equations of the IPM synchronous motor on the rotating d-q rotating coordinates are

$$\begin{bmatrix} v_d \\ v_q \end{bmatrix} = \begin{bmatrix} R + pL_d & -w_{re}L_q \\ w_{re}L_d & R + pL_q \end{bmatrix} \begin{bmatrix} i_d \\ i_q \end{bmatrix} + \begin{bmatrix} 0 \\ w\lambda_f \end{bmatrix} \quad (1)$$

$$T = \frac{3P\lambda_s}{4L_dL_q} \left[2\lambda_f L_q \sin\delta - \lambda_s \left(L_q - L_d \right) \sin 2\delta \right] \quad (2)$$

where

R stator armature resistance, Ω:

L_d, L_q direct and quadrature inductance, H;

ω_{re} electrical rotor speed, rad/s;

T electromagnetic torque, Nm;

P number of pole pairs;

λ_s, λ_f stator and rotor flux linkage;

p differential operator;

$v_d,\ v_q,\ i_d$ and i_q are voltages and currents in the rotor reference frame.

III. DIRECT TORQUE AND FLUX CONTROL (DTFC)

The basic principle of DTFC is to directly select stator voltage vectors according to the differences between the reference and actual torque and stator flux linkage. The current controller followed by a pulse width modulation (PWM) comparator used in vector control is not required in DTC systems, and the parameters of the motor are also not used, except the stator resistance. Therefore, DTFC has advantages of lesser parameter dependence and fast torque response when compared with the torque control via PWM current control [7].

Fig. 2 shows the system diagram of a modified DTFC IPM drive integrated with space vector modulation (SVM), which is used in this paper for application of high frequency signal injection for estimation of rotor position and speed. It includes flux and torque estimators, Proportional + Integral (PI) torque and flux controllers and a space vector modulation block. A DC bus voltage sensor and two output current sensors are needed for the flux and torque estimation. Speed sensor is not necessary for the torque and flux control.

In order to estimate the stator flux and torque using the voltage model, an integrator is used in (3), of which the scalar form is (4). The voltage vector can be obtained by multiplying the DC bus voltage with the selected voltage space vector and the current vector can be obtained by measuring any 2 of the 3 phase currents. The torque is estimated by (5).

$$\lambda = \int (v - Ri)dt + \lambda_0 \quad (3)$$

$$\lambda_\alpha = \int (v_\alpha - Ri_\alpha)dt + \lambda_{\alpha 0}$$
$$\lambda_\beta = \int (v_\beta - Ri_\beta)dt + \lambda_{\beta 0} \quad (4)$$

$$T = \frac{3}{2}p(\lambda \times i) = \frac{3}{2}p(\lambda_\alpha i_\beta - \lambda_\beta i_\alpha) \quad (5)$$

At very low speeds, the back EMFs are too small to be used to obtain accurate stator flux estimation. The effects of forward voltage drop and dead-time of power devices become very significant at very low speeds. None of the voltage compensators in the existing literature have been able to estimate the stator flux and torque accurately down to zero speed, mainly due to the non-linear characteristics of IGBTs and diodes. The current model of the IPM machine can be used for flux and torque estimation at very low speeds once accurate rotor position information is obtained. This is done using high frequency signal injection method described in the next section. The stator flux linkages and torque are then estimated using the current model as in (6) and (7).

$$\lambda_d = L_d i_d + \lambda_f$$
$$\lambda_q = L_q i_q \quad (6)$$

$$T = \frac{3}{2}P(\lambda_d i_q - \lambda_q i_d) \quad (7)$$

IV. HF SIGNAL INJECTION

Several high frequency injection methods can be found in the literature. These methods can be classified into α-β frame rotating injection [21], d-q frame pulsating injection [22] and d-q frame rotating injection [23]. In this paper, the d-q frame persistent HF rotating carrier injection is implemented, where an alternating voltage is used for injection. A carrier excitation signal fluctuating at angular frequency ω_c, and having amplitude V_c, as shown in (8), is superimposed on the d component of the stator voltage in the estimated rotor reference frame.

$$V_{hf} = V_c \cos(\omega_c t) \quad (8)$$

The frequency of the injected voltage carrier, ω_c, should be high enough to ensure sufficient spectral separation between itself and the fundamental excitation to reduce the requirements of the band-pass filters. An alternating HF current response is detected in the q direction of the estimated rotor reference frame with its amplitude modulated by the rotor position estimation error. The method of the demodulation process is shown in Fig. 2. The high frequency component of measured current in the q direction, i_{qc}, is obtained by band-pass filtering (BPF) of the q component of measured current. The HF current signal is then demodulated and low-pass filtered (LPF) to extract an error signal given by (9).

$$\varepsilon = LPF\{i_{qc} \sin(\omega_c t)\} \quad (9)$$

This error signal is ideally [24]

$$\varepsilon = K_\varepsilon \sin(2\widetilde{\theta}_m) \qquad (10)$$

where $K_\varepsilon = \dfrac{V_c}{\omega_c}\dfrac{L_q - L_d}{4L_dL_q}$. K_ε is the signal injection gain and $\widetilde{\theta}_m = \theta_m - \widehat{\theta}_m$ is the estimated error of the rotor position.

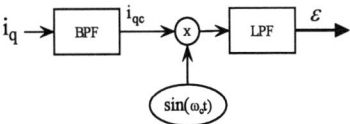

Fig. 2 Demodulation scheme used to obtain error signal

The HF signal injection method to obtain information of rotor position becomes difficult to apply at high speed mainly because the alternating exciting frequency must be completely off a fundamental wave frequency as a prerequisite. The voltage model (back EMF integration) is used for flux calculation at high speed and the speed of the stator flux vector is then used to complete the speed loop. The change of algorithms at low and high speeds has been realized by using weight coefficients W_1 and W_2 to determine the input to the flux controller. In this case, W_1 represents the weighting for flux obtained from the rotor reference frame (current model) during high frequency signal injection while W_2 is the weighting of flux obtained in the stationary reference frame (voltage model). This is shown in Fig. 3 where W_1 fully dominates below 100 rpm and W_2 above 500 rpm. The crossover of algorithms does not affect the estimated position as $W_1 + W_2 = 1$.

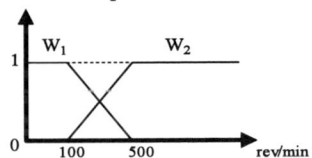

Fig. 3   Weighting coefficients for changeover of algorithms

V. EXPERIMENTAL RESULTS

The effectiveness of the proposed sensorless DTFC scheme was tested experimentally. A DS1104 DSP card was used to carry out the real-time algorithm. A three-phase insulated gate bipolar transistor (IGBT) intelligent power-module is used for an inverter. Coding of real-time control software was done using C language. The experimental setup is shown in Fig. 4. The PWM signals were generated on the DS1104 board. In experiments, the sampling period of the DTFC was set to 200µs. A dc machine whose armature current is separately regulated is used to emulate the load. An incremental encoder was used to obtain the position signal which was solely used for comparison and not for control purposes. The reference flux value was selected according to the maximum torque per ampere (MTPA) trajectory to increase the efficiency of the overall drive system. The block diagram of the sensorless drive is shown in fig. 5.

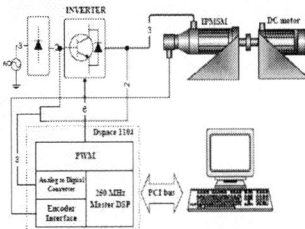

Fig. 4 Experimental Setup.

Fig. 6 illustrates the speed, torque and flux responses of the sensorless DTFC drive at a zero speed due to nominal load torque steps. The reference, estimated and actual speed signals are shown in blue, red and green respectively. The estimated and reference torque and flux signals are shown in red and blue respectively. The load was initially applied to the machine and was subsequently removed. We can observe that the estimated speed tracks the actual speed very well and the estimation error is very small during the transients and steady-state. Hence, the sensorless DTFC drive is capable of zero speed operation with full-load.

The low-speed dynamic response of the sensorless drive is shown in fig. 7. The machine is reversed from -5 rpm to 5 rpm without load. The estimated speed follows

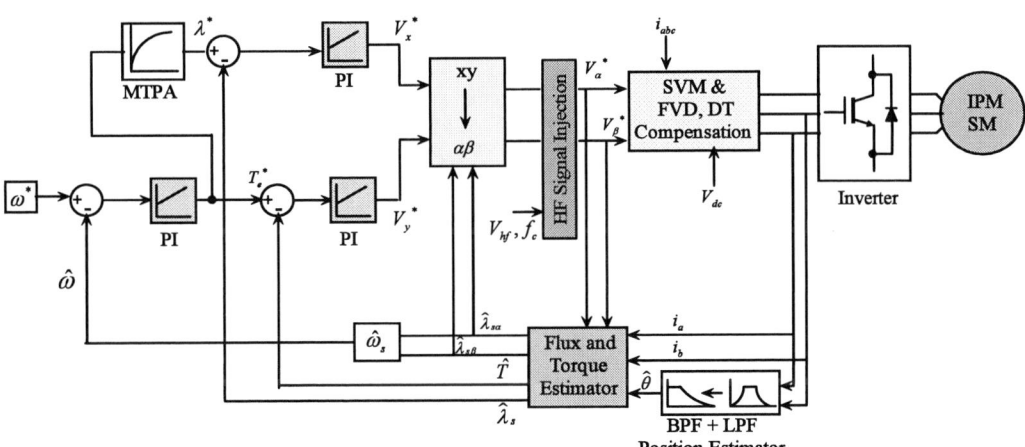

Fig. 5. Block diagram of the sensorless SVM-DTFC scheme.

2271

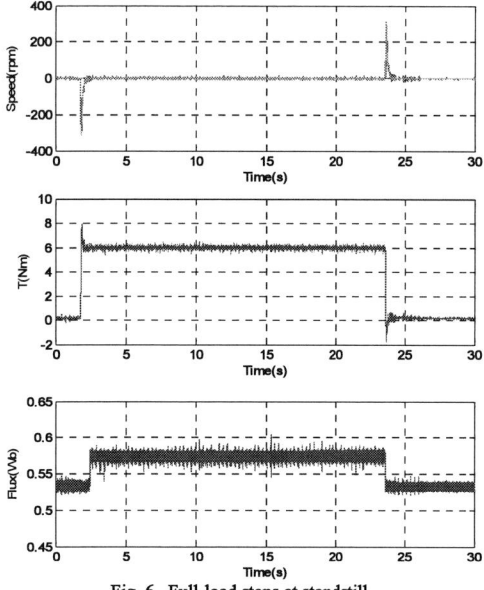

Fig. 6. Full-load steps at standstill.

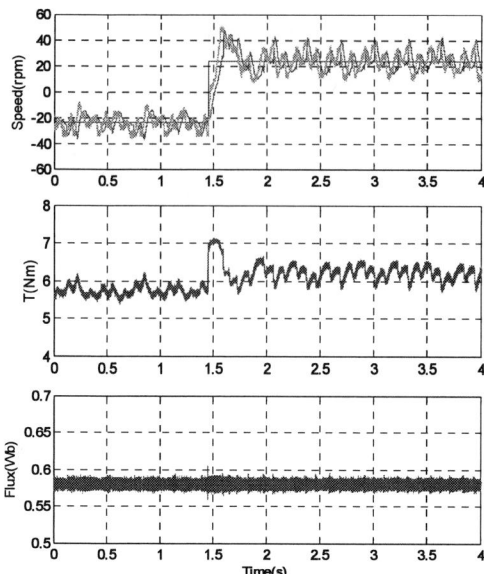

Fig. 8 Full-load speed reversal from -24rpm to 24rpm.

the actual speed very well during the reversal. Speed reversal at nominal load from -24rpm to 24rpm can be visualized from fig. 8. Even under load condition, the estimated speed follows the actual speed very well during the reversal and the position estimation error is very small, confirming the effectiveness of the high frequency injection algorithm.

The high-speed dynamic response of the DTFC drive is depicted in fig. 10. The machine is accelerated from -1000 rpm to 1000 rpm. Again, we can see that the current-voltage model flux estimation handover is more than satisfactory. The estimated speed follows the actual speed very closely in the transient interval.

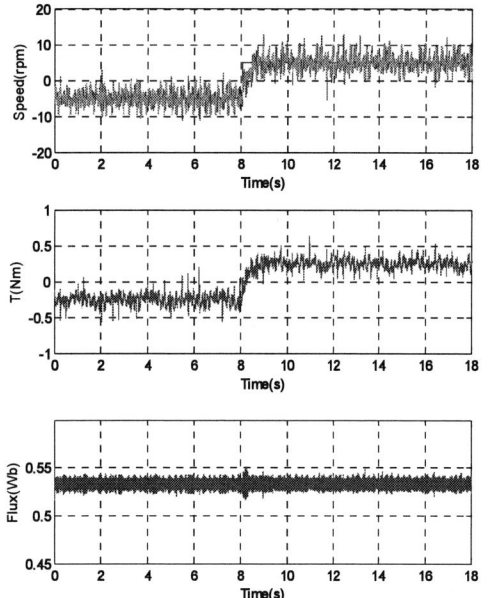

Fig. 7. No-load low speed reversal from -5rpm to 5rpm.

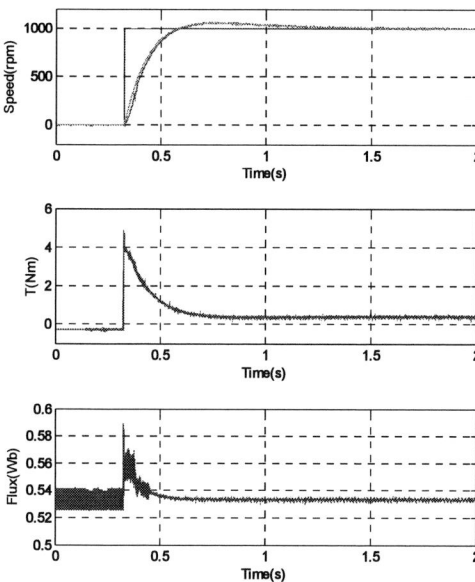

Fig. 9. Start-up performance of the sensorless DTFC drive.

Fig. 9 portrays the start-up profile of the sensorless DTFC under no-load condition. The IPMSM is accelerated from standstill to 1000rpm. Therefore, there exists a transition from the current model-based observer to the voltage model-based observer. The speed, torque and flux responses in fig. 9 indicate a smooth transition between both observers.

VI. CONCLUSION

In this paper, a sensorless DTFC scheme based on SVM for low speed and standstill operation of IPMSMs has been introduced. The SVM-DTFC utilizes two PI controllers to regulate the torque and stator flux linkage independently. The reference voltage vector is synthesized with the SVM technique. A high frequency signal injection technique has been adopted to extract the rotor

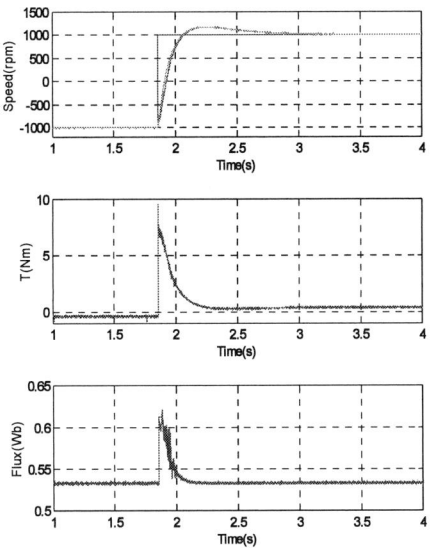

Fig. 10. No-load high speed reversal.

position at low speeds. A hybrid voltage-current model flux observer with a smooth changeover algorithm enables stator flux estimation over the entire speed range.

The proposed SVM-DTFC method is capable of handling full-load over the entire speed range. Experimental results presented confirm the veracity of the sensorless DTFC scheme for speed and position control of IPM synchronous motors.

TABLE I
PARAMETERS OF IPMSM USED IN THIS PAPER

Number of poles	P	4
Stator resistance	R_s	5.8 Ω
Magnet flux linkage	λ_f	0.533 Wb
d-axis inductance	L_d	0.0447 H
q-axis inductance	L_q	0.1024 H
Phase voltage	V	132 V
Line current	I	3 A
Base speed	ω_b	1260 rpm
Rated torque	T	6 Nm

REFERENCES

[1] M. Depenrock, "Direct self-control of inverter-fed machine," *IEEE Trans. Power Electron.*, vol. 3, pp. 420-429, Oct. 1988.

[2] I. Takahashi and T. Naguchi, "A new quick-response and high efficiency control strategy of an induction motor," *IEEE Trans. Ind. Applicat.*, vol. IA-22, pp. 820-827, Sept./Oct. 1986.

[3] L. Zhong, M. F. Rahman, W. Y. Hu, and K. W. Lim, "Analysis of direct torque control in permanent magnet synchronous motor drives," *IEEE Trans. Power Electron.*, vol. 12, pp. 528-536, May 1997.

[4] L. Zhong, M. F. Rahman, W. Y. Hu, and K. W. Lim, "A direct torque controlled interior permanent magnet synchronous motor drive incorporating field weakening," *IEEE Trans. Ind. Applicat.*, vol. 34, pp. 1246-1253, Nov./Dec. 1998.

[5] C. French and P. Acarnley, "Direct torque control of permanent magnet drives," *IEEE Trans Ind. Applicat.*, vol. IA-32, pp. 1080-1088, Sept./Oct. 1996.

[6] M. F. Rahman, L. Zhong and K. W. Lim, "A direct torque controlled interior permanent magnet synchronous motor drive incorporating field weakening," in *Conf. Rec. IEEE-IAS Annu. Meeting*, vol. 1, 1997, pp. 67-74.

[7] L. Tang, L. Zhong, M. F. Rahman and Y. Hu, "A Novel Direct Torque Control for Interior Permanent Magnet Synchronous Machine Drive System with Low Ripple in Flux and Torque and Fixed Switching Frequency," *IEEE Trans. Power Electron.*, vol. 19, pp. 346-354, March 2004.

[8] C. G. Mei, S. K. Panda, J. X. Xu and K. W. Lim, "Direct Torque Control of Induction Motor – Variable Switching Sensors," *Conf. Rec. IEEE-PEDS*, July 1999, Hong Kong, pp. 80-85.

[9] C. Martins, X. Roboam, T. A. Meynard and A. S. Caryalho, "Switching Frequency Imposition and Ripple Reduction in DTC Drives by Using a Multilevel Converter," *IEEE Trans. Power Electron.*, vol. 17, issue 2, pp. 286-297, March 2002.

[10] C. Lascu, I. Boldea and F. Blaabjerg, "A Modified Direct Torque Control for Induction Motor Sensorless Drive," *IEEE Trans. Ind. Applicat.*, vol. 36 no.1, pp.122-130, Jan/ Feb 2000.

[11] G. S. Buja and M. P. Kazmierkowski, "Direct torque control of PWM inverter-fed AC motors – a survey," *IEEE Trans. Ind. Electron.*, vol. 51, pp. 744-757, 2004.

[12] M. F. Rahman, L. Zhong, K. W. Lim, "A direct torque controlled interior permanent magnet synchronous motor drive without a speed sensor," *IEEE Trans. Energy Conv.*, vol. 18, pp. 17-22, March 2003.

[13] M. E. Haque, L. Zhong, M. F. Rahman, "A sensorless initial rotor position estimation scheme for a direct torque controlled interior permanent magnet synchronous motor drive," *Applied Power Electron. Conf. and Expo., APEC 2001*, 16th Annual IEEE, vol. 2, pp. 879-884, March 2001.

[14] S. Bolognani, R. Oboe, M. Zigliotto, "Sensorless full digital PMSM drive with EKF estimation of speed and position," *IEEE Trans. Ind. Electron.*, vol. 46, issue 2, pp. 184-191, Feb. 1999.

[15] Z. Xu, M. F. Rahman, "Direct Torque and Flux Control of an IPM synchronous motor drive using variable control approach," in *30th Annu. Conf. IEEE Ind. Electron. Soc., IECON 2004.*

[16] H. Kim, K.-K. Huh, R. D. Lorenz, and T. M. Jahns, "A novel method for initial rotor position estimation for IPM synchronous motor drives," *IEEE Trans. Ind. Applicat*, vol. 40, no. 5, pp 1369-1378, Sep./Oct. 2004.

[17] Y. Jeong, R. D. Lorenz, T. M. Jahns, and S. Sul, "Initial rotor position estimation of an interior permanent magnet synchronous machine using carrier-injection methods," in *Proc. IEEE Int. Elect. Machines Drives Conf. (IEMDC'03)*, Madison, WI, Jun. 2003, vol.2, pp. 1218-1223.

[18] H. Kim, M. C. Harke, and R. D. Lorenz, "Sensorless control of interior permanent magnet drives with zero-phase lag position estimation," in *Conf. Rec. IEEE-IAS Annu. Meeting*, 2002, pp. 86-91.

[19] C. I. Pitic, G.-D. Andreescu, F. Blaabjerg, I. Boldea, "IPMSM Motion-Sensorless Direct Torque and Flux Control," in *Conf. Rec. IEEE-IECON*, 32nd Annual IEEE.

[20] I. Boldea, C. I. Pitic, C. Lascu, G.-D. Andreescu, L. Tutelea, F. Blaabjeerg, P. Sandholdt, "DTFC-SVM Motion Sensorless Control of a PM-Assisted Reluctance Synchronous Machine as Starter-Alternator for Hybrid Vehicles," *IEEE Trans. Power. Electron.*, vol. 21, no.3, pp 711-719, May. 2006.

[21] Spiteri, C. Cilia, J. Michallef, B. and Apap, M. "Sensorless Vector Control of Surface Mount PMSM using High Frequency Injection", *Power Electronics Machines and Drives*, 16-18 April 2002, Conference Publication No. 487, IEE2002

[22] Jung-Ik Ha, and Seung-Ki Sul, "Sensorless Field-Orientation Control of an Induction Machine by High-Frequency Signal Injection", *IEEE Transaction on Industry Applications*, Vol. 35, No. 1, January/February 1999.

[23] C. Caruana, G.M. Asher, K.J. Bradley and M. S. Woolfson, "Flux Position Estimation in Cage Induction Machines using Synchronous Injection and Kalman Filtering, "*IEEE Transactions on Industry Applications*, Vol. 39, No. 5, September/October 2003, pp 1372-1378.

[24]   M. Corley and R.D. Lorenz, "Rotor position and velocity estimation for a salient-pole permanent magnet synchronous machine at standstill and high speeds," *IEEE Trans. Ind. Applications*, vol. 43, no. 4, pp 784-789, July/Aug. 1998.

[25]   Piippo, A. Hinkkanen, M. Luomi, J. "Sensorless Control of PMSM drives using a combination of voltage model and HF signal injection," *Industry Applications Conference*, 2004, 39th IAS Annual Meeting, vol. 2, 3-7 Oct. 2004, pp 964-970.

Sensorless Control of PM Synchronous Motors Using a Predictive Current Controller with Integrated INFORM and EMF Evaluation

Manfred Schrödl, Christian Simetzberger

Vienna University of Technology/Institute of Electrical Drives and Machines, Vienna, Austria,
e-mail: _manfred.schroedl@tuwien.ac.at_

Abstract— The paper describes a new current control algorithm with integrated measuring algorithms used for sensorless control of permanent magnet (PMSM-) synchronous motor drives. The measuring information is evaluated to detect the rotor position at low speed range using the INFORM method and at high speed range using an EMF-based model. No extra test pulses are necessary for sensorless position detection. A further advantage of the current controller with ingrated INFORM/EMF measurement is that only few parameters are necessary for control.

Keywords—**INFORM, current control loop, sensorless control, permanent magnet synchronous motor.**

I. INTRODUCTION

For some years, permanent magnet synchronous machines (PMSMs) have got a wide application range in the field of speed-variable AC drives because they offer a lot of advantages compared to other motor types like induction motors. They offer

- higher efficiency due to low rotor losses
- compact construction and higher torque per volume ratio
- inherent positioning capability due to synchronous operation
- simple sensorless flux detection using either back EMF or saliency effects.

A disadvantage of PMSMs (without damper windings) compared to induction motors is that they cannot be controlled feed-forward with high quality.

Hence, PMSMs are usually operated in a closed control loop. Therefore, position information is necessary according to field-oriented control theory.

This position sensor reduces the robustness of the drive, increases the costs and needs some extra space. Hence, for low-cost applications, substituting the sensor by mathematical models yields an economic solution with an additional benefit of better efficiency then induction motor solutions. A further cost reduction is possible if the current measurement is performed in the DC link using a simple ohmic shunt instead of expensive transfo-shunts in the motor phases.

For about 20 years, various attempts have been made to eliminate the sensor by using mathematical models based on electrical measurement quantities, like phase or DC link currents and voltages.

The "sensorless" algorithm may either evaluate the PMSM's behavior passively without influencing the inverter control or may be combined with active measuring sequences or special operating conditions, like short circuits, zero crossing etc.

The main effects which are used for evaluation are

- The back EMF (induced voltage) when the rotor rotates at a certain speed [1-9]
- Saturation or geometric saliency at low speed (signal injection [10-14], INFORM method [15-17])

Obviously, at zero speed, no back EMF occurs in the PMSM. Hence, at low speed a saliency-based model must be used. However, in this case, the inductance difference in d- and q-axis of the PMSM must be sufficiently large.

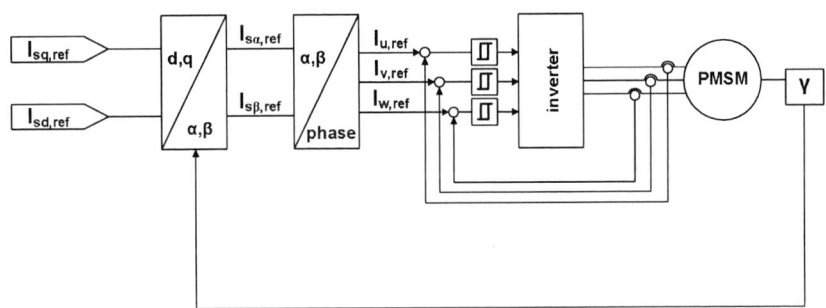

Fig. 1. Current control in a stator-oriented reference frame using phase current hysteresis controllers

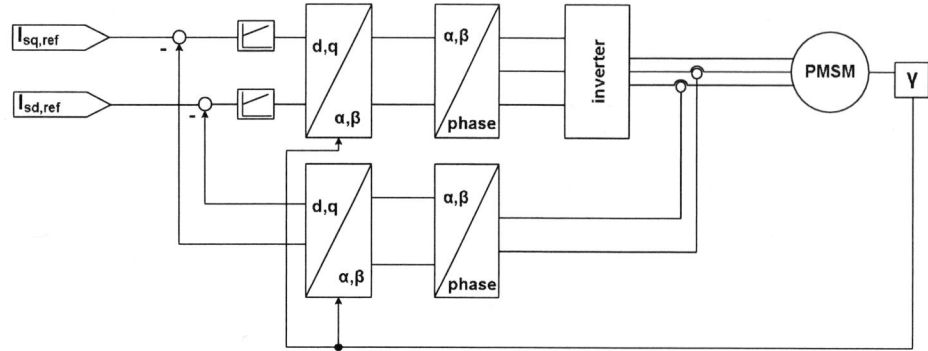

Fig. 2. Current control in a rotor oriented reference frame using PI controllers in d- and q-axis

This can be a problem, especially in surface-mounted magnet motors with almost cylindric construction and hence position-independent electrical behaviour.

Good static and dynamic performance at low speed and standstill has been achieved by the INFORM method presented in chapter IV. In normal INFORM evalutation, test signals are applied to the inverter with a certain disturbance of the PWM pulse pattern defined by the current controllers. One goal of this paper is to eliminate extra test pulses by using the inverter switching states calculated by the presented predictive current controller.

In this way, current change information evoked by the current control algorithm is used for the whole speed range using an EMF- and an INFORM-evaluation at the same time.

II. CURRENT CONTROL OF THE PMSM

A classical field-oriented PMSM control uses an outer speed control loop and inner current control loops. Several current control methods have been used for a long time. Two typical methods (stator oriented current control using hysteresis controllers and rotor-oriented current control using PI controllers) are mentioned briefly. Then, a new predictive current control method is presented in detail. All controllers need the information about the rotor orientation.

A. Stator-oriented current controllers

Fig. 1 shows the structure of a typical stator-oriented current control with hysteresis controllers for phase current control. The reference current information is obtained by the field-oriented control – the desired torque defines the reference q-current component, the flux level can be influenced by the d-current component. From the d- and q-reference current components, the reference phase current components are calculated using the actual position information. Since these reference values vary more or less sinusoidal in time (at steady-state operation), a PI-controller cannot be used because it produces unacceptable phase shifts between reference and actual currents. Hence, nonlinear controllers like simple hysteresis controllers are very fast and can directly define the inverter switching states. Typical problems of controllers with integral parts like wind-up phenomena

are avoided in this way. The new controller presented in chapter III is also a stator-oriented approach without integral control parts.

B. Rotor-oriented current controllers

In highly dynamic applications with PWM generation, (PI-) current controllers in the rotor-oriented reference frame are frequently used. A typical structure is shown in Fig. 2. Coupling effects between d- and q-control loop are managed either by the integral controller parts or by special decoupling algorithms.

III. NEW PREDICTIVE CURRENT CONTROLLER WELL-SUITED TO SENSORLESS POSITITON DETECTION

Predictive current controllers have been well-known for many years [18], [19]. The principle is to predict the behaviour of the controller during the next sampling period(s). This is achieved by calculating the change of the current during the next period based on the stator voltage equation. This prediction is performed for several inverter states according to the controller type. In the presented new controller, a cyclic change of the voltage space phasor angular position is carried out (voltage space phasor directions change after each interval Δt according to the sequence u-v-w-u-v-w-..) and only the sign of the respective voltage is calculated depending on the predicted current deviation. The following subsections explain the current measurement, the current prediction and the control law.

A. Current measurement

In order to build a very cost-efficient hardware structure of the inverter, the current is measured only in the DC link, e.g. using a simple ohmic shunt with low parasitic inductance (Fig. 8). A reconstruction of the phase currents is carried out by measuring the DC link current and utilizing the actual switching states of the inverter. As a consequence of the cyclic (120°-) switching of the voltage space phasor direction (explained in subsection B), all phase currents are mapped cyclically into the DC link current and hence it is possible to reconstruct all phase currents within short time. As shown in Fig. 3, the current is measured at begin and end of each interval Δt. That means that two phase currents are measured within a very short time before and after each switching event. Hence, it is admissible to calculate the missing third phase currents by Kirchhoff's law, as shown in Fig. 3. The current difference between the measurements at beginning and end of interval Δt is stored in the so called "current rise table" (Fig. 3). The current changes at the 6 different possible active voltage space phasors are stored separately.

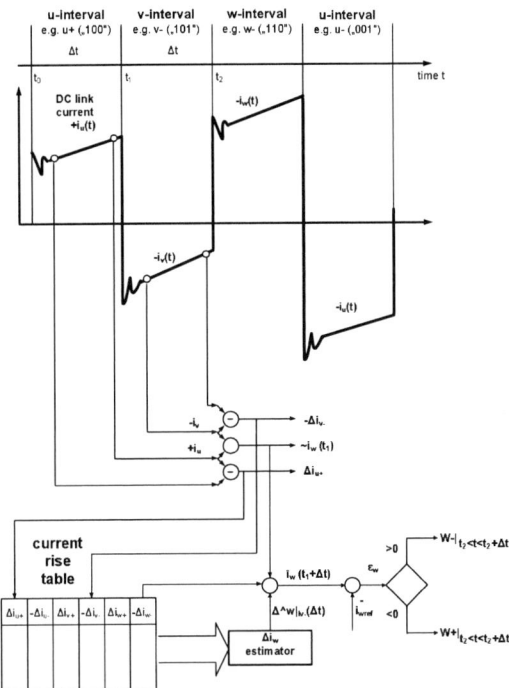

Fig. 3.   Measuring DC-link current and data processing using a current rise table for current control and sensorless position and speed detectionof

B. Prediction of the current

During interval $\Delta t(k)$, the behavior of the current in the phase which will be controlled in the next interval $\Delta t(k+1)$ is predicted. This can be achieved using the estimation at beginning of interval $\Delta t(k)$ (using Kirchhoff's law) and using the current rise table, which has information about the expected current change during interval $\Delta t(k)$ (Fig. 3).

C. The control law

As already stated, during each interval Δt, only one phase current is controlled. The controlled phases are changed cyclically (Fig. 4). Furthermore, Fig. 4 shows the influence of the back EMF (induced voltage phasor $\underline{u}_i$).As a consequence, the current change space phasors are not parallel to the applied voltage space phasors. To pass a full control cycle, three intervals Δt are necessary.

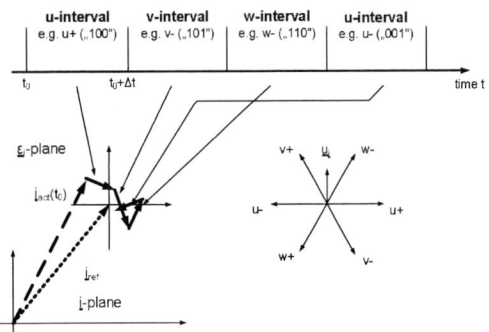

Fig. 4.   Cyclic change of the controlled phases and behaviour of the current control considering the induced voltage $\underline{u}_i$.

We realized the simple strategy of using the sign of the (estimated) phase current deviation from its reference value at beginning of interval Δt. A more detailed explanation of the decision process of the cyclic phase controller is shown in Fig. 5.

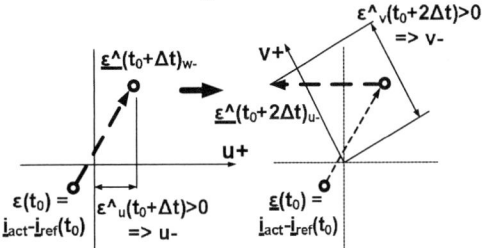

Fig. 5.   Illustration of the cyclic phase control process using current prediction

D. Using current controller data for sensorless control

The information in the current rise table can be used not only for predictive control but also for sensorless rotor position and speed detection using the INFORM and a special back-EMF method ("short circuit model"). In Fig.6, the structure of this evaluation is depicted. As can be seen from basic INFORM and "short circuit model" theory, current changes according to a positive and a negative phase voltage (stored in the current rise table) can be added (then only the back EMF has a resulting influence and an EMF-based evaluation is possible) or subtracted (then the back EMF is eliminated and only the saliency, evaluated by INFORM, has influence). For more information about INFORM see chapter IV. The Back-EMF based short circuit model is explained in chapter V.

Both the INFORM and the EMF model produce information about the rotor angular position which is used as input into an observer (based on a simple model of mechanical system), as shown in Fig. 6.

The back-EMF model works fine at high speed but provides poor information at low speed, so we have to use relative weight factors to disable the EMF at low speed, see Fig. 7. Compared to classical INFORM evaluations, the INFORM weight can be high within the whole speed range because even at high speed the information from the current rise table yields high quality

2277

information (Normally, when using special INFORM measurement pulses, which disturb the current controller, the INFORM method is restricted to about 20% of rated speed).

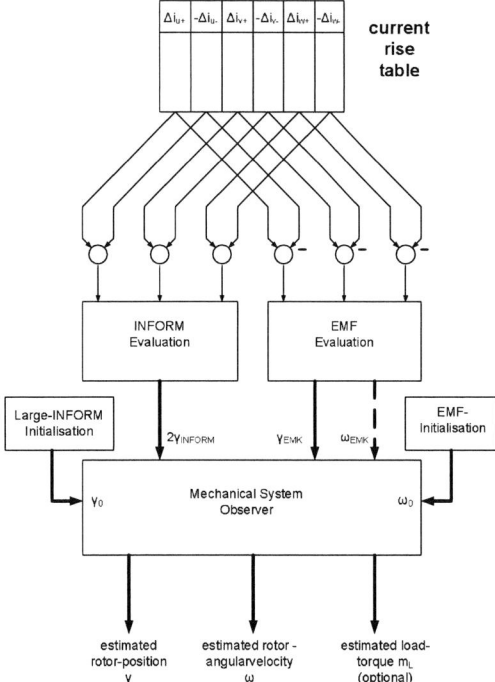

Fig. 6. Data analysis and calculation of rotor orientation

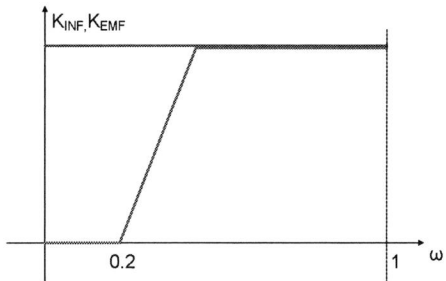

Fig. 7. Relative weights between INFORM and EMF evaluation used in the observer feedback loop (INFORM is used in full speed range)

E. Current controller realization

The whole block diagram of the predictive controller with integrated sensorless position detection capability is displayed in Fig. 8. The measurement is reduced to a simple DC-link current measurement with intelligent data processing via the current rise table. This information is processed for predictive current control in the block "cyclic phase current controller" and for rotor position detection in the block "INFORM and EMF evaluation". The outer control loops generating the reference values of d- and q-current components (speed- and possibly position control loop) are well-known and not depicted here.

The explained current controller was implemented on a low-cost 16 bit digital signal processor (TMS320LF2403A from Texas Instruments).

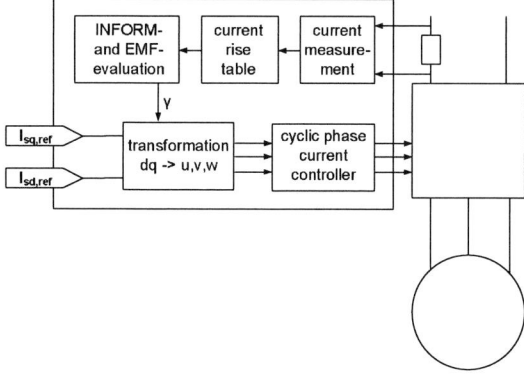

Fig. 8. Block diagram of the implemented current controller

IV. THE INFORM METHOD

A very brief overview about the installed INFORM Method (Indirect flux detection by online reactance measurement) [16] is given in the following subsection.

A. Description

As already mentioned, the INFORM method uses test algorithms by utilizing voltage steps and measuring the current response, either using a special measuring sequence interrupting the PWM control or, as shown here, is integrated into an intelligent current control loop. For detection of the axis with maximum (or minimum) magnetic conductance, a sequence of voltage space phasors $\underline{u}_S$ is applied to the PSM via the inverter and the current reaction $d\underline{i}_S / d\tau$ is measured. We define

$$\underline{y}_{\text{INFORM}} := \frac{d\underline{i}_S/d\tau}{\underline{u}_S} \qquad (1)$$

By the reason of saliency effects $\underline{y}_{\text{INFORM}}$ is a 180°-periodic function, which can be modelled as a circle in the complex plane using the parameters $\arg(\underline{u}_S) = \gamma_U$ and rotor angular position γ (Fig. 9; details are given in [17]):

$$\underline{y}_{\text{INFORM}} = y_0 - \Delta y \cdot \exp\left[j(2\gamma - 2\gamma_U)\right] \qquad (2)$$

For measuring $\underline{y}_{\text{INFORM}}$ explicitly, test voltage space phasors $\underline{u}_S$ are applied to the motor during operation. The respective current changes $d\underline{i}_S / d\tau$ are measured. Turning the rotor from POS1 to Position POS3 in Fig. 9, a half circle of $\underline{y}_{\text{INFORM}}$ with radius Δy and offset y_0 is produced. Applying a positive flux-parallel current i_{Sd} increases the circle radius and can be used to improve the INFORM capability of a drive. Fully symmetric PSMs with almost no saturation yield a very small radius Δy.

2278

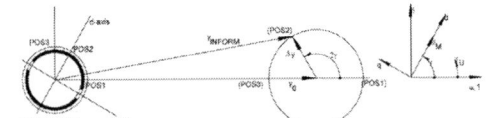

Fig. 9. Complex function $\underline{y}_{INFORM}$ at two different saturation levels and given voltage test phasor depending on the rotor position.

Combining measurements in different test directions of $\underline{u}_S$ transforms the information of $\underline{y}_{INFORM}$ into an offset-free circle $\underline{C}_{INFORM}$ with

$$\underline{C}_{INFORM} = \Delta y \cdot \exp\left[j(2\gamma)\right] \tag{3}$$

B. Initialisation

The ability of the flux-parallel current component to change the radius Δy can be used to distinguish between the two possible solutions ($\gamma_1 = \gamma$, $\gamma_2 = \gamma + 180°$) at initialization state. A modified INFORM method [20], called "Large INFORM initialisation" in Fig. 6, eliminates the 180°-uncertainty by a short test at startup. The basic idea is to shift the magnetic set point by a relatively large stator current phasor and perform an INFORM evaluation in this modified set point (Fig. 10) Then, the observer stores this information about the most likely solution.

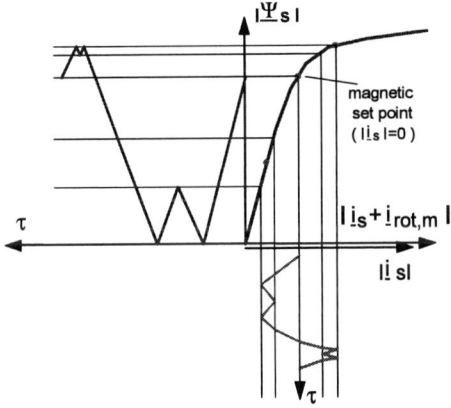

Fig. 10. Elimination of the 180 degrees ambiguity; basic principle of shifting the magnetic set point

The large INFORM measurement is repeated in different directions, yielding directly the rotor position. The procedure needs typically about 1 millisecond to initialize the position. Then, the normal INFORM procedure is started, providing the signal γ_{INFORM}.

V. THE EMF METHOD

A. Classical EMF model with integration

The stator flux linkage is obtained by integration of the stator voltage after subtracting the ohmic voltage drop:

$$\underline{\Psi}_s = \int_0^\tau \left(\underline{u}_s(\tau') - \underline{i}_s(\tau') \cdot r_s\right) d\tau' + \underline{\Psi}_s(0) \tag{4}$$

However, the feedforward integration is unpracticable due to drift effects, parameter uncertainties and measuring errors. Therefore, a weak stabilizing feedback (Fig. 12) is used. The integrator is hence modelled by a first order lag:

$$\frac{d\underline{\Psi}_s}{d\tau} = \underline{u}_s(\tau) - \underline{i}_s(\tau) \cdot r_s - K \cdot \underline{\Psi}_s \tag{5}$$

This model is well suited in the high frequency range. However, at zero and low frequencies, the model works completely wrong. This can be seen in the Bode diagram (Fig. 11).

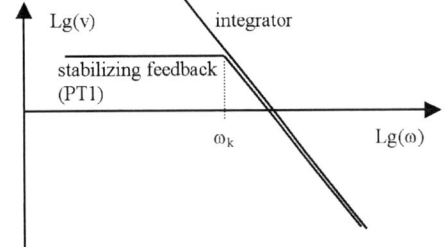

Fig. 11. Bode diagram of integrator and approximating first order lag.

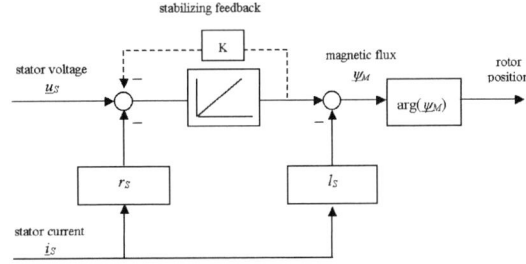

Fig. 12. Integrating EMF model for position detection at higher speed

The permanent magnet flux linkage is obtained by subtracting the flux part evoked by stator current and self inductance from the stator flux linkage. Then, the angular position of the permanent magnet flux linkage provides the rotor angular position.

B. Short circuit model

Consider all machine terminals connected to the positive or the negative dc-link voltage, which represents a more or less ideal short circuit of the armature windings. We get the following space phasor equation:

$$\underline{u}_s = 0 = r_s \underline{i}_s + l_s \frac{d\underline{i}_s}{d\tau} + \frac{d\underline{\Psi}_M}{d\tau} \tag{6}$$

The rotor flux linkage $\underline{\Psi}_M$ is represented in a stator-oriented reference frame by

$$\underline{\Psi}_M = |\underline{\Psi}_M| \cdot \exp\left(j \cdot \gamma\right) \tag{7}$$

Neglecting the voltage drop caused by the stator resistance, we get from (6) and (7)

$$\frac{di_S}{d\tau} = -j \cdot \frac{|\underline{\Psi}_M| \cdot \omega}{l_S} \cdot \exp(j \cdot \gamma) : \quad \omega := \frac{d\gamma}{d\tau} \qquad (8)$$

Calculating the arguments of (9) yields the desired flux angular position:

$$\gamma = arg \frac{di_S}{d\tau} + \frac{\pi}{2}\Big|_{u_S=0} \qquad \omega > 0 \qquad (9)$$

$$\gamma = arg \frac{di_S}{d\tau} - \frac{\pi}{2}\Big|_{u_S=0} \qquad \omega < 0 \qquad (10)$$

The major advantages are that no PMSM parameters influence the measurement result and that $\underline{\Psi}_m$ can be detected without integration. Furthermore, the measurement is independent of the DC link voltage.

A certain disadvantage is that, at higher angular velocities or in current control strategies without using short circuits like in the presented paper, (almost) no evluable short-circuit inverter states occur, so that the short circuit has to be inserted additionally.

C. Extension of the short circuit model to arbitrary inverter states – "Virtual circuit model"

To overcome the disadvantage mentioned above, it is shown how to use arbitrary inverter states for detecting the flux angular position γ [21]. The following voltage equations are valid for two inverter states (the stator resistance is again neglected) within a short time period ($\gamma_I \approx \gamma_{II}$):

$$u_{S,I} = l_S \cdot \frac{di_{S,I}}{d\tau} + \frac{d\underline{\Psi}_{M,I}}{d\tau} \qquad (11)$$

$$u_{S,II} = l_S \cdot \frac{di_{S,II}}{d\tau} + \frac{d\underline{\Psi}_{M,II}}{d\tau} \qquad (12)$$

With

$$u_{S,II} = u_{S,I} \cdot \exp(j \cdot \Delta\alpha_N) \quad \Delta\alpha_N \neq 0. \qquad (13)$$

In (11) and (12), $d\Psi_{M,I}/d\tau \approx d\Psi_{M,II}/d\tau = d\Psi_M/d\tau$ is assumed. Multiplying (12) by $\exp(-j\Delta\alpha_N)$ and subtracting from (11) yields

$$0 = l_S \cdot \left(\frac{di_{S,I}}{d\tau} - \frac{di_{S,II}}{d\tau} \cdot \exp(-j\Delta\alpha_N)\right) +$$

$$j\omega|\underline{\Psi}_M| \cdot \exp(j \cdot \gamma) \cdot (1 - \exp(-j\Delta\alpha_N)) \qquad (14)$$

Equation (14) is a generalization of the basic idea. By a linear combination of two active inverter states [(11) and (12)], a mathematical zero state is constructed. The proposed calculation of the permanent magnet flux is possible without waiting for zero states or generating auxiliary zero states, respectively. The desired flux angle is again obtained by using the arguments of (14) [17]:

$$\gamma = arg \left(\frac{di_{S,I}}{d\tau} - \frac{di_{S,II}}{d\tau} \cdot \exp(-j\Delta\alpha_N)\right) +$$

$$\frac{\pi}{2}\text{sign}(\omega) - arg\,(1 - \exp(-j\Delta\alpha_N)). \qquad (15)$$

For the special case that $\Delta\alpha_N = \pi$, we get from equation (13):

$$u_{S,II} = -u_{S,I} \qquad (16)$$

With equation (14) and (15) we get:

$$0 = l_S \cdot \left(\frac{di_{S,I}}{d\tau} + \frac{di_{S,II}}{d\tau}\right) + j2\omega|\underline{\Psi}_M| \cdot \exp(j \cdot \gamma) \qquad (17)$$

$$\gamma = arg \left(\frac{di_{S,I}}{d\tau} + \frac{di_{S,II}}{d\tau}\right) + \frac{\pi}{2}\text{sign}(\omega) \qquad (18)$$

The desired current changes are available in the current rise table. This strategy enables a very simple calculation without any machine parameters.

In Equation (15) it is assumed that the ohmic voltage drop and the change of the rotor flux within the measurement time can be neglected. It is also assumed that the DC link voltage is constant during evaluation.

VI. MEASUREMENT RESULTS

In the real time DSP-based realization, a classical field oriented control structure (current control with superimposed speed- and optional position control loop) was used. However, the position information is obtained by the described "current rise table" of the current controller structure (Figs. 3, 6 and 8).

The measured complex INFORM circle $\underline{C}_{INFORM}$ of the tested motor (DC link voltage 48 V, Irms=10 A) at full load (rated torque is normally worst-case) is shown during operation at 10% and 25% of rated speed in Figs. 13 and 14, respectively (left-hand sides). The desired angular information $2\gamma_m$ is obtained by calculating the argument of $\underline{C}_{INFORM}$. Thus, eliminating the saturation-depending radius Δy . As expected, the INFORM curve $\underline{C}_{INFORM}$ is slightly noisy. (standard deviation of angular error is about 6°). A good comparison of the INFORM and EMF measurement quality can be seen from the calculated permanent magnet flux space phasor based on the virtual short circuit model (right-hand sides of Figs. 13 and 14). In both cases, the angular positon is calculated using the arctan-function of the complex curves ($\underline{C}_{INFORM}$ and ($\omega_m \cdot \underline{\psi}_M$), respectively. It can be seen from Figs 13 and 14 that the EMF-model noise content is reduced at increased speed.

In Figs. 15 and 16, the angular posititon (observer output) and a controlled phase current (phase u) at full load torque and at 10% and 25% of rated speed is shown. According to the relative weights of INFORM and EMF mesurements (Fig. 7), the observed position at 10% of rated speed is obtained only by INFORM information (EMF measurement input of the observer is inactive). At 25% of rated speed, both INFORM and EMF measurements are used as observer input quantities.

As an example of the dynamic behaviour of the predictive current controller, the first milliseconds of a startup from standstill at nominal torque is presented (Fig. 17). Starting angle is -90°, hence, the q-axis points in the direction of phase axis u and the q-current is more or less identical to the phase current i_u for the first milliseconds after reference torque step (from 0 to 1, upper curve in Fig. 17). The middle curve in Fig 17 shows the actual q-current in the DSP reconstruced from DC link current measurement and estimated rotor position). The lower curve shows the measured phase current i_u. It can be seen that the predictive current controller performs a step in the motor torque from 0 to 1 within 1.5 milliseconds.

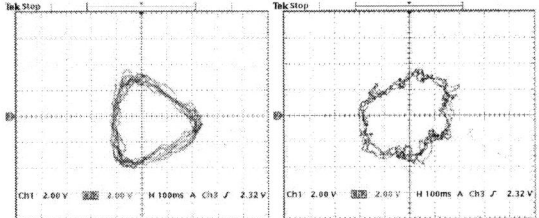

Fig. 13. INFORM (left) and EMF(right) curve at w=0.1 and nominal torque

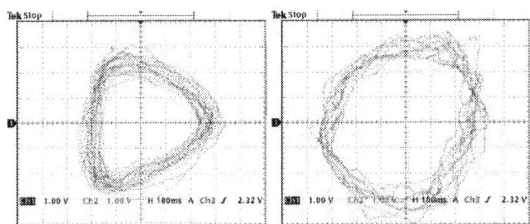

Fig. 14. INFORM (left) and EMF(right) curve at w=0.25 and nominal torque

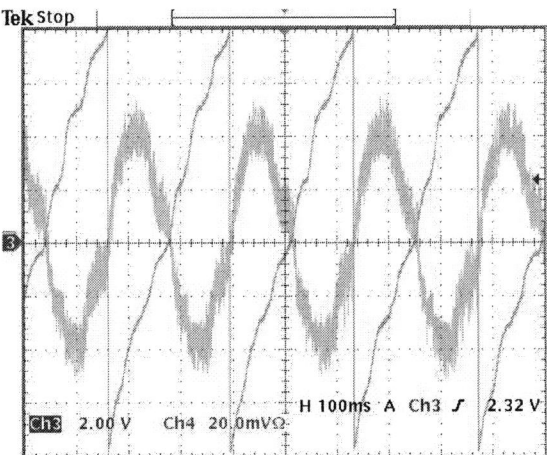

Fig. 15. Observer angle and phase U current at w=0.1 and nominal torque

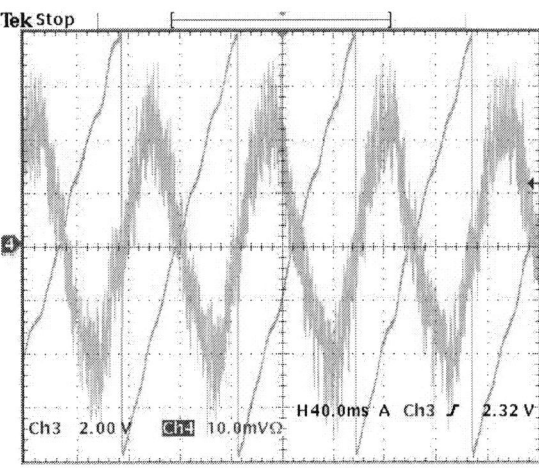

Fig. 16. Observer angle and phase u current at w=0.25 and nominal torque

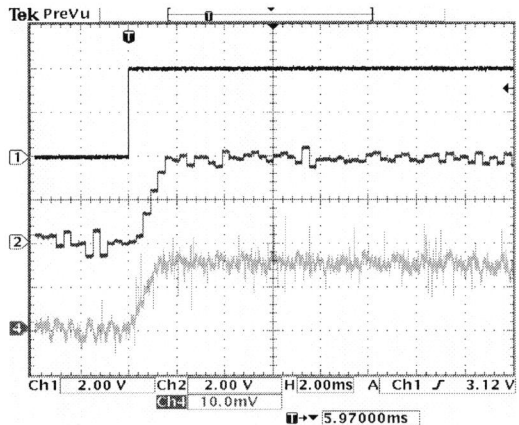

Fig. 17. Dynamic behaviour of current controller (response to a torque reference step within 1.5 ms). Upper curve: reference torque, middle curve: actual q-current, lower curve: measured phase current i_u.

VII. CONCLUSION

In the paper, a new predictive current controller with integrated sensorless position detection was presented. The only used measurement quantity was the DC link current. In the low (and even high) speed range, the INFORM method was implemented, in the high-speed range, the EMF-based virtual short circuit model is active. Both models use current changes, stored in a "current rise table". A simple mechanical observer yields improved speed and position information. Measurement results showed that a highly-dynamic current- (and hence, torque-) control was achieved.

ACKNOWLEDGMENT

The authors are very much indebted to the Austrian Science Foundation (FWF) which generously supports the work at the Department of Electrical Drives and Machines at the Vienna University of Technology - Faculty of Electrical Engineering and Information Technology.

REFERENCES

[1] Erdman, D.M., Harms, H.B., Oldenkamp, J.L., *Electronically commutated DC motors for the appliance industry*, IEEE-IAS conference record, 1984, pp. 1339-1345.

[2] Iizuka, K., Uzuhashi, H., Kano, M., Endo, T., Mohri, K., *Microcomputer control for sensorless brushless motor*, IEEE Trans. Ind. Appl., 1985, No. 4, pp. 595-601.

[3] Becerra, R.C., Jahns, T.M., Ehsani, M., *Four-quadrant sensorless brushless ECM drive*, IEEE APEC conference record, 1991, pp. 202-209.

[4] Moreira, J.C., *Indirect sensing for rotor flux position of permanent magnet AC motors operating in a wide speed range*, IEEE IAS conference record, 1994, pp. 401-407.

[5] Shen, J.X., Zhu, Z.Q., Howe, D., *Sensorless flux-weakening control of permanent magnet brushless machines using third-harmonic back-EMF*, IEEE Conference record, 2003, pp. 1229-1235.

[6] Consoli, A., Musumeci, S., Raciti, A., Testa, A., *Sensorless vector and speed control of brushless motor drives*, IEEE Trans. On Industrial Electronics, 1994, Vol. 41, No 1, pp. 91-96.

[7] Ertugrul, N., Acarnley, P., *A new algorithm for sensorless operation of permanent magnet motors*, IEEE Trans. On Ind. Appl., 1994, Vol. 30, No. 1, pp. 126-133.

[8] Wu, R., Slemon, G., *A permanent magnet drive without a shaft sensor*, IEEE IAS conference record, 1990, pp. 553-558.

[9] Matsui, N., Shigyo, M., *Brushless DC motor control without position and speed sensors*, IEEE Trans. On Ind. Appl., 1992, vol. 28, No. 1, pp. 120-127.

[10] Binns, K., Jabbar, M, *A high-field self-starting permanent magnet synchronous motor*, Proc. IEE vol. 128, pp. 157-160, 1981.

[11] Binns, K., Shimmin, D., Al-Aubidy, K, *Implicit rotor-position sensing for permanent magnet self-commutating machine drives*, Proc. ICEM, Boston (USA), 1990, Vol. 3, pp. 1231-1236.

[12] Jansen, P.L., Lorenz, R.D, *Transducerless position and velocity estimation in induction and salient AC machines*, IEEE Trans. Ind. Appl., 1995, Vol. 31, pp. 240-247

[13] Jansen, P.L., Corley, M., Lorenz, R.D, *Flux, position and velocity estimation in AC machines at zero speed via tracking of high frequency saliencies*, Proc. EPE, 1995, pp. 154-160

[14] Linke, M., Kennel, R., Holtz, J.: *Sensorless speed and position control of synchronous machines using alternating carrier injection*, IEEE Int., Electric Machines and Drives Conference, Madison (USA), 2003, pp. 1211-1217

[15] M. Schrödl, *Sensorless Control of AC Machines at Low Speed and Standstill Based on the INFORM Method*, 31st IEEE Industry Applications Society, San Diego California, 1996, Vol. 1, pp. 270-278

[16] M. Schrödl, *Detection of the Rotor Position of a Permanent Magnet Synchronous Machine at Standstill*, International Conference on Electrical Machines, ICEM, Italy, 1988, pp. 195-197

[17] M. Schrödl, *Sensorless Control of AC machines*, Habilitationsschrift, VDI-Fortschrittsberichte No. 117, 1992, Reihe 21

[18] J. Holtz, S. Stadtfeld, *A PWM Inverter Drive System with On-line Optimized Pulse Patterns*, Proceedings of European Conference on Power Electronics, Brüssel, 1985, pp. 3.21-3.25

[19] M. Schrödl, *Ein prädiktives Stromregelverfahren für hochdynamische Drehstromantriebe*, Elektrotechnik und Informationstechnik (E&I), 1988, pp. 62-65

[20] M. Schrödl, Sensorless *Control of permanent-magnet synchronous machines at arbitrary operating points using a modified INFORM flux model*, European Transactions on Electrical Power Engineering (ETEP), Vol. 3, No. 4, 1993, pp. 277-283

[21] M. Schrödl, R. Wieser, *EMF-based rotor flux detection in Induction motors using virtual short circuits*, IEEE Trans. On Ind. Appl., 1998, Vol. 34, No. 1, pp. 142-147

Torque Sensorless Control of Induction Motor

Karel Jezernik[*] and Miran Rodič[+]

[*]University of Maribor / FERI, Maribor, Slovenia, *karel.jezernik@uni-mb.si*
[+]University of Maribor / FERI, Maribor, Slovenia, *miran.rodic@uni-mb.si*

Abstract- **In this paper the torque and speed sensorless induction motor (IM) control is presented. The main idea consists in utilizing the best qualities of the both nonlinear structures of the motor observer and the current controller. Motor observer is based on flux error between modeled and measured quantities. In addition, the rotor flux observer is using combination of the feedforward and feedback terms in order to enhance the sensitivity of the observer. Torque and stator current control as well as the rotor flux observer are based on the sliding mode theory, which results in high degree of robustness towards parameters variation and the external disturbances. Proposed control scheme is implemented on DSP system extended with FPGA where event-driven current controlled modulator is realized. DSP serves for time discrete speed control and observer, while time critical current control is implemented on FPGA. Results demonstrate high efficiency of the proposed estimation and control method.**

Keywords—**sliding mode control, sensorless control, induction motor**

I. INTRODUCTION

Recent theoretical advances in the field of hybrid and discrete event-systems, and significant increase of the computational power available for the control of the power electronic systems are inviting both the control and the power electronics communities to adopt traditional control schemes associated with power electronics applications. In order to raise the performance and efficiency of the drive applications, faster and more sophisticated current control schemes are required. The conventional current control scheme consisting of discrete-time current controller and pulse-width modulator is replaced with the new sequential switching current control strategy. Inherent switching operation of the three phase bridge requires adopted control principles. Hysteresis controllers can be a good alternative for such applications. They are robust to system parameter variations, exhibit very good dynamics, require simple implementation and enable direct control of the bridge transistors without special modulators. Their main drawback is a limited control of transistor switching frequency. [1],[2].

In this paper by combining the variable structure system and Lyapunov design [3] a novel sliding mode algorithm of controller/observer for induction motor is developed. This control method is based on estimation of the rotor flux and speed of IM and is due to use of sliding mode principle robust against variation of load torque,

machine parameters and external disturbances. For both, controller and observer, there are used nonlinear control principles, namely estimated EMF and machine terminal voltage built a nonlinear feedforward control, sliding mode principle based on state variable errors are used as feedback to guarantee stability of control system. The proposed method is investigated and verified in hardware in the loop simulation experimentally.

II. DYNAMIC MODEL OF INDUCTION MOTOR

A. Machine Dynamics

Control of induction motor (IM) is still a challenging problem due to its nonlinear dynamics, limited possibility to measure or estimate necessary state variables and presence of the switching converter with its own nonlinearity as a power modulator in control loop. The dynamics of IM consist of mechanical motion (4), dynamics of stator electromagnetic system (1) and the dynamics of the rotor electromagnetic system (2):

$$\frac{di_s^s}{dt} = \frac{1}{\sigma L_s}\left(u_s^s - R_s i_s^s - \frac{L_m}{L_r}\frac{d\Psi_r^s}{dt}\right), \qquad (1)$$

$$\frac{d\Psi_r^s}{dt} = R_r\frac{L_m}{L_r}i_s^s + \left(-\frac{R_r}{L_r}\begin{bmatrix}1 & 0\\ 0 & 1\end{bmatrix} + p\omega_r\begin{bmatrix}0 & -1\\ 1 & 0\end{bmatrix}\right)\Psi_r^s, \qquad (2)$$

$$T_e = \frac{2}{3}p\frac{L_m}{L_r}\left|\Psi_r^s \times i_s^s\right|, \qquad (3)$$

$$\frac{d\omega_r}{dt} = \frac{1}{J}\left(T_e - T_L\right), \qquad (4)$$

where ω_r is mechanical rotor angle speed, the two dimensional complex space vectors $\Psi_s = \left[\Psi_{sa}, \Psi_{sb}\right]^T$, $\Psi_r = \left[\Psi_{ra}, \Psi_{rb}\right]^T$, $u_s = \left[u_{sa}, u_{sb}\right]^T$, $i_s = \left[i_{sa}, i_{sb}\right]^T$ are stator and rotor flux, stator voltage and current, respectively, T_e is motor torque, T_L is load torque, J is inertia of the rotor and p is the number of pole pairs.

One of the most important issues in implementing direct torque control (DTC) or field oriented control (FOC) strategies for IM is to obtain real-time instantaneous flux level and orientation with sufficient accuracy for the entire speed range, from almost standstill to high speed level. The difficulty in flux estimation lies with the non-linear induction machine model, which is characterized by speed dependent and time varying parameters. In order to illustrate this non-linear behavior of IM control let us express the derivation of developed electrical torque of IM from (3). This yields for torque variation:

978-1-4244-1741-4/08/$25.00 ©2008 IEEE

$$\frac{dT_e}{dt} + \left(\frac{R_s}{\sigma L_s} + \frac{R_r}{\sigma L_r} \right) T_e = \frac{2}{3} \frac{p}{\sigma L_s} \frac{L_m}{L_r} \left(\boldsymbol{\Psi}_r \times \boldsymbol{u}_s - p\omega_r \boldsymbol{\Psi}_s \bullet \boldsymbol{\Psi}_r \right), (5)$$

where $\times$ indicates cross product and $\bullet$ indicates dot product. It can be recognized from (5), that torque variation is the sum of two terms. The first term depends on the stator (R_s) and rotor (R_r) resistance and reduces the absolute value of the torque (T_e). The second term represents the effect of the applied control voltage vector ($\boldsymbol{u}_s$) on the torque and is dependent on the operating condition of IM. It can be noted that some switching voltage vectors may cause positive torque variation at low dynamic EMF value and negative torque variation at high value of back induced voltage.

The crucial point in control of IM is to make the electromagnetic torque and the flux of IM independently controllable. Similarly to torque variation, the rotor flux variation can be described from (1) and (2) as:

$$\frac{d}{dt} |\boldsymbol{\Psi}_r| = \frac{L_r}{L_m} \frac{1}{|\boldsymbol{\Psi}_r|} \left(\boldsymbol{\Psi}_r \bullet \boldsymbol{u}_s - R_s \boldsymbol{i}_s \bullet \boldsymbol{\Psi}_r - \sigma L_s \frac{d\boldsymbol{i}_s}{dt} \bullet \boldsymbol{\Psi}_s \right). \quad (6)$$

The variation of the rotor flux is determined with the dot product between rotor flux and applied input voltage vector and depends mostly on stator parameters variation $R_s, \sigma L_s$. Both torque and flux variations are highly nonlinear in applied control voltage $\boldsymbol{u}_s$ regarding IM rotor flux of machine $\boldsymbol{\Psi}_r$. The conventional control method of IM is based in case of FOC on simplification of rotor flux components $\boldsymbol{\Psi}_r = \left[\boldsymbol{\Psi}_{rd}, \ \boldsymbol{\Psi}_{rq} \right]^T = \left[\boldsymbol{\Psi}_{rd}, \ 0 \right]^T$. The DTC method replaces the IM coupling with hysteresis control.

In real IM control rotor flux in q-axis will not be zero and FOC method is due variations of mostly parameters, inappropriate in sensorless drives applications. The DTC method is in principle speed sensorless, but due to use of voltage stator model of IM and approximation of stator resistance $R_s \sim 0$ in the flux model, current and torque variation by low speed are slightly higher then in case of nominal speed [4][5].

B. Control Procedure

The design of sliding mode system consists generally of two procedures: design of the switching surface and design of the sliding mode controller [5]. The switching surface is designed to obtain a design performance for the system output variables. In VSS control, the goal is to keep the system motion on the manifold S, which is defined as:

$$S = \left\{ \boldsymbol{y} : \sigma(\boldsymbol{y},t) = G \boldsymbol{y} \right\} = \boldsymbol{0}; \quad \sigma = \boldsymbol{y}^d - \boldsymbol{y}, \quad (7)$$

where $\boldsymbol{y}^d, \boldsymbol{y}$ are state variables of desired and estimated value and σ is control error.

The sliding mode control [5] should be chosen such that the candidate Lyapunov function satisfies the Lyapunov stability criteria. This can be assured for

$$V = \sigma^T \sigma / 2 > 0 \quad \text{and} \quad \dot{V} = \sigma^T \dot{\sigma} < 0, \quad \dot{\sigma} = \Phi(\sigma). \quad (8)$$

There can be many different ways of selecting function $\Phi(\sigma)$. For the application in electrical drive control two particular forms may be of interest:

$$\dot{V} = \sigma^T \Phi(\sigma) < 0 \Rightarrow \begin{cases} \Phi(\sigma) = -\Gamma sign(\sigma) \\ \Phi(\sigma) = -D\sigma; D > 0 \end{cases}, \quad (9)$$

where Γ is diagonal or matrix with predominant diagonal, $sign(\sigma) = \left[sign(\sigma)_1, \cdots, sign(\sigma_m) \right]^T$ and D is positive definite matrix.

If vector function $\Phi(\sigma)$ is selected to be $\Phi(\sigma) = -\Gamma sign(\sigma)$ then resulting control will be discontinuous and manifold (7) will be reached in finite time. If $\Phi(\sigma)$ is selected as $\Phi(\sigma) = -D\sigma; D > 0$ then resulting control is expected to be continuous and sliding mode manifold will be reached in infinite time. The asymptotic stability of the solution $\sigma = \sigma(x^r, x) = 0$ will be guarantied in this case.

Both solutions will be applied to the electrical motor control. Implementing multiple loop control of a drive, most of computational time is denoted to the inner current control loop calculation and generation of space vectors for proper drive current signal. For this time critical operation of current control, the FPGA implementation for new event-driven current controlled modulator (EDCCM) based on discontinuous control $\Phi(\sigma) = -\Gamma sign(\sigma)$ of current error will be used. Continuous sliding mode control in form $\Phi(\sigma) = -D\sigma; D > 0$ will be implemented in rotor flux observer. The continuous control function will then be

$$\boldsymbol{u}(t) = \boldsymbol{u}_{equ} + \left(GB \right)^{-1} \dot{\sigma}, \quad (10)$$

where $\boldsymbol{u}_{equ}$ is continuous function [5]. For implementation the approximation of error function $\sigma(x^r, x)$ will be written in discrete time form after applying Euler's approximation

$$\frac{\sigma((k+1)T_s) - \sigma(kT_s)}{T_s} = GB \left(\boldsymbol{u}_{eq}(kT_s) - \boldsymbol{u}(kT_s) \right) \quad (11)$$

Here T_s is the sampling time and $k = Z^+$. By discretizing (10) and combining it with (11) one can eliminate the $\boldsymbol{u}_{eq}(kT_s)$ from (10) and arrive to the following discrete time version of the control (10)

$$\boldsymbol{u}_k = \boldsymbol{u}_{k-1} + \left(GBT_s \right)^{-1} \left(\left(DT_s + I \right) \sigma_k - \sigma_{k-1} \right). \quad (12)$$

Proposed algorithm ensures the sliding mode existence in manifold (7) and thus ensures the robustness of the closed loop system behavior against external disturbance and the parameters' changes. The control algorithm (12) has a feedforward term expressed by $\boldsymbol{u}_{k-1}$ and feedback term determined with control error dynamics and will be used in speed controller of IM.

III. Proposed VSC Torque and Flux Control Scheme

A. An Event Driven Induction Motor Current Control

Consider an IM together with the voltage source inverter (Fig. 1). Based on the transistor switch pattern, instantaneous control input is determined and thus a distinctive structure of the IM as dynamical system is determined. Applying proper switching among possible values, the system can be forced to follow desired transitions among possible structures and thus exhibit

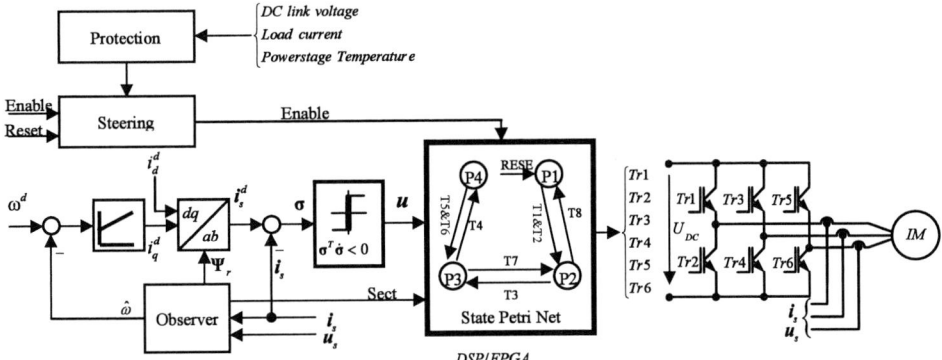

Fig. 1. Block diagram of event-driven sliding mode control of IM.

desired motion [7]. This allows application of the discrete-event driven approach [8], [9] to the IM control system. Each transistor switch pattern is considered as *discrete state* of the system and the change of the transistor switch pattern is considered as a *transition* of the system among discrete states. The transition of the discrete system among discrete states can be considered as occurrence of an *event*. To control the transitions of the system among the discrete states, some additional *conditions* are introduced. The switching matrix plays the role of switching elements determining the power exchange between energy storage elements, introducing change in the structure of the system and has making design in the framework of switching control a natural choice. The aim of this paper is to present an application of switching control in switching power inverters which is embedded in FPGAs environment.

Considering the drive current error as conditions for the transitions, a *discontinuous current control* is achieved which is fast, robust and simple for implementation. To achieve safe and manageable drive operation, monitoring and protection functions should be included. They are event-driven inherently, as they react on the change of logic conditions. Space vector representation of the inverter output voltages v_i (i=0,..,7) actual and in stationary frames of coordinates, are depicted in Fig. 2. The inverter output voltage vectors are stationary while the stator voltage vector is rotating with the stator frequency. Six active switching vectors of the three phase transistor inverter result in six output voltage vectors denoted $v_1 ... v_6$ and the zero vector denoted v_7 or v_0 depending on the connection of the switches. According to the signs of the projection of the stator voltage Us on the phase voltages ports u_{s1}, u_{s2} and u_{s3}, the phase plane is divided into six sectors denoted Sect 1 ... Sect 6. When the stator voltage vector is in particular sector a subset of inverter vectors is selected for the realization of the control. In Fig. 2, the stator voltage space vector u_s is in sector 1. In this sector voltage vectors v_0, v_7, v_2 and v_6 are selected for the IM current

control. v_0, v_7 are two zero vectors, while v_2, v_6 are two nearest adjacent live output voltage vectors to this . With the use of the discrete event system theory, four output voltage vectors v_0, v_7, v_2 and v_6 are recognized as discrete states of the system.

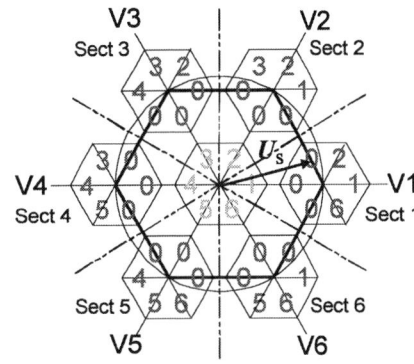

Fig. 2. Stator voltage V_s sector allocation.

The design of the current control system involves the selection of the transition between system structures such that the desired tracking of the motor current is obtained. The transition involves the selection of the appropriate output voltage vectors and thus it is very suitable for the application of the sliding mode framework. By defining current loop control error as $\sigma_i = i_s^d - i_s$ where i_s^d, i_s are desired and actual stator current of motor. The selection of the voltage vector that will ensure the stability of the current tracking can be realized by analyzing the derivative of the candidate Lyapunov function.

$$V = \frac{1}{2}\sigma^T\sigma = \frac{1}{2}\left(i_s^d - i_s\right)^T\left(i_s^d - i_s\right), \qquad (13)$$

Derivatives of current control error may be expressed as

$$\frac{d}{dt}\left(i_s^d - i_s\right) = \frac{d}{dt}i_s^d - \frac{1}{L_s}\left(u_s - R_s i_s - e_r\right), \qquad (14)$$

where u_s is voltage control input, $R_s i_s$ is resistive voltage drop and e_r is EMF of the motor. Rewriting (16) in the flux oriented frame of references it is easy to prove that the selection of the control inputs as discontinuous control having sign opposite to the sign of the corresponding current error guaranty the tracking of the desired current. The actual implementation of such algorithm requires that one determine the changes of the ON OFF state of the inverter switches. The mapping of the control from the rotor flux vector oriented frame of references to the stationary frame of references may be done by projecting the errors in the current loop to the orts of the phase vectors as depicted by the following relations:

$$S_R = \left(1 - \text{sign}(A)\right)/2, \ S_S = \left(1 - \text{sign}(B)\right)/2,$$
$$S_T = \left(1 - \text{sign}(C)\right)/2 \tag{15}$$

where

$$A = \left(i_{sa}^d - i_{sa}\right)$$
$$B = -\left(i_{sa}^d - i_{sa}\right)/2 - \sqrt{3}\left(i_{sb}^d - i_{sb}\right)/2, \tag{16}$$
$$C = -\left(i_{sa}^d - i_{sa}\right)/2 + \sqrt{3}\left(i_{sb}^d - i_{sb}\right)/2$$

Notice that if S_R, S_S, S_T equal to zero simultaneously, no current is delivered to the motor.

B. Petri Net Discrete-event Based Controller

Events represent allowed transition among the discrete states i.e. allowed switching. The structure of the proposed strategy is represented by Petri Net graph (Fig. 3) [10],[11].

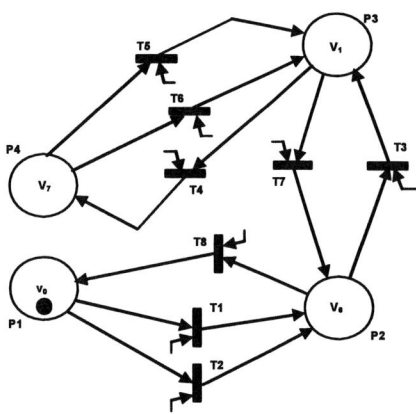

Fig. 3. PN-graph of the switching sequence in Sector 1.

The PN-graph from Fig. 3 can be described in the matrix based recursive form. The proposed design has inputs u, events x, discrete states m and outputs y denoted by logical vectors. m_o denotes initial discrete state. Their components take values 0 and 1, where 1 means that the appropriate input is set, an event has occurred, the state is active or output is set, while 0 means the opposite. The structure of the DES, describing the relations among the particular variables u, x, m and y is denoted using

Boolean matrices C, S, F and R. The recursive, matrix-based formalism for description of event-driven systems and algorithms is illustrated in Fig. 4 and the corresponding equations are denoted by:

$$u = \begin{bmatrix} S_R \\ S_S \\ S_T \\ \text{sign}(u_{s1}) \\ \text{sign}(u_{s2}) \\ \text{sign}(u_{s3}) \end{bmatrix}, \ x = \begin{bmatrix} T_1 \\ T_2 \\ T_3 \\ T_4 \\ T_5 \\ T_6 \\ T_7 \\ T_8 \end{bmatrix}, \ m = \begin{bmatrix} P_1 \\ P_2 \\ P_3 \\ P_4 \end{bmatrix}, \ y = \begin{bmatrix} Tr_1 \\ Tr_2 \\ Tr_3 \\ Tr_4 \\ Tr_5 \\ Tr_6 \end{bmatrix}, \tag{17}$$

$$m_0 = \begin{bmatrix} 1 \\ 0 \\ 0 \\ 0 \end{bmatrix}$$

$$C = \begin{array}{c} \\ T_1 \\ T_2 \\ T_3 \\ T_4 \\ T_5 \\ T_6 \\ T_7 \\ T_8 \end{array}\begin{bmatrix} u_1 & u_2 & u_3 & u_4 & u_5 & u_6 \\ 1 & x & x & 1 & 0 & 0 \\ x & x & 0 & 1 & 0 & 0 \\ x & 1 & 0 & 1 & 0 & 0 \\ 0 & x & 1 & 1 & 0 & 0 \\ 1 & x & x & 1 & 0 & 0 \\ x & x & 0 & 1 & 0 & 0 \\ 1 & 0 & x & 1 & 0 & 0 \\ 0 & x & 1 & 1 & 0 & 0 \end{bmatrix}, \ S = \begin{array}{c} \\ T_1 \\ T_2 \\ T_3 \\ T_4 \\ T_5 \\ T_6 \\ T_7 \\ T_8 \end{array}\begin{bmatrix} P_1 & P_2 & P_3 & P_4 \\ 0 & 1 & 0 & 0 \\ 0 & 1 & 0 & 0 \\ 0 & 0 & 1 & 0 \\ 0 & 0 & 0 & 1 \\ 0 & 0 & 1 & 0 \\ 0 & 0 & 1 & 0 \\ 0 & 1 & 0 & 0 \\ 1 & 0 & 1 & 0 \end{bmatrix}, \tag{18}$$

$$F = \begin{array}{c} \\ T_1 \\ T_2 \\ T_3 \\ T_4 \\ T_5 \\ T_6 \\ T_7 \\ T_8 \end{array}\begin{bmatrix} P_1 & P_2 & P_3 & P_4 \\ 1 & 0 & 0 & 0 \\ 1 & 0 & 0 & 0 \\ 0 & 1 & 0 & 0 \\ 0 & 0 & 1 & 0 \\ 0 & 0 & 0 & 1 \\ 0 & 0 & 0 & 1 \\ 0 & 0 & 1 & 0 \\ 0 & 1 & 0 & 0 \end{bmatrix}, \ R = \begin{array}{c} \\ Tr_1 \\ Tr_2 \\ Tr_3 \\ Tr_4 \\ Tr_5 \\ Tr_6 \end{array}\begin{bmatrix} P_1 & P_2 & P_3 & P_4 \\ 0 & 1 & 1 & 1 \\ 0 & 1 & 0 & 1 \\ 0 & 0 & 1 & 1 \\ 1 & 0 & 0 & 0 \\ 1 & 0 & 1 & 0 \\ 1 & 1 & 0 & 0 \end{bmatrix}$$

$$u_c(k) = \overline{C \times \overline{u}(k)}, \tag{19}$$
$$x(k) = \overline{F \times \overline{m}(k)} \, \& \, u_c, \tag{20}$$
$$m(k+1) = m(k) + M^T x(k), \tag{21}$$
$$y(k) = R \times m(k). \tag{22}$$

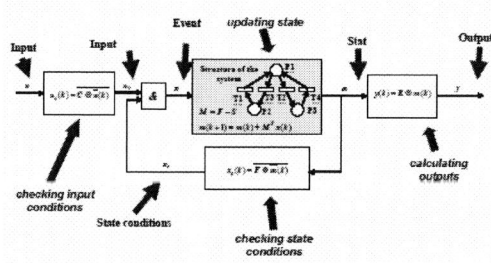

Fig. 4. Recursive description of event driven systems.

Inputs u denotes switching conditions S_R, S_S, S_T of the Lyapunov stability criteria and signs of the drive stator phase voltages u_{s1}, u_{s2} and u_{s3}, where sign(x) = 1 denotes positive and sign(x) = 0 negative values. Events x denote conditions for occurrence of transitions among discrete states. Discrete states m correspond to the appropriate output voltage vectors and outputs y denote the transistor gate signals for the generation of corresponding output voltage vectors (Fig. 4).

Proposed algorithm ensures the current tracking. The components of the reference current are generated by the outer loop controllers (flux and torque/velocity controllers) and the reference current vector can be then written as:

$$i_s^d = \frac{\left|\hat{\mathbf{\Psi}}_r\right|}{L_m} + j\frac{3}{2p}\frac{L_r}{L_m}\frac{T_e^d}{\left|\hat{\mathbf{\Psi}}_r\right|}. \tag{23}$$

IV. ROTOR FLUX OBSERVER

This rotor flux observer, is based on the stator equation, where the derivative of the estimated stator flux is calculated from measured stator voltage and current. The observer equation (24) represents the first order vectorial differential equation. The stator voltage u_s and current i_s serve as control input to the estimated stator flux $\hat{\mathbf{\Psi}}_s$. The measured value of the stator voltage is used instead of the commonly used reference voltage, in order to avoid voltage error influence due to power-stage non-linear behavior:

$$\frac{d}{dt}\hat{\mathbf{\Psi}}_s = \hat{u}_s - \hat{R}_s i_s + \hat{u}_k, \quad \hat{u}_k = \hat{\mathbf{\Psi}}_r\left(u_m + ju_p\right). \tag{24}$$

Non-modeled dynamic is set as a remaining signal $\hat{u}_k$, calculated from the magnitude error of the rotor flux $\left|\Delta\mathbf{\Psi}_r\right|$.

The stator parameters of the IM appear in the rotor flux observer; i.e. stator resistance $\hat{R}_s$ and stator inductance $\hat{L}_s$. The influence of the stator inductance variation is small, but stator resistance changes significantly during the operation. The variation of the stator resistance $\Delta\hat{R}_s$ impacts on the estimated rotor flux and, due to this, on the variation of the IM's torque.

The influence of the rotor flux variation's magnitude is compensated by introducing a non-linear magnitude and orientation feedback compensator in the observer. The switching function of the VSC flux magnitude controller C_m is set to the error between the reference and estimated rotor flux magnitude:

$$\frac{d\sigma_m}{dt} = -D_m\sigma_m = 0, \quad \sigma_m = \mathbf{\Psi}_r^d - \left|\hat{\mathbf{\Psi}}_r\right|. \tag{25}$$

The discrete part of the resulting unknown offset voltage u_m is:

$$C_m : u_m(k) = \frac{K_m}{T_s}\left(\left(1 + T_s D_m\right)\sigma_m(k) - \sigma_m(k-1)\right). \tag{26}$$

The variation of the stator resistance $\Delta\hat{R}_s$ impacts to the

torque variation of the IM, expressed by the variation of the rotor flux and torque:

$$u_p = \frac{1}{J}\int_0^t \Delta T_e\, d\tau, \quad \Delta T_e = \sigma_m\frac{T_e^d}{\mathbf{\Psi}_r^d}. \tag{27}$$

The switching function of the VSC orientation controller C_p takes into account the torque variation:

$$\frac{d\sigma_p}{dt} = -D_p\sigma_p = 0, \quad \sigma_p = \Delta T_e\, \text{sign}\left(\hat{\omega}T_e^d\right). \tag{28}$$

The source of the connection between the torque error and the rotor flux error is the influence of the stator resistance error, when the torque is applied. The discrete form of the orientation correction signal becomes:

$$C_p : u_p(k) = u_p(k-1) + \frac{K_p}{T_s}\left(\left(1 + T_s D_p\right)\sigma_p(k) - \sigma_p(k-1)\right) \tag{29}$$

Correction input signals, influence the magnitude and orientation of the magnetic flux error, make the proposed observer robust to the parameter uncertainties variations. The resulting diagram of both, an amplitude and phase-controlled, variable frequency two phase oscillator is shown in Fig. 5, which is suitable for providing the estimated stator and rotor fluxes of the IM.

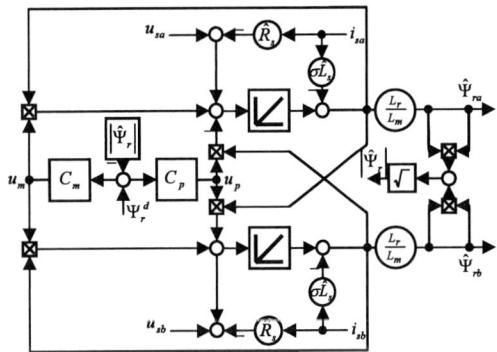

Fig. 5. Block diagram of rotor flux observer.

The estimated synchronous speed will be now expressed with cross product:

$$\hat{\omega}_s = \frac{1}{\left|\hat{\mathbf{\Psi}}_s\right|^2}\frac{d\hat{\mathbf{\Psi}}_s}{dt}\times\hat{\mathbf{\Psi}}_s. \tag{30}$$

The estimated speed is computed entirely from the estimated torque, rotor flux and it's time derivative:

$$\hat{\omega}_r = \frac{\hat{\mathbf{\Psi}}_r\times\dfrac{d\hat{\mathbf{\Psi}}_r}{dt}}{p\left|\hat{\mathbf{\Psi}}_r\right|^2} - \frac{2R_r}{3p^2}\frac{\hat{T}_e}{\left|\hat{\mathbf{\Psi}}_r\right|^2}. \tag{31}$$

The resulting estimated speed is subjected to high noise levels due to the derivative term that can be reduced by employing low pass filters.

V. RESULTS

The simulation program in Matlab/Simulink environment of the sensorless sliding mode IM control was developed. The sliding mode torque and flux algorithm was implemented on the in house developed DSP/FPGA board [11]. The DSP/FPGA board contains Texas Instruments TMS 320C32 digital signal processor and Xilinx Spartan family field programmable gate array. DSP serves for A/D conversion and generating of the reference current. Replacing usual sequential calculation of algorithms on the DSP by parallel executable FPGA hardware, increases the calculation speed. A/D conversion is the most critical operation regarding time and takes 5 μs. According to the fact, that A/D conversion takes most of the calculation time, switching frequencies up to 200 kHz are theoretically possible. The DSP remains for the implementation of less critical speed control and observer algorithm. Experiments confirm the applicability of the proposed algorithm as well as achieving of higher switching frequencies in current control task.

Fig. 6, Fig. 7 and Fig. 8 shows the transients response of phase motor currents, active torque of IM during low speed operation and motor speed .

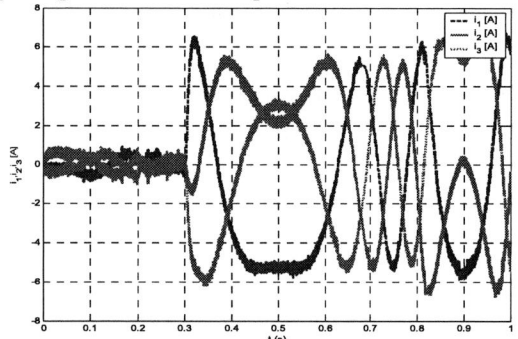

Fig. 6. Phase current during low speed operation

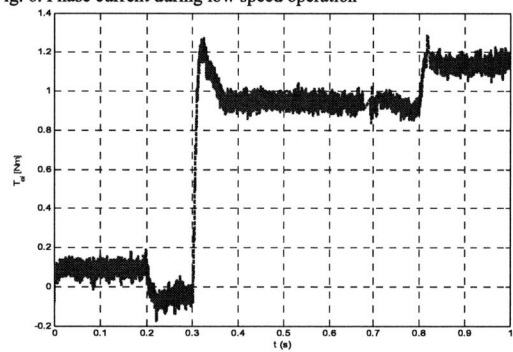

Fig. 7. Calculated produced torque during low speed operation.

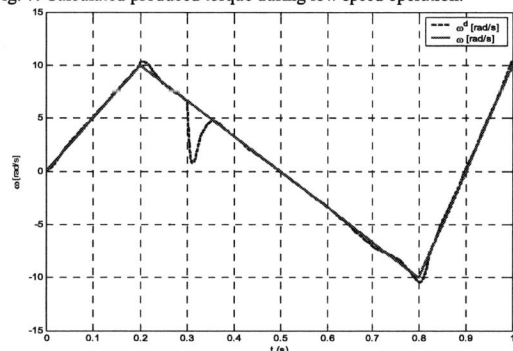

Fig. 8. Speed reference tracking at low speed operation.

VI. CONCLUSION

A full digital controller for switching converters has been proposed. The most important difference from previously proposed digital controllers is that it is based on specific FPGA hardware instead of the common DSP solutions. The main advantage of this method is that the all the logic is executed continuously and simultaneously (concurrent operation) and new high speed algorithms can be used in this way. However, complex arithmetic operations should be avoided whenever possible, because they need many resources. The other main characteristic of the switching controller is: simplicity.

Implementing the control algorithm in a hardware description language (HDL) allows high flexibility and technology independence. The same controller can be directly synthesized into any other FPGA or even in an ASSIC, or it can also be added to other logic blocks forming a more complex multi-task system in a single chip. Solutions based on specific hardware, that allows high concurrency, are suitable to be used in power electronics and motion control applications.

REFERENCES

[1] Z.H. Akpolat, G.M. Asher, J.C. Clare, "Dynamic Emulation of Mechanical Loads Using a Vector Controlled Induction Motor-Generator Set", *IEEE Trans. on Industrial Electronics*, vol. 46, no. 2, 1999, pp. 370-379.

[2] M. Rodič, K. Jezernik, M. Trlep, "A Feedforward Approach to the Dynamic Emulation of Mechanical Loads", Proceedings of the 35th Annual IEEE Power Electronics Specialists Conference (PESC'04), pp. 4595-4601, 2004.

[3] V.I. Utkin, *Sliding modes in control and optimization*, Springer Verlag, Berlin, 1992.

[4] G.S. Buja, and M.P. Kazmierkowski, "Direct Torque Torque Control of PWM Inverter-Fed AC Motors-A Survey". *IEEE Transactions on Industrial Electronics*, vol. 51, no. 4, 2004, pp., 744-757.

[5] W. Leonhard, *Control of Electric Drives*, Springer Verlag, Berlin, Germany, 3rd edition, 2001.

[6] V.I. Utkin, "Sliding Mode Control Design Principles and Applications to Electric Drives", *IEEE Transactions on Industrial Electronics*, vol. 40, 1993, pp. 23-36.

[7] K. Zhou, and D. Wang, "Relationship Between Space-vector Modulation and Three-phase Carrier-based PWM: a Comprehensive analysis," *IEEE Trans. Industrial Electronics*, vol. 49, no. 1, pp. 186-196, Feb. 2002.

[8] A. Tilli, and A. Tonielli, "Sequential Design of Hysteresis Current Controller for Three-phase Inverter," *IEEE Trans. Industrial Electronics*, vol. 45, no. 5, pp. 771-781, Oct. 1998.

[9] A. Polic, and K. Jezernik, "Matrix based Event-driven Approach for Current Control Design of VSI," in Proc. of Int. Symp. Industrial Electronics - ISIE, Dubrovnik, 2005, pp.917-922.

[10] R. Zurawski, and M.C. Zhou, "Petri Nets and Industrial Applications: A Tutorial," *IEEE Trans. Industrial Electronics*, vol. 41, no.6, pp. 567-5830, Dec. 1994.

[11] A. Polic, and K. Jezernik, "Closed-loop Matrix based Model of Discrete Event Systems for Machine Logic Control Design," *IEEE Trans. Industrial Informatics*, vol. 1, no. 1, pp. 39-46, Feb. 2005.

Application of the induction motor torque - observer to the control of turbo - machines

Andrzej Dębowski, Daniel Lewandowski

Technical University of Lodz/Institute of Automation Control, Łódź, Poland

e-mail: *andrzej.debowski@p.lodz.pl, daniel.lewandowski@p.lodz.pl*

Abstract—This paper describes the results of the experiment with an application of the outer torque observer to the radial fan driven by induction motor fed from the voltage inverter. The observer is designed to be an additional stand-alone device connected between the induction motor and the power supply, separated from the inner signals of the inverter controller. The inputs for the observer algorithm are the three-phase currents and voltages measured in the conductors connected to the motor terminals. The output signals of this observer estimating the stator flux and motor electro-magnetic torque are used to the compressor supervisory control.

Index Terms—induction motor drive, induction motor torque-observer, turbo - compressor anti-surge control

I. INTRODUCTION

The flow machines are commonly used in many fields of the industry. An example of such a machine is the radial compressor which can be used to compress and transport various kinds of gases. The flow machine can be driven by steam turbine or induction motor. To keep the working point of the flow machine in the safe region and to prevent undesirable effects (like the compressor surge), is very useful to apply the driving torque as the control variable [1], [2]. In existing high-power compression systems access to the instantaneous information about the torque is practically possible by the installation of the appropriate strain gauge sensors on the shaft coupling. However the maintenance of such a torque sensor is very inconvenient. The installation of this torque sensor can be omitted in the machine driven by the steam turbine by calculating the turbine power [3].

In this paper it is proposed, that in case when the compressor is driven by induction motor supplied from the any voltage inverter, the outer torque observer should be added to estimate the driving torque. This observer calculates the motor internal state from measured electrical quantities like three-phases currents and voltages supplying the motor and can be integrated with the inverter supervisory control. In many cases, especially in high power compression systems when the voltage inverters have simple construction and any vector state control, the instantaneous electromagnetic torque, produced by the

This paper describes a part of research work supported by the Polish Ministry of Science and Higher Education, grant No. N510 3063 33.

motor is unknown and not accessible for the automatic supervisory control system. This comment is obviously true when the motor is supplied directly from the power net. In both cases the information about the motor driving torque is very useful for the accurate compressor control. In comparison to the strain gauge sensor the observer can also calculate the stator flux amplitude, the stator current amplitude and angle between ψ_s and i_s. These signals allow to achieve the induction motor drive quality improvements also in case if the simple U/f = const control scheme is used.

Coming up to the expectation concerning the increased precision in controlling parameters of compression in many industrial applications, a new control algorithm is proposed by one of the authors [2]. The algorithm was tested during the simulations and it consists of two parts. First part of this algorithm is responsible for realisation the desired control task when the machine state remains in the safe region of the working area. The second part of the control algorithm is responsible for protecting the compressor - its goal is to keep the machine state in the safe region.

Proposed algorithm is based on approximation of static characteristic of compression ratio π for the compressor or static pressure p for the fans as the function of mass flow q and shaft speed ω. The shape of characteristic of static pressure in the working area can be approximated by quadratic function defined by coefficient matrix $\mathbf{A}$ of dimension 3 x 3 and written in quadric form transcription (1) where $\mathbf{Q}$ and $\mathbf{\Omega}$ are given by (2).

$$p(q, \omega) = \mathbf{Q}^T \mathbf{A} \mathbf{\Omega} \qquad (1)$$

$$\mathbf{Q} = \mathbf{E}(q) \qquad \mathbf{\Omega} = \mathbf{E}(\omega)$$
$$\mathbf{E}(x) = \begin{bmatrix} x^2 \\ x \\ 1 \end{bmatrix} \qquad (2)$$

The static pressure characteristic of the fan is shown on the figure 1. The formula (1) is very convenient in calculation of the pressure gradient which can be written in form (3).

$$\nabla p(q, \omega) = \begin{bmatrix} \frac{d\mathbf{Q}^T}{dq} \mathbf{A} \mathbf{\Omega} \\ \mathbf{Q}^T \mathbf{A} \frac{d\mathbf{\Omega}}{d\omega} \end{bmatrix} \qquad (3)$$

978-1-4244-1741-4/08/$25.00 ©2008 IEEE

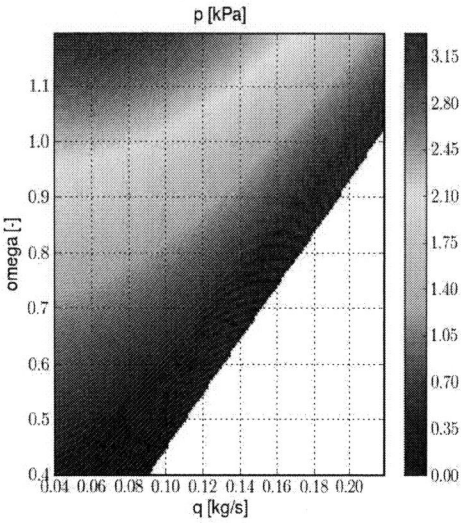

Fig. 1: The fan $p(q, \omega)$ characteristic

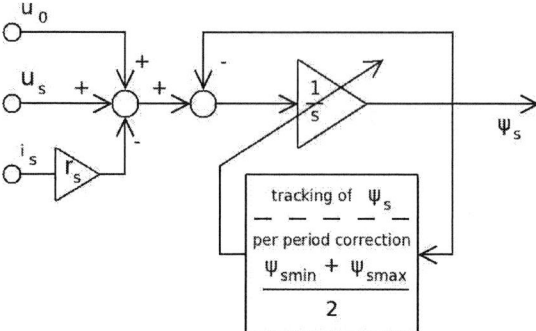

Fig. 2: Block scheme of the modified voltage model.

The proposed algorithm takes into consideration the "surge effect" [4]. This name describes a complex phenomenon which can occur during usage of the turbocompressor. This phenomenon occurs in situation when the total torque on impeller blades is less then pressure between outlet and inlet. This situation enables the fluid to flow backwards for a moment what causes rapid fluctuations in speed and power on the shaft. These fluctuation can even lead to the compressors damage. The surge phenomenon occurs when machine state approaches the maximum of its working characteristic. Because the efficiency of the system is highest in this area then it is obvious that machine state should be kept as close to the maximum as it is possible.

II. TORQUE OBSERVER CONSTRUCTION

It is well-known that the electromagnetic torque t_e depends upon the stator magnetic flux vector ψ_s and the stator current vector i_s, and can be calculated from the equation (4).

$$t_e = \psi_s \times i_s \qquad (4)$$

The i_s can be easily calculated by Clarke transformation on the basis of phase currents measurement. However the stator flux vector cannot be directly measured and has to be estimated. One of the method of solving the problem is to use a reduced order observer based on voltage and current flux model equations [5], [6], [7]. However this requires significant computational power and parameter identification of the induction motor. The authors look for a simpler but still useful flux estimation model:

- integration of current model
- integration of voltage model (5)

It is well known that current equation provide valid results only for low speed of motor [8]. With increasing

speed the error of current model disqualifies this solution. On the contrary the voltage equation is valid in higher ranges of motor speed which is more suitable in compressor system. Some authors proposed use of both models switched by speed signal [8]. Interesting results were received with application of voltage model with inverse current model [9]. As it is already mentioned, application of current model is bounded with more number of induction motor parameters. This enforces full identification and leads to problem with parameter variation during normal control system operation. The current model also requires more computational resources. However the flow machines works with high rotary speeds and the accuracy in these conditions is required. This allows for application of voltage model with stator resistance r_s as only one parameter. The equation (5) is also independent from motor speed signal or its estimation.

$$\omega_b^{-1} \dot{\psi}_s = u_s - r_s \cdot i_s \qquad (5)$$

On the other hand the voltage model requires integration in open loop. This introduces the problems with dc drift and integrator saturation. Moreover measurement of u_s is troublesome when motor is fed by voltage inverter.

III. INTEGRATION IN OPEN LOOP

The main expectation for the outer observer is that it will be working without any knowledge of control algorithm or any control signal inside inverter. The only signals available for the observer are u_s and i_s calculated on the basis of measurements of phase voltages and phase currents. Rapid changes of u_s requires integration of equation (5) with as small time step as possible. The block scheme of proposed algorithm is shown on figure 2.

The voltage equation can be properly solve only when u_s and i_s have the average value equal zero. Measurement process introduces u_0 vector into right side of voltage model and transforms equation (5) into (6).

$$\omega_b^{-1} \dot{\psi}_s = u_s - r_s \cdot i_s + u_0 \qquad (6)$$

In case where input signal has nonzero average value the pure integrator cannot be implemented. The use of low pass filter (LPF) instead of pure integrator was proposed

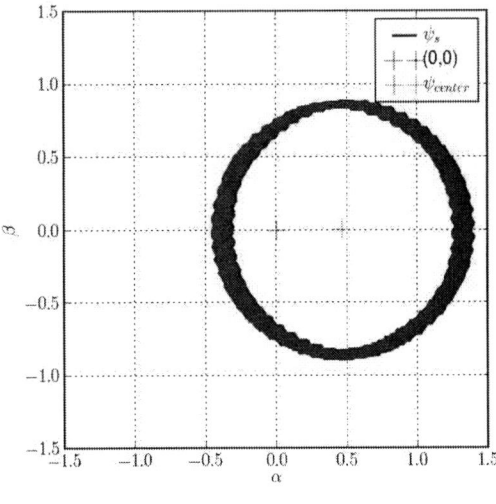

(a) Rotation of ψ_s

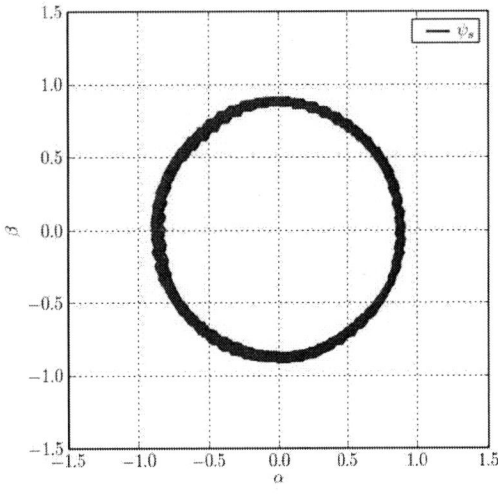

(a) Rotation of ψ_s

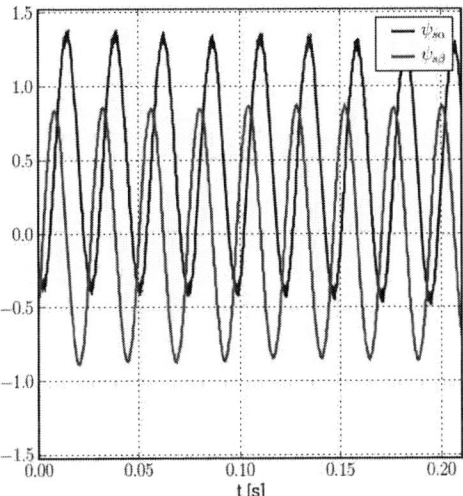

(b) Components of ψ_s

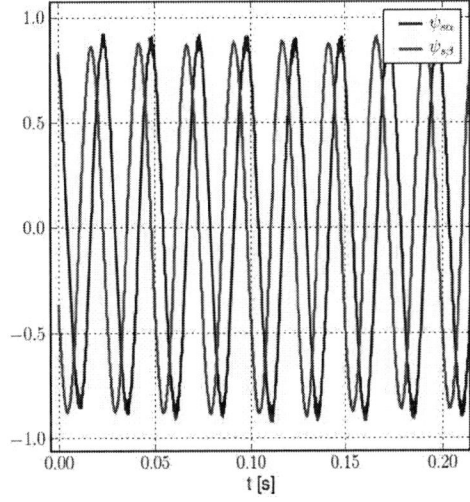

(b) Components of ψ_s

Fig. 3: Experimental results of LPF without correction algorithm. Rotation of the ψ_s is shifted right to the beginning of coordinates system. The $u_0 \neq 0$ component is introduced by measurement process

Fig. 4: Experimental results for LPF with correction algorithm. Rotation of ψ_s in steady state is achieved with center in beginning of coordinates system.

in [10]. The output of LPF contains the ψ_s with offset vector ψ_0 as it is shown of figure 3a. The ψ_0 which can be approximated by (7) and it is mean to be constant vector determined by starting conditions and u_0.

$$\psi_{center} = \frac{1}{2}\left(\psi_{max} + \psi_{min}\right) \qquad (7)$$

To eliminate this error there was proposed regulator which corrects the integrator of LPF once per rotation by ψ_{center} vector which can be calculated according to

equation (7). The ψ_{max} and ψ_{min} are minimum and maximum value of ψ_s in last period. The correction of the integrator is performed once per rotation which leads to minimization of ψ_{center} in long term and ensures zero average value of ψ_s in steady state. Figure 5 presents centering process of LPF output. The properly corrected integration process can observed as circular rotation of ψ_s in steady state as shown on figure 4a. Circular rotation is equivalent to sinusoidal waveform shifted in phase by $\frac{\pi}{2}$ of ψ_s coordinates - figure 4b.

Similar correction method was proposed in [11] but it

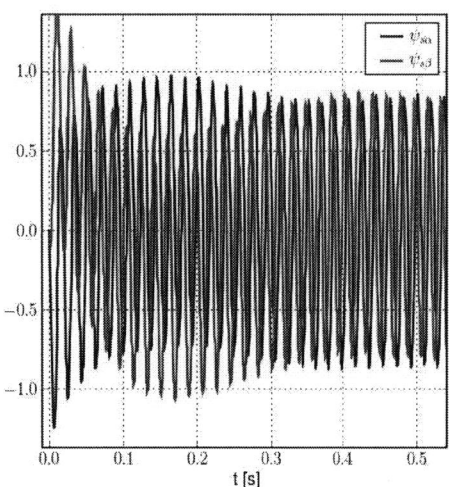

Fig. 5: Experimental results with LPF with proposed correction algorithm. Convergence of ψ_{center} toward zero.

was used in inner observer compensation block.

IV. EXPERIMENTS

The schematic of the stand is presented on figure 6. Experiments were conducted with laboratory stand based on the radial fan driven by 0.75kW induction motor. The fan was put in the tank which outlet flow can be regulated by throttle. The throttle simulated different pneumatic load of the installation and allowed test the plant in various operating conditions. The induction motor was fed by 5kW voltage inverter with implemented indirect flux and torque control algorithm (IDTFC) [12]. The only measurements available for the torque observer were phase voltages and phase currents.

The observer was implemented on the LPC2214 ARM processor with external simultaneous, 8-channel A/D converter. The sampling rate of proposed algorithm was 128kHz. The observer and the inverter were communicating by RS232 link with dedicated communication module which provide connection with PC computer. The PC computer provide higher level control of fan system with data acquisition. Because high bandwich was required, the link was established by Ethernet Network. The latencies introduced by general network in point to point configuration are about milliseconds and they can be neglected in comparison of fan time constant on the rate of single seconds [13].

The exemplary experiment results with outer observer are shown on figure 7. The estimated torque $\tilde{t}$ reflects dynamic changes of setvalue torque t. However the static accuracy is better when ω is close to 0.5. With increasing speed the $\tilde{t}$ varies more from t. This is the effect of inaccuracy in measurement of phase voltages.

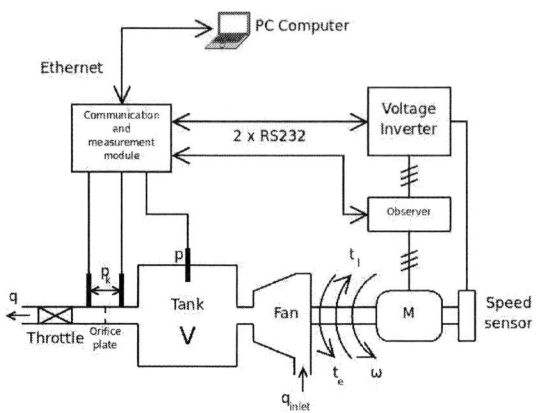

Fig. 6: Schematic of laboratory stand

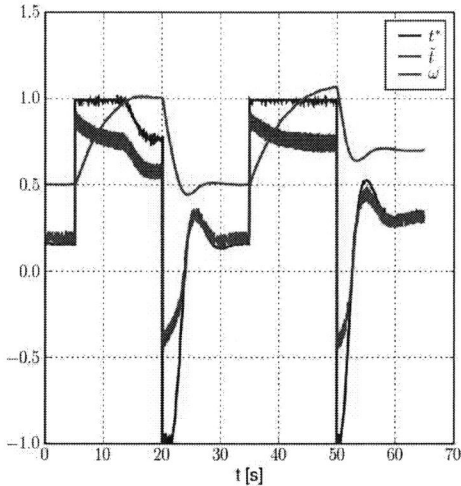

Fig. 7: Comparison of the outer observer output $\tilde{t}$ with the inverter torque setvalue t^*. The speed setpoint (not shown) changes (1.0 at 5s, 0.5 at 20s, 1.1 at 35s, 0.7 at 50s), the drive speed ω is controlled by PI regulator

V. CONCLUSIONS

The electromagnetic torque estimation is well known problem with multiple solutions. The most of the researchers explores problem of the inner observer and estimation of ψ_r. It is related to popularity of field oriented control methods of the induction motor. However switching frequency of the high power and high voltage inverters is limited to houndreds of hertzs. Of course this practically excludes current controllers and field oriented control method. On the other hand simple U/f = const control method is commonly used.

The authors proposed the outer observer to improve torque control of existing U/f inverters. The voltage model based on equation (5) was implemented in general purpose microprocessor. Experimental results have shown

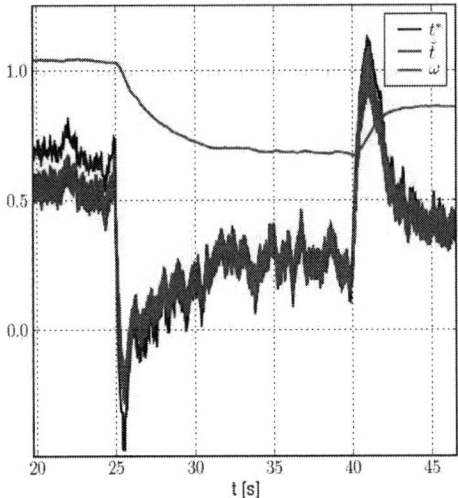

Fig. 8: Comparison of the outer observer output $\tilde{t}$ and the inverter torque setvalue t^* with setpoint $p = 1.0$kPa (not shown)

(figure 7) that estimated torque has significant error depend on speed. The main source of estimation error is inaccuracy in measurement of phase voltages in the voltage inverter fed environment. On the other hand the outer observer is able to recreate the dynamic changes of the electromagnetic torque - figure 8.

REFERENCES

[1] T. Gravdahl, O. Egeland, and S. Vatland, "Active surge control of centrifugal compressors using drive torque," in *Proc. of the 40th IEEE Conference on Decision and Control*, Orlando, 2001, pp. 1286 – 1291.

[2] D. Lewandowski, "Controlling of the machines with quadratic performance characteristic," in *Proceedings of the VII Domestic Conference "ELEKTROWNIE CIEPLNE. Eksploatacja – Modernizacje – Remonty"*, Słok k/Bełchatowa, 2005, pp. 193 – 200, in Polish.

[3] A. Dębowski, W. Błasiński, and A. Potapczyk, "Integrated control system of the turbo-compressor driven by the steam turbine," in *Proceedings of the XV Domestic Conference in Automatic Control*, vol. 3, Warszawa, 2005, pp. 263 – 268, in Polish.

[4] E. Tuliszko, *Compressors, blowers, fans*. Warszawa: WNT, 1976.

[5] T. Orłowska-Kowalska, P. Wojsznis, and C. Kowalski, "Comparative study of different flux estimators for sensorless induction motor drive," in *Proceedings of International Conference ELECTRIMACS'99*, vol. 2, 1999, pp. 2.51 – 2.56.

[6] Z. Krzemiński, "A new speed observer for control system of induction motor," in *Power Electronics and Drive Systems, 1999. PEDS '99. Proceedings of the IEEE 1999 International Conference on*, vol. 1, Hongkong, 1999, pp. 550–560.

[7] R. Nilsen and M. Kaźmierkowski, "Control theory and applications," in *IEE Proceedings D*, vol. 136, 2005, pp. 35 – 43.

[8] H. Tunia and M. Kaźmierkowski, *Automatyka napędu przekształtnikowego*. Warszawa: PWN, 1987.

[9] S. Kuhne and U. Riefenstahl, "A new torque calculator for ac induction motor drives that improves accuracy and dynamic behaviour," in *Industrial Electronics, 1999. ISIE '99. Proceedings of the IEEE International Symposium on Industrial Electronics ISIE '99*, vol. 2, 1999, pp. 498 – 503.

[10] J. Hu and B. Wu, "New integration algorithms for estimating motor flux over a wide speed range," in *Transactions of Power Electronics*, vol. 13, 1998, pp. 969 – 977.

[11] J. Holtz and J. Quan, "Drift- and parameter-compensated flux estimator for persistent zero-stator-frequency operation of sensorless-controlled induction motors," in *IEEE Transactions on Industry Applications*, vol. 44. IEEE Industry Applications Society, 2003.

[12] A. Dębowski, "Pośrednie sterowanie w napędzie elektrycznym przy wykorzystaniu stymulatora stanu," in *Scientific bulletin of Łódź Technical University*, vol. 552. Łódź: Wydawnictwo Politechniki Łódzkiej, 1991, in Polish.

[13] D. Lewandowski and M. Morawski, "Experiments with actuator's control via the general purpose networks," in *Proceesings of the XII International conference - System Modelling and Control SMC'2007*, 2007.

Observer of induction motor speed based on exact disturbance model

Zbigniew Krzemiński

Gdansk University of Technology/Deparment of Electrical and Control Engineering, Gdansk, Poland,
e-mail: *z.krzeminski@ely.pg.gda.pl*

Abstract—A simplified speed observer was proposed previously by the author to estimate the rotor speed needed in control system of a drive with induction motor. Small error of estimated rotor speed was observed but errors of the stator current and rotor flux vector components appear in dependence of rotor speed. A new structure of the speed observer is proposed now with exact disturbance model and gain coefficients depending on stator frequency. The new structure was derived from the previous one by analysis of dependencies appearing in differential equations. Explanation of observer stability is presented. Properties of the proposed observer are presented using simulations of full control system.

Keywords—Adjustable speed drive, induction motor, non-linear control, sensorless control.

I. INTRODUCTION

Control systems for the induction motor based on nonlinear transformation of variables and nonlinear feedback were proposed by the author more then 20 years ago [1] and further developed and extended in [2] and [3]. The idea of nonlinear transformation was used in [4 – 10]. At presence the control algorithm looks simple and may be technically realized using low cost floating point digital signal processors. If the rotor speed is measured, a Luenberger observer may be used to estimate remaining unmeasured variables, usually components of a rotor flux vector. Recently standard requirements for industrial drives with induction motors include sensorless speed control which means that no encoder is applied in the system. The rotor speed have to be estimated and many methods are proposed for this purpose such as adaptive observers, sliding mode based observers, model reference adaptive systems (MRAS) and Kalman filters. A new kind of observer based on estimation of disturbances defined as additional variables was proposed by the author [2] to estimate the rotor speed. Such speed observer was applied to the control system with nonlinear feedback and was a basis for high performance drive with induction motor. The sensorless system with speed observer was compared with systems based on adaptive observer and MRAS system and results were presented in [3]. The conclusion was that nonlinear feedback requires exact estimation of the rotor speed and application of the speed observer results in better drive performances than systems based on both other methods.

Although the speed observer presented in [2] acts with very low speed error, a few drawbacks result from its simple structure. Errors of the rotor flux and stator current components, limited speed range and instabilities near zero speed cause main problems. To overcame the above mentioned drawback a modified structure of the speed observer is proposed in this paper.

II. MODEL OF THE INDUCTION MOTOR

The vector model of the induction motor is expressed as differential equations for the stator current vector and the rotor flux vector of the following form:

$$\frac{d\mathbf{i_s}}{d\tau} = a_1\mathbf{i_s} + a_2\mathbf{\psi_r} - j\omega_a\mathbf{i_s} - ja_3\omega_r\mathbf{\psi_r} + a_4\mathbf{u_s}, \tag{1}$$

$$\frac{d\mathbf{\psi_r}}{d\tau} = a_5\mathbf{i_s} + a_6\mathbf{\psi_r} - j(\omega_a - \omega_r)\mathbf{\psi_r}, \tag{2}$$

$$\frac{d\omega_r}{d\tau} = \frac{1}{J}\left(\frac{L_m}{L_r}\text{Im}\left|\mathbf{\psi_r^*i_s}\right| - m_0\right), \tag{3}$$

where $\mathbf{u_s}$, $\mathbf{i_s}$, $\mathbf{\psi_r}$ are the stator voltage, stator current and rotor flux vectors, is m_0 the motor load, τ is the relative time, ω_r is the angular rotor velocity, ω_a is the angular velocity of the reference frame and

$$a_1 = -\frac{R_sL_r^2 + R_rL_m^2}{wL_r}\;;\;\; a_2 = \frac{R_rL_m}{wL_r}\;;\;\; a_3 = \frac{L_m}{w}\;;$$

$$a_4 = \frac{L_r}{w}\;;\;\; a_5 = \frac{R_rL_m}{L_r}\;;\;\; a_6 = -\frac{R_r}{L_r}\;;\;\; w = L_sL_r - L_m^2\;;$$

R_s, R_r are the stator and rotor resistances, L_s, L_r, L_m are the stator, rotor and mutual inductances.

All the variables and parameters used in the paper are expressed in p.u. system.

III. A SIMPLE SPEED OBSERVER FOR THE INDUCTION MOTOR

Estimation of a rotor speed in a simple speed observer was proposed in [11]. Properties of the speed observer were analyzed in [12] and many applications to sensorless control systems were described in [13], [3]. The structure of the speed observer was derived on a basis of the Luenberger observer for vector components determined in separate axes of an unmoving frame of references. The main idea was based on assumption that for low speed components of an electromotive force vector may be interpreted as small disturbances in two separated observers for current and flux vector components. Anyway the speed observer has a property that the rotor speed is exactly estimated in wide range.

The differential equations of the simple speed observer are as follows [2]:

$$\frac{d\hat{i}_{s\alpha}}{d\tau} = a_1\hat{i}_{s\alpha} + a_2\hat{\psi}_{r\alpha} + a_3\zeta_\beta + a_4u_{s\alpha} \tag{4}$$
$$+ k_3\left(k_1\left(i_{s\alpha} - \hat{i}_{s\alpha}\right) - \hat{\omega}_r\zeta_\alpha\right),$$

$$\frac{d\hat{i}_{s\beta}}{d\tau} = a_1\hat{i}_{s\beta} + a_2\hat{\psi}_{r\beta} - a_3\zeta_\alpha + a_4u_{s\beta} \tag{5}$$
$$+ k_3\left(k_1\left(i_{s\beta} - \hat{i}_{s\beta}\right) - \hat{\omega}_r\zeta_\beta\right),$$

$$\frac{d\hat{\psi}_{r\alpha}}{d\tau} = a_5\hat{i}_{s\alpha} + a_6\hat{\psi}_{r\alpha} - \zeta_\beta - k_2\left(\hat{\omega}_r\hat{\psi}_{r\beta} - \zeta_\beta\right), \tag{6}$$

$$\frac{d\hat{\psi}_{r\beta}}{d\tau} = a_5\hat{i}_{s\beta} + a_6\hat{\psi}_{r\beta} + \zeta_\alpha + k_2\left(\hat{\omega}_r\hat{\psi}_{r\alpha} - \zeta_\alpha\right), \tag{7}$$

$$\frac{d\zeta_\alpha}{d\tau} = k_1\left(i_{s\beta} - \hat{i}_{s\beta}\right), \tag{8}$$

$$\frac{d\zeta_\beta}{d\tau} = -k_1\left(i_{s\alpha} - \hat{i}_{s\alpha}\right), \tag{9}$$

$$\hat{\omega}_r = S\left(\sqrt{\left(\zeta_\alpha^2 + \zeta_\beta^2\right)/\hat{\psi}_r^2} - k_4\left(V - V_f\right)\right), \tag{10}$$

where k_1, k_2, k_3, k_4 are gain coefficients,

$$\hat{\psi}_r^2 = \hat{\psi}_{r\alpha}^2 + \hat{\psi}_{r\beta}^2, \tag{11}$$

$$V = \zeta_\alpha\hat{\psi}_{r\beta} - \zeta_\beta\hat{\psi}_{r\alpha}, \tag{12}$$

$$S = \text{sign}\left(\zeta_\alpha\hat{\psi}_{r\alpha} + \zeta_\beta\hat{\psi}_{r\beta}\right) \tag{13}$$

and V_f is the filtered signal V.

The above observer is based on replacement of the rotor flux vector components multiplied by the rotor speed in differential equations for stator current and rotor flux vector components by disturbances defined as new variables as follows:

$$\zeta_\alpha = \omega_r\psi_{r\alpha}, \tag{14}$$

$$\zeta_\beta = \omega_r\psi_{r\beta}. \tag{15}$$

In this way the motor model is split into two parts, each for one axis. One of the parts is formed by (4), (6) and (9). The second one is formed by (5), (7) and (8). The variables defined by (14) and (15) may be interpreted as disturbance and general theory of the Luenberger observers may be applied. In accordance with this theory the disturbances are estimated in additional integrators forming the disturbance model. The rotor speed is calculated from variables determined in the model of the induction motor and in the disturbance model.

The disturbances are estimated by pure integration of errors between actual and estimated current vector components in unmoving frame of references. As all variables are sinusoidal in this frame of references in steady states, the errors of current vector components different from zero are needed to receive the values of disturbances on the outputs of integrators. The errors of the rotor flux vector components appear as consequence of the current errors.

The consequence of nonzero errors of the current vector components is form of feedbacks in (4) and (5).

Two separated observer are coupled together by feedbacks introduced into (6) and (7). These feedbacks are main idea of the simple observer. Because the rotor flux vector components are not measurable in the control

system, the feedbacks are formed using variables from two parts of the observer. In this way these feedbacks provide proper dependencies in inner structure of the observer, in contrary to the feedbacks containing errors of the current vector components which provide matching of the observer with the motor. In this way two different kinds of feedback appear in the proposed speed observer. The first one will be called the inner feedback, as only variables from the observer are included into expressions for errors. The second one will be called the external feedback, as errors for feedback are calculated from observer variables and machine variables.

IV. SPEED OBSERVER WITH EXACT MODEL OF DISTURBANCES

Drawbacks of the simple speed observer may be removed by introducing changes resulting from analysis of observer structure and transformation of algebraic dependences and differential equations. Exact differential equations may be used for disturbances in the observer structure instead of simplified model. The differential equations for disturbances take the following form resulting from differentiations of (14) and (15) :

$$\frac{d\zeta_\alpha}{d\tau} = a_5\hat{\omega}_r\hat{i}_{s\alpha} + a_6\zeta_\alpha - \hat{\omega}_r\zeta_\beta + \hat{\psi}_{r\alpha}\frac{d\hat{\omega}_r}{d\tau} + k_1\left(i_{s\beta} - \hat{i}_{s\beta}\right), \tag{16}$$

$$\frac{d\zeta_\beta}{d\tau} = a_5\hat{\omega}_r\hat{i}_{s\beta} + a_6\zeta_\beta + \hat{\omega}_r\zeta_\alpha + \hat{\psi}_{r\beta}\frac{d\hat{\omega}_r}{d\tau} - k_1\left(i_{s\alpha} - \hat{i}_{s\alpha}\right). \tag{17}$$

The derivative of estimated rotor speed $\hat{\omega}_r$ may be replaced by approximated value calculated in a digital realization of the speed observer.

Application of exact dynamical model of disturbances in (16) and (17) results in elimination of disturbance errors appearing in the simplified observer. Additionally the current errors are reduced in steady states and the rotor flux vector component are estimated without errors. The external feedbacks in (4) and (5) depend only on errors of the stator current vector components.

Instead of calculating the rotor speed from (10) the following expression is used:

$$\hat{\omega}_r = \left(\hat{\psi}_{r\alpha}\zeta_\alpha + \hat{\psi}_{r\beta}\zeta_\beta\right)/\hat{\psi}_r^2. \tag{18}$$

Validity of (18) is obvious taking into account definition of disturbance vector. The main advantage of (18) in comparison with (10) is simple form and no need of square root and additional determination of the speed sign. The other difference is that more information is used in (18) because an angle between the disturbance vector and the rotor flux vector is included. In contrary, only amplitude of the disturbances and rotor flux is used in (10).

Feedbacks used in (6) and (7) may be simplified using (18) without damping as follows:

$$\hat{\omega}_r\hat{\psi}_{r\alpha} - \zeta_\alpha = -\hat{\psi}_{r\beta}\left(\zeta_\alpha\hat{\psi}_{r\beta} - \hat{\psi}_{r\alpha}\zeta_\beta\right)/\hat{\psi}_r^2, \tag{19}$$

$$\hat{\omega}_r\hat{\psi}_{r\beta} - \zeta_\beta = \hat{\psi}_{r\alpha}\left(\zeta_\alpha\hat{\psi}_{r\beta} - \hat{\psi}_{r\alpha}\zeta_\beta\right)/\hat{\psi}_r^2. \tag{20}$$

Additional damping introduced into (10) may be now added to (19) and (20). As errors of the estimated rotor flux and disturbance vector are equal to zero, the signal V is equal to zero in steady states and filtered value V_f may be omitted. Anyway the signal V is multiplied by sign S in (10). Expression (13) depends of sign of the rotor speed but from more exact analysis it results that the sign have to be determined from sign of angular velocity of the rotor flux vector calculated from:

$$\hat{\omega}_\psi = \hat{\omega}_r + a_5 \hat{x}_{12} / \hat{\psi}_r^2, \tag{21}$$

and

$$S_\psi = \text{sign}\left(\hat{\omega}_\psi\right), \tag{22}$$

where

$$\hat{x}_{12} = \hat{\psi}_{r\alpha} \hat{i}_{s\beta} - \hat{\psi}_{r\beta} \hat{i}_{s\alpha} \tag{23}$$

is the multiscalar model variable proportional to the motor torque.

Adding the damping terms and taking into account the above considerations the following expressions are received:

$$\hat{\omega}_r \hat{\psi}_{r\alpha} - \zeta_\alpha = \left(-\hat{\psi}_{r\beta} \left(\zeta_\alpha \hat{\psi}_{r\beta} - \zeta_\beta \hat{\psi}_{r\alpha}\right) \right. \\ \left. - k_4 S_\psi \hat{\psi}_{r\alpha} \left(\zeta_\alpha \hat{\psi}_{r\beta} - \zeta_\beta \hat{\psi}_{r\alpha}\right)\right)/\hat{\psi}_r^2, \tag{24}$$

$$\hat{\omega}_r \hat{\psi}_{r\beta} - \zeta_\beta = \left(\hat{\psi}_{r\alpha} \left(\zeta_\alpha \hat{\psi}_{r\beta} - \zeta_\beta \hat{\psi}_{r\alpha}\right) \right. \\ \left. - k_4 S_\psi \hat{\psi}_{r\beta} \left(\zeta_\alpha \hat{\psi}_{r\beta} - \zeta_\beta \hat{\psi}_{r\alpha}\right)\right)/\hat{\psi}_r^2. \tag{25}$$

Denominator in (24) and (25) changes slowly and has an influence on observer gain if omitted. Additionally damping feedbacks may be added to equations (4) and (5). After rearranging subscripts at observer gains, the following equations are received taking the above under consideration:

$$\frac{d\hat{i}_{s\alpha}}{d\tau} = a_1 \hat{i}_{s\alpha} + a_2 \hat{\psi}_{r\alpha} + a_3 \zeta_\beta + a_4 u_{s\alpha} \\ + k_1 \left(i_{s\alpha} - \hat{i}_{s\alpha}\right) + k_2 V \hat{\psi}_{r\alpha} - k_3 S_\psi V \hat{\psi}_{r\beta}, \tag{26}$$

$$\frac{d\hat{i}_{s\beta}}{d\tau} = a_1 \hat{i}_{s\beta} + a_2 \hat{\psi}_{r\beta} - a_3 \zeta_\alpha + a_4 u_{s\beta} \\ + k_1 \left(i_{s\beta} - \hat{i}_{s\beta}\right) + k_2 V \hat{\psi}_{r\beta} + k_3 S_\psi V \hat{\psi}_{r\alpha}, \tag{27}$$

$$\frac{d\hat{\psi}_{r\alpha}}{d\tau} = a_5 \hat{i}_{s\alpha} + a_6 \hat{\psi}_{r\alpha} - \zeta_\beta - k_4 V \hat{\psi}_{r\alpha} + k_5 S_\psi V \hat{\psi}_{r\beta}, \tag{28}$$

$$\frac{d\hat{\psi}_{r\beta}}{d\tau} = a_5 \hat{i}_{s\beta} + a_6 \hat{\psi}_{r\beta} + \zeta_\alpha - k_4 V \hat{\psi}_{r\beta} - k_5 S_\psi V \hat{\psi}_{r\alpha}, \tag{29}$$

$$\frac{d\zeta_\alpha}{d\tau} = a_5 \hat{\omega}_r \hat{i}_{s\alpha} + a_6 \zeta_\alpha - \hat{\omega}_r \zeta_\beta + \hat{\psi}_{r\alpha} \frac{d\hat{\omega}_r}{d\tau} + k_6 \left(i_{s\beta} - \hat{i}_{s\beta}\right), \tag{30}$$

$$\frac{d\zeta_\beta}{d\tau} = a_5 \hat{\omega}_r \hat{i}_{s\beta} + a_6 \zeta_\beta + \hat{\omega}_r \zeta_\alpha + \hat{\psi}_{r\beta} \frac{d\hat{\omega}_r}{d\tau} - k_6 \left(i_{s\alpha} - \hat{i}_{s\alpha}\right). \tag{31}$$

Taking into account that derivative of estimated rotor speed may be replaced by its value calculated in the control system as follows:

$$\frac{d\hat{\omega}_r}{d\tau} = \frac{\Delta\hat{\omega}_r}{\Delta\tau}, \tag{32}$$

where $\Delta\hat{\omega}$ is an increment of the rotor speed during the time period $\Delta\tau$ and coefficients k_3 and k_5 may be defined as nonlinear functions of the angular speed of the

rotor flux vector, the following equations of the speed observer are:

$$\frac{d\hat{i}_{s\alpha}}{d\tau} = a_1 \hat{i}_{s\alpha} + a_2 \hat{\psi}_{r\alpha} + a_3 \zeta_\beta + a_4 u_{s\alpha} \\ + k_1 \left(i_{s\alpha} - \hat{i}_{s\alpha}\right) + k_2 V \hat{\psi}_{r\alpha} - k_3 \left(\omega_\psi\right) V \hat{\psi}_{r\beta}, \tag{33}$$

$$\frac{d\hat{i}_{s\beta}}{d\tau} = a_1 \hat{i}_{s\beta} + a_2 \hat{\psi}_{r\beta} - a_3 \zeta_\alpha + a_4 u_{s\beta} \\ + k_1 \left(i_{s\beta} - \hat{i}_{s\beta}\right) + k_2 V \hat{\psi}_{r\beta} + k_3 \left(\omega_\psi\right) V \hat{\psi}_{r\alpha}, \tag{34}$$

$$\frac{d\hat{\psi}_{r\alpha}}{d\tau} = a_5 \hat{i}_{s\alpha} + a_6 \hat{\psi}_{r\alpha} - \zeta_\beta - k_4 V \hat{\psi}_{r\alpha} + k_5 \left(\omega_\psi\right) V \hat{\psi}_{r\beta}, \tag{35}$$

$$\frac{d\hat{\psi}_{r\beta}}{d\tau} = a_5 \hat{i}_{s\beta} + a_6 \hat{\psi}_{r\beta} + \zeta_\alpha - k_4 V \hat{\psi}_{r\beta} - k_5 \left(\omega_\psi\right) V \hat{\psi}_{r\alpha}, \tag{36}$$

$$\frac{d\zeta_\alpha}{d\tau} = a_5 \hat{\omega}_r \hat{i}_{s\alpha} + a_6 \zeta_\alpha - \hat{\omega}_r \zeta_\beta + \hat{\psi}_{r\alpha} \frac{\Delta\hat{\omega}_r}{\Delta\tau} + k_6 \left(i_{s\beta} - \hat{i}_{s\beta}\right), \tag{37}$$

$$\frac{d\zeta_\beta}{d\tau} = a_5 \hat{\omega}_r \hat{i}_{s\beta} + a_6 \zeta_\beta + \hat{\omega}_r \zeta_\alpha + \hat{\psi}_{r\beta} \frac{\Delta\hat{\omega}_r}{\Delta\tau} - k_6 \left(i_{s\alpha} - \hat{i}_{s\alpha}\right), \tag{38}$$

$$V = \zeta_\alpha \hat{\psi}_{r\beta} - \zeta_\beta \hat{\psi}_{r\alpha}, \tag{39}$$

$$\hat{\omega}_r = \left(\hat{\psi}_{r\alpha} \zeta_\alpha + \hat{\psi}_{r\beta} \zeta_\beta\right)/\hat{\psi}_r^2, \tag{40}$$

$$\omega_\psi = \hat{\omega}_r + R_r L_m \left(\hat{\psi}_{r\alpha} i_{s\beta} - \hat{\psi}_{r\beta} i_{s\alpha}\right)/\left(L_r \hat{\psi}_r^2\right). \tag{41}$$

Gains of the simple and modified observers depend on the working point, specially on angular velocity of rotor flux.

V. ANALISYS OF STABILITY OF THE SPEED OBSERVER

Analysis of stability of the simple observer is difficult because nonlinearities resulting from expression for the rotor speed are complicated and no equivalent motor model is defined. The steady state errors depend on working point of the motor and gain coefficients. Up to now no results of stability investigations of simple speed observer were presented.

The more advanced speed observer was presented in [14] with disturbances estimated in an simplified model of harmonic oscillator:

$$\frac{d\hat{\zeta}_\alpha}{d\tau} = -\hat{\omega}_\zeta \hat{\zeta}_\beta - k_6 \left(i_{s\beta} - \hat{i}_{s\beta}\right), \tag{42}$$

$$\frac{d\hat{\zeta}_\beta}{d\tau} = \hat{\omega}_\zeta \hat{\zeta}_\alpha + k_6 \left(i_{s\alpha} - \hat{i}_{s\alpha}\right), \tag{43}$$

$$\frac{d\hat{\omega}_\zeta}{d\tau} = \gamma\left(\hat{\psi}_{r\alpha} \hat{\zeta}_\beta - \hat{\psi}_{r\beta} \hat{\zeta}_\alpha\right), \tag{44}$$

where ω_ζ is the synchronous angular velocity of the disturbance model.

Although the model of disturbances is simplified in (42) - (44), the above adaptive observer works in wide rotor speed range up to three times rated speed. On the other hand some difficulties were observed for very low rotor speed. range.

Other advantage of the adaptive observer is possibility to proof its stability using the second Lapunov method.

2296

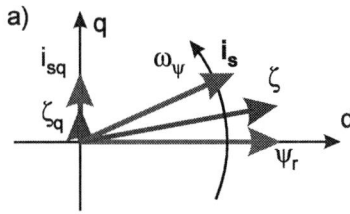

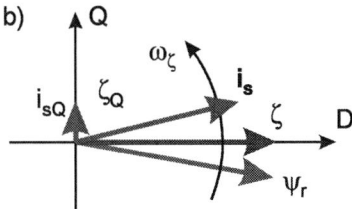

Fig. 1. The rotor flux and disturbance vectors in flux a) and disturbance b) oriented frames of references.

Details of the proof were presented in [15] and are prepared for publication.

In contrary an extended motor model may be defined by introducing additional variables. Equations for the observer errors may be derived for the modified observer base on exact model of disturbances. Anyway the equations for observer errors remain nonlinear and complicated.

Analysis the observer stability will be done here in a simplified way. The main property of the observer is estimation of the rotor speed with very small error. This property results from equality of the angular velocity of the rotor flux vector and the angular velocity of the disturbance vector. Additionally the angle between the these two vectors has to be equal to zero.

To analyze the angular velocity of the rotor flux vector the following equations my written in the rotating frame of references:

$$\frac{d\hat{\psi}_{rd}}{d\tau} = a_5 \hat{i}_{sd} + a_6 \hat{\psi}_{rd} - \zeta_q + \omega_\psi \hat{\psi}_{rq} \\ - k_4 V \hat{\psi}_{rd} + k_5 (\omega_\psi) V \hat{\psi}_{rq} , \quad (45)$$

$$\frac{d\hat{\psi}_{r\beta}}{d\tau} = a_5 \hat{i}_{s\beta} + a_6 \hat{\psi}_{r\beta} + \zeta_\alpha - \omega_\psi \hat{\psi}_{r\alpha} \\ - k_4 V \hat{\psi}_{r\beta} - k_5 (\omega_\psi) V \hat{\psi}_{r\alpha} , \quad (46)$$

where ω_ψ is the angular velocity of the rotor flux vector and d,q denotes the axis of rotating frame of references.

Assuming that the q component of the rotor flux vector is equal 0, the angular velocity of the rotor flux vector may be calculated from (46) as follows:

$$\omega_\psi = a_5 \hat{i}_{sq} / \hat{\psi}_{rd} + \zeta_d / \hat{\psi}_{rd} + k_5 (\omega_\psi) \zeta_q \hat{\psi}_{rd} . \quad (47)$$

For the angular velocity of the vector of disturbances the following equations in the rotating frame of references my be written:

$$\frac{d\zeta_D}{d\tau} = a_5 \hat{\omega}_r \hat{i}_{sD} + a_6 \zeta_{D\alpha} - \hat{\omega}_r \zeta_{Q\beta} + \omega_\zeta \zeta_Q \\ + \hat{\psi}_{rD} \frac{\Delta \hat{\omega}_r}{\Delta \tau} + k_6 (i_{sQ} - \hat{i}_{sQ}) , \quad (48)$$

$$\frac{d\zeta_Q}{d\tau} = a_5 \hat{\omega}_r \hat{i}_{sQ} + a_6 \zeta_Q + \hat{\omega}_r \zeta_D - \omega_\zeta \zeta_D \\ + \hat{\psi}_{rQ} \frac{\Delta \hat{\omega}_r}{\Delta \tau} - k_6 (i_{sD} - \hat{i}_{sD}) , \quad (49)$$

where ω_ζ is the angular velocity of the disturbance vector and D,Q denotes the axis of frame of references.

Assuming that the Q component of the disturbance vector is equal 0, the angular velocity of the disturbance vector may be calculated from (49) as follows:

$$\omega_\zeta = a_5 \hat{\omega}_r \hat{i}_{sQ} / \zeta_D + \hat{\omega}_r - k_6 (i_{sD} - \hat{i}_{sD}) / \zeta_D . \quad (50)$$

If the angular velocities ω_ψ i ω_ζ are not equal, their values change in accordance to (47) and (50). As shown in Fig. 1a) if the angle of the rotor flux lags the disturbance vector, positive q components of disturbance vector appears and the angular velocity of the rotor flux vector increases. On the contrary, as seen in Fig. 1a) and Fig. 1b) Q component of the current vector decreases and, as follows from (50), the angular velocity of the disturbance vector decreases. In this way the angle between the flux and disturbance vectors decreases to zero. Similar effects appears if the disturbance vector lags the rotor flux vector.

The above considerations show not only stability of the observer but explain principle of operation.

The above explanation my be applied to the other versions of the speed observer, specially for the simple observer with the rotor speed calculated from (18).

VI. THE CONTROL SYSTEM

The presented observers were applied in the control system based on the multiscalar model of the induction motor. The following variables were controlled:

$$x_{11} = \omega_r , \quad (51)$$

$$x_{12} = \psi_{r\alpha} i_{s\beta} - \psi_{r\beta} i_{s\alpha} , \quad (52)$$

$$x_{21} = \psi_r^2 , \quad (53)$$

$$x_{22} = \psi_{r\alpha} i_{s\alpha} + \psi_{r\beta} i_{s\beta} . \quad (54)$$

Interpretation of variables (51) - (54) were discussed in [1], [2], [3].

The nonlinear feedback is applied and two independent linear subsystems appear. The differential equations of the subsystems are as follows:

- mechanical subsystem

$$\frac{dx_{11}}{d\tau} = \frac{L_m}{J L_r} x_{12} - \frac{1}{J} m_0 , \quad (55)$$

$$\frac{dx_{12}}{d\tau} = \frac{1}{T_v} (-x_{12} + m_1) , \quad (56)$$

- electromagnetic subsystem

$$\frac{dx_{21}}{d\tau} = -2 \frac{R_r}{L_r} x_{21} + 2 \frac{R_r L_m}{L_r} x_{22} , \quad (57)$$

2297

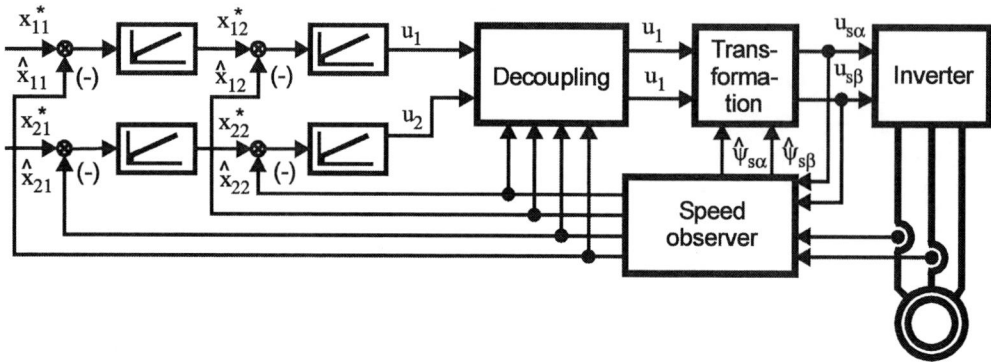

Fig. 2. Scheme of the control system.

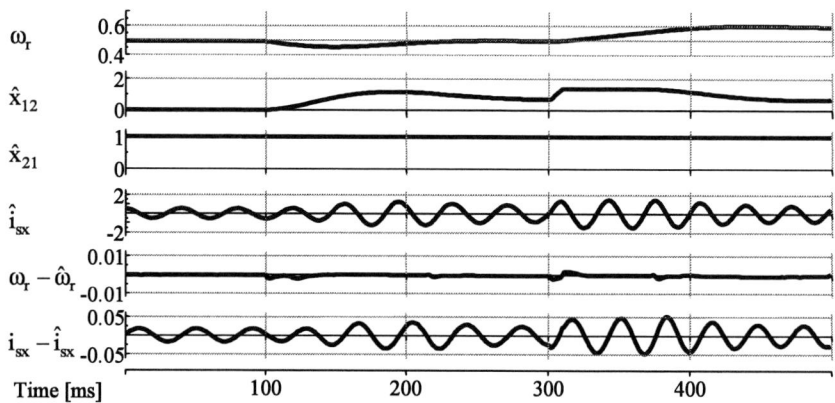

Fig. 3. Transients in the control system with observer based on exact model of disturbances.

$$\frac{dx_{22}}{d\tau} = \frac{1}{T_v}(-x_{22} + m_2),\qquad(58)$$

where T_v is time constant and J moment of inertia.

Cascade controllers were used to control two independent linear subsystems resulting from application of nonlinear feedback. The control system is shown in Fig. 2. Properties of the nonlinearly controlled induction motor were discussed in previous papers [1], [2]. Similar settings of controllers were used for system with exact observer and with simplified observer. The stator current was limited by limiting the motor torque.

VII. RESULTS OF SIMULATIONS

To show properties of the speed observer based extended model of the induction motor transient of selected variables and speed and current errors are presented in Fig. 3. In instant 100 ms a nominal load was applied for the rotor speed equal to 0.5 p.u. The set value for the rotor speed was applied in instant 300 ms. Very small speed errors, less than 0.003 p.u. appear in transients. The current errors are less than 0.05 p.u. but, as it can be seen in Fig. 4, these errors are caused mainly by errors of current phase.

Transients during speed reversal from −2.5 to 2.5 in p.u. in the system with observer based on exact model of disturbances are presented in Fig. 4. Errors of estimated speed are small and the control system is stable. Small errors of the estimated stator current appear for high speed region. For the speed less than nominal the speed error is less than 0.002 in transients. There is only small region of low speed where the speed error is greater but less than 0.01.

For comparison transients during speed reversal from − 1.18 to 1.18 in p.u. in the system with observer based on simplified model of disturbances are presented in a first part of Fig. 5. It can be seen that coupling between variables appears in field weakening region. For speeds higher than 1.18 p.u. the system was unstable. The reason is that the errors of the stator current vector components depend on the rotor speed and increase for speed higher than 1 p.u. Higher errors of the estimated speed appear. If the rotor speed is not greater than 1 p.u., as shown in the second part of Fig. 5, the control system works properly although high error of current amplitude appears.

VIII. CONCLUSION

The proposed speed observer with exact model of disturbances estimates the rotor speed with very small errors in the wide range of the rotor speed. High exactness of the proposed speed observer is observed in transients and in steady states. The variables estimated using the proposed observer may be used in the control system with nonlinear decoupling feedback.

ACKNOWLEDGMENT

Scientific work financed as development project from resources on science in years 2007 – 2009.

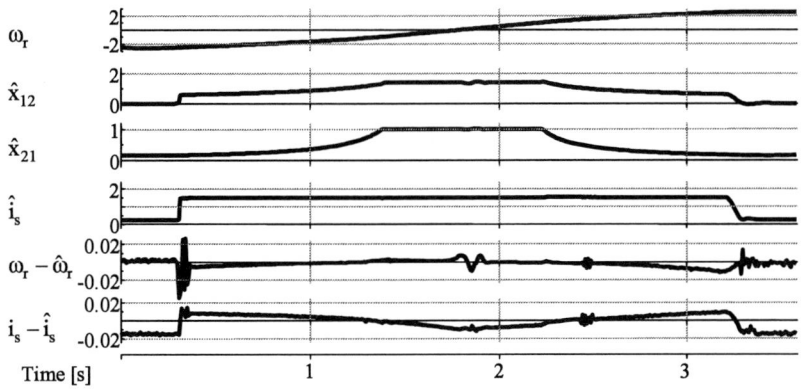

Fig. 4. Transients during speed reversal in the system with observer based on exact model of disturbances.

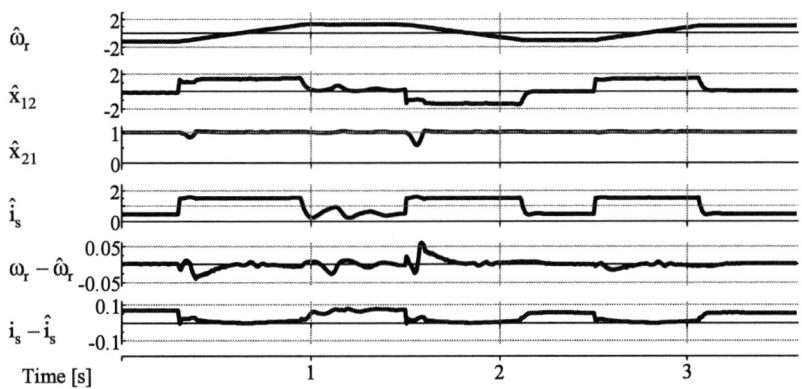

Fig. 5. Transients during speed reversal in the system with observer based on simplified model of disturbances.

REFERENCES

[1] Krzeminski Z., "Nonlinear control of induction motor," *Proceedings of the 10th IFAC World Congress*, Munich, 1987, pp. 349-354.

[2] Krzemiński Z., "Nonlinear feedback and control strategy of the induction motor," *Proceedings of the IFAC Nonlinear Control System Design Symposium*, Bordeaux, France, 1992.

[3] Krzemiński Z., Lewicki A., Włas M., "Properties of sensorless control system based on multiscalar models of the induction motor," *Special Issue of COMPEL "Selected Papers from the 18th Symposium on Electromagnetic Phenomena in Nonlinear Circuits"*. COMPEL Vol 25, no 1, pp. 195 – 206.

[4] Kim G. S., Ha I. J., and Ko M. S., "Control of induction motors for both high dynamics performance and high-power efficiency," *IEEE Trans. on Industrial Electronics*, vol. 39, 1992.

[5] Bellini A., "An adaptive control for induction motor drives based on a fully linearized model," *5th European Conference on Power Electronics and Applications*, Brighton, 1993.

[6] Luckjiff G., Wallace I., Divan D., "Feedback Linearization of Current Regulated Induction Motors." *Power Electronics Specialists Conference, 2001, PESC*, 2001 IEEE 32nd Annual Vol. 2, 17-21 Junew 2001 Pages: 321 – 326.

[7] Marino R., Peresada S., Valigi P., "Adaptive input-output linearizing control of induction motors," *IEEE Trans. on Automatic Control*, vol. 38, 1993.

[8] Chung-Hyuk Yim, Gyu-Sik Kim, Chang-Hwan Kim, "Decoupling control of induction motors with motor parameter identification," *Industry Application Conference, 1996. Thirty-First IAS Annual*

Meeting, IAS '96, Conference Record of the 1996 IEEE. Vol 1. Pages 221 –228.

[9] Pavlov A., Zaremba A., "Direct Torque and Flux Regulation in Sensorless Control of an Induction Motor," *Proceedings of the American Control Conference, Arlington, VA June, 25-27, 2001*.

[10] Ojo O. and Dong G., "Efficiency optimizing control of induction motor using natural variables," *Applied Powert Electronics Conf. And Exposition, 2004, APEC '04*, Nineteenth Annual IEEE Volume 3, @004 Pages: 1622 – 1627

[11] Krzemiński Z., "A new speed observer for control system of induction motor," *Proc. of IEEE Int. Conf. on Power Electronics and Drive Systems, PESC'99, Hong Kong*, pp. 555 – 560.

[12] Krzemiński Z., "Estimation of rotor speed for nonlinear control of the induction motor," *Int. Conf. EPE-PEMC'2002, Cavtat & Dubrownik, 2002*.

[13] Wlas M., Krzeminski Z., Guzinski J., Abu-Rub H., Toliyat H.A.: Artificial Neural Network Based Sensorless Nonlinear Control of Induction Motors, *IEEE Trans. on Energy Convertion*, Vol 20 no 3, 2005, pp. 520 – 528.

[14] Adamowicz M., Krzemiński Z.: Novel Adaptive Flux Observer for Wide Speed Range Sensorless Control of Induction Motor. The 12 European Conference on Power Electronics and Applications, Aalborg, Denmark, September 2-5, 2007.

[15] Adamowicz M.: Control system for the induction motor with with reduced flux in the air gap. Doctoral thesis, Gdansk, 2008. (in polish)

Experimental Performance Evaluation for Low Speed and Regenerating Operation of Sensor-less Vector Control System of Induction Motor Using Observer Gain Tuning

Kazuhiro Ohyama*, Greg Asher** and Mark Sumner**

| *Department of Electrical Engineering | **Department of Electrical and Electronic Engineering |
| Fukuoka Institute of Technology | University of Nottingham |

Corresponding Author: Kazuhiro Ohyama
Email: ohyama@ee.fit.ac.jp

Abstract – The design method of adaptive rotor flux observer gain to improve stability at low speed and regenerating mode was proposed in [1]. The method is based on the stability analysis. The stability analysis utilizes the linearized model considering all systems including each control loop. Therefore the proposed method considers the effects of motor constants and control circuit constants. The stability analysis using the transfer function for rotor speed considers the arrangement of pole and zero and the steady state error. The rotor flux observer gain which improves the stability for each operating condition was clarified. Also the real time tuning method of observer gain was proposed in [1]. The validity of the proposed method was confirmed on the simulation using Matlab Simulink and the experiment. Then this paper evaluates the experimental performance for low speed and regenerating operation of sensor-less vector control system of induction motor using observer gain tuning.

1. Real Time Observer Gain Tuning

The proposed adaptive rotor flux observer is shown in Fig. 1. The OGT included in Fig. 1 is shown in Fig. 2. In the OGT, the k is a function for the speed and electric torque. However, the speed reference is used as the speed, since the speed is not detected. And, the electric torque is estimated using following equation.

$$T_e^{est} = \frac{P}{2}\frac{M}{L_r}\phi_{rd}^{est} i_{sq}^{ref} \qquad (1)$$

In the OGT, the value of k is constant over 10rpm speed intervals and the 10% torque intervals. Therefore, for an operating point in the middle of interval, the error increases. Also k changes as a step, when the operating point changes. Then, the output value of k that smoothes the step is proposed. Using the speed reference and torque estimate in real time, 4 points of k adjoining the operating point are identified. Fig. 3 shows the 4 points that surround the operating point. The 4 points are $k(x,y)$, $k(x-10,y)$, $k(x,y-10)$, $k(x-10,y-10)$. The x [rpm] value is obtained by rounding up the first figure of

speed reference ω_r^{ref}, and y [%] value is obtained by cutting the first figure of the electric torque estimate T_e^{est}. The k applied to the adaptive rotor flux observer is estimated according to linear approximation style shown in following equations.

$$k = k_1 - \frac{k_2 - k_1}{10}\left(T_e^{est} - y\right) \qquad (2)$$

where,

$$k_1 = k(x,y) - \frac{k(x-10,y) - k(x,y)}{10}\left(\omega_r^{ref} - x\right)$$

$$k_2 = k(x,y-10) - \frac{k(x-10,y-10) - k(x,y-10)}{10}\cdot\left(\omega_r^{ref} - x\right)$$

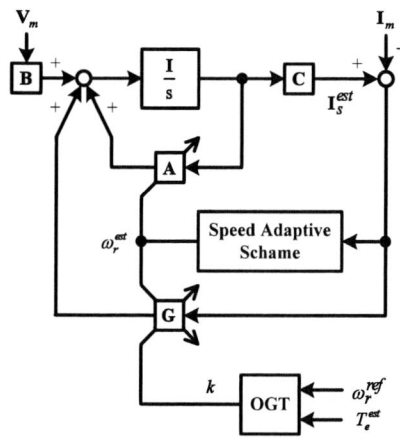

Fig. 1. Adaptive rotor flux observer.

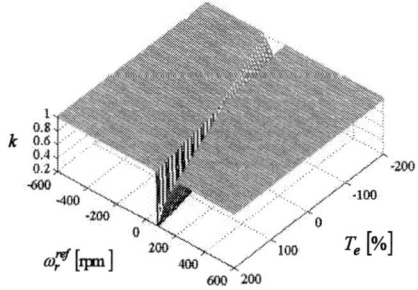

Fig. 2. Observer gain table

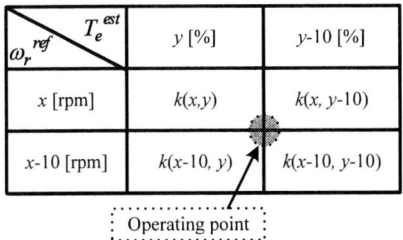

Fig. 3. 4 points that surround an operating point.

2. Experimental Performance Evaluation

2.1 Experimental equipment

Figure 4 shows the experimental equipment. The induction motor 1 (IM1) is the 1.5kW test machine and IM2 is the load machine. The inverter (INV1: MWINV-4R222) is driven according to the PWM signal which is sent form the control system (DS1103). The control period is 100 [μs], and the PWM switching frequency is 10 [KHz]. The IM2 is torque controlled by the inverter (INV2: VARISPEED616-G5). The loading test of ±150 [%] rated torque for IM1 can be carried out with the experimental equipment.

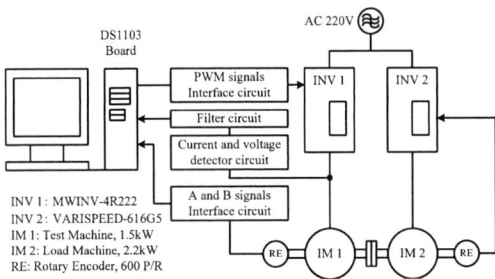

Fig. 4. Experimental equipment.

2.2 Sensorless Speed –Torque Characteristics

Figure 5 shows the simulated unstable torque-speed region for various load torques. The unstable regions are distinguished with calculated poles and zeros. The unstable region exists around the zero frequency line ($\omega_e = 0$ [Hz]) for not using observer gain tuning ($k = 1$). Its range becomes wider in proportion to the load torque. Although the unstable region exists for using observer gain tuning, the range is minimized and lies on the zero frequency line.

Figure 6 shows the speed-torque characteristics in simulation using Matlab Simulink. Fig. 7 shows the speed-torque characteristics in experiment. In this test, IM1 was held under speed control at the speeds shown. IM2 was changed from +150% motoring mode to -150% regenerative mode at -5 % per second. The gain k was changed in real time according to the machine torque derived from the i_{sq}^{ref}. The

experiment is carried out after the on-line tuning of stator resistance and rotor resistance using the method of [2].

However in practice the tuned resistance parameters are not perfect. Therefore the rotor speed changes depend on the load torque. The rotor speed control accuracy depends on the accuracy of estimated rotor resistance. The maximum rotor speed control error is about 15rpm at 400rpm of the rotor speed reference. This error is 0.83% for 1800rpm of the rated rotor speed.

The results of Fig. 7 show the improvement when using the on-line OGT. When using the OGT, the system is instability occurs at 50% only for negative (regen) loads greater than 100% rated. This is a clear improvement compared with not using the OGT. In the experiment, the rotor speed could not cross the zero frequency.

The experimental unstable torque-speed region for various load torque is shown in Fig. 8. This figure clearly shows the difference between not using OGT and using OGT. Also the unstable region agrees quite well with the results of Fig. 5.

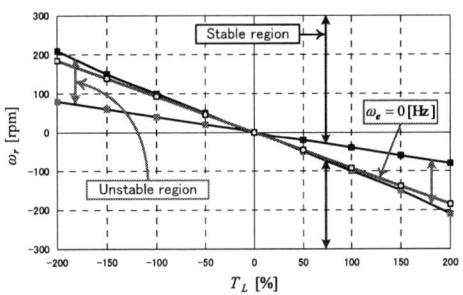

(a) not using observer gain tuning ($k = 1$)

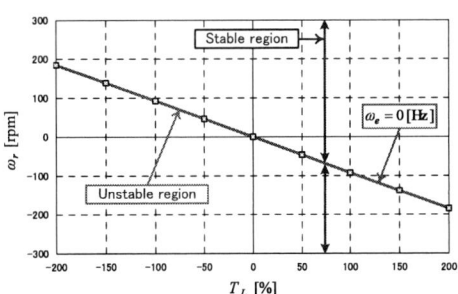

(b) using observer gain tuning

Fig. 5. Simulated unstable torque-speed region for various load torque.

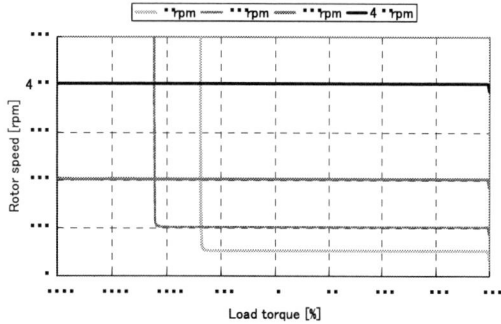

(a) not using observer gain tuning (k= 1)

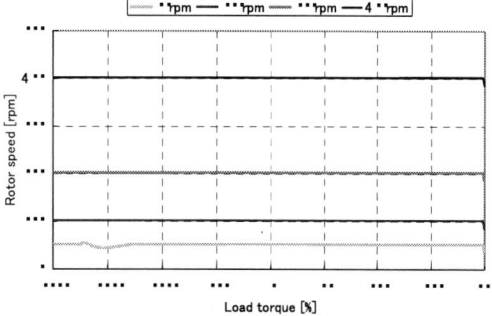

(b) using observer gain tuning

Fig. 6. Simulated Speed-torque characteristics.

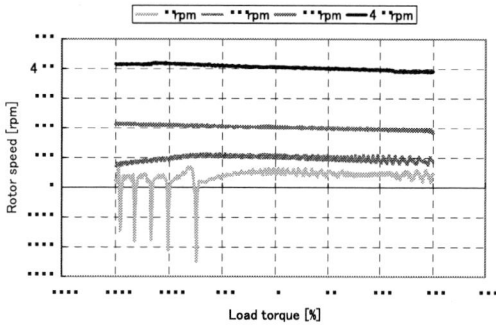

(a) not using observer gain tuning (k= 1)

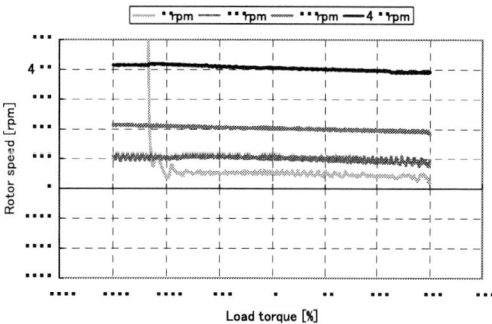

(b) using observer gain tuning

Fig. 7. Experimental Speed-torque characteristics.

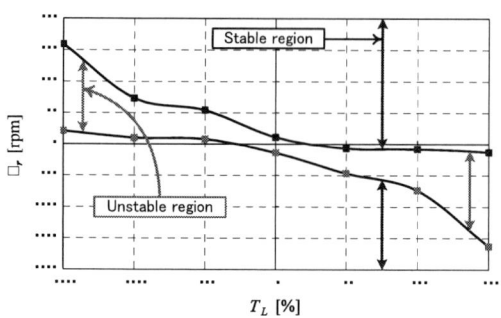

(a) not using observer gain tuning ($k = 1$)

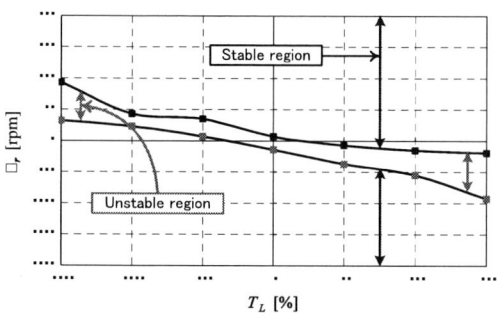

(b) using observer gain tuning

Fig. 8. Experimental unstable torque-speed region for various load torque.

2.3 Sensorless Transient Responses

Figures 9 and 10 show the simulation results for rated regenerating operation without and with the OGT respectively. The speed reference changes from 300rpm to -300rpm, and then from -300rpm to 300rpm. The load torque is kept at 100%. Motoring and Regenerative operations and passage through zero frequency at 50rpm/s acceleration of rotor speed reference occurs when using the real time tuning of observer gain. When it is not used, the system fails when the rotor speed reference changes from 0rpm to -300rpm and stator frequency crosses zero. In Fig. 9, this occurs at approximately t = 13s. At this moment, the rotor speed estimation fails and keeps same value because the stator frequency is zero (See Fig. 5). As a result, the q-axis stator current keeps same value, and the rotor speed control is lost. Of course, even the system using observer gain tuning becomes unstable when the smaller acceleration rate of rotor speed reference is used.

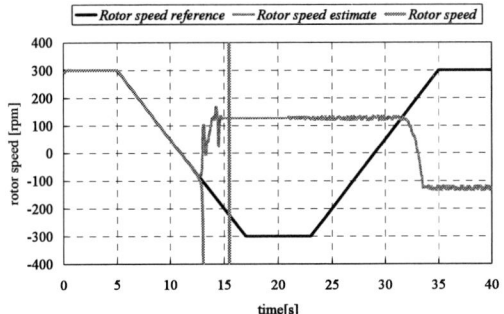

Fig. 9. Simulated responses to 300 to -300rpm ramps under rated motoring and regeneration, without OGT (k = 1).

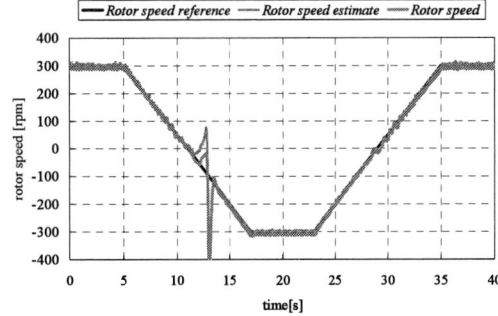

Fig. 11 Experimental responses to 300 to -300rpm ramps under rated motoring and regeneration, without OGT ($k = 1$).

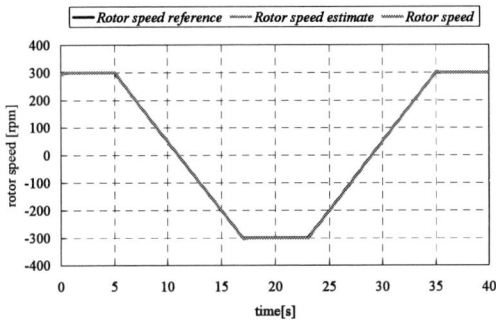

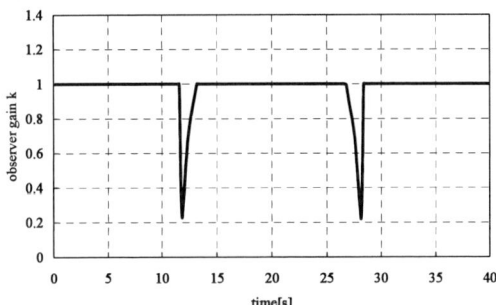

Fig. 10. Simulated responses to 300 to -300rpm ramps under rated motoring and regeneration, with OGT.

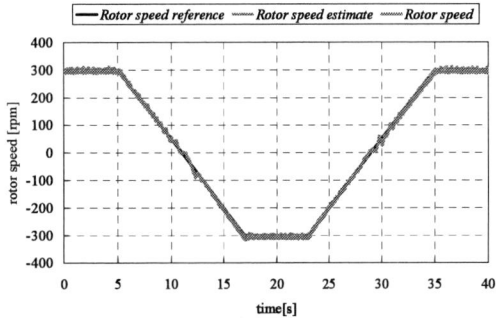

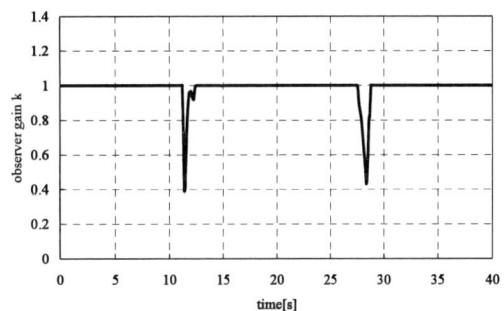

Fig. 12. Experimental responses to 300 to -300rpm ramps under rated motoring and regeneration, with OGT.

Figures. 11 and 12 shows the experimental results corresponding to the simulated results of Figs. 9 and 10. The OGT system is capable of regenerative operation and passage through zero frequency at 50rpm/s acceleration of rotor speed reference. The effect of the observer gain tuning can be confirmed in the real machine operation. The 50rpm/s acceleration of rotor speed reference is the smallest value to cross the zero frequency even for using OGT.

Figures 13 and 14 show the experimental responses to 300 to -300rpm ramps with 50% rated motoring and regeneration torque. Both experimental results have the resistance parameters detuned by means of changing the resistance parameters in the controller. As described in [1], the OGT, which improves the stability for the resistance parameters detuning, cannot be obtained. Then the same OGT is used for Figs. 13 and 14 and it is seen that using the OGT does not appear to make a great difference.

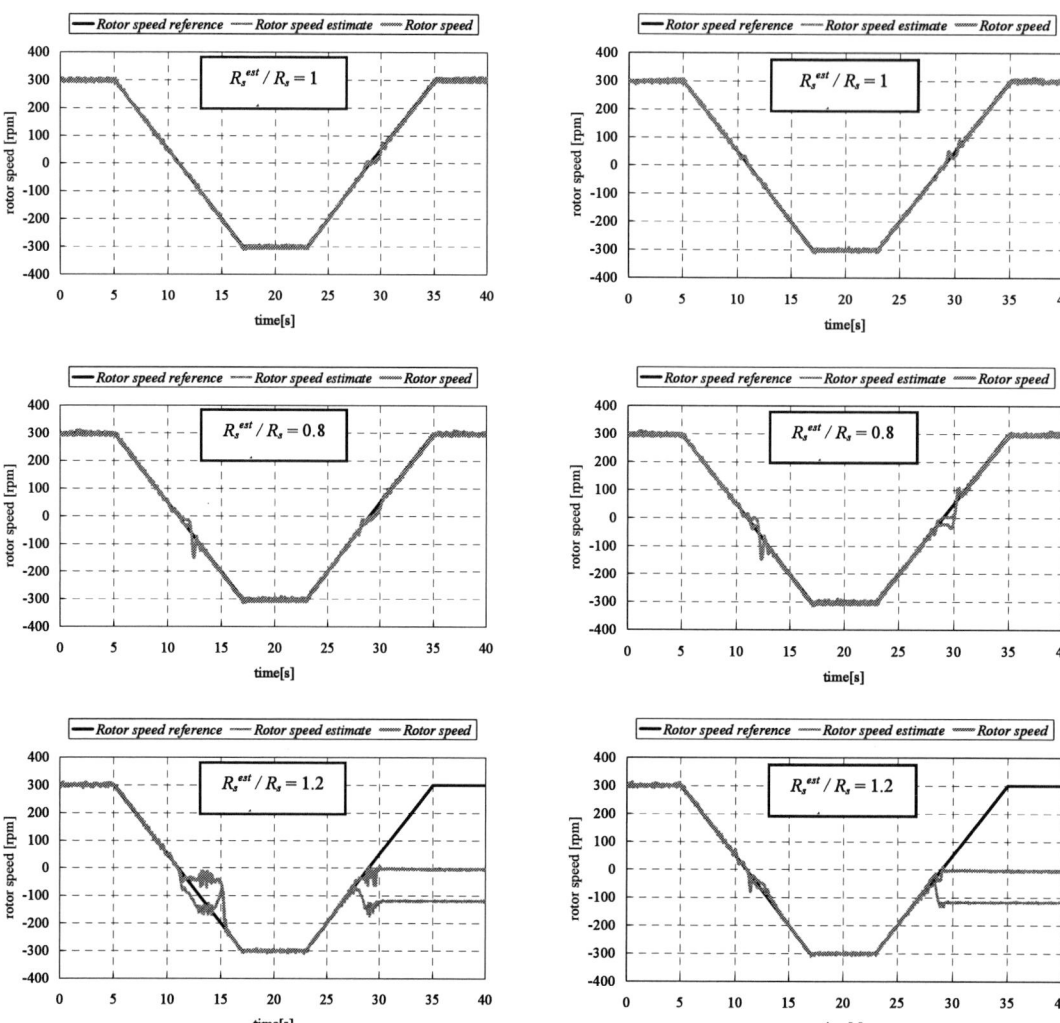

Fig. 13. Experimental responses to 300 to -300rpm ramps under 50 % rated motoring and regeneration, without OGT (k=1.0).

3. Conclusion

The observer gain design method, a form of gain scheduling and which can consider the effect of motor parameters and control parameters has been introduced and investigated. The observer gain tuning method can be used for applying the designed observer gain in the sensorless vector control drive. Both simulation and real machine test have been carried out in order to confirm the effect of proposing observer gain design method and observer gain tuning method. By using the optimized observer gain and the control parameters, designed with the rotor flux of twice rated value, it was confirmed that the stability of the low speed and regenerating operation was improved and that this was consistent with simulation.

Fig. 14. Experimental responses to 300 to -300rpm ramps under 50 % rated motoring and regeneration, with OGT.

The use of the Observer Gain tuning table was shown to increase the stability margin for the case when the motor parameters are tuned. A reasonable fit with simulation is obtained. It was found that using the OGT did not improve the stability when resistance parameters were detuned. This was expected for the variations of the stator resistance since Figs 13 and 14 shows there is not much scope for improvement. The stability improvements for rotor resistance variation, although slightly improve in transients, was not very great. Further work is required to investigate improvements under resistance variation further. Note that although the unstable region for using OGT became smaller, it still exists around the zero frequency line.

Reference

[1] K. Ohyama, T. Hamaoka, G. M. Asher and M. Sumner, "Stability Improvement of Sensorless Vector Control System of Induction Motor Using Real Time Tuning of Adaptive Rotor Flux Observer Gain", in Proc. Int. Conf. IEEE IECON, pp. 1475-1480, (2006-11).

[2] H. Kubota and K. Matsuse, "Speed Sensorless Field-Oriented Control of Induction Motor with Rotor Resistance Adaptation", IEEE Transaction Industry Application, vol.30, no.5, pp. 1219-1994 (1994-9/10).

Application of the Stator Current-based MRAS Speed Estimator in the Sensorless Induction Motor Drive

Mateusz Dybkowski, Teresa Orlowska-Kowalska, *Senior Member IEEE*

Wroclaw University of Technology, Institute of Electrical Machines, Drives and Measurements, Wroclaw, Poland
E-mail: mateusz.dybkowski@pwr.wroc.pl, teresa.orlowska-kowalska@pwr.wroc.pl

Abstract - The paper deals with the analysis of the vector controlled induction motor drive with a novel MRAS-type rotor speed estimator. The stability analysis method of this novel MRAS estimator is proposed. The influence of equivalent circuit parameter changes of the induction motor as well as coefficients of the adaptation algorithm of the MRAS scheme to the pole placement of the estimator transfer function and stability of the whole drive system is analysed and tested. The allowable range of motor parameter changes is determined, which guarantees the stable operation of the sensorless field oriented induction motor drive with this speed and flux estimator. Dynamical performances of the vector control system with the current-type MRAS estimator are tested in the laboratory set-up.

Keywords - Induction motor, variable speed drive, vector control, sensorless control, estimation technique

I. INTRODUCTION

In the recent years, remarkable efforts have been made to the development of state variables reconstruction of the induction motor (IM), such as: rotor or stator flux vectors, motor electromagnetic torque and rotor speed, to obtain speed sensorless drive systems [1] – [4].

The rotor speed for sensorless drives can be estimated by various techniques [1], [3], [4]. The simplest method is based on the angular velocity of rotor flux vector and slip calculation, based on the rotor flux vector coordinates obtained using the IM model. This method is quite popular and simple to implement, but the obtained accuracy is not very good due to a great sensitivity to motor parameter uncertainties. The other methods are based on the extended Luenberger observers or extended Kalman filters [3], [5], which are more robust to the IM parameter changes or identification errors, but are much more complicated in technical realization. Another solution for speed estimation is based on the model reference adaptive system (MRAS) principle, in which an error vector is formed from the outputs of two models, both dependent on different motor parameters. The error is driven to zero through adjustment of a parameter that influences one of models. The MRAS approach has the advantage in the simplicity of used models, which are simulators of chosen electromagnetic state variables of the IM, in comparison with nonlinear or extended state observers or Kalman filters. Thus they are easy in implementation and have direct physical interpretation.

MRAS-based speed estimators, developed so far, can be divided into three groups:

- the rotor flux error-based MRAS scheme (MRAS^F), developed by Tamai [6] and Schauder [7], called a classical MRAS speed estimator, is one of the most popular methods. This solution is based on the rotor flux error calculated using two different IM models, namely voltage and current models of the rotor flux. The speed is determined using the closed loop signal from the output of PI controller operated by the flux error signal.

- the back-EMF error-based MRAS schemes, where the error vector used for the rotor speed correction is obtained from the comparison of the measured and calculated back-EMF of the induction motor [8], [9].

- the stator current error-based MRAS schemes, where the stator current is estimated by suitable stator current model and compared with the measured value, to obtain the speed-error correction signal [10], [11]. These concepts are not widely known and in this paper their comparison and advantage will be analysed and discussed.

It is worth saying that sometimes the widely known nonlinear full-order flux observer with speed adaptation loop (NFO^A), proposed firstly in [12], is considered as a kind of stator-current MRAS schemes [13]. But this concept does not uses state variables simulators, like in original MRAS schemes and thus require much more effort on the design stage, because not only parameters of the speed adaptation loop must be chosen, but also the gain matrix of the full-order observer must be designed.

One of the most popular algorithmic method to the rotor speed and flux reconstruction is the MRAS^F estimator introduced by [6] and then modified by other authors [7], [14], [15]. This estimator is highly sensitive to motor parameter changes due to the high sensitivity of rotor flux current and voltage models used for the rotor flux vector estimation [13], [16].

In this paper the novel MRAS^{CC} estimator is proposed, where the measured stator current of the induction motor – used as a reference system, is compared with the stator current estimated using the stator voltage-current model, adjusted with the estimated rotor speed calculated by the adaptation algorithm. Such stator current model requires the information on the rotor flux vector. In the proposed MRAS^{CC} estimator this state variable is calculated based on the speed dependent current model of the rotor flux [1], [3], on the contrary to the solution with the voltage flux model, proposed in [10]. Moreover the adaptation algorithm is also modified in comparison with the classical MRAS^F solution [6].

978-1-4244-1741-4/08/$25.00 ©2008 IEEE

The concept of this estimator was firstly presented in [11], but only some simulation and experimental tests were shown there. In this paper the stability analysis method of such estimator is proposed. The sensitivity to motor parameters changes of the field-oriented IM drive with the proposed MRAS^{CC} rotor speed and flux estimator is tested. The theoretical analysis is confirmed by experimental tests of the sensorless drive system under different operation conditions.

II. MATHEMATICAL MODEL OF THE MRAS^{CC} ESTIMATOR

The new MRAS^{CC} estimator is based on the comparison between the measured stator current of the induction motor and the estimated current obtained from the stator current model. This mathematical model of the stator current can be calculated from the combined voltage and current flux models [1], [3] and is described by the following equation:

$$\frac{d}{dt}\mathbf{i}_s^e = -\frac{r_r x_m^2 + x_r^2 r_s}{\sigma T_N x_s x_r^2}\mathbf{i}_s^e + \frac{1}{\sigma T_N x_s}\mathbf{u}_s + \frac{x_m r_r}{\sigma T_N x_s x_r^2}\mathbf{\Psi}_r^i - j\omega_m^e \frac{x_m}{\sigma T_N x_s x_r}\mathbf{\Psi}_r^i \quad (1)$$

where: ω_m^e – estimated rotor angular speed, r_s, r_r, x_s, x_r, x_m – stator and rotor resistances, stator, rotor and magnetizing reactances, $\mathbf{u}_s$, $\mathbf{i}_s^e$, $\mathbf{\Psi}_r^i$ – stator voltage, estimated stator current and rotor flux vectors respectively, $\sigma=1-x_m^2/x_s x_r$, $T_N=1/2\pi f_s$.

Such stator current model requires the information about the rotor flux vector. In the proposed MRAS^{CC} estimator this state variable is calculated based on the speed dependent current model of the rotor flux [1], [3]:

$$\frac{d}{dt}\mathbf{\Psi}_r^i = \left[\frac{r_r}{x_r}(x_M \mathbf{i}_s - \mathbf{\Psi}_r^i) + j\omega_m^e \mathbf{\Psi}_r^i\right]\frac{1}{T_N} \quad (2)$$

Both stator current model (1) and rotor flux model (2) are adjusted by the estimated rotor speed, according to the schematic diagram presented in Fig. 1.

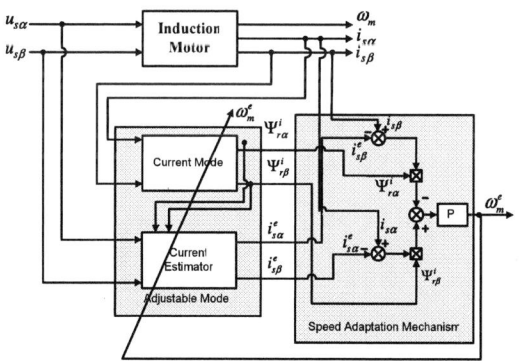

Fig. 1 Rotor speed reconstruction using MRAS^{CC} system

In the classical MRAS speed estimator, a PI controller is used as an adaptation block [6], [7], which calculates the rotor speed based on the difference between the rotor flux vectors estimated by two rotor flux models (voltage and current models). In the MRAS^{CC} estimator the used adaptation algorithm is different from the classical solution and is based on the error between estimated and measured stator current, according to the formula used in the full-order flux observer with speed adaptation, developed in [12]:

$$\omega_m^e = K_P\left(e_{i_{sa}}\Psi_{r\beta}^i - e_{i_{s\beta}}\Psi_{ra}^i\right) + K_I\int\left(e_{i_{sa}}\Psi_{r\beta}^i - e_{i_{s\beta}}\Psi_{ra}^i\right)dt \quad (3)$$

where $e_{i_{s\alpha,\beta}} = i_{s\alpha,\beta} - i_{s\alpha,\beta}^e$ - error between estimated and measured stator current.

The obtained rotor speed value is used in the current model of rotor flux and stator current estimator as changeable parameter, as shown in the Figure 1.

In the classical MRAS^F rotor speed estimator the voltage model is used as a reference model and the current model is an adjustable model [6]. In the proposed MRAS^{CC} estimator the induction motor is used as a reference system and the current flux model together with current estimator are adjustable models. Estimated stator current (1) is compared with its measured value and the signal $\mathbf{e}_{is}$ is used in the speed adaptation mechanism (3).

The similar solution was proposed in [10], but the difference consists in the application of the voltage model for the reconstruction of rotor flux, which is used in the stator current estimator. As it will be shown latter, such solution is less robust to motor parameter uncertainties than the proposed here.

III. STABILITY ANALYSIS OF THE MRAS^{CC} ESTIMATOR

The stability analysis of the novel MRAS^{CC} speed and rotor flux estimator is tested from the point of view of the induction motor and PI controller parameters changes, based on the estimator transfer function.

From the point of view of the rotor speed estimation, the MRAS^{CC} estimator can be analysed as a system controlled by the signal $e_\Psi^{i\Psi}$, which is the combination of the stator current error and the rotor flux, used in the adaptation loop (Fig. 1):

$$e_\Psi^{i\Psi}(s) = \left(i_{sa}(s) - i_{sa}^e(s)\right)\Psi_{r\beta}(s) - \left(i_{s\beta}(s) - i_{s\beta}^e(s)\right)\Psi_{ra}(s). \quad (4)$$

This can be illustrated as in Fig. 2.

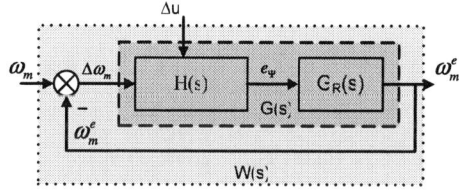

Fig. 2. Simplified block scheme of the MRAS^{CC} estimator used for stability analysis

At the output of G_R controller (PI type), the estimated rotor speed is thus obtained:

$$\omega_m^e(s) = G_R(s)e_\Psi^{i\Psi}(s), \quad (5)$$

where the transfer function of the PI controller is presented by equation:

$$G_R(s) = K_P\left(1 + \frac{1}{sT_I}\right). \quad (6)$$

For such stator current based MRAS speed estimator, the global asymptotic stability can be proved based on the second Lyapunov method, as in case of NFO^A estimator [12]. Thus the estimator stability to changes of motor and PI controller parameters can be checked using small signal perturbation analysis. The difference between the actual and the estimated rotor speed $\Delta\omega_m = \omega_m - \omega_m^e$ is used in the adjustable current model of the rotor flux (2) and in the stator current estimator (1). These equations can be written, using Laplace transform, in the following way:

$$i_{s\alpha}^e(s) = \frac{x_m r_r \Psi_{r\alpha}^i(s) + x_m x_r \Psi_{r\beta}^i(s)\omega_m + x_r^2 u_{s\alpha}}{r_r x_m^2 + r_s x_r^2 + s\sigma T_N x_s x_r^2}$$

$$i_{s\beta}^e(s) = \frac{x_m r_r \Psi_{r\beta}^i(s) - x_m x_r \Psi_{r\alpha}^i(s)\omega_m + x_r^2 u_{s\beta}}{r_r x_m^2 + r_s x_r^2 + s\sigma T_N x_s x_r^2} \quad , \quad (7)$$

$$s\Psi_{r\alpha}^i(s) = \frac{1}{T_N}(\frac{r_r}{x_r}(x_m i_{s\alpha}(s) - \Psi_{r\alpha}^i(s)) - \omega_m \Psi_{r\beta}^i(s)) \quad , \quad (8)$$

$$s\Psi_{r\beta}^i(s) = \frac{1}{T_N}(\frac{r_r}{x_r}(x_m i_{s\beta}(s) - \Psi_{r\beta}^i(s)) + \omega_m \Psi_{r\alpha}^i(s))$$

After substituting equations (7) and (8) into (4) and assuming, that during system response to the change of one of its input signals Δu (depending on the change of the speed error value $\Delta\omega_m$ – Fig. 2), f.e. the stator current $i_{s\alpha}$, changes of other variables (stator current $i_{s\beta}$ and stator voltage $u_{s\alpha}, u_{s\beta}$) are equal zero, and $\Delta\omega_m$ tends to zero ($\omega_m^e \rightarrow \omega_m = \omega_0$, what is evident for globally stable estimator), the signal $e_\Psi^{i\psi}(s)$ is described by the equation:

$$e_\Psi^{i\psi}(s) = \frac{r_r x_m x_r^3 \omega_0(r_s + sT_N \sigma x_s)}{((r_r + x_r sT_N)^2 + \omega_0^2 x_r^2)(r_r x_m^2 + x_r^2(r_s + sT_N \sigma x_s))} i_{s\alpha}^2(s) = H(s)u(s), \quad (9)$$

where:

$$H(s) = \frac{r_r x_m x_r^3 \omega_0(r_s + sT_N \sigma x_s)}{((r_r + x_r sT_N)^2 + \omega_0^2 x_r^2)(r_r x_m^2 + x_r^2(r_s + sT_N \sigma x_s))}, \quad (10a)$$

$$u(s) = f(\Delta\omega_m, i_{s\alpha}). \quad (10b)$$

Transfer function of the open loop MRAS^{CC} estimator is described by the equation:

$$G(s) = H(s)G_R(s) = \frac{(K_P + K_P sT_I)r_r x_m x_r^3 \omega_0(r_s + sT_N \sigma x_s)}{sT_I((r_r + x_r sT_N)^2 + \omega_0^2 x_r^2)(r_r x_m^2 + x_r^2(r_s + sT_N \sigma x_s))} \quad (11)$$

So the transfer function of the closed loop is as follows:

$$W(s) = \frac{\omega_m^e(s)}{\omega_m(s)} = \frac{r_r \omega_0(K_P + K_P sT_I)x_m x_r^2 b}{r_r^3 sT_I x_m^4 + sT_I(\omega_0^2 + s^3 T_N^2)x_r^4 b + r_r^2 sT_I x_r(r_r x_s + sT_N(2x_m^2 + \sigma x_s x_r)) + }{r_r x_r^2(\omega_0^2 sT_I x_m^2 + K_P \omega_0(1 + sT_I)x_m x_r b + s^3 T_I T_N(2r_r x_s + sT_N(x_m^2 + 2\sigma x_s x_r)))} \quad (12)$$
$$\text{with } b = r_r + sT_N \sigma x_s$$

The transfer function (12) of the MRAS^{CC} speed estimator $W(s)$ depends on IM parameters, the actual rotor speed, and on PI controller coefficients (K_P, T_I).

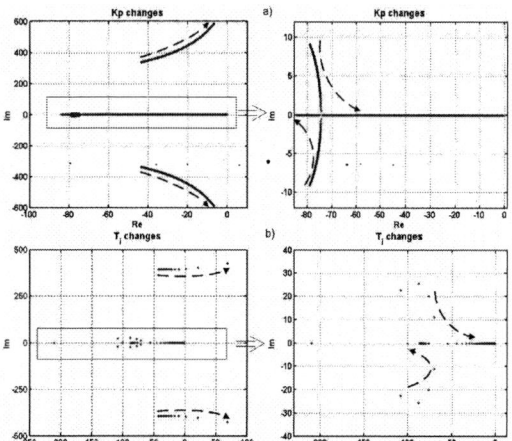

Fig. 3. Pole placement of the transfer function of the MRAS^{CC} speed estimator depending on coefficients of PI controller in the speed adaptation loop: K_P change (a), T_I change (b)

In Fig. 3 – Fig. 5 the pole placement of the MRAS^{CC} transfer function is presented, depending on IM parameters and on PI coefficients (arrows shows the change of pole placement with the increase of chosen parameter).

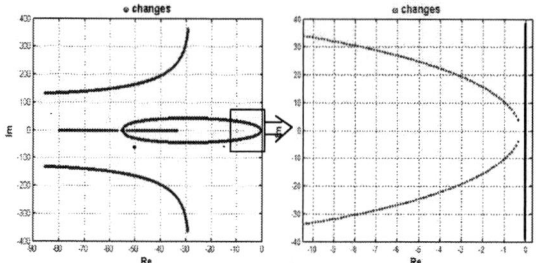

Fig. 4. Pole placement of the transfer function of the MRAS^{CC} speed estimator depending on the speed value ω_m in the range $0 \rightarrow \omega_{mN}$

The MRAS^{CC} estimator is stable for both PI controller parameter changes: K_P (Fig. 3a) and T_I. (Fig. 3b). The pole placement of the estimator is tested for parameter $K_P = (0.01 - 5000)$ [p.u.], making it possible to obtain very good dynamic performance of the MRAS^{CC} estimator. Further increasing this parameter does not improve the dynamical performance of the estimator and does not cause loosing its stability.

The change of the T_I parameter has much bigger influence on the estimator stability. It is changed in the range $T_I = (0,001 - 5)$s. Only T_I values between $(3 - 5)$s make loss of stability of the tested speed estimator (Fig. 3b).

The pole placement of the MRAS^{CC} estimator in the function of the rotor speed changes, in the range $0 \rightarrow \omega_{mN}$, is presented in Fig. 4. The change of the rotor speed does not cause the loss of the estimator stability.

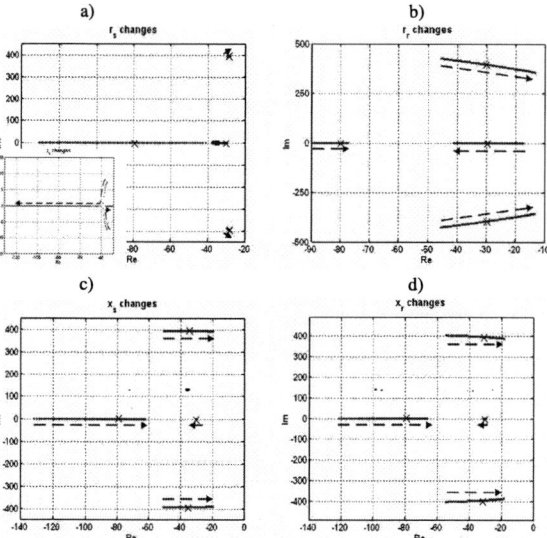

Fig. 5. Pole placement of the transfer function of the MRAS^{CC} speed estimator depending on the stator (a), rotor (b) resistance, stator (c) and rotor (d) reactance changes from -50% to +50% of their nominal values

Basic advantage of the elaborated rotor flux and speed estimator is its small sensitivity to the IM parameter changes. Results of the sensitivity tests of the MRAS^{CC} estimator to incorrect identification of stator and rotor resistances and reactances, in the range of ±50% of their nominal values are presented in Fig. 5 (points corresponding to nominal values of parameters are marked with cross).

Incorrect identification of the stator resistance to a small extent is influencing for moving poles of characteristic equation; they are located, even for the extreme mismatch of this parameter, in a left-hand half of the *s*-plane. Poor estimation of the rotor resistance causes bigger influence for moving poles of the characteristic equation in (12), however does not bring them for the right-hand half of the *s*-plane, even for very big identification errors, in the range of +50%. Similarly, the incorrect identification (or changes) of the stator and rotor reactances in the range of ±50% of their nominal value does not involve loss of the estimator stability.

IV. ROBUSTNESS OF THE DFOC SYSTEM WITH MRAS^{CC} ESTIMATOR TO PARAMETER UNCERTAINTIES

The robustness of the MRAS^{CC} speed estimator to the induction motor parameter changes is also tested in the closed-loop operation, in the field-oriented vector controlled drive. The estimation errors of the rotor speed and flux are calculated from simulated transients of the DFOC drive equipped with the MRAS^{CC} estimator. In the rotor flux and speed control loops the state variable values obtained from the MRAS^{CC} estimator are used. The sensitivity analysis is performed for constant coefficients of PI controller in the speed adaptation mechanism of the MRAS^{CC} estimator, under steady state operation of the drive system, with chosen reference speed and nominal load torque.

The rotor speed and flux estimation errors are calculated as follows:

$$\Delta_{\Psi_r} = \frac{\sum_{i=1}^{N} \frac{\Psi_r - \Psi_{rest}}{\Psi_r}}{N} 100\%, \quad \Delta_\omega = \frac{\sum_{i=1}^{N} \frac{\omega_m - \omega_{est}}{\omega_m}}{N} 100\% \quad (13)$$

where: $\Psi_r, \Psi_{rest}, \omega_m, \omega_{est}$ – actual and estimated rotor flux magnitudes and speed in each numerical step, N – number of calculation steps .

To show clearly the advantages of the developed MRAS^{CC} estimator, the similar sensitivity analysis is performed for the current-based MRAS -type speed estimator [10], which uses the voltage model for the rotor flux vector calculation (referred here as MRAS^{CV}) and for the classical, rotor flux model-based MRAS^{F} estimator [6]. The sensitivity results are presented in Fig. 6 – Fig. 8.

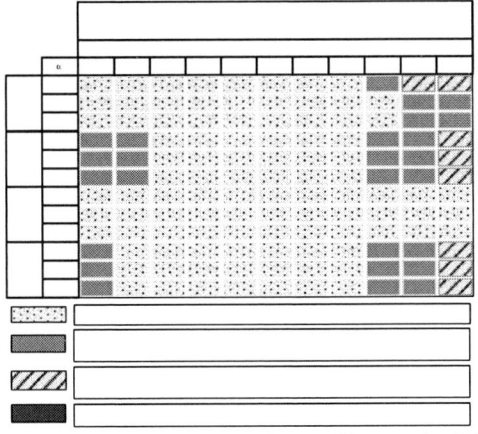

Fig. 6. Sensitivity to motor parameter changes of the sensorless field oriented control with MRAS^{CC} estimator

It is seen distinctly from the comparison of tables in Fig. 6 – Fig. 8, that the new MRAS^{CC} speed estimator based on the stator current estimator and current model of rotor flux is much less sensitive to the IM parameter changes or identification errors than MRAS^{CV} and the classical MRAS^{F} estimators, and works well for different speed levels.

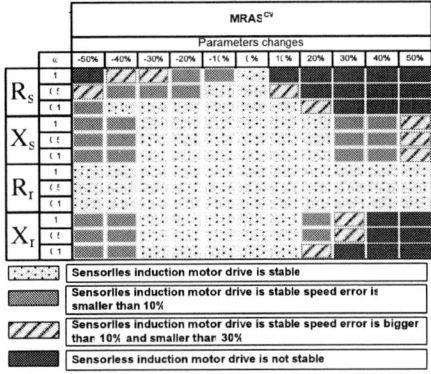

Fig. 7. Sensitivity to motor parameter changes of the sensorless field oriented control with MRAS^{CV} [10] estimator

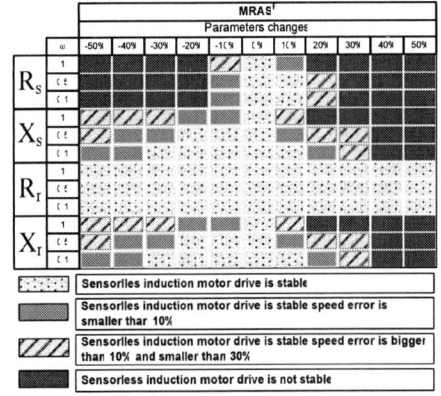

Fig. 8. Sensitivity to motor parameter changes of the sensorless field oriented control with MRAS^{F} [6] estimator

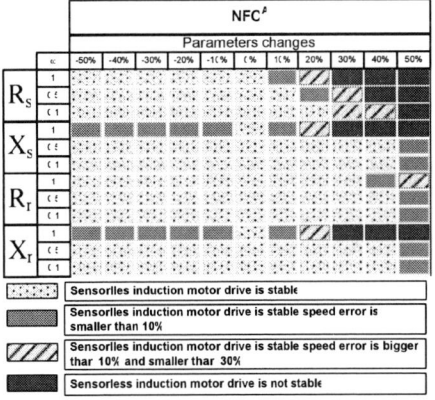

Fig. 9. Sensitivity to motor parameter changes of the sensorless field oriented control with NFO^{A} scheme [12]

Moreover it is much robust to parameter uncertainties than the well known nonlinear full-order observer with speed adaptation (NFO^{A}) [12], which is sometimes considered also as a current error-based MRAS scheme [13], what is shown in Fig. 9.

2309

The MRAS^{CC} estimator can reconstruct the rotor flux and speed with much lower errors for misaligned motor parameters as well for the dynamical as for steady-state drive system operation. For bigger parameter errors, the drive system with the MRAS^{CV} and MRAS^F estimators, as well as with the NFO^A scheme, can even lose the stability (Fig. 7 – Fig. 9), whereas the new MRAS^{CC} estimator works stable in the DFOC drive system even for ±50% parameter changes, as is shown in Fig. 6.

V. CHOSEN EXPERIMENTAL RESULTS

The sensorless induction motor drive with MRAS^{CC} estimator is implemented in the laboratory set-up with PC computer using the dSPACE software. The schematic diagram of the experimental test bench is shown in Fig. 10.

The experimental set-up is composed of the IM motor fed by the SVM voltage inverter. The motor is coupled to a load machine (AC motor supplied from a inverter).

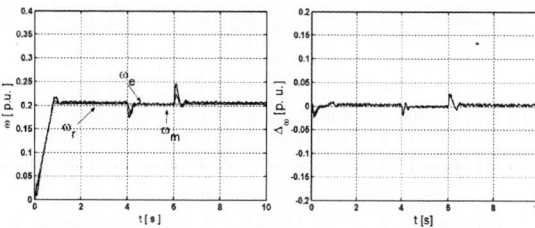

Fig. 10. Schematic diagram of the laboratory test bench

The driven motor has the nominal power of 1.5 kW. The speed and position of the drive are measured by the incremental encoder (36000 imp./rev), only for the comparison with the estimated speed in the sensorless drive system. The control and estimation algorithms are implemented in DS1103 card. In Fig. 11 to Fig. 15 chosen drive system transients, in different operation regimes are presented. In Fig. 11 transient of the estimated and measured speed for start-up of induction motor drive to 20% of the nominal speed is presented. For $t_1 = 4$s the nominal load torque is applied. Error between the measured and estimated speeds is equal zero in a steady-state; during load change transients, very small dynamical errors occur.

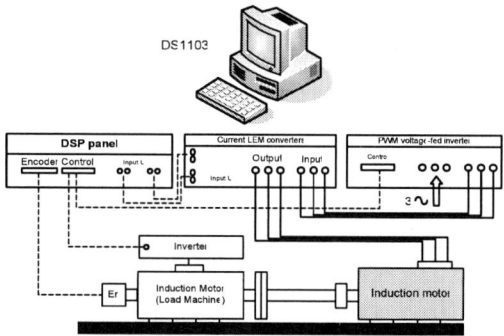

Fig. 11. Experimental transients of the IM drive with MRAS^{CC} estimator for start-up to $\omega_{ref} = 0.2\omega_{mN}$ and load torque changes $m_o=m_N$

Next in the Fig. 12 the fast reverse operation of the sensorless DFOC drive, in the low speed range is demonstrated ($\omega_{ref} = \pm 0.04 \omega_{mN}$). The similar reverse operation in the low speed region, but with simultaneous changes of

the load torque, is presented in Fig. 13. Also in this case the speed error (calculated between the estimated and measured rotor speed) is equal zero, except transients caused by the fast speed reference or load torque changes.

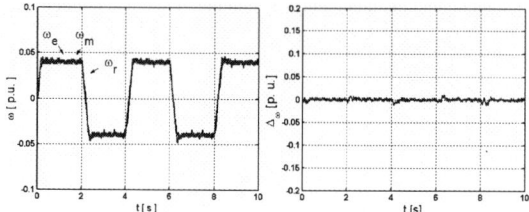

Fig. 12. Experimental transients of the IM drive with MRAS^{CC} estimator for reverse operation at low speed $\omega_{ref} = \pm 0.04\,\omega_{mN}$ and $m_o=0.2m_N$

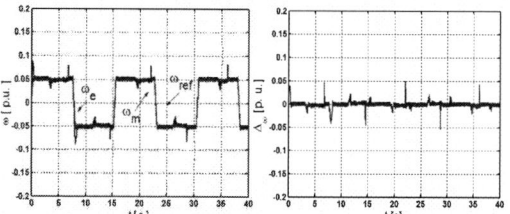

Fig. 13. Experimental transients of the IM drive with MRAS^{CC} estimator for the reverse operation at low speed reference (±0.05ω_{mN}) and load torque changes $m_o = m_N$

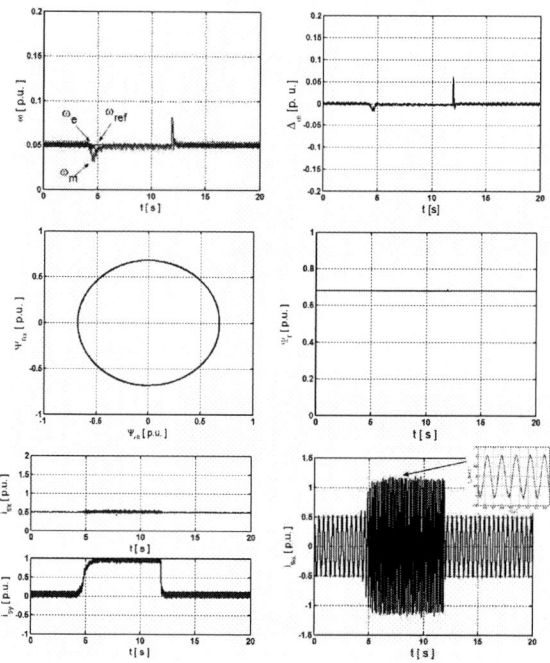

Fig. 14. Experimental transients of the IM drive with MRAS^{CC} estimator (reference, measured and estimated speed (a), speed difference (b), rotor flux hodograph (c) and magnitude (d), stator current components (e), phase current (f) for constant speed reference $\omega_{ref}=0.05\,\omega_{mN}$, and load torque change from $m_o=0$ to $m_o=m_N$

To show the proper operation of the DFOC structure in the low speed region, under full load torque, not only speed but also internal control variables transients are shown in Fig. 14. It can be seen, that the rotor flux magnitude is constant (nominal value, in [p.u.]) and stator current components in

2310

the flux and torque control loops, in rotating field coordinate system *x-y*, as well as the phase stator current, react correctly to the step changes of the load torque.

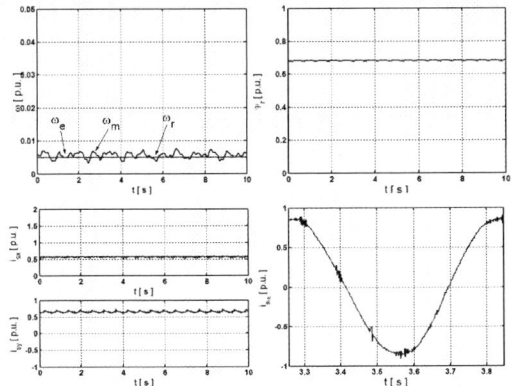

Fig. 15. Experimental transients of the IM drive with MRAS^{CC} estimator for constant speed reference $\omega_{ref}=0.005\,\omega_{mN}$, $m_o=0.9m_N$

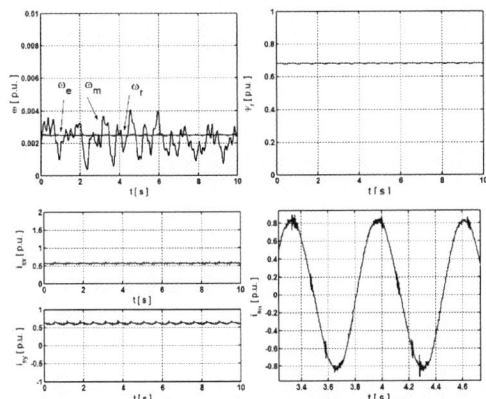

Fig. 16. Experimental transients of the IM drive with MRAS^{CC} estimator for constant speed reference $\omega_{ref}=0.0025\,\omega_{mN}$, $m_o=0.9m_N$

In Fig. 15 and Fig. 16 the very low speed operation is demonstrated for $\omega_m=0.005\,\omega_{mN}$ (0,5%≈14rpm≈1,46rad/s) and $\omega_m=0.0025\,\omega_{mN}$ (0,25%≈7rpm≈0,73rad/s), respectively. It is seen that the sensorless drive system works properly, without any changes in the design parameters (K_P, K_I) of the new MRAS^{CC} estimator and controllers of the DFOC structure. The transient speed reconstruction error is close zero.

In the analyzed sensorless drive system, both speeds, estimated and measured, are equal each other not only for the steady state operation but also under speed reference and load torque changes the speed estimation error tends to zero very fast (Fig. 11 to Fig. 16). The motor speed, measured only for comparison, tunes the estimated value in all drive system operation conditions. The developed current-based MRAS^{CC} speed estimator performs very well in the whole speed range, including very low speeds.

VI. CONCLUSION

The new speed estimator based on the MRAS concept, which uses the rotor flux current-based model and stator current estimator, performs very well in the wide range of the speed reference, in the sensorless DFOC drive system. The application of the real induction motor as a kind of a reference object in this MRAS estimator concept, as well as the adjustable stator current estimator and current flux

model, improves the robustness of the speed reconstruction to motor parameters uncertainties. It works stably even for relatively big parameters' mismatch (±50%), in contrary to the classical rotor flux error-based MRAS^{F} estimator [6], other type of stator current error-based MRAS^{CV} [10] or even NFO^{A} [12] estimators.

The proposed speed estimator can be easily implemented in the microprocessor system and can be used not only for the speed but also for the simultaneous rotor flux vector and stator current reconstruction. The experimental verification confirmed the good behaviour of the rotor flux and speed estimation in the closed-loop operation of the vector controlled drive, in the wide speed range. Drive system works well for very low speed reference with load torque. The stability analysis confirms very good dynamical and robustness performance of the proposed MRAS^{CC} rotor speed and flux estimator.

ACKNOWLEDGMENT

Research work financed by The Ministry of Sciences and Higher Education (Poland) under Grant N510 023 32/2345 (2007-2009)

APPENDIX

MOTOR DATA		RATED VALUES	
$P_N = 1.5$	[kW]	$n_N = 2820$	[rpm]
$U_N = 230/400$	[V]	$f_N = 50$	[Hz]
$I_N = 5.9/3.4$	[A]	$p_b = 3$	

Parameters: $T_M=0.0188$ [s]

R_s	R_r	X_s	X_r	X_m	
3.68	4.033	119.93	119.93	115.77	[Ω]
0.0543	0.0595	1.769	1.769	1.7076	[p.u.]

REFERENCES

[1] Vas P., *Sensorless vector and direct torque control*. Oxford University Press, New York, 1998.

[2] Kazmierkowski M. P., Blaabjerg F., Krishnan, *Control in Power Electronics – Selected Problems*, Academic Press, USA, 2002

[3] Orlowska-Kowalska T., *Sensorless induction motor drives*, Wroclaw University of Technology Press, Wroclaw, 2003.

[4] Holtz J., Sensorless Control of Induction Machines - with or without Signal Injection, *IEEE Trans. on Ind. Electr.*,vol.53, No.1, 2006, pp.7-30

[5] Orlowska-Kowalska T., Application of the Extended Luenberger Observer for Flux and Rotor Time Constant Estimation in Induction Motor Drives, *IEE Proceedings, Part D*, 1989, vol.136, pp.323-330

[6] Tamai S., Sugimoto H., Masao Y., Speed sensorless vector control of IM with Model Reference Adaptive System, *Proc. of IEEE/IAS*, pp.189-195, 1987

[7] Schauder C., Adaptive speed identification for vector control of induction motors without rotational transducers, *IEEE Trans. on Ind. Appl.*, vol. 28, No. 5, pp. 1054-1061, 1992

[8] Peng F.Z., Fukao T., Robust speed identification for speed-sensorless vector control of induction motors, *IEEE Trans. on Industry Applications*, vol. 30, no. 5, 1994, pp.1234-1240

[9] Rashed M., Stronach A.F., A stable back-EMF MRAS-based sensorless low-speed induction motor drive insensitive to stator resistance variation, *IEE Proc.-Electr. Power Applic.*, vol. 151, No.6, 2004, pp.685-693

[10] Sobczuk D.L., Application of ANN for control of PWM inverter fed induction motor drives, *Ph.D. Thesis, Warsaw UT*, Poland, 1999.

[11] Orlowska-Kowalska T., Dybkowski M., Dynamical properties of induction motor drive with novel MRAS estimator, *Electrical Review* (Poland), R. 82, no. 11, 2006, pp. 35-38.

[12] Kubota H., Matsuse K., Nakano T., New adaptive flux observer for wide speed range motor drives, *Proc. IEEE-IECON Conf. 1990*, pp.921-926.

[13] Ohyama K., Asher G. M., Sumner M., Comparative Analysis of Experimental Performance and Stability of Sensorless Induction Motor Drives, *IEEE Trans. on Ind. Electr.*, vol.53, No.1, 2006, pp. 178–186.

[14] Tajima Y, Hori Y., Speed Sensorless Field Oriented Control of the Induction Machine, *IEEE Trans. Ind. Appl.*, vol.29, No.1, 1993, pp.175-180

[15] Orlowska-Kowalska T., Dybkowski M., Improved MRAS-type speed estimator for the sensorless induction motor drive, *Proc. of XIX Symp. EPNC'2006*, Maribor, Slovenia, 2006, pp.105-106.

[16] Orlowska-Kowalska T., Wojsznis P., Kowalski C.T., Dynamical performances of sensorless induction motor drive with different flux and speed observers, *Proc. of 10th Int. Conf. EPE'2001*, Graz, Austria, 2001.

State and Parameter Estimation in Induction Motors using Sliding Modes

Sachit Rao*, Martin Buss*, Vadim Utkin†

*Institute of Automatic Control Engineering, TU-Muenchen, Munich, Germany, e-mail: *sachit.rao,mb@tum.de*
†Department of Electrical Engineering, Ohio State University, Colmbus, Ohio, USA, e-mail: *utkin.2@osu.edu*

Abstract— We present two observer based solutions to the popular problem of sensorless speed control of induction motors. The observers which provide simultaneous motor speed and flux estimation are based on the basic principles of sliding mode control theory. We present an observer, which incorporates rotor flux dynamics, that can also be used to simulataneously estimate rotor resistance (if unknown). We propose a second observer design, defined by stator flux dynamics, that is independent of the unknown rotor resistance. As for the controller of the induction motor itself, which now relies on the outputs of each observer, we propose designs again based on sliding mode control theory.

I. INTRODUCTION

Discontinuous control offered by sliding mode control theory makes it a natural choice for controlling inverter driven induction machines. Apart from this advantage, this theory provides simplicity in control design and robustness against unknown or time-varying plant parameters. However, irrespective of the control algorithm, information on motor speed, flux linkages etc. are required. Recent advanced developments in data acquisition and processing allow the implementation of the sensorless control idea, this paper presumes the availability of such devices and offers a sliding mode based observer-controller system.

Control of an induction machine has been a subject of much study over the past several decades. Sliding mode control has also been presented abundantly in technical literature, both from a theoretical and implementation perspective. Hence, in this paper, background details are kept to the minimum and the interested or novice control engineer is directed to [1], [2], and [3] for fundamentals of sliding mode control theory and induction machines.

A. Two Sets of Machine Equations

Controlling motor speed by controlling torque and flux independently, but without the use of speed or flux sensors is the main focus of this paper. To solve this fundamental problem, we study the induction machine at the equation level–using the system of equations derived in [4]. In the standard (α, β) frame, the induction machine equations can be written in terms of the stator currents i_α, i_β and

---

¹This research was supported in part by the Deutsche Forschungsgemeinschaft (German Research Foundation), for which the authors are grateful.

---

the rotor fluxes $\psi_{r\alpha}, \psi_{r\beta}$ as

$$
\begin{aligned}
\dot{i}_\alpha &= \beta\eta\psi_{r\alpha} + \beta\omega\psi_{r\beta} - \gamma i_\alpha + \frac{1}{\sigma L_S}u_\alpha \\
\dot{i}_\beta &= \beta\eta\psi_{r\beta} - \beta\omega\psi_{r\alpha} - \gamma i_\beta + \frac{1}{\sigma L_S}u_\beta \\
\dot{\psi}_{r\alpha} &= -\eta\psi_{r\alpha} - \omega\psi_{r\beta} + \eta L_M i_\alpha \\
\dot{\psi}_{r\beta} &= -\eta\psi_{r\beta} + \omega\psi_{r\alpha} + \eta L_M i_\beta \\
J\dot{\omega} &= \frac{PL_M}{L_R}T_i - T_L; \quad T_i = (\psi_{r\alpha}i_\beta - \psi_{r\beta}i_\alpha)
\end{aligned}
\tag{1}
$$

where u_α, u_β are the input stator voltages, ω is the rotor speed, and the positive constants

$$
\eta = \frac{R_R}{L_R}, \quad \sigma = 1 - \frac{L_M^2}{L_S L_R}
$$

$$
\beta = \frac{PL_M}{\sigma L_S L_R}, \quad \gamma = \frac{R_S + (L_M^2/L_R^2)R_R}{\sigma L_S}
$$

are functions of the self and mutual inductances $L_{S,R,M}$ and the stator and rotor resistances $R_{S,R}$; J, P, and T_L are the rotor mass moment of inertia, number of pole-pairs, and the load torque respectively.

By making the reasonable assumption that the magnetic circuits are linear, we are allowed to use the relations

$$
\begin{aligned}
\psi_{s\alpha} &= L_S i_{s\alpha} + L_M i_{r\alpha} \\
\psi_{s\beta} &= L_S i_{s\beta} + L_M i_{r\beta} \\
\psi_{r\alpha} &= L_M i_{s\alpha} + L_R i_{r\alpha} \\
\psi_{r\beta} &= L_M i_{s\beta} + L_R i_{r\beta}
\end{aligned}
\tag{2}
$$

where $i_{r\alpha,\beta}$ are the rotor currents, to represent the induction motor in terms of stator currents and stator fluxes $\psi_{s\alpha,\beta}$ as

$$
\begin{aligned}
\dot{i}_\alpha &= -\gamma_s i_\alpha - P\omega i_\beta + \frac{1}{\sigma L_S}(\eta\psi_{s\alpha} + P\omega\psi_{s\beta}) \\
&\quad + \frac{1}{\sigma L_S}u_\alpha \\
\dot{i}_\beta &= P\omega i_\alpha - \gamma_s i_\beta - \frac{1}{\sigma L_S}(P\omega\psi_{s\alpha} - \eta\psi_{s\beta}) \\
&\quad + \frac{1}{\sigma L_S}u_\beta \\
\dot{\psi}_{s\alpha} &= -R_S i_\alpha + u_\alpha \\
\dot{\psi}_{s\beta} &= -R_S i_\beta + u_\beta \\
J\dot{\omega} &= PT_i - T_L; \quad T_i = (\psi_{s\alpha}i_\beta - \psi_{s\beta}i_\alpha) \\
\gamma_s &= \frac{R_S L_R + R_R L_S}{\sigma L_S L_R}
\end{aligned}
\tag{3}
$$

The unknown quantities in both the models of the induction machine are the speed ω, rotor and stator fluxes $\psi_{r,s\alpha,\beta}$, and finally, the rotor resistance R_R, which can vary significantly, albeit slowly, with temperature. The problem statements can be described as follows: first, estimate the unknown variables and second, use these estimates to design a speed control strategy. An observer is designed to solve the first problem and a controller based on the Field-Oriented Control (FOC) technique [2], [4], is used for the second; sliding mode control theory is applied in solving both problems. For more ideas on observers: [5] which provides an exhaustive survey on sliding mode based observer designs for AC machines and [6], [7] which offers a basic version of the observer proposed in this paper.

B. Sensorless Control

If the induction motor is used as an actuator, as it usually is, then an important objective could either be to maintain a constant rotor speed (speed control) or to rotate the rotor by a pre-specified angle (position control). We only consider the speed control problem, the position control problem can be solved by a simple extension and adding a position sensor. To control rotor speed, we use from the mechanical equation in (1) and (3)

$$J\dot{\omega} = k_{r,s}T_i - T_L, \quad k_r = (PL_M/L_R), k_s = P$$

the torque function

$$T_i = (\psi_{s,r\alpha}i_{s\beta} - \psi_{s,r\beta}i_{s\alpha}) \quad (4)$$

as control. For T_i to be used as control, all the states which it is dependent on should be available for measurement. The flux variables, $\psi_{r,s\alpha,\beta}$, can be measured only in the air-gap, and to place sensors in that location would drive up the cost of the motor considerably, there is also the reliability of the sensor output to be considered. Moreover, to maintain speed, at say, a constant value $\omega = \omega_{\text{ref}}$, ω needs to be measured. Again, the reasons of cost are sufficient to motivate a search for other techniques that allow for its sensor-less estimation. Of course, we have to presume that the stator currents $i_{\alpha,\beta}$ can be easily measured with the help of robust, fast, and inexpensive current sensors.

These issues have led to the term 'sensor-less' control. The problem statement can be summed up thus: control speed without measuring it. In order to solve this problem, we use observers. The reasons are three-fold:

- to estimate ω,
- to measure $\psi_{r,s\alpha,\beta}$ without adopting integration techniques which would require exact knowledge of their initial values, and
- to reduce chattering when using a sliding mode speed controller.

II. CONTROL LAWS AND THEIR REQUIREMENTS

Since we proposed to use the input torque T_i to control speed, a possible intermediate control law is

$$T_i = \frac{(-\alpha J(\omega - \omega_{\text{ref}}) + T_L)}{k_{r,s}} = T_{\text{ref}}, \quad \alpha > 0 \quad (5)$$

Since $T_i = (\psi_{r,s\alpha}i_\beta - \psi_{r,s\beta}i_\alpha)$, we would need to control $i_{\alpha,\beta}$ and $\psi_{r,s\alpha,\beta}$ as well using the real control inputs $u_{\alpha,\beta}$ so that $T_i = T_{\text{ref}}$; [8] provides details on implementation of assigning desired $u_{\alpha,\beta}$ using inverters.

To design a flux and current controller, we adopt the Field Oriented Control (FOC) technique. The main philosophy of this technique is the independent control of current and flux through decoupling. In the new frame of reference, (d, q) which is the result of applying this technique, the currents and fluxes defined in the (α, β) frame are aligned with the respective flux vector, leading to a reduced order and decoupled set of equations. This order reduction is possible through the use of the transformations

$$i_{r,sq} = \frac{i_\beta \psi_{r,s\alpha} - i_\alpha \psi_{r,s\beta}}{|\psi_{r,sd}|}$$

$$i_{r,sd} = \frac{i_\beta \psi_{r,s\beta} + i_\alpha \psi_{r,s\alpha}}{|\psi_{r,sd}|}$$

$$|\psi_{r,sd}| = \sqrt{\psi_{r,s\alpha}^2 + \psi_{r,s\beta}^2}, \; |\psi_{r,sq}| = 0 \quad (6)$$

which leads to the rotor flux model

$$\dot{i}_{rq} = -\gamma i_{rq} - \omega i_{rd} - \beta\omega\psi_{rd} - \eta L_M \frac{i_{rq}i_{rd}}{\psi_{rd}} + \frac{1}{\sigma L_S}u_q$$

$$\dot{i}_{rd} = -\gamma i_{rd} + \omega i_{rq} + \beta\omega\psi_{rd} + \eta L_M \frac{i_{rq}^2}{\psi_{rd}} + \frac{1}{\sigma L_S}u_d$$

$$\dot{\psi}_{rd} = -\eta\psi_{rd} + \eta L_M i_{rd}$$

and

$$\dot{i}_{sq} = -\gamma_s i_{sq} + P\omega i_{sd} - \frac{P\omega}{\sigma L_S}\psi_{sd} + R_S\frac{i_{sd}i_{sq}}{\psi_{sd}}$$

$$\quad + (\frac{1}{\sigma L_S} - \frac{i_{sd}}{\psi_d})u_q$$

$$\dot{i}_{sd} = -\gamma_s i_{sd} - P\omega i_{sq} + \frac{\eta}{\sigma L_S}\psi_{sd} + \frac{1}{\sigma L_S}u_d$$

$$\quad + \frac{i_{sq}u_q}{\psi_{sd}} - \frac{R_S}{\psi_{sd}}i_{sq}^2$$

$$\dot{\psi}_{sd} = -R_S\psi_{sd} + u_d$$

as the stator flux model in the (d, q) frame.

The input torque is now given by $T_i = \psi_{r,sd}i_{r,sq}$. Thus, $\psi_{r,sd}, i_{r,sq}$ can be controlled independently using the real control inputs $u_{\alpha,\beta}$, which in the new frame are transformed into $u_{d,q}$ using

$$u_q = \frac{u_\beta \psi_{r,s\alpha} - u_\alpha \psi_{r,s\beta}}{|\psi_{r,sd}|}$$

$$u_d = \frac{u_\beta \psi_{r,s\beta} + u_\alpha \psi_{r,s\alpha}}{|\psi_{r,sd}|} \quad (9)$$

We design sliding mode controllers for $\psi_{r,sd}$ and $i_{r,sq}$ under the assumption that all states and machine parameters are known. We will then identify the states that can or cannot be measured and design an observer for their estimation. As stated in the abstract, we propose two observers based on (1) and (3).

A. Sliding mode speed controller

A sliding mode controller is designed using a 2 step design procedure; we illustrate this procedure, for brevity, only for the stator flux model, [9] offers more design details for designing a controller based on the rotor flux model. First, choose as surfaces

$$s_{sq} = i_{sq}^* - i_{sq}$$
$$s_{sd} = \psi_{sd}^* - \psi_{sd} \tag{10}$$

such that $i_{sq}^* \psi_{sd}^* = T_{\text{ref}}$. Second, choose discontinuous controls

$$u_q = M_q \text{sign}(s_{sq})$$
$$u_d = M_d \text{sign}(s_{sd}) \tag{11}$$

For simplicity, we assume that i_{sq}^*, ψ_{sd}^* are constants. To show that sliding mode occurs, calculate

$$\dot{s}_{sq} = \gamma_s i_{sq} - P\omega i_{sd} - \frac{P\omega}{\sigma L_S}\psi_{sd} - R_S \frac{i_{sd}i_{sq}}{\psi_{sd}}$$
$$- \left(\frac{1}{\sigma L_S} - \frac{i_{sd}}{\psi_{sd}}\right)M_q \text{sign}(s_{sq})$$
$$\dot{s}_{sd} = -R_S\psi_{sd} - M_d \text{sign}(s_{sd}) \tag{12}$$

so that for high enough $M_{q,d} > 0$, $i_{sq} \to i_{sq}^*$ and $\psi_{sd} \to \psi_{sd}^*$ in finite time [1]. The closed-loop motion equations are of reduced order (by 2) and is indeed the dynamics of the direct component of the current i_{sd}. To analyse the stability of this equation and to show that $\left(\frac{1}{\sigma L_S} - \frac{i_{sd}}{\psi_{sd}^*}\right) \neq 0$, calculate equivalent controls $u_{d,qeq}$ by solving for them from $\dot{s}_{sd,q} = 0$ and substitute them in the dynamics of i_{sd} in (3). For simplicity, we perform the transformation $x = \left(\frac{\psi_{sd}^*}{L_S} - i_{sd}\right)$ which leads to

$$\dot{x} = -\frac{1}{x}\left(\gamma_s x^2 - \frac{\psi_{sd}^*}{L_S}\gamma_s x + \frac{\eta}{\sigma}i_{sq}^{*2}\right) - \frac{\psi_{sd}^*}{\sigma L_S}(\eta + R_S) \tag{13}$$

and whose linearized version is stable about $x \neq 0$. This procedure is valid as the current component i_{sd} is non-zero when the stator and rotor coils are magnetized, let this steady state value be i_{sd}^*. Hence, our proposed sliding mode control laws (5) and (10) will ensure that $\omega = \omega_{\text{ref}}$.

B. Requirements for control

Irrespective of the model chosen for the induction machine, we would need information on $\psi_{r,s\alpha,\beta}$ and ω to evaluate discontinuous controls, similar to (11), for the enforcement of sliding mode and to implement the intermediate law (5). To estimate the unknown states, we propose sliding mode based flux and speed observers, whose design and convergence properties will be presented in the remainder of the paper.

III. FLUX AND SPEED OBSERVERS

A. Models

After having identified the requirements of control, we satisfy them by proposing sliding mode based observers for either model of the induction machine that the user prefers to choose. Hence, for the rotor flux model, where

we also claimed that the rotor resistance R_R is unknown, we use from (1)

$$\dot{\hat{i}}_\alpha = \beta\hat{\eta}\hat{\psi}_{r\alpha} + \beta\hat{\omega}\hat{\psi}_{r\beta} - \gamma i_\alpha + \frac{1}{\sigma L_S}u_\alpha - \beta\hat{\psi}_{r\alpha}\mu$$
$$\dot{\hat{i}}_\beta = \beta\hat{\eta}\hat{\psi}_{r\beta} - \beta\hat{\omega}\hat{\psi}_{r\alpha} - \gamma i_\beta + \frac{1}{\sigma L_S}u_\beta - \beta\hat{\lambda}_\beta\mu$$
$$\dot{\hat{\psi}}_{r\alpha} = -\hat{\eta}\hat{\psi}_{r\alpha} - \hat{\omega}\hat{\psi}_{r\alpha} + \hat{\eta}L_M i_\alpha + C\hat{\psi}_{r\beta}\mu \tag{14}$$
$$\dot{\hat{\psi}}_{r\beta} = -\hat{\eta}\hat{\psi}_{r\beta} + \hat{\omega}\hat{\psi}_{r\alpha} + \hat{\eta}L_M i_\beta - C\hat{\psi}_{r\alpha}\mu$$
$$\hat{\eta}(t) = \frac{\hat{R}_R(t)}{L_R}$$

where $\hat{\eta}(t)$ is the estimate of η. For the stator flux model, from (3), we propose

$$\dot{\hat{i}}_\alpha = -\gamma_s i_\alpha - P\hat{\omega}i_\beta + \frac{1}{\sigma L_S}(\eta\hat{\psi}_{s\alpha} + P\hat{\omega}\hat{\psi}_{s\beta})$$
$$+ \frac{1}{\sigma L_S}u_\alpha + \hat{\psi}_{s\alpha}\mu$$
$$\dot{\hat{i}}_\beta = P\hat{\omega}i_\alpha - \gamma_s i_\beta - \frac{1}{\sigma L_S}(P\hat{\omega}\hat{\psi}_{s\alpha} - \eta\hat{\psi}_{s\beta})$$
$$+ \frac{1}{\sigma L_S}u_\beta + \hat{\psi}_{s\beta}\mu$$
$$\dot{\hat{\psi}}_{s\alpha} = -R_S i_\alpha + u_\alpha - \hat{\psi}_{s\beta}\mu$$
$$\dot{\hat{\psi}}_{s\beta} = -R_S i_\beta + u_\beta + \hat{\psi}_{s\alpha}\mu \tag{15}$$
$$\gamma_s = \frac{R_S L_R + R_R L_S}{\sigma L_S L_R}$$

$\hat{\omega}, \mu$ are inputs for each observer, which have to be designed separately. $\hat{\psi}_{r,s\alpha,\beta}, \hat{i}_{\alpha,\beta}$ are the estimates of the respective fluxes and stator currents.

We further justify our separate sliding mode observer designs:

1) We can prove the adaptive capability of the rotor flux observer (14) by showing that $\hat{\eta} \to \eta_{\text{real}}$: thus ensuring that the observer will yield accurate estimates of flux and speed.

2) We can completely ignore the effect of unknown η by using the stator flux observer (15). This proposition can be justified by closely observing the machine equations in the (d, q) frame for each model. In (7), $\psi_{rd} \to \psi_{rd}^*$ can be achieved only by using i_{rd} as fictitous control, and this control law would require exact knowledge of η_{real}, which would make observer implementation without this knowledge or without adaptation inaccurate. However, in (8), the real control control u_d can be directly selected to ensure $\psi_{sd} \to \psi_{sd}^*$ without direct dependence on η_{real}.

B. Performance

We follow the 2-step procedure outlined before for the design of a sliding mode controller to select the observer inputs $\hat{\omega}, \mu$ by, first, selecting as surfaces

$$s_{1r,s} = \tilde{i}_\beta\hat{\psi}_{r,s\alpha} - \tilde{i}_\alpha\hat{\psi}_{r,s\beta}$$
$$s_{2r,s} = \tilde{i}_\alpha\hat{\psi}_{r,s\alpha} + \tilde{i}_\beta\hat{\psi}_{r,s\beta} \tag{16}$$

2314

where $\tilde{i}_{\alpha,\beta} = \hat{i}_{\alpha,\beta} - i_{\alpha,\beta}$ are the current estimate errors; we similarly define $\tilde{\psi}_{r,s\alpha,\beta} = \hat{\psi}_{r,s\alpha,\beta} - \psi_{r,s\alpha,\beta}$ as the flux estimate errors. In order to enforce sliding mode and eventually select $\hat{\omega}, \mu$, we evaluate, from the rotor flux model

$$\dot{s}_{1r} = \beta\eta\left(\hat{\psi}_{r\alpha}\tilde{\psi}_{r\beta} - \tilde{\psi}_{r\alpha}\hat{\psi}_{r\beta}\right) - \beta\hat{\omega}\left(\hat{\psi}_{r\alpha}^2 + \hat{\psi}_{r\beta}^2\right)$$
$$+ \beta\omega\left(\hat{\psi}_{r\alpha}\psi_{r\alpha} + \hat{\psi}_{r\beta}\psi_{r\beta}\right)$$
$$- \hat{\eta}\left(s_{1r} + L_M\left(i_\beta\tilde{i}_\alpha - i_\alpha\tilde{i}_\beta\right)\right) + (C\mu - \hat{\omega})s_{2r}$$
$$\dot{s}_{2r} = \beta\left(\hat{\psi}_{r\alpha}^2 + \hat{\psi}_{r\beta}^2\right)(\tilde{\eta} - \mu) + \beta\eta\left(\hat{\psi}_{r\alpha}\tilde{\psi}_{r\alpha} + \hat{\psi}_{r\beta}\tilde{\psi}_{r\beta}\right)$$
$$+ \beta\omega\left(\psi_{r\alpha}\hat{\psi}_{r\beta} - \hat{\psi}_{r\alpha}\psi_{r\beta}\right)$$
$$- \hat{\eta}\left(s_{2r} + L_M\left(i_\alpha\tilde{i}_\alpha + i_\beta\tilde{i}_\beta\right)\right) + (\hat{\omega} - C\mu)\,s_{1r} \tag{17}$$

and, from the stator flux model

$$\dot{s}_{1s} = P\tilde{\omega}\hat{i}_{sd}\hat{\psi}_{sd} - \frac{P}{\sigma L_S}\hat{\omega}\hat{\psi}_{sd}^2 + \frac{\eta}{\sigma L_S}\tilde{\psi}_{sq}\hat{\psi}_{sd}$$
$$+ \frac{P}{\sigma L_S}\omega\left(\hat{\psi}_{sd}^2 - \tilde{\psi}_{sd}\hat{\psi}_{sd}\right) + \left(\tilde{i}_\beta\dot{\hat{\psi}}_{s\alpha} - \tilde{i}_\alpha\dot{\hat{\psi}}_{s\beta}\right)$$
$$\dot{s}_{2s} = -P\tilde{\omega}\hat{i}_{sq}\hat{\psi}_{sd} + \frac{\eta}{\sigma L_S}\tilde{\psi}_{sd} - \frac{P}{\sigma L_S}\omega\tilde{\psi}_{sq}\hat{\psi}_{sd}$$
$$+ \mu\hat{\psi}_{sd}^2 + \left(\tilde{i}_\alpha\dot{\hat{\psi}}_{s\alpha} + \tilde{i}_\beta\dot{\hat{\psi}}_{s\beta}\right) \tag{18}$$

where $\hat{\psi}_{sd}^2 = \left(\hat{\psi}_{s\alpha}^2 + \hat{\psi}_{s\beta}^2\right) > 0$ and

$$\tilde{\psi}_{sq} = \left(\hat{\psi}_{s\alpha}\tilde{\psi}_{s\beta} - \hat{\psi}_{s\beta}\tilde{\psi}_{s\alpha}\right)/\left(\hat{\psi}_{sd}\right)$$
$$\tilde{\psi}_{sd} = \left(\hat{\psi}_{s\alpha}\tilde{\psi}_{s\alpha} + \hat{\psi}_{s\beta}\tilde{\psi}_{s\beta}\right)/\left(\hat{\psi}_{sd}\right) \tag{19}$$

to choose as discontinuous controls

$$\hat{\omega} = \omega_0\mathrm{sign}(s_{1r,s}) \text{ and } \mu = \mu_0\mathrm{sign}(s_{2r,s}) \tag{20}$$

so that

$$s_{1,2r,s} \to 0 \text{ and } \dot{s}_{1,2r,s} \to 0 \tag{21}$$

Proof of existence of sliding mode and method of selection of control magnitudes ω_0 and μ_0 can be found in [1]. With the occurrence of sliding mode

$$\tilde{i}_{\alpha,\beta} \to 0 \tag{22}$$

1) Flux Estimation: To show that the flux estimate errors $\tilde{\psi}_{r\alpha,\beta} \to 0$ as well with the discontinuous controls as chosen above, we have to evaluate the equivalent controls $\hat{\omega}_{\mathrm{eq}}, \mu_{\mathrm{eq}}$. Thus, for the rotor flux model, these are

$$\hat{\omega}_{\mathrm{eq}} = \frac{\omega\left(\hat{\psi}_{r\alpha}\psi_{r\alpha} + \hat{\psi}_{r\beta}\psi_{r\beta}\right)}{\hat{\psi}_{rd}^2} + \frac{\eta\left(\hat{\psi}_{r\alpha}\tilde{\psi}_{r\beta} - \tilde{\psi}_{r\alpha}\hat{\psi}_{r\beta}\right)}{\hat{\psi}_{rd}^2}$$
$$\mu_{\mathrm{eq}} = \tilde{\eta} + \frac{\omega\left(\hat{\psi}_{r\alpha}\psi_{r\beta} - \hat{\psi}_{r\alpha}\psi_{r\beta}\right)}{\hat{\psi}_{rd}^2}$$
$$+ \frac{\eta\left(\hat{\psi}_{r\alpha}\tilde{\psi}_{r\alpha} + \hat{\psi}_{r\beta}\tilde{\psi}_{r\beta}\right)}{\hat{\psi}_{rd}^2} \tag{23}$$

where $\hat{\psi}_{rd}^2 = \left(\hat{\psi}_{r\alpha}^2 + \hat{\psi}_{r\beta}^2\right) > 0$ and $\tilde{\eta} = \hat{\eta} - \eta$. Similarly, for the stator flux model, the equivalent controls are

$$\hat{\omega}_{\mathrm{eq}} = \omega - \omega\frac{\tilde{\psi}_{sd}}{\psi_{sd}^*} + \tilde{\omega}\sigma L_S\frac{i_{sd}^*}{\psi_{sd}^*} + \eta\frac{\tilde{\psi}_{sq}}{P\psi_{sd}^*}$$
$$\mu_{\mathrm{eq}} = -P\omega\frac{\tilde{\psi}_{sq}}{\psi_{sd}^*} + \tilde{\omega}\sigma L_S\frac{i_{sq}^*}{\psi_{sd}^*} - \eta\frac{\tilde{\psi}_{sd}}{P\psi_{sd}^*} \tag{24}$$

Again, for brevity, we only present the proof of convergence of the flux estimates to their real values only for the stator flux model, results are similar for the rotor flux model and can be found in [9].

Once sliding mode occurs, the sliding mode equations, now of reduced order (by 2), have to be derived and analyzed. These equations are the flux estimate error dynamics

$$\dot{\tilde{\psi}}_{s\alpha} = -\hat{\psi}_{s\beta}\mu_{\mathrm{eq}}$$
$$\dot{\tilde{\psi}}_{s\beta} = +\hat{\psi}_{s\alpha}\mu_{\mathrm{eq}} \tag{25}$$

with μ_{eq} substituted from (24). In order to perform a complete error analysis, the results of the sliding mode speed controller, described in the previous section, have be used. This will satisfy two goals, it will: 1. make the flux estimate error analysis easier, and 2. lead to a complete stability analysis of the controller-observer system in the closed-loop, the second goal obviously being more important. Indeed, if the control action, now evaluated based on $\hat{\psi}_{s\alpha,\beta}$ and $\hat{i}_{s\alpha,\beta}$, ensures that $\hat{i}_{sq,d} \to i_{sq,d}^*$ and $\hat{\psi}_{sd} \to \psi_{sd}^*$ which we assumed to be constants, then we can use them to analyse the dynamics of $\tilde{\psi}_{s\alpha,\beta}$. The transformations (19) lead to the closed loop system

$$\dot{\tilde{\psi}}_{sq} = A_s\tilde{\psi}_{sd} - \frac{R_S i_{sd}^*}{\psi_{sd}^*}\tilde{\psi}_{sq}$$
$$\dot{\tilde{\psi}}_{sd} = -A_s\tilde{\psi}_{sq} - \frac{R_S i_{sd}^*}{\psi_{sd}^*}\tilde{\psi}_{sd} \tag{26}$$

where $A_s = (R_S i_{sq}^* - u_q - \psi_{sd}^*\mu)/\psi_{sd}^*$. Since the magnetizing (or direct) component of current and flux $i_{sd}^*, \psi_{sd}^* > 0$, stability of the above system is evident and hence $\tilde{\psi}_{s\alpha,\beta} \to 0$.

2) Speed Estimation: From (24) and (26), it is evident that the equivalent control $\hat{\omega}_{\mathrm{eq}} \to \omega$. Now, $\hat{\omega}_{\mathrm{eq}}$ is essentially the slow component of the fast switching control input $\hat{\omega}$, thus it can be extracted by passing $\hat{\omega}$ through a low-pass filter [1]. Thus, to implement the speed control law (5), we can directly use $\hat{\omega}_{\mathrm{eq}}$. A similar result is valid for the rotor flux model as well.

3) Adaptive Parameter Estimation: This feature is a part of the rotor flux based observer alone, for which it can be proved that $\tilde{\psi}_{r\alpha,\beta} \to 0$. Thus, from (23), the equivalent control $\mu_{\mathrm{eq}} \to \tilde{\eta}$. Since, $\eta = \hat{\eta} - \mu_{\mathrm{eq}}$, the rotor resistance $R_R = L_R * \hat{\eta}$ can be found. But, this means that the initial estimate of $\hat{\eta}$ should be very close to the true η, which is unknown. Moreover, η varies with the motor temperature, so a static calculation definitely leads to an error in parameter estimation. Therefore, a dynamic law should be used so that the estimate converges to the true value. Proposition: the law

$$\dot{\hat{\eta}} = -\gamma \mu_{\text{eq}} \tag{27}$$

in tandem with the observer and where the gain $\gamma > 0$ will ensure parameter convergence. The proof is quite simple. Indeed, if η varies very slowly in time, $\dot{\tilde{\eta}} = \dot{\hat{\eta}}$. Hence, from (27)

$$\dot{\tilde{\eta}} = -\gamma \tilde{\eta} \tag{28}$$

and for $\gamma > 0$, $\hat{\eta} \to \eta$.

IV. SIMULATION RESULTS

We studied the performance of our proposed observers by simulating the controller-observer system for 2 different motors. The simulation results are as shown in the figures that follow for each observer.

For the rotor flux observer (RFO), we used a motor having the following constants [10]: $L_S = L_R = 0.47\text{H}$, $L_M = 0.44\text{H}$, $R_S = 8\Omega$ and a nominal value of the rotor resistance $R_R = 3.6\Omega$ ($\eta = (R_R/L_R) = 7.7$), $P = 2$, and $J = 0.05\text{Nms}^2$. The desired flux and speed values were 1.5Wb and 100rad/s respectively. A constant load torque of $T_L = 5\text{Nm}$ was applied.

For the stator flux observer (SFO), we used a motor having the following constants: $L_S = L_R = 5.9e - 4H$, $L_M = 5.5e - 4H$, $R_S = .0106$, $R_R = 0.0118$ (in Ω), $P = 1$, $J = 0.05\text{Nms}^2$, desired speed $\omega^* = 100\text{rad/s}$, desired flux $\psi_{r,sd} = 0.1\text{Wb}$, and $T_L = 5\text{Nm}$.

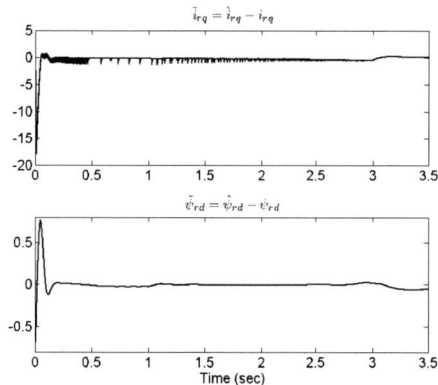

Fig. 1. RFO: Flux and Current Estimate Errors

V. CONCLUSIONS

RFO: By performing the transformations

$$e_1 = \hat{\psi}_{r\alpha}\tilde{\psi}_{r\alpha} + \hat{\psi}_{r\beta}\tilde{\psi}_{r\beta}$$
$$e_2 = \hat{\psi}_{r\alpha}\tilde{\psi}_{r\beta} - \tilde{\psi}_{r\alpha}\hat{\psi}_{r\beta} \tag{29}$$

we can arrive at the closed-loop system in the case of the RFO, the closed-loop system is of the form

$$\dot{e}_1 = -\eta e_1 + A_1 e_2$$
$$\dot{e}_2 = -(A_1 + \delta_1)e_1 - \delta_2 e_2 \quad A_1 = (F_1 + F_2) \tag{30}$$

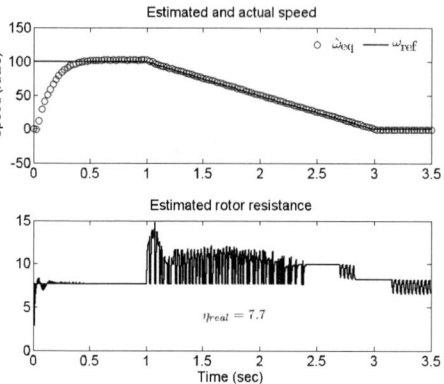

Fig. 2. RFO: $\hat{\omega}_{\text{eq}}$, ωref; Estimated R_R

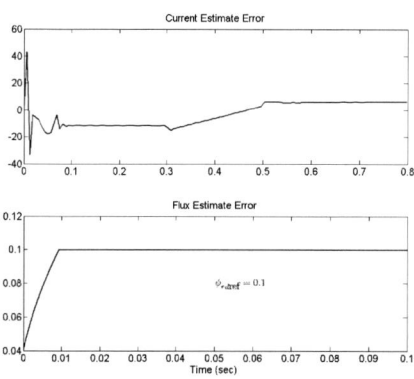

Fig. 3. SFO: Flux and Current Estimate Errors

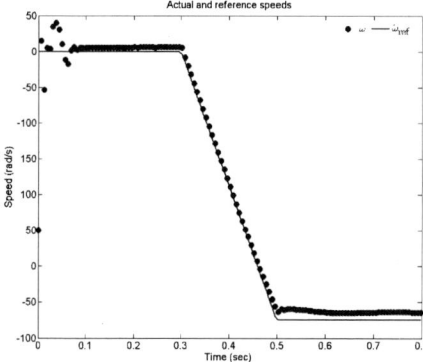

Fig. 4. SFO: ω, ωref; Estimated R_R

which is similar to (26) and where

$$F_1 = \frac{1}{\hat{\psi}_{rd}^2}(-\omega e_1 + \eta e_2), F_2 = (L_M \hat{\eta} i_{sq}^*)/\hat{\psi}_{rd}$$

$$\delta_1 = \frac{L_M \eta i_{rq}^*}{\hat{\psi}_{rd}} + \omega, \delta_2 = \frac{L_M \eta i_{rq}^* \omega}{\hat{\psi}_{rd}}$$

By performing a Lyapunov function based stablity analysis–details of which can be found in [9]–the system (30) proves to be stable only when $T_i \omega > 0$, i.e. only when the induction machine acts as a motor.

A common and valid objection that can be raised to this result is that the induction machine need not act as a motor at all times, i.e. during braking. We offer a simple solution to this problem: switch off the adaptation process and use the latest value of $\hat{\eta}$ to continue with the estimation of the flux and speed even during braking. This solution is not very restrictive as during braking, the emphasis is more on stopping the motor than on its speed control. For more mathematical details on braking and the regenerative mode, refer to [11] and [12].

The equivalent controls $\mu_{eq}, \hat{\omega}_{eq}$ were extracted as the output of a filter with a cut-off frequency of 200Hz and 40Hz respectively. As the filtered component $\mu_{eq} \neq 0$, there is no need of an external signal having the persistency of excitation property for adaptation, even at standstill. The estimation yields the exact value for $\hat{\eta}$ at constant speed.

SFO: To show that speed estimation can be performed at all speeds, a speed reversal experiment was simulated for the induction motor. The speed estimate $\hat{\omega}_{eq}$ was extracted as the output of a filter with a cut-off frequency of 10 kHz.

To conclude, we have presented two observer designs that the user can select based on the favoured mathematical description of the induction machine. The performance capabilities and limitations of each observer has been exhibited.

REFERENCES

[1] V.I.Utkin, J.G.Guldner, and J.Shi, *Sliding Mode Control in Electromechanical Systems.* Taylor & Francis, 1999.

[2] W. Leonhard, *Control of Electric Drives.* Springer-Verlag, 2001.

[3] P. Krause, *Analysis of Electric Machinery.* McGraw-Hill, 1986.

[4] R. Marino, S. Peresada, and P. Valigi, "Adaptive input-output linearizing control of induction motors," *Automatic Control, IEEE Transactions on,* vol. 38, no. 2, pp. 208–221, Feb 1993.

[5] Z. Yan and V. Utkin, "Sliding mode observers for electric machines - an overview," in *IECON 02 [28th Annual Conference of the Industrial Electronics Society],* vol. 3, no. 2, November 2002, pp. 1842 – 1847.

[6] A. Derdiyok, Z. Yan, M. Guven, and V. Utkin, "A sliding mode speed and rotor time constant observer for induction machines," in *IECON 01 [27th Annual Conference of the Industrial Electronics Society],* vol. 2, November-December 2001, pp. 1400–1405.

[7] Z. Yan, C. Jin, and V. Utkin, "Sensorless sliding-mode control of induction motors," *IEEE TRANSACTIONS ON INDUSTRIAL ELECTRONICS,* vol. 47, no. 6, pp. 1286–1297, December 2000.

[8] V.I.Utkin, "Sliding mode control design principles and applications to electric drives," *IEEE TRANSACTIONS ON INDUSTRIAL ELECTRONICS,* vol. 40, pp. 23–36, 1993.

[9] S. Rao, M. Buss, and V. Utkin, "An adaptive sliding mode observer for induction machines," in *2008 American Control Conference,* Seattle, Washington, USA, June 2008 2008, pp. 1947–1951.

[10] B.Aloliwi, H.K.Khalil, and E.G.Strangas, "Robust speed control of induction motors: application to a benchmark example," *INTERNATIONAL JOURNAL OF ADAPTIVE CONTROL AND SIGNAL PROCESSING,* vol. 14, pp. 157–170, 2000.

[11] E. G. Strangas, H. K. Khalil, B. Oliwi, L. Lauhinger, and J. M. Miller, "A robust torque controller for induction motors without rotor position sensor: Analysis and experimental results," *IEEE Transactions on Energy Conversion,* vol. 14, no. 4, pp. 1448–1458, December 1999.

[12] J. Jiang and J. Holtz, "An efficient braking method for controlled ac drives with a diode rectifier front end," *IEEE Transactions on Industry Applications,* vol. 37, no. 5, pp. 1299–1307, 2001.

Torque Transient Alleviation in Fixed Speed Wind Generators by Indirect Torque Control with STATCOM

Marta Molinas, Jon Are Suul and Tore Undeland
Norwegian University of Science and Technology
Department of Electrical Power Engineering, Trondheim, Norway,
marta.molinas@elkraft.ntnu.no

Abstract— Gearboxes for wind turbines must ensure high reliability over a period of 20 years withstanding cumulative and transient loads. Grid disturbances affecting wind turbines with induction generators directly connected to the grid generates electromagnetic torque transients that will result in significant stresses and fatigue of the gearbox. This paper presents a technique by which the transient torques during recovery after a grid fault can be smoothed for fixed speed wind turbines with induction generators directly connected to the grid. The technique labeled Indirect Torque Control (ITC) is suggested in order to reduce the drive train mechanical stresses caused by the characteristics of the induction machine when decelerating through the maximum torque region. The basis of the approach consists of controlling the induction generator terminal voltage by the injection/absorption of reactive current using a STATCOM. By controlling the terminal voltage with the STATCOM, the electromagnetic torque of the generator is indirectly smoothed. The control concept is shown by simulations on a wind generation system model built in PSCAD, where the smoothing effect of the proposed technique is seen during the recovery after a three phase to ground fault condition. The influence of shaft stiffness and mutual damping on the proposed control is further investigated on a two mass model of the wind generation system.

*Keywords—*Wind energy, Static Synchronous Compensator STATCOM; Voltage Source Converter

I. INTRODUCTION

During the last decades, extensive research has been focused on operation of variable speed wind turbines. In addition to the possibility for increased energy capture, one of the main motivations for investigating variable speed wind turbine topologies has been to partly or completely decouple the mechanical transients of the wind turbine from the electrical transients of the power system. One clear benefit from variable speed operation is the possibility to control the electromagnetic torque directly and by that being able to smooth mechanical stresses on the wind turbine drive train. Since variable speed wind generators has been the subject of research for several years, extensive literature can be found on control and operation of different configurations and power electronics interfaces [1], [2]. However, induction generators directly connected to the grid are still used in

many modern wind turbine installations today. By being locked to the grid voltage and frequency, operating at approximately fixed speeds, they are vulnerable to contingencies such as faults and disturbances that might appear in the nearby grid. In the case of induction generators directly connected to the grid, the electromechanical torque can not be directly controlled by the stator currents, and the stress on the drive train will be determined by the input mechanical torque and the electrical connection to the grid. However, if a fast acting controllable source of reactive current is implemented at the stator terminals of the directly connected wind generators, indirect control of the electromagnetic torque can be possible by controlling the induction generator terminal voltage with the injection/absorption of reactive power. During recovery after a grid fault the drive train of the turbine can experience high torque caused by the torque-speed characteristics of induction machines, and such condition is the topic of investigation in this paper. The basis for the approach is a standard STATCOM control intended for improving the Low Voltage Ride Through (LVRT) capability of the wind turbine as investigated in [5],[10] and in that context an additional control feature for reducing the transient torque during the voltage recovery is suggested.

The additional control feature suggested in this paper is labeled Indirect Torque Control (ITC), since the torque of the induction machine is influenced indirectly by utilizing a STATCOM to modulate the flow of reactive current and by that the voltage on the terminals of the induction machine. Several studies have confirmed how reactive compensation can increase the torque capability and by that the stability limit of induction generators [11],[12], but as shown in [5], [10] this will also increase the maximum torque that occurs during the recovery process. The basic idea of the ITC is therefore to reduce the compensation level of the STATCOM during the recovery process after a fault, after stability has been ensured but before the voltage is completely recovered and the speed has returned to the pre-fault value. Depending on the case, also the inductive region of STATCOM operation can be exploited to reduce the voltage at the machine terminals and limit the torque of the induction generator during the recovery process, on the cost of reactive power flow from the grid.

978-1-4244-1741-4/08/$25.00 ©2008 IEEE

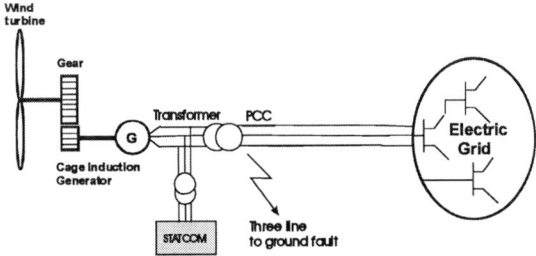

Fig. 1. Schematic configuration of the system under study: directly connected induction generator with STATCOM

II. INDIRECT TORQUE CONTROL BY USE OF STATCOM

Figure 1 shows the schematic configuration of the wind generation system used to investigate the indirect torque control (ITC) [10]. The system consists of an induction generator directly connected to the grid and driven by a wind turbine through a gear box to convert the low speed of the turbine shaft into a high speed that matches the rated rotational speed of the induction generator. The induction generator is connected to the grid through a transformer, and a STATCOM is connected at the generator terminals to control the voltage level at the generator by injection of reactive current [4]. As reported in [5], the STATCOM can be used to improve the transient stability and the critical clearing time of the wind generator and by that increase the LVRT capability. High levels of reactive compensation to improve the fault ride-through capability of the system will however increase the maximum torque of the generator during the recovery process [3]. In this context, the control system of the STATCOM could be expanded, as indicated in the block diagram of Fig. 2, to include the ITC control in addition to the normal STATCOM control, to allow for torque transient alleviation during the recovery process after a grid fault. This can be achieved by reducing the voltage reference of the STATCOM control system after reclosing and by that the reactive compensation when stability is ensured but before the grid voltage and the speed of the generator has returned to the pre-fault values[10]. In this way the STATCOM can improve the torque capability of the induction generator when this is needed to keep the system stable, but also reduce the maximum torque during recovery once stability has been ensured and by that limit the strain on the drive train. This assumes relevance especially in the context of LVRT where wind turbines cannot just disconnect from the grid to protect the installation from mechanical damage that might be caused by the stresses of repeated peak torque transients.

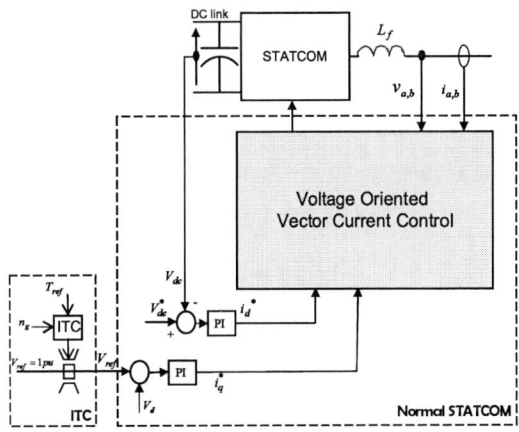

Fig. 2. Block diagram of the control system including ITC and normal STATCOM

At the reclosing instant of a fault sequence and afterwards, transients of the electromagnetic torque will result in significant stresses for the wind turbine mechanical system and can have harmful effects on the fatigue life of drive train sensitive components such as the gearbox [6][7][8]. Gearbox fatigue is caused by stressing of the gearbox teeth in response to torque overloads. For an input torque in excess of the gearbox rating, the fatigue damage increases in the extent to which the rating is exceeded and also as the length of the time the overload persists [9]. In addition to that, lifetime of the gearbox is reported to be influenced by the load-duration distribution. The influence of the mean wind speed on the lifetime of gearbox is identified and found that the accumulated duration of torque levels significantly influences the fatigue load on the gearbox and therefore its lifetime [13]. Taking this into account, not only the high transient torques will represent stresses on the gearbox but also the cumulative torque stresses under normal operation by adding up to the high transient torques during recovery after a fault. The short circuit initial torque transients are not the target of the proposed ITC control. This paper focuses on the recovery process after breaker re-closing operation for being one of the cases that represents high transient torques for induction machines. A three-phase grid failure is used as example to put in evidence the torque transients that appear at the reclosing.

The concept of the Indirect Torque control can be explained on basis of the equivalent circuit of Figure 3 [10]. This assumes the simplifications corresponding to quasi-static operation as discussed in [5], and can be considered reasonable when limiting the torque and giving

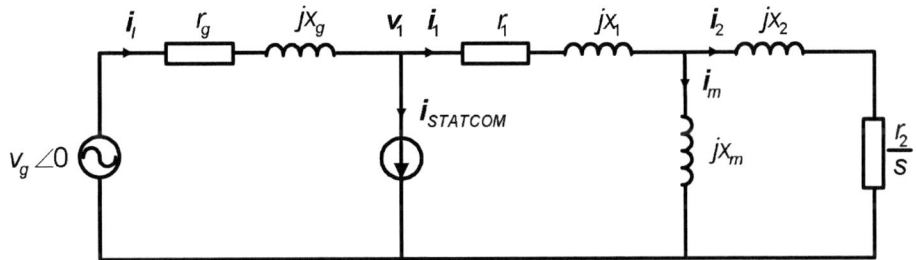

Fig. 3. Quasi stationary equivalent circuit for the system under study, consisting of the traditional induction machine equivalent, the STATCOM modelled as a current source and a grid equivalent

2319

TABLE I.
MACHINE AND GRID PARAMETERS USED IN SIMULATIONS

Asynchronous machine	
Power : 2.26 MVA Generator inertia constant : J=2H = 6 s Self Damping : 0.008 pu	r_1 = 0.008553 pu x_1 = 0.07365 pu x_m = 4.376 pu r_2 = 0.01045 pu x_2 = 0.1787 pu τ_m = 1 pu
Two mass model	**STATCOM**
H_{WT} = 2.5 s H_G = 0.5 s K_{shaft} = 50 D_{mutual} = 10	Stationary current reference limitation of 1.0 pu reactive current. Vector control with inner current loop.
Grid	**Fault type and duration**
x_g = 0.1126 pu r_g = 0.01126 pu	350 ms three phase to ground

as a result slower deceleration of the generator. On this basis, the derivation of the control concept can be understood starting from the equation of the electromagnetic torque as given in (1). This equation can provide the required rotor current i_2, from a reference value for the torque, that must be set to a slightly higher value than the mechanical torque (1.15 pu is used as reference in this case), after stability has been ensured and the recovery process has started. From the equivalent circuit of Figure 2 the current i_1 can be obtained as given in equation (2). Current i_1 in equation (3) will give the voltage v_1 at the terminals of the STATCOM that will be used as reference voltage for the control of the STATCOM. In this way it is possible to indirectly control the torque of the induction generator by controlling the voltage at its terminals. As seen by this equation, the control structure of the ITC will require the information about the generator speed for calculation of the reference value. The system control structure is sketched in Fig. 2, where the signals needed for the ITC are indicated and the ITC part is enclosed in the dashed block on the left. The rest of the control structure corresponds to the normal STATCOM control.

Equation (4) which is derived from the circuit of Figure

3 by fulfillment of the condition of a specific torque based on quasi-static considerations as presented in [10] will determine the required reactive current injection by the STATCOM when implementing the indirect torque control. From this criterion the required reactive current to ensure stability or to limit the torque below a specified level can be derived in the form of curves of required rating of reactive compensation as a function of speed at clearing of the induction generator. The value of this current is influenced by the selected setting of the reference of electromagnetic torque whose necessary condition is to be slightly higher than mechanical torque to ensure stability. The length of the recovery process will be affected by the difference between the mechanical torque and the selected value for T_{em}.

$$\tau_{em} = \frac{r_2}{s}\left|i_2\right|^2 \tag{1}$$

$$i_1 = \frac{\dfrac{r_2}{s} + j\left(x_2 + x_m\right)}{jx_m}i_2 \tag{2}$$

$$v_1 = i_1\left(r_1 + r_{eq,r} + j\left(x_1 + x_{eq,r}\right)\right) \tag{3}$$

III. ILLUSTRATION OF CONCEPT

Simulations performed with PSCAD/EMTDC on the model in Figure 1 provide an indication and illustration of the effect of the ITC technique on the recovery part of the transient torques for one set of generator and grid parameters as shown in Table I. The contingency investigated is breaker re-closing after a three phase to ground fault at the point of connection of the STATCOM.

Figure 4a shows the STATCOM current with the normal STATCOM control and with the ITC control. It can be noticed that some time after the reclosing, during the recovery process, the STATCOM current with ITC goes from injection to absorption of reactive current to reduce the torque. With the normal STATCOM control,

$$\left|i_{STATCOM}\right|^2 + \frac{2 \cdot i_2\left(r_{eq,i2} \cdot r_{eq,STATCOM} + x_{eq,i2} \cdot x_{eq,STATCOM}\right)}{r_{eq,STATCOM}^2 + x_{eq,STATCOM}^2}\left|i_{STATCOM}\right| + \frac{\left(r_{eq,i2} \cdot i_2\right)^2 + \left(x_{eq,i2} \cdot i_2\right)^2 - \left|v_g\right|}{r_{eq,STATCOM}^2 + x_{eq,STATCOM}^2} = 0 \tag{4}$$

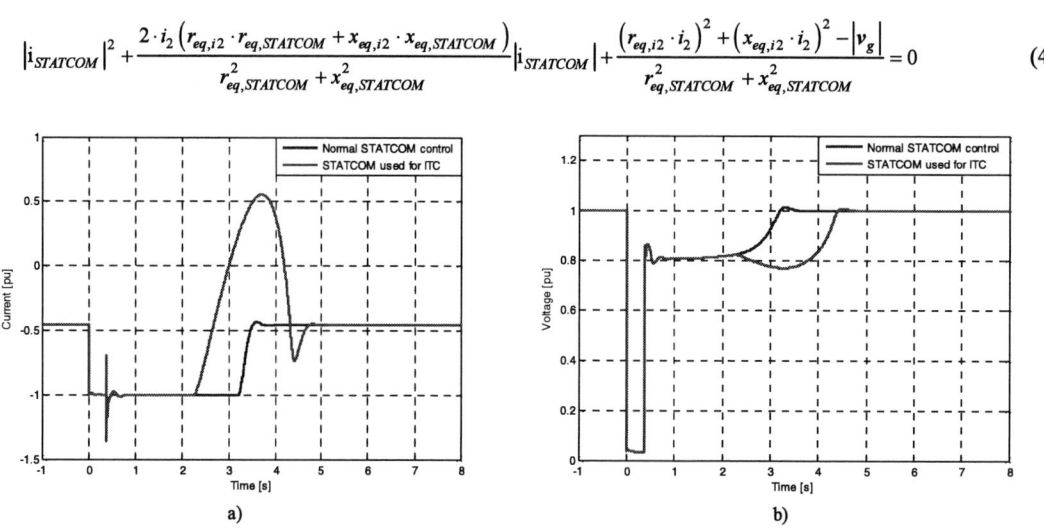

Fig. 4. Time responses of terminal voltage with normal STATCOM control and ITC control, and time responses of STATCOM reactive current with normal STATCOM control and ITC control

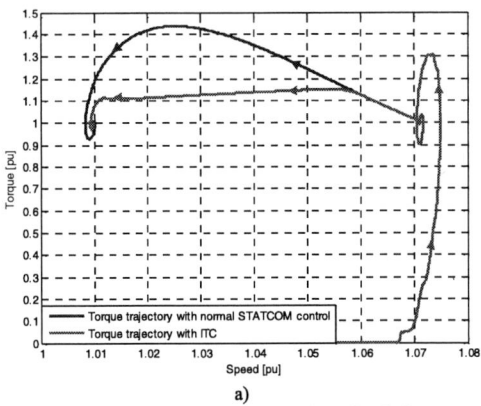
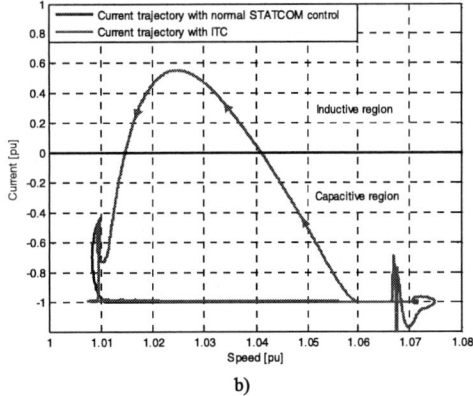

a)   b)

Fig. 5. Torque and current trajectories during recovery process with normal STATCOM control and with ITC control

there will be only capacitive operation of the STATCOM, and the speed of recovery of the system is the fastest possible. The recovery process with the ITC is longer than with the normal STATCOM control, because the decelerating torque is limited. Fig. 4b shows the difference between the terminal voltages with the normal STATCOM control and the ITC control. The terminal voltage remains below rated value for longer time due to the slower deceleration introduced by the ITC control. The value of the voltage depth depends on the system parameters and the reference torque selected when implementing the ITC. In the simulations presented in this paper the torque reference is chosen to be 1.15 pu.

In Fig. 5a the torque trajectories for both cases; normal STATCOM and ITC control shows the difference in peak torques during the recovery process. With ITC control as expected by the torque reference setting, the maximum torque amplitude is of 15% at the beginning of the recovery process being reduced as it gets close to the rated speed and rated voltage. This reduction in torque is caused by the dynamics of the machine when decelerating as described in [5]. As a result of the almost constant torque of the ITC, there is a characteristic linear change of speed of the generator. With the normal STATCOM control and for parameters used in this study, the maximum torque amplitude that appears during the recovery process is about 43% above the rated torque. The propagation of these torque transients in the drive train is determined by its torsional characteristics, which is investigated in section IV with a two mass model. Figure 5b also shows the reactive current trajectory of the STATCOM as function of the generator speed. The part of the trajectory that corresponds to the recovery process is the same as the results from solving (4) for i_q as a function of speed.

Both types of control give maximum reactive current compensation during the fault and immediately after the fault clearing. With the normal control objective of the STATCOM to keep the voltage at the rated value, the compensation current is kept at the maximum value until the speed of the generator is reduced to the pre-fault value and the STATCOM is able to bring the terminal voltage back to the reference value. With the ITC, the compensation is reduced, and the STATCOM current even goes into the inductive region to limit the terminal voltage of the generator, and by that to limit the torque as clearly shown by the torque trajectory.

The torque transients after a grid fault as analysed in this paper, are a source of stresses in the gearbox that will be on top of the stresses during normal operation originated by the variability of wind speeds. The combined effect of these disturbances on the fatigue life of the gearbox needs to be investigated as well as detailed investigation of the performance of the new proposed approach on a multi-mass model for concluding on the validity of the proposed technique. In order to further investigate the effect of the ITC on the electromagnetic torque of the generator when the torsional torque is taken into consideration, a two mass model of the wind generation system is introduced in the next section [14].

IV. INFLUENCE OF TWO MASS MODEL WIND GENERATION SYSTEM ON ITC

When the wind turbine and the wind generator are modeled as one mass lumped model with a combined inertia constant, stability analysis based on such model may give significant error when compared to a multi-mass model [15],[16]. The effects of inertia constants, shaft stiffness, self-damping of the individual masses, and mutual damping of the adjacent masses, is taken into consideration when investigating the performance of the ITC in real wind turbine generation systems. A two-mass drive train model as indicated in Fig. 6 is built and simulated in PSCAD with the parameters given in Table I. The effects of shaft stiffness and damping factor on the ITC performance is investigated by considering a 3 phase to ground fault at the PCC in Fig. 1.

Simulation results shown in Fig. 7 using the parameters indicated in Table I show the influence of stiffness and mutual damping on the performance of the ITC control. Time responses of turbine and generator speed show the

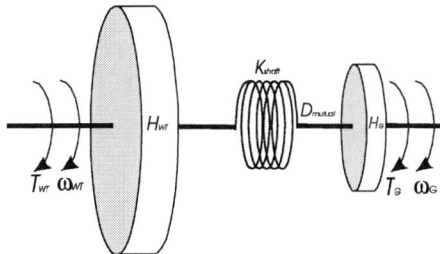

Fig. 6. Two mass model of the wind energy generation system

2321

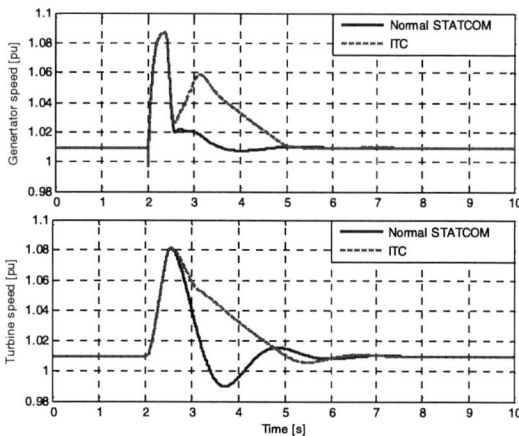

Fig. 7. Time responses of turbine and generator speed with normal STATCOM and ITC

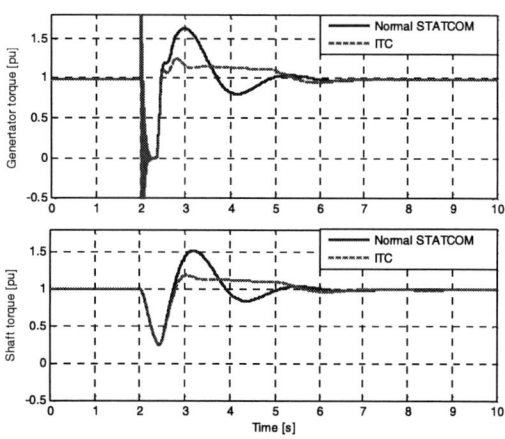

Fig. 8. Time responses of turbine shaft and generator torque with normal STATCOM and ITC

characteristic linear speed of the ITC after reclosing.

Figure 8 shows the time responses of the turbine shaft and generator torques. The effect of the ITC can be seen by the smoothed torque compared to the normal STATCOM control as a result of the torque limitation

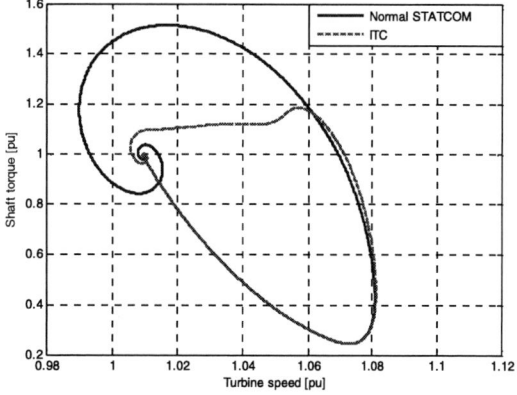

Fig. 9. Turbine shaft torque trajectory with normal STATCOM and ITC

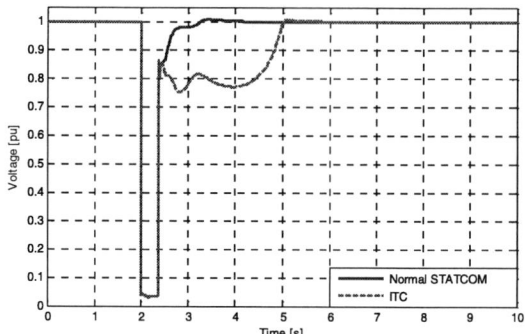

Fig. 10. Time responses of voltage at the PCC with normal STATCOM and ITC

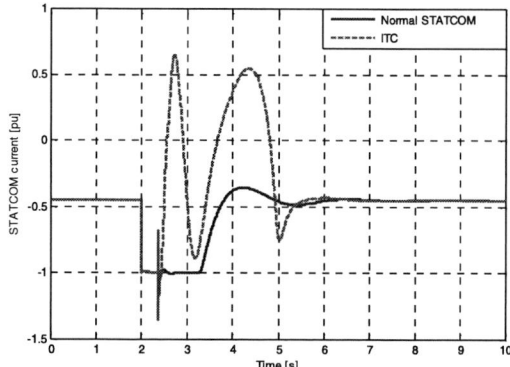

Fig. 11. Time responses of STATCOM currents with and without ITC

imposed by the ITC controller. Figure 9 shows the turbine shaft torque trajectory with reduced torsional torque as a result of the ITC control. This gives an indication on the reduced torque stresses for the gearbox compared to the stresses with normal STATCOM control. The voltage at the PCC is shown in Fig. 10 with the characteristic reduced voltage with ITC control and slower recovery process compared to the normal STATCOM control. It should be noted that the reduced stresses in the turbine shaft is at the price of a slower recovery process and lower terminal voltage after reclosing. The STATCOM current during the fault and recovery process is shown in Fig. 11. During the fault, both controllers provide the maximum amount of reactive current but at the reclosing instant the current from the ITC control responds to the voltage calculated according to the algorithm proposed in this paper, and the STATCOM goes from capacitive to inductive operation depending on the voltage reference value.

The critical mutual damping that makes the system to be on the stability limit with the ITC control is investigated and the simulation results corresponding to that case are shown in Fig. 12. Time responses of generator and turbine speeds are shown, where the ITC influence is seen by the oscillations compared to the normal STATCOM control. Reducing the value of the mutual damping below $D_{mutual} = 5$ will make the system to become unstable. Figure 13 shows the time responses of generator and turbine shaft torques in the stability limit for

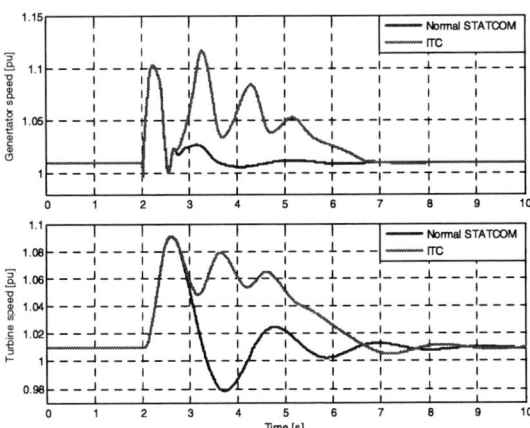

Fig. 12. Time responses of generator and turbine speeds with and without ITC

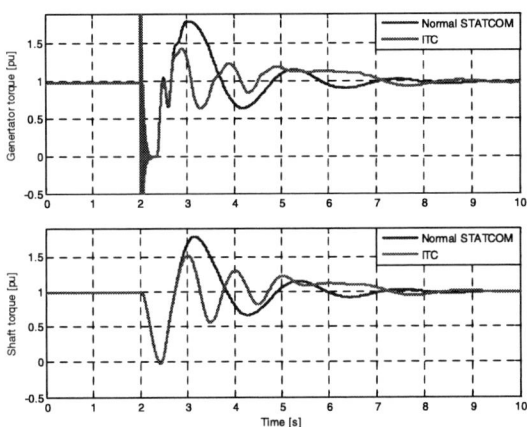

Fig. 13. Time responses of turbine shaft and generator torque with normal STATCOM and ITC control

normal STATCOM control and ITC showing the tendency to oscillatory instability with the ITC. Figure 14 shows the shaft torque trajectory with a clearly reduced shaft torque with ITC but with considerable more oscillations compared to the normal STATCOM control. Figures 15

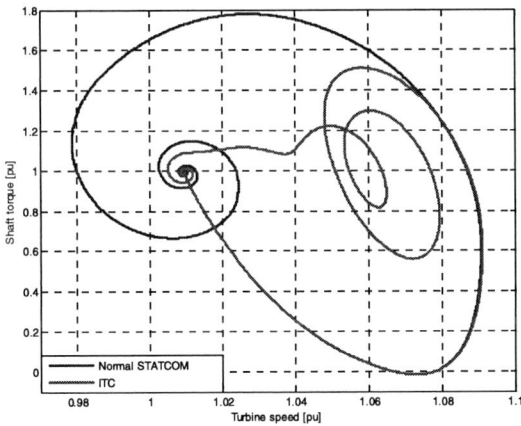

Fig. 14. Turbine shaft torque trajectory with normal STATCOM and ITC control

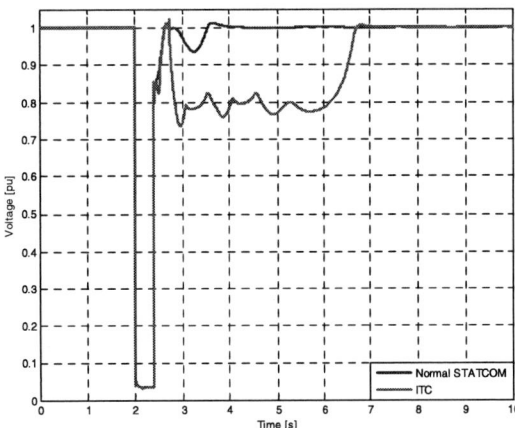

Fig. 15. Time responses of PCC voltage with and without ITC

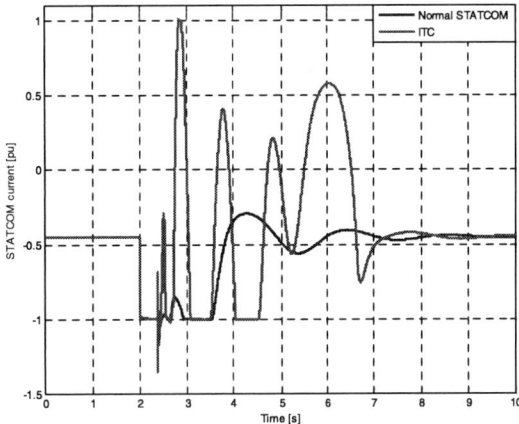

Fig. 16. Time responses of STATCOM currents with and without ITC

and 16 show the time responses of the PCC voltage and STATCOM current with and without the ITC control. The responses become more oscillatory as a result of the typical speed oscillations originated by the soft shaft of the two mass model and the recovery process lasts until the current oscillations are well damped.

The results with reduced damping factor indicate that the wind generation system can become oscillatory unstable and the ITC does not perform as expected due to the soft shaft and little damping [17]-[19]. For having a realistic indication of the validity of the ITC control, parameters of mutual damping of wind turbines in the rating range of interest need to be tested by simulations. To improve performance for low damping factors, the equations of the two mass model should be included in the derivations of the ITC control logic and damping control could be included in the control structure of the ITC to suppress the vibration caused by mechanical resonance [18].

V. CONCLUSIONS

Torque transients developed in asynchronous generators during the recovery process after a grid failure has been investigated for a fixed speed wind generation system. These torque transients represent stresses for the gearbox and other mechanical components. To alleviate these torque transients an indirect torque control technique has been proposed and implemented by extending the control capabilities of a STATCOM. Simulation results indicate how the STATCOM can allow for the implementation of such control strategy to reduce the mechanical stresses on the drive train of a wind turbine during recovery after a fault, but at the expense of a reduced voltage level and a longer recovery time. The terminal voltage of the generator is reduced by limiting the reactive compensation from the STATCOM.

Simulation results on a one mass model provides a qualitative indication of the effect that torque transients would have on the gearbox, at the same time of providing a good proof of the effectiveness of the ITC in alleviating such torque transients. Simulation results on a two mass model system give an indication of the limitations of the ITC in a real system when the controller does not take into account the complete model of the system. The effect of mutual damping on the performance of the ITC is indicated by the oscillatory behavior that is typical of a two mass model system. Low damping factor will affect the proposed control in a negative way, requiring more sophisticated considerations for attaining a good performance in such conditions. Further work is mandatory for designing a controller that can take into account the two mass model into the controller structure. Controller parameter sensitivity and dependency on speed measurement are the other two factors which should be taken into consideration for further improving the performance of the ITC.

A trade-off between minimum required voltage level in the grid, allowed flow of reactive power depending on grid code requirements, and smoothed transient torque should be attempted when implementing the ITC control.

REFERENCES

[1] F.D. Kanellos, S.A. Papathanassiou, N.D. Hatziargyriou, "Dynamic Analysis of a Variable Speed Wind Turbine equipped with a Voltage Source AC/DC/AC Converter Interface and a Reactive Current Control loop," in *10th Mediterranean Electrotechnical Conference*, vol. 3, pp. 986-989, 2000

[2] S.A. Papathanassiou, M.P. Papadopoulos, "Dynamic Behavior of Variable Speed Wind Turbines under Stochastic Wind," in *IEEE Transactions on Energy Conversion*, vol. 14, No. 4, pp. 1617-1623, Dec. 1999

[3] M. Molinas, J. A. Suul, T. Undeland, "Wind farms with increased transient stability margin provided by a STATCOM", in Proc. *CES/IEEE 5th International Power Electronics and Motion Control Conference*, IPEMC2006, 13-16 Aug. 2006, vol. 1, pp. 63-69

[4] N.G.Hingorani, L.Gyugyi, *Understanding FACTS: Concepts and Technology of Flexible AC Transmission Systems*, Piscataway, NJ: IEEE Press, 2000, Chap. 5

[5] M. Molinas, J.A. Suul, T. Undeland, "A Simple Method for analytical Evaluation of LVRT in Wind Energy for Induction Generators with STATCOM or SVC," in *Proceedings of the 12th European Power Electronics Conference EPE 2007*, Aalborg, Denmark, September 2007

[6] M. Papadopoulus, P. Malatestas, J. Tegopulos, "Stresses of self excited induction generators during abnormal supply conditions," in *Proc. of International Conference on Electrical Machines ICEM 1992*, vol 3, pp. 1072-1076, 1992

[7] J. Faiz, M. Ghaneei, A. Keyhani, "Performance analysis of fast reclosing transients in induction motors," in *IEEE Trans. on Energy Conversion*, vol. 14, no. 1, pp. 101-107, March 1999.

[8] S. Papathanassiou, M. Papadopoulos, "Mechanical Stresses in Fixed-Speed Wind Turbines due to Network Disturbances," in

IEEE Trans. on Energy Conversion, vol. 16, no. 4, pp. 361-367, December 2001.

[9] W.E. Leithead, S. de la Salle, D. Reardon, "Role and Objectives of Control of Wind Turbines," in *IEE Proceedings –C*, vol. 138, No. 2, pp. 135-148, March 1991

[10] J. A. Suul, M. Molinas, T. Undeland, "Indirect Torque Control of Induction Machines during Voltage Recovery after Grid Faults," unpublished.

[11] S. K. Salman, A. L. J. Teo, "Investigation into the Estimation of the Critical Clearing Time of a Grid Connected Wind Power Based Embedded Generator", in *Proc. IEEE/PES Transmission and Distribution Conference and Exhibition*, vol. 2, pp. 975-9806, 10 Oct. 2002

[12] M. Molinas, J. A. Suul, T. Undeland, "Low Voltage Ride Through of Wind Farms with Cage Generators: STATCOM versus SVC," in *IEEE Transactions on Power Electronics*, vol. 23, No.3, pp. 1104-1117, May 2008

[13] B.Niederstucke, A. Anders, P. Dalhoff, R. Grzybowski, "Load Data Analysis for Wind Turbine Gearboxes," Germanischer Lloyd WindEnergy GmbH, www.gl-group.com/pdf/nst_paris.pdf, accessed on March 2007

[14] Y.Shima, R. Takahashi, T. Murata, J. Tamura, Y. Tomaki, S.Tominaga, A. Sakahara,"Transient Stability Simulation of Wind Generator Expressed by Two-Mass Model," in *IEEJ Trans. of the Institute of Electrical Engineers of Japan B*, vol. 125-B, no. 9, pp. 855-864, 2005

[15] S.M. Muyeen, Md. Hasan Ali, R. Takahashi, T. Murata, J. Tamura, Y. Tomaki, A. Sakahara, E. Sasano, "Comparative Study on Transient Stability Analysis of Wind Trubine Generator System Using Diferent Drive Train Models," in *IET Renewable Power Generation Journal*, vol. 1, no. 2, pp. 131.141, 2007

[16] S.M. Muyeen, M.H. Ali, R. Takahashi, T. Murata, J. Tamura, Y.Tomaki, A. Skahara, E.Sasano, "Transient Stability Analysis of Wind Generator by Using Six-Mass Drive Train Model," in *The 2006 International Conference on Electrical Machines and Systems (ICEMS 2006)*, Reference No. 00082, Nagasaki, Japan, 2006.

[17] Y.Hori, H. Iseki, K. Sugiura, "Basic Consideration of Vibration Suppression and Disturbance Rejection Control of Multi-inertia System using SFLAC (State Feedback and Load Acceleration Control)," in IEEE Transactions on Industry Applications, Vol. 30, no. 4, pp. 889-896, July/August 1994

[18] K. Sugiura, Y. Hori, "Vibration Suppression in 2-and 3-Mass System Based on the Feedback of Imperfect Derivative of the Estimated Torsional Torque," in IEEE Transactions on Industrial Electronics, Vol. 43, no. 1, pp. 56-64, February 1996

[19] Y. Hori, H. Sawada, Y. Chun, "Slow Resonance Ratio Control for Vibration Suppression and Disturbance Rejection in Torsional System," in IEEE Transactions on Industrial Electronics, Vol. 46, no. 1, pp. 162-168, February 1999

Flicker Study on Variable Speed Wind Turbines with Permanent Magnet Synchronous Generator

Weihao Hu[*†], Zhe Chen[†], Yue Wang[*] and Zhaoan Wang[*]

[*] School of Electrical Engineering, Xi'an Jiaotong University, Xi'an, China, e-mail: *whu@iet.aau.dk*
[†] Institute of Energy Technology, Aalborg University, Aalborg, Denmark, e-mail: *zch@iet.aau.dk*

Abstract—Grid connected wind turbines are fluctuating power sources that may produce flicker during continuous operation. This paper presents a simulation model of a MW-level variable speed wind turbines with a permanent magnet synchronous generator (PMSG) and a full-scale converter developed in the simulation tool of PSCAD/EMTDC. Flicker emission of this system is investigated during continuous operation. The dependence of flicker emission on wind characteristics (mean speed, turbulence intensity), 3p torque oscillations due to wind shear and tower shadow effects and grid conditions (short circuit capacity, grid impedance angle) are analyzed. Flicker mitigation is realized by output reactive power control of the variable speed wind turbines with PMSG. Simulation results show the output reactive power control is an effective measure to mitigate the flicker during continuous operation of grid connected wind turbines.

Keywords—Flicker, wind energy, renewable energy systems.

I. INTRODUCTION

Because of energy shortage and environment pollution, the renewable energy, especially wind energy has become more considerable all over the world. As the wind power penetration into the grid is increasing quickly, the influence of wind turbines on the power quality is becoming an important issue. One of the important power quality aspects is flicker.

Flicker is defined as "an impression of unsteadiness of visual sensation induced by a light stimulus, whose luminance or spectral distribution fluctuates with time" [1], which can cause consumer annoyance and complaint. Flicker is induced by voltage fluctuations, which are caused by load flow changes in the grid. The flicker emission produced by grid connected variable speed wind turbines with PMSG and full-scale converter during continuous operation is mainly caused by fluctuations in the output power due to wind speed variations, the wind shear and the tower shadow effects [2]. As a consequence, an output power drop will appear three times per revolution for a three bladed wind turbine.

There are many factors that affect flicker emission of grid connected wind turbines during continuous operation, such as wind characteristics and grid conditions [3], [4]. Variable speed wind turbines have shown better performance related to flicker emission in comparison with fixed speed wind turbines [5]. But the flicker study becomes necessary and imperative as the wind power penetration level increases quickly.

Variable speed wind turbines with multipole PMSG and full-scale converter are becoming more popular worldwide, because of some advantages such as no gearbox, high power density and easy to control [6]. In this paper, a simulation model of a MW-level variable speed wind turbine with a PMSG and a full-scale converter is developed in PSCAD/EMTDC [7], [8]. Base on the wind turbine model, flicker emission of variable speed wind turbines with PMSG and full-scale converter is investigated during continuous operation. The factors that affect flicker emission of wind turbine, such as wind characteristics, 3p torque oscillations and grid conditions are analyzed. Flicker mitigation is realized by output reactive power control of the back-to-back full-scale converter.

II. WIND TURBINE MODEL

The wind turbine considered here applies a PMSG, using a back-to-back full-scale PWM voltage source converter connected to the grid. Variable speed operation of the wind turbine can be realized by appropriate adjustment of the rotor speed and pitch angle.

A complete wind turbine model includes the wind speed model, the aerodynamic model of the wind turbine, the mechanical model of the transmission system and models of the electrical components, namely the PMSG, PWM voltage source converters, transformer, and the control and supervisory system. Fig. 1 illustrates the main components of a grid connected wind turbine.

Wind simulation plays an important task in the wind turbine modeling. A wind model used in this paper has been developed in [9]. The wind model provides an equivalent wind speed for each wind turbine, which is conveniently used as input to an aerodynamic model of the wind turbine.

The relation between the wind speed and aerodynamic torque may be described by the following equation:

$$Tw = \frac{1}{2} \rho \pi R^3 v_{eq}^2 \frac{C_P(\theta, \lambda)}{\lambda} \qquad (1)$$

where Tw is the aerodynamic torque extracted from the wind (Nm); ρ is the air density (kg/m³); R is the wind turbine rotor radius (m); v_{eq} is the equivalent wind speed (m/s); θ is the pitch angle of the rotor (deg), $\lambda = \omega R / v_{eq}$ is the tip speed ratio; ω is the wind turbine rotor speed (rad/s); and C_p is the aerodynamic efficiency of the rotor.

978-1-4244-1741-4/08/$25.00 ©2008 IEEE

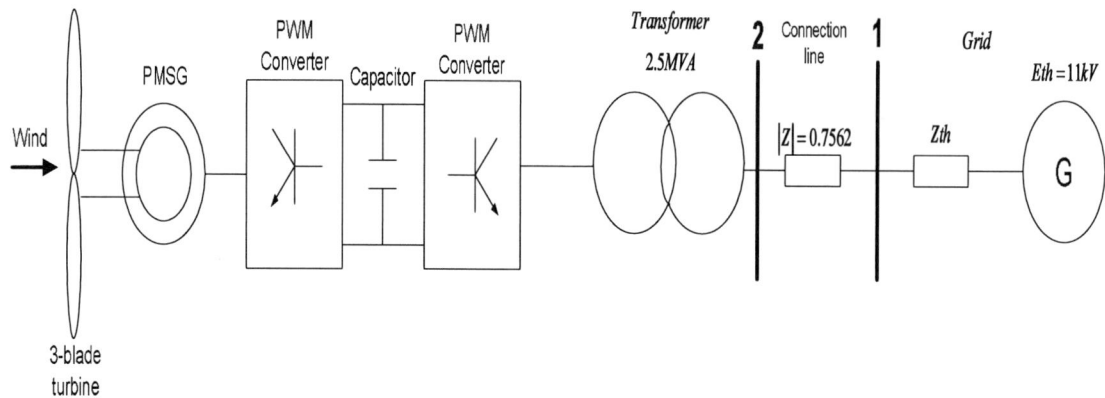

Fig. 1. Block diagram of a grid connected wind turbine with a PMSG and a full-scale converter.

3p torque oscillations are important parts in the aerodynamic model. A comprehensive yet pragmatic model of 3p torque oscillations due to wind shear and tower shadow effects has been developed for a three-blade wind turbine [10], which is applied in this paper. Based on this model, the equivalent wind speed (v_{eq}) will have three components. The first (v_{eq_0}) is the hub height wind speed, the second ($v_{eq_{ws}}$) is due to the wind shear, and the third ($v_{eq_{ts}}$) is due to the tower shadow. Therefore, the equivalent wind speed can be expressed as (2) whose components are shown as (3)-(5)

$$v_{eq} = v_{eq_0} + v_{eq_{ws}} + v_{eq_{ts}} \tag{2}$$

$$v_{eq_0} = V_H \tag{3}$$

$$v_{eq_{ws}} = V_H \Big[\frac{\alpha(\alpha-1)}{8} (\frac{R}{H})^2 $$
$$ + \frac{\alpha(\alpha-1)(\alpha-2)}{60} (\frac{R}{H})^3 \cos 3\beta \Big] \tag{4}$$

$$v_{eq_{ts}} = \frac{mV_H}{3R^2} \sum_{b=1}^{3} \Big[\frac{a^2}{\sin^2 \beta_b} \ln \left(\frac{R^2 \sin^2 \beta_b}{x^2} + 1 \right) $$
$$ - \frac{2a^2 R^2}{R^2 \sin^2 \beta_b + x^2} \Big] \tag{5}$$

where V_H is the wind speed at hub height (m/s), α is the empirical wind shear exponent, H is the elevation of rotor hub (m), β is the azimuthal angle of the blade (deg), β_b is respectively the azimuthal angle of each blade (deg), a is the tower radius (m), x is the distance from the blade origin to the tower midline (m), and

$$m = \Big[1 + \frac{\alpha(\alpha-1)R^2}{8H^2} \Big]$$ is a coefficient of the wind turbine.

When modeling the transmission system, it is a common practice to neglect the dynamic of the mechanical parts, as their responses are considerably slow in comparison to the fast electrical ones. The transmission system may be modeled by a single equation of motion

$$J_{WG} \frac{d\omega}{dt} = T_W - T_G - D\omega \tag{6}$$

where J_{WG} is the wind turbine mechanical inertia plus generator mechanical inertia (kg•m^2), T_G is the generator electromagnetic torque (Nm), and D is the friction coefficient (Nm/rad).

PSCAD/EMTDC software library provides dedicated model of PMSG and an average model without switches is used for power electronics converters so that the simulation can be carried out with a larger time step resulting in a simulation speed improvement.

III. CONTROL SCHEMES

Variable speed wind turbines may have two different control goals, depending on the wind speed. In low to moderate wind speeds, the control goal is maintaining a constant optimum tip speed ratio for maximum aerodynamic efficiency. In high wind speeds, the control goal is to keep the rated output power in order not to overload the system.

Two control schemes are implemented in the wind turbine model: speed control and pitch control. The speed control can be realized by adjusting the generator power or torque. The pitch control is a common control method to regulate the aerodynamic power from the turbine.

A. Speed Control

Vector-control techniques have been well developed for PMSG using back-to-back PWM converter. Two vector-control schemes are designed respectively for the generator-side and grid-side converters, as shown in Fig. 2.

The objective of the vector-control scheme for the grid-side PWM converter is to keep the DC-link voltage constant regardless of the magnitude of the generator power, while keeping sinusoidal grid currents. The objective of the vector-control scheme for the generator-side PWM converter is to control the optimal power tracking for maximum energy capture from the wind by adjusting the speed of the wind turbine.

Normally, the reference values of both generator-side and grid-side converters, Q_{s_ref} and Q_{g_ref} are set to zero to ensure unity power factor operation and reduce currents of both generator-side and grid-side converters.

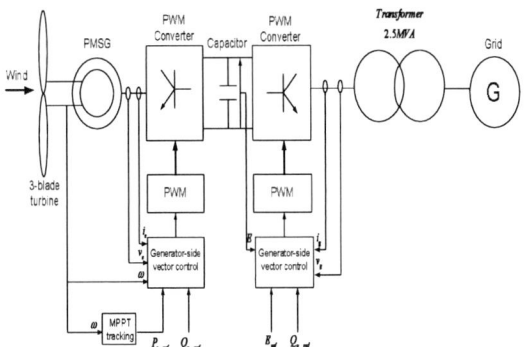

Fig. 2. Block diagram for the vector-control schemes of PMSG.

B. Pitch Control

The aerodynamic model of the wind turbine has shown that the aerodynamic efficiency is strongly influenced by variation of the blade pitch with respect to the direction of the wind or to the plane of rotation. Small changes in pitch angle can have a dramatic effect on the power output. For wind speeds above the rated value, the pitch control scheme takes over the wind turbine control to limit the output power. The relationship between the pitch angle and the wind speed is shown in Fig. 3 [5].

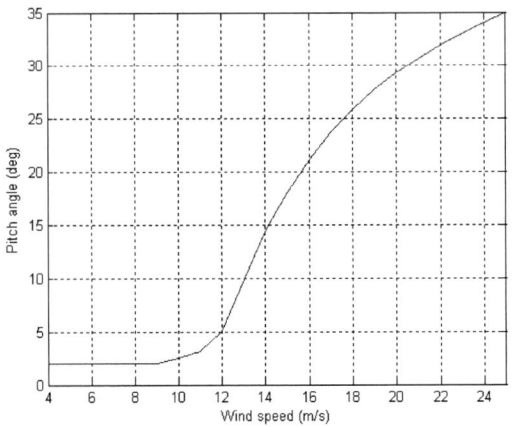

Fig. 3. The relationship between the pitch angle and the wind speed.

IV. FLICKER EMISSION

The level of flicker is quantified by the short-term flicker severity P_{st}, which is normally measured over a ten-minute period. According to IEC standard IEC 61000-4-15 [11], a flickermeter is built to calculate the short-term flicker severity P_{st}. The P_{st} of bus 2, the PCC, is calculated on the basis of the voltage variation.

The basic parameters of the flicker emission are shown in Table I. The wind turbine generates 2 MW active power during rated operation, while the output reactive power is normally controlled as zero to keep unity power factor.

TABLE I.
BASIC PARAMETERS

Parameter	Value
Rated power	2 MW
Rated speed	15.5 r/min
Rated voltage	0.69 kV
Stator resistance	0.03 p.u.
Xd	0.775 p.u.
Xq	0.775 p.u.
Stator leakage inductance	0.064 p.u.
Magnetic strength	1 p.u.

Flicker emission of grid connected wind turbines depends on many factors, such as mean speed (v), turbulence intensity ($In= \Delta v /v$), short circuit capacity ($SCR=S_k/S_n$) and grid impedance angle (ψ_k), where Δv is the wind speed standard deviation, S_k is the short circuit apparent power of the grid where the wind turbines are connected, S_n is the rated apparent power of the wind turbines.

The wind applied in this investigation is produced according to the wind model and the 3p oscillation model due to wind shear and tower shadow effects, which are introduced in Section II. The equivalent wind speed and the output active power of the wind turbine are shown in Figs. 4-5, respectively.

The variation of short-term flicker severity P_{st} with mean wind speed is shown in Fig. 6. In low wind speeds, P_{st} is very low due to a small output power. Then P_{st} increases with the mean wind speed. For higher wind speeds, where the wind turbine reaches rated power, the flicker level decreases. The reason is that the pitch angle control and the variable speed control.

The variation of short-term flicker severity P_{st} with turbulence intensity is shown in Fig. 7. P_{st} increases with the turbulence intensity. The more turbulence in the wind results in larger flicker emission.

An approximately inversely proportional relationship between the short-term flicker severity P_{st} and the short circuit capacity ratio in Fig. 8. The wind turbine would produce greater flicker in weak grids than in stronger grids.

The variation of short-term flicker severity P_{st} with grid impedance angle is shown in Fig. 9. When the grid impedance angle increases, the short-term flicker severity P_{st} decreases. When the grid impedance angle approaches 90 degrees, the flicker emission is minimized.

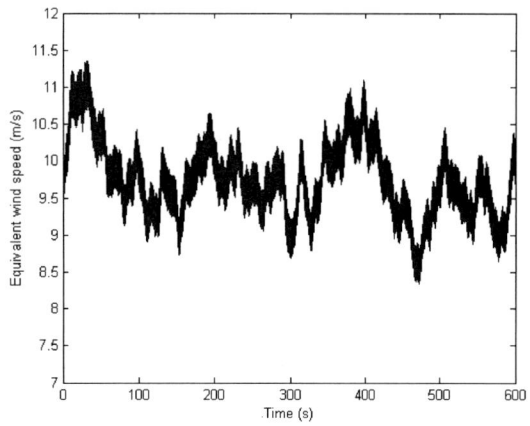

Fig. 4. The equivalent wind speed

(v =10 m/s, In=0.1, SCR=10, $\psi_k = 60^o$).

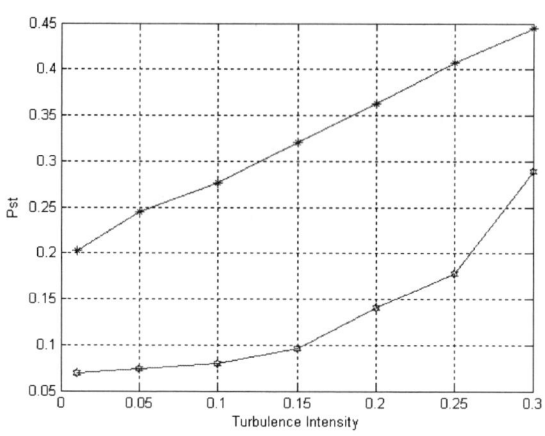

Fig. 7. P$_{st}$ variation with turbulence intensity

(v=9m/s (upper), v=16m/s (lower), SCR=10, $\psi_k = 60^o$).

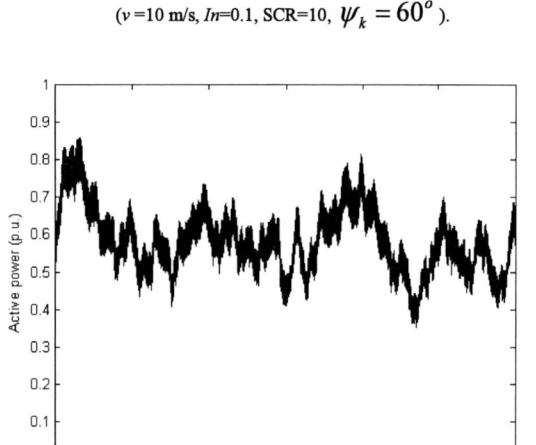

Fig. 5. The output active power

(v =10 m/s, In=0.1, SCR=10, $\psi_k = 60^o$)

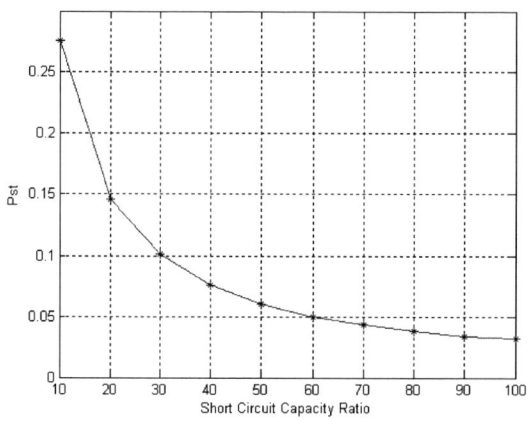

Fig. 8. P$_{st}$ variation with short circuit capacity

(v=9m/s ,In=0.1, $\psi_k = 60^o$).

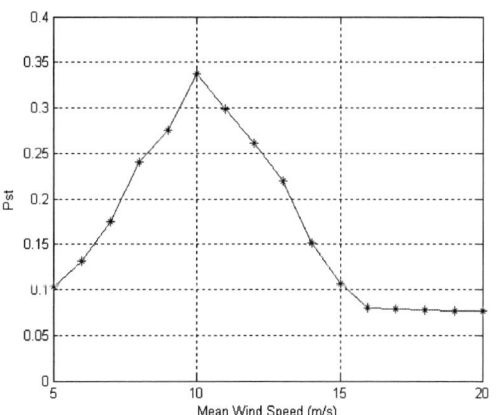

Fig. 6. P$_{st}$ variation with mean wind speed

(In=0.1, SCR=10, $\psi_k = 60^o$).

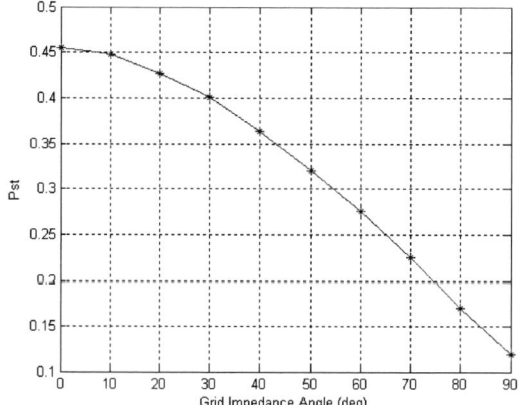

Fig. 9. P$_{st}$ variation with grid impedance angle

(v=9m/s, In=0.1, SCR=10).

V. FLICKER MITIGATION

The voltage change across a power line may be approximately calculated with the following equation:

$$\Delta V = \frac{PR+QX}{V} \qquad (7)$$

Define:

$$\tan \psi_k = \frac{X}{R}$$

$$\tan \psi = \frac{Q}{P} \qquad (8)$$

With the definition of equation (8), (7) can be rewritten as:

$$\Delta V = \frac{PR \bullet \cos(\psi - \psi_k)}{V \bullet \cos \psi \bullet \cos \psi_k} \qquad (9)$$

As shown is (9), when the difference between the grid impedance angle (ψ_k) and the power factor angle (ψ) approaches 90 degrees, the voltage variation as well as the corresponding flicker level is reduced. It is possible that the wind turbine output reactive power is regulated to vary with the output active power by the grid-side converter control. It behaves similarly to a STATCOM at the PCC. But the grid-side converter is already there without any additional cost.

Figs. 10-12 illustrate the flicker levels with output power factor angle control and without such control. The comparison is done with different parameters, such as mean wind speed, turbulence intensity, and short circuit capacity ratio. These results are shown that the wind turbine output reactive power control provides an effective means for flicker mitigation regardless of mean wind speed, turbulence intensity and short capacity ratio.

VI. CONCLUSION

This paper presents a simulation model of a MW-level variable speed wind turbines with a PMSG and a back-to-back full-scale PWM voltage source converters developed in the simulation tool of PSCAD/EMTDC. On the basis of the developed wind turbine model and corresponding control schemes, flicker emission of this system is investigated during continuous operation. The dependence of flicker emission on wind characteristics (mean speed, turbulence intensity), 3p torque oscillations due to wind shear and tower shadow effects and grid conditions (short circuit capacity, grid impedance angle) are analyzed. Simulation results show that the wind characteristics and grid conditions have significant effects on the flicker emission of the variable speed wind turbine with PMSG.

Flicker mitigation is realized by output reactive power control of the variable speed wind turbines with PMSG. The wind turbine output reactive power is regulated to vary with the output active power by the grid-side converter control, which leads to reduced flicker levels. Simulation results show the output reactive power control is an effective measure to mitigate the flicker during continuous operation of grid connected wind turbines.

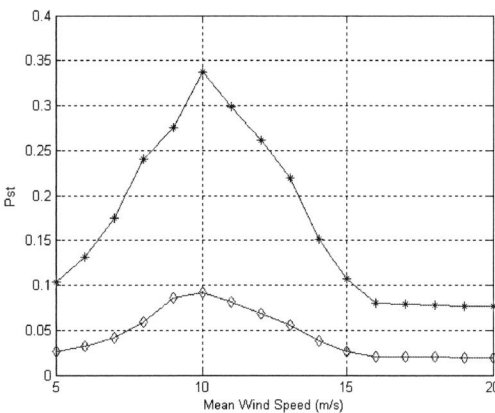

Fig. 10. P_{st} variation with mean wind speed (In=0.1, SCR=10, $\psi_k = 60^o$, without Q control (asterisk), with Q control (diamond)).

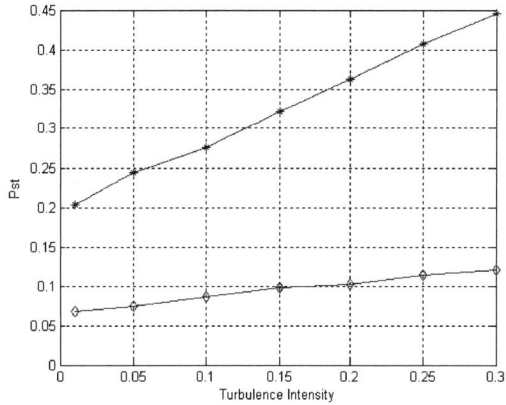

Fig. 11. P_{st} variation with turbulence intensity (v=9m/s, SCR=10, $\psi_k = 60^o$, without Q control (asterisk), with Q control (diamond)).

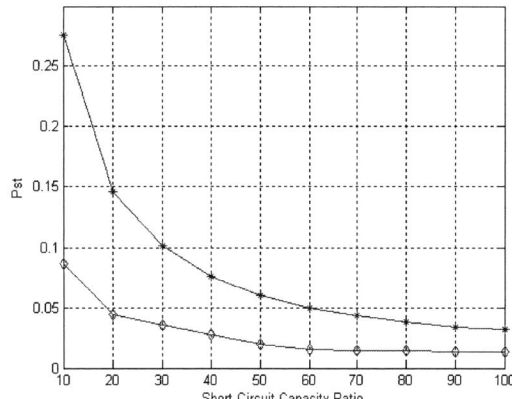

Fig. 12. P_{st} variation with short circuit capacity (v=9m/s ,In=0.1, $\psi_k = 60^o$, without Q control (asterisk), with Q control (diamond)).

REFERENCES

[1] L. Rossetto, P. Tenti, and A. Zuccato, "Electromagnetic compatibility issues in industrial equipment," *IEEE Ind. Appl. Mag.*, vol. 5, no. 6, pp. 34–46, Nov./Dec. 1999.

[2] Å. Larsson, "Flicker emission of wind turbines during continuous operation," *IEEE Trans. Energy Convers.*, vol. 17, no. 1, pp. 114–118, Mar. 2002.

[3] H. Sharma, S. Islam, T. Pryor, and C. V. Nayar, "Power quality issues in a wind turbine driven induction generator and diesel hybrid autonomous grid," *J. Elect. Electron. Eng.*, vol. 21, no. 1, pp. 19–25, 2001.

[4] M. P. Papadopoulos, S. A. Papathanassiou, S. T. Tentzerakis, and N. G. Boulaxis, "Investigation of the flicker emission by grid connected wind turbines," in *Proc. 8th Int. Conf. Harmonics Quality of Power*, vol. 2, Athens, Greece, Oct. 14–16, 1998, pp. 1152–1157.

[5] Tao Sun, Zhe Chen and Frede Blaabjerg, "Flicker study on variable speed wind turbines with doubly fed induction generators" *IEEE Trans. Energy Convers.*, vol. 20, no. 4, pp. 896–905, Dec. 2005.

[6] M´onica Chinchilla, Santiago Arnaltes and Juan Carlos Burgos. "Control of Permanent-Magnet Generators Applied to Variable-Speed Wind-Energy Systems Connected to the Grid". *IEEE Trans. Energy Convers.*, Vol. 21, No. 1, Mar. 2006, pp. 130–135.

[7] *PSCAD User's Guide*, Manitoba HVDC Research Centre Inc., Winnipeg, MB, Canada, 2003.

[8] *EMTDC User's Guide*, Manitoba HVDC Research Centre Inc., Winnipeg, MB, Canada, 2003.

[9] P. A. C. Rosas, P. Sørensen, and H. Bindner, "Fast wind modeling for wind turbines," in *Proc. Wind Power 21st Century: EUWER Special Topic Conf. Exhibit.*, Kassel, Germany, Sep. 25–27, 2000, pp. 184–187.

[10] Dale S. L. Dolan and Peter W. Lehn, "Simulation Model of Wind Turbine 3p Torque Oscillations due to Wind Shear and Tower Shadow" *IEEE Trans. Energy Convers.*, Vol. 21, No. 3, Sep. 2006, pp. 717–724.

[11] *Electromagnetic Compatibility (EMC)—Part 4: Testing and Measurement Techniques—Section 15: Flickermeter—Functional and Design Specifications*, IEC Std. 61 000-4-15, Nov. 1997.

Power Output Characteristics Analysis of Wind Energy Converter Control Methods

Bingchang Ni, Constantinos Sourkounis

Research Group for Power Systems Technology and Power Mechatronics,
Faculty of Electrical Engineering and Information Sciences, Ruhr-University Bochum, Bochum, Germany
e-mail: *ni@eele.rub.de*
e-mail: *sourkounis@eele.rub.de*

Abstract—The stochastically fluctuating offer of wind energy limits the supply capacity in electrical grids by means of technical and economical reasons. This paper analyses the influence of the control method on the power output characteristic of wind energy converters. Conventional controls, like optimal rotor speed control, and a new control, the stochastic dynamic optimization, are compared in regard of their grid impact. The results show the superior performance of the stochastic dynamic optimization. The power gradient and the deviation from the mean power output value are significantly lowered, reducing the wind energy converter's impact on the grid stability.

Keywords—adaptive control, optimal control, stochastic optimisation, wind power generation

I. INTRODUCTION

Over the past decades environmental issues, as well as the ambition of autonomy are gaining importance in global power generation. The goal of CO_2 reduction has led to an increasing amount of energy provided by renewable sources, like wind, water, solar and geothermal. Wind power generation plays a vital role in achieving this goal. The share of electricity generated by wind power converters is world wide steadily growing. By the end of 2006 the world wide installed power of wind energy converters reached 74 GW, with an increase rate of 28% annually since 1995. In 1990 wind power converters in the EU generated 0.8 TWh of electric energy; by 2006 this number reached over 80 TWh and this growth is expected to continue. Additionally, the increasing energy and commodity prices, the oil price in May 2008 was about five times a high as five years ago, has added to the attractiveness of alternative power sources.

Wind energy conversion is subject to two major restrictions from the power source; the wind mass's kinetic energy. The first restriction is that the availability of wind energy can not be controlled; furthermore wind energy underlies a stochastic fluctuating behaviour, where the location has great influence on the fluctuation characteristics. These fluctuations cause dynamic power oscillations in the wind energy converter's power drain. Especially at locations with turbulent wind characteristics, critical load peaks might occur. They propagate from the wind rotor to the connected electrical grid and cause premature damage in mechanical components, thermal overloads in electrical components, as well as voltage variations in the mains power supply.

In principle, variable speed wind energy converters provide the technical possibility to reduce the cumulative load in the power drain and the influence on the mains power supply. Therefore, specially designed automatic controls and operation management are required to run the appliance. The automatic controls are reference variable controls, which set the system to the optimal operating point. The main task of the operation management is to force a power or rotor speed set point for the momentarily wind condition.

In this paper different control methods are analysed under special focus on their influence on the resulting power output characteristic of the wind energy converter. The control methods considered in this investigation are optimal rotational speed control, torque control, average power control and stochastic dynamic optimisation. These control methods are described in chapter three. In chapter four the influence of the controls on the output power characteristic are compared. The final chapter summarises the investigation results. Foremost, the second chapter gives a short overview of the system behaviour of Wind Energy Converters (WEC).

II. WIND ENERGY CONVERSION SYSTEM

The WEC can be described by the energy conversion chain. First of all, the wind rotor converts the air mass's kinetic energy into mechanical (rotating) energy. The drive shaft transfers this rotating energy to the electrical generator. Most WEC use gear boxes to match the operating speed of the wind rotor and the generator, usually synchronous machines or doubly fed induction machines. The electrical energy is then fed into the

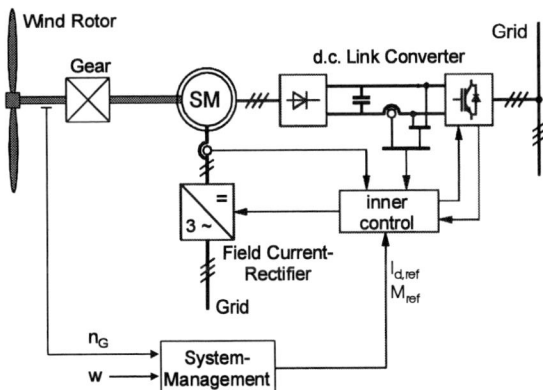

Fig. 1.   Exemplary configuration of a variable speed WEC [8]

connected grid. The generator can be directly connected to the mains power supply or power converter systems provide the coupling of the generator and the electrical grid (figure 1). [1], [8]

From the system side of view, the requirements are high system efficiency, long life cycle and high appliance availability. In addition the application specific requirements are maximum energy yield, low influence on the grid and low strain/stress in the power drain.

The air mass's kinetic energy has a specific power P_{wind} that is characterized by the air mass density ρ_A, the surface area of the wind flow A_R and the wind speed w. it can be described by the following formula:

$$P_{Wind} = \frac{1}{2}\rho_A A_R w^3. \quad (1)$$

Keep in mind the cubic influence of the wind speed on the wind power.

To gain the maximum energy yield, the wind rotor has to operate at highest efficiency. The aerodynamic efficiency is expressed by the c_P coefficient, which determines the amount of power extracted (P_{Rotor}) from the wind power crossing the rotor area (P_{Wind}):

$$c_P = \frac{P_{Rotor}}{P_{Wind}}. \quad (2)$$

According to Betz [2] the theoretically maximum c_p value is $c_{P,Betz} = 0.59$. The efficiency of wind rotors strongly depends on the airfoils angle of incidence. This angel results from the rotational speed of the wind rotor and the wind speed, that are perpendicular to each other. The ratio between the circumferential velocity of the rotor tip and the wind speed is described by λ, which is named the tip speed ratio:

$$\lambda = \frac{2\pi R n_R}{w}. \quad (3)$$

where R is the rotor radius and n_R is the rotor speed.

Wind rotor performance is usually described by the c_P over λ curve. For easy understanding, the c_P - λ curve can be transformed into a rotor power over rotational speed curve. From this curve (figure 2), the rotor power at a given rotational speed can be readout according to the wind speed. The wind rotor's operation point in the characteristic power field changes with each change of wind speed, e.g. from 1 to 2 in figure 2. Therefore, the controller has to adapt the rotor speed for maximum power output from 2 to 3. Depending on the dynamic behavior of the applied control method this might cause additional undesirable torque and power alterations.

In order to gain advantage of this special characteristic the WEC must be able to operate at variable speed. It demands a special structure of the generators grid coupling. The main task is the decoupling of the generators rotational speed from the grids voltage frequency. In figure 1 an example of a variable speed WEC's structure was shown, where a dc-link connects the generator to the grid. [6]

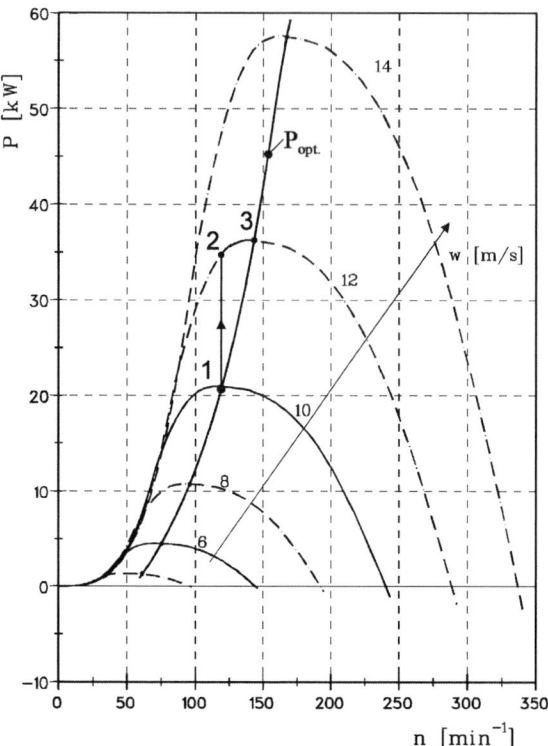

Fig. 2. Power field example of a wind rotor

III. CONTROL METHODS FOR WEC'S

The control can be divided into two stages. The inner control loop sets the plant to the desired operation point given by the outer loop, the system management. The influence variable of the inner control loop is the generator torque counteracting the rotor torque. By the torque difference a change in rotational speed of the drive train results. A rotor torque higher than the generator torque leads to an acceleration of the drive train, and a lower leads to a slow down.

The controlled variable of the system management depends on the executed control method. Generally single variable control methods follow different goals, like maximum power yield, low torque oscillations or low power output fluctuations. The system management is also responsible for overall operation of the plant, like the run-up procedure, emergency shut down, etc. In the following, the four different control methods analysed in this paper are described.

A. Optimal Rotor Speed Control

The optimal rotor speed controller aims to always operate the wind rotor at the rotor speed with maximum power yield. The correlation between wind speed and optimal rotor speed can be derived from the wind rotor's characteristic power field (figure 2). Figure 3 shows the rotor speed control loop structure.

The disadvantage of this simple speed control is that for fast increasing wind speeds w, the rotor speed increases slower than the reference speed n_{ref} due to high system inertia. Entering a high input value into the rotor speed controller causes a sudden drop of the output power since

2332

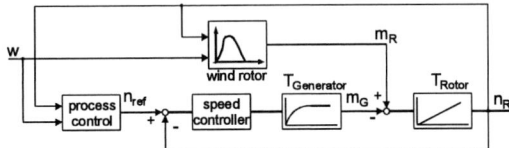

Fig. 3. Block diagram of the optimal rotor speed control

the speed controller tries to rapidly accelerate the rotor by decreasing the generator's counter torque. For fast decreasing wind speeds, the controller reacts in the opposite direction by setting the generator to the maximum power output, creating a high counter torque to slow down the wind rotor. This results in strong fluctuations of the power output at unstable wind conditions (cp. section IV).

A measure for reducing the undesired output power fluctuations is to limit the control deviation. Thus, the maximum speed controller reaction is restricted, leading to a fast control characteristic for small control deviation and a greater time constant with less power fluctuation for large deviations. Therefore, a lower control deviation limit also results in an increased deviation from the maximum power factor due to slower speed adaptation.

B. Torque control

The torque control was designed to avoid load peaks in the WEC drive train, like they occur with the optimal rotor speed control. It works, like the optimal speed control with a wind rotor model. From the measured wind speed and rotor speed the control computes the actual torque generated by the rotor (M_R). This value is passed on to the generator as the reference torque. Therefore the generator torque follows the rotor torque for the momentarily wind speed. Through this, peaks in the power output are avoided. The rotor torque can calculated with the following formula:

$$M_R = \frac{P_\omega}{\omega_R} = \frac{\frac{1}{2}c_P\rho\pi R^2 w^3}{2\pi n_R} = \frac{2c_P\rho\pi^3 R^5}{\lambda^3}\cdot n_R{}^2 \quad (4)$$

Based on the request that $\lambda = \text{constant} = \lambda_{opt}$ for maximum power yield, the generator torque (M_G) can be set by applying following formula:

$$M_G = \eta_G\eta_{Gear}\frac{2c_p\rho\pi^3 R^5}{\lambda^3}\cdot n_R^2. \quad (5)$$

where η_G and η_{Gear} are the efficiency of the generator and the gear box.

C. Average Power Control

Another approach for controlling the WEC is to determine an output power reference value. From figure 2 and the measured values w and n_R, the controller determines if the appliance is running at the maximum power factor. Otherwise, the controller generates a designated power time sequence to converge to the desired state. The reference power output is computed using the mean wind speed value. The best mean wind speed calculation method showed to be a recursive mean value calculation. Hereby each newly measured value is

weighted for example with a factor of 7% and it is added to the old mean value, which is weighted with a factor of 93%.

$$\overline{w}_{new} = (1-\varepsilon)\cdot\overline{w}_{old} + \varepsilon\cdot w_{actual} \quad (6)$$

where $\varepsilon = 0.07$ in the given example, but generally $0 \leq \varepsilon \leq 1$. This addition is carried out once a second. Old values constantly loose their influence in the mean value calculation, but are never discarded completely. An increased ε value would lead a faster adaptation of the rotor speed to changing wind speeds, resulting in a better wind power utilization, the trade-off however, being a stronger fluctuation of the power output with the disadvantage of more stress on the electrical and mechanical subsystems.

D. Stochastic Dynamic Optimization (StoDO)

A new approach for WEC control is the stochastic dynamic optimization [4] with iterative adaptive wind speed probability distribution, named "iterative self-adapting system management" (ISSM) [5], [8]. Here, the WEC is understood to be a stochastic process where the wind speed is subject to a stochastic disturbance z. The aim of the control is to find the operation state with the minimum output power fluctuation and the maximum wind power utilization.

Figure 4 schematically shows this approach with the system management determining the set points of the WEC. The state values of the process are the wind speed w and rotational speed n_R, of which only the rotational speed is influenced by the controller. The dc-link current $I_{d,ref}$ serves as the controllable variable determined by applying the control algorithm on the measured state values. This control algorithm is adapted to the changing wind speed distribution, which is location specific, and dc-link current frequency distributions, which results from the WEC operation. The control structure is not static, but

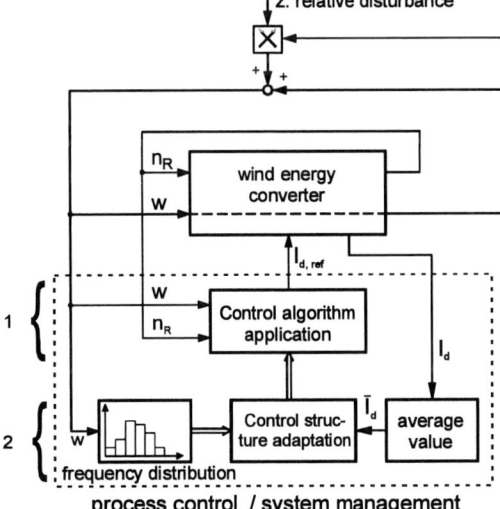

process control / system management

1: execution each computation cycle

2: e. g. once every 100 computation cycles

Fig. 4. Layout of stochastic dynamic optimization control

2333

continuously updated, noticeably slower than the power set point. Therefore this control method is able to adapt to changing wind conditions at the WEC installation site.

The optimal control algorithm is implemented in the appliance's first execution level. It consists of a look-up table for the dc-link current reference values. The input variables of the look-up table are w and n_R. Meanwhile statistical data of the wind speed and dc-link are gathered during operation, which are stored in the wind speed frequency distribution and the average generator current respectively.

For each operation state the control calculates two values which indicate the deviation from the mean power output value (P-index)

$$P\_index = \frac{\sum_{i=1}^{\infty}\left|P_i - \overline{P}\right|}{\overline{P}}, \qquad (7)$$

which is also an indication for the output power fluctuation rate, and the deviation from the rotor's optimal power coefficient (c_P-index),

$$c_P\_index = \frac{\sum_{i=1}^{\infty}\left|c_{P,i} - c_{P,opt}\right|}{c_{P,opt}}, \qquad (8)$$

which indicates the aerodynamic efficiency of the wind rotor operation.

The control aim is to keep the power output fluctuation as low as possible and, at the same time, let the wind rotor operate at its highest efficiency. Therefore, both index values have to be as small as possible. To express both objectives in one formula, the P_index and the c_P_Index are put into one formula called the cost function φ:

$$\varphi = g \cdot \frac{\left|P_i - \overline{P}\right|}{\overline{P}} + (1-g) \cdot \frac{\left|c_{P,i} - c_{P,opt}\right|}{c_{P,opt}}, \qquad (9)$$

where $0 \leq g \leq 1$ is a weighting factor. Now, the goal of the StoDO controller is to minimise this cost function throughout operation.

For every possible operation state at the next time step (t+1) the actual cost can be calculated according to (9). With any given transition of w_t to w_{t+1} the next operation state can be computed with the rotor torque, the actual rotor speed and the assumed generator torque for each possible $I_{d,ref}$ value. From the torque difference and the system inertia the next operation state, depending on the applied $I_{d,ref}$ value, can be calculated. For WECs, the transition of w is stochastic, and therefore the expected cost is calculated by applying the frequency distribution of wind speed transitions in the calculation. The goal of the StoDO control is not to only minimise the cost of the next state but to minimise all future operation states. Therefore the principle of optimal control [4] is implemented by computing the residual cost of each operation state. Then the StoDO control chooses the $I_{d,ref}$ value that makes the WEC approach the state with minimum transitional and residual costs. These operation state depending $I_{d,ref}$ values are put into a look-up table and passed on to the control algorithm at each control structure adaptation

cycle. With each alteration of the wind speed frequency distribution and mean value of I_d, the residual costs change too and are recalculated. Thereby, the wind energy converter is able to adapt to changing conditions by updating the wind turbulence frequency distribution $P(z|w)$.

IV. COMPARISON OF POWER OUTPUT

The control methods were tested by simulations with a model of a 5MW WEC and on a test plant, where a variable speed wind energy converter with a nominal electrical power of 22 kVA was modulated under consideration of the subsystems individual non-linear nature. For this purpose power grid simulation software was used to modulate the individual behaviours of the subsystems, rotor, gearbox/shaft, synchronous machine and electric power converter. The wind speed was read from a file containing the dynamics of different geographical locations (mountains, flat country), created prior with help of pseudo random-number generators. The wind characteristic used for these analyses was of highly fluctuating nature. To analyse the investigation results, the wind speed w, wind rotor power P_R, rotational speed n_R, dc-link current I_d, the electrical power output P_{el}, and the power coefficient c_P were logged.

In practice the speed control (figure 5) causes fluctuations in the electrical output, which eventually

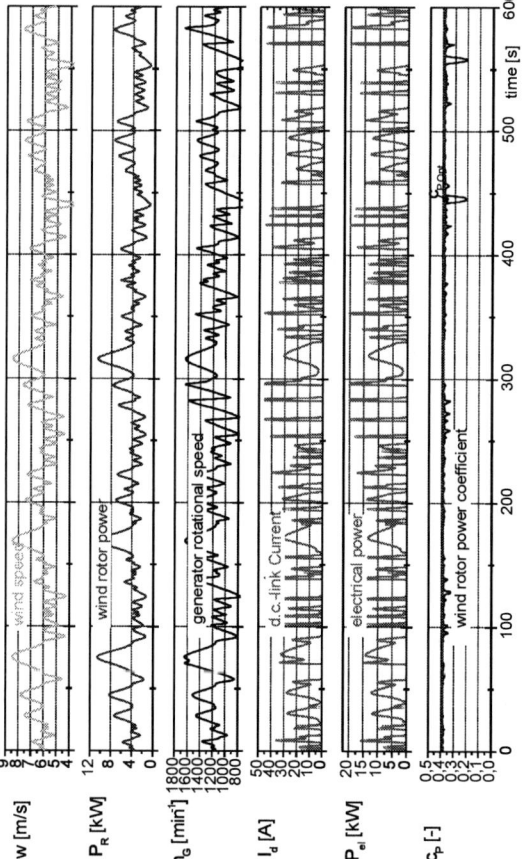

Fig. 5. Power flow in the power drain by conventional speed control method with stochastic coastline wind speed

result in disturbances in the grid (e.g. flicker effect) and a heavy work load on the drive train. Especially transient dynamic fluctuations of wind speed trigger transitional processes. As described in section III, during highly turbulent periods the controller causes the generator to suddenly switch to maximum or zero power output in order to quickly match the rotor speed with the strongly altered optimal rotor speed. Additionally, the controller does not try to keep the rotor speed within its given limits. When the rotor reaches its upper or lower speed limits, the system management overrules the speed controller and takes action to return the rotor speed back to its valid range by switching the generator output to maximum or zero power. For short time periods, even a fast adaptation of the rotational speed results in an insignificant increase of energy yield only, but causing strong disturbances on the grid.

Compared to the speed control the StoDo control shows a significant improvement regarding the power output fluctuation (figure 6). The diagram shows almost no variation of the power output value. For most of the time it stays constant at about $4 - 5$ kW. The frequency and the size of the peaks are greatly improved in term of a minimal grid impact of the WEC's power delivery to the mains supply.

For a better analysis of the power output diagram characteristics, two frequency distributions were created. Figure 7 shows the relative frequency distribution of the power output deviations from the average power output over the whole analyzed time period. From figure 5 one can see that, with the optimal rotor speed control, the power output is equals to zero for an extended time period. The frequency of the zero power output can be read from figure 7, indicated by the peak at a deviation of approximately -0.9 MW. For the rest of the deviation the frequency stays below 0.5% most of the time. Therefore no significant power output value is maintained for a relevant time period.

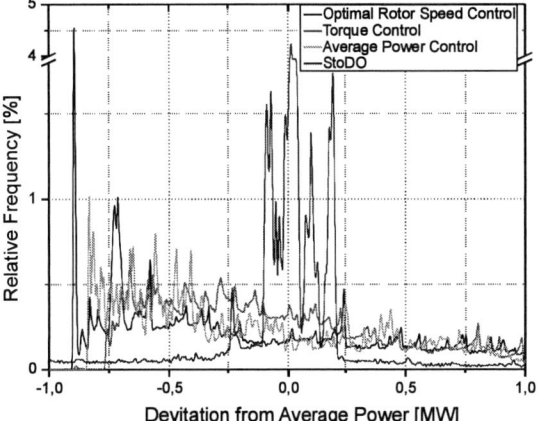

Fig. 7. Simulation of a 5 MW WEC: Frequency of output power deviation from the mean value

The torque control shows no significantly increased frequency at any deviation. The most frequent deviation is just above 0.5 %. Therefore the power output covers a wide range with almost constant frequency. The average power control shows a similar behaviour like the torque control with a slight bias to lower power output, with the zero power output having a slightly increased frequency.

The StoDO curve shows the desired behaviour of a control that maintains the power output constantly near the mean value. The frequency of the deviation from the mean power value (deviation of 0 MW) is extremely high. Very close to it, the frequency stays at high value and rapidly drops to a dispensable value of under 0.1 %.

The second frequency distribution describes the occurrences of power gradients. The grid is not only influenced by the range of the power fluctuation, but the speed of the fluctuation has an even more important role for understanding the impact on the mains power supply.

Figure 8 shows the relative frequency distribution of the power gradients. The optimal rotor speed control has a small peak close to 0 kW/s, with a relevant bump in the frequency distribution from 0 kW/s to 125 kW/s. The torque control shows a significant bias to almost a normal distribution with a bias towards -50 kW/s. The average power control shows a very good behaviour with a major peak at 0 kW/s and a relatively fast decrease of the frequency distribution for positive and negative gradients. The best performance of all controls shows the StoDO control. The peak at 0 kW/s is even higher and the decrease of the frequency around it is even steeper than those of the average power control.

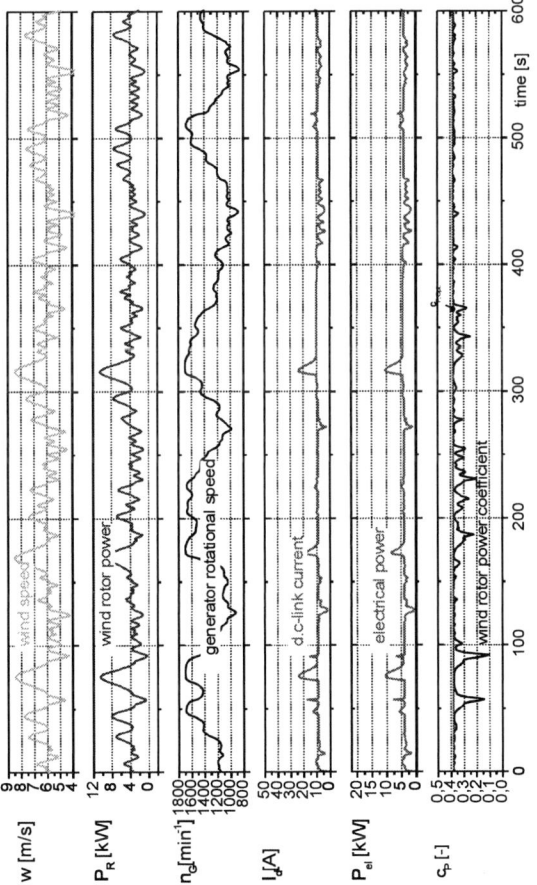

Fig. 6. Power flow in the power drain by iterative self-adapting control method with stochastic coastline wind speed

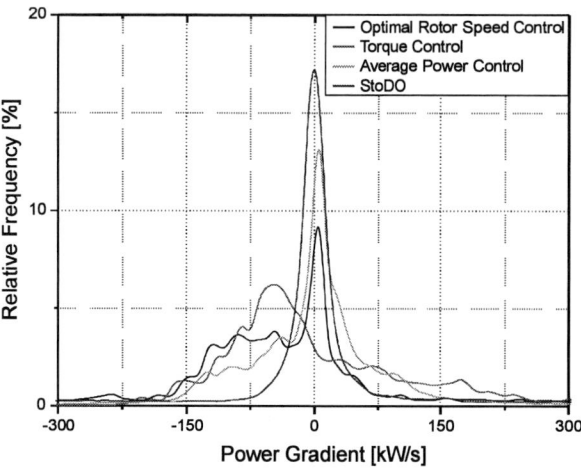

Fig. 8.   Simulation of a 5 MW WEC: Frequency of power gradient

V. RÉSUMÉ

The analysis shows that the control methods of WEC have great influence on their power output characteristics and therefore on their grid impact. To evaluate the grid impact, the deviation of the power output is of great influence, but even more important is the power gradient. The controls show very different behaviour regarding their grid impact. The common control methods are not optimised in this direction. Otherwise, the stochastic dynamic optimisation has a very smooth power output characteristic without sacrificing the overall power yield. Therefore it is recommended to put more effort on reducing the grid impact of WECs, especially regarding the fact of an increasing number being connected to the mains power supply.

REFERENCES

[1] R. Gasch, J. Twele, "Wind Power Plants", James & James, April 2005

[2] A. Betz, "Windenergie und ihre Ausnutzung durch Windmühlen", Vandenhoekk und Rupprecht, Göttingen 1926

[3] S. Heier. "Grid Integration of Wind Energy Conversion Systems", 2nd ed., Wiley & Sons, 2006.

[4] M. Papageorgiou, "Optimierung", 2. Auflage. Oldenbourg, Muenchen, Wien, 1996.

[5] C. Sourkounis, B. Ni, "Drive Train Control for Wind Energy Converters Based on Stochastic Dynamic Optimisation", Proceedings IEEE Industrial Electronics Conference 2007, Paris, France

[6] C. Sourkounis, "Windenergiekonverter mit maximaler Energieausbeute am leistungs-schwachen Netz", Dissertation, TU Clausthal, 1994. Clausthal-Zellerfeld: Papierflieger, 1995.

[7] J.-P. Molly, "Windenergie: Theorie - Anwendung - Messung", 2. Auflage. Mueller, Karlsruhe, 1990.

[8] C. Sourkounis. Drehzahlelastische Antriebssysteme unter stochastischen Belastungen. Habilation, TU Clausthal, 2004. Clausthal-Zellerfeld: Papierflieger, 2004.

A Cooperative Control Method for Output Power Smoothing and Hydrogen Production by Using Variable Speed Wind Generator

Rion Takahashi[1], Hirotaka Kinoshita[1], Toshiaki Murata[1], Junji Tamura[1]
Masatoshi Sugimasa[2], Akiyoshi Komura[2], Motoo Futami[2], Masaya Ichinose[2], Kazumasa Ide[2]
[1]Kitami Institute of Technology, 165 Koen-Cho, Kitami, Japan.
[2]Hitachi, Ltd., 7-1-1 Omika-Cho, Hitachi, Japan.
E-mail: rion@pullout.elec.kitami-it.ac.jp

Abstract — This paper presents a combination system of wind energy conversion and hydrogen production. Hydrogen is expected as alternative energy sources in the future, and it is the best way to produce it from renewable energy like wind energy. On the other hand, the output of a wind generator, in general, fluctuates greatly due to wind speed variations, and thus the output fluctuations can have a serious influence on the power system. In the proposed system, a variable speed wind generator is adopted, and an electrolyzer is installed in parallel with it for hydrogen production. Output power from the wind generator is smoothed and supplied to the power system as well as to the electrolyzer based on the cooperative control method. The performance of the proposed system is evaluated by simulation analyses, in which simulations are performed by using PSCAD/EMTDC.

Keywords — Wind energy conversion, Power smoothing, Variable speed wind generator, Hydrogen production

I. INTRODUCTION

In recent years, hydrogen has attracted much attention as an alternative energy to oil due to the problems of global warming and exhaustion of fossil fuel. Hydrogen can be produced from electrical power and water by using an electrolyzer. However, if the electrical power generated from thermal power plants is used for hydrogen production, environmental problems stated above cannot be solved. In this paper, hydrogen production by wind power, which is one of the renewable energy sources, is proposed. Since the wind power is a clean energy source without emission of carbon dioxide, it is the best electrical power source for hydrogen production.

In this paper, a cooperative control method of the combination system composed of a variable speed wind generator and High-purity Hydrogen and Oxygen Generator (HHOG) system is presented. Because output power from the variable speed wind generator also fluctuates, a power smoothing control using the HHOG system with variable power consumption is presented. In general, however, electrolyzers constituting the HHOG system should be operated under constant input power, because electrodes in the electrolyzer can be deteriorated under variable current and voltage conditions. In this paper, a new method is proposed in which the total power supply to the HHOG is varied in stepwise manner by

changing the number of activated electrolyzers but consumed power in each electrolyzer is maintained constant (full-load or no-load). Since in such control method, however, power output to the grid can be varied in the stepwise manner, another smoothing control of the stepwise output is performed by the flywheel operation of variable speed wind generator. Consequently, in the proposed wind generator system, the smoothed power can be supplied to the power system as well as hydrogen is generated in the electrolyzers with constant power consumption. The effectiveness of the proposed system is confirmed by simulation analyses by PSCAD/EMTDC.

II. MODEL SYSTEM

The model system used in the simulation analyses is shown in Fig.1. DFSG, the variable speed wind generator, is connected to an infinite bus through a transformer and a transmission line. With detecting the active power output to the grid, P, and the terminal voltage, V_G, at point G, output power smoothing and constant terminal voltage control at point G are performed. HHOG is connected to the grid through a diode rectifier, a DC-DC converter, and a transformer. HHOG consists of several electrolyzers, producing hydrogen gas by consuming a part of output power from DFSG, in which the number of activating electrolyzers is determined according to the output from DFSG.

III. MODEL OF WIND TURBINE

The mathematical relation for the mechanical power

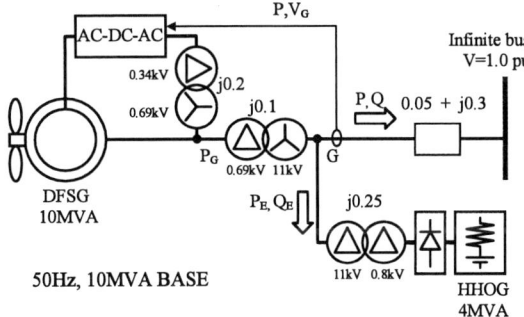

Fig. 1. Model system

extraction from wind can be expressed as eq.(1), where, P_M is the extracted power from the wind, ρ is the air density (kg/m^3), R is the blade radius (m), V_w is the wind speed (m/s) and C_p is the power coefficient which is a function of both tip speed ratio, λ, and blade pitch angle, β (deg). C_p is expressed in eq.(2) [1], where ω_m is the rotational speed of the wind turbine (rad/s).

$$P_M = 0.5\rho C_p\left(\lambda,\beta\right)\pi R^2 V_W^3 \left[W\right] \tag{1}$$

$$C_p\left(\lambda,\beta\right) = 0.5\left(\Gamma - 0.02\beta^2 - 5.6\right)e^{-0.17\Gamma} \tag{2}$$

$$\lambda = \frac{\omega_m R}{V_W}, \quad \Gamma = \frac{R}{\lambda}\cdot\frac{3600}{1609} \tag{3}$$

$$\omega_{m\_op} = 0.0775 V_W \left[pu\right] \tag{4}$$

The characteristic of C_p will change depending on the wind speed greatly. Eq.(4) is a relation between the optimal rotational speed, ω_{m_op}, corresponding to the maximum power coefficient, and the wind speed, V_W. This expression was obtained by differentiating C_p with respect to ω_m with assuming $\beta = 0$. Fig.2 shows the characteristics between the wind turbine power P_M and the rotor speed ω_m for various wind speed as well as the maximum turbine power, P_{max}. A dashed line shows a locus for the operation of the wind generator. In addition, the wind turbine is driven within a speed range from 0.7 to 1.3 pu in this study. Reference values for the wind generator output and the power consumption of HHOG are determined according to P_{max} and ω_{m_op} by the method shown later.

IV. MODEL OF HYDROGEN PRODUCTION

In general, output power of the wind generator fluctuates greatly due to wind speed variations. In this paper, a new combination system to overcome this prob-

lem is proposed, which consists of the wind generator and High-purity Hydrogen and Oxygen Generator (HHOG) system [2] composed of electrolyzers. Each electrolyzer is modeled in this paper as an electric circuit shown in Fig.3, and HHOG consists of 6 electrolyzers as shown in Fig.4. In the proposed system, operation shifting among these electrolyzers is adopted in order to average the operation time of each electrolyzer as mentioned later.

The number of electrolyzers being connected to the electrical supply can be changed by ON/OFF control of a chopper circuit in each electrolyzer. Consequently the total consumed power in HHOG can be varied, while input power to each electrolyzer is controlled constant. Such operation is adopted here because varying load operation of electrolyzer can have a serious influence on its service life. In addition, the soft-start operation of electrolyzer, in which supplied voltage is increased gradually by the chopper, is also possible. A simple model composed of a resistance and internal voltage source [2] has been used for the electrolyzer in the simulation analyses. Parameters of the electrolyzer are shown in Table I, and it is assumed each electrolyzer has the same parameters in this study.

V. MODEL OF WIND GENERATOR

Doubly-Fed Synchronous Generator (DFSG) [3] is used as the variable speed wind generator in this study. DFSG can control active power output and reactive

Table I. Parameters of the electrolyzer

Rated capacity of one block	0.66 MW
Internal resistance R_0	0.5 Ω
Internal voltage V_0	1407.5 V

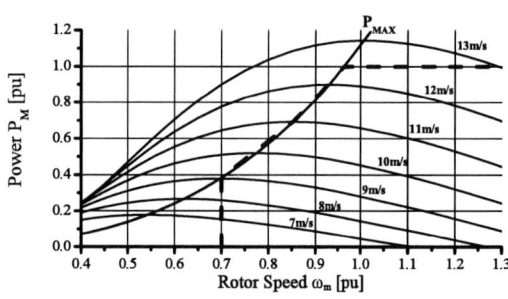

Fig. 2. P_M-ω_m curves for different wind speed

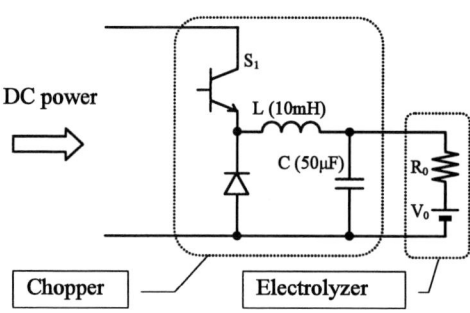

Fig. 3. Electrolyzer circuit

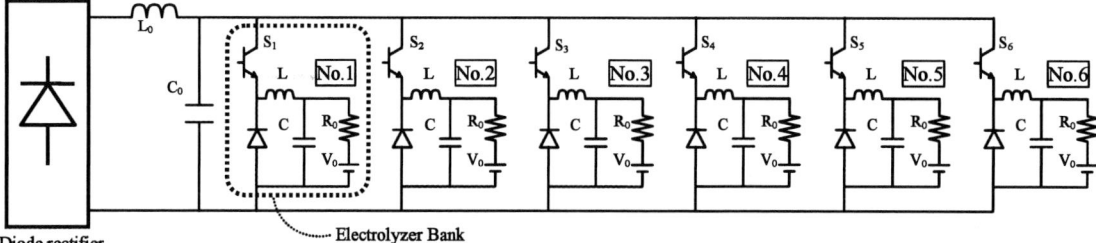

Fig. 4. Hydrogen generator circuit.

power output independently by AC excitation control in the rotor circuit. In addition, the construction of DFSG is basically the same as that of a wound rotor induction machine and the required power capacity of a power converter for secondary (rotor circuit) excitation can be decreased to be about several ten percent of the machine rated capacity depending on the speed range of DFSG. Therefore the total cost can be decreased.

Fig.5 shows a brief of the secondary excitation control system for DFSG. This consists of the generator output control (inverter side) and the excitation power control (converter side). The active power output, P, and the terminal voltage, V_G, at the grid are regulated by the generator output control. The method to determine the reference value, P_{REF}, for P will be shown later. The terminal voltage, V_G, is controlled to be the rated value (1.0 pu). The DC link voltage, E_{dc}, and reactive component of the excitation current, I_q', are regulated by the excitation power control to be constant (E_{dc}=1.0 pu, I_q'=0 pu). Ratings and parameters of the DFSG are shown in Table II. As stated before, it is assumed in this study that the rotor speed takes a value within the range from 0.7 to 1.3 pu, that is, ±30% of the synchronous speed.

VI. PROPOSED CONTROL METHOD

If some part of fluctuating component of the wind generator output can be consumed in the HHOG, it is expected that smoothed power can be supplied to the power system. In order to achieve this control strategy, the reference value for the output to the power system and the number of active electrolyzers in the HHOG are very important. The decision methods for them are described in the following.

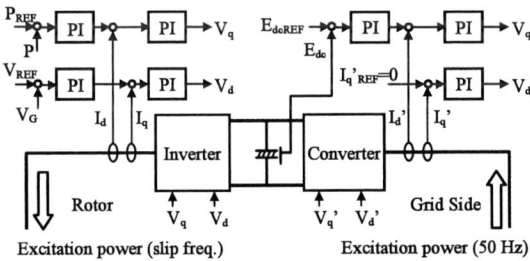

Fig. 5. Secondary excitation control system

Table II. Ratings and parameters of the DFSG and its excitation system

Rated Power	10 MVA
Rated Voltage	690 V
Stator resistance	0.01 pu
Rotor resistance	0.01 pu
Excitation reactance	3.5 pu
Stator leakage reactance	0.15 pu
Rotor leakage reactance	0.15 pu
Inertia constant (H)	1.5 sec
DC link voltage	700 V
DC link capacitor	25000 μF
PWM carrier frequency	1000 Hz

A. Active power output reference (P_{REF})

The reference value of the DFSG output, P_{REF}, is calculated to smooth the power flow into the power system. Fig.6 shows a basic concept of the calculation [4], in which P_{max} denotes the wind generator output obtained by the Maximum Power Point Tracking (MPPT) operation. P_{max} is smoothed by using a low pass filter (first order delay unit) at first, and then P_{max_LPF} is obtained. Next, standard deviation of the difference, P_{max_σ}, between P_{max} and P_{max_LPF} is calculated, and then P_{REF} can be obtained by subtracting P_{max_σ} from P_{max_LPF} as shown in Fig.6 and Block A in Fig.7. Eventually, the fluctuating component and smoothed component of the wind generator output can be separated.

In Fig.6, P_c is the difference between P_{max} and P_{REF}, and it is needed to be consumed in the HHOG as mentioned above. According to P_c, the number of active electrolyzers is determined as explained later. P_c is almost positive in whole operation period as can be seen in Fig.6. Thus the smoothed power, P_{REF}, can be supplied to the power system because the fluctuating component P_c is consumed in the HHOG. However when P_{max} becomes less than P_{REF}, P_c becomes negative and then P_{REF} cannot be achieved only by the HHOG. In this period, the difference between P_{REF} and P_{max} is transiently supplied by discharging the kinetic energy of the wind turbine. At this time, as the turbine speed ω_m decreases from ω_{m_op} at MPP, P_{REF} is decreased so as to keep the turbine speed at ω_{m_op}. This control is performed in Block B of Fig.7.

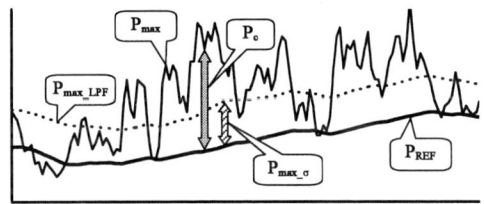

Fig. 6. Example of P_{REF} and P_c

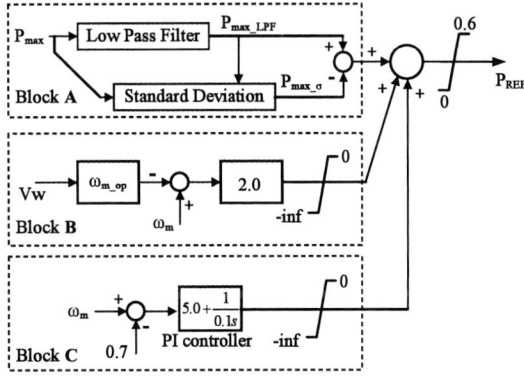

Fig. 7. Block diagram to determine P_{REF}

In addition, another control block shown in Block C of Fig.7 works to keep the rotor speed to be over 0.7 pu because it is the minimum speed in the wind generator. On the other hand, a blade pitch angle of the wind turbine is adjusted so that the rotor speed does not exceed the maximum limit of 1.3 pu. In this paper, the blade pitch actuator is modeled by using a first order time delay with a time constant of 5 second, and the rate of pitch angle changing is limited to 10 degrees per second to avoid excess mechanical stress. Finally P_{REF} is determined to be the sum of output signals from block A to C in Fig.7, and it is limited under the maximum output of 6MW (0.6 pu) because the HHOG consumes 4MW (0.4 pu) at the rated condition.

B. Switching operation of electrolyzers in the HHOG

The number of the active electrolyzers in the HHOG is determined according to P_c. In this paper, the following method is adopted for the operation of electrolyzers, that is, one electrolyzer does not activate under a partial load condition, in other words, it operates in either full load or no load (disconnected) condition. The rated power capacity of each electrolyzer is 0.66 MW, and the number of active electrolyzers among 6 electrolyzers is changed according to P_c in the stepwise manner. However, if a simple changing strategy is applied, the electrolyzers can be operated with excessive ON/OFF switching because the wind generator output varies greatly due to the wind speed fluctuation.

In this study, the following strategy is adopted to avoid the excessive switching; that is, the controller generates ON signal when the power, P_c, becomes greater than the present power of the HHOG by 1.32MW (double of the capacity of one electrolyzer). After that, if the ON signal continues for 5 seconds, then one electrolyzer is activated as shown in Fig.8. But if the ON signal does not continue for 5 seconds, then the controller resets the ON signal. Thus another electrolyzer is not activated. When P_c becomes less than the present power of the HHOG (for example, 1.32 or 0.66 (MW) in Fig.8), one of the activated electrolyzers is switched off immediately. Due to the switching method, excessive switching of electrolyzers can be avoided.

C. Operation shift of the electrolyzers

In the changing strategy of the number of electrolyzers explained above, the operation shift of electrolyzers is also adopted to avoid a biased operation period of electrolyzers. If the rule explained above is adopted as it is, it is expected that the activation rate of the first electrolyzer becomes very large and that of the last electrolyzer becomes very small as shown in the case of "simple operation" in Fig.9. Such biased operation of electrolyzers can shorten the survice life of particular electrolyzers. Therefore the operation shift is introduced as shown in tha case of "operation shift" in Fig.9. When one of the electrolyzers being OFF condition needs to be switched ON, the electrolyzer which was turned off

firstly among the OFF electrolyzers is switched ON. On the other hand, when one of the electrolyzers being operated needs to be switched OFF, the electrolyzer which has been in operation for the longest time is switched OFF. As a result, the operation time of each electrolyzer can be averaged.

D. Flywheel effect of the wind generator

Because power consumption of the HHOG can be controlled only in the stepwise manner as explained above, the fluctuating power component, P_c, cannot be consumed completely in the HHOG. In this paper, it is proposed to use the flywheel effect of the wind turbine generator to compensate the residual power of P_c; that is, the difference between P_c and the consumed power of the HHOG is stored into the wind turbine rotor as kinetic energy. Although the wind turbine speed increases beyond the MPP speed in this situation, it can be returned back to the MPP speed by extending the activating time of electrolyzers. The electrolyzer of which activating time has been extended is turned OFF when P_c becomes less than the power consumption of the HHOG, and then the excess kinetic energy of wind turbine, E_{WG}, becomes zero at this instant. E_{WG} can be calculated from eq.(5), where J

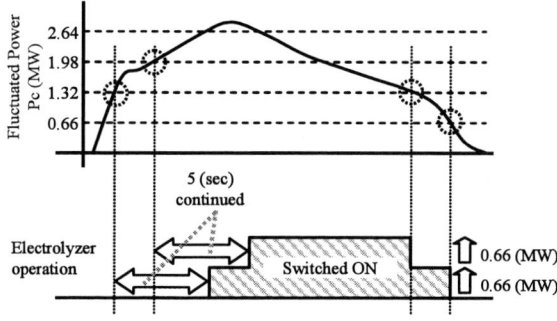

Fig. 8. Switching of hydrogen generator

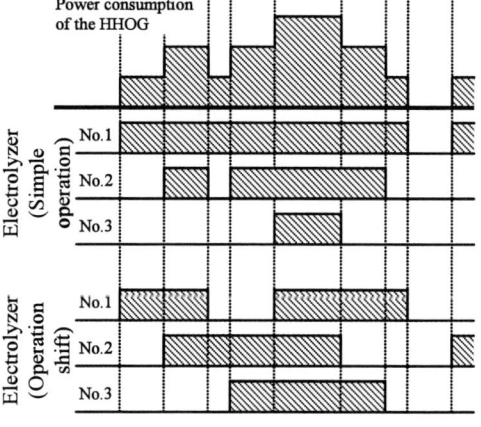

Fig. 9 Concept of the operation shift

Table III. Operation rule of electrolyzers

The number of activated blocks	Condition of ON signal	Condition of OFF signal
1	$P_c \geq 1.32$ (MW)	$P_c \leq 0.69$ (MW) AND $E_{WG} = 0$
2	$P_c \geq 1.98$ (MW)	$P_c \leq 1.32$ (MW) AND $E_{WG} = 0$
3	$P_c \geq 2.64$ (MW)	$P_c \leq 1.98$ (MW) AND $E_{WG} = 0$
4	$P_c \geq 3.31$ (MW)	$P_c \leq 2.64$ (MW) AND $E_{WG} = 0$
5	$P_c \geq 3.97$ (MW)	$P_c \leq 3.31$ (MW) AND $E_{WG} = 0$
6	$P_c \geq 4.62$ (MW)	$P_c \leq 3.97$ (MW) AND $E_{WG} = 0$

$$E_{WG} = \frac{1}{2} JP \left(\omega_m^2 - \omega_{m\_op}^2 \right) \tag{5}$$

is inertia constant (sec), P is the rated power of the wind generator (W). When the wind turbine operates at the MPP speed, E_{WG} is zero. These conditions of electrolyzer operation are shown in Table.III.

VII. SIMULATION RESULTS

Fig.10 shows the wind speed data used in the simulation analyses. Responses of P_{max} and P_{REF} are shown in Fig.11. P_{max} is the maximum output power of the wind generator, which is equivalent to the output power to the grid in the case of MPPT operation without the HHOG. Then the reference of smoothed power, P_{REF}, can be obtained using the method of section VI (A). P_c is calculated by subtracting P_{REF} from P_{max}. Then P_E and the number of operating electrolyzers are determined based on P_c. Fig.12 shows responses of the DFSG output, P_G, and supplied power to the grid, P, from which it can be seen that P_{REF} and P are almost the same. Fig.13 shows responses of the actual speed ω_m and the optimal speed ω_{m_op} of the wind turbine. Due to the proposed stepwise switching operation of electrolyzers, it is seen that the surplus energy from the wind is stored as kinetic energy in the rotor and thus the rotor speed increases over the optimal speed ω_{m_op}. Though the operating speed of the generator deviates from the optimal (MPP) speed as shown in Fig.13, output to the power system can be smoothed effectively. Moreover, as can be seen from Fig.13, the rotor speed is returned back to near the ω_{m_op} due to the absorption of excess kinetic energy, E_{WG}, by the extension of activation period of electrolyzers. Fig.14 shows blade pitch angle of the wind turbine. It is adjusted so that the wind turbine speed does not exceed 1.3pu. Fig.16 shows the reactive power response, and the voltage at point G, V_G, is maintained almost constant due to the reactive power control as shown in Fig.15.

Table. IV. Rate of operation of each electrolyzer.

Number of electrolyzer	Consumed power in each electrolyzer [MJ]	Rate of operation [%]
1	205.00	51.68
2	231.87	58.45
3	231.04	58.24
4	202.71	51.10
5	194.29	48.97
6	208.58	52.57

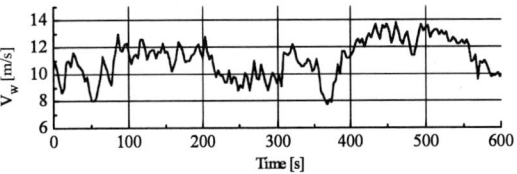

Fig. 10. Wind speed

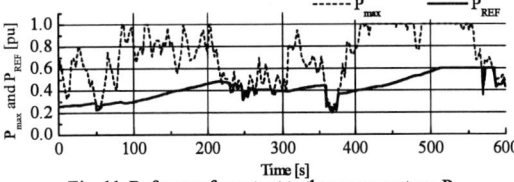

Fig. 11. Reference for output to the power system, P_{REF} and generator output at MPP speed, P_{max}

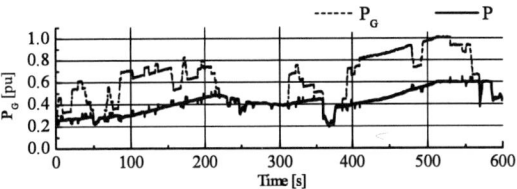

Fig. 12. Output from the DFSG, P_G, and output power to the power system, P

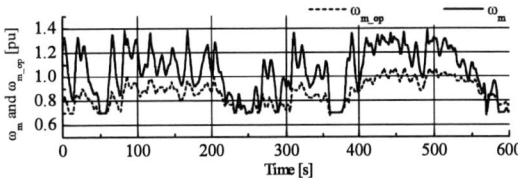

Fig. 13. Actual speed, ω_m, and the optimal speed, ω_{m_op}, of DFSG

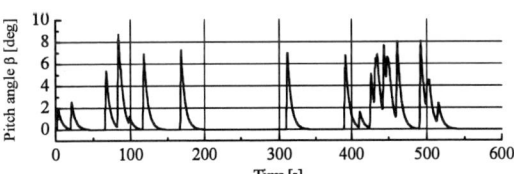

Fig. 14. Blade pitch angle of wind turbine

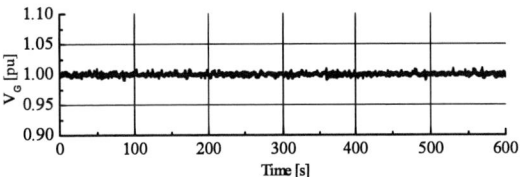

Fig. 15. Terminal voltage at point G

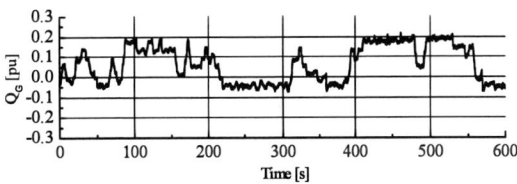

Fig. 16. Reactive power output of the DFSG

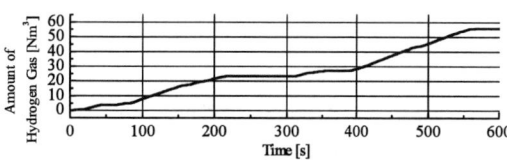

Fig. 17. Amount of the hydrogen gas

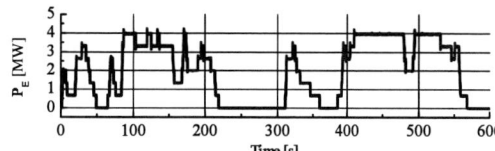

Fig. 18. Total consumed power of the HHOG

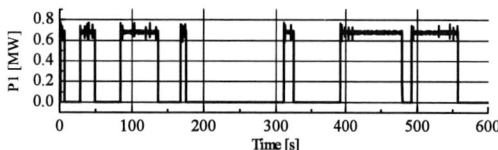

Fig. 19. Consumed power of the electrolyzer No.1

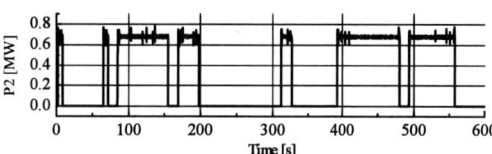

Fig. 20. Consumed power of the electrolyzer No.2

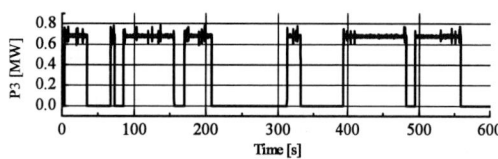

Fig. 21. Consumed power of the electrolyzer No.3

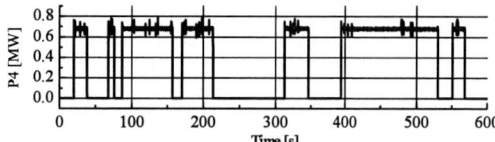

Fig. 22. Consumed power of the electrolyzer No.4

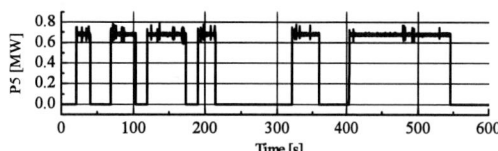

Fig. 23. Consumed power of the electrolyzer No.5

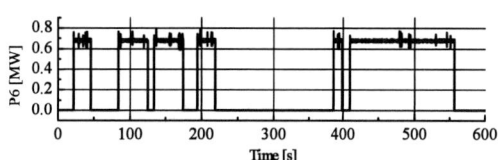

Fig. 24. Consumed power of the electrolyzer No.6

Figs.19 to 24 show consumed power in each electrolyzer and Fig.18 shows the total consumed power in the 6 electrolyzers (i.e., consumed power of the HHOG). It is clear that excessive switching cannot be seen in each electrolyzer and the operation time of each electrolyzer can be averaged effectively as indicated in Table IV. Fig.17 shows the amount of the produced hydrogen gas.

VIII. CONCLUSION

In this paper, a cooperative control of wind energy conversion system composed of a variable speed wind generator and a High-purity Hydrogen and Oxygen Generator (HHOG) for hydrogen production has been proposed, in which smoothed power can be supplied to the power grid as well as hydrogen can be generated in the HHOG. In the proposed system, HHOG is composed of parallel connected several electrolyzers and each electrolyzer is operated only under full-load or no-load condition by switching control because variable load operation can have a bad influence on the service life of electrolyzers. Though the total consumed power of HHOG can be controlled only in a stepwise manner, smoothed power can be supplied to the power grid by cooperating the speed control of the wind generator using its flywheel effect. In addition, a new switching method and the operation shift of electrolyzers are proposed in order to avoid the excessive ON/OFF switching of electrolyzers in the HHOG system.

The effectiveness of the proposed system has been confirmed by simulation analyses using PSCAD/EMTDC. As a result, it can be concluded that the proposed wind energy conversion system is very effective for supplying smoothed power to the power grid as well as generating hydrogen.

As a future study, we are going to examine the annual total efficiency of the proposed system.

IX. REFERENCES

[1] O.Wasynczuk et.al. ,:DYNAMIK BEHAVIOR OF A CLASS OF WIND TURBINE GENERATORS DURING RANDOM WIND FLUCTUATI-ONS IEEE Transactions on Power Apparatus and Systems, Vol.PAS-100 ,No.6, pp.2837-2854,1981.

[2] Jun Hirose, Tatsuhiko Isagawa: "A High-Purity Hydrogen and Oxygen Generator (HHOG) for Chemical Industry" Technical document of Shinkou Pantetuku, Vol. 40, No.2, pp. 48-56, 1997.

[3] Rion Takahashi, Junji Tamura, Motoo Futami, Mamoru Kimura, Kazumasa Ide: "A New Control Method for Wind Energy Conversion System Using a Doubly-Fed Synchronous Generator" IEEJ Trans. PE, Vol.126, pp.225-235, No.2, 2006.

[4] Tomonobu Senju, Ryosei Sakamoto, Naomitsu Urasaki, Toshihisa Funabashi, Hideki Fujita, Hideomi Sekine: "Output power leveling of wind turbine generator for all operating regions by pitch angle control", IEEE Trans. on Energy Conversion, Vol.21, No.2, pp.467-475, June 2006.

A new interconnecting method for wind turbine/generators in a wind farm and basic characteristics of the integrated system

Shoji Nishikata* and Fujio Tatsuta[†]

* Department of Electrical and Electronic Engineering, School of Engineering,
Tokyo Denki University, 2-2 Kanda-Nishiki-cho, Chiyoda-ku Tokyo, Japan, e-mail: *west@cck.dendai.ac.jp*
[†]Department of Information Systems and Multimedia Design, School of Science and Technology for Future Life,
Tokyo Denki University, 2-2 Kanda-Nishiki-cho, Chiyoda-ku Tokyo, Japan, e-mail: *tatsuta@im.dendai.ac.jp*

Abstract—A new interconnecting method for a cluster of wind turbine/generators is proposed, and some examples of the basic characteristics of the integrated system are shown. This method can be achieved with wind turbine generating system using a shaft generator system. A group of wind turbine-generators can be interconnected easily with the proposed method, and high reliability and electric output power with high quality are also expected. Moreover, since this method enables to send generated power through a long-distance DC transmission line, the optimum site for wind turbines can be selected so as to acquire the maximum wind energy.

Keywords— Adjustable speed generation system, Current source inverter (CSI), Renewable energy systems, Thyristor, Windgenerator systems.

I. INTRODUCTION

It is needed to construct a large-scale wind farm for generating a large amount of electric power by kinetic energy of wind, and such a site should be far distant from urban areas or off shore. Also, the interconnecting method for the wind turbine/generators is essential in forming the wind farms.

We already proposed a DC-link type wind turbine generating system using a shaft generator system [1], [2], which is widely used for power sources in a ship [3], and revealed the usefulness of the system. In this type of generating system a large-scale smoothing capacitor is unnecessary because current-source inverter is adopted. Also, since this type of inverter uses in general thyristors as switching devices, the system scale can be easily enlarged and system reliability is improved greatly when compared with IGBTs in the voltage-source inverters. Furthermore, as in [1], [2], output electricity with high quality is obtained with this generating system.

In this paper a new interconnecting method for a group of wind turbine/generators, which are based on the above-mentioned system, is proposed. In this method, the outputs of each AC generator coupled with the wind turbine are rectified, respectively, and these rectified outputs are connected in series and integrated in the DC link. The resultant output DC power is converted again to AC power with the thyristor inverter. Thus, only one inverter is enough, making the system very simple, and the inverter can be placed everywhere without restraint. Thereby, we can select the optimum site for wind turbine/generators with a long-distance DC transmission line. A basic idea of control method for the proposed system is also discussed, and some examples of system characteristics are shown in this paper.

II. BASIC EQUATIONS OF THE SYSTEM

Let us first show the basic equations for a wind turbine/generator which are necessary for the following discussion. As a wind turbine generator, permanent magnet synchronous generator (PMSG) is used here. In Fig. 1, mechanical energy is acquired from kinetic energy possessed by wind through a wind turbine, and the PMSG converts it to electrical energy. The output of PMSG is converted to DC power through a thyristor converter. The output power of the wind turbine P_t, which is equal to converted DC power if the losses in the generator and converter are neglected, is given by

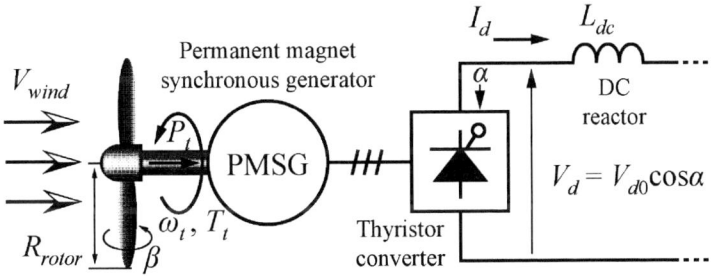

Fig. 1. Wind turbine generator and thyristor converter.

978-1-4244-1741-4/08/$25.00 ©2008 IEEE

$$P_t = \frac{1}{2} C_p(\lambda, \beta) \rho \, A_W V_{wind}{}^3 = V_d I_d \qquad (1)$$

where, ρ, A_w, V_{wind}, V_d and I_d are air density, rotor swept area, wind velocity flowing into wind turbine, output voltage and current of the converter, respectively. Also C_p is performance coefficient which is the conversion ratio of turbine output to wind power [4].

Here, it should be noted that C_p is expressed as a function of tip speed ratio λ and blade pitch angle β. The tip speed ratio λ is defined as

$$\lambda = \frac{\omega_t R_{rotor}}{V_{wind}} \qquad (2)$$

where, ω_t is angular velocity of the wind turbine and R_{rotor} is blade radius.

In general, the wind turbine should be driven to keep λ a constant value to obtain C_p as large as possible. So, λ and C_p are assumed to be constant for a given pitch angle β in this discussion for the sake of simplicity. In this case, the angular velocity ω_t is proportional to the wind velocity V_{wind}. On the other hand, it should be recognized that the non-control output voltage of the converter V_{d0} is almost proportional to ω_t since PMSG is used as the wind turbine generator.

When the coefficient of proportion between V_{d0} and ω_t is assumed to be K_d, we have Eq. (3) for controlled output voltage of the converter.

$$V_d = V_{d0} \cos \alpha = K_d \omega_t \cos \alpha = \frac{K_d \lambda}{R_{rotor}} V_{wind} \cos \alpha \qquad (3)$$

where, α is control angle of converter and $V_{d0} = K_d \omega_t$.

As to the torque of the wind turbine T_t, we obtain Eq. (4) from Eqs. (1), (2), and (3).

$$T_t = \frac{P_t}{\omega_t} = \frac{C_p \rho \, A_W R_{rotor} V_{wind}{}^2}{2\lambda} = I_d K_d \cos \alpha \qquad (4)$$

III. SYSTEM CONFIGURATION

Fig. 2 shows the configuration of the proposed wind turbine generating system composed of two or more sets of wind turbine and PMSG. As in the figure, the output of each thyristor converter is connected in series and the unified DC output is fed to the current-source thyristor inverter through a DC transmission line. The synchronous compensator connected to the inverter through a duplex reactor provides reactive power needed for commutation of the inverter thyristors and that required in the AC output as well [1], [2].

It should be noticed that the output voltage distortion in the inverter caused by the commutation of thyristors is completely compensated with the well-designed duplex reactor so as to cancel the subtransient inductance of the synchronous compensator [3].

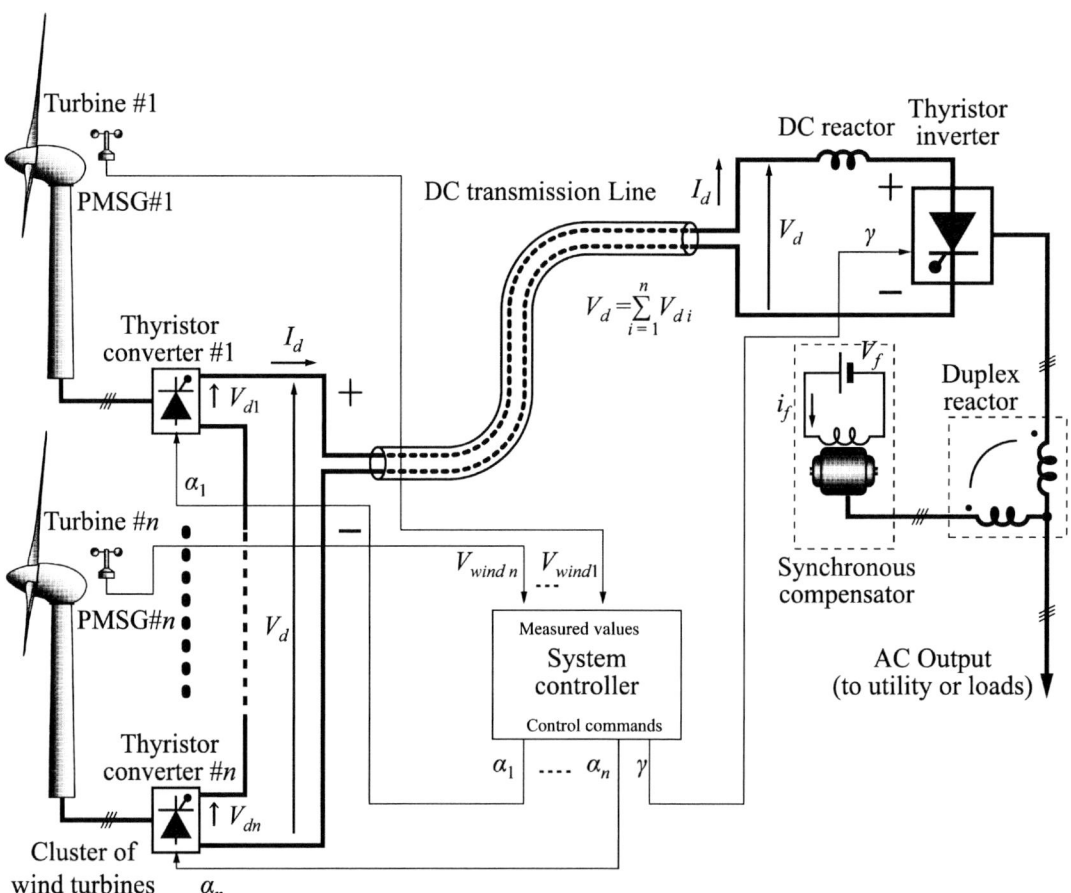

Fig. 2. Proposed interconnecting method for wind turbine/generators and system configuration.

The system controller shown in the figure collects observed data on the wind velocities $V_{wind\,i}$ ($i = 1, 2, ..., n$) at each wind turbine. In this controller DC link current I_d and then leading angle for commutation of the inverter γ are calculated based on $V_{wind\,i}$ (i. e., total output power gained by the wind farm, ΣP_{ti}), and the control angles of converter α_i ($i = 1, 2, ..., n$) for the individual converter (output voltage V_{di} ($i = 1, 2, ..., n$)) are also determined.

When there exists only one wind turbine, it is the case that we reported in [1] and [2]. In the following, we discuss how to control the system.

IV. OPERATING METHOD FOR THE SYSTEM CONSISTING OF ARBITRARY NUMBER OF WIND TURBINE GENERATORS

A. Loads connected through the thyristor inverter

We discuss here the way of controlling α_i for individual thyristor converter when the system loads are connected in the DC link through the thyristor inverter as in Fig. 2.

For this case, let $V_{wind1} \sim V_{windn}$ be the wind velocities flowing into #1~#n wind turbines, respectively, and $V_{w\,max}$ be the maximum value among these wind velocities. That is,

$$V_{w\,max} = \max\left(V_{wind1}, V_{wind2}, \cdots V_{windn}\right) \tag{5}$$

We set the control angle for the converter at which wind velocity is $V_{w\,max}$ as 0. Then, the output power and output DC voltage for this wind turbine become maximum as:

$$P_{t\,max} = \frac{1}{2}C_p \rho A_w V_{w\,max}^{\;3} \tag{6}$$

$$V_{d\,max} = \frac{K_d \lambda}{R_{rotor}} V_{w\,max} \tag{7}$$

As a result, DC link current turns into:

$$I_d = \frac{P_{t\,max}}{V_{d\,max}} = \frac{C_p \rho A_w R_{rotor} V_{w\,max}^{\;2}}{2 K_d \lambda} \tag{8}$$

Besides the wind turbine with the maximum wind velocity, the output power P_{ti} and output DC voltage V_{di} for #i wind turbine are given as:

$$P_{ti} = \frac{1}{2}C_p \rho A_w V_{wind\,i}^{\;3} = V_{di} I_d \tag{9}$$

$$V_{di} = \frac{K_d \lambda}{R_{rotor}} V_{wind\,i} \cos\alpha_i \tag{10}$$

Since DC link current I_d is the same for all the converters because of DC link, the following relationships are obtained:

$$\frac{P_{ti}}{P_{t\,max}} = \frac{V_{di}}{V_{d\,max}} = \frac{V_{wind\,i}^{\;3}}{V_{w\,max}^{\;3}} = \frac{V_{wind\,i}\cos\alpha_i}{V_{w\,max}} \tag{11}$$

Hence, control angle α_i for converter #i should be controlled as:

$$\cos\alpha_i = \frac{V_{wind\,i}^{\;2}}{V_{w\,max}^{\;2}}, \quad \alpha_i = \cos^{-1}\frac{V_{wind\,i}^{\;2}}{V_{w\,max}^{\;2}} \tag{12}$$

The total DC link voltage V_d and the total output power P_{fTotal}, which are the input voltage and input power, respectively, to the inverter, become :

$$V_d = \frac{K_d \lambda}{R_{rotor}} \sum_{j=1}^{n}\left(V_{wind\,j}\cos\alpha_j\right) \tag{13}$$

$$P_{fTotal} = \frac{1}{2}C_p \rho A_w \sum_{j=1}^{n}\left(V_{wind\,j}^{\;3}\right) \tag{14}$$

B. Resistive load connected in the DC link

As another primitive investigation, we discuss the method of controlling output voltage for each converter for the case of a load of constant resistance connected in the DC link instead of the thyristor inverter shown in Fig. 2.

The total power derived from all the wind turbines P_{fTotal} is given with Eq. (14), and DC link voltage V_d applied to the load resistance turns into:

$$V_d = \sqrt{R_L \cdot P_{fTotal}} = \sqrt{\frac{1}{2}R_L C_p \rho A_w \sum_{j=1}^{n}\left(V_{wind\,j}^{\;3}\right)} \tag{15}$$

Since the same DC current I_d flows in all of the converters because of DC link, the output DC voltage for the individual converters has to be controlled depending on the wind conditions for the turbines in order to provide a stable operation of the system. That is, the control angles should be determined based on the relationships between the power obtained by the wind and the consumption power.

Hence, the contribution of each wind turbine to the total power is assigned as:

$$\frac{P_{ti}}{P_{fTotal}} = \frac{V_{wind\,i}^{\;3}}{\sum_{j=1}^{n}\left(V_{wind\,j}^{\;3}\right)} \tag{16}$$

Thus, the output voltage for #i wind turbine converter should be:

$$V_{di} = V_d \frac{V_{wind\,i}^{\;3}}{\sum_{j=1}^{n}\left(V_{wind\,j}^{\;3}\right)} = \sqrt{\frac{R_L C_p \rho A_w}{2\sum_{j=1}^{n}\left(V_{wind\,j}^{\;3}\right)}} V_{wind\,i}^{\;3}$$

$$= \frac{K_d \lambda}{R_{rotor}} V_{wind\,i} \cos\alpha_i \tag{17}$$

As a result, the control angle for #i converter can be calculated as:

$$\alpha_i = \cos^{-1}\left(\frac{R_{rotor} \cdot V_{wind\,i}^{\;2}}{} \cdot \sqrt{R_L C_p \rho A_w}\right) \tag{18}$$

V. BASIC CHARACTERISTICS FOR THE CASE OF TWO WIND TURBINE GENERATORS

On the basis of the system equations derived above, we explore here the system characteristics for the case of two sets of wind turbine/generator as a basic investigation.

In this case, $n=2$ and the system is given by Fig. 3 for the inverter load. The whole load is shared between two generators and the control angles of each converter should be properly controlled depending on the wind velocity. It ought to be reminded here that the output voltage and frequency of the inverter can be kept constant by controlling field current of the synchronous compensator shown in Fig. 2 as well as leading angle of commutation of inverter [1], [2].

When $V_{wind1} > V_{wind2}$, then α_1 is set to be 0 according to foregoing discussion and α_2 is calculated with Eq. (12). Here, let us investigate the steady-state characteristics of the system having the constants shown in Table 1 [2].

Figs. 4 and 5 give examples of calculated results for the cases when α_1 is fixed to be 0 and V_{wind1} = 11 m/s = constant; meanwhile, V_{wind2} changes in the range of from 0 to 11m/s.

In Fig. 4, the characteristics of angular velocities of the wind turbines ω_{t1}, ω_{t2}, and α_2 versus V_{wind2} are shown. It is clarified that although ω_{t2} increases with V_{wind2}, ω_{t1} is kept constant because of a constant tip speed ratio λ, and that control angle α_2 decreases with an increase in V_{wind2} according to Eq. (12).

TABLE I.
PARAMETERS USED IN THE CALCULATIONS.
(COMMON IN BOTH #1 AND #2 TURBINE/GENERATOR)

Blade radius	R_{rotor}	7.5 [m]
Tip speed ratio	λ	4.0 = constant
Performance coefficient	C_p	0.3 = constant
Air density	ρ	1.225 [kg/m³]
Generator coefficient	K_d	40.0 [V·s/rad]
Rated wind velocity		11 [m/s]
Rated output power		43 [kW]

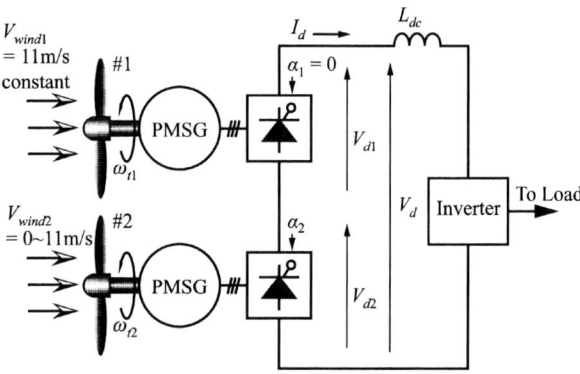

Fig. 3. Two wind turbine/generators connecting an inverter load.

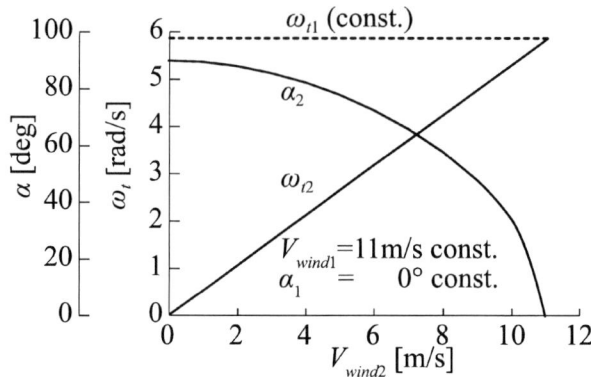

Fig. 4. Angular velocities of wind turbines ω_{t1}, ω_{t2} and control angle of converter α_2 versus V_{wind2}.

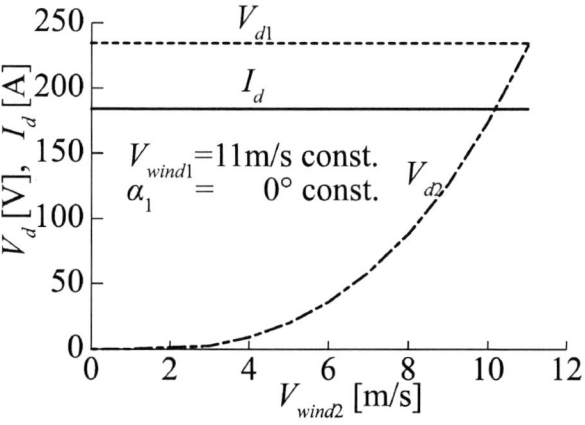

(a) I_d, V_{d1}, V_{d2} versus V_{wind2}

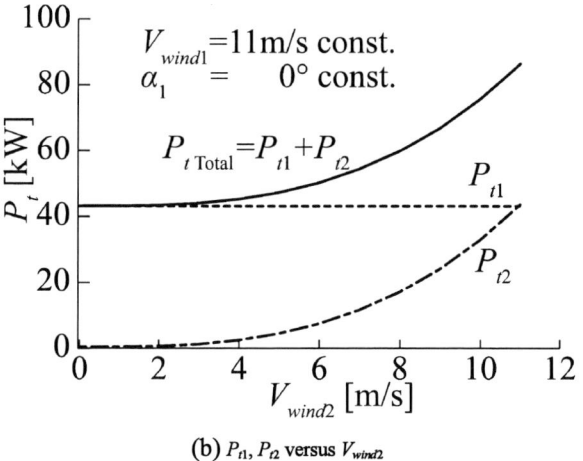

(b) P_{t1}, P_{t2} versus V_{wind2}

Fig. 5. Characteristics of output powers and DC link voltages and current versus wind velocity at turbine #2 V_{wind2}.

Fig. 5 shows the characteristics of DC link current I_d, DC voltages of the converters V_{d1} and V_{d2} (in (a)) and system output P_{tTotal}, which is equal to $P_{t1} + P_{t2}$ (in (b)), when V_{wind2} changes. It can be seen from this figure that DC voltage V_{d2} and output power P_{t2} increase with an increase in V_{wind2}, while V_{d1} and P_{t1}, as well as I_d are kept

power when the proposed system is used, resulting in an optimum operation of a wind farm.

Next, the steady-state characteristics of the system are discussed for the case of a load of constant resistance connected in the DC link as in Fig. 6.

Based on Eqs. (15) – (18) the characteristics of α_1 and α_2 can be calculated for the case of constant DC link resistance load (R_L=1Ω, in this case), and those of DC link voltages and current can also be clarified.

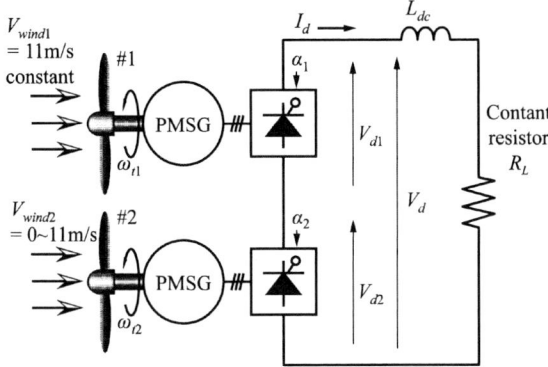

Fig. 6. Operation with the resistor load.

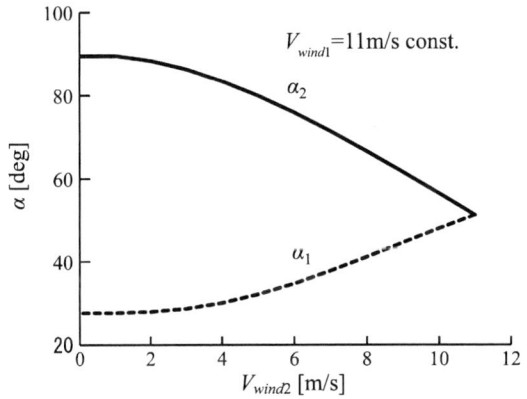

Fig. 7. Control angles α_1, α_2 versus V_{wind2} for the case with constant load resistance.

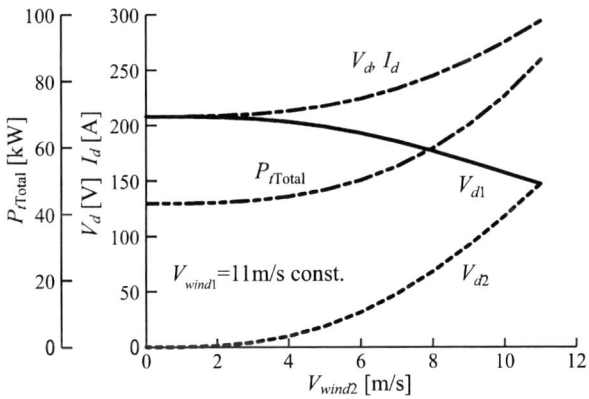

Fig. 7 shows the characteristics of α_1 and α_2 for a change in V_{wind2} when V_{wind1} = 11 m/s = constant. In this calculation the parameters shown in Table 1 are used again. It is noted that the control angles have to be controlled to realize a stable operation of the system.

Fig. 8 shows steady-state characteristics of P_{fTotal}, V_{d1}, V_{d2}, V_d (= $V_{d1} + V_{d2}$), and I_d for the case when V_{wind2} changes in the range of from 0 to 11m/s. It is noticed that I_d is directly proportional to V_d because the load resistance is constant in this case, while for the system in Fig. 3 I_d is kept constant as in Fig. 5(a).

VI. SYSTEM PERFORMANCES
FOR THREE WIND TURBINE GENERATORS

In this section we explore the system performances for the case of three wind turbine/generators feeding the loads through the inverter (i.e., n =3 in Fig. 2). Since the wind turbines are usually driven by natural wind, we investigate here performance characteristics of the system for the case when the wind velocities running into the turbines fluctuate with respect to time as in Fig. 9.

Figs. 10 to 13 show various performances of the system for the wind conditions given in Fig. 9. The constants used for these calculations are the same as in Table I for three turbine/generators.

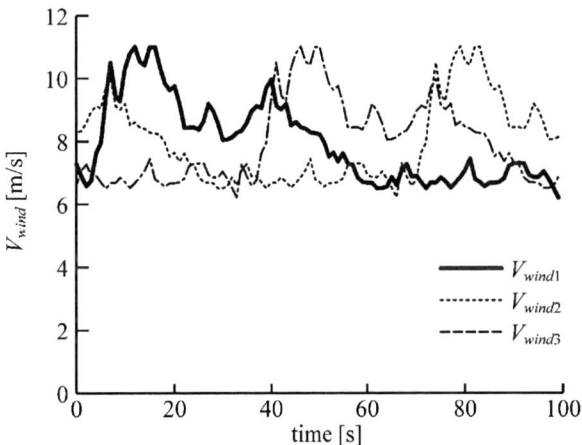

Fig. 9. Changes in wind velocities.

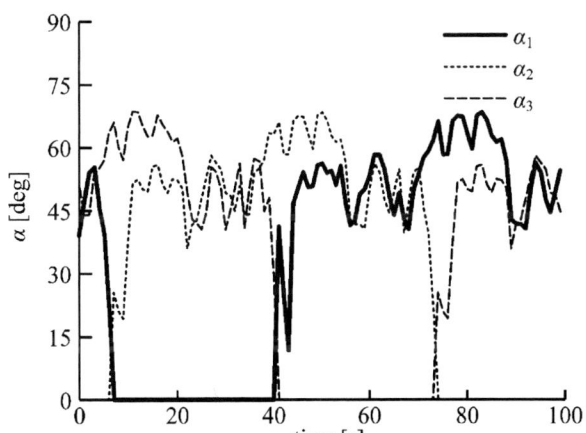

2347

In Fig. 10 the changes in the control angles of the converters are shown. Since this angle is controlled to be 0 for the converter at which wind velocity of the turbine is maximum, the converter with α of 0 is replaced depending on the wind velocity as in the figure.

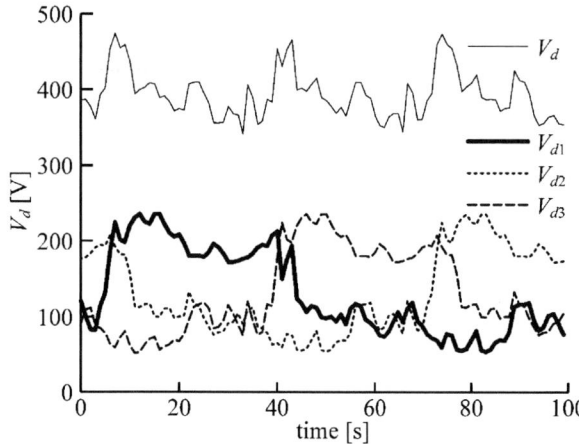

Fig. 11. Controlled output voltages for the converters.

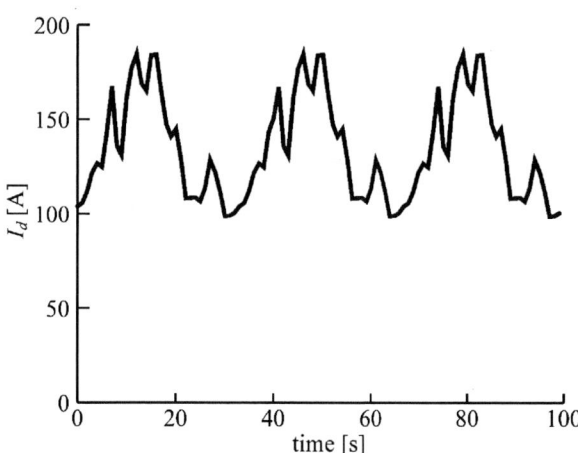

Fig. 12. DC link current.

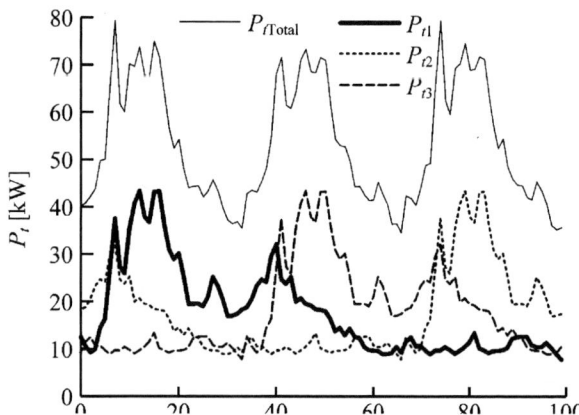

Fig. 11 gives the changes in the converter output voltages. It is clarified from this figure that the maximum voltage alters among these three converters being dependent on the wind velocity. It is also noticed that the change in the resultant output voltage, which is DC link voltage, becomes smoother than those in the individual output voltage.

As to DC link current, since it is proportional to the square of maximum wind velocity due to Eq. (8), the change in I_d is relatively large as in Fig. 12.

In Fig. 13 the output powers of the individual generators and the resultant output power generated by the whole system are shown. It can be seen that although the output power obtained from each of the generators changes considerably, the change in total output power is reduced remarkably, showing the usefulness of the proposed system.

It should be recognized here that the calculation results discussed so far are based on the steady-state equations introduced here to clarify the fundamental characteristics of the system, and more detailed investigations including dynamic performances of the system are necessary to obtain more precise results.

VII. CONCLUSION

In this paper, a new interconnecting method of two or more sets of wind turbine/generator used in a wind farm has been proposed, and basic characteristics of the integrated wind turbine generating system have been discussed.

In the proposed system, only one externally commutated thyristor inverter is required for a cluster of wind turbines, and output voltage without distortion can be achieved with ease, realizing a very simple configuration of wind farm with high quality of output power as well as high reliability.

In addition to these advantages, only one DC link is used and the optimum site for wind turbines, such as off shore, can be readily selected in order to obtain more power from wind because DC transmission system is entirely appropriate for the proposed system.

ACKNOWLEDGMENT

This work is partially supported by Research Institute for Science and Technology of Tokyo Denki University Q06E-03/Japan.

REFERENCES

[1] F. Tatsuta, and S. Nishikata.: Studies on a Wind Turbine Generator System using a Shaft Generator System, Proc. of 8th Int. Conf. on Electrical Machines and Systems (ICEMS 2005), Nanjing (China), pp.931-936, Sept. 2005.

[2] F. Tatsuta, S. Nishikata: Performance Characteristics of a Practical Scale Wind Turbine Generating System using a Shaft Generator System: European Conference on Power Electronics and Applications (EPE) 2007, Sept. 2007

[3] S. Nishikata, Y. Koishikawa, F. Mita, and T. Kataoka.: A Shaft Generator System without Output Voltage Distortion, Trans. IEE Japan, vol.119-D, No.12, pp.1549-1555, Dec. 1999 (in Japanese).

[4] Siegfried Heier: Grid Integration of Wind Energy Conversion Systems, second edition, p.44, John Wiley & Sons, Ltd. 2006.

Educational aspects of mechatronic control course design for collaborative remote laboratory

Andreja Rojko*, Darko Hercog* and Karel Jezernik*

* University of Maribor, Faculty of Electrical Engineering and Computer Science, Maribor, Slovenia, e-mail:
andreja.rojko@uni-mb.si, darko.hercog@uni-mb.si, karel.jezernik@uni-mb.si

Abstract—Paper addresses educational aspects of a distance learning mechatronic control course, which is one of the 18 courses that compound an international collaborative remote laboratory designed for students of electrical engineering. The work is undertaken in the frame of European Leonardo da Vinci project EDIPE "E-learning Distance Interactive Practical Education" with a goal to offer Web based experimental courses to the students from 11 countries that participate in the project. Course presented in this paper is motion control of mechatronic device with nonlinear dynamics. Addressed topics are modelling of dynamics of mechatronic device, design of linear and nonlinear motion controllers, implementation and optimization of motion controllers for real mechatronic system. Three remote experiments are available as a part of the course: motion control of mechatronic device with cascade controller, motion control with PD controller and computed torque motion controller. For the course two educational strategies are developed; one for geographically distant students and one for local students. For geographically distant students the course is executed completely remote. For local students some parts of remote course are combined with traditional learning in the classroom and laboratory sessions in order to improve course's flexibility and to achieve best educational outcome.

Keywords—Education methodology, Mechatronics, Motion control, Virtual instrument.

I. INTRODUCTION

Nowadays, the traditional educational methods used at the academic institutions are due to the society and lifestyle changes challenged in many ways; educational, organizational and technological. Therefore many new educational methodologies and supporting education tools are being developed. Remote curses supported by remote laboratories are also one of new, very popular tools, as they offer to the students and to the tutors much more flexibility as in-the-classroom lectures and conventional laboratory experiments [2], [3], [4], [6]. However design and implementation of such courses is very time consuming and demanding, especially from technical point of view.

Collaborative solutions for building of remote courses with remote experiment such as it is collaborative remote laboratory "E-learning Distance Interactive Practical Education" (EDIPE) [1] that comprehends remote courses designed at 11 European technical faculties, can significantly lessen technical and other problems concerning design of remote laboratories. Especially problems as organization and design of Web learning environment, design of booking system for the remote

experiments and Web server maintenance can be solved centrally for all project partners. Beside this it is also important that collaborative remote laboratories offer a lot of easy accessible learning resources, usually much more as a single institution can provide.

However when designing the course for the collaborative remote laboratory, which will also be available to the students from all other in the project involved faculties, some additional requirements must be meet. First the course topics must address up to date problem interesting for wide range of engineering students. Second, the course must be prepared in the way that the students with diverse knowledge backgrounds, prior knowledge and experience and different problem solving approaches are able to successfully finish the course autonomously or with minimal intervention of the teacher. Further the course structure should preferably be flexible so that the teachers from other countries and also local teachers can adapt it to their needs.

In the paper educational aspects of developing mechatronics control course for EDIPE collaborative remote laboratory are presented. Course was developed at the Institute of robotics, Faculty of Electrical Engineering and Computer Science, University of Maribor, Slovenia. Topics include modelling, simulations and applied linear (cascade position, velocity and current controller) and nonlinear motion control (model based computed torque controller) of the mechatronics devices with highly nonlinear dynamics of the mechanical part. Target students group are mechatronics students, students of the electrical and mechanical engineering and other engineering students with adequate background.

The paper is organized as follows. Second section presents teaching objectives and course documentation. Third section describes educational strategy for the students from other faculties and fourth section presents strategy developed for local students. Last section draws some conclusions.

II. COURSE OBJECTIVES AND DOCUMENTATION

Objectives of the course design include following:

- *Modelling of the mechatronic device with highly nonlinear dynamics.* Major task is derivation of the differential equations of motion for the mechanical part of mechanism with spring, Figure 4, by using physics knowledge and by inspecting photos, live Web picture, schemes and description from documentation. Calculation of the necessary model's parameters by using given dimensions of the mechanism and materials' densities data is also included.

- *Design and application of the linear motion controllers.* Linear motion controller is a cascade of PI current controller, PI velocity controller and P position controller. Students design few different cascade controllers; full cascade, cascade of position and current controller, cascade of position and velocity controller. Remote experiments include optimization of parameters for the position and velocity controller.

- *Design and application of the model based nonlinear motion controller.* As nonlinear controller computed torque position controller is implemented. Students design the computed torque controller for the device according to the derived dynamical equation. After that remote experiments are executed.

- *Understanding of a discrepancy between applicability of the linear and nonlinear control methods.* Based on measured results, the efficiency of all applied controller schemes is evaluated and compared.

For the course an extensive documentation in English and Slovene is prepared. For each of the topics first theoretical background is presented. Citations to other relevant literature enable the learner to find further information. After each theoretical part, an exercise for the students is given. Difficulty level of those exercises is progressing from simple repeating of the simulations which are in details described in the materials, to the autonomous execution of more complex simulations and finally to the execution of remote experiments.

Documentation is prepared in two versions, one for the students and other one for the teachers from remote faculties. Documentation for teachers additionally includes the expected results and the suggestions how the course could be included in the local educational process. There is also a short documentation that describes technical details concerning booking of the remote experiments, access to the remote experiments and their execution.

III. EDUCATIONAL STRATEGY FOR GEOGRAPHICALLY DISTANT STUDENTS

The structure of the course enables the tutors from other faculties to apply it in their educational process in different manners. One option is that the lecturer includes only the remote experiments as an example shown in the regular control lectures. Second option is to apply the whole course or some of its components as the seminary work that is autonomously executed by the students and replaces one part of the regular lectures and laboratory work. This case is described in the continuation.

Educational strategy for autonomous execution of course by geographically distant students is based on three phases; pre-laboratory, remote experiments and summary. Structure of the course is shown in Figure 1.

1) Pre-laboratory
First the students are provided with the access to the Moodle based course Web page, Figure 2. Next the students study the introductory part of the Web materials to get familiar with the problem, to clarify terms and get known with the outline and requirements of the course. If

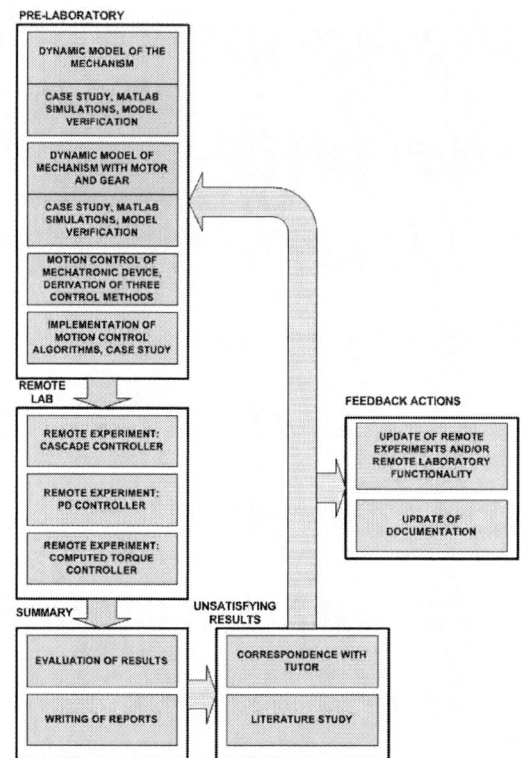

Figure 1: Course structure

their knowledge is insufficient they are instructed to study the literature cited in the documentation or other relevant materials.

First discussed topic in pre-laboratory part is dynamic analysis of the mechatronic device. Theory is followed by the case study. After studying theoretical background the students examine the experimental device which is mechanism with spring (Figure 3) with DSP-2 control system (Figure 4). For this they use photos, schemes and live Web picture of mechanism. Next step is derivation of the mechanism's dynamic model, which is presented with nonlinear differential equations of second order. The model is then verified by the MATLAB/Simulink simulations which are executed locally by the students on their personal computers. Basic simulation model can be downloaded from the course web page, but the students have to complete it to obtain fully functional simulation model of mechanism's dynamics.

Second, the problem of building an active dynamic model that is dynamic model of mechanism together with motor drive and gears is discussed. First model of DC motor is derived. MATLAB/Simulink simulation model of motor is built by using derived equations and motor data sheet. Next the role of gears in the mechatronic devices is discussed. As the case study dynamic model of mechanism with spring with motor and gear is derived. After studying the model the students have to perform some simulations for checking if the motor with gear has sufficient torque for producing desired motion of mechanism. Also influence of lower and higher gear ratios is examined by simulations.

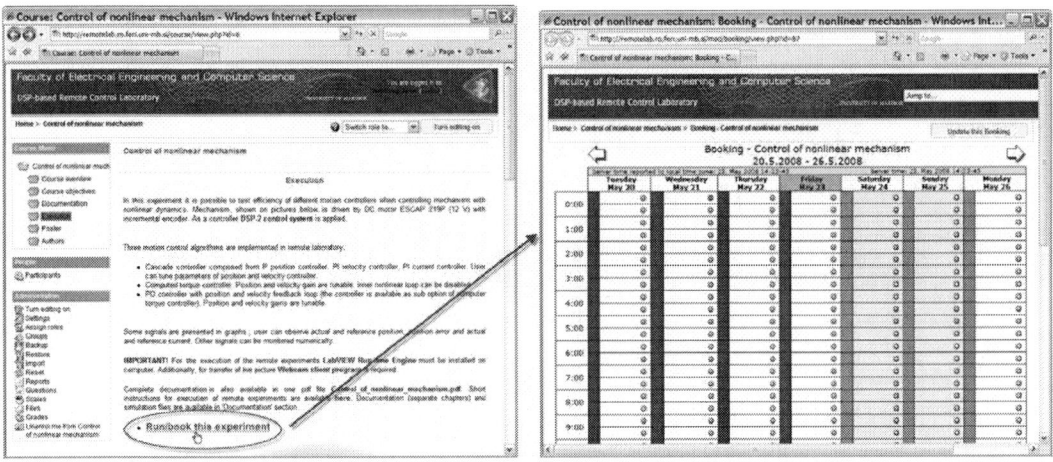

Figure 2: Course Web page with booking system

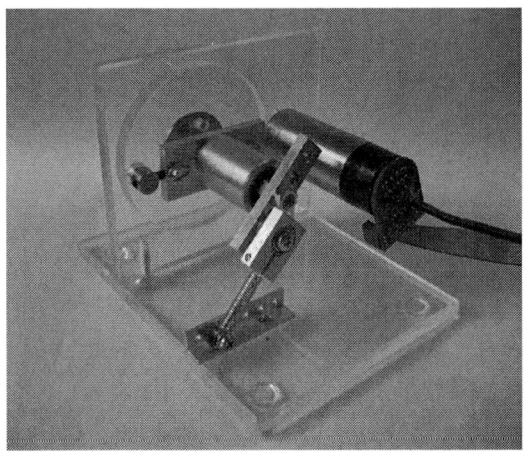

Figure 3: Mechanism with spring

Figure 4: DSP-2 control system

Next the problem of motion control of the mechanisms with nonlinear dynamics is discussed. At the beginning the computed torque controller is derived. From this basic nonlinear control method also two other controllers that is PD controller with gravitation compensation and simple PD controller with position and velocity loop are derived. Last derived motion control method is cascade position and velocity control which is also frequently used in industrial controllers. All controllers are compared from different aspects as efficiency, computational requirements, necessary knowledge concerning dynamic model and necessary measured and reference signals. Stability conditions are also discussed. Finally the adequate controllers for motion control of the mechanism with spring are designed.

In the final part of the pre-laboratory implementation details concerning control of the mechanism with spring are covered. Topics as sampling time, calculation of reference trajectory, influence of noise in measured signals for different controllers are discussed.

2) Remote lab

In the remote lab three motion controllers are implemented: cascade controller, PD controller and computed torque controller. Details and implemented schemes are presented and discussed in the course literature.

For performing remote experiments the students have to install programs LabVIEW Run-time Engine and WebCam Client on their personal computer. Both programs can be downloaded from the course web page. Next the students book the experiment by using booking table for the experiment, Figure 2. It is possible to reserve only one time slot in advance. After the reservation expires it is possible to book another time slot. When time slot is valid the temporary web page with user interface as shown in Figure 5 is created and the user automatically gains control over experiment. Live picture of experiment can be seen with using Webcam client, Figure 6.

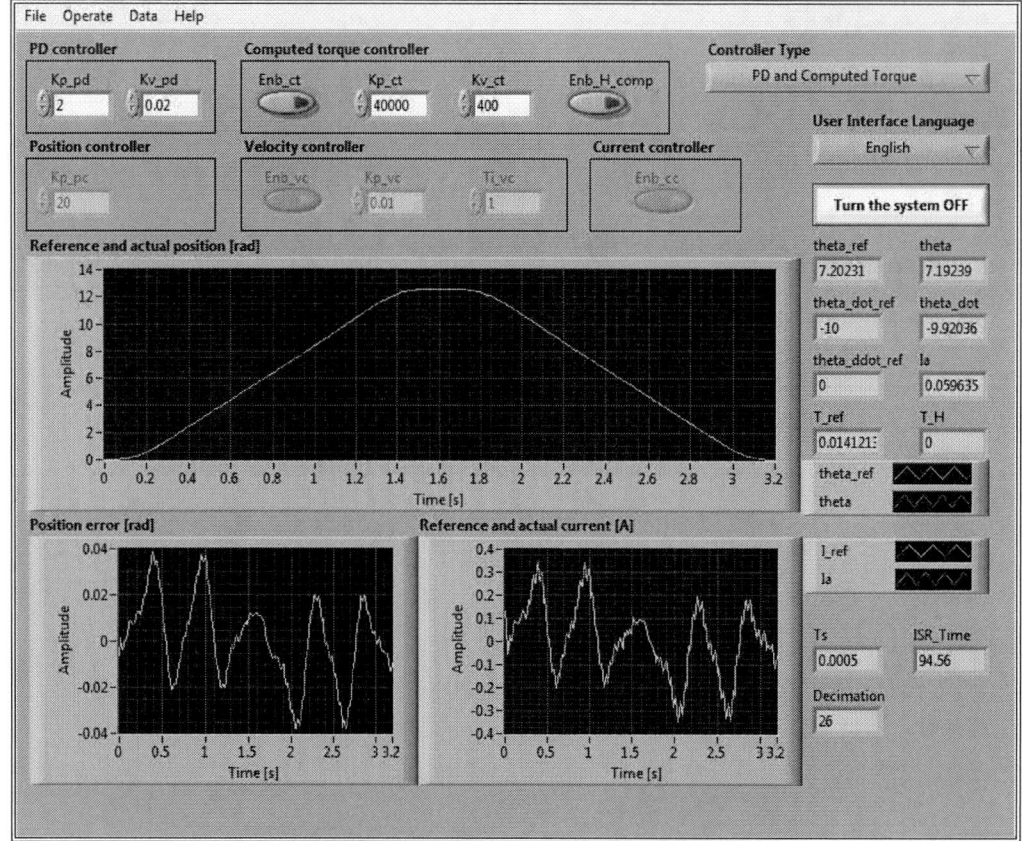

Figure 5: User from end for execution of remote experiments

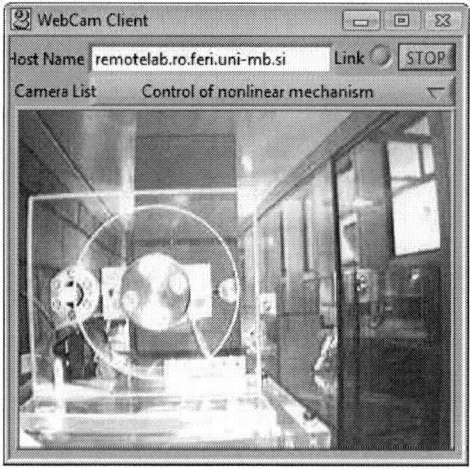

Figure 6: Webcam live picture

User interface, Figure 5, enables the choice between three basic controllers and tuning of the controller's parameters. Parameter values that are allowed are limited to the values for which the motion control is stable. In the case when user writes lower or higher value of the parameters that is allowed, minimum respectively maximum allowed values will be applied. In this way instability of motion controllers and consequently excessive tear and wear and overloading of the motor is avoided. In the case of the computed torque controller user can also change the structure of the controller by including or excluding compensation of spring torque and friction.

Monitored signals that are presented in user interface in graphical form are reference and actual position of mechanism, actual and reference motor current and position error. Other signals as are reference and actual velocity, calculated torque and reference acceleration can be monitored numerically.

Students' task is to explore the influence of high and low values of position and velocity gains for all three controllers and find optimal parameters for which lowest position error is achieved. After execution of remote experiments the efficiency of the controller should be compared and results commented.

3) Summary tasks

As a course conclusion the students write report where all simulated and measured results are commented and evaluated. Students hand the reports to their local teacher, who also does the grading.

The students are also asked to write general comments to the staff of the faculty that had prepared the course. The comments should describe overall experience and suggestions for improving the course, its documentation and remote experiments together with the booking system.

IV. EDUCATIONAL STRATEGY FOR LOCAL STUDENTS

Locally the course is implemented as a part of the regular course in robot control for the students of mechatronics and automation. The students must first attend the lectures. Next, for laboratory work, two options are available. The students with weak knowledge are advised to attend conventional laboratory sessions. At the conventional laboratory sessions the course is executed under supervision of the tutor and the experiments are done locally in the laboratory. Students work in groups of two to three. At the end of laboratory sessions they are aimed to perform also some remote Web based experiments, so that each student also gets an opportunity to work autonomously and update lab report with hers or his individual results.

Second group of students that are students with good background knowledge and good practical experience at handling of mechatronic devices and measurement equipment can instead of combining local and remote experiment execute only remote experiments.

This course is also often offered to the foreign students that study at the faculty, since the course documentation is in English. The students are then free to perform course autonomously and completely remote under supervision of the tutor in the local laboratory.

V. CONCLUSION

In the paper mechatronic control course for an international collaborative remote laboratory is presented. Described course structure includes pre-laboratory part with locally performed simulations, remote experiments and evaluation of results. The course is given with complete on line documentation which enables individual work with the students with different knowledge levels and from different engineering fields. Remote experiments are supported by the user-friendly booking system. Continuous improvement of the course, including technical solutions of the remote lab and course documentation is based on the feedback information provided from the students. The complete course, especially remote experiments with an efficient booking system can be benefit in the case of large number of the students and limited quantity of the available experimental equipment.

ACKNOWLEDGMENT

This work has been performed within the project "E-learning Distance Interactive Practical Education (EDIPE)". The project was supported by the European Community within framework of Leonardo da Vinci II programme (project No CZ/06/B/F/PP-168022). The opinions expressed by the authors do not necessarily reflect the position of the European Community, nor does it involve any responsibility on its part.

REFERENCES

[1] P. Bauer, J. Dudak, D. Maga, V. Hajek, "Distance Practical Education for Power Electronics," *International Journal of Engineering Education*, vol. 23, no. 6, 2007.

[2] T. Chang, D. Chang, "A Hands-on Graduate Real-time Control Course: Development and Experience," *International Journal of Engineering Education*, vol. 21, no. 6, pp. 1083-1092, 2005.

[3] A. R. S. Castellanos, L. Hernandez, I. Santana, E. Rubio, "Platform for distance development of complex automatic control strategies using MATLAB," *International Journal of Engineering Education*, vol. 21, no. 5, pp. 790-797, 2005.

[4] B. Duan, K. V. Ling, H. Mir, M. Hosseini, R. K. L. Gay, "An online laboratory framework for control engineering courses," *International Journal of Engineering Education*, vol. 21, no. 6, pp. 1068-1075, 2005.

[5] D. Hercog, K. Jezernik, "Rapid control prototyping using MATLAB/Simulink and a DSP-based motor controller," *International Journal of Engineering Education*, vol. 21, no. 4, pp. 596-605, 2005.

[6] D. Hercog, B. Gergič, S. Uran, K. Jezernik, "A DSP-based Remote Control Laboratory," *IEEE Transactions on Industrial Electronics*, vol. 54, no. 6, pp. 3057-3068, 2007.

PEMCWebLab – Distance and Virtual Laboratories in Electrical Engineering: Development and Trends

Pavol Bauer*, Viliam Fedák†, Otto Rompelman*

* Delft University of Technology, Delft, The Netherlands, e-mail: *P.Bauer@TUDelft.NL*
† Technical University, Košice, Slovak Republic, e-mail: *viliam.fedak@tuke.sk*

Abstract—**The paper deals with basic philosophy and structure of remote controlled laboratory for experimentation in Electrical Engineering. The laboratory collects experiments from fields of Power Electronics, Electrical Machines, Electro-Mechanical and Motion Control Systems. The workbenches in the real laboratory are used over internet. The real experiments of the remote controlled laboratory are placed at various universities. A special care is devoted to preparation of the learners to experimentation, including examining their knowledge before joining to the experiment. To have a feeling of participation in the experiment, except of the measurement data, also the signal from web camera is transferred to a distance operator.**

Keywords—**e-learning, remote controlled experiments, engineering education.**

I. INTRODUCTION

Although modelling and dynamic simulation are basic tools for understanding and verifying theoretical subjects, the experimentation with a real system plays a fundamental role that cannot be replaced. Analysis of these tools from view of modern education is given in [1].

Practical education needs to be based understanding phenomena that occur in real systems. Remote control of the experiments trough Internet comes as a solution to these problems, allowing students to control them, without leaving their normal workplace.

Rapid development of ICT technologies since beginning of 90ties enabled expansion of online distance laboratories. Their utilisation presents the latest trends in education – to get practical experience by experimentation with measurements, verifying properties of complex equipment and well as analysis of the equipment in various operation points. Tacking advantage of the internet and development of related technologies, an increasing number of remote access solutions are being developed. A remote hardware experiment are adapted in such a way that it can also be accessed from the Web that enables distance sharing of the experiment by other individuals and/or institutions.

Currently we can find numerous solutions for remote controlled experiments in various fields of practical education and also there are running numerous projects focused to their developments that are presented at special conferences, e.g. in [3], [4] and special sessions of the scientific conferences [1], [8], [9], [11], [13], [22], etc.

An overview of some typical solutions is presented in the reference [1]. Within the PEMCWebLab a more complex system of virtual laboratories is developed containing not only single experiments but the whole set of the experiments [2] that are distributed across Europe (Fig.1) and the experiments allocated there cover basic fields of electrical engineering. They are placed in different laboratories at universities. The system to be developed presents an open system enabling later expansion. The virtual (distance) laboratories are not any web-based simulation. They present real electro-technical experiments conducted in the laboratory, but they are remotely controlled and monitored by web-based tools with visualization of measuring apparatus, electronic components and many more factors.

Design of such a system except of technical solutions dealing with access and sharing of remote experiments requires solving number of other tasks that is pointed out in the following.

II. REMOTE PEMCWEBLAB, ITS FUNCTIONALITY AND SUBSYSTEMS

To support distance learning in electrical engineering a set of remotely controlled real experiments from fields of electrical engineering mainly from Power Electronics and Electrical Drives and motion control has been developed, so that they create the *PEMCWebLab* (Fig. 3). It can be approached via webpage www.PEMCWebLab.com (Fig.2) where there is also a direct link to the booking

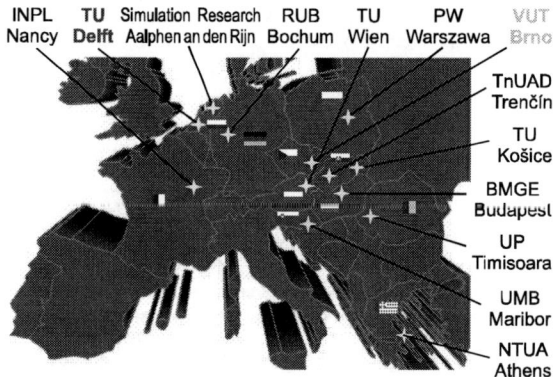

Fig. 1. PEMCWebLab project partners.

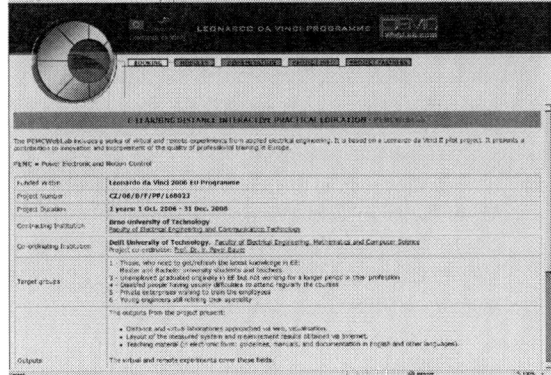

Fig. 2. Web page of the PEMCWebLab (www.PEMCWebLab.com).

system and more information about the project, dissemination results and description of different modules (measurements).

The *PEMCWebLab* creates an integrated learning platform. Several learning issues are addressed such as:

- Learning objectives
- Education
- Animation
- Simulation
- Experiment

In the first part the learning objectives of each experiment are addressed. In the part education a theoretical background of the each individual experiment is given. Interactive animations developed in the previous project are addressed further. The last educational method before experiment is the simulation.

A. Learning objectives

The experiments are not only analysis oriented (to measure and see the results) but also synthesis oriented. They should involve a design aspect. Therefore the measurements are designed as a project with educational philosophy.

The learning objectives for each single experiment are defined with the goal that:

- the objectives should be described in terms of knowledge and/or skills to be acquired by students,
- the objectives should be realistic given the attributes of the target group (prior knowledge, skills) and the time available for the students (credit points),
- the outcomes should be testable; if not, they should be left out or reformulated,
- there is an ability to analyze the function and working of electrical power (electronic) systems,
- there is an ability to design, implement and verify electrical power (electronic) systems.

Since both analysis and synthesis are key competences of engineers, the main objectives of the project are highly appropriate in an engineering education environment. A few remarks should be made with respect to these issues. Analyzing existing systems implies measurements. It is, however, very important, that students learn to conceive measurements as a means, rather than a goal. Measurements are a means to verify a hypothesis of the functioning of the system under analysis. The modules should indeed be structured such, that this concept indeed becomes clear. Therefore, it is proposed, that any measurement is embedded in the following structure of consecutive steps:

- Define the function of the system
- Decompose into subsystems with sub-functions
- Formulate hypotheses on characteristics of subsystems
- Voltages, currents
- Design measurements for verification of hypotheses
- Carry out measurements
- Compare results with hypotheses
- Evaluate learning

As far as synthesis is concerned, all modules will lead to a common design methodological approach for the student.

The main function of *PEMCWebLab* is to provide a web-based remote control for designed experiments. The learning process includes several, specially designed, experimental tasks. However, for safety reasons no one will be allowed to perform any experiment until he, or she, has shown adequate knowledge of the experiment.

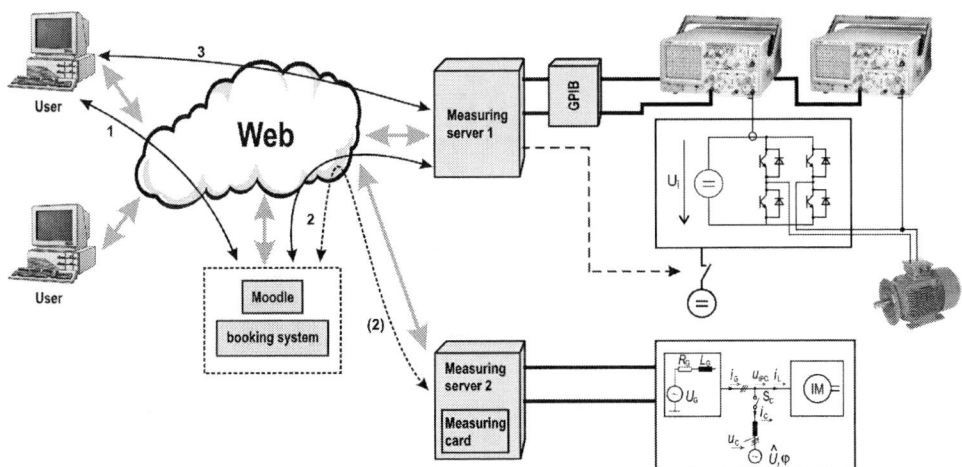

Fig. 3. Principal structure of the *PEMCWebLab* distance laboratory.

Entering wrong input parameters, due to insufficient knowledge of the experiment, may also lead to improper operation of the experiment. Therefore, a learning routine is designed for learners to gain the prerequisite knowledge which is required before attempting the experiment.

After completion of the online experiment, the learners are given an opportunity to take a simple questionnaire or alternatively to submit their report through the available feedback subsystem for its final evaluation (depending on the requirement enforced by the instructor). All learning procedures are recorded for future reference and analysis.

III. MANUALS

The quality of a product is largely determined by the quality and completeness of accompanying documents. Documents are meant for communication. It can be very helpful to contemplate on the four main issues of communication. These issues can be expressed as questions, viz:

1) *What* do I want to tell it?
2) *Why* do I want to tell it?
3) *To whom* to I want to tell it?
4) Once these questions are adequately answered, the answer to the final question can be formulated:
5) *How* do I tell it?

The order in which the questions are answered, it is a logical one, if we start from scratch. In the PEMCWebLab project we have already chosen for an educational structure and the overall goals are clear: better and easier learning of complex matters in electrical power engineering. Therefore let us start with the third question: *To whom* do I want to tell it? The answer immediately leads to the so called target groups. If we turn to the first questions in somewhat more detail, it will be clear, that the target groups will need different information. Furthermore, the same information may be presented in different ways for the different target groups.

The main target groups are:

A. Students
B. Teachers
C. Technical staff responsible for maintenance

A. *Students*

The goal of the document is to make students familiar with the educational objectives of the module. It should provide them with sufficient information to work with the module. It should be clear what students are expected to do and how their results will be assessed. In order to achieve this, the document could contain the following items (chapters, sections):

1) Educational objectives
2) Assumed entrance competencies
3) Structure of the module (block diagrams)
4) Instructions how to use the module
5) Assignments/tasks
6) Instructions how to communicate with teachers and fellow students
7) How to communicate suggestions for improvements

B. *Teachers*

The document for teachers is aimed at understanding the different (technical) functions available in the module and enabling the creation of an educational environment for students. The document should at least contain the following items (chapters, sections):

1) Goal of the module: the educational objectives as envisaged by the developers of the module.
2) How to create experiments for the students.
3) How to create assignments.
4) Suggestions for assessment.
5) How to create a structure for communication with students.
6) How to communicate suggestions for improvements to the developers and/or their successors.

C. *Technical staff responsible for maintenance*

As everything in life, things need maintenance. This holds for the PEMCWebLab - modules as well. Though initially the maintenance may be taken care of by the developers of the modules, after some time the attention of these developers will be deviated to other issues, and the maintenance will be left to other people or disregarded. Unfortunately, the latter is more often a rule rather than an exception. An analysis afterwards of this undesired development usually leads to the conclusion, that things were not sufficiently documented.

The contents of the document(s) for maintenance are largely dependant on the form of the modules. It should contain at least:

1) Functional block diagrams,
2) Input and output specifications of different modules,
3) Detailed diagrams of individual blocks including key measurement points with their normal voltages, waveforms etc.,
4) Structural diagrams of software modules (source codes only on CD-ROM or DVD).

IV. EXPERIMENT ADMINISTRATION

A central booking system is available at the project page PEMCWebLab.com Booking system is provided through Moodle software. Layout of the Moodle pages for all experiments is uniform. This page will contain menu with the following submenus:

1) Learning objectives
2) Education
3) Animation
4) Simulation
5) Experiment

All the submenus at the booking system are to be accessed without restriction of number of students. The actual booking is provided in the *submenu Experiment*. The experiments can be booked one week ahead, the length of the offered time window for the experiment varies from 5 to 30 min. Before the experiment becomes available online, it should be tested to verify the correctness of the experiment results as well as the stability of the experimental set-up. The power to some experiment is available 24 hours a day; some experiments are available for safety reasons in the working hours only.

An administrator of each experiment can restrict the use if the experiment for his purposes during some days or hours only. Supervisors have to routinely check the status of each experiment to make sure that each of them is functionally correct and is available for use.

Several clients can connect to PEMCWebLab.com simultaneously. However, Internet bandwidth becomes extremely limited when too many remote users request to use this system. Several concurrent, remote users are allowed via an Internet connection for each experiment. However, each experiment in the *PEMCWebLab* can be operated only by a single remote user at a time. The system thus considers each experiment as a "resource", and remote users who wish to operate a specific experiment should first get permission to operate the experiment. Once the resource is in use, other remote users cannot access that resource, because it is then marked as "locked." All the remote users without access permission can see only the online, real-time video of that experiment.

Server Site Administration

As already said every experiment has its own server and it is located at the different location. Remote users first logged onto a main booking server, after which they will be directed to the specific server for actually performing the experiment get into the page of the experiment itself.

V. SET OF REMOTE EXPERIMENTS

A Leonardo da Vinci EU project titled "*E-learning Distance Interactive Practical Education - EDIPE*" [2] is suggested and approved to create a full set of distance laboratories. Twelve universities with the span across the EU are participating in the project (Fig. 1).

The expected specific results of the project are:

- Learning objectives for the distance experimental education,

- The guidelines for project oriented measurements with the learning objectives for distance and /or virtual practical education,

- Synthesis oriented experimental measurements,

- Technology and technical documentation for distance practical education and measurements via the Internet,

- Different designed measurements each with its own philosophy.

The outputs from the project will present:

- teaching material (in electronic form; guidelines, manuals, documentation in English and other languages),

- distance and virtual laboratories approached via web,

- vizualization and layout of the measured system, and

- measurement results obtained via Internet.

The following modules are proposed (Table 1, grouped into sets of modules) in such a way that they cover fundamentals and basic applications of the EE and advance topics including the application "as well".

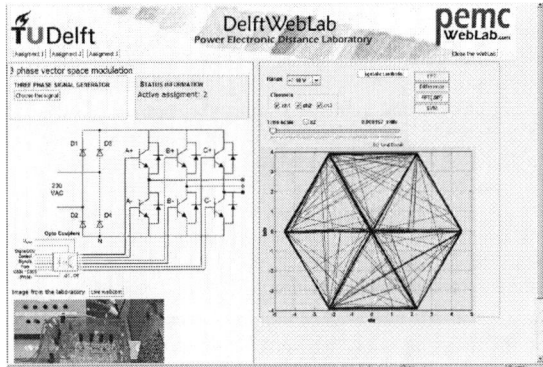

Fig. 4. Remote experiment for three-phase space vector modulation.

TABLE 1
LIST OF MODULES WITH REMOTE CONTROLLED EXPERIMENTS IN THE EDIPE PROJECT

Group N°	Group of specialized subjects	Module N°	Module title
1	Fundamentals of Electrical Engineering	1.1	Single Phase and Three Phase Rectifier Circuits
		1.2	DC Circuit Measurements and Resonant AC Circuits
2	Power Electronics	2.1	Power Converters
		2.2	Power Factor Correction
		2.3	PWM Modulation
		2.4	DC-DC Converter for Renewable Energy Sources and Microgrid
		2.5	Power Quality and Active Filters
		2.6	Power Quality and/or Electromagnetic Compatibility
3	Electrical Machines	3.1	Basic Electrical Machinery – Synchronous Generator
		3.2	DC Machines
		3.3	Basic Electrical Machinery – DC Motor
		3.4	Basic Electrical Machinery – Asynchronous Motor
4	Electro-Mechanical and Motion Control Systems	4.1	Basic Elements of Internet based Tele-manipulation
		4.2	Mechatronics, HIL (Hardware in the Loop) Simulation
		4.3	High Dynamic Drives - Motion Control
		4.4	Automotive Electrical Drive
		4.5	Complex Control of a Hoist Equipment by a SLC
		4.6	Intelligent Gate Control by a Small Logic Controller (SLC)

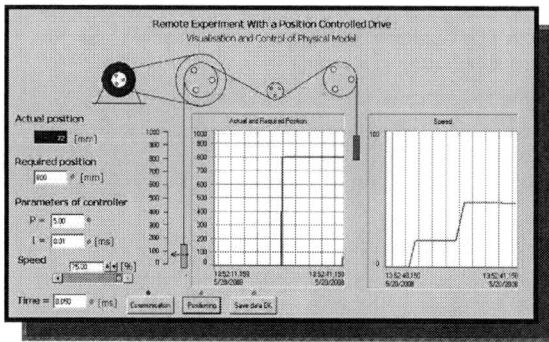

Fig. 5. Remote experiment with a position controlled drive.

For illustration figures with examples from two modules are shown (Fig. 4, Fig. 5). Both modules are treated in detail in one of the previously published papers [22], [23].

VI. CONCLUSIONS
AND EXPECTED DEVELOPMENT TRENDS

These hands-on remote experiments aim to repeat prior knowledge or lecture demonstrations and to invite to active participation. Without taking risks the students can manipulate materials and discover results. The exercises turned out to be useful for self study, if instructions and assignments were given. Although the study results of the students that used the learning platforms were better, the transfer of this knowledge and skills in new real life applications are still doubtful. Students report that they find it difficult to bridge the sub domains to integrated wholes.

Nowadays education is changing towards paying attention to integration of knowledge areas and to development of skills for learning. The market expects universities to deliver professionals at an academic level, with skills for cooperation, communication, problem solving in new situations. The curricula should be adapted to competence development and active participation1. The students are internet-minded: they work and study and have many social contacts. Teachers could use this behavior as a quality and develop courses with virtual assignments, cooperation tasks, simulations and just-in-time theory instructions.

Departments of EE are challenged to work out these characteristics in their educational programs. Embracing competences development in academic programs teaching staff might consider using the 'real life' professional situations as ingredients. Real life situations are cases, problems, practices, projects from industry, consultancy and research. While working with authentic material the students apply knowledge in an integrated way. They experience the meaning behind the knowledge and they have to deal with the limitations of the contexts. Students learn how to solve problems of future professional life and are challenged to develop autonomy and responsibility. The next trend consists therefore integration of different labs within the mentioned groups but also beyond.

ACKNOWLEDGMENT

This work has been performed within the project „E-learning Distance Interactive Practical Education (EDIPE)", (project No CZ/06/B/F/PP-168022) supported by the EC within framework of Leonardo da Vinci II programme and partly by the project No 229930-CP-1-2006-1-RO-MINERVA-IDENTITY "Individualized Learning Enhanced by Virtual Reality". The opinions expressed by the authors do not necessarily reflect the position of the EC, nor does it involve any responsibility on its part.

REFERENCES

[1] P. Bauer, V. Fedak, V. Hajek, and I. Lampropoulos, "Survey of distance laboratories in oower electronics," *IEEE 39th Annual Power Electronics Specialists Conference*, Rhodos, 2008, ISBN 978-1-4244-1668-4.

[2] E-learning Distance Interactive Practical Education – EDIPE, Pilot Project of the EU Leonardo da Vinci Vocational Training Programme. Project No CZ/06/B/F/PP-168022, duration 10/2006 - 09/2008, at http://www.powerweblab.com/.

[3] F. Coito, L. Gomes, and A. Costa, "Simulation, Emulation and Remote Experiments," Proceedings of the *Workshop on using VR in Education*, Lisboa, March 2007, pp. 99-110, ISBN 978-989-20-0715-1.

[4] Proceedings of *International Conference on Remote Engineering and Virtual Instrumentation* (REV'07), Porto, Portugal, June 25-27, 2007, www.i-joe.org/ojs/.

[5] F. G. Lerro and M. D. Protano, "Web-based remote Semiconductors Devices Testing Laboratory," *International Journal of Online Engineering (iJOE)*, 2007, pp. 161-164.

[6] K. Yeung and Huan, J., "Development of a Remote-Access Laboratory: a DC Motor Control Experiment." *Computers in Industry*, Vol. 52, Issue 3, Dec. 2003, pp. 305-311. (http://www3.mae.cuhk.edu.hk).

[7] V. Silva, V. Carvalho, R. M. Vasconcelos., and F. Soares, "Remote PID control of a DC motor," *iJOE*, 2007, p. 147, www.i-joe.org.

[8] D. Maga, J. Sitar, and J. Dudak, "Measurement of electrical machines in virtual laboratory". 10th *Int. Symposium on Mechatronics „Mechatronika 2007"*, June 6-8, 2007, Trenčianske Teplice, Slovakia.

[9] E. Ernest; R. Sztylka; B. Ufnalski; and W. Koczara, "Methods in teaching modern ac drives: inverter-fed motor system with internet-based remote control panel," *12th International Power Electronics and Motion Control Conference*, EPE-PEMC 2006, Portorož, 2006, pp. 2130 – 2133.

[10] M. P. Kazmierkowski and R. Bracha, "Virtual laboratory of power electronics pulse width modulation in three-phase converters " www.isep.pw.edu.pl/icg/vla.

[11] A. Rojko, D. Hercog, and K. Jezernik, "Advanced control course with teleoperation in the mechatronics study", *16th Int. Conf. on Electrical Drives and Power Electronics, EDPE 2007*, The High Tatras, Sept. 2007.

[12] D. Gillet; A. V. Nguyen Ngoc, and Y. Rekik, "Collaborative web-based experimentation in flexible engineering education", *IEEE Transactions on Education*. Vol. 48, 2005, No. 4. p. 696-704.

[13] K. W. E.Cheng, C. L. Chan, N. C. Cheung, and D. Sutanto, "Virtual laboratory development for teaching power electronics," Int. Conf., *EPE-PEMC 2004*, Riga, Latvia. Sept. 2004.

[14] Z. Yi., J. Jian-jun, and F. S. Chun, "A LabVIEW – based, interactive virtual laboratory for electronic engineering," Education. *Int. J. Engng Ed.*, Vol. 21, 2005, No. 1, pp. 94-103.

[15] A. B. Buckman, "VI-based introductory electrical engineering laboratory course", *Int. J. Engng Ed.*, Vol. 16, 2000, No. 3, pp. 212-217.

[16] C. S. Peek, O. D. Crisalle, S. Deapraz, and D. Gillet, "The virtual control laboratory paradigm: architectural design

requirements and realization through a dc-motor example," *Int. J. Engng Ed.*, Vol. 21, 2005, No. 6, pp. 1134-1147.

[17] C. Fernandez, M. A. Vicente, and L. M. Jimenez, "Virtual laboratories for control education: a combined methodology;" *Int. J. Engng Ed.*, Vol. 21, 2005, No. 6, pp. 1059-1067.

[18] N. Ertugrul. "New era in engineering experiments: an integrated and interactive teaching/learning approach, and real-time visualisations," *Int. J. Engng Ed.*, Vol. 14, 1998, No. 5, pp. 344-355.

[19] V. G. Agelidis, "A laboratory-supported power electronics and related technologies undergraduate curriculum for aerospace engineering students," *Int. J. Engng Ed.* Vol. 21, 2005, No. 6, pp. 1177-1188.

[20] P. H. Gregson and T. A. Little, "Designing contests for teaching electrical engineering design," *Int. J. Engng Ed.*, Vol. 14, 1998, No. 5, pp. 367-374.

[21] Ertugrul N., "Towards Virtual laboratories: a survey of LabVIEW-based teaching/ learning tools and future trends,: *Int. J. Engng Ed.*, Vol. 16, 2000, No. 3, pp. 171-180.

[22] V. Fedak, T. Balogh, P. Bauer, and S. Jusko, "Virtual and Remote Experimentation in Motion Control", Int. Conference on Mechatronics, AD University of Trencin, Faculty of Mechatronics, May 2008, ISBN 978-80-8075-305-4

[23] P. Bauer, J. Dudak, and D. Maga, "Distance practical education with DelftWebLab," Int. Conf *EPE-PEMC 2006*, **Portoroz**, August 30- September 1, ISBN 1-4244-0121-6, pp.1528-1535

Integrated multimedia educational program of a DC servo system for distant learning

Gábor Sziebig, István Nagy, Rafael Kálmán Járdán, Péter Korondi
Budapest University of Technology and Economics,
Department of Automation and Applied Informatics, Budapest, Hungary,
e-mail: {nagy, jrk, korondi}@elektro.get.bme.hu

Abstract— The paper presents a complete (animation, simulation and internet based measurement) multimedia educational program of DC servo system for distant learning. The animation program describes the basic operation of a DC motor, derives the torque-speed characteristics and explains the basic steps of control design. The animation program includes screens for teaching in class and for individual study as well. With the guidance of this animation program the students can simulate the control of a given servo system. The final and most important step in any kind of education in the field engineering is the measurement. This is the most challenging step in distant learning. The students should enter to a web page to access the experimental set up. The experimental set up includes a DC servo motor with 4 quadrant drive. The drive can be switch on and off via internet. The students can write a simple PI or sliding mode controller program which is inserted into the communication frame program to operate the servo motor. The measurement result is generated in such form which is compatible to the simulation results the students can compare their simulation and measurement results.

Keywords—Electrical engineering education, distant learning, e-learning, sliding mode control.

I. INTRODUCTION

The e-learning materials described in this paper are presented in B.Sc. and M.Sc. education of integrated and mechatronics engineers. They learn it in several different subjects in an unified way, which enables continues refreshment of their knowledge. The students learn about the basis of electric motors, control and software engineering, power electronics first and later about motion control which integrates their previous study. The main advantages of this approach is that it can be used in adult vocational education and training (life-long education) as well.

To meet the competitively and environmentally challenges in the manufacturing industry, automation and robotization is the most important measure hence turning the manual work power from tiring repetitive tasks into complex tasks where knowledge and human skills are required. This is both due to the increasing complexity level of products and the focus on improved working environment within EU. However, advanced robotic systems requires advanced knowledge within many classical fields of engineering and surely the small and medium sized enterprises, so typical in EU industry, do not cover all these areas of expertise. One usual problem is the different backgrounds of the attendees. In case of an advanced motion control course, the professor can refer to the internet for the necessary background. The project course to combine the new technology and new learning methods. The interactive multimedia applications (animations, simulations, virtual actuality sections) combined with the Web-based laboratory tests results a Personal Learning Environment.

The whole educational program has not available yet in e-learning form. Here, the DC motor related examples are selected. The animation program can be down loaded from http://dind.get.bme.hu/animation/.

Within the framework of the Leonardo da Vinci program of the European Union (EU), a project called INETELE incorporating eight Universities from eight member countries is aimed at developing multimedia software for teaching the subject of Electrical Engineering (EE) [1] [2] [3] [4] [5] [6]. The present paper is concerned only with a small fraction of the program developed by the team at the Budapest University of Technology and Economics.

The DC motors have a special historical role in the field of industrial electronics since all industrial servo drives used DC motors in the past and the first microprocessor controlled drive [7] also applied DC motor. Even if they have several drawbacks they are used in recent applications [8][9]. The main advantages of a DC servo motor drive is that it is simple from the point of view of control. Before the advent of micro controllers they were the only solutions for servo drive systems. It is easy to adopt any control method for a DC servo system. That is why the newly proposed control methods are frequently applied for a DC servo system first. On the other hand, there is a trend to control all kind of servo drives (field oriented induction motor drives [10] and brushless drives [11]) like a DC servo drive.

PID controller is still the most common controller method in the industrial applications [12]. The other popular method is the sliding mode control which was introduced in the late 1970's [13] but it is used recently in high-performance motion control systems [14]. In recent applications, the sliding mode control is combined with different soft computing methods [15][16].

Sliding mode control of variable structure systems has a special role in the field of robust control. On one hand, the exact description of sliding mode needs advanced mathematics, which was established by [17] [18] in the early sixteen's. On the other hand, it is quite easy to implement in most engineering systems, a simple relay is necessary in most cases.

The organization of the paper is as follows: Section II describes the concept of the animation and presents some selected parts of the animation program. Section III summaries simulations. Section IV presents the internet based laboratory measurements. Section V concludes the paper.

II. ANIMATION OF A DC SERVO MOTOR AND DRIVE

It is difficult to describe a dynamic and interactive animation by static screens. The problem is the same when the operation of a motor is explained by static figures.

A. Concept and structure of animation

Explaining the sophisticated processes of servo systems and their control methods is a real challenge by the traditional methods. The static figures shown by books or the computer projector with power point are not well suited for individual distant learning. On the other hand the possibilities by the modern multimedia methods can optimally be adapted in this area [19]. Utilizing the opportunities of the technique, the sequence of topologies and transient processes of the system can well be shown by animated figures and understood easier. The basics of the method in most cases is the application of simulation techniques to obtain the results representing the complex processes. The advantages of animated representation are obvious in the study of both simple and complex units.

There are basically two kinds of screens [20]: Main screens; Screens for teaching in class room. They are designed for projection in classroom. Only main screens are presented in this paper. Large letters, figures, tables are used. The information content is limited. They are explained by the professor. Sub screen; Screens containing Supporting Text for distant study at home. They are designed for using them out of class-room for individual distant learning. They apply smaller letters, figures, tables and contain substantial amount of information to. It makes possible to study and learn the material of screens at home for students without teacher. The Problems/Questions and Interactive Study are presented in the form of sub screens.

B. Content of animation

1) Physical model of DC motor: The physical model of the DC motor is introduced (see in Fig.2) including its construction and its equivalent circuit. The chapter contains 2 subchapters. Both are animated.

2) Time-domain equations: This chapter focuses on the equivalent circuit of the DC motor and the time-domain equations. The animation contains 9 identical frames, which are represented by Fig. 3.

The static torque-speed characteristic and the working point is derived (see in Fig. 4). The animation contains 6 individual (and 42 intermediate) frames. The intermediate frames necessary to float (ride) the equations or variables into another equation smoothly. This "floating equation

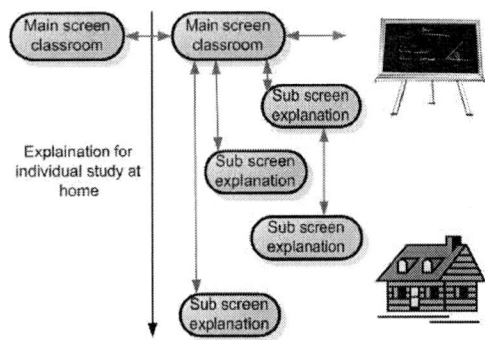

Fig. 1. Structure of the screens

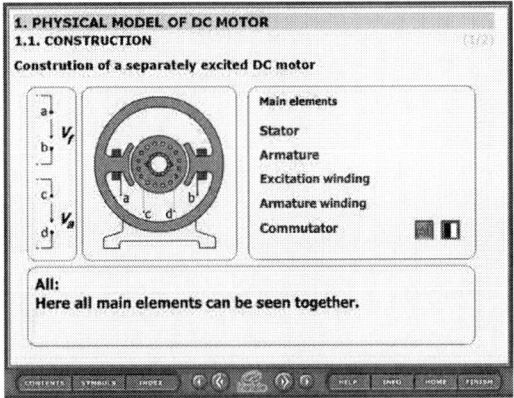

Fig. 2. Construction of a DC motor

technic" is used in several parts of this animation. Four parameters can be modified. This figure is on-the-fly type the actual characteristic is calculated on-line according to the actual parameters. In this case the sophisticated system of equations can cause the problem because of the limited processor power. It is accurate and fast enough in this case.

3) Transfer functions of a DC motor: It explains the derivation of the transfer function of the DC motor in a detailed form step-by-step. The animation contains 13 individual (and 132 intermediate for floating equations technic) frames. The derivation of the transfer function is represented by one figure (see in Fig. 5).

This animation uses a very special way to explain the derivation. On the first page there are the initial equations. More initial equations appear than they slide into the result equations. This system repeats itself until the desired derivation ends. By clicking the two PLAY buttons the animation can go forward (right play button) or backward (left play button). The directions are different the backward way is not animated. The progress bar graphically shows in which phase the derivation is. This way student can follow how many percentage of the derivation has elapsed. Teacher can vary the speed of the presentation according to the progress bar.

4) Block diagram of a DC motor: The block diagram changes from Fig. 6 to Fig. 8 step-by-step. The animation

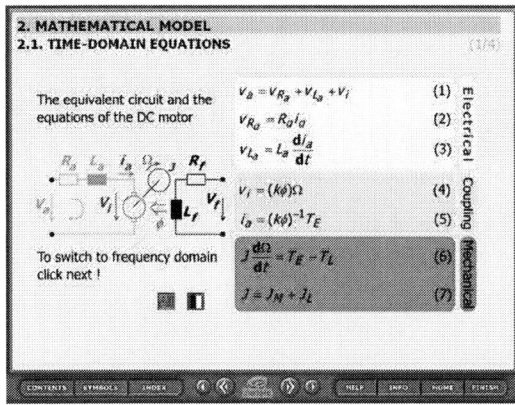

Fig. 3. Time-domain equations

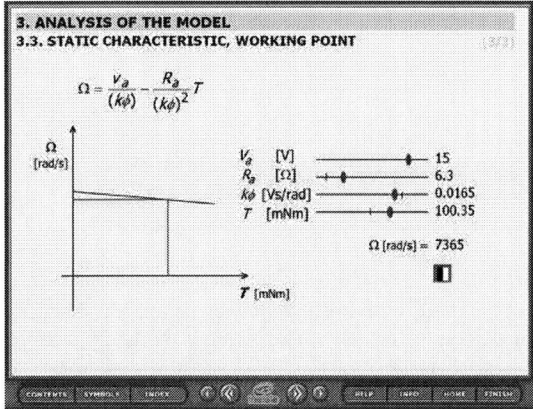

Fig. 4. Interactive figure of the static torque-speed characteristics

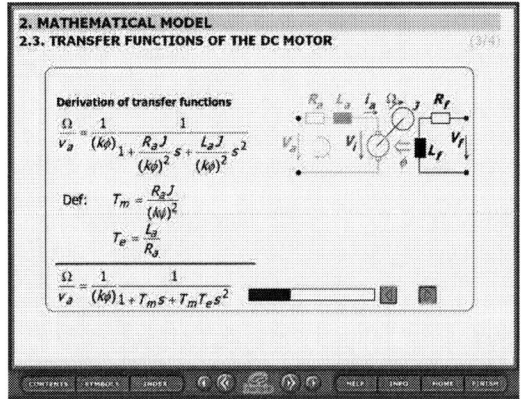

Fig. 5. A stage of the derivation of transfer function

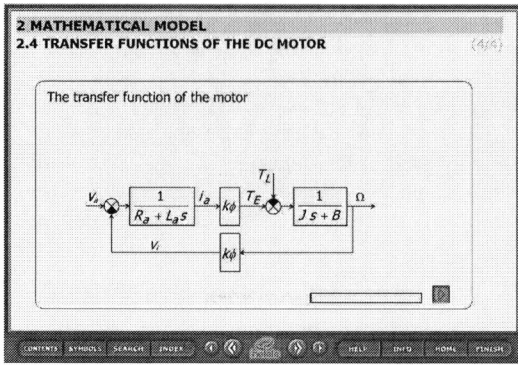

Fig. 6. First screen of DC motor block diagram

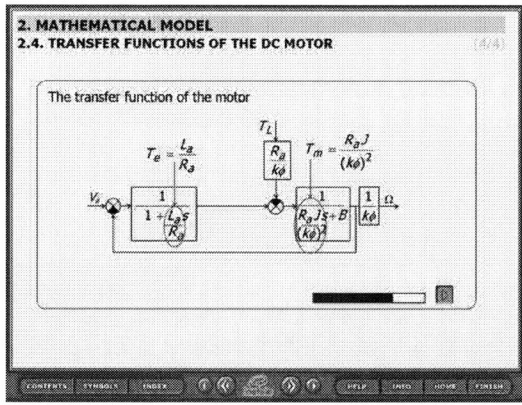

Fig. 7. A stage of the animation

contains 4 individual (and 127 intermediate) frames. In this animation the progress bar is applied as well as in the previous subchapter can be seen.

5) PID controller: The task is introduced briefly by static slide shown in Fig 9. The colours help the student to identify the motor and the two control loops. The steps of control design method are explained in a detailed form. Here only two screens are presented.

Fig. 10 explains the meaning of phase margin. The animation shows how to determine A_p if $30...60°$ of remaining phase margin (for stable and smooth working) is suggested. It contains 7 individual frames.

The aim of the animation to explain how to design the phase margin in a reversed non-analytical way where the Bode plots let to see the phase margin. Student can vary A_p between 0.5 and 3.5 and can check the effect of different amplification. The animation uses the preliminary simulated results and graphs. A_p can be varied from 0.5 to 3.5 in 7 steps (0.5; 1.0; 1.5; 2.0; 2.5; 3.0; 3.5) so the flash file is small enough but the number of the frames are enough to see the difference.

The static slide of Fig. 11 helps to remember the student to the open and the close loop transfer functions.

6) Sliding Mode Control: Several advanced control methods are discussed. The sliding mode control is presented here. The controlled plant is a DC motor, the

2362

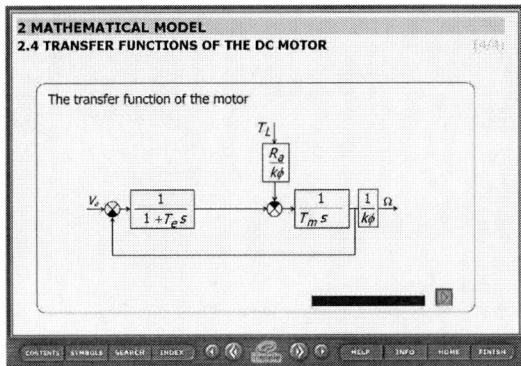

Fig. 8. Last screen of DC motor block diagram

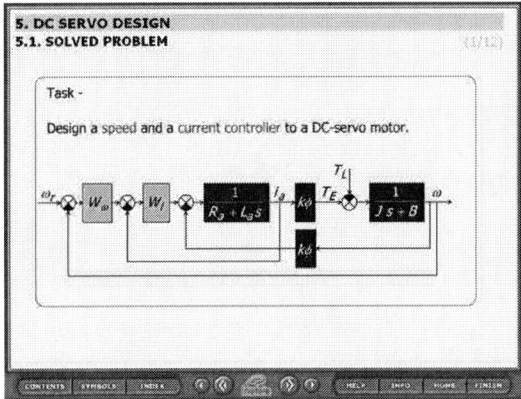

Fig. 9. Cascade control loops

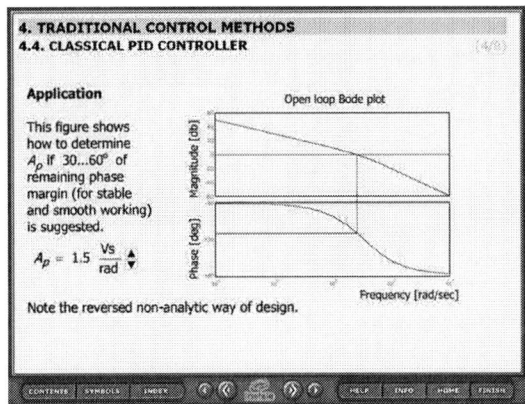

Fig. 10. Calculation of phase margin by the Bode-diagram

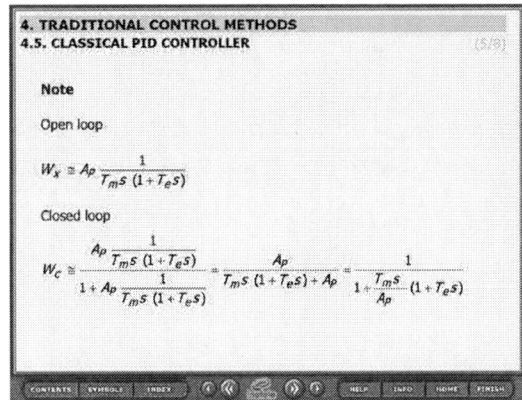

Fig. 11. Open and closed loop transfer functions

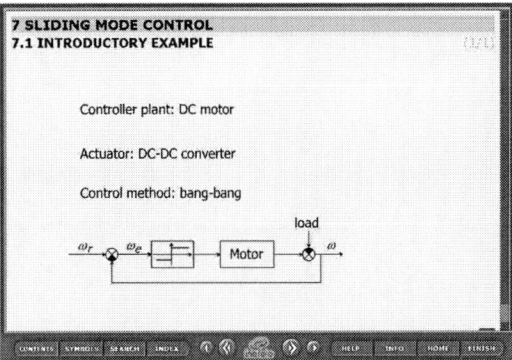

Fig. 12. Simplified system

actuator is a DC-DC converter and the control method is bang-bang type.

This means that the DC motor is controlled by a 2 state relay which can accelerate or decelerate the motor (see in Fig. 12).

The derivation of the simplified per unit equations of the error is shown in Fig. 13.

It is a second order differential equation with zero damping (since the armature resistor is ignored, see the bottom line of Fig. 13), which means that the error trajectory moves along a circle. The center of that error trajectory circle depends on the switching state of the DC-DC chopper as shown in the animated Fig. 14 and Fig. 15.

The suppressing process of the error is shown in the animated Fig. 16. Assuming that the system starts from standstill zero position and the reference signal ω_r is constant and it is a half of the no load speed ω_{nl} than $\omega_e = \omega_r$ and $\dot{\omega}_e = 0$ at $t = 0$ time instant. Defining a switching line as $\omega_e + \lambda\dot{\omega}_e = \sigma = 0$, the DC-DC chopper is switched on when the system trajectory is above the switching line ($\sigma > 0$), it means the trajectory moves along a circle with a center of $\omega_e = \omega_r - \omega_{nl}$, $\dot{\omega}_e = 0$ and the DC-DC chopper is switched off when the system trajectory is below the switching line ($\sigma < 0$), it means the trajectory moves along a circle with a center of $\omega_e = \omega_r$,

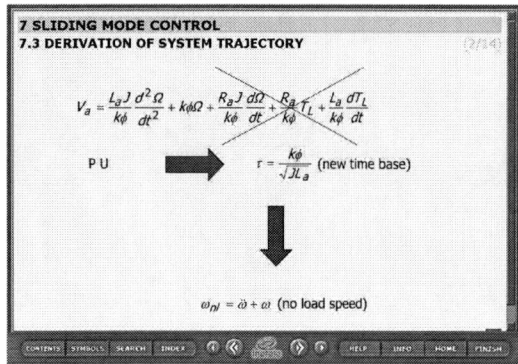

Fig. 13.   Derivation of error trajectory

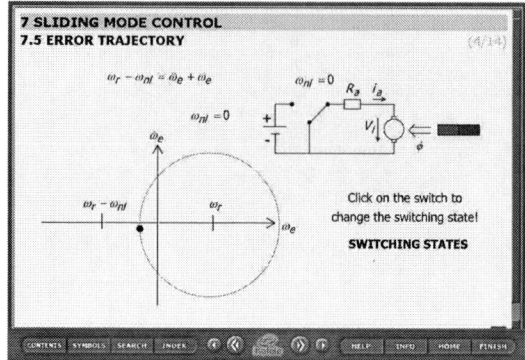

Fig. 15.   Error trajectory when the DC motor is switched off

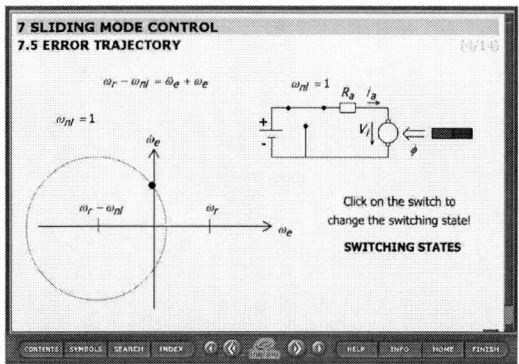

Fig. 14.   Error trajectory when the DC motor is switched on

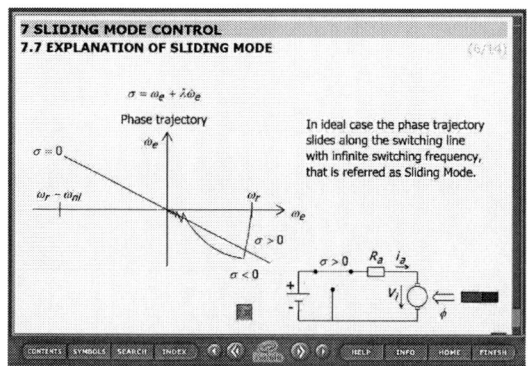

Fig. 16.   Sliding mode

$\dot{\omega}_e = 0$. The first part of the trajectory is a circle with center of $\omega_e = \omega_r - \omega_{nl}$, $\dot{\omega}_e = 0$. When the trajectory crosses the sliding line a switch occurs. Since the error trajectory must be continuous, it is continued from the position where the switching occurred along a circle with a new center. An interesting phenomenon occurs after the second switch when the error trajectory moves above the switching line. A little delay is shown in the animated figure. If this delay is eliminated then the error trajectory returns back immediately below the switching line which initiates an other switch. This phenomenon repeats itself in the opposite direction and finally the error trajectory remains on the switching line and the switching frequency is infinite. This is referred to as the siding mode. It is easy to explain the robustness of sliding mode by this animation. If the DC voltage of the DC-DC chopper is changing a little than the center of trajectory circle is moving a little. If the armature resistor is not ignored than the error trajectory moves along a spiral with decreasing radius. If that modifications are within certain limits than they have no effect on the sliding mode i.e. the error trajectory always returns immediately back to the side of the switching line where it was before the previous switching. Fig. 16 shows the final state of this animation.

III. SIMULATION

The first step from theory to application is simulation. The students use Matlab-Simulink software. The simulation model of a DC servo motor and drive are shown in Fig. 17 and 18. The students can check the torque-speed characteristics by simulation and they can study the performance (overshot, settle down time, oscillation, robustness) of the cascade speed controller with different parameters and phase margins.

IV. E-LABORATORIES

Sometimes a simple web browser is not enough in remote laboratories. The strength of a web browser is

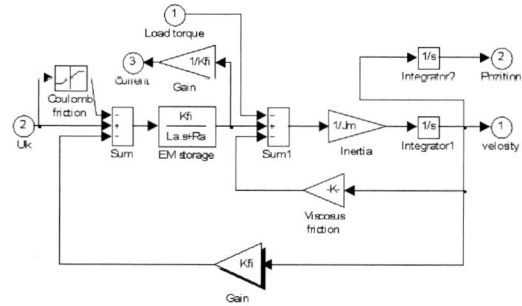

Fig. 17.   Simulink model of a DC motor

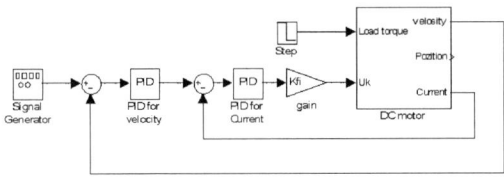

Fig. 18. Simulink model of a DC drive

simplicity and existence on every PC. But most of the remote laboratories based only on a web browser [21], [22], [23], [24], [25] allow only control parameter modification or a simple choose in control laws. An add-on for a web browser (ex.: Java or Flash) provides more capabilities for the remote experiments, but not a full control. Instead of using a special program [26], [27], [28] to access remote laboratory, the students learning at Department of Automation and Applied Informatics at Budapest University of Technology and Economics (BUTE) can choose from two equivalent solutions. These students are usually fulfilling their B.Sc. studies in mechanical engineering and not well trained in computer programming. The most important contributions in the following two solutions are that, after only a short learning period (less than an hour), the students are able to create a controller, independently to their previous programming skills. All the controllers (P, PI, PID, Sliding mode, etc.) can be visualized, tested, observed and every little change can be seen in the results. The students can even see (via a webcam), what are they doing in the laboratory. The first solution is based on a service provided by most of the PCs located in traditional laboratories, the Remote Desktop Connection (RDC). By giving access to students to these PCs, the laboratory can be accessed 24h a day, which results cost-effective education. The second solution is a standard web browser based remote laboratory, but without any limitation in controller design and execution. The website does not need any add-on, because all the logics and intelligence are located on server side, which results thin client application. From both solutions, remote experiments can be run on DC servo motor.

A. System architecture

In Fig. 19. the architecture of the proposed solutions can be seen. The solutions only differ in the way of connection to the remote experiment. The DC servo motor is turned on and off (with a PSU, controlled by a D/A card), every time a student logs in to the system. This is required, because of security rules, which declare that experimental laboratory equipment can not be turned on 24h a day.

B. Remote Experiment Framework

The solution equality is achieved by using a software framework, which can be used the same way by the two different solutions. The framework separates the controller and the communication part of the DC servo

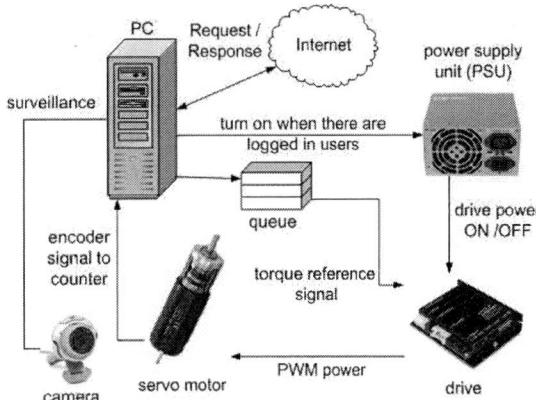

Fig. 19. System architecture

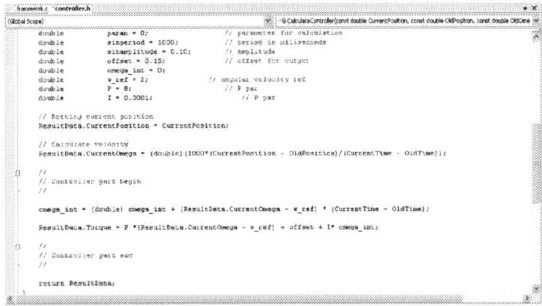

Fig. 20. Controller file

program. This split in the framework allows students to write their own controller, independently on the solution they use. In case of the first solution (RDC), after connecting to one of the PC in the laboratory, only the file that contains the controller function has to be filled in. This file already contains the input parameters and the parameters that are required for calculations. A sample controller file can be seen in Fig. 20. After error checking and compiling the measurement is run. The results are saved in comma separated files, which can be read by any simulation program (ex.: Matlab).

In case of the second solution (web browser based), the student logs in to the website, where only the controller itself has to be typed in to the form. There are predefined parameters and variables, listed on the website, which can be used in the controller. A screenshot of the website can be observed in Fig. 21. After posting the form, the framework compiles the code and runs the measurement. The results are sent to the student via e-mail, or can be downloaded from the server.

C. Distant learning

In Section II. the animation part of the learning process was already presented. The knowledge, earned by watching the animations and interacting with the Flash movies, is used in the remote laboratory. A step by step learning program helps students to understand everything.

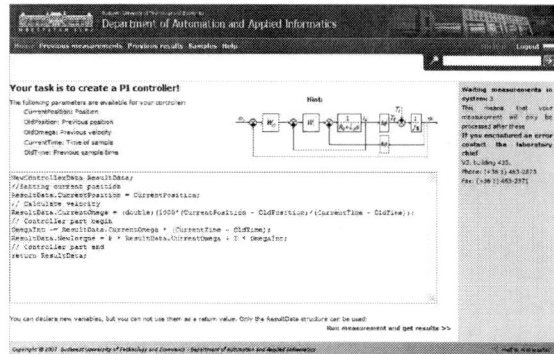

Fig. 21. E-laboratory website

From the basis (rotate the DC motor) to the upper level knowledge (Sliding mode control). The students decide, which education solution they want to use (RDC or web browser based). RDC is similar, what and how they do in traditional laboratory classes, but use more resources at students PCs. The standard web browser based remote laboratory can be used from anywhere, anytime, even from palmtops, mobile phones, etc. In this section a short overview of the learning program will be presented. The step by step learning program consist the following lessons: Introduction to programming languages
Variable definition, creation, value assign, properties
Introduction to the Remote Experiment Framework
Usage and predefined variables

1) Measurement
   Turning On/Off the experimental system
   Rotate the DC motor (without control)
2) Measurement
   Rotate the DC motor with a given load
3) Measurement
   Design and implement a P controller
4) Measurement
   Design and implement a PI controller
5) Measurement
   Design and implement a PID controller
6) Measurement
   Design and implement a Sliding mode controller

All the results can be visualized and the running of experiments can be observed through a web camera.

V. CONCLUSION

A complete distance learning program is presented starting from the animation and ending at internet based measurement. It is difficult to describe a dynamic and interactive animation by static screens. The problem is the same when the operation of a motor must be explained by static figures. Because of the separation of the communication frame program and the controller program, all students can carry out this simple measurement even if they are not good at programming. Since the actual measurement when the motor is allocated to one student takes only a few seconds, several students can measure the same motor virtually in the same time. The development of such educational program needs a complete new way of thinking, some elements of this new educational approach were presented in this paper.

ACKNOWLEDGMENT

This work was supported by the European Community program Leonardo da Vinci no 2002-CZ/02/B/F/PP-134009, the Hungarian Research Fund (OTKA TO46240 and K62836), Control Research Group and Janos Bolyai Research Scholarship of Hungarian Academy of Science for their financial support.

REFERENCES

[1] J. Hamar, R. K. Jardan, P. Korondi, I. Nagy, Z. Sepa, Z. Suto, K. Zaban, and H. Weiss, "Teaching and learning nonlinear dynamics by multimedia," in *Proc. Electrimacs 2005*, Hammamet, Tunisia, Apr. 2005.

[2] P. Bartal, J. Hamar, R. K. Jardan, P. Korondi, I. Nagy, Z. Sepa, Z. Suto, K. Zaban, H. Funato, E. Masada, and S. Ogasawara, "Learning multimedia software for teaching nonlinear dynamics," in *Proc. Control in Power Electronics and Motion Control (IPEC'05)*, Niigata, Japan, Apr. 2005.

[3] V. Fedak, P. Bauer, V. Hajek, H. Weiss, B. Davat, S. Manias, I. Nagy, P. Korondi, R. Miksiewicz, P. Duijsen, and P. Smektal, "Interactive e-learning in electrical engineering," in *Proc. EDPE'03*, High-Tatras, Slovakia, Sept. 2003, pp. 368–373.

[4] P. Bauer, "E-learning for better understanding of power quality problems and compensators," in *Proc. 11th International Power Electronics and Motion Control Conference (EPE-PEMC'04)*, Riga, Latvia, Sept. 2004.

[5] B. Davat, P. Bauer, and P. V. Duijsen, "Teaching of power electronics: From graphic representation to animation," in *Proc. 11th International Power Electronics and Motion Control Conference (EPE-PEMC'04)*, Riga, Latvia, Sept. 2004.

[6] P. Bauer and V. Fedak, "Educational visualization of different aspects for power circuits and electrical drives," in *Proc. 11th International Power Electronics and Motion Control Conference (EPE-PEMC'04)*, Riga, Latvia, Sept. 2004.

[7] K. Saito, K. Kamiyama, T. Ohmae, and T. Matsuda, "A microprocessor-controlled speed regulator with instantaneous speed estimation for motor drives," *IEEE Trans. Ind. Electron.*, vol. 35, no. 1, pp. 95–99, Feb. 1988.

[8] C. Chan, S. Hua, and Z. Hong-Yue, "Application of fully decoupled parity equation in fault detection and identification of dc motors," *IEEE Trans. Ind. Electron.*, vol. 53, no. 4, pp. 1277–1284, June 2006.

[9] F. Betin, A. Sivert, A. Yazidi, and G.-A. Capolino, "Determination of scaling factors for fuzzy logic control using the sliding-mode approach: Application to control of a dc machine drive," *IEEE Trans. Ind. Electron.*, vol. 54, no. 1, pp. 296–309, Feb. 2007.

[10] M. Boussak and K. Jarray, "A high-performance sensorless indirect stator flux orientation control of induction motor drive," *IEEE Trans. Ind. Electron.*, vol. 53, no. 1, pp. 41–49, Feb. 2006.

[11] J. Moreno, M. Ortuzar, and J. Dixon, "Energy-management system for a hybrid electric vehicle, using ultracapacitors and neural networks," *IEEE Trans. Ind. Electron.*, vol. 53, no. 2, pp. 614–623, Apr. 2006.

[12] R.-E. Precup, S. Preitl, and P. Korondi, "Fuzzy controllers with maximum sensitivity for servosystems," *IEEE Trans. Ind. Electron.*, vol. 54, no. 3, pp. 1298–1310, Apr. 2007.

[13] V. Utkin and K. Young, "Methods for constructing discontinuous planes in multidimensional variable structure systems," vol. 31, no. 10, pp. 1466–1470, Oct. 1978.

[14] K. Abidi and A. Sabanovic, "Sliding-mode control for high-precision motion of a piezostage," *IEEE Trans. Ind. Electron.*, vol. 54, no. 1, pp. 629–637, Feb. 2007.

[15] F.-J. Lin and P.-H. Shen, "Robust fuzzy neural network sliding-mode control for two-axis motion control system," *IEEE Trans. Ind. Electron.*, vol. 53, no. 4, pp. 1209–1225, June 2006.

[16] C.-L. Hwang, L.-J. Chang, and Y.-S. Yu, "Network-based fuzzy decentralized sliding-mode control for car-like mobile robots," *IEEE Trans. Ind. Electron.*, vol. 54, no. 1, pp. 574–585, Feb. 2007.

[17] A. G. Filippov, "Application of the theory of differential equations with discontinuous right-hand sides to non-linear problems in automatic control," in *1st IFAC congress*, 1960, pp. 923–925.

[18] ——, "Differential equations with discontinuous right-hand side," *Ann. Math Soc. Transl.*, vol. 42, pp. 199–231, 1964.

[19] U. Mnz, P. Schumm, A. Wiesebrock, and F. Allgwer, "Motivation and learning progress through educational games," *IEEE Trans. Ind. Electron.*, vol. 54, no. 6, pp. 3141–3144, 2007.

[20] P. Bartal, P. Bauer, J. Hamar, R. K. Jardan, P. Korondi, I. Nagy, Z. Suto, K. Zaban, H. Funato, and S. Ogasawara, "Multimedia course for power electronics, nonlinear dynamics and motion control," in *Proc. IEEE Power Electronics Education Workshop*, Recife, Brasil, June 2005.

[21] C. C. Ko, B. M. Chen, J. P. Chen, J. Zhang, and K. C. Tan, "A web-based laboratory on control of a two-degrees-of freedom helicopter," *International Journal of Engineering Education*, vol. 21, no. 6, pp. 1017–1030, 2005.

[22] M. Casini, D. Prattichizzo, and A. Vicino, "The automatic control telelab - a web-based technology for distance learning," *IEEE Control Syst. Mag.*, vol. 24, no. 3, pp. 36–44, 2004.

[23] J. Sanchez, S. Dormido, R. Pastor, and F. Morilla, "A java/matlab-based environment for remote control system laboratories: illustrated with an inverted pendulum," *IEEE Trans. Educ.*, vol. 47, no. 3, pp. 321–329, 2004.

[24] A. R. S. Castellanos, L. Hernandez, I. Santana, and E. Rubio, "Platform for distance development of complex automatic control strategies using matlab," *International Journal of Engineering Education*, vol. 21, no. 5, pp. 790–797, 2005.

[25] B. Duan, K. V. Ling, H. Mir, M. Hosseini, and R. K. L. Gay, "An online laboratory framework for control engineering courses," *International Journal of Engineering Education*, vol. 21, no. 6, pp. 1068–1075, 2005.

[26] J. Henry and C. Knight, "Modern engineering laboratories at a distance," *International Journal of Engineering Education*, vol. 19, no. 3, pp. 403–408, 2003.

[27] T. Kikuchi, S. Fukuda, A. Fukuzaki, K. Nagaoka, K. Tanaka, T. Kenjo, and D. A. Harris, "Dvts-based remote laboratory across the pacific over the gigabit network," *IEEE Trans. Educ.*, vol. 47, no. 1, pp. 26–32, 2004.

[28] D. Hercog, B. Gergic, S. Uran, and K. Jezerniks, "A dsp-based remote control laboratory," *IEEE Trans. Ind. Electron.*, vol. 54, no. 6, pp. 3057–3068, 2007.

Electromechanical Actuators WEB-lab

Dusan Maga[1], Jan Sitar[1], Juraj Dudak[1], Rene Hartansky[1], Peter Siroky[1], Jan Halgos[1],
Pavol Bauer[2]

[1] Faculty of Mechatronics TnUAD, Studentska 1, Trencin, Slovakia
tel.: +421 – 32 – 7417501, fax +421 – 32 – 7417515 e-mail: maga@yhman.tnuni.sk
[2] Delft University of Technology, Postbus 5, Delft, the Netherlands

Abstract - **Faculty of Mechatronics (FM TnUAD), as one of EDIPE (Leonardo da Vinci) project partners, has built a laboratory of electromechanical actuators, controlled and measured by available web services. The preliminary results have been presented at important world-wide conference proceedings or journals. The recent state in electromechanical actuators Web-Lab will be presented at paper. The used control technology, as well as the measuring equipment and technologies, will be described and discussed. Also the very first experience with designed technology will pre presented with special attention paid to viewpoint of end-user and laboratory staff. Also the basic ides of booking systems, developed by University of Maribor, and its connection to laboratories in Trencin will be presented.**

Keywords - **Education methodology, Education tool, Electrical machine, Measurement**

I. INTRODUCTION

New approaches and new technologies are used mainly in the science, industrial working process and in education. With influences of evolution in automatization, information technologies and computers technologies is possible monitoring and controlling of industrial process by virtual equipment with distant access. The measurement process has several stages: data collection, data analyze, data presentation and data storage. These functions are usually implemented in data acquisition systems - DAQ. The cost behind measurement equipment or DAQ system is increasing together with his capabilities. In present time the computers are most suitable solution for analysis, data collection and presentation. With this method can be decreasing the cost and the measurement system is possible very easy modified. Designed virtual laboratory have some possibilities between LabView, ControlWeb and DasyLab [1]. The new tendency in technologies and science are mirrored in education too. The main areas of implementation are the information technologies and e-learning. E-learning is education process, which is using the information and communication technologies for training and distribution of research and learning materials, student active communication with teacher and for study inspection. In our case the connection between virtual laboratory with remote control and measurement with learning process can be obtained the powerful tool in educational process [2]. In contrast to [3, 4], where the measurement is based on previous FE solution (simulation), the finite element models will be used only to demonstrate the flux distribution in the measured electric machinery.

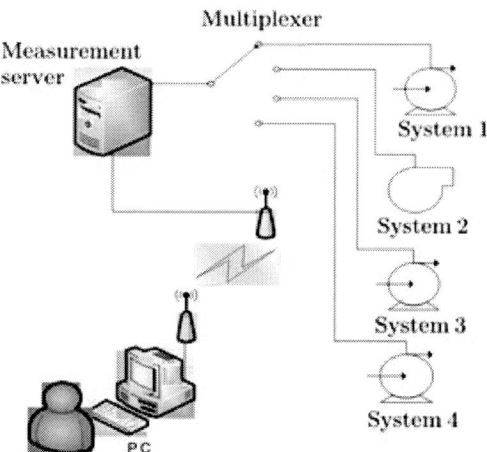

Fig. 1 Measurement system based on multiplexer technology

With the virtual laboratory is possible for students acquire the new knowledge's from everywhere. They need only high speed internet for data transfer. After server connection is possible for student repeating his knowledge and reduplicates his measurement on real equipment. In measurement only one observation station will be used. This state can be seen in Fig. 1, where the measurement on several workplaces is presented. The actual values in all measured workplaces are not desired in the same time. Measurement will be realized on individual workplaces independently (after measurement and data processing from one measurement can be switch through the next measurement) [5, 6]. The multiplexers are used for switching between various measurement places. They are cheep and very suitable solution for various data measurement. In solution can be used from two to several hundreds connection points. Multiplexers are worked on many switching principles. The most important principles are eg electromechanical principle and/or mechanical principle.

II. WEB-LABORATORY EQUIPMENT

Control of electrical machines is realized by Lucas-Nülle measurement workplace (Fig. 2). With this measurement system can be controlled and measured the torque and revolution in several electrical machines (asynchronous and synchronous motor, DC motor and generator) [7]. The workplace consists of controlled servomotor, measured electrical machines as asynchronous and synchronous motor, DC motor and generator, DC and AC power sources and from evaluation

978-1-4244-1741-4/08/$25.00 ©2008 IEEE

and control electronics (rectifier, frequency converter). An important part of the measurement is the three axis position system, used to provide a physical switch between different measuring equipment. All equipments are connected by GPIB and RS232 buses. Measured values for current, voltage, frequency and speed are transferred cross GPIB and RS232 buses directly into the LabView server in the real time.

Fig. 2 View on measurement workplace at laboratories of FM TnUAD

III. DESIGN OF AUTOMATED MEASUREMENT SYSTEM

The final design of automated measurement in virtual laboratory is focused mainly on measurement of direct current electrical motors and alternating current electrical motors. The block diagram for shunt motor measurement can be seen in Fig 3. Individual values for currents and voltages are acquired from multimeters HP 34401A. These multimeters are connected with the PXI server by GPIB bus. Power source regulation is realized directly by stepper motor with RS-232 bus. Finally, the RS-232 bus and control board is used to control of Lucas-Nülle measurement system (frequency converter for synchronous servo drive/brake unit control) and the values for various voltages (speed – 1 V/1000 rpm and torque – 1 V/1 Nm) are obtained and stored. Complete measurement workplace is controlled by LabView server (PXI computer with DAQ boards). The same computer is used as the web server too.

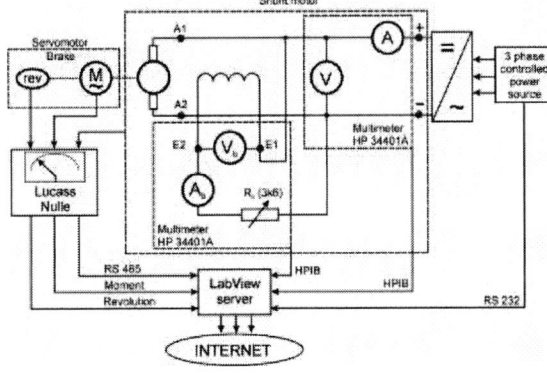

Fig. 3 Block diagram for one of designed measurements

Final data collection and complete control system is realized by PXI computer and DAQ measurement and control boards. PXI computer consist of CPU unit with various input and output busses (USB, GPIB, RS-232), operating system WinXP, LabView programming software, web server and DAQ measurement boards (frequency generator, multimeter and oscilloscope). Complete virtual laboratory workplace is programmed in LabView program. Web server is built from PXI computer too and its applications are programmed in Lab View. This industrial computer provides the real time measurement, calculations (based on math scripts files and block diagrams), motor control and web publishing (the measurement can be realized via internet on real equipments) for complete distance laboratory measurement system. All result of the measured circuits, electrical machines and power electronic systems are presented by internet education system (e-learning) on department web page. After completing the experiment, the learners can simple send an experiment report through the feedback subsystem. All learning procedures are recorded for the future references and analysis.

The internet web-page is build on LabView programming environment. The PXI server is used as web server and observation station too. The internet pages with active access for real measurement control and evaluation are placed on this server. The control, measurement and data transfer connections are created in the LabView program with the assistance of internal libraries (libraries for RS-232, RS-485 and GPIB buses). More deatiled measurement description, measurement schemes and technology description can be found in [8].

Individual measurement can be controlled by web page interface cross internet. One of the integrated parts of the measuring systems is the Moodle technology based booking procedure, developed by one of Edipe partners [9]. The approach of automated measuring systems, especially from viewpoint of minimizing the human errors, is presented in [10].

IV. BOOKING SYSTEM

In Spite of used web-design and software, a real experiment (not only simulation processes) will be done at the laboratories. This requires a unique access to the end-point of the system – laboratory. Since the measurement process will be done from different places (even from different countries), it is impossible to coordinate these accesses. Theoretically, one operator (student) is able to increase the speed of controlled equipment, while at the same time the other one should decrease it. This general access should lead to a lot of unpredictable emergency situations. To prevent these, a restricted access to measurement point should be applied. Only one operator should really operate the measurement, the others should only watch, what is going on. One approach is based on first-come first-win logic. This means, that the one, who will reach the laboratory as the first, will operate and control the equipment. This can not be used in real laboratory, since the students needs to know, when they will be able to run the exercises. The solution is in prepared booking system [9]. The basic idea of used booking system is based on keeping the internet address of measurement page classified and do not present it to end-user. The connection to this page will be done by redirection from booking system page, where the user must register them first. The end-user will not know the direct address to measurement web page neither after the successful (or unsuccessful) measurement. The booking system will administer the accesses and does not allow the user to book undesirable duplicities or waste the laboratory time. At the same time, the booking system does not allow the other user nor control neither watch the measurement. The layout of booking system internet page can be seen at Fig. 4.

A mirror of main web page has been created to allow all users to watch the processes in web-laboratory. The address of this web page will be known and this page will be accessible to unlimited number of user without restrictions. The only and the mayor restriction will be the impossibility to control the processes – these users will only be able to watch. In this case, students will be able to watch their colleagues; however, this does not mean that they will not be able to book the suitable space for themselves.

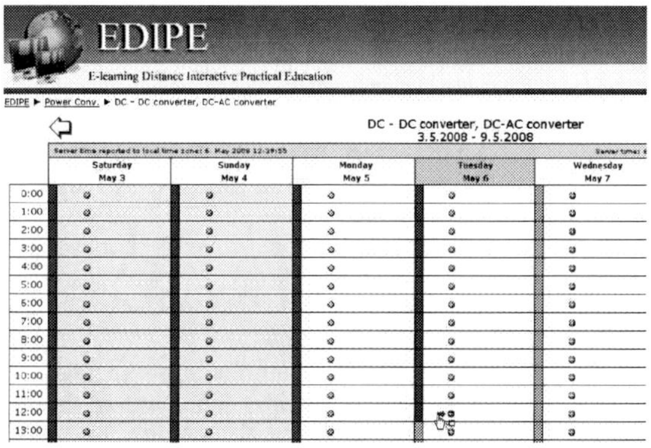

Fig. 4 View on Booking System Internet Page

V. CONCLUSIONS

A part of wide-ranged education based programme, designed and solved by Faculty of Mechatronics (FM), Alexander Dubcek University of Trencin (ADUT), Slovakia, has been presented in the paper. The presented laboratory exercises are pointed at electrical machinery (electromechanical actuators). The measurement is based on technology and techniques used at FM ADUT, where also the technology required to hardware-web connection and control has been implemented and applied. The paper presents the basic ideas of this web laboratory and the recent state of laboratory development. Also the basic ideas of used booking systems, preventing the possible crossing of active users, supplemented by systems allowing to follow the processes has been presented. The booking systems, in this case, can be understood as a "meeting" point of 12 project partners with different parts of the project, based on electrical, respectively technical education.

Present paper proves that even partners from different geographical dispersion and cultural (historical background) can work together under a collaborative engineering strategy [11].

ACKNOWLEDGEMENT

This work has been performed within the project "Elearning Distance Interactive Practical Education (EDIPE)". The project was supported by the European Community within framework of Leonardo da Vinci II programme (project No CZ/06/B/F/PP-168022). The opinions expressed by the authors do not necessarily reflect the position of the European Community, nor does it involve any responsibility on its part.

REFERENCES

[1] D. MAGA, J. SITAR, J. DUDAK. Measurement of Electrical Machines in Virtual Laboratory. In Mechatronika 2007. Trencin: TnUAD Trencin, 2007. ISBN 978-80-8075-209-5. pp. 107-112.

[2] P. BAUER, et al. Distance Practical Education in Power Electronics. International Journal of Engineering Education. Vol. 23, Nr. 6., 2007. ISSN 0949-149X.

[3] T. ORLOWSKA-KOWALSKA, J. LIS, K. SZABAT. Identification of the induction motor parameters at standstill using soft computing methods. COMPEL-The international journal for computation and mathematics in electrical and electronic engineering. Volume: 25, issue: 1, 2006. pp. 181-194. ISSN 0332-1649

[4] A. REINAP, et al. Simulation and experimentation of a single-phase claw-pole motor. COMPEL-The international journal for computation and mathematics in electrical and electronic engineering. Volume: 25, issue: 2, 2006. pp. 379-388. ISSN 0332-1649

[5] P. BAUER, J. DUDAK, D.MAGA. Distance practical education with DelftWebLab. In EPE-PEMC 2006. Maribor: University of Maribor, 2006. ISBN 1-4244-0121-6. pp.1528-1535

[6] P. BAUER, et al. PEMCWebLab-Distance practical education for Power Electronics and Electrical Drives. In 38th Annual IEEE Power Electronics Specialists Conference Power Electronics Education Workshop (PEEW07). Los Angeles: IEEE Power Electronics Society, 2007.

[7] D. MILJAVEC, P. JEREB. Electrical Machines (In Slovenian: Elektricni stroji). Ljubljana: [s. n.], 2005. 428 pages. ISBN 961-236-851-1

[8] J. SITAR et al. Distance Laboratory Of Electromechanical Actuators Measurement. In Mechatronika 2008. Trencin: TnUAD Trencin, 2008. In press.

[9] S. URAN, D. HERCOG, K. JEZERNIK. Remote Control Laboratory with Moodle Booking System. In Proceedings of the IEEE International Symposium on Industrial Electronics (ISIE). [s.l.]: [s.n.], 2007. pp. 2978-2983.

[10] N. H. FUENGWARODSAKUL, S. E. BAUER, R. W. De DONCKER. Characteristic Measurement System for Automotive Class Switched Reluctance Machines. EPE Journal. Volume 16, no. 3, 2006. pp 44+. ISSN 0939-8368

[11] E. GONZALEZ, et al. Cross Cultural Issues on Globally Dispersed Design Team Performance: The PACE Project Experiences. International Journal of Engineering Education. 2008, vol 24, Nr. 2. pp. 328-335. ISSN 0949-149X.

Power Quality and Active Filters as Web-Controlled Experiment in the frame of PEMC WebLab

Volker Staudt*, Andreas Steimel*, Pavol Bauer† and Vítězslav Hájek‡

* Ruhr-Universität Bochum, Bochum, Germany, e-mail: *staudt@eele.rub.de, steimel@eele.rub.de*
† Delft University of Technology, Delft, The Netherlands, e-mail: *p.bauer@tudelft.nl*
‡ Brno University, Brno, Czech Republic, e-mail: *hajek@feec.vutbr.cz*

Abstract— PEMC WebLab features online experiments sponsored by the EU. Several real-world experiments aim at students and engineers of electrical engineering and mechatronics who want to extend or refresh knowledge. The Power Quality and Shunt Active Filters experiment is presented. The paper gives the theoretical background concerning Power Quality, describes the compensation method and the active filter, the example loads which are compensated and the learning objectives for the students and engineers. Based on understanding Power Quality and its implications the possibilities, limitations and side effects of a shunt active filter, featuring extended static compensator (STATCOM) functionality, are discussed. The experiment can be found via *http://pemcweblab.com.*

Keywords— Education tool, Active filter, Power Quality, STATCOM.

I. INTRODUCTION

Power-electronics devices have changed electrical engineering, especially electrical power engineering, considerably. Very fast control of energy flow, e.g., for the control of electrical drives, led to many new or functionally enhanced applications. However, while loads once used to be more or less linear, today many loads are nonlinear, generating harmonics and interharmonics. Power theory once was centered on fundamental-frequency reactive power with the fundamental-frequency displacement factor $\cos(\varphi)$ being identical with the power factor λ. This is no longer so. Devices causing considerable current harmonics are – correctly – labeled with $"\cos(\varphi) = 0.99"$ without giving the power factor, which may be as low as 0.5. Strongly nonsinusoidal currents are relevant, causing, e.g., grid-voltage distortion and unexpected overheating of transformers and cables.

New technologies may cause problems, but they also offer solutions. Active filters, based on, e.g., shunt-connected voltage-source inverter systems, eliminate reactive currents (STATCOM functionality) and even undesired harmonics (active filter functionality). Automatic control algorithms adapt to actual load conditions and eliminate parameter dependencies.

Many engineers in industry and utilities are not yet aware of the new situation, because it was not part of their education. Students have to be trained and to acquire experience to use these new technologies adequately.

They have to be aware of advantages and drawbacks of the enabling technologies.

It is not possible for university institutes to provide a whole set of power-electronics-related experiments. With this background and the support of the Leonardo da Vinci Program of the European Union the Power Electronics Motion Control Web Labarotory (PEMC WebLab, *http://pemcweblab.com*) uses web-browser-based internet access to combine highly relevant experiments and adequate documentation for the training of students and engineers wishing to update their knowledge [1 – 3]. This paper describes the properties of a remote-controlled shunt active filter experiment hosted at Ruhr-Universität Bochum, integrated into the PEMC WebLab.

Within the scope of the experiment, Power Quality and its relation to power theory and compensation issues is explained. In addition to theory, experimental results based on a shunt active filter illustrate one method for practical compensation of current harmonics and reactive power. The method presented [4 – 6] is known and applied in practice, it is very economical with low computational effort, making it appealing for practical application.

Finally, the connection of the experiment to the Internet via a standard PC equipped with a Meilhaus® data acquisition board is described.

II. POWER QUALITY AND POWER THEORY

A. Power Quality

Power Quality is a generic term encompassing many aspects, partly covering reliability and quality of service provided by the utility, partly defining limits concerning harmonics, interharmonics and flicker imposed on the consumer [7 – 9]. In the following the paper concentrates on the situation of a consumer having to meet requirements for grid access or having to mitigate problems within his own internal grid. This is also the focus of the experiment and the associated educational material.

B. Power Theory

Judging Power Quality and defining compensation needs clearly requires a concise reference. Optimal conditions must be known in all conceivable situations, also for non-sinusoidal voltages and currents in multi-

978-1-4244-1741-4/08/$25.00 ©2008 IEEE

conductor circuits [9 – 12]. Further information on power theory can be found, e.g., in [13, 14].

In short, power theory bases on the following: All currents of a circuit always sum up to zero because of Kirchhoff's law. Matching voltages: "One voltage per conductor, all these voltages summing up to zero" are defined by introducing the virtual star point (Fig. 1).

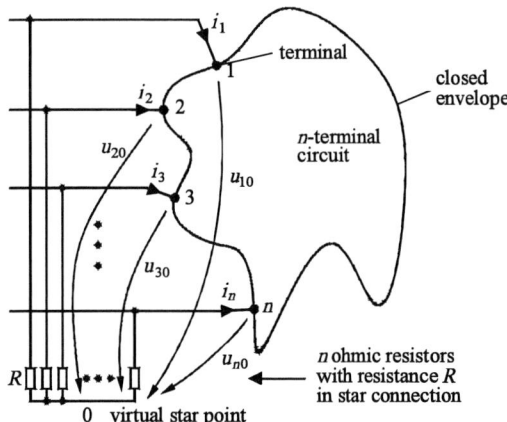

Fig. 1. Definition of the virtual star point.

By this, all currents and voltages of an *n*-terminal circuit are clearly defined.

Power theory demands that in the optimal case all currents should be proportional – with constant factor of proportion – to the respective voltages against the virtual star point, leading to a power factor of unity.

In practice, this optimal condition is often disregarded. In most practical cases current harmonics are limited, making symmetrical sinusoidal currents the desired waveform. With low voltage distortion, both requests are similar. With noticeable voltage distortion, currents proportional to the voltage (and therefore non-sinusoidal) reduce undesired voltage components by damping them [4, 14], leading to improved grid characteristics.

From an economical point of view a consumer should not invest more money than needed to meet the requirements imposed by the grid operator. In this context, cost and benefits of active filters have to be seen in comparison to usual filter banks.

While the drawback of active filters is mainly the cost factor, the main advantages are:

- automatic reaction to changes in load behaviour
- resonance problems eliminated
- no parameter sensitivity (ageing of components)
- flexibility (can be easily re-used elsewhere)
- properties controlled by software

III. SHUNT ACTIVE FILTER (PARALLEL COMPENSATOR)

A. General

A shunt active filter (parallel compensator) is able to impress – within limits given by its design – any desired set of currents into the internal point of connection (IPC) between source and load (Fig. 2). Usually three conductors (not the neutral) are used to connect the

compensator, so space vector notation is used efficiently to describe voltages and currents.

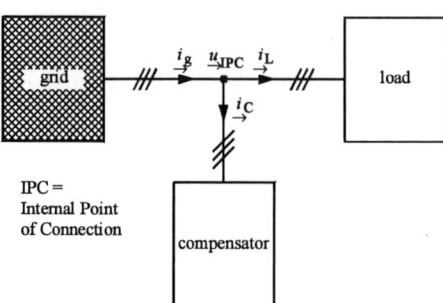

Fig. 2. Compensator connected in parallel to the load, quantities in space vector notation.

Undesired load-current components are absorbed by the compensator and kept away from the source. Considering economical constraints only those current components which exceed limits or directly cause costs (like, sometimes, reactive power) should be compensated. In addition, the voltage at the IPC can – within limits – be stabilized by impressing additional reactive currents causing a suitable voltage drop at the source impedance.

B. Types of shunt active filters

Shunt active filters impressing mainly reactive currents are commonly known as static (synchronous) compensators (STATCOM) [15, 16]. Normally they have to provide high power (several MVA) and therefore only relatively low switching frequencies are possible, allowing no compensation of harmonics. By quickly adjusting reactive currents with the help of suitable closed-loop control algorithms they stabilize the source voltage and reduce losses of energy transmission.

Shunt active filters in general provide low to medium power (up to some hundred kVA) with switching frequencies in the range of 2 kHz to 8 kHz, allowing the compensation of harmonics in addition to that of reactive currents. Closed-loop control schemes are usually employed, making these devices independent of load behavior and parameters.

C. The shunt active filter used in the experiment

The structure of the experimental set is shown in Fig. 3, its physical realization in Fig. 4.

The experimental set-up mainly consists of the compensator section, the load section and a section for analysis and visualization. This last section is for stand-alone demonstrations only, it cannot be remote-controlled. The load section contains an ohmic resistor fed by a grid-connected six-pulse thyristor rectifier with capacitive smoothing featuring strongly nonlinear currents and an idling induction machine giving reactive currents. The compensator section contains the voltage-source IGBT converter and a T-type filter for grid connection (Fig. 3). The filter mitigates switching-frequency harmonics. A simplified structural diagram used on the web page of the experiment is shown in Fig. 5.

2372

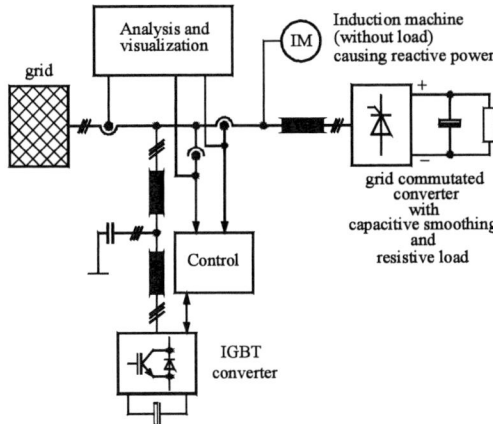

Fig. 3. Structure of the experimental set.

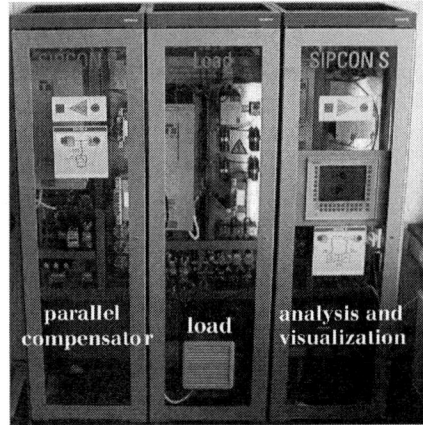

Fig. 4. Shunt active filter used in the experiment.

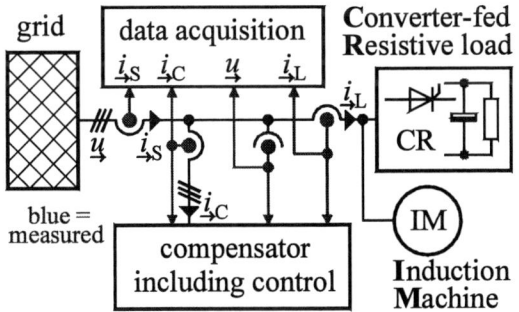

Fig. 5. Simplified structural diagram for use on web page.

Data acquisition means here the connection to the PC which links the experiment to the Internet and acquires the measurement data to be displayed. Details on this aspect can be found below.

D. Control of the shunt active filter

The shunt active filter used in the experiment is designed to provide reactive power compensation and compensation of selected low-order harmonics with minimal effort. Minimal effort means minimal voltage and

current rating of the converter and in the sense of minimal calculation power of the closed-loop compensator control.

In load currents, usually only certain sets of harmonics with certain sequence characteristics have considerable amplitude. These are +1, −1, −5, +7, −11, +13, e.g., where "− " refers to negative sequence and "+" refers to positive sequence. The control of the filter detects and eliminates only these selected sets of harmonic sequence components (except, of course, +1 active which is the desired set of currents). Depending on the application, the sequence components to be compensated may be selected differently – in the experiment the selection detailed above is fixed.

The control of the active filters utilizes rotating coordinates. For each harmonic-sequence component μ (e.g., $\mu = -5$) the following equivalent scheme is valid:

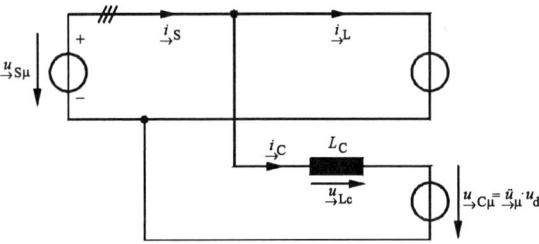

Fig. 6. Equivalent circuit for a harmonic-sequence component.

The source voltage u_g may contain a certain harmonic voltage of order μ which is disregarded in the following. The load current is impressed by the load and remains unchanged. The compensator generates a harmonic voltage system $u_{C\mu}$ of order μ described by a space vector. It is generated based on the DC link voltage u_d and a reference space vector $\ddot{u}_\mu$. The voltage system causes a voltage drop at the T-type filter (simplified, here, by an inductor L_C) leading in turn to a harmonic current sequence component i_C of compensator currents. Currents and voltages at the inductor have a phase shift of 90°.

The voltages could be calculated directly from phase and amplitude of the load current system to be compensated, in practice they are determined by an I-type controller. In the tutorial of the experiment both options are explained: Calculation of the needed voltage is used in one simulation part supporting the educational objectives, automatic I-controller-based control is used in a second simulation and the experiment itself. Fig. 7 gives the structure of the control for the elimination of a harmonic current system of order μ with $\omega_\mu = \mu\, 2\pi f_g$, where f_g is the grid frequency (50 Hz in Europe):

Fig. 7. Structure of the control for harmonic current-system elimination.

2373

On the left-hand side the source current space vector enters the controller. It is transformed into rotating coordinates, turning the selected harmonic system into a DC component. The sliding integration eliminates all harmonics but not the DC portion, interharmonics are damped but cannot be eliminated completely. The result is the complex Fourier Series coefficient of the source current of harmonic system μ. An I-type controller and a 90°-rotation give the voltage space vector needed to force the compensation current with regard to the total inductance of the T-type filter dominating in the compensation-frequency range. This result is still in rotating coordinates. A back-transformation into original coordinates gives the associated voltage space vector in original coordinates, which is then one part of the voltage set-point value of the compensator. Due to the I-type controller, the source current harmonic is eliminated completely if no source voltage harmonic of the same frequency exists.

All voltage components estimated by the controllers are summed up and give the voltage set-point value for the converter of the active filter via the reference space vector $\vec{u}_R$ as discussed above for single harmonic systems:

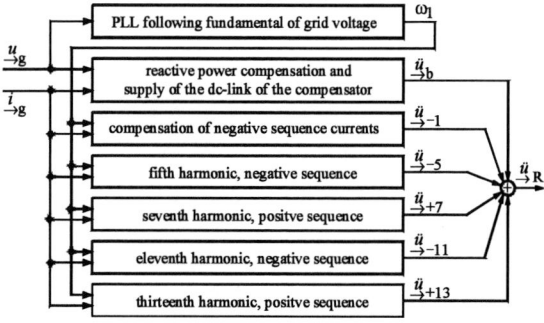

Fig. 8. Separate controllers generating compensator voltage command

In addition to the current system elimination discussed above, Fig. 8 shows controllers taking care of the DC-link voltage and reactive power compensation. Only these two tasks need information about the source voltages.

E. Practical considerations

Power theory demands currents proportional to the voltages for unity power factor. In case of distorted voltages, currents should also be distorted.

Standards do not go thus far – they set limits for the allowed distortion of source voltages and also for the harmonics contained in the currents taken by a load. It is sufficient to meet these limits and an engineering task to do this with minimal effort and cost.

Therefore the concept presented here eliminates harmonics without consideration of the voltages, reducing calculation effort and thereby cost of the hardware components. Only selected harmonic-sequence components are eliminated – so that the power and consequently the cost of the compensator are also reduced. This allows to comply with standards efficiently while reducing the cost of the active filter.

With regard to the educational objectives of the experiment this shows that usually not the technically best

solution is chosen. Instead it is an engineering task to reach the needed functionality at minimal cost.

IV. SIMULATION

A simulation structure is supplied to make the students familiar with the system, allowing to change parameters and reducing the time needed for the experiment itself. The simulation is based on the Caspoc® simulator by Simulation Research [17]. This full-scale simulator suitable for demanding simulation of power-electronics circuits has been developed with special regard to education. A freeware viewer, which allows to change all parameters (but not the structure) gives many possibilities of modifications "playing around with the system" to interested students. Fig. 9 shows the source current harmonics (upper part) and the compensator current harmonics (lower part) for the compensated case. It can be seen that the selected control structure of the simulation (Fig. 10) effectively eliminates fifth, seventh, eleventh and thirteenth harmonic. The **Harmonic Compensator (HC)** block was developed for this simulation and contains the control structure already shown in Fig. 7. Parameters, like harmonic order and amplification, can be set. Amplitude and phase of the selected harmonic system are displayed.

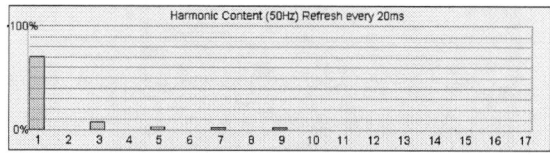

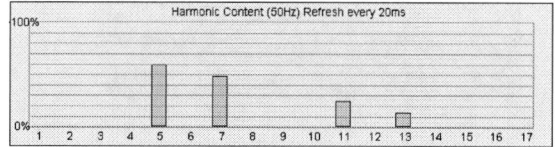

Fig. 9. Simulation results: Harmonics in source current (upper figure) and compensator current (lower figure). 100% equals 10 A

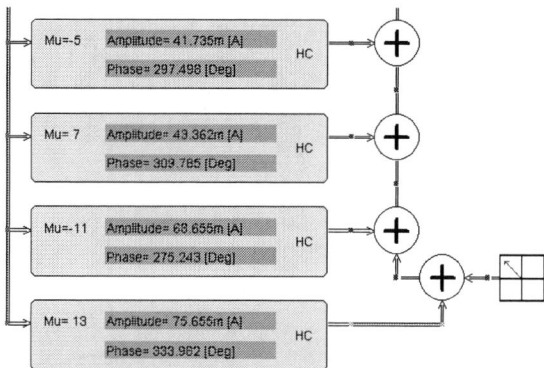

Fig. 10. Control section of the simulation (Mu means μ)

Time functions resulting from simulation are displayed in Fig. 11. The simulation mainly supports the understanding of harmonic compensation, therefore four identical HC blocks are used but no reactive power compensation. It becomes obvious that compensating only four sets of harmonic current systems improves the source

2374

current considerably – although remaining harmonics are clearly visible.

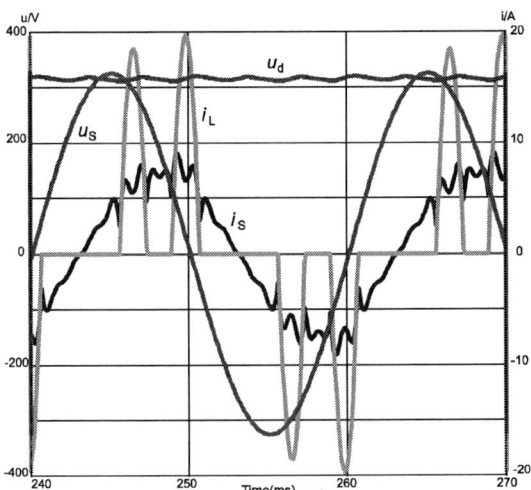

Fig. 11. Simulation result (-5th, 7th, -11th and 13th compensated)

The experiment itself also compensates reactive current, which is quite large because the thyristors are fired with an angle of about 70°. In addition to the quantities measured at the experiment, the DC link voltage u_d is displayed in Fig. 11. The idling induction machine is not modeled in the simulation.

V. LEARNING OBJECTIVES

A. Theoretical part

Similar to the structure of this paper up to this point, but with increased redundancy and partly in more detail, the students are introduced to the meaning of Power Quality and the basics of power theory. From theory and economical constraints a practical solution is derived: The compensator treated in the experiment.

The learning material is structured into the sections

- Introduction
- Basic Definitions
- Power Quality and Compensation
- The Experiment
- Control Aspects

Based on Adobe® Presenter, questions have to be answered correctly to access the next section and finally gain access to the experiment itself. In this way the learning outcome is continuously monitored.

The section "Control Aspects" is positioned behind the section "The Experiment", because learning about control in rotating coordinates is an add-on for advanced students.

B. Use of Supporting Simulation

Simulations are supplied to enhance the understanding:

The first simulation is dedicated to compensator current generation. The students perform the task of calculating suitable compensator voltages "manually", on the basis of given load current and inductivity of the T-type filter. The desired compensation current (frequency, amplitude,

phase) is extracted from the simulated load current. The values calculated for the compensator voltage (one harmonic only) are then inserted into the simulation, the resulting source current should no longer contain the harmonic under consideration. Simulation results are displayed in time domain and by a bar graph showing the Fourier series components.

A second version of the simulation contains the I-type controllers for selective harmonic-sequence-system compensation. These can be switched on and off, demonstrating that the controllers are – mainly – independent of each other. It can also be seen in which way the selection of harmonics to be compensated influences waveform- and Fourier series representation of the source current. Advanced students use this simulation to get acquainted with the dynamics of the control scheme, they may change the parameters of the controllers, for example.

C. The Experiment

A first version of the web page controlling the experiment is shown in Fig. 12. Depending on the passcode entered when first accessing the page, stationary mode (for the basic experiment) or dynamic mode (for advanced students) is enabled. While on the one hand tuning the same experimental set-up to two groups of students, basic and advanced, getting "dynamic control" is certainly a motivation for the students to learn more.

The upper left part of the web page displays a simplified structural diagram. The upper right part controls the functions of the experiment and starts a measurement. The lower right part displays the results in time domain (conductor quantities or space-vector quantities) or gives the Fourier series components. The lower left part gives one of two web-cam streams: One showing the set-up as a whole and a second one the induction machine which is one of the loads which can be switched on and off.

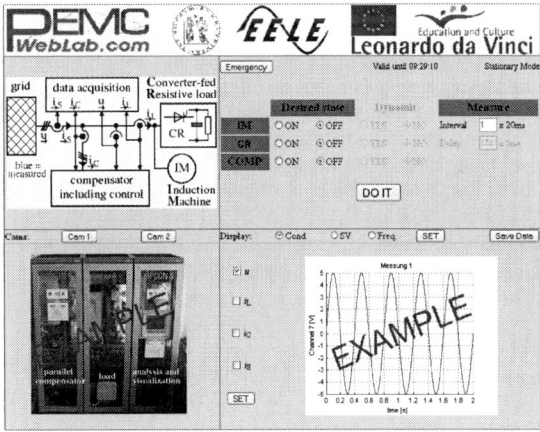

Fig. 12. First version of web page controlling the experiment

Measurement results can also be downloaded for post-processing. The students select the different loads and compare time-domain and frequency-domain representation of voltages and currents before and after compensation.

D. Comparison between experiment and simulation

Having finished the experimental part, the students are asked to compare simulation results and experimental results. The simulation of the compensator is based on ideal voltage sources not comprising the switching of the voltage-source inverter employed in the experiment. Also, the simulation does not contain source voltage harmonics. These differences are analyzed by the students and point out that in power electronics experiments are needed to verify theory and simulation.

When compensating the sinusoidal reactive current of the induction machine with this type of compensator, it becomes obvious that while removing the reactive currents the compensation introduces a set of harmonics. This type of operation is known as "Static Compensator" (STATCOM) operation. The students learn that also in this case advantages and drawbacks have to be accounted for.

VI. Experimental Results

The following figure (Fig. 13) shows experimental results under stationary conditions.

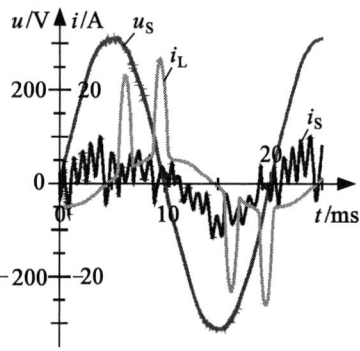

Fig. 13. Measured voltage and currents

The load current i_L shows on the one hand reactive current caused by the induction machine and on the other hand the typical spikes of a grid-commutated rectifier with capacitive smoothing. Usually these current spikes would be in phase with the voltage, but here a thyristor bridge with a firing angle of about 70° is used to produce increased harmonic currents at reduced active power and to introduce additional reactive power. The source current i_S contains considerably less spikes and practically no reactive current – but is still far from being sinusoidal due to the compensation strategy selected. However, the improvement is considerable. This can also be seen when evaluating the Fourier series components of load current and source current shown in the Figures 14 and 15.

Fundamental frequency current is considerably reduced, mainly due to reactive current compensation. The compensated harmonics are nearly eliminated. Harmonics above the 14th are not shown here – but can be analyzed by the students in the experiment.

The result seen in Fig. 15 certainly meets usual limits for grid currents, verifying the suitability of the compensation concept.

The results are close to those displayed in the section on simulation.

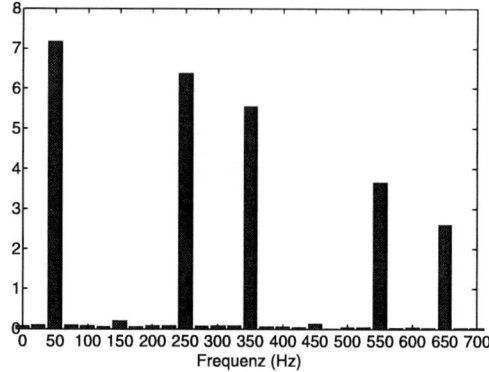

Fig. 14. Load current harmonics

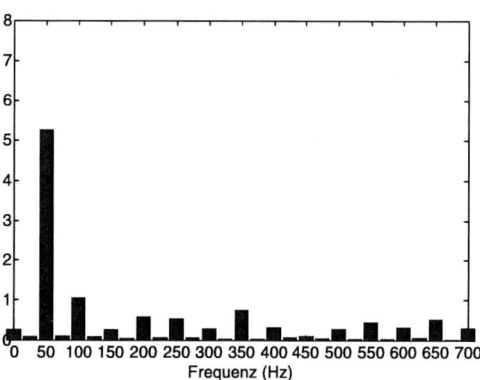

Fig. 15. Source current harmonics

VII. IT Hardware Used

While the automatic control of the compensator is included in the test set-up seen in Fig. 2, appropriate hardware is needed to control the functions of the experiment and measure currents and voltages. This is solved by a Meilhaus® data acquisition card and Matlab®-based software. The general structure of this set-up is shown in Fig. 16.

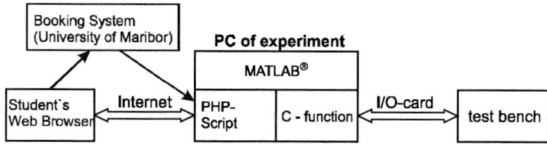

Fig. 16. Source current harmonics

To access the experiment, the student's web browser has to connect to the booking system set-up and administrated by the University of Maribor. If the student has access rights, a dedicated php-script on the PC connecting to the experiment is generated and the student is redirected to this script. Based on the input made by the student, the script calls up Matlab® which in turn uses C-functions accessing the I/O card. The I/O card features A/D conversion for measuring purposes as well as digital outputs for control. It initiates the experiment and performs the measurement, passing data over to Matlab® which generates the requested diagrams.

VIII. CONCLUSION

A web-controlled experiment explaining Power Quality and Active Filters is presented. The basic concept of Power Quality and the practical considerations leading to the solution realized in the experimental set are discussed. A clear structure with integrated learning assessment guides the student through tutorial material to the experiment itself. Simulations support the learning outcome. Basic and advanced version provide two levels of education and enhance motivation for advanced students. The design of the web page is described.

Experimental results of the experiment itself and a discussion of the way in which it is connected to the internet complete the paper.

ACKNOWLEDGMENT

This work has been performed within the project "E-learning Distance Interactive Practical Education (EDIPE)". The project was supported by the European Community within framework of Leonardo da Vinci II programme (project No CZ/06/B/F/PP-168022). The opinions expressed by the authors do not necessarily reflect the position of the European Community, nor does it involve any responsibility on its part.

REFERENCES

[1] Bauer, P.; Dudak, J.; Maga, D., "Distance practical education with DelftWebLab"; EPE-PEMC 2006, Portoroz, August 30 – September 1, ISBN 1-4244-0121-6, pp.1528–1535

[2] Bauer, P.; Maga, D.; Sitar, J.; Dudak, J.; Hartansky, R., "PEMCWebLab – Distance practical education for Power Electronics and Electrical Drives", 38th Annual IEEE Power Electronics Specialists Conference Power Electronics Education Workshop (PEEW07)

[3] Bauer, P.; Staudt, V., "Remote controlled practical education for Power Electronics", 12th European Conference on Power Electronics and Applications, 2007, Aalborg

[4] Sonnenschein, M.; Weinhold, M., "Comparison of time-domain and frequency-domain control schemes for shunt active filters", ETEP Vol. 9 (1999), No. 1, pp. 5–16

[5] Sonnenschein, M.; Weinhold, M.; Zurowski R., "Shunt-Connected Power Conditioner for Improvement of Power Quality in Distribution Networks", 7th International Conference on Harmonics and Quality of Power (ICHQP VII), 1996, Las Vegas, pp. 27–32

[6] Depenbrock, M.; Staudt, V., "The FBD-Method as tool for compensating total non-active currents", 8th International Conference on Harmonics and Quality of Power (ICHQP VIII), Oct. 1998, Athens, pp. 320–324.

[7] IEEE Standard 519-1992: "Recommended Practices and Requirements for Harmonic Control in Electrical Power Systems"

[8] IEC 1000-2: "Electromagnetic Compatibility" Part 2: Environment. Section 2: Compatibility levels for low-frequency conducted disturbances and signalling in public low-voltage power supply systems. Section 4: Compatibility levels in industrial plants for low-frequency conducted disturbances

[9] "IEEE trial-use standard definitions for the measurement of electric power quantities under sinusoidal, nonsinusoidal, balanced, or unbalanced conditions" IEEE Std. 1459–2000, Jan. 30, 2000.

[10] Depenbrock, M., "The FBD-Method, a Generally Applicable Tool for Analysing Power Relations", IEEE Trans. Power Syst., Vol. 8, 1993, No. 2, pp. 381–386

[11] DIN 40110 Teil 1, "Wechselstromgrößen; Zweileiter-Stromkreise" ("AC quantities, two-conductor circuits", in German language), Beuth-Verlag, March 1994, Berlin

[12] DIN 40110 Teil 2, "Mehrleiter-Stromkreise" ("Multi-conductor circuits", in German language) Beuth-Verlag, Nov. 2002, Berlin

[13] Ferrero, A., "Definition of Electrical Quantities Commonly Used in Non-Sinusoidal Conditions", ETEP Eur. Trans. Elect. Power Eng., Vol. 8, 1998, No. 4, pp. 235–240

[14] Staudt, V., "Fryze – Buchholz – Depenbrock: A time-domain power theory" ISNCC tutorial, Lagow, June 2008, in press

[15] Sobrink, K. H.; Pedersen, J. K.; Pedersen, K. O. H.; Schettler, F.; Bergmann, K.; Jenkins, N., "Power compensation of 24 MW wind farm using a 12-pulse voltage-source converter", International Conference on Large High Voltage Electric Systems, Vol. 4, August/September, 1998

[16] Bijlenga, B.; Gruenbaum R.; Johansson, T., "SVC Light – a powerful tool for power quality improvement". ABB Review 6/1998

[17] www.simulation-research.com

Distant learning of Pulse Width Modulation Techniques for Voltage Source Converters

Bartlomiej Kamiski*, Dariusz Sobczuk[†]

*Warsaw University of Technology, Warsaw, Poland, e-mail: *kaminskb@isep.pw.edu.pl*
[†]Warsaw University of Technology, Warsaw, Poland, e-mail: *darso@isep.pw.edu.pl*

Abstract—Paper presents distant learning of PWM techniques using java applets and remotely controlled and monitored experimental setup. The scope of presented material as well as exercises details are discussed. Moreover, the topology and components of the experimental setup are also presented. .

Index Terms—Teaching, Pulse Width Modulation, Software, Voltage Source Inverters.

I. INTRODUCTION

Today, vast majority of modern electrical drives uses some kind of modulation scheme, to allow adjustable speed operation.

Since its introduction to power converters PWM theory was a subject to extensive research over the years. As a result, there is a number of techniques used, ranging from simple carrier based to a more complex space vector. Each of these, having its advantages and disadvantages.

Therefore, it is essential to educate students as well as engineers with those methods, in order to improve their knowledge of drive operation. As a result they would better understand modern drive characteristics and performance. Although, good theory explanation is essential, students get the most from practical experience. Nothing can replace manual modification of parameters and observation of the results on a scope. On the other hand access to the hardware, which would allow to experiment with different schemes and especially inverter topologies maybe difficult.

The paper presents laboratory setup intended to remotely teach PWM techniques, via internet.

II. CONTENTS OF THE THEORY EXPLANATION WEBPAGES

Before any excercises can be done effectively, the theory must be presented first. The explanation starts with presentation of inverter topologies, then a purpose of modulation is discussed. As it is shown on fig. II, the explanation is achieved by means of a standard textbook.

A. Inverter topologies

The theory of PWM schemes covered by the course relates only to three phase, two-level and three-level NPC [3], [4] inverters supplying symmetrical RLE type loads. The textbook shows schematics of inverters, discusses their principles of operation and provides basic equations.

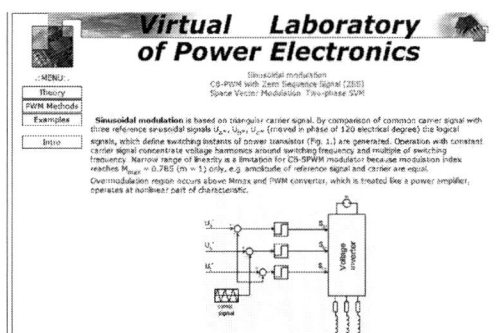

Fig. 1. Screenshot of PWM theory explanation webpage

B. Carrier based sinusoidal modulation for two level inverter

Sinusoidal modulation is based on a triangular carrier signal. By comparison of common carrier signal with three reference sinusoidal signals U_a^*, U_b^*, U_c^* (moved in phase of $2/3\pi$) the logical signals, which define switching instants of the power transistors, are generated (fig. 2).

Operation with constant carrier signal frequency, concentrate voltage harmonics around switching frequency and its multiples. Narrow range of linearity is a limitation for CB-SPWM modulator because modulation index reaches $m_{max} = 1$ ($M_{max} = \pi/4 = 0,785$) only, e.g. amplitude of reference signal and carrier are equal. Overmodulation region occurs above m_{max}. The AC side voltage is expressed as:

$$U_{LN} = m \frac{U_{DC}}{2} \qquad (1)$$

where m is a modulation index.

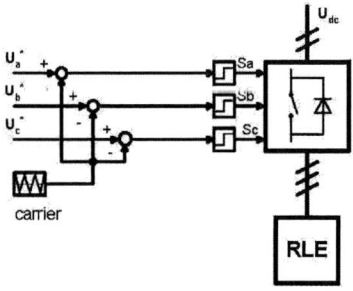

Fig. 2. Sinusoidal modulator for two level inverter

978-1-4244-1741-4/08/$25.00 ©2008 IEEE

The waveforms for carrier based method are presented on figure 3.

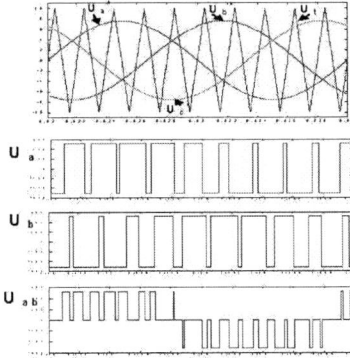

Fig. 3. Sinusoidal modulation for two level inverter

C. Space Vector Modulation (SVM) for two-level inverters

The SVM strategy, based on space vector representation (Fig.4), becomes very popular due to its simplicity. A

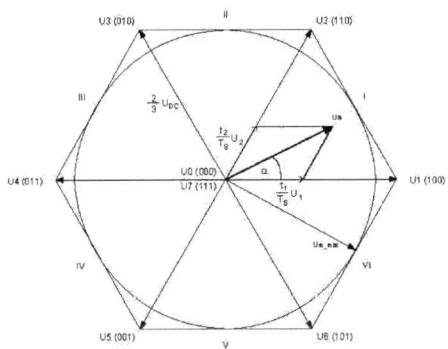

Fig. 4. Space vector modulation for two level inverter

three-phase, two-level converter provides eight possible switching states, made up of six active and two zero switching states. Active vectors divide plane into six sectors, where a reference vector U^* is obtained by switching (for proper time) two adjacent vectors.

It can be seen, that vector U^* (4) is possible to obtain by the different turn on/off sequence of U_1 and U_2, and the zero vectors. Allowable length of the U^* vector, for any angle, is equal to $U*_{max} = \frac{U_{DC}}{\sqrt{3}}$ Contrary to CB-PWM, in the SVM there is no separate modulators for each phase. Reference vector U^* is sampled with a fixed clock frequency $2f_s = 1/T_s$, and next $U^*(T_s)$ is used to solve equations, which describe switching times t_1, t_2, t_0 and t_7 (Fig. 5). Microprocessor implementation is described with the help of simple trigonometric relationship for first sector, and, recalculated for the next sectors (n)

$$t_1 = \frac{2\sqrt{3}}{\pi}MT_s sin(\pi/3 - \alpha) \qquad (2)$$

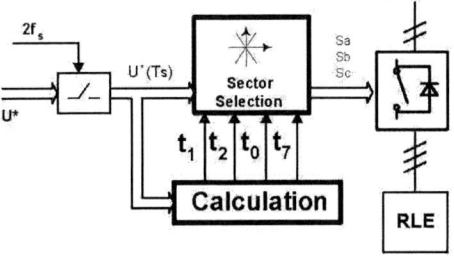

Fig. 5. Space vector modulator for two level inverter

$$t_2 = \frac{2\sqrt{3}}{\pi}MT_s sin(\alpha) \qquad (3)$$

After calculation of t_1 and t_2, the residual sampling time is reserved for zero vectors U_0, U_7 with condition $t_1 + t_2 < T_s$. The equations (2), (3) are identical for all variants of SVM. The only difference is in different placement of zero vectors $U_0(000)$ and $U_7(111)$. It gives different equations defining t_0 and t_7 for each of the method, but total duration time of zero vectors must fulfill condition:

$$t_{0,7} = T_s - t_1 - t_2 = t_0 + t_7 \qquad (4)$$

Then, the voltage between DC link midpoint and load midpoint will be equal to:

$$U_{N,0} = \frac{1}{T_s}(\frac{-U_{DC}}{2}t_0 - \frac{U_{DC}}{6}t_1 + \frac{U_{DC}}{6}t_2 + \frac{U_{DC}}{2}t_7) \quad (5)$$

D. Two carrier Level Shifted Modulation (LSM) for three level inverter

Two carrier level shifted modulation is based on two triangular carrier signals, which are shifted adjacently on the vertical axis. By comparison of the carrier signals with three reference sinusoidal signals U_a^*, U_b^* and U_c^* (shifted by $2/3\pi$) the logical signals, which define switching instants of power transistors are generated. The control algorithm of the inverter is very simple.

If the reference is

- highter than the carriers the positive state is chosen
- between both carriers zero state is chosen
- lower than the carriers the negative state is selected

as a result output voltage waveforms are obtained as presented on figure 6

E. Three-level Space Vector Modulation

A three-phase three-level inverter provides twenty seven vectors (Fig. 7): 3 zero (000, 111, 222), 12 internal (100, 211, 110, 221, 010, 121, 011, 122, 001, 112, 101, 212), 6 middle (210, 120, 021, 012, 102, 201) and 6 external (200, 220, 020, 022, 002, 202). External vectors divide plane for six sectors and each sector is divided into four triangular regions (Fig. 8) in respect to the modulation indexes as shown in Fig. 9 and Table I ,[2]. Reference vector U_{ref} is obtained by switching on (for proper time) three adjacent vectors in each region, but it can be realized by the different on/off sequence, what has influence on performance of the modulator.

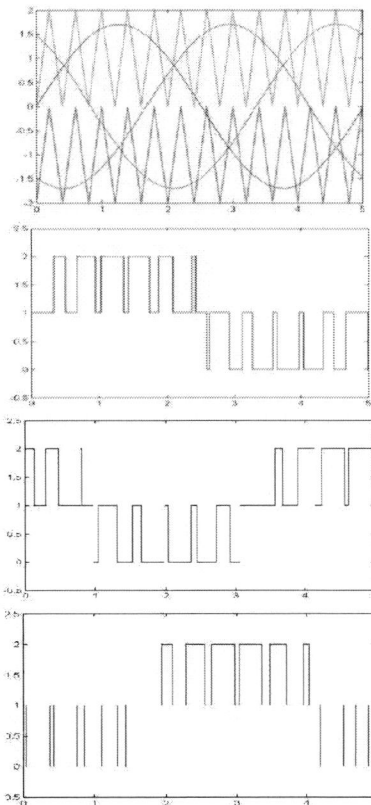

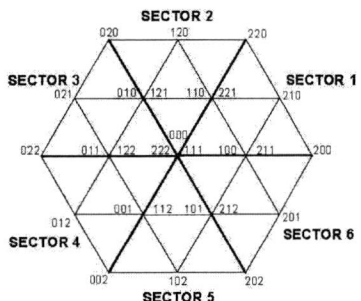

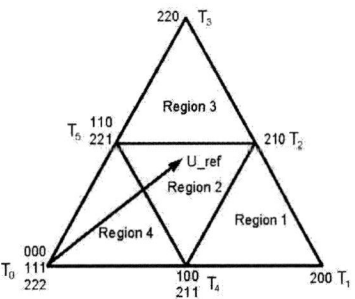

Fig. 6.   Level shifted PWM for three-level inverter

SECTOR 2
SECTOR 3
SECTOR 1
SECTOR 4
SECTOR 6
SECTOR 5

Fig. 7.   Voltage plane of three phase three-level inverter

Fig. 8.   Division of the sector into regions

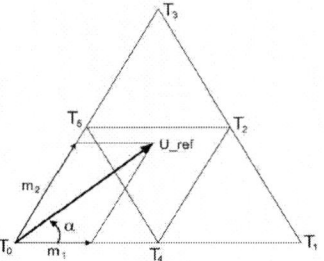

Fig. 9.   Selection of the vectors according to modulation indices

	Region	Duration of the switching state
$m_1 > 1$	1	$T_1 = m_1 - 1$ $T_2 = m_2$ $T_4 = 2 - m_1 - m_2$ $T_0 = T_3 = T_5 = 0$
$m_1 < 1, m_2 < 1, m_1 + m_2 > 1$	2	$T_2 = m_1 + m_2 - 1$ $T_4 = 1 - m_2$ $T_3 = 1 - m_1$ $T_1 = T_5 = T_0 = 0$
$m_2 > 1$	3	$T_2 = m_1$ $T_3 = m_2 - 1$ $T_5 = 2 - m_1 - m_2$ $T_0 = T_1 = T_4 = 0$
$m_1 < 1, m_2 < 1, m_1 + m_2 < 1$	4	$T_4 = m_1$ $T_5 = m_2$ $T_0 = 1 - m_1 - m_2$ $T_1 = T_2 = T_3 = 0$

where: $M = \dfrac{U\_ref}{\pi \cdot U_{dc}}$, $m_1 = M\left(\cos\alpha - \dfrac{\sin\alpha}{\sqrt{3}}\right)$, $m_2 = 2 \cdot M \dfrac{\sin\alpha}{\sqrt{3}}$

Fig. 10.   TABLE I. Selection of region with respect to the modulation indices

III. INTERACTIVE JAVA APPLETS

Understanding of the PWM theory can be greatly improved by interactive simulations. Course website contains series of interactive Java applets [5], [6], [7], allowing student evaluate different schemes by changing modulation parameters and observe resulting waveforms in the real time. Since Java applet is a mini program executed under control of a web browser, there are no access restrictions neccesary. Moreover, since Java is hardware independent, simulation part of the course can be viewed and executed on machines working under any of the web browsers and operating systems.

Figure 11 shows Java applet, which presents the idea of Space Vector Modulation of the two level inverter. The user can interact with the applet by placing mouse cursor over the voltage plane. When the mouse is left-clicked, the reference vector is plotted. At the same time switching times are calculated and presented on the charts.

Similar functionality is provided by Java applet explaining SVM for the three-level circuit (Fig. 12).

IV. EXPERIMENTAL SETUP

In order to complement obtained theoretical knowledge about PWM techniques, it is very important to learn about issues related to the modulator implementation in the real hardware. Therefore, during experimental setup design, a list of main objectives was created. The main objective was related to remote access to the hardware

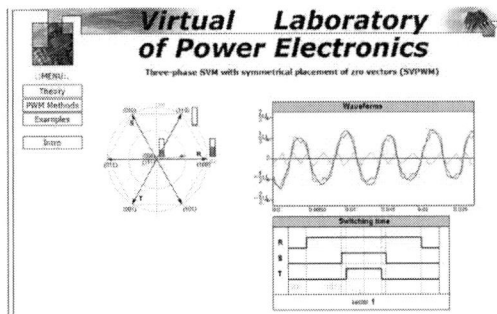

Fig. 11. Screenshot of Java applet explaining two level space vector modulation

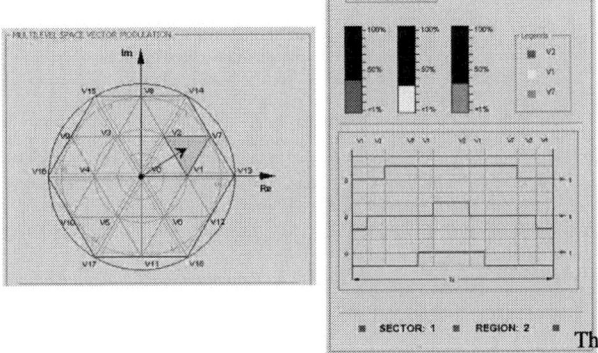

Fig. 12. Screenshot of Java applet explaining three level space vector modulation

via internet. Therefore, it was required that all changes in the power circuit or modification of the modulator would be able only, by means of software. As an example, any change of inverter topology between two and three level, or any load change was achieved by sending adequate command to the DSP controller. Additionally, all signals and quantities related to inverter performace were required to be accessible remotely. Nonetheless, important was to provide access restrictions to ensure safe operation of the system. Basically, only one user is allowed to modify system structure and parameters at a given period. The rest is able only to observe the circuit.

For selected power circuit topology, it was needed to provide easy way to change modulation schemes along with its parameters, in order to compare methods performance. Additionally, it was intended to allow access to hardware implementation specific parameters like for example blanking time delay.

A. Hardware setup

On the basis of presented requirement a hardware setup was constructed, which structure is shown on figure 13. The three phase voltage source inverter allows to change its topology from conventional two-level to the NPC three-level, by means of setting appriopriate bit in its IGBT driver board register. Similarly, the driver board can be set with different values of the blanking time delay.

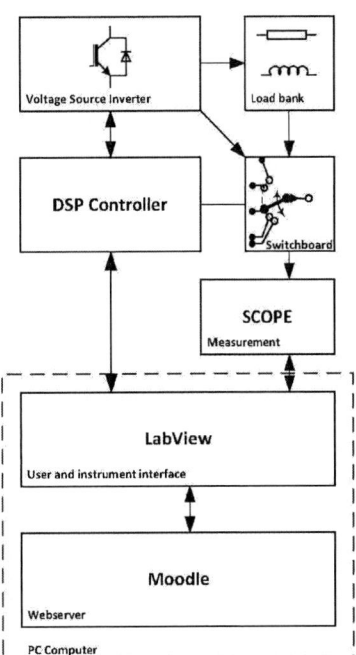

Fig. 13. Block diagram of remote PWM teaching setup

The inverter supplies fixed, three phase RLE type load. The reprogrammable modulator is implemented on high performance DSP/FPGA board [1], so that, apart from change of the modulation technique, all of the parameters can be easily modified. Additionally, the DSP controller can set the cofiguration of the signal switchboard, which collects all measurement signals from the system and directs them to the scope. The hardware setup is shown on figure 14.

Fig. 14. Hardware setup

B. User Interface

The control of the system topology, modulation schemes as well as routing and acquisition of measurement signals is realized by application created in LabView environment. The application integrates communication protocol with the DSP board as well as with the scope. As a result, easy to use visual interface to experiment is provided (fig. 15). Selected inverter topology is shown on

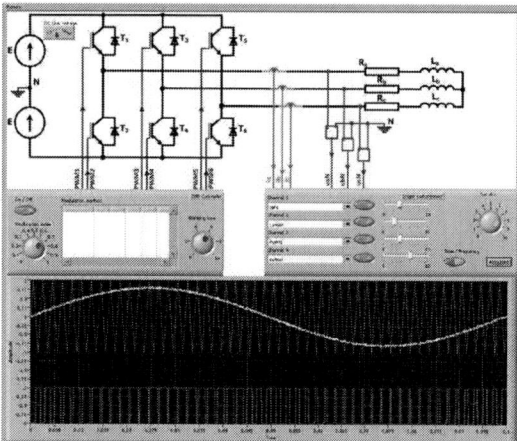

Fig. 15.   Screenshot of laboratory control panel

the top of the experiment webpage. Apart from the circuit schematic, signal path are also presented.

Modulation method along with its parameters can be set using DSP control panel on the left side, below the circuit schematic.

The control of the measurement signal routing is obtained by a panel located on the right side of the webpage. Additionaly, the panel provides setting measurement data acquisition parameters like channel amplification or a time scale. Due to limited bandwidth of the internet connection, data acquisition is not performed in the real time. After setup of the acquisition parameters, the measurement is started manually, by pressing a button labeled "'ACQUIRE"'. Then after a delay, acquired data is downloaded and presented on a chart, located at the lower side of the screen.

The master control of the theory explanation and experiments webpages is governed by Moodle Course Management system. The software provides access restrictions by means of booking system.

V. Conclusions

Proposed learning tool will allow remote teaching of various PWM techniques. Apart from detailed theory presentation, through interactive simulations to remote experiments on real circuit, the student will be able to gain knowledge and understand existing modulation techniques.

VI. Acknowledgement

This work has been performed within the Project "E-learning Distance Interactive Practical Education". The Project was supported by the European Community within Framework of Leonardo da Vinci II programme (Project

No CZ/06/B/F/PP-168022). The opinions expressed by the authors do not necessarily reflect the position of the European Community, nor does it involve any responsibility on its part.

References

[1] B. Kaminski, K. Wejrzanowski, W. Koczara, "An application of PSIM to rapid prototyping of DSP controlled power electronics converters" *Proceeding of PELINCEC conference, Warsaw, Poland 2005*

[2] W. Kolomyjski, M. Malinowski, M. P. Kazmierkowski Adaptive Space Vector Modulator for Three-Level NPC PWM Inverter-Fed Induction Motor *AMC2006, Istanbul, Turkey, pp.523-528*

[3] A. Nabae, I. Takahashi, and H. Akagi, A new neutral-point clamped PWM inverter, *IEEE Trans. on Ind. Applicat., vol. IA-17, pp518-523, Sept/Oct., 1981*

[4] J. Rodiguez, J.-S. Lai and F.-Z. Peng, Multilevel Inverters: A Survey of Topologies, Controls and Applications, *IEEE Trans. on Ind. Electronics, vol. 49, no. 4, pp. 724-738, August 2002.*

[5] M. P. Kazmierkowski, R. Bracha, M. Malinowski Web Based Teaching of Pulse Width Modulation Methods for Three-Phase Two-Level Converters *EPE-PEMC2006, Portoroz, Slovenia, pp.2134-2139*

[6] Sobczuk D. Internet Based Teaching of Pulse Width Modulation for Three-Level Inverters *in Proceedings of EUROCON 2007*

[7] Sobczuk D. E-Learning Presentation of Pulse Width Modulation Methods for Two and Three-Level Converters *in Proceedings of Electrical Drives and Power Electronics Conference, Slovakia 2007*

Modern design optimisation exploiting field simulation

Jan K. Sykulski

School of Electronics and Computer Science, University of Southampton
Southampton, SO17 1BJ, UK, e-mail: *jks@soton.ac.uk*

Abstract—**The presentation will review some of the new developments in optimisation techniques and their relevance to the design of electrical machines and drive systems. Cost effective algorithms will be explored for computationally expensive modelling processes such as encountered when field simulation techniques are employed in CAD aided design. Surrogate modelling, kriging-assisted methods, pareto-optimality and design sensitivity will be emphasised.**

Keywords—**Design, modelling, simulation, optimisation.**

I. INTRODUCTION

This review builds on previous publications by the author, in particular the overview presented at the last EPE-PEMC conference in Portoroz [1]. It was argued then that significant advances had taken place in the field of Computational Electromagnetics and demonstrated how numerical field simulation could aid the design of electrical machines and drive systems. Harnessing primarily the versatile finite element approach, the commercially available software – including general purpose packages – already offers a mature tool for performance prediction, optimisation and general design. Two particular challenges were also identified as a potential road map to successful design: modelling of the multi-physics problems and applying multi-objective optimisation. This review focuses on the latter and will relate to another review published last year addressing the emergence of new optimisation techniques for tackling computationally expensive modelling processes [2]. Both publications together contain a vast selection of relevant references and web addresses which the reader is encouraged to consult for details.

II. HIERARCHICAL DESIGN

There are many definitions of design, for our purposes the most appropriate is probably the one which describes design as *a process of searching for a device or structure which satisfies a set of requirements*. This is certainly an inverse problem where the requirements may be expressed in terms of the physical sizes, the inputs and/or the outputs, as well as possibly some special characteristics or properties.

A traditional ('trial and error') design typically consists of the following steps: (1) guess a solution, (2) build the device and measure its performance, (3) modify the device to more nearly match the requirements, (4) carry on improving until specification is met within acceptable tolerance. The modifications are usually performed on the basis of simple models, design expertise and "know-how".

A design engineer is expected to have an appreciation of how a change in a particular parameter will affect the device performance. In other words, he/she has a mental picture of how small changes in any parameter will affect each aspect of the desired performance. This is in fact a concept of sensitivity, which incidentally has a much wider application than just guiding a search, as it can be used to determine the effects of manufacturing errors (robustness of a design), it provides basis for finding performance parameters (e.g. force calculation) and it can be formulated as an optimisation method, an approach which will be pursued later in this talk.

Alternately, if no experience or models exist, random variations can be tried, the performance measured and appropriate models developed. However, the notion of 'virtual prototyping' becomes relevant here as it may be cheaper and more efficient to explore design space using computational (software) models rather than real (hardware) counterparts. Different levels of complexity of these computational models may be appropriate at different design stages, ranging from primitive equivalent circuits, through simple magnetic circuit representations, full numerical 2D and 3D formulations, to integrated field-circuit multi-physics system models. This leads us nicely into the concept of *hierarchical design*.

It is increasingly argued that the most efficient approach to design is to combine all available tools, methods and approaches – from very simple to advanced – and use them in a logical way, the simplest of which is the top-down approach. Thus we start with a very large design space and use approximate but very fast solutions (e.g. equivalent circuits, semi-empirical, design sheets, etc) taking full advantage of extensive knowledge base available and conducting extensive optimisation. The design space is then progressively reduced so that more accurate, but computationally more expensive, models can be used, such as 2D finite element, static or steady-state. Some constrained optimisation may accompany, perhaps coupling to circuit models as well. Finally, for most devices, full 3D finite-element, often transient, simulation needs to be conducted, effectively providing a fine tuning of the design. Under this hierarchical structure all models of varied complexity have their important role to play, although the most important stage in the process appears to be the middle tier, where major decisions are taken and where the geometric and topological structure of the device is being shaped up. Although the models used here may not be the most accurate, they certainly offer better accuracy than the simplistic treatment of the top tier, although there may be a certain overlap between the first and the second stage in terms of models used.

978-1-4244-1741-4/08/$25.00 ©2008 IEEE

The essential part of the whole design process is optimisation and its most distinctive aspect is that it is computationally expensive as each calculation of the objective function typically requires a full finite element field solution, and often several such solutions (to find an average torque for example), or a solution to a coupled problem. Thus methods of optimisation relevant to this type of design must be efficient in requiring as few function calls as possible.

III. SINGLE- AND MULTI-OBJECTIVE OPTIMISATION

Central to this discussion is the so called 'no free lunch' (NFL) theorem which prohibits the existence of an algorithm which would outperform all other optimisation algorithms, when averaged over all possible problems. In other words, when averaged over an infinite number of all types of problems, every algorithm performs the same. However, design engineers are only interested in a subset of problems, thus – consistent with the NFL theorem – it is possible to identify a set of algorithms which outperform others over a particular domain of interest.

Single-objective optimization problems (SOOP) are well researched and various techniques are in abundance. Performance criteria used for comparison of algorithms include: best function value found, CPU time, number of function evaluations, accuracy, success rate and stopping criteria. In multi-objective optimization problems (MOOP) the aim is to simultaneously minimize a number of different objectives. This may be achieved using *scalarizing methods*, which combine the multiple objectives of the MOOP using some function, and then use one of the methods for single-objective optimization. *Non-scalarizing methods*, on the other hand, consider each objective function individually.

It is helpful to introduce the notion of *pareto-optimality*. Rather then a single optimal solution, we arrive at a set of possible designs forming a pareto-optimal front. For a multi-objective problem it is very unlikely that all objectives can achieve optimum simultaneously (the so called 'utopia point'), thus the final design is necessarily a compromise, but rather then making *a priori* decisions by combining objectives through for example weighting functions, we leave the final choice until after *a posteriori* assessment of all possible solutions.

Recent years have witnessed significant research effort in developing new techniques for optimisation of computationally intensive problems. These are mainly relying on *surrogate modelling* with various basis functions selected to model the reality. The surrogate model which stands out, due to its solid statistical foundations, is *kriging*, which is essentially as a Gaussian process characterized by its mean and its covariance function. In kriging-assisted SOOPs, the method of selecting search points may be *two-stage*, where first the surrogate model is fitted to the observed points and then a utility function is constructed to determine the next search point, or *one-stage*, where a design vector is determined which would yield the most credible response surface. Almost all existing algorithms are two-stage; however one-stage algorithms have been successfully constructed using both kriging and other surrogate models.

A detailed review of these new methods may be found in [2], and attention is drawn to the following recent SOOP techniques: Efficient Global Optimisation (EGO),

Generalized Expected Improvement (GEI), Weighted Expected Improvement (WEI), Credibility of a Hypothesis (CH) and Minimizer Entropy (ME) criterion.

The scalarizing methods for converting MOOP into SOOP include: ε-constraint (ε-C), Weighting method (W), Weighted Metrics (WM) (including the Tchebycheff metric) method, Achievement scalarizing Function approach (AF), Lexicographic Ordering approach (LO) and Value Function method (VF). The main purpose of a scalarizing function is to combine the multiple objectives of a MOOP in such a way that the contours of the resulting function are able to capture every point on the Pareto-optimal front.

It is interesting to note that there are a large number of selection criteria for SOOP and a large number of methods for transforming a MOOP to a SOOP, thus creating a huge number of scalarizing MOOP Algorithms (made possible with kriging), only very few of which have already been investigated in terms of their efficiency as practical tools.

Finally, all the techniques mentioned address the crucial aspect of any optimisation algorithm, that is striking a careful balance between *exploitation* and *exploration*.

IV. SENSITIVITY STUDIES

Special attention needs to be paid to a class of methods based on sensitivity analysis, which offer the advantage of having computation times independent of the number of design variables, thus making them particularly useful in topology optimisation, where the parameterization enables all feasible shapes of electromagnetic devices to be explored. These algorithms are still at early stages of development and not versatile enough to be considered as competitive against surrogate modelling, but their performance is very impressive.

The sensitivity formula needs to be derived for both the *primary system* and the *adjoint system*. By exploring the analogy between the two formulations the geometric and material properties of the adjoint system are found to be the same as those of the primary system and sources may be recognised as electric current or permanent magnet. Thus the adjoint system – despite having been introduced as a mathematical derivation – is physically well based. An added benefit is that if the objective function is energy related the system becomes *self-adjoint*.

V. CONCLUDING REMARKS

The talk will focus on new developments in methods of optimisation relevant to the design of electrical machines and drive systems. Many efficient new approaches have recently been proposed, in particular for computationally intensive problems, and optimisation continues to be a very active area of research. Most existing design systems are based on algorithms developed in the 80s and 90s, while the new techniques offer exciting new opportunities.

REFERENCES

[1] J. K. Sykulski, "Field Simulation as an Aid to Machine Design: the State of the Art," *Proceedings of the 12th International Power Electronics and Motion Control Conference EPE-PEMC 2006, Portoroz, Slovenia,* IEEE Catalog Number: 06EX1282C, Library of Congress: 2005938592, pp. 1937–1942, September 2006.

[2] J. K. Sykulski, "New trends in optimization in electromagnetics," *Przeglad Elektrotechniczny,* ISSN 0033-2097, vol. 83, pp. 13–18, June 2007.

Transmission-Line Modelling of Wave Propagation Effects in Machine Windings

Herbert De Gersem[*], Olaf Henze[†], Thomas Weiland[†], Andreas Binder[‡]

[*]Katholieke Universiteit Leuven, Faculteit Wetenschappen, Kortrijk, Belgium,
e-mail: *Herbert.DeGersem@kuleuven-kortrijk.be*

[†]Technische Universität Darmstadt, Institut für Theorie Elektromagnetischer Felder, Darmstadt, Germany,
e-mail: *Henze/Weiland@temf.tu-darmstadt.de*

[‡]Technische Universität Darmstadt, Institut für Elektrische Energiewandlung, Darmstadt, Germany,
e-mail: *ABinder@ew.tu-darmstadt.de*

Abstract— In this paper, the parameters of a transmission-line model for electrical machine windings are calculated on the basis of a finite-element model of the winding cross-section. Three finite-element formulations are compared: a low-frequency approximation, a high-frequency approximation and an eddy-current formulation. Combining wires into windings and the connection of parts in series or parallel is treated on the level of the transmission-line model. The application to a particular winding configuration indicates the validity ranges of the low- and high-frequency approximations in comparison with the eddy-current formulation. The propagating modes are examined.

Keywords— finite element methods, electrical machines, wave propagation, transmission line models, windings.

I. INTRODUCTION

Inverter-fed electrical drives may suffer from overvoltage effects when a relatively long cable is applied between the power-electronic converter and the electrical machine [1], [2]. Due to a mismatch of the winding input impedance and the cable characteristic impedance, reflections occur, which cause a substantially overvoltage at the motor terminals. These effects have been studied both experimentally as on the basis on numerical models [3]. Commonly, transmission-line models (TLMs) are applied, both for the cable [4], [5] as for the machine windings [6]. The lumped parameters inserted in the TLMs are determined on the basis of analytical formulae or numerical simulations. The accuracy of these lumped parameters is of primordial importance in order to guarantee relevant results, especially if a broad frequency range is needed. In this paper, three different finite-element formulations are described and compared. It is shown that a more accurate, but also more expensive, eddy-current formulation closes the gap between a low-frequency approximation and a high-frequency approximation. The influence of the resolution of the finite-element mesh is studied.

II. TLM MODEL FOR A WIRE CONFIGURATION

A. Multi-conductor transmission-line model

The wire configuration is modelled as a multi-conductor transmission-line model (TLM) (Fig. 1). The TLM model consists of a *resistance matrix* **R**, an *inductance matrix* **L**, a *capacitance matrix* **C** and a *conductance matrix* **G**, which are all expressed per unit length,

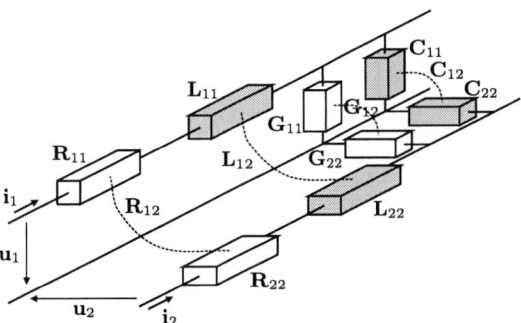

Fig. 1. Multi-conductor transmission-line model for the case of 2 conductors and 1 return path. The resistance matrix **R**, the inductance matrix **L**, the conductance matrix **G** and the capacitance matrix **C** are indicated per unit length and, therefore, should be multiplied by dx to obtain a valid circuit representation. The off-diagonal elements of the matrices are indicatively represented by dashed connection lines.

and therefore carry the units Ω/m, H/m, F/m and S/m, respectively. These matrices are combined in the *impedance matrix* $\mathbf{Z} = \mathbf{R} + \jmath\omega\mathbf{L}$ and the *admittance matrix* $\mathbf{Y} = \mathbf{G} = \jmath\omega\mathbf{C}$ where ω denotes the angular frequency. The outer shield or one of the wires is considered as a reference conductor. The voltages between the other conductors and the reference conductor are collected in the vector $\mathbf{u}(x)$, whereas the currents through all conductors except the reference conductor are gathered in the vector $\mathbf{i}(x)$. The voltage and currents depend on the unidirectional spatial coordinate x measured along the transmission line.

The TLM model is described by the *telegraph equations* in matrix form, i.e.,

$$\frac{d\mathbf{u}(x)}{dx} = -(\mathbf{R} + \jmath\omega\mathbf{L})\,\mathbf{i}(x)\,; \tag{1}$$

$$\frac{d\mathbf{i}(x)}{dx} = -(\mathbf{G} + \jmath\omega\mathbf{C})\,\mathbf{u}(x)\,. \tag{2}$$

The simultaneous solution of (1) and (2) is simplified by a so-called *modal decomposition*. The voltages, and currents are transformed into the *modal voltages* $\mathbf{u}_\mathrm{m}$ and the *modal currents* $\mathbf{i}_\mathrm{m}$ by the orthonormal matrices $\mathbf{Q}_\mathrm{u}$ and $\mathbf{Q}_\mathrm{i}$ such that the *modal impedance matrix* $\mathbf{Z}_\mathrm{m}$ and the *modal admittance matrix* $\mathbf{Y}_\mathrm{m}$ become diagonal:

$$\mathbf{u}_\mathrm{m} = \mathbf{Q}_\mathrm{u}^{-1}\mathbf{u}\,; \tag{3}$$

978-1-4244-1741-4/08/$25.00 ©2008 IEEE

$$\mathbf{i}_{\mathrm{m}} = \mathbf{Q}_{\mathrm{i}}^{-1}\mathbf{i} \; ; \tag{4}$$

$$\mathbf{Z}_{\mathrm{m}} = \mathbf{Q}_{\mathrm{u}}^{-1}\mathbf{Z}\mathbf{Q}_{\mathrm{i}} \; ; \tag{5}$$

$$\mathbf{Y}_{\mathrm{m}} = \mathbf{Q}_{\mathrm{i}}^{-1}\mathbf{Y}\mathbf{Q}_{\mathrm{u}} \; . \tag{6}$$

This decomposed is calculated by solving the eigenvalue problem

$$(\mathbf{YZ})\,\mathbf{Q}_{\mathrm{i}} = \mathbf{Q}_{\mathrm{i}}\,(\mathbf{Y}_{\mathrm{m}}\mathbf{Z}_{\mathrm{m}}) \tag{7}$$

and putting $\mathbf{Q}_{\mathrm{u}} = \mathbf{Z}\mathbf{Q}_{\mathrm{i}}$. In the transformed telegraph equations,

$$\frac{d\mathbf{u}_{\mathrm{m}}(x)}{dx} = -\mathbf{Z}_{\mathrm{m}}\mathbf{i}_{\mathrm{m}}(x) \; ; \tag{8}$$

$$\frac{d\mathbf{i}_{\mathrm{m}}(x)}{dx} = -\mathbf{Y}_{\mathrm{m}}\mathbf{u}_{\mathrm{m}}(x) \; , \tag{9}$$

the modes are decoupled. Hence, the differential equations can be solved for each mode separately. For the k-th mode, the modal voltage and modal current are

$$\mathbf{u}_{\mathrm{m},k} = \mathbf{u}_{\mathrm{m},k,0}^{+}e^{-\jmath\beta_k x} + \mathbf{u}_{\mathrm{m},k,0}^{-}e^{+\jmath\beta_k x} \; ; \tag{10}$$

$$\mathbf{i}_{\mathrm{m},k} = \frac{\mathbf{u}_{\mathrm{m},k,0}^{+}}{Z_{\mathrm{ch},k}}e^{-\jmath\beta_k x} - \frac{\mathbf{u}_{\mathrm{m},k,0}^{-}}{Z_{\mathrm{ch},k}}e^{+\jmath\beta_k x} \; , \tag{11}$$

where $\beta_k = \sqrt{\mathbf{Z}_{\mathrm{m},k,k}\mathbf{Y}_{\mathrm{m},k,k}}$ is the *propagation constant* for mode k and

$$Z_{\mathrm{ch},k} = \sqrt{\frac{\mathbf{Z}_{\mathrm{m},k,k}}{\mathbf{Y}_{\mathrm{m},k,k}}} \tag{12}$$

is the *characteristic impedance* associated with mode k. The k-th column of $\mathbf{Q}_{\mathrm{u}}$ and the k-th column of $\mathbf{Q}_{\mathrm{i}}$ correspond to a certain pattern for the voltages and the currents, called a *propagation mode*. Each mode has a particular propagation constant β_k, propagation velocity $v_k = \omega/\beta_k$ and characteristic impedance $Z_{\mathrm{ch},k}$. In symmetric three-phase systems, the three occurring modes are called the *direct mode*, *inverse mode* and the *common mode*. In this paper, at first, we consider the winding cross-section as a system of a large number of conductors. Besides a common-mode pattern, a large variety of modes arises.

B. Propagation matrices

The propagation of the k-th mode through the multi-conductor configuration over a length ℓ is described by

$$
\begin{bmatrix} \mathbf{u}_{\mathrm{m},k}(\ell) \\ \mathbf{i}_{\mathrm{m},k}(\ell) \end{bmatrix}
= \underbrace{\begin{bmatrix} \cosh(\beta_{\mathrm{m},k}\ell) & -Z_{\mathrm{ch},k}\sinh(\beta_{\mathrm{m},k}\ell) \\ -Y_{\mathrm{ch},k}\sinh(\beta_{\mathrm{m},k}\ell) & \cosh(\beta_{\mathrm{m},k}\ell) \end{bmatrix}}_{\mathbf{A}_{\mathrm{m},k}(\ell)}
\begin{bmatrix} u_{\mathrm{m},k}(0) \\ i_{\mathrm{m},k}(0) \end{bmatrix} \tag{13}
$$

where $Y_{\mathrm{ch},k} = 1/Z_{\mathrm{ch},k}$. The *modal propagation matrices* $\mathbf{A}_{\mathrm{m},k}(\ell)$ are concatenated in the 2-by-2 block matrix $\mathbf{A}_{\mathrm{m}}(\ell)$. Then, the propagation of the untransformed voltages and currents can be calculated from

$$
\begin{bmatrix} \mathbf{u}(\ell) \\ \mathbf{i}(\ell) \end{bmatrix}
= \underbrace{\mathbf{Q}_{\mathrm{ui}}\mathbf{A}_{\mathrm{m}}(\ell)\,\mathbf{Q}_{\mathrm{ui}}^{-1}}_{\mathbf{A}(\ell)}
\begin{bmatrix} \mathbf{u}(0) \\ \mathbf{i}(0) \end{bmatrix} \tag{14}
$$

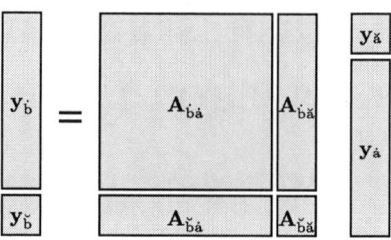

Fig. 2. Schematic representation of the partitioning of the propagation formula for a multi-conductor system.

with the transformation matrix $\mathbf{Q}_{\mathrm{ui}} = \mathrm{diag}\{\mathbf{Q}_{\mathrm{u}},\mathbf{Q}_{\mathrm{i}}\}$. A shorthand for (14) which will be used below, is $\mathbf{y}_{\mathrm{b}} = \mathbf{A}\mathbf{y}_{\mathrm{a}}$.

An alternative formulation expresses the input and output currents in function of the input and output voltages:

$$
\begin{bmatrix} i_{\mathrm{m},k}(0) \\ i_{\mathrm{m},k}(\ell) \end{bmatrix}
$$
$$
= \underbrace{\frac{Y_{\mathrm{ch},k}}{\sinh(\beta_{\mathrm{m},k}\ell)} \begin{bmatrix} \cosh(\beta_{\mathrm{m},k}\ell) & -I \\ I & -\cosh(\beta_{\mathrm{m},k}\ell) \end{bmatrix}}_{\mathbf{B}_{\mathrm{m},k}(\ell)}
$$
$$
\begin{bmatrix} u_{\mathrm{m},k}(0) \\ u_{\mathrm{m},k}(\ell) \end{bmatrix} . \tag{15}
$$

The *modal TLM admittance matrix* $\mathbf{B}_{\mathrm{m}}$ collects all 2-by-2 matrices $\mathbf{B}_{\mathrm{m},k}$. The *TLM admittance matrix* $\mathbf{B}$ follows from

$$\mathbf{B} = \mathbf{Q}_{\mathrm{ii}}\mathbf{B}_{\mathrm{m}}\mathbf{Q}_{\mathrm{ii}}^{-1} \tag{16}$$

where $\mathbf{Q}_{\mathrm{ii}} = \mathrm{diag}\{\mathbf{Q}_{\mathrm{i}},\mathbf{Q}_{\mathrm{i}}\}$.

C. Connections of several multi-conductor TLM models

Machine windings are complicated arrangements of different winding parts. The TLM-model of an entire, heterogeneous machine winding arises by connecting the TLM-models of individual winding parts.

1) Winding connection: Until now, a winding part embedded in a slot is considered as a multi-conductor system with the associated TLM model $\mathbf{y}_{\mathrm{b}} = \mathbf{A}\mathbf{y}_{\mathrm{a}}$. The series connection of all conductors introduces additional constraint on the currents and the voltages. When the conductors are connected in the same order as they are listed in the vectors $\mathbf{u}$ and $\mathbf{i}$, the voltage and the current at the beginning of the $(n+1)$-th conductor is equal to the voltage and the current at the end of the n-th conductor, i.e.,

$$
\begin{bmatrix} \mathbf{u}_{\mathrm{a},n+1} \\ \mathbf{i}_{\mathrm{a},n+1} \end{bmatrix}
= \begin{bmatrix} \mathbf{u}_{\mathrm{b},n} \\ \mathbf{i}_{\mathrm{b},n} \end{bmatrix} \tag{17}
$$

or, in short, $\mathbf{y}_{\mathrm{a},n+1} = \mathbf{y}_{\mathrm{b},n}$. The only remaining degrees of freedom are $\mathbf{y}_{\mathrm{a},1}$ and $\mathbf{y}_{\mathrm{b},n_{\mathrm{wire}}}$ where n_{wire} stands for the number of wires. The propagation formula $\mathbf{y}_{\mathrm{b}} = \mathbf{A}\mathbf{y}_{\mathrm{a}}$ can be partitioned accordingly (Fig. 2)

$$
\begin{bmatrix} \mathbf{y}_{\breve{\mathrm{b}}} \\ \mathbf{y}_{\check{\mathrm{b}}} \end{bmatrix}
= \begin{bmatrix} \mathbf{A}_{\breve{\mathrm{b}}\breve{\mathrm{a}}} & \mathbf{A}_{\breve{\mathrm{b}}\dot{\mathrm{a}}} \\ \mathbf{A}_{\check{\mathrm{b}}\breve{\mathrm{a}}} & \mathbf{A}_{\check{\mathrm{b}}\dot{\mathrm{a}}} \end{bmatrix}
\begin{bmatrix} \mathbf{y}_{\breve{\mathrm{a}}} \\ \mathbf{y}_{\dot{\mathrm{a}}} \end{bmatrix} . \tag{18}
$$

Here, the subscripts $\breve{a}$ and $\breve{b}$ indicate the first and last entry, respectively. The subscripts $\dot{a}$ and $\dot{b}$ indicate the

2386

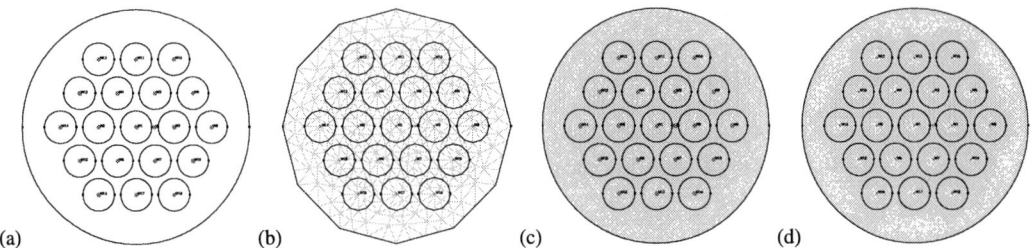

Fig. 3.   (a) Geometry and (b-d) FE meshes of the cross-section of a single slot.

remaining entries. The connection constraint reads $\mathbf{y}_{\mathrm{\acute{a}}} = \mathbf{y}_{\mathrm{\grave{b}}}$. The propagation matrix for the winding with all conductors are connected in series, is found by some algebraic manipulation:

$$\mathbf{y}_{\mathrm{\breve{b}}} = \left(\mathbf{A}_{\mathrm{\breve{b}\breve{a}}} + \mathbf{A}_{\mathrm{\breve{b}\acute{a}}}\left(\mathbf{I} - \mathbf{A}_{\mathrm{\grave{b}\acute{a}}}\right)^{-1}\mathbf{A}_{\mathrm{\grave{b}\breve{a}}}\right)\mathbf{y}_{\mathrm{\breve{a}}} . \quad (19)$$

2) Series connections: The propagation matrix of the series connection of two TLM-models with propagation matrices $\mathbf{A}_1$ and $\mathbf{A}_2$ is $\mathbf{A}_{\mathrm{series}} = \mathbf{A}_1\mathbf{A}_2$.

3) Parallel connection: When two TLM models described by the TLM admittance matrices $\mathbf{B}_1$ and $\mathbf{B}_2$ are connected in parallel, the resulting TLM model is represented by $\mathbf{B} = \mathbf{B}_1 + \mathbf{B}_2$.

III. FINITE-ELEMENT MODELS

The cross-sectional geometry of a machine winding is commonly too complicated to allow for an analytical computation of the impedance matrix $\mathbf{Z}$ and the admittance matrix $\mathbf{Y}$. Here, the parameters of the transmission-line model are calculated using a 2D finite-element (FE) model. The model consists of a cross-section of a single slot in which the contours of the individual wires are resolved by the FE mesh (Fig. 3). The entire slot cross-section is denoted by Ω_{slot}. The wire cross-sections and their boundaries are denoted by Ω_q and Γ_q, respectively. The index $q = 1, \ldots, n_{\mathrm{wire}}$ counts the wires. The slot boundary is denoted by Γ_0. The computational domain Ω_{slot} is triangulated (Fig. 3). The set of nodes of the FE mesh that are strictly inside insulation, is denoted by $\mathcal{S}_{\mathrm{c}}$. The set of nodes of the FE mesh that are not on the boundary Γ_0 of the computational domain is denoted by $\mathcal{S}_{\mathrm{e}}$. The set of wires is denoted by $\mathcal{S}_{\mathrm{d}}$. The same subscripts are used to partition matrices and vectors. The indices i and j are typically used for addressing members of $\mathcal{S}_{\mathrm{c}}$ or $\mathcal{S}_{\mathrm{e}}$, whereas the indices p and q are used for counting wires, hence, addressing elements of $\mathcal{S}_{\mathrm{d}}$. During the solution of the problem, adaptive mesh refinement is applied to enhance the accuracy of the results, especially for eddy-current computations [7].

IV. ADMITTANCE MATRIX

The admittance matrix $\mathbf{Y}$ represents the relation between a voltage distribution $\mathbf{u}$ at the wires and the currents $\mathbf{i}$ between the wires. The admittance matrix is computed from the electroquasistatic approximation of the Maxwell equations in terms of the electric scalar potential φ, i.e.,

$$-\nabla \cdot \left((\sigma + \jmath\omega\varepsilon)\nabla\varphi\right) = 0 \quad (20)$$

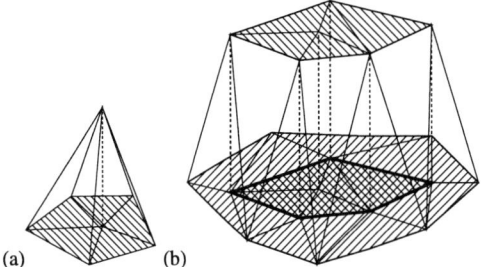

Fig. 4.   (a) linear FE shape function $N_{\mathrm{c},j}$ and (b) agglomerated shape function $N_{\mathrm{d},q}$ (the supports are indicated by hatching).

where σ is the conductivity, ε is the permittivity and ω is the angular frequency. Below, the material parameters will be represented by the single, complex-valued conductivity $\underline{\sigma} = \sigma + \jmath\omega\varepsilon$. A reference potential of 0 V is assigned to the boundary Γ_0 of the slot. Hence, the stator laminations and housing are considered at a return path for the current with a zero impedance. This assumption is troublesome if a iron yoke is laminated. It is assumed that the cross-sections of the wires are equipotential surfaces.

The problem is discretised on the FE mesh. Standard, linear FE shape functions $N_{\mathrm{c},j}(x,y)$, $j \in \mathcal{S}_{\mathrm{c}}$ are associated with the nodes which are strictly inside the insulation material (Fig. 4). An *agglomerated* shape function $N_{\mathrm{d},q}(x,y)$, $q \in \mathcal{S}_{\mathrm{d}}$ is constructed for each wire cross-section Ω_q (Fig. 4). The functions $N_{\mathrm{d},q}$ enforce all mesh nodes at Ω_q to carry the same potential [8]. The electric scalar potential is written as a linear combination of both shape functions, i.e.,

$$\varphi = \sum_{j\in\mathcal{S}_{\mathrm{c}}}\varphi_{\mathrm{c},j}N_{\mathrm{c},j}(x,y) + \sum_{q\in\mathcal{S}_{\mathrm{d}}}\varphi_{\mathrm{d},q}N_{\mathrm{d},q}(x,y) . \quad (21)$$

A system of algebraic shape functions is obtained by weighing (20) with the same FE shape functions:

$$\begin{bmatrix} \mathbf{K}_{\mathrm{cc}}^{(\underline{\sigma})} & \mathbf{K}_{\mathrm{cd}}^{(\underline{\sigma})} \\ \mathbf{K}_{\mathrm{dc}}^{(\underline{\sigma})} & \mathbf{K}_{\mathrm{dd}}^{(\underline{\sigma})} \end{bmatrix}\begin{bmatrix} \varphi_{\mathrm{c}} \\ \varphi_{\mathrm{d}} \end{bmatrix} + \begin{bmatrix} 0 \\ \mathbf{g}_{\mathrm{d}} \end{bmatrix} = \begin{bmatrix} 0 \\ 0 \end{bmatrix} \quad (22)$$

or in short

$$\mathbf{K}^{(\underline{\sigma})}\varphi + \mathbf{g} = \left(\mathbf{K}^{(\sigma)} + \mathbf{K}^{(\varepsilon)}\right)\varphi + \mathbf{g} = 0 . \quad (23)$$

All blocks in the system matrix of (22) are assembled as

$$\mathbf{K}_{\dagger\ddagger,i,j}^{(\alpha)} = \int_\Omega \alpha\nabla N_{\dagger,i}\cdot\nabla N_{\ddagger,j}\,\mathrm{d}\Omega . \quad (24)$$

Here the superscript (α) indicates the material property incorporated in the diffusion matrix and the subscript †‡ can form any combination of the subscripts c and d. The matrix $\mathbf{K}^{(\varepsilon)}$ is positive definite. The matrix $\mathbf{K}^{(\sigma)}$ is positive semi-definite due to the possible presence of nonconductive parts in the model. The additional term reads

$$\mathbf{g}_{\mathrm{d},p} = -\int_{\partial\Omega} J_n N_{\mathrm{d},p}\, \mathrm{d}\Gamma \ . \tag{25}$$

Here, J_n is component of the current density which leaves the computational domain $\partial\Omega$ through its boundary $\partial\Omega$. Hence, the boundary-integral term $\mathbf{g}_{\mathrm{d},p}$ corresponds to the current $-I_p$ entering the computational domain through $\partial\Omega_p$. The electric scalar potentials φ_{d} are equal to voltages applied to the individual wires. Hence, the elimination of the degrees of freedom φ_{c} associated with nodes which are strictly inside the computational domain leads to the *Schur complement system*

$$\left(\mathbf{K}_{\mathrm{dd}}^{(\sigma)} - \mathbf{K}_{\mathrm{dc}}^{(\sigma)}\left(\mathbf{K}_{\mathrm{cc}}^{(\sigma)}\right)^{-1}\mathbf{K}_{\mathrm{cd}}^{(\sigma)}\right)\varphi_{\mathrm{d}} = -\mathbf{g}_{\mathrm{d}} \ . \tag{26}$$

Hence, the admittance matrix equals

$$\mathbf{Y} = \mathbf{K}_{\mathrm{dd}}^{(\sigma)} - \mathbf{K}_{\mathrm{dc}}^{(\sigma)}\left(\mathbf{K}_{\mathrm{cc}}^{(\sigma)}\right)^{-1}\mathbf{K}_{\mathrm{cd}}^{(\sigma)} \ . \tag{27}$$

The conductance matrix $\mathbf{G}$ and the capacitance matrix $\mathbf{C}$ can be extracted using the relation $\mathbf{Y} = \mathbf{G} + \jmath\omega\mathbf{C}$. $\mathbf{G}$ and $\mathbf{C}$ inherit the properties of $\mathbf{K}^{(\sigma)}$ and $\mathbf{K}^{(\varepsilon)}$. The symmetry of $\mathbf{G}$ and $\mathbf{C}$ is guaranteed by construction. The eigenvalues of $\mathbf{K}^{(\sigma)}$ have a positive real part and a strictly positive imaginary part, thanks to the positive (semi-)definiteness of $\mathbf{K}^{(\sigma)}$ and $\mathbf{K}^{(\varepsilon)}$, respectively. This property is maintained in the Schur complement $\mathbf{Y}$. Hence, $\mathbf{Y}$ is decomposable into a symmetric semi-positive definite conductance matrix $\mathbf{G}$ and a positive definite capacitance matrix $\mathbf{C}$.

In many cases, the conductivity of the insulation material can be neglected. Then, (27) directly gives an expression for the capacitance matrix:

$$\mathbf{C} = \mathbf{K}_{\mathrm{dd}}^{(\varepsilon)} - \mathbf{K}_{\mathrm{dc}}^{(\varepsilon)}\left(\mathbf{K}_{\mathrm{cc}}^{(\varepsilon)}\right)^{-1}\mathbf{K}_{\mathrm{cd}}^{(\varepsilon)} \ . \tag{28}$$

V. IMPEDANCE MATRIX

Several approaches for calculating the impedance matrix can be distinguished. They differ in the assumption granted on the skin effect inside the wires. In DC operation, a homogeneous current distribution is assumed. In high-frequency operation, the current is concentrated in an infinitesimally small layer at the wire boundary. In the mid-frequency range, a finite redistribution of the current over the wire cross-section is considered. In all approaches proposed here, the outside boundary of the wire configuration is assumed to form a zero impedance for the returning current. This assumption is valid for a shielded cable but may become doubtful for an unshielded cable or for windings embedded in a lamination. For the low- and high-frequency cases, the resistance matrix $\mathbf{R}$ and the inductance matrix $\mathbf{L}$ can be calculated separately. The impedance matrix $\mathbf{Z}$ then follows by combining them, i.e., $\mathbf{Z} = \mathbf{R} + \jmath\omega\mathbf{L}$.

A. Low-frequency approximation

In low-frequency operation, the current is assumed to be homogeneously distributed over the wire cross-section. The resistance matrix is calculated assuming DC conditions, i.e.,

$$\mathbf{R}_{p,p} = \frac{1}{\sigma S_p} \tag{29}$$

where σ is here the conductivity of the wire material and S_p is the cross-section of wire p. Because all TLM parameters are here computed per unit length, the length of the wire is not present in (29).

The inductance matrix is computed as a reduction of the magnetostatic formulation of the Maxwell equations:

$$\nabla \times (\nu\nabla \times \mathbf{A}) = \mathbf{J}_{\mathrm{s}} \tag{30}$$

where $\mathbf{A}$ is the magnetic vector potential, $\nu = 1/\mu$ is the reluctivity, μ is the permeability and $\mathbf{J}_{\mathrm{s}}$ is the applied current density. In this model, only currents perpendicular to the slot cross-section and fluxes flowing in the cross-section are relevant. Then, the applied current density and the magnetic vector potential have only a z-component, i.e., $\mathbf{J}_{=}(0,0,J_{\mathrm{s},z})$ and $\mathbf{A} = (0,0,A_z)$ and the vectorial partial differential equation (30) is reduced to a scalar one:

$$-\frac{\partial}{\partial x}\left(\nu\frac{\partial A_z}{\partial x}\right) - \frac{\partial}{\partial y}\left(\nu\frac{\partial A_z}{\partial y}\right) = J_{\mathrm{s},z} \ . \tag{31}$$

The slot boundary serves as the reference conductor and is equipped with a homogeneous Dirichlet boundary condition, i.e., $A_z = 0$ at Γ_0. A piecewise constant function $\mathbf{t}_q = (0,0,t_{z,q})$ where $t_{z,q} = 1$ in Ω_q and $t_{z,q} = 0$ outside Ω_q, is assigned to each wire cross-section Ω_q. The currents $\mathbf{i}$ applied to the wires are related to the source current density by

$$J_{\mathrm{s},z} = \sum_{q=1}^{n_{\mathrm{wire}}} \frac{t_{z,q}}{S_q}\mathbf{i}_q \ . \tag{32}$$

The flux per unit length is directly related to the z-component of the magnetic flux density, i.e., the flux per unit length flowing between two arbitrary points (x_1,y_1) and (x_2,y_2) in the 2D cross-section equals $\phi = A_z(x_2,y_2) - A_z(x_1,y_1)$. Hence, the flux per unit length coupled with a loop formed by a single wire and the slot boundary equals the magnetic vector potential, averaged over the wire cross-section:

$$\phi_p = \frac{1}{S_p}\int_{\Omega_p} A_z\, \mathrm{d}\Omega \ ; \tag{33}$$

$$= \int_{\Omega} A_z t_{z,p}\, \mathrm{d}\Omega \ . \tag{34}$$

The formulation is discretised on the FE mesh. The discretisation of A_z reads

$$A_z = \sum_{j\in\mathcal{S}_{\mathrm{e}}}^{n_{\mathrm{c}}} \mathbf{a}_{\mathrm{e},j} N_{\mathrm{e},j}(x,y) \ . \tag{35}$$

The algebraic system of equations is obtained by applying the Ritz-Galerkin weighting approach as before, i.e.,

$$\mathbf{K}_{\mathrm{ee}}^{(\nu)}\mathbf{a}_{\mathrm{e}} = \mathbf{F}_{\mathrm{ed}}\mathbf{i} \tag{36}$$

where $\mathbf{K}_{ee}^{(\nu)}$ is constructed as in (24). The components of the righthandside term read

$$\mathbf{F}_{ed,i,q} = \int_\Omega N_{e,i} t_{z,q} \, d\Omega . \quad (37)$$

From (34), it follows that the coupled fluxes are

$$\phi = \mathbf{F}_{ed}^T \mathbf{a}_e . \quad (38)$$

Hence, the FE model gives a relation between the coupled fluxes per unit length and the applied currents, i.e.,

$$\phi = \underbrace{\mathbf{F}_{ed}^T \left(\mathbf{K}_{ee}^{(\nu)}\right)^{-1} \mathbf{F}_{ed}}_{\mathbf{L}} \mathbf{i} , \quad (39)$$

from which the inductance matrix $\mathbf{L}$ follows directly. The symmetry and positive definiteness of $\mathbf{L}$ is given by construction.

In low-frequency operation, the magnetic flux lines penetrate the wires (Fig. 5b). It is obvious that this approximation becomes doubtful for higher frequencies.

B. High-frequency approximation

For very high frequencies, it can be assumed that all magnetic flux is expelled out of the wires. When calculating the resistance matrix $\mathbf{R}$, the skin effect is taken into account, i.e.,

$$\mathbf{R}_{p,p} = \frac{1}{\sigma S_p} \chi \quad (40)$$

where χ is a correction factor dependent on the shape of the wire cross-section, the material parameters and the frequency. For the case of a circular cross-section,

$$\chi = \frac{\xi r}{2} \frac{I_0\left(\xi r\right)}{I_1\left(\xi r\right)} \quad (41)$$

where $\xi = \sqrt{\jmath \omega \mu \sigma}$ is the Helmholtz constant, r is the radius and I_0 and I_1 are modified Bessel functions of the 0th and 1st order respectively. The correction factor for a rectangular cross-section is given in [6].

The inductance matrix is also computed as a reduction of the magnetostatic formulation (30). The fact that all flux is expelled from the wires and stator laminations is enforced by applying floating-potential boundary conditions to all wires. Again, the slot boundary serves as a reference, which is enforced by the boundary condition $A_z(\Gamma_0) = 0$. The flux per unit length coupled with wire Ω_p equals the constant value $A_{z,p}$ for the magnetic vector potential at Γ_p. The boundaries of the wires are flux walls, which is introduced in the model by using the agglomerated shape functions $N_{d,,q}$ for discretising A_z at all wire boundaries. The discretisation of A_z reads

$$A_z = \sum_{j \in \mathcal{S}_c} \mathbf{a}_{c,j} N_{c,j}(x,y) + \sum_{q \in \mathcal{S}_d} \mathbf{a}_{d,q} N_{d,q}(x,y) . \quad (42)$$

The algebraic system of equations is obtained by applying the Ritz-Galerkin weighting approach as before, i.e.,

$$\begin{bmatrix} \mathbf{K}_{cc}^{(\nu)} & \mathbf{K}_{cd}^{(\nu)} \\ \mathbf{K}_{dc}^{(\nu)} & \mathbf{K}_{dd}^{(\nu)} \end{bmatrix} \begin{bmatrix} \mathbf{a}_c \\ \mathbf{a}_d \end{bmatrix} = \begin{bmatrix} 0 \\ \mathbf{i}_d \end{bmatrix} \quad (43)$$

where all $\mathbf{K}_{pq}^{(\nu)}$ are constructed as in (24). The components of the righthandside term read

$$\mathbf{i}_{d,p} = \int_{\partial\Omega} J_{s,z} N_{d,p} \, d\Gamma \quad (44)$$

which equals the current $\mathbf{i}_p$ through wire Ω_p. The elimination of $\mathbf{a}_c$ from (43) leads to the *Schur complement* system

$$\left(\mathbf{K}_{dd}^{(\nu)} - \mathbf{K}_{dc}^{(\nu)} \left(\mathbf{K}_{cc}^{(\nu)}\right)^{-1} \mathbf{K}_{cd}^{(\nu)}\right) \mathbf{a}_d = \mathbf{i}_d . \quad (45)$$

Hence, the inductance matrix reads

$$\mathbf{L} = \left(\mathbf{K}_{dd}^{(\nu)} - \mathbf{K}_{dc}^{(\nu)} \left(\mathbf{K}_{cc}^{(\nu)}\right)^{-1} \mathbf{K}_{cd}^{(\nu)}\right)^{-1} . \quad (46)$$

The inductance matrix is symmetric by construction. The positive definiteness follows from the property that any Schur complement of a positive definite matrix is positive definite as well.

When the conductivity of the insulation material can be neglected and when, moreover, the slot is filled with a homogeneous material with permittivity ε and permeability μ, $\mathbf{L}$ and $\mathbf{C}$ are related by

$$\mathbf{L}\mathbf{C} = \varepsilon \mu \mathbf{I} \quad (47)$$

where $\mathbf{I}$ is the unit matrix. The relation follows directly from (46), (28) and (24). The relation is only valid for the case that the wires and the stator lamination can be considered as perfect conductors and the slot filling material is perfectly insulating. In this particular case, the eigenvalues of $\mathbf{L}\mathbf{C}$ equal $\varepsilon\mu$ and, hence, all wave propagate at the velocity of light associated with the insulating material. For very high frequencies and with the restrictions mentioned above, all electromagnetic fields propagate through the wire configuration with the same velocity.

In high-frequency regime, the magnetic flux lines are completely inside the insulating material (Fig. 5c).

C. Eddy-current formulation

In the mid-frequency range, an arbitrary current redistribution dependent on the frequency and the inhomogeneities of the geometry have to be considered. Then, the impedance matrix is computed as a reduction of a magnetoquasistatic formulation of the Maxwell equations, reduced to the 2D cross-section:

$$-\frac{\partial}{\partial x}\left(\nu \frac{\partial A_z}{\partial x}\right) - \frac{\partial}{\partial y}\left(\nu \frac{\partial A_z}{\partial y}\right) + \jmath\omega\sigma A_z = J_{s,z} . \quad (48)$$

The source current density is related to the voltages per unit length $\mathbf{u}_q$ applied to each of the wires $q \in \mathcal{S}_d$. The relation between $J_{s,z}$ and $\mathbf{u}$ can be formalised in similar way as the relation (32) between $J_{s,z}$ and $\mathbf{i}$ in the low-frequency case, i.e.,

$$J_{s,z} = \sum_{q=1}^{n_{wire}} \sigma t_{z,q} \mathbf{u}_q . \quad (49)$$

The total current density in the wires is given by

$$J_z = J_{s,z} - \jmath\omega\sigma A_z . \quad (50)$$

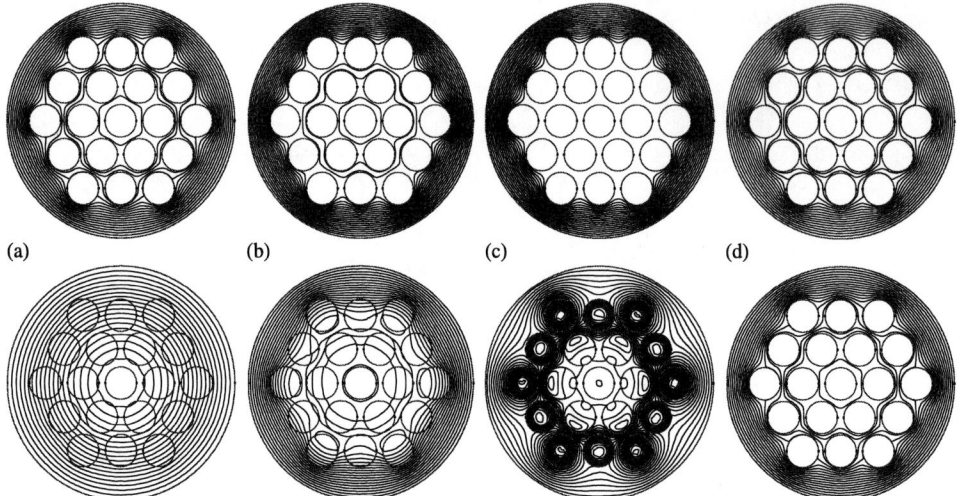

Fig. 5. Common-mode patterns (above, the equipotential lines of the electric scalar potential, below, the magnetic flux lines): (a) $f = 1$ kHz (low-frequency approximation); (b) $f = 100$ kHz (low-frequency approximation); (c) $f = 100$ kHz (eddy-current formulation); (d) $f = 100$ kHz (high-frequency approximation)

The z-component of the magnetic vector potential is discretised as in (35). The weighting procedure leads to the algebraic system of equations

$$\left(\underbrace{\mathbf{K}_{ee}^{(\nu)} + \jmath\omega\mathbf{M}_{ee}^{(\sigma)}}_{\mathbf{H}}\right)\mathbf{a}_e = \mathbf{M}_{ee}^{(\sigma)}\mathbf{D}_{ed}\mathbf{u} \qquad (51)$$

where $\mathbf{M}_{ee}^{(\sigma)}$ is given by

$$\mathbf{M}_{ee,i,j}^{(\sigma)} = \int_\Omega \sigma N_{e,i} N_{e,j}\,\mathrm{d}\Omega\ . \qquad (52)$$

and the coupling matrix for the wire voltages is

$$\mathbf{D}_{ed,i,q} = \int_\Omega t_{z,q}\,\mathrm{d}\Omega\ . \qquad (53)$$

The currents through the wires are found by integrating (50) in space:

$$\mathbf{i} = \mathbf{D}_{ed}^T\left(\mathbf{M}_{ee}^{(\sigma)}\mathbf{D}_{ed}\mathbf{u} - \jmath\omega\mathbf{M}_{ee}^{(\sigma)}\mathbf{a}_e\right)\ . \qquad (54)$$

Now, $\mathbf{a}_e$ is solved from (51) and introduced in (54). Then, the relation between the wire currents $\mathbf{i}$ and the wire voltages per unit length becomes

$$\mathbf{i} = \mathbf{D}_{ed}^T\left(\mathbf{M}_{ee}^{(\sigma)} - \jmath\omega\mathbf{M}_{ee}^{(\sigma)}\mathbf{H}^{-1}\mathbf{M}_{ee}^{(\sigma)}\right)\mathbf{D}_{ed}\mathbf{u}\ . \qquad (55)$$

The first matrix factor is

$$\mathbf{D}_{ed}^T\mathbf{M}_{ee}^{(\sigma)}\mathbf{D}_{ed} = \mathbf{R}_{dd}^{-1} \qquad (56)$$

and represents the DC conductance matrix. The second matrix term represents a combined resistive-inductive effect dependent of the frequency of the excitation and the geometry and material distribution of the configuration. The impedance matrix reads

$$\mathbf{Z} = \left(\mathbf{D}_{ed}^T\left(\mathbf{M}_{ee}^{(\sigma)} - \jmath\omega\mathbf{M}_{ee}^{(\sigma)}\mathbf{H}^{-1}\mathbf{M}_{ee}^{(\sigma)}\right)\mathbf{D}_{ed}\right)^{-1}\ . \qquad (57)$$

TABLE I

DATA OF THE TRIANGULATIONS USED IN THE FE CALCULATIONS: THE LENGTH OF THE EDGES AT THE WIRE SURFACE SERVES AS AN INDICATIVE MESH SIZE h, THE FREQUENCY $f_{\max}$ GIVES A THEORETICAL VALUE FOR THE FREQUENCY AT WHICH EDDY CURRENTS ARE STILL RESOLVED BY THE FE GRID.

	# nodes	h	$f_{\max}$
mesh 1 (Fig. 3b)	406	260 μm	64 kHz
mesh 2 (Fig. 3c)	7623	44 μm	2.3 MHz
mesh 3 (Fig. 3d)	45329	8.7 μm	58 MHz

The resistance matrix $\mathbf{R}$ and the inductance matrix $\mathbf{L}$ are found as the real and imaginary parts of $\mathbf{Z}$. The resistance matrix is no longer diagonal. The off-diagonal elements of $\mathbf{R}$ represent the losses introduced in a wire by the neighbouring wires due to proximity effects. Both the resistance and inductance matrices depend on the frequency. The symmetry of both matrices is guaranteed by the form of the expression. It is possible to show that $\mathbf{R}$ inherits the properties of $\mathbf{M}_{ee}^{(\sigma)}$, whereas $\mathbf{L}$ inherits the properties of $\mathbf{K}_{ee}^{(\nu)}$.

In the mid-frequency range, the magnetic flux lines penetrate the wires up to a certain extent (Fig. 5d).

VI. BENCHMARK WINDING

The simulation techniques developed in this paper have been applied to a benchmark winding. The winding consists of 19 wires arranged as shown in Fig. 3a. The winding passes through a slot of a ferromagnetic yoke with a total length of 2 m. The ferromagnetic material is assumed to have a high conductivity, such that the return path for the current can be assumed to be infinitely conductive. Three triangulations of different resolution are used for the FE calculations (Table I and Fig. 3b-d). The results for the low- and high-frequency approximations of the capacitance and inductance matrices are almost

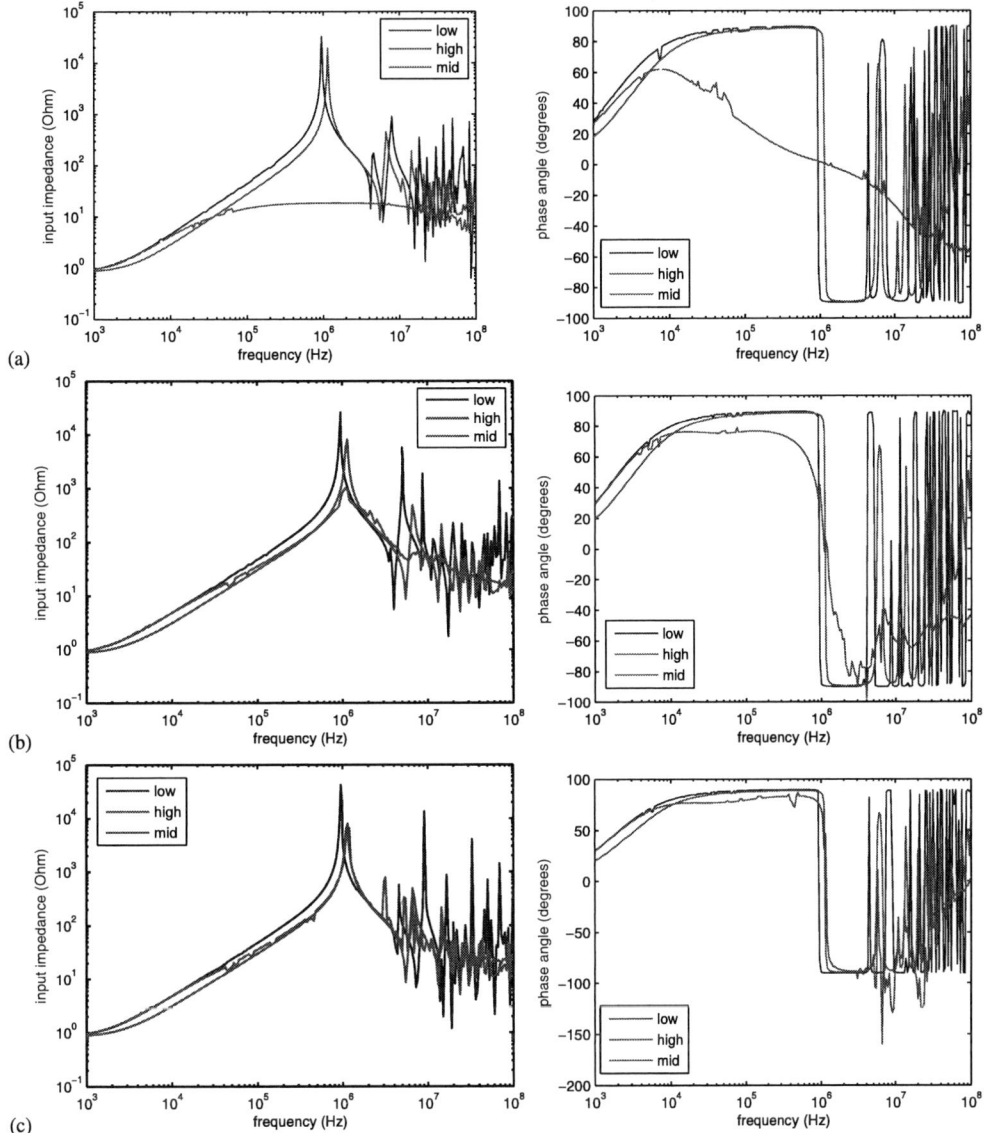

Fig. 6. Magnitudes and phases of the input impedances of the machine winding in function of frequency: TLM parameters calculated on the basis of (a) FE mesh 1, (b) FE mesh 2 and (c) FE mesh 3), low-frequency and high-frequency approximation and eddy-current formulation.

the same for the coarse and fine mesh. Notice, that these approximations do not require the skin depth to be resolved in the FE mesh.

The field patterns for the common mode are shown in Fig. 5. The pattern depends on the frequency of the excitation and on the formulation used for the FE calculation. The common-mode patterns obey the symmetries in the benchmark wire configuration. If the wire configuration would be completely symmetric, the patterns would be independent of the frequency and the calculation approach. Notice that the electric and magnetic field pattern is the same for the high-frequency approximation, as explained above.

The frequency dependence of the input impedance is

clear from Fig. 6. The low-frequency and high-frequency approximation show some similarities, especially for low frequencies. The first resonance is, however, different. In the high-frequency region, no agreement is found anymore. The eddy-current formulation, accounting for skin and proximity effects with a finite thickness, is similar to the low-frequency approximation in the low-frequency range. It is clear that, with increasing frequency, the mid-frequency results tend to the results for the high-frequency approximation. This is not observed in the results for the coarse mesh (Fig. 6), which is easily explained by the insufficient resolution of the FE mesh (Fig. 3b). When the mesh size of the elements at the surface of the wire is larger than the expected skin depth, one should doubt

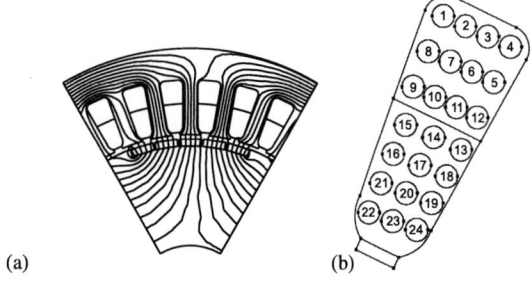

(a)   (b)

Fig. 7. (a) Magnetic flux in one pole of the PMSM and (b) geometry of the slot cross-section with ordering of the turns.

TABLE II

DATA OF THE PERMANENT-MAGNET SYNCHRONOUS MACHINE.

rated power	3.0	kW
rated speed	1930	rpm
rated torque	15.3	Nm
rated current	10.6	A
number of slots per pole	6	
number of turns per slot	24	
iron stack length	100	mm

the accuracy of the FE results. It is clear from the figures that one should prefer to use the low-frequency and high-frequency approximations, if one is not able to guarantee a sufficiently fine FE mesh. The TLM parameters calculated on the finer mesh of Fig. 3c are sufficient to predict the first resonance frequency (Fig. 6b). The magnitude of the resonance is, however, not captured. The finest mesh (Fig. 3d) delivers TLM parameters which give accurate results beyond the first resonance (Fig. 6). The 2D FE model, however, already has more than 40000 nodes.

VII. PMSM WINDING

The developed simulation method has been applied to the stator winding of a six-pole permanent-magnet synchronous machine (PMSM) (Fig. 7). The data of the machine are collected in Table II. The PMSM is equipped with a double-layer winding with 24 turns per slot. The input impedance of one phase of the winding is shown in Fig. 8. Again, the eddy-current formulation clearly shows the transition between the low-frequency and the high-frequency regime. The results have been obtained assuming an infinitely conductive return path. This is, however, not the case for a laminated stator yoke. The return currents experience a rather long return path closing through the laminations and the housing [9], [10]. This can be modelled by an additional impedance in the TLM model or by a Robin boundary condition in the FE models [11].

VIII. CONCLUSIONS

The paper clearly distinguishes three formulations for the calculation of transmission-line parameters on the basis of a finite-element model. The eddy-current formulation gives accurate results in the frequency range in which both the low-frequency and high-frequency

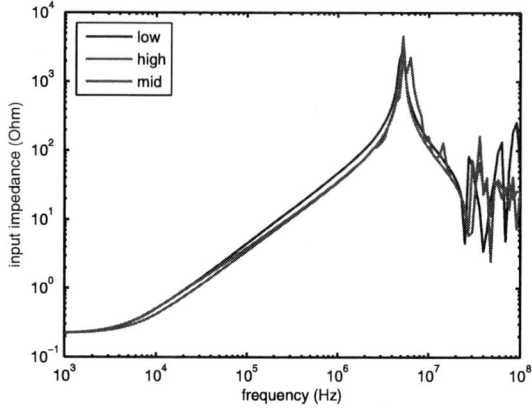

Fig. 8. Input impedance of one phase of the PMSM winding.

approximations are not applicable, as long as a finite-element mesh with a sufficient resolution is used.

ACKNOWLEDGMENT

This work has been carried out in the collaborative research group (Forschergruppe 575) "High-frequency parasitic effects in inverter-fed electrical drives" funded by the German Research Foundation (Deutsche Forschungsgemeinschaft, DFG).

REFERENCES

[1] A. Binder, "Armature insulation stress of low voltage AC motors due to inverter supply," in *Proc. ICEM*, vol. 2, Paris, France, Sept. 1994, pp. 431–436.

[2] M. Kaufhold, H. Auinger, M. Berth, J. Speck, and M. Eberhardt, "Electrical stress and failure mechanism of the winding insulation in PWM-inverter-fed low-voltage induction motors," *IEEE Trans. Ind. Elek.*, vol. 47, no. 2, pp. 396–402, Apr. 2000.

[3] A. Moreira, T. Lipo, G. Venkataramanan, and S. Bernet, "High-frequency modeling for cable and induction motor overvoltage studies in long cable drives," *IEEE Trans. Ind. Applicat.*, vol. 38, no. 5, pp. 1297–1306, September/October 2002.

[4] G. Skibinski, R. Kerkman, D. Leggate, J. Pankau, and D. Schlegel, "Reflected wave modeling techniques for PWM AC motor drives," in *IEEE Applied Power Electronics Conference*, Anaheim, CA, USA, Feb. 1998, pp. 1021–1029.

[5] B. Bolsens, K. De Brabandere, J. Van den Keybus, J. Driesen, and R. Belmans, "Transmission line effects on motor feed cables," in *Proc. IEMDC*, Madison, Wisconsin, USA, June 2003, pp. 1866–1872.

[6] O. Henze, Z. Çay, O. Magdun, H. De Gersem, T. Weiland, and A. Binder, "A stator coil model for studying high-frequency effects in induction motors," in *Proc. SPEEDAM*, Ischia, Italy, June 2008.

[7] U. Pahner, R. Mertens, H. De Gersem, R. Belmans, and K. Hameyer, "A parametric finite element environment tuned for numerical optimization," *IEEE Trans. Magn.*, vol. 35, no. 5, pp. 2936–2939, Sept. 1998.

[8] P. Dular, W. Legros, and A. Nicolet, "Coupling of local and global quantities in various finite element formulations and its application to electrostatics, magnetostatics and magnetodynamics," *IEEE Trans. Magn.*, vol. 34, no. 5, pp. 3078–3081, Sept. 1998.

[9] P. Maeki-Ontto and J. Luomi, "Common-mode flux calculation of ac machines," in *Proc. ICEM*, Brugge, Belgium, Aug. 2002, paper No. 549, 6 pages on CDROM.

[10] A. Muetze, H. De Gersem, and T. Weiland, "Influence of teeth and cooling ducts on the HF common mode flux of inverter-fed AC machines," in *Proc. IEEE Ind. Appl. Soc. 40th Ann. Meet.*, Kowloon, Hong Kong, Oct. 2005, p. 7 pages on CD ROM.

[11] S. Yuferev and N. Ida, "Selection of the surface impedance boundary conditions for a given problem," *IEEE Trans. Magn.*, vol. 35, no. 3, pp. 1486–1489, May 1999.

An efficient field-circuit coupling method by a dynamic lumped parameter reduction of the FE model

F. Henrotte, E. Lange, and K. Hameyer

Institute of Electrical Machines, RWTH Aachen University, Aachen, Germany, fh@iem.rwth-aachen.de

Abstract—A field-circuit coupling method is presented, whose basic idea is to extract from the FE model, on basis of energy balance considerations, a temporary lumped parameter representation of the electrical machines, to be used in the circuit simulator model of the power electronic supply. The dynamic coupled model of the complete drive obtained this way can be iterated over a limited period of time, with a time step adapted to the high frequency of electronic commutations. When the temporary representation of the machine has come, or is expected to have come under a given accuracy threshold, a new FE simulation is performed in order to generate an updated set of lumped parameters, and the process is repeated. This method allows decoupling the time constants of the field problem from that of the circuit problem, which is typically one or two orders of magnitude smaller. This yields a considerable saving of computation time with a controllable, at least a posteriori, loss of accuracy. The method presented in this paper is also characterized by the fact that only known quantities of the nonlinear FE method are used in the identification process, which can therefore be done systematically and in an automated way.

I. INTRODUCTION

THIS paper presents a method for coupling efficiently the computational time expensive 2D or 3D Finite Element (FE) model of an electrical machine with the external power electronics circuit describing the supply system.

Over the past decades various approaches have been developed to simulate such systems supplied through high frequency switching components. Numerically strongly coupled approaches, e.g. [1]- [3], ensure consistency but suffer from long simulation times. Numerically weakly coupled approaches on the other hand, e.g. [4]- [5], are able to make benefit from the difference between the time constant of the field problem and that of the power electronics circuit, which can be of several orders of magnitude. Weak coupling allows as well to work with two specialized software, with a great benefit in flexibility and reliability.

The idea of the coupling is to extract from the FE model of the machine a set of slowly varying lumped parameters (inductances, resistances, electromotive forces) that represent the machine seen from stator terminals, and to use them in the circuit equations. The field problem is thus temporarily

represented in the circuit simulator equation by a small set of lumped parameters, and time-consuming finite element simulations are done only from time to time to update this set at a rate much slower than the rate of switching of the electronic components in the supply system. The overall computation time of the coupled problem is thus considerably reduced. Additional lumped parameters for machines with rotor terminals connected to an external circuit could be treated the same way. This will however not be considered in this paper.

The updating of the lumped parameters set, is done either on a regularly basis (on the 20th circuit simulator iteration for instance), or on basis of an error estimator that triggers the generation of a new set of linearized data when the error of the present set exceeds a given threshold.

The proposed model is applied to a transformer, a permanent magnet synchronous machine (PMSM) and a three phase claw-pole alternator. The simulation results are compared with measurements and with numerical results obtained with other software packages by numerically strongly coupled approaches.

II. LUMPED PARAMETER EXTRACTION

A. Energy Balance in Magnetodynamics

The lumped parameters that represent an electrical machine seen from its stator terminals are the phase voltages ΔU_r, the phase resistances R_r, the inductance matrix L_{rs}, and the electromotive forces E_r induced in stator windings by the rotation of the machine, $r = 1, \ldots, m$, where $m = 3$ is the number of phases. One has

$$\Delta U_r = R_r I_r + \partial_t \varphi_r \tag{1}$$

$$= R_r I_r + E_r + L_{rs} \partial_t I_s, \tag{2}$$

where φ_r is the flux linkage in phase r.

The method presented in this paper is characterized by the fact that only known quantities of the nonlinear FE method are used for the identification process, which can therefore be done systematically and in an automated way. Those FE quatities are : the FE system matrix M_{ij}, the Jacobian matrix J_{ij} which represents the 1st order linearization of the the nonlinear field equations around the system state at time t, and finally the computed torque T. All lumped parameters characterizing the FE model, as it is seen by the external circuit at time t, can be systematically extracted from them.

A relation between the field state variable (the vector potential **a**) and the circuit state variables (the phase fluxes φ_r) is required for making this extraction. It is obtained on basis of energy considerations. Multiplying (1) with I_r and summing over all phases gives

$$\sum_r I_r \, \partial_t \varphi_r = -\sum_r R_r I_r^2 + \sum_r I_r \Delta U_r. \tag{3}$$

This equation has a counterpart in the field domain

$$\int_\Omega \mathbf{j} \cdot \mathcal{L}_\mathbf{v}\, \mathbf{a} = -\int_\Omega \frac{\mathbf{j}^2}{\sigma} - \int_\Omega \mathbf{j} \cdot \mathrm{grad}\, u, \tag{4}$$

where $\mathcal{L}_\mathbf{v}\, \mathbf{a}$ denotes the material derivative of **a**, i.e. a time derivative accounting for movement [6]. Equations (4) and (3) must be identifiable term by term to ensure that the energy balance in the field domain and in the lumped parameter domain are equivalent.

The principle of the identification between field and circuit variables follows now from the observation that the current density in stranded conductors can be written

$$\mathbf{j} = \sum_r I_r \mathbf{w}_r, \tag{5}$$

where the auxiliary field $\mathbf{w}_r$ can be regarded as the shape functions of the phase currents I_r. In 2D, $\mathbf{w}_r = \pm \mathbf{e}_z P_p n_r / (\sum_k S_{rk})$, with S_{rk} are the cross section areas of all the slots carrying the current of phase r, n_r the number of turns and P_p the number of parallel paths. In 3D, $\mathbf{w}_r$ is a vector field whose field lines follow the threads of the stranded coil, Fig. 1. This vector field can be computed beforehand by an electric vector potential formulation, $\mathrm{rot}\, \mathbf{t} = \mathbf{j}$.

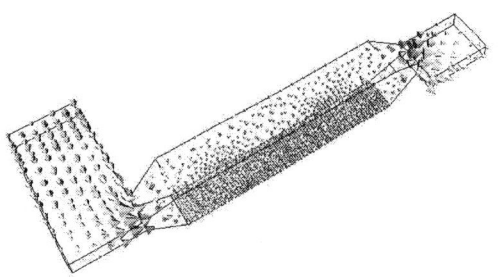

Figure 1. Current shape function of a single stator phase.

Substituting (5) in (4) yields readily

$$R_r = \int_\Omega \sigma^{-1} |\mathbf{w}_r|^2 \quad , \quad \Delta U_r = -\int_\Omega \mathbf{w}_r \cdot \mathrm{grad}\, u. \tag{6}$$

On the other hand, substituting (5) in

$$\int_\Omega \mathbf{j} \cdot \mathcal{L}_\mathbf{v}\, \mathbf{a} = \sum_r I_r \partial_t \varphi_r. \tag{7}$$

yields

$$\partial_t \varphi_r = \int_\Omega \mathbf{w}_r \cdot \mathcal{L}_\mathbf{v}\, \mathbf{a}. \tag{8}$$

If the current distribution functions $\mathbf{w}_r$ are assumed not to depend on time (which is exactly the case in stranded coils), the sought mapping between **a** and φ_r is

$$\varphi_r = \int_\Omega \mathbf{w}_r \cdot \mathbf{a}, \tag{9}$$

which, in a mesh, can be expressed by the rectangular matrix

$$\varphi_r = W_{ri} a_i \quad \text{with} \quad W_{ri} = \int_\Omega \mathbf{w}_r \cdot \boldsymbol{\alpha}_i \tag{10}$$

where $\boldsymbol{\alpha}_i$ are the shape functions of the **a** field, and a_i the corresponding coefficients.

B. Extraction of the inductance matrix

Let

$$M_{ij}(\mathbf{a})\, a_j = b_i, \tag{11}$$

with

$$b_i = \int_\Omega \mathbf{j} \cdot \boldsymbol{\alpha}_i = I_r \int_\Omega \mathbf{w}_r \cdot \boldsymbol{\alpha}_i = I_r W_{ir}, \tag{12}$$

be the nonlinear FE equations describing the electrical machine subjected to given phase and excitation currents.

Now, let I_r^* be the phase currents at time t, and $b_i^* = I_r^* W_{ir}$ the corresponding right-hand sides. Solving (11) with $b_i \equiv b_i^*$ and a fixed rotor angular position $\delta\Theta = 0$ gives a_j^* and a first order linearization around this particular solution writes

$$M_{ij}(a_j^* + \delta a_j) = M_{ij}(a_j^*) a_j^* + J_{ij}(a_j^*) \delta a_j = b_i^* + \delta b_i \tag{13}$$

with the Jacobian matrix $J_{ij} \equiv \left(\partial_{a_j} M_{ik}(a_j^*)\right) a_k^*$. Since $M_{ij}(a_j^*) a_j^* = b_i^*$, one has

$$J_{ij}(a_j^*) \; \delta a_j |_{\delta\Theta=0} = \delta b_i. \tag{14}$$

One can now repeatedly solve (14) with the right-hand sides $\delta b_i = \delta I_r\, W_{ir}$ obtained by perturbating one after the other m phase currents I_r and obtain m solution vectors for $\delta a_j |_{\delta\Theta=0}$. Since (14) is linear, the magnitude of the perturbations δI_r is arbitrary. One can so define by inspection the tangent inductance matrix L_{rs}^∂ of the electrical machine seen from terminals as

$$\delta\varphi_r |_{\delta\Theta=0} = W_{rj} \; \delta a_j |_{\delta\Theta=0}$$
$$= W_{rj} J_{ji}^{-1}(a_j^*) W_{is} \; \delta I_s, \equiv L_{rs}^\partial \; \delta I_s \tag{15}$$

with

$$L_{rs}^\partial = W_{rj} J_{ji}^{-1}(a_j^*) W_{is}. \tag{16}$$

Similarly, one can identify the secant inductance matrix L_{rs} and by solving (11) repeatedly with linearly independent phase currents I_r to obtain

$$\varphi_r = W_{rj}\, a_j = W_{rj} M_{ji}^{-1}(a_j^*) W_{is}\, I_s \equiv L_{rs}\, I_s \qquad (17)$$

with

$$L_{rs} = W_{rj} M_{ji}^{-1}(a_j^*) W_{is}. \qquad (18)$$

It is practical for these identifications to use a solver capable of efficiently dealing with multiple right-hand sides.

C. Extraction of the motion induced voltage

One can now complement (15) to account for the electromotive forces. One has

$$\delta\varphi_r = L_{rs}^{\partial}\delta I_s + \partial_\Theta\varphi_r\,\delta\Theta. \qquad (19)$$

The direct computation of the Θ derivative requires to slightly shift the rotor, remesh, solve the FE problem, evaluate new fluxes and calculate a finite difference. In order to avoid this tedious process, one can again call on the energy principles. One has

$$\partial_\Theta\varphi_r = \partial_\Theta\partial_{I_r}\Psi_M = \partial_{I_r}\partial_\Theta\Psi_M = \partial_{I_r}T \qquad (20)$$

where T is the torque and Ψ_M is the magnetic energy of the system. During the identification process described above, it is thus easy to calculate additionally the torque corresponding to the perturbed solutions $\delta a_j|_{\delta\Theta=0}$, and to evaluate the motion induced voltage

$$E_r \equiv \partial_\Theta\varphi_r\,\partial_t\Theta \qquad (21)$$

of each phase as the variation of torque with the perturbation of the corresponding phase current I_r.

Beware however that, as the torque is a nonlinear function of the fields, the perturbations needs in this case be small. Because of the linearity of (14), one may scale the perturbation currents in (20) which yields:

$$\partial_\Theta\varphi_r = \frac{T(a_j^*) - T(a_j^* + \lambda\delta a_j|_{\delta\Theta=0})}{\lambda\delta I_r} \text{ with } \lambda = \kappa\frac{||a_j^*||_2}{||\delta a_j||_2}. \qquad (22)$$

The scale factor is chosen between $0.01 \leq \kappa \leq 0.05$. Both the direct calculation of the Θ derivative by finite differences and the proposed energy-based approach (22) have been implemented.

III. REALIZATION OF COUPLING

The numerically independent solution process of the circuit and the field problem requires a time stepping scheme synchronizing the Modified Nodal Analysis (MNA) used by the circuit simulator and the FE Analysis. A basic scheme consists in holding the time step ΔT_{FE} of the FE system constant. At the beginning of the transient simulation, the initial values of the tangent inductance matrix $L_{rs}^{\partial 0}$ and the induced voltages E_r^0 are

calculated by a magnetostatic FE calculation, and incorporated into the equation system of the circuit simulator. If the circuit simulator reaches $t >= t_{FE}^{k-1} + \Delta T_{FE}$, a new set of phase currents is imposed and the magnetostatic FE model is solved, yielding an updated set of lumped parameters, $L_{rs}^{\partial k}$ and E_r^k. This basic scheme is compatible with an adaptive time stepping circuit simulator. The latter can freely adapt the time step to accurately account for topological changes occuring in the circuit, caused by e.g. switching semi-conductor components.

Any circuit simulator in combination with a magnetostatic FE-solver can be used provided both packages have proper interface capabilities. In this paper, the circuit simulator *Simplorer* [7] and the in-house FE library *i*MOOSE [8] have been used. *Simplorer* provides a C-Interface, giving access to state variable information at different stages of the time stepping scheme.

For the field and circuit simulators run on different operating systems, and in order to avoid tedious and error-prone data exchange via files, a network based data exchange has been implemented. To reduce the implementation effort to a minimum while preserving maximum flexibility, the communication is based on the free implementation omniORB [9] of the CORBA standard. All communication is done via the standard network. The CORBA standard defines a platform independent interface definition language, by which remote procedure calls are made transparent to the program designer, who need thus not worry about the implementation of a complete network stack [10].

IV. APPLICATION EXAMPLES

A. Transformer

The poposed method is first validated on a 2D single phase transformer model. The primary side ($n_{pri} = 900$ turns) of the transformer is fed by the voltage source $E = 325\,\text{V}\cos(2\pi\cdot\frac{t}{T})$, while the secondary side ($n_{sec} = 50$ turns) is connected to the load resistance $R_L = 3\,\Omega$.

The simulation is carried out with a fixed time step $\Delta t_c = 5\cdot 10^{-5}\,\text{s}$ for the circuit simulator and different time steps Δt_{FE} for the field domain resulting in 38, 95, 356 or 670 FE time steps, for a total simulation time of $2T = 40$ ms.

For the sake of comparison, the system is also solved by applying a numerically strong coupling with the 2D transient FE solver of the *i*MOOSE package [8]. The relative errors $e_I = (I_{ref} - I)/\max(I_{ref})$ between numerically weakly and strongly coupled approaches on the primary and secondary currents are shown in Fig. 2 and Fig. 3 respectively. Both figures reveal an error accumulation over time at a rate depending on Δt_{FE}: the smaller Δt_{FE}, the smaller the rate of amplification of the error.

The ripple observed in Fig. 3 can be ascribed to the stepwise approximation of the inductances, Fig. 4. By decreasing Δt_{FE}, the variation in time of the inductances is smoothened, which yields a more accurate result.

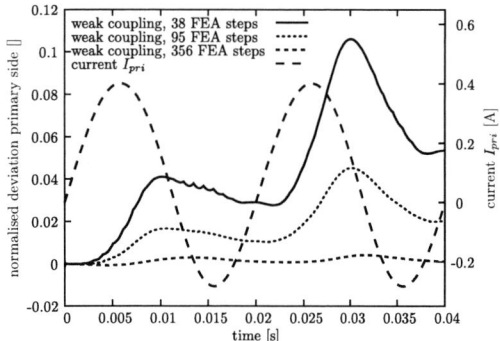

Figure 2. Relative error on the primary side current I_{pri}.

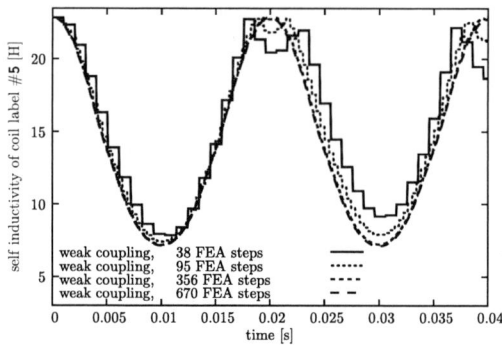

Figure 4. Linearised self inductivity of coil label #5.

In summary, the proposed method gives results comparables to those obtained by the strong coupling approach. Choosing an appropriate time step Δt_{FE} is however crucial to obtain accurate results.

B. Permanent Magnet Generator

The method has also been applied to a three phase synchronous permanent magnet generator connected via B6-bridge rectifier to the DC voltage source E1. The coupled problem has been simulated with both *Maxwell* [7] and the *iMOOSE /Simplorer* coupling described above. The topology of the system is shown in Fig. 5. The block iMOOSECon-nector_3 in Fig. 5 contains the lumped parameters L_{rs}^{∂} and E_r. All circuit parameters except those lumped parameters are held constant throughout the simulation.

The comparison of the simulation results are shown in Fig. 6 and Fig. 7. The currents in phase u and the total current I_{tot} can be seen in Fig. 6. They are in good accordance, except at the beginning of the simulation. Having a closer look at the terminal voltage depicted in Fig. 7, one recognizes however overshoots at the peak values as well as around the singular

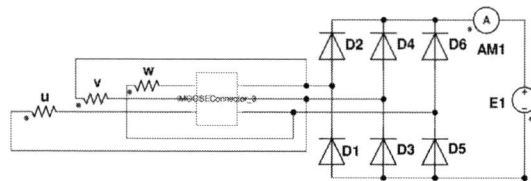

Figure 5. Generator connected to a b6 rectifier working connected to a DC voltage source.

switching points of the diodes in the calculation results of *Maxwell*. This behaviour indicates an inaccurate zero crossing detection in the the diodes of the bridge. The simulation results of the proposed lumped parameter coupled method are in good accordance with the simulation results of *Maxwell* with a significant time saving of approximately 25%.

C. Claw pole generator

The proposed coupling method is finally validated on a three phase claw-pole alternator which is connected via a B6-bridge D1-D6 to the constant voltage source E1 representing the battery in series with the resistance R4, Fig. 10. Due to its irreducible 3D flux path structure and the connected bridge rectifier, the claw pole generator is a challenging field-circuit coupled system. The winding resistances are labelled R1 to R3. The excitation current is assumed to be constant in this simulation, and concerns thus only the FE system.

The inverter topology is given Fig. 10. The coupled model is simulated with the electromotive forces calculated both by finite differences and by the energy-based approach (Sec. II-C). Additionally, the software package JMAG [11] is used for the sake of comparison, see Tab. I.

The alternator is analysed at different rotation speeds. Tab. II shows the mean output currents of the alternator. Due to license issues, the simulation with JMAG at 6000 rpm could not be carried out. The simulated and measured output currents

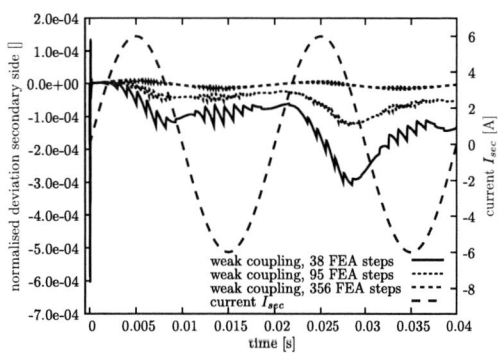

Figure 3. Relative error the secondary side current I_{sec}.

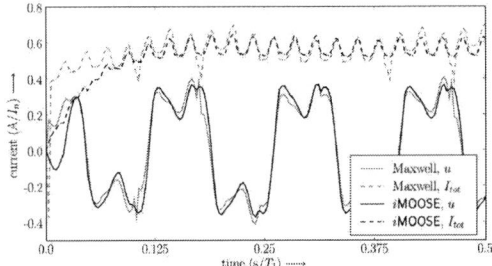

Figure 6. Total and phase u current forms at $f_1 = \frac{5}{6}T_n^{-1}$.

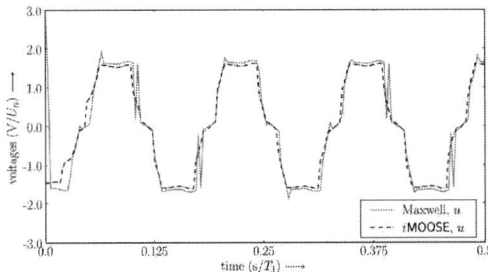

Figure 7. Terminal voltage of phase u at $f_1 = \frac{5}{6}T_n^{-1}$.

are shown in Fig. 8. While implementation a) performs well at low rotation speeds, the simulated current for the highest speed clearly exceeds the measurements. Implementation b) appears to have a small offset compared to the measurement. The implementation c) performs worst at low rotation speeds but simulates the output current very accurately at medium rotation speeds. The relative error is shown in Tab. III as well as in Fig. 9.

The induced voltages calculated by implementation a) are very sensitive to the chosen shift of the rotor. The offset of the implementation b) might stem from the calculation of the torque by the Maxwell stress tensor. A verification and comparison with different approaches of the calculation of the torque will provide useful additional information. All simulations deliver an output current larger than the measured

Table I
OVERVIEW OF THE NUMERICAL APPROACHES CHOSEN TO SIMULATE THE CLAW POLE ALTERNATOR.

	Simulation approach	Comment
a)	iMOOSE $\partial_\Theta \varphi_r$	The induced voltages are calculated by finite differences similar to postprocessing.
b)	iMOOSE $\partial_{I_r} M$	The induced voltages are calculated by the variational approach as proposed in II-C.
c)	JMAG	Commercial software package. Implementation and exact approach unknown.

Table II
COMPARISON OF THE MEAN OUTPUT CURRENTS OF THE CLAW POLE ALTERNATOR AT DIFFERENT ROTOR SPEEDS.

rotation speed	measured	iMOOSE $\partial_\Theta \varphi_r$	iMOOSE $\partial_{I_r} M$	JMAG
1500 rpm	45.73 A	58.80 A	65.03 A	80.04 A
1800 rpm	79.09 A	84.30 A	97.79 A	105.78 A
3000 rpm	128.76 A	145.18 A	139.12 A	131.80 A
6000 rpm	149.01 A	194.93 A	156.95 A	n.a.

Table III
RELATIVE ERROR OF THE OUTPUT CURRENT.

rotation speed	iMOOSE $\partial_\Theta \varphi_r$	iMOOSE $\partial_{I_r} M$	JMAG
1500 rpm	29 %	42 %	75 %
1800 rpm	7 %	24 %	34 %
3000 rpm	13 %	8 %	2 %
6000 rpm	31 %	5 %	n.a.

one.

The phase and the output current waveforms simulated by implementation b) are depicted in Fig. 11. The flux density of the modeled claws is illustrated in Fig. 12.

V. DISCUSSION

The proposed field-circuit coupling method is applicable to simulate complex problems, 2D or 3D, with or without motion. Two approaches to calculate the motion induced voltages have been implemented. Despite the good agreement obtained between simulated and measured currents, the energy based approach is numerically stable. Though the calculation of the torque is crucial for this approach, it appears to be more reliable than the calculation of the electromotive forces by finite differences. Further investigations on the calculation of the torque are expected to give more information about the offset in the simulated mean output current.

The communication between *Simplorer* and iMOOSE via network avoids error prone file locking mechanisms necessary for synchronizing the solution process and additionally bridges the different operating systems.

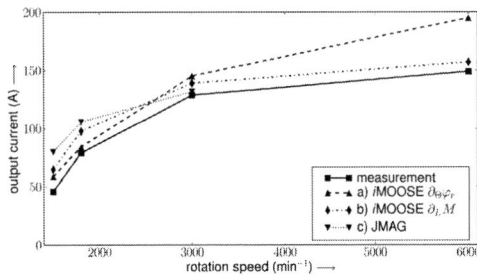

Figure 8. Comparison of the mean output currents of the claw pole alternator.

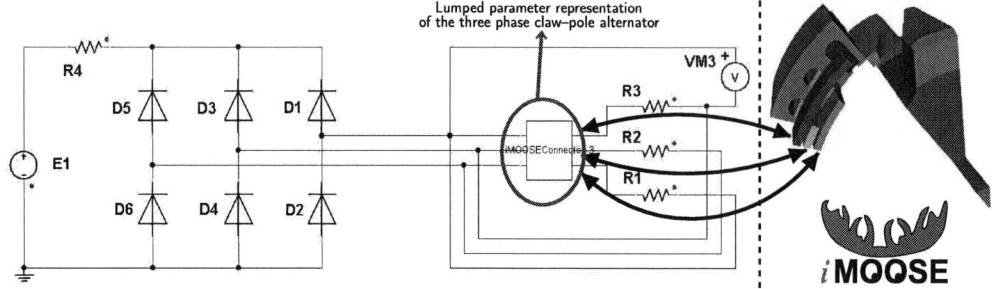

Figure 10. Application of the proposed coupling: three phase claw pole alternator connected to a rectifier working on a constant voltage source.

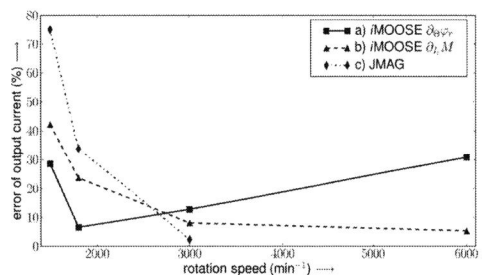

Figure 9. Relative deviation of output current through the battery E1.

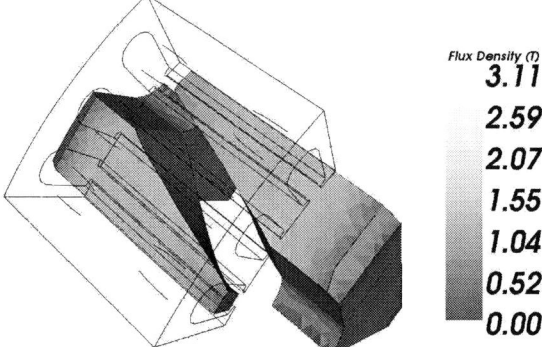

Figure 12. Midvalue of flux density at $n = 1500\,\text{rpm}$ with an excitation of $I_f = 1760$ Aturns.

Furthermore, the weak coupling reduces the computation time compared to numerically strongly coupled approaches, still providing good and accurate results. During a standard simulation cycle the circuit simulator performs 10 times more transient steps before a new FE extraction is calculated. Thus, compared to numerically strongly coupled approaches, a great saving of time is achieved.

REFERENCES

[1] T. Dreher, and Gérard Meunier, "3D Line Current Model of Coils and External Circuits", *IEEE Transactions on Magnetics*, vol. 31, no. 3, May

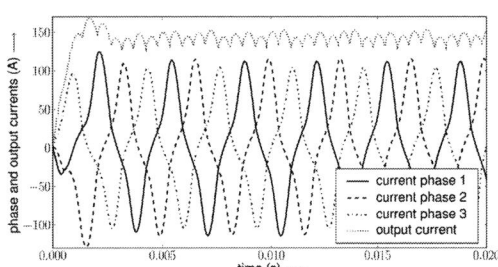

Figure 11. Phase and output current waveforms of the alternator.

1995.

[2] P.J. Leonard, and H. C. Lai, "Treatment of Symmetry in Three Dimensional Finite Element Models of Machines Coupled to External Circuits", *IEEE Transactions on Energy Conversion*, vol. 14, no. 4, December 1999.

[3] A. Canova, M. Ottella, and D. Rodger, "Coupled Field–Circuit Approach to 3D FEM Analysis of Electromechanical Devices", *Proceedings of the 9th Int. Conf. on Electrical Machines and Drives*, pp. 71–75, 1999.

[4] P. Zhou, W. N. Fu, D. Lin, S. Stanton, and Z. J. Cendes, "Numerical Modeling of Magnetic Devices", *IEEE Transactions on Magnetics*, vol. 40, pp. 1803–1809, July 2004.

[5] P. Zhou, D. Lin, W. N. Fu, B. Ionescu, and Z. J. Cendes, "A General Cosimulation Approach for Coupled Field–Circuit Problems", *IEEE Transactions on Magnetics*, vol. 42, no. 4, pp. 1051-1054, April 2006.

[6] F. Henrotte and K. Hameyer, "The Structure of EM Energy Flows in Continuous Media", *IEEE Transactions on Magnetics*, vol. 42, no. 4, pp. 903-906, April 2006.

[7] Ansoft Corporation, "Simplorer", *http://www.ansoft.com*, [online], visited on May 9th, 2008.

[8] G. Arians, T. Bauer, C. Kaehler, W. Mai, C. Monzel, D. van Riesen and C. Schlensok, "Innovative modern object–oriented solving environment – iMOOSE", *http://www.imoose.de*, [online], visited on February 14th, 2008.

[9] Duncan Grisby, Apasphere Ltd, "omniORB – a robust high performance CORBA ORB for C++ and Python", *http://omniorb.sourceforge.net*, [online], visited on July 14th, 2008.

[10] E. Lange, M. van der Giet, F. Henrotte, K. Hameyer, "Circuit coupled simulation of a claw pole alternator by a temporary linearization of the

3D-FE model", Submitted to the XVIIIth International Conference on Electrical Machines, ICEM'08, September 2008.

[11] JRI Solutions, Ltd., "JMAG", *http://www.jri.co.jp/pro-eng/jmag/e/jmg/*, [online], visited on July 11th, 2008.

[12] H. Bai, S. Pekarek, J. Tichenor, W. Eversmann, D. Buening, G. Holbrok, M. Hull, R. Krefta, and S. Shields, "Analytical Derivation of a Coupled-Circuit Model of a Claw-Pole Alternator With Concentrated Stator Windings", *IEEE Transactions on Energy Conversion*, vol. 17, no. 1, pp. 32–38, March 2002.

[13] N. A. Demerdash and T. W. Fello, "Electric Machinery Parameters and Torques by Current and Energy Perturbations from Field Computations – Part I: Theory and Formulation", *IEEE Transactions on Energy Conversion*, vol. 14, no. 4, pp. 1507–1513, December 1999.

Coupled field-circuit-mechanical model of an electromagnetic actuator operating in error actuated control system

Lech Nowak

Poznań University of Technology, Institute of Electrical and Electronic Engineering, Poznań, POLAND
e-mail: lech.nowak@put.poznan.pl

*Abstract–*An algorithm of coupled field-circuit simulation of the dynamics of an electromagnetic linear actuator operating in error actuated control system is presented. The software consists of three main parts: (a) numerical model of the actuator dynamics which includes equations of a transient electromagnetic field in a non-linear conducting and moving medium, (b) discrete model of electric circuit and (c) optimization solver. Numerical implementation is based on the finite elements. The influence of the PID controller settings on the actuator operation is shown. In order to find optimal parameters of the system the genetic algorithm is applied. The simultaneous optimization of both: actuator structure and regulator settings has been carried out.

*Keywords–*Electromagnetic actuators, electromagnetic field, transient analysis, feedback, optimisation.

I. INTRODUCTION

In order to simulate the operation of electromagnetic actuators, the precise mathematical models of electromagnetic, thermal and mechanical phenomena should be applied. The most common approach is equivalent circuit description, with lumped parameters. But more accurate simulation requires the field description of the electromagnetic phenomena – so called distributed parameters model.

The phenomena of different nature are often coupled mutually. The coupling arise due to material or state variables reciprocal connections [1, 3, 6]. For example, in the devices containing non-linear, conducting and moving parts, the waveforms of currents which excite the field are not known in advance, i.e prior to the field computation, but the field depends on the exciting currents, in turns. This is because usually only the supply voltages waveforms are explicitly known. The exciting currents depend on induced electromotive forces, i.e depend on transient magnetic flux linkages and on the armature movement, as well. Therefore, the complex, mathematical model of coupled field, circuit and mechanical phenomena must be applied [2, 3]. The additional equations of the actuator circuit, the electronic supply and feedback [5, 6] must be included into set of equation which describe the electromagnetic field. All equation of such model are solved simultaneously.

Therefore, there is need to link different softwares: the software for electromagnetic field computation, the software for electric and electronic circuits simulation [2]. Moreover, since the mechanical movement significantly influences on transient field distribution and

transient circuit state variables, the software for mechanical computation must be attached. In some case it is necessary to include the software for thermal computation. In case of design process, the software for CAD or for the optimisation should be included, as well.

In this paper the combined computer enviroment for simulation of: (a) transient electromagnetic field, (b) electric and electronic circuits and (c) mechanical motion dynamics is presented. These three units are interconnected by the integral quantities such as: winding currents, electromotive forces, electromagnetic driving force and mechanical displacement or velocity [6, 7].

Finally, the procedure of optimal design, based on the genetic algorithm, has been added. The optimisation solver has been elaborated regardless of the "combined" computer software for the system dynamics simulation. Both units are connected through the transformation of the design variables used in model of the system and their dimensionless counterparts used in optimisation procedure.

On the other hand, the measurement of certain variables in automatic control systems, in particular mechanical quantities such as displacement or rotational speed, is inconvenient and often leads to considerable errors what lessening the reliability of the systems. Therefore the sensorless systems are recently developed. In these systems, some of the variables are computed on the basis of a mathematical model of the electromechanical converter (electrical machine, linear actuator). The accuracy of the model is very important. Therefore , the proposed field-circuit models are being increasingly applied to the estimation of state variables. In the near future these types of models may be used "on line" in automatic control systems.

II. LINEAR MOVEMENT ACTUATORS

Linear movement converters such as electromagnetic actuators and linear motors constitute a significant part of electromechanical energy converters. Such systems have many advantages in comparison to classical electric drives with rotational machines. The structure of the system is simplified because the transmission and coupling mechanisms are unnecessary. Also, the reliability and the robustness of the system improve thanks to the elimination of clearances [4, 5, 6]. The advantages are specifically evident in the case of fast systems, which require great precision of positioning.

The paper presents an algorithm for field-circuit simulation of the dynamics of a multi-stable electromagnetic linear movement actuator working in an error actuated control system (including feedback). In such types of actuators, a linear control characteristic (i.e. proportionality between electric input signal and output mechanical quantity) is available. They are also used in magnetic levitation and magnetic bearing systems.

The most common used electromagnetic actuators have so called plunger-type structure – Fig. 1 [7]. A roller-shaped mover plunges into a cylindrical coil positioned inside the ferromagnetic core. Such actuators are widely used as servos in automation systems or as driving devices of hydraulic and pneumatic electro-valves. The reverse movement is caused by a return spring.

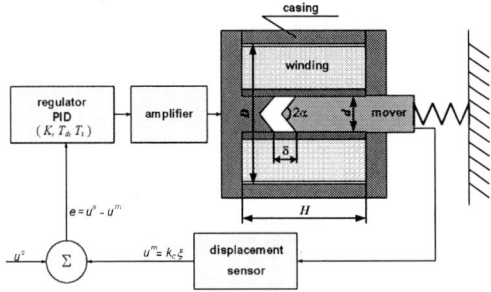

Fig. 1 Control system with plunger-type electromagnetic actuator

Electromagnetic plunger actuators have compact construction and the largest force-to-mass ratio. Usually, they operate bistably; the plunger is pulled in or released. Unfortunately, it is difficult to obtain multi-stable or continuously variable performance due to the very steep force-displacement steady-state characteristic $F_e(\delta)$. This characteristic should possibly be flattened. This is usually achieved by tapering the front surface of the plunger or magnetic shunts. Good results can also be obtained using saturated ferromagnetic cores [6]. But in order to obtain precisely linear control characteristic the closed loop error actuated control system must be applied [5, 6].

In Figure 2 steady state characteristics of two actuators are compared: in (a) the actuator with a plunger with a flat front surface is supplied with DC rated voltage $U_N = 27\,\text{V}$. In (b) the same actuator, this time equipped with a tapered plunger, is supplied with much higher voltage $U = 60\,\text{V}$. Due to the conical plunger shape and strong core saturation the characteristic in (b) is almost horizontal. The dotted line illustrates the characteristic of the loading linear spring. As it can be observed, stable position of the mover can be obtained in (b). It corresponds to the air gap equal to 5 mm.

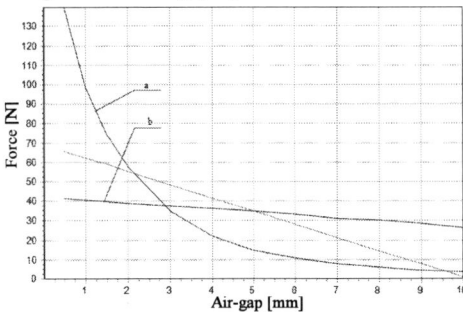

Fig. 2. Steady state characteristics: (a) flat plunger (U = 27 V), (b) tapered plunger (U = 60 V)

III. FIELD-CIRCUIT MODEL OF THE ACTUATOR

The non-linear and conducting properties of the solid ferromagnetic core have been taken into account. The field equation has been coupled to the electric circuit equations and to the equation of mechanical motion. Equations describing feedback, i.e. the displacement sensor and PID controller, has also been included.

Plunger actuators are characterized by axial symmetry. Therefore, the 2D cylindrical co-ordinate system r, z can be applied. The auxiliary magnetic potential $\phi(r,z,t) = \rho\, A_\varphi(r,z,t)$ is introduced, where A_φ is the tangential component of the magnetic vector potential, $\rho = 2\pi r$. The complete model of electromagnetic and mechanical phenomena comprises [3, 6, 7]:

(a) the equation describing a transient electromagnetic field in a non-linear conducting and moving environment:

$$\vec{\nabla}\cdot\left(\nu\rho^{-1}\vec{\nabla}\Phi\right) = -J + \gamma\rho^{-1}\left(\frac{\partial\Phi}{\partial t} + v\frac{\partial\Phi}{\partial z}\right), \qquad (1)$$

where J is the current density; v is the velocity; ν, σ are the reluctivity and conductivity, respectively;

(b) Kirchhoff's equation describing the actuator winding:

$$d\psi/dt + R\cdot i = u, \qquad (2)$$

where: i, u are the winding current and the supply voltage, respectively, R is the winding resistance; ψ is the magnetic flux linkage;

(c) the equation of mechanical motion:

$$m\frac{d^2\xi}{dt^2} + k_f\frac{d\xi}{dt} + F_{lo}(\xi) = F_e(\xi, \Phi), \qquad (3)$$

where m is the mass of moving parts, ξ is the mover displacement (linear co-ordinate), k_f is the friction coefficient, F_e, F_{lo} are the electromagnetic and loading forces, respectively;

(d) the equation describing the proportional-plus-integral-plus-derivative controller (PID controller):

$$u = K\left(e(t) + \frac{1}{T_i}\int_0^t e\,dt + T_d\frac{de}{dt}\right), \qquad (4)$$

$$e = u^s - u^m, \qquad u^m = k_{sen}\xi, \qquad ($$

where u is the controller output signal; $e = u^s - u^m$ is the voltage error, which is equal to the difference between the set value u^s proportional to the set value ξ^s of the mover position, and the measured sensor output voltage u^m proportional to the actual position ξ; k_{sen} is the sensor constant, K is the controller amplification; T_d, T_i are the controller differentiation and integration constants.

IV. NUMERICAL IMPLEMENTATION OF THE ALGORITHM

Numerical implementation is based on the finite element method (FEM). In order to solve the non-linear system of coupled field-circuit equations the Newton-Raphson iterative process has been adopted.

To solve transient and dynamic problem the time-stepping Cranck-Nicolson procedure has been used. As a result, for the n-th instant, the following non-linear set of equations is obtained:

$$\mathbf{M}_n(\mathbf{\Phi}_n, \xi) \cdot \mathbf{\Phi}_n = \mathbf{N} i_n + \mathbf{G} \mathbf{E}_{n-1} \qquad (5)$$

where: $\mathbf{M}_n(\mathbf{\Phi}_n, \xi) = \mathbf{S}_n(\mathbf{\Phi}_n, \xi) + 2(\Delta t)^{-1} \mathbf{G}$,

$$\mathbf{E}_{n-1} = 2(\Delta t)^{-1} \mathbf{\Phi}_{n-1} + \frac{d\mathbf{\Phi}}{dt}\Big|_{n-1} .$$

Here, $\mathbf{S}$ is the stiffness matrix, $\mathbf{\Phi}$ is the vector of nodal potentials, $\mathbf{G}$ is the matrix of conductances of rings formed by 2D mesh (in the region with eddy currents), $\mathbf{N}$ is the matrix of turn numbers associated with the nodes.

If the current values $i_n = i(t_n)$ are enforced then equation (5) may be solved directly at each time-step. Otherwise, the circuit equations must be taken into account [2, 3, 7]. According to the assumed difference scheme, the electric circuit equation takes the following discrete form:

$$2(\Delta t)^{-1} \mathbf{N}^T \mathbf{\Phi}_n + R\, i_n = V_n \qquad (6)$$

where: $V_n = u_n + \mathbf{N}^T \left\{ 2(\Delta t)^{-1} \mathbf{\Phi}_{n-1} + d\mathbf{\Phi}/dt\big|_{n-1} \right\}.$

Similarly the discrete form of the motion equation can be written as follows:

$$m\left(4(\Delta t)^{-2}(\xi_n - \xi_{n-1}) - 2(\Delta t)^{-1} v_{n-1} - dv/dt\big|_{n-1}\right) + $$
$$+ k_f\left(2(\Delta t)^{-1}(\xi_n - \xi_{n-1}) - v_{n-1}\right) + F_{lon} = F_{en} \qquad (7)$$

wherefrom the formula for the displacement ξ_n in successive time-steps may be derived:

$$\xi_n = \xi_{n-1} + \frac{(2m + \Delta t\, k_f)\Delta t\, v_{n-1} + (F_{en} - F_{lon} + m\, a_{n-1})(\Delta t)^2}{4m + 2\Delta t k_f} \qquad (8)$$

The position ξ_n at the n-th time-step is not known in advance, hence the forces $F_{en} = F_e(\mathbf{\Phi}_n, \xi_n)$ and $F_{lon} = F_{lo}(\xi_n)$ in the right-hand side of (6) are not known explicitly. Therefore, in order to determine ξ_n, iterative calculations are required. Each "superior" iteration associated with the calculation of the position ξ_n includes internal iterations concerning non-linearity. Hence, field distribution is computed dozens of times at each time-step. For this reason it is very important to restrict the number of "superior" iterations to maximum 2 or 3. A very effective procedure is proposed. At the beginning of each time-step the following ordinary differential equation of motion is solved:

$$m\frac{d^2\zeta}{d\tau^2} + k_f\frac{d\zeta}{d\tau} + F_{lo}(\zeta) = \widetilde{F}_e(t_{n-1} + \tau), \quad \tau \in (0, \Delta t) \qquad (9)$$

where $\tau \in (0, \Delta t)$, $\Delta t = t_n - t_{n-1}$ is the considered n-th time period, $\widetilde{F}_e(\tau)$ is the polynomial extrapolation of the force of order $r \leq n$-1 constructed on the basis of the points (F_{ei}, t_i), where $F_{ei} = F_e(t_i)$; $i = n-r, n-(r-1), \dots, n-1$. The initial conditions are: $\xi(\tau = 0) = \xi_{n-1}$, $d\xi/d\tau\big|_{\tau=0} = v_{n-1}$, $F_e(\tau = 0) = F_{en-1}$.

After solving equation (9) the tentative (approximate) displacement and force values: $\xi(\tau = \Delta t) = \widetilde{\xi}_n$ and $\widetilde{F}_{lon} = F_{lo}(\widetilde{\xi}_n)$ are calculated. Next, the grid is re-meshed accordingly to the change of the position $\widetilde{\xi}_n$ and the system of field-circuit equations is solved and the electromagnetic force F_{en} is calculated. Finally, after using (8), the new value ξ_n is obtained. If the difference between $\widetilde{\xi}_n$ and ξ_n is greater than the permissible incorrectness ε_ξ, then the calculation is repeated with a new value of $\widetilde{\xi}_n$.

Finally, according to assumed scheme, the discrete equation describing the PID controller has been formed:

$$u_n = Ke_n + KT_d\left\{2(\Delta t)^{-1}(e_n - e_{n-1}) - de/dt\big|_{n-1}\right\} + $$
$$+ \frac{K}{T_i}\left\{J_{n-1} + 0.5(e_n + e_{n-1})\Delta t\right\} \qquad (10)$$

Here J_{n-1} is the value of the integral at the previous instant, $e_n = u^s - k_{sen}\widetilde{\xi}_n$ is the error for $t = t_n$ and $\xi = \widetilde{\xi}_n$. The displacement $\widetilde{\xi}_n$ is determined iteratively and so is the error e_n.

V. DYNAMIC OPERATION OF THE ACTUATOR WITHOUT FEEDBACK

Dynamic operation of the tapered plunger electromagnetic actuator, after the application of the DC supply voltage $U = 60$ V has been considered. The steady state characteristic and the loading characteristic are shown in Fig. 2. The air gap $\delta(t)$ has changed from the initial stable value $\delta_b = 8$ mm to the final stable value $\delta_f = 2$ mm. The displacement $\xi(t) = \delta_b - \delta(t)$ has changed in interval $(0, 6\text{mm})$. The actuator has been loaded with a spring having the linear characteristic $F_{lo}(\xi) = F_{lo}(0) + k_s \xi$, where initial force $F_{lo}(0) = 14.4$ N and coefficient of elasticity $k_s = 6800$ N/m.

First, a very small friction coefficient $k_f = 0.004$ Ns/m has been assumed. As it can be observed in Fig. 2, there is a stable point of equilibrium for $\delta_s = 5$ mm ($\xi_s = 3$ mm), though due to mechanical and magnetic inertia this point has not been reached. The mover attains the final position $\delta = \delta_f$, i.e. it strikes on the stopper. In order to obtain the point $\delta_s = 5$ mm strong mechanical or electrical damping is necessary.

Figure 3b shows the dynamic operation of the same system but supplemented with a mechanical damper of linear force-velocity characteristic. Large value of coefficient $k_f = 0.4$ Ns/m has been assumed. The time-variations of the winding current $i(t)$ related to the steady-state value, the electromagnetic force $F_e(t)$ related to the maximum value and the relative displacement $x(t) = \xi(t)/(\delta_b - \delta_f)$ are shown. The point $\delta_s = 5$ mm is attained but unfortunately significant oscillations occur and the response time increases significantly.

2402

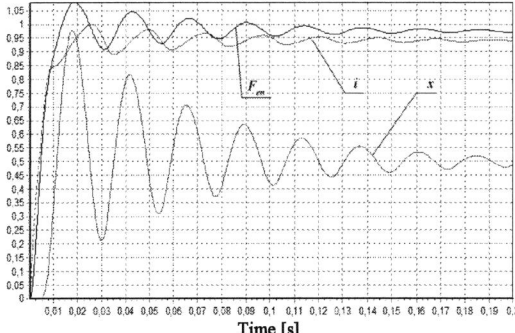

Fig. 3. Dynamic operation of the actuator for U = 60 V, k_f = 0,4 Ns/m

VI. DYNAMIC OPERATION OF THE ACTUATOR IN A CLOSED-LOOP AUTOMATIC CONTROL SYSTEM

As it has been noticed in the previous sections, in order to attain a stable position inside the range (δ_b, δ_f), a strong damping is necessary. Therefore, an error actuated control system with a PID controller has been used. The differentiating term plays role of "electric damper".

The set value ξ^s = 3 mm (δ^s =5mm, x^s =0.5) has been imposed. It corresponds to the set voltage signal $u^s = k_{sen}\xi^s = 0.3$ V. The dynamic response for two different PID controller settings is presented in Fig. 4a, b.

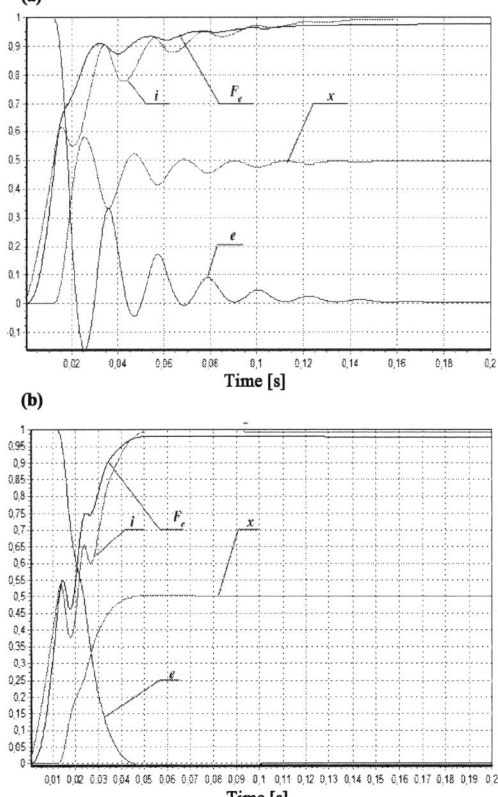

Fig. 4. Dynamics of the system for two different controller settings:
(a): K =100, T_d = 0,0029 s, T_i = 0,012 s,
(b): K = 100, T_d = 0,014 s, T_i = 0,012 s

The time -variations of the relative values of the current $i(t)$, the force $F_e(t)$ the displacement $x(t)$ and the error $e(t)$ are shown. As it can be observed, when the constant T_d is too small considerable oscillations occur and the response time increases; in the presented example – from 0.045 s to 0.125 s.

Figures 5a, b illustrate the operation of the same system for different dynamic disturbances. The controller settings are the same as in the previous examples. At $t = 0$ the set value $x^s = 0.5$ has been imposed. Next, at $t = 0.2$ s the loading force $F_{lo}(0)$ has been stepwise decreased – from 14.4 to 8.0 N. Subsequently, at $t = 0.35$ s the previous value of force has been reinstated. As it can be observed, the system operates stably.

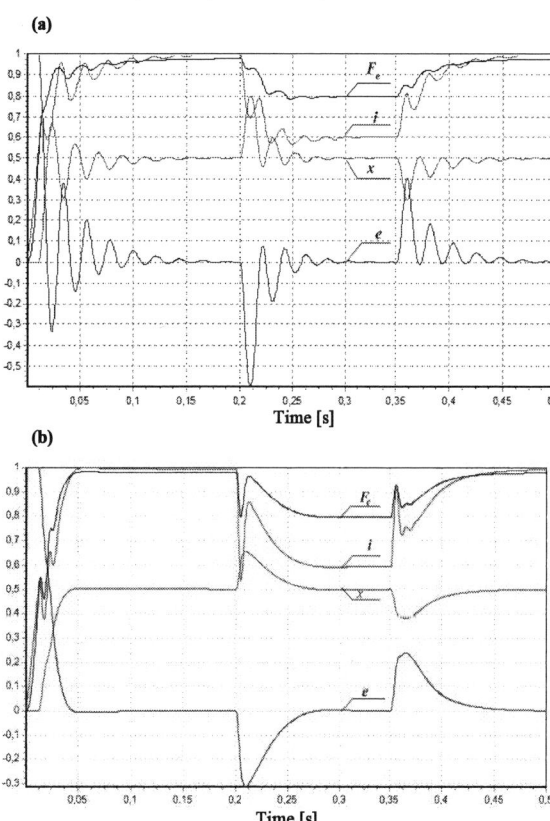

Fig. 5. Behavior of the system during disturbances

Figures 6a, b illustrate comparison of the response of two systems (a) with standard amplifier and (b) with PWM amplifier with the pulse frequency $f = 1$ kHz. The following regulator settings has been assumed: $K = 100$, $T_d = 12$ ms, $T_i = 14$ ms.

VII. OPTIMIZATION OF THE ACTUATOR DIMENSIONS AND THE REGULATOR SETTINGS

The optimisation solver has been elaborated regardless of the software for field circuit simulation of the actuator and feedback. Both units are connected through the transformation of the real design variables s_i used in the model of the actuator and their dimensionless counterparts x_i used in the optimisation procedure. The set of the data must include the lower s_{imin} and upper s_{imax} limits (expected values) of each variable s_i.

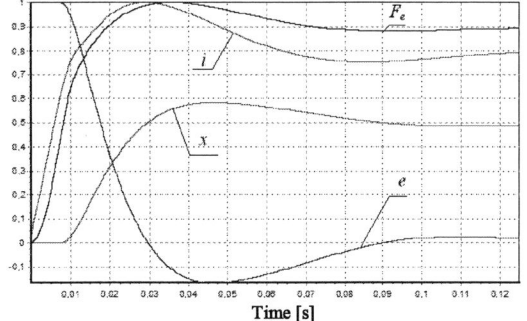

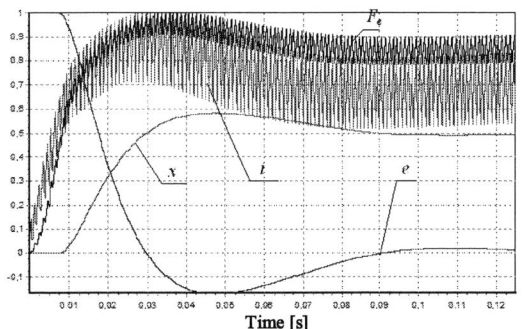

Fig. 6. Dynamic response of the system with (a) standard and (b) PWM amplifier

The variable transformations are given by the relations:

$$x_i = (s_i - s_{imin})/(s_{imax} - s_{imin}), \qquad (11)$$

If $s_i \in \langle s_{imin}, s_{imax} \rangle$ then $x_i \in \langle 0,1 \rangle$.

In the elaborated genetic procedure objective function $f(\mathbf{X})$ (individuals adaptation) is maximised. But the natural criterion of the optimisation of the dynamic system under the consideration is to minimize the set-up time T_{set} of the error actuated system. Therefore, the criterion has to be modified. Assuming the time of reference T_{ref} the objective function is constructed as follows:

$$f(\mathbf{X}) = 1 - T_{set}/T_{ref} \qquad (12)$$

Value of T_{ref} should be of the order $2(T_{set})_{opt}$, where $(T_{set})_{opt}$ is the expected value of the setting time for optimal individual. In such case, the conversion (12) practically doesn't change the measure of the differences between individuals, what ensures unchanged probability of individuals reproduction.

A. Optimisation of the Actuator Dimensions

The set value $\xi^s = 3$ mm ($\delta^s = 5$mm, $x^s = 0.5$) has been imposed. It has been assumed, that mover position is set up when the relative error decreases below 5%. The set-up time T_{set} was minimised.

First, the relatively simple problem was being solved. Initially, two design variables: relative angle of the tapered plunger $s_1 = \alpha/90^\circ$ and its relative diameter $s_2 = d/\delta_b$ (see Fig. 1) have been assumed. These parameters have the crucial influence on the winding

inductance, the electromagnetic force and mass of the moving part. Therefore, they have significant influence on dynamic operation of the actuator. The outer dimensions (Fig. 1) had the constant values $D = 26$ mm, $H = 22$ mm during the optimisation process. Number of individuals $n_{inv} = 200$ has been assumed. The following regulator settings have been assumed: $K = 100$, $T_d = 12$ ms, $T_i = 14$ ms.

After 28 generations the optimal design variables $\alpha = 30.3^\circ$, $d = 11.01$ mm and optimal set-up time $T_{set} = \mathbf{31}$ ms have been obtained, what is by 42 % less than for best variant drawn during the initiation stage. The dynamic response for the optimal variant is presented in Fig. 7.

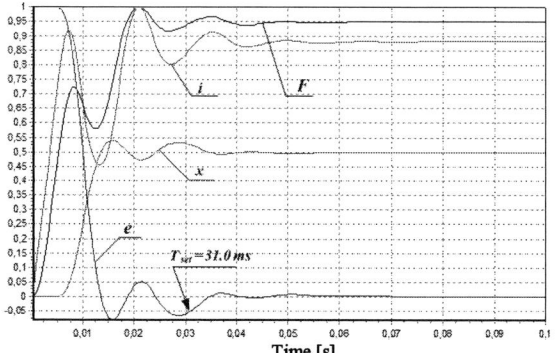

Fig.7 Dynamic response after structure optimisation

B. Optimisation of the PID regulator settings

This time, as the design variables the regulator three settings: $s_1 = K/100$, $s_2 = T_d/T_{cr}$, $s_3 = T_i/T_{cr}$, have been assumed. Here $T_{cr} = 0.025\,s$ is so called critical (reference) value. The set-up time T_{set} was minimised. The actuator structure optimal parameters $\alpha = 30.3^\circ$, $d = 11.01$ mm calculated in previous section are taken into computation. As a result of optimisation the optimal PID regulator setting have been obtained $K = 127$, $T_d = 7.5$ ms, $T_i = 9.275$ ms. For these optimal settings, the set-up time T_{set} decreased to $\mathbf{27.5}$ ms. The dynamic response for this case is shown in Fig. 8.

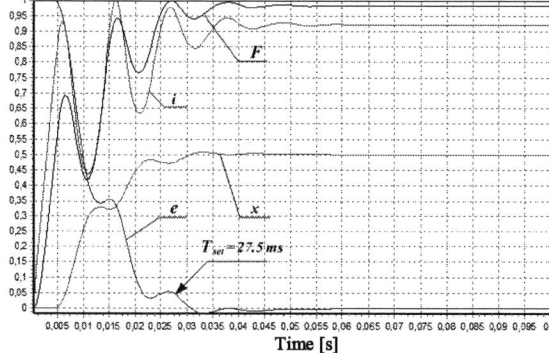

Fig.8 Dynamic response after optimisation of the regulator settings

Such iterative procedure, i.e. alternately optimisation of the actuator structure and regulator settings may lead to the optimal comprehensive variant of the whole system, i.e system optimal in relation to the both group of variables. As an example, the results obtained after renewed optimisation of the actuator structure (assuming obtained optimal regulator settings) are shown in Fig. 9.

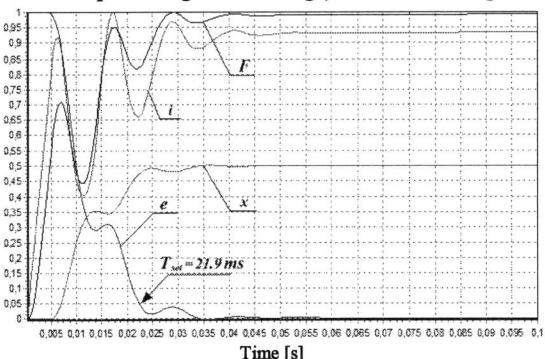

Fig. 9. Dynamic response after second optimization of the structure for optimal regulator settings

Due to this additional step the set-up time T_{set} decreased by 25.5 % – from **27.5** to **21.9** ms.

However, such iterative procedure is very time consuming. Much more effective is comprehensive optimisation of the system, assuming five design variables. This is because the computation time in genetic algorithm doesn't depends on the number of variables as strongly as in case of deterministic methods.

C. Complex optimisation of the actuator and regulator

In this section, the results of simultaneous optimisation of both: the actuator parameters and regulator settings are presented. First, the simpler problem has been solved. <u>Five</u> design variables: $s_1 = \alpha/90^\circ$, $s_2 = d/\delta_b$, $s_3 = K/100$, $s_4 = T_d/T_{cr}$, $s_5 = T_i/T_{cr}$ have been assumed. The following optimal values of these variables have been obtained: $\alpha = 28.33^\circ$, $d = 10.17$ mm, $K = 140.80$, $T_d = 7.47$ ms, $T_i = 5.58$ ms. Substantial improvement of the dynamic properties has been achieved. The set-up time desreased to **19.6 ms**. The response for this variant is presented in Fig. 10.

Next, the outside dimensions (see Fig. 1) have been also included into the optimisation process. The extended complex problem with <u>seven</u> variables has been solved. The variables are assumed as follows: $s_1 = \alpha/90^\circ$, $s_2 = d/\delta_b$, $s_3 = D/d$, $s_4 = H/\delta_b$; $s_5 = K/100$, $s_6 = T_d/T_{cr}$, $s_7 = T_i/T_{cr}$ –.

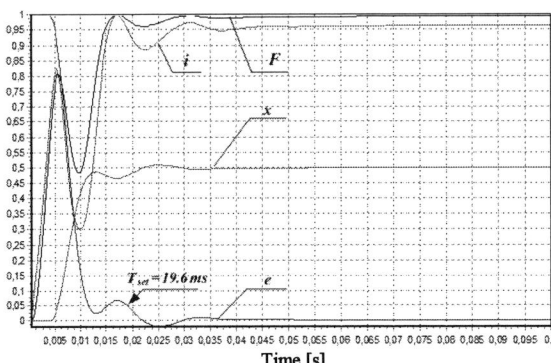

Fig. 10. Dynamic response after comprehensive optimization (5 variables)

The following optimal values of actuator and regulator parameters has been obtained: $\alpha = 35.23^\circ$, $d = 10.52$ mm, $D = 15.15$ mm, $H = 38.80$ mm, $K = 154.9$, $T_d = 6.28$ ms, $T_i = 6.67$ ms.

As a result of such comprehensive optimisation, the lowest time $T_{set} = \mathbf{10.78}$ ms has been attained. Results are presented in Fig. 6. The comprehensive, simultaneous optimisation of all main actuator dimensions and regulator settings enabled to reduce the set-up time almost by 65% in comparison to previous optimal system, presented in Fig. 11.

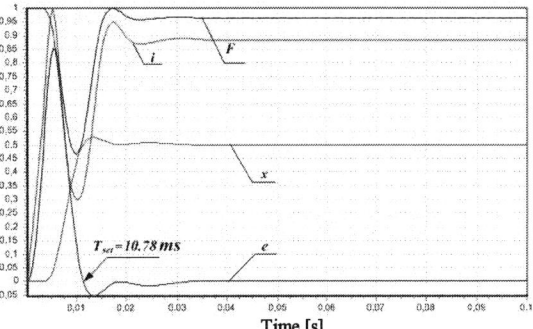

Fig. 11. Dynamic response after comprehensive optimisation (7 variables)

VIII. CONCLUSIONS

The elaborated algorithm and the computer code can be an effective tool for field-circuit simulation of the dynamics of an electromagnetic linear actuator that operates in an automatic control system.

The influence of PID controller settings on the operation of the actuator can be investigated. Different dynamic disturbances such as step change of the set value of mover position or change of loading force can be analysed.

Due to the complex optimisation of the regulator settings and actuator parameters, the cost of design and production the prototypes can be essentially decreased.

REFERENCES

[1] Benderskaya G., Clemens M., De Gersem H,., Weiland T., Embedded Runge-Kutta Methods for Field Circuit Coupled Problems With Switching Elements, IEEE Trans. Magn., vol.41, No. 5, May 2005, pp.1612-1615.

[2] Benderskaya G., De Gersem H,., Weiland T., Interpolation Technique for Effective Determination of Switching Time Instant for Field Circuit Coupled Problems With Switching Elements, Compel, Vol. 25, No.1, 2006, pp 64-70.

[3] De Gersem, H.; Mertens, R.; Lahaye, D.; Vandewalle, S.; Hameyer, K.; Solution Strategies for transient, field-circuit coupled systems, IEEE Trans. Magn., Vol.36, No. 4, July 2000, pp.1531-1534.

[4] Ho S.L., Li Y., Lin X., Xu J.Y., Lo Wc., Wong H.C., Design and Dynamic Analyses of Permanent Magnetic Actuator for Vacuum Circuit Breaker, 14th Conference on the Computation of Electromagnetic Fields, 2003, Saratoga Springs, New York USA

[5] Kiam Heong A., Chong G., Yun Li, PID control system analysis, design, and technology, IEEE Transactions on Control System Technology,Vol.13, 2005, pp. 559 - 576

[6] Nowak L., Radziuk K., Dynamics of an Electromagnetic Linear Actuator Operating in Error Actuated Control System, COMPEL, Vol. 26, No. 4, 2007, pp. 941-951.

[7] Nowak L., Radziuk K., Transient analysis of PWM-excited electromagnetic actuators, IEE Proceedings-Science, Measurement and Technology, Vol. 149, 2002, pp. 199 - 202

Simulation and Investigation of Magnetorheological Fluid Brake

Wiesław Łyskawiński, Wojciech Szeląg and Cezary Jędryczka

*Poznań University of Technology / Institute of Electrical Engineering and Electronics, Poznań, Poland, e-mail:
Wieslaw.Lyskawinski@put.poznan.pl, Wojciech.Szelag@put.poznan.pl, Cezary.Jedryczka@doctorate.put.poznan.pl

Abstract — The paper deals with coupled electromagnetic, hydrodynamic and mechanical motion phenomena in magnetorheological fluid (or simply MRF) brakes. The governing equations of these phenomena are presented. The numerical implementation of the mathematical model is based on the finite element method. Elaborated computer program is used to simulate the operation of MRF brake prototype. In order to verify elaborated algorithm and program, the investigation of the prototype of MR brake is carried out. The dynamic properties as well as mechanical and control characteristics of the brake are determined. The influence of MR fluid properties on functional parameters of the brake is also considered. Chosen results of simulations and measurements are presented.

Keywords — simulation, design, transducer, measurement, test bench.

I. INTRODUCTION

The characteristic feature of MRF is a dependence of their viscosity on magnetic field. A change in viscosity is inseparably connected with a change of yield stress in the fluid. The relationship between the yield stress τ_0 and the magnetic flux density B for MRF-132LD produced by Lord Corporation is shown in Fig. 1. The stress changes during the increase and decrease of magnetic flux density occur in microseconds [3, 6, 15]. The fluids retain their properties in the temperature range from $-40°C$ to $150°C$. The *B-H* characteristic for MRF 132LD is shown in Fig. 2. Relative magnetic permeability of the fluid is small, $\mu_r < 10$.

Owing to their properties, magnetorheological fluids are useful for the efficient control of the transmission of torques and forces. They are used, among others, in rotary brakes, clutches, and rotary and linear dampers. The working principle of magnetorheological electromagnetic transducers is based on the fact that viscosity changes when the fluid is exposed to magnetic field. The viscosity and the stresses increase with the growth of the field and so does the yield strength counteracting the motion of moveable parts in the transducers.

The paper proposes a mathematical model of coupled electromagnetic, hydrodynamic and mechanical motion phenomena that can be applied to simulate steady and transient states of the magnetorheological fluid brake. In order to verify elaborated method of analysis of the MR brake, the prototype of the brake with cylindrical rotor have been designed and tested at Poznań University of Technology. The influence of MR fluid properties on the brake parameters has been investigated. The tests have been performed for two types of MR fluids produced by

the Lord Corporation: MRF 132LD, MRF 132AD. Selected measurement results are shown in the paper.

II. COUPLED PHENOMENA MODEL

The phenomena observed in electromagnetic brakes with magnetorheological fluid need to be analysed in terms of fields. In the brake, the velocity field of the fluid depends on the angular velocity of the rotor and distribution of the yield stress in the fluid. These stresses are function of the magnetic field distribution. The yield strength counteracts the motion of the rotor in the brake. Therefore, the velocity field of the fluid and the mechanical stress field are coupled with the electromagnetic field. The field coupling makes the analysis of the transients in the brake highly complicated. What makes it even more intricate is the changing character of those fields and the non-linear character of the equations describing them.

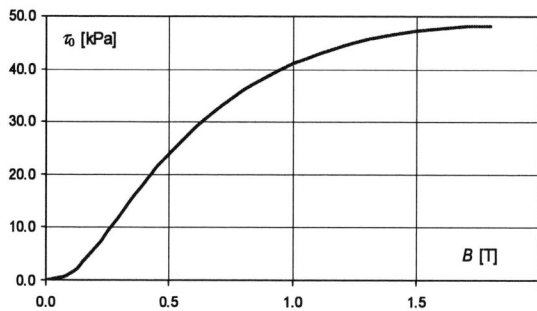

Fig. 1 The yield stress $\tau_0 = f(B)$ for MRF 132LD

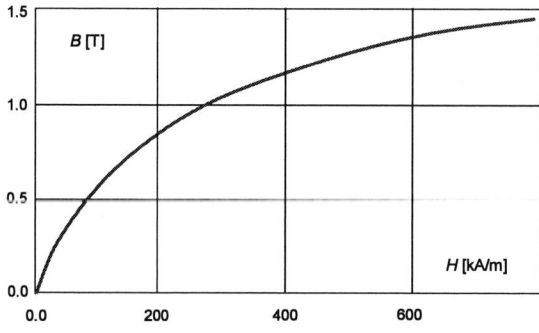

Fig. 2. *B-H* curve for MRF 132LD at 20°C

In the paper a model of coupled phenomena in magnetorheological transducers is presented. The

978-1-4244-1741-4/08/$25.00 ©2008 IEEE

attention was concentrated on electromagnetic and hydrodynamic phenomena and also on the dynamics of movable elements in the brake. The problems pertaining to heat, ventilation and vibration have been disregarded.

In the paper, a magnetorheological brake with axial symmetry is considered – Fig. 3. A cylindrical coordinate system r, z, ϑ was applied. In this case, the equation describing the transient electromagnetic field in the brake can be expressed as [9, 11, 13].

$$\frac{\partial}{\partial r}\left(\frac{1}{\mu l}\frac{\partial \varphi}{\partial r}\right)+\frac{\partial}{\partial z}\left(\frac{1}{\mu l}\frac{\partial \varphi}{\partial z}\right)=J-\frac{\gamma}{l}\frac{\partial \varphi}{\partial t}. \qquad (1)$$

Here $l = 2\pi r$, $\varphi = 2\pi r A_\vartheta$, where A_ϑ is the magnetic vector potential, $J = i/s$ is the current density in the winding, i is the winding current, s is the cross-section area of the conductor, μ is the magnetic permeability and γ is the conductivity of the region with eddy currents. For the MRF $\gamma = 0$.

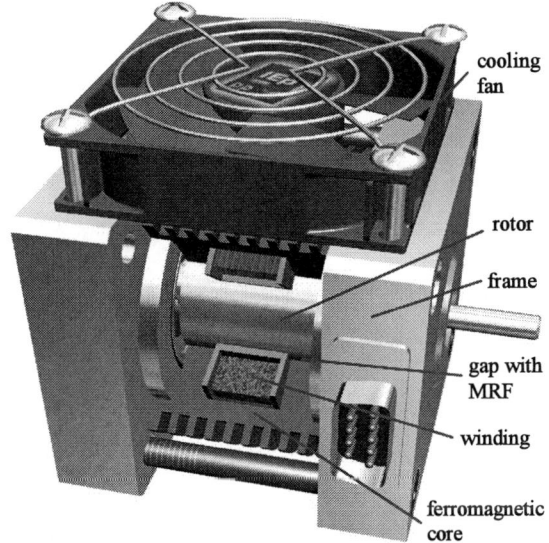

Fig. 3. The brake with MRF

In general, the transient electromagnetic field in MRF devices is voltage-excited. This means that the currents i in the windings are not known in advance, i.e. prior to the electromagnetic field calculation [1, 5, 12]. Therefore, it is necessary to consider the equations of the electric circuit of the device. The set of independent loop equations may be written as

$$\boldsymbol{u} = \boldsymbol{R}i + \frac{d}{dt}\boldsymbol{\varPsi}, \qquad (2)$$

where $\boldsymbol{u}$ is the vector of supply voltages, $\boldsymbol{i}$ is the vector of loop currents, $\boldsymbol{R}$ is the matrix of loop resistances, $\boldsymbol{\varPsi}$ is the flux linkage vector. The vector $\boldsymbol{\varPsi}$ is calculated by means of the field model.

The phenomenological approach was used to describe fluid dynamics. In this approach, the fluid is treated as a non-conducting continuum of properties determined by density ρ, dynamic viscosity υ and magnetic permeability μ [2, 10, 14]. In the hydrodynamic model, the laminar

flow of a non-compressible fluid is investigated [2, 4, 8]. The motion of the liquid in the ϑ-direction is caused by the motion of the rotor. It is assumed that the internal energy and temperature of the fluid are constant. For such conditions, the differential equation of the motion of the fluid may by written as [11]

$$\frac{\partial}{\partial r}\left(\frac{\upsilon_z}{l}\frac{\partial \phi}{\partial r}\right)+\frac{\partial}{\partial z}\left(\frac{\upsilon_z}{l}\frac{\partial \phi}{\partial z}\right)=\frac{\rho}{l}\frac{\partial \phi}{\partial t}. \qquad (3)$$

Here $\phi = 2\pi r v_\vartheta$, where v_ϑ is the component of velocity v in the ϑ-direction, ρ is the fluid density, and υ_z is the equivalent dynamic viscosity of the fluid.

The description of the problem (3) should be completed by non-slip boundary conditions $v_\vartheta = r\omega$ and $v_\vartheta = 0$ on the surface of the rotor and the frame, respectively, where ω is the angular velocity of the rotor.

In order to determine the equivalent dynamic viscosity of the fluid, physical properties of the fluid were taken into account. MRFs belong to the non-Newtonian group of fluids. The properties of such fluids can be described by the Bingham model [2, 8]. For Bingham model the equivalent dynamic viscosity of the fluid may be determined as [11]

$$\upsilon_z = \eta_p + \tau_0(B)/\|\boldsymbol{D}\| \quad for \quad \|\boldsymbol{\tau}\| > \tau_0(B), \qquad (4.a)$$

$$\upsilon_z = \infty \quad for \quad \|\boldsymbol{\tau}\| \le \tau_0(B). \qquad (4.b)$$

where η_p is a plastic viscosity of the fluid, $\boldsymbol{D}$ is the deformation tensor, and τ is the stress tensor.

The yield stress $\tau_0(B)$ in (4) is determined from characteristic shown in Fig. 1, on the basis of the distribution of the magnetic flux density B obtained from (1) and (2). The norms $\|\boldsymbol{D}\|, \|\boldsymbol{\tau}\|$ of the deformation tensor $\boldsymbol{D}$ and of the stress tensor τ can be expressed as [8, 11, 12]

$$\|\boldsymbol{D}\| = \left(\frac{1}{2}\sum_{i=1}^{2}\sum_{j=1}^{2}D_{i,j}^2\right)^{1/2}, \quad \|\boldsymbol{\tau}\| = \left(\frac{1}{2}\sum_{i=1}^{2}\sum_{j=1}^{2}\tau_{i,j}^2\right)^{1/2}, \quad (5.a)$$

where

$$D_{i,j} = 0 \quad for \quad \|\boldsymbol{\tau}\| \le \tau_0(B), \qquad (5.b)$$

$$\mathbf{D} = 0.5\left[\nabla\mathbf{v} + (\nabla\mathbf{v})^T\right] \quad for \quad \|\boldsymbol{\tau}\| > \tau_0(B), \qquad (5.c)$$

$$\tau_{i,j} = \left(\eta_p + \tau_0(B)/\|\mathbf{D}\|\right)D_{i,j} \quad for \quad \|\boldsymbol{\tau}\| > \tau_0(B). \quad (5.d)$$

The fluid behaves like an elastic body for $\|\boldsymbol{\tau}\| \le \tau_0(B)$ and like a body of plastic viscosity η_p for $\|\boldsymbol{\tau}\| > \tau_0(B)$. Moreover, from (5.d), stress $\tau_{i,j}$ has two components. The first one depends on plastic viscosity η_p, while the second one is connected with yield stress $\tau_0(B)$, induced in the fluid by magnetic field.

During the simulation of phenomena in a MRF electromechanical brake, equations (1), (2), (3) describing the electromagnetic and hydrodynamic phenomena must be solved together with the equation of dynamics of brake's movable elements. This equation of mechanical equilibrium has the following form

$$J_i \frac{d\omega}{dt} + T_0 + T_B = T_d, \qquad (6)$$

where J_i is the moment of inertia, $T_B = T_B(B, \omega)$ is the braking torque associated with the rotation of the rotor and the occurrence of magnetic field in the brake, T_0 is the braking torque produced in the packing and bearings, and T_d is the external driving torque. The braking torque T_B can be determined as an integration of stress tensors along closed surface s placed in the MRF and contained rotor using the equation

$$T_B = \oiint_s r(\tau_{\vartheta} + \tau_{e\vartheta}) ds. \qquad (7)$$

The vectors $\tau_{\vartheta}, \tau_{e\vartheta}$ in this equation describe the stress in the fluid and the electromagnetic stress acting in the direction ϑ at a tangent to the external surface of the brake's rotor.

III. FINITE ELEMENT FORMULATION

The equations (1), (2), (3), (6) are coupled through the viscosity function $v_z = v_z(B, \|D\|)$, the braking torque $T_B = T_B(B, \omega)$ and through the boundary condition $v_{\vartheta} = r\omega$. Therefore, these equations should be solved simultaneously.

In order to solve coupled equations, the finite element method and a "step-by-step" procedure were used [5, 12]. The finite element and time discretization lead to the following system of non-linear algebraic matrix equations

$$\begin{bmatrix} S_n + (\Delta t)^{-1} G & -w \\ -w^T & -\Delta t R \end{bmatrix} \begin{bmatrix} \varphi_n \\ i_n \end{bmatrix} = \begin{bmatrix} (\Delta t)^{-1} G \varphi_{n-1} \\ -\Delta t u_n - w^T \varphi_{n-1} \end{bmatrix}, \quad (8)$$

$$[S'_n + (\Delta t)^{-1} G'] \phi_n = (\Delta t)^{-1} G' \phi_{n-1}, \qquad (9)$$

where n denotes the number of time-steps, Δt is the time-step length, S, S' are the magnetic and hydrodynamic stiffness matrices, φ, ϕ are the vectors of the nodal potentials φ and ϕ respectively, w^T is the matrix that transforms the potentials φ into the flux linkages with the windings, G is the matrix of conductances of elementary rings formed by the mesh and G' is the matrix whose elements depend on the dimensions of the elementary rings and fluid density ρ.

Motion equation (6) is approximated by difference formula [11]

$$J_i(\alpha_{n+1} - 2\alpha_n + \alpha_{n-1})/(\Delta t)^2 = T_{d,n} - T_{B,n} - T_0, \quad (10)$$

where α is the angular position of the rotor, $T_{d,n} = T_d(t_n)$, $T_{B,n} = T_B(t_n)$.

The angular velocity ω of the rotor may be calculated according to the expression

$$\omega(t_n + 0.5\Delta t) = (\alpha_{n+1} - \alpha_n)/\Delta t. \qquad (11)$$

The braking torque $T_{B,n}$ is described by formula (7). In the considered brake, the component B_{ϑ} of the magnetic flux density B is equal to zero. Therefore, in (7) the component $\tau_{e\vartheta}$ of Maxwell stress tensor is equal to zero.

The main difficulty in obtaining a numerical solution of magnetorheological fluid flow problem, given by equation (3), is the existence of a surface separating the regions of sheared and non-sheared fluid [2]. The position of this surface is not known in advance, i.e. prior to the velocity field calculation [6]. Utilization of the previously described equivalent dynamic viscosity formulations eliminates the need to track the surface separating these two flow regions and simplifies the solution. It leads, however, to singularities since the equivalent dynamic viscosity v_z attains an infinite value in the regions where $\|D\| = 0$, i.e. in the regions where the fluid behaves like a solid body. In order to avoid such a problem, equations (4.a) and (4.b) are replaced by following equation proposed in [6]

$$v_z = \eta_p + \frac{\tau_o}{\|D\|}\left(1 - e^{-m\|D\|}\right), \qquad (12)$$

where m is an exponential growth parameter.

Extensive numerical experiments led to the establishment of $m = 100$ as high enough to obtain accurate solutions [6].

IV. SIMULATION AND INVESTIGATION RESULTS

The presented algorithm for solving the equations of coupled phenomena model was implemented in a computer program. The program was used for simulation of the magnetorheological brake operation. The transients and the steady state in the prototype of the electromagnetic brake built at Poznań University of Technology, shown in Fig. 3, were considered. This is a cylindrical-rotor brake system. Magnetic field is excited by a ring coil in a stator. The diameter and the length of the rotor are about 26 and 27 mm, respectively. The maximum braking torque is about 1.7 Nm. One of the advantages of the brake is low electric power consumption of the winding, not exceeding a couple of watts.

First, the elaborated program was used to determine the electromagnetic field and the velocity field of the fluid when constant voltage is applied to the winding of the brake. It was assumed that the rotor's angular velocity ω equals 150 rad/s and that in the brake the fluid of MRF 132 LD type was used. Selected examples of the magnetic field lines' distributions and the respective distributions of lines connecting the points with identical velocity values are shown in Fig. 4.

a)

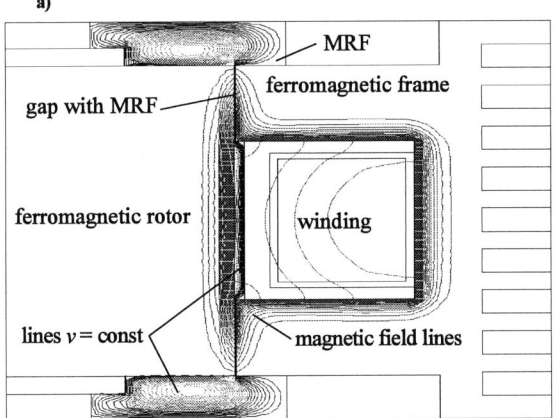

b)

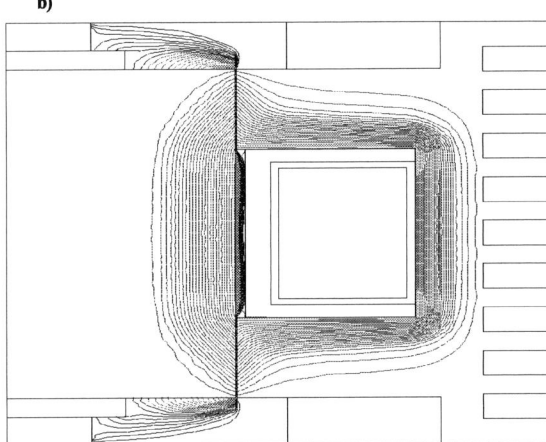

c)

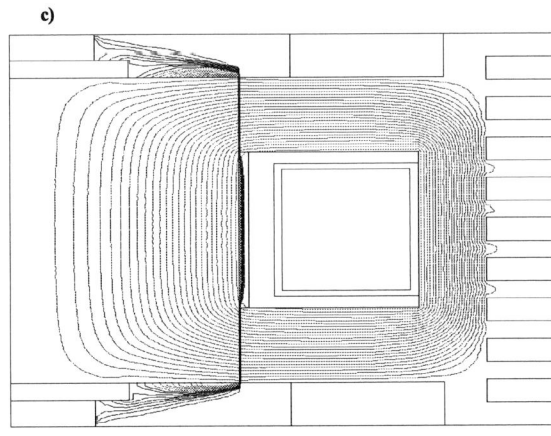

Fig. 4. Distribution of the magnetic field lines and the lines $v =$ const for
a) $t = 0.004$s, b) $t = 0.05$s, c) steady state

Next, dynamic states in a brake driven by an induction motor were analysed. The motor was supplied from an inverter. The characteristic $T_d = f(\omega)$ of the motor and the braking torque $T_0(\omega)$ from equation (10) associated with friction were measured. A bench for magnetorheological brake test is shown in Fig. 5. The test bench can be used to measure and register time characteristics of torque,

rotational speed, supply voltage, and current in the excitation winding [7].

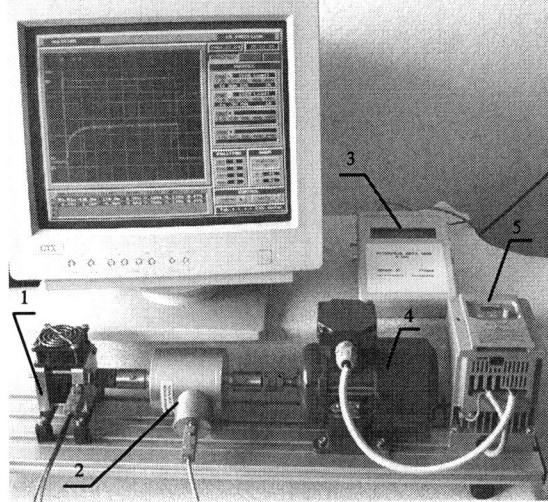

Fig. 5. Test bench for the MRF brake: brake (1), interface BETA (2), measuring head Mt5Nm (3), induction motor (4), converter (5)

A transient state associated with the supply of voltage on the field-exciting coil in the brake was considered. It was assumed that prior to the voltage supply the angular velocity ω of the motor had been equal to 50 rad/s. The calculated torque-time $T(t) = J_r d\omega/dt + T_B + T_0$, current-time $i(t)$ and the velocity-time $\omega(t)$ characteristics are shown in Fig. 6.

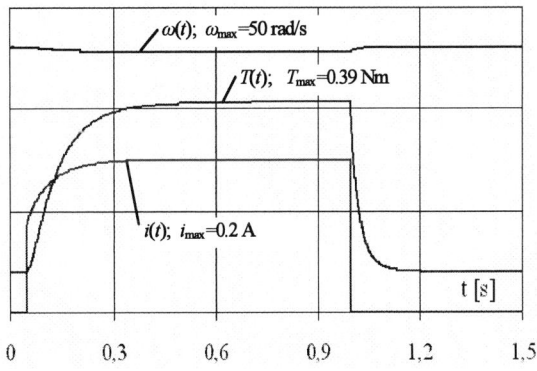

Fig. 6. Calculated torque-time $T(t)$, current-time $i(t)$ and the velocity-time $\omega(t)$ characteristics for the length of gap $g = 0.6$ *mm*

In order to verify the calculations, the driving torque $T_d(t)$, the current $i(t)$ and the angular velocity $\omega(t)$ were measured on the prototype of the brake. The results are shown in Fig. 7. The investigation proved that the braking torque distribution at constant voltage supply to the excitation winding is multiexponential.

The steady states of the MRF brake were also investigated. A prototype brake was tested with two MR fluids: MRF 132AD and MRF 132LD. The influence of properties of MR fluids (Table I) and the length g of gap between stator and rotor on the parameters of the brake were examined.

2409

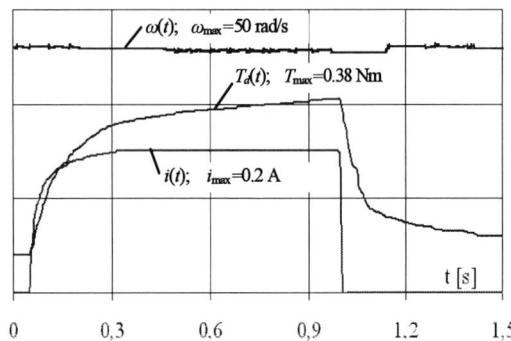

Fig. 7. Measured torque-time $T_d(t)$, current-time $i(t)$ and the velocity-time $\omega(t)$ characteristics, $g = 0.6$ mm

TABLE I.
SELECTED PROPERTIES OF TESTED MRF FLUIDS

Type of fluid	Carrier fluid	$\tau_0(B)$ for $B = 1T$	Viscosity for $B=0$, $D=100$ 1/s
MRF 132LD	Mineral oil	41 kPa	0.30 Pa s
MRF 132AD	Mineral oil	38 kPa	0.18 Pa s

Selected measured relationships between the braking torque and the rotational speed of the rotor at a constant exciting current are shown in Fig. 8. It follows from the figure that for both types of MR fluids, the braking torque does not change significantly at rotational speeds exceeding 250 rpm.

In order to determine the control characteristics of the brake, the relationship between the braking torque and the exciting current at a constant rotational speed was tested. The measurements were repeated for different values of rotational speed, in both rotational directions of the rotor. Before each test, the magnetic circuit of the brake had been demagnetised. It was found that the direction of rotation does not have any influence on the torque value. The control characteristics for both fluids are shown in Fig. 9. The total braking torque for the current in the exciting coil $I = 0$ and the demagnetised magnetic circuit of the brake is equal to the sum of the torque T_0 associated with friction in the seals and bearings and torque $T_B(0,\omega)$. The component T_0 has been measured for the brake without magnetorheological fluid. The total torque is higher for MRF 132LD because of its higher viscosity. Higher torques were achieved for this fluid at identical values of the exciting current.

Fig. 9 shows also characteristics $T(I)$ calculated by means of the elaborated computer program for analysis of coupled phenomena in the brake. Good agreement between calculations and measurements has been achieved.

Influence of the length g of the gap between stator end rotor on breaking torque T has been investigated. The length of the gap has been changed in the range from 0.1 to 0.8 mm (by change of the rotor diameter). The characteristics $T(I)$ have been measured for several constant rotational speeds. The results obtained for $n = 500$ rpm are shown in Fig. 10. For small currents I, the influence of the gap g on the torque T was insignificant. For greater currents, the torque was increasing along with decrease of the length of the gap.

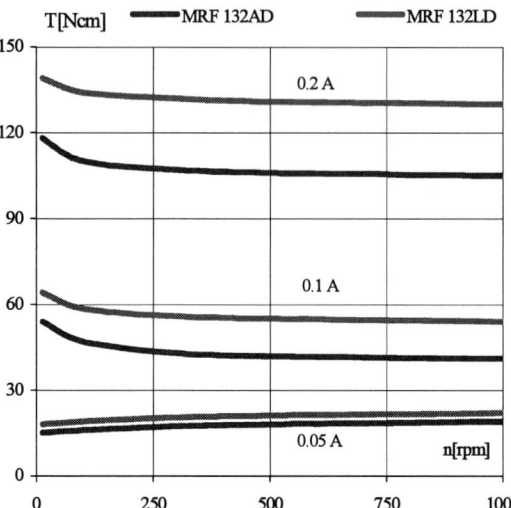

Fig. 8. Mechanical characteristics at constant current values in the excitation winding of the brake, $g = 0.1$ mm

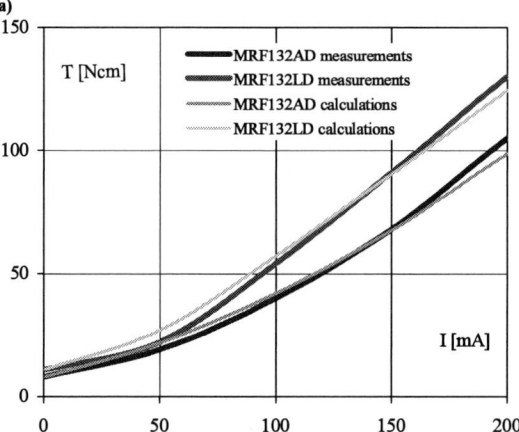

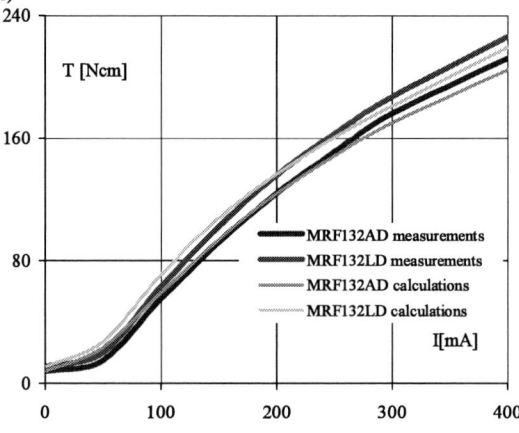

Fig. 9. Relationship between the braking torque and the current in the excitation winding for two different MR fluids at constant rotational speed (a) 1000 rpm (b) 12 rpm; $g = 0.1$ mm

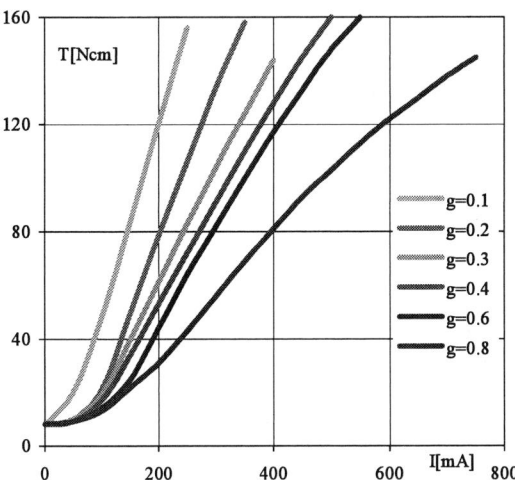

Fig. 10. Characteristics $T(I)$ for the brake filled by MRF 132 LD; g = const, n=500 rpm

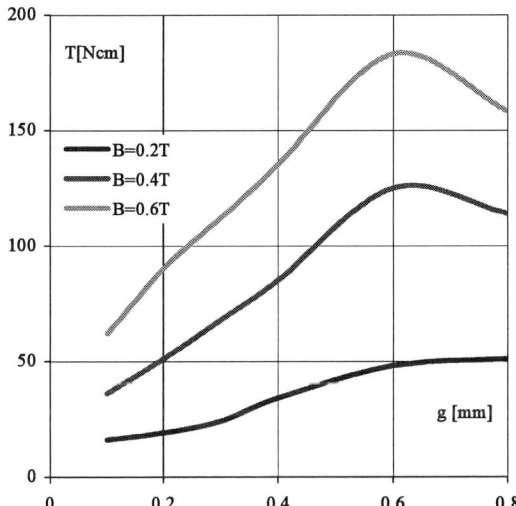

Fig. 11. Characteristics T(g) for B = const and n = 500 rpm

Elaborated program has been also used to establish the dependence of the magnetic flux density B in the gap on exciting current I, for several gaps length g. Next, on the basis of obtained relationships $B(I)$ and measured breaking torques $T(I)$, the characteristics $T(g)$ for B = const were determined, as shown in Fig. 11.

From Fig. 11, we can see that optimal length g of the gap is equal to 0.6 mm. For this length, and given magnetic flux density, maximum breaking torque T is obtained. Relatively big gap length $g = 0.6$ mm is also good from the brake feasibility point of view.

V. CONCLUSIONS

In the paper, a field model of coupled field phenomena in an electromechanical brake with magnetorheological fluid was presented. The algorithm for solving the equations of the model was described. In the consideration, the non-linear properties of materials, the eddy currents induced in solid elements, flows of the fluid and the rotor movement were taken into account.

On the basis of this algorithm a computer program was elaborated. The program proved to be useful in simulating the transients and the steady states of magnetorheological brake. Good concordance between the calculations and measurements was achieved. The influence of the properties of the MR fluid and the length of the gap between stator an rotor on the parameters of the MRF brake was investigated. Slightly greater breaking torque was obtaine

d using fluid of type MRF 132 LD. It has been determined that from the view point of the greatest braking torque, gap length 0.6 mm is optimal.

The presented model of coupled phenomena, elaborated software and results of investigations are very useful in MRF brake design process.

REFERENCES

[1] M. Besbes, Z. Ren, A. Razek, Finite element analysis of magneto-mechanical coupled phenomena in magnetostrictive materials, IEEE Trans. Magn., 1996, Vol. 32 No. 3, pp. 1058-1061.

[2] R. B. Bird., W.E. Stewart, E.N. Lightfoot, *Transport phenomena*, John Wiley & Sons, New York 1960.

[3] J.D. Carlson, D.M. Catanizarite, K.A. Clair, Commercial magnetorheological fluid device. *Proc. of 5th Int. Conf. On ER Fluids, MR Suspensions and Associated Technology*, Singapore 1996, pp. 20-28.

[4] T.J. Chung, *Finite Element Analysis in Fluid Dynamics*, McGraw-Hill Inc. 1978.

[5] A. Demenko, *Time stepping FE analysis of electric motor drives with semiconductor converter*, IEEE Trans. Magn. 1994, Vol. 30 No. 5, pp. 3264-3267.

[6] K.J. Hammand, *The effect of hydrodynamic conditions on heat transfer in a complex viscoplastic flow field*, International Journal of Heat and Mass Transfer, 2000, No. 43, pp. 945-962.

[7] W. Łyskawiński, W. Szeląg Magnetorheological fluid brake test stand. *Computer Applications in Electrical Engineering, Electrical Engineering Comeete of Polish Academy of Sciences*, Poznań University of Technology, Poznań, 2004, pp. 555-665.

[8] C. Nouar, I.A Frigaard., *Nonlinear stability of Poiseuilla flow of Bingham fluid: theoretical results and comparison with phenomenological criteria*, Journal Non-Newtonian Fluid Mechanic 2001, No. 100, pp. 127-149.

[9] L. Nowak, Simulation of the dynamics of electromagnetic driving device for comet ground penetrator, *IEEE Trans. Magn.* 1998, Vol. 34 No.5, pp. 3146-3149.

[10] R.E. Rosensweig, Ferrohydrodynamics, *Cambridge University Press*, 1985.

[11] W. Szeląg, Finite element analysis of coupled magnetorheological fluid devices, *COMPEL – The International Journal for Computation and Mathematics in Electrical and Electronic Engineering*, Emerald Group Publishing Limited, 2004, Vol. 23, No. 3, pp. 813-824.

[12] W. Szeląg, P. Sujka, R. Walendowski, Field-circuit transient analysis of a magnetorheological fluid brake, *COMPEL – The International Journal for Computation and Mathematics in Electrical and Electronic Engineering*, Emerald Group Publishing Limited, 2004, Vol. 23, No. 4, pp. 986-992.

[13] W. Szeląg, Simulation of coupled phenomena in magnetoreolorheological electromechanical transducers. *Electromotion*, Vol. 9, No. 4, October-December 2002, pp. 181-186.

[14] S.L. Verardi, J.R. Cardoso, A solution of two-dimensional magnetohydrodynamic flow using the finite element method, *IEEE Trans. Magn.* 1998, Vol. 34 No. 5, pp. 3134-3137.

[15] www.lord.com.

Field and Field-Circuit Description of Electrical Machines

Andrzej Demenko*, Kay Hameyer[†]

* Institute of Electronics and Electrical Engineering, PUT, Poznań, Poland, e-mail: *andrzej.demenko@put.poznan.pl*
† Institute of Electrical Machines, RWTH, Aachen University, Germany, e-mail: *Kay.Hameyer@iem.RWTH-Aachen.DE*

Abstract—The field and coupled field–circuit models of electrical machines are presented. The field model consists of: (a) finite element (FE) equations for the magnetic field, (b) equations describing eddy-currents and (c) equations, which describe the currents in the machine's windings. Moreover the FE equations are coupled by the electromagnetic torque to the differential equations of motion. In the presented field-circuit model the flux linkages with the windings are expressed by two components. One component with inductances and the other described by edge or nodal values of the magnetic potential. The FE equations are derived by using the notation of circuit theory. The approach to consider the differential equation of motion in the simulation is discussed in the paper.

Keywords—electrical machines, finite element method, magnetic field, eddy currents, field-circuit coupling.

I. Introduction

Since several years now, numerical methods to simulate the electromagnetic field are applied in the analysis and synthesis of electrical machines. The classical equivalent circuit methods are supported by additional procedures of field analysis. As a result we obtain the so called field-circuit model [31, 33]. The machine parameters and characteristics can be calculated on the basis of field formulation with omission of circuit equations. This description is treated as in fully field.

In the paper we try to systemize the applied models of electrical machines and actuators. We will consider the electrical machine models which are used to describe the electromagnetic phenomena and characteristics for the steady states and transients operation of the device.

It seems that the most distinctive feature of the discussed models is the method of flux linkage calculation. In the typical approaches the flux linkage with the winding is determined by inductances. The inductances represent coefficients or functions that express dependence between the flux linkages with windings or parts of it and the field exciting winding currents. For example, the dependence between the flux associated with the end of the qth winding and the current in the pth winding expresses mutual inductance of the winding ends.

The models, described by the systems of ordinary differential equations with inductances, will be considered as the circuit models. In the field models winding inductances do not exist. Flux linkages and electromagnetic torques and forces are calculated using field quantities, e.g. the magnetic vector potential. The field equations are then coupled to the equations that describe the winding connections and contain terms

defined by field quantities and lumped parameters [1, 10, 24]. The models of this type can be considered as the field-circuit [7, 8, 21, 28].

In the paper particular attention is paid to the field and field-circuit models of typical electromechanical converters (electrical machines and actuators). Therefore, it is assumed that the winding voltages and speed are the input quantities. However, winding currents and electromagnetic torque are the output quantities.

The field models of electrical machine consist of the following equations: (a) finite element equations of magnetic field; (b) finite element equations that describe eddy currents distribution, (c) equations that describe currents in the winding. Moreover the field and circuit equations are coupled via the electromagnetic torque or force to the equation of motion.

Most often, field models are formed using magnetic and electric potentials. Therefore the FE equations (a) describe magnetic field in terms of magnetic scalar potential Ω or magnetic vector potential A. Moreover, equations (b) express eddy currents by electric scalar potential V or electric vector potential T. Equations (c) are related to the loops of multiply connected conductors. In the notation of field theory these equations describe magnetic vector potential T_0 [2]. Edge values of T_0 represent loop currents in the loops around 'holes' in multiply connected conductors, e.g. the loop currents in the winding composed of stranded conductors or the loop currents around the holes in regions with eddy currents [15].

Because magnetic field and electric field of conducting currents can be expressed by scalar or vector potentials there are three formulation of equation (a)–(b)–(c). We can apply Ω–T–T_0 formulation, or A–V–T_0 formulation, or description using A–T–T_0. Of course, in specific case field model can be simplified, e. g. when eddy currents are negligible we consider only equations (a)–(c) that are represented by formulation Ω–T_0 or A–T_0.

The presented field model is formed with the intention to join it with equation of supply and control circuit. Therefore, it is advantage to express equations (a)–(b)–(c) using the language of circuit theory. First the FE equations of magnetic field will be presented.

II. Finite Element Equations of Magnetic Field

Two most popular FE formulations are discussed: (α) the formulation using scalar potential Ω and nodal elements, and (β) the formulation using vector potential and edge elements. In the formulation (α) polynomial interpolating function $\Omega(x,y,z)$ is constructed on the nodal values of Ω, i.e. on the nodal potential. However, the

978-1-4244-1741-4/08/$25.00 ©2008 IEEE

formulation (β) applies edge value of A. For oriented edge P_1P_2 the edge value of A is equal to the line integral of A on P_1P_2 [15]. The edge value of A for edge P_1P_2 can be considered as a loop flux in the loop around P_1P_2. Thus in the formulation (α) magnetic field is defined by the nodal magnetic potentials and in the case of (β) formulation magnetic field is described by thy loop fluxes.

It has been shown [14, 15] that FE equations represent nodal and loop equations of two types of networks: 'edge networks' (EN) where branches are associated with edges of the elements, and 'facet networks' (FN) with branches connecting the centers of the relevant facets with the centre of the element volume. FE model composed of 8-node hexahedrons is used to illustrate these networks - see Figs. 1, 2. The structural matrices of the networks are the FE representation of differential operators. For example, the transposed nodal incidence matrix k_n of EN represent 'grad' operator and the transposed loop matrix k_e for FN is the network representation of 'curl' operator. The nodal equations for EN are equivalent to the nodal finite element formulation using scalar potential Ω, whereas loop equations for FN refer to the edge element formulation based on vector potentials A.

Table I summarises the equations for both models and shows: (a) equations that describe branch fluxes ϕ_b in EN and node-to-node magnetic 'voltages' $u_{\Omega f}$ for FN, and (b) FE equations using the notation of equivalent networks.

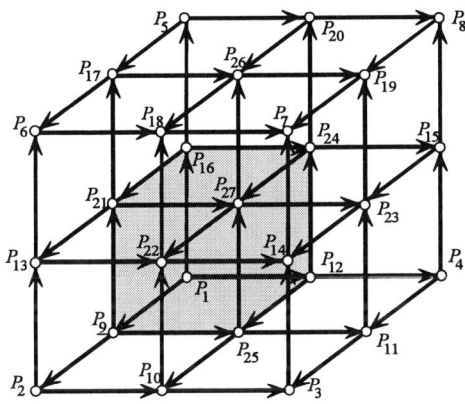

Fig. 1. Edge graph of 8 hexahedrons.

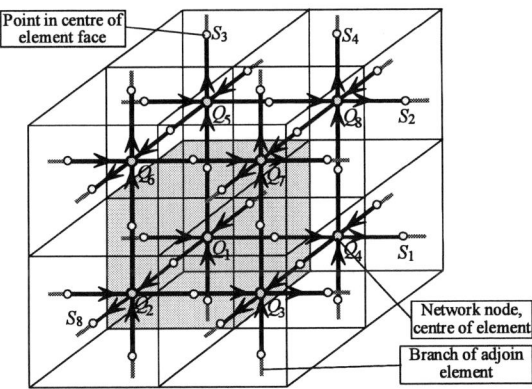

Fig. 2. Facet graph of 8 hexahedrons.

TABLE I.
EQUATIONS OF EQUIVALENT MAGNETIC NETWORKS

Network	Branch equation	Substitutions	FE equations
Edge	$\phi_b = \Lambda(u_\Omega + \Theta_{be})$	$u_\Omega = k_n\Omega$	$k_n^T\Lambda k_n^T\Omega = -k_n^T\Lambda\Theta_{be}$ *
Facet	$u_{\Omega f} = R_\mu\phi_f - \Theta_{bf}$	$\phi_f = k_e\phi_e$	$k_e^T R_\mu k_e\phi_e = k_e^T\Theta_{bf}$ **
Comments: Ω is the vectors of nodal potentials, Λ is the matrices of branch permeances, Θ_{be}, Θ_{bf} are the vectors of branch *mmf*s, ϕ_e is the vectors of loop fluxes; R_μ is the matrices of branch reluctances, *nodal equations of EN, **loop equations of FN			

It should be noted that in equivalent models of element there exist inter-branch couplings [14, 15].The matrices of branch parameters are not diagonal.

In the presented equations the branch *mmf*s are establish from loop currents (ampere-turns) in the loops around branches. However, when using the FN it is not necessary to know the branch sources, instead the loop sources are needed. The loop *mmf*s are represented by the current passing through the loops of FN. Thus, for scalar potential formulation we should define loop currents to determine the right hand side (RHS) vector of FE equations. Whereas, for vector potential method the RHS vector can be calculated using the current passing through the loops.

In general, we can consider three categories of currents that are the source of magnetic field: (i) conducting currents, (ii) magnetizing current in the region with permanent magnets, (iii) displacement current. In the discussed model displacement current are neglected. Magnetizing currents are assumed to be known. These currents are determined by magnetization vector H_m. The edge value of H_m represents the magnetizing current in the loop around edge. The most complicated is to establish the *mmf*s caused by the conducting currents. It is advantage to separate two kinds of conducting currents: eddy currents and currents inducted in windings. The eddy currents are calculated using FE method.

III. FINITE ELEMENT EQUATIONS OF EDDY CURRENTS

System containing region with eddy currents field may be described using electric scalar potential V or electric vector potential T. FE equations for these formulations are constructed in the way similar to the FE for magnetic field. The FE equations for scalar potential formulation and nodal elements represent nodal equations of edge electric network. Whereas, the FE equations for vector potential T and edge elements are equivalent to the loop equations of facet electric network with loops around element edges. The edge value of T represents loop current that may be considered as the eddy current.

The equations of electric network are shown in Table II. The branch equations express the branch currents i_b in EN and the node to node potentials u_{Vf} in FN.

In the electric equivalent networks, similar to magnetic ones, inter-branch coupling exists. For example, the current in the ith branch of the edge element model depends on the voltage across the conductance of the jth branch, whereas the voltage of the branch q in the facet model is linked to the flux in the branch p.

When formulating equations presented in Table II, branch *emf*s are expressed by time derivative of magnetic flux in the loops around the network branches. Therefore, the branch *emf*s in the EN are calculated using the loop fluxes in the facet magnetic network. In the case of the FN

2413

TABLE II.
EQUATIONS OF EQUIVALENT ELECTRIC NETWORKS

Network	Branch equation	Substitutions	FE equations
Edge	$i_b = G(u_V + e_{be})$	$u_V = k_n V$	$k_n^T G k_n^T V = -k_n^T G e_{be}$ *
Facet	$u_{Vf} = R i_f - e_{bf}$	$i_f = k_e i_e$	$k_e^T R k_e i_e = k_e^T e_{bf}$ **

Comments: V is the vectors of nodal potentials, Λ, is the matrices of branch conductances, e_{be}, e_{bf} are the vectors of branch *emf*s, i_e is the vectors of loop currents; R is the matrices of branch resistances, *nodal equations of EN, **loop equations of FN

that is analysed using the loop method it is not necessary to establish the branch *emf*s e_{bf}. The RHS vector e_{mf} in loop equations is represented by the loop *emf*s, $e_{bf} = k_e^T e_{bf}$. Thus, we need the loop *emf*s. The loop *emf*s in FN are expressed by time derivative of flux passing through the loops, i.e. flux associated with branch of EN.

The important disadvantage of formulation using vector potential T is that the method can only be applied to simply connected conductors, e.g. solid parts of a core with no 'holes'. Whereas, electrical machine windings must be considered as a multiply connected regions. The FE equations for the classical T formulation refer to loops around the element edges, see loops with current i_{mi}, i_{mk} in Fig. 3. The loops around the 'holes' do not exists. It is thus necessary to modify classical T method and to introduce additional equations describing the loop currents flowing around the 'holes', e.g. currents i_{ci} ($i=1,2,3$) in Fig. 3. These currents are circuit representation of the edge values of T_0 introduced in [2, 4]. The edge values of T_0 describe also currents in the windings composed of thin stranded conductors with negligible skin effect.

IV. EQUATIONS OF WINDING CURRENTS

In the presented approach the winding currents represent loop currents in the determined loops of multiply connected conductors, see currents i_{ci} in Figs. 3, 4. The winding outlets are considered to be out of the region divided into the FEs. In general the winding equations can be connected with the equations of supply system. Here, to simplify description it is assumed that the terminal voltages are given and the loop *emf*s e produced by external sources are known.

The procedure of winding equation formulation starts with the description of winding loops in the finite element space. As a result we get matrices that transform winding currents into magnetic field sources and transform fluxes ϕ_b, ϕ_e (see Table I) into the flux linkages with the winding loops. There are two methods of winding description in the finite element space. The method based on the definition of intersection points between the winding loops and the finite element facets [13, 16]. This method describes windings in the facet element space. The more universal is method that relies on the calculation of intersection points between the finite element edges and the surfaces of winding loops. This method describes the winding in the edge element space [13].

The path of the ith winding loop L_i can be defined by the parametric equations of oriented curves, $r=r_i(t)$. For a real winding these equations have a very complicated form. Therefore, it is suggested each loop L_i be replaced by a set of m_i closed plane curves (sub-loops) $L_{i,j}$, e.g. by triangles or parallelograms (Fig. 5). In order to describe sub-loops in the edge element space we should define the oriented surfaces $S_{i,j}$ of $L_{i,j}$ – see Fig. 5.

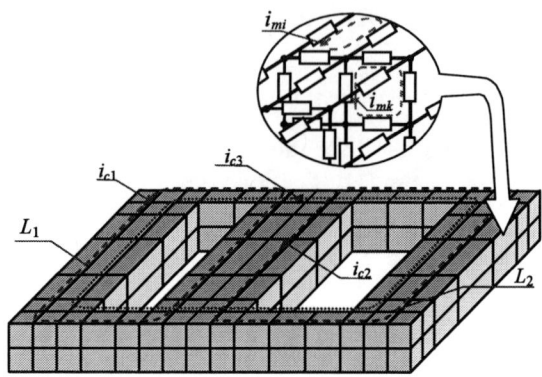

Fig. 3. Multiply connected conducting regions with eddy currents i_m.

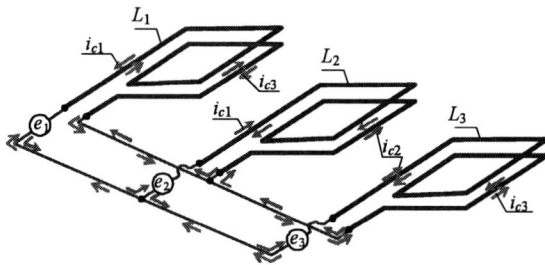

Fig. 4. Loops of 3-phase winding composed of stranded conductors.

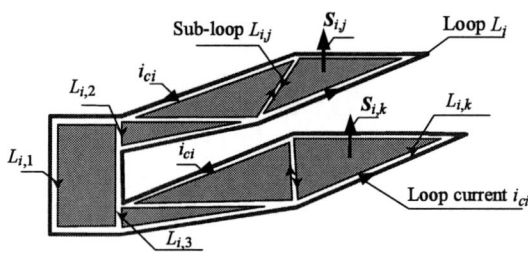

Fig. 5. Sub-loops $L_{i,j}$ of loop L_i with current i_{ci}.

As a result the winding loops are represented by a set of closed oriented plane curves of parametric equations $r=r_{i,j}(t)$ and by plane oriented open surfaces $S_{i,j}$ of parametric equations $r=r_{i,j}(u,v)$. Usually, the description of winding loops may be simplified by ignoring very small sub-loops. For example, the loop L_i in Fig. 5 may be represented by 5 loops only. The flux that penetrates loops $L_{i,2}$, $L_{i,3}$ is very small and these loops may be ignored.

The presented representation of winding by sub-loop makes easy the process of forming matrix N_e that describes windings in the edge element space. The number of intersection points between the edge $K_{p,q}$ (going from P_p to P_q) and surfaces $S_{i,j}$ of L_i represent the entry of N_e that is related to the loop L_i and to the edge $K_{p,q}$. In the calculation of intersection points the sense of $K_{p,q}$ in relation to the sense of $S_{i,j}$ should be taken into account. The entry $N_{eK_{p,q},i}$ is equal to the difference between the numbers of intersection point of negative and positive scalar product $S_{i,j} K_{p,q}$ [13]. Therefore, in Fig. 6, $N_{eK_{4,5},i}=0$.

The product of matrix N_e and vector i_c is equal to the ampere-turns θ_{be} around the edges. The ampere-turns around the edges represent the branch *mmf*s in magnetic EN. Thus, the vector Θ_{be} can be defined as follows

$$\Theta_{be} = \theta_{be} = N_e i_c \,. \tag{1}$$

The vector θ_{be} of ampere-turns in the loops around the branches of EN can be transformed into the ampere-turns θ_{bf} in the loops around the branches of FN [15]. Fig. 7 explains the transformation in the case of network model of hexahedra. The transformation matrix K consists of weighted average factors. The product of matrix K and vector θ_{be} gives the vector θ_{bf} and the branch *mmf*s in FN,

$$\Theta_{bf} = \theta_{bf} = K\Theta_{be} = K\Theta_{be} \,. \tag{2}$$

In the formulations using potential A it is not necessary to know the branch sources. The loop sources should be determined. The loop *mmf*s Θ_{mf} in FN are equal to the ampere-turns (total currents) θ_{mf} that penetrate the loops of FN. The ampere-turns θ_{mf} are obtained by multiplication of the loop matrix k_e^T and the vector θ_{bf},

$$\Theta_{mf} = \theta_{mf} = k_e^T \theta_{bf} = k_e^T K N_e i_c \,. \tag{3}$$

The total currents that pass through the loops of FN may also be calculated from the currents (ampere-turns) θ_{me} that cross the element faces, i.e. that cross the loops of EN and represent the facet values of current density. Matrix k_e transposes the currents in the loops around the edges into the currents in the branch of FN, i.e. into the total currents θ_{me}. Thus, using (1), we can find that the vector θ_{me} of ampere-turns that cross the element facets is

$$\theta_{me} = k_e \theta_{be} = k_e N_e i_c \,. \tag{4}$$

The matrix product $k_e N_e$ is equal to the matrix N_f that describes winding loops in the facet element space.

$$N_f = k_e N_e \tag{5}$$

The transposition matrix N_f can be easy determined by the calculation of intersection points between the loops $L_{i,j}$ and element facets F_q. Fig. 8 illustrates the method of matrix N_f calculation. The winding loop L_i intersects two times facet F_q. The scalar products of F_q and edges of $L_{i,j}$, $L_{i,k}$ are negative. Therefore the entry $N_{fq,i}$ is equal to -2.

When matrix N_f and currents i_c are given it is easy to calculate the vector θ_{me}. Then, using matrix K, the vector θ_{mf} and loop *mmf*s Θ_{mf} in facet network can be found [15].

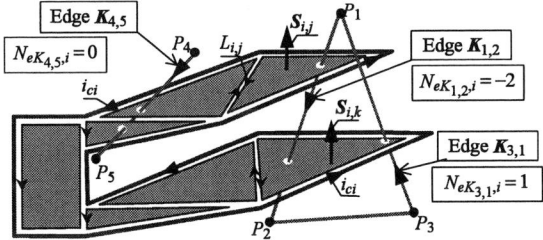

Fig. 6. Winding loop in the edge element space.

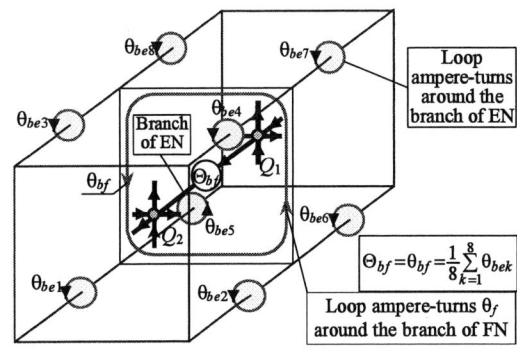

Fig. 7. Transformation of ampere-turns around the branches of EN into the ampere-turns around the branches of EN.

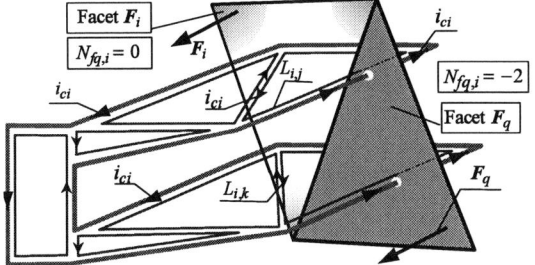

Fig. 8. Winding in the facet element space.

For given matrix N_f the vector Θ_{mf} may be expressed in the following form

$$\Theta_{mf} = \theta_{mf} = K^T \theta_{me} = K^T N_f i_c \,. \tag{6}$$

It is interesting to notice that there are two methods of forming the loop *mmf*s in FN: (a) the method using description of winding by the matrix N_e – see (3), and (b) the method using matrix N_f– see (6).

The presented above descriptions of winding can be used in the calculations of flux linkages with the loops L_i. For EN of branch fluxes ϕ_b the vector Ψ of flux linkages is

$$\Psi = N_e^T \phi_b \,. \tag{6}$$

The description of winding loops in the edge element space is not unique. Thus matrix N_e is not unique either. The set of surfaces $S_{i,j}$ with the total boundary L_i is not unique. However, the results of vector Ψ calculation are independent of the choice of $S_{i,j}$.

In the case of the magnetic vector potential method and the facet magnetic network two formulas can be applied

$$\Psi = N_e^T K^T k_e \phi_e \,, \tag{7a}$$

$$\Psi = N_f^T K \phi_e \,. \tag{7b}$$

A comparison of equations (7) with respect to (5) leads to the identity $K^T k_e = K k_e^T$.

The presented above descriptions of branch and loop mmfs and flux linkages are summarized in Table III. Formulas shown in Table III have been used in the formulation of winding equations, i.e. equations that describe loop currents i_c.

TABLE III.
DESCRIPTIONS OF MMFS AND FLUX LINKAGES ELECTRIC NETWORKS

Network	Branch *mmfs*	Loop *mmfs*	Flux linkages
Edge	$\Theta_{be} = N_e i_c$	$\Theta_{me} = N_f i_c$	$\Psi = N_e^T \phi_b$
Facet	$\Theta_{bf} = KN_e i_c$	$\Theta_{mf} = K^T N_f i_c$	$\Psi = N_f^T K \phi_e$

Comments: matrix N_e describes windings in the edge element space, matrix N_f describes windings in the edge element space i_c is the vector of currents in winding loops $K^T k_e = Kk_e^T$, $N_f = k_e N_e$

The winding equations can be described in the following unified form

$$R_m i_c + p\Psi = e \,, \qquad (8)$$

where R_m is the matrix of loop resistances, p=d/dt and e is the vector of external *emfs*, see Fig. 4. It should be notice that (8) also relates to the loop around the holes in multiply connected conductors with eddy currents, i.e. to the loops with currents i_{ci} in Fig.3. In equations for system in Fig. 3 the vector e is equal the voltages produces by eddy currents; i.e. $e = -N_e^T k_e^T R k_e i_e$.

V. EQUATIONS OF COUPLED MAGNETO-ELECTRIC MODEL

In the preceding section it has been discussed the coupling between the FE equations of magnetic field and the equations of winding currents. In order to form the complete field model of machine the links between the magnetic field and eddy current field must be consider. It has been shown [15] that branch sources in FN are established from loop quantities in EN, and – by symmetry – branch sources in EN are found from loop quantities in FN. Branch *mmfs* Θ_{be} in EN correspond to loop currents. Branch *emfs* e_{be} in EN are found as time derivatives of loop fluxes ϕ_e in FN. Using the symbols in Tables I, II, the branch sources of EN can be written as follows

$$\Theta_{be} = i_e \,, \quad e_{be} = -d\phi_e/dt \,. \qquad (9)$$

In the analysis of FN we need loop sources. The loop *mmf* is equivalent to the current passing through the loop of the magnetic network, thus loop *mmfs* Θ_{mf} in the FN correspond to the branch currents i_b in the EN, see Table II. In the FN models of eddy current regions the loop *emfs* may be found by taking time derivatives of branch fluxes in the magnetic network passing through the loops of the electric network, i.e. fluxes ϕ_b in the branches of EN, see Table I. Thus, the loop sources in FN can be express in the following form

$$\Theta_{mf} = i_b = G(k_n V - p\phi_e) \,, \qquad (10a)$$

$$e_{mf} = -p\phi_b = -p(\Lambda(k_n \Omega + \Theta_{be})) \,. \qquad (10b)$$

The field sources in FN can be also calculated using the relations presented in section IV. The winding loops should be assumed to be eddy current loops, i.e. $i_c = i_e$ and $N_e = 1$. As a result we obtain

$$\Theta_{mf} = k_e^T K i_e \,, \quad e_{mf} = -pK^T k_e \phi_e \,. \qquad (11)$$

On the basis of the presented above equations the field model of electrical machine may be constructed. Further

on we will show the equations of the model for the formulations described in Section I, i.e. for Ω–T–T_0 formulation, for A–V–T_0 formulation, and for description using A–T–T_0.

The FE equation for formulation Ω–T–T_0 are represented by nodal equations of edge magnetic network coupled with the loop equations that describe eddy currents in electric FN and currents in winding loops. These equations can be written in the following matrix form

$$\begin{bmatrix} k_n^T \Lambda k_n & k_n^T \Lambda & k_n^T \Lambda N_e \\ p\Lambda k_n & R_e + p\Lambda & (R_e + p\Lambda)N_e \\ pN_e^T \Lambda k_n & N_e^T(p\Lambda + R_e) & R_m + N_e^T p\Lambda N_e \end{bmatrix} \begin{bmatrix} \Omega \\ i_e \\ i_c \end{bmatrix} = \begin{bmatrix} k_n^T \Lambda \theta_b \\ 0 \\ e \end{bmatrix} .$$

$$(12)$$

Here R_e is the matrix of loop resistances for loops with eddy currents, $R_e = k_e^T R k_e$, and θ_b is the vector of additional branch *mmfs* in the permanent magnet region. These *mmfs* represent the edge values of magnetization vector. In the above equation vector i_c describes the winding currents and the currents in the loops around the 'holes' in the region with eddy currents. For systems without eddy currents equation (12) becomes simpler and have the following matrix form

$$\begin{bmatrix} k_n^T \Lambda k_n & k_n^T \Lambda N_e \\ pN_e^T \Lambda k_n & R_m + N_e^T p\Lambda N_e \end{bmatrix} \begin{bmatrix} \Omega \\ i_c \end{bmatrix} = \begin{bmatrix} k_n^T \Lambda \theta_b \\ e \end{bmatrix} . \qquad (13)$$

It seems that for eddy currents calculation the most convenient is A–V–T_0 formulation. This formulation is equivalent to the loop analysis of facet magnetic network, coupled with nodal analysis of EN for eddy currents and with the loop description of winding with stranded conductors. The FE equations for A–V–T_0 formulation are

$$\begin{bmatrix} k_e^T R_\mu k_e + Gp & -Gk_n & -K^T N_f \\ -pk_n^T G & k_n^T Gk_n & 0 \\ pN_f^T K & 0 & R_m \end{bmatrix} \begin{bmatrix} \phi_e \\ V \\ i_c \end{bmatrix} = \begin{bmatrix} \theta_m \\ 0 \\ e \end{bmatrix} , \qquad (14)$$

where θ_m is the vector of loop *mmfs* in the regions with permanent magnets; $\theta_m = k_e^T K \theta_b$.

In the analysis of electrical machines we can also apply A–T–T_0 formulation. The FE equations for this formulation represent the loop equations for magnetic and electric FNs. These equations can be expressed as follows

$$\begin{bmatrix} k_e^T R_\mu k_e & -k_e^T K & -K^T N_f \\ pK^T k_e & R_e & k_e^T RN_f \\ pN_f^T K & N_f^T Rk_e & R_m \end{bmatrix} \begin{bmatrix} \phi_e \\ i_e \\ i_c \end{bmatrix} = \begin{bmatrix} \theta_m \\ 0 \\ e \end{bmatrix} . \qquad (15)$$

In the case of electrical machine of negligible eddy currents both (14) and (15) are simplified into the following matrix equations

$$\begin{bmatrix} k_e^T R_\mu k_e & -K^T N_f \\ pN_f^T K & R_m \end{bmatrix} \begin{bmatrix} \phi_e \\ i_c \end{bmatrix} = \begin{bmatrix} \theta_m \\ e \end{bmatrix} . \qquad (16)$$

These equations are the FE representation of A-T_0 formulation and seems to more simpler than (13) where there is the time derivative of permeance matrix Λ.

VI. MOVEMENT SIMULATION

Movement simulation is one of the most important problem in the FE analysis of electrical machines. There are a lot of techniques of modelling the changes of rotor position in the FE description of electrical machines [6, 11, 26, 30].

The FE methods taking into account the movement can be divided into two categories: (a) techniques with the fixed grid independent of the moving region position (FG), and (b) the techniques with the moving grid (MG) [26, 29, 33]. The FG methods have been successfully applied in the analysis of the systems with the homogenous moving part and constant speed. In the specific cases, the problems with the changes of speed can also be solved by the fixed grid methods. The FG method needs the suitable discretization of the region considered and the close correlation between the value of rotor speed and the value of element side dimensions. For example, in the case of time step method application, the time step length must be controlled to get the displacement "for element to element" in each time step.

The moving grid methods are more universal. In these techniques the grid is divided into two parts: the moving part associated with the moving elements, i.e. with the rotor of electric machine, and the fixed part associated with the other elements, e.g. with the stator of electric machine. Between these parts, the interconnecting band or slip surface is created. The most popular methods of coupling the fixed and moving part through the band or the slip surface can be easy explain using the notation of equivalent edge and facet networks.

In the scalar potential method, i.e. method equivalent to the nodal analysis of EN, the changes of rotor position are modelled by the changes of nodal permeances matrix $k_n^T \Lambda k_n^T$. For vector potential method that corresponds to the loop analysis of FN the changes of rotor position are represented by the changes of loop reluctance matrix $k_e^T R_{\parallel} k_e$. At first we will explain the moving grid method for the formulation using magnetic vector potential A. In this formulation the changes can be related to the following matrices of product $k_e^T R_\mu k_e$: (a) the structural matrix k_e, (b) the matrix R_μ of branch reluctances or (c) both the matrix k_e and R_μ. In the methods (a) we consider discrete position of rotor that differs on the distribution of nonzero elements of loop matrix k_e. As a result we obtain a set of matrices k_e for successive rotor positions. This set represents the data points for interpolation functions that express the dependence of k_e on the angle α that describes the rotor position [11]. It can be proved that if the trigonometric interpolation is applied then the method is equivalent to the method of harmonic weighting functions presented in [9]. When a suitable interpolation of matrix k_e is applied the method of category (a) guarantees high accuracy. However due to the increase of matrix $\{k_e(\alpha)\}^T R_\mu k_e(\alpha)$ density the procedure of solving the FE equations becomes complex and time-consuming. Also, in the case of methods of category (b) with $R_\mu = R_\mu(\alpha)$, the matrix of loop reluctances is dense what results in the increase of computation time. The most representative method of category (b) is the air-gap element method.

The simplest methods of movement simulation belong to category (c). A typical representative of these methods is moving band method with remeshing of FE network. In this method, at each time step of the rotor displacement, the finite elements are distorted and $R_\mu = R_\mu(\alpha)$ or, if the distortion is large, the band is remeshed, i.e. the nonzero elements of matrix k_e change their position [6]. In the matrix product $k_e^T R_\mu k_e$, function $R_\mu(\alpha)$ is a piecewise continuous and the matrix k_e is represented by discrete set of its values. As a result the derivative of matrix $k_e^T R_\mu k_e$ entries with respect to angle α is not continuous that produces the ripple in calculated quantities [11]. However, due to the simplicity, the moving band method belongs to the most popular methods of movement simulation.

The presented method of movement simulation for vector potential formulation can be easy adopted for scalar potential method. Changes of matrices R_μ, k_e are represented by changes of matrices Λ, k_n in the scalar potential methods

VII. ELECTROMAGNETIC TORQUE

One of the most important advantages of the FE analysis of rotating electrical machines is ability to calculate the electromagnetic torque with high accuracy. The relationships that describe electromagnetic torque are formed on the basis of virtual work principle. The obtained formulas can be divided into 2 categories: (a) the force density formulas, e.g. Lorenz formula, the method of magnetizing currents, and (b) stress tensor formulas, e.g. the Maxwell stress tensor formula [25]. Of course, for the exact solution of Maxwell's equations the formulas of both categories give the identical result. However, in the case of FE models the results of different methods of torque calculation are not identical. Moreover, the position of integration surface has an effect on the result of stress tensor method. Therefore, very often in the FE models the electromagnetic torque is calculated using the formulas that are obtained from virtual work principle implemented to discrete systems [5, 12, 25, 26].

In accordance with virtual work principle, for scalar potential method, the torque is equal to the magnetic coenergy derivative versus the virtual moving. An interpolation function can be applied to describe this derivative. The interpolation gives

$$T(\alpha) = \left.\frac{\partial W_c(\alpha + \Delta\alpha)}{\partial(\Delta\alpha)}\right|_{\Delta\alpha=0} = \frac{W_c(\alpha+\beta) - W_c(\alpha+\beta)}{2\beta}, \quad (17)$$

where $W_c(\alpha\pm\beta)$ is the magnetic coenergy for rotor position $\alpha\pm\beta$, see Fig 9. From (17), using symbols in Table 1 we obtain

$$T(\alpha) = \frac{1}{2\beta}\Omega^T\left[\left.(k_n^T \Lambda k_n)\right|_{\alpha+\beta} - \left.(k_n^T \Lambda k_n)\right|_{\alpha-\beta}\right]\Omega. \quad (18)$$

For vector potential method the magnetic energy derivative versus the virtual moving is considered [12]. The derivative can be approximated by the finite differences and as a result we have

$$T(\alpha) = -\frac{1}{2\beta}\phi_e^T\left[\left.(k_e^T R_\mu k_e)\right|_{\alpha+\beta} - \left.(k_e^T R_\mu k_e)\right|_{\alpha-\beta}\right]\phi_e. \quad (19)$$

In the literature, there are also other methods of work principle representation. Particularly noteworthy are the

2417

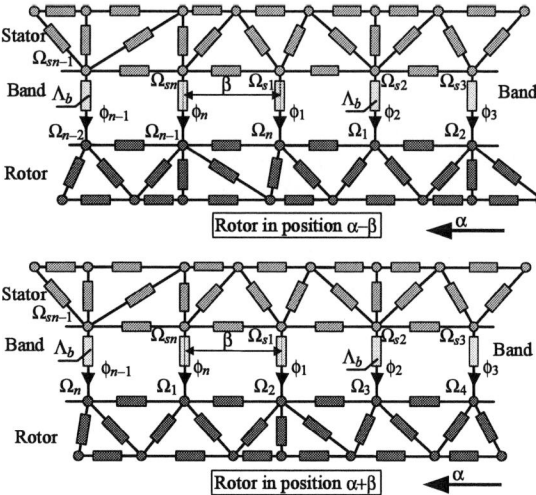

Fig. 9.   A part of magnetic EN for 2 discrete rotor positions

methods adapted to the applied techniques of movement simulation [5, 11]. For example, if in the procedure of movement simulation, $R_\mu = R_\mu(\alpha)$ then

$$T(\alpha) = -0.5\phi_e^T \{k_e^T (dR_\mu/d\alpha)k_e\}\phi_e. \qquad (20)$$

This formula and (18), (19) can be considered as the FE representation of Maxwell stress formula with integration surface related to the band between the stator and rotor.

VIII. FIELD-CIRCUIT MODEL

The presented below field-circuit (F-C) model of electrical machine should not be mixed up with the field-circuit model of the electric drive. The field-circuit model of electric drive is constructed by joining FE equations given in section V with equations that describe supply system [20, 23, 27]. Moreover in the F-C model of drives the field and circuit equations are coupled through the expressions that define the electromagnetic and opposite torque to the equation of motion [3, 10, 12].

In this paper the term field-circuit model of electrical machine relates to the approaches that express the flux linkages Ψ with the windings by two components: (a) components defined by field quantities and (b) component represented by inductances. Thus, expressions in Table III become

$$\Psi = N_f^T K\phi_e + L_e i_c, \quad \Psi = N_e^T \phi_b + L_e i_c, \qquad (21a,b)$$

where L_e is the matrix of equivalent inductances. The equivalent inductances relate to this part of winding that is out of the region divided into the finite elements.

The F-C model is applied when the magnetic field is assumed to be 2D independent of the coordinate parallel to the rotor shaft. In the 2D model the matrix L_e describes the inductances of winding ends.

It can be seen that equations of F-C model are similar to the equations presented in section V. To obtain equations of F-C model matrix R_m in (12) – (16) should be replaced by sum $R_m + pL_e$.

The F-C model with the 2D description of magnetic field is simpler than 3D. The number of unknowns in FE

equations is considerably smaller. Therefore, although the software for 3D calculation is commonly available the 2D F-C models are still in wide use.

Some authors use the phrase 'field-circuit' for the circuit equivalent model with the FE calculations of circuit parameters, mostly inductances. The method based on the FE calculations of circuit parameters is effective in specific cases only. Three factors determine the method efficiency: (a) number of coupled windings, (b) saturation ratio, (c) influence of eddy currents on the flux linkages [16, 18, 22]. The circuit model with the FE calculation of circuit parameters can be successfully used in the analysis of systems without eddy currents with non saturated core. The circuit model aided by the FE method can also be effectively applied in the description of system with saturated core but with no-mutual coupling between the windings, e.g. to simulate switch reluctance motor. To form the model of this motor we should define self inductance only. The FE calculation self inductance only are not time consuming, even if it is necessary to take into account that the inductance is a function of current and rotor position.

In the FE calculations of self and mutual inductances we may apply presented above equations of field model. The vector of self and mutual inductances of the ith winding is equal to the vector Ψ of flux linkages produced by elementary current in this winding. The ith component of Ψ represents the self inductance, the jth component is equal to the mutual inductance between the ith and j^{th} winding. In the calculation of inductances for machine of saturated core we should take into consideration that matrices R_μ, Λ depend on the currents i_c. Therefore, in the calculations of inductance for given values of currents i_c the FE equations should be solved two times: first for given vector i_c to determine the values of entries of matrix R_μ or Λ, and then for elementary currents to find the vectors of self and mutual inductances.

It can be calculate that to describe the inductances of six coupled windings for 20 values of currents and 20 rotor position the FE equations should be solved more than billion times. Thus, the FE calculations of circuit parameters are very time consuming.

Usually, in the description of electrical machines the equivalent circuits are applied. The equivalent circuits are formed by the application of current transformation; e.g. Clarke and Park transformation. The inductances of equivalent transformed system can be directly calculated using the presented above field model. In the algorithm the transposed currents i_T are assumed to be known. For elementary values of currents i_T the currents i_c in winding loops are calculated, $i_c = k^T i_T$, where k is the transformation matrix. Then, for the currents i_c the FE equations are solved and the vectors Ψ_T of flux linkages for transformed system are determined, $\Psi_T = k\Psi$. These vectors represent the inductances of transformed circuit model.

It should be noticed that the described algorithm differs from the classical approaches. In the classical approaches at first the inductances of real system are determined and then these inductances are transformed [19]. The calculations using the algorithm with direct calculations of equivalent inductances are much less time consuming [3, 17]. In has been shown [17] that in the case of 3-phase balanced system that produces rotating magnetic field in saturated core the procedure of equivalent inductances calculations needs to solve FE equations $6m$ times only, where m is the number of given rms values I_s of phase current to form function $L(I_s)$.

IX. CONCLUSIONS

The paper presents the so called field and the field-circuit models of electrical machines. The particular approaches have been elaborated to simulate the machine's behavior with the presented models [7, 8, 21]. The developments in these procedures and in the methods to solve large systems of equations enable to apply the presented model in many practical applications of technical importance.

In the discussion on field and field-circuit models practicability the three directions of applications should be separated, namely: (a) design, (b) diagnostics and (c) control. The field methods are commonly used in the design stage of electrical machines. With the application of high performance field models the designing engineer can avoid the costs of building expensive prototypes of the device under study.

Field methods are also applied for diagnostic purpose [32]. However, in diagnostics the field models are not as popular as for the machine design. Recently, the field models become more and more popular in the analysis and synthesis of electrical machine control, even though the control methods are based on the classical circuit approaches. Mostly, the field methods are used to calculate the parameters of the equivalent circuit of considered control system [3, 18]. The field methods can be especially helpful in the case of sensorless control when an accurate description of machine parameters is required. It seems that untill now, due to the high computational costs, the exact 3D field model has not been directly applied in the control systems.

REFERENCES

[1] A. Arkkio A., "Time stepping finite element analysis of induction motors," *Proceedings of ICEM'88*, Piza 1988, pp. 275-280.

[2] S. Bouissou, F. Piriou, "Study of 3D formulations of model electromagnetic devices," *IEEE Trans. Magn.*, Vol. 30, No.5, 1994, pp. 3228-31.

[3] S. Brulé S., A. Tounzi, "Comparison between FEM and the Park's models to study the control of an induction machine," *Proceedings of ICEM'2000*, Helsinki Vol. 1, 2000, pp.548–552.

[4] V. P. Bui, Y. Le Floch, G. Meunier, J. L. Coulomb, "A new three-dimensional (3-D) scalar finite element method to compute T_0," *IEEE Trans. on Magn.*, Vol. 42, No.4, 2006, pp. 1035–38.

[5] J. Coulomb, G. Meunier, "Finite element implementation of virtual work principle for magnetic or electric force and torque computation," *IEEE Trans. on Magn.*, Vol. 20, No. 5, 1984, pp.1894–96.

[6] S. Davat, Z. Ren, M. Lajoie-Mazenc, "The movement in field modelling," *IEEE Trans. Magn.*, Vol.21, No.6, 1985, pp. 2296–98.

[7] H. De Gersem, R. Mertens, D. Lahaye, S. Vandewalle, K. Hameyer, "Solution strategies for transient, field-circuit coupled systems," *IEEE Trans. on Magn.*, Vol. 36, No. 4, 2000, pp.1531–34.

[8] H. De Gersem, R. Mertens, U. Pahner, R. Belmans, K. Hameyer, "A topological method used for field-circuit coupling," *IEEE Trans. on Magn.*, Vol. 34, No.5, 1998, pp. 3190–93.

[9] H. De Gersem, M. Ion M., T. Weiland T., A. Demenko, "Trigonometric interpolation at sliding surfaces and moving band of electrical machine models, " *COMPEL*, Vol. 25, No 1, 2006, pp. 31-42.

[10] A. Demenko, "Time-stepping FE analysis of electric motor drives with semiconductor converters", *IEEE Trans. on Magn.*, Vol. 30, No.5, 1994, pp. 3264–67.

[11] A. Demanko, "Movement simulation in finite element analysis of electric machine dynamics," *IEEE Trans. on Magn.*, Vol. 32, No. 3, 1996, pp. 1553–56.

[12] A. Demenko, "3D edge element analysis of permanent magnet motor dynamics", *IEEE Trans. on Magn.*, Vol.35, No.5, pp. 3220–23, 1998.

[13] A. Demenko, "Representation of windings in the 3-D finite element description of electromagnetic converters," *IEE Proc. Sci. Meas. Technol.*, Vol. 149, No. 5, 2002, pp. 186 – 189.

[14] A. Demenko, J. Sykulski, "Network equivalents of nodal and edge elements in electromagnetics," *IEEE Trans. on Magn.*, Vol. 38, No.2, pp. 1305-08, 2002.

[15] A. Demenko, Sykulski J. "Network models of three-dimensional electromagnetic fields," *ICS Newsletter*, Vol. 13, No 3, 2006, pp.3-13.

[16] A. Demenko, L. Nowak, W. Pietrowski, "Calculation of end-turn leakage inductances of electrical machines using the edge element method," *COMPEL*, Vol. 20, No. 1, 2001, pp. 132–139.

[17] A. Demenko, L. Nowak, W. Pietrowski, "Calculation of magnetization characteristic of a squirrel cage machine using edge element method," *COMPEL*, Vol. 23 No. 4, 2004, pp. 1110–18.

[18] A. Di Napoli, S. Santini, "FEM identification of parameters of vector control of induction motors," *Proceedings of ICEM'2000*, Helsinki Vol. 1, 2000, pp. 71–75.

[19] Ch. Jones, The *Unified Theory of Electrical Machines*, Butterworth & Co. (Publishers) LTD, London, 1967.

[20] Y. Kawase, Y. Hayashi, T. Yamaguchi, "3D finite element analysis of permanent magnet motor excited from square pulse voltage source, " *IEEE Trans. Magn.*, Vol. 32, No. 3, 1996, pp. 1537 – 40.

[21] D. Lahaye, S. Vandewalle, K. Hameyer, "An algebraic multilevel preconditioner for field-circuit coupled problems," *IEEE Trans. on Magn.*, Vol. 38, No. 2, 2002, pp. 413–416.

[22] E. Lange; F. Henrotte, K. Hameyer, "A circuit coupling method based on a temporary linearization of the energy balance of the finite element model," *IEEE Trans. on Magn.*, Vo. 44, No.6, 2008, pp. 838–841.

[23] F. Piriou, A. Razek, "Coupling of saturated electromagnetic system to nonlinear power electronic devices," *IEEE Trans. Magn.*, Vol. 24, No. 1, 1988, pp. 274-277.

[24] F. Piriou, A. Razek, "Finite element analysis in electromagnetic systems accounting for electric circuits, *IEEE Trans. Magn.*, 1993, Vol. 29, No. 2, pp. 1669–75.

[25] Z. Ren, "Comparison of different force calculation methods in 3D finite element modelling," *IEEE Trans. on Magn.*, Vol. 30. No.5, 1994, pp. 3471–75.

[26] N. Sadowski, Y. Laverve, M. Lajoie-Mazenc, J. Cros, "Finite element torque calculation in electrical machine while considering the movement," *IEEE Trans. Magn.*, Vol.28, No.2, 1992, pp. 1410–13.

[27] N. Sadowski, Y.Lavevre, M.Lajoie-Mazenc, S.Astier, "Finite element simulation of electrical motors fed by current inverters," *IEEE Trans. Magn.*, Vol. 29, No. 2, 1993, pp. 1683-88.

[28] E. G. Strangas, "Coupling the circuit equations to the non-linear time dependent field solution in inverted driven induction motors," *IEEE Trans. Magn.*, Vol. 21, No. 6, 1985, pp. 2408–11.

[29] C.W. Trowbridge, J.K. Sykulski, "Some key developments in computational electromagnetics and their attribution," *IEEE Trans. Magn.*, Vol. 42, No. 4, 2006, pp.503–508.

[30] I. A. Tsukerman, "Overlapping finite elements for problems with movement," *IEEE Trans. Magn.*, Vol.28, No.5, 1992, pp. 2247–49.

[31] I. A. Tsukerman, A. Konrad, G. Meunier, J. C. Sabonnadiere, "Coupled field -circuit problems; trends and accomplishment," *IEEE Trans. Magn.*, Vol. 29, No. 2, 1993, pp. 1701-04.

[32] L. Weili, X. Ying, S. Jiafeng, L. Yingli, "Finite-element analysis of field distribution and characteristic performance of squirrel-cage induction motor with broken bars," *IEEE Trans. Magn.*, Vol. 43, No. 4, 2007, pp. 1537-40.

[33] S. Williamson, "Induction motor modelling using finite elements," *Proceedings of ICEM'94*, Paris, Vol. 1, 1994, pp. 1–8.

Interaction between Thermal Impedance and Parasitics in Power Sections

Stefan Förster*, Andreas Lindemann†

*Otto-von-Guericke-University Magdeburg, Magdeburg, Germany, e-mail: *stefan.foerster@ovgu.de*
†Otto-von-Guericke-University Magdeburg, Magdeburg, Germany, e-mail: *andreas.lindemann@ovgu.de*

Abstract—**Considerations refer to thermal optimisation and its coherence with the appearance of electrical parasitics in power electronic systems. Analytical and finite element simulation assisted analysis of thermal resistance depending on geometry of layered assemblies is discussed. Resulting approach provides a basis for considerations on parasitics applying partial element equivalent circuit method (PEEC).**

Keywords—**Packaging, System Integration, Thermal Design**

I. INTRODUCTION

The thermal optimisation process of layered assemblies with integrated insulation leads to geometrical reshaping of layers having essential influence on thermal resistance. Regarding desired heat spreading effects this will determine geometrical design of electrically conducting parts. Thus the appearance of inductive and capacitive parasitics will be influenced. Analysis is intended to retrieve the different parasitics separately as parts of a closed formulation for overall electrical behaviour. Therefor partial element equivalent circuit (PEEC) method as a very powerful tool is applied. Lumped electrical circuit elements can be extracted and possibly integrated in a common circuit simulator.

II. THERMAL RESISTANCES

A. Reference and Improved Calculation

Irregular geometry and discontinuous thermal conductivity λ of layered structure [1] results in difficult estimation of overall thermal resistance of assemblies mentioned. Heat spreading of the well thermally conductive circuit layer under chip designates the effectively used cross sectional area for heat flux passage through insulation. This passage with poor thermal conductivity has essential influence on thermal resistance. Thus analysis is intended to determine heat flux density distribution within layers up to heat sink. With this knowledge it is possible to mark out significant parts of the geometry. Also minimum mounting distance needed between heat dissipating elements on substrate may be retrieved.

According to consideration of systems aforementioned Fig. 1 gives an example with a silicon chip soldered on a DCB substrate, used for later calculations. All layers (indicated later by index n) are for reference first assumed to contribute with the same cross sectional area equal to that of chip. For reference calculation the overall thermal resistance rectangular to substrate layers is determined.

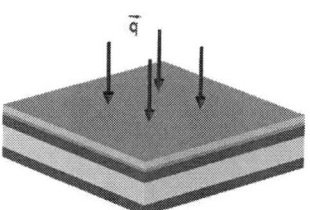

Fig. 1. Silicon chip of $6\text{mm} \cdot 6\text{mm} \cdot 200\mu m$ soldered with $100\mu m$ lead on DCB substrate ($300\mu m$ Cu, $630\mu m$ Al$_2$O$_3$, $300\mu m$ Cu)

Because temperature is assumed to depend only on z-coordinate, heat flux density q

$$\vec{q} = -\lambda_n \cdot grad T \tag{1}$$

with $T = f(z)$ is constant within all layers. Calculation of the example given by Fig.1 with

$$R_{th,n} = \frac{s_n}{\lambda_n \cdot A} \tag{2}$$

$$R_{th,ges} = \sum_n R_{th,n} \tag{3}$$

with A: cross sectional area, s: layer thickness
leads to $R_{th,ges} = 1,034\text{K/W}$.

According to direction of heat flux, first principal heat spreading layer is considered, which is DCB top copper layer. In [2] the analytical solution of temperature distribution $T = f(r, z)$ for cylindrical symmetry is given. A corresponding assembly with dimensions given in Fig. 2 will be used for following calculations assuming a uniform heat flux density to go through a circular area equal to chip area. As additional boundary conditions, the circular bottom plane equal to DCB area is assumed to be at fixed temperature thus describing an ideal heatsink.

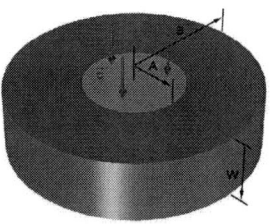

Fig. 2. Cylindrical symmetrical equivalent of DCB top copper layer and a circular area $\pi \cdot A^2$ equal to chip area yielding heat flux

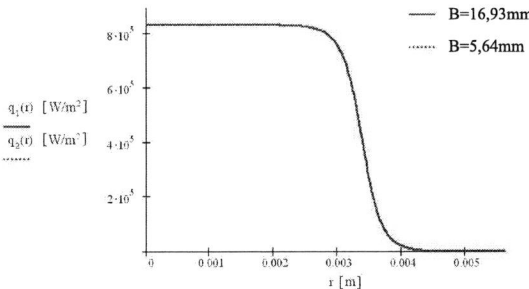

Fig. 3. Heat flux density on the bottom of DCB top copper layer depending on radius r relative to assembly centreline after conversion to the cylindrical equivalent with constant value $A = 3,39\text{mm}$

TABLE I

R_{th} VALUES OF THE SUBSEQUENT LAYERS

Calculation	Reference [K/W]	Improved [K/W]
Chip	$3,75 \cdot 10^{-2}$	$3,75 \cdot 10^{-2}$
Solder	$5,56 \cdot 10^{-2}$	$5,56 \cdot 10^{-2}$
DCB circuit	$2,11 \cdot 10^{-2}$	$2,11 \cdot 10^{-2}$
DCB insulation	$8,75 \cdot 10^{-1}$	$4,95 \cdot 10^{-1}$
DCB base	$2,11 \cdot 10^{-2}$	$1,19 \cdot 10^{-2}$
Σ	$1,03$	$0,62$

The maximum thermal resistance is calculated by use of the difference of highest temperature appearing (at centre of top plane) and assumed heat sink temperature.

Heat flux that exits the assembly on bottom area will be non-uniform. It can be calculated and described as a function of radius r relative to centreline:

$$\vec{q_2}(r) = -\lambda \cdot \frac{\partial T(r,z)}{\partial z}\bigg|_{z=z_{\max}} \quad (4)$$

Fig. 3 gives the result for DCB top copper layer of the assembly given in Fig. 2 with $P = 30\text{W}$ applied and for two different values of radius B resulting in $\pi \cdot B^2 = 100\text{mm}^2$ and $\pi \cdot B^2 = 900\text{mm}^2$. No significant change in heat flux density distribution can be determined. Thus the assembly can be reduced by means of substrate dimensions with low repercussion on thermal behaviour. The cross sectional area effectively used for passage of heat flux can be determined. This method can be applied to each layer. Fig. 3 indicates neglectable heat flux for approximately $r \geq 4,5\text{mm}$. Using $r = 4,5\text{mm}$ for simplified calculation of thermal resistance of DCB insulating and bottom copper layer according to eq. 3 leads to $R_{th,ges} = 0,621\text{K/W}$. For comparison of the results of the reference calculation according to Fig. 1 and the improved calculation according to Fig. 2 and Fig. 3 the R_{th} values of the subsequent layers are given in Tab. I.

B. FEM Simulation

Results of last section are validated using finite element simulation [3]. For this purpose the thermal behaviour of a component frequently needed in applications is

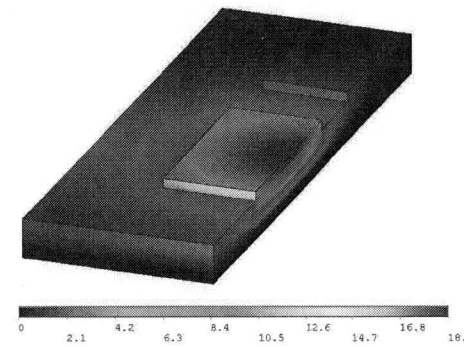

Fig. 4. Cut-off view of temperature difference as result of finite element simulation where surface of DCB bottom copper layer is fixed.

investigated. The discrete IGBT IXER35N120D1 is characterised with $R_{thjc} = 0,6\text{K/W}$ for the IGBT chip and is extensively considered in [4]. The IXER35N120D1 features the same thickness by means of substrate, chip and solder layers as the assembly given in Fig. 1. Also the IGBT chip is dimensioned with $(6 \cdot 6)\text{mm}$.

Fig. 4 gives the cut-off view of the temperature distribution when when a heat generation of $P = 30\text{W}$ is applied to the IGBT chip. Thermal resistance is determined by the difference of maximum chip temperature and fixed bottom temperature of the assembly:

$$
\begin{aligned}
R_{th,ges} &= 18,85\text{K/30W} \\
&= 0,628\text{K/W}
\end{aligned}
$$

The approximately circular shape of isothermal lines considered before is confirmed by Fig. 4. An abstraction to a cylindrical symmetrical assembly seems appropriate as used before.

The following considerations of electrical behaviour refer to these results. Especially the circular shape of isothermal lines within all thermally conducting layers is of interest.

III. IDENTIFICATION OF ELECTRICAL PARASITIC ELEMENTS

A. PEEC model approach

In this section the determination of the electrical behaviour of a complexly shaped geometry is studied. The inductances, capacitances and ohmic resistances at defined connection points are of special interest because they represent the behaviour as easily interpretable lumped parameters.

Because optimisation processes lead to changes in geometry it seems appropriate to extract parameters of parts of the assembly independently from all the rest.In [5] the influence of inductive parasitics including the backside baseplate of a module thus involving backside eddy-currents on switching of IGBTs is investigated. Therefor PEEC method (Partial Element Equivalent Circuit) is used giving solution of exact inductances for the example geometry considered. In [6] similar considerations are made with a conducting loop above an ideally conducting plate. This method provides the possibility to partially

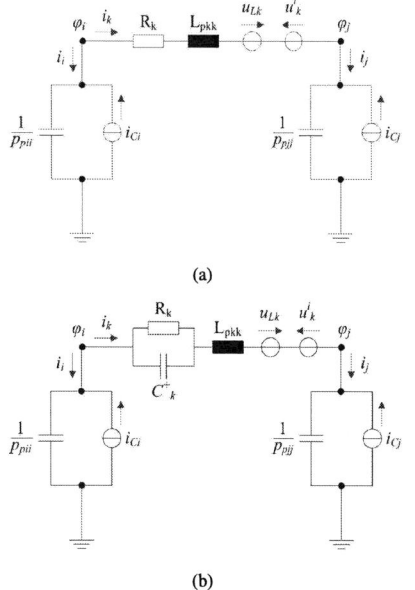

(a)

(b)

Fig. 5. Equivalent circuits for partial elements of a) conducting and b) dielectric regions

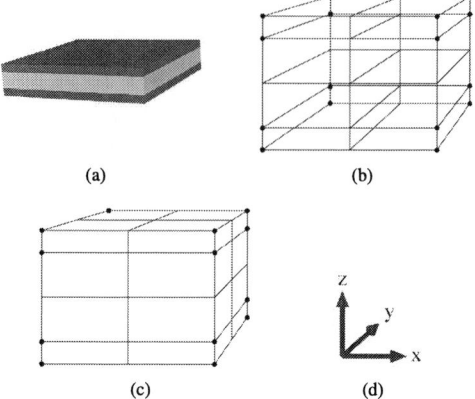

(a)　　　　　　　　(b)

(c)　　　　　　　　(d)

Fig. 6. a) conductor geometry for PEEC model parameter extraction, b)current cells, c)potential surface cells and d) coordinate directions

extract inductances and capacitances and integrate them into a common circuit simulator. So it seems appropriate to use this strategy to receive information of influence of thermal optimisation on appearance of parasitics.

PEEC method depends on volumetric decomposition of an electrically conducting or dielectric geometry into cells which are advantageously arranged rectangularly or in parallel (partial elements) [7]. Also two- and three-dimensional extraction is possible e.g. when conductors with several access points are considered [6], [8].

The maximum geometric partial element length has to be chosen in accordance to the minimum wavelength of interest. A sufficient relation is given with eq. 5 where c is the speed of light and χ should as far as possible be chosen greater than 10 [9].

$$l \leq \frac{c}{\chi \cdot f_{\max}} \qquad (5)$$

From the electrical field integral equation EFIE the complete equivalent circuits for the partial elements from orthogonal discretisation can be obtained [9]. Fig. 5 gives the proposed equivalent circuits of one PEEC cell for conductors Fig. 5(a) and for dielectric regions Fig. 5(b). The main difference is due to the effects with dielectric materials and is represented by the excess capacitance C^+. The partial inductance L_{pkk} for each dimension x, y and z can be determined exactly through volumetric integration. But calculation effort becomes extensive when dimensions for integration of the cells vary. Thus approximation formulas which are based on exact integration can be used but must be treated carefully. With subdivision of one cell into subcells and the combined usage of simplified self and mutual formulas an advantageous way for approximation of the exact value for complex structures can be found.

As with inductances it is also possible to advantageously determine capacitances by use of PEEC-method. Charges are assumed to only be stored on surfaces of conductors and between materials having different dielectric constants [10]. Volume decomposition as aforementioned leads to PEEC elements each represented by a volumetric part and surface cells which lead to a formulation of "coefficients of potential". As with partial inductances also self (p_{pii}, p_{pjj}) and mutual coefficients of potential are defined. Unit is $1/\text{F}$.

For an example in the following an assembly geometrically equivalent to a parallel plate capacitor with dielectric medium is considered with conductors on top and bottom plane which have a finite thickness h. Thus a 3-dimensional discretisation can be realised that enables this assembly to serve as a subpart of a complexly shaped assembly including current cells for all dimensions of the cartesian coordinate system and at least one potential surface cell on each node. Fig. 6 gives the steps from real assembly Fig. 6(a) to the discretised model for partial element extraction with the current cells for dimension y Fig. 6(b) and the surface cells placed on every boundary surface between materials having different dielectric constants Fig. 6(c). With respect to all dimensions x, y and z a number of 28 current cells and 40 surface cells are retrieved. According to the definition of partial inductances and coefficients of potential [11] the formulas eq. 12 and eq. 13 can be used for calculation of the coefficients of potential for laminar and filament like conductors when the permittivity ϵ of the surrounding media is constant. Therefor results of partial self and mutual inductances are divided by $(\mu_0 \cdot \mu_r \cdot \epsilon_0 \cdot \epsilon_r \cdot l^2)$, where $\mu_0 \cdot \mu_r$ is the permeability, $\epsilon_0 \cdot \epsilon_r$ the permittivity and l the conductor's length. In practical applications the assumption of constant ϵ will not apply. Thus calculation method must be modified. An advantageous method for simulation purposes is to additionally discretise all volumes of $\epsilon \neq \epsilon_0$ according to Fig. 5(b) and Fig. 6 [9]. Thus dielectric PEEC cells are included providing the

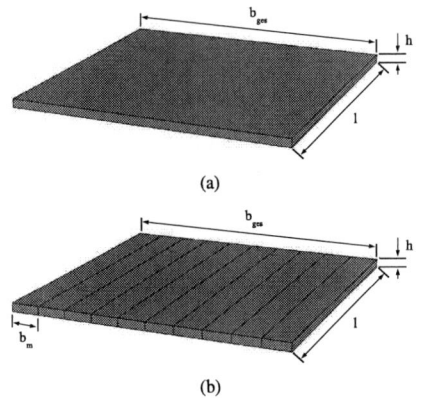

(a)

(b)

Fig. 7. Conductor geometry for PEEC model parameter extraction

TABLE II

MUTUAL INDUCTANCE OF PARTIAL ELEMENTS WITH FOOTPRINT OF
$(10 \cdot 10)\text{mm}^2 \cdot 0, 3\text{mm}$ UNDER CHIP

Element	1-2	1-3	1-4	1-5	1-6
$L_{p,ij}$[nH]	4,186	2,985	2,350	1,940	1,651

Element	1-7	1-8	1-9	1-10	
$L_{p,ij}$[nH]	1,435	1,268	1,134	1,025	

possibility to assume a "free space" problem with constant $\epsilon = \epsilon_0$ again.

As a calculation example inductive parasitics formulated as partial inductances of a more simple geometry are considered in the following. Fig. 7 gives a part of an assembly considered in section II. It represents the copper metallisation of the circuit layer of DCB substrate. For a simplified consideration current is first assumed only in dimension l. Applying the generalised formulation of eq. 12 [7], [10] also considering height h on Fig. 7(a) with $b - 10\text{mm}$, $h - 300\mu\text{m}$ and $l = 10\text{mm}$ leads to

$$L_P = 2,913\text{nH}$$

Fig. 7(b) gives the same overall geometry but subdivided in 10 geometrically equal conductors arranged in parallel. For direct current and homogeneous current density over cross sectional area of metalisation Fig. 7(a) and Fig. 7(b) are equivalent. Only with volumetric decomposition as given with Fig. 7(b) effects of alternating current — e.g. skin effect — can be considered. With eq. 13 the values of mutual inductances given with Tab. II follow where i and j indicate the conductor segments considered. With $m = 10$ segments for the self inductance of every distributed component $L_{P,ii}$ follows according to the generalised formulation of eq. 12 as used before.

$$L_{P,ii} = 6,534\text{nH} \tag{6}$$

With eq. 7 [7] the total self inductance can be calculated.

$$L_{P,\Sigma} = \frac{1}{m^2} \cdot \sum_{i=1}^{m} \sum_{j=1}^{m} L_{P_{ij}} \tag{7}$$

With $m = 10$ and thus $b_m = 1\text{mm}$ follows

$$L_{P,\Sigma} = 2,868\text{nH}$$

The resulting total partial self inductance of Fig.7(b) from partial self and mutual inductances of the distributed components of the conductor equals the calculated value of Fig.7(a) for infinite number of filaments over conductors cross sectional area. The difference in comparison with partial self inductance calculated for Fig. 7(a) mainly results from simplified assumption of only ten current filaments for mutual inductance calculation (Tab. II). Depending on whether skin- and proximity effects are included or not the number of subdivisions of the cross sectional area of a conductor must be chosen [12].

When current is assumed to go through bottom plane of a chip and solder, the current density in copper metalisation of the circuit layer of DCB substrate is inhomogeneous even in case of direct current. In this case at least two-dimensional PEEC models are needed. *FastHenry* for this purpose provides two dimensionally conductive planes which can be contacted on arbitrary position only restricted by the mesh size.

B. FEM Simulation

In supplement to PEEC method inductances can also be determined using finite element method (FEM) simulation. As this method requires to model conductor and surrounding media the calculation effort becomes high when outer magnetic field is considered. Thus the following example is intended to also demonstrate the advantages of PEEC method for the tasks of this work. For this purpose a rectangular conductor according to qualitative view of Fig. 8 of width b, height h and length l is embedded in a cylindrical volume with radius r_L. Dimension l of conductor and centerline of cylinder are parallel to y-axis of the cartesian coordinate system, thus oriented in direction of current. The length of the cylinder equals the length l of the conductor. The difference between conductor's volume and cylinder's volume represents the ambient air. Thus r_L has to be chosen appropriately to assure flux-parallel (Dirichlet) boundary condition being applicable. Symmetry to x-y and y-z planes is not used for explanation purposes.

With $b = 1\text{mm}$, $h = 300\mu\text{m}$, $l = 10\text{mm}$ and $r_L = 20\text{mm}$ the conductor's internal inductance $L_{P,int}$ is retrieved

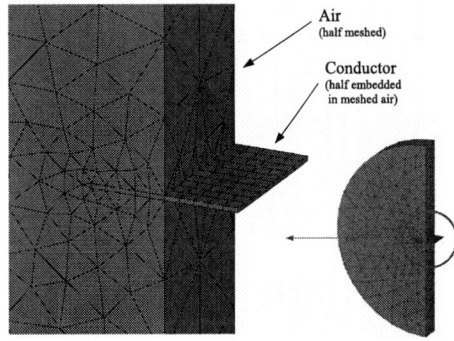

Fig. 8. Finite Element Model (incompletely and coarse meshed) having a reduced value of r_L for explanation purposes

2423

from stored energy in magnetic field:

$$L_{P,int} = 0,3359\text{nH}$$

It is also possible to determine the total partial self inductance of the conductor. It is the in-plane total magnetic flux related to the current through conductor. The integration area for magnetic flux is bounded by the conductor centerline, its length and infinity normal to the conductors centerline. The external inductance then can be calculated by use of:

$$L_P = L_{P,int} + L_{P,ext} \qquad (8)$$

The PEEC method based result of eq. 6 with eq. 8 applied leads to

$$
\begin{aligned}
L_{P,ext} &= 6,534\text{nH} - 0,3359\text{nH} \\
&= 6,1981\text{nH}
\end{aligned}
$$

While internal inductance is constant and proportional to length l of conductor with direct current, dominating outer inductance is a nonlinear function of l for partial elements. This is due to absence of return conductor with respect to inductance as property of closed loops.

C. PEEC Extraction Software

As with finite element method also PEEC parameter extraction for complexly shaped geometries can only be advantageously done by the help of computers employing effective methods for calculation and storing of partial elements with inductive and capacitive couplings. The PEEC cell structure as given with Fig. 5 is extracted from a given arrangement intended to serve for simulation of the electrical behaviour by means of lumped parameters along with a complex electrical circuit with SPICE-like simulators. Often the determination of only parts of the PEEC model — e.g. partial inductance, capacitance or ohmic resistance — at defined connection points is sufficient to predict the electrical behaviour.

Numerous software tools were developed since PEEC method was proposed. Noncommercial tools are of special interest because of their open source and availability. Thus *FastHenry* for partial inductance and ohmic resistance extraction and *FastCap* for capacitance calculation are suitable software tools. Both were developed at the Massachusetts Institute of Technology M.I.T for the solution of Maxwell equations and extraction of circuit parasitics and are accessible with *FastModel* editor available at *www.fastfieldsolvers.com* .Since *FastCap* and *FastHenry* are independent the additional tool *ConvertHenry* automates the conversion of geometrical structures prepared as *FastHenry* input file to the definition of potential plates requested by *FastCap*. Some restrictions — e.g. absence of direct conversion of 2-dimensional conducting ground planes — have to be observed.

The tools above mentioned are used in the following to investigate the electrical behaviour of the electrically conducting structure of the IGBT IXER35N120D1 cohesive with the thermal optimisation results for circuit layer structuring from section II.

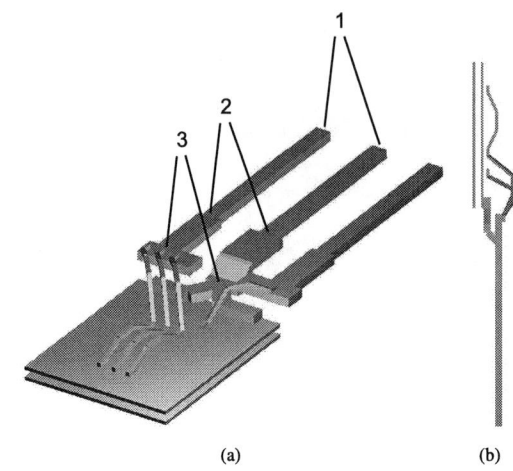

(a)   (b)

Fig. 9. a) 3-dimensional view and b) side view of IGBT IXER35N120D1 model for PEEC parameter extraction

IV. INDUCTANCES

For investigation of the influence of thermal optimisation on electrical behaviour, first principal characteristic of the IXER35N120D1 original structure is determined. Fig. 9 gives the model prepared as a *FastHenry* input file with three representative connection point pairs. Depending on which current path is to be calculated the end nodes of all bond wires geometrically related to one chip are electrically shorted to DCB top copper layer. Al dimensions equal the original dimensions of the discrete component. Fig. 10 and Fig. 11 give the frequency dependent partial inductance at the connection point pairs given in Fig. 9. The decrease of inductance with higher frequency is due to the skin effect. Thus in accordance with the magnetic field distribution inner inductance decreases while outer inductance remains. Connection point pair 1 mostly is unacceptable for in-circuit conduction because of the significant inductance parts of the connection pins. Connection point pair 3 features the lowest inductance values but is not accessible

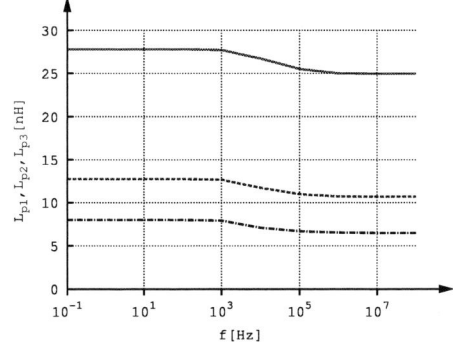

Fig. 10. Partial inductance of IXER35N120D1 model with the diode chip current path closed at connection point pairs 1(red), 2(blue) and 3(black)

2424

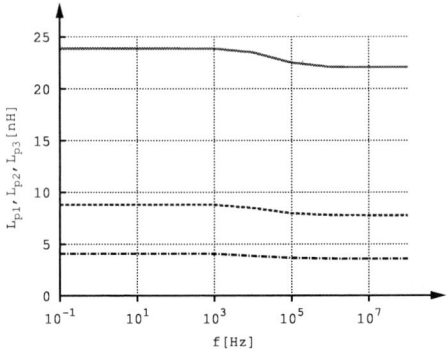

Fig. 11. Partial inductance of IXER35N120D1 model with the IGBT chip current path closed at connection point pairs 1(red), 2(blue) and 3(black)

because of the packaging. Thus the discrete component should as far as possible be connected near connection point pair 2.

Without restructuring the in-plane shape of the circuit layer of the DCB substrate, improvement of thermal behaviour can be achieved by

- an increase of the thickness of the copper layers and
- a decrease of the insulation layer thickness.

Dependency of partial inductance at the connection point

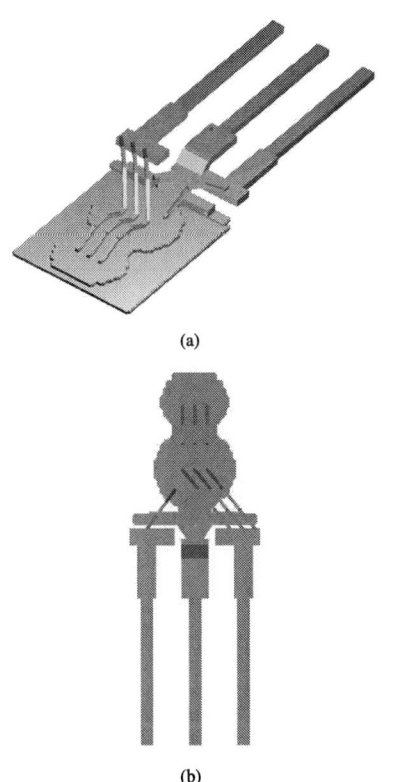

(a)

(b)

Fig. 12. a) 3-dimensional view and b) top view of IGBT IXER35N120D1 model for PEEC parameter extraction with thermally optimised circuit layer of DCB substrate

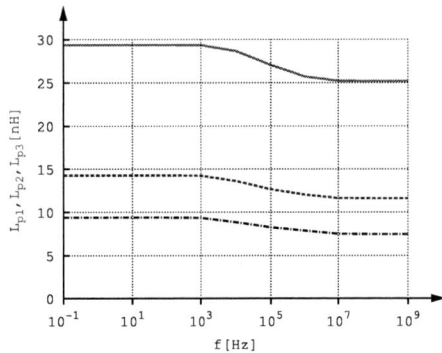

Fig. 13. Partial inductance of the reshaped IXER35N120D1 model with the diode chip current path closed at connection point pairs 1(red), 2(blue) and 3(black)

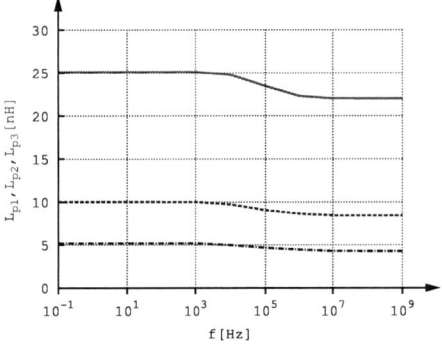

Fig. 14. Partial inductance of the reshaped IXER35N120D1 model with the IGBT chip current path closed at connection point pairs 1(red), 2(blue) and 3(black)

pairs given in Fig. 9 on variation of the above mentioned parameters has been investigated by calculating the frequency dependent partial inductance. Results show no significant change of inductance values due to the short and flat current path structure on DCB substrate of IXER35N120D1. Partial inductance with the current path of the IGBT chip closed occurs with lower values due to the situation closer to the external connection pins than the diode chip. Fig. 14 gives the results.

Thermal optimisation through circuit layer reshaping discussed in section II in most simple case leads to a circular circuit layer design around the silicon chip on substrate with common center. Fig. 12(a) gives the model of IXER35N120D1 prepared as a *FastHenry* input file where the circuit layer of DCB substrate is reduced to radius $r_I = 4, 8\text{mm}$ around the IGBT chip and $r_D = 4\text{mm}$ around the diode chip. The top view with DCB bottom layer removed for explanation purposes is given in Fig. 12(b). The circuit layer area for the diode chip is flattened due to the dimensions of the substrate of the original model. Fig. 13 gives the results for the partial inductance with the diode chip current path closed. Although the model with circularly optimised circuit layer drastically reduces the circuit layer area for current conduction in comparison with the original model Fig. 9, no significant discrepancy in the frequency dependent partial inductance at each connection point pair occurs. This is due to the flat structure of the conducting parts of the DCB which even remains after thermal optimisation when interconnecting parts of the layer are kept at an adequate width. The main part of the inductance values anyway results from the external connection pins.

V. CAPACITANCES

Since *FastHenry* and *FastCap* are independent it is not possible to extract partial inductances and coefficients of potential in common. With the software version of *ConvertHenry* used for this work it is not possible to convert the ground planes used for the two dimensionally conducting DCB copper layers from *FastHenry*. Thus all surfaces of the conductors subdivided into quadrilateral or triangular potential plates must be set-up manually resulting in extensive modelling effort. The principle of the extraction of the coefficients of potential and calculation of capacitances is discussed in the following by means of a more simple example.

Two conductors according to Fig. 7(b) are considered with length $l = 10$mm, width $b_{ges} = 10$mm and $m = 10$ subdivisions for each cell. This assumption results in two surface cells each subdivided in ten subcells. Applying eq. 12 and dividing the result by $(\mu_0 \cdot 1 \cdot \epsilon_0 \cdot 1 \cdot l^2)$ leads to the partial self coefficient of potential of one subcell:

$$P_{P,self,sub} = 6,3428 \; 1/\text{pF}$$

The values for mutual coefficients of potential between the subcells of one conductor directly result from the values given in Tab. II by dividing the values by $(\mu_0 \cdot 1 \cdot \epsilon_0 \cdot 1 \cdot l^2)$. Applying

$$P_{P,self} = \frac{1}{m^2} \cdot \sum_{i=1}^{m} \sum_{j=1}^{m} P_{P_{ij}} \tag{9}$$

according to section III leads to the partial self coefficient of potential of one conductor:

$$P_{P,self} = 2,6249 \; 1/\text{pF}$$

The partial mutual coefficient of potential between the two conductors depends on their geometrical arrangement. A parallel-plate capacitor assembly is considered in the following with a distance of $d = 630\mu$m of the plates corresponding to the insulation of the DCB substrate of IXER35N120D1. The volume between the two plates is filled with aluminium oxide with $\epsilon_r = 8,5$. It is discretised as aforementioned, thus influence on resulting capacitance will be discussed later. The mutual coefficients of potential between the subcells of the conductors can be calculated as done before with respect to the now appearing distances between the filaments. For calculation of the total partial mutual coefficient of potential between the conductors eq. 9 can be used [7] where i now indicates the subcells of conductor one and j that of conductor two. For an arrangement as given this leads to

$$P_{P,mutual} = 2,3764 \; 1/\text{pF}$$

In case of no dielectric media and thus free space present between the plates, capacitance can be retrieved from coefficients of potential by assigning an electrical potential of $\varphi_1 = 1$V to one conductor and $\varphi_2 = 0$V to the other. Then two coupled equations can be formulated for the electrical potential,

$$\varphi_1 = P_{11} \cdot Q_1 + P_{12} \cdot Q_1 \tag{10}$$
$$\varphi_2 = P_{21} \cdot Q_2 + P_{22} \cdot Q_2 \tag{11}$$

where Q_1 and Q_2 are the charges of the conductors, $P_{12} = P_{21} = P_{P,mutual}$ and $P_{11} = P_{22} = P_{P,self}$. Solving eq. 10 and eq. 11 for Q_1 and Q_2 for an arrangement as given this leads to

$$Q_1 = 2,0354\text{pC}, \; Q_2 = -1,8350\text{pC}$$

Applying

$$C = Q_2/(\varphi_2 - \varphi_1)$$

leads to the capacitance between the conductors.

$$C = 1,8350\text{pF}$$

With *FastCap* instead the mutual coefficients of potential are calculated for quadrilateral plates not only for filaments. The different result $C = 1,461$pF is obtained for the same arrangement. Because of inhomogeneous charge distribution on the plates both results are inaccurate. The plates must be subdivided in two dimensions with attention to a refinement in border regions. Calculation with *FastCap* using 2704 subcells each plate results in

$$C = 1,582\text{pF}$$

For the DCB substrate of the IXER35N120D1 the assumption of infinitesimally thin plates is unusable because conductor thickness of 300μm is quite half of the insulation thickness which is 630μm. Analytical calculation is very extensive because all conductor faces must be included. With *FastCap* using 6032 subcells each conductor with a thickness of the conductors of 300μm is obtained:

$$C = 1,655\text{pF}$$

When dielectric media is considered and all parameters of the PEEC model for all directions of the coordinate system are determined its volumetric discretisation for each direction of the cartesian coordinate system is needed when an orthogonal 'Manhattan Discretisation' is used [9]. According to the aforementioned parallel plate capacitor arrangement the dielectric volume between the plates with length $l = 10$mm, width $b = 10$mm and height $h = 630\mu$m is considered. For each dimension x, y and z one partial element equivalent circuit cell is defined. The surfaces of the dielectric cuboid carry charges and thus are represented by plane cells as already considered before in case of two conductors with no dielectric media between. For simplified analytical consideration one volume cell in rectangular direction to the capacitor plate's surfaces is assumed and its 'excess capacitance' partially representing the dielectric behaviour is calculated [9]:

$$C^+ = \frac{\varepsilon_0 \cdot (\varepsilon_r - 1) \cdot b \cdot l}{d}$$
$$= 10,5407\text{pF}$$

Simplified summation of calculated capacitance between the plates and dielectric capacitance leads to $C_\Sigma = 12,0805\text{pF}$.

For comparison the formula for a parallel-plate capacitor assuming ideal conditions — e.g. homogeneous charge distribution on the plates — with area A and distance d of the plates is applied and because of the definition of the excess capacitance leads to the similar result $C = 11,9461\text{pF}$ which is well in accordance.

With *FastCap* the potential plates of the dielectric region can be included in calculation same as the thickness of the conductors of $300\mu\text{m}$ by modelling all conductor faces. With 13116 subcells for the whole arrangement is obtained: $C_\Sigma = 12,30\text{pF}$

Considerations demonstrate the possibility to calculate capacitances of any structures if they are subdivided into sufficiently small PEEC volume and potential surface cells. If the arrangement of cells is appropriately chosen the set-up of an PEEC model for inclusion in a simulation with SPICE-like simulators is possible.

VI. CONCLUSION AND OUTLOOK

This paper gives a basic approach for future designs of power electronic systems. Estimation of thermal resistance can be done using a cylindrical abstraction. Cross sectional area of insulation for heat flux passage essentially influences resulting thermal resistance. With help of PEEC method and FEM simulation the influence of thermal optimisation on formation of parasitics can be analysed. Combined considerations provide the possibility to avoid extension of parasitics when thermal aspects are regarded. Further work will have to deal with methodical structuring of optimisation processes and measurements for validation.

APPENDIX

A. Partial Self Inductance

Partial self inductance of a conductor as given with Fig. 7 can be calculated applying eq. 12 when an infinitesimal small value h is assumed:

$$
L_P = \frac{\mu \cdot l}{6 \cdot \pi} \cdot \left[3 \cdot \ln\left(\frac{l}{b} + \sqrt{\left(\frac{l}{b}\right)^2 + 1} \right) \right. \quad (12)
$$
$$
+ \frac{b}{l} + 3 \cdot \frac{l}{b} \cdot \ln\left(\frac{b}{l} + \sqrt{\left(\frac{b}{l}\right)^2 + 1} \right)
$$
$$
\left. - \sqrt{\left(\left(\frac{l}{b}\right)^{\frac{4}{3}} + \left(\frac{l}{b}\right)^{-\frac{2}{3}} \right)^3 + \left(\frac{l}{b}\right)^2} \right]
$$

B. Partial Mutual Inductance

Assuming conductors as filaments as given in Fig. 15 eq. 13 can be applied for partial mutual inductance.

Fig. 15. Mutual inductance of two conductor filaments

$$
M_P = \frac{\mu}{4 \cdot \pi} \cdot \left[\alpha \cdot \ln\left(\frac{\alpha}{d} + \sqrt{\left(\frac{\alpha}{d}\right)^2 + 1} \right) \right. \quad (13)
$$
$$
- \beta \cdot \ln\left(\frac{\beta}{d} + \sqrt{\left(\frac{\beta}{d}\right)^2 + 1} \right)
$$
$$
- \gamma \cdot \ln\left(\frac{\gamma}{d} + \sqrt{\left(\frac{\gamma}{d}\right)^2 + 1} \right)
$$
$$
+ \delta \cdot \ln\left(\frac{\delta}{d} + \sqrt{\left(\frac{\delta}{d}\right)^2 + 1} \right)
$$
$$
- \sqrt{\alpha^2 + d^2} + \sqrt{\beta^2 + d^2}
$$
$$
\left. + \sqrt{\gamma^2 + d^2} - \sqrt{\delta^2 + d^2} \right]
$$
$$
\alpha = l + m + \delta , \quad \beta = l + \delta , \quad \gamma = m + \delta
$$

REFERENCES

[1] S. Förster and A. Lindemann, "Thermal optimisation in low voltage - high power applications using SMT components," in *Proceedings of the IYCE 07*, Budapest, Hungary, 2007.

[2] D. P. Kennedy, "Spreading Resistance in Cylindrical Semiconductor Devices," *Journal of Applied Physics*, vol. 31, no. 8, August 1960.

[3] J. V. den Keybus, T. Nobels, and R. Belmans, "Thermal Design of converters using discrete power components incorporating an IGBT and a freewheeling diode," in *Proceedings of the EPE2005*, Dresden, Germany, 2005.

[4] S. Förster and A. Lindemann, "Cooling of Insulated Assemblies," in *Proceedings of the CIPS 2008*, Nürnberg, Germany, 2008.

[5] B. Gutsmann, P. Mourick, and D. Silber, "Exact Inductive Parasitic Extraction for Analysis of IGBT Parallel Switching including DCB-Backside Eddy Currents," in *Proceedings of the Power Electronics Specialists Conference PESC 00*, vol. 3, Galway, Ireland, June 2000, pp. 1291 – 1295.

[6] S. Thamm, S. V. Kochetov, and G. Wollenberg, "Modellierung flächenhafter Verbindungsstrukturen für leistungselektronische Anwendungen mittels der Methode der partiellen Elemente," in *Proceedings of the EMC 06*, Düsseldorf, Germany, 2006.

[7] A. E. Ruehli, "Inductance Calculations in a Complex Integrated Circuit Environment," *IBM Journal on Research and Development*, pp. 470–481, September 1972.

[8] P. A. Brennan, N. Raver, and A. E. Ruehli, "Three-Dimensional Inductance Computations with Partial Element Equivalent Circuits," *IBM Journal on Research and Development*, vol. 23, no. 6, pp. 661–668, November 1979.

[9] M. L. Zitzmann, "Fast and Efficient Methods for Circuit-based Automotive EMC Simulation," Dissertation, Universität Erlangen-Nürnberg, Erlangen-Nürnberg, Germany, February 2007.

[10] S. P. Weber, "Effizienter Entwurf von EMV-Filtern für leistungselektronische Geräte unter Anwendung der Methode der partiellen Elemente," Dissertation, Technische Universität Berlin, Berlin, Germany, May 2007.

[11] J. Ekman, G. Antonini, A. Orlandi, and A. E. Ruehli, "Impact of Partial Element Accuracy on PEEC Model Stability," *IEEE Transactions on Electromagnetic Compatibility*, vol. 48, no. 1, February 2006.

[12] G. Antonini, A. Orlandi, and C. R. Paul, "Internal Impedance of Conductors of Rectangular Cross Section," *IEEE Transactions on Microwave Theory and Techniques*, vol. 47, no. 7, July 1999.

Discussion of Internal and External High Frequency Common Mode Noise Current on a Chopper Circuit

Tetsuya Mitani*, Keiji Wada*, Toshihisa Shimizu*, Hiromichi Ohashi[†]

*Tokyo Metropolitan University / Department of Electrical and Electronics Engineering
1-1 Minami Osawa, Hachioji, Tokyo, Japan
e-mail: *shimizut@tmu.ac.jp*
[†]Advanced Institute of Science and Industry

Abstract— The mechanism of common-mode noise current in control and gate-drive circuits on a chopper circuit is discussed. For high power density configurations, common-mode noise current flows not only in the main circuit, but also in the control and gate-drive circuits. This noise current increases the risk of control circuit misoperation, and might flow out from control and gate-drive circuits and interfere in a utility line, etc. The factors for the generation of noise currents and the propagation paths of these noise currents are examined. The influence of this noise current on the EMI noise is shown from the experimental results.

Keywords— EMC/EMI, High frequency power converter, High power density systems, Noise.

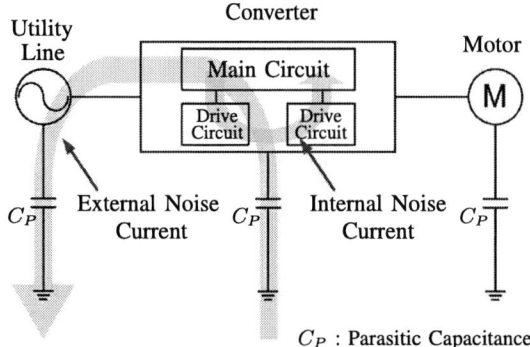

C_P : Parasitic Capacitance

Fig. 1.   Definition of internal and external noise current.

I. INTRODUCTION

In recent years, the power density of power converters has been increasing, i.e., the power density increases by a factor of 10 every 15 years, so that the power density has increased 100-fold in the last 30 years [1]. In order to increase the power density of converters, the reduction in losses is essential, and also the reduction in the volume of components used in converters. In general, the volume of passive components, such as inductors and transformers, can be reduced by increasing the switching frequency of the power converters. Fortunately, the switching characteristics of power semiconductors have been improved, and the resultant switching frequency can be raised to around 10 to 100 kHz. In recent years, the power density of power converters has been increased to around 3 to 10 W/cc. However, the volume of power converters is now again dominated by the volume of heat sinks and passive components, so that the increase in power density has come to an abrupt stop. Fortunately, progress in the research and development of wideband gap semiconductors, such as SiC, has enabled low on-state resistance and ultra-high-speed switching of power semiconductor devices, and the switching frequency of converters can be increased to more than 100 kHz using these devices [2].

On the other hand, high-speed switching results in an increase of electromagnetic noise problems in communication or measurement equipment, and in utility lines, etc. In order to prevent such interference, much research has been undertaken in the last 10 years, such

as the analysis of common-mode noise currents flowing into utility lines through the parasitic capacitance of the power converter main circuit [3]. For high power density configurations, the control circuit and gate-drive circuits have to be located very close to the main circuit, and also the power devices on the main circuit are operated under very high dv/dt and di/dt conditions. Noise currents generated by the switching devices then flow into the control and gate-drive circuits, and cause the misoperation of the converter control circuit. The experimental results of noise currents, and the expected misoperation of the control circuit caused by these currents are presented [4]. Moreover, this noise current might flow out from the control and gate-drive circuits and interfere in the utility line, etc. Therefore, in order to suppress EMI noise from high power density configurations with high-frequency switching, the study of noise currents flowing through the control and gate-drive circuits has become a very important issue. However, no paper has discussed and analyzed noise currents in this respect.

In this investigation, common-mode noise currents flowing through control and gate-drive circuits were measured, and the characteristics are discussed with respect to the suppression of noise currents in high power density power converters. Fig. 1 depicts the two kinds of common-mode noise current in converter systems. One is the common-mode noise current that flows out to the utility line through the main terminal of the power

978-1-4244-1741-4/08/$25.00 ©2008 IEEE

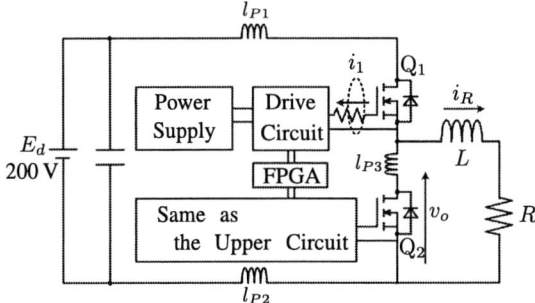

Fig. 2. Circuit configuration.

TABLE I
CIRCUIT PARAMETERS.

Device		Specification
MOSFET	Type	2SK3936 (TOSHIBA)
	V_{DSS}	500 V
	I_D	23 A
	C_{iss}	4250 pF
	C_{oss}	420 pF
	C_{rss}	10 pF
DC Voltage	E_d	200 V
Inductor	L	500 μH
Output Resistance	R	5 Ω
Gate Resistance	R_g	4.7 Ω~50 Ω

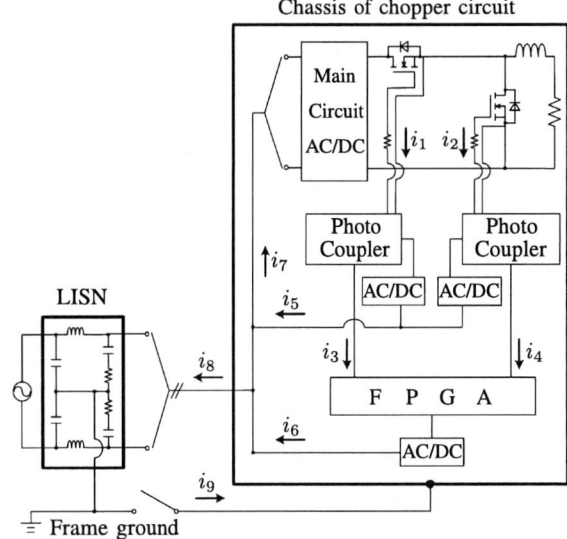

Fig. 3. The detailed connection of each component.

converter, and is defined as "external noise current". Another is the common-mode noise current that circulates in the power converter chassis through the control and gate-drive circuit, and is referred to as "internal noise current". The internal and external noise currents are measured under various conditions in this study, and the factors of noise current generation and the propagation paths of these noise currents are examined.

II. CIRCUIT CONFIGURATION

Fig. 2 shows the circuit configuration of the buck-chopper circuit used under test, and Table. I shows the circuit parameters. The DC bus voltage is set to 200 V and two high voltage MOSFETs (2SK3936,Toshiba) are used as the switches. MOSFET Q_1 is driven by the gate-drive circuit with double pulse sequences, and MOSFET Q_2 is kept in the turned-off condition, because the body diode of Q_2 is used as a free-wheeling diode. The pulse width of the Q_1 gate signal is varied in order to obtain the required current flow through MOSFETs Q_1 and Q_2. Three kinds of gate resistance R_g are used in order to vary the switching speed of the MOSFETs.

Fig. 3 shows the detailed connection of each component, including the gate-drive circuits, power supplies, a field programmable gate array (FPGA) for the control component, a line impedance stabilizing network (LISN), and the chassis for the chopper circuit. A photo-coupler is used in order to isolate the gate signals between the two MOSFETs and the FPGA. Isolated-type AC/DC converters are used for both the control circuit and the

main circuit in order to isolate each portion, and also a transformer and rectifier are used for gate-drive power supply. A LISN is used to standardize the input impedance from the converter input. The chopper circuit is assembled on a chassis. The noise currents are measured in the case where the chassis is connected to the frame ground of the LISN and that where it is not connected to the frame ground, referred to as "grounding syste" and "non-grounding system", respectively.

Fig. 3 also shows the measured points of the common-mode noise currents on the experimental circuit. i_1 to i_6 are the common mode noise currents that flow through each part of the control and gate-drive circuits. i_7 is the common mode noise current that flows toward the main circuit, i_8 is the common mode noise current that flows out to the utility line side from the chopper circuit, and i_9 is the common mode noise current that flows from the frame ground of the LISN to the chassis. In this paper, i_1 to i_7 are defined as the internal noise current, and i_8 to i_9 are defined as the external noise current. A digital oscilloscope (DPO4104,Tektronix) and AC current transformers (MODEL2877 (300 Hz to 200 MHz),Pearson) were used for the measurements.

III. FACTORS FOR GENERATION OF INTERNAL NOISE CURRENT

When internal noise currents flow in the control and gate-drive circuits, serious misoperation of the power converter can occur. Therefore, in order to examine the factors of internal noise current generation, the internal noise current i_1 is measured under various conditions with respect to the switching operation of Q_1, and the maximum amplitude I_1 is examined. The non-grounding system is used for this measurement, in order to reduce the influence of the external noise current.

Fig. 4 shows three experimental waveforms of i_1 at

the instant of the MOSFET Q_1 turn off, when the load currents I_R are 5, 10 and 20 A, respectively. The gate resistor R_g is selected to be 10 Ω. It is clear that each of the common-mode current waveforms of i_1 has a damping oscillation. The resonant frequency of these currents coincide with the oscillation frequency of the turn off voltage on the MOSFET. The maximum amplitude of i_1 is varied depending on the amplitude of the load current, I_R.

Figs. 5 and 6 show the maximum amplitude of the common-mode current i_1 vs. the load current I_R at the time when Q_1 is turned off and turned on, respectively. The maximum amplitude I_1, increases linearly as the load current increases, as shown in Fig. 5. The reason for these results is given as follows. Since Q_1 is turned off at this time, the capacitance C_{DS}, between drain and source of Q_1, is charged by I_R. When I_R becomes large, the charge speed of C_{DS}, which is equivalent to dv/dt, increases in proportion to I_R. The maximum amplitude I_1 is expected to be determined by dv/dt, so that the resultant I_1 increases as the load current I_R increases. In contrast, I_1 decreases as the load current I_R increases, as shown in Fig. 6. The reason for these results is given as follows. At the time when Q_1 is turned on, I_R is flowing through the body diode of Q_2. Since the body diode of Q_2 must be turned off, the reverse recovery current flows through the body diode of Q_2 before turning the body diode off. In the case when I_R is small, the reverse recovery time becomes short, and the resultant dv/dt of the output voltage, v_o, becomes large, because a change in v_o occurs during the reverse recovery time. Therefore, I_1 decreases when I_R becomes large.

Fig. 7 shows the relation between the maximum amplitude I_1 and the dv/dt of v_o. The maximum amplitude of the common-mode current increases in proportion to the value of dv/dt. This means that the previous results shown in Fig. 5 and Fig. 6 reflect the dv/dt value of v_o. Since the amplitude of I_1 in Fig. 5 is much larger than that in Fig. 6, it is important to determine the factors of dv/dt at the turn off transients.

Fig. 8 shows the measured value of dv/dt at the turn off of Q_1, when the load current I_R is changed. Here, the gate resistance R_g, is changed between 10 and 50 Ω. In the region where I_R is small, the change of dv/dt almost follows a straight line determined by the following equation.

$$\frac{dv}{dt} = \frac{I_R}{C_{DS}} \qquad (1)$$

However, if I_R becomes large, the change in dv/dt does not follow Eq. (1), but the value is limited to a saturated value that is determined by the gate resistance, R_g. This phenomenon reflects the influence of the miller effect at the charge period of C_{DS}. In other words, in the case when I_R is increased to the saturated value, dv/dt does not depend on I_R, but is determined by the gate resistance. It is concluded that the amplitude of the load current and the gate resistance are the factors determining dv/dt.

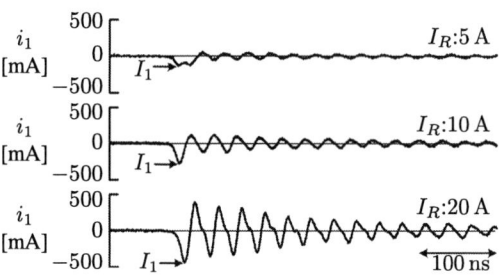

Fig. 4. Experimental waveforms of i_1.

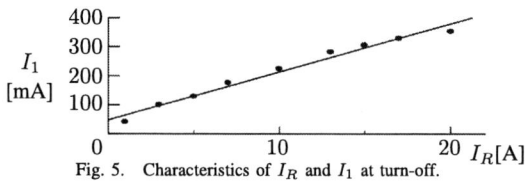

Fig. 5. Characteristics of I_R and I_1 at turn-off.

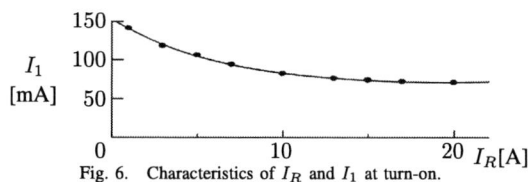

Fig. 6. Characteristics of I_R and I_1 at turn-on.

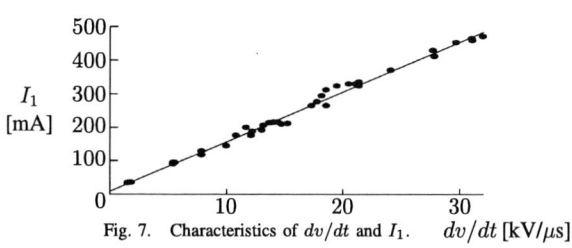

Fig. 7. Characteristics of dv/dt and I_1.

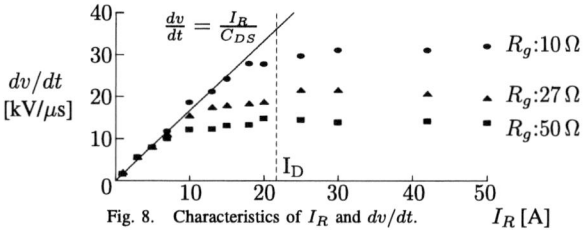

Fig. 8. Characteristics of I_R and dv/dt.

IV. ANALYSIS OF NOISE CURRENTS IN THE NON-GROUNDING SYSTEM

In order to analyze the propagation path of the noise currents, the noise currents of each point were measured in the non-grounding system. Fig. 9 shows the experimental waveforms of the output voltage v_o, and each of the common-mode noise currents measured at the points shown in Fig. 3 for the turn off transition of Q_1. The load current I_R is set to $20\,A$, and the gate resistance R_g is $4.7\,\Omega$. When the output voltage v_o is changed from $200\,V$ to $0\,V$, v_o oscillates with a frequency around $50\,MHz$, as shown in Fig. 9. The oscillation frequency is determined by the parasitic inductance l_{P1}, l_{P2} and l_{P3} of the chopper circuit, and the output capacitances of MOSFETs Q_1 and Q_2. The major frequency component of the internal noise currents i_1 to i_6 is $50\,MHz$, which corresponds to the oscillation frequency of v_0. On the other hand, the major frequency components of the noise currents i_7 and i_8 are $50\,MHz$ and $9\,MHz$. The $9\,MHz$ component is the noise current that flows out to the utility line from the main circuit of the chopper; however, the noise current generating factors and the propagation path of this component are not clear at this stage.

The propagation path of the $50\,MHz$ component of the noise current is expressed as follows. i_1 is the internal noise current that flows into the high side gate-drive circuit from the high side MOSFET Q_1, and the amplitude of i_1 is the largest in all the current waveforms of Fig. 9. i_2 is the internal noise current that flows into the low side gate-drive circuit from the low side MOSFET Q_2. The experimental waveforms of i_1 and i_2 show that the phase difference between i_1 and i_2 is approximately 180 degrees and the maximum amplitude of i_2 is $1/4$ that of i_1. Therefore, a part of the internal noise current i_1 flows into the low side MOSFET Q_2 from the high side MOSFET Q_1. This means that the internal noise current circulates between the high and low side MOSFETs through the parasitic capacitance of the gate-drive circuits.

The other part of the internal noise current i_1 is divided into the noise current i_5, which flows out from the power supplies for the gate-drives, and the noise currents i_3 and i_4, which flow into the control circuit from the photo-couplers for the gate-drives. The amplitudes of the noise currents i_3 and i_4 are very small compared to i_5, but are sufficient to cause misoperation of the control circuit. Furthermore, the waveform of i_6 shows that a part of the noise currents i_3 and i_4 flows out from the power supply for the FPGA.

Fig. 10 shows the experimental waveforms of $i_1 + i_2$ and $i_5 + i_6$. The amplitude of $i_5 + i_6$ is a little smaller than that of $i_1 + i_2$. Therefore, most of the noise current $i_1 + i_2$ flows out to $i_5 + i_6$, but the difference between these two current may flow through uncertain current path.

Moreover, i_7 is the internal noise current that flows into the power supply for the main circuit, and i_8 is the external noise current that flows out to the utility line side, respectively. It is clear from a comparison of waveforms i_5, i_7 and i_8 that, most of the internal noise current i_5

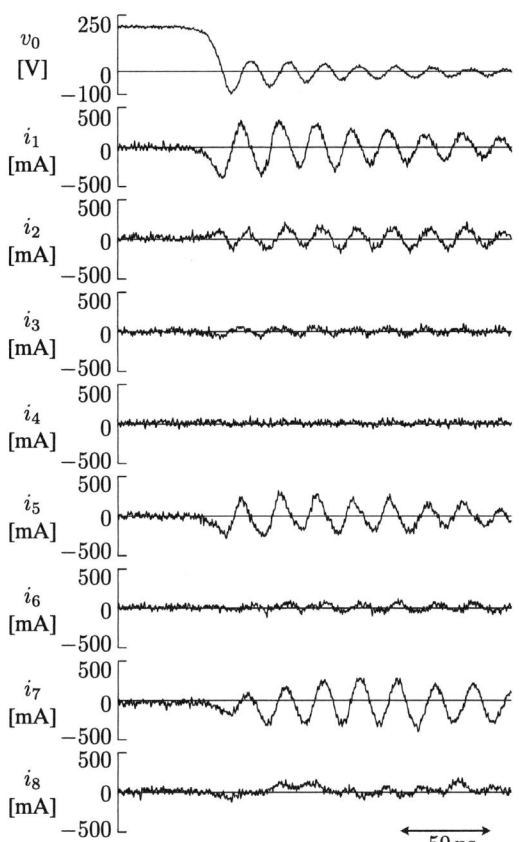

Fig. 9. Experimental waveforms of v_o and i_1 to i_8 in the non-grounding system.

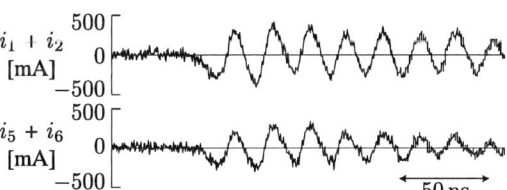

Fig. 10. Experimental waveforms of $i_1 + i_2$ and $i_5 + i_6$ in the non-grounding system.

flows toward the main circuit of the chopper, but part of i_5 flows out to the utility line side.

From the above discussion, most of the internal noise current circulates in the chopper circuit. Since this noise current might cause misoperation of the control and gate-drive circuits, the noise current must be suppressed. Moreover, part of the internal noise current flows out to the utility line side from the chopper circuit, even if a non-grounding system is used. It is thought that the reason for this is that the chopper circuit is not absolutely isolated from the grounding potential.

V. ANALYSIS OF NOISE CURRENTS IN THE GROUNDING SYSTEM

The experimental results for the non-grounding system show that part of the internal noise current flows out to the utility line side. For the grounding system, this noise current might become even more significant. In order to evaluate this, the noise currents of each point were also measured in the grounding system.

Fig. 11 shows the experimental waveforms of the output voltage v_o, and each of the common-mode noise currents measured at the points shown in Fig. 3 for the turn off transition of Q_1 in the grounding system. Similar to the case of the non-grounding system experiment, the load current I_R is set to 20 A, and the gate resistance R_g is 4.7 Ω.

The experimental waveforms of i_1 to i_6 in Fig. 11 are almost the same as the waveforms in Fig. 9 with respect to the oscillation frequency and the amplitude. Therefore, the internal noise current has three kinds of propagation paths. The 1st is the path which circulates between the high and low side MOSFETs through the gate-drives, the 2nd is the path which flows into the control circuit, and the 3rd is the path which flows out from the power supplies for the gate-drives.

Similar to the case of the non-grounding system, the noise current i_5, which flows out from the power supplies for the gate-drives, is divided into the internal noise current i_7 and the external noise current i_8. However, the amplitude of the noise current i_8, which flows out to the utility line side, in Fig. 11 is much larger than the amplitude of i_8 in Fig. 9. This indicates that the noise current, which flows out to the utility line from the control and gate-drives, becomes much larger than that in the case of the non-grounding system.

The reason for this result is explained as follows. Fig. 12 shows the experimental waveforms of $i_1 + i_2$ and $i_5 + i_6$ in the grounding system. The waveform of $i_1 + i_2$ is almost the same as the waveform in Fig. 10, but the amplitude of $i_5 + i_6$ is a little larger than that in Fig. 10. Therefore, unlike the case of the non-grounding system, the amplitude of $i_5 + i_6$ is almost the same as that of $i_1 + i_2$. At this time, it is thought that the noise current, which flows through uncertain current path, decreases compared to the non-grounding system. And the external noise current i_8 increases, because the impedance of the external noise current path becomes small in the grounding system.

Moreover, i_9 is the external noise current that flows from the frame ground of the LISN to the chassis of chopper circuit. The experimental waveforms of i_8 and i_9 show that the noise current i_8 flows through the LISN, and is transmitted to the chassis. It is thought that this noise current flows through the parasitic capacitance between the chassis and the heat sinks of the MOSFETs.

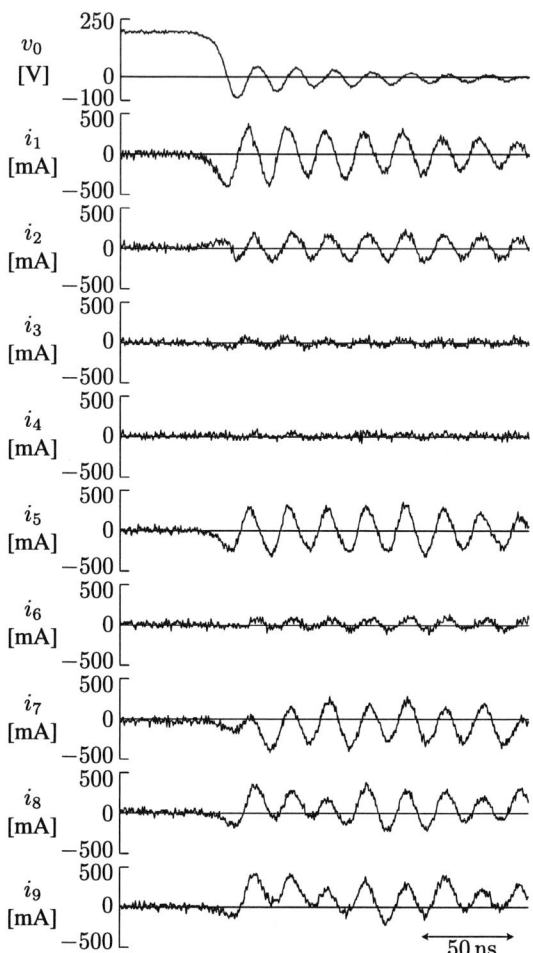

Fig. 11. Experimental waveforms of v_o and i_1 to i_9 in the grounding system.

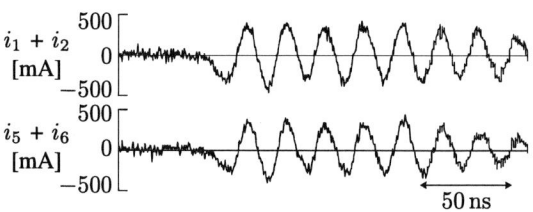

Fig. 12. Experimental waveforms of $i_1 + i_2$ and $i_5 + i_6$ in the grounding system.

On the other hand, the external noise currents i_7, i_8 and i_9 contain a 9 MHz component. This 9 MHz component is the noise current, which flows out to the utility line through the main circuit of chopper, the power supply of the main circuit and the parasitic capacitance of the main circuit, without passing the control and gate-drive circuits.

From the above discussion, a part of the noise current that flows out from the power converter is the noise current that flows out from the control and gate-drive circuits. Moreover, the amplitude of the noise current that flows out to the utility line from the control and gate-drive circuits is large enough compared to that which flows out from the main circuit of the power converter. Consequently, with respect to the problem of EMI, not only the noise current that flows out from the main circuit, but also the noise current that flows out from the control and gate-drive circuits must be suppressed.

VI. CONCLUSION

The internal noise current and external noise current of a chopper circuit was discussed. The factors for the generation of noise currents, and the propagation paths of these noise currents were described. Moreover, the influence of the noise current that flows out to the utility line side from the control and gate-drive circuits on the EMI noise was shown from the experimental results.

ACKNOWLEDGMENT

This research project was supported by KAKENHI (Grant-in-Aid for Scientific Research)(No.19360132). The Authors thanks Prof. Yukihiko Sato and Dr. Yusuke Hayashi for their discussion.

REFERENCES

[1] H.Ohashi, IEEJ, Vol.102, No.3, pp.168-171, 2002 (in Japanese).
[2] M.Tsukuda, I.Omura, T.Domon, W.Saito, T.Ogura, "Demonstration of High Output Power Density (30 W/cc) Converter using 600 V SiC-SBD and Low Impedance Gate Driver", IEEJ IPEC Niigata, pp.1184-1189, 2005.
[3] M.Shoyama, G.Li, T.Ninomiya, "Balanced Switching Converter to Reduce Common-Mode Conducted Noise", IEEE Transactions on Industrial Electronics, Vol.50, No.6, pp.1095-1099, 2003.12.
[4] K.Wada, K.Shirakawa, T.Shimizu, "Discussion of Internal Noise Currents in a Control Circuit on a 200-kHz Switching PWM Inverter", PCC Nagoya, LSI-3-3, 2007.

A Novel Digital Control Method for DC-DC Converter

Fujio Kurokawa[†,††], Masashi Okamatsu[††], Yuichi Sumida[††],
Yasuhiro Mimura[†††] and Masahiro Sasaki[†††]

† Department of Electrical and Electronic Engineering, Nagasaki University ,1-14 Bunkyo-machi,
Nagasaki, 852-8521 Japan, Tel/Fax: +81 95 819 2553, E-mail:fkurokaw@nagasaki-u.ac.jp
†† Graduate School of Science and Technology, Nagasaki University,1-14 Bunkyo-machi,
Nagasaki, 852-8521 Japan, Tel/Fax: +81 95 819 2553, E-mail:d707121c@cc.nagasaki-u.ac.jp
††† Shindengen Electric Manufacturing Co.,Ltd, New-Ohtemachi Bldg., 2-1 Ohtemachi 2-chome, Chiyoda-ku,
Tokyo, 100-0004 Japan, TEL: +81-3-3279-4431, FAX: +81-3-3279-6478, E-mail: yasuhiro@shindengen.co.jp

Abstract— The purpose of this paper is to present a novel control method for the digitally controlled dc-dc converter and to improve the dynamic characteristics using the proposed simple control circuit. The feature is that the derivative control in the time-domain is combined to the FIR filter in the frequency-domain. As a result, it is clarified that the transient time and overshoot of the reactor current are suppressed to within 1.43ms and 1.4A and that these result is approximately 60% smaller than the conventional minimum phase FIR filter method, respectively. Moreover, it is revealed that the number of parts of the digital control circuit is 15% smaller than that of the analog control circuit and the cost is also 5% smaller.

Keywords— Converter control, High frequency power converter, Pulse Width Modulation (PWM), Switched-mode power supply.

I. INTRODUCTION

The power supply system plays the important role of Electrical and Electronic systems reliability. The high controllability and monitoring function are required in the switching power supply to improve the reliability of these systems [1]-[6]. In this case, the digitally controlled switching power supply is useful because it has the advantage of realizing both control and monitoring tasks.

Furthermore, a key distinguishing feature of digital control circuit is able to combine with the other digital IC in the electronics system. Conventionally, since the conventional control circuit is analog, the controller of the switching converter is extra parts in the digital system. When the digital control circuit is applied to practical use, the digital control circuit can lead to decrease both the size and cost of not only the power supply system but also digital electronics system.

We have already reported the FIR filter control method for digitally controlled dc-dc converter [7]. This method has superior feature, that is, the control algorithm and the control circuit are simple. In this case, prior attempts to improve the transient response have been inconclusive because this digital filter has no phase-lead compensation in time-domain.

Generally, the IIR filter control method is used to improve the dynamic characteristics of the digitally controlled converter [8]. However, the control parameters are complex and the control circuit needs the DSP

(Digital Signal Processor). Therefore, the cost is expensive and the limitation of switching frequency depends on the DSP.

This paper presents a novel simple control method of the switching dc-dc converter and the improved dynamic characteristics. Usually, the digital P-I-D control is used in the time-domain and the FIR, the IIR and the other digital control method is designed in the frequency-domain. Although the former control circuit is simple, it differs essentially from the digital signal processing method that the domain is different to the latter. The latter control circuit is not simple. So, this paper takes a different approach toward the digital control circuit. In the proposed method, the novel idea is introduced. The feature is that the derivative control in the time-domain is combined to the FIR filter in the frequency-domain. As a result, both the simple control circuit and the superior dynamic characteristics are obtained in the proposed method.

First, the operation principles of each method are described. Next, the dynamic characteristics are compared. Furthermore, the number of parts and cost are discussed.

II. OPERATION PRINCIPLE

Figure 1 shows the digitally controlled dc-dc converter. This digital control circuit senses the output voltage E_o. T_r is the main switch, D is the fly wheel diode, L is the energy storage reactor, C is the output smoothing capacitor and R is the load. i_L is a reactor current. Figure 2 shows the circuit configuration of the general digital control circuit with DSP.

The output voltage e_o is sent to the A-D converter through the anti-aliasing filter and is converted into digital amount N_n. The relation between the input and output values of the A-D converter is given by equation (1) when it approximately shows the linear expression by considering the width of the quantization to be small.

$$N_n = G_{AD}e_o \qquad (1)$$

where n denotes an n-th switching cycle, and the digital amount N_n is a positive integer number. G_{AD} is a gain of the A-D converter. The digital amount N_n is sent to DSP. In DSP, the numerical value N_{Ton} that corresponds to the on-time interval T_{on} is calculated.

978-1-4244-1741-4/08/$25.00 ©2008 IEEE

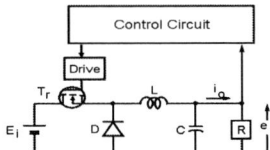

Fig. 1 Digitally controlled dc-dc converter

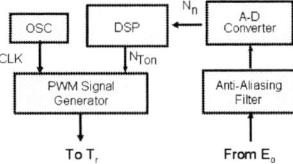

Fig. 2 General digital control circuit with DSP.

The relation between the on-time interval T_{on} and the numerical value N_{Ton} is shown as follows;

$$\frac{T_{on,n+1}}{T_S} = \frac{N_{Ton,n+1}}{N_{Ts}} \qquad (2)$$

where N_{Ts} is a numerical value corresponding to the switching period T_s ($=1/f_s$). N_{Ts} is calculated in the PWM signal generation circuit which is composed of a digital comparator or a counter. According to the relation between the on-time interval T_{on} and the numerical value N_{Ton}, T_{on} is generated. This T_{on} regulates the output voltage e_o.

Usually, the digital P-I-D control method and the digital filter method is applied to the digital control circuit for dc-dc converter. As purpose of this paper is concerned, it is not necessary to discuss the digital P-I-D control method. So, the digital filter method is discussed in conventional method.

The minimum phase FIR filter is implemented by low-pass digital filters, and are represented as N_{min} for the output voltages by the following equations [7];

$$N_{min} = \sum_{i=0}^{q} h_i N_{n-i} \qquad (3)$$

Especially, the moving average which is one of the FIR filter is represented as N_{ma} for the output voltages by the following equations[7], [9];

$$N_{ma} = \frac{1}{m} \sum_{i=0}^{m-1} N_{n-i} \qquad (4)$$

This method is performed by a very simple algorithm as shown in Fig. 3. The function devices are just the adder, shift-register and multiplier. m is the number of the sampling point of the moving average.

The on-time interval $N_{Ton,n+1}$ of each is represented by the following equations.

$$N_{Ton,n+1} = N_B - K(N_{min} - N_R) \qquad (5)$$

$$N_{Ton,n+1} = N_B - K(N_{ma} - N_R) \qquad (6)$$

where N_B, N_R and K_D are the bias, the control reference values and the differential coefficient as follows

$$N_B = N_{Ts}(1 + r/R)\frac{e_o}{E_i} \qquad (7)$$

$$N_R = \sum_{i=0}^{q} h_i G_{AD} e_o^* \qquad (8)$$

where h_i denotes the digital filter coefficients and e_o^* is the desired output voltage of the dc-dc converter.

On the other hand, the IIR filter is designed using the double primary conversion. The converted digital filter becomes the second-order IIR filters as shown in the next equation [8];

$$y_n = a_o x_n + a_1 x_{x-1} + a_2 x_{n-2} - b_1 y_{n-1} - b_2 y_{n-2} \qquad (9)$$

where a_0, a_1, a_2, b_1 and b_2 are coefficients of IIR filter, x_n is a difference between the digital value N_n and the desired value N_R. x_n is shown in the following equation.

$$x_n = N_n - N_R \qquad (10)$$

Then, the on-time interval N_{Ton} of IIR filter control circuit is represented by the following equation.

$$N_{Ton,n+1} = N_B - y_n \qquad (11)$$

where N_R is given by the following equation.

$$N_R = G_{AD} e_o^* \qquad (12)$$

Figure 4 shows the proposed digital control circuit and Fig. 5 shows the circuit configuration of moving average and differential part. It is seen in Fig. 5 that the derivative control in the time-domain is combined to the moving average as the FIR filter in the frequency-domain and that the proposed circuit is very simple and low cost.

The on-time interval $N_{Ton,n+1}$ is represented by the following equations.

$$N_{Ton,n+1} = N_B - K(N_{ma} - N_R) - K_D(N_n - N_{n-1}) \qquad (13)$$

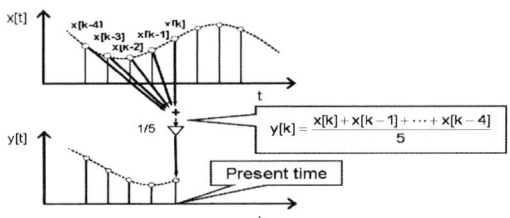

Fig. 3 Operation principle of moving average.

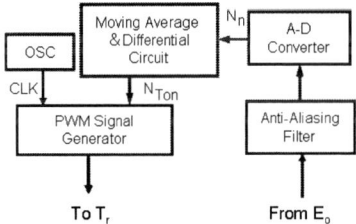

Fig. 4 Proposed digital control circuit.

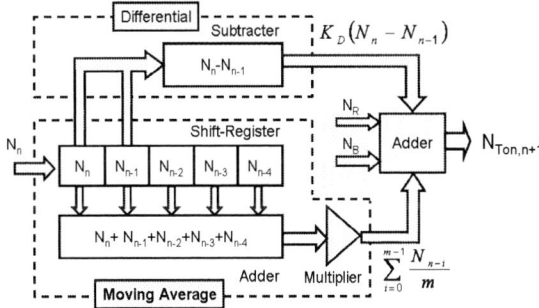

Fig. 5 Circuit configuration of moving average and differential part.

III. TRANSIENT RESPONSE

A. FIR Filter Control Method

Figure 6 shows the simulated transient response of the dc-dc converter with conventional minimum phase FIR filter control method in step change of the load resister R from 50Ω to 10Ω. The simulator is PSIM. The switching frequency is 100kHz. The main circuit parameters are E_i=20V, E_o*=10V, L=0.247mH, C=950µF and r=0.165Ω. The control parameters of FIR filter is shown in Table 1. The undershoot of output voltage e_o, overshoot of reactor current i_L and transient time are 245mV, 1.6A and 2.7ms, respectively.

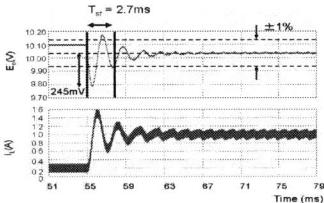

Fig. 6 Transient response of conventional FIR filter method.

Table 1 Control parameters of FIR filter.

Passage area Frequency (kHz)	5
Transition area (kHz)	15
Passage area Ripple (dB)	0.08
Quantity of decrement (dB)	13

B. IIR Filter Control Method

Figure 7 shows the transient response in case of using the IIR filter control method to improve the previous characteristics. Table 2 shows the control parameters. The undershoot of output voltage e_o, overshoot of reactor current i_L and transient time are 150mV, 1.4A and 1.12ms, respectively, and these characteristics are improved compared with FIR filter method. However, the coefficients of IIR filter as shown in Eq. (9) are complex. So, this method needs the DSP and then it is difficult to combine the other digital IC.

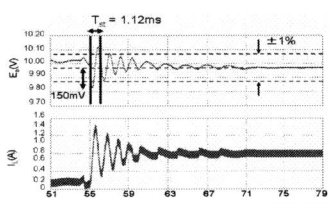

Fig. 7 Transient response of conventional IIR filter method.

Table 2 Control parameters of IIR filter.

First Gain G_1 (dB)	40
Second Gain G_2 (dB)	15
First Cut off Frequency f_a (Hz)	1
Second Cut off Frequency f_b (Hz)	10
Third Cut off Frequency f_c (Hz)	40k

C. Proposed Control Method

Figures 8(a) through (d) show the transient response of the proposed circuit in case of K_D=8, taking K_C as parameter. Also, Fig. 9 shows the undershoot and transient time against K_C. In these figures, the oscillation occurs when K_C is large as shown in Fig. 8(d). Therefore, the superior characteristics are obtained when K_C is selected at 0.005 as shown in Fig. 8(b) and Fig. 9 because both the undershoot and transient time are not variable in from 0.003 to 0.008.

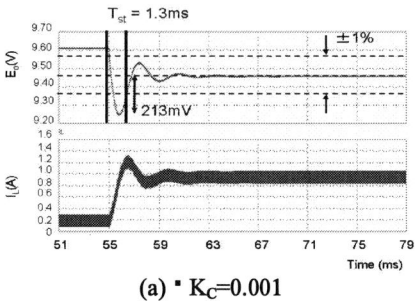

(a) ˙ K_C=0.001

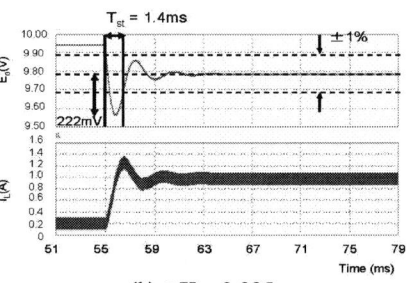

(b) ˙ K_C=0.005

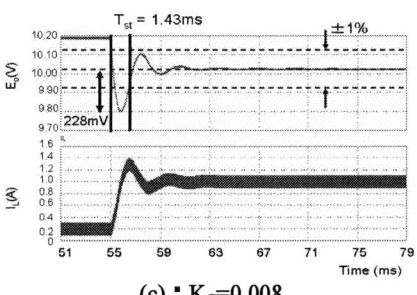

(c) ˙ K_C=0.008

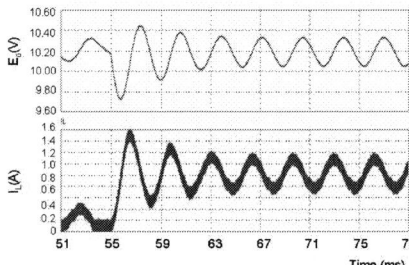

(d) ˙ K_C=0.011

Fig. 8 Transient response of proposed digital control method, taking K_C as a parameter (K_D=8).

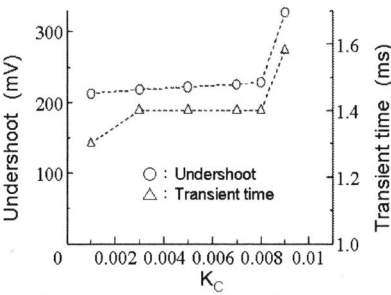

Fig. 9 Undershoot and transient time in proposed control method against K_C ($K_D=8$).

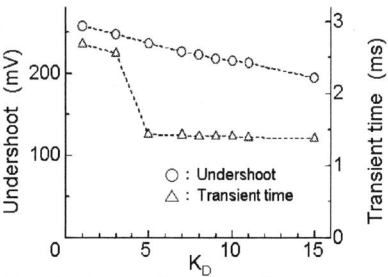

Fig. 11 Undershoot and transient time in proposed control method against K_D ($K_C=0.005$).

Figures 10(a) through (c) show the transient response of the proposed circuit in case of $K_C=0.005$, taking K_D as parameter. Furthermore, Fig. 11 shows the undershoot and transient time against K_D. The transient response has good characteristics when K_D is large over 5 as shown in Fig. 10 and Fig. 11. Therefore, K_D is selected at 8.

As a result, it is seen that the undershoot, overshoot and transient time of the dc-dc converter with a new control method are 228mV, 1.4A and 1.43ms, respectively. In this case, M is equal to 50 and K_D is equal to 8, respectively. Although this result is worse than that of the IIR filter method, it is superior than that of the conventional minimum phase FIR filter method. Especially, the overshoot of the reactor current and the transient time of the proposed method are quite similar to that of IIR filter method. Therefore, it is revealed that the transient time and overshoot of the reactor current are suppressed to within 1.43ms and 1.4A and that these result is approximately 60% smaller than the conventional minimum phase FIR filter method, respectively.

In other words, it is clarified that both the simple control circuit and the superior dynamic characteristics are obtained in the proposed method. Moreover, the results indicate that the number of parts of the digital control circuit is 15% smaller than that of the analog control circuit and the cost is also 5% smaller.

IV. CONCLUSION

It is seen that the dc-dc converter with a new simple control circuit has a superior transient response compared with the conventional one of the FIR filter method. Although the simulation result of the proposed circuit is worse than that of the IIR filter method, it is superior than that of the conventional minimum phase FIR filter method. Especially, the overshoot of the reactor current and the transient time of the proposed method are quite similar to that of IIR filter method.

On the other hand, the circuit configuration of the proposed novel control circuit is very simple compared with the FIR and IIR filters, however, the transient time and overshoot of the reactor current are suppressed enough to within 1.43ms and 1.4A, respectively. These results are approximately 60% smaller than the conventional minimum phase FIR filter method.

Moreover, it is revealed that the number of parts of the digital control circuit is 15% smaller than the analog control circuit and the cost is also 5% smaller.

We confirm that these results are useful to realize the next generation model of the switching power supply.

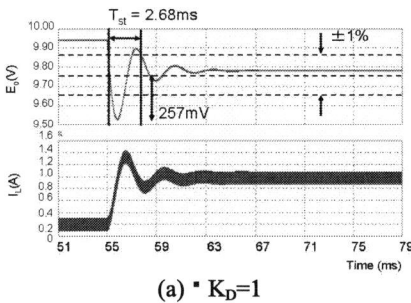

(a) $K_D=1$

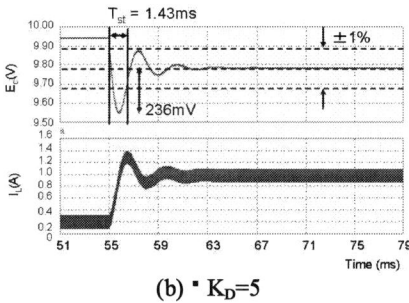

(b) $K_D=5$

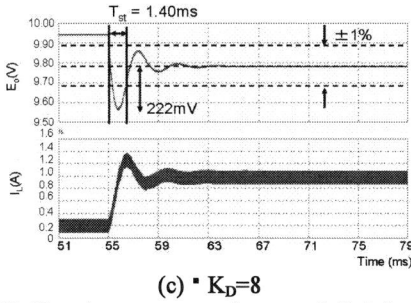

(c) $K_D=8$

Fig. 10 Transient response of proposed digital control method when changing K_D ($K_C=0.005$).

ACKNOWLEDGMENT

This work is supported in part by the Grant-in-Aid for Scientific Research (No.18310117) of JSPS (Japan Society for the Promotion of Science) and the Ministry of Education, Science, Sports and Culture.

REFERENCES

[1] D. Maksimovic, R. Zane and R. Erickson: "Impact of digital control in power electronics", Proceedings of-2004-International Symposium on Power Semi-conductor Devices & ICs, Kitakyushu, pp. 13-22, May 2004.

[2] F. Kurokawa, M. Sasaki, S. Hiura and H. Matsuo: "1 MHz high-speed digitally controlled dc-dc converter", Invited Paper, IEICE Trans. on Commun., Vol. E87-B, No. 12, pp. 3437-3442, Dec. 2004.

[3] W. Stefanutti, S. Saggini, E. Tedeschi, P. Mattavelli and P. Tenti: "Simplified model reference tuning of PID regulators of digitally controlled dc-dc converters based on crossover frequency analysis" IEEE PESC '07 Record, pp.785-791, June 2007

[4] F. Kurokawa, K. Tanaka and H. Eto: "Performance Characteristics of Switching DC-DC Power Converter with Static Model Reference", Proceedings of the ICEMS '06, pp.1-5, Nov. 2006.

[5] V. Yousefzadeh, A. Babazadeh, B. Ramachandram and E. Alarcon: "Proximate time-optimal digital control for dc-dc converters," IEEE PESC '07 Record, pp.124-130, June 2007.

[6] I. Voss, S. Schroder and R. W. De Doncker: "Predictive digital current control using advanced average current sampling algorithm for multi-phase 2-quadrant dc/dc converters", Proc. of IEEE APEC '07, pp. 8-13, Feb 2007.

[7] F. Kurokawa, W. Okamoto and H. Matsuo: "A Comparison of Steady-State Characteristics of Buck-Type DC-DC Converter Using DSP," Invited Paper, Wiley Periodicals, Inc., Electronics and Communications in Japan, Part 1, Vol. 90, No. 5, pp.1-10, May 2007.

[8] F. Kurokawa, W. Okamoto, K. Tanaka and K. Nakajima: "Improvement of Dynamic Characteristics of Digitally Controlled Switching Power Converter," Proceedings of the IEEJ&IEEE Power Conversion Conference, Vol.4, pp.1147-1153 (2007.4)

[9] B. Mulgrew, P. Grant and J. Thompson: "Digital signal processing", Macmillan Press Ltd., London, 1999.

A Novel Single/Three-phase Matrix Converter For High Power Integration

Makoto Saito[*]

[*] Shibaura Institute of Technology /Dep. of Electrical Engineering, Tokyo, Japan,
e-mail: *saitom@sic.shibaura-it.ac.jp*

Abstract — This paper deals with a three-phase to single-phase power conversion system using a matrix converter topology. In order to compensate for instantaneous power unbalance occurring between three-phase circuits and single-phase circuits, a small reactor is installed into a matrix converter. The proposed circuit and control diagrams as well as some important simulation results are described herein.

Keywords — Matrix converter, Single / three-phase power conversion.

I. INTRODUCTION

We are surrounded by electric railways, electric home appliances, and many other facilities and equipment that run on single-phase power. Yet, single-phase power is rarely readily available in that form. It is, instead, normally supplied to users after conversion from three-phase power sources. Among the largest consumers of three-phase to single-phase power are electric railways, many of which operate individual machines with power appetites that exceeds several MW. Such consumers normally receive single-phase power after it has been converted from three-phase power via a Scott-transformer, after which the three-phase power source must be balanced.

This is because, unless the single-phase power obtained from the transformer is balanced, a Scott-transformer generates unbalance or harmonics in a three-phase power supply current. However, since the M-T power is rarely balanced in practice, SVC should be prepared on the three-phase side of the Scott- transformer, or a power balancer should be attached between M and T to compensate for the imbalance or harmonics that often occur in a three-phase power supply current. For a large-capacity single-phase load, a power electronics device is eventually necessary even when a Scott-transformer is used.

Use of a power electronics device enables single-phase to three-phase conversion without a Scott-transformer by combining a three-phase pulse width modulation rectifier and a single-phase inverter. However, since power efficiency is reduced, and a large installation area is required due to the necessity of installing a large-capacity capacitor or inductor in the middle of a direct current, this configuration doe not have sufficient advantages over the use of a Scott-transformer to be feasible if used alone.

And, while many three-phase to single-phase matrix converters have ever been proposed, such converters are aimed at conversion from commercial three-phase power to high-frequency single-phase power, and are not suitable for converting commercial three-phase power to commercial single-phase power.

In response to this, the authors have devised the three-phase to single-phase matrix converter with a power decoupling reactor shown in Figure 1. This design marks the first time efficient and direct three-phase to single-phase power conversion has been achieved with a power electronics device alone.

The proposed circuit basically consists of a matrix converter and a reactor Lc. The matrix converter directly generates commercial single-phase voltage from commercial three-phase power, while the small-capacity reactor Lc compensates for any momentary power unbalance occurring between three-phase power and single-phase power, thus making the power-supply current sinusoidal.

This paper describes the configuration and control method of the proposed circuit and the results of basic performance verification simulations. The power efficiency, device volume, and other system characteristics will be introduced in a separate report.

II. CIRCUIT CONFIGURATION

Figure 1 shows the main circuit configuration of the proposed three-phase to single-phase matrix converter with power decoupling reactor. This circuit can roughly be divided into a matrix converter (MC) and a reactor Lc. While the MC directly generates single-phase voltage from three-phase voltage, the small-capacity reactor Lc compensates for the momentary power unbalance occurring between the three and single phases. The MC consists of nine two-way switches, a snubber capacitor Sc, and a surge clamp circuit.

To prevent a harmonic outflow to the three-phase power input and the single-phase power output, the reactor Ls is connected in series with the three-phase power input and Lf and Cf are connected to the single-phase power output, respectively. The resistor Rs, which is connected in parallel with Ls suppresses the resonance of Ls and Cs. The necessary capacity of Ls will be mentioned later.

III. CONTROL METHOD

A. Basic Control Theory

Regarding the system in Figure 1, the momentary power of the power decoupling reactor Lc is S_c. If the power loss of the main circuit is ignored, the following power equation can be obtained for the entire system:

$$v_r i_r + v_s i_s + v_t i_t = S_c + v_o i_o \qquad (1)$$

978-1-4244-1741-4/08/$25.00 ©2008 IEEE

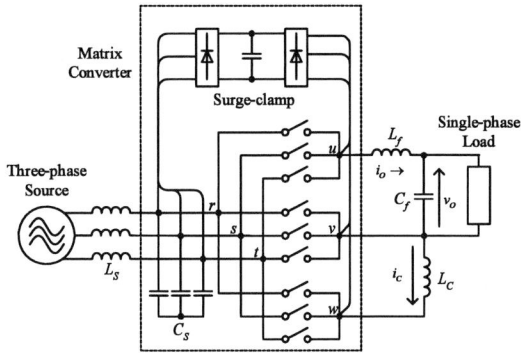

Figure.1 Three/single phase MC with power decoupling reactor.

The authors devised "Single-phase d-q transformation using a Hilbert transformer"[1]. With this usage, the single-phase momentary power $v_o i_o$ is defined as a new quantity.

Based on Hilbert transform H [ ], v_o and i_o are first extended to the following vector quantity:

$$\left. \begin{array}{l} \vec{v}_o = v_o + jH\left[v_o\right] \\ \vec{i}_o = i_o + jH\left[i_o\right] \end{array} \right\} \quad (2)$$

From the above equation, the momentary phase angle ε of v_o can be calculated as follows:

$$\varepsilon = \tan^{-1} \frac{H\left[v_o\right]}{v_o} \quad (3)$$

Into the d-q coordinate system whose d-axis reference angle is ε, the vectors $\vec{v}_o$, and $\vec{i}_o$ of the equation are transformed (see Figure 2).

$$\left. \begin{array}{l} \vec{v}_o{}^{dq} = \vec{v}_o e^{-j\varepsilon} \equiv v_{od} + j0 \\ \vec{i}_o{}^{dq} = \vec{i}_o e^{-j\varepsilon} \equiv i_{od} + ji_{oq} \end{array} \right\} \quad (4)$$

When these d-q coordinate quantities are used, v_o and i_o become as follows:

$$\left. \begin{array}{l} v_o = v_{od} \cos\varepsilon \\ i_o = i_{od} \cos\varepsilon - i_{oq} \sin\varepsilon \end{array} \right\} \quad (5)$$

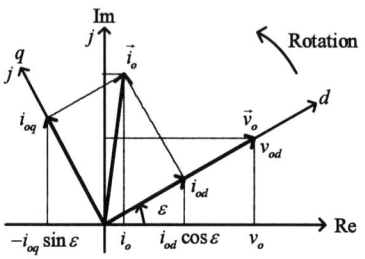

Figure.2 Single-phase d-q transformation.

Therefore, the single-phase momentary power $v_o i_o$ can be expressed as follows:

$$v_o i_o = \frac{1}{2} v_{od} i_{od} \left(1 + \cos 2\varepsilon\right) - \frac{1}{2} v_{od} i_{oq} \sin 2\varepsilon \quad (6)$$

If S_c is controlled by Equations (1) and (6) as follows:

$$S_c = -\frac{v_{od}}{2}\left(i_{od} \cos 2\varepsilon - i_{oq} \sin 2\varepsilon\right)$$

$$= \frac{v_{od}}{2}\sqrt{i_{od}^2 + i_{oq}^2}\, \sin\left(2\varepsilon + \psi\right) \quad (7)$$

Where, $\psi = \tan^{-1}\dfrac{-i_{od}}{i_{oq}}$ (8)

The three-phase momentary power $v_u i_u + v_v i_v + v_w i_w$ is defined uniquely with v_{od} and i_{od} as follows:

$$\frac{1}{2} v_{od} i_{od} = v_u i_u + v_v i_v + v_w i_w \quad (9)$$

The ic necessary for controlling S_c by Equation (7) is $i_c = I_c \sin\left(\varepsilon + \psi\right)$. If the differential values of I_c and ψ are ignored:

$$S_c = L_C \frac{di_c}{dt}i_c = \frac{\dot{\varepsilon} L_C I_c^2}{2}\sin\left(2\varepsilon + 2\psi\right)$$

$$= \frac{v_{od}}{2}\sqrt{i_{od}^2 + i_{oq}^2}\,\sin\left(2\varepsilon + \psi\right) \quad (10)$$

From the above equation:

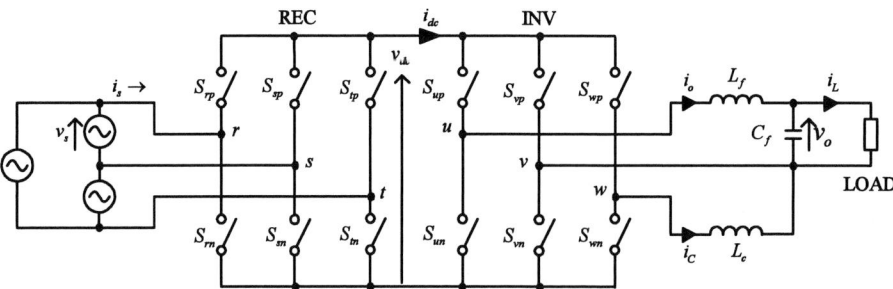

Figure.3 Virtual REC/INV circuit.

$$i_e = \sqrt{\frac{v_{od}\sqrt{i_{od}^2 + i_{oq}^2}}{\dot{\varepsilon}L_C}} \sin\left(\varepsilon + \frac{\psi}{2}\right) \quad (11)$$

In order to allow the differential values of I_e and ψ to be ignored, high-frequency components are eliminated from i_{od} and i_{oq} with a low-pass filter of about several kHz before the applications of Equations (8) and (11).

B. PWM Control of MC by Virtual REC/INV Circuit

In the proposed system, a PWM rectifier using two-way switches and the PWM inverter circuit (hereinafter, called the virtual REC/INV circuit) shown in Figure 3 are utilized to provide control.

The I-O voltage relationship of the matrix converter shown in Figure 1 can be defined by the following switching function:

$$\begin{bmatrix} v_u \\ v_v \\ v_w \end{bmatrix} = [\mathbf{C}_M]\begin{bmatrix} v_r \\ v_s \\ v_t \end{bmatrix} \quad (12)$$

Where, $[\mathbf{C}_M]$ is the switching function of the matrix converter.

$$[\mathbf{C}_M] = \begin{bmatrix} S_{ur} & S_{us} & S_{ut} \\ S_{vr} & S_{vs} & S_{vt} \\ S_{wr} & S_{ws} & S_{wt} \end{bmatrix} \quad (13)$$

The I/O relationship of the virtual REC/INV circuit shown in Figure 3 can be expressed as follows:

$$\begin{bmatrix} v_u \\ v_v \\ v_w \end{bmatrix} = [\mathbf{C}_I][\mathbf{C}_R]\begin{bmatrix} v_r \\ v_s \\ v_t \end{bmatrix} \quad (14)$$

Where, $[\mathbf{C}_I]$ and $[\mathbf{C}_R]$ are the switching functions of the virtual inverter and the virtual rectifier, respectively.

$$[\mathbf{C}_I] = \begin{bmatrix} S_{up} & S_{un} \\ S_{vp} & S_{vn} \\ S_{wp} & S_{wn} \end{bmatrix} \quad (15)$$

$$[\mathbf{C}_R] = \begin{bmatrix} S_{rp} & S_{sp} & S_{tp} \\ S_{rn} & S_{sn} & S_{tn} \end{bmatrix} \quad (16)$$

In Figures 1 and 3, the three-phase supply voltage, reactor Lc voltage, and single-phase output voltage are equal. Therefore, the switching function of the matrix converter can be defined from that of the virtual REC/INV circuit as follows:

$$[\mathbf{C}_M] = [\mathbf{C}_I][\mathbf{C}_R] \quad (17)$$

As Figure 4 shows, PWM modulators are prepared for the virtual rectifier and the virtual inverter. Switching functions (pulses) available from the modulators are synthesized by the above equation to drive the MC.

C. Virtual rectifier control

Figure 5 shows the virtual rectifier. The three-phase PWM rectifier circuit consisting of two-way switches has

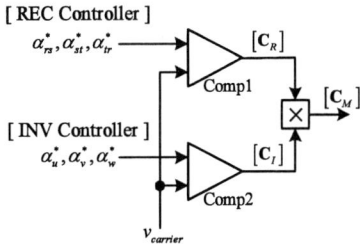

Figure. 4 MC control based on virtual REC/INV circuit.

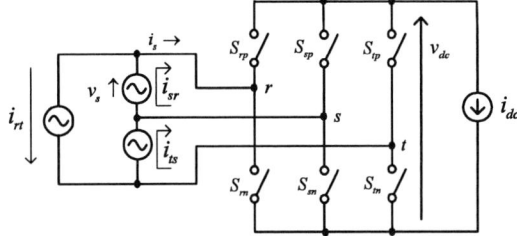

Figure. 5 Virtual rectifier.

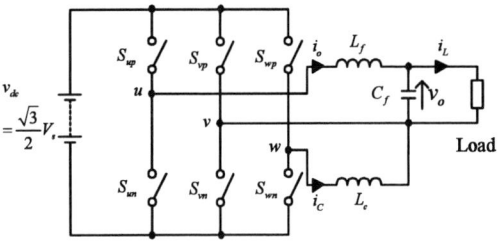

Figure. 6 Virtual inverter.

a three-phase voltage source connected to r, s, and t. The DC current source i_{dc} connected to the DC bus simulates the virtual inverter circuit viewed from the DC bus in Figure 4.

Considering the supply voltage as $v_s = V_s \cos\phi$, this rectifier circuit performs triangular modulation with the following modulation factor α_{rt}^*, α_{sr}^*, α_{ts}^*:

$$\left.\begin{aligned} \alpha_{rt}^* &= 1.1547\sin\phi + 0.1667\sin 3\phi \\ \alpha_{st}^* &= 1.1547\sin(\phi - 2\pi/3) + 0.1667\sin 3\phi \\ \alpha_{ts}^* &= 1.1547\sin(\phi - 4\pi/3) + 0.1667\sin 3\phi \end{aligned}\right\} \quad (18)$$

This controls the three-phase power side to power factor =1. For your reference, ϕ is easy to detect with the "systematic phase PLL circuit using single-phase d-q transformation [2]" devised by the authors.

D. Virtual inverter control

In the virtual REC/INV circuit, the virtual rectifier is equivalent to the DC power source. v_{dc} when viewed from the DC bus, the virtual inverter circuit shown in Figure 6 can be obtained. This inverter consists of a three-phase bridge inverter where the single-phase output vo is connected between u and v and the reactor Lc between v and w. The purpose of control by this inverter is to make

2441

the vo and the current ic of the reactor Lc agree with each other. The target values of both vo and ic are sinusoidal and it is difficult to suppress stationary deviation by mere proportional integral control. Therefore, with the "single-phase d-q transformation using a Hilbert transformer" devised by the authors, stationary deviation is suppressed by feedback control in the d-q coordinate system. Figure 7 shows the control block for this purpose.

1) vo control system

If vo is once extended to a complex number $\vec{v}_o$ based on the Hilbert transform and then converted into d-q coordinate quantities vod and voq with the vo phase command ε^* as the d-axis reference angle, it can be expressed as follows:

$$v_o = v_{od}\cos\varepsilon^* - v_{oq}\cos\varepsilon^* \qquad (19)$$

If the vo target value is $v_o^* = v_{od}^*\cos\varepsilon^*$, a desired vo value can be obtained by control to vod = vod* and voq = 0.

If Hilbert transform is executed ideally on vp, proportional integral control is permitted on vod and voq in the d-q coordinate system. In the actual Hilbert transform, however, the phase characteristic will deviate away from ideal as the frequency becomes higher. By noting the fact that the frequency characteristic changes neither in the rest frame nor in the d-q coordinate system, proportional control is executed in the rest frame and integral control in the d-q coordinate system. This allows the construction of a control system that is not affected by the high frequency characteristic of Hilbert transform.

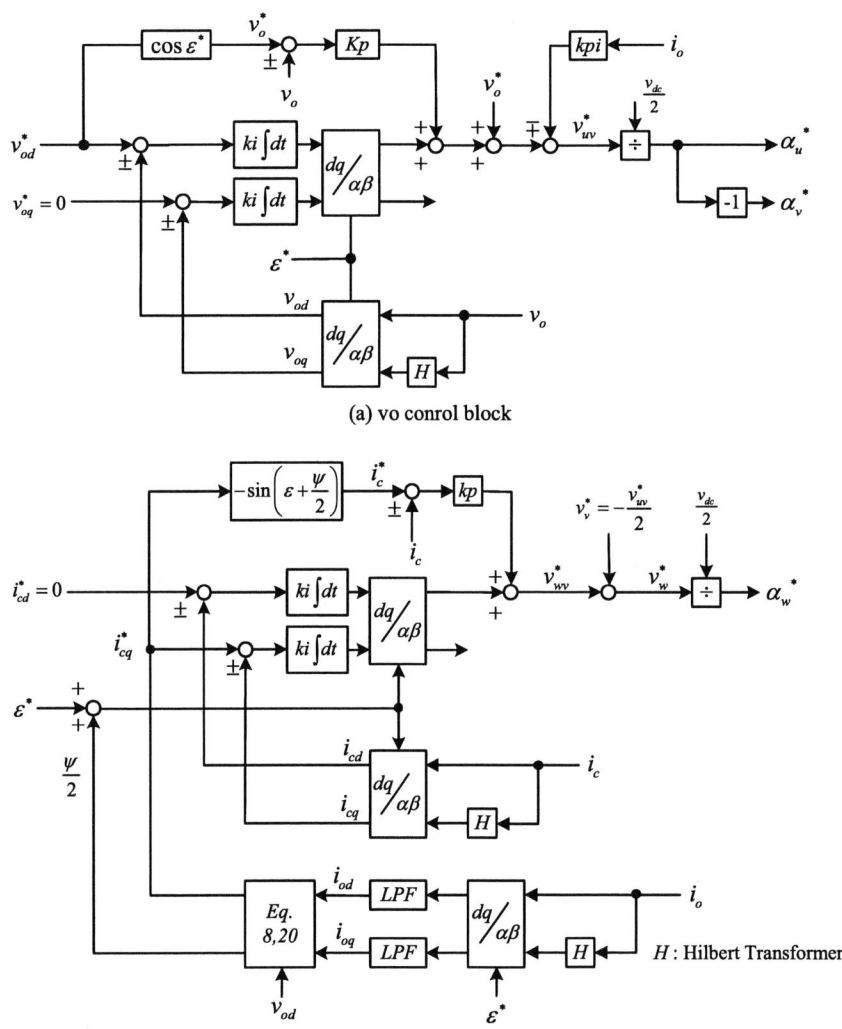

(a) vo conrol block

(b) ic control block

Figure.7 Control diagram of virtual PWM inverter

Figure 7(a) shows the control block. First, vo is extended to a complex number based on the Hilbert transform and then converted into d-q coordinate quantities vod and voq with ε^* as the reference angle. Next, integral control is executed on the quantities. The integrator outputs the d and q coordinate command values of the virtual inverter u-v voltage. These values are transformed with ε^* in the rest frame. Since the imaginary part of this rest frame is a fictitious quantity, only the real part is used as a command value.

In the rest frame, the command value vo* of vo is generated from vod* and ε^* and its proportional control with vo is executed. This adjuster output is synthesized with the aforementioned integrator output in the rest frame in order to generate the u-v phase voltage command value. One half of the value is used as the u-phase voltage command value and the other half as the v-phase voltage command value.

2) ic control system

If ic is first extended to a complex number $\vec{i}_c$ based on Hilbert transform, and then converted into d-q coordinate quantities icd and icq with $\varepsilon^* + \psi/2$ as the d-axis reference angle, it can be expressed as follows:

$$i_c = i_{cd}\cos\left(\varepsilon^* + \psi/2\right) \ i_{cq}\sin\left(\ ^* \ /2\right) \quad (20)$$

Where, ε^* can be regarded as matching its actual value ε under vo control. If icd and icq follow the following values, Equation (20) will match Equation (11).

$$\left.\begin{array}{l} i_{cd} = 0 \equiv i^*_{cd} \\ \\ i_{cq} = -\sqrt{\dfrac{v_{od}\sqrt{i^2_{od} + i^2_{oq}}}{\dot{\varepsilon}L_c}} \equiv i^*_{cq} \end{array}\right\} \quad (21)$$

Figure 7(b) shows the control block. As in the vo control system, icd and icq receive integral control in the d-q coordinate system and proportional control in rest frame to avoid the influence of the high-frequency characteristic of the Hilbert system. In this process, ic is first extended to a complex number based on Hilbert transform, then converted into d-q coordinate quantities icd and icq with $\varepsilon^* + \psi/2$ as the reference angle, after which it receives integral control. i^*_{cq} and ψ are calculated from iod and ioq after the high-frequency components are removed with a low-pass filter. The integrator outputs the d and q coordinate command values of the virtual inverter u-v voltage. These received values are transformed with $\varepsilon^* + \psi/2$ in the rest frame. Since the imaginary part of this rest frame is a fictitious quantity, only the real part is used as a command value.

In the rest frame, the command value ic* of ic is generated from icq* and $\varepsilon^* + \psi/2$ and its proportional control with ic is executed. This adjuster output is synthesized with the aforementioned integrator output in the rest frame to generate the w-v phase voltage command value. The v-phase voltage command value generated by the vo control system is then subtracted from this

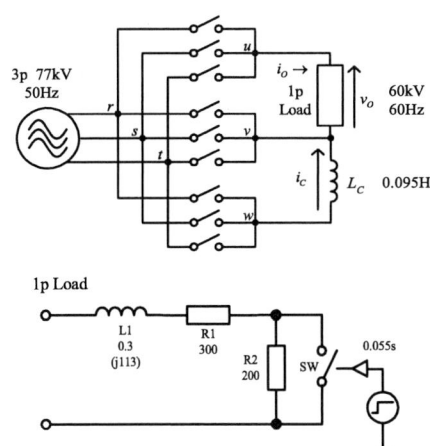

Figure.8 Simulation circuit.

command value to calculate the w-phase voltage command value.

IV. SIMULATION RESULTS

A. Purpose and Conditions

In order to confirm the voltage and current basic characteristics of the proposed system, a simulation was executed with the circuit shown in Figure 8. Table 1 lists the simulation parameters. To verify the circuit performance, the power-side LC filter (Ls, Cs) and load-side LC filter (Lf, Cf) were removed in order to eliminate the influence of their resonance. In addition, ideal two-way switches were used without any special commutation sequence assigned. Since the output LC filter was removed, the output voltage was subject to open-loop control ($\alpha_u = -\alpha_{\bar{v}} \ 0.9\cos\varepsilon^*$).

As for the voltage and current basic characteristics, the stationary and transient characteristics at a step-wise change of load were focused this time. As Figure 8 shows, the loads of R and L circuits were varied by opening or closing the switch SW connected to R2 in parallel. The SW was opened for an impedance of 512 ohm and a power factor of 0.975 and closed for an impedance of 321 ohm and a power factor of 0.936. When the SW was closed, the time constant of the load was 1 ms.

B. Results and Discussion

Figure 9 shows the simulation results. From the top, the waveforms are those of the power-supply voltage vr, the power-supply currents ir, is, and it, the output voltage vo and current io, the output current d-axis and q-axis components iod and ioq, the power-compensating reactor current command value ic* and current actual value ic. The power-supply current and output voltage waveforms are then measured after processing with a 5 kHz low-pass filter. In this simulation, the SW is closed at 0.055 s.

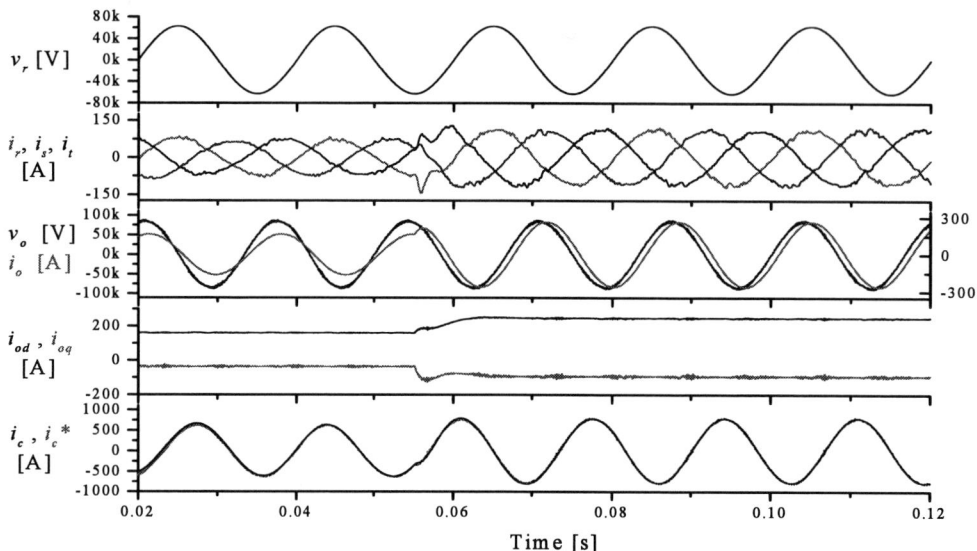

Figure 9 Simulation results.

Because of the open loop of modulation factor 0.9, the amplitude of the output voltage vo is theoretically 0.9 x power line voltage amplitude x $\sqrt{3}/2$ = 84.9 kV, matching the simulated waveform. This voltage becomes vod.

When the SW is open, the impedance is 512 ohm and the power factor is 0.975. The amplitude of the output current io is theoretically 166 A where the d-axis component iod is 166 x cos (power factor angle) = 162 A and the q-axis component ioq is 166 x sin (power factor angle) = 36.9 A, matching the corresponding simulated waveforms, respectively. Furthermore, when the SW is closed, the theoretical values by similar calculations match the simulated waveforms. The power-compensating reactor current ic and its target value ic* are controlled with no stationary deviation, whether the SW is open or closed. The target value of ic* is calculated by Equation (11). For example, when the SW is open, the theoretical amplitude is 628A, matching the simulated waveform. From these results, we can say that the simulation matches the theory.

The most important point of this system is its ability to make a power-supply current sinusoidal. Whether the SW is open or closed, the simulated waveforms of ir, is, and it are three-phase sinusoidal with some stationary distortion. Since the power-supply r-phase voltage vr and r-phase current ir are controlled in the same phase, the power-supply current is found to be synchronous with the power source. Judging from these factors, the purpose of this system has been achieved. The stationary power-supply current distortion may be attributable to the PWM pattern of the matrix converter.

The waveforms of ir and it show spikes for about 2 ms after a sudden change of load. As Equation (10) shows, this system provides control based on the assumption that the load does not fluctuate greatly, and by ignoring sudden changes of iod and ioq. The aforementioned spikes may

be attributable to this assumption. For about 10 ms from a sudden change of load, iod and ioq change with some distortion in ir, is, and it. The iod waveform changes like the first-order delay system with a time constant of about 5 ms and the ioq waveform changes like the second-order delay system with an overshoot. Since the load time constant is 1 ms and the iod and ioq detection LPF is 3 kHz, the iod and ioq responses may be attributable, not to their influence, but to the mutual interference between the d and q axes. This mutual interference might have left some distortion in ir, is, and it at a sudden change of load. This is a subject that will be discussed in the future.

V. CONCLUSION

This paper proposed a circuit system using a matrix converter to, for the first time, realize direct three-phase to single-phase power conversion exclusively with a power electronics device. By simulation, the proposed circuit was verified to be capable of compensating for the momentary power unbalance that occurs during three-phase and single-phase conversion by use of a small-capacity reactor that makes the three-phase power supply current sinusoidal without a single-phase load power factor.

Experiments are now being conducted on the system to verify the simulated results. In the future, the voltage and current control characteristics of various single-phase loads will be clarified.

REFERENCES

[1] M. Saitou, et al. "Generalized theory of instantaneous active and reactive powers in single-phase circuits based on Hilbert Transform", Proc. on PESC'02, pp. 1419-1424, (2002).

[2] M. Saitou, et al." A control strategy of single-phase active filter using a novel d-q transformation", Peoc. on IAS'03, Vol.2, pp.1222-1227,(2003).

An Effective Design Method for High Power Density Converters

Yusuke Hayashi[*], Kazuto Takao[**], Toshihisa Shimizu[***] and Hiromichi Ohashi[*]

[*] National Institute of Advanced Industrial Science and Technology, Tsukuba, Japan,
e-mail: yusuke-hayashi@aist.go.jp
e-mail: h.ohashi@aist.go.jp
[**]Electron Devices Laboratory, Corporate Research & Development Center, Toshiba Corporation, Kawasaki, Japan,
e-mail: kazuto.takao@toshiba.co.jp
[***] Tokyo Metropolitan University, Hachioji, Japan,
e-mail: shimizut@tmu.ac.jp

Abstract— Energy Semiconductor Electronics Research Laboratory in AIST has developed an integration design methodology for high power density converters. A part of this work, a novel converter loss estimation method based on Power Converter Platform Concept was proposed. The proposed method estimates the converter loss exactly under real circuit operation condition, by taking the correlation among converter parameters into account. The availability of the proposed method was shown, and the design consideration was carried out for the power conditioner of the photovoltaic solar generation.

Keywords— Design, High Power Density Systems, High Frequency Power Converter

I. INTRODUCTION

High output power density (OPD) power converters have been essential for future growing markets such as electric vehicles and power supply systems for information and communication networks. Figure 1 shows the output power density roadmap for power electronics converters [1]. The OPD of the commercially available power converters has been lineally increasing by a factor of 2 figures in the last 30 years, and is now at a level of about 5 W/cm3. Intensive studies are under way [2]-[4], and the OPD of more than 30 W/cm3 will be realized around 2015.

There are two key issues to realize high OPD power converters. One is the utilization of ultra low loss and high speed semiconductor power devices such as the silicon (Si) super junction (SJ) MOSFET or the silicon carbide (SiC) MOSFET to downsize the volume of the heat sink. The other is high frequency operation to shrink the volume of passive components.

Influences of circuit stray parameters on the switching loss of power devices increase under high speed switching operation. Power loss of the passive filter increases under high frequency operation. They must be estimated exactly to realize high OPD power converters, and a novel converter loss estimation method is indispensable.

Researches about the development of the integration design methodology for high OPD converters are under way in Energy Semiconductor Electronics Research Laboratory [5]. A part of this work, a novel converter loss estimation method based on Power Converter Platform

Concept has been proposed, taking converter parameters such as device parameters, circuit stray parameters, passive filter parameters and control parameters into account.

This paper presents a novel converter loss estimation method based on Power Converter Platform Concept mainly. The integration design methodology for high OPD converters is introduced at first. Then, the detail of the novel converter loss estimation method is shown. The design consideration for the power conditioner of the photovoltaic solar generation is carried out, after the availability of the proposed method is confirmed.

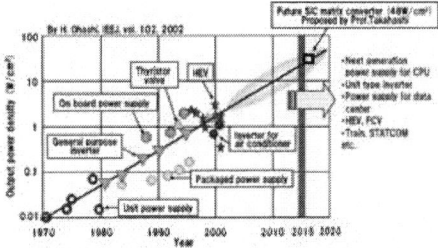

Fig. 1. Output power density roadmap for PE converters [1].

II. INTEGRATION DESIGN METHODOLOGY FOR HIGH OPD CONVERTERS

A. Basic concept for the high OPD converter design

A basic concept for the high OPD converter design is illustrated in Fig. 2. The concept consists of three parts mainly; they are a circuit parameter optimization part, a converter structure optimization part and an assembly core technology part.

In the circuit parameter optimization part, power device parameters, circuit stray parameters, passive filter parameters and control parameters are extracted ideally to minimize the total converter loss under the real circuit operation condition. Dimensions of a chip, a circuit board, passive components and a heat sink are designed in the converter structure optimization part. The component layout is considered to minimize a converter volume, operating an electro-magnetic analysis and a thermal analysis. In the assembly core technology part, materials for the packaging are selected and the reliability is considered. All information is stored in a design database, and optimal parameters are extracted to fabricate the converter prototype.

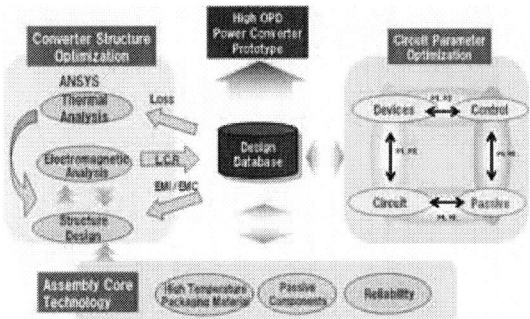

Fig. 2. A basic concept for the design of high OPD integrated Converters [5].

A power converter consists of a large number of components. To realize a high OPD converter, optimal component layout in the structure design becomes important. To realize the optimal component layout for the thermal diffusion, the power loss caused by power devices, circuit stray parameters, passive components and the control strategy must be estimated exactly. The exact converter loss estimation becomes a key technology, because the high OPD converter causes the high heat density.

The converter loss is estimated in the circuit parameter optimization part shown in Fig. 2. Power Converter Platform Concept has been proposed to develop the converter loss estimation method. In the next section, Power Converter Platform Concept is introduced.

B. Power converter platform concept

Novel semiconductor power devices such as Si - SJ MOSFET and SiC MOSFET are very attractive because of the low specific on-resistance. The volume of the heat sink can be reduced, and it can be designed smaller under the high temperature operation in the case of SiC devices.

High speed switching operation is necessary to realize high OPD power converters. High speed switching enables to reduce the switching loss of power devices, and high frequency operation can be carried out. As a result, volume of the passive components can be reduced.

However, converter parameters, especially device and circuit stray parameters, affect on the switching behavior largely under high speed switching operation. In this case, device and circuit stray parameters become dominant factors of the power device loss. On the other hand, high frequency operation causes the increase of the passive component loss. The core loss, especially the hysteresis and eddy current loss, increases in the passive filter inductor, and the core loss depends on the control parameters largely. Therefore, the correlation among converter parameters, such as device parameters, circuit stray parameters, passive filter parameters and control parameters must be quantified to estimate the total converter loss exactly.

Power converter platform concept has been proposed to resolve the issue [6]. Figure 3 shows the Power Converter Platform Concept. Characteristics of the proposed concept are summarized as follows.

- A power converter is classified into four factors mainly. They are a power device factor, a converter circuit factor, a passive filter factor and a control factor.

- Relations between two factors are quantified. Then, models are developed to describe the power loss. Then, developed models are evaluated under real circuit operation condition.

- Developed models are bundled. Finally, the total converter loss is described by the converter parameters, and the design database is created to extract parameters to minimize the converter loss.

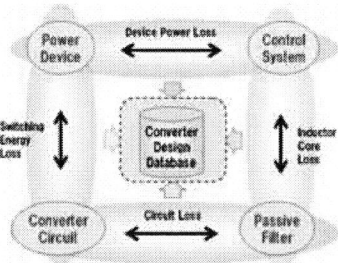

Fig. 3. Power converter platform concept for the optimal parameter design.

An experimental apparatus has been constructed to realize the platform concept. Figure 4 shows the overview of the apparatus. Correlations between two factors in converter factors are quantified here, and the converter design database is created. Specifications of the experimental apparatus are summarized as follows.

- Experimental apparatus consists of parameter adjustable circuit boards, a high speed digital controller and passive components.

- Circuit configuration of the apparatus is a three-phase three-leg hard switching inverter. Output power of the inverter is 3kW and DC voltage is 300V. This is assumed for the power drive systems and the distributed generators.

- Converter parameters are variable. Especially, circuit stray parameters such as circuit stray inductances and circuit stray capacitances are adjustable on the parameter adjustable circuit board [7].

Using the apparatus, a power loss limit model has been proposed to estimate switching energy under the real circuit condition. This model quantifies the correlation between the power device factor and the converter circuit factor. One of features of the model is that the switching energy of power devices can be described by the accumulation of the stored energy in device parameters and circuit stray parameters. Nonlinear transient voltage and current waveforms, which depend on the device parameters, are not required. Therefore, the power loss limit model enables exact switching energy estimation of power devices.

A loss map has been reported to estimate the core loss exactly under a real PWM operation condition [8] [9]. The loss map is the database of dynamic minor loop's areas which represent the energy loss of the magnetic core. The loss map is described using the bias magnetic field Ho and the magnetic flux density ripple ΔB. The correlation between the passive filter factor and the control factor can be quantified by using the loss map.

The dependency of the switching energy on circuit stray parameters is shown in Fig. 4 (a). The switching energy is calculated by using the power loss limit model. In high speed switching, the switching energy depends on the circuit stray parameters largely. The power loss limit model is indispensable to estimate the device switching loss under real circuit operation condition.

The dependency of core loss energy on the bias magnetic field and the magnetic flux density is also shown in Fig. 4 (b). The core loss increases in the high frequency operation and the loss changes drastically by the bias magnetic field and the magnetic flux density. The loss map is available to estimate exact core loss under real control condition.

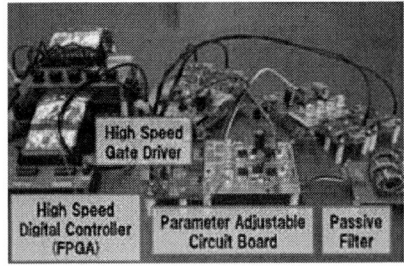

Fig. 4. The experimental apparatus for the power converter platform concept.

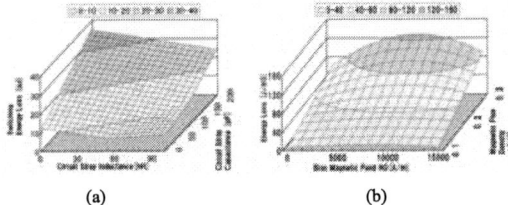

(a)   (b)

Fig. 5. Examples of energy loss estimation (a) based on the power loss limit model and (b) based on the loss map.

C. Converter Loss Estimation Method based on the power converter platform concept

Power losses originating from power devices and passive components occupy a large percentage of the total converter loss [10]. The switching loss of the power device and the core loss of the passive filter are now estimated exactly by the power loss limit model and the loss map respectively.

However, the power loss limit model and the loss map have been proposed independently. These two models

have to be integrated to estimate the total converter loss exactly. Figure 6 shows the flowchart to estimate the total converter loss including the power loss limit model and the loss map.

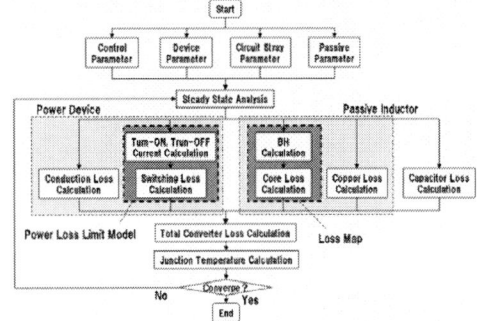

Fig. 6. Flowchart of the converter loss estimation based on the platform concept

The total converter loss consists of the power device loss and the inductor loss mainly. The power device loss consists of the conduction loss and the switching loss. The inductor loss consists of the copper loss and the core loss.

The switching loss of the power device is calculated by using the power loss limit model. The power loss limit model requires device parameters, circuit stray parameters, turn on and turn off currents mainly. Device parameters and circuit stray parameters are determined initially. Switching currents are obtained in the steady state analysis, taking the effect of the filter inductor into account. The conduction loss of the power device is calculated by using the conventional formula.

The core loss of the inductor is calculated by using the loss map. The database of the minor loop's areas, the induced voltage, the induced time and the average current through the inductor are required to estimate the core loss mainly. The database of the minor loop's area is initially selected in the passive filter parameter. The induced voltage and the average current are obtained in the steady state analysis, taking the effect of the switching time of power devices into consideration.

The flowchart shown in Fig. 6 is operated repeatedly for various parameters, and the design database is constructed. Optimal parameters to minimize the converter loss are extracted by using the database.

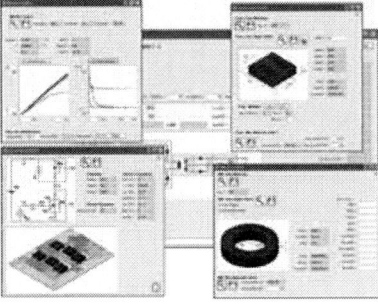

Fig. 7. Graphical user interface of the converter loss simulator based on the platform concept

The simulator has been developed to estimate the converter loss under real circuit operation condition. Figure 7 shows the interface of the developed simulator. Device parameters, circuit stray parameters, passive component parameters and control parameters are set up arbitrarily or selected from the database. The converter loss is estimated, using the power loss limit model and the loss map based on the flowchart shown in Fig. 6.

D. Availability of the converter Loss Estimation Method

The accuracy of the proposed converter loss estimation method has been investigated for 1kW 100kHz power conditioner for PV solar generation [6]. The circuit configuration of the power conditioner is shown in Fig. 8. The power losses originating from MOSFETs, diodes, inductors and DC capacitor in both the boost chopper and the full bridge single-phase inverter are estimated using the developed converter loss simulator.

Figure 9 shows the comparison between estimated power loss and the measured converter loss. Output power of the converter is controlled, changing the load resistance. The bar graph shows losses originating from MOSFETs, diodes, inductors and DC capacitors in both the boost chopper and the full bridge single-phase inverter. Green dots mean the measurement results. The estimation results are in good agreement with the measured results, and the accuracy of 95% has been obtained.

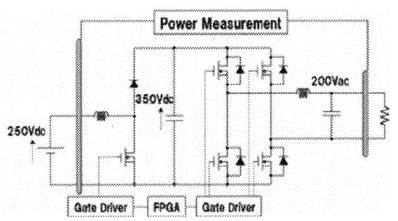

Fig. 8. The equivalent circuit of the power conditioner for PV solar generation.

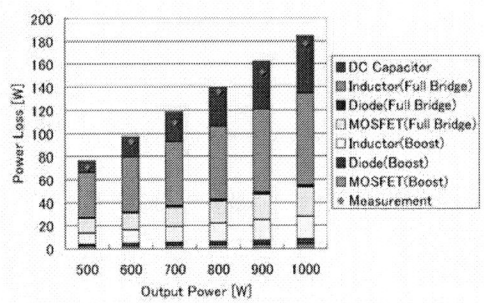

Fig. 9. The measured converter loss and the estimated power loss for the power conditioner.

III. DESIGN CONSIDERATION FOR HIGH OPD POWER CONVERTERS

The availability of the proposed converter loss estimation method has been confirmed in the previous chapter. The design consideration is carried out for the power conditioner of PV solar generation here. The specification of the designed power conditioner is shown in Table 1. A Si - SJ MOSFET and a SiC – SBD hybrid pair is assumed. The possibility of OPD is estimated, under the condition that the efficiency is 96%.

TABLE I. PARAMETERS OF THE DESIGNED POWER CONDITIONER

Output Power	3000W
DC Input Voltage	250V
DC Bus Voltage	350V
AC Output Voltage	200V (RMS Value)
Isolation	Transformer less
Semiconductor Devices	Si-SJMOSFET(600V) Si-SBD(30V) SiC-SBD(600V)
Magnetic Material	Sendust
Control	PWM Hard Switching
Efficiency	96%
Si-SJ MOSFET	Vbr = 600V
	0.68m Ωcm2 at 125 deg. C.
Si-SBD	Vbr = 30V
	0.35V at 15A at 125 deg. C.
SiC-SBD	Vbr = 600V
	1.4V at 15A at 125 deg. C.

A. Converter loss estimation in circuit parameter optimization part

Total power loss of the power conditioner is estimated by the developed simulator. A Si - SJ MOSFET of 68mΩcm2 at 125 degrees Celsius and a SiC - SBD of 1.4V at rated current at 125 degrees are utilized here. Figure 10 shows the calculation result of the power loss originating from single MOSFET in the full bridge inverter of the power conditioner. The junction temperature of 125 degrees Celsius is assumed. The ideal circuit condition (a) in Fig. 10 and the real circuit condition (b) in Fig. 10 are shown. Power loss consists of conduction loss and the switching loss. The loss is calculated for the chip area of the MOSFET and the switching frequency. This calculation is operated for various circuit stray parameters, and the design database is created.

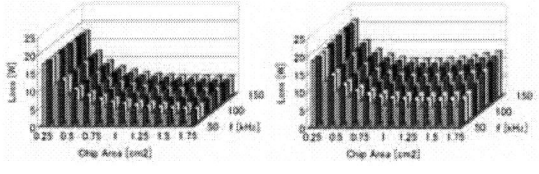

(a) Ideal Circuit Condition (b) Real Circuit Condition
(a) Lst=0nH, Lsc=0nH, CsH-0pF, CsL=0pF, Rg=0ohm
(b) Lst=9nH, Lsc=3nH, CsH-8pF, CsL=28pF, Rg=2.2ohm

Fig. 10. Power loss originating from single MOSFET in the full bridge inverter of the power conditioner.

The sendust core is applied for the magnetic materials here. The filter inductors at the DC input side and the AC output side are designed to keep the current ripples lesser than 20% of the inductor peak current. Figure 11 shows the power loss originating from the filter inductor at the AC output side of the power conditioner.

The power loss in the boost chopper can be also calculated. The converter efficiency can be now estimated, and the result is shown in Fig. 12. The efficiency is estimated for the switching frequency and the chip area of

2448

the MOSFET. The ideal circuit condition and the real circuit condition are illustrated. Parameters of the inductor are fixed at 4000A/m of the bias magnetic field. The switching frequency and the chip area are extracted ideally to satisfy the efficiency of 96% from Fig. 12.

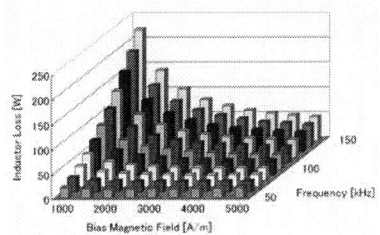

Fig. 11. Inductor loss originating from the filter inductor at the AC output side of the power conditioner.

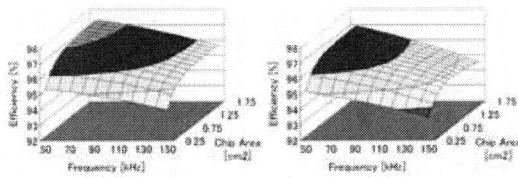

(a) Ideal Circuit Condition   (b) Real Circuit Condition
(a) Lst=0nH, Lsc=0nH, CsH-0pF, CsL=0pF, Rg=0ohm
(b) Lst=9nH, Lsc=3nH, Lsc=3nH, CsH-8pF, CsL=28pF, Rg=2.2ohm

Fig. 12. Converter efficiency estimation of the 3kW power conditioner.

B. Structure design and OPD estimation

The structure design of the circuit board is carried out here. Chip areas of the MOSFET and the SBD are selected, and the circuit board is designed at first. Then, the thickness of the board is determined, taking the isolation into account. Circuit stray parameters are estimated by the electro-magnetic analyses, and the parameters to satisfy the result in the circuit parameter optimization part is extracted.

The specifications of power devices are shown in Table 1. The Si-SBD of 30V and 0.3cm2 is utilized to block the body diode of the MOSFET. The SiC-SBD of 0.04cm2 is applied as the free wheel diode. The MOSFET of 68mΩcm2 and 0.77cm2 is utilized in this consideration. These chips are commercially available, and the MOSFET of 0.77cm2 realizes the high frequency operation as shown in Fig. 10 and Fig. 12.

The circuit board of the power conditioner is designed by using the selected power devices. Figure 13 shows the illustration of the board and the electro-magnetic analysis result. The switching energy of the MOSFET and the circuit stray parameters are shown in Fig. 13. The circuit board is the double sided copper board with ALN insulation of 15kV/mm. Thickness of the board is determined in terms of the power loss and the isolation problem. Based on the database from the circuit parameter optimization part, the thickness of the 0.6mm is selected.

Figure 14 shows the converter loss and the efficiency for the 3kW power conditioner. Circuit stray parameters are that the circuit stray inductance Ls is 9nH, the

common stray inductance Lsc is 3nH, high side stray capacitance CsH is 8pF, low side stray capacitance CsL is 28pF and gate resistance Rg is 2.2Ω. Filter inductors are designed at bias magnetic field of 4000A/m. The efficiency of 96% is realized at 90 kHz.

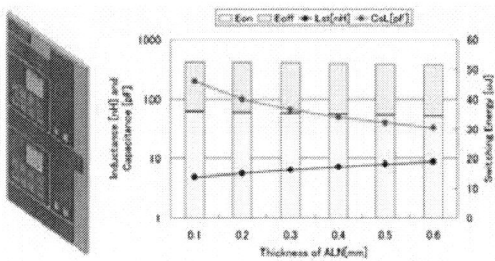

Fig. 13. Circuit stray parameters and switching energies of single MOSFET on the circuit board of 1 leg part of the inverter in the power conditioner.

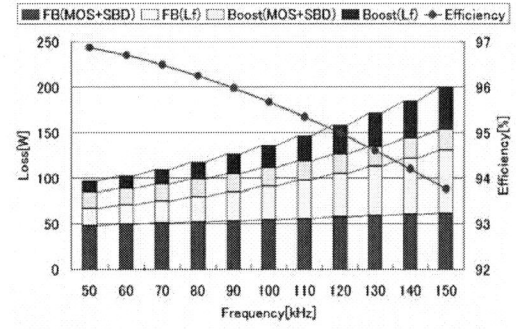

Fig. 14. Converter loss and the efficiency estimation for 3kW power conditioner under real circuit condition.

The thermal analysis is carried out to estimate the heat sink volume. The calculation result of power loss shown in Fig. 14 is utilized for this analysis. Natural cooling is usually employed to the power conditioner for PV solar generation. The thermal analysis is operated based on the following equation.

$$Q = h \cdot S \cdot (Tj - Ta)$$

Where, "Q" means total loss of the power conditioner. "S" means the area for the heat transfer. "Tj" and "Ta" are the junction temperature of the power device and ambient temperature respectively. Tj is 125 degrees Celsius, and Ta is 27 degrees Celsius. "h z is the heat transfer coefficient, and 8 W/m2K is applied.

Finally, OPD of the power conditioner is estimated. Volume of the circuit board for the boost chopper and the full bridge inverter, the input filter, the output filter and the heat sink are considered. The gate driver and the power supply for the driver are included in the circuit board. Here, the 3kW power conditioner for PV solar generation of 7.8 W/cc with 96% efficiency will be accomplished by using the Si - SJ MOSFET and the SiC – SBD hybrid pair.

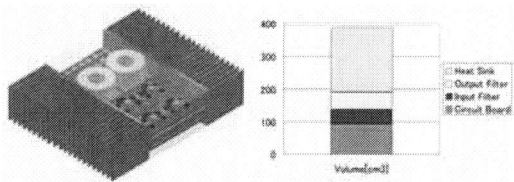

Fig. 15. Volume estimation of the 3kW power conditioner for PV solar generation.

The possibility of higher OPD of the 3kW power conditioner is discussed. OPD of 7.8W/cc is estimated under the condition of the designed circuit board and the efficiency of 96%. Ideal OPD of the power conditioner can be also investigated using the proposed method.

Figure 16 shows OPD estimation results of the 3kW power conditioner. The chip area of the MOSFET and the switching frequency are varied. This figure is drawn based on Fig. 10, and the inductor is assumed to operate at 4000A/m of bias magnetic field. In Fig. 16 (b), OPD of 8W/cc is possible when the designed structure shown in Fig. 15 is applied. Figure 16 (a) shows OPD of about 10W/cc is possible at 150 kHz with the chip area of 1.75 cm2 under the ideal circuit condition. The management of circuit stray parameters has the possibility to increase OPD of the power converter.

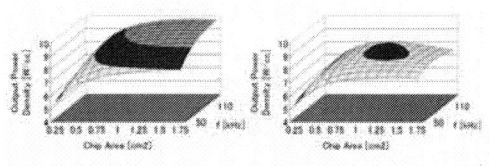

(a) Ideal Circuit Condition   (b) Real Circuit Condition
(a) Lst=0nH, Lsc=0nH, CsH=0pF, CsL=0pF, Rg=0ohm
(b) Lst=9nH, Lsc=3nH, CsH=8pF, CsL=28pF, Rg=2.2ohm

Fig. 16. OPD of the 3kW power conditioner for PV solar generation.

IV. POWER CONVERTERS DESIGNED BY USING THE NOVEL CONVERTER LOSS ESTIMATION METHOD

Three types of power converters have been demonstrated in Energy Semiconductor Electronics Research Laboratory. They are the power conditioner for the photovoltaic solar generation, the SiC power converter for 400W DC motor drive and the three-phase inverter of Si SJMOSFET and SiC-SBD hybrid pair.

The power conditioner and the three-phase inverter have been constructed to confirm the validity of the proposed methods. The three-phase inverter with 10W/cm3 of output power density has been achieved with more than 100 kHz switching operation. The power converter for the DC motor drive has been constructed by using the SiC-IEMOSFET and SiC-SBD made in Energy Semiconductor Electronics Research Laboratory. The effectiveness of the SiC power devices and the availability of the design methods have been demonstrated.

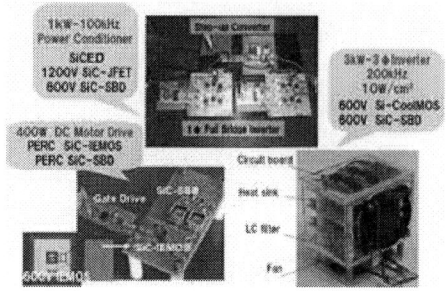

Fig. 17. Demonstrated Power Converters using the Design Method based on the Platform Concept

V. 4. CONCLUSION

A novel converter loss estimation method based on Power Converter Platform concept has been proposed to realize high output power density integrated converters. An experimental apparatus has been fabricated to realize the concept. The detail of the proposed converter loss estimation method has been shown, and the converter loss simulator based on the concept has been introduced. The availability of the novel converter loss estimation method has been confirmed, and three types power converter fabricated in Energy Semiconductor Electronics Research Laboratory has been introduced.

Researches about the development of the power converter integration design methodology are under way to realize high OPD converters in Energy Semiconductor Electronics Research Laboratory.

ACKNOWLEDGMENT

A part of this work was supported by New Energy and Industrial Technology Development Organization (NEDO) of Japan. We greatly appreciate fruitful comments of Prof. Y. Sato at Chiba University for this study.

REFERENCES

[1] H. Ohashi, "Recent Power Devices Trend", IEEJ, Vol.12, No.3, pp.168-171, 2002 (in Japanese).

[2] J. W. Kolar, U. Drofenik, J. Biela, M. L. Heldwein, H. Ertl, T. Friedli and S. D. Round, "PWM Converter Power Density Barriers", Proceedings of the Fourth Power Conversion Conference PCC Nagoya 2007, Nagoya, Japan, Apr. 2-5, 2007.

[3] Z. Liang, B. Lu, J. D. van Wyk and F. C. Lee, "Integrated CoolMOS FET/SiC-Diode Module for High Performance Power Switching", IEEE Transactions on Power Electronics, Vol.20, Issues 3, pp.679-686, May, 2005.

[4] I. Omura, M. Tsukuda, W. Saito and T. Domon, "High Power Density Converter using SiC-SBD", Proceedings of the Fourth Power Conversion Conference PCC Nagoya 2007, Nagoya, Japan, Apr. 2-5, 2007.

[5] H. Ohashi, "Research Activities of the Power Electronics Research Center with Special Focus on Wide Band Gap Materials", in Proc. CD-ROM, 4th International Conference on Integrated Power Systems (CIPS2006), 2006.

[6] Y. Hayashi, K. Takao, T. Shimizu and H. Ohashi, "High Power Density Design Methodology", Proceedings of the Fourth Power Conversion Conference PCC Nagoya 2007, Nagoya, Japan, Apr. 2-5, 2007.

[7] K. Takao, Y. Hayashi and H. Ohashi, "Study on High Frequency Limitation of SJ-MOSFET/SiC-SBD Pair in Comparison with Normal MOSFET/SiC-SBD Pair", Proceedings of the Fourth

Power Conversion Conference PCC Nagoya 2007, Nagoya, Japan, Apr. 2-5, 2007.

[8] S. Iyasu, T. Shimizu and K. Ishii, "A Novel Inductor Loss Calculation Method on Power Converters Based on Dynamic Minor Loop", IEEJ Transactions on Industry Applications, Volume 126-D Number 7, pp. 1028-1034, 2006.

[9] T. Shimizu and K. Ishii, "An Iron Loss Calculating Method for AC Filter Indcutors Used on PWM Inverters", in Proc. CD-ROM, 37th IEEE Power Electronics Specialists Conference (PESC2006), 2006.

[10] A. Lidow, D. Kinzer, G. Sheridan and D. Tam, "The Semiconductor Roadmap for Power Management in the New Millennium", PROCEEDINGS OF THE IEEE, VOL. 89, NO. 6, JUNE 2001.

Power Devices in Polish National Silicon Carbide Program

Mariusz Sochacki*, Andrzej Kubiak[†], Zbigniew Lisik[†] and Jan Szmidt*

* Warsaw University of Technology/Institute of Microelectronics and Optoelectronics, Warsaw, Poland, e-mail:
msochack@elka.pw.edu.pl
[†] Technical University of Lodz/Department of Semiconductor and Optoelectronics Devices, Lodz, Poland, e-mail:
akubiak@p.lodz.pl

Abstract—Polish Government Program "New technologies based on silicon carbide and their applications in high frequency, high power and high temperature electronics" covers an project package that consists of three general tasks. The contribution presents the overview of projects in the field dealing with the design and manufacturing of power SiC semiconductor devices.

Keywords—silicon carbide, Schottky diode, PiN diode, MOSFET, JFET

I. INTRODUCTION

For several years, the investigations dealing with the features of different silicon carbide polytypes as well as the technology of silicon carbide device manufacturing are more and more intensive. It is caused by the extremely attractive material parameters of silicon carbide in comparison to silicon ones, which is characterized by the larger band gap, the better heat conductivity, the larger electron saturation velocity and the larger critical electric field strength. These parameters, potentially, allow manufacturing silicon carbide devices characterized by ratings unattainable or attainable very difficult for the devices manufactured from silicon or other semiconductors. The large critical electric field strength allows getting high voltage p-n junctions with the breakdown voltage larger than 10 kV. The large band gap makes possible manufacturing low noise devices or devices working at high temperatures exceeding even 700°C whereas the large saturation electron velocity makes silicon carbide an excellent candidate for high frequency devices with the possible maximal frequency reaching THz. The good thermal conductivity is very crucial from the reliability point of view (thermal stresses) and thermal management problems.

The basic obstacle in practical use of silicon carbide in electronics consists in different, very often extreme, demands concerning the technology processes in comparison to the processes of silicon technology. It causes that the technology processes and characterization procedures well know in the silicon technology occur often to be inappropriate in the case of silicon carbide. The Polish National Silicon Carbide Program is aimed at solving some of these problems and it is expected that the results of its run will create the room for further activities towards the wider introduction of silicon carbide into electronics industry. It covers several research programs gathered in three tasks, and a few of them, dealing with

the silicon carbide power devices, is more detailed presented in the paper.

II. PRESENTATION OF THE PROGRAMS

The program consists of three thematic tasks aimed at SiC in bulk and substrate form, SiC device manufacturing and SiC devices application, respectively. The first task is dominating in the program. Its main project is devoted to work out the technology of 6H:SiC and 4H:SiC monocrystal bulks produced from silicon carbide powder using the methods basing on Lelly approach. Other projects of this task concern next steps necessary to obtain 6H:SiC and 4H:SiC polished substrates with and without an epilayer, characterized by the quality satisfying device producers.

The second one covers projects dealing with different aspects of device manufacturing. Since the well-characterized substrate is a base for any research activity aimed at semiconductor device manufacturing, a few projects is concentrated on the characterization methods specific for SiC structures, which are essential for the final results of technological processes. One of them concerns e.g. the study of electrical, optical and photoelectrical characterization methods for MIS silicon carbide structures. Other projects deal with the technology problems like the technology of selective ion implantation, which is the key process for source and drain doping in MISFET technology as well as the termination in high voltage devices, the technologies of electrical contacts for silicon carbide devices and packaging technology taking also into account the problems resulting from silicon carbide application in high temperature electronics. In this task, the separate group is formed by the projects aimed at the manufacturing technology of particular devices with the final goal to deliver the demonstrators of these devices. They deal with Schottky diodes, PiN diodes, MISFET and JFET transistors manufactured on SiC substrates as well as with high-frequency HFET transistors and Schottky diodes manufactured as $A^{III}N$/SiC heterostructures. Some of these projects, related to the power SiC devices are presented more detail below in the successive chapters.

The third task is the smallest one in the program and covers two projects only, concentrated on the investigations of advantages and disadvantages resulting from the application of SiC devices in power electronics circuits. Basing on the commercial available elements like Schottky diodes and MESFET transistors mainly, they cover the investigations of device features from the point of view of their influence on design process and work conditions in the real power electronics equipment.

III. SCHOTTKY DIODES

Many power electronics circuits consist of a semiconductor switch connected with an anti-parallel diode, which parameters have a strong influence on power losses in the circuit. Such diode parameters as its reverse recovery time and recovery charge determine the final energy efficiency of the circuit and the replacement of Si PiN freewheeling diodes by SiC ones can results in orders of magnitude less storage charges leading to reduction of the turn-off energy lossess in the circuit [1]. Application of silicon carbide Schottky diodes improves energy efficiency, power density and energy quality simultaneously due to high breakdown voltage with low specific resistance, very little charge storage and fast recovery. The absence of recovery current peak indeed results in noise and EMI reduction during switching what is an important aspect of high power converter design. All these aspects decided that the development the proper technology for SiC Schottky diode was an element of a few projects undertaken in the program. The up to day carried out activities are shown briefly below.

A. Electrical simulations

The electrical characteristics were simulated using ATLAS device simulation software provided by SILVACO International Inc. Four basic Schottky diode structures were analyzed to find out the main advantages of different techniques of edge termination: diode without any contact termination, diode with floating rings termination, diode with field plate termination and diode with ion implantation beneath contact edge. The epitaxial layer thickness was tuned between 5 μm and 10 μm and doping level was changed between 5×10^{15} cm^{-3} and 1×10^{16} cm^{-3} because breakdown voltage of 1200 V is proposed in the project. The influence of Schottky barrier height on diode parameters was verified within the range of 4.6 – 5.8 eV. The lower value corresponds to the typical barrier of titanium on SiC and the higher one is related to iridium oxide. Both types of metallization will be deposited on SiC surface.

Forward characteristics were simulated within temperature range of 20°C–500°C. Reverse characteristics were calculated for elevated temperature only because the accuracy of the simulation at room temperature is unsatisfied and there are a lot of troubles with algorithms convergence. The simulations of reverse characteristics were focused on determination of breakdown voltage rather than leakage current calculation. The impact ionization rate model proposed by Selberherr was applied for this purpose with the default coefficients adjusted for SiC. The obtained solution confirmed that the contacts without any termination breaks down at around 150 – 200 V. The most efficient way of termination appears the ion implantation beneath the contact edge. The calculated characteristics at temperature of 500°C were presented in Fig. 1 for 10 μm-thick drift region doped up to 5×10^{15} cm^{-3} and 1×10^{16} cm^{-3}, respectively. The breakdown voltage of 1200 V is achievable without any additional operations. The presented result could be obtained by 80 μm-width p-type ring implantation with total doping of 10^{17} cm^{-3}.

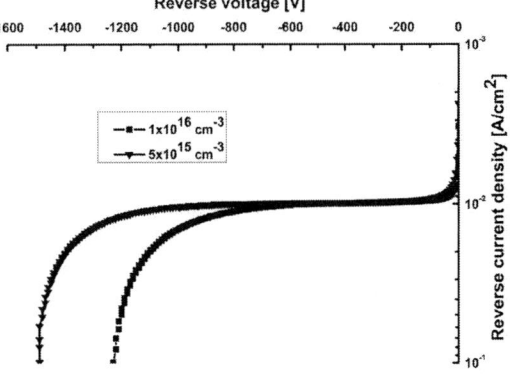

Fig. 1 Simulated reverse characteristics of Schottky diode at 500°C.

B. Device fabrication

Circular Schottky diodes were fabricated on high quality 4H-SiC (0001) wafers purchased from SiCrystal AG using Ni, Ir and IrO$_2$ as Schottky barrier. Both the epitaxial layer and the heavily doped substrate were n-type doped with concentration of $N_D = 5 \times 10^{15}$ cm^{-3} and $N_D = 5 \times 10^{18}$ cm^{-3}, respectively. The epitaxial layers were fabricated at Institute of Electronic Materials Technology. The entity is involved in the technology of SiC crystallization in the project. An ohmic contacts were formed on the back side by rapid thermal annealing (RTA) at 1000°C in Ar of a 200 nm thick nickel film deposited in sputtering chamber. 50 nm-thick Ir and Ni Schottky contacts were deposited by DC magnetron sputtering from Ir and Ni target in Ar ($P_{DC} = 50$ W, $p_{Ar} = 1 \times 10^{-2}$ mbar). Thin films of iridium oxide (50 nm-thick) were fabricated by RF reactive magnetron sputtering from Ir target in O$_2$/Ar plasma at oxygen to argon pressure ratio of 0.1.

The rare earth metal and its oxide was used in this work for comparison purpose because the Schottky barrier height values of widely investigated metals with high work functions (Ni, Pt, Au, Pd) were lower than those predicted by the Schottky-Mott theory [3-5]. Additionally, most of the metals interdiffuses into SiC within temperature range of 450°C-500°C leading to instability of Schottky barrier height [6-8]. The abovementioned problems could be overcome by application of the rare earth metals and their conducting oxides which have higher work function (Ir: 5.6 eV, IrO$_2$: 5.8 eV) and create more thermally stable metal/semiconductor interface [9-10].

In our work, the I-V characteristics were analyzed to find a correlation between extracted parameters of Ni/4H-SiC, Ir/4H-SiC, IrO$_2$/4H-SiC Schottky diodes and initial surface treatment. Therefore, prior to Schottky contact deposition around ten different methods of surface cleaning including reactive ion etching (RIE) were performed to find out the optimal way of surface treatment.

C. Characterization

The static I-V characteristics of Schottky contacts were measured using Keithley 251 Source Meter. Three methods were applied to extract Schottky barrier height (SBH) from forward I-V characteristics (proposed by Rhoderick [11], Cheung [12] and Norde [13]). The forward and reverse current density versus voltage measured at room temperature is shown in Fig. 2 and Fig. 3, respectively.

The forward current density is linear over more than four orders of magnitude at low voltages for every diode. The linearity of the characteristic for selected diodes is even more than nine orders of magnitude. It confirms very strong influence of the applied method of surface preparation on diode properties. The current density tends to saturate at higher voltages due to the series resistance of the diode.

Schottky barrier height ϕ_b, ideality factor η and specific resistance R_{sp_ON} of the diodes were extracted from I-V measurements using three different methods which are based on thermionic emission theory. According to the mentioned theory the I-V relationship for Schottky junction is given by the following equation:

$$I = A^{**}AT^2 \exp\left(-\frac{q\phi_{b0}}{kT}\right)\left(\exp\frac{qV}{\eta kT}-1\right) \qquad (1)$$

where A^{**} is effective Richardson constant, A is the diode area, ϕ_{b0} is apparent barrier height at zero-bias while η is ideality factor. For our calculations the theoretical value of A^{**} was used for 4H-SiC (146 Acm^{-2}K^{-2}) [14] which was determined by using a value of the effective mass of 0.2m$_0$ [15]. The voltage (V) in (1) is often replaced by V-$R_{sp\_ON}I$ to take into account the specific resistance (R_{sp_ON}) of the diode. Three methods of extraction are applied in this work to minimize the specific errors caused by the series resistance and to find a general trend of influence of cleaning methods on diode parameters. The extracted parameters are dependent on the method used and must be considered with care in order to get the most reliable and accurate evaluation of R_{sp_ON}, ϕ_{b0} and η. Rhoderick's model fails when ideality factor is voltage dependent which can be caused by a strong recombination current contribution and by the strong influence of series resistance [16]. It is then necessary to introduce other methods to evaluate the diode parameters, specifically, Norde's and Cheung's method. These methods uses plots of auxiliary functions to evaluate the diode parameters more accurately. However, the methods have their own weaknesses as well.

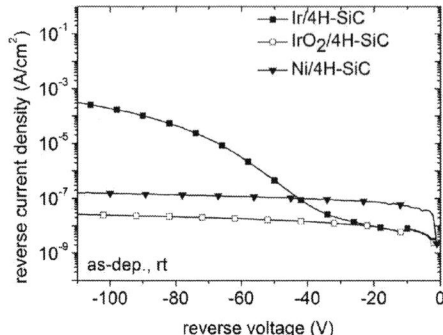

Fig. 3 Reverse characteristics of as-deposited Schottky diodes at room temperature

The main disadvantages of Norde's method are that the ideality factor is assumed to be very close to unity which is not always true for a real diode and that only a single point of I-V characteristic is used to calculate the barrier height. These problems often lead to the overestimation of the series resistance value. Good linearity of $log\,(I)$–V characteristic is required for accurate extraction of parameters by Cheung's method but this one seems to be the most reliable and accurate among the discussed methods [17]. The summarized results for different methods are presented in Tab. I.

Forward current density of 200 A/cm^2 was achieved for small area devices (3×10^{-4} cm^2). The forward current capability is still affected by too high value of specific contact resistance and it is a real challenge to improve the technology. Rare earth metals and its oxides are tested at elevated temperature to demonstrate its superiority over a classical Schottky contacts. Hitherto, the operating temperature increase of at least 100°C seems to be possible by application of the rare earth compounds.

IrO$_2$ metallization is clearly superior to all the other applied materials assuring the highest Schottky barrier, the lowest leakage currents and the ideality factor close to unity. However, the specific resistance of the diode must be decreased to improve current capability simultaneously.

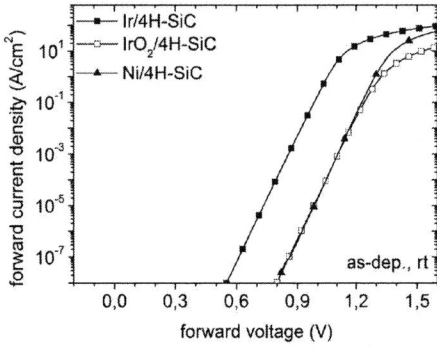

Fig. 2 Forward characteristics of as-deposited Schottky diodes at room temperature

TABLE I. EXTRACTED PARAMETERS OF SCHOTTKY DIODES

Ir/4H-SiC					
ϕ_b (eV)			η		R_{sp_ON} (mΩcm^2)
Rhoderick	Cheung	Norde	Rhoderick	Cheung	Cheung
1.44	1.46	1.47	1.03	1.00	4.6
IrO$_2$/4H-SiC					
1.66	1.64	1.72	1.04	1.07	19.6
Ni/4H-SiC					
1.50	1.63	1.69	1.17	1.03	2.8

IV. PIN DIODES

The Schottky barrier diodes are unattractive for blocking voltage higher than 2 kV due to the increase in specific resistance with the increase of drift layer thickness and the decrease of its doping. This obstacle can be avoided when the bipolar PiN diodes is taken into account. In the bipolar diodes, due to the conductivity modulation, it is possible to reduce the drift layer resistance as the result of its flooding by excess carriers when the forward current flows. This way, one can obtain the high voltage semiconductor switch with a very attractive blocking voltage/forward voltage rate if the quality of SiC substrates used to bipolar diodes manufacturing is sufficiently high. Any defects can be crucial for the features of manufactured device and e.g. its forward voltage drift can be essentially affected by stacking faults related to basal plane screw dislocation [2].

The mentioned above role of SiC bipolar diodes caused that the technology processes allowing their manufacturing as a high voltage power device are the subject of research in a few projects. They are aimed at doping technologies, both the ion implantation that is already used in SiC devices fabrication and the thermal diffusion that is still under research, dielectric and passivation layers technologies that are necessary among others to the high voltage junction termination and general, high temperature processing technologies that are the specific new additionally but inherent element of the majority of SiC technologies. In some projects, an effort is also made to touch the aspects of high temperature electronics application in te field of ohmic contacts technologies and packaging issues. The up to day carried out activities in some of above topics are shown briefly below.

A. High temperature processes

All the doping technologies that can be used for SiC PiN diodes fabrication require the high temperature processes. In the case of ion implantation doping, an additional annealing step is necessary after the main process of implantation to activate the introduced acceptor or donor atoms Its temperatures are significantly large exceeding sometimes even 1600°C. In the case of diffusion doping, the process can run successfully only when the temperature exceeds 1600°C. Unfortunately, the temperatures exceeding 1600°C evoke the sublimation process leading to destruction of the surface layer of processed SiC structure and in consequence, it can result in the damage of manufactured device. Therefore, it is of crucial importance to minimize the sublimation effects during the high temperature processes.

This problem is the subject of carried our investigations with the goal to develop the procedure that allows blocking the sublimation during the SiC annealing in the temperatures reaching 2000°C. All the investigations concerns the high temperature Gegussa furnace that allows getting the temperatures till 2500°C in the work chamber under low pressure conditions and controlled gas flow. The investigations that has been done till now, one can split into experimental one and theoretical one.

The theoretical investigations consisted in two-dimensional (2-D) numerical analysis of thermal phenomena taking place in the work chamber of the furnace under different work conditions resulting among other from the chamber design changes. Numerical model of chamber, work out on the ANSYS software platform, was the basic tool for the numerical analysis. It was complex model, included all the essential parts of the furnace, the heat and mass flow inside the chamber as well as all the mechanisms of heat conduction, radiation and heat abstraction caused by water cooling the chamber.

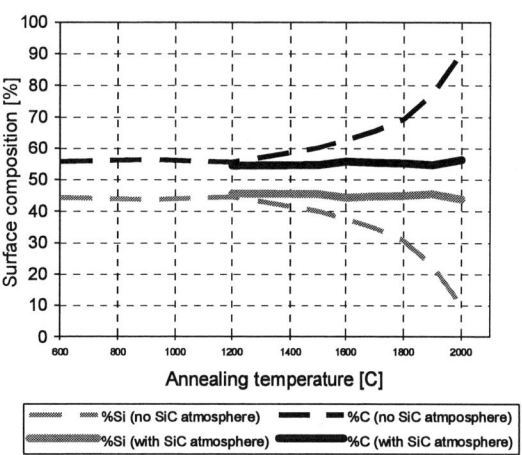

Fig.4. Influence of "SiC atmosphere" on sublimation process

The discrepancy between simulation results and the real values was less than 5% within the most interesting temperature range of 1000°C-2000°C. The experimental investigations consisted in the furnace modifications, among other, according to the results of numerical simulation and in the test processes with the artificially generated SiC atmosphere surrounding the processed SiC substrate. These experiments allowed defining the process conditions blocking the sublimation effect on the surface of processed substrate.

It is shown in Fig.4 that covers the comparison of the percentage changes of the Si and C content in the surface layer of SiC structure being annealing with and without the protected "SiC atmosphere". The measurements of the surface compositions have been made using the XPS technique.

B. Ohmic contacts to SiC

Long-term thermally stable ohmic contacts to n-type and p-type SiC with low specific contact resistance are extremely important, since parasitic resistances limits forward current capability. Pure Ni contacts to n-type SiC were fabricated by electron beam evaporation. Ni/TaSi and Ni/Si stacks were deposited by DC magnetron sputtering in Ar plasma from Ni, Ta_5Si_3 and Si targets, respectively. Post-metallization annealing was carried out within temperature range of 600°C-1100°C under Ar or N_2 flow in rapid thermal processing (RTP) chamber. Current-voltage characteristics and circular transmission line method (c-TLM) were applied for contact resistance extraction. The lowest value of 2×10^{-5} Ωcm^2 was obtained for pure Ni metallization. It means that the improvement of the ohmic contact technology is necessary because the presented results are not satisfied and specific contact resistance is still too high for high-current power devices. Ohmic contacts to p-type SiC will be developed as well. The project coves electrical contacts technologies for silicon carbide widely. Especially, it is concentrated on the packaging problems when the appropriate electrical connections between the pads on silicon carbide structure and the package contact areas are needed. Wire bonding technology and flip-chip bonding will be investigated for high-power electronics purposes.

V. FIELD-EFFECT TRANSISTORS (FETs)

FET transistors have been designed and developed in the project. Lateral structures were chosen for evaluation because such devices are more attractive for integration in power ICs. The meaningful progress in MISFETs technology is limited by insufficient quality of dielectric/semiconductor interface which disturbs inversion channel mobility. In spite of more than 20 years of research, difficulties surrounding thermal oxidation of SiC surface have not yet been overcome and MOSFETs are not yet commercially available. It is expected that when the basic problem is solved the lateral MISFETs will be one of the most attractive power device for moderate breakdown voltage. Simultaneously, the technology of lateral JFETs has been developed in the project and advantages of both technologies have been investigated. Reduced surface field (RESURF) design is analyzed for both structures to increase breakdown voltage.

Currently, layouts with different channel lengths and RESURF zones lengths were electrically simulated. The photolithography masks were designed and fabricated. The source and drain areas of MISFETs and gate area of JFETs will be defined in an ion implantation process. It means that the ion implantation process and post-implantation annealing is a key factor of the final success. Multiple implantations with different energies and doses were simulated and performed. The sample temperature was maintained at 500°C during the implantation process. Optical methods like spectroscopic elipsometry and micro-Raman scattering studies of Al^+ ion-implanted SiC substrates were performed. The methods indicated the formation of post-implantation amorphous regions in a subsurface of as-implanted SiC substrates. An efficient ways of recrystallization and dopant activation will be studied.

Different dielectric materials and their deposition techniques will be evaluated for gate and passivation technology besides thermal oxidation. The most promising layers of SiO_2, Si_3N_4, Al_2O_3, AlN, $BaTiO_3$, BN, C_xN_y will be fabricated by atomic layer deposition (ALD), impulse plasma deposition (IPD) and plasma-enhanced chemical vapor deposition (PE-CVD).

VI. SUMMARY

Main goals of Polish National Silicon Carbide Program were presented in this paper. Different approaches used for the design, optimization, fabrication, characterization and challenges associated with the technology were discussed in more detail. Simultaneously, the executed and evaluated part of the research was shown. Verified demonstrators of Schottky diodes, PiN diodes, MOSFETs and JFETs are expected as a final result of the program.

ACKNOWLEDGMENT

The research is supported by Polish Ministry of Science and High Education under grant 1/0-PBZ-MEiN-6/2/2006.

REFERENCES

[1] P. Alexandrov, J. H. Zhao, W. Wright, M. Pan, H. Weiner, "Inductively-loaded half-bridge inverter characterization of 4H-SiC merged PiN/Schottky diodes up to 230A and 250°C", *Electron. Lett.*, vol. 37, pp. 1261-1262, 2001.

[2] R. E. Stahlbush, K. X. Liu, M. E. Twigg, "Effects of dislocations and stacking faults on the reliability of 4H-SiC PiN diodes", Reliability Physics Symposium Proceedings, 2006, 44th Annual., IEEE International, pp. 90-94.

[3] L. M. Porter, R. F. Davies, "A critical review of ohmic and rectifying contacts for silicon carbide", Mater. Sci. Eng., vol. B34, pp. 83-105, 1995.

[4] A. Itoh, H. Matsunami, "Analysis of Schottky barrier heights of metal/SiC contacts and its possible application to high-voltage rectifying devices", Phys. Status Solidi A 162, pp. 389-408, 1997.

[5] C. Raynaud, K. Isoird, M. Lazar, C. M. Johnson, N. Wright, J. Appl. Phys., "Barrier height determination of SiC Schottky diodes by capacitance and current-voltage measurements", vol. 91, pp. 9841-9847, 2002.

[6] D. E. Ioannou, N. A. Papanicolaou, P. E. Nordquist, Jr., "The effect of heat treatment on Au Schottky contacts on β-SiC", IEEE Trans. Electron Dev., vol. 34, 1694-1699, 1987.

[7] N. A. Papanicolaou, A. Christou, M. L. Gipe, "Pt and $PtSi_x$ Schottky contacts on n-type β-SiC", J. Appl. Phys., vol. 65, pp. 3526-3530, 1989.

[8] N. Lundberg, M. Östling, C. -M. Zetterling, P. Tägström, U. Jansson, "CVD-based tungsten carbide Schottky contacts to 6H-SiC for very high-temperature operation", J. Electron. Mater. Vol. 29, 372-375, 2000.

[9] L. Krusin-Elbaum, M. Wittmer, "Conducting transition metal oxides: Possibilities for RuO_2 in VLSI metallization", J. Electrochem. Soc., vol. 135, pp. 2610-2614, 1988.

[10] M. L. Green, M. E. Gross, L. E. Papa, K. J. Schnoes, D. Brasen, "Chemical vapor deposition of ruthenium and ruthenium dioxide films", J. Electrochem. Soc., vol. 132, pp. 2677-2685, 1985.

[11] E. H. Rhoderick, R. H. Williams, "Metal-semiconductor contacts", University Press, Oxford, 1988.

[12] S. K. Cheung, N. M. Cheung, "Extraction of Schottky diode parameters from current-voltage characteristics", Applied Physics Letters, vol. 49, pp. 85-87, 1986.

[13] H. Norde, "A modified forward I-V plot for Schottky diodes with high series resistance", J. Appl. Phys., vol. 50, pp. 5052-5053, 1979.

[14] A. Itoh, T. Kimoto, H. Matsunami, "High performance of high-voltage 4H-SiC Schottky barrier diodes", IEEE Electron Device Lett., vol. 16, pp. 280-282, 1995.

[15] W. Götz, A. Schöner, G. Pensl, W. Suttrop, W. J. Choyke, R. Stein, S. Leibenzeder, "Nitrogen donors in 4H-silicon carbide", J. Appl. Phys., vol. 73, pp. 3332-3338, 1993.

[16] J. H. Werner, "Schottky barrier and I-V plots – small signal evaluation", Appl. Phys. A, vol. 47, pp. 291-300, 1998.

[17] V. Aubry, F. Meyer, "Schottky diodes with high series resistance: Limitations of forward I-V methods", J. Appl. Phys., vol. 76, pp. 7973-7984, 1994.

SiC Power Semiconductor Devices for new Applications in Power Electronics

Dominique Planson*, Dominique Tournier, Pascal Bevilacqua, Nicolas Dheilly, Herve Morel, Christophe Raynaud, Mihai Lazar, Dominique Bergogne, Bruno Allard, Jean-Pierre Chante

*Ampere Lab INSA Lyon, F-69621 Villeurbanne Cedex, France, e-mail: *dominique.planson@insa-lyon.fr*

Abstract—**This paper addresses the benefits of SiC semiconductor, owning excellent physical properties able to fulfill new scope of applications in terms of high temperature, high voltage and for more specific applications. Devices and applications developed at Ampere laboratory are detailed.**

Keywords—**SiC-device, High temperature electronics, Power semiconductor device, High voltage Device, Power integrated circuit.**

I. INTRODUCTION

The first unipolar power silicon carbide devices (Schottky diodes) were commercialized in 2001 [1], [2]. Nowadays power JFETs are available as engineering samples [3], [4] . These two devices enable to build a large number of power switching converters. It is worth listing the scope of applications for which silicon carbide devices can advantageously replace classical silicon power semiconductor devices. The first kind of applications for SiC power devices are applications for which the use of silicon is impossible or restricted. The second kind of applications for SiC devices correspond to systems in which mass reduction is a key issue, e.g. for embedded systems like in transport.

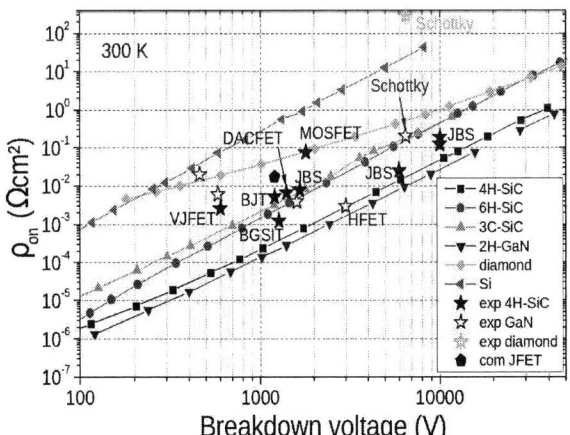

Fig. 1. On state resistance for unipolar devices for various semiconductor materials.

Figure 1 illustrates the on-state resistance for different semiconductor materials. It yields the optimal on-resistance limit versus breakdown voltage of an infinite plane junction for different semiconductor materials. These are theoretical limits. Stars correspond to power

device demonstrators. It is clear that there are over the theoretical limit of Si and also from the SiC limits. From this figure, it is clear that there is still a great margin to get the optimal performances of SiC devices.So, clearly, SiC power device applications are high temperature and/or high voltage applications. However, most of recent studies have shown that the use of SiC devices does not consist in a simple replacement of silicon devices by silicon carbide devices. Indeed a new design of power electronic systems must take advantage of the properties of the new SiC power devices.

II. HIGH VOLTAGE APPLICATIONS

Such applications are medium term applications because high voltage devices are not ready for industrial applications. However 10 kV demonstrators have been developed, $15kV$ demonstrators are under development even $30kV$ and further may soon be obtained. The interest in such devices is huge since they enable to :

- replace heavy $50Hz$ high voltage electrical transformers by high voltage static power electronics.
- replace AC-current distribution networks by DC-current distribution networks and eliminating numerous associated problems like instability (wind-power farm ...) even blackout. This enables an easier integration of renewable energy sources (Photovoltaic, Wing power...) and removes the need of $50Hz$ electric transformers.
- improve the high voltage system protection with very fast active protection systems.

The main issues of these new applications are :

- availability of high bipolar diodes and switches (BJT, Thyristors, ...),
- electrical insulation of the high voltage converters (packaging, passivation, ...),
- thermal cooling of the high voltage converters, highly insulated driver for high voltage converters even multi-level converters.

Ampere lab works are divided in several tasks:

A. *Design of high voltage periphery protection*

Design of efficient peripheral protection is fundamental for high voltage devices. This protection should have an optimum geometry in order to avoid local critical electric field areas. There are various possible protections such as JTE (Junction Termination Extension), Mesa, field plates

978-1-4244-1741-4/08/$25.00 ©2008 IEEE   2457

or guard rings [5]. Design of classical JTE needs an accurate control of the doping dose to reach the complete depletion of the termination well for a given reverse voltage (V_{BR}). The peripheral protection efficiency is the ratio between breakdown voltage of the protected diode and of the ideal diode (plane parallel junction), this last being named theoretical breakdown voltage afterwards. JTE process for SiC device is most often an ionic implantation followed by a thermal annealing. Figure 2 shows the breakdown voltage of a bipolar diode against the termination doping dose for several JTE lengths.

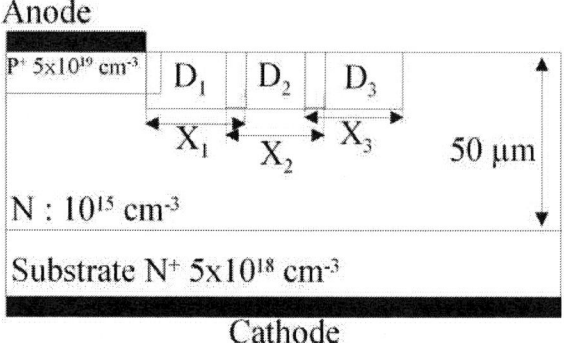

Fig. 3. Vertical cross-section of the triple JTE (D1, D2 and D3) protected diode.

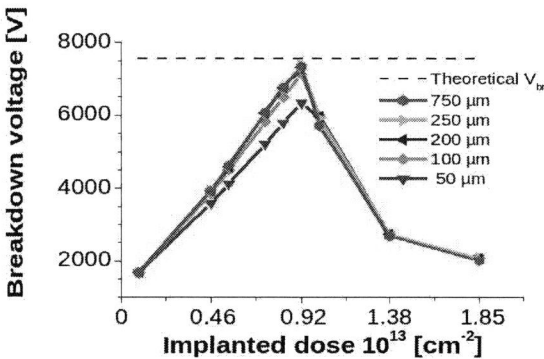

Fig. 2. Breakdown voltage against implanted dose for a constant lateral doping profile and for several JTE lengths.

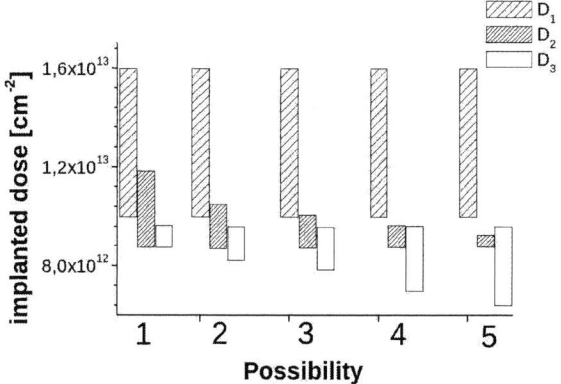

Fig. 4. Five possible dose configurations and their variation ranges to obtain a 99% protection efficiency.

With an epitaxial layer thickness of $50\mu m$ and a doping level of $10^{15} cm^{-3}$, the diode is able to sustain $7.5kV$. The JTE depth is $0.8\mu m$. As can be observed from Figure 1, breakdown voltage is very sensitive to the implanted dose. Besides this voltage is almost independent from the well length if the JTE is longer than $100\mu m$. The breakdown voltage is calculated by numerical bi-dimensional simulation using $MEDICI^{TM}$ software [6] and impact ionization coefficients published by Konstantinov [7]. V_{BR} is defined as the voltage involving a leakage current of $1\mu A.\mu m^{-1}$. JTE with variable lateral doping is a good way to get an efficient protection valid for a wide range of doping concentration. Thanks to the diffusion process, it is possible to realise this kind of JTE in silicon. However, due to the lack of dopant diffusion in SiC, variable doping JTE is very difficult to obtain. A good alternative is a JTE with multiple constant doping zones to approximate the optimum lateral profile. Figure 3 presents a SiC bipolar diode protected by a three-zone JTE that has been optimized to achieve a breakdown voltage that is as close as possible to the theoretical breakdown voltage.

Each zone is defined by its doping dose D_i and its length X_i. Figure 4 shows 5 possible configurations of doping variation of the three zones enabling a V_{BR} of 7500 V i.e. 99% of the theoretical breakdown voltage. The highest doped zone is the first one and its variation range is the widest. For the two other zones, dose choices are more critical, because of the wider $D2$ variation range, and of the narrower $D3$ one.

A 7500 V breakdown voltage can be obtained using simple or triple JTE. Nevertheless the latter option (which is a discretization of the ideal variable lateral doping profile) enables a larger flexibility on doping concentrations. The only drawback of this technique is a little more complicated technological process.

B. Fabrication of high voltage bipolar devices

The aim of this study is to show potentialities of silicon carbide for high voltage power systems. Test structures are SiC bipolar diodes able to sustain $1.3kV$ and $5kV$. Emitter and periphery protection (JTE) are obtained by ion implantation of Aluminium in epitaxial n-type layer grown on commercial substrate. The number of devices which exhibit efficiency $> 84\%$ has been improved during the last four years. Indeed 70% of the $1.3kV$ diodes with an emitter implanted at $300\degree C$ and a long JTE (120 m) exhibit a breakdown voltage higher than $1100V$ (up to $1300V$) without any impact of the ambient conditions (air or SF6 gas or dielectric liquid). The ambient conditions impact breakdown voltage only for diodes with short JTE length (50m) [8]. These results show in the same time that post-implantation annealing realized in the lab furnace [8] allows a full activation of dopants in JTE. In forward state, current densities and efficiency have been improved

(200 A/cm2 @ 5 V at room temperature) also due to the improvement of metallization step and subsequent annealing at IMM Bologne [9]. Results of acceptable ohmic contacts on highly doped P+ zones are comparable with other results already published in literature [10]. As it can be observed in Figure 5, Ampere-lab is now able to present results of breakdown voltage in the range of 4 kV, with a maximum of 4.8 kV with low reverse current density in fluoride ambient (Galden and SF6) [11].

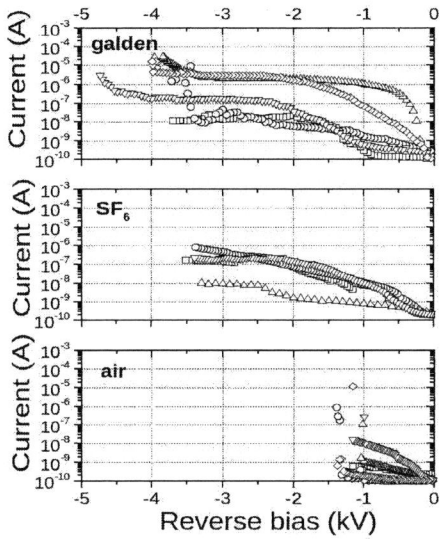

Fig. 5. Reverse I-V characteristics of $5kV$ bipolar diodes measured in three different ambient conditions: air, galden and SF6. Active area $A = 0.05mm^2$.

Without passivation, breakdown voltage is only $1.3kV$ for these devices with a JTE length of $250m$ (optimized with respect to the semiconductor properties). This clearly proves the need for passivation in order to fully take benefit of SiC properties for high voltage applications. Theoretical interest to use passivation materials with high dielectric permittivity (ϵ) in order to reduce electric field outside SiC have been shown using simulation software MEDICI on a 5 kV bipolar diode protected by JTE. For a given structure, only a sufficiently high ϵ value allows to reduce electric field peaks both at emitter and JTE extremities, by stretching space charge region outside the JTE [12]. Another advantage to use a high ϵ material is to significantly reduce the sensitivity on breakdown voltage with respect to the optimal dose of JTE. From the technological point of view, choice of such material (with high ϵ) and a compatible deposition method remains to be solved before obtaining experimental results. Among available and practical insulating materials, polyimide materials specifications exhibit good dielectric together with good thermal properties under usual conditions. Thus an experimental characterization using MIM and MIS structures has been launched in order to quantify their effective characteristics within an extended temperature range. For high voltage (up to $50kV$) and high temperature (up to $500\,^\circ C$) measurements, a specific set-up has

to be developed.

1) Fabrication of high voltage thyristor: Since silicon based technology is reaching its physical limits concerning blocking and power handling capability, GTO thyristors based on SiC are under investigation for compact future pulsed power systems [13]. Reaching breakdown voltages of $19kV$ [14], the potential of SiC based devices appears very interesting. According to primary device simulations using the finite element code MEDICITM, the developed GTO-thyristors should be able to block voltages up to $6kV$. For the device realization, a n-type 4H-SiC wafer material is obtained from Cree ResearchTM, including a $PP - NP+$ (from the wafer up to the top) epitaxial layer structure (6). Anode P+ type

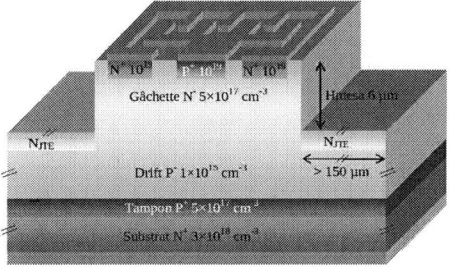

Fig. 6. Thyristor structure with planar electrodes for anode and gate and combined periphery protection: MESA and JTE

and gate N+ type layers were obtained by Aluminium (Al) and Nitrogen (N) ion implantation doping followed by a high temperature annealing in the $1700\,^\circ C/30min$ range in order to activate the dopants. Mesa structures were fabricated by Reactive Ion Etching with Ni masks with a particular attention to avoid micromasking formation and preserving the initial roughness of the surface. Ohmic contacts are formed by deposing layers composed on $Al/Ni/Ti$ alloys followed by a RTA annealing at $1000\,^\circ C$ during 1 to 2 min. Measurements have been performed on chip, in oil and as shown in Figure 7 a breakdown voltage of nearly $5kV$ has been reached.

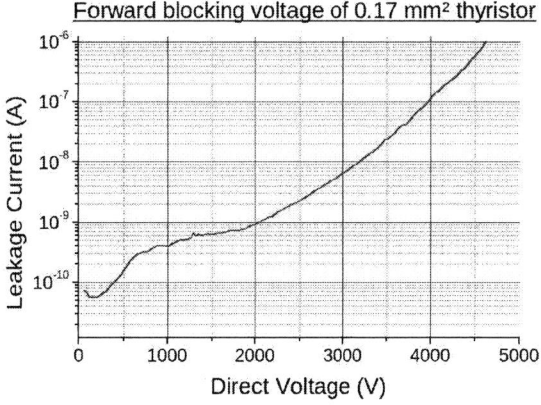

Fig. 7. Forward blocking characteristics of the device.

III. SPECIFIC APPLICATIONS USING SiC DEVICES

The ability for SiC devices to operate in the high temperature range, 250 ° C up to 600 ° C allows applications not possible with conventional silicon devices, especially for mains-operated systems.

A. Current limiting devices

To protect power electronic circuits against over-current or over-voltage either serial protection (SP) or a parallel protection (PP) can be used. For the serial protection (8, usually used components are Serial Protection Device (SPD) or Current Limiting Device (CLD), being either fuses, mechanical contactors, superconductors or polymers.

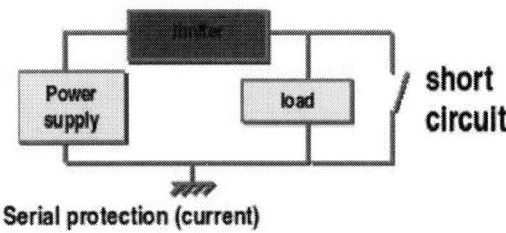

Fig. 8. Circuit using the protecting device

In all cases it is necessary to dissipate the overload energy through the protection device which has to sustain high voltage and high current simultaneously. These constraints imply a very fast temperature increase. So, when the protection device is active it has to sustain high junction temperature. All these constraints will globally define the electrical specifications of this kind of device. Up to now, only few semiconductor current limiter structures have been described in the literature [17], [16]. Although Current Regulative Diode components already exist, their voltage and current capabilities ($V_{BR} = 100V$, $I_{MAX} = 10mA$), do not allow to use them in power systems. A promising application of SiC-based devices is current limitation for power system protection, which benefits from its high thermal conductivity and wide band gap. In the steady state operating mode (or passive state), the voltage drop across the component must be as low as possible. In the active state, (limiting phase), the current limiter must sustain a high current, under high voltage bias. The resulting high power density must not cause the component failure. Two specific current limiting devices (CLD) have been studied [18], [20], [19] taking into account previous considerations.

Figure 9 shows a cross section schema of the VJFET and the main parameters to adjust. This device has a channel divided in two parts: a vertical one and a lateral one. The source is grounded and current flows from drain to source. P-buried layers are designed so that the VJFET is normally-on and presents a low specific resistance. When the drain voltage rises, the current saturates at a voltage corresponding to the pinch-off of both vertical and horizontal channels. In the saturation mode, the device

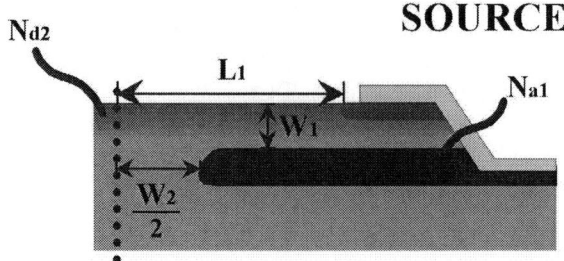

Fig. 9. VJFET parameters to be optimized.

presents an important on resistance (R_{ON}), resulting from the serial resistance of both parts of the channel and the drift resistance of the epitaxial layer. Due to self heating, current decreases as the voltage increases (since electron mobility decreases and induces a current reduction). This effect is amplified while increasing the limiting current density. When a negative bias is applied between gate and source, the PN junction formed by the Pwell and the epitaxial layer is reverse biased leading to current modulation. The device optimization job consists in the design of buried layer (the gate electrode of the devices, which is formed by either high energy implantation or epitaxial process), and the tuning of horizontal channel parameters. A trade off between specific resistance and blocking capabilities was investigated by means of analytical model and finite element simulations.

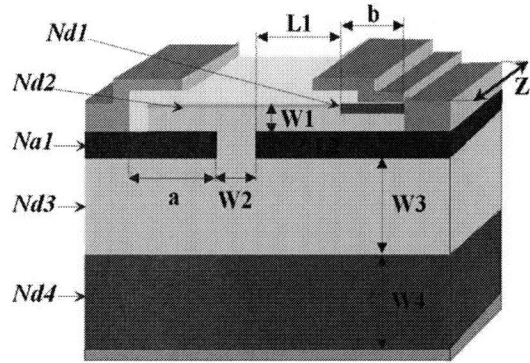

Fig. 10. JFET structure with its parameters.

Figure 10 presents the vertical JFET structure, 3 terminals device, with all parameters to be optimised by simulation with ISE TCAD. An AccuMOSFET, 2 terminals device, has also been designed and fabricated.

Fabrication process (7 level masks) complexity is similar to silicon power device one. The samples were manufactured at Centro Nacional de Microelectronica, in Barcelona. Key points of the device fabrication are high energy implantation, annealing activation and ohmic contact formation. Based on CNM- SiC technology, specific steps have been optimized: horizontal channel definition (implanted doses value as well as geometrical parameters) for both the VJFET and the AccuMOSFET. An additional step has been tuned for the AccuMOSFET: the MOSFET channel oxide layer. First batch of small area devices has been fabricated. Figure 11 presents the

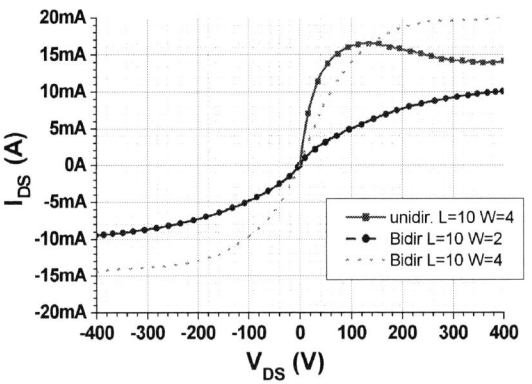

Fig. 11. Electrical measured characteristics for unidirectional and bi-directional limiter.

measured electrical characteristics for an unidirectional VJFET. Measurements were done by applying $0.5s$ pulse ($f = 2Hz$), with drain biased up to 400 V. The specific on-resistance varies from $176m\Omega.cm^2$ up to $237m\Omega.cm^2$. The maximum pulsed power density dissipated by the CLD in the limiting mode is $160kW/cm^2$. Limiting capabilities have also been measured for a bi-directional device made of two unidirectional devices connected head to tail (i.e. drain connected). This one exhibits a specific on-resistance of $700m\Omega.cm^2$. Highest breakdown in current limiting state were measured to be $810V$, corresponding to a high pulsed power density of $140kW/cm^2$. As it can be observed from Figure 12, the current limiting device is able to react in a very short time (less than $1\mu s$).

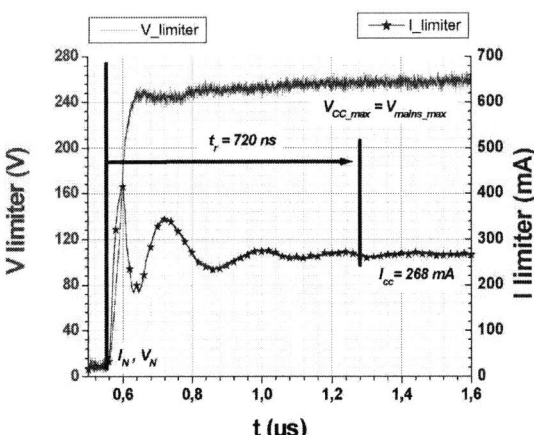

Fig. 12. Switching waveforms of the current limiter with gate control.

Based on small area devices optimization work, larger area AccuMOSFET ($S = 0.3cm^2$) have been fabricated.

Static characterizations have been performed using a specific pulsed power measurement setup. Depending on the layout of the device (geometrical variation), saturation current in the range of $300A$ to $450A$ has been measured (Figure 13), with a pinch-off voltage around $V_P = 12V$.

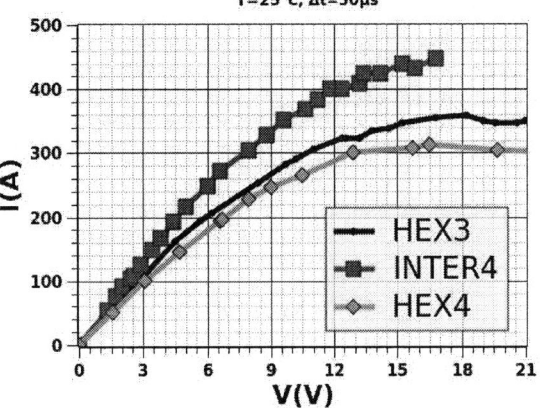

Fig. 13. Electrical measured characteristics for unidirectional high current liming AccuMOSFET device.

Self heating effect for short time pulses ($t_P = 100\mu s$) has been estimated by an analytical model. Mobility, i.e. current, reduction of 9% for a $100\mu s$ pulse gives an internal temperature increase estimation of $10\,^\circ C$. Based on experimental measurements, analytical modeling and electro-thermal finite element simulations, the short circuit energy capability of an AccuMOSFET CLD has been estimated to be $ECC_{max} = 35J$. We report on both simulation work, fabrication process and experimental results of a unidirectional and bi-directional current limiter structure based on a VJFET and ACCUMOSFET, with improved current capabilities (specific resistance as low as $40m\Omega$, and a saturation current between $200A < I_{SAT} < 450A$). The design was made using the ISE software with respect to the technological limitations. The first demonstrator exhibits optimistic dynamics characteristics. Bi-directional components were measured by associating two devices head to tail. Short circuit demonstration was done using $5W - 240V$ bulbs as a load on low current rating devices. Very low response time to short circuit has been measured (as low as $t_R = 1\mu s$). Using such a component with increased current ratings should permit a complexity reduction of classical mechanical device. Next challenges consist in both current sensing and current modulation in limiting state using the gate electrode so that to reduce short circuit power losses.

B. Monolithic converters

Hybrid approaches suffer from interconnection reliability. Integration appears to be a more effective approach providing capable power devices and passive components. Due to their physical properties, semiconductors with large band-gap are very interesting in this case. The integration allows to reduce considerably the switching time and to reduce losses and parasitics related to the device interconnections. This integrated system based on the design of SiC lateral JFET with a breakdown voltage of 900V is presented in Figure 14.

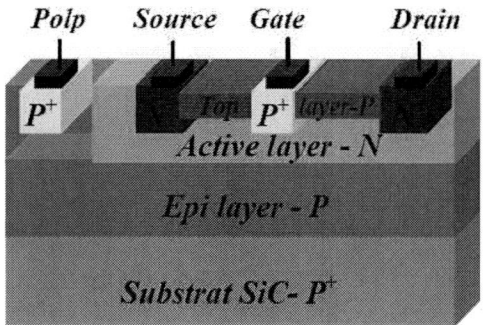

Fig. 14. Schematic presentation of the SiC lateral JFET with a double RESURF structure.

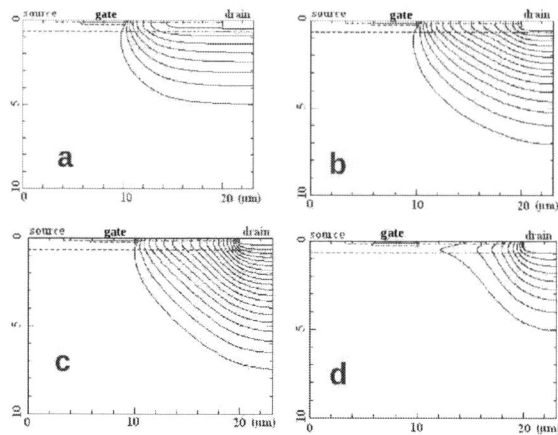

Fig. 15. Potential distribution for the lateral JFET with a doping concentration of the Ptop layer : $5.10^{16}cm^3(a), 1.10^{17}cm^3(b), 2.10^{17}cm^3(c) and 5.10^{17}cm^3(d)$

In order to improve the blocking voltage and to reduce the Ron-resistance, the RESURF technique [15] is applied in this work to design SiC lateral JFETs. The P^+type region (gate), N^+type region (source and drain), N type active layer and $P - type$ top layer are realized by ion implantation. The epitaxial layer at the top of the substrate is defined by a $P - type$ region with a uniform doping concentration. A double-junction is created in the top and the bottom of the active $n - type$ layer, improving its lateral depletion and allowing increasing of its doping. The $n - type$ active layer constituting the JFET channel, this technique allows to obtain a JFET with higher channel doping and so a lower on-resistance by keeping a high V_{BR}. A relationship between the doping concentration and the thickness of layers must be respected to vertically deplete both the $p - type$ top layer and $n - type$ channel layer before the lateral junction breakdown. The structure of these devices was studied with $MEDICI^{TM}$. In each area of the device structure, different parameters are defined such as profile and concentration of doping impurities which were estimated by Monte-Carlo simulations taking into account the 4H-SiC crystalline structure. The lateral JFET is a normally-on device. A negative gate voltage is required to turn the device off. The channel doping is $2.10^{17}cm^{-3}$ and its thickness is $0.40\mu m$ which is smaller than the thickness of the depleted region produced by the applied drain-to-source voltage. The thickness of the $p - type$ top layer is $0.12\mu m$ The potential line distributions in the double RESURF SiC JFET in forward blocking mode are shown in Figure 15 for different doping concentrations of the top layer varying from $5.10^{16}cm^{-3}$ to $5.10^{17}cm^{-3}$.

The V_{BR} variation is also presented in Figure 16. The potential lines are distributed mainly in the p-type epilayer due to the low doping concentration of this layer. The presence of the top layer with a low doping concentration ($5.10^{16}cm^{-3}$ -Figure 15-a) generates a potential line distribution transversally at the device surface. In this

case, the value of blocking voltage is equal to $385V$. In the same way, a high doping concentration of the top layer with a high doping concentration ($5.10^{17}cm^3$ - Figure 15-d) makes difficult to deplete this layer. The maximum value of electric field is obtained at the drain side. In this case, the potential lines are concentrated that involves the device to reach prematurely the blocking voltage.

The potential distribution is optimum, when the doping concentration of the top layer ranges between $10^{17}cm^{-3}$ and $2.10^{17}cm^{-3}$, thus obtaining a trapezoidal electric field profile between the drain and the gate. In this case, the Space Charge Region (SCR) surface is wider when he VBR of the JFET is increased up to $1kV$, as presented in Figure 16.

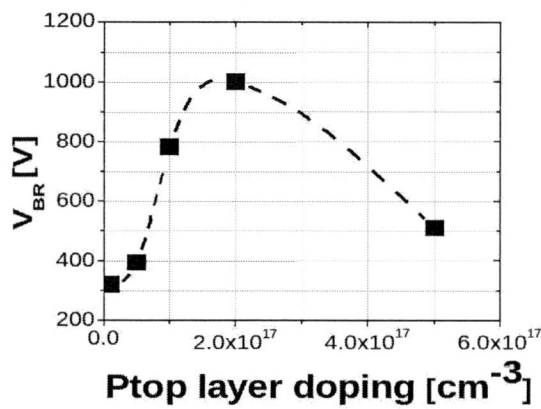

Fig. 16. V_{BR} versus top layer doping for the JFET structure.

The lateral JFETs run is performed (Figure 17). The first results show that in blocking state, the RESURF structures are validated in terms of voltage rating. Breakdown voltage values as high as $600V$ were obtained which is conform to the specific requirements for which these

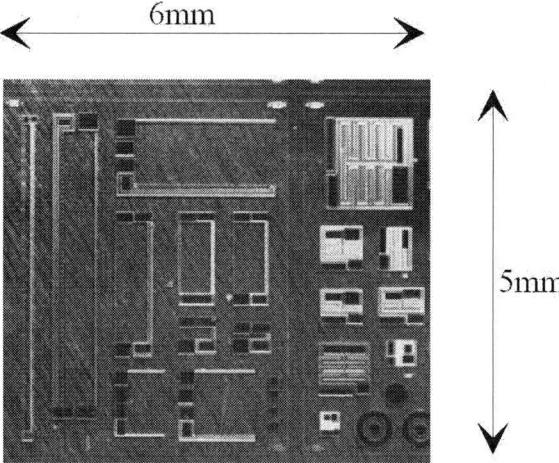

6mm

5mm

Fig. 17. Top view of the SiC lateral JFETs with different inter-digitated structures and different sizes.

devices have been designed.

IV. CONCLUSION

This paper shows the already existing devices and applications using SiC devices. Higher theoretical values could be reached but the environment of the devices remains a critical issue. Dealing with Current Limiting Devices, although we demonstrate they are able to sustain high current densities, the packaging and interconnections are limiting their performances. It seems to be clear that for high voltage devices, the periphery protection must now combine several classical techniques, and must take into account the passivation features (dielectric permittivity, thicknesses and interface states density). There is still works to be done to reach the real performances of SiC semiconductor material, for the fabrication of high voltage, high temperature and integrated power systems.

ACKNOWLEDGMENT

The authors would like to thank all the financial support from different institutions and companies, namely DGA, CNRS, ISL, Schneider-Electric, ST micro-electronics, Alstom and Hispano-Suiza. Authors also gives special thanks to Philippe Godigon of the "Centro Nacional de Microelectronica" of Barcelona for clean-room SiC-process support.

REFERENCES

[1] D. Stephani, Status, prospects and commercialization of SiC power devices. *Device Research Conference*, in proceedings, p14, 2001.
[2] http://www.cree.com
[3] S. Round, M. Heldwein, J. Kolar I. Hofsajer P. Friedrichs, A SiC JFET Driver for a 5 kW, 150 kHz Three-Phase PWM Converter, *IAS Conference*, p 410-416, 2005.
[4] http://www.siced.de
[5] B.J. Baliga, "Modern Power Devices", New York : Wiley, 1987
[6] Technology Modeling Associates, Inc. MEDICI: Two dimensionnal semiconductor device simulation, Version 2005.2.0, Vol. 1 et Vol. 2, 2005
[7]]A. O. Konstantinov, Q. Wahab, N. Nordell, U. Lindefelt. Ionization Rates and Critical Fields in 4H-SiC Junction devices, *Appl. Phys. Lett. 71, 1, 1997, p. 90-92*

[8] M. Lazar, G. Cardinali, C. Raynaud, A. Poggi, D. Planson, R. Nipoti, J.P. Chante, The role of the ion implanted emitter state on 6H-SiC power diodes behavior. A statistical study, 2004 - *Materials Science Forum, vol. 457-460, p. 1025-1028*
[9] R. Nipoti, F. Moscatelli, A. Scorzoni, A. Poggi, G. C. Cardinali, M. Lazar, C. Raynaud, D. Planson, M-L. Locatelli, J-P. Chante, Contact Resistivity of Al/Ti Ohmic Contacts on p-Type Ion Implanted 4H- and 6H-SiC, *MRS Proceedings vol. 742, 2002 Fall, p. K6.2*
[10] J. Crofton, S. E. Mohney, J.R. Williams, T. Isaacs-Smith, Finding the Optimum Al-Ti Alloy Composition for Use as an Ohmic Contact to P-Type SiC, 2002, *Solid State Electronics, vol. 46, No. 1, pp 109-113,*
[11] C. Raynaud, M. Lazar, D. Planson, J.P. Chante, Z. Sassi Design, fabrication and characterization of 5 kV 4H-SiC p+n planar bipolar diodes protected by junction termination extension *2004 Materials Science Forum, vol. 457-460, p. 1033-1036.*
[12] M.L. Locatelli, K. Isoird, S. Dinculescu, V. Bley, T. Lebey, D. Planson, E. Dutarde, M. Mermet-Guyennet, "Study of suitable dielectric material properties for high electric field and high temperature power semiconductor environment", *EPE'03 conference, 2-4 Sept 2003,* Toulouse, France.
[13] S. Scharnholz, V. Zorngiebel, P. Brosselard, and E. Spahn; *in IEE Proc. of the 1st European Pulsed Power Symposium (EPPS),* 22-24 October 2002, Saint Louis, France; 2002; p. 14/1-14/6.
[14] Y. Sugawara, D. Takayama, K. Asano, R. Singh, J. Palmour, T. Hayashi 1219 kV 4H-SiC pin diodes with low power loss, *Proceedings of 2001 International Symposium on Power Semiconductor Devices & Ics,* Osaka, p. 27-30
[15] Adriaan W. Ludikhuize "A Review of RESURF Technology" *ISPSD'2000* Toulouse. France, p. 11-18
[16] Siemens Corp. Current limiter circuit, *Patent n W09727657, 30/07/97,*
[17] J.L. Sanchez et al., Design and fabrication of a new high voltage current limiting device for serial protection applications, *Proc. 8th International Symposium on Power Semiconductor Devices and ICs, ISPSD96, Hawai,* USA, 1996, pp. 201-205.
[18] D. Tournier et. al. Current limitation with SiC devices, *in proceedings of 2002 EPE conference,* Toulouse 2002.
[19] Nallet, F. et. al. ,Very low R/sub ON/ measured on 4H-SiC accu-MOSFET high power device, *Power Semiconductor Devices and ICs, 2002. Proceedings of the 14th International Symposium on,* 2002, pp. 209- 212
[20] D. Tournier et. al., Optimal layout for 6HSiC VJFET controlled current limiting device, *Diamond and Related Materials Volume 12, Issues 3-7, March-July 2003, Pages 1220-1223*

Silicon carbide Schottky diodes and MOSFETs: solutions to performance problems

Owen J. Guy*, Michal Lodzinski*, Ambroise Castaing*, P. M. Igic*, Amador Perez-Tomas#, Michael R. Jennings†, Philip A. Mawby†

*School of Engineering, Swansea University, Singleton Park, Swansea, UK, SA2 8PP e-mail: *o.j.guy@swansea.ac.uk*
#Centro Nacional de Microelectrónica, Campus Universidad Autónoma de Barcelona, 08193 Bellaterra, Barcelona, SPAIN.
† School of Engineering, University of Warwick, Coventry CV4 7AL, UK

Abstract— **Silicon carbide has long been hailed as the successor to silicon in many power electronics applications. Its superior electrical and thermal properties have delivered devices that operate at higher voltages, higher temperatures and with lower on-resistances than silicon devices. However, SiC Schottky diodes are still the only devices commercially available today. Though SiC Schottkys are now being used with silicon IGBTs in 'hybrid' inverter modules, the real advantages will be seen when silicon switching devices can be replaced by SiC. This paper describes the current state of SiC diode and MOSFET technology, discussing possible solutions to making these devices commercially viable.**

Keywords— **Silicon Carbide, Power semiconductor device, New switching devices, MOSFET, High temperature electronics.**

I. INTRODUCTION

Developments in silicon carbide growth technology have led to the release of 100 mm diameter wafers with zero micropipes (ZMP®) and significant reductions in the densities of a number of other defects and dislocations. In addition to the progress in crystal quality and wafer diameter, made by Cree [1], technology for increasing growth rates [2] using chlorine gas chemistry should also contribute to lower wafer prices. Currently, cost is still high but there are signs from manufacturing giants such as Toyota that the improved performance of SiC over silicon devices could eventually make SiC a financially viable product [3]. Despite these encouraging noises, commercially available SiC devices have been restricted to Schottky diodes, from Cree, Infineon and more recently Bipolar Junction Transistor for power applications from the new Swedish manufacturer, TRANSIC [4].

The other area where SiC has a clear advantage over silicon is in High Temperature Electronics (HTE). Wide bandgap semiconductors, such as SiC, will be critical in enabling applications above 300°C. Few components specified for operation in ambient above 125°C are currently available, though several silicon and SOI high temperature ICs and sensors are on the market. The world market for high temperature electronics (HTE) has been forecast to reach $376.8 million by 2003 and is expected to grow to $887.1 million by 2008 [5]. The excellent thermal conductivity and high temperature capability of SiC devices suggest that SiC is a realistic option for HTE.

This paper discusses the current state of the art of SiC Schottky diodes and MOSFETs highlighting the technical problems that have held back commercialization of these devices. Some of the innovative solutions to these issues are discuss in terms of improved device performance.

II. SCHOTTKY BARRIER DIODES

Silicon carbide Schottky Barrier Diodes (SBDs), rated up to 1.2 kV [1, 7, 8] are already commercially available rated at up to 20 A. SiC One of the main concerns regarding SiC Schottky diodes has been in obtaining larger area diodes and good device yields. This has been influenced largely by the quality and defect density of SiC material. Lower micropipe densities have enabled devices with areas greater than 8.2 mm by 4 mm in size, with current ratings up to 50 A now possible [6].

Schottky diodes, supplied by Cree, with a 20 A, 1200V rating have reverse leakage currents of around 200 μA at $V_R = 1200V$, at room temperature. Infineon 9 A, 600V SiC Schottky diodes have leakage currents of around 100 μA at $V_R = 600V$ [7]. These leakage currents are slightly higher than those reported for Infineon's 9 A, 600V and 1200V silicon diodes ($I_R = 50$ A at $V_R = 600V$) and ($I_R = 100$ A at $V_R = 1200V$), respectively [8]. However, the SiC and silicon diodes exhibit very similar leakage currents at higher junction temperatures (silicon: 750 A where $Tj = 150°C$; SiC:) both at $V_R = 600V$.

Leakage currents of Schottky diodes have tended to be higher than desired, especially at higher voltages, but better edge termination and the use of merged PiN Schottky (MPS) or Junction Barrier Schottky (JBS) diodes have improved these characteristics.

Higher voltage devices are currently in development. These include 3.3 kV Schottky barrier diodes from the "Establish SiC for Applications in Power Electronics in Europe ESCAPEE" project [9]. The highest blocking rating obtained for ESCAPEE diodes was 4.7 kV (Fig. 1).

The advantages of Schottky barrier diodes are their relatively simple structure, combined with low on-state resistances, compared to Si PiN diodes

1) Switching: Another major advantage of SiC Schottky diodes is their faster switching speed and almost zero reverse recovery. The superior switching characteristics of SiC Schottky diodes mean they can switch around 5 times faster than an equivalently rated Si PiN diode, and have virtually no reverse recovery [10] (Fig. 2).

978-1-4244-1741-4/08/$25.00 ©2008 IEEE

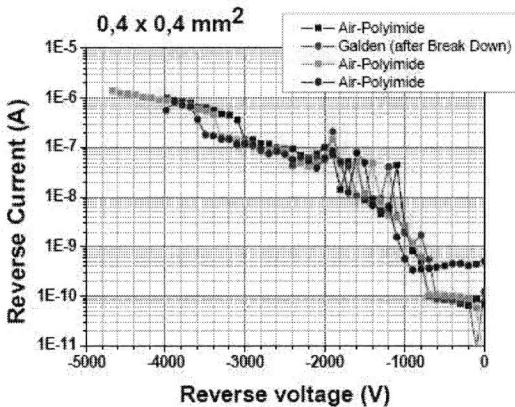

Fig. 1 Reverse leakage current of 0.4 mm x 0.4 mm 4 kV rated area Schottky barrier diode.

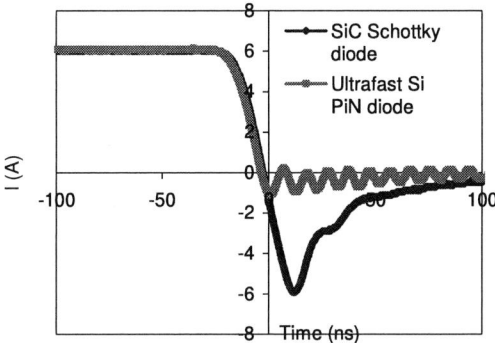

Fig. 2 Switching behaviour of SiC Schottky diode compared to an ultra-fast silicon PiN diode of similar blocking rating.

2) On-state performance: Schottky barrier diodes (SBDs) are often assumed to conform to thermionic emission theory with current density / voltage (J(V)) characteristics described by (1) and (2).

$$J_T = J_0 \left[\exp\left(q \frac{V - JR_{on}}{\eta kT} \right) - 1 \right] \quad (1)$$

where

$$J_0 = A^{**}T^2 \exp\left(\frac{-q(\phi_{bn} - \Delta\phi)}{kT} \right) \quad (2)$$

and V is the applied voltage, R_{on} is the specific on resistance, J_0 is the saturation current density, η the ideality factor, A^{**} is the effective Richardson's constant for 4H-SiC (146 Acm^{-2} K^{-2}), k is Boltzmann's constant, T the temperature in Kelvin and ϕ_{bn} is the barrier height and $\Delta\phi$ is the parameter due to image-force barrier-lowering at the Schottky interface. The term $\phi_{bn} - \Delta\phi$ in (2) can be denoted as ϕ_b, the effective barrier height. The image-

force lowering of the Schottky barrier is actually very small under forward bias (around 0.01 eV) and can usually be neglected.

Provided the forward bias is not large enough for the series resistance of the device to be significant, (1) can be written as (3).

$$J_T = J_0 \left[\exp\left(\frac{qV}{\eta kT} \right) - 1 \right] \quad (3)$$

The ideality factor η can then be extracted from the slope of the linear portion of a ln J against V plot, and, provided that this value is not too far removed from unity ($\eta = 1$), ϕ_b can be obtained from the intercept of the same ln J against V plot.

3) Ideality factor: The ideality factor of good quality Schottky diodes is generally between 1.0 and 1.1. Generally, metals which form a higher Schottky barrier to SiC, such as nickel, yield more ideal behaviour. The barrier height ϕ_b of the Schottky diode is strongly influenced by the work function of the metal. A higher work function metal like nickel yields a higher values of ϕ_b (around 1.5 eV) than a lower work function metal like titanium ($\phi_b \approx 1.0$ eV).

Annealing of the metal / SiC contact in the temperature range of 300 to 500ºC can improve the ideality of the contact but usually results in a slight increase (approximately up to 0.2 eV) in ϕ_b.

This annealing may have several effects including the disruption or deterioration of a thin interfacial oxide layer between the metal and the semiconductor, which is reported start to occur rapidly around this temperature [11]. The 500°C Schottky anneal may also stabilize the metal Schottky interface – creating a more stable and reliable contact.

Annealing at temperatures greater than 500ºC can result in a degradation of the contacts rectifying properties. Larger barrier heights give improved blocking characteristics – including lower levels of leakage current. However, the larger barrier height also increases the threshold or turn-on voltage of the diode, resulting a larger forward voltage drop V_f.

4) On-resistance: The other important parameter, in terms of the Schottky diode performance in forward bias mode, is the specific on resistance $R_{on,sp}$. This can be extracted from the inverse of the slope of the linear portion of the J against V plots (4).

$$R_{on,sp} = \frac{1}{dJ / dV} \quad (4)$$

The specific on-resistance $R_{on,sp}$ of Schottky diodes (is mainly determined by the resistance of the SiC material epitaxial layer R_{epi} (10).

$$R_{epi} = \frac{W_D}{q\mu_n N_D} = \frac{4V_B^2}{\varepsilon_s E_c^3 \mu_n} \quad (5)$$

2465

where W_D is the drift region width, q is the electron charge, μ_n is the electron mobility, N_D is the donor impurity concentration, V_B is the breakdown voltage, ε_s is the permittivity of SiC, and E_c is the critical electric field. To sustain the high breakdown voltages required for the high voltage diodes, reduced doping concentrations and increased depletion widths are required, leading to an increase in the serial resistance R_{on}. There are additional resistances due to the SiC substrate and ohmic contact resistances, but these are much smaller relative to the resistance of the epi-layer.

5) Heterojunction (HJ) diodes: One innovative method for producing lower on-resistance diodes is to use heterojunction pn structures.

By replacing existing all Si solutions with their SiC counterparts, designers now have the ability to extract more power output from the drive for the same given size packaging. Monolithic integration of Si and SiC devices should be considered as an attractive solution in itself for devices or for a broad range of applications including smart power integrated circuits with the control part implemented in silicon.

Ultra low on-resistances have been reported for this type of device [12].

6) High temperature performance; High temperature operation of devices is key in the switch from silicon to SiC technology, as the size and weight of packaging for SiC devices can be significantly reduced at increased operating temperatures [13]. The target operating temperature of the ESCAPEE diodes was 225°C, some 50°C higher than that recommended for other commercial SiC diodes. The reliable performance of diodes up to this temperature is, therefore, critical to the production of commercial quality devices.

Temperature has an important effect on the forward characteristics of Schottky diodes [14, 15, 16, 17, 18, 19]. Several factors affect the determination of ϕb and η and their temperature-dependent behaviour. In general ϕb is assumed to decrease with temperature, because the expansion of the lattice causes a decrease in the band-gap energy Eg [20], according to (6) [21]. Though there is little literature information available on the parameters for 4H-SiC, Eg has been calculated to vary from 3.25eV at 273K to 3.15eV at 573K using EgT0 = 3.265eV, α = 3.3e-4 eV K-1 and β = 0 K; a decrease of 0.1eV.

$$E_g(T) = E_g^{T_0} - \alpha^{E_g} \frac{T^2}{\beta^{E_g} + T} \qquad (6)$$

However, ϕ_b for 4H-SiC has been found to be strongly dependent on the metal work function, and has been reported by Itoh [23] to vary according to (7).

$$\phi_b = 0.7\phi_M - 1.95 \qquad (7)$$

In practice, the assumptions that ϕ_M and χ_{SiC} do not change when the metal and semiconductor are brought into contact are not entirely valid. Surface states, surface impurities, surface dipole layers, contributions from electronic interactions with the crystal and with other electrons, adsorbed surface atoms, and a non-uniform

Schottky interface, can all affect both χ_{SiC} and ϕ_M [20]. Adsorption of atoms on the metal surface can lead to a change in the dipole layer and hence ϕ_M. Surface states can affect the charges at the interface and can cause an increase in the band bending and consequently ϕ_M. Many of these contributing factors may be temperature-dependent and influence changes in ϕ_b.

Several of the parameters used to extract the barrier height from I(V) curves, may also be dependent on temperature. These include the ideality factor η and the Richardson constant A^{**}, which may vary as a function of the temperature dependence of electron energy and the maximum electric field. Errors in η and A^{**} will consequently affect the precise determination of ϕb. Roccaforte et al [22] extracted values for A^{**} of 256 ± 51 and 196 ± 60 A/cm^2 K^2 for Ti/SiC and Ni$_2$Si diodes respectively. They also argued that A** is very sensitive to the condition of the interface, in spite of the conventional predictions of a constant value, independent of the metal.

We have identified a linear relationship between ϕ_b and η, for the Ni and Ti Schottky diodes up to 200°C. The increase in ϕ_b was accompanied by a corresponding decrease in η when the temperature was raised, strongly suggesting that the two parameters are not independent of one another. A similar, approximately linear correlation between ϕ_b and η, has been reported for Si and GaAs Schottky contacts [23]. This finding was attributed to inhomogeneous interfaces, resulting in non-uniform Schottky contacts [24]. The relationship between ϕb and η, for both Ni and Ti Schottky contacts observed in the present study suggests that the theory of non-uniform Schottky interfaces also applies to 4H-SiC, and that inhomogeneities play an important role in experimental characteristics. The presence of an inhomogeneous contact can be associated with current flowing via two pathways; over a lower barrier and over a higher barrier [25]. Fig. 3 compares the ideality factor and barrier height of Ni diodes as a function of temperature. These results are similar to those reported by Roccaforte *et al* [22], who also obsereved a temperature dependent relationship between the ideality factor and barrier height of Ni /SiC Schottky diodes despite their nearly ideal behaviour.

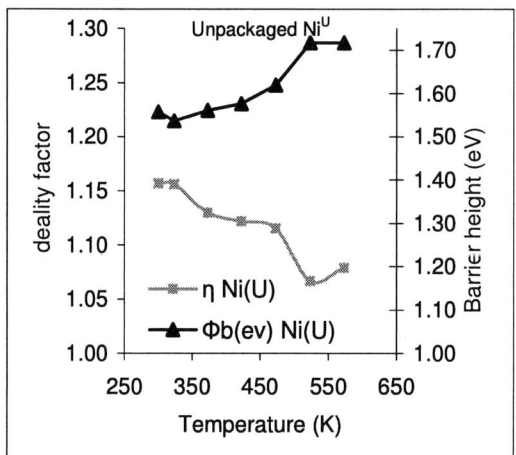

Fig. 3 Schottky barrier diode ideality factor η and barrier height ϕ_b as a function of temperature for Ni diodes.

Temperature also has a significant influence on the on-resistance of SiC Schottky diodes. Both Ti and Ni and Ni diodes, were observed to have temperature dependencies of around $T^{2.4}$ for $R_{on,sp}$ - attributed to decreases in electron mobility in SiC with increasing temperatures [26]. As the temperature is raised, most of the impurities are ionized and the effect of phonon scattering becomes significant. The electron mobility decreases and consequently the resistance increases. The temperature dependence of the electron mobility in SiC is reportedly similar to the $T^{-2.4}$ dependence of Si [27]. In a study on SiC and GaAs Schottky diodes by Luo *et al*, the increase in $R_{on,sp}$ was assigned a $T^{1.89}$ dependence for 6H-SiC [27]. Kimoto, Urushidani *et al.* have reported temperature dependences of T^2 for 6H-SiC and $T^{2.1}$ for 4H-SiC Schottky diodes [27], while Chilukuri and Baliga reported a $T^{2.44}$ relationship for 4H-SiC diodes [28].

7) Reverse bias: The reverse leakage current J_r, of all Schottky diodes, increases significantly at higher temperatures due to the increased number of carriers. The experimental leakage current is much higher than predicted by simulations performed in this work, and has a linear dependence on the reverse voltage (Fig. 4). In the simulation model, the generation current is proportional to the depletion region width, and thus shows a square root dependence on the reverse voltage V_r according to (11).

$$W = \sqrt{\frac{2\varepsilon V_r}{q N_D}}$$
(8)

The high levels of leakage current are suggested to be due to additional leakage mechanisms such as defect related conduction [29, 30]. Higher leakage currents, relative to simulation data, have also been reported by Park et al [31], who attribute the variation from $V^{1/2}$ dependence, to the effects of trap assisted tunneling. Traps are active generation and recombination centres and yield higher leakage currents as the electric field increases. Strong electric fields increase the free carrier density in the depletion region, increasing the probability of carriers tunneling to traps in the band-gap [32], in addition to Schockley-Read-Hall (SRH) thermal emission. Electric fields, caused by high voltages, can be particularly high around stressed regions, defects, dislocations, and at the curved regions at the perimeter of the device [32]. Increased intrinsic carrier concentrations at these regions can lead to increases in leakage currents and localized breakdown [32]. A simple leakage current model shows that the square root voltage dependence does not adequately describe the experimentally observed leakage current of the SiC diodes. It is suggested that additional tunneling mechanisms such as Trap Assisted Tunneling (TAT) might be responsible for the higher observed leakage currents.

We have observed that 1kV rated Schottky diodes perform reliably at temperatures of up to 200°C. However, at 300°C, though there is little effect detectable in the forward bias characteristics, high voltage reverse bias conditions can cause an irreversible degradation of the blocking performance of Schottky diodes. Specifically, this results from the appearance of a low Schottky barrier conduction pathway, which allows higher levels of reverse leakage current. This degradation occurs only on application of both high temperature (300°C) and high voltage.

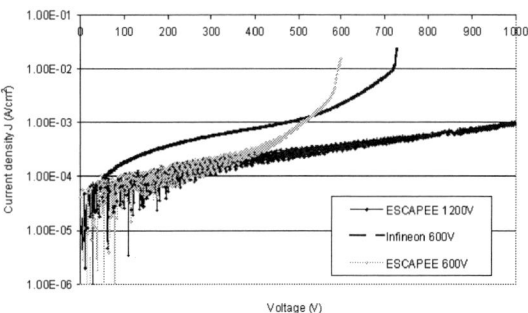

Fig. 4 Comparison of reverse blocking voltage characteristics for ESCAPEE 600 V and 1200 V Ni^U Schottky diodes (active area: 0.4 mm x 0.4 mm) and a commercial Infineon 600 V Schottky (active area: estimated 1.1 mm^2).

This indicates that 4H-SiC carbides could be capable of reliable operation at temperatures of up to 300°C, though further reliability testing under standardized conditions must be performed.

8) Schottky diode design: The reverse blocking voltage of any diode is limited primarily by the onset of avalanche breakdown, resulting from impact ionization. Impact ionization is mainly dependent on the peak electric field inside the device. In order to utilise the potential blocking capability of SiC, Schottky diodes must be designed in a way so that localised build-up of high electric fields is avoided.

The simplest vertical SiC Schottky diode consists of a metal contact on top of a low-doped SiC epilayer and Ohmic contact at the backside of the wafer. In practice, high electric fields tend to occur at the periphery of the Schottky metal contact (*E*-Field crowding at the edge of Schottky metal contact) – resulting in premature breakdown. This build-up can be dissipated by using edge termination techniques such field plates [33], mesa structures [34], Junction Termination Extension (JTE) [35], guard rings [36], and highly resistive surface regions formed by argon implants [37].

The effect of utilizing a JTE is shown in Fig. 5. The Schottky diode without edge termination shows a breakdown voltage of approximately 200 V, with leakage current density lower than 0.15 A/cm^2. The Schottky diode with JTE demonstrates a soft breakdown at around 450 V (Fig. 6).

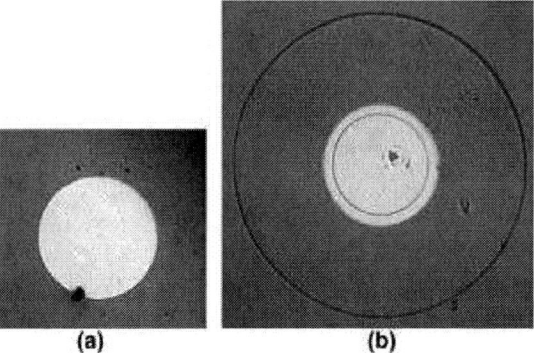

Fig. 5. Optical microscope images of the damaged devices (a) Schottky diode without JTE (b) Schottky diode with JTE.

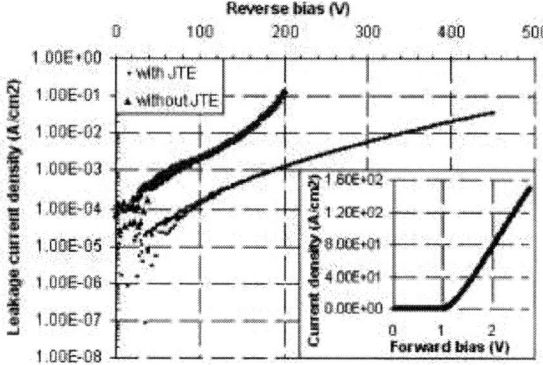

Fig. 6. Experimental *I–V* characteristics of Schottky diodes.

Post test, optical microscope examinations conducted on each damaged devices exhibited clear evidence of surface flashover failure near metal periphery for non-terminated devices. Evidence of destructive failure was seen in the JTE Schottky diodes was confined to the near central region of the metal contact.

The Schottky diode with JTE shows a breakdown voltage more than two times higher than that of the device without the JTE structure. Using more sophisticated device designs the blocking rating attainable can be almost the same as the theoretical breakdown capability.

Despite being able to achieve blocking voltages of up to around 5 kV, Schottky diodes do tend to high marginally higher reverse leakage currents than PiN diodes. This is due primarily to leakage over the Schottky barrier – especially at higher temperatures.

To counter this, alternative diode structures such as the Junction Barrier Schottky (JBS) or Merged PiN Schottky (MPS) diode have been investigated. They utilize adjacent pn junctions to prevent the electric field from rising at the Schottky junctions. SBDs as well as JBS diodes operating with a forward voltage lower than the turn-on voltage of the pn junctions exhibit faster switching than PiN diodes due to the lack of minority carrier injection.

The diode layout (the ratio between Schottky contact length, L_N, and bipolar contact length, L_P) has a great impact on the forward and turn-off losses. Thus, with a suitable layout, one can tailor the switching losses of a JBS diode, which is an important advantage for power electronics.

Thus JBS diodes have similar on-state compared to Schottky diodes, but superior reverse characteristics. JBS diodes have also been reported to be more reliable at higher ambient temperatures.

The switching behaviour of the JBS diode produces an increase of the reverse current peak and recovery time with temperature compared to the Schottky switching behaviour. This is due to the bipolar component of JBS. However, the switching losses of the SiC JBS remain much lower than that of a Si or a SiC pin diode, especially at high temperature.

The JBS structure, and edge termination structures like JTEs and guard rings are all achieved by ion implantation. Electrical activation of implants in SiC requires annealing at temperatures of above 1500°C. However, this has a side-effect of producing severe damage to the SiC

substrate. The surface roughness can be increased by up to ten times that of the original SiC surface (Fig. 7).

Schottky diodes fabricated on rough surfaces can display severely degraded rectifying performance (Fig. 8a). This surface damage can be minimised by using a protective carbon-cap on the surface of SiC samples, during annealing. The carbon-cap can then be removed after the high temperature anneal using an oxygen plasma etching process.

After removal of the carbon-cap, the surface roughness is reduced by up to an order of magnitude, compared to an unprotected sample annealed under the same conditions. Ideality factors, barrier heights and specific on state resistances of Ni Schottky diodes, fabricated on protected samples (after oxygen plasma etch removal of the capping layer), were found to be similar to those for Schottky diodes fabricated on samples that had not undergone any annealing (Fig. 8b). In contrast, the un-protected surface morphology is significantly altered by annealing at 1600°C, producing a lowered Schottky barrier, and non-rectifying behaviour (Fig. 8a). This work shows that the surface damage caused by high temperature annealing of 4H-SiC can be greatly reduced by the utilisation of a protective carbon-cap, and that Schottky diodes fabricated on a carbon-cap protected surface (after oxygen plasma etch removal of the capping layer), display similar electrical characteristics to diodes on un-annealed samples.

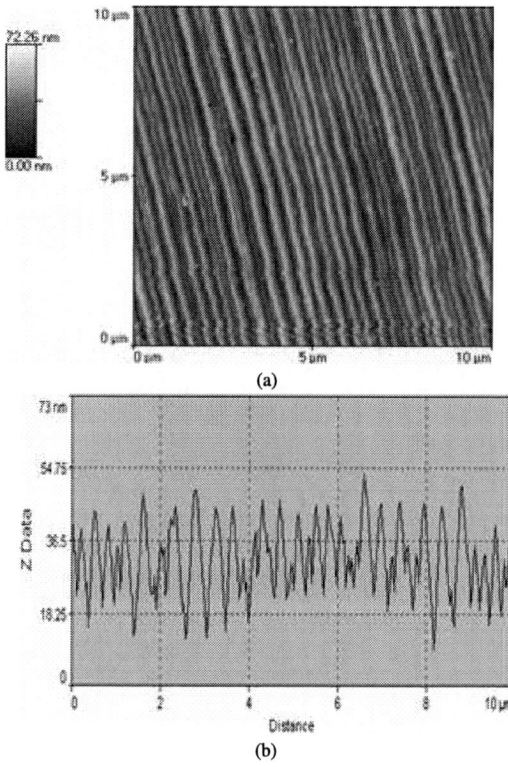

Fig. 7 (a) AFM micrograph and (b) line profile of sample annealed without any protective graphite cap.

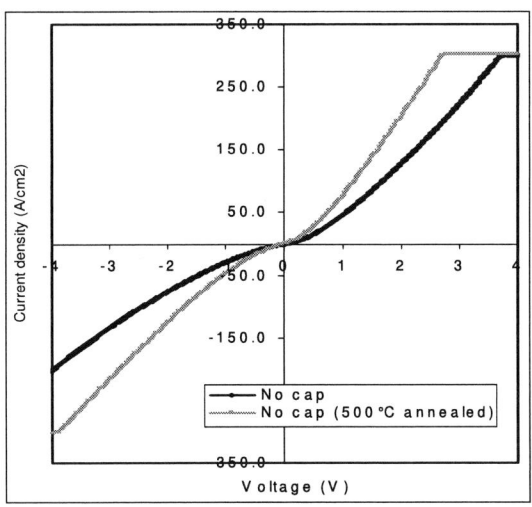

(a)

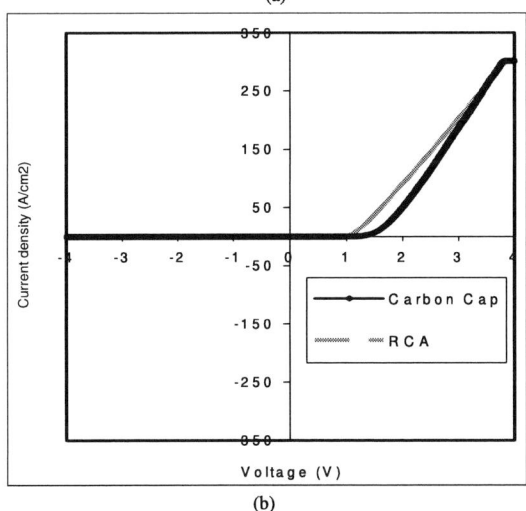

(b)

Fig. 8 I(V) curves for Schottky diodes fabricated on (a) un-annealed RCA cleaned samples and 1600°C annealed graphite capped samples and (b) uncapped samples annealed at 1600°C.

III. SiC MOSFETS

The carbon cap method of surface protection could also produce improvements in the quality of the semiconductor-oxide interface in MOS devices. MOS capacitors fabricated on protected and damaged surfaces (as illustrated in Fig. 9). Indeed, MOS capacitors fabricated on undamaged surfaces show lower interface state densities than MOS capacitors on damaged surfaces (Fig. 10).

9) MOSFET performance: SiC switching devices are still some way from being commercially available. The development of silicon carbide MOSFETs has also been problematic, primarily due to low channel mobilities, caused by high SiO_2/SiC interface state densities. Thermally grown oxidation of SiC often leads to a disordered interfacial region, containing silicon oxycarbides [38, 39, 40], and a high density of interface states that can trap electron or hole carriers, or act as recombination centres or scattering sites. This results in low MOS channel inversion layer mobilities and

consequently, higher on-resistances than those which could theoretically be achieved with lower interface state densities. High interface state densities (D_{it}) near the conduction band in SiC MOS devices are a primary inhibitor to the development of commercial SiC MOSFETs. These problems have prompted research into the oxidation process and the effects of oxidation temperature, post oxidation annealing (POA) and metal enhanced oxidation techniques. Some of these techniques yield improvements in channel mobilities from around 5 $cm^2 V^{-1} s^{-1}$ up to today's reported state of the art peak mobilities [41] of 50 $cm^2 V^{-1} s^{-1}$ to 150 $cm^2 V^{-1} s^{-1}$ [42]. However the value of mobility obtained by Olafsson *et al* has recently be reported to be attributed to the effect of sodium – which is highly undesirable in a MOS device. Even taking in to account recent improvements in interface trap densities, D_{it} values are still high and channel mobilities are still far below expectations.

Perhaps because of this, attention is now turning to alternative unipolar switching devices, like VJFETs and SITs. Indeed SiCED have developed a mature structure lateral-vertical JFET concept [43].

Despite the continued delay of SiC MOSFETs, there have been significant improvements in power MOSFET development. Cree, for example have produced MOSFETs in 4H-SiC with low specific on-resistance values (8 $m\Omega cm^2$) and superior switching performance [44]. AIST have also produced very low on-restistance MOSFETs using an epitaxial re-growth technique [45].

Research is now also concerned with the reliability of MOSFET gate oxides [46]

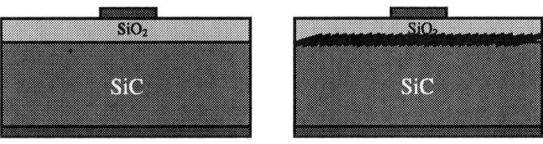

Fig. 9 Schematic of MOS capacitors fabricated on smooth and damaged surfaces (resulting from high temperature annealing)

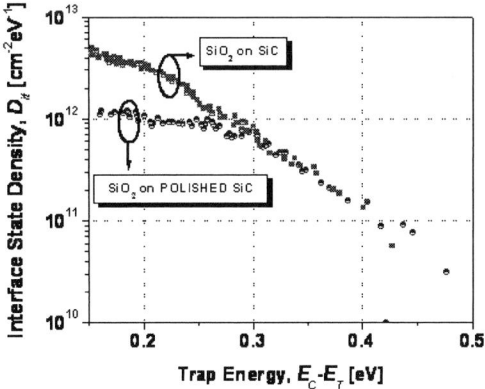

Fig. 10 Comparison of Dit values for MOS capacitors fabricated on smooth and damaged surfaces respectively.

2469

Though the major SiC device producers are very guarded with regard to their device processing techniques, there are several possible solutions to SiC MOSFET problems under investigation.

An alternative to direct thermal oxidation of SiC, is deposited silicon dioxide via chemical vapour deposition (CVD), sputtering or MBE. However, MOSFETs using CVD oxides deposited on SiC showed poor channel mobilities [47]. In contrast, very good results have been obtained using oxidation of a UHV deposited thin silicon film on to an atomically clean surface, with very low interface state densities reported [48].

A variation on this method is a partial oxidation of a silicon layer, deposited on a SiC substrate. In this structure, a thin silicon layer remains between the SiO_2 and SiC layers, so that the oxide interface is formed between silicon and SiO_2, rather than SiC and SiO_2. This should yield the advantage of much lower densities of interface states for the SiO_2/Si interface of silicon MOS structures, compared to interface state densities of conventional SiO_2/SiC structures. This partial silicon oxidation process could result in an improved oxide interface, with higher channel mobilities yielding a lower on-resistance MOSFET. This research is the focus of the UK INTRINSiC project [49]. The silicon layer can be fabricated using several techniques including; Chemical Vapour Deposition (CVD), Molecular Beam Epitaxy (MBE) [8], Electron Beam Evaporation under UHV conditions (EBE-UHV) and Layer Transfer (LT) of a thin silicon layer via wafer bonding. A major problem with silicon growth on SiC is the large lattice mismatch, which prevents growth of a high-quality single crystalline layer.

Another approach is to use a combination of a thin thermally grown oxide with a thicker TEOS oxide layer deposited on top. This has produced improved results when compared with thermally grown oxides [50].

Alternative dielectric materials may also provide solutions to MOSFET reliability issues. Both HfO_2 and Al_2O_3 have been used as a potential replacement for SiO_2 in the development of Metal Insulator Semiconductor or MISFET devices in SiC [51]. Hino et al have reported excellent performance from Al_2O_3 structure [52]. Stacked gate structures are also being investigated, an example being high-k Ta_2Si stacked on SiO_2 [53].

ACKNOWLEDGMENT

The authors acknowledge the support of the UK Department of Trade, under the Technology Programme scheme, project No: TP/3/OPT/6/I/17311.

REFERENCES

[1] Cree Inc., 4600 Silicon Drive, Durham, NC 27703, 919-313-5300, 919-313-5452 FAX. Available at http://www.cree.com/Products/pwr_index.asp

[2] H. Pedersen, S. Leone, A. Henry, V. Darakchieva, E. Janzén, 'Very high epitaxial growth rate of SiC using MTS as chloride-based precursor', Surface and Coatings Technology 201 (2007) 8931-8934.

[3] http://compoundsemiconductor.net/cws/article/news/31473

[4] http://www.transic.com/

[5] http://compoundsemiconductor.net/cws/article/magazine/11663

[6] http://compoundsemiconductor.net/cws/article/news/27036

[7] Infineon SiC data sheets available at: http://www.infineon.com/cms/en/product/channel.html?channel=ff80808112ab681d0112ab6a50b304a0

[8] Infineon silicon data sheets available at: http://www.infineon.com/cms/en/product/channel.html?channel=ff80808112ab681d0112ab6a527f04a6

[9] 'Establish Silicon Carbide Applications in Power Electronics in Europe (ESCAPEE)'. Available at www.siliconcarbide.org.

[10] Morisette, D.T., Cooper Jr., J.A., Melloch, M.R., Dolny, G.M., Shenoy, P.M., Zafrani, M., and Gladish, J.: 'Static and dynamic characterization of large-area high-current-density SiC Schottky diodes', IEEE Transactions on electron devices, 2001, 48, (2), pp. 349-352

[11] Luo, J., Chung, K.-J., Huang, H., and. Bernstein, J.B.: 'Temperature Dependence of Ron, sp in Silicon Carbide and GaAs Schottky' in Proc. of 40th annual IEEE International Reliability Physics Symposium, (New Jersey) 2002.

[12] Tanaka, H.; Hayashi, T.; Shimoida, Y.; Yamagami, S.; Tanimoto, S.; Hoshi, M.; Power Semiconductor Devices and ICs, 2005. Proceedings. ISPSD '05. The 17th International Symposium on 23-26 May 2005 pp. 287 - 290

[13] Ozpineci, B., Tolbert, L. M., Islam, S.K., Hasanuzzaman, M.: 'Effects of silicon carbide (SiC) power devices on HEV PWM inverter losses', in Proc. Industrial Electronics Society, 2001. IECON '01. The 27th Annual Conference of the IEEE , 2001, 2, (29), pp. 1061-1066

[14] Lee, S.-K., Zetterling, C.-M., and Ostling, M.: 'Schottky diode formation and characterization of titanium tungsten to n- and p-type silicon carbide', J. Appl. Phys., 2000, 87 (11), pp. 8039-8044

[15] Itoh A., and Matsunami, H.: Phys. Stat. Sol. 162, (1997) pp. 389

[16] Treu, M., et al., Mat. Sci. Forum 353-356 (2001) pp. 679-682.

[17] Hatakeyama, T., and T., Takashi, T. Shinohe, Materials Science Forum vol. 389-393, p.1169 (2002) pp. 1169-1172.

[18] Hatakeyama, T., Kushibe, M., Watanabe, T., Imai, S., Shinohe, T.: "Optimum design of a SiC Schottky barrier diode considering reverse leakage due to a tunneling process" Mater. Sci. Forum 433-436 (2003) pp. 831-834.

[19] Blascuic-Diitru, C., Horsfall, A.B., Vassilevski, K.V., Johnson, C.M., Wright, N.G., O'Niel, A.G.: "Characterisation of the high temperature performance of 4H-SiC Schottky barrier diodes", Mater. Sci. Forum 433-436 (2003) pp. 823-826.

[20] Itoh, A., and Matsunami, H.: 'Single crystal growth of SiC and Electronic Devices' Critical Reviews in Solid State and Materials Sciences, 1997, 22, (2), pp. 111-197

[21] C.M Zetterling Process technology for Silicon Carbide devices, EMIS Processing Series 2, Published by INSPEC, The Institution of Electrical Engineers, London, UK 2002.

[22] Roccaforte, F. La Via, V. Raineri, R. Pierobon, E. Zanoni, J. Appl. Phys., Vol. 93, No. 11, 1, 2003, pp. 9137-9144

[23] Kampen, T. U. and Monch, W.: 'Lead contacts on Si(111):H-1 × 1 surfaces' Surf. Sci., 1995, 331-333, pp. 490-495.

[24] Schmitsdorf, R.F., Kampen, T.U., and Monch, W., 'Correlation between barrier height and interface structure of Ag/Si(111) Schottky diodes', 1995, Surf. Sci., 324, (2-3), pp. 249-256.

[25] Schmitsdorf, R.F., Kampen, T.U., and Monch, W.: 'Explanation of the linear correlation between barrier heights and ideality factors of real metal-semiconductor contacts by laterally nonuniform Schottly barriers', J. Vac. Sci. Technol. 1997, B 15, (4), pp. 1221-1226.

[26] Barrett, D.L., and Campbell, R.B., 'Electron Mobility Measurements in SiC Polytypes', J. Appl. Phys., 1967, 38, pp. 53-55.

[27] Kimoto, T., Urushidani, T., Kobayashi, S., Matsunami, H.: 'High-voltage (>1 kV) SiC Schottky barrier diodes with low on-resistances' IEEE Electron. Dev. Lett., 1993, 14, pp. 548-550.

[28] Chilukuri, R.K., and Baliga, B.J.: 'High voltage Ni/4H-SiC Schottky rectifiers', *in Proc. of the 11th International Symposium on Power Semiconductor Devices and ICs*, ISPSD '99, May 1999, 26-28, pp.161 – 164.

[29] Miller, E.J., Schaadt, D.M., Yu, E.T.: 'Reduction of reverse-bias leakage current in Scottky diodes on GaN grown by molecular-beam epitaxy using surface modification with an atomic force microscope,' *Journal of Appl. Phys.*, 2002, 91, (12), pp. 9821-9826.

[30] Miller, E.J., Dang, X.Z., Yu, E.T.: 'Gate leakage current mechanisms in AlGaN/GaN heterostructure field-effect transistors;' *Journal of Appl. Phys.*, 2000, 88, pp.5951-5958,

[31] Park, J.E., Shields, J., Schroder, D.K., 'Non-volatile memory disturbed due to gate and junction leakage currents,' *Solid-State Electronics*, 2003, 47, pp. 588-864.

[32] Hurkx, G.A.M, de Graafg, H.C., Kloosterman, W.J., Knuvers, M.P.G.: 'A new analytical diode model including tunneling and avalanche breakdown', *IEEE Trans. Electron Dev.*, 1992, 39, (9), pp. 2090-2098

[33] V. Saxena, J.N. Su and A.J. Steckl, High-voltage Ni– and Pt–SiC Schottky diodes utilizing metal field plate termination, *IEEE Trans Electron Dev* 46 (1999) (3), pp. 456–464.

[34] P.G. Neudeck, D.J. Larkin, J.A. Powell and L.G. Matus, 2000 V 6H–SiC p–n junction diodes grown by chemical vapour deposition, *Appl Phys Lett* 64 (1994) (11), pp. 1386–1388.

[35] K.J. Schoen, J.M. Woodall and J.A Cooper Jr., Design considerations and experimental analysis of high-voltage SiC Schottky barrier rectifiers, *IEEE Trans Electron Dev* 45 (1998) (7), pp. 1595–1604.

[36] Ueno, T. Urushidani, K. Hashimoto and Y. Seki, The guard-ring termination for the high-voltage SiC Schottky barrier diodes, *IEEE Electron Dev Lett* 16 (1995), pp. 331–332.

[37] D. Alok, B.J. Baliga and P.K. McLarty, A simple edge termination for silicon carbide devices with nearly ideal breakdown voltage, *IEEE Electron Dev Lett* 15 (1995), pp. 394–395

[38] Hirofumi Kurimoto, Kaoru Shibata, Chiharu Kimura, Hidemitsu Aoki, Takashi Sugino, 'Thermal oxidation temperature dependence of 4H-SiC MOS interface', *Appl. Surf. Sci.* 253 (2006) pp. 2416-2420.

[39] V.V. Afanas'ev, M. Bassler, G. Pensl, M. Schulz, 'Analysis of the electron traps at the 4H-SiC/SiO2 interface using combined CV/thermally stimulated current measurements', *Phys. Stat. Sol.* A 162 (1997) pp. 321-337.

[40] O.J. Guy, T.E. Jenkins, M. Lodzinski, A. Castaing, S.P. Wilks, P. Bailey T.C.Q. Noakes, 'Ellipsometric and MEIS Studies of 4H-SiC/Si/SiO2 and 4H-SiC/SiO2, Interfaces for MOS Devices' *Mater. Sci. Forum* 556-557 (2007) pp. 509-512

[41] C.-Y. Lu, J.A. Cooper Jr., T. Takashi, G. Chung, J. R. Williams, K. McDonald, L.C. Feldman, 'Effect of Process Variations and Ambient Temperature on Electron Mobility at the SiO2/4H-SiC Interface', IEEE Trans. Electron Devices, 50 (2003) pp. 1582-1588.

[42] H.O. Olafsson, G. Gudjonsson, F. Allerstam, E.O. Sveinbjornsson, T. Rodle, R. Jos, 'Stable operation of high mobility 4H-SiC MOSFETs at elevated temperatures' Electron. Lett., 41 (2005) 825-826.

[43] http://www.siced.de/hp527/Device-Concepts.htm

[44] S-H. Ryu, S. Krishnaswami, B. Hull, B. Heath, M. Das, J. Richmond, A. Agarwal, J. Palmour, J. Scofield, 'A Study on the Reliability and Stability of High Voltage 4H-SiC MOSFET Devices', Materials Science Forum 527-529 (2006) 1313-1316

[45] Shinsuke Harada, Mitsuo Okamoto, Tsutomu Yatsuo, Kenji Fukuda, Kazuo Arai, 'Analysis of Low On-Resistance in 4H-SiC Double-Epitaxial MOSFET' Materials Science Forum Vols. 483-485 (2005) 813-816.

[46] 'Reliability of High Voltage 4H-SiC MOSFET Devices', Sumi Krishnaswami, Sei-Hyung Ryu, Bradley Heath, Anant Agarwal, John Palmour, Aivars Lelis, Charles Scozzie, James Scofield, Mater. Res. Soc. Symp. Proc. Materials Research Society, 911 (2006) 0911-B13-01.

[47] A. Perez-Tomas, P. Godignon, N. Mestres, R. Pérez, J. Millan, 'A study of the influence of the annealing processes and interfaces with deposited SiO2 from tetra-ethoxy-silane for reducing the thermal budget in the gate definition of 4H–SiC devices', Thin Solid Films 513 (2006) 248-252.

[48] D. Ziane, J.M. Bluet, G. Guillot, P. Godignon, J. Monserrat, R. Ciechonski, M. Syväjärvi, R. Yakimova, L. Chen, P. Mawby, 'Characterizations of SiC/SiO2 interface quality toward high power MOSFETs realization', Mater. Sci. Forum, 457-460 (2004) 1281-1286.

[49] The INTRINSIC project:

http://www.siliconcarbide.org/intrinsic/

[50] C. Blanc, D. Tournier, P. Godignon, D.J. Brink, V. Soulière, J. Camassel, 'Process Optimisation for <11-20> 4H-SiC MOSFET applications', Mater. Sci. Forum 527-529 (2006) 1051-1054

[51] C.M. Tanner, J. Choi, J.P. Chang, 'Experimental and First-Principles Studies of the Band Alignment at the HfO2/4H-SiC (0001) Interface' Materials Science Forum 527-529 (2006) pp. 1071-1074.

[52] S. Hino, T. Hatayama, N. Miura, T. Oomori, E. Tokumitsu, 'Fabrication and Characterization of 4H-SiC MOSFET with MOCVD-grown Al2O3 Gate Insulator', Mater. Sci. Forum (2008) In press.

[53] A. Perez-Tomas, M.R. Jennings, P.M. Gammon, G.J. Roberts, P.A. Mawby, J. Millan, P. Godignon, J. Montserrat, N. Mestres, 'SiC MOSFETs with thermally oxidized Ta2Si stacked on SiO2 as high-k gate insulator', Microelectronic Engineering 85 (2008) pp. 704–709.

Characterization of the Static and Dynamic Behavior of a SiC BJT

M.M.R. Ahmed*, N-A.Parker-Allotey*, P.A. Mawby* Muhammed Nawaz[†] and Carina Zaring[†]

* Warwick University/Electrical & Electronics, Coventry, UK, e-mail: Mohamed.ahmed@*warwick.ac.uk*

[†] TranSiC AB, Electrum 207, Isafjordsgatan 22 C5, SE-16440 Kista, Sweden

Abstract— Silicon carbide (SiC) bipolar junction transistors (BJTs) are interesting candidates for high temperature and for high power applications primarily due to their low conduction losses and fast switching capability. The aim of this paper is to test and evaluate both the static and dynamic characteristics of SiC bipolar junction transistor (developed by TranSiC) rated at 600 V and 6 A at different temperatures. The high power curve tracer 371B has been used to test the DC output characteristics of the device in a temperature range of – 40 °C to 175 °C. A single pulse switching inductive load circuit has been used to test the dynamic characteristics of the SiC BJT. The experimental results with a normal gate drive circuit, show that the device has a turn on time of less than 0.5 μs and turn off time of less than 0.35 μs under test condition of 300 V, 10 A in an ambient temperature of range – 40 °C to 125 °C. In addition, the experimental data were analyzed to obtain the device performance parameters like the turn on, off time, transistor gain and switching losses.

Keywords—Power electronics, Silicon Carbide (SiC), Bipolar Junction Transistor (BJT), Characterization.

I. INTRODUCTION

Power semiconductor devices play an important role in improving electric aircraft performance and often dictate the efficiency, cost and size of these systems. In these applications, a growing demand for devices being capable of operating at high temperatures is developing. High temperature semiconductor devices can be placed "on the engine", in very harsh operating conditions with an ambient temperature which could vary from – 40 °C to 200°C. Silicon carbide bipolar junction transistors (SiC BJTs) are the most attractive candidate device for high voltage and high temperature switching applications because of their low specific on-resistance, wide bandap, high breakdown field, high thermal conductivity and temperature independent switching characteristics [1-9]. At the same time, SiC BJTs do not suffer from issues pertaining to MOS mobility, gate oxide reliability like the SiC power MOSFETs. The SiC BJT development resulted in record current and power dissipation densities of $180kA/cm^2$ and $6 MW/cm^2$ [10], operation up to 500°C and radiation hardness up to 1.6 Mrad [11]. The current gain is typically between 30 and 60 and was measured to reduce to 60% of its room temperature value at 200°C and to 40% at 500°C [12]. While excellent room temperature characteristics in terms of high breakdown voltage, low on-resistance and high gain have been reported, very limited data is available for their assessment at high

temperature under different bias condition. This work demonstrates the suitability of SiC based BJTs at different temperature.

II. DEVICE CONSTRUCTION

A schematic cross-section of 4H-SiC based NPN BJT is shown in fig. 1. A three layer epitaxy (NPN type structure) has been grown in a single continuous growth step. The emitter layer is composed of two steps with different doping concentration. The emitter and base mesa structures were defined by ICP (inductively coupled plasma) etching using SiO_2 as a mask. Aluminum ions were implanted to form the low resistance base contact. Another aluminum implantation was introduced to define the junction termination extension (JTE) to suppress the surface electric field. The implants were activated at 1650 °C. A thermal oxide under N_2O environment was grown for passivation. The emitter and base contact metals were Ni and Ni/Ti/Al respectively, while bottom collector contact was based on Ni/Au. The top pad metallization was aluminum.

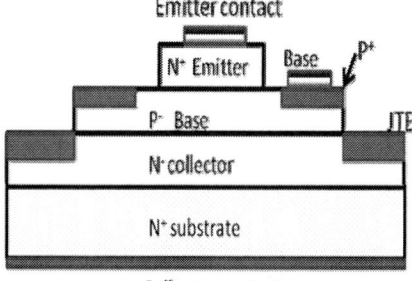

Fig. 1: Schematic cross-section of SiC BJT (top) and top view of bonded SiC BJT (bottom) indicating emitter and base contact pads.

III. DC CHARACTERISTICS

This section describes the characterization of SiC BJT and its temperature dependency through dc measurements. The I-V characteristics of the SiC BJT for the forward conduction condition and blocking voltage were captured by a Tektronix 371B high power curve tracer. Figs. 2 -5 show the dc output characteristics at ambient temperature of -40 °C, 25 °C and 175 °C respectively. Fig. 6 shows the on state and specific on state resistances versus ambient temperature. Both forward voltage drop and specific on state resistance increases with increasing temperature as a result of the decreased current gain and increased drift region resistance. It can be seen that. Fig. 5 shows the blocking voltage of the device which is around 800 V (defined at collector of 100 μA). The blocking voltage capability depends on the base and collector layer thickness and doping concentrations in the appropriate layers. JTE (junction termination extension) has been employed to suppress the surface electric field and hence improving the breakdown voltage of the device.

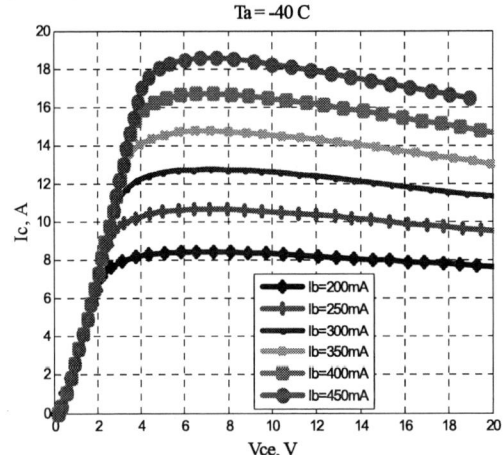

Fig. 2. DC output characteristics @ Ta = - 40 °C

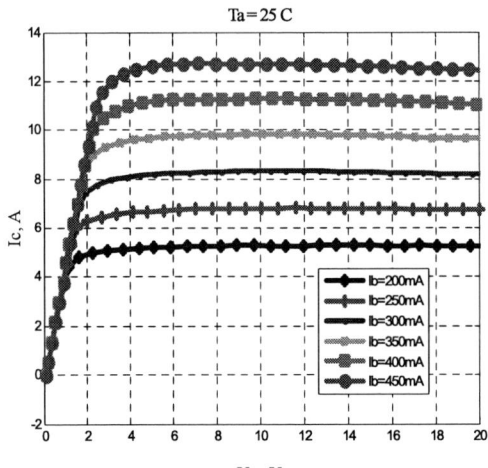

Fig. 3. DC output characteristics @ Ta = 25 °C

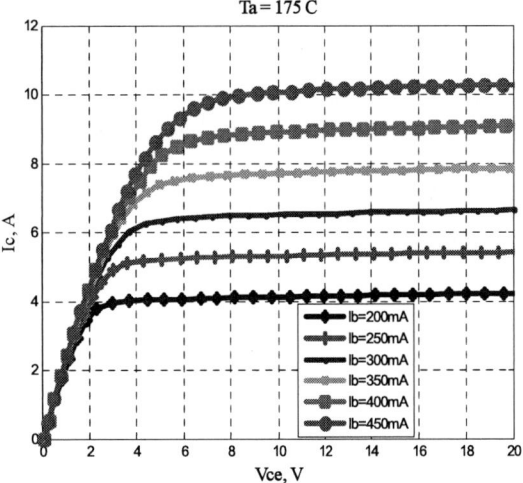

Fig. 4. DC output Characteristics @ Ta = 175 °C

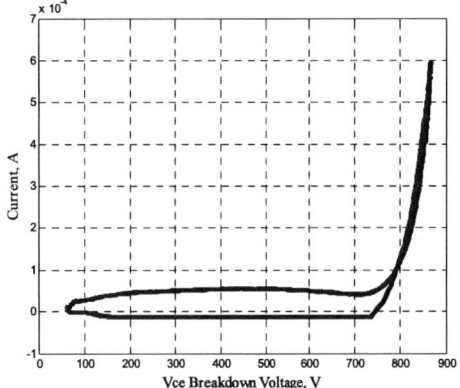

Fig. 5. Collector current as a function of collector-emitter voltage with open base BJT

A. Specific on-state resistance

The specific on-state resistance for a given base current, on-state resistance is calculated by the slop of the Ic versus Vce curve in the linear region of device operation. Reducing on-state resistance and forward voltage drop is one figure of merit of the device. The specific on-state resistance is mainly dependent on the contact resistance (I.e., improved annealing condition of metal layers of base, emitter and collector terminals) and doping concentration and the thickness of the base and collector layers. A BJT can be made with a lower specific on-state resistance device than that of unipolar devices such as MOSFET, and JFET due to the junction voltage cancellation, the absence of the channel region and conductivity modulation. The conductivity modulation can be used in BJT with light doped and thick drift layer. For high voltage applications conductivity modulation is needed to obtain a sufficiently low forward voltage drop. The temperature dependencies of the on state and specific on-state resistances measured on a 600 V, 6 A device are shown in Fig. 6.

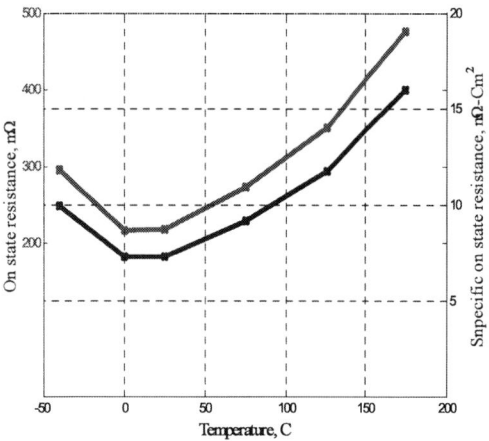

Fig. 6. On-state and specific resistance of SiC BJT

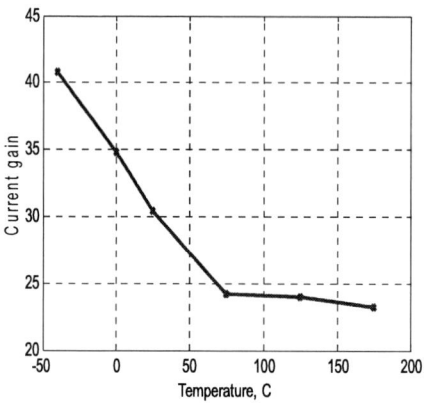

Fig. 8. Current gain of SiC BJT

B. Current Gain (β)

The vital parameter for SiC BJT applications is the Common Emitter Current Gain. The current gain of the SiC BJT is defined as ratio between base current and collector current (β). The current gain of SiC power transistor is normally investigated by pulse measurement to avoid self heating of device [13,14]. Figs. 7&8 show Gummel plot and current gain respectively of SiC BJT at various temperatures. Typical values of β vary from 5 to 29, with maximum current gain (β_{max}) of about 30.4 at room temperature at a base current 200mA. The current gain initially decreases with increasing temperature. The drop in gain is due to decreased current injection efficiency (i.e., increasing ionization level of Al acceptor atoms and hence providing more hole current to the emitter). The decrease in gain at high temperature is compensated by the competing process of increased minority carrier life time in the base. The maximum collector current at room temperature was about 10 A but reduces to about 6 A at 175 °C.

IV. DYNAMIC CHARCTERISTICS OF SiC BJT

In aerospace power electronics applications the switching characteristics of semiconductor switch play an important role in the overall performance (losses, size and cost) of these devices. Figs 9 & 10 show the schematic and experimental test circuit. The overall circuit is based on a classical inductive switch circuit, which consists of the power supply, an inductive load, a freewheeling diode, SiC BJT (DUT) and environmental chamber which control ambient temperature. The experimental test circuit parameters are: Vdc=300V, Vgg=18V, Rb=11Ω, C = 4.7mF and L = 8.39 mH.

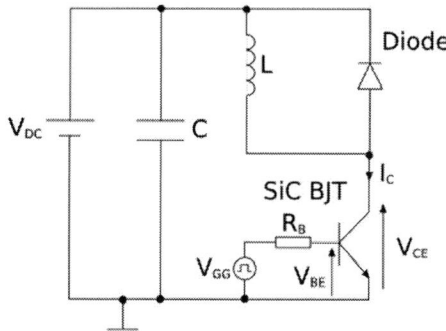

Fig. 9 Schematic diagram of test circuit

Fig. 10 Experimental set up

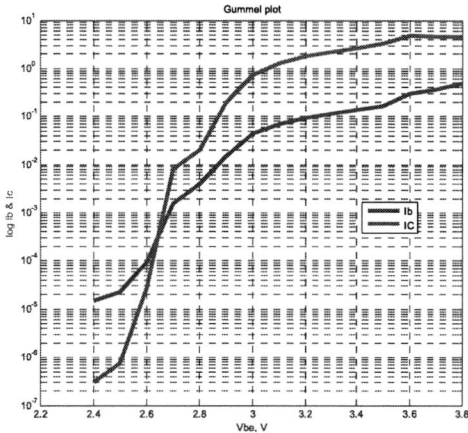

Fig. 7. Gummel plot of SiC BJT

To evaluate the inductive switching performance of the SiC BJT, a programmable function generator was used to provide a double-pulse signal to the gate drive circuit of the BJT under test, which reduces the energy consumption and prevent device self heating. The base drive in the inductive test rig is a MOSFET gate drive chip (HCPL-3120) that is capable of delivering 2.0A of current to the base of the DUT. It contains an opto-coupler for isolating the low voltage turn-on signal from the high voltage side of operation. The switching characteristics of SiC BJT were tested under various ambient temperature ranging from -40 °C to 125 °C. During these tests, temperature of the free-wheeling diode is not varied to investigate the performance temperature dependence of the SiC BJT alone.

A. Turn on

The SiC BJT is tested under bus voltage 300 V and a load current of 10 A. The turn-on waveforms of, common-collector voltage, collector and base currents at ambient temperature of -40 C, 25 C and 125 C are shown in Figs. 11, 12 and 13 respectively.

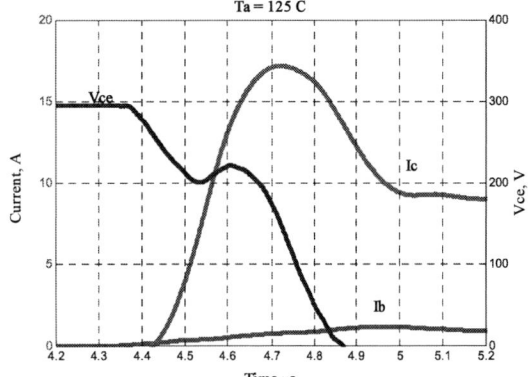

Fig. 13 Detailed SiC BJT turn-on waveforms @ Ta= 125 °C

B. Turn off

The SiC BJT is tested under bus voltage 300 V and a load current of 10 A. The turn-off waveforms of, common-collector voltage, collector and base currents at ambient temperature of -40 C, 25 C and 125 C are shown in Figs. 14, 15 and 16 respectively. It can be seen from the previous results that both rise and fall times are approximately independent on the temperature and in the range of microseconds respectively. The slow speed of turn on and turn off process are due to the normal gate drive used in these tests.

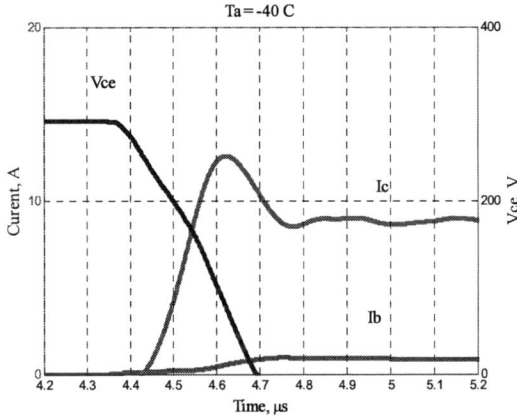

Fig. 11 Detailed SiC BJT turn-on waveforms @ Ta= -40 °C

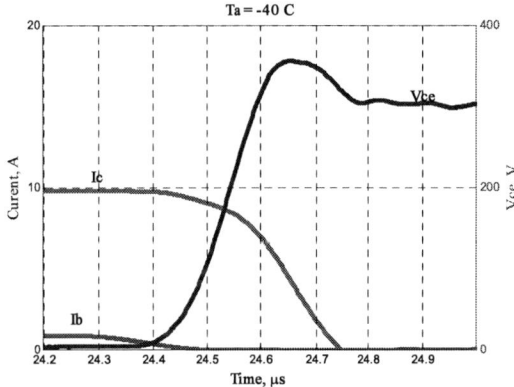

Fig. 14 Detailed SiC BJT turn-off waveforms @ Ta= -40 °C

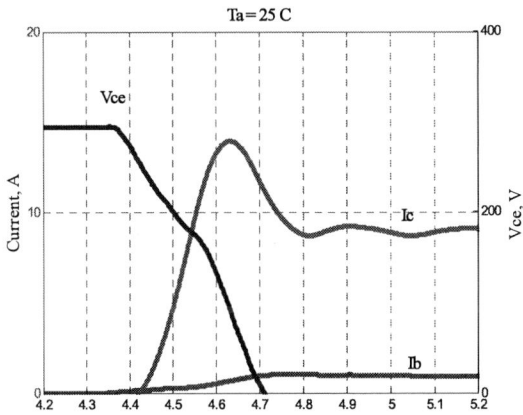

Fig. 12 Detailed SiC BJT turn-on waveforms @ Ta= 25 °C

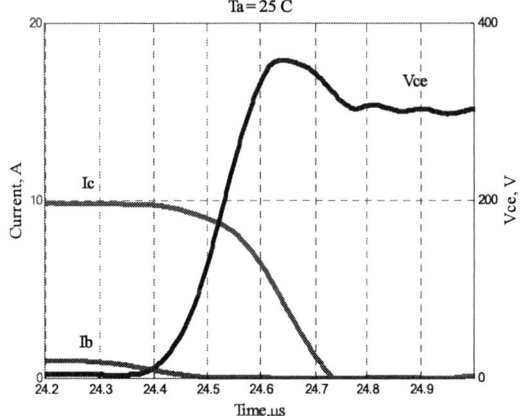

Fig. 15 Detailed SiC BJT turn-off waveforms @ Ta= 25 °C

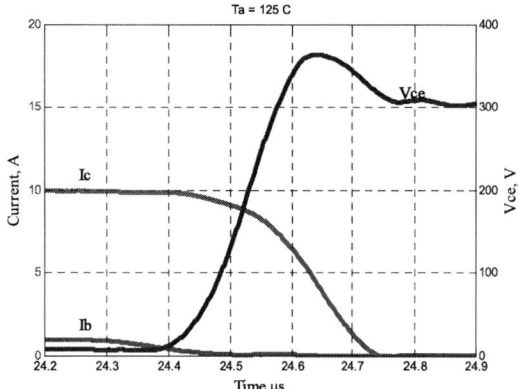

Fig. 16 Detailed SiC turn-off waveforms @ Ta= 125 °C

C. *Turn-on and turn-off switching losses*

The SiC BJT turn-on and turn-off power loss were obtained by multiplying both collector current and collector-common voltage. Figs 17 & 18 show turn-on and turn-off power losses versus temperatures respectively. Figs 19 & 20 show that both the turn-on time and energy are dependent on the temperature. While Fig. 18 shows that both turn-off time and energy are independent on temperature and are in range of 300 ns and 300 μJ.

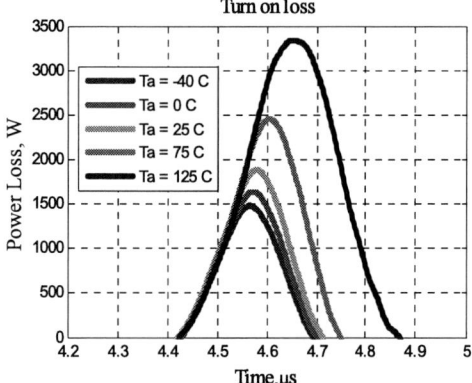

Fig. 17 SiC BJT turn-on power loss vs temperature

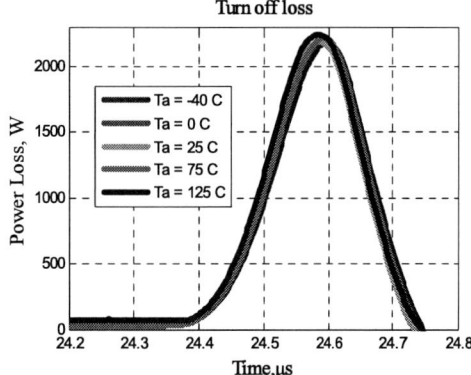

Fig. 18 SiC BJT turn-off power loss vs temperature

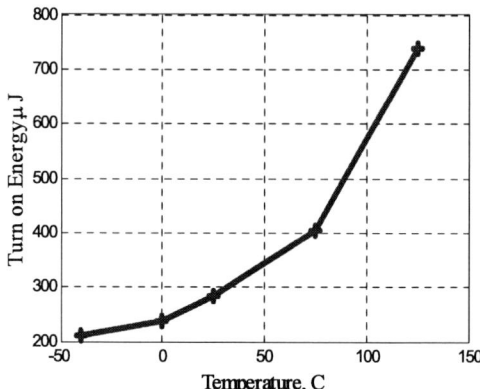

Fig. 19 Tunn-on energy vs temperature

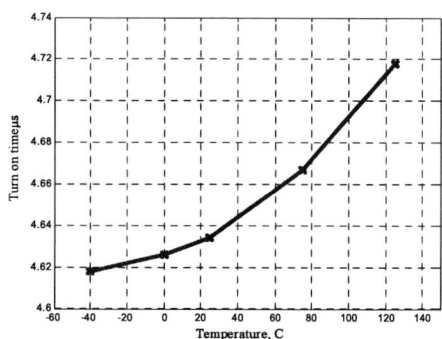

Fig. 20 Tunn-on time vs temperature

V. CONCLUSION

In this paper, static and dynamic characterization of a novel 600 V / 6 A SiC BJT developed by Transic has been analyzed and presented.

- The dc output characteristics have been performed under various junction temperatures vary from – 40 o°C to 175 °C using Tektronix curve tracer 370B. The dynamic characteristics have been examined under the following condition 300 V, 10 A and ambient temperature vary from -40 °C to 125°C.

- The extracted parameters from the measured results were presented, which are suitable for modeling the device.

- The experimental results show that this device is a promising switch with good static characteristics, low switching loss and robust turn off capability of high power density without applying negative base current. In addition, the overall switching speed of the tested SiC BJT remained almost unchanged at different temperatures making this device an attractive candidate switch for aerospace applications.

Further work needed to test the device under high temperatures up to 300 °C to validate the complete model of the fabricated SiC BJT.

ACKNOWLEDGMENT

The author would like to thank Stevens Martin, Mike Baily and Adrain Shiply from GE Aviation for their comments and discussions on this paper. The author would also like to thank the EPSRC for their generous financial assistance.

REFERENCES

[1] A.T. Bryant, P.R. Palmer, E. Santi, J.L. Hudgins and P.A. Mawby, "Review of Advanced Power Device Models for Converter Design and Simulation". Invited paper, ICTES Conf. Rec., Chennai, India, December 2007.

[2] A. K. Agarwal, S. Ryu and K. A. Jones, "Recent progress in SiC bipolar junction transistors," Proceedings of The IEEE International Symposium on Power Semiconductor Devices and ICs, ISPSD 2004, pp. 361–364.

[3] A. Johnson and D. James, "Characteristics of high-speed silicon carbide (sic) power Transistors", Instrumentation and Measurement Technology Conference- IMTC 2007, pp. 1-6.

[4] C. Weitzel, J. Palmour, and M. Bhatnagar, "Silicon Carbide High-Power Devices," IEEE Transactions on Electron Devices, Vol. 43, No. 10, October 1996, pp. 17321741.

[5] G. Yan, Q. Alex,and Z Qingchun, "Analysis of Operational Degradation of SIC BJT Characteristics", Proceedings of the 19th International Symposium on Power Semiconductor Devices & ICs May 27-30, 2007, pp. 121-125.

[6] G. Yan, Q. Alex,and K.Anant, "Comparison of Static and Switching Characteristics of 1200V 4H-SiC BJT and 1200V Si-IGBT", IEEE Transactions on Industry Applications May 2008, Vol. 44, No. 3, pp. 887-893.

[7] J. Asumadu, and J. Scofield, "Characteristics of High-Speed Silicon Carbide (Sic) Power Transistors", Proceedings of IEEE Conference on Instrumentation and Measurement Technology 2007, pp. 1-6.

[8] H.-S. Lee, M. Domeij, C.M. Zetterling, M. Östling, F Allerstam, and E.Ö. Sveinbjörnsson, IEEE Electron Device Lett, Vol. 28, No. 11, p 1007 (2007).

[9] A. Claudio, W. Hongfang and A. Huang, "Static and dynamic characterization of silicon carbide bipolar junction transistor", Proceedings of the 29th Annual Conference of the IEEE Industrial Electronics Society, 2003, Vol. 2, pp. 1173- 1178.

[10] Y. Luo, L. Fursin and J.H. Zhao, "Demonstration of 4H-SiC power bipolar junction transistors", IEE Electronics Letters, 2000, Vol. 36, No.17, p.1496-1497.

[11] G. Yan, Q. Alex and S. Charles, "Analysis of SiC BJTs RBSOA", Proceedings of The ieee International Symposium On Power Semiconductor Devices and ics, ISPSD 2006, pp. 281 – 284.

[12] X. Xiaojun, G. Yan and D. Zhong, "400khz, 300W sic BJT based high power density PFC Converter", Proceedings of The 37th IEEE Power Electronics Specialists Conference, 2006, pp. 1-5.

[13] Z. Jianhui and A. Petre Alexandrov, 4H-SiC Power Bipolar Junction Transistor With a Very Low Specific ON-Resistance of 2.9 mΩ · cm2", IEEE Transactions on ELECTRON DEVICE LETTERS, 2006, Vol. 27, No. 5, pp. 368-370.

[14] C.-F. Huang and J. A. Cooper, Jr., "High current gain 4H-SiC NPN bipolar junction transistors," IEEE Electron Device Lett., vol. 24, no. 6, pp. 396–398, Jun. 2003.

An active network control method using distributed energy resources in microgrids

Takayuki Tanabe, Yoshinobu Ueda, Toshihisa Funabashi
Meidensha Corporation
Think Park Tower, 2-1-1, Osaki, Shinagawa-Ku, Tokyo,
141-6029 Japan

Shigeo Numata, Kimio Morino, Eisuke Shimoda
Shimizu Corporation
3-4-17, Ecchu-jima, Koto-Ku, Tokyo,
135-8530 Japan

Abstract— An active network control method using distributed energy resources in microgrids is shown. Some features of the system were outlined. And also, power system stabilizer to compensate for the fluctuation of active power in microgrid is proposed. The microgrid with these new techniques can contribute to efficient operation of DGs and stable power supply for customers.

I. INTRODUCTION

In June 2002, the "Energy Policy Basic Law" was enacted and the basic guidelines for energy supply and demand were released.

- In regard to the assurance of a stable supply of energy, the introduction of new energy resources is positively promoted so that the rate of self-reliance in energy can be raised in Japan.

- In regard to the compatibility with the environment, the Kyoto Protocol was concluded to reduce the emission of greenhouse effect gases (CO_2).

- In regard to the active utilization of the market mechanism, deregulation of electric utilities is promoted through the introduction of power wholesale markets and so on.

The introduction of new energy contributes not only to the improvement of the rate of self-reliance in energy but also to the reduction of greenhouse effect gases.

A concept of microgrid was proposed and is being developed in many laboratories and organizations [1-2]. Microgrid is defined as a small grid in which distributed energy resources and loads are placed together and controlled efficiently in an integrated manner. In many microgrids, new energy is used as a core power source. We believe that the development of technologies relating to this energy supply system contributes to the accelerated introduction of new energy resources. Microgrid also contributes to utility grid's load leveling by controlling power flow between utility grid and microgrid according to a predetermined power flow pattern. Also, it will be able to contribute to an efficient operation of distributed generations by operation planning considering grid economics and energy efficiency. In addition, microgrid is useful for the area with no power system or weak

power system in rural area. It can be operated in an islanded manner with appropriate control scheme.

Shimizu Corporation and Meidensha Corporation recently developed an active network control system and the Power Systems Stabilizer (PSS) using Electric Double Layer Capacitors (EDLCs). In this paper, a microgrid control system is proposed. In this control system, operation planning is realized based on generation and load forecasting by using neural network and fuzzy systems. It includes multi-objective evaluation of generation cost and CO_2 gas emission with some constraints. Unit commitment of generations includes start/stop of power generations and storages. Load following function is accomplished based on PI control scheme. Power Systems Stabilizer (PSS) for microgrids has been developed with an Electric Double Layer Capacitor (EDLC). Rapid fluctuations of renewable generations are compensated with PSS. It enables rapid and frequent charge/discharge and realizes a long lifetime. PSS compensates both active and reactive power and it can also realize constant reactive power supply and consequently it can decrease reactive power of generators in microgrids.

According to a study for microgrid written in this paper, Shimizu Corporation and Meidensha Corporation are jointly executing a field test.

II. OUTLINE OF MICROGRID FIELD TESTS IN JAPAN

The NEDO (New Energy and Industrial Technology Development Organization) started three research projects, which deal with new energy integration to local power system field test in 2004. The sites are in Aomori, Aichi and Kyoto [3].

In Aomori project, field tests were started to develop a distributed energy supply system, in which some loads in special districts are supplied by this supply system with private power lines and makes no influence to utility power system with which the energy supply system is connected at one point. In this project gas engine generator and battery system are used to control variable power output of photovoltaic and wind turbine generators. Electric power is supplied with private power lines to four school and some buildings. Heats are supplied to water treatment works. Total DER capacity is 710kW, in which PV: 80kW, Wind: 20kW, Biomass gas engine: 510kW and Battery: 100kW.

In Aichi project, fuel cells are used as main generations.

The site was World Expo 2005. Fuel cell systems are MCFC: 350kW, 370kW, PAFC: 800kW, and SOFC: 50kW. Total DER capacity is 2400kW. PV: 330kW and NAS battery: 500kW are also used. In Kyoto project, capacity of each resources are wind: 50kW, Biogas: 400kW, Fuel cell: 250kW, and PV: 50kW.

The Shimizu Corporation has built a Microgrid at its research labs in Tokyo, Japan [5-10]. It includes two gas engines (90kW and 350kW) as well as extra storage in form of a 400kWh NiMH battery (50kW×8h, with 200kW inverter) and an 100kW EDLC.

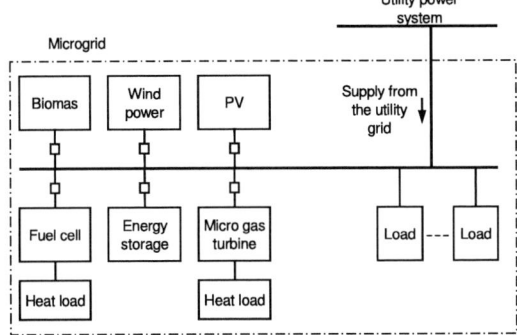

Fig.1. An example of microgrid.

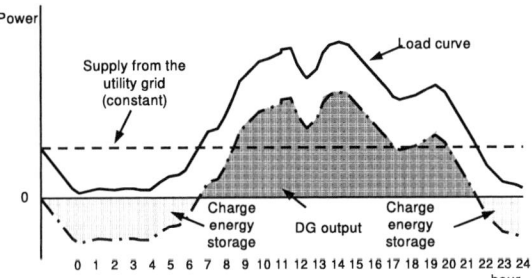

.Fig.2. An example of supply and demand control.

III.   SUPPLY AND DEMAND CONTROL SYSTEM

By making the supply power from the utility grid to the microgrid constant or a specified pattern, the SDC system contributes to the load leveling of the utility power system. Fig.1 shows an example of microgrid and Fig.2 shows an example of supply and demand control. By the control based on the operation plan of the distributed generations considering their economics and energy efficiency, the SDC system realizes the efficient operation of the distributed generations in the microgrid.

The SDC system consists of the control system that executes the operation planning and supervisory control of the total microgrid and the supervisory control terminals that executes supervisory control of distributed generations and loads.

Generator output control is executed according to the operation plan that is made based on the pattern of supply

power from the grid (In this system it is constant as shown in Fig.2) and the load demand pattern forecasted in advance.

For making the supply power coincident to the planned pattern at high accuracy, the output control of DG dispatched to load following is done. Control is carried out to realize both control modes of the constant power flow control (to assure kW following accuracy) by which the power flow at the interconnected point is maintained as constant and the power balancing control (to assure simultaneity in terms of kWh). The output control value is determined from the planned output value and real-time collected total supply and total demand. Output control signal is determined considering DG's response time and data transmission delay time in the information network.

IV.   POWER SYSTEM STABILIZER WITH EDLC

A.   Features of EDLCs

In order to compensate for sudden load fluctuations occurring in the microgrid, the EDLCs are used as the charge/discharge elements because they are capable of high-speed charging and discharging.

The features of equipment are described below.

- High-speed response: Compared with storage batteries, the EDLCs are capable of high-speed charging and discharging.

- Long operational life after frequent charge/discharge actions: Compared with storage batteries, the EDLCs can withstand frequent charge/discharge duty cycles. In addition, the operational life is long.

- Generation of active power in high preference: When fluctuations occur in active power reactive power at the same time, active power is output in higher performance.

- Generation of a constant reactive power: According to the reactive power command value sent from the supply and demand control system, it is possible to generate a constant level of reactive power burden to be shouldered by the generator.

B.   Equipment Configuration

Fig.3 shows the equipment configuration. The power is 100kVA. The EDLC manufactured by Meidensha Corporation comes in a module of 45F in capacitance. Six units of such modules are connected into a capacitor bank consisting of three-serial and two parallel modules. Thus, the resultant capacitor bank has a total capacitance of 30F. In order to cope with load switching (closure and tripping), the operating DC voltage is defined based on the rated voltage (middle voltage) of 336 volts and this voltage is practically used within a voltage range of 240 – 400 volts according to the charging and discharging cycles. The time needed for charging and discharging is 2 seconds for charging and 2 seconds for discharging, respectively. This is based on the assumption that this time is consumed from the middle voltage to the maximum or minimum voltage with a +/-100kW output in the charging or discharging phase.

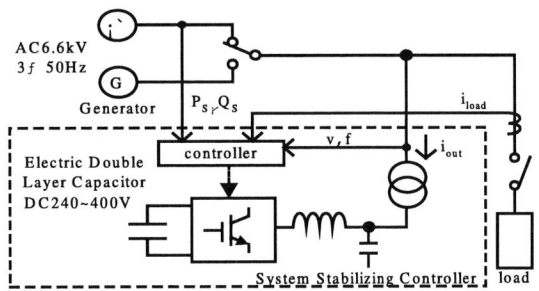

Fig.3 A configuration of power system stabilizer

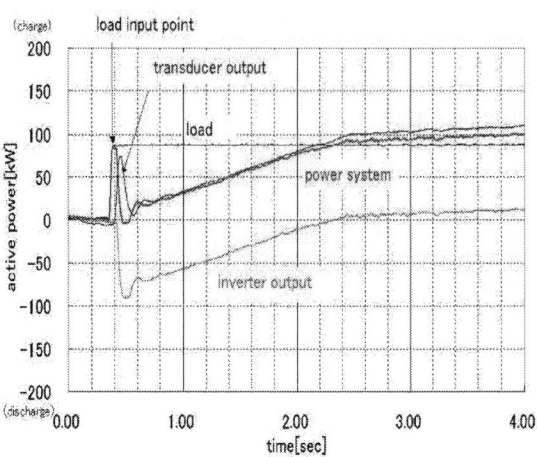

Fig.4 Power flow control of interconnected operation (laboratory)

C. Operation of Power System Stabilizer

Fig.2 shows the operation of the power system stabilizer. During interconnected operation with the power system, the power system stabilizer performs power flow control to compensate for the power flow fluctuation at the point of common coupling (PCC). By this control active or reactive power at the PCC is used to determine the power flow fluctuation at the PCC. The Power system stabilizer then generates an output to compensate for the determined power flow fluctuation.

When the microgrid stays in its islanding operation independent of the utility power network system, the power system stabilizer continues its control after selecting either power flow control to compensate for the load power fluctuation or a CVCF control to maintain the same levels of voltage and frequency in the microgrid system.

In the power flow control for islanding operation, the load current is used to determine the load power fluctuation and the power system stabilizer generates an output to compensate for this power fluctuation. In the CVCF control mode, the fluctuation components of system voltage and frequency ar determined. Based on these components, frequency fluctuations are controlled by the active power and voltage fluctuations are controlled by the reactive power.

D. Evaluation Test

The evaluation test has been carried out in the factory . Test circuit is shown in Fig.7. Fig.8 shows an example, the result of the active power fluctuation control under the interconnected condition. Due to the resistor load insertion to the load insertion point, power at the PCC fluctuates rapidly. But the power system stabilizer compensates for this in 0.1 seconds or so.

V. CONCLUSIONS

This paper introduced a microgrid control system. Some features of the system were outlined. And also, power system stabilizer to compensate for the fluctuation of active power in microgrid is proposed. The microgrid with these new techniques can contribute to efficient operation of DGs and stable power supply for customers.

REFERENCES

[1] T.Funabashi and R.Yokoyama, "Microgrid Field Test Experiences in Japan," IEEE PES General Meeting, 2006

[2] S.Morozumi and K.Nara, "Recent Trend of New Type Power Delivery System and its Demonstrative Project in Japan," IEEJ Trans. PE, Vol.127, No.7, 2007, pp.770-775

[3] M., G.Ventakaramanan, J.Kondoh, R.Lasseter, H.Asano , N.Hatziagyriou, J.Oyarzabal, and T.Green, "Real-World MicroGrids-An Overview," 2007 IEEE International Conference on System of Systems Engineering, San Antonio, USA., April 16-18, 2007

[4] N.Hatziargyriou, H. Asano, R.Iravani, C. Marnay: "Microgrids," IEEE power & energy magazine, vol. 5, no. 4 (2007)

[5] A.Denda, "Shimizu's microgrid research activities," Symposium on Microgrids, Montreal, June 2006

[6] T.Tanabe, Y.Ueda, S.Suzuki, T.Ito, N.Sasaki, T.Tanaka and T.Funabashi and R.Yokoyama, "Optimized Operation and Stabilization of Microgrids with Multiple Energy Resources," ICPE 2007.

[7] T.Ito, T.Tanabe and N.Sasaki, "Development of Microgrid Control Systems," Meiden Review, No.140, No.2 2007, pp.4-7

[8] S.Suzuki and Y.Ueda, "System Stabilizing Controller by Electric Double Layer Capacitor," Meiden Review, No.140, No.2 2007, pp.8-11

[9] S.Suzuki, Y.Ueda, S.Numata and A.Denda, "Combined Power Supply Control of Distributed Generators in Micro Grid (Par.2) – Development of System Stabilizing Controller by Electric Double Layer Capacitor -," IEE Japan Power and Energy Society annual meeting, September 2006 (in Japanese)

[10] Y. Ueda, S.Suzuki, T. Tanabe, T. Ito, S. Numata, K. Morino, and E. Shimoda, "Power System Stabilizer with an EDLC for Microgrids," IEEJ Workshop in Power Engineering, Power Systems Engineering and Semiconductor Power Conversion, January 2008 (in Japanese)

Energy Management in Solar Photovoltaic Plants based on ESS

M. Lafoz, L. García-Tabarés, M. Blanco

Centro de Estudios del Transporte. Centro de Estudios y Experimentación de Obras Públicas (CEDEX).
Madrid. Spain.

Julián Camarillo, 30. 28037 Madrid (Spain).
Ph: +34 913357194   Fax: +34 913357197
Marcos.Lafoz@cedex.es

Abstract – **Taking into account the increasing penetration of renewable energy sources in the electric grid it has become important to manage these resources to improve the availability, reliability and quality of the electric power. In order to connect to the electric grid a solar photovoltaic (PV) plant with the characteristic of Energy Management it is required a set of conditions established by the Transmission System Operator (TSO). Among those, there are important points as the reliability in the power prediction, obligation to provide energy reserve and support to voltage disturbances. In order to provide a reserve of energy, some possibilities are proposed in this paper, all of them based on the Energy Storage System concept. A hybrid batteries and flywheels solution will be analysed for a specific application of 1MW, 1MWh solar photovoltaic plant. Only technical evaluation has been developed.**

Keywords - Solar Cell System, Renewable energy systems, Energy system management, Energy Storage, Battery management system, Flywheel, voltage regulator modules.

I. INTRODUCTION

With the increasing penetration of renewable energy sources in the Spanish electric grid the Transmission System Operator (TSO) is concerned about the consequences of a massive introduction of a non-dispatchable electric power sources. Currently, the renewable energy generation plants work supplying the maximum power obtained from their primary energy source meanwhile the rest of power plants are able to regulate the power supply to fit the power demand. A new specific and applicable norm is being performing to fix new requirements for renewable energy plants in order to improve the availability and controllability of the electric grid ensuring the stability. One proposal is to define the concept of Energy Managed Power Generation Plant [1]. This concept is related to the possibility of the power plant to provide the regulation power required in the intraday electric market. In the particular case of Spain, the TSO is the company *Red Eléctrica de España*.

In addition to providing stability to the grid it could be interesting the concept of time-shifting in the case of PV plants, consisting on shifting loads to periods of high supply [2].

The concept of energy management could be considered as a more extended idea, implying three capabilities: solar prediction, power regulation and energy storage reserve.

Energy Management =
Solar Prediction + Power Regulation + Energy Storage Reserve

A. Solar Prediction

Solar photovoltaic generation plants have the best success when a good prediction can be achieved. The natural non predictable character of solar energy is a handicap for its use in the electric system. Moreover, one other disadvantage occurs during the winter, when the maximum power consumption peak at the grid coincides with the moment of sunset. TSO would like to ensure a high level of reliability to achieve a good quality level in the electric supply. In fact, considering the energy management requirements, the program reliability must be of 90% during 24 hours and 95% during 6 hours, including the failure possibility in both cases. Moreover, any appropriate mechanisms must be provided to correct the program deviations.

Based on historic files it is possible to adapt the obtained energy more and more accurately to the real situation by means of predictive models and studies [3][4].

B. Power Regulation

The electric grid requirements and the TSO determine different regulation levels [1]: Primary Regulation is related to the generation plant (supplying power during 15 seconds to compensate any unbalance between supplied and demanded powers), Secondary and Tertiary Regulations are related to the external voltage and power regulation and depends on the grid behaviour.

Providing the solar photovoltaic generation plant with some kind of energy reserve or energy storage system (ESS) it is possible to compensate electrical problems in the electric grid such as: power regulation, static compensation (STATCOM) and voltage disturbances like voltage sags.

978-1-4244-1741-4/08/$25.00 ©2008 IEEE

TSO requirements in terms of power capability are: To supply at least two times the rated current of the connected group during the voltage sag; to provide the system with an automatic voltage regulation, considering the voltage previous to the failure as a command.

C. C. Energy Storage Reserve

In order to obtain the accreditation of Energy Management for the generation plant, several requirements have to be verified related to Energy Storage Reserve:

1. The plant has to be able to increase the power program between 0% and 30% of the maximum power of the plant.

$$\Delta P_{max} = 0{,}30 \ P_{max}$$

2. The increase of the power program implies that the sumatory of the inicial power ($P_{initial}$) and the power increment (ΔP) will be lower than the maximum power of the plant

$$P_{initial} + \Delta P \leq P_{max}$$

For instance, if P_{max} = 3MW, a power reserve of 1MW has to be provided at the generation plant.

About the energy considerations, due to the fact of the intraday electrical market has hourly program, a power supply during 1 hour is required for the system. Therefore, for the same example of a 3MW plant, an energy reserve of 1MWh has to be provided.

Different energy storage systems (ESS) can be evaluated and each one has more appropriate characteristics for every application [5][6][7]. For middle and large scale developments (>100MW and >100MWh) the most obvious storage systems at this moment are compressed air, melted salts and pumped hydro storage. These systems require specific geography and not very good efficiency is got. Therefore some other types may be found as an alternative for middle and large scale electricity storage. Ultracapacitors, Batteries Energy Storage Systems (BESS) and Kinetic Energy Storage Systems (KESS) could be candidates to carry out the requirements. In general terms, batteries are more suitable for low power and energy, ultracapacitors are more appropriate for high power but low energy and are still not mature enough to achieve the requirements of an energy generation application. Finally, kinetic energy storage results very convenient for high power and energy because of the energy density.

Three energy storage technologies have been compared in terms of volume and mass energy and power densities. Table I presents the results for the following systems:

Ultracapacitors: (390V Maxwell module), E = 1,35 MJ, P = 58 kW, V = 0,2158 m³, m = 165 Kg
KESS: (Omega Plus by CEDEX [ref x3]), E = 200 MJ, P = 300 kW, V = 2 m³, m = 10.000 Kg
Batteries: (Exide-Tudor OPzS1200), E= 3600 MJ, P = 1MW, N=850, V=52 m³· m = 73.000 Kg)

TABLE I.
ENERGY AND POWER DENSITIES FOR THE MAIN ENERGY STORAGE TECHONOLOGIES

	Energy		Power	
	$e_v = E/V$	$e_m = E/m$	$p_v = P/V$	$p_m = P/m$
Ultracapacitors	6,25 MJ/m³	8 KJ/Kg	268,7 kW/m³	0,35 kW/Kg
Battery OPzS	97 MJ/m³	49 KJ/Kg	27 kW/m³	0,0136 kW/Kg
KESS	100 MJ/m³	20 KJ/Kg	180 kW/m³	0,03 kW/Kg

Batteries and flywheel will be analysed in the following chapter, considering the characteristics of both systems to provide energy reserve at the solar photovoltaic plant to get energy management.

II. BATTERY ENERGY STORAGE SYSTEM

Batteries have been historically the most conventional energy storage system from the beginning of the 20th century. More recently, beginning of 1980s, the evolution of the materials and the power electronics have led to renew the interest in Battery Energy Storage Systems (BESS). Battery technologies can cover a wide spectrum of purposes resulting more interesting for high energy application. During the last years many projects have been developed in large-scale systems [8] using different batteries technologies. As grid applications can be pointed out: spinning reserve, load levelling, peak saving, voltage-frequency stabilization and reactive control (VAR).

The characteristics of cost and cycle life from several technologies are shown in Table I [9].

TABLE II.
CHARACTERISTICS OF DIFFERENT BATTERIES

	Lead-Acid	NiCd	NaS	VRF
Storage cost, €/Kwh	200-500	500-1300	400-1500	200-1800
Cycle life	1000-2000	2000	2500	3000

Considering the solution of lead-acid batteries and commercial data of EXIDE OPzS batteries, several results have been analysed to check the adaptability to the current application. For a 1MWh and 1MW energy reserve specifications, a number of 850 units of OPzS1200 batteries are required divided into 2 strings connected to 2 power converters.

Depending on the operation parameters, the behaviour of the batteries will vary. Increasing the discharge time, a much higher capacity can be used at the batteries, as figure 1 shows. This effect is more intense for bigger batteries.

Another important parameter is the minimum voltage chosen for the discharge. This type of batteries has a nominal voltage of 2 volts and a set of them are associated in series to get a minimum 650V at the worse case. The lower is the minimum

2482

discharging voltage the more capacity can be got from the system, as figure 2 presents. On the other hand, deep discharges produce a reduction of the battery life.

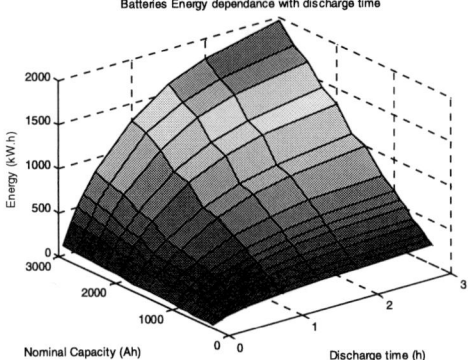

Figure 1. Battery energy dependence with discharge time.

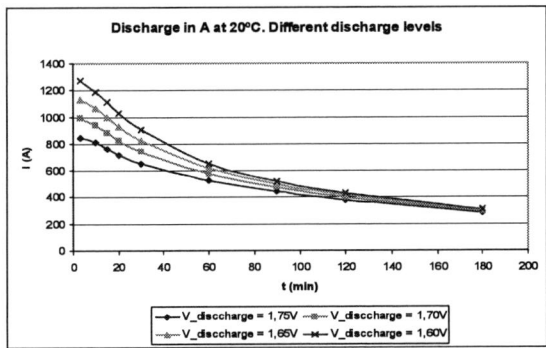

Figure 2. Discharge current considering different discharge voltage.

Figure 3 shows that the shorter the discharge time, the lower the discharge current. This result is translated in a better behaviour of the battery if the energy is supplied at a lower power rate. This phenomenon is due to the chemical efficiency of the battery.

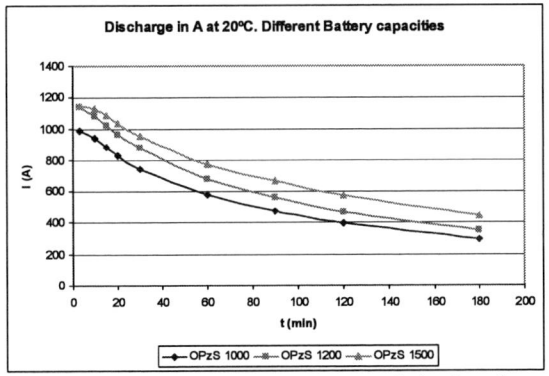

Figure 3. Discharge current considering different discharging voltage.

About the batteries cost, some results are presented in figure 4. Different curves have been displayed for different discharge time. For the same amount of energy, the cost is increasing with the power but is decreasing if the discharge time is lower.

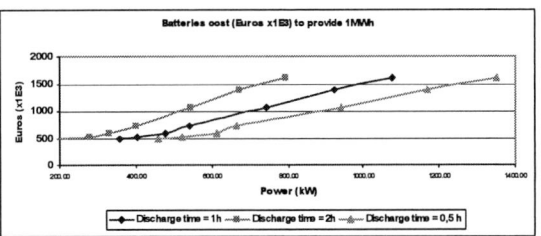

Figure 4. Batteries cost for 1MWh varying the power and the discharge time

One of the most important drawbacks of the storage systems based on batteries is that the recharge can not be produced at a rated current but a C/5 level (being C the battery capacity) which implies that if the discharge time is 1h at rated power, the recharging time will be almost 5 hours. The system is only suitable to cover power demand sceneries where the maximum power is not higher than 1MW and the discharge-recharge period is higher than 6 hours.

As a summary, batteries results as a good solution for massive energy storage due to the good energy density, well known technology and low cost but they have the main drawbacks:

- Their limited cycle life
- Recharge cycle very slow
- The low power density
- Environmental problems with Pb.

Therefore, some other energy storage systems should be analysed as an alternative. That is the case of the kinetic energy storage, based on flywheels, described in the next chapter.

III. KINETIC ENERGY STORAGE SYSTEMS

The decision of selecting a particular energy storage system depends on many factors, mainly the energy and power requirements, the type of installation, available dimensions to install the system and maintenance necessities. In a KESS the flywheel and the electrical machine can be selected separately to cover a wide range of energy and power. Other considerations like the energy concentration increase the advantages of the kinetic energy storage over batteries and ultracapacitors for this specific application.

A description of the KESS proposed for the PV solar plant is presented in this chapter.

Different prototypes with different power and energy levels have been developed in CEDEX [10]. Machine, flywheel and power electronics design have followed the target to get a robust final product with the minimum power losses.

A. Machine and Flywheel Technology Description

The part of the system where the energy is physically stored is the flywheel which is made of a high strength steel alloy, forged and machined to an almost-cylindrical shape avoiding stress concentration at any point and reducing them to the maximum theoretical value which is located in the axis of the cylinder. The shaft of the flywheel is machined from the same initial piece. Its lower part holds the rotor of the electrical machine and the bearings are placed at each end of this shaft. In order to reduce the load of the bearings, it is magnetically levitated using a hybrid system of permanent magnets and one electromagnet which adjusts the net weight to the required optimum value. Guidance of the flywheel is achieved by conventional high speed ceramic bearings. The rotating mass is working in a vacuum in order to reduce frictional losses. A 6/4 Switched Reluctance Machine is used to rotate the flywheel.

B. Power Electronics and Control

A double electronic converter is used to drive the KESS, composed of a machine-side converter (MSC) and a grid-side converter (GSC) sharing a common DC link. The MSC is in charge of driving the switched reluctance machine [11][12] and the GSC is in charge of controlling the DC link voltage and connecting the system to the electric grid.

From 0 to 4500 rpm the machine is supplied with 650V and the current level, due to the reduced electromagnetic force, is controlled by hysteresis band strategy. Equation 1 presents the current evolution in a saturated machine. The usual operation range is between 4500 and 6600 rpm whilst the DC voltage is controlled so as to achieve a single pulse current in the machine phases. As a result, the commutation losses at the converter and the hysteresis losses in the machine are being reduced to their maximum.

$$\frac{di}{dt} = \frac{V_{dc} - R \cdot i - w \cdot \frac{d\lambda}{d\theta}}{\frac{d\lambda}{dt}} \quad \text{(Eq.1)}$$

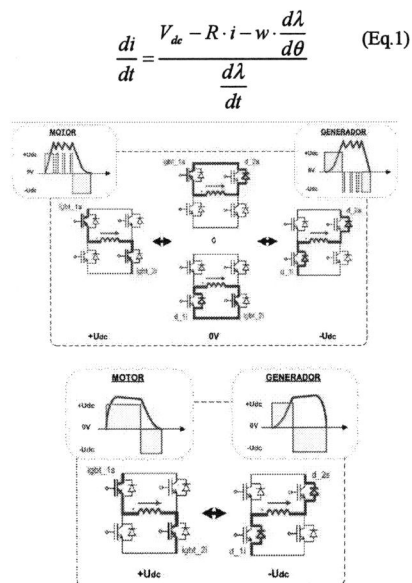

Figure 6. MSC topology and switches selection for low speed and high speed operation modes.

Figure 6 shows the different commutation modes for the MSC, hysteresis-band and single-pulse. Current and

voltage are shown in the schemes with IGBTs status so as to share the electrical stress. During hysteresis band operation, +Vdc, -Vdc and 0 volts are applied to the machine phases to control the current. During single-pulse operation, just +Vdc and –Vdc is sufficient to control the current whilst the commutation frequency is reduced to the value of *w(rpm)/30*.

The control architecture has been specifically proposed for this application. Reproducing the idea to achieve a totally independent operation (electric and electronic), a decentralised control architecture based on microcontroller boards connected through a 1Mbps CAN network is used. The on-board basic circuit is a 16-bit single-chip Renesas H8s2623 microcontroller working at clock frequency of 20 MHz. Modularity and decentralisation allows to distribute tasks in order to achieve very short cycle times, which are needed to deal with the high di/dt managed at the machine-side ($f_{sample}>50kHz$). Besides, this scheme allows the controller to grow larger in a loose manner, adding additional boards for controlling or monitoring other possible subsystems and for interconnecting with the solar plant operation, TSO command, etc.

The first complete prototype is a 120 kVA, 5MJ KESS. Figure 5 shows some details of the electrical machine and flywheel.

Figure 5. 120 kVA, 5MJ flywheel prototype.

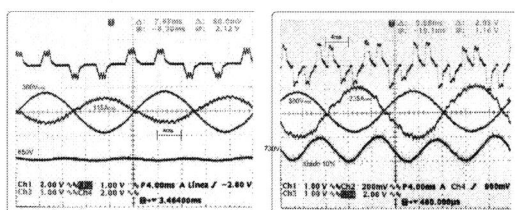

Figure 7.a. SRM charging operation (motor mode) at different speeds

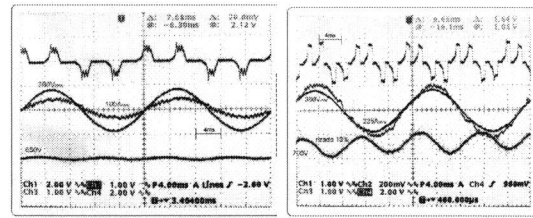

Figure 7.b. SRM discharging operation (generator mode) at different speeds

Figure 7 presents some experimental results obtained with the 120kVA-5MJ KESS prototype.

2484

A bigger prototype of 350kVA, 200MJ KESS, presented in figure 8, has been also developed and is being testing in CEDEX.

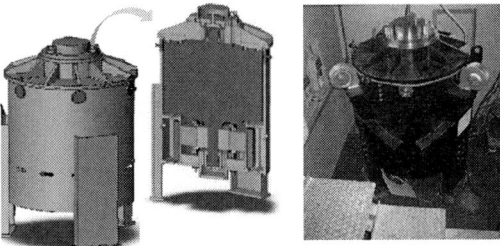

Figure 8. 350kVA, 200MJ KESS prototype

Some experimental results are included in figure 9 during charging and discharging operations.

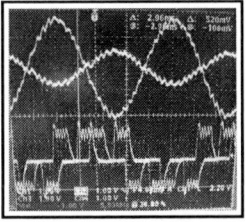 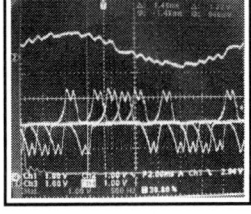

Figure 9. Network and machine waveforms of the 350 kVA, 200 MJ machine during motor and generator operation.

C. Efficiency Considerations

A KESS will spin continuously waiting for the power demand and just recovering the maximum speed when if gets an established value named recovery speed. So as the most important value to calculate the efficiency of the system is the amount of energy that is required to maintain the flywheel around the maximum speed, ready to supply the stored energy.

This target will be got as long as the aerodynamic losses are reduced with an appropriate vacuum level.

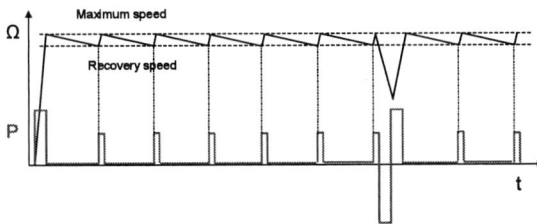

Figure 10. Speed and power profile during KESS operation.

Several experimental tests have been developed with the 120kVA machine at different vacuum levels inside the machine. Results are presented in figure 11. Taking into account the losses reduction by decreasing the power and the limitation due to the electric isolation (Paschen Curve) at lower pressure, 10mbar has been selected. An average power of 0,88 kW is required at these conditions.

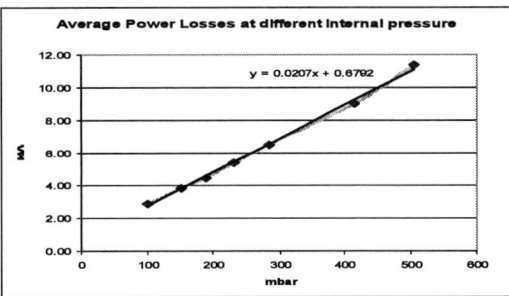

Figure 11. Evaluation of the average losses at different pressures.

As a summary, evaluating if a KESS is suitable for this energy and power specifications, energy requirement is the most restrictive. For instance, for an energy purpose of 1MWh, 36 big machines of 200MJ would be required (considering a discharge rate of 50%), meanwhile to cover the 1MW purpose would be achieved just with 3 machines of 350kVA. Therefore, the required KESS for this application would have a big flywheel and a very small machine. That is not an appropriate design since a much higher machine can be include at the system without an important increase in volume, weight and cost.

IV. HYBRID BATTERIES-FLYWHEELS ENERGY STORAGE SYSTEM

Considering the BESS solution for the energy management of the PV solar plant and the limitations reflected in chapter II, only a few sceneries can be afforded with it, as figure 12 presents. Maximum output power of 1MW can be supplied and in case of providing more power, the energy requirement is not achieved. Conventionally the BESS was oversized to get this target, increasing the total cost.

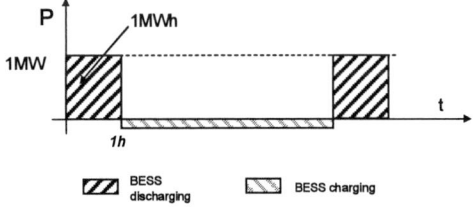

Figure 12. Operation with a BESS.

A more flexible capability can be provided by means of a hybrid BES-KES system as figure 13 displays. Since the batteries are operating with a lower maximum power, a larger energy amount can be supplied as figures 2 and 3 have demonstrated. Moreover, if the power demand takes no longer than 1h, the rest of power can be used to charge the KESS providing later an extra energy available when the batteries are completely discharge. Also if the power demand requires a high dP/dt, KESS can supply this faster than BESS does. Another improvement that can be achieved is the possibility to get power peaks higher than the maximum power of the plant during short periods of time.

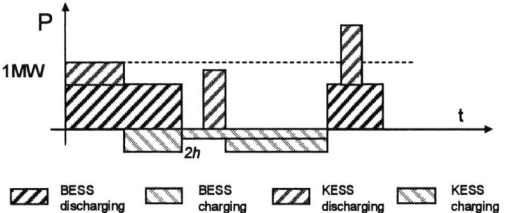

Figure 13. Operation with the hybrid BES-KES system.

Considering the TSO requirements, more and more selective, it is worth the effort to work in obtaining the most management level at the generation plant.

The global control operation of the plant will be integrated with the control of the solar inverter but over an independent platform. A control of the hybrid BES-KES system is being developed based on the following parameters: electric grid command, batteries voltage and flywheels energy. Since the flywheels operation is more flexible in terms of charge-discharge frequency, the operation criterion is to carry out the grid command by protecting the batteries from any operation that damage them, using the KESS to absorb the short-term power demands.

Figure 14 present the scheme of the PV solar plant integrating the hybrid BES-KES system proposed.

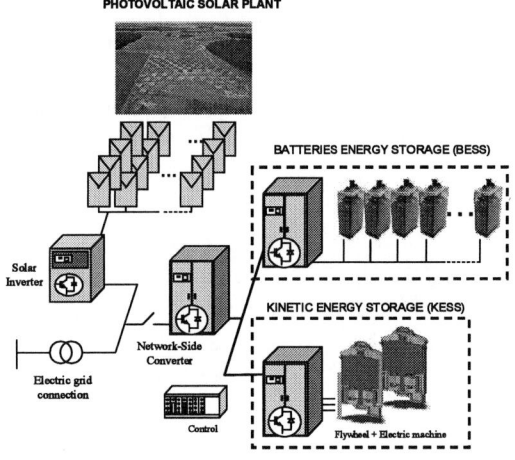

Figure 14. PV plant scheme including a hybrid BES-KES system.

V. CONCLUSIONS

Depending on the power and energy values proposed for the solar photovoltaic plant, different ESS may be selected to result in a better set of conditions to connect to the electric grid, guarantying the power quality and the electric grid stability. Among the different ESS, batteries energy storage (BES) and kinetic energy storage (KES) systems has been technically evaluated to provide an energy reserve of 1MWh, with a maximum power supply of 1MW. Advantages and drawbacks for both systems

were pointed out and it was learned that they can be complementary for some energy and power specifications.

A hybrid system based on BES-KES is proposed, able to afford with more flexibility the different energy-power sceneries that the TSO can propose to the generation plant.

REFERENCES:

[1] Valoración de Gestionabilidad de Centrales Termosolares. Red Eléctrica Española. Anexo de la Solicitud de acceso a red de transporte para centrales de generación fotovoltaica. Nov. 2007.

[2] Tyson Lawrence, Robert A. Zogg (TIAX Cambridge, Massachusetts, USA) " Distributed Energy Storage: Role in building-Based PV System". Procs. EESAT 2007.

[3] M. S. L. Florens. Data-driven solar wind model and prediction of type II bursts. GEOPHYSICAL RESEARCH LETTERS, VOL. 34, L04104, doi:10.1029/2006GL028522, 2007.

[4] R. Montenegro, G. Montero, J.M. Escobar, E. Rodríguez, J.M. González-Yuste, E. Rodríguez-Jiménez, A. González y F. Cerezo. "Predicción Eólica y Solar Combinando MM5 y Mallas de Elementos Finitos Adaptativas". Inst. Univ. de Sistemas Inteligentes y Aplicaciones Numéricas en Ingeniería, Universidad de Las Palmas de Gran Canaria, España.

[5] Ter-Gazarian, A.; "Energy Storage for Power Systems". London, UK: Peter Pergrinus Ltd., Redwood Books, Trowbridge, Wiltshire, UK. 1994.

[6] Ribeiro, P.F.; Johnson, B.K.; Crow, M.L.; Arsoy, A.; Liu,Y.; "Energy Storage Systems for Advanced Power Applications". Proc. of the IEEE, Dec 2001, Vol 89, pp. 1744-1756.

[7] "Investigation on Storage Technologies for Intermittent Renewable Energies: Evaluation and Recommended R&D strategy". Investire Network. Storage Technology Report WP-ST6-Flywheel. 2003

[8] Carl D. Parker "Lead-acid battery energy-storage systems for electricity supply networks". Journal of Power Sources, 2001.

[9] A. Oudalov, D. Chartouni, C. Ohler, G. Linhofer. "Value Analysis of Battery Energy Storage Applications in Power Systems". PSCE 2006 Proc. Pages 2206-2211.

[10] C.Vázquez, M.Lafoz, D. Ugena, L.G. Tabarés, Control System for the Switched Reluctance Drive of a Flywheel Energy-Storage Module, European Transactions on Electrical Power, 2007.

[11] Barnes, M.; Pollock, Ch.; "Power Electronics Converters for Switched Reluctance Drives". IEEE Transactions on Power Electronics, Vol. 13, N° 6, pp. 1100-1111, November 1998.

[12] Krishnan, R.; "Switched Reluctance Motor Drives: Modelling, Simulation, Analysis, Design and Applications", CRC Press LLC, New York, 2001.

A Method of Three-Phase Balancing in Microgrid by Photovoltaic Generation Systems

Masahide Hojo*, Yuta Iwase*, Toshihisa Funabashi[†] and Yoshinobu Ueda[†]

* The University of Tokushima, Tokushima, Japan, e-mail: *hojo@ee.tokushima-u.ac.jp*
[†] Meidensha Corporation, Tokyo, Japan, e-mail: *funabashi-t@mb.meidensha.co.jp*

Abstract—Recently, a small-scale power system has been able to perform autonomous or islanding operation with some distributed generators. But in such a microgrid, it is afraid that serious voltage imbalance may be caused by unbalanced loads during the islanding operation. Fortunately, grid interactive inverters of the distributed generators often have some amount of surplus capacity. This paper deals with a three-phase balancing method utilizing the surplus capacity. By the proposed method, the inverters output negative sequence currents to suppress the voltage imbalance within their surplus capacities. Operating characteristics and its effectiveness are confirmed by some simulation results.

Keywords—Power quality, distribution of electrical energy, renewable energy systems, three-phase system, photovoltaic.

I. INTRODUCTION

Power quality is one of issues in an electrical power distribution system. With rapid development of power electronics or related technologies, various types of power converters play a role of power quality enhancement [1]. In terms of three-phase system, load balancing is a critical issue and it is inherent in the system. A direct solution is to replace the electrical loads but it is difficult to be accomplished. As an alternative solution, installation of power converters has the potentials to realize the load balancing [2], if the converters have enough capacities to compensate the unbalances.

On the other hand, various types of distributed generators (DG) become available in an electrical power distribution network. Nowadays, a small-scale power system can be operated autonomously with such DGs. When the total capacity of the DGs is enough to support the all loads, the power system can move to an islanding operation. Such a small-scale power system can be referred to as a microgrid. There are some papers about the possibilities of the microgrid and its useful application [3], [4].

From the viewpoint of power quality in the islanding microgrid, it is afraid that serious voltage imbalance may be caused by insufficient load balancing [5]. In the conventional distribution system, most of the unbalanced load can compensate each other because many loads are installed. The easiest way to overcome the imbalance is to replace the load [6] as mentioned before. Of course it can be a good and simple answer on the first stage of the system design of the microgrid. However it may be difficult in most of practical cases in which the microgrid starts its operation. A large voltage imbalance may affect proper operation of other rotating type generators or motors in the microgrid. Although employing a compensator for the voltage imbalance can be also another solution, it will require some additional cost.

On the other hand, a photovoltaic generation system (PV system) can be one of the major generation systems in the microgrid. It means that there are many inverters in the microgrid as the utility interactive inverters for the PV systems. As the PV output power widely varies during a day, there is some surplus capacity of the inverter in accordance with the variations. Considering that many PV systems are installed and each inverter has surplus capacity, the surplus capacities are widely distributed in the microgrid. Therefore, it is useful to utilize the surplus capacities of the utility interactive inverters for the line voltage management. Especially, it is expected that the inverter contribute largely to improve power quality in the night time when the inverters of the PV systems are completely idle. Therefore, it is useful to apply some additional functions such as power quality enhancement to the inverters. Authors have some challenges to the power quality enhancement by the grid interactive inverters [7].

However, as the inverter must be operated within its rated capacity, a suitable control system should be considered in order to realize an effective compensation of voltage imbalance within the capacity. In case that the operation is restricted by an output current limiter, effective voltage balancing may not be achieved because of the output current distortion by the hard limitation. In order to overcome this difficulty, authors have developed an adjustable compensation method.

This paper deals with a three-phase balancing method utilizing the surplus capacity. By the proposed method, the inverters output additional currents of negative sequence to reduce unbalanced load currents. As a result, the voltage imbalance which is caused by the unbalanced load can be suppressed by the grid interactive inverters within their surplus capacities. As the method directly calculates the compensation current, the intensity of the compensation can easily regulated. It is confirmed by some simulation studies that the method can suppress the voltage imbalance effectively.

II. CONTROL SYSTEM OF THE GRID INTERACTIVE INVERTER

Fig. 1 shows a simple microgrid model with a grid interactive inverter which employs the three-phase balancing control. A three-phase voltage source E_s represents a high-voltage distribution line. It is assumed that voltage source E_s is balanced but the inductive load (R_l, X_l) is unbalanced. In this case, the grid interactive

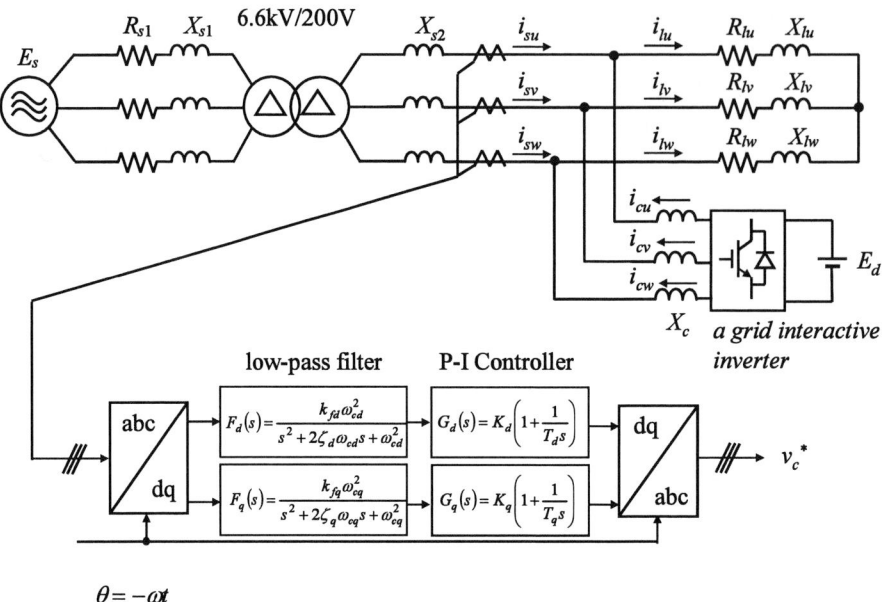

Fig. 1. A grid interactive inverter with the three-phase balancing control in a microgrid model.

inverter compensates the unbalanced load current so as to balance the three-phase source current.

A block diagram of the inverter control system is shown in Fig. 2. The inverter detects the line currents at the source side and converts them to two-axis components. The two-axis transform can be formulated as

$$
\begin{bmatrix} i_{sd} \\ i_{sq} \\ i_{s0} \end{bmatrix} = \frac{2}{3} \begin{bmatrix} \cos\theta & \cos\left(\theta - \frac{2}{3}\pi\right) & \cos\left(\theta + \frac{2}{3}\pi\right) \\ \sin\theta & \sin\left(\theta - \frac{2}{3}\pi\right) & \sin\left(\theta + \frac{2}{3}\pi\right) \\ \frac{1}{2} & \frac{1}{2} & \frac{1}{2} \end{bmatrix} \begin{bmatrix} i_{su} \\ i_{sv} \\ i_{sw} \end{bmatrix} . \quad (1)
$$

This transform is executed by the phase angle of the negative sequence as represented by

$$ \theta = -\omega t . \quad (2) $$

By the calculations, the variables i_{sd} and i_{sq} include both ac and dc components. The ac components correspond to the positive sequence current. Therefore, the negative sequence components of the line current can be extracted from the observed source current as a dc quantity of them through a low-pass filter. The cut-off frequency of the filter should be carefully designed because it affects transient responses of the controller as well as depend on the fundamental frequency of the system.

By this method, the intensity of the balancing control can be easily modified with regulating the dc component by a control gain k. In this summary, the control gain is assumed to be fixed for the simplicity.

The output voltage of the grid interactive inverter for the compensation is determined by the conventional proportional-integral controller in each component.

Finally, three-phase voltage reference v_c^* which is used for PWM control of the inverter is calculated from the two-axis components. The inversed transform is formulated as

$$
\begin{bmatrix} v_{cu}^* \\ v_{cv}^* \\ v_{cw}^* \end{bmatrix} = \begin{bmatrix} \cos\theta & \sin\theta & 1 \\ \cos\left(\theta - \frac{2}{3}\pi\right) & \sin\left(\theta - \frac{2}{3}\pi\right) & 1 \\ \cos\left(\theta + \frac{2}{3}\pi\right) & \sin\left(\theta + \frac{2}{3}\pi\right) & 1 \end{bmatrix} \begin{bmatrix} v_{cd} \\ v_{cq} \\ v_{c0} \end{bmatrix} . \quad (3)
$$

These calculations are summarized in Fig. 1.

In this study, the dc voltage of the grid interactive inverter is supplied by an ideal dc voltage source as shown in Fig. 1.

The control strategies can easily employ the intensity regulation of the balancing compensation. One can easily regulate the intensity by modifying the both low-pass filter gains simultaneously.

III. SIMULATION STUDIES

The balancing abilities are confirmed by following simulation studies. The system shown depicted in Fig. 1 is used and the parameters are summarized in Table 1.

A. Examples of Unbalanced Load Compensation

Fig. 2 (a) shows the unbalanced load current in this case study. The output current of the three-phase inverter is shown in Fig. 2 (b), which includes negative sequence component for the balancing control. The source currents are also depicted in Fig. 2 (c). By these figures, it is confirmed that the inverter can realize the load current balancing.

TABLE I.
SYSTEM PARAMETERS

Parameters of a microgrid
System frequency f_s = 50Hz Voltage source, E_s = 200V, line to line, rms Line impedance, (R_{s1}, X_{s1}) = (0.37, 0.05) Ω Line reactance, X_{s2} = 0.63 Ω Inductive load, (R_{lv}, X_{lv}) = (R_{lw}, X_{lw}) = (0.4, 0.063) Ω [pf=0.99] $\qquad\qquad (R_{lu}, X_{lu})$ = (0.37, 0.31) Ω [pf=0.69]
Parameters of an inverter
Switching frequency : 10kHz, the triangular wave PWM Dc voltage, E_d = 300V Interconnecting reactance, X_e = 0.63 Ω Gains of the low-pass filter, $k_{fd} = k_{fq}$ = 0.001 Cut-off frequencies of the low-pass filter, $f_{cd} = f_{cq}$ = 10Hz Control gains of the P-I controller, K_d = –0.03, K_q = –0.03 Time constants of the P-I controller, T_d = 0.1ms, T_q = 1ms

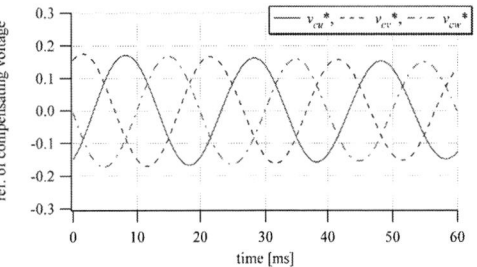

Fig. 3. Reference signal for the inverter output voltage.

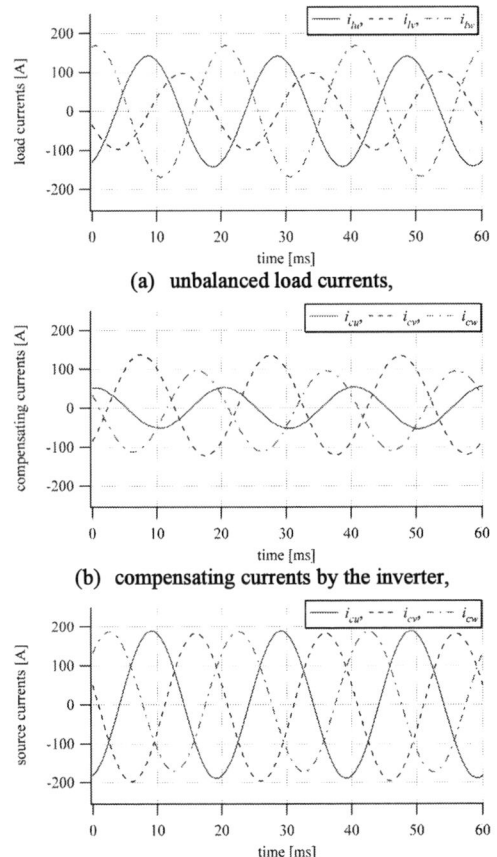

(a) unbalanced load currents,

(b) compensating currents by the inverter,

(c) compensated source current.

Fig. 2. Line currents with the balancing method (case 1).

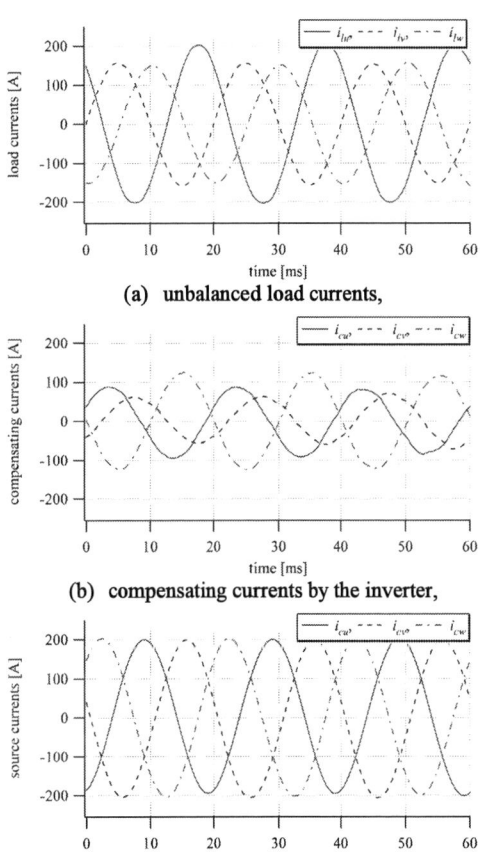

(a) unbalanced load currents,

(b) compensating currents by the inverter,

(c) compensated source current.

Fig. 4. Line currents with the balancing method (case 2).

Fig. 3 shows the reference signal for the output voltage of the inverter. It should be noted that the waveforms are the negative components.

Fig. 4 shows the other test case where the load is replaced to high power factor load. The load on the phase u is replaced to (R_{lu}, X_{lu}) = (0.20, 0.04) Ω whose power factor is 0.98. When the low impedance with high power

factor load is connected only on the phase u, the grid interactive inverter can compensate the load unbalance successfully as shown in Fig. 4.

B. *Intensity Regulation of the Load Balancing Compensation*

In the next results, the intensity regulation of the load balancing compensation is tested by a simulation study. Fig. 5 shows the results with same parameters used in the case 2, which is displayed in Fig. 4, except that the gains of the low-pass filter is decreased to the half of their original values. As shown in Fig. 5, Compensating current is reduced and the source current unbalance remains. As

2489

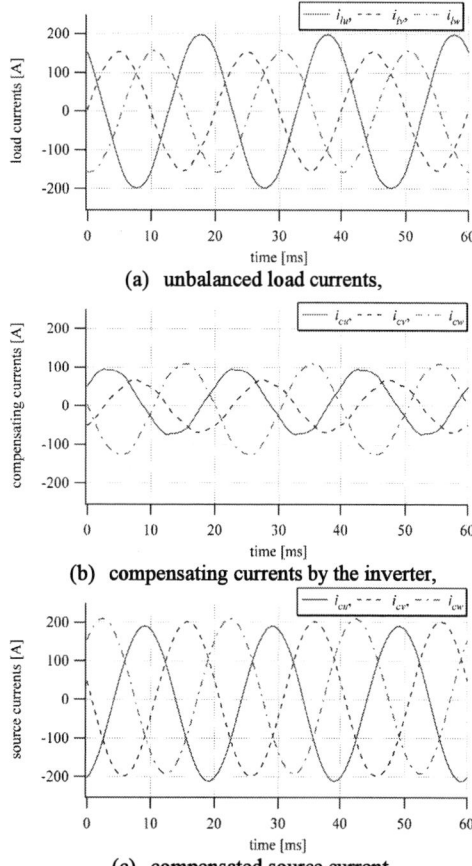

(a) unbalanced load currents,

(b) compensating currents by the inverter,

(c) compensated source current.

Fig. 5. Line currents with the regulated balancing method.

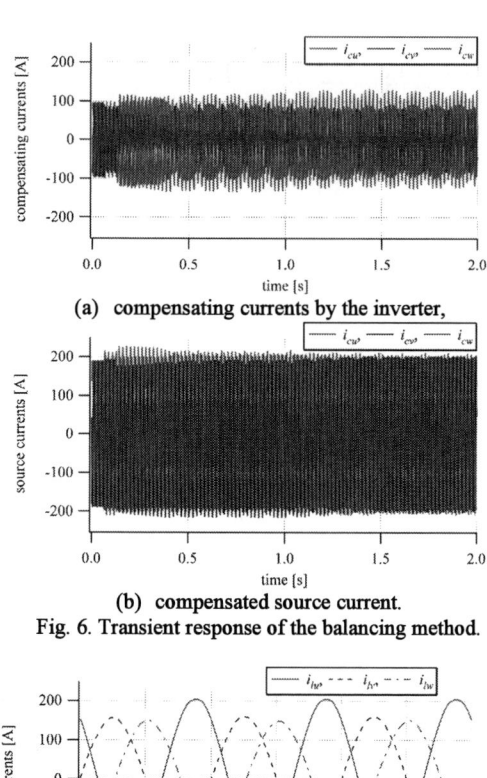

(a) compensating currents by the inverter,

(b) compensated source current.

Fig. 6. Transient response of the balancing method.

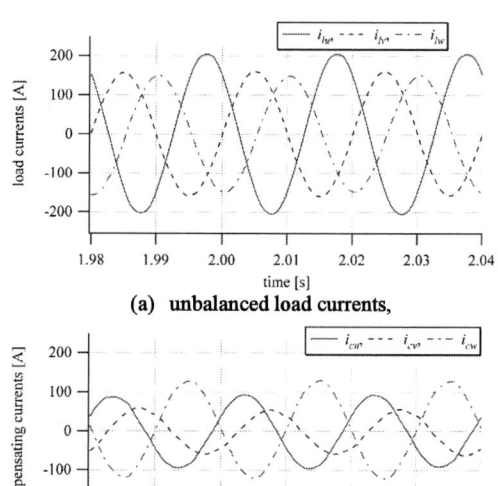

(a) unbalanced load currents,

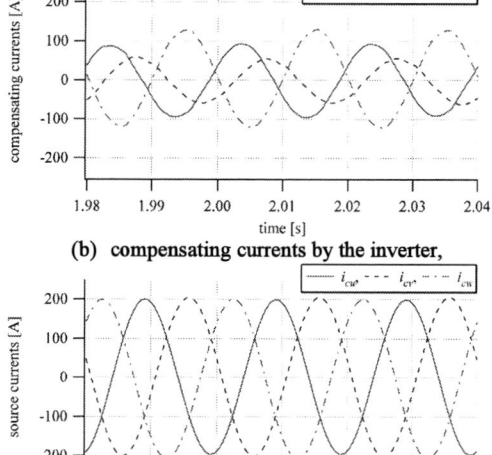

(b) compensating currents by the inverter,

(c) compensated source current.

Fig. 7. Line currents after the load varied.

compared to the Fig. 4, it can be seen that both of the effect and compensating current are reduced. As the regulation can be accomplished only by modifying the low-pass filter gains, the regulation of the compensation can be applied as an online function along with the surplus capacity of the inverter.

C. Transient Response against a Sudden Load Change

Finally, transient response of the compensating method is investigated in case of load variation. The load was initially balanced but the load on the phase u was switched to its half value at the time of t=0.06s. This means that the grid interactive inverter must start its compensating function. All system parameters are same as the first study except that the load started from the balanced load. That is, $(R_{lu}, X_{lu}) = (R_{lv}, X_{lv}) = (R_{lw}, X_{lw}) = (0.4, 0.063)$ Ω. As the method employs low-pass filters, the response can be affected by the cut-off frequency of the low-pass filter.

Fig. 6 shows the transient response of the system. Before the load variation, the load current was balanced well as shown in Fig. 6 (a). The load current changed at t=0.06s and the three-phase current became unbalanced current. However, just after the load variation, the grid interactive inverter started to compensate the unbalanced load current. Then, the compensation is well accomplished around t=2.0s. Fig. 7 shows the current waveforms around the time. It can be confirmed that the balancing the source current is achieved. From these results, the load current balancing can be realized in about two seconds.

IV. CONCLUSIONS

This paper deals with a three-phase balancing method based on direct determination of the negative sequence line current. This method features that its intensity of the compensation can be easily regulated by online. This possibility is suitable for utilizing the surplus capacity of the photovoltaic generation systems in a microgrid.

The system operation is verified under various situations by some simulation studies. First, its proper operation against unbalanced loads which have different power factor is confirmed. Secondly, regulation of the intensity to compensate the unbalanced load is investigated. It is proved that the intensity can be easily regulated by modification of the low-pass filter gains. Finally, transient response is also tested because the system is subjected to frequent load variations under the practical situation. When the load was suddenly switched from balanced to unbalanced load, the grid interactive inverter successfully started the compensation and accomplished in two seconds.

From these results, the grid interactive inverter is expected to provide a useful solution to enhance the load current balancing in the microgrid. Especially, three-phase voltage sourced inverters can play an important role as the electrical power generation system in the microgrid. This method has a potential to enhance the power quality in the microgrid.

As the future work, these operations should be verified by an experimental setup. In addition to this, management method of multiple inverters should be also considered under some practical situation.

REFERENCES

[1] A. Ghosh and G. Ledwich, *Power Quality Enhancement Using Custom Power Devices*, Kluwer Academic Publishers, Dordrecht, 2002.

[2] A. Ghosh and A. Joshi, "A new approach to load balancing and power factor correction in power distribution system," *IEEE Trans. on Power Delivery*, vol. 15, no. 1, pp. 417–422, January 2000.

[3] S. Abu-Sharkh, R.J. Arnold, J. Kohler, R. Li, T. Markvart, J.N. Ross, K. Steemers, P. Wilson and R. Yao, "Can microgrids make a major contribution to UK energy supply?" *Renewable and Sustainable Energy Reviews*, vol.10, no. 2, pp.78-127, April 2006.

[4] T. Kato, Y. Kondo, Y. Suzuoki and T. Funabashi, "Conceptual study of microgrid for electricity supply during urban disaster," *Proc. of Intl. Conf. on Advanced Power System Automation and Protection*, no. P560, April 2007.

[5] J. Sumita, K. Nishioka, T. Shimakage, Y. Noro, Y. Ito and M. Yabuki, "Study of measures to reduce voltage fluctuation and influence of unbalance loads during isolated operation of a microgrid," *The 2007 Annual Meeting Record IEE Japan*, no. 6-033, March 2007 (in Japanese).

[6] H. Maejima, T. K. Uchida, K. Temma, K. Kuroda and M. Shimomura, "A demonstrative microgrid project in Hachinohe city part 3: Study of system unbalance under islanding operation," *The 2007 Annual Meeting Record IEE Japan*, no. 6-031, March 2007 (in Japanese).

[7] M. Hojo and T. Ohnishi, "Adjustable Harmonic Mitigation for Grid-Connected Photovoltaic System Utilizing Surplus Capacity of Utility Interactive Inverter," *IEEE Power Electronics Specialists Conference–2006 Record*, pp.1734-1739, June 2006.

Development of HILS(Hardware In-Loop Simulation) System for MMS(Microgrid Management System) by using RTDS

Jin-Hong Jeon[*], Jong-Yul Kim[*], Seul-Ki Kim[*], ong-Bo Ahn[*], and JuneHo Park[†]

[*] Korea Electrotechnology Research Institute, Chang-won, Korea, jhjeon@keri.re.kr
[†] Pusan National University, Busan, Korea

Abstract— This paper proposes a structure of an HILS based development System for MMS. This system is composed of RTDS and its hardware interface devices, MMS, communication emulator modules. The RTDS is operated for real-time simulation for both microgrid and microsource models. By using this HILS system, the secondary regulation action of MMS during an islanded operation is evaluated. The secondary regulation action of MMS is designed to control the microsource's power output set-point in order to make the BESS power output zero. The test results show that the secondary regulation action of MMS can regulate the frequency and voltage while minimizing the consumption of the stored energy of BESS.

Keywords— Microgrid, MMS, RTDS, HILS, Islanded Operation

I. INTRODUCTION

Though the penetration of distributed generations to the electric power system is limited due to the lack of economical benefits, it will be accelerated for various reasons, such as environmental friendliness, the diverse needs of the end user for high quality power, the restructuring of the electric power industry, and restrictions on the extension of power transmission and distribution facilities. The increase in DGs' (Distributed Generators) penetration depth and the presence of multiple DGs in electrical proximity to one another have brought about the concept of the microgrid [1-3], which is a cluster of interconnected DGs, loads and intermediate energy storage systems. As usual, the microgrid operates in grid-connect mode, but when a fault occurs in the upstream grid the microgrid should disconnect and shift into islanded operation mode. In grid-connect mode, the frequency of the microgrid is maintained within a tight range by the main grid. In an islanded operation, however, which has relatively few microsources, the local frequency control of the microgrid is not straightforward. The frequency of the power system has a strong coupling with the active power in the network. If demand increases, the frequency will fall unless there is a matching increase in generation and vice versa. The rate of change of frequency depends on the inertia of the systems (i.e., the larger the inertia, the smaller the rate of change). During a disturbance, the frequency of the microgrid may change rapidly due to the low inertia present in the microgrid. Therefore, local frequency control is one of the main issues in islanded operation [4]. To overcome this challenge, the cooperative control of the microsources and the energy storage system is very important in maintaining the frequency of the microgrid during islanded operation.

By the cooperative control strategy, the ESS handles the frequency and voltage as a primary control. Then, the secondary regulation control of MMS returns the current power output of ESS into a pre-planned value. In this paper, the secondary regulation action of MMS is evaluated by HILS test.

II. CONFIGURATION OF MICROGRID

A microgrid is a cluster of interconnected microsources that are referred to as distributed generators, loads and intermediate energy storage systems that co-operate with each other to be collectively treated by the grid as a controlled load or generator [5]. Fig. 1 shows a typical configuration of a microgrid. It comprises, in addition to loads, RES (Renewable Energy Sources) as well as a diesel engine, gas engine, micro turbine and an energy storage system as support systems to fulfill the stable operation of a facility. The microgrid disconnects from the utility grid, however, and transfers into islanded operation mode when a fault occurs in the upstream grid. The conventional microsources existing in the distribution network mainly use rotating machines. They are directly connected the grid to supply electric power, but new technologies, such as the micro turbine, PV, Wind, and Fuel Cells, that are proposed for use in the microgrid are not suitable for supplying energy to the grid directly. They have to be interfaced with the grid through an inverter. In terms of controllability of power output, there are two types of microsources in the network: controllable and non-controllable sources. Controllable sources mean that the power output of each microsource can be controlled by the command signal arbitrarily. On the other hand, non-controllable sources, such as RES, do not have this manageable nature since their output power depends on the availability of the primary source (wind, sun, etc.). The loads are either critical or non-critical. Critical loads require a high reliability and quality of power without any short-term interruptions in both normal and emergency situations. Non-critical loads may be shed during emergency situations. The intermediate ESS (Energy Storage System) is an inverter-interfaced battery bank (BESS), SMES, supercapacitor or flywheel. The storage device in the microgrid is analogous to the spinning reserve of large generators in the conventional grid. They ensure the balance between energy generation and consumption, especially during islanded operation [6].

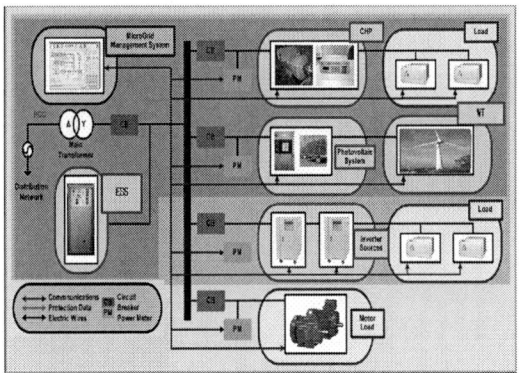

Fig.1. Typical configuration of a microgrid.

III. MICROGRID MANAGEMENT

A. Hierarchical Control Structure of Microgrid

The microgrid has a hierarchical control structure, as shown in Fig. 2. It has two control layers: MMS (Microgrid Management System) and LC (Local Controller). The MMS is a centralized controller that deals with management functions such as disconnection and re-synchronization of the microgrid along with the load shedding process. In addition to this function, the MMS is responsible for the supervisory control of microsources and the energy storage system. Using collected local information, the MMS generates a power output set point and provides it to the LCs. An LC is a local controller that is located at each microsource and controls the power output according to the power output set from the MMS [7].

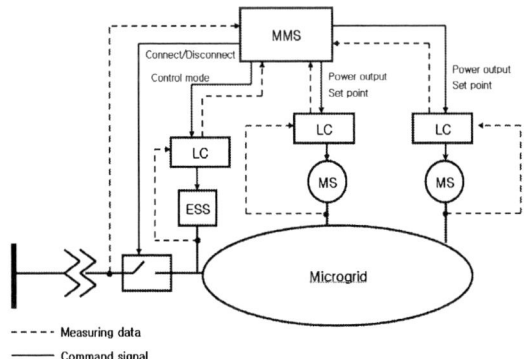

Fig. 2. Hierarchical control structure of a microgrid.

B. Management of Microgrid

The microgrid can operate in both grid-connected and islanded mode. Its functionalities as well as its control structure depend on the mode of operation. In the following section, these modes of operation and the transitions from one mode of operation to the other are analyzed in detail.

a) Islanded Mode

In islanded mode, due to the unavailability of the utility grid, two requirements must be fulfilled: the power balance between the generation and the consumption and the control of the main parameters of the installation (Voltage amplitude and frequency). Due to the non-controllable nature of RES generators, the controllable microsources such as engine generator, fuel cell, and energy storage system are responsible for ensuring the power balance by means of absorbing and injecting the power difference between the renewable generation and the local load. In case of an energy excesses, the management system has to decrease the output power of the controllable microsources to maintain the frequency and voltage of the microgrid. On the other hand, if all available power is not enough to feed the local loads, the management system will detach non-critical loads.

b) Grid-Connected Mode

In grid-connected mode, the balace between generation and the consumption as well as the control of the parameters of the system is guaranteed by the utility grid. Thus, generators are regulated with the criterion of optimized economic exploitation of the installation.

For the RES generators, it is considered more economic to generate energy locally than to buy it from the grid. Therefore, their references are calculated with the objective of extracting the maximum power from the primary sources. Concerning the controllable microsource, the objective of the control is to optimize the microgrid performance in order to contribute to the grid operation and, in this way, receive as much a bonus as possible and avoid possible penalties:

- By means of the exact fulfillment of the production program, economic operation of the microgrid can be achieved.
- By controlling the power at the point of common coupling(PCC), the microgrid operates as a constant power generator or a constant power load.
- By filtering the active power injected to the grid or absorbed from the grid, the impact on power quality can be decreased.

IV. HILS TEST SYSTEM

The RTDS is a fully digital electromagnetic transient power system simulator capable of continuous, sustained real-time operation[8]. That is, it can solve the power system equations fast enough to continuously produce output conditions that realistically represent conditions in the real network. The RTDS has been widely accepted as an ideal tool for the design, development and testing of power system protection and control schemes. Fig. 3 is a schematic diagram of the HILS system for a real-time simulation and a management function test of MMS. The HILS system consists of a microgrid model realized by RTDS, PC-based MMS, and communication emulator. The microgrid model, including microsources, loads, and distribution line is represented by RSCAD software and plays a real time simulation. For the simulation, needed modules of the RTDS system are a WIF card for a communication of system, four 3PC cards with optic ports for system interface, a GPC card for a network solution, two 16 bit DA for output signal, which transmit a monitoring data for MMS, two 16 bit AD cards for input signal, which come from the MMS to control the power output of the microsources, and a DIO card for digital input and output signal, as shown in Fig. 4.

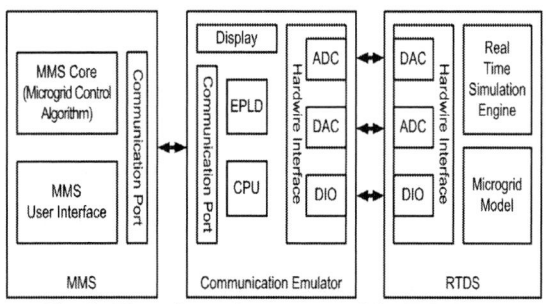

Fig. 3. The schematic structure of the HILS system

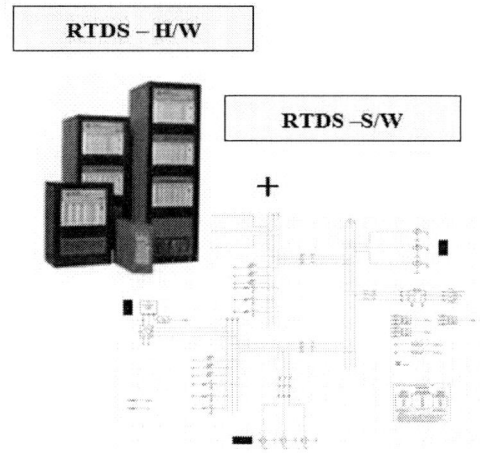

Fig. 4. A picture of implemented RTDS system

A PC-based MMS consists of an MMI module and a management function algorithm module. The MMI module gives operating status information of the microgrid and trend view of monitored data. Management function algorithm is implemented by C++ language, as shown in Fig. 5. In a real microgrid system, the MMS should communicate with every component. However, RTDS doesn't have its own communication interface device; it just gives a hardwire interface with H/W.

Therefore, a communication emulator is required to mimic the communication between MMS and RTDS, modeling the microgrid. The communication emulator connects with the RTDS by a hardwire interface and with MMS by a RS-485 serial communication port.

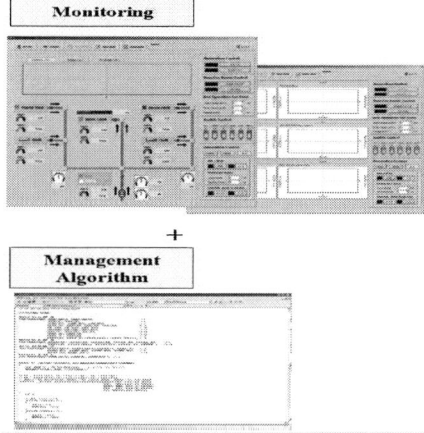

Fig. 5. A structure of MMS

V. SECONDARY REGULATION FUNCTION OF MMS DURING ISLANDED OPERATION

A. Control of Frequency and Volatge by ESS [4]

In grid-connect mode, the frequency of the microgrid is maintained within a tight range by the main grid, but, in islanded mode, which has relatively few microsources, the local frequency control of the microgrid is not straightforward. During islanding, the power balance between supply and demand does not match at the moment. As a result, the frequency of the microgrid will fluctuate, and the system can experience a blackout unless there is an adequate power balance matching process. The rate of change of frequency depends on the inertia of the systems (i.e., the larger the inertia, the smaller the rate of change). During a disturbance, the frequency of the microgrid may change rapidly due to the low inertia present in the microgrid. The energy storage system inverter controller responds in milliseconds. Otherwise, because of the nature of the control of some microsources such as the diesel generator, gas engine, and fuel cell, these microsources have a relatively slow response time. Obviously, the ESS should play an important role in maintaining the frequency and voltage of the microgrid during islanded operation. In grid-connect operations, all of the microsources and ESS adopt the fixed power control mode, which means that the microsources and ESS generate constant active and reactive power. Usually, power output of the ESS may be fixed at zero when the microgrid is operated in grid connect mode, but, since the fixed power control of ESS supplies constant power, it cannot provide proper frequency and voltage controlling ability in islanded operation. Therefore, the control scheme of the ESS has to be switched from fixed power control to droop or constant frequency/voltage control during islanded operation. Otherwise, other controllable microsources are still fixed power control. In this paper, frequency/voltage control is considered. Fig. 6 shows the detailed control block of an ESS, as previously mentioned.

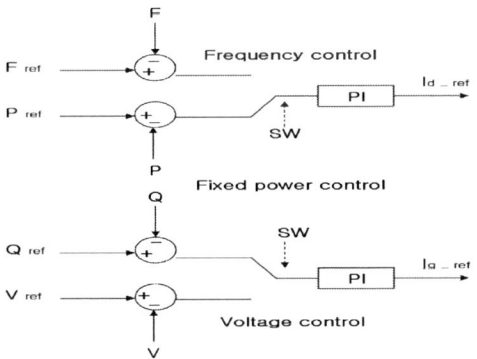

Fig. 6. Control scheme of the ESS.

B. Secondary Regulation Control

Though the frequency and the voltage of microgrid in islanded operation can be effectively controlled by applying a droop control scheme in the ESS, the control capability of the ESS may be limited by its available energy storage capacity. Therefore, the power output of the ESS should be brought back to zero as soon as possible. The secondary regulation of MMS is in charge of returning the current power output of the ESS to the pre-planned value, which is usually set at zero. When operating in islanded mode, the ESS regulates the frequency and the voltage in local control actions, and then the MMS calculates the proper power outputs of each microsource to make the power output of the ESS equal zero, as shown in Fig. 7. The calculation procedure of MMS is as follows:

a) *Calculate the power deviation of ESS:*

$$\Delta P_{ESS} = P_{ESS}^{mea} - P_{ESS}^{sch} \qquad (1)$$

where ΔP_{ESS} is the power deviation, P_{ESS}^{mea} is the measured active power output, and P_{ESS}^{sch} is the pre-planned power output.

b) *Calculate the power output of each microsource:*

$$\Delta P_{ref,i} = pf\_P_i \times \Delta P_{ESS} \qquad (2)$$

where $\Delta P_{ref,i}$ is the change of power output at ith microsource, and pf_P_i is the participant factor of ith microsource.

Therefore, the final active power output set point of the microsource is determined by the summation of base values and change of power output, as shown in Fig. 8.

The base set point is updated every minute, and the change of power output value is calculated every a few seconds by the MMS. The final reactive power output set point produced can also be calculated in the same way as active power.

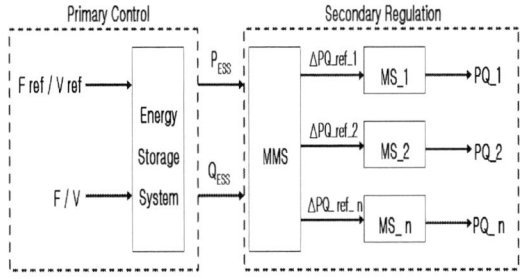

Fig. 7. Cooperative control structure.

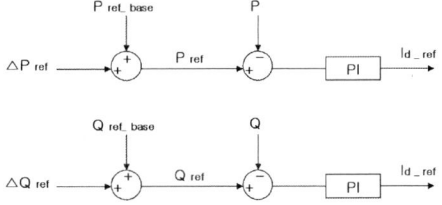

Fig. 8. Microsource power output set point calculation.

VI. CONFIGURATION OF TEST SYSTEM

A test system under the RTDS environment was developed to evaluate the dynamic behavior of the microgrid. Fig. 9 shows the studied microgrid model in RTDS software. The PV/wind Hybrid and diesel generation system has been connected to 380V low voltage lines, which connects to a 22.9kV distribution feeder through a pole transformer. The BESS is connected to a 380V busbar, which is near the PCC. The 380V line is connected to microsources and two loads through the overhead line. In the RTDS model, the RES sources and BESS are modeled as an equivalent current source model for analysis convenience. A typical synchronous generator model in the RTDS library is used to represent the diesel generator. The upstream grid is modeled by an equivalent voltage source with thevenin impedance, and the load and line impedance are represented by constant impedance models, R and X. The detailed aspects of the test system are as follows:

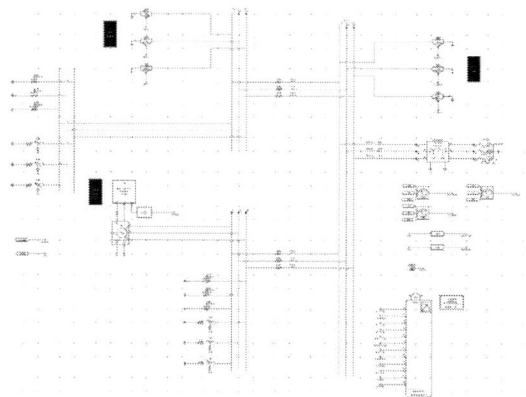

Fig. 9. RTDS model of test microgrid system.

a) Test System Configuration
 - No. of Sources: 3 (Hybrid, Diesel generator, BESS)
 - No. of Loads: 2

b) Generation Capacity of Microsources
 - Hybrid 25kVA
 - Diesel generator 25kVA
 - BESS system 15kVA

c) Total Load
 - 40kW+j19kVar
 - Constant impedance model (R/X)

d) Transformer
 - 3phase 22.9/0.38kV 100kVA
 - Leakage impedance %Z = 6%

e) Line impedance
- R= 0.1878ohm/km, X=0.0968ohm/km

VII. TEST RESULTS

The management function of the MMS during islanded operation is evaluated. In this operating mode, the BESS controls the frequency and voltage in primary, and then the secondary regulation control action is performed by the MMS. The initial condition is characterized by a total load of 25kW+j12kVar and the generation of PV/Wind hybrid 10kW+j6kVar, and diesel generator 15kW+j6kVar.

The two consecutive events are applied. Total load increases from 25k+j12kVar to 30kW+j12kVar and then decreases back to initial state. For first and second load changing, the power output of BESS increases or decreases from zero to 5kW or from zero to -5kW very quickly. The power output of BESS changes from zero to a certain value to control the frequency and the voltage as soon as the disturbance occurs, and it is returned to zero due to secondary regulation control, which is performed in the MMS as shown in Fig. 10. The power output of microsources is changed from an initial constant value to a new set point calculated by (1)-(2). This secondary regulation control can reduce the consumption of the stored energy of BESS without lowering the control performance. Fig. 11 shows the feature of tested MMS, which is monitoring and conducting secondary regulation control action.

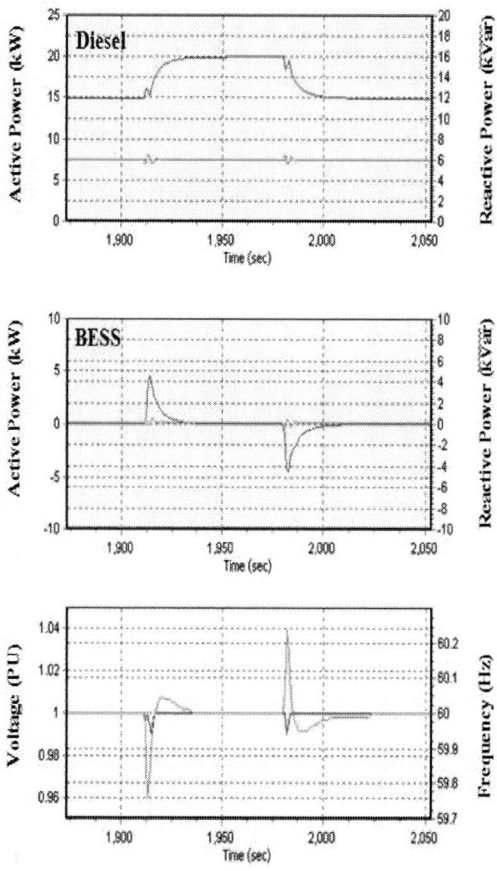

Fig. 10. Test results.

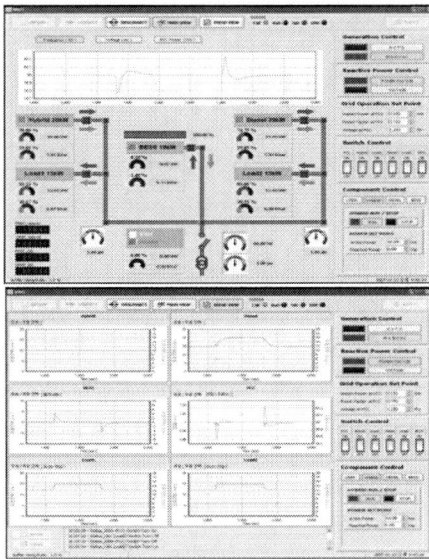

Fig. 11. Feature of tested MMS.

VIII. CONCLUSIONS

In this paper, a structure of an HILS based development system for an MMS is presented. This system is composed of RTDS and its hardware interface devices, MMS, communication emulator modules. By using this HILS system, the secondary regulation action of MMS during islanded operation is evaluated. The secondary regulation action of an MMS is designed to control the microsource's power output set-point in order to make the BESS power output zero. The test results show that the secondary regulation action of MMS can regulate the frequency and voltage while minimizing the consumption of the stored energy of BESS.

REFERENCES

[1] CERT, "Integration of Distributed Energy Resources: The CERTS MicroGrid Concept", LBNL-50829, 2003.

[2] N. D. Hatziargyriou, A.P. Meliopoulos, "Distributed Energy Sources: Technical Challenges", *IEEE Power Engineering Society Winter Meeting*, pp. 1017-1022, 2002.

[3] F. Katiraei, M. R. Iravani, P. W. Lehn, "Micro-Grid Autonomous Operating During and Subsequent to Islanding Process", *IEEE Trans. Power Delivery*, vol. 20, pp. 248-257, 2005.

[4] EU Microgrid Project, "Microgrids Large Scale Integration of Microgeneration to Low Voltage Grids", ENK-5-CT-2002-00610, 2004.

[5] R.H. Lasseter, "MicroGrids", *IEEE Power Engineering Society Winter Meeting*, pp. 146-149, 2001.

[6] Paolo Piagi, Robert Lasseter, "Autonomous Control of Microgrids", *IEEE Power Engineering Society Meeting*, 2006.

[7] J. A. Pecas Lopes, C. L. Moreira, A. G. Madureira, "Defining Control Strategies for MicroGrids Isladed Operation", *IEEE Trans. Power Syst.*, vol. 21, pp. 916-924, 2006.

[8] P. Rorsyth, T. Maguire, and R. Kuffel, "Real-time digital simulation for control and protection system testing", in Proc. 35th Annual IEEE Power Electronics Specialists Conference, Aachen, Germany, June 20-25, 2004, pp. 329-335.

Power Quality Analysis of Jeju Island Power System with Wind Farm and HVDC System

Jae-Hong Kim[1], Eel-Hwan Kim[1], Se-Ho Kim[1], Jaeho Choi[2], Gil-Soo Jang[3], Seung-Ho Song[4]

[1]Dept. of Electrical Eng., Cheju National Univ., Jeju, Korea, *ehkim@cheju.ac.kr*
[2]School of Electrical & Computer Eng., Chungbuk National Univ., Cheongju, Korea, *choi@chungbuk.ac.kr*
[3]Dept. of Electrical Eng., Korea Univ., Seoul, Korea, *gjiang@Korea.ac.kr*
[4]Dept. of Electrical Eng., Kwangwoon Univ., Seoul, Korea, *ssh@kw.ac.kr*

Abstract— This paper presents the modeling and power quality analysis of Jeju island power system connected with wind farm, current source type HVDC system, and thermal power plant. It is for indicating the influence of wind farm operation in steady and transient state in Jeju island power system. For the computer simulation, four kinds of main items are modeled, which are 83(MW) wind farm, 300(MW) HVDC system, thermal power plant, and Jeju power load. To analyze the influence of the wind power generation to the Jeju power system, two kinds of simulations are carried out by using the PSCAD/EMTDC program. One is the steady state operation under the different wind speeds, and the other is the transient state operation when all of wind farms in Jeju island are disconnected from the Jeju power grid instantaneously due to the over wind speed than the rated value. These comparative studies are useful to analyze the influence of wind farms on the Jeju island power system stability.

Keywords—PSCAD/EMTDC, wind power generation, power quality

I. INTRODUCTION

In recent years, the wind power generation is very attractive to the renewable energy field because it is effectively possible to improve the capturing wind energy capability with the development of power electronics, aero dynamics and computer technology. Specially, with the effectuation of the Kyoto Protocol, many countries are concerned about the wind power generation. So, it is a trend in wind power generation to combine the large scaled wind turbine generation system (WTGS) and build a grid connected wind farm [1][2]. The location for wind farm is usually specified depending on the wind speed, and therefore the wind farm is concentrated in the special area. The wind in Jeju island is strongest in Korea and it has the best condition for wind power generation. The annual wind speed in the east and west area of Jeju island is about 7~8[m/s]. This is why so many companies want to invest a lot of funds to construct the wind farm in that area [3].

But there are some constrained conditions for the penetration of wind power generation to the power system in Jeju island due to the distribution of wind velocity and weak power load. First of all, total power generation capacity in Jeju island including HVDC

system is only about 730[MW]. And the capacity to install the wind power generation should be limited which can guarantee the power system stability. In 2008, 34[MW] power capacity of wind power generation system is being operated and about 50[MW] systems are under construction. And there are two times more applications for wind power generation so far. Then, it is one of the hot issues in Jeju island how much capacity of wind farm is acceptable without any harmful effects on the stability of Jeju power system. It is strongly recommended to analyze the influence of wind power generation system to the Jeju power grid before penetrating the wind farms to the grid.

In this paper, two kinds of simulations are carried bout for the analysis of its influence by using the PSCAD/EMTDC program which is widely favored for the analysis of power system including power electronics devices [4][5]. One is the steady state operation under the different wind speeds, and the other is the transient state operation when all of wind farms in Jeju island are disconnected from the Jeju power grid instantaneously caused by the over wind speed than the rated value. It is very useful to analyze the variation of grid voltage and frequency under the transient state for the understanding of the influence of wind power generation in Jeju island power system. Finally, the extensive results will be described and discussed.

II. CHARACTERISTICS OF POWER SYSTEM AND WIND VELOCITY IN JEJU ISALND

A. Power system in Jeju island

Although small, the power system in Jeju island has many kinds of power source components, including thermal power plant generation, transmission, distribution, HVDC and renewable energy system. It shows that the possibility of electric fault is high and it will be increased as the new install of renewable energy sources. And this means that power system stability in Jeju island will be debased. Also, the rate of annual load increase is almost about 8~9[%], which is faster than that of the main land of Korea. As the Jeju island to become a free-trade international province, the need for reliable supply of electric energy sources is increasing urgently. Not only the supply energy, but also the quality of power is becoming an important issue to meet the expectation of international standard in building the infrastructures. Therefore, the expansion of generation facilities is an essential requirement for securing power system stability and reliability in Jeju island. Furthermore, since the island is located in a hurricane path and is known to have frequent lightings, there has been frequent contingencies because of the system heavily depending on HVDC tie lines from main land Korea and sporadic variable output from many wind farms located in the east and west side of Jeju island. The total installed capacity of Jeju island in 2008 is 730[MW], which includes 150[MW] transfer through the HVDC system from the mainland. The power transfer through the HVDC is about 40[%] of the minimum demand power, which is almost 310[MW] in 2007.

B. Wind characteristics of Jeju island

Figure1 shows the wind distribution map of Jeju island. Considering the map, the mean wind speed in the east and west area is almost 6~8[m/s], but north and south is 4~6[m/s]. This means that the east and west area have the very good conditions for wind power generation site than the north or south area, respectively. Now a days, there are many wind farms under constructions in the east and west area and about 223[MW] systems are applied accumulatively.

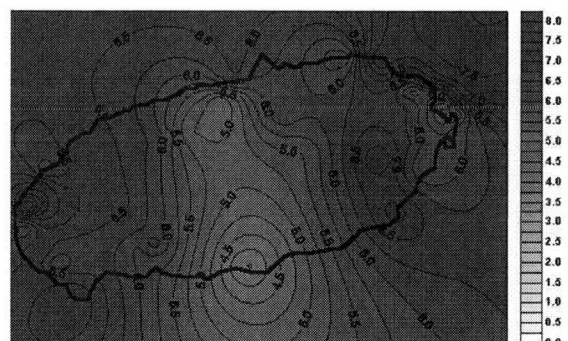

Fig. 1. Wind map of Jeju island.

III. MODELING AND SIMULATION

To verify the feasibility of the proposed scheme, computer simulations have been carried out. 1.5[MW] NEGMICON WTGS named NM72C, current source type HVDC system, thermal power plant included diesel engine and steam turbine and Jeju power grid load are modeled and used in the simulation.

A. Wind turbine modeling

Figure 2 shows the model scheme of WTGS in the PSCAD program.

On the process of modeling the WTGS, it is very difficult to know all of the real parameters because WTGS is very complex system composed of a lot of mechanical, electrical, and aero dynamic systems. This means that the results of computer simulation can be incorrect. So using the real output data of model machine located in HanGeong wind farm established in 2006 in the west area of Jeju island, the power performance curve versus wind speed from cut-in to cut-off is obtained and made a look-up-table. In the modeling process, the interval of wind speed in the power curve is 0.1[m/s], and output data is mean value, which is derived using more than 15 data numbers according to the wind speed.

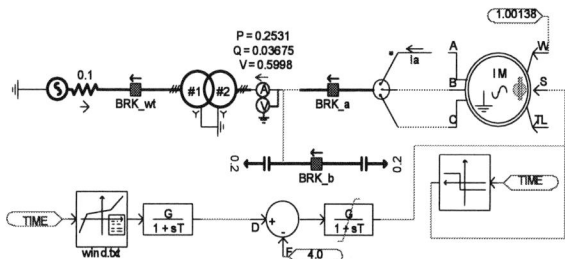

Fig. 2. WTGS model.

Figure 3 shows the simulation results of model system and the real mean output data measured from Jan. 2007 to Feb. 2007 in HanGeong wind farm. In Fig.3, P_{wt} and Q_{wt} represent the real active and reactive power of model WTGS, respectively, and also P_{wt_s} and Q_{wt_s} are simulation results using PSCAD/EMTDC program. Comparing with two results, these are almost same and we can say that the simulation model of WTGS is good and can be used to analyze the power system analysis.

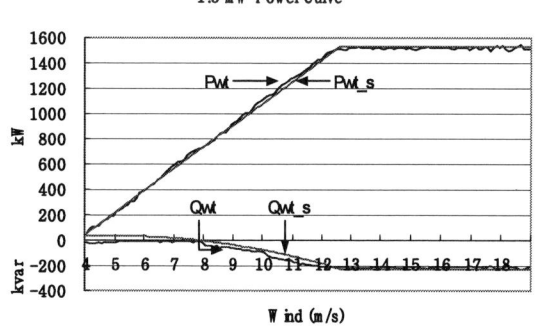

Fig.3 Power curve of model system

C. Power system modeling of Jeju island

Figure 4 shows the PSCAD/EMTDC model of Jeju island power system. There are five mainly kinds of modeling items, which are thermal power plant, current type HVDC, wind farms, transmission line impedance and

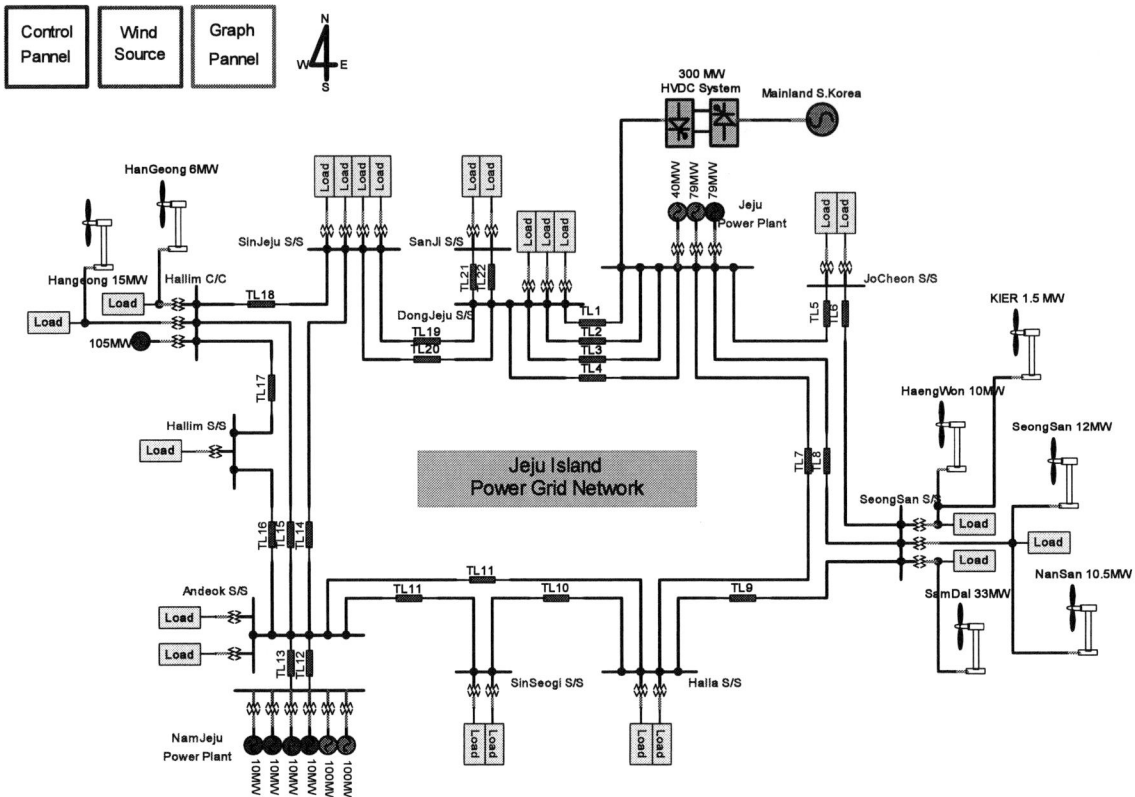

Fig.4 . PSCAD model of Jeju island power system.

load. On the assumption of 320[MW] base load in Jeju island, the supply of electric power is from three kinds of power generation system. First, one is thermal power plant composed of a 40[MW] diesel engine, one 79[MW] and two 100[MW] steam turbine generation systems. Second, HVDC system supplies the power capacity from 40[MW] to 150[MW] on condition of frequency control mode.

Third, wind farms located in seven different sites supply maximum 83[MW] under variable wind speed.

Fourth, impedance value in the 154[kV] transmission line composed of steady state and zero sequence are modeled by using the actual data.

Fifth, the number of load connected in 60[KVA] rated transformer is 23. And also, power ratings of one load is from 13 to 15[MW]. While making the model items, as can as possible actual data are applied. In the actual power system of Jeju island, it is possible to supply the nearly 150[MW] by HVDC system during the constant power control mode. But, in the frequency control mode, the supply power capacity is from 40[MW] to 150[MW] depending on the change of load and power generation value.

D. Simulation results and discussion

Let us assume that there are seven sited wind farms which has 1.5[MW], 6[MW], 10[MW], 10.5[MW],

12[MW], 15[MW], and 33[MW] power ratings respectively in Fig. 4, and all of those systems generate the power according to the variable wind speed under mean wind valued 10[m/s]. And also, four thermal power plants and HVDC system supply the power. At this time, supposing the base load rated 320[MW] is modeled, and also the component ratio of power supply is almost thermal power 170[MW] versus HVDC 150[MW] operated frequency control mode. This means that HVDC system can absorb the output variation of wind power generation according to variable wind speed. It is very useful for promoting the stability of Jeju island power system and will be able to increase the making capacity of wind power generation.

To verify the effectiveness of proposed modeling system, five different variable wind velocities depending on the wind farm sites are activated such as Fig. 5. Figures 6 and 7 show the active and reactive power outputs of wind farm located in Fig. 4 according to the variable wind velocities, respectively. In this simulation results, the output of wind farms are revealed in proportion to the 3 power of variable wind speed.

And also, to analyze the transient characteristics of voltage and frequency in Jeju island power system, all of wind farms shut down at the same time over the rated wind speed at 32[sec]. In Figs. 6 and 7, the total active output power of wind generation is changed from almost

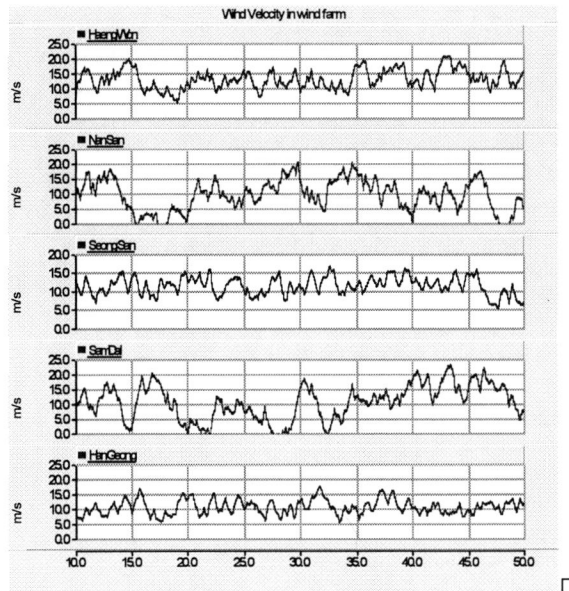

Fig. 5. Wind velocities in each wind farm.

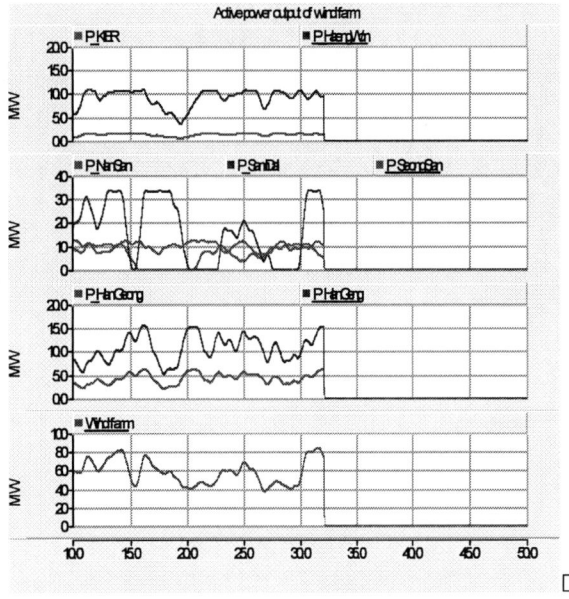

Fig. 6. Active power outputs of each wind farm.

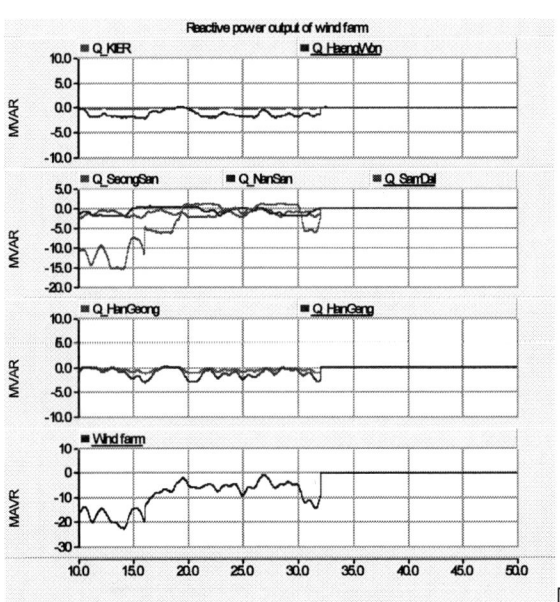

Fig. 7. Reactive power outputs of each wind farm.

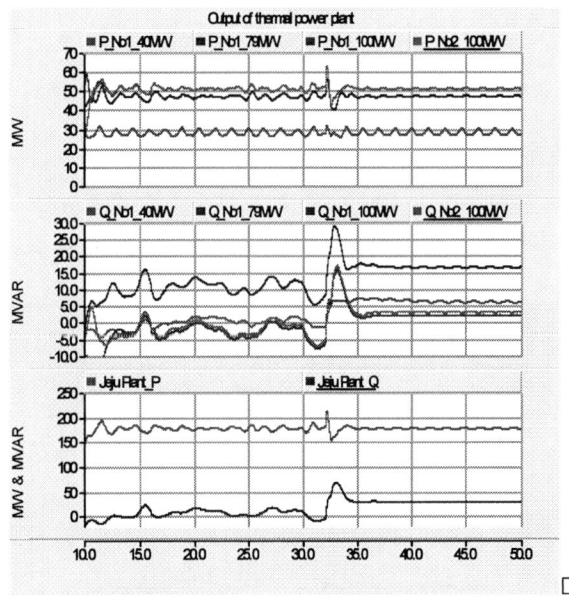

Fig. 8. Power outputs of each thermal power plant.

83[MW] to 0 and reactive power is changed from 14[MVAR] to 0.

At the same time, thermal power plant and HVDC system are replied in response to lacking of wind power generation. Figure 8 shows the output power of each thermal power plant. In this figure, 40[MW] diesel engine which has the 6.71[sec] response time as the turbine-generator combined inertia constant generates the oscillating power ripple such as ± 5% of rated value, and almost no change of output power at 32[sec]. Both of 100[MW] steam turbine generators which have the 5.91[sec] response time as the turbine generator combined

inertia constant have the faster response than that of 79[MW] steam turbine generator

The inertia constant of the latter is 6.9[sec], so this is why the former system has the faster response. And values of peak to peak ripple in Fig. 8 are almost 3, 18, and 24[MW], respectively.

Supposing the same operating conditions of Fig.6, the active and reactive output variations of all of the Jeju thermal power plant, wind farms, and HVDC system in Jeju island are shown in Fig. 9.

In Fig.9, the output of HVDC system is replied in inverse proportion to the wind farm power generation

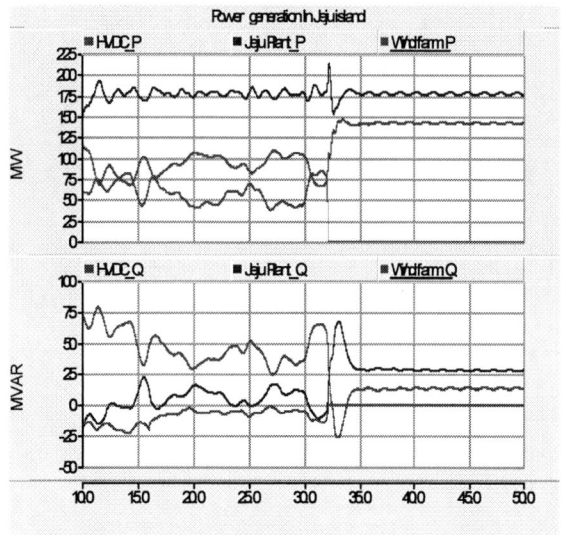

Fig. 9. Active power and reactive power variation of Jeju thermal power plant.

simultaneously. But total output power of thermal power plant have a very little effect on the variation of wind power output except for the period of transient state at 32 [sec]. According to this figure, the variation of wind power generation output can be more absorbed than the thermal power plant. This means that HVDC system which has the faster inertia constant than the thermal power plant will be able to play an important role to ensure the stability of Jeju island power system

Figure 10 shows the characteristics of voltage and frequency in the transmission line TL6, TL9, and TL18. All of the former transmission lines are linked with the SeongSan S/S. In the simulation results, the frequency has almost constant value except at 32[sec]. At this time, the value of frequency change is 0.3[Hz] for almost 2 cycles.

In the transmission line, the variation of voltage magnitude has small ripple according to the variation of the wind power generation and the oscillation of thermal power plant output. During the steady state, the variation of voltage magnitude is almost below 0.01[pu]. But, in the transient state, the voltage drop for 2[sec] is almost 0.05[pu]. This means that the sudden change of wind power generation in Jeju island will be able to produce an effect on the power quality in the transmission line.

Figure 11 shows the characteristics of voltage and frequency in the load connected in SeongSan S/S under the same condition in Fig. 10. The characteristics of simulation results are almost same. But the value of frequency drop in the load is lager than that of the transmission line. According to this simulation results, it is well verified that the influence of wind farms connected to the distributed lines on the power quality can be severed than that of transmission line.

IV. CONCLUSIONS

Modeling and power quality analysis of thermal power plant, current source type HVDC system, seven sited wind farms, transmission lines and loads in Jeju island power system have been proposed and simulated under

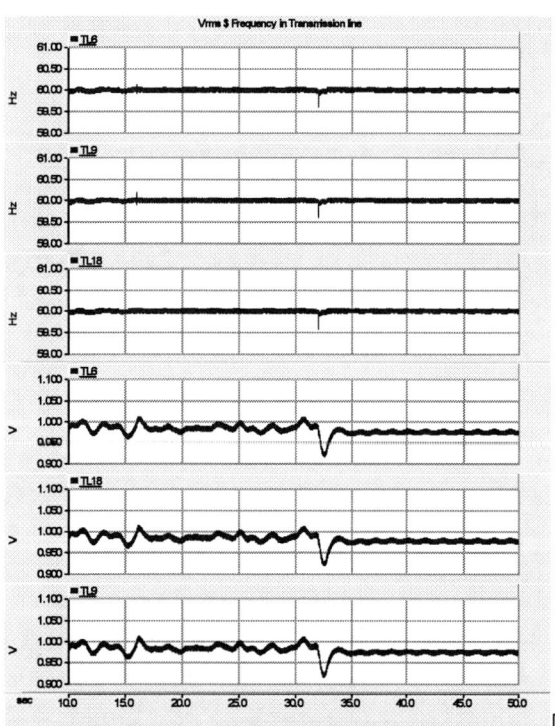

Fig. 10. Voltage and frequency in the transmission lines.

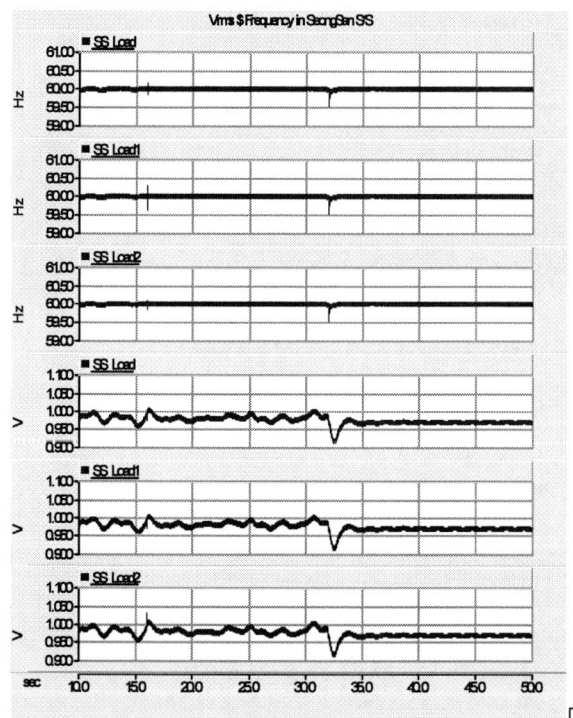

Fig. 11. Voltage and frequency in the load.

the variable wind speed using PSCAD/EMTDC program

To show the validity of the proposed computer simulation, the comparative simulations were carried out based on the seven wind farms. The results of simulation using the mainly five model items under the steady and transient state at 32[sec] are presented, respectively. Analyzing the simulation results, we can be expected that the proposed schemes and method are very useful for constructing the wind farms and for strengthening the stability of power system.

REFERENCES

[1] Thomas Ackermann, *Wind Power in Power Systems*, John Wiley & Sons, 2004.

[2] M. Godoy Simoes and Felix A. Farret, *Renewable Energy Systems*, CRC Press, 2004.

[3] Kwang Y. Lee, Se-Ho Kim, Eel-Hwan Kim, and Ho-Chan Kim, "Fuel Cell as an Alternative Distributed Generation Source under Deregurated Power System," *KIEE Trans. B*, Vol. 55A, No. 12, pp. 561-568, 2006.

[4] Oplimpo Anaya-Lara and F. Michael Hughes, "Influence of Windfarms on Power System Dynamic and Transient Stability," *Wind Engineering,* Vol. 30, No. 2, pp. 107-123, 2006.

[5] Akmatov, V., "Variable Speed wind turbines with doubly fed induction generators," Wind Engineering , Vol. 26, No. 2, pp. 85-108, 2002.

[6] *PSCAD/EMTDC Mannual*, 2006.

[7] *HVDC Light Topology*, ABB Handbook, 1999.

A New Control Method for
Power Turbine Generators Using an Accurate
Ship Plant System Model

Nobumasa Matsui[†, ††], Fujio Kurokawa[†, †††] and Keiichi Shiraishi[††††]

† Graduate School of Science and Technology, Nagasaki University, Japan

††Choryo Control System Co., Ltd., Japan, e-mail: *nobumasa_matsui@ryousei.co.jp*

†††Department of Electrical and Electric Engineering, Nagasaki University, Japan,
e-mail: *fkurokaw@nagasaki-u.ac.jp*

††††Mitsubishi Heavy Industries, Ltd., Nagasaki Shipyard & Machinery Works, Japan,
e-mail: *keiichi_shiraishi@mhi.co.jp*

Abstract— The purpose of this paper is to present a new control method for a power turbine generator (PTG) system using an accurate ship plant system model that ascertains startup characteristics. The PTG system is a waste heat recovery type generation system making use of exhaust gas from the main shipboard diesel engines. However, with the PTG systems, control during startup is not entirely clear. Startup characteristics from power turbine startup through synchronization to the system bus are ascertained using an accurate simulation model, and startup sequence control is clarified for a combination of gas valve operation and load bank. A prototype constructed on the basis of the simulation. As a result, in the startup process, so as to limit speed overshooting that is 0.83 p. u. at 100% output from main diesel engine as the severest condition. The synchronization to the system bus is attained, and highly satisfactory results are achieved.

Keywords— energy system management, generation of electrical energy, modeling, ship, simulation, systems engineering.

I. INTRODUCTION

On-board power requirements for large commercial vessels such as container ships and oil tankers are typically met by systems composed of a diesel generator (DG), a steam turbine generator (TG), and a shaft generator (SG) installed on the propeller shafts. Propulsion can be by means of main diesel engines, steam turbines, gas turbines, electrical propulsion, and so forth, although the main diesel engines are used for the majority of the world's vessels.

Conventionally, the main diesel engine exhaust gas has been subject to rotational energy recovery using a turbocharger and heat energy recovery using an exhaust gas economizer, so as to boost on-board plant efficiency.

Even greater improvements in fuel economy are now urgently required due to recent rises in fuel prices, and one arrangement that has drawn attention lately is a power turbine that extracts part of the exhaust gas from the main diesel engines [1]. An STG system using a power turbine in combination with a steam turbine to produce the electrical power has been proposed [2], but not all marine plants make use of the steam turbines. Also, given demand

for a lower power output than would be provided by the STG system, development of a power turbine based PTG system has been initiated to enable generation of several hundred kW with a power turbine alone as initial development of the PTG system.

In the STG systems, the power source for the generator is composed of a steam turbine and a power turbine, with the speed control applied to the steam turbine [3]. After reaching the rated speed following STG system startup, the system enters a state of synchronizing operation.

However, with the PTG systems, the speed control during startup is difficult because of four reasons. First, the speed control during the PTG system startup is performed using valves arranged on the exhaust gas system that drives the system. Since output of the main diesel engine is affected by vessel speed demands, the state of exhaust gas is not constant. Secondly, an exhaust gas contains considerable soot, and valves on the exhaust gas system for the power turbine are subject to deposits of soot and ash. Thirdly, stable speed control is quite difficult because the valve actuators require between 0.5 and 1.0 second operating from fully closed to fully open. Final factor is that a certain amount of the exhaust must be reserved for the requirements of the main diesel engine itself. So, the control method during startup is not established in these various issues and constraints.

This paper presents a new control method for startup of the PTG system using an accurate ship plant system model. At first, the valves that are placed along the exhaust gas system are operated so as to be ascertained the startup characteristics for the speed control. This is to be evaluated in the operating range 40% through 100% output of the main diesel engine with the PTG system. Next, arrangement of a valve and a load bank can be represented in the simulation for startup operation. The proposed control method of startup is clarified for a combination of the valve operation and load bank. Furthermore, a prototype system is constructed on basis of the simulation results in order to verify the proposed control method.

Accordingly, the design accuracy during the planning stage is improved with respect to the system layout and control, while the risk can also be avoided in prototype testing, and tuning time of the controller can be reduced.

978-1-4244-1741-4/08/$25.00 ©2008 IEEE

II. POWER TURBINE SYSTEM

The power system for a marine plant is shown in Figs. 1(a) and (b). BUS represents the power system bus, CB is the circuit breaker, L is the total load requirement and G is the generators, that is, DG is the diesel generator, TG is the steam turbine generator SG is the shaft generator in Fig 1. A block diagram for conventional on-board power supply is shown in Fig. 1(a). DG would be the primary power source on most ships, but, being fed by the fossil fuel, a ship owner might also select from energy recovery using SG, TG, and so forth for improved power supply efficiency. However, not all vessels are equipped with TG or SG, such that arrangements for power supply vary with the type of vessel under consideration.

The composition of proposed PTG system is shown in Fig. 1(b), representing an energy efficient measure for the marine power. When the power requirements are the low in the marine plant, the PTG system would be capable of serving as the sole electric power source. The PTG location along the gas system is shown in Fig. 2. The main diesel engine is used for propulsion of the vessel, and the exhaust gas from the main diesel engine is collected at the manifold. In existing ships with the diesel plant, the exhaust gas collected at the manifold is routed through the turbocharger and then released into the atmosphere via the exhaust gas economizer and a funnel. The turbocharger

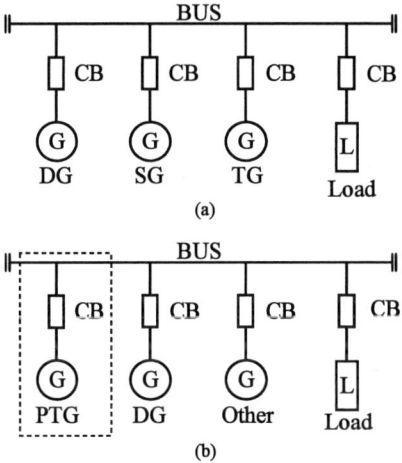

Fig. 1. Outline of electrical system bus in the ship plant. (a) Conventional system. (b) Proposed system with PTG.

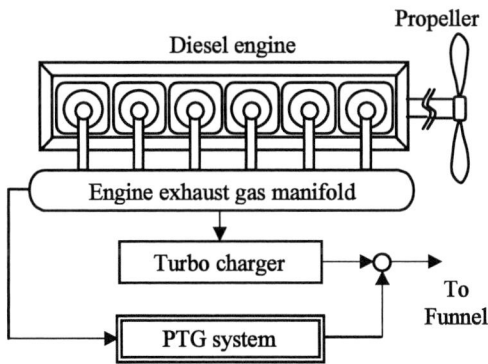

Fig. 2. Outline of diesel engine and PTG system.

drives a compressor by converting the energy obtained from the exhaust gas, and directs compressed air into the cylinders of the main diesel engine [2]. In the system using a power turbine, part of the exhaust gas at the manifold is extracted to drive a power turbine, and the energy recovered thereby is converted to electrical power by a generator in this new power supply arrangement.

III. PLANT MODEL

It is generally difficult to build an accurate model of a total plant simulation, requiring much time and effort. In database simulation system (DBSS) [3], equipment arrangement and characteristics as in an actual plant are given by the database, thus enabling easier construction of thermal power and marine plant models [4], [5].

Figure 3 shows the constitution of the plant model for the PTG system in Fig. 2. B_1 and B_2 are the boundary elements in this simulation. B_1, as the power source, is the exhaust gas manifold of the main diesel engine from the main diesel engine cylinders. B_2 is the exhaust gas element in this system. V_1 is the gas valve for the in-service state of generation using the power turbine, V_2 is the gas valve for the out-of-service state, PT is the power turbine, RG is the reduction gear, G is the generator and L is the total load in the ship. BUS is the electrical system bus, with a frequency of 60Hz and voltage of 440V in the simulation. When the power turbine is in the in-service state, the exhaust gas flow is supplied from B_1 to PT via V_1. PT is driven by the exhaust gas energy with gas temperature of approx. 400°C. PT speed reaches approx. 20000 rpm. G is driven by PT, but rated speed of the generator is 1800 rpm at 60Hz. The differential speed is high enough for the system to require a reduction gear.

The paragraphs A through E describe the equation a part of plant model for the PTG system. The paragraph A describes the boundaries of the plant model with respect to gas fluid. The paragraphs B and C describe an equation of gas fluid model. The paragraph D describes an equation of electrical model and E describes an equation of speed model using reduction gear.

A. Gas condition of engine output

In the exhaust of the main diesel engine, the gas temperature is approx. 400 °C, and the gas pressure is approximately 3.5 bar. This condition is for 100% of the main diesel engine output. The simulation model is given both temperature and pressure as design data from the engine's manufacturer. When the engine output is constant, the gas temperature and pressure are constant. Also, the pressure of B_2 is constant.

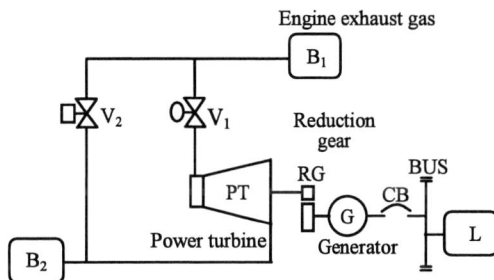

Fig. 3. Diagram of the power turbine generator system model.

B. Valve model

The exhaust gas at flow rate W is generally in the proportion to the differential pressure of the valve ΔP as indicated by equation (1) below. W is the gas flow rate, and ΔP is the differential pressure of the valve. The valve model is simulated by following equation (2). Where W_V is the gas flow rate, P_{e_V} is the inlet pressure of valve, P_{a_V} is the outlet pressure of valve, ξ is a constant as the coefficient of pressure loss, v_{e_V} is the specific volume of gas at valve inlet and φ is the position of the valve. In Fig. 3, both V_1 and V_2 are simulated by Eq. (2). The capacity of V_1 decides according to the specification of power turbine. The capacity of the V_2 is decided that can be bypass all of gas flow as same as the PT.

$$W \propto \Delta P \tag{1}$$

$$W_V = \sqrt{\frac{P_{e\_V} - P_{a\_V}}{\xi v_{e\_V}}} \cdot \varphi^2 \tag{2}$$

C. Power turbine model

The gas flow rate corresponding to the characteristics of the power turbine is given by the following equation of the diffuser, where W_{PT} is the gas flow rate, P_{e_PT} is the inlet pressure of the power turbine, P_{a_PT} is the outlet pressure of the power turbine, v_{e_PT} is the specific volume of gas at the power turbine inlet, η is the efficiency of the power turbine, and κ is the ratio of specific heat.

$$W_{PT} = \sqrt{2g \frac{m}{\eta(m-1)} \frac{P_{e\_PT}}{v_{e\_PT}} \left(\psi^{\frac{2}{m}} - \psi^{\frac{m+1}{m}} \right)} \tag{3}$$

where

$$m = \frac{\kappa}{\kappa + \eta(1 - \kappa)} \tag{4}$$

$$\psi = \frac{P_{a\_PT}}{P_{e\_PT}} \tag{5}$$

When Eq. (3) is used in the simulation, it is necessary to consider η as a parameter because η changes in the gas condition with both the temperature and pressure of the gas. When the power turbine model employs equation (6) below, the flow rate corresponding to the characteristics of the power turbine can be easily obtained in this simulation for approximation of the design data. The parameter α is the constant of flow coefficient. β is nondimensional flow as following equation (7), where W_{PT_0} is rated gas flow.

$$W_{PT} = \alpha \left(P_{e\_PT} - P_{a\_PT} \right)^{0.58} \tag{6}$$

$$\beta = \frac{W_{PT}}{W_{PT\_0}} \tag{7}$$

The exhaust temperature of the power turbine is given by the following equation under the adiabatic change, where θ_{e_PT} is the inlet temperature of the power turbine, and θ_{a_PT} is the outlet temperature of the power turbine. κ is given as a constant value that is 1.35 in this simulation.

$$\theta_{a\_PT} = \theta_{e\_PT} \cdot \left(\frac{P_{a\_PT}}{P_{e\_PT}} \right)^{\frac{\kappa-1}{\kappa}} \tag{8}$$

Output of the power turbine is given by the equation below, where H_{e_PT} is the inlet enthalpy of the power turbine, H_{a_PT} is the outlet enthalpy of the power turbine and L_{PT} is the output power of power turbine. The enthalpy of gas is given by the function of gas temperature and gas contents, where W_{PT} is according to Eq. (6). Here, gas contents are the same condition at the power turbine inlet/outlet, the gas contents are constant and the function uses the gas temperature in the simulation.

$$L_{PT} = W_{PT} \cdot \left(H_{e\_PT} - H_{a\_PT} \right) \tag{9}$$

where

$$H_{e\_PT} = f\left(\theta_{e\_PT} \right), \quad H_{a\_PT} = f\left(\theta_{a\_PT} \right) \tag{10}$$

D. Generator model

The effective power L_G and reactive power Q_G for the generator model are according to equation (11) and (12) [6], where E is electromotive force, V_t is terminal voltage, X_d is direct axis synchronous reactance, X_q is quadrature axis reactance and δ is the internal phase angle.

$$L_G = \frac{EV_t}{X_d} \sin\delta + \frac{\left(X_d - X_q\right)V_t^2}{X_d X_q} \sin\delta \cos\delta \tag{11}$$

$$Q_G = \frac{EV_t \cos\delta - V_t^2}{X_d} - \frac{\left(X_d - X_q\right)V_t^2}{X_d X_q}\left(1 - \cos^2\delta\right) \tag{12}$$

E. Reduction gear model

The reduction gear model is the model of speed connecting between the power turbine and generator. The speed of the power turbine is approximately 20000 rpm, while that of the generator is 1800 rpm. This is difference is not inconsequential. So, when the element is considered as a rigid body, the reduction gear model is given by the equation below, where τ_{PT} is the torque of power turbine, τ_G is the torque of generator, g is acceleration of gravity, J_{PT} is the inertia of the power turbine, J_G is the inertia of the generator and ω is an angular velocity. GD^2 is 2nd moment of the power turbine or generator. Here, the GD^2 of the reduction gear is ignored because it is smaller than the power turbine or generator.

$$\left(J_{PT} + \frac{1}{\gamma^2} J_G \right) \frac{d}{dt} \omega = g \cdot \left(\tau_{PT} - \tau_G \right) \tag{13}$$

where

$$\gamma = \frac{\omega_{PT}}{\omega_G} \tag{14}$$

$$J = \frac{GD^2}{4} \tag{15}$$

$$\tau_{PT} = \frac{L_{PT}}{\omega_{PT}} \tag{16}$$

$$\tau_G = \frac{L_G}{\omega_G} \tag{17}$$

χ is nondimensional speed by following equation (19), where the speed N is according to Eq. (18), N_0 is the rated speed.

$$N = \frac{60}{2\pi} \cdot \omega \tag{18}$$

$$\chi = \frac{N}{N_0} \tag{19}$$

IV. PTG STARTUP CHARACTERISTICS

Output from the main diesel engine, i.e., the drive source for the propulsion, is affected by demands for the vessel speed as well as the need to balance propeller speed against relevant sea conditions, meaning that such output is not constant. In this respect, the quantity of state for the exhaust gas collected in the manifold after combustion in the cylinders changes in accordance with engine output. This output fluctuation is an extremely important issue during startup of the PTG system, which is driven by exhaust gas extracted from the manifold. It would be desirable for the PTG system startup to be independent of the state of output from the main diesel engine, but, since the PTG system makes use of diesel exhaust gas, some

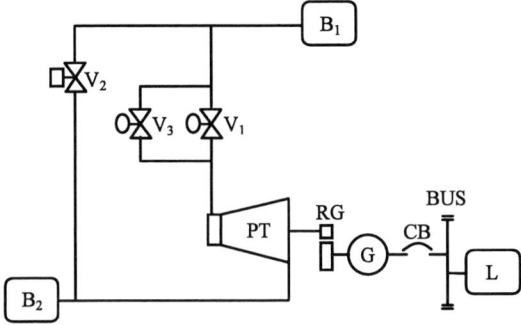

Fig. 4. Diagram of power turbine generator system model for improvement.

minimum level of exhaust gas is necessary. Here, the case is considered where the minimum quantity of state for the exhaust gas from the main diesel engine to enable the PTG system startup is 40%, and synchronization to the system bus after the PTG system startup is reached rated speed at either 40% or 100% output.

In the PTG system shown in Fig. 3, the sequence from the PTG system startup to synchronize to the system bus is as follows: open the PT inlet valve V_1 to allow flow of exhaust gas to PT, and close the bypass valve V_2. The speed of PT increases during this process, with the exciter of generator in-serviced upon reaching the rated speed, and so as to attain synchronicity with the electrical system bus. However, V_1 is difficult to control the speed because exhaust gas contains contaminants such as soot. Moreover, the operation time between the fully closed and fully open state is at least 0.5 seconds. When V_1 is fully open at the rated flow of exhaust gas, it must have the capacity to allow the full flow, but this capacity is excessive at startup.

Figure 4 shows the addition of small-valve V_3 for use during startup. Small-valve V_3 has 15% of the capacity of V_1, and simulation was undertaken for the PTG system startup at 40% and 100% of main diesel engine output. The simulation result is shown in Figs 5. and 6. using the model that describes paragraphs A through E in chapter III.

Figures 5(a) and (b) show the simulation result of PT speed and PT gas flow at 40% output from the main diesel engine. □is according to Eq. (19). □is according to Eq. (7). The valve positions before 0 second are 0% for V_1, 100% for V_2, and 0% for V_3. At 0 second, small- valve V_3 is fully open to allow PT speed to rise to the rated speed.

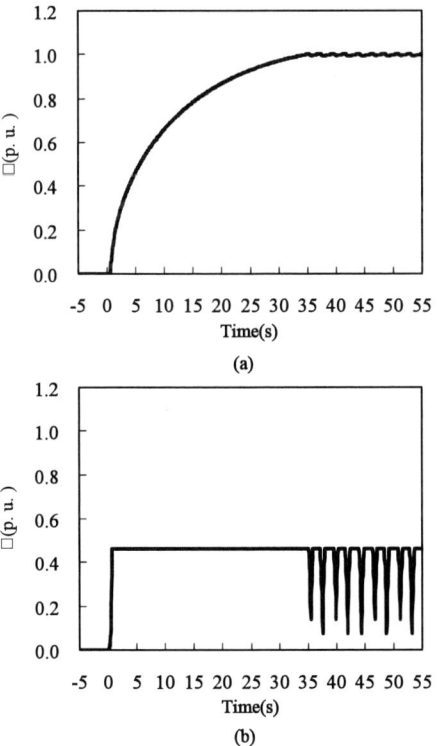

Fig. 5. PT speed response in 40% diesel engine output. (a) PT speed. (b) PT gas flow.

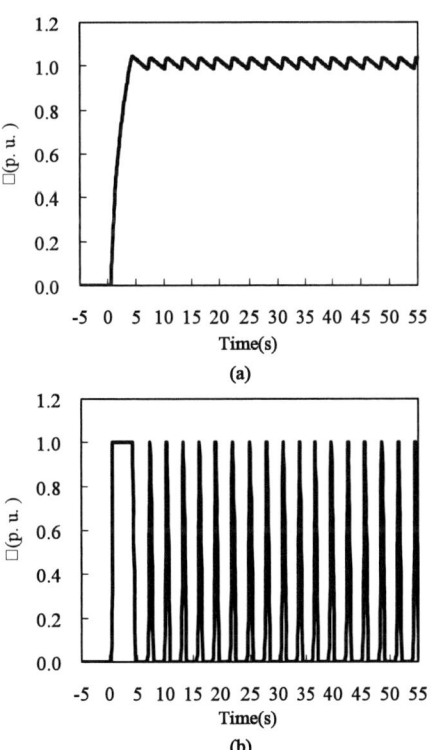

Fig. 6. PT speed response in 100% diesel engine output. (a) PT speed. (b) PT gas flow.

The rated speed is achieved about 35 seconds after fully opening V_3, which is subsequently operated so as to maintain speed. The PT speed is controlled between 0.995 and 1.002 p. u. at about the point when rated speed is reached, and synchronization to the system bus can be in-serviced to be smoothly accomplished. However, the gas flow indicated in Fig. 5(b) fluctuates by about 0.3 p. u., such that requirements for a constant amount of exhaust gas are not satisfactory.

Figures 6(a) and (b) show the PT speed and gas flow at 100% output from the main diesel engine. The valve positions before 0 second are the same as in the previously noted case of 40% output: 0% for V_1, 100% for V_2, and 0% for V_3. At 0 second, small-valve V_3 is fully open to allow PT speed to rise to the rated speed. The rated speed is achieved about 5 seconds after fully opening V_3, which is subsequently used to maintain speed. The PT speed is controlled between 0.989 and 1.037 p. u. at approx. the point when rated speed is reached, and it is difficult to synchronize to the system bus. Also, since the inlet pressure is high for V_3, even though the valve capacity is 15% of V_1, β becomes about 0.5 p. u. in full open of V_3.

Consequently, it is clarified that, while valve-based startup could be managed for the PTG system at the lower gas state, such speed control would be extremely difficult at high output from the main diesel engine that would be employed. Furthermore, considering the fluctuations occurring in the amount of PT gas flow, such fluctuations would be substantial regardless of the main diesel engine output region, and the influence on the main diesel engine would not be negligible. As a result, it is determined that the required conditions can not be satisfied by means of valve manipulation.

V. PROPOSAL CONTROL METHOD

Valve-based control of the PTG system startup was considered in the previous section, in which it was concluded that the overall system requirements could not be met solely by means of valves.

Figure 7 shows an improved the PTG system diagram in Fig. 4, with a load bank newly placed between generator G and breaker CB. The load bank is capable of consuming a load equivalent to the output of the generator. Figure 8 shows the startup sequence for the PTG system in Fig. 7. t_0 through t_6 show operating event time. χ is the expected value of speed in Fig. 8(a). I_{EX} is exciter current of the generator in Fig. 8(b). As before t_0, V_1 and V_3 are at 0% and V_2 is at 100% prior to startup, after which PT inlet valve V_3 is opened so that exhaust gas flows to PT,

and PT speed is raised from t_0 to t_1. Next, V_2 closes at t_1, the exciter of generator is in-serviced at t_2 when the speed is 0.5 p. u., and the load bank is also in-serviced, such that overshooting of the initial speed is limited at t_3. The load bank is adjusted in accordance with the speed setting from t_3 to t_4, and the PT rated speed is achieved in a controlled manner at t_4. Upon attaining the rated speed, the circuit breaker is closed and the system is synchronized to the bus system at t_5. Finally, V_1 opens at t_6. Furthermore, for improvement of starting torsional stress of the shaft, since the load bank controls the PT speed, the open time of the startup valve V_3 can be slowly from t_0 to t_1. The open time of V_3 is shown in Fig. 9. The doted line is the open time of V_3 in Fig. 4. The solid line is the open time of V_3 in Fig. 7. And so the valve capacity of V_3 is changes to 3% from 15% of V_1.

Figure 10 shows simulation results for the sequence of PT speed control in Fig. 8. χ is according to Eq. (19). The main diesel engine output of 100% is indicated by a solid line, an output level of 75% is indicated by a uniform dotted line, and a level of 40% is denoted by a dotted line. In each case, the generator field current and load bank is in-serviced at t_2 after closing V_2 at t_1 in the startup process, so as to limit speed overshooting at t_3, these are 0.81 p. u. at 100% output, 0.75 p. u. at 75% output and 0.58 p. u. at 40% output. Also, by operating the load bank, a rise to rated speed is achieved in accordance with speed demands, demonstrating effective speed control in all cases. Moreover, the shaft of torsion stress is improved because the PT speed confirms that reaches to 0.2 p. u. by approx. 40 seconds after operating of V_3 in Fig. 10. Based on the simulation results, it is clarified that on-board system requirements could be sufficiently met with a sequence by which the load bank is placed between the generator and the breaker, and, after the generator reaches the proper condition for exciter, the PTG system control is entrusted to the load bank.

Figure 11 shows the speed test results at startup for a prototype constructed on the basis of the simulation

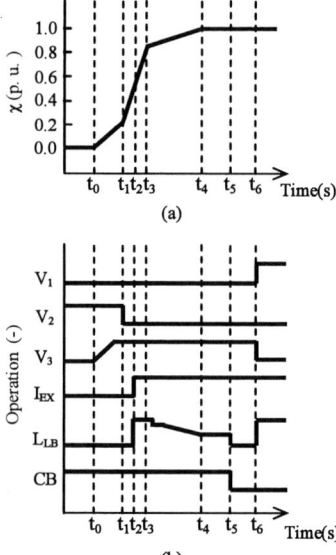

(a)

(b)

Fig. 8. Time chart of operation. (a)PT speed. (b) Operation.

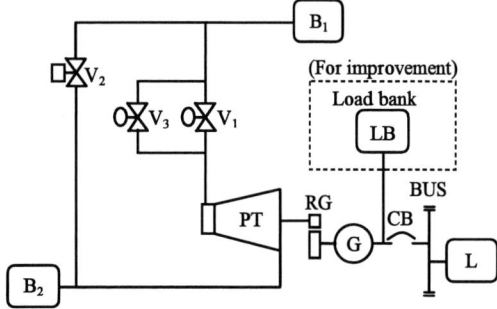

Fig. 7. Diagram of power turbine generator system model. for improvement.

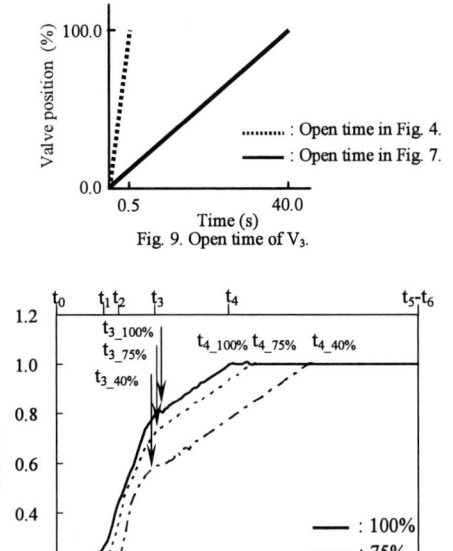

Fig. 9. Open time of V_3.

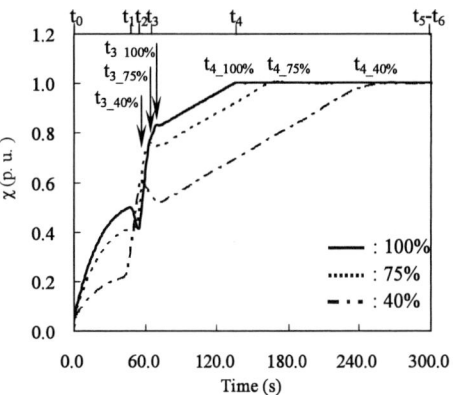

Fig. 11. PT speed response in prototype plant when the power turbine starts up by using a new proposal sequence.

Fig. 10. PT speed response in simulation when the power turbine starts up by using a new proposal sequence.

results using the load bank and the sequence in Fig. 8. The main diesel engine output of 100% is indicated by a solid line, an output level of 75% is indicated by a uniform dotted line, and a level of 40% is denoted by a dotted line. In each case, the generator field current and load bank is in-serviced at t_2 after closing V_2 at t_1 in the startup process, so as to limit speed overshooting at t_3, these are, 0.6 p. u. at 40% output and 0.75 p. u. at 75% output, furthermore, 0.83 p. u. at 100% output as the severest condition for startup. During t_0 through t_2, the speed raises faster than simulated because each capacity of V_2 and V_3 is larger than simulated that for the final design. However, the prototype allowed smooth control of PTG speed in all of cases. The synchronization to the system bus is attained, and highly satisfactory results are achieved. There is no space here for the accurate model that the valves are operated t_0 through t_3 will be present near in feature after the model is identified with measured.

VI. CONCLUSION

The PTG system startup characteristics are ascertained using an accurate plant model, and a new proposal control method is presented in this paper. This result is as follows.

(1) In the case of valve placement along the exhaust gas system for speed control during the PTG system startup, fluctuations in the gas flow utilized at startup are substantial. The influence on the main diesel engine from the exhaust gas used should be minimized, and gas fluctuations are undesirable. Furthermore, the speed of PTG is controlled between 0.989 and 1.037 p. u. at approx. the point when rated speed is reached, and it is difficult to synchronize to the system bus. As a result, it is determined that control is difficult using only valves.

(2) The startup characteristics from the power turbine startup through synchronization to the system bus are ascertained using the accurate simulation model. Moreover, the control method of startup is clarified for the combination of gas valve operation and load bank. As a result, in the startup process, so as to limit speed overshooting that is 0.81 p. u. at 100% output from main diesel engine as the severest condition for startup.

(3) A prototype is constructed on basis of the results of the simulation. Startup sequence control using the valve and load bank arrangement under consideration facilitated safe and smooth PTG system startup. As a result, in the startup process, so as to limit speed overshooting that is 0.83 p. u. at 100% output from main diesel engine. The synchronization to the system bus is attained, and highly satisfactory results are achieved, while the risk can also be avoided in prototype testing.

PTG is an extremely effective system for fuel cost reduction in the face of rising fuel prices, and systems capable of providing several thousand kW are being considered.

REFERENCES

[1] K. Imakiire, Y. Iwanaga, T. Ohashi and E. Matsuo: "Development of Energy-Efficient MET Turbocharger into Super-Turbogenerationg System (STG system) for Maximum Waste Energy Exploitation," Proc. The CIMAC World Congress on Combustion Engine Technology, 1987.

[2] K. Shiraishi, M. Kimura and T. Teshima: "Development and application of MET-MA turbochargers," Proc. The CIMAC World Congress on Combustion Engine Technology, no. 30, pp.1-12, May 2007.

[3] T. Kishikawa: "Development and the present conditions of DBSS," Technical Report of Cyouryou Control System co., ltd., vol. 5, pp.53-59, Apr. 2005.

[4] N. Matsui and F. Kurokawa: "A New Approach of Thermal Power Plant Model Using VVVF Inverter for Accuracy Improvement," Proc. The KIEE International Conference on Electrical Machines and Systems 2007, pp.32-37, Oct. 2007.

[5] N. Matsui and F. Kurokawa: "Improvement of Transient Response of Thermal Power Plant Using VVVF Inverter," Proc. The IEEE International Conference on Power Electronics and Drive Systems, pp.1209-1214, Nov. 2007.

[6] K. Aratame: "Application of Electric Power System Calculations," Denki-syoin, 2004.

Voltage profile support in distribution networks – influence of the network R/X ratio

B. Blažič* and I. Papič*
* University of Ljubljana, Faculty of Electrical Engineering, Ljubljana, Slovenia

Abstract—One of the problems emerging with a rising share of distributed generation (DG) is also the regulation of voltage levels in different operating conditions. Distribution network voltage profile can be controlled with the use of active compensators and also with converters which are used for connection of DG. Both devices are usually based on a voltage-source converter with a capacitor or dc source connected to the dc side. The paper presents the influence of the network R/X ratio on compensation effectiveness of shunt compensators. This ratio influences the ability to control voltage and must be taken into account at device design.

Keywords: Active filter, voltage source converter, distributed generation, power quality.

I. INTRODUCTION

The foreseen large scale integration of distributed generation (DG) into distribution networks brings new challenges for the development of distribution systems. A large number of sources make it difficult to maintain an adequate level of power quality and to keep voltage within defined limits [1], [2].

In nowadays distribution systems power electronic devices are already present and are used to increase the quality of power supplied to the customers [3]–[5]. For active compensation, devices based on a voltage-source converter (VSC) are used. They can, for example, provide active filtering of harmonics, load balancing, power factor correction and voltage regulation. A similar approach that is used for the design of VSC-based compensators can be adopted also for the design of converters which are used for the connection of various DG (e.g. photovoltaic modules) to the network. Such converters may offer additional functionality (beside the delivery of active power) and provide some power quality enhancement. Clearly, such sources should be rewarded by the distribution system operator for providing ancillary services.

This paper deals with voltage regulation in distribution networks with the use of shunt connected active compensators or the use of converters for DG connection which provide the additional functionality of voltage control. We concentrate on a special aspect of voltage regulation, i.e. its effectiveness depending on the network R/X ratio. The R/X ratio in distribution and transmission networks may differ substantially. In transmission networks which are composed mainly of overhead lines the ratio is usually lower than 0.5. In distribution networks both, overhead lines and cables are used, resulting in a higher R/X ratio ranging from 0.5 to as high as 7, where higher values are typical for low-voltage networks.

The basic structure of a shunt active filter converter is presented first. The influence of the R/X ratio on voltage compensation effectiveness is evaluated with the use of a simple network model and a practical case is presented by means of simulations in PSCAD.

II. ACTIVE FILTER AND THE DC/AC CONVERTER

A basic connections scheme of a shunt active filter [3] is shown in Fig. 1. Active compensators are typically based on VSC-s. They usually operate as a source of reactive power and enable, for example, power factor correction, voltage regulation, harmonics compensation… If an active filter has a source on the DC side (e.g. batteries) it is capable also of active power exchange.

The DC/AC VSC is the main building block of active compensators and is also used for connecting DC sources the ac network (Fig. 2). The VSC basically generates ac voltage from dc voltage, where the magnitude, the phase angle and the frequency of the output voltage can be

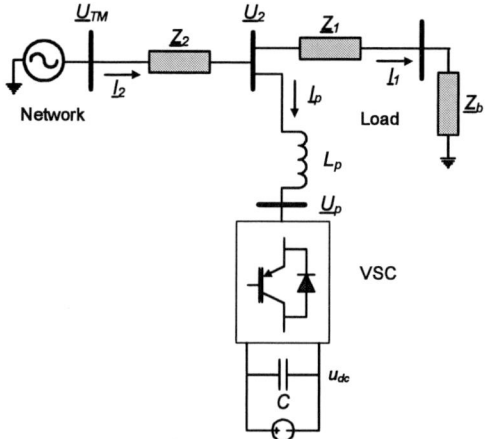

Fig. 1: Shunt connected active filter

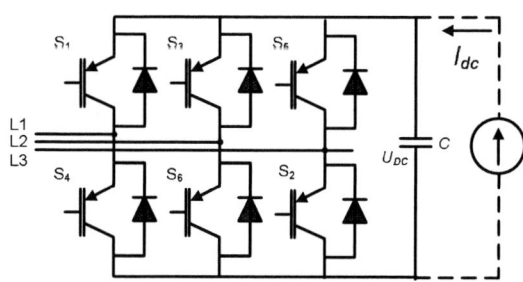

Fig. 2: Basic structure of a voltage source converter (VSC)

controlled. The converter consists of connected semiconductor valves, has a capacitor on the DC side and is shunt connected to the network through a coupling inductance.

A. Operation

Basic operation of a VSC is presented in Fig. 3 [6]. In the case there is no power source on the DC side, the device can exchange only reactive current (I_p) with the network and the generated voltage U_p is in phase with the voltage at the point of connection (U_2). However, if there is a power source connected to the DC side, the converter can exchange also active power with the network (Fig. 3). With an appropriate control algorithm decoupled control of active and reactive power can be achieved. By proper control of output voltage the desired current can be forced to flow through the coupling reactance.

III. VOLTAGE COMPENSATION

Relatively high R/X ratios in distribution networks affect also the ability of compensators to regulate voltage. The effectiveness of voltage regulation depending on the R/X ratio will be evaluated next [7].

A. Network

The evaluation will be based on the simple network showed already with the single-line diagram in Fig. 1. Load Z_b is connected to a stiff voltage source U_{TM} through a line (cable) which is split into two parts with impedances Z_1 and Z_2. A compensator is shunt connected at an arbitrary point between the two sections. The equivalent circuit of the proposed network is shown in Fig. 4. Network elements data are given in Table I.

TABLE I
NETWORK DATA

pu system	Network	Line (cable)	Load
S_{pu}=1 MVA	U_{TM}=1 pu∠°0	Al 240 mm²	S_b=5 pu
U_{pu}=20 kV		l=10 km	P_b/Q_b=1
		R/X=0,833	
		z=0.234 Ω/km	

The shunt compensator was modelled as a voltage source U_p connected to the network through an inductance $X_p = 0.3$ pu. Compensator current I_p is calculated with (1), where m is the voltage modulation factor and δ is the phase angle between the generated voltage U_p and the network voltage U_2. We can also say the compensator is modelled as a constant current source. The current is defined with the voltage drop on the coupling inductance which is determined with the modulation factor m and phase angle δ.

$$ I_p = \frac{U_2 - U_p}{jX_p\sqrt{3}} = \frac{U_2 - mU_2 e^{j\delta}}{jX_p\sqrt{3}} \quad (1) $$

The voltage at the point of connection U_2 is calculated with (2). With a known voltage U_2 also U_1 can be calculated.

$$ U_2 = \left(\frac{1}{Z_2} + \frac{1 - me^{j\delta}}{X_p} + \frac{1}{Z_1 + Z_b} \right)^{-1} \frac{U_{TM}}{Z_2} \quad (2) $$

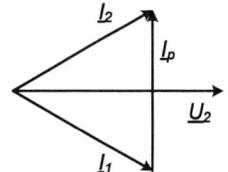

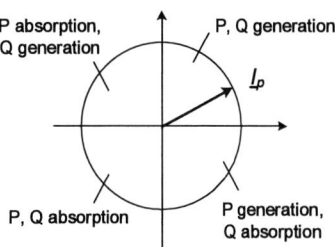

Fig. 3: Active compensator operation

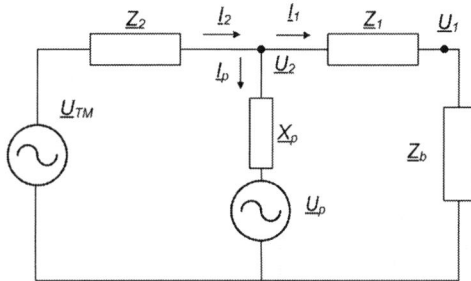

Fig. 4: Equivalent circuit of the observed network

B. Results

In the first case the compensator was connected at a distance 0.2 km from the load. The distance from the network was therefore 9.8 km. Fig. 5 shows the absolute voltage at the point of compensator connection (U_2) as a function of the R/X ratio of the line. Different curves represent different levels of reactive power injection (in pu), where a negative sign means reactive power generation and a positive sign means reactive power consumption. The load power factor was $\cos\varphi = 0.95$ (i.e. $P_b/Q_b = 3$). Fig. 6 shows the results for the same case but with a different load power factor, i.e. $P_b/Q_b = 1$. Fig. 7 shows voltage U_2 variation if compared to the voltage without compensation (for the same case).

The results indicate that voltage compensation with reactive power is more effective at low R/X ratios and almost ineffective at high R/X ratios. It can be also observed that a shunt compensator has a relatively small effect on voltage level. With a series connected compensator the same voltage increase could be achieved with less power injection. On the other hand we must take into account that a series compensator influences just voltage after the point of connection whereas a shunt compensator affects the voltage profile in a large network area.

In the second case we assumed that the compensator is capable also of active power exchange. The absolute value

2511

of voltage $\underline{U_2}$ was again calculated for different ratios of active and reactive power injection and for the case without compensation. The compensator rated power is fixed at 4 pu. The results are shown in Fig. 8 and 9, for load power factors $P_b/Q_b = 3$ and $P_b/Q_b = 1$ respectively.

As it follows from the results active power injection has far more impact on voltage level at high R/X ratios and reactive power injection has more impact at low R/X ratios. At $R/X = 1$ the effectiveness of active and reactive power compensation is equal. It can be concluded that for R/X values between 0.4 and 1.8 a combination of active and reactive power compensation is most effective.

Fig. 10 shows the influence of compensator distance from the load on voltage at load connection point ($\underline{U_1}$). The graph shows the dependency of the U_1 absolute value on the Z_1/Z_2 ratio. The ratio of absolute impedances also equals to the ratio of lengths l_1/l_2. The voltage U_1 is calculated for different operating points of the compensator. All the results show that the compensator has a larger influence on load voltage level when connected closer to the load. In this particular case with relatively large R/X ratio ($R/X = 2$) better results are achieved with active power injection.

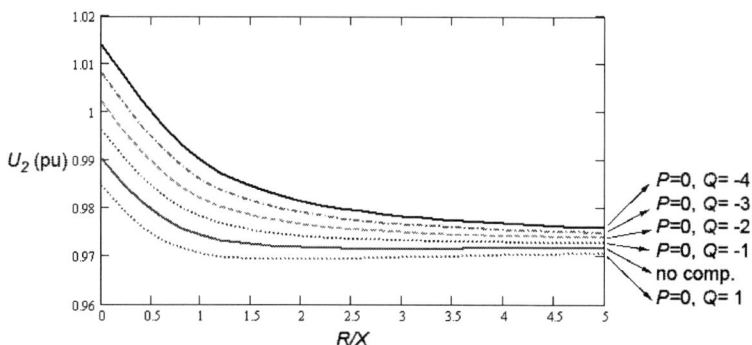

Fig. 5: Dependence of absolute value of voltage U_2 on the line R/X ratio (P_b/Q_b=3).

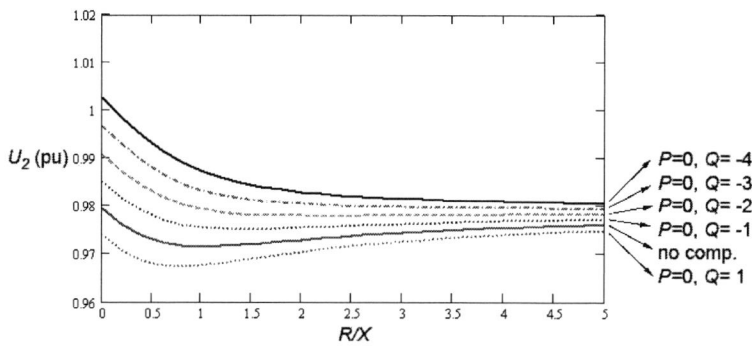

Fig. 6: Dependence of absolute value of voltage U_2 on the line R/X ratio (P_b/Q_b=1).

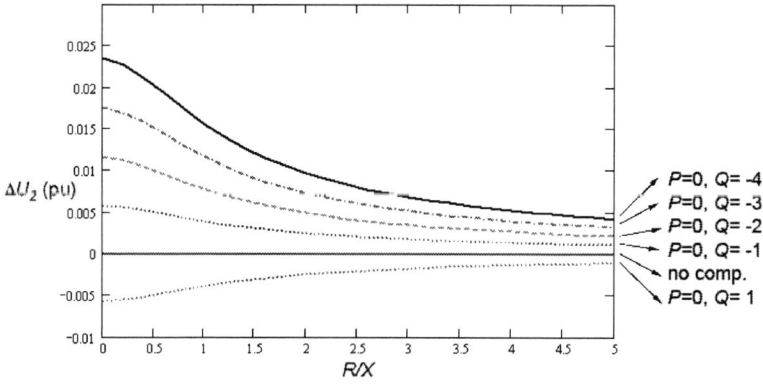

Fig. 7: Voltage U_2 variation depending on the line R/X ratio (P_b/Q_b=3).

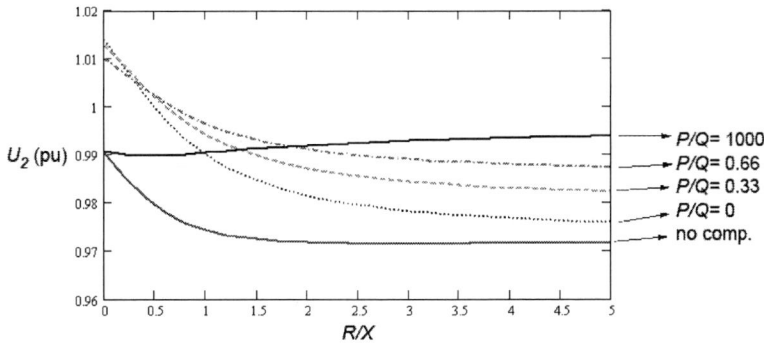

Fig. 8: Dependence of absolute value of voltage U_2 on the line R/X ratio, S_p = 4 pu (P_b/Q_b=3).

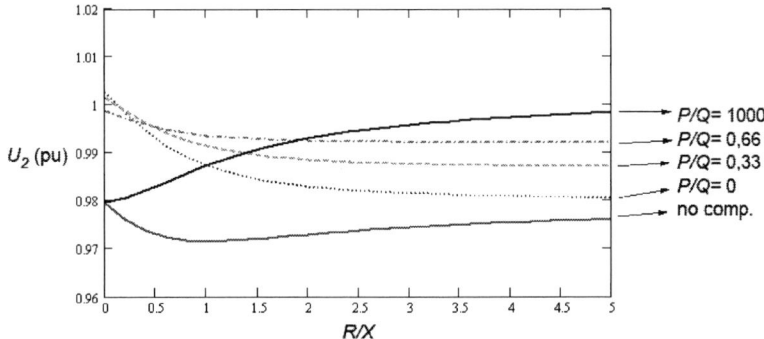

Fig. 9: Dependence of absolute value of voltage U_2 on the line R/X ratio, S_p = 4 pu (P_b/Q_b=1).

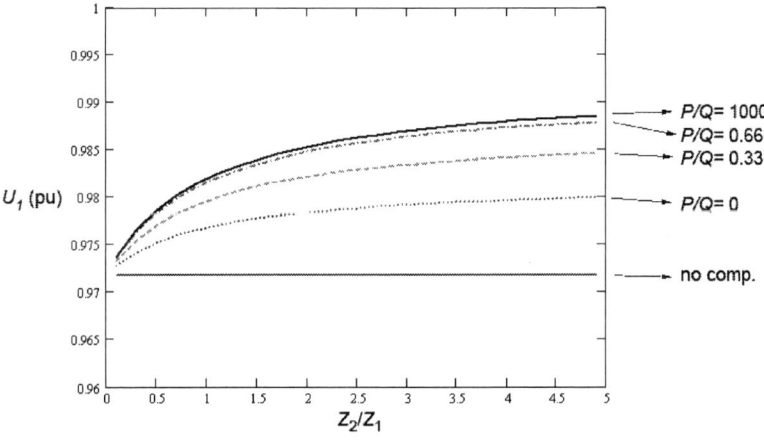

Fig. 10: Dependence of absolute value of voltage U_1 on the compensator location, S_p = 4 pu, R/X = 2, (P_b/Q_b=3).

IV. SIMULATION CASE

To illustrate some practical implications of voltage regulation in distribution networks and the influence of the R/X ratio a simulation case is presented next. Simulations were carried out in the PSCAD programme. The simulated network is shown in Fig. 11. A 20 kV substation is fed from the 110 kV transmission network with a short circuit power of 2000 MVA and an R/X ratio of 0.1. The 110/20 kV and the 20/0.4 kV transformers data are given in Fig. 11. For both transformers an R/X ratio of 0.1 was used. Impedances Z_1 and Z_2 represent line impedances. Same line data as in the calculation example were used and were given in Table I. A load was connected at the 0.4 kV level and was composed of a fixed load (P_b = 3 MW, Q_b = 1 MVAr) and a motor drive (P_M = 400 kW). An active filter with a source on the DC side was connected to the 20 kV bus. The filter was capable of active and reactive power exchange. The basic operation of the compensator was already introduced in section II. For firing pulses generation hysteresis switching was employed. The device was used for voltage regulation at the connection point.

The active filter task was to compensate the voltage drop that was the consequence of motor drive starting. Two cases were simulated. The reference voltage level was set to the value before the motor started for each case. In the first case the line (impedances Z_1 and Z_2) R/X ratio

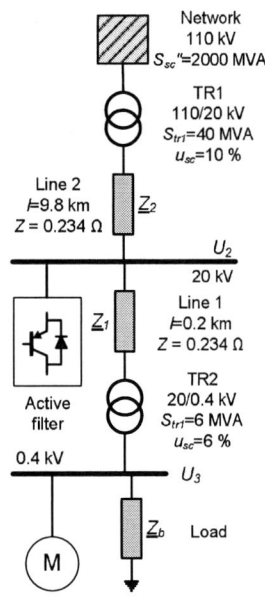

Fig. 11: Simulated network (PSCAD)

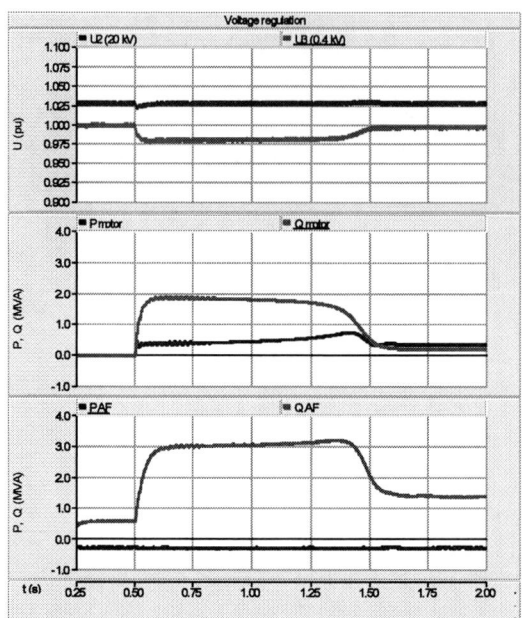

Fig. 13: Voltage compensation with reactive power, R/X = 4

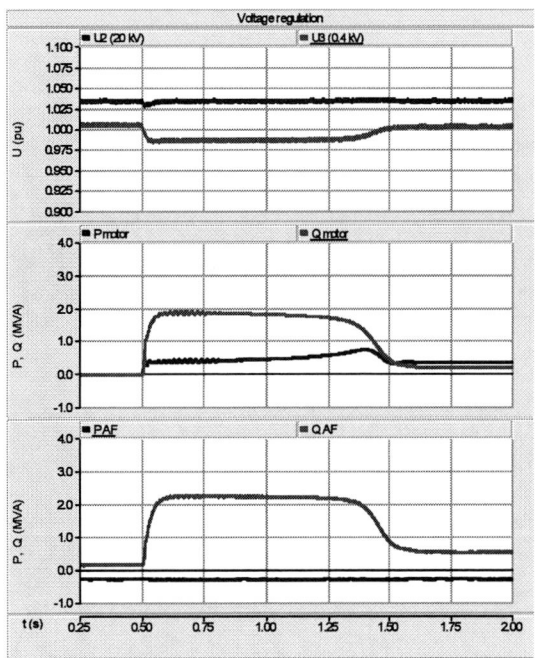

Fig. 12: Voltage compensation with reactive power, R/X = 0.5

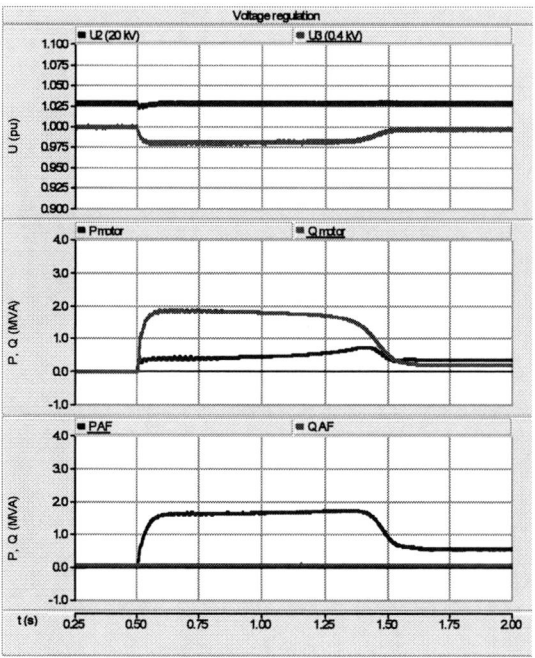

Fig. 14: Voltage compensation with active power, R/X = 4

was set to 0.5 while in the second case the ratio was 4. The absolute impedance value was the same in both cases. The results are shown in Fig. 12 – 14. The graphs show voltages at the 20 kV and 0.4 kV bus (U2 and U3, in pu), active and reactive power of the motor (P motor, Q motor) and active and reactive power of the active filter (P AF, Q AF).

Results for the first case (R/X = 0.5) are shown in Fig. 12. Only reactive power is used for voltage compensation. The compensator is able to maintain a constant voltage at

20 kV level also during motor start with a maximum reactive power injection of approximately 2.2 MVAr. Results for the second case (R/X = 4) and reactive power compensation are shown in Fig. 13. This time the required reactive power is approximately 3.2 MVAr which is 45 % more than in the first case. Fig. 14 also shows the results for the second case but with compensation with active power. A 1.8 MW active power injection was needed to keep the voltage at the level before the motor started. Of course, for such compensation an energy source is needed, whereas in the form of energy storage or a DG source.

2514

V. CONCLUSIONS

The *R/X* ratios in distribution networks may have high values when compared to the ratios in transmission networks. This fact influences also the compensation effectiveness of shunt connected compensators or DG connected to the network through converters that enable compensation. The results have shown that at low *R/X* ratios (lower than 0.4) reactive power is most effective for voltage regulation. At high *R/X* ratios (higher than approx. 1.8) active power has the largest impact on voltage. At medium R/X ratios a combination of active and reactive power is most effective. Although such results are expected, clearly identified limits enable proper design of compensator's control algorithms.

To present the discussed issue in a realistic system a small distribution network was simulated in PSCAD. An active filter was used to compensate the voltage drop cause by the starting of a motor. The results confirmed that reactive power compensation is more effective at low *R/X* ratios while active power compensation is more efficient at high ratios.

The calculations and simulations results suggest that if voltage regulation is used in networks, attention has to be paid also to the network *R/X* ratio. For example, replacement of an overhead line with an underground cable will have an impact also on the compensation characteristics of active compensation devices. Voltage regulation is an important topic also in networks with a high share of distributed generation, usually connected to low- and medium-voltage levels. If the connected generators participate at network voltage regulation the network *R/X* ratio has to be taken into account. The results also show that shunt compensation has a relatively small impact on voltage levels but, on the other hand, has an influence on voltage profile in a large part of a network.

REFERENCES

[1] N. Jenkins, et al., "Embedded generation", IEE, London, UK, 2000.

[2] R. O'Gorman, M. Redfern, "The impact of distributed generation on voltage control in distribution systems", 18th International Conference on Electricity Distribution, Turin, 6-9 June 2005.

[3] Ghosh A., Ledvich G., "Power Quality Enhancement Using Custom Power Devices", Kluwer Academic Publishers, USA, 2002.

[4] E. Acha, et al., "Power Electronic Control in Electrical Systems", Newnes Power Engineering Series, 2002.

[5] N.G. Hingorani and L. Gyugyi, "Understanding FACTS: concepts and technology of flexible AC transmission systems", New York: IEEE Press, 2000.

[6] Song Y. H., Johns A. T., "Flexible ac Transmission Systems (FACTS)", IEE, London, 1999.

[7] I. Papi☐ B. Blaži☐ "FACTS technology efficiency in distribution networks with high DG penetration", Digesec CRIS Workshop 2006, December 6th to 8th 2006, Magdeburg, Germany.

Modeling, Simulation and Analysis of Conducted Common-Mode EMI in Matrix Converters for Wind Turbine Generators

S. Zhang and K.J. Tseng

Centre for Smart Energy Systems, Nanyang Technological University, Singapore, e-mail: *ekjtseng@ntu.edu.sg*

Abstract—This paper introduces an equivalent circuit analysis method to investigate common-mode effects in matrix converter applied in wind energy systems. Transfer characteristics of common-mode current against emission sources are analyzed in order to identify the coupling of transfer functions due to absence of DC link. The comprehensive analysis results provide insight into the common-mode current generation, coupling, and transmission. In addition, the matrix converter time-domain model is established with accurate parasitic parameters from experimental measurements. The validity and effectiveness of analysis method has been verified by time-domain simulation.

Keywords—Matrix Converter, Wind Turbine Generator, EMI, Common-Mode

I. INTRODUCTION

It is generally known that common-mode (CM) conducted electromagnetic interference (EMI) is caused by common-mode current flowing through the parasitic capacitance of transistors, diodes, and transformers to ground in the power circuit. The conducted common-mode current is mainly dependent on the dv/dt of power devices and on the common-mode paths in switching power converters [1, 2]. In addition, common-mode current is also the main cause of machine bearing lifetime degradation.

Consequently, conducted common-mode currents in power converter-based wind generation system at the machine and at the grid sides would be a serious issue for machine operation reliability and EMI compliance. Various research findings have been reported on the reduction of common-mode current using common-mode EMI filters [3-5]. Improved modulation strategies have also been proposed to reduce output common-mode voltage of corresponding converters [6-8]. Especially, the zero vector modification of AC-DC-AC converter is helpful in eliminating total common-mode voltage which is causing common-mode current problems [8].

Since last decade, there has been strong interest in matrix converter which can realize AC-AC direct conversion without intermediate DC electrolytic capacitors [9]. However matrix converters are susceptible to EMI related issues due to the complex power devices architecture [10]. A method to design an integrated filter of matrix converter for drive application was reported which shown that it could meet strict EMI regulations with a small value of filter volume [11]. Additionally,

several proposed technologies were reported to eliminate peak value of common-mode voltage in matrix converters [12-15]. However detailed studies on common-mode current caused by common-mode voltage on coupling paths in matrix converters are lacking. It is important to investigate common-mode current issues directly related to system operation reliability and harmful EMI effects [3-5, 16, 17]. Therefore, this paper introduces a method to investigate the common-mode issues together with common-mode emission sources and coupling paths.

The paper firstly presents the investigation of parasitic parameters of matrix converter associated with common-mode paths and their characterization from repeated experimental measurements. Then accurate time-domain simulation models of matrix converter system have been established in order to investigate the conducted common-mode current issue. Equivalent circuit is useful in modeling common-mode emission effects along with different coupling paths. Consequently, the virtual DC-link equivalent circuit of matrix converter was adapted and modified to analyze its common-mode behavior. This is a new and significant contribution. The paper then provides the analysis of the common-mode coupling characteristics of input and output impedances of matrix converter. The comprehensive analysis is helpful in evaluation of EMI associated with conducted common-mode currents, and in the design of filters for its mitigation. Finally, the verification of the equivalent circuit analysis method by accurate time-domain simulations is presented.

II. WIND GENERATION SYSTEM CONFIGURATION BASED ON MATRIX CONVERTER

AC-AC matrix converter has been reported in the application of wind generation based on doubly-fed induction generator (DFIG) [18, 19]. It has several characteristics which do not appear in back-to-back converter, such as the absence of DC electrolytic capacitors, and the bidirectional power flow ability. Fig. 1(a) shows the fundamental configuration of matrix converter. The matrix converter consists of nine bidirectional switches, arranged in a manner such that any input phase may be connected to any output phase at any time. Each individual switch is capable of rectification and inversion. The matrix converter is controlled using double space vector modulation (SVM) [9], employing the use of input current and output voltage SVM with virtual DC-link technology as shown in Fig.1 (b).

978-1-4244-1741-4/08/$25.00 ©2008 IEEE

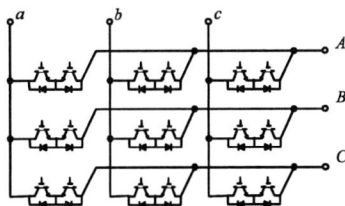

(a) Fundamental configuration of matrix converter

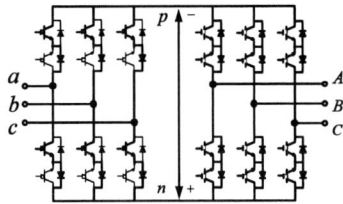

(b) Equivalent virtual DC-link structure of matrix converter

Fig. 1. Matrix converter topology

Fig.2 shows the typical wind turbine generator (WTG) system configuration based on matrix converter and doubly fed induction generator (DFIG). A matrix converter is connected directly between the rotor and the AC grid, providing a single-stage AC-AC power conversion of two voltage systems. The grid-side connection is made via a three-phase LC filter to suppress high-order harmonics in the currents. The use of a stator-flux oriented control is employed on the rotor matrix converter. The d-axis current is aligned with the stator-flux linkage vector. Simple PI controllers can be employed to control the d-q axis currents. The regulation of the d-axis current allows for control of the stator-side reactive power flow, where as the q axis current helps regulate the stator-side active power [18]. Details of this method are beyond the scope of this paper but can be referenced in [19].

In order to analyse the common-mode current in power electronics system, it is well known that parasitic capacitances play an important role in providing common-mode paths. Therefore, accurate analysis results should be based on characterization of the corresponding parameters in a suitable manner. Hence, the characterization method applied in this study is explained in the next section.

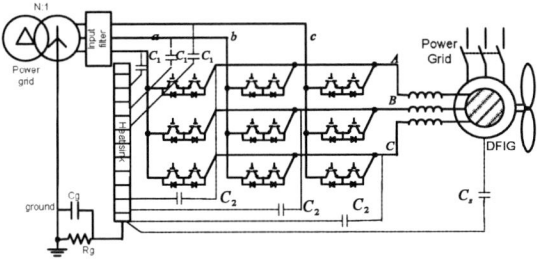

Fig.2: Wind generation system with matrix converter

III. TIME-DOMAIN SYSTEM MODEL SETUP

A simulation using the ANSOFT SIMPLORER package has been developed which enables the entire system including high-frequency effects to be modelled in the time domain. This simulation together with experimental measurements allows the dominant high-frequency current paths to be identified. Consequently, the simulation model would be divided into several segments such as power semiconductor devices, induction machine and matrix converter distributed configuration. Moreover, the input low-frequency harmonics filter is included in the setup to form full common-mode paths. The cable is assumed to be short enough for its contributions to be neglected.

A. Power semiconductor devices

Several semiconductor physics phenomena need to be considered with high accuracy in modeling power devices, because these phenomena dominate the high-frequency switching characteristics. For high-power IGBT devices, the important effects that need to be modeled are conduction modulation, charge storage, metal–oxide–semiconductor (MOS) capacitances, electro–thermal interaction, and breakdown phenomenon. Challenges in developing such models for circuit simulation arise from the need to simultaneously fulfill the contradicting requirements of high quantitative accuracy and low computation overheads.

Another challenge in modeling the IGBT device is the model parameters extraction. The parameters should be obtained through some experiments without too much difficulty or complexity. Various parasitics exist inside the IGBT module and they play a very important role in the EMI generations. However, modeling of parasitic elements has been a very challenging task because of: [20, 21, 22]

    1)   The parasitic elements are difficult to identify;

    2)   The parasitic inductances or capacitances are usually too small to be measured using regular impedance analyzer;

    3)   The parasitic elements may be physically inaccessible inside a package.

To overcome these difficulties, nonlinear switching characteristics of IGBT can be modeled by curve fitting method so as to follow the experimental waveform of applied IGBTs [23]. Fortunately, this kind of approach can be easily realized in SIMPLORER which has custom-defined function to reproduce our prediction model for common-mode conducted analysis. Fig. 3 shows the simulation results of IGBT modules which follow the actual operation profile of power devices.

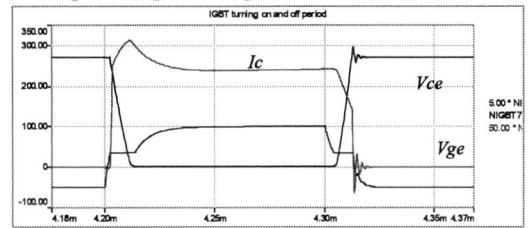

Fig.3: Simulation results of IGBT turning on and off waveforms

B. Machine model

The low-frequency behavior of the induction machine is modeled using standard differential equation representation, which is easily implemented in SIMPLORER. This model accurately reproduces the fundamental current and the currents due to the low-frequency PWM harmonics. The low-frequency inductance and resistance can be worked out from the winding turns and cross-sectional structure. The accurate values should be measured from the winding terminals to the machine neutral point with impedance analyzer (HEWLETT PACKARD 4194A) over low frequency range (10~500Hz) in this study.

However, parasitic inductance and capacitance of the winding to ground would significantly contribute to common-mode paths. The inductance and capacitance values for every winding are determined by measuring the leakage inductance and the winding to frame capacitance with an impedance analyzer over the frequency range 100Hz to 30MHz [3]. Fig. 4 shows the measured characteristic of the machine common-mode impedance. From these measurements, it was determined that the motor has a stray capacitance of 1.76nF in the frequency range of 100Hz – 100 kHz. Therefore, the equivalent circuit for high frequency characteristics is estimated and also confirmed with the corresponding component values of RLC structure associated with high frequency impedance.

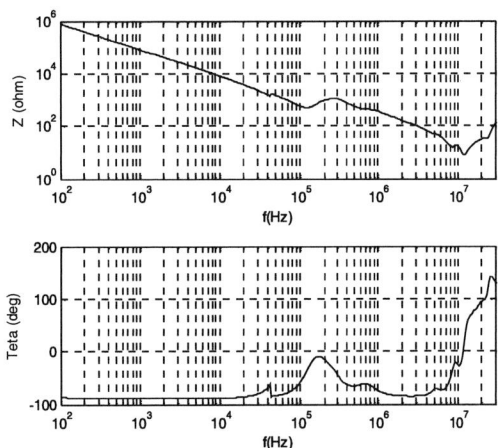

Fig. 4: Measured common-mode impedance of the test induction machine

Fig. 5 shows the common-mode equivalent circuit and corresponding RLC parameters of the induction machine which include both low-frequency and high frequency effects. From a theoretical point of view, the common-mode inductance is one-third of the stator leakage inductance per phase at a line frequency of 50 Hz. However, this is not applicable in a frequency range of 1 kHz – 1 MHz because of the high-frequency effects of non-negligible parasitic components in the motor.

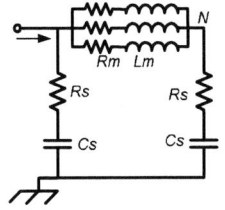

Parameter	Value
Rm	19 ohm
Lm	200 mH
Rs	1 ohm
Cs	1.76 nF

Fig. 5: Common-mode equivalent circuit and electrical parameters of the test induction machine

C. Distributed parameters of matrix converter

The parasitic inductance and capacitance are related to various design considerations, like circuit components layout, PCB design, device selection and thermal dissipation structures. Due to these factors, even two products based on the same system design may have quite different parasitic parameters due to different physical layout.

The best way to extract the distributed parameters would be measurements by wide-response signal analyzer which has a wide frequency analysis range and accuracy. In another way, equivalent circuit method can be used to estimate key parameters. The former method should be carried out after a prototype is established; the latter can be chosen as an approximation before the prototype is established.

The study considered the parasitic effects of topology arising from power devices and ignored the layout effects. Generally, switches are equipped with heat sink as in Fig. 2 and it is inevitable to have parasitic capacitances between them. In addition, the heat sink is connected to ground. With this assumption, we calculated the parasitic capacitance using the definition of capacitance given in Equation (1). Here, S is the area of the heat sink and d is the thickness of the insulator between the sink and the switches.

$$C_s = \frac{K\varepsilon_0 S}{d} \tag{1}$$

Where $\varepsilon_0 = 8.854 \times 10^{-12}$ is the dielectric constant of vacuum and K is the relative dielectric constant of the insulator.

Using Equation (1), key parameters can be extracted via time-domain simulations in SIMPLORER. Initially, a phase voltage is applied to the parasitic capacitances. When it is changed in a step-wise shape due to switching, a high frequency leakage current is generated along a common mode route, from the switch, through the capacitor and to the ground. This may cause EMI problems. The details are discussed in following section.

IV. COMMON-MODE ANALYSIS IN MATRIX CONVERTER

The conducted common-mode currents would be present in common-mode sources and EMI paths. In this paper we consider the dv/dt of switches as the main EMI sources. The common-mode paths in matrix converter-based WTG are made up of common-mode parasitic circuits of the generator, the converter and the grid. The equivalent circuit analysis method [24] to analyse the common-mode current problem is presented in this section. The grid and machine side common-mode currents are analysed separately due to their different effects on system reliability. Then the issues of common-mode effects on potential applications can be anticipated.

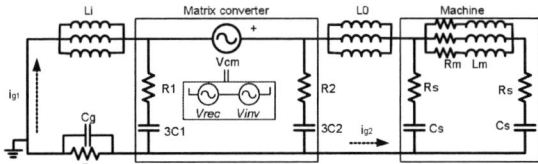

Fig. 6: The common-mode equivalent circuit model of wind turbine generator based on matrix converter

A. Common-mode equivalent circuit of matrix converter

The parasitic nature of matrix converter is quite different from the back-back converter because of the absence of DC link stage. There is only one common-mode source in matrix converter due to the direct AC-AC conversion. The corresponding common-mode paths and emission sources in matrix converter are illustrated in Fig. 6. The common-mode emission source in machine side in matrix converter system can be deduced from input grid side voltage and the PWM switching functions of the three phases [9].

$$V_{cm}(t) = \frac{V_{Ag} + V_{Bg} + V_{Cg}}{3} = \sum_{\substack{i=a,b,c \\ j=A,B,C}} \frac{v_{ig}(t) \times S_{ij}}{3} \quad (2)$$

Where $v_{ag}(t)$, $v_{bg}(t)$ and $v_{cg}(t)$ are the input phase sources with respect to ground, as shown in Fig 2. S_{ij} is the switching function of matrix converter. We have adapted the dual space vector modulation of matrix converter. This common-mode voltage is more complicated to analyze because of tighter coupling between input and output sides in matrix converter based system. Fig. 7 shows the time domain simulation results of common-mode voltage sources in matrix converter.

Using the principle of dual space vector modulation, the matrix converter is divided into a virtual rectifier and a virtual inverter in order to model the common-mode conducted paths. A virtual DC stage without any capacitor separates the virtual rectifier and inverter. The mid neutral point O of the virtual DC-link is taken as the potential to reference the virtual divided common-mode emission sources at the input and at the output terminals of the converter [8]. Therefore, the common-mode voltage can be modeled as two separate emission sources as shown in Fig. 5:

$$V_{cm} = \frac{V_{Ao} + V_{Bo} + V_{Co}}{3} - \frac{V_{ao} + V_{bo} + V_{co}}{3} + \frac{V_{ag} + V_{bg} + V_{cg}}{3} \quad (3)$$

Where, V_{Ao}, V_{Bo}, V_{Co} are voltages between output terminals and virtual mid-point; V_{ao}, V_{bo}, V_{co} are voltages between input terminals and virtual mid-point; and V_{ag}, V_{bg}, V_{cg} are the input phase to ground voltages. If symmetrical input voltage sources are applied to the matrix converter wit h their neutral points tied to ground, equation (3) can be simplified as

$$V_{cm} = V_{inv} - V_{rec} + 0 \quad (4)$$

Where, V_{inv} and V_{rec} are the equivalent virtual common-mode sources of the matrix converter.

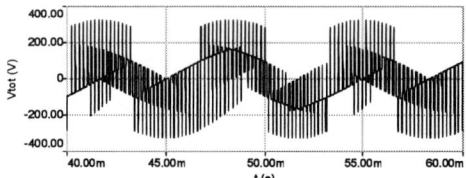

Fig.7: Direct common-mode emission source in matrix converter

The common-mode voltage is associated with switching regulation. The equivalent DC link voltage in matrix converter is shown in Fig. 8(a). Due to the virtual DC voltage, the virtual output side common-mode voltage would be as shown in Fig. 8(b). The maximum voltage envelope would appear to have low frequency fluctuations. The input side here is considered as controlled rectifier with current SVM, and the corresponding common-mode voltage presented in Fig.8(c) would be much larger than the one in back-back converter.

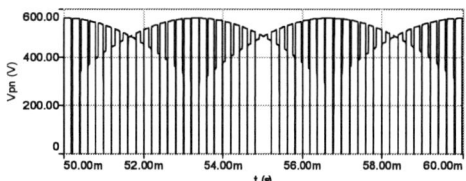

(a) Virtual dc link voltage

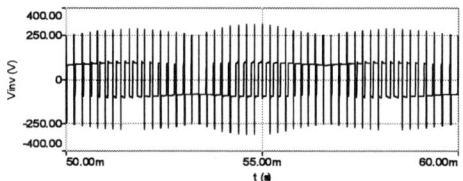

(b) Equivalent inverter side source

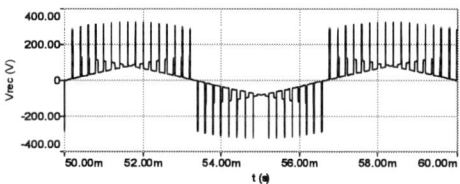

(c) Equivalent rectifier side source

Fig. 8: Virtual common-mode sources in matrix converter

Note that the most dominant frequency component present in the common-mode voltage is 5 kHz or the carrier frequency. Both equivalent common-mode emissions of the matrix converter would spread around with 5 kHz; contributing to high frequency effects in the grid side and machine side common-mode paths.

B. Common-mode transfer gain analysis of matrix converter

Based on the principle of dual space vector modulation, the equivalent input current vector modulation and output voltage vector modulation have to be adapted so that the input and output side common-mode voltage sources would be PWM modulated. Consequently, the dominant frequency components present in the common-mode voltage is at the carrier frequency in both input and output side. This would be a major difference in equivalent analysis method for matrix converter compared to back-back converter. Therefore, the two equivalent common-mode sources would contribute to the generation of common-mode currents resulting in serious EMI problems. The respective grid and machine side common-mode current transfer gains $G_{in}(j\omega) = \dfrac{I_{g1}}{V_{cm}}$

and $G_{out}(j\omega) = \dfrac{I_{g2}}{V_{cm}}$ can be expressed by:

$$G_{in}(j\omega) = \frac{(R_1 + \frac{1}{3j\omega c_1})(R_2 + \frac{1}{3j\omega c_2} + z_m)}{(R_2 + \frac{1}{3j\omega c_2})z_m(R_1 + \frac{1}{3j\omega c_1} + z_g) + (R_1 + \frac{1}{3j\omega c_1})z_g(R_2 + \frac{1}{3j\omega c_2} + z_m)}$$

(5)

$$G_{out}(j\omega) = \frac{(R_2 + \frac{1}{3j\omega c_2})(R_1 + \frac{1}{3j\omega c_1} + z_g)}{(R_2 + \frac{1}{3j\omega c_2})z_m(R_1 + \frac{1}{3j\omega c_1} + z_g) + (R_1 + \frac{1}{3j\omega c_1})z_g(R_2 + \frac{1}{3j\omega c_2} + z_m)}$$

(6)

Where, z_m is the common-mode equivalent impedance of the induction machine, and z_g is the common-mode impedance of front-end route. From Equations (5) and (6), several parasitic parameters: c_1, c_2, z_m and z_g would contribute to the transfer functions of common-mode currents in both grid and machine sides.

Fig. 9 presents the characteristics of $G_{in}(j\omega)$ and $G_{out}(j\omega)$ in the frequency range of 1 kHz to 30MHz. The grid side common-mode current can be expressed as

$$i_{g1} = G_{in}(j\omega) \times (V_{rec} + V_{inv})$$

(7)

The transfer function would be more serious in the intermediate frequency range which would result in high transfer ratio of common-mode machine current.

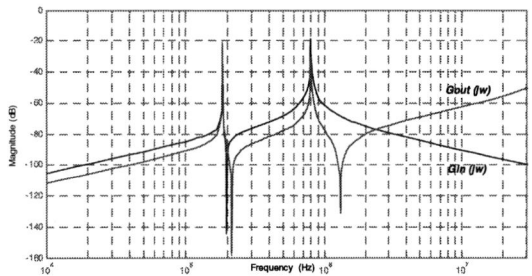

Fig. 9: Common-mode transfer functions $G_{in}(j\omega)$ and $G_{out}(j\omega)$ of matrix converter

In order to understand parasitic contributions to common-mode currents, corresponding transfer functions $G_{in}(j\omega)$ and $G_{out}(j\omega)$ with different parasitic capacitances C_1 and C_2 have been plotted as shown in Fig. 10. In practice, C_1 and C_2 can be assumed to be equal. From the transfer characteristics, it can be seen that increase in parasitic capacitance would introduce more common-mode currents in the grid side in the lower frequency spectrum. This is because the parasitic

common-mode paths of matrix converter would have lower impedance resulting in higher common-mode currents. It is essentially caused by the inductance impedance and capacitor impedance along with operation frequency. In grid side common-mode path, the inductance would play a dominant role on common-mode current, thus the stray capacitance of the machine can contribute to the spectrum of high-frequency interval as shown in Fig. 10(b). From the theoretical viewpoint, at the high frequency domain, the capacitance will play a role of short circuit function. The common-mode current caused by common-mode voltage sources will be directly shorted by the parasitic capacitors without flowing through grid side common-mode paths. Moreover, both V_{recs} and V_{invs} working at carrier frequency associated with the common-mode current transfer characteristics can result in more serious common-mode currents.

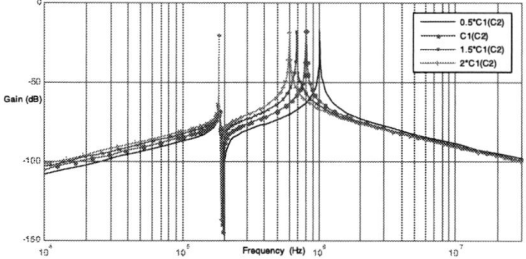

(a) Grid side $G_{in}(j\omega)$

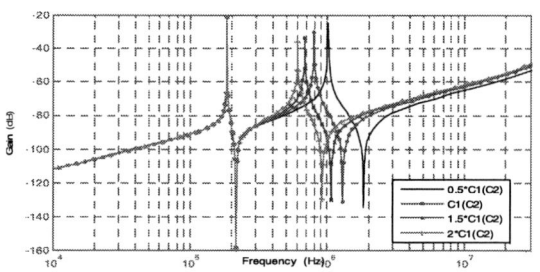

(b) Machine side $G_{out}(j\omega)$

Fig. 10: Common-mode transfer characteristics with different parasitic capacitances

C. Coupling effects on transfer gains

Grid side common-mode impedance z_g usually includes the ground impedance and the inductance of the input filter as shown in Fig. 1. Input EMI filter is a general method to eliminate the serious common-mode emission problem in order to satisfy prevailing EMI standards. There is no DC stage common-mode parasitic capacitor in the matrix converter which would augment the coupling between the grid side and the machine side [25]. Hence, we will analyze whether the input inductance of front-end impedance would introduce significant influences on $|G_{in}(j\omega)|$ and $|G_{out}(j\omega)|$ in this section. Any parasitic resistance shall be ignored here in order to focus on the dominant components of inductance and capacitance. Equations (5) and (6) can be simplified to:

2520

$$G_{in}(j\omega) = \frac{1 + j\omega \cdot 3c_2 \cdot z_m}{z_m + z_g + j\omega \cdot 3(c_1 + c_2)z_g z_m} \quad (8)$$

$$G_{out}(j\omega) = \frac{1 + j\omega \cdot 3c_1 \cdot z_g}{z_m + z_g + j\omega \cdot 3(c_1 + c_2)z_g z_m} \quad (9)$$

Substituting z_g with $j\omega L_i$ as shown in Fig. 5, and considering invariant resistance with frequency, we can express gain magnitudes as

$$\left| G_{in}(j\omega) \right| = \frac{\left| 1 + j\omega \cdot 3c_2 \cdot z_m \right|}{\left| z_m(1 - 3\omega^2(c_1 + c_2) \cdot L_i) + j\omega \cdot L_i \right|} \quad (10)$$

$$\left| G_{out}(j\omega) \right| = \frac{\left| 1 - 3\omega^2 c_1 \cdot L_i \right|}{\left| z_m(1 - 3\omega^2(c_1 + c_2) \cdot L_i) + j\omega \cdot L_i \right|} \quad (11)$$

The increasing input inductance would reduce the transfer gain of $\left| G_{in}(j\omega) \right|$ which is the principle of common-mode EMI filter. However, the L_i effect on $\left| G_{out}(j\omega) \right|$ is difficult to gauge from Equation (11) directly. Hence, the relationships of $\left| G_{in}(j\omega) \right|$ and $\left| G_{out}(j\omega) \right|$ against grid side inductance and frequency are presented graphically in Fig. 11.

From theoretical viewpoint, the inductance would effectively reduce the grid side common-mode current in the high-frequency domain. The admittance magnitude of $\left| G_{in}(j\omega) \right|$ can be seen from Fig. 11 (a). The transfer gain $\left| G_{in}(j\omega) \right|$ is decreasing with the increase in inductance. Consequently, the corresponding common-mode current would be reduced.

Although the input inductance affects the $\left| G_{in}(j\omega) \right|$ significantly, its effect on the output side $\left| G_{out}(j\omega) \right|$ is almost negligible, as shown in Fig. 11 (b). The frequency on the peak value of $\left| G_{out}(j\omega) \right|$ would be decreasing with the increase in grid side inductance. Hence, the coupling characteristics of $\left| G_{in}(j\omega) \right|$ and $\left| G_{out}(j\omega) \right|$ is not as severe as expected, despite the absence of DC link. This has implications on the design of input filter which should be considered along with the output side of matrix converter.

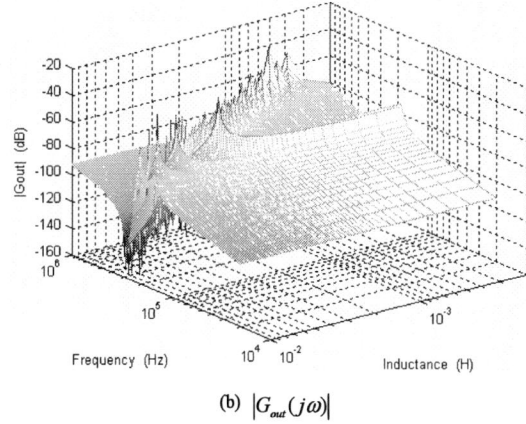

(b) $\left| G_{out}(j\omega) \right|$

Fig. 11: Magnitudes of $\left| G_{in}(j\omega) \right|$ and $\left| G_{out}(j\omega) \right|$ as functions of frequency and inductance L_i

V. SIMULATION RESULTS

To verify the above analysis method, time-domain simulation has been carried out based on the system in Fig. 2 with the parameters in Table 1, using SIMPLORER.

TABLE.1 Main parameters in simulation

Parameters	Value
Input voltage sources	230V AC 50Hz
Output frequency	10 Hz
Switching frequency	5 KHz

Fig. 12 shows the time domain output side results of matrix converter. The line to line voltage is given in Fig. 12(a) and the load 3-phase currents are shown in Fig.12(b). From these two figures, the fundamental characteristics of matrix converter based machine system can be easily seen in terms of the line-line voltage in both topologies.

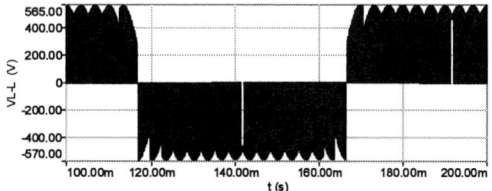

(a) Output line-line voltage

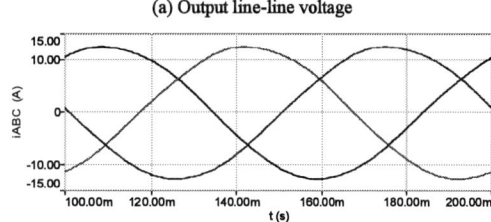

(b) 3-phase output current

Fig.12: Simulation of matrix converter

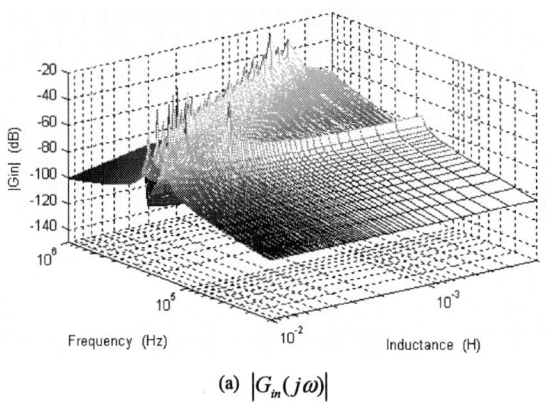

(a) $\left| G_{in}(j\omega) \right|$

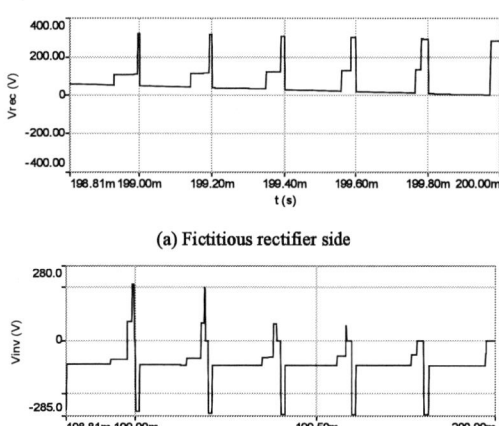

(a) Fictitious rectifier side

(b) Fictitious inverter side

Fig.13: Common-mode voltage of matrix converter

Fig. 13 (a) and (b) show the grid side conducted common-mode currents associated with the corresponding common-mode sources in matrix converter. The common-mode current is generated by the dv/dt of the common-mode voltage presented in Fig.13. The generation of common-mode current can be estimated through the profile of common-mode voltage. The analysis based on equivalent circuits is in agreement with the results in time domain model. In order to illustrate the corresponding results between analysis method and time-domain simulation, the spectral analysis of common-mode currents is shown in Fig.14 (c). The difference during low-frequency is mainly due to absence of parasitic line resistances and inductances. However, the analysis method would be effective to analyze the conducted common-mode current issue in frequency domain range of interest. It is obvious that the corresponding analysis method for common-mode current is in agreement with circuit simulation. It can be seen that the equivalent circuit method is useful to analyse the common-mode current problem without having time domain simulation. Moreover, the virtual rectifier and inverter model of matrix converter is effective for analysing the common-mode issues so that the composite switching modes in matrix converter can be easily equated.

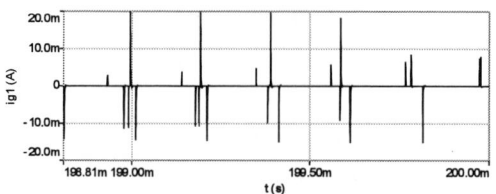

(a) Equivalent circuit analysis method

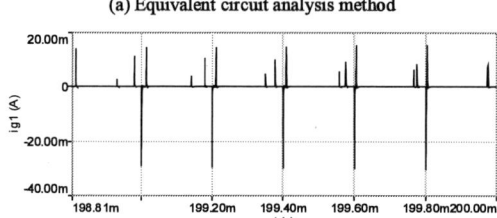

(b) Time-domain model

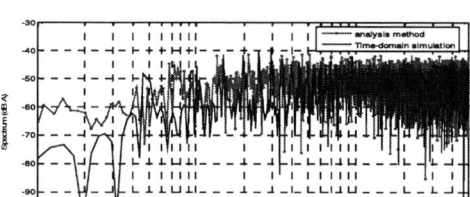

(c) Current spectral diagram

Fig.14: Grid side common-mode simulation results of matrix converter

VI. CONCLUSION

This study presents a method to evaluate the conducted common-mode effects in matrix converter based wind turbine generation system. The method is based on equivalent circuit model of the matrix converter with a virtual DC link. The common-mode models of the input filter and the induction generator are included as well. The conducted common-mode current variation with parasitic parameters has been presented graphically. For verification, a time domain simulation using SIMPLORER with full models of the entire matrix converter based WTG was carried out. It has been found that it can be used to analyze the common-mode currents with reasonable accuracy. Moreover, there is negligible common-mode coupling between the input and the output sides of the matrix converter despite the absence of a DC stage.

REFERENCES

[1] Li Ran, Sunil Gokani, Jon Clare, Keith John Bradley, and Christos Christopoulos, Conducted Electromagnetic Emissions in Induction Motor Drive Systems Part I: Time Domain Analysis and Identification of Dominant Modes IEEE TRANSACTIONS ON POWER ELECTRONICS, VOL. 13, NO. 4, JULY 1998

[2] Li Ran, Sunil Gokani, Jon Clare, Keith John Bradley, and Christos Christopoulos, Conducted Electromagnetic Emissions in Induction Motor Drive Systems Part II: Frequency Domain Models IEEE

TRANSACTIONS ON POWER ELECTRONICS, VOL. 13, NO. 4, JULY 1998

[3] Hirofumi Akagi, Takafumi Doumoto, An Approach to Eliminating High-Frequency Shaft Voltage and Ground Leakage Current From an Inverter-Driven Motor, IEEE TRANSACTIONS ON INDUSTRY APPLICATIONS, VOL. 40, NO. 4, JULY/AUGUST 2004

[4] Nobuyoshi Mutoh, Mitsukatsu Ogata, Kayhan Gulez, and Fumio Harashima, New Methods to Suppress EMI Noises in Motor Drive Systems, IEEE TRANSACTIONS ON INDUSTRIAL ELECTRONICS, VOL. 49, NO. 2, APRIL 2002

[5] Satoshi Ogasawara, Hideki Ayano, and Hirofumi Akagi, An Active Circuit for Cancellation of Common-Mode Voltage Generated by a PWM Inverter, IEEE TRANSACTIONS ON POWER ELECTRONICS, VOL. 13, NO. 5, SEPTEMBER 1998

[6] M. R. Baiju, K. K. Mohapatra, R. S. Kanchan, and K. Gopakumar, A Dual Two-Level Inverter Scheme With Common Mode Voltage Elimination for an Induction Motor Drive IEEE TRANSACTIONS ON POWER ELECTRONICS, VOL. 19, NO. 3, MAY 2004

[7] Amit Kumar Gupta,, Ashwin M. Khambadkone, A Space Vector Modulation Scheme to Reduce Common Mode Voltage for Cascaded Multilevel Inverters IEEE TRANSACTIONS ON POWER ELECTRONICS, VOL. 22, NO. 5, SEPTEMBER 2007

[8] Hyeoun-Dong Lee, Seung-Ki Sul, A Common Mode Voltage Reduction in Boost Rectifier/Inverter System by Shifting Active Voltage Vector in a Control Period, IEEE TRANSACTIONS ON POWER ELECTRONICS, VOL. 15, NO. 6, NOVEMBER 2000

[9] Huber L, Borojevic D. Space Vector Modulated Three-Phase to Three-Phase Matrix Converter with Input Power Factor Correction [J], IEEE Trans on Industry Applications, 1995, 31(6): 1234-1246.

[10] Aten M., Towers G., Whitley C., and et al. Reliability comparison of matrix and other converter topologies, IEEE Transactions on Aerospace and Electronic Systems, 2006, 42(3):867-875.

[11] Kume T., Yamada K., Higuchi T., and et al. Integrated Filters and Their Combined Effects in Matrix Converter, IEEE Transactions on Industry Applications, 2007, 43(2):571-581.

[12] Han Ju Cha, Prasad N. Enjeti. An approach to reduce common mode voltage in matrix converter [J].IEEE Transactions on Industry Applications, 2003, Vol.39(4):1151-1159

[13] Hua Lin, Bi He, Xiaofeng Zhang, A Modulation Strategy to Reduce Common-Mode Voltage for Current-Controlled Matrix Converters. IECON 2006-32nd Annual Conference on Industrial Electronics, Nov. 2006 Page(s):2775 – 2780

[14] Yue Fan, Wheeler Patrick W., Clare Jon C., Mitigation of Common-mode Voltage in Matrix Converter, The 3rd IET International Conference on Power Electronics, Machines and Drives, 2006. Mar. 2006 Page(s):296 - 300

[15] Rzasa J., Control of a matrix converter with reduction of a common mode voltage, IEEE Compatibility in Power Electronics, 2005. June 1, 2005 Page(s):213-217

[16] Meng Jin, Ma Weiming, Pan Qijun, Kang Jun, Zhang Lei, and Zhao Zhihua, Identification of Essential Coupling Path Models for Conducted EMI Prediction in Switching Power Converters, IEEE TRANSACTIONS ON POWER ELECTRONICS, VOL. 21, NO. 6, NOVEMBER 2006

[17] Meng Jin and Ma Weiming, A New Technique for Modeling and Analysis of Mixed-Mode Conducted EMI Noise, IEEE TRANSACTIONS ON POWER ELECTRONICS, VOL. 19, NO. 6, NOVEMBER 2004

[18] Zhang, L., Watthanasarn, C., Shepherd, W., Application of a Matrix Converter for the Power Control of a Variable-Speed Wind-Turbine Driving a Doubly-Fed Induction Generator, 23rd International Conference on Industrial Electronics, Control and Instrumentation, Vol. 2, pp. 906-911, Nov. 1997.

[19] Keyuan, H., Yikang, H., "Investigation of a Matrix Converter-Excited Brushless Doubly-Fed Machine Wind-Power Generation System," The 5th International Conference on Power Electronics and Drive Systems, Vol. 1, pp. 743-748, Nov. 2003.

[20] Huibin Zhu, Allen R. Hefner, Jr., and Jih-Sheng (Jason) Lai, Characterization of Power Electronics System Interconnect Parasitics Using Time Domain Reflectometry, IEEE

TRANSACTIONS ON POWER ELECTRONICS, VOL. 14, NO. 4, JULY 1999

[21] Huibin Zhu, Jih-Sheng Lai, Allen R. Hefner, Yuqing Tang, and Chingchi Chen, Modeling-Based Examination of Conducted EMI Emissions From Hard- and Soft-Switching PWM Inverters, IEEE TRANSACTIONS ON INDUSTRY APPLICATIONS, VOL. 37, NO. 5, SEPTEMBER/OCTOBER 2001

[22] K. Xing, F. C. Lee, and D. Borojevic, Extraction of parasitics within wire-bond IGBT modules, in Proc. IEEE Applied Power Electronics Conf., 1998, pp. 497–503.

[23] Meng Jin, Ma Weiming, Power converter EMI analysis including IGBT nonlinear switching transient model. Proceedings of the IEEE International Symposium on Industrial Electronics, Volume 2, 20-23 June 2005 Page(s):499 - 504 vol. 2

[24] A. E. Ruehli, Equivalent circuit models for three-dimensional multiconductor systems, IEEE Trans. Microwave Theory Tech., vol. 22, pp.216–221, Mar. 1974.

[25] J. L. Schanen, E. Clavel, and J. Roudet, Modeling of low inductance busbar connections, IEEE Ind. Applicat. Mag., pp. 39–43, Sept./Oct. 1996.

Design of Frequency Shift Acceleration Control for Anti-islanding of an Inverter-based DG

Seul-Ki Kim*, Jin-Hong Jeon*, Heung-Kwan Choi and Jonng-Bo Ahn*

* Korea Electro-technology Research Institute, Changwon, Korea, e-mail: *blksheep@keri.re.kr*

Abstract—the paper proposes frequency shift acceleration control (FSAC) for anti-islanding of an inverter-based distributed generator. The key concept is to disturb an islanded system frequency by controlling reactive current output in d-q frame, and accelerate the resulting frequency shift using an acceleration gain. Design method of the acceleration gain is proposed using analytical approaches, based on small-signal stability and step input response. Simulation and experimental results are presented.

Keywords—frequency shift acceleration control, anti-islanding, inverter-based distributed generators, small-signal analysis

I. INTRODUCTION

The purpose of active anti-islanding methods is to inject additional disturbance into a distributed generator (DG) output according to pre-set schemes, and force system parameters such as voltage magnitude, frequency, etc. to deviate from the anti-islanding threshold values in islanding condition. There have been various active schemes proposed and implemented [1-4].

This paper proposes a useful option for active anti-islanding of an inverter-based DG, which is frequency shift acceleration control (FSAC) algorithm with an acceleration gain. The FSAC disturbs an islanded system frequency by controlling reactive current of a grid connection inverter, and accelerate the resulting frequency shift using an analytically designed accelerating gain. The proposed scheme is conceptually based on application of positive feedback control in *dq*-frame [5, 6]. FSAC includes real power component of the grid inverter in reactive current control loop, so that the frequency shift acceleration gain, say FSAC gain, is consistently effective regardless of magnitude of real power generation of the DG. Such an idea makes FSAC different from the conventional positive feedback method [5, 6]. Analytical approaches using small-signal stability and step input analysis are presented to determine the optimal range of the acceleration gain. The proposed control is intended to achieve successful anti-islanding with minimized non-detection zone, non-compromised power quality, and no additional equipments required. Simulation study was conducted with PSCAD/EMTDC package, a power system transient analysis tool. Hardware test was performed using a grid inverter and RLC passive loads. The simulation and experimental results verify that the proposed algorithm is very effective and practical in real DG applications.

II. FREQUENCY SHIFT ACCELERATION CONTROL

A. Frequency Positive Feedback in DQ control

According to DQ frame based instantaneous power control theory [7], real and reactive power of a grid inverter can be independently controlled by regulating the q-axis and d-axis current at the terminal, respectively.

$$P_{inv} = \frac{3}{2} e_q i_q \qquad (1)$$

$$Q_{inv} = -\frac{3}{2} e_q i_d \qquad (2)$$

where e_q is instantaneous voltage at the terminal and equals $\sqrt{2}\,V$. Fig. 1 presents block diagram of the real and reactive power controller.

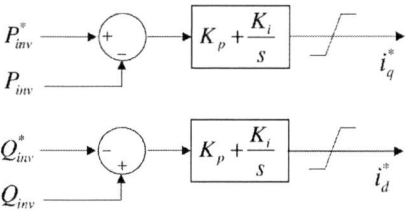

Fig. 1. Power controller (P-I controller)

Frequency positive feedback can be applied in DQ implementation [5, 6]. It does not distort the current waveform and has little impact on harmonics. Fig. 2 presents the frequency positive feedback mechanism in DQ implementation. When the measured frequency increases, the positive feedback commands the reactive power of the DG to be decreased so that the frequency increases more. The frequency will keep moving in the positive direction for its initial variation and be eventually out of the normal ranges.

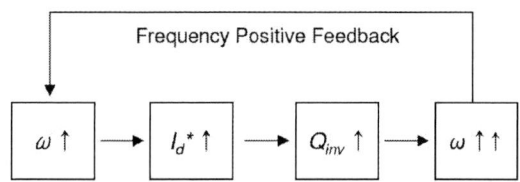

Fig. 2. Frequency positive feedback in dq-frame

B. Frequency Positive Feedback in DQ control

The FSAC basically adopts frequency positive feedback in DQ implementation. The reactive power controller equation is expressed as

$$\left(Q_{inv} - Q^*\right)\left(K_P + \frac{K_i}{s}\right) + \left(\omega\frac{\omega_f}{s+\omega_f} - \omega_0\right)K_{pf} = i_d^* \quad (3)$$

where Q^*, ω_0, i_d^*, ω_f, K_p, K_i, and K_{pf} are the reactive power command, nominal angular frequency, the d-axis current command, filtering frequency, proportional and integral gains of PI controller and positive feedback gain, respectively.

The main difference in FSAC from the normal dq-based positive feedback is to set the positive feedback gain as a frequency shift acceleration gain multiplied by the q-axis current command, as shown in (2).

$$K_{pf} = i_q^* \cdot K_a \quad (4)$$

where i_q^* is the q-axis current command and K_a is a frequency shift acceleration gain [s/radian]. The relationships that shows non-detection zone of under/over frequency can be expressed as below [8].

$$Q_f \cdot \left(1 - \left(\frac{f_0}{f_{min}}\right)^2\right) \le \frac{\Delta Q}{P_{inv}} \le Q_f \cdot \left(1 - \left(\frac{f_0}{f_{max}}\right)^2\right) \quad (5)$$

where f_0, f_{min}, and f_{max} are the nominal frequency and under/ over frequency thresholds, respectively. The (5) can be described as (6)

$$P_{inv} \cdot Q_f \cdot \left(1 - \left(\frac{f_0}{f_{min}}\right)^2\right) \le \Delta Q \le P_{inv} \cdot Q_f \cdot \left(1 - \left(\frac{f_0}{f_{max}}\right)^2\right) (6)$$

In (4) the reactive power mismatch thresholds are influenced by the real power generation of a DG: that is, the reactive power mismatch required for shifting frequency beyond the anti-islanding thresholds is in proportion to the real power output. It is reasonable that the positive feedback gain should reflect the real power component as shown in (4). By doing so the acceleration gain does not need to be readjusted for different levels of real power of the DG. The q-axis current command is proportional to real power of the DG and so the amount of reactive power disturbance is automatically adjusted by the q-axis current command.

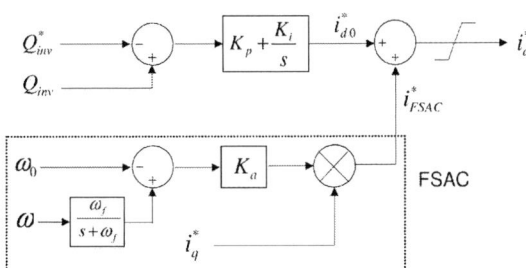

Fig. 3. Proposed FSAC controller

FSAC can be readily implemented into reactive power control loop of the conventional DG inverter just by adding the frequency acceleration component of equations (3) and (4), as shown in Fig. 3. In the proposed scheme, no additional device or accessory is needed.

III. DESIGN OF ACCELERATION GAIN

A. Small Signal Analysis for Lower Limit of K_a

A small signal analysis [9, 10] was used to determine the lower limit of the acceleration gain. Normal lower-level current controller has very fast dynamics, i.e., less than a few mili-seconds, and in (3), the d-axis current command may be replaced with the actual current i_d. So by linearizing (3), a small-signal equation (11) can be obtained.

$$\left(K_P + \frac{K_i}{s}\right)\Delta Q_{inv} + K_{pf}\frac{\omega_f}{s+\omega_f}\Delta\omega = \Delta i_d \quad (11)$$

A small-signal form of (9) can be written as below.

$$\Delta Q_{inv} = -\frac{3}{2}e_q\Delta i_d \quad (12)$$

Reactive load may be described by (13). In islanding condition ω can be approximated as ω_0. (13) can be expressed into its small signal representation (14) with ω replaced with ω_0.

$$Q_{load} = \frac{3}{2}e_q^2\left(\frac{1}{\omega L} - \omega C\right) \quad (13)$$

$$\Delta Q_{load} = -\frac{3}{2}e_q^2\left(\frac{1}{\omega_0^2 L} + C\right)\Delta\omega \quad (14)$$

In islanding condition, reactive power is balanced with reactive load as below.

$$Q_{inv} = Q_{load} \quad (15)$$

Quality factor may be written as (16) [8].

$$Q_f = \omega_0 R C = \frac{R}{\omega_0 L} \quad (16)$$

Substitution of (15) and (16) into (14) leads to (17)

$$\Delta Q_{inv} = -3e_q^2\left(\frac{Q_f}{\omega_0 R}\right)\Delta\omega \quad (17)$$

Also, (18) can be obtained from (12) and (17).

$$\Delta i_d = 2e_q\left(\frac{Q_f}{\omega_0 R}\right)\Delta\omega \quad (18)$$

Finally, characteristic equation (19) can be obtained by substituting (17) and (18) into (11) and developing it in terms of $\Delta\omega$.

$$s^2 + \left[\frac{e_q \left\{ 2 + 3 e_q \left(K_p + \frac{K_i}{\omega_f} \right) \right\} \left(\frac{Q_f}{\omega_0 R} \right) - K_{pf}}{2 e_q \left(\frac{Q_f}{\omega_0 R} \right) \left(1 + \frac{3}{2} e_q K_p \right)} \right] \omega_f s \qquad (19)$$

$$+ \frac{3}{2} e_q \left(\frac{K_i \omega_f}{1 + \frac{3}{2} e_q K_p} \right) = 0$$

where V_n is nominal voltage of inverter terminal (rms). For the system of (19) to be unstable, the positive feedback gain should meet the below inequality condition.

$$K_{pf} > \left\{ 2 + 3 e_q \left(K_p + \frac{K_i}{\omega_f} \right) \right\} \cdot \left(\frac{Q_f}{\omega_0} \right) \cdot \left(\frac{e_q}{R} \right) \qquad (20)$$

The q-axis current i_q can be described as (21).

$$i_q = e_q / R \qquad (21)$$

Accordingly, (22) can be obtained by substituting (8) and (21) into (20).

$$K_a > \left\{ 2 + 3 e_q \left(K_p + \frac{K_i}{\omega_f} \right) \right\} \cdot \left(\frac{Q_f}{\omega_0} \right) \qquad (22)$$

e_q can be approximated as $\sqrt{2} \, V_n$, which denotes inverter nominal voltage (rms), and (22) can be rewritten as (23).

$$K_a > \left\{ 2 + 3\sqrt{2} V_n \left(K_p + \frac{K_i}{\omega_f} \right) \right\} \cdot \left(\frac{Q_f}{\omega_0} \right) \qquad (23)$$

In (20) and (21), it is observed that the lower limit of K_{pf} is dependent on the q-axis current. This implies that the setting of the frequency positive feedback gain should be changed according to real power output of a DG inverter. Accordingly, the proposed FSAC can eliminate the real power dependency of the control gain, as shown in (23), by setting it as the acceleration gain multiplied by the q-axis current command, as presented in (8).

In (23), the acceleration gain K_a mainly depends on the quality factor of the load, and the proportional and integral gains of its upper level controller. Greater Q_f requires greater acceleration gain. K_a also is influenced by frequency of the measuring filter. A filter with higher frequency requires larger value of the acceleration gain. This is considered as reasonable since lower filtering frequency indicates that noises of system frequency are bigger.

B. Step Response Analysis for Upper Limit of K_a

Larger value of K_a means that greater disturbance is added to reactive power of a DG inverter for the same size of frequency deviation. When a system is islanded, the larger K_a will positively contribute to anti-islanding, whereas in grid-tied operation the greater gain negatively impacts on the grid. So it is desirable to impose an upper threshold on the acceleration gain to limit the resulting reactive disturbance. To do so, frequency step input analysis is used.

Applying the Laplace transform into (7) gives

$$\Delta i_d (s) = \left(K_P + \frac{K_i}{s} \right) \Delta Q_{inv} (s) + K_{pf} \Delta \omega(s) \qquad (24)$$

Equation (6) may be expressed in its small signal equation.

$$\Delta Q_{inv}(s) = -\frac{3}{2} e_q \Delta i_d(s) \qquad (25)$$

By substituting (25) into (24) and replacing the frequency deviation with a step function, the relationship between the frequency shift and reactive power disturbance (26) can be obtained.

$$\left| \Delta Q_{inv}(s) \right| = \frac{K_{pf}}{K_p + 2/(3 e_q) + K_i / s} \left| \frac{\Delta \omega}{s} \right| \qquad (26)$$

Inverse Laplace transformation of (26) can be described as below.

$$\Delta Q_{inv}(t) = \frac{K_{pf}}{K_p + 2/(3 e_q)} \left| \Delta \omega \right| \exp[st] \qquad (27)$$

Inequality (28) can be obtained from (27).

$$\Delta Q_{max} > \frac{K_{pf}}{K_p + 2/(3 e_q)} \left| \Delta \omega_{max} \right| \qquad (28)$$

where ΔQ_{max} is the maximum allowable reactive power disturbance, and $\Delta \omega_{max}$ is the maximum step change of system frequency in grid-tied operation. Here, it is assumed that the preset threshold for reactive power disturbance is given as a ratio of the maximum reactive disturbance to real power output. Then, upper limit condition (29) can be obtained by substituting (5) and (8) into (28).

$$K_a < \left(1 + \frac{3}{2} e_q K_p \right) \frac{1}{\left| \Delta \omega_{max} \right|} \cdot \eta_{preset} \qquad (29)$$

where η_{preset} is the preset ratio of the maximum allowable reactive power disturbance to the real power and equals ($\Delta Q_{max} / P_{inv}$). The e_q can be approximated as V_n and hence (29) can be rewritten as (30).

$$K_a < \left(1 + \frac{3\sqrt{2}}{2} V_n K_p \right) \frac{1}{\left| \Delta \omega_{max} \right|} \cdot \eta_{preset} \qquad (30)$$

The upper limit is closely influenced by the K_p of the upper level controller.

IV. SIMULATION STUDY

PSCAD/EMTDC simulations were made to validate the proposed FSAC. A simulation test circuit was configured as shown in Fig. 1. A grid connection inverter was represented by a detailed digital-time simulation model that included inner current loop, outer power loop, and even PWM switching block. Islanding detection conditions were in accordance with IEEE Std. 1547 [11]. Settings of DG inverter generation and RLC load followed the relevant guidelines specified in IEEE std. 929-2000 and UL 1741 [12,13]. The simulation settings are summarized as follows.

- $P_{inv} = P_{load} = 20$kW
- $Q_{inv} = Q_{load} = 0$kVar
- $V_n = 220$V, $V_{max} = 110\% * V_n$, $V_{min} = 88\% * V_n$
- $f_0 = 60$ Hz , $f_{max} = 60.5$ Hz, $f_{min} = 59.3$ Hz
- Disconnection : within two seconds of the formation of an island
- $Q_f = 2.5$
- Filtering frequency = 100 Hz
- R = 7.26Ω , L = 7.7mH, C = 913.24μ F
- $\eta_{preset} = 0.1$, which means that the maximum allowable reactive power variation is 2kVar, 10% of real power generation 20kW.
- $|\Delta\omega_{max}| = 2\pi *(0.3\text{Hz})$

The proportional gain, K_p, and integral gain K_i was 10 and 50, respectively. A proper range of K_a was calculated as below from (23) and (30).

$$0.076 < K_a < 0.30 \qquad (31)$$

It would be reasonable to choose a medium value within the range. So the acceleration gain was set to 0.15. Fig. 4 shows voltage and frequency variations in case of no active anti-islanding scheme applied. Mismatches between generation and load of real and reactive power were zero and there was no actual shift in both voltage and frequency. In case of the proposed FSAC implemented, the frequency was shifted at the moment of islanding occurrence and its shift was accelerated to go beyond the detection threshold, as shown in Fig. 5.

Fig. 6 shows frequency variation in case with $K_a = 0.078$, which was the lowest gain that started destabilizing the islanding system. The simulated value was very close to the calculated 0.076.

In bulk electric power system since, a great number of large scale generators are well in coordination for synchronism of entire system, and there is little change in frequency in normal condition. In this study, 0.3Hz was assumed as the maximum step-change in frequency during grid-tied operation, and η_{preset} was set to 0.1. Hence the maximum reactive disturbance owing to 0.3Hz of frequency change was designed not to exceed 2kVar. The upper limit of Ka was estimated as 0.3 by (30). Fig. 7 shows the resulting reactive power variations owing to the step frequency change with Ka equal to 0.3. The simulation result showed good agreement with the calculated result. Fig. 8 compares current harmonics for cases without and with FSAC ($K_a = 0.3$). It should be noted that there were little discrepancies between two cases. The proposed FSAC gives little negative impact on power quality of grid, so long as its frequency shift acceleration gain is appropriately designed by (23) and (30).

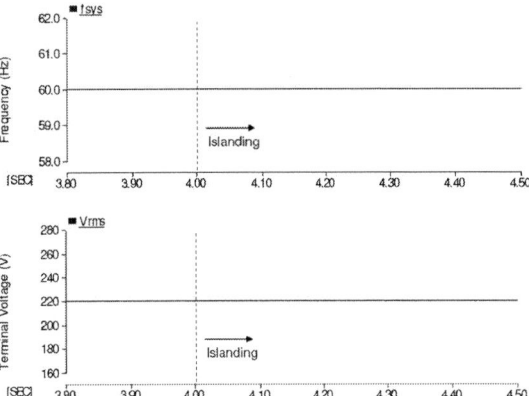

Fig. 4 Frequency and voltage in case without anti-islanding

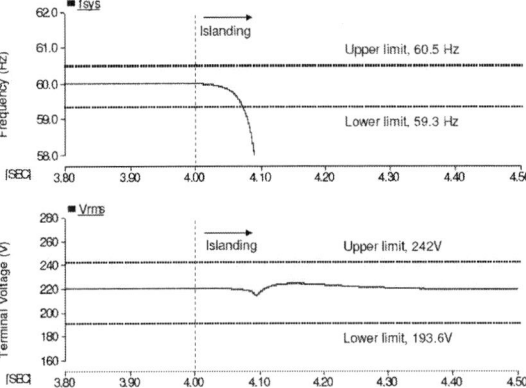

Fig. 5 Frequency and voltage in case with FSAC

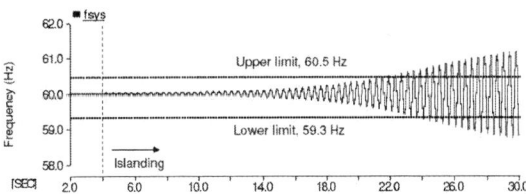

Fig. 6 Frequency variation with $K_a = 0.078$

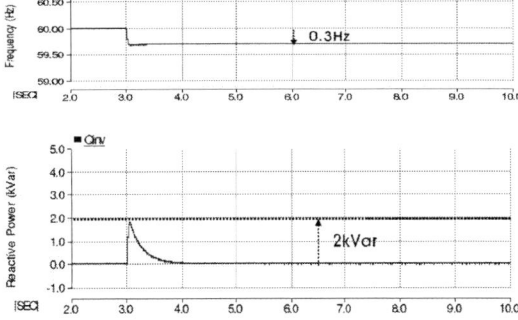

Fig. 7 Reactive power disturbance for step change in frequency during grid-connected operation

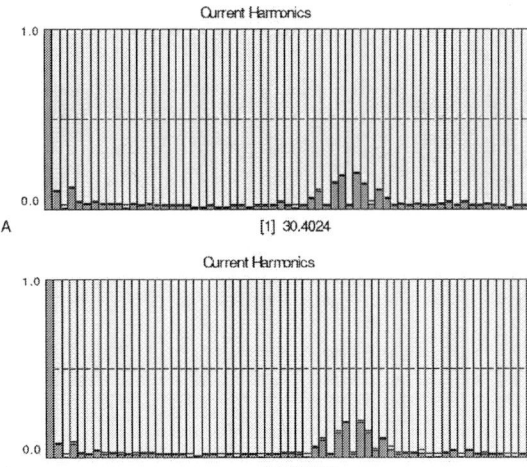

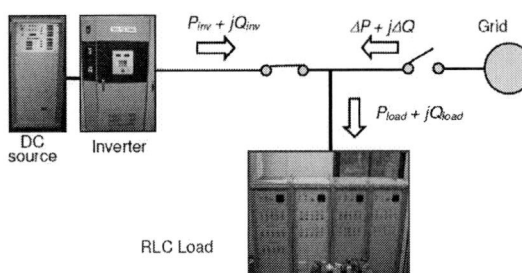

Fig. 8 Comparison of current harmonic spectra: (top) before FSAC (bottom) after FSAC

V. EXPERIMENTAL TEST

The proposed scheme was experimentally tested as configured in fig. 9. A three phase grid-connection inverter and RLC passive load were used. Test conditions and parameters were as follows.

- P_{inv} = 4.0kW, Q_{inv} = 0kVar
- K_p = 10, K_i = 5 (for outer power loop)
- P_{load} = 4.0kW, Q_{load} = 0kVar
- Q_f of RLC load = 2.5

Grid-side breaker was opened after the inverter generation and RLC load were made balanced, i.e., in Fig. 9, $\Delta P = \Delta Q = 0$. Fig. 10 shows the results of islanding operation with no active anti-islanding scheme. Inverter did not stop operation and continued energizing the RLC load. Inverter voltage and current were still alive.

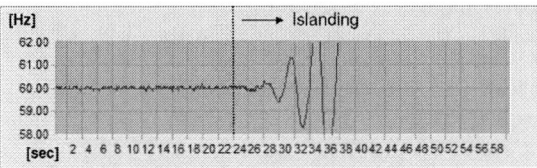

Fig. 9 Experimental test configuration

The proposed FSAC was implemented into the grid connection inverter with K_a=0.1 in its outer power control loop. After the grid-side switch was opened, inverter detected its islanding operation and stopped supplying electric power, as shown in fig. 11. Experimental result for the lower limit of K_a was 0.057. Fig. 12 presents the frequency variation when K_a of 0.057 was applied. The swinging magnitude gradually increased to system collapse. Even if the experimental lower limit was not as much close to the calculated result as the simulated result,

it was considered as close enough to validate the proposed gain design of (23) and (30).

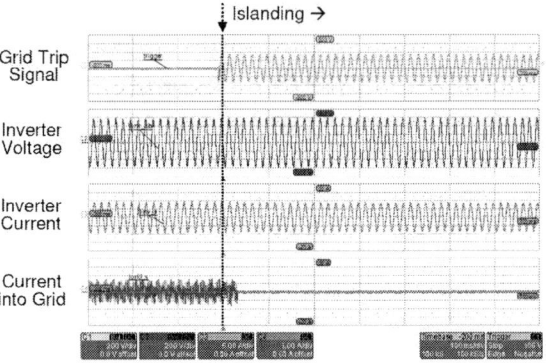

Fig. 10 Anti-islanding test with no active scheme

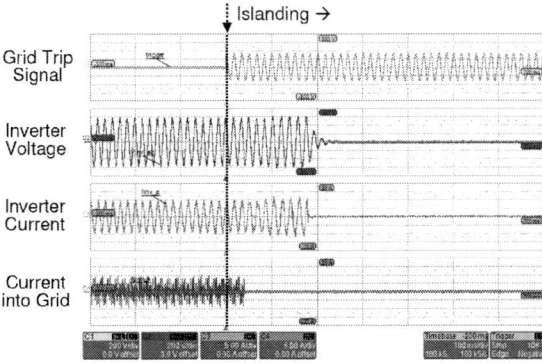

Fig. 11 Test results for power controlled DG with FSAC (K_a = 0.1)

Fig. 12 Frequency of power controlled DG (K_a=0.057)

VI. CONCLUSION

Frequency shift acceleration control was proposed for anti-islanding of an inverter-based DG. Analytical approaches were presented to design the effective range of the frequency shift acceleration gain based on small-signal analysis and step-input response. Simulation and experimental tests were carried out to validate the proposed FSAC scheme and gain designing method. Simulation and experimental results showed satisfactory performance and its feasibility in industrial applications. The FSAC was originally focused on the 3-phase DG, but m y be extended into a single phase DG.

REFERENCES

[1] I. Suh, and L. Chang, ICPE'07 Tutorial 3 Power Electronic Converters for Distributed Generation, The 7th International Conference on Power Electronics, October 2007.

[2] Luiz A. C. Lopes, and Huili Sun, "Performance Assessment of Active Frequency Drift Islanding Detection Methods", IEEE Trans. Energy Conversion, Vol. 21, No. 1, pp. 171-180, March 2006.

[3] G. A. Smith, P. A. Onions and D. G. Infield, "Predicting islanding operation of grid connected PV inverters", IEE Proc.-Electr. Power Appl., Vol. 147, No. 1, pp. 1-6, January 2000.

[4] G.-K. Hung, C.-C. Chang, and C.-L. Chen, "Automatic Phase-Shift Method for Islanding Detection of Grid-Connected Photovoltaic Inverters", IEEE Trans. Energy Conversion, Vol. 18, No. 1, pp. 169-173, March 2003.

[5] Z. Ye, R. Walling, L. Garces, R. Zhou, L. Li and T. Wang, Study and Development of Anti-Islanding Control for Grid-Connected Inverters, NREL/SR-560-36243, May 2004.

[6] Z. Ye, L. Li, L. Garces, C. Wang, R. Zhang, M. Dame, R. Walling, and N. Miller, "A New Family of Active Anti-Islanding Schemes Based on DQ Implementation For Grid-Connected Inverters", 2004 35th Annual IEEE Power Electronics Specialist Conference, Aachen Germany, 2004.

[7] H. Akagi, E. H. Watanabe, and M. Aredes, Instantaneous Power Theory and Applications to Power Conditioning", IEEE PRESS, 2007.

[8] Z. Ye, A. Kolwalkar, Y. Zhang, P. Du, and R. Walling, "Evaluation of Anti-Islanding Schemes Based on Nondetection Zone Concept", IEEE Trans. Power Electronics, Vol. 19, No. 5, pp. 1171-1176, September 2004.

[9] E. Coelho, P. Cortizo, and P. Garcia, "Small-Signal Stability for Parallel-Connected Inverters in Stand-alone AC Supply Systems", IEEE Trans. Industry Applications, Vol. 38, No. 2, pp. 533-542, March/April 2002.

[10] Prabha Kundur, Power System Stability and Control, McGrawHill, 1993.

[11] IEEE, Std 1547.1-2005 IEEE Standard Conformance Test Procedures for Equipment Interconnecting Distributed Resources with Electric Power Systems.

[12] IEEE, Standard 929-2000, IEEE Recommended Practice for Utility Interface of Photovoltaic (PV) Systems, 2000.

[13] UL 1741 Inverters, Converters, and Controllers for Use in Independent Power Systems, 2001.

Integrated Power Converter for Photovoltaic and Fuel Cell Systems in Home.

Yasuyuki NISHIDA [1], Shinichiro SUMIYOSHI [2] and Hideki OMORI [2]

[1] Energy Electronics Lab., College of Engineering, Nihon University; Tokusada, Tamura, Kouriyama, 963-8642, JAPAN, nishida@ee.ce.nihon-u.ac.jp

[2] Matsushita Electric Industrial Co., Ltd; 2-3-1-2, Noji-higashi,Kusatsu, Shiga, 525-8555, JAPAN

Abstract – Renewable energy has been a great concern to obtain ecological society that can continuously grow. As an attractive solution, the Fuel Cell cogeneration system has been attracted and some first products have been developed in Japan and in some other centuries/areas. A product in Japan, which is the first developed system and few hundreds of the sets are under operation in Japanese homes, has been investigated and practically evaluated in this digest. The novel topology consisting of a high-frequency transformer isolated soft-switching dc-dc converter and a utility interactive PWM inverter are introduced and practically evaluated.

Keywords—Converter, photovoltaic, fuel cell, home.

I. INTRODUCTION

The R&D of distributed and renewable energy systems has been becoming essential to obtain ecological energy systems. The authors have been dealing with photovoltaic systems and fuel-cell cogeneration systems. This paper descries the converter system applied on the applications.

II. FUEL-CELL SYSTEM

A. Overall System

Fig. 1 (a) shows a whole appearance of the products while Fig. 1 (b) shows an appearance of the converter set. As shown in Fig. 2, the converter set consists of a quasi-resonant type soft-switching dc-dc converter with high-frequency isolation transformer and a following utility interactive PWM inverter of hard-switching single-phase bridge. How to realize a high efficiency under a lower input voltage (e.g., 40V, output of the fuel-cell) in the DC-DC converter and an utility current with a high waveform quality (e.g., THD = 2%) are the dominant considerations in the developing works. Some details of the DC-DC converter and the inverter are described in the following chapters III and IV, respectively

Fig. 1. Appearance of Converter Set (1kW, 330 x 212 x 110 mm, 7.7 little, 5.4 kg) of Fuel-Cell Cogeneration Product.

B. DC-DC Converter
i) Topology of DC-DC Converter

The dc-dc converter, shown in Fig. 2, consists of a full-bridge high-frequency (HF) inverter, a HF isolation transformer T and a HF diode rectifier. The full-bridge inverter has four switching arms, each of which consists of a self-turn-off device (Q1 – Q4) and a freewheeling diode (D1 – D4). A capacitor (C1a – C1d) is connected in parallel in each the arm to realize Zero-Voltage-Switching (or Zero-Voltage turn-off) and other some inductors and capacitors are employed to perform a quasi-resonant soft-switching operation. The dc-dc converter and the HF transformer T converters the low voltage (e.g., 40V) obtained from the fuel-cell into a high voltage (e.g., 400V) with a high efficiency by means of a specially arranged quasi-resonant soft-switching topology.

ii) Topology of PWM Inverter

The utility in Japanese home consists of single-phase three-wire system (i.e., $100V_{RMS}+100V_{RMS}$ system), and the outer two wires offering higher voltage (i.e., $200V_{RMS}$) are connected to the inverter output through a harmonic trap LC filter as shown on right-hand-side of Fig. 2.

iii) Control of DC-DC Converter

The output voltage of the full-bridge type dc-dc converter can be controlled by adjusting the operating frequency and/or the operating phase-displacement-angle of the two half-bridges (i.e., left-hand-side one consisting of Q1 and Q2 and right-hand-side one consisting of Q3 and Q4). To regulate the output voltage of the proposed dc-dc converter, only the operating frequency is adjusted in medium to full load condition while both the operating frequency and the

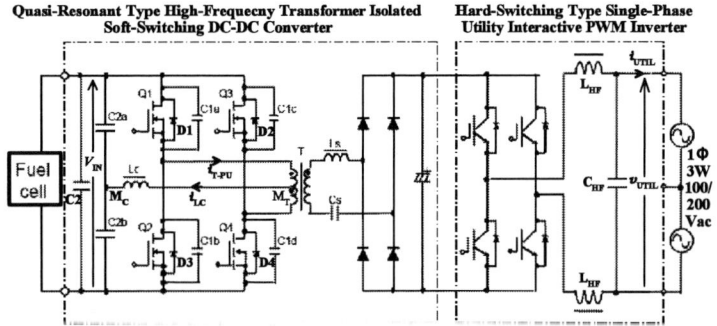

Fig. 2. Whole Topology of 1kW Converter Set of Fuel-Cell Cogeneration System

phase-displacement-angle are adjusted in light load condition, as shown in Fig. 3.

The phase-displacement-angle is set to the maximum (i.e., 180[deg]) under full to medium load condition so that the input voltage is fully utilized in the output under the condition. Due to this arrangement, the secondary voltage offers a full-width rectangular waveform as shown in Fig. 3.

The operating frequency of the DC-DC converter is set to be close to the resonant frequency of the series LC circuit (i.e., L_S and C_S in Fig. 2) under full and heavy load condition so that the impedance of the resonant circuit remains in a very low value and a higher power can be fed to the output. As a result, the secondary current offers a sinusoidal waveform as shown in Fig. 3.

The operating frequency is increased to decrease the output power while the phase-displacement-angle is kept to the maximum (i.e., 180 deg) under full to medium load condition. In a medium load condition, the secondary current obtains a distorted waveform with lagging fundamental component against the voltage since the operating frequency is higher than the resonant frequency, as seen in Fig. 3.

iv) Switching Operation in DC-DC Converter

Heavy to Medium Load Condition

Under heavy to medium load condition, the dc-dc converter performs soft-switching by means of quasi-resonant and ZVS schemes. The resonant inductor L_S and resonant capacitor C_S connected in series on secondary circuit are for the quasi-resonant switching while the capacitors (C1a etc.) connected in parallel to the switching arms are for the Zero-Voltage turn-off. The soft-switching technique is an ordinary one and the detail is omitted in the digest but described in detail in the final conference paper.

Light Load Condition

To achieve the ZVS, the parallel capacitor (C1a etc.) has to be fully discharged by the current flowing through the capacitor and the primary winding of the HF transformer T during in a certain switching transition period. The current is maintained by the magnetizing inductance L_{MAG} of the transformer T and the inductor L_S on the secondary. Under a light load condition, the output current (i.e., the current flowing through the inductor L_S) is low in amplitude. Additionally, the magnetizing current

is also low in amplitude in a light load condition since the applied voltage on the primary winding of the transformer T is low in amplitude (i.e., narrow width of rectangular voltage) due to the phase-shift control of the full-bridge dc-dc converter to obtain lower output voltage. Thus, the current to discharge the parallel capacitor (C1a etc.) has not enough amplitude and the parallel capacitor does not achieve full-discharged condition during in the current transition. As a result, the ZVS is not achieved under a light load condition.

To overcome this problem, an additional current source, i.e., the inductor L_C, is connected between the center tap M_T of the primary winding and center point M_C of the input dc rails. Under the full (or a heavy) load condition, the voltage potential of the center tap M_T is kept to be identical to that of the center point M_C of the input at all (or almost all) the time. Thus, no voltage (or a voltage with a very low amplitude) is applied on the auxiliary inductor L_C in this condition. Therefore, the current i_{LC} flowing through the auxiliary inductor L_C is none (or very low in amplitude) and additional losses due to the current are none (or very low) under full to medium load condition. As the output power decreases the amplitudes of the applied voltage on the auxiliary inductor L_C increases due to the phase-shifting control of the full-bridge dc-dc converter. Hence, the amplitude of the auxiliary current i_{LC} to assist discharging the parallel capacitor (C1a etc.) increases as the output power decreases. As a result, the ZVS can be achieved under a light load condition too with this topology.

v) Realization of Controller of DC-DC Converter

The controller is realized by a micro-controller (MCU) based system as shown in Fig. 4. The output voltage is sensed and fed to an A/D converter in the MCU through an Isolation Amp. The MCU executes the processing of the feedback control for the output voltage by referring to the sensed and reference voltages. The obtained data is

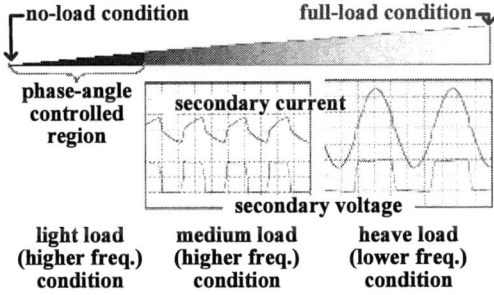

Fig. 3. Output Voltage Control

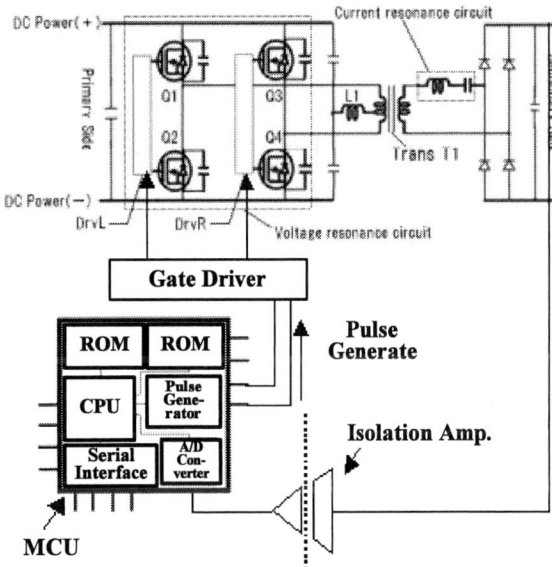

Fig. 4. Block Diagram of Controller

2531

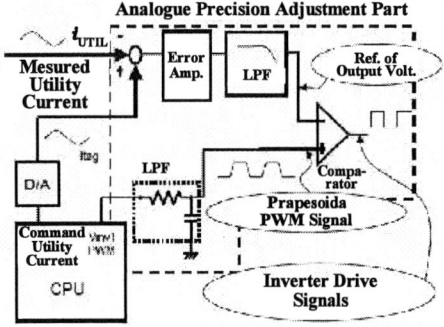

Fig. 5. Utility Current Controller

set to the Pulse Generator that feeds the drive signals to the switch through the Gate Driver.

C. Utility Interactive Inverter

The Utility interactive inverter consists of a smoothed dc input voltage, single-phase standard full-bridge, a harmonic trap filter consisting of a series inductor L_{HF} and parallel capacitor C_{HF} as seen in Fig. 2. Although the topology is a standard one, a precision output current (i.e., Utility current) is applied to obtain the output current (i.e., the Utility current) i_{UTIL} of a fine waveform quality.

Fig. 5 shows a block diagram of the precision output current control system that consists of digital signal processing part with a CPU and analogue precision adjustment part. Although the detail of the controller is omitted in this digest, it will be described in the final conference paper.

D. Practical Evaluation of Product

To verify the ZVS performance of the new dc-dc converter and whole of the converter set including the utility interactive inverter, one of the products has been tested. TABLE I show the specification of the converter set.

i) Operating WaveformsDC-DC Converter

DC-DC Converter (Soft-Switching Operation)

Fig. 6 (a) and (b) show voltage and current waveforms of the switches Q2 and Q4 for a full-load condition (P_O=1 [kW]) and a light-load condition (P_O= 100 [W]), respectively, obtained from a product. As seen from the

operating waveforms, the switches perform soft-switching in both the turn-on and turn-off.

Utility-Interactive PWM Inverter

Fig. 7 shows Utility voltage v_{UTIL} and Utility current i_{UTIL} (i.e., inverter output current) obtained from a product. By applying the precision control described in the previous chapter, the Utility current draws a sinusoidal waveform with a unity displacement factor as seen in the figures.

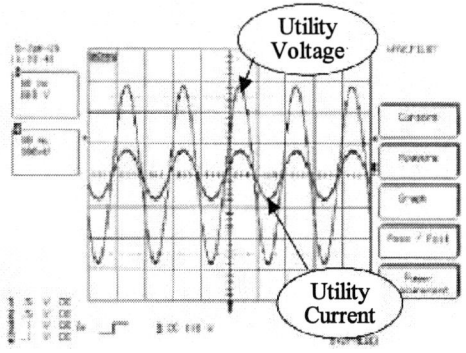

Fig. 7. Utility Voltage and Output/Utility Current (Experiment)

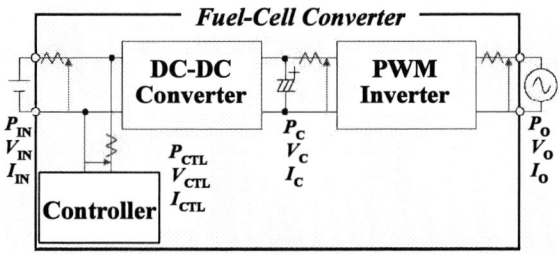

Fig. 8. Efficiency Measuring Arrangement

TABLE I
Specification of Product Converter Set

Rated Output Power of Inv. Po-RAT	1.0 kW
DC Input Voltage Range V$_{IN}$	34 – 60 V
Rated AC Output Voltage V$_{AC-RAT}$ (Utility/Mains RMS Voltage)	200 V$_{RMS}$ (1ϕ 3-Wire System)
Cooling	Forced Air
Size	330 x 212 x 110 mm (7.7 little)
Weight	5.4 kg

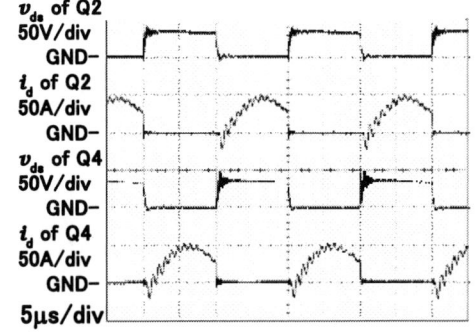

(a) Full-Load (1kW) Condition	(b) Light-Load (100W) Condition

Fig. 6. Voltage and Current Waveforms of Switches Q2 and Q4 for Full-Load (1kW) Condition (experiments)
(Input Voltage = 40V, Utility Voltage = 200V$_{RMS}$)

2532

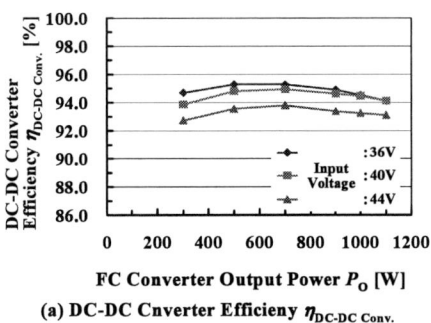

(a) DC-DC Cnverter Efficieny $\eta_{\text{DC-DC Conv.}}$

(b) PWM Inverter Efficieny η_{INV}

(b) FC Converter Overall Efficiency $\eta_{\text{FC-Conv.}}$

Fig. 9. Efficiency of Converter Set

Efficiency

Since the products have been developed as a renewable energy and ecological system, the overall efficiency of the system is essential. In this section, efficiency of the dc-dc converter, the utility interactive inverter, whole the converter set and overall the FC system obtained from one of the products are shown.

Fig. 8 shows an arrangement to measure efficiency of the converter where the powers, voltage and current of the DC-DC converter input (P_{IN}, V_{IN} and I_{IN}), the PWM inverter input (P_C, V_C and I_C), the PWM inverter output (P_O, V_O and I_O) and the Controller input (P_{CTL}, V_{CTL} and I_{CTL}) are measured. Fig. 9 shows measuring results of the efficiency.

Since the above-mentioned fine soft-

switching operation is applied on the product, the DC-DC converter offers a high efficiency $\eta_{\text{FC-Conv.}}$ for a wide load range of output power P_O under a severe low input voltage condition as seen in Fig. 9 (a). The utility inverter is an ordinal one and it offers a very high efficiency for a wide output power range. The total efficiency $\eta_{\text{FC-Conv\_all}}$ of the DC-DC converter and PWM inverter cascade set is shown in Fig. 9 (c). Due to the efficiency improvement of the DC-DC converter, the total efficiency of the converter set offers a high $\eta_{\text{FC-Conv.}}$ value for a wide output power range as seen in the figure.

According to experimental data obtained from a product, when both of the electrical output and thermal output are fully utilized, the total energy efficiency reaches a high value, i.e., 80% in the products. This is an advantage of the cogeneration system.

III. PHOTOVOLTAIC SYSTEM

A. Power Circuit Topology

Fig. 10 shows the proposed utility interactive photovoltaic system. The power circuit consists of solar-cells (the output voltage is V_{DC1}), a two-quadrant dc-dc converter, a suppressed capacitor C_2 (film capacitor), a single-phase inverter, an ac filtering inductor L_2, a noise-filter NF and a utility of single-phase 240 [V_{RMS}]. The diode D_S in the dc-dc converter is to modify the topology and it has been omitted here. The operation with the diode D_S is discussed later.

The two-quadrant dc-dc converter consists of an electrolytic capacitor C_1, a dc-inductor L_1 and two reverse-conducting switching arms. This converter feeds the power produced by the solar-cells to the utility via the inverter with the boost-mode operation by means of the switch SW_F and the reverse conducting diode D_B. While the converter performs the buck-mode operation by means of the switch SW_B and the reverse conducting diode D_F to feed the power buck from the utility to the

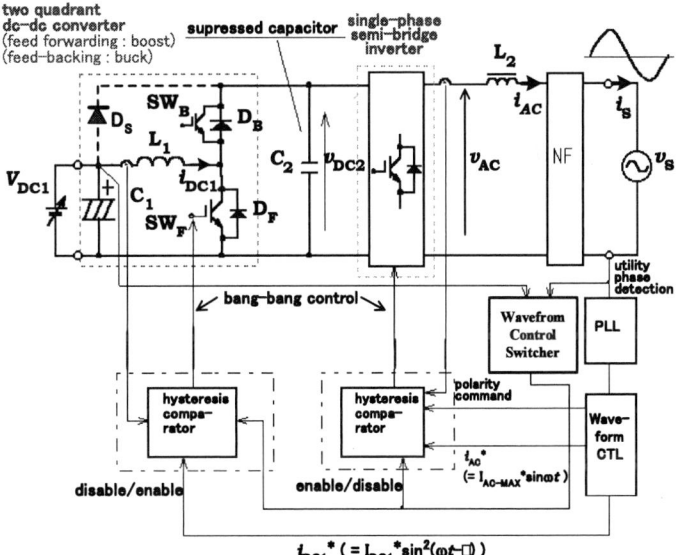

Fig. 10. Proposed Utility Interactive Photovoltaic System.

2533

capacitor C_1.

Comparing to the conventional system with the similer topology, the dc-capacitor C_2 on the input-side of the inverter is suppressed significantly as introduced below.

B. Utility Current Control

The control of the utility current i_S in the proposed system is unique and it's achieved by means of both the dc-dc converter and the inverter.

Operation and Current Control of DC-DC Converter

This dc-dc converter is a two-quadrant type, and it feeds the power obtained from the solar-cells to the inverter with boost-mode operation. On the other hand, it feeds the energy back from the utility to the dc capacitor C_1 with buck-mode operation. The energy feedback with the buck-mode operation is rarely utilized to control the power factor of the utility to reduce the utility voltage. Since this case is occurred very rarely and is not focus here, the detail is omitted in this paper.

Since the voltage produced by the solar-cells is not high enough to obtain certain level of ac-voltage through the inverter, a boost-mode dc-dc converter is necessary to connect between the solar-cells and the inverter in the transformer-less system. In this project, however, the solar-cells unit is designed to produce the output voltage of approximately 200 [V] in the maximum output power condition (i.e., 4.5kW).

A higher voltage can be obtained depending on solar-cells arrangement but it affects reasonable system designing. On the other hand, the inverter needs approximately 400 [V] or more in the conventional system since the utility voltage is up to 280 [V_{RMS}] and the maximum value reaches almost 400 [V].

When the solar-cells voltage V_{DC1} is greater than the absolute value of the utility voltage $|v_{AC}|$, the inverter operates only as a polarity switcher, i.e., it connects the positive/negative dc-rail to the higher/lower potential rail on the ac-side. This period is called "Period-A" hereafter. Since the inverter does not perform any switching in this period, the absolute value of the ac-voltage $|v_{AC}|$ appears on the dc-side (i.e., $v_{DC2} = |v_{AC}|$). However, the inductance of the ac-inductor L_2 is very low (its %-reactance in the products is only 2.5%) and the voltage drop is negligible. Thus, the inverter dc-input voltage v_{DC2} equals the absolute value of the utility voltage $|v_S|$ (i.e., $v_{DC2} = |v_S|$). As a result, the solar-cells voltage (i.e., dc-dc converter input voltage) V_{DC1} is lower than the output voltage v_{DC2}. Therefore, the dc-dc converter can perform a boost-mode operation in this period.

If the PFC is achieved somehow, the utility current i_S draws a sinusoidal waveform in-phase with the utility voltage v_S. Thus, the instantaneous power p_S fed from the utility is expressed as;

$$p_S = 2V_{S\text{-RMS}} I_{S\text{-RMS}} \sin^2 \omega t, \qquad \ldots(1)$$

where $V_{S\text{-RMS}}$ and $I_{S\text{-RMS}}$ are the RMS-values of the utility voltage and current, respectively.

Since the series and parallel impedance of the noise-filter NF is enough low and high, respectively, its series voltage drop and parallel by-pass current are negligible. Further, if drop-voltages, bypass-currents and losses of the inverter, capacitor C_2 and the dc-inductor L_1 are all neglected, the instantaneous power p_S fed to the utility equals the instantaneous dc-power p_{DC1} ($=V_{DC1} i_{DC1}$). Thus, we obtain;

$$p_S = p_{DC1} = V_{DC1} i_{DC1} . \qquad \ldots(2)$$

From equations (1) and (2), we obtain;

$$
\begin{aligned}
i_{DC1} &= p_S / V_{DC1} \\
&= \left(2V_{S\text{-RMS}} I_{S\text{-RMS}} \sin^2 w t \right) / V_{DC1} \qquad \ldots(3) \\
&= I_{DC1} \sin^2 w t,
\end{aligned}
$$

where

$$I_{DC1} = 2V_{S\text{-RMS}} I_{S\text{-RMS}} / V_{DC1}. \qquad \ldots(4)$$

However, if we can force the dc-current i_{DC1} to produce the waveform shown in Eq. (4), the PFC is achieved. The dc-current i_{DC1} is the input current of the boost dc-dc converter, and its waveform is controllable when the input voltage V_{DC1} is lower than the output voltage v_{DC2}, i.e., in the discussing condition in Period-A. To realize the waveform control, the following current reference $i_{DC1}*$ is generated in the part "Waveform CTL" shown in Fig. 1.

$$i_{DC1}* = I_{DC1}* \sin^2 \left(\omega t + \phi \right), \qquad \ldots(5)$$

where

$$I_{DC1}* = 2V_{S\text{-RMS}} I_{S\text{-RMS}}* / V_{DC1}. \qquad \ldots(6)$$

The RMS-value and average-value of voltages (i.e., $V_{S\text{-RMS}}$ and V_{DC1}) on the right-hand-side in Eq. (6) are detected through sensors in the system. It notes that the capacitor voltage V_{DC1} can be regarded as constant since capacitance of C_1 is large enough. $I_{S\text{-RMS}}*$ on the right-hand-side in Eq. (6) is the reference of the RSM-value of the utility current i_S, and it determines the power fed from the solar-cells to the utility. Thus, this value is determined in the "Maximum Power Tracking" control part (not shown in Fig. 1) in the control system and fed to "Waveform CTL" shown in Fig. 1. The "Maximum Power Tracking" control in the system is based on a conventional one and its detail is omitted in this paper.

In practice, however, the capacitor C_2 produces a phase displacement between the dc-inductor current i_{DC1} and the utility current i_S, and it results in decrease of the power-factor. To take the displacement into account in the control, a displacement phase-angle ϕ is set in the reference $i_{DC1}*$ as shown in Eq. (5). The phase-angle ϕ is adjusted in the "Waveform CTL" by referring to the detected displacement phase-angle between the utility voltage v_S and current i_S. This is the principle how to control the instantaneous input current i_{DC1} of the dc-dc converter to achieve the PFC in Period-A. The actual current i_{DC1} is controlled to follow the reference with so called Bang-Bang control by means of a "histeresis controller." This PFC theory is based on the instantaneous control for the energy stored in the dc-inductor L_1 and is applicable to difference topologies[2].

ii) Operation and Current Control of Inverter

As shown in Fig. 11, the inverter consists of two half-bridges; the half-bridge on the left-hand-side is with two high-frequency type switches (SW_{HF-P} and SW_{HF-N}) while the other half-bridge on the right-hand-side is with two low-frequency switches (SW_{LF-P} and SW_{LF-N}). As mentioned, this inverter (i.e., all the four switches) operates without switching in the Period-A so that the positive/negative dc-rail is connected to the higher/lower potential ac-rail at all the time in the Period-A. In this period, the boost dc-dc converter plays the role to achieve the PFC. On the other hand, the inverter plays the role in the remaining period (called "Period-B" hereafter), and the high-frequency switches (i.e., SW_{HF-P} and SW_{HF-L}) operate alternately each other with a high-frequency. This switching is controlled based on the Bang-Bang control by means of the "histeresis controller" as shown in Fig. 12. The waveform (i.e., "$\sin\omega t$") of the reference $i_{AC}{}^*$ is generated in "Waveform CTL" referring to the detected utility voltage v_S while the amplitude $I_{AC-MAX}{}^*$ is determined based on "Maximum Power Tracking" control. By means of the information (i.e., "$\sin\omega t$" and $I_{AC-MAX}{}^*$) the reference $i_{AC}{}^*$ is synthesized in the "Waveform CTL."

Since the dc-voltage v_{DC2} is greater than the absolute value $|v_{AC}|$ of the ac-voltage v_{AC} in this Period-B, the inverter can control the ac-current i_{AC} to follow the reference $i_{AC}{}^*$.

iii) Summing up the operation and comparing to conventional scheme

As discussed in Section A and B, only one of the two power converters (i.e., the boost dc-dc converter in Period-A or the inverter in Period-B) operates in high-frequency switching and other one interrupts the switching in this proposed scheme as shown in Fig. 12. On the other hand, both the power converters operate with high-frequency switching at all the time in conventional schemes. Thus, the switching losses are significantly reduced in the proposed one. Additionally, a part of conduction losses produced in the inverter is

reduced since switches with lower forward-voltage-drop are applicable for the low-frequency half-bridge. As a result, the losses and the size of the power unit (including the two power converters and a heat-sink) decrease. Further, the bulky electrolytic capacitor with a large capacitance connected in the input of the inverter in the conventional products is replaced by a small film capacitor C_2 in the new products. Due to the new topology and control, the proposed system becomes advantageous for the conventional one in size, weight, cost and maintenance.

However, it is difficult to compress both the capacitor C_1 and C_2 at the same time but C_1 can be compressed instead of C_2. In this case, the total size and cost of the two capacitors is almost the same to the proposed one but the inverter input voltage v_{DC2} becomes smooth. Thus, the boost dc-dc converter cannot control the ac current i_{AC} or the utility current i_S. As a result, both the dc-dc converter and the inverter must operate PWM at all the time. Therefore, the switching loss reduction and the utilization of cheep and low forward-voltage-drop switches are no longer applicable in this case. From the practical view of point, the proposed topology and control scheme is reasonable to realize a utility interactive inverter system with cheap, low losses, light weight and compact size.

This inverter is, however, designed as a home appliance, and acoustic noise must be low. Thus, the air-forced-cooling is difficult to apply. In such the condition, low loss property is essential and the proposed loss reducing technique is desirable.

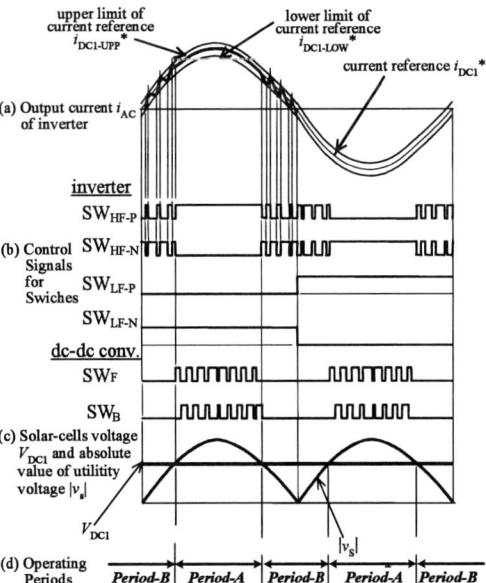

Fig. 12. Operating principle of Inverter and DC-DC converter.

Only the inverter (in Period-A) or the dc-dc converter (in Period-B) is operated with high-frequency PWM to waveshape the utility current and thus, losses caused by PWM is reduced significantly.

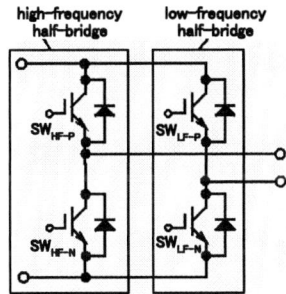

Fig. 11. Inverter.

It consists of a high-frequency half-bridge with fast IGBTs and a low-frequency half-bridge with slower IGBTs.

TABLE II. Specification of Product

Items	Data
Rated Output Power	4.5 [kW]
DC Voltage Fed from Solar-Cells	80 – 350 [V_{AVE}]
Rated Utility Voltage	220 [V_{RMS}] (1-phase 3-wire)
Utility Frequency	50 or 60 [Hz]
Cooling	Natural Air Cooling
Size	550 x 300 x 124 [mm] (20 [litters])
Weight	20 [kg]

TABLE III Specifications of IGBTs

Items	Specifications	
	GT60J321 (Fast Type)	GT60j322 (Slow Type)
Coll.-Emit. Voltage (V_{CES})	600 [V]	
Collector Current (I_C)	60 [A] @ DC, 120 [A] @ 1[ms] pulse	
Collector-Emitter Saturation Voltage ($V_{CE(sat)}$)	1.2[V] @ I_C=10[A] 1.55[V]@I_C=60[A]	0.9[V] @ I_C=10[A] 1.25[V]@I_C=60[A]
Turn-ON time (t_{ON})	0.40[μs]	0.3[μs]
Turn-OFF time (t_{OFF})	0.65[μs]	1.40[μs]
Emit.-Coll. Voltage (V_F)	1.5[V] @ I_F=60[A]	1.2[V] @ I_F=60[A]
Reverse-recovery time (t_{rr})	0.1[μs] @ I_F=60[A]	0.6[μs] @ I_F=60[A]

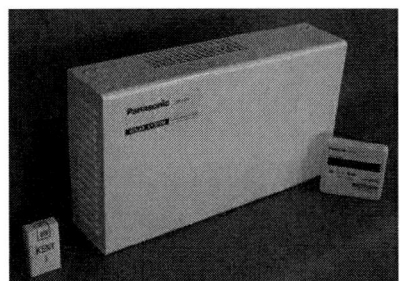

Fig. 13. Appearance of Product.

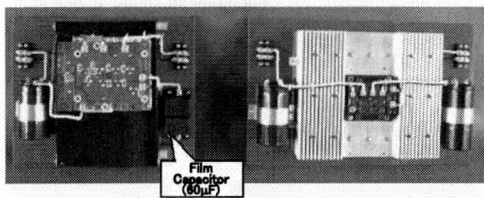

Fig. 14. Power Units in New (left) and Conventional (right) Products.

The size in the new product is 2/3 of the conventional one, and the left-hand-side electrolytic capacitor in conventional product has been replaced by a film capacitor in the new product.

B. Products and Practical Evaluation

To confirm the validity of the theory, the performance of new product has been evaluated. TABLE II and III show the specification of the products and the employed IGBTs, respectively. Figures 13 and 14 show the appearance of the new product and the power units of the new and conventional products, respectively.

Table II shows that the dc voltage fed from the solar-cells varies in wide range (i.e., 80 to 350 [V_{AVE}]) due to the Maximum Power Tracking. Although the rated utility voltage is 220 [V_{RMS}], it can reach much higher value in practice and a case of 280 [V_{RMS}] has been considered in the design. The IGBTs (with the ratings of 600 [V] and 60 [A] and package of TO3-P) are employed in the all switches but those in the high-frequency half-bridge and dc-dc converter are fast-type while those in the low-frequency half-bridge are slow-type, as shown in Table II. The forward voltage drops (or collector-emitter saturation voltages) of the fast- and slow-type IGBTS are 1.55 [V] and 1.25 [V] (@ I_C=60[A]), respectively.

As seen in Fig. 5, the power unit of the new product is compact and the size is approximately 2/3 of that of the conventional one. This size-reduction is achieved due to less size of the heat sink (i.e. less power losses) and less

of one electrolytic capacitor.

Fig. 15 and 16 show operating waveforms of the new product while Table IV shows circuit condition and measured data. The waveforms shown in Fig.15 and 16 have been obtained under the condition shown in Table

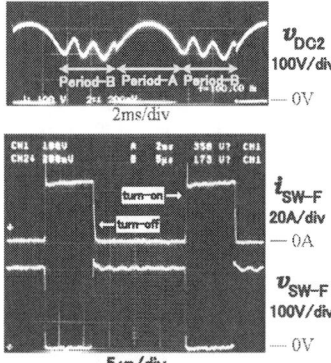

Fig. 15. Operating Waveforms.

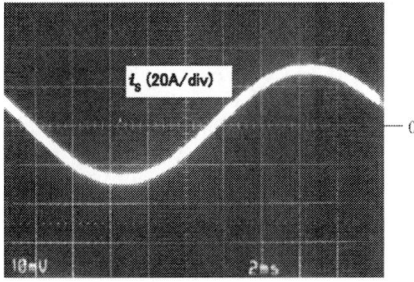

Fig. 16. Operating Waveform of Utility Current i_S.

TABLE IV
Circuit Condition and Measured Data for Operating Waveforms in Fig. 4 and 5.

Items	Data
DC-Inductors (L_1 and L_2)	1.0 mH
DC-Capacitor (C_1)	2,000 μF
DC-Capacitor (C_2)	60 μF
Output Power (P_{O-RAT})	4.5 kW
DC Voltage Fed from Solar Cell (V_{DC1})	208 V_{AVE}
Utility Voltage (V_S)	220 V_{RMS}
Utility Frequency (f_S)	50 Hz
Efficiency (η)	95.2 %
Total Harmonic Distortion of Utility Current i_S (THD-i_S)	1.6 %

IV. The waveform of the intermediate dc voltage v_{DC2} (top trace in Fig. 15) shows that the voltage equals the absolute value of the utility voltage v_S (220 [V_{RMS}] and the maximum value is 311 [V] for this case) in Period-A, as discussed in the theory. The middle and bottom traces in Fig.15 show the waveforms of the current i_{SW-F} and voltage v_{SW-F} of the boost switch SW_F in the Period-A. It can be seen from these switching waveforms that the IGBT offers very fast switching in turn-on and turn-off. The spike current occurring at the turn-on instant of the switch SW_F is due to reverse recovery of the reverse conduction diode D_B.

On the other hand, the dc voltage v_{DC2} is higher than the absolute value of the utility voltage v_S in the remaining period (i.e., Period-B). In this period, the boost dc-dc converter cannot operate and it interrupts the operation. Thus, the dc-capacitor C_1 is connected to the dc-capacitor C_2 though the dc-inductor L_1 and reverse conducting diode D_B. As a result, the intermediate dc voltage v_{DC2} produces an oscillation caused by L_1 and C_2 as shown in Fig. 6, where the effect of C_1 on the oscillation is negligible since the capacitance is large enough compared with C_2. This oscillation, however, is not so significant and no damage is produced in this system. In this period, the inverter operates with high frequency switching to perform the Bang-Bang control for the ac current i_{AC}. However, an alternative topology to avoid the oscillation is introduced later.

Fig. 16 shows the waveform of the utility current i_S. From this waveform, we can recognize that the waveshaping by the boost dc-dc converter and the inverter with the Bang-Bang control has been operated ideally. The Total-Harmonic-Distortion THD-i_S of the current is only 1.6 [%].

Fig.17 shows efficiency curve s for the power P_S fed to the utility where the dissipated power of 30 [W] in the controller has been taken into account. The efficiency has marked very high value and is more than 95.5 [%] and 95 [%] at the medium and full load condition, respectively, under the condition of $V_S = 240$ [V_{RMS}]. Thus, the effect of the novel current modulation scheme (i.e., alternation of high-frequency switching operation between the boost converter in Period-A and the inverter in Period-B) to reduce the switching losses has been confirmed. TABLE V shows a loss analysis for the condition of $P_S = 4.5$ [kW] and $V_S = 208$ [V_{RMS}] where the efficiency is 94.5 [%] and the total loss is 261 [W]. It can be known from the table that the power dissipation of the power devices is low and it results in the high efficiency. The efficiency in the maximum power condition increases by 0.67 [%] if the power dissipation of the controller (30 [W]) is omitted.

By the way, the oscillation caused by L_1 and C_2 (refer to the top trace in Fig. 15) may cause acoustic noise produced by L_1. By employing the additional diode D_S shown in Fig. 10, the oscillation can be eliminated as shown in Fig. 18. The dc capacitor C_2 is connected to C_1 through the diode D_S in the modified topology, and thus the oscillation does not occur. The performance of the system with the modified topology with D_S is almost the

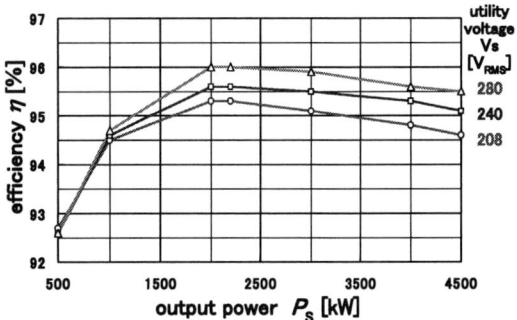

Fig. 17. Measured Efficiency.
A very high Efficiency has been obtained especially in medium load condition.

TABLE V
Loss Analysis for Condition of $P_S = 4.5$ [kW] and
$V_S = 208[V_{RMS}]$

Items	Losses [W]
Power Devices in DC-DC Converter (SW_B & SW_F)	72
DC-Inductor (L_1)	52
DC-Capacitor (Film Capacitor: C_2)	5
Power Devices in Inverter	53
DC-Capacitor (Electrolytic Capacitor: C_1)	5
AC-Inductor (L_2)	28
Control Circuitry	30
Others	16
Total	261

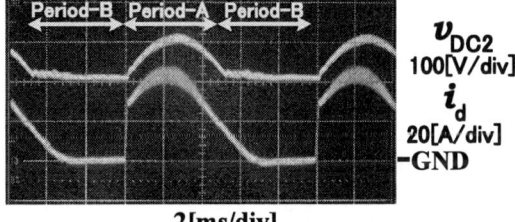

2[ms/div]

Fig. 18. Operating Wafeforms of Modified Toplogy with Diode D_S.
The oscillation in Period-B has been disappeared. The current i_d of the dc-inductor L_1 flows even in the beginning half of the Period-B to discharge the stored energy.

same to that of the original topology without D_S.

III. CONCLUSIONS

A novel utility-interactive inverter for photovoltaic system and a high-frequency transformer isolated soft-switching DC-DC converter and a utility interactive PWM inverter cascade converter set for Fuel-Cell cogeneration system has been described. Several techniques to realize integrated power converter in the products are described in detail.

A Comparison of Position Control Structures for Ironless Linear Synchronous Motor

Martin HRASKO*, Pavol MAKYS*, Marek FRANKO*, Jozef KUCHTA*

* Electrotechnical Research and Projecting Company, j.s.c. /CNKEA R&D, Nova Dubnica, Slovakia,
e-mail: hrasko@evpu.sk, makys@kves.uniza.sk, franko@evpu.sk, kuchta@evpu.sk

Abstract— Comparison of two position control systems for ironless linear synchronous motor with permanent magnets is presented. The first control system is based on classic cascade structure, comprising inner current and speed loop with PI controllers, and an outer position control loop with controller with variable parameter P. The other control system is similar as the first one. Furthermore, its outer loop is based on sliding mode control principles. The paper compares abilities of these two structures for high precise position control focused on high steady state accuracy at full load range. Simulations results and experimental verifications are presented as well.

Keywords— linear motor, permanent magnet motor, position control, sliding mode, robust control.

I. INTRODUCTION

In many industrial applications of linear electrical drives it is specified that the moving part (primary part) should be controlled to a constant or variable reference position or that machine part, possibly the tool of milling machine, should follow a prescribed trajectory, so this translational movements may be part of a multi-dimensional motion. Depending on the application, there are different requirements with regard to accuracy or dynamic response.

II. CASCADE VECTOR CONTROL

In generally, the most know position system is a cascade structure including several superimposed control loops for force, speed and position. An acceleration control loop is occasionally added to eliminate the effects of load force. It is no substitute for force control, however, because it cannot prevent static overload, for instance when the drive is mechanically jammed. Main advantages of cascade structure and its nested control loops are [1]

- transparent structure,
- step-by-step design, beginning with the innermost loop thereby solving the stability problem,
- use of standard controllers P, PI, PID etc.,
- commissioning is greatly simplified by closing one control loop after other, from inside out,
- opening of outer loops permits simple procedure for diagnostics and field test.

Theoretically there is only one serious drawback of the cascade control structure as seen in Fig. 1 that is caused by the fact that the response to the reference input becomes progressively slower as more outer loops added.

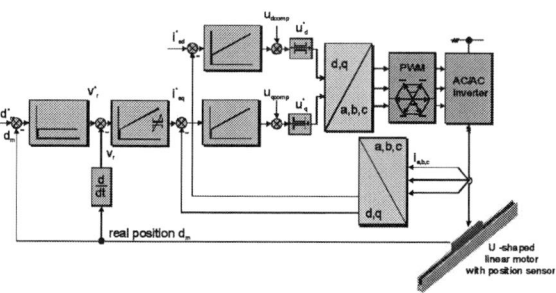

Fig. 1. The cascade vector control of linear motor.

Practically, another terrific drawback occurs. Cascade structure it not able to eliminate steady state error due to discreteness or resolution of variables (demand value for inner loop) for digital control.

Let assume at first that position error is very small value but bigger than particularly resolution. Although, this position error is multiplied by P constant of position controller, results may by smaller than resolution for next inner loop, in our case it is speed loop. Consequently, speed controller is not producing any request for current controller thereby position error is not minimised.

A. Position controller with adaptive constant

Position controller is proportional with P constant. It is necessary to scale allowance in witch is represented position error e and controller output (in this case there is demanded speed) if is it implemented in fixed point arithmetic. Position error is defined as difference between demanded and real (measure) mechanical position of ironless linear synchronous motor with permanent magnets. Controller output (demanded speed) is represented as real physical value in case when floating point arithmetic is used.

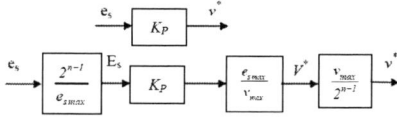

Fig. 2. Block diagram of demanded speed calculation
a) floating point arithmetic, b) fixed point arithmetic

When fixed point is used, controller output and also input is represented as an integer number. See Fig. 2 [2]. When P controller is implemented in fixed point arithmetic it is necessary to compute output value with minimal loss of accuracy. Computing accuracy is depended on:

- number of bits for output controller
- mathematic and arithmetic technique of computation

Real implementation with fixed point arithmetic of position controller for ironless linear synchronous motor with permanent magnets is presented on Fig. 3.

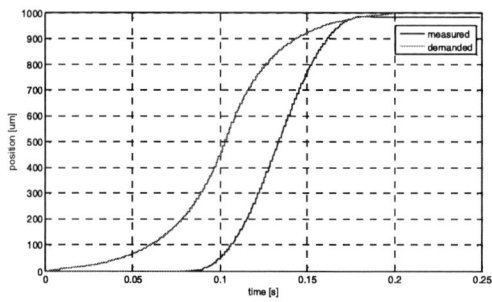

Fig. 3. Demanded and real (measured) mechanical position as a function of time (demanded position was set on 1000μm)

We can see position offset on Fig. 3 when fixed point arithmetic is used on position controller without P constant adaptation. This offset is caused by:

- speed and current controller quantization (controller output non – sensitivity for small demanded value of speed and also q- axis current)
- small gain of position controller for low regulation error

This offset can be adapted by higher gain of position controller in low regulation error. Controller gain depending on regulation error (absolute value) is presented on Fig. 4 (measured and implemented in fixed point arithmetic).

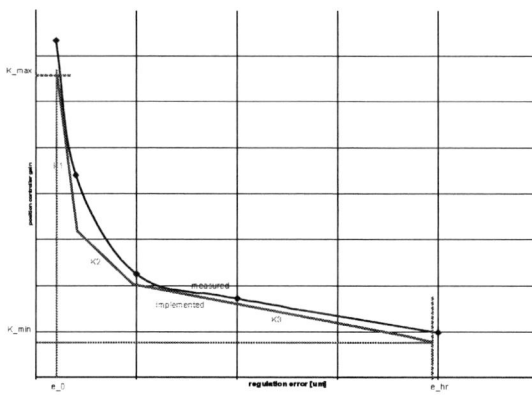

Fig. 4. Position controller gain as a function of regulation error

Where K is position controller gain, K_{max} is gain for minimal regulation error K_{min} is a nominal gain (K_p for position controller), e_0 is regulation error with constant – maximum controller gain, e_hr is maximum regulation error. Mechanical position of ironless linear synchronous motor with permanent magnets with gain controller adaptation is presented on Fig. 5.

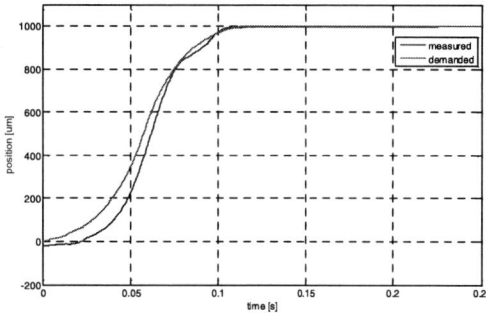

Fig. 5. Demanded and real (measured) mechanical position as a function of time with gain controller adaptation
(demanded position was set on 1000 μm)

As we can see on Fig. 5 the regulation error was minimized by using gain controller adaptation. Some vibration on real (measured) mechanical position are caused by switching between adaptation lines K1, K2 and K3 (see Fig. 4). Best way, how to reach smooth response of mechanical position is used continuous line for gain adaptation. On the other hand this line is difficult to implement in fixed point arithmetic (with DSP).

III. SLIDING MODE CONTROL

The basic form of sliding mode control is well known [3],[4],[5],[7]. This system has great robustness even when the plant parameters are changed or external disturbances are applied. Fig. 6 shows an overall block diagram of the position control system.

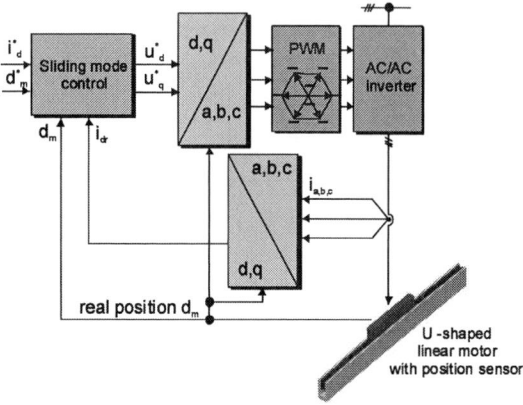

Fig. 6. Sliding mode control of linear motor

Fig. 7 and Fig. 8 show inner loops separately for d and q axis. In sliding mode control, only the plant rank n has to be known (position loop, n=3), so, constant for q-axis can be derived as follows:

2539

$$F(s) = \frac{d_m(s)}{d_m^*(s)} = \left[\frac{\omega_0}{s+\omega_0}\right]^n = \left[\frac{1}{1+\frac{s}{\omega_0}}\right]^n = \tag{1}$$

$$\left[\frac{1}{1+s\frac{T_s}{1,5(1+n)}}\right]^{n=3} = \frac{1}{1+s\frac{T_s}{2}+s^2\frac{T_s^2}{12}+s^3\frac{T_s^3}{216}}$$

where T_s is the settling time. Constant for d-axis can be derived similar way (only current loop, plant rank n=1) as follows:

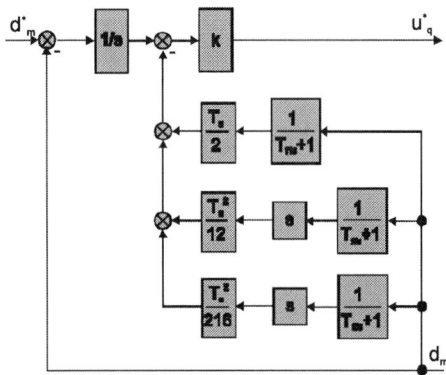

Fig. 7. Sliding mode control structure of q- axis

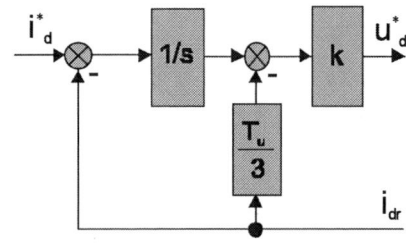

Fig. 8. Sliding mode control structure of d- axis

$$F(s) = \frac{i_r(s)}{i_d^*(s)} = \left[\frac{\omega_0}{s+\omega_0}\right]^n = \left[\frac{1}{1+s\frac{T_s}{1,5(1+n)}}\right]^n \overset{n=1}{=} \frac{1}{1+s\frac{T_s}{3}} \tag{2}$$

Substituting Fig. 7 and Fig. 8 to Fig. 6, yielding equations:

$$i_d^*(s) = i_r(s)\left(1+s\frac{T_s}{3}\right) \tag{3}$$

$$d_m^*(s) = d_m\left(s\left(1+s\frac{T_s}{2}+s^2\frac{T_s^2}{12}+s^3\frac{T_s^3}{216}\right)\right) \tag{4}$$

The transformation equations (3) and (4) to the time domain system yields:

$$i_d^* = i_r + \frac{T_s}{3}\frac{d}{dt}i_r \tag{5}$$

$$d_m^* = d_m + \frac{T_s}{2}\frac{d}{dt}d_m + \frac{T_s^2}{12}\frac{d^2}{dt^2}d_m + \frac{T_s^3}{216}\frac{d^3}{dt^3}d_m \tag{6}$$

Expressing demanded voltages u_q^* and u_d^* from Fig. 7 and Fig. 8 by using equations (5),(6) yields:

$$u_d^* = \int k\left[i_d^* - i_r - \frac{T_s}{3}\frac{d}{dt}i_r\right]dt = k\left[\int(i_d^* - i_r)dt - \frac{T_s}{3}i_r\right] \tag{7}$$

$$u_q^* = \int k_1\left[d_m^* - d_m - \frac{T_s}{2}\frac{d}{dt}d_m - \frac{T_s^2}{12}\frac{d^2}{dt^2}d_m - \frac{T_s^3}{216}\frac{d^3}{dt^3}d_m\right]dt =$$

$$= k_2\left[\int(d_m^* - d_m)dt - \frac{T_s}{2}d_m - \frac{T_s^2}{12}\frac{d}{dt}d_m - \frac{T_s^3}{216}\frac{d^2}{dt^2}d_m\right] \tag{8}$$

A. Verification by Simulation

The simulation results of SMC for 1mm position step are presented on Fig. 9 - 12. The computation step is h=1e-4s. The setting time for position control is 0.3s and Fig.9-12 shows a good agreement with this theoretical predictions. The main aim to apply this control method was eliminate to steady state error and Fig. 10 shows that steady state error is on the limit of the sensor resolution (1μm).

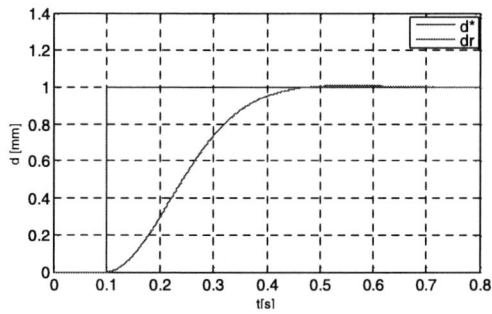

Fig. 9. Simulation results for sliding mode position control of LMST, position as a function of time

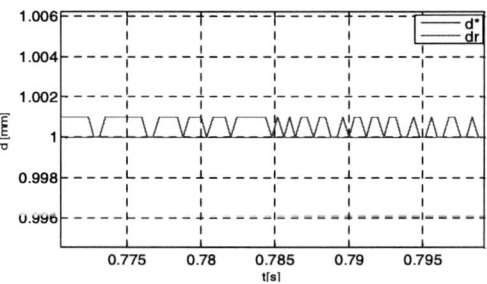

Fig. 10. Simulation results for sliding mode position control of LMST position as a function of time (detail), steady state error

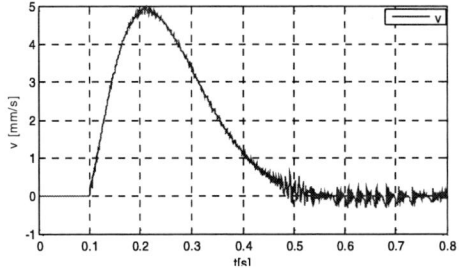

Fig. 11. Simulation results for sliding mode position control of LMST speed as a function of time

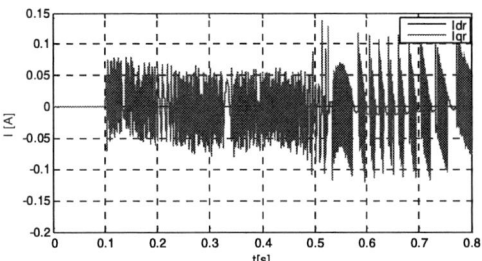

Fig. 12. Simulation results for sliding mode position control of LMST d and q component of stator current as a function of time

IV. EXPERIMENTAL RESULTS

Experiments with the cascade vector control and sliding mode control of position have been performed. They were integrated within a variable structure vector controlled drive with direct typing of mechanical position. The block diagram of the linear drive is presented on Fig. 1.

The drive (see Fig. 13) employs a 7.5 kVA inverter with TI DSP TMS LF2812 and synchronous linear motor with permanent magnets (LMST) developed in EVPU J.s. Company [9]. Parameters of LMST are: $F_t = 95$ N, $U_{sN} = 91$ V, $f_{sN} = 266$Hz, $2p = 8$, $R_f = 440$ mΩ, L_d=157 µH, L_q=141,3 µH. A linear optical position sensor was used from RENISHAW Company with 1µm resolution in case with cascade vector control. Sampling frequency of current loop was set on 10 kHz.

Fig. 13. Synchronous linear motor with permanent magnets (LMST), developed in EVPU j.s.c

The experimental results of cascade vector control of position with position gain controller adaptation are presented on Fig. 14 and Fig. 15. Demanded mechanical position was given by "S" curve master algorithm.

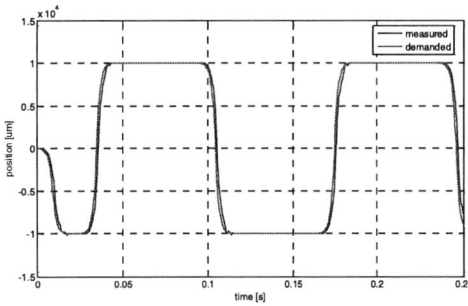

Fig. 14. Experimental results for cascade vector control of LMST mechanical position as a function of time (±10000 µm)

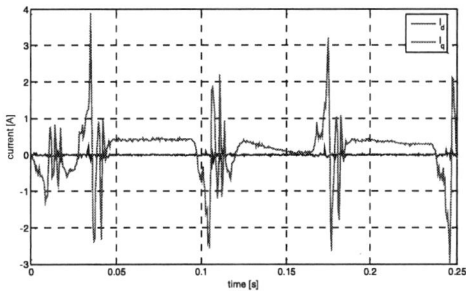

Fig. 15. Experimental results for cascade vector control of LMST d and q component of stator current as a function of time

Experimental results from sliding mode control of LMST are shown on Fig. 16, Fig. 17 and Fig. 18.

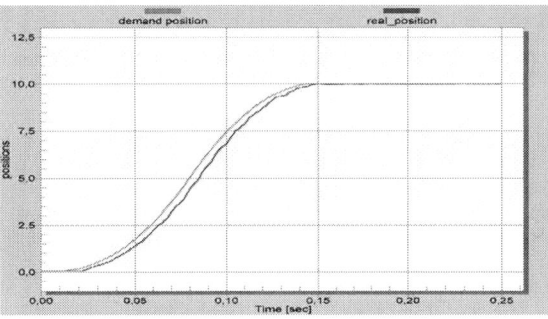

Fig. 16. Experimental results for sliding mode control of LMST mechanical position as a function of time (10000 µm)

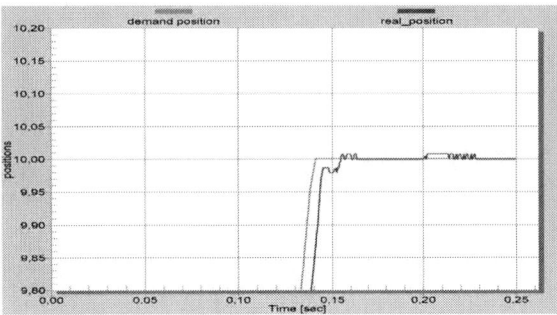

Fig. 17. Experimental results for sliding mode control of LMST mechanical position as a function of time (detail of steady state error)

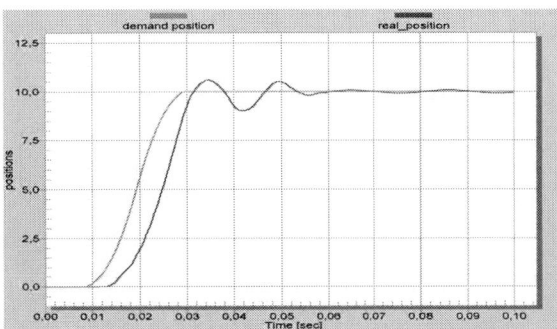

Fig. 18. Experimental results for sliding mode control of LMST mechanical position as a function of time (faster dynamic response)

In case when sliding mode control was applied, position sensor with 10µH resolution has been used. Fig. 16 shows behavior of position sliding mode control where control system is following demanded S-curve similarly to cascade control on Fig. 14 for better comparison with this one.

Although dynamic position error between demanded and real position is bigger for sliding mode control, steady state error is minimized under sensor resolution (see Fig. 17). Experiments to reach the faster dynamic response incurred unwanted oscillations as shown on Fig. 18.

V. CONCLUSIONS

This paper deals with cascade vector position control with gain controller adaptation. Sliding mode control for synchronous linear motor with permanent magnets was presented and verified by simulation. Experimental results from cascade vector position with gain controller adaptation and position sliding mode control are given in last. The dynamic behavior was better with cascade control against the better static features of sliding mode control. However, sliding mode control has a capability to enhance dynamic behavior but these improvements require implementation the key algorithms by 32 bit arithmetic or implement all control system into 32 bit DSP. Control algorithm includes in the some equations with very small and large numbers and the 16 bits arithmetic is not suitable to get appropriate accuracy in all range. Moreover implementation on fixed point DSP makes obvious complications with scaling all variables and equations.

ACKNOWLEDGMENT

This work was supported by the Slovak Research and Development Agency under the contract No. APVV-99-031205 and Slovak Grant Agency VEGA, project No.4087/07 *"Servosystems with Rotational and Linear Motors without Position Sensor"*.

REFERENCES

[1] W. Leonard, *"Control of electrical drives,"* Springer Berlin 2001, ISBN 3-540-41820-2IEEE.

[2] Kuchta, J. et al.: *"Research and Development of New Generations Electric Linear Drives with High Resolution of Position"*. Stage report of project APVV – 99 – 031205, January 2008. EVPU a.s. V025/08

[3] Utkin, V. I.: *'Sliding Modes in Control and Optimisation'*, Springer-Verlag, Berlin 1992

[4] Arellano-Padilla, J., Asher, G. M., Summer, M.: *'Robust Fuzzy-Sliding Mode Position Controller for Motor Drives Operating Variable Loads'*, proc. of EPE-PEMC 2002 conference, Cavtat, Croatia, CD-Rom

[5] Gyeviki, J., Toth, I. T., Rozsahegyi, K. : *'Sliding mode control and ist application on a servo-pneumatics positioning system'*, Transactions on Automatic Control and Computer Science, University of Timosoara, vol. 49, Romania, 2004

[6] Perdukova, D., Fedor, P., Timko, J.:'Modern method of complex drives control', Acta Technica CSAV vol. 49, Czech Republic, 2004, pp.31-45

[7] Rivkin, S., Izosimov, D., Palomar-Lever, E. : *'Digital sliding mode based references limitation law for sensorless control of an electromechanical systems'*, Journal of Physics, IOP publ. Ltd, series 23, pp. 192-201, 2005

[8] Urlep, E., Jezernik, K.: *'Speed Sensorless Variable Structure Control of PMSM'*, proc. of IEEE IECON 2006 conference, Paris, France, 2006, pp. 1194-1199.

[9] Kuchta, J., Franko, M., Hrasko, M., Fulier, M.: *"Design of Ironless Synchronous Linear Motor with Permanent Magnets and Experimental Verification of Field Distribution for Drive with High Resolution of Position"*, proc. of LDIA 2007 conference Lille, France 2007, pp. 195-196

[10] Vittek, J., Briš, P., Štulrajter, M., Makyš, P., Comnac, V., Cernat, M.: *Chattering Free Sliding Mode Control Law for the Drive employing PMSM Position Control*, OPTIM 2008, Brasov, RO, 2008, 05, 22. - 23., AFC, str.: 115 – 120

[11] Hrasko, M., Makys, P., Franko, M., Kuchta, J.:*" Improved Control Structure for Linear High Precise Positioning System"*, proc. of ELEKTRO2008 conference, Zilina, Slovakia 2008,pp. 112 - 115

A Comparison of Sliding Mode Approaches to a Nanometre Position Control Application

Paul Andreas Stadler*, Stephen James Dodds†

* Oerlikon ESEC, Cham, Switzerland, e-mail: *paul.stadler@oerlikon.com*
† University of East London/School of Computing and Technology, London, United Kingdom,
e-mail: *stephen.dodds@spacecon.co.uk*

Abstract—A vacuum air bearing based linear drive is presented that is capable of precision positioning with a relative accuracy of a few nanometres. Three different control systems embodying Sliding Mode Control (SMC) are compared by simulation and experiment regarding their individual positioning performance. The three control designs investigated are high gain SMC, a first order and a second order version of outer loop SMC where the inner loop consists of a Forced Dynamic Control law (FDC, a special form of state space controller) applied to the linear drive to be positioned.

Keywords— Sliding mode control, active damping, electrical drive, modelling, motion control, robustness.

I. INTRODUCTION

Linear drives are used in many different applications such as machine tools, coordinate measuring machines, bonding machines, inspection systems for the semiconductor industry and many more. They are able to be positioned to very high accuracy [1]. In this work, a linear drive actuated by a voice coil motor and based on a vacuum air bearing, as shown in Fig. 1, is to be positioned to accuracies in the region of nanometres. The presence of a significant vibration mode in the vacuum air bearing under consideration has been discovered [2] and the mass-spring model of this is shown in Fig. 1. The noise of the air cushion of the vacuum air bearing stimulates its eigenmode thereby impairing the positioning accuracy when using conventional control techniques such as PID or state control. In this paper, three versions of SMC control techniques are investigated with the aim of minimising the influence of the vibration mode. Sliding mode control (SMC) [3], is a robust control technique and here it is aimed at achieving a specified non-overshooting position control dynamics independent of the vibration mode. The first control system is directly based on SMC. The second and third systems are based on an outer SMC loop to enhance the robustness of an inner model based control loop, one giving a first order response and the other giving a second order response. In both these cases, the inner loop is based on a forced dynamic control (FDC) law leading to nominal prescribed closed loop behaviour [4] and employs active modal damping.

A model exhibiting the aforementioned vibration mode has been developed in [2] together with simulations and experimental tests with FDC but the experimental performance was found to be highly dependent on the

observer settling time, indicating modelling errors, as discussed in section VII. The approach taken to overcome this problem was to increase the robustness against modelling uncertainty by adding an outer sliding mode control loop and the first results were presented in [5]. The existing state feedback control algorithm based on FDC supported by an observer nominally yields a critically damped second order dynamics with a specified settling time. This is augmented by an outer control loop using a special form of sliding mode control (SMC) where a high gain is used instead of a switching element to avoid chattering. This outer SMC loop will not add an additional state to the closed loop system. In fact, in the sliding mode, the state closely follows the sliding boundary which has one dimension less than that of the state space and therefore the closed loop system is only of first order. Two versions of the outer SMC loop are investigated yielding, respectively, first and second order dynamics (FDCSMC1, FDCSMC2). The latter is achieved by closing an additional outer integral control loop around the sliding mode to achieve the same nominal closed loop dynamics as the inner FDC loop alone with appropriately chosen sliding boundary coefficients. So the new system will be robust against parameter variations and external disturbances. In this paper, a comparison is made between the performances of the SMC, FDCSMC1 and FDCSMC2 systems.

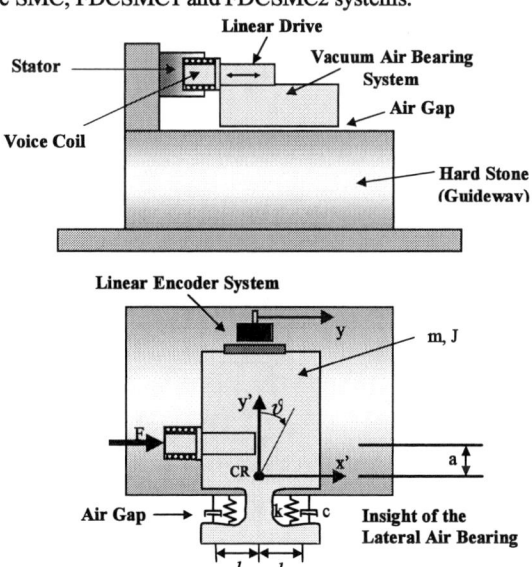

Fig. 1 View of the vacuum air bearing based linear drive.

An introduction to the high gain version of SMC is given in section III. Then its direct application to the vacuum air bearing based linear drive [6] will be compared with its application in the two outer loop variants. The accuracies predicted by simulations and the accuracies achieved experimentally will be compared with one another.

With reference to Fig. 1, a voice coil actuator produces a horizontal translational displacement of the table which is measured by a linear encoder, whose output, y, is required to follow a demand, y_{ref}, with the aid of closed loop position control. Fig. 2 shows the hardware components incorporated in the closed loop and a dSPACE system is used as the real time environment to run the control algorithm code produced using Matlab/Simulink.

II. STATE SPACE MODEL

The state space model was set up using the force balance and the torque balance equations, yielding

$$
\begin{cases}
\dot{x}_1 = x_2 \\
\dot{x}_2 = \dfrac{k_a k_m}{m}(u-d) \\
\dot{x}_3 = x_4 \\
\dot{x}_4 = \dfrac{1}{J}\left(k_a k_m (u-d)a - 2l^2 k x_3 - 2cl^2 x_4\right) \\
y = x_1 + R x_3 :
\end{cases}
\quad
\begin{array}{l}
\text{State} \\
\text{differential} \\
\text{equations} \\
\\
\text{Output equation}
\end{array}
\quad (1)
$$

where u is the control variable producing a proportional output current via an amplifier constant k_a. The voice coil actuator force constant is k_m. The constants c, l and k are defined in Fig. 1. The state variables are $x_1 = x$, $x_2 = \dot{x}$, $x_3 = \vartheta$ and $x_4 = \dot{\vartheta}$, where x is the translational displacement of the centre of rotation of the moving platform and ϑ is the angular displacement of the platform.

The plant parameters were identified as follows:
- Mass: m = 3.3 kg
- Moment of inertia: J = 30e-3 kg*m²
- Distance: a = 0.03 m
- Distance: l = 0.12 m
- Motor constant: k_m = 11.1 N/A
- Amplifier constant: k_a =0.8 A/V
- Spring constant: k =30e5 N/m
- Damping constant: c =53 N/(m/s)
- Radius: R = 0.15m

This exhibits a resonance frequency at 280Hz.

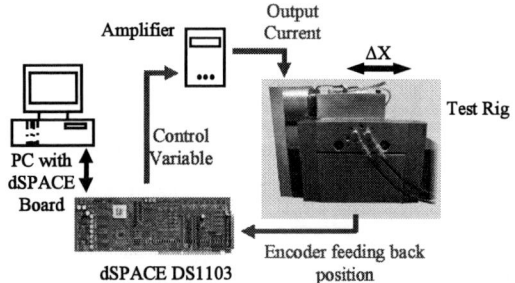

Fig. 2. Hardware components showing basic control loop.

III. THE SETTLING TIME FORMULA

The heuristic formula [8] derived by the second author for the settling time (5% criterion) of linear closed loop systems with coincident poles at $s_{1,2,\ldots,n} = -1/T_c$ is used frequently throughout this paper. This is defined as the time taken for the step response to reach 95% of its steady state value and is given by

$$ T_s = \frac{3}{2}(1+n)T_c \qquad (2) $$

IV. SLIDING MODE CONTROL

A. Assumptions and Advantages

It is assumed that the rank of the plant is known but the advantage of SMC over classical methods is that a specific closed loop dynamics is guaranteed while allowing considerable errors in the values of the plant parameters. In the vacuum air bearing application, this would allow the object carried by the platform to be changed, thereby changing the mass and moment of inertia of the moving part, without any need for controller readjustment.

B. Basic Concept

Sliding mode control [3] is a form of bang-bang state feedback control in which the switching boundary in the state space is designed such that the state trajectories of the phase portrait within the operational envelope of the control application are directed towards the boundary from both sides. Then, with an arbitrary initial state, after the state trajectory first crosses the boundary, the control variable switches, theoretically at infinite frequency and with a continuously changing mark-space ratio, so that the state trajectory is held on the boundary and appears to slide in it. This is a sliding mode. If the chosen state variables are the output, y, and its derivatives, then in the sliding mode, the closed-loop performance is entirely determined by the equation of the switching boundary and is independent of the plant parameters and any external disturbance, indicating extreme robustness.

The maximum order of output derivative used by the sliding mode control law is equal to $r-1$, where r is the rank of the plant. This is also the order of the closed-loop system.

C. Boundary Layer

The basic bang-bang form of sliding mode control suffers, in practice, from control chatter in which $u(t)$ switches at a finite frequency determined by any sensor or actuator lags and the sampling period of the digital implementation. The signum function implementing the switching function in the state space is responsible for this, but the control chatter is eliminated in the control system presented by replacement of the signum function by a high gain element with saturation limits, $\pm u_{max}$, as shown in Fig. 3.

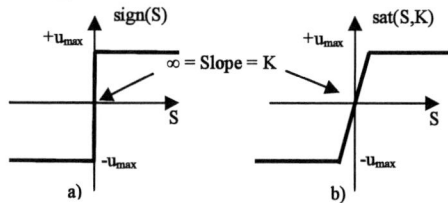

Fig. 3. a) Signum function and b) Saturation function

This means that a boundary layer replaces and straddles the switching boundary in the phase plane [2]. If the value of K is reduced, the boundary layer becomes broader and the robustness against plant parameter uncertainties and external disturbances becomes significantly less than that of the ideal sliding mode controller.

It is worthy of note that within the boundary layer, the saturation function is equivalent to a high gain, K, and if the switching function is linear, the controller is equivalent to a standard classical one, but the robustness becomes apparent when viewing the root locus w.r.t. K for a linear plant. As $K \to \infty$, the root loci (giving the poles of the closed-loop system) move onto the zeros of the open loop transfer function, which are the roots of the characteristic equation of the linear closed-loop system in the ideal sliding mode.

D. Sliding Mode Control of a Vacuum Air Bearing Based Linear Drive

Since the plant (1) is of rank, 2, the closed loop system is just of first order using the basic sliding mode control law. With the finite gain introduced by the boundary layer, however, electrical cables and pneumatic tubes connected to the moving platform cause significant steady state errors on the nanometre scale, and so an outer integral control loop, with anti-windup, is added to eliminate this steady state error. A block diagram of the resulting system is shown in Fig. 4.

Note that a low-pass filter with time constant, T_f, has been added to avoid amplification of measurement noise at high frequencies which would risk unsatisfactory performance due to control saturation. In theory, with $T_f = 0$, it is fairly straightforward to show that as $K \to \infty$, the closed-loop system has the following ideal critically damped transfer function:

$$G_{ID}(s) = \left(\frac{1}{1+\frac{2}{9}T_s s}\right)^2 = \frac{1}{\frac{4}{81}T_s^2 s^2 + \frac{4}{9}T_s s + 1} \qquad (3)$$

where T_s is the settling time given by (2) with $n = 2$ if

$$\left\{ T_c = T_s/9, \quad K_i = 9/(4T_s) \right. \qquad (4)$$

In practice, for sufficiently small T_f, and a finite but sufficiently large K, there are two dominant closed loop poles very nearly equal to $s_{1,2} = -9/(2T_s)$ in value and therefore the closed loop dynamics closely approximates that of (3). A sketch of the root locus for the system of Fig. 4 is shown in Fig. 5.

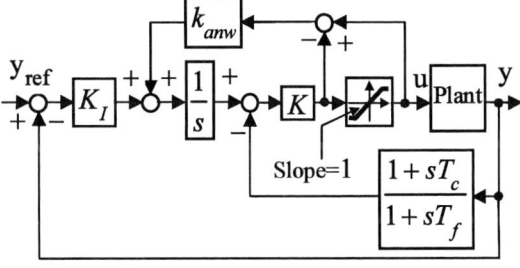

Fig. 4. Closed loop block diagram with high gain SMC controller.

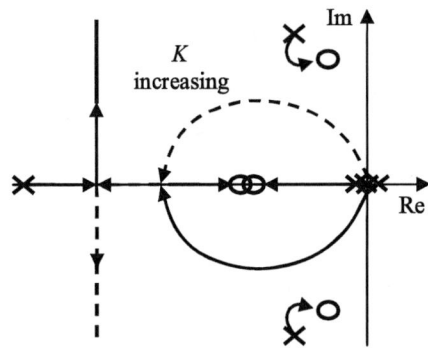

Fig. 5. Root locus sketch for the high gain SMC controller.

Three poles start at the origin (two from the plants rigid body and one from the additional integrator avoiding the steady state error) and two start from the complex conjugate poles due to the oscillatory mode and move on towards the two zeros near the imaginary axis as K increases. In fact, it is clear, that a sufficiently high gain K is necessary to ensure that no oscillation in the output occurs, i.e., the poles of the closed loop system have to be real and/or close to and nearly cancelled by the zeros. Further, by magnifying the root locus around the origin, two of the three poles are seen to move into the right half plane when K is increased from zero causing instability before K is sufficiently high to return the poles to the left half plane meaning that the system is conditionally stable.

It is clear from the above that when K is too small there will be oscillation of the step response. This has been confirmed by simulation and K can be increased until the oscillation vanishes and the predicted step response is nearly the same as the ideal one that can easily be simulated using (3). Satisfactory performance has been predicted by Matlab/Simulink simulation with the controller parameters of Table 1.

TABLE I. HIGH GAIN SMC SIMULATION PARAMETERS

$K = 30e3$	$K_{anw} = 1$
$T_s = 100ms$	$K_I = 9/4/Ts$
$T_f = 0.1ms$	

V. FDC AND OUTER SLIDING MODE CONTROL LOOP

The main focus in this section is on a version of SMC where an outer loop is closed around a forced dynamic control (FDC) law positioning the linear drive. The FDC control method is introduced briefly in the following subsection before continuing with the two versions of the outer SMC control loop.

A. Forced Dynamic Control

Forced dynamic control [4] is a form of state control but this differs from conventional linear state control in a) compensating external disturbances if they are included in the plant model, b) catering for nonlinear plants and c) cancelling any zeros of a linear plant. Since the complex conjugate zero pair of the vacuum air bearing plant are close to the imaginary axis of the s-plane and two of the closed loop poles coincide with them upon applying FDC, an oscillatory mode occurs with too small a damping ratio. To overcome this problem, a fictitious output equation is used with the real time state space plant model of the observer [2] to yield an artificially larger damping ratio, η_1. This FDC law is presented in [2] for the vacuum air bearing but departures from the nominal dynamics were observed in the experimental results, attributed to plant modelling errors, better results being obtained with SMC [6]. SMC alone, however, has to 'do all the work' to achieve the desired closed loop dynamics. This encouraged the authors to design a new control system in which an outer SMC loop is applied just to *assist* the FDC loop in applying the correct control to achieve the required closed loop dynamics when the real plant is mismatched with respect to the plant model on which the FDC is based [4, 5]. Then, in view of the slight reduction in robustness of the SMC that would be expected with the use of the finite gain of the saturation element of Fig. 3, better results might be expected than with SMC alone for the same gain.

B. Basic Concept of Outer Sliding Mode Control

As mentioned, the inner loop comprises an FDC control law [2] applied to the plant, supported by an observer for state and disturbance estimation. This is the 'virtual plant' with demanded position, y'_{ref}, as the control input from the sliding mode control law, as shown in Fig. 6.

The output derivative is estimated independently of the observer in Fig. 6 together with measurement noise filtering. This is shown to clarify the principle of operation but in the real implementation, the velocity estimate is taken from the observer.

In the following, the filtering time constant, T_f, will be assumed sufficiently small to have no substantial effect on the closed-loop dynamics and will be ignored. Since the rank of the virtual plant is 2, the switching boundary of the sliding mode controller is one dimensional [4] and given by

$$y - y_{ref} + T_c \dot{y} = 0 \tag{5}$$

where T_c is the time constant of the first order closed loop dynamics in the sliding mode. This boundary becomes the centre of the boundary layer when implementing the high gain saturation element of Fig. 3 (b). Then the sliding mode control law is:

$$\begin{cases} S = y - y_{ref} + T_c \dot{y} \\ u = u_{max} sat(S, K) \end{cases} \tag{6}$$

This drives the state trajectory towards the boundary (5) from both sides over a finite region of the boundary that increases with the control saturation limits. These limits will be such that this condition for sliding motion holds

over the range of practical values of the output, y, and $\dot{y}$ that will occur. Then once the state trajectory enters the boundary layer, it will be 'trapped' by it and appear to 'slide' along a path in a close neighbourhood of the boundary while the output, y, approaches the reference input, y_{ref}, if it is constant. Boundary equation (4) is then the first order differential equation of the closed loop system and the corresponding closed loop transfer function is as follows:

$$G_{ID}(s) = \frac{1/T_c}{s + 1/T_c} \tag{7}$$

In this first order example, it is well known that the settling time (also given by (2) with $n = 1$), is $T_s = 3T_c$.

The real plant is of rank two and the virtual plant is the closed loop transfer function yielded by the FDC, designed to have a critically damped transfer function:

$$G_{VP}(s) = \left(\frac{1/T_{c,VP}}{s + 1/T_{c,VP}} \right)^2 \tag{8}$$

where the time constant is chosen according to (2) yielding $n = 2$ $T_{c,VP} = T_s/4.5$. Fig. 7 shows a simplified version of Fig. 6 in the sliding mode. The closed loop transfer function is:

$$G_{cl}(s) = \frac{K\left[\left(1/T_{c,VP}\right)/\left(s + 1/T_{c,VP}\right) \right]^2}{1 + K\left(1 + sT_c\right)\left[\left(1/T_{c,VP}\right)/\left(s + 1/T_{c,VP}\right) \right]^2} \tag{9}$$

from which:

$$\lim_{K \to \infty} G_{CL}(s) = \frac{1}{1 + sT_c} \tag{10}$$

which yields the behaviour predicted by (5). Again, using (2), the value of the overall closed loop time constant T_c as a function of the settling time with $n = 1$ is:

$$T_c = \frac{1}{3}T_s \tag{11}$$

Since, however, the transfer function (10) in the sliding mode, differs from transfer function (8) of the virtual plant, the SMC law will have to change the dynamic response in addition to compensating for the plant modelling errors in the inner loop, although the settling times of (8) and (10) are made the same. This additional demand on the SMC, however, can be removed by closing an integral control loop around the SMC loop and then

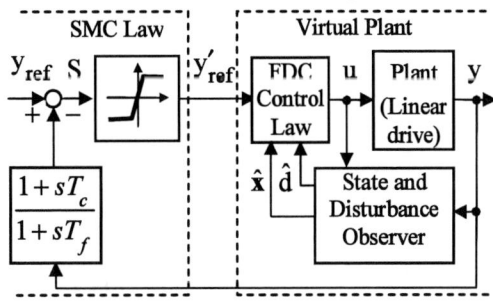

Fig. 6. Introduction of SMC outer control loop.

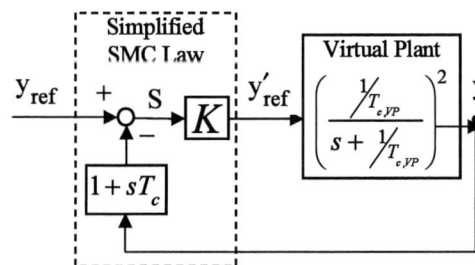

Fig. 7: Simplified SMC block diagram (FDCSMC1).

adjusting the integral gain and T_c to yield the same closed loop transfer function as (8). The implemented system, including integral anti-windup is shown in Fig. 8.

Replacing the inner SMC loop by the ideal transfer function (7) and setting $T_f = 0$ for calculating K_I and T_c yields Fig. 9.

Here, the closed loop transfer function is:

$$G_{cl}(s) = \frac{K_I/T_c}{s^2 + (1/T_c)s + K_I/T_c} \qquad (12)$$

The desired closed loop transfer function is chosen to be the same as (7). Again applying the settling time formula (2) with $n = 2$:

$$G_{cl}(s) = \left(\frac{9/2T_s}{s + 9/2T_s}\right)^2 = \frac{\frac{81}{4T_s}}{s^2 + \frac{9}{T_s}s + \frac{81}{4T_s}} \qquad (13)$$

Comparing equations (11) and (12) yields:

$$\left\{ T_c = \frac{1}{9}T_s, \quad K_I = \frac{9}{4T_s} \right. \qquad (14)$$

The two different outer sliding mode control loops show significant differences in the control variable with the same defined settling time, T_s, as will be seen in the next section.

VI. SIMULATION RESULTS

All the simulations show responses to a step position reference input together with a step disturbance force equivalent to a 25 mV step i.e. 20mA in the coil current applied at $t = 0.5\,s$ which is for the purpose of comparing

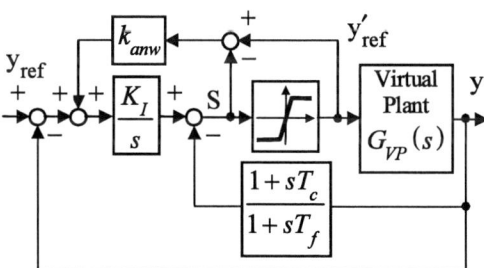

Fig. 8: Closure of outer integral control loop.

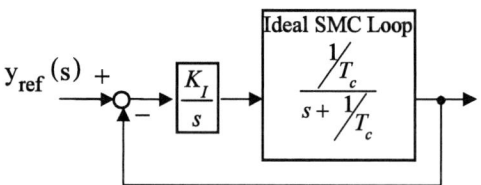

Fig. 9: Simplified block diagram for parameter determination (FDCSMC2).

the disturbance rejection properties and rather larger than would be experienced in practice.

The vacuum air bearing, despite its lack of friction, is believed to produce significant plant noise due to the random movement of air molecules within the air cushion, which will continually excite the oscillatory mode. Such excitation is evident in the linear encoder output of the experimental rig (peaking around $\pm 20\,nm$) with the control actuation turned off. To simulate this, a random signal is added to x_3 and the part of the control variable acting on the oscillatory mode at a level yielding similar amplitude in y to that observed experimentally without control.

Fig. 10 shows the simulated step response of the basic SMC together with the ideal step response, for $K = 10^4$. The maximal position deflection (due to the disturbance step of d=25mV) of the slider is $0{,}64\,\mu m$.

The controller parameters for the outer SMC loop versions are given in Table II.

Fig. 11 and Fig. 12 show the simulated step and disturbance responses for FDCSMC1 and FDCSMC2 respectively. The deflection in both cases is greater than with SMC on its own and shows an undershoot in the case of the second order version of the outer sliding mode control loop.

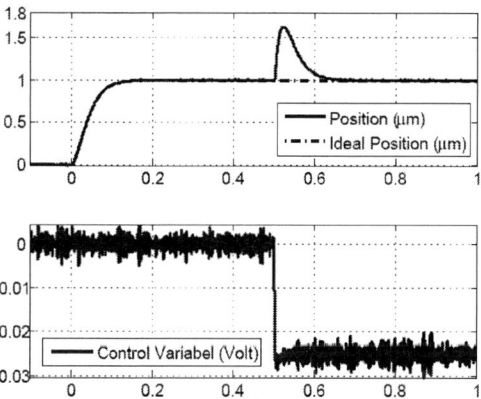

Fig. 10. Simulation of step and disturbance response using SMC.

TABLE II. CONTROLLER PARAMETERS

FDCSMC1	$K = 30$	$K_{anw} = 1e-9$
	$T_{s\,smc} = 100\,ms$ (Inner loop settling time)	$\eta_1 = 0.2$ (Inner loop damping ratio)
	$T_{s\,fdc} = 200\,ms$ (Inner loop settling time)	$T_{so} = 8\,ms$ (FDC observer settling time)
	$T_f = 0.1\,ms$	
FDCSMC2	$K = 100$	$K_{anw} = 1e-9$
	$T_{s\,smc} = 100\,ms$	$T_{so} = 8\,ms$
	$T_{s\,fdc} = 200\,ms$	$T_{so} = 8\,ms$
	$T_f = 0.1\,ms$	

VII. EXPERIMENTAL RESULTS

Fig. 13 shows the experimental step response for SMC.

The experimental disturbance response of Fig. 14. agrees well with the deflection predicted through the simulation.

The experimental results found for FDCSMC1 and FDCSMC2 also show good agreement with the simulations.

To assess the relative position control accuracy, Fig. 19 to Fig. 21, inclusive, show the experimental position measurements with zero reference input together with the standard deviation.

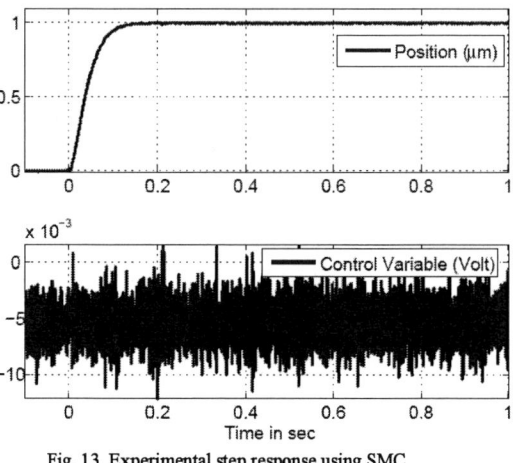

Fig. 13. Experimental step response using SMC.

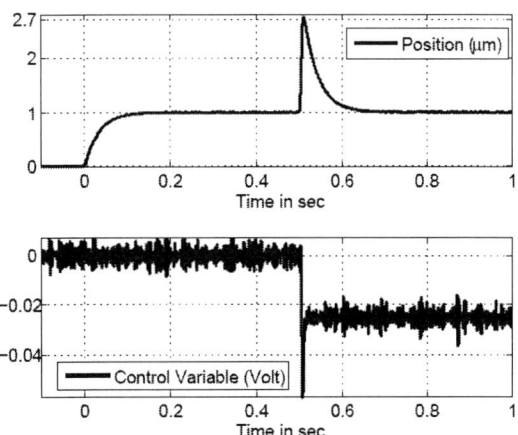

Fig. 11. Simulated step and disturbance response using FDCSMC1.

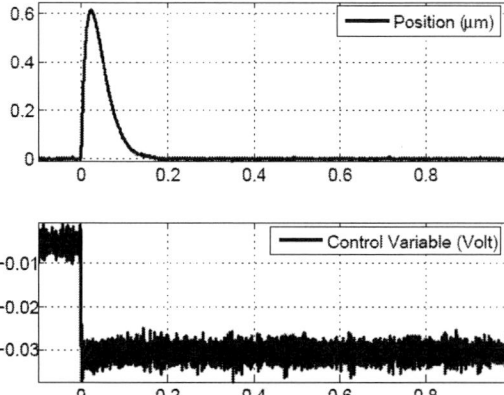

Fig. 14. Experimental disturbance response using SMC.

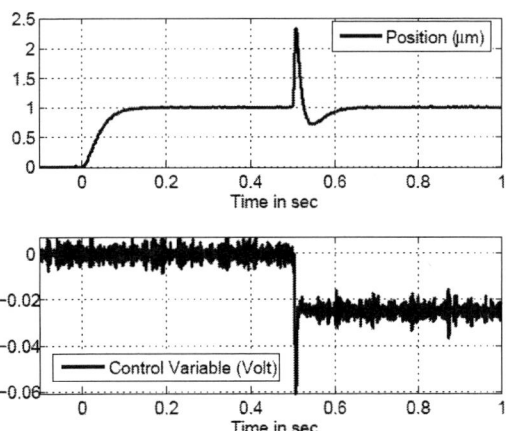

Fig. 12. Simulated step and disturbance response using FDCSMC2.

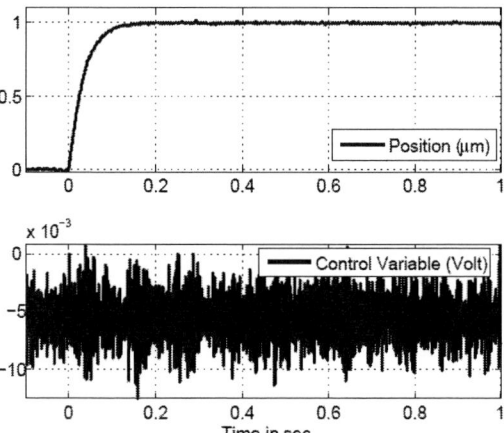

Fig. 15. Experimental step response using FDCSMC1.

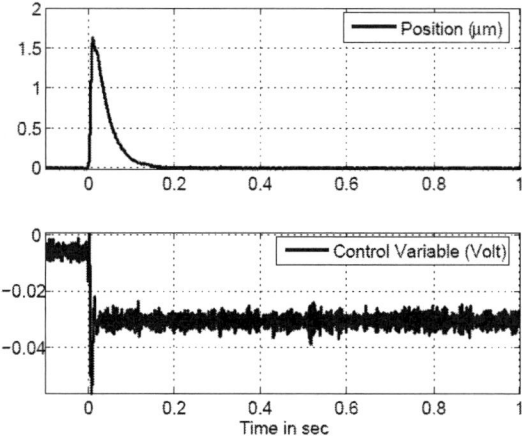

Fig. 16. Experimental disturbance response using FDCSMC1.

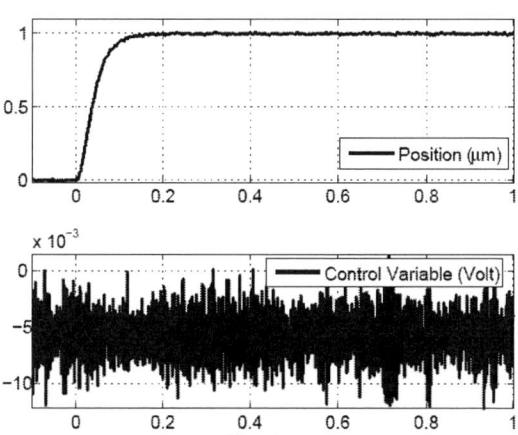

Fig. 17. Experimental step response using FDCSMC2.

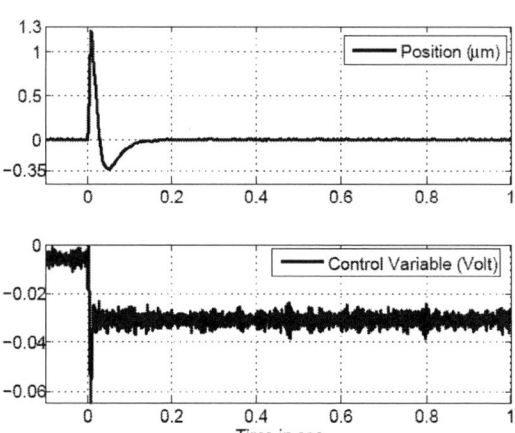

Fig. 18. Experimental disturbance response using FDCSMC2.

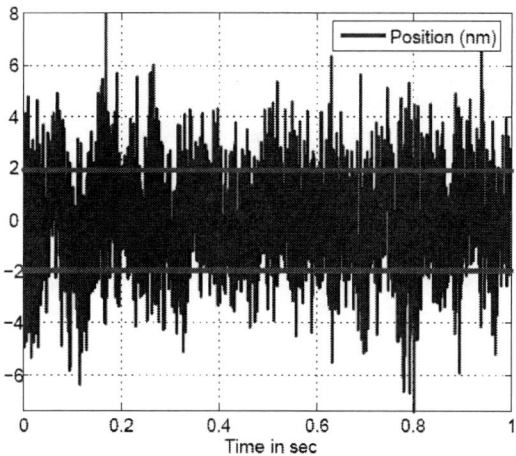

Fig. 19. Position measurement with zero reference input using SMC.

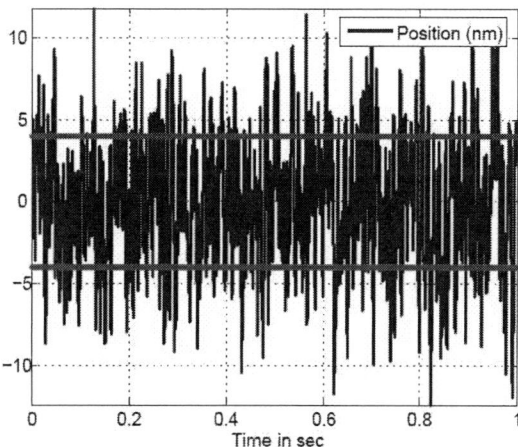

Fig. 20. Position measurement with zero reference input using FDCSMC1.

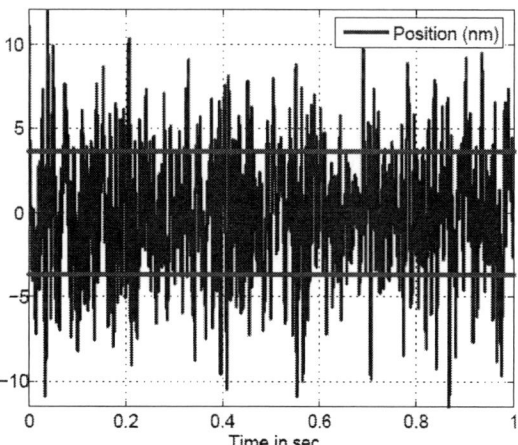

Fig. 21. Position measurement with zero reference input using FDCSMC2.

The results are summarised in Table III to give an overview of the three different control techniques investigated and their individual performances regarding the following criteria:

- Reference tracking capability (transient response)
- Compliance, defined as the transient peak position error in μm per 0.222N step disturbance force (since a step of 0.222N was applied in the laboratory)
- Robustness against parameter variations/model order uncertainties
- Complexity of the design procedure
- Applicability for practical implementation
- Computational demand
- Position stability for zero reference input (standard deviation)

VIII. CONCLUSIONS

Clearly SMC has the advantages of less complexity than FDCSMC1/2 without the need for an accurate plant model (except for knowledge of its rank) and better external disturbance rejection. The major advantage of SMC is that the gain, K, can be increased to improve robustness without altering the settling time. On the other hand, SMC has to be handled with care when applying the necessary high gain in the real world and reliable simulations must still be used, including noise sources, to predict performance and cautious experimental procedures are recommended.

Both versions of the outer sliding mode control loop, FDCSMC1 and FDCSMC2, strongly improved the performance when compared with FDC alone [2,5], which was impaired by a lower limit on the observer settling time, below which uncontrollable vibration modes were observed experimentally, thereby limiting the bandwidth of the closed loop system. The measured disturbance rejections show a very good agreement with the predicted results found by the simulations. Unfortunately, however, increasing the gain, K, beyond the relatively low limits shown in Table 1 in an attempt to increase the control loop stiffness to improve the disturbance rejection (i.e., minimise the compliance) caused stimulation of the aforementioned uncontrollable vibration modes experimentally, thereby limiting the performance. This offset the potential advantage of the SMC in the outer Performance Comparisons

TABLE III. CONTROLLER PERFORMANCE COMPARISONS

	SMC	FDCSMC1	FDCSMC2
Transient Response	Good	Acceptable	Acceptable
Compliance	0.6 μm	1.62 μm	1.25 μm
Robustness	Yes	Yes	Yes
Complexity	Low	High	High
Applicability	Yes	Yes	Yes
Computational Demand	Low	Moderate	Moderate
Standard Deviation	2nm	4nm	3.7nm

loop over SMC alone by forcing the gain, K, to be reduced from 3×10^4 to only 30 (FDCSMC1) and 100 (FDCSMC2). These vibration modes were audible and attributed to elasticity of the mechanical framework supporting the voice coil motor. It is therefore suggested that higher gains might be permitted by increasing the stiffness of this mechanical structure. The second order version of the outer SMC loop might also be improved by refinement of the anti-windup loops for the observer disturbance estimation integrator and the additional integrator in the outer SMC-loop.

Additionally, FDCSMC1 shows an undesirable high peak in the control variable demand for larger step position responses which increases as K is increased.

One significant advantage of SMC is its simplicity meaning little effort needed to programme the real time system and the simulations together with relatively small computational power demands.

The overall conclusion has to be that with the vacuum air bearing hardware as it stands, SMC alone is the best solution but it would be worth stiffening the mechanical structure of the voice-coil actuator support in an attempt to alleviate the vibration mode problem, thereby allowing the gain, K, to be increased for FDCSMC1 and FDCSMC2.

REFERENCES

[1] A. Cassat, N. Corsi, R. Moser, and N. Wavre. "Direct Linear Drives: Market and Performance Status", *Proceedings of the 4th International Symposium on Linear Drives for Industry Applications*, LDIA 2003, SP-01, (2003).

[2] P. A. Stadler, S. J. Dodds, and H. G. Wild, "Simultaneous High Precision Control of the Position and an Oscillatory Mode of a Vacuum Air Bearing Linear Drive", *Proceedings of the 5th International Symposium on Linear Drives for Industry Applications*, LDIA 2005.

[3] V. I. Utkin, "Sliding Modes in Control and Optimisation", *Springer Verlag*, 1992.

[4] Vittek, J. and Dodds, S. J.: *Forced Dynamics Control of Electric Drives*, EDIS, Žilina University Publishing Centre, 2003, ISBN 80-8070-087-7.

[5] P. A. Stadler, S. J. Dodds, and H. G. Wild, "Robustness Improvement for Positioning a Vacuum Air Bearing Linear Drive Using an Outer Sliding Mode Control Loop", *Proceedings of AC&T 2006*, SCOT, University of East London, UK, ISBN 0-9550008-1-5, London, Jan 2007.

[6] P. A. Stadler, S. J. Dodds, and H. G. Wild, "Comparison of nanometer motion control techniques for vacuum air bearings," *Proceedings of Control 2006, Glasgow, UK*, 2006.

[7] S. J. Dodds, "Forced Dynamic Control: A Model Based Control Technique Illustrated by a Road Vehicle Control Application", *Proceedings of AC&T 2006*, SCOT, University of East London, UK, ISBN 0-9550008-1-5.

[8] S. J. Dodds, "Settling time formulae for the design of control systems with linear closed loop dynamics", *Proceedings of AC&T 2008*, SCOT, University of East London, UK, ISBN 0-9550008-5-8.

Sliding Mode Control of PMSM Drives Subject to Torsion Oscillations in the Mechanical Load

Stephen J. Dodds[*], Jan Vittek[†]

[*]School of Computing and Technology, University of East London, UK, e-mail: *stephen.dodds@spacecon.co.uk*
[†]Faculty of Electrical Engineering, University of Žilina, Žilina, Slovak republic, e-mail: *Jan.Vittek@fel.uniza.sk*

Abstract— A control system for permanent magnet synchronous motor electric drives with a significant torsion vibration mode is presented based entirely on sliding mode principles to achieve robustness against external load torques and parametric modelling uncertainties in the motor and/or the driven mechanical load. The sliding mode control law respects the vector control condition by keeping the direct axis current component approximately zero as well as controlling either the rotor or load position to follow the demanded position with prescribed closed loop dynamics. When controlling the load position, the torsional oscillations are completely eliminated. To avoid control chatter, a boundary layer is introduced by replacing the relay control switching transfer characteristic (signum function) by a high gain with the same control saturation limits. The user is only required to provide the demanded position and specify the settling time of the step response, no controller tuning being necessary. The simulations predict that the desired robustness will be achieved.

Keywords—active damping, control of drive, electric drive, highly dynamic drive, motion control, robustness, sliding mode control, synchronous motor, vector control.

I. INTRODUCTION

A. Overview

The main contribution of this paper is a relatively simple sliding mode controller that simultaneously achieves vector control and position control with prescribed closed loop dynamics for a permanent magnet synchronous motor (PMSM) drive with a mechanical load that may contain a significant vibration mode, and without the need for any knowledge of the motor or load parameters. This is achieved through the important property of robustness offered by sliding mode control (SMC) [1]. The definition of robustness taken here is as follows. A control system is robust if it maintains specified closed loop dynamic responses to its reference inputs when a) the plant parameters are changed and b) external disturbances are applied. SMC is well known to achieve this form of robustness and several PMSM drive control systems have been designed using SMC principles. For drives assumed to have rigid coupling and load dynamics, an SMC based outer loop has been applied to a model based forced dynamic control system in [2] to provide robustness, and other forms of SMC are applied directly in [4] to [7]. The combination of the elasticity and the inertia of the materials used in controlled mechanisms, however, gives rise to vibration modes that, if ignored, can, impair control accuracy and some progress on their active control has already been made. For PMSM drives containing a single torsion

vibration mode, SMC has been applied in an outer loop in [8], directly in [9] and even extended to cater for model order uncertainty in [10]. The sliding mode drive control system offered here, however, is unique in completely eliminating the torsional oscillations while controlling the load mass position to respond to changes in the demanded position with prescribed closed loop dynamics, aided by the Dodds settling time formula [11].

B. An Introduction to Sliding Mode Control

Sliding mode control (Utkin, 1992) is a well known technique for achieving robustness. It not only guarantees stability but can achieve a specified closed-loop dynamic response to the reference input that does not change significantly in the presence of parametric changes or external disturbances. It is also applicable to nonlinear plants.

Fig. 1 shows the general block diagram of a sliding mode control system designed to yield a precisely defined closed-loop dynamic performance for a single input, single output (SISO) plant.

Here, $\mathbf{x} = \begin{bmatrix} x_1 & x_2 & \cdots & x_n \end{bmatrix}^T$, is the state vector, where n is the order of the plant, and $\mathbf{y} = \begin{bmatrix} y & \dot{y} & \cdots & y^{(r-1)} \end{bmatrix}^T$ is the vector of output derivatives. The rank (or relative degree) of the plant is r, such that the rth output derivative is $y^{(r)} = h_r(\mathbf{x}, u)$ and is *not* a state variable because of its dependence on the control input, u. The elements of $\mathbf{y}$ are *all state variables* and the set of equations for the derivatives of $\mathbf{y}$ shown in Fig. 1 constitute a transformation to a new set of state variables. As can be

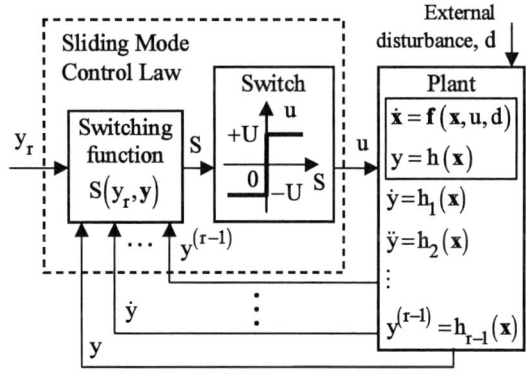

Fig. 1. Basic SISO sliding mode control.

978-1-4244-1741-4/08/$25.00 ©2008 IEEE

seen, the sliding mode control law is a bang-bang control law in which u switches between maximum and minimum values of $\pm U$ whenever S passes through zero. The switching function, $S(y_r, \mathbf{y})$, is designed such that over the normal range of operating states, u is automatically switched to the value that drives $\mathbf{y}$ towards the switching boundary,

$$S(y_r, \mathbf{y}) = 0 \Rightarrow S\left(y_r, y, \dot{y}, \cdots y^{(r-1)}\right) = 0 \qquad (1)$$

from both sides as illustrated in Fig. 2 for $r = 2$. Here, the infinite continuum of trajectories shown in red for $u = +U$ together with that shown in blue for $u = -U$ are referred to as the closed loop phase portrait. The term 'phase' is used to describe a set of state variables consisting of a chosen variable and its derivatives. With an arbitrary starting point, a, the green trajectory at first moves towards the switching boundary under control saturation and upon crossing the boundary at point, b, $\mathbf{y}$ is held on the boundary due to u rapidly switching between $+U$ and $-U$, in theory at an infinite frequency and with a continuously varying mark-space ratio. Under these circumstances, the point, $\mathbf{y}$, in the output derivative sub state space trajectory zigzags about the switching boundary but remains in a close neighbourhood of the boundary and on a larger scale appears to *slide* in the surface and the system is said to be operating in a *sliding mode*. Also, during this *sliding motion*, the switching boundary equation (1) is, in theory, obeyed and becomes the closed loop differential equation. If this is linear, i.e.,

$$y_r - \left(y + q_1\dot{y} + q_2\ddot{y} + \ldots + q_{r-1}y^{(r-1)}\right) \qquad (2)$$

where q_i, $i = 1, 2, \ldots, r-1$, are constant coefficients that may be chosen to yield the desired closed loop dynamics, then the closed loop system is linear with transfer function

$$\frac{y(s)}{y_r(s)} = \frac{1}{1 + sq_1 + s^2q_2 + \ldots + s^{r-1}q_{r-1}} \qquad (3)$$

which is *independent of the plant parameters and the external disturbance*. It must be realised that this performance is only attained while in sliding motion. The aforementioned condition for sliding motion that the point, $\mathbf{y}$, in the r dimensional sub-state space is driven back towards the switching surface (1) from both sides by the control law may be expressed mathematically as

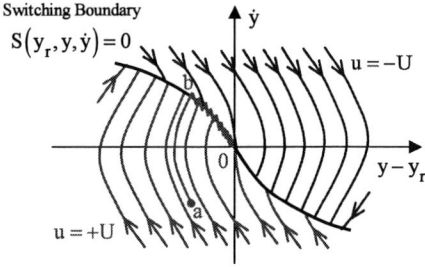

Fig. 2. Illustration of closed loop phase portrait and trajectory

$$S\dot{S} < 0 \qquad (4)$$

This condition will only be satisfied over a finite region of the switching boundary but, in general, this region may be increased in size by increasing the maximum control level, U.

The sliding mode control system described above is a state feedback control system. If the plant with order, n, is of full rank, then $r = n$ and $\mathbf{y} = \left[y \ \dot{y} \cdots y^{(r-1)}\right]^T$ is a complete state vector, enabling *complete* control of the plant according to standard control theory. If, however, the plant is not of full rank, i.e., $r < n$, which is the case when controlling the rotor position of the PMSM drive with the flexible coupling, then the sliding mode control law can only control a subsystem of the plant with the state variables, $y_r, y, \dot{y}, \cdots y^{(r-1)}$. In this case, there exists an uncontrolled subsystem of order, $n - r$. The dynamics of this uncontrolled subsystem is referred to as the *zero dynamics*. In fact, a linear plant with transfer function, $y(s)/u(s) = N(s)/D(s)$, which is not of full rank has $n - r$ zeros and the poles characterising the zero dynamics are roots of $N(s) = 0$, i.e., the plant *zeros*. Zero dynamics will be seen in the rotor angle control considered later.

II. THE MODEL OF THE CONTROLLED PLANT

Although the parameters of the motor and its mechanical load are not needed for the control system design, the complete model is given here as it is used for the simulations and the relative degree determinations. The PMSM and its mechanical load will first be modelled separately and then considered together for the relative degree determination.

The state differential equations of the PMSM are

$$di_d/dt = -Ai_d + B\omega_r i_q + Fu_d \qquad (5)$$

$$di_q/dt = -\left(Ci_d + E\right)\omega_r - Di_q + Gu_q \qquad (6)$$

$$d\omega_r/dt = M\left(\Gamma_c - \Gamma_{Lr}\right) \qquad (7)$$

$$d\theta_r/dt = \omega_r \qquad (8)$$

in the synchronously rotating d-q co-ordinate system, where i_d, i_q are the stator current vector components, u_d, u_q are the voltage vector components, *which are also the control variables*, ω_r and θ_r are the rotor angular velocity and angle, Γ_c is the electromagnetic torque given by

$$\Gamma_c = J_r\left(H + Ki_d\right)i_q \qquad (9)$$

and Γ_{Lr} is the total load torque acting on the rotor given by

$$\Gamma_{Lr} = \Gamma_{Lre} + \Gamma_{Ls} \qquad (10)$$

where Γ_{Lre} is an external load torque applied directly to the rotor and Γ_{Ls} is the load torque presented by the driven mechanical load, which in this case is the torsion torque of the flexible shaft driving the load.

The constant coefficients are:

$$\begin{cases} A=R_s/L_d; & B=pL_q/L_d; & C=pL_d/L_q; \\ D=R_s/L_q; & E=p\Psi_{PM}/L_q; & F=1/L_d; & G=1/L_q; \\ H=3p\Psi_{PM}/(2J); & K=3p\left(L_d-L_q\right)/(2J); & M=1/J_r \end{cases}$$

where Ψ_{PM} is the permanent magnet flux, R_s is the stator resistance, L_d and L_q are the direct and quadrature axis inductances, p is the number of pole pairs and J_r is the rotor moment of inertia. It is evident from (8) that the control torque delivered by the motor is:

The driven mechanism is a balanced mass with moment of inertia, J_L, coupled to the motor shaft via a torsion spring with spring constant, K_s, as shown in Fig. 3(a). The corresponding torque balance equations are as follows:

$$J_r\ddot{\theta}_r = \Gamma_c - \Gamma_{Lre} + K_s\left(\theta_L - \theta_r\right) \qquad (11)$$

$$J_L\ddot{\theta}_L = K_s\left(\theta_r - \theta_L\right) - \Gamma_{Le} \qquad (12)$$

where J_r is the rotor moment of inertia, θ_r is the rotor angle, θ_L is the load mass angle, Γ_{Lre} and Γ_{Le} are the external load torques applied, respectively, to the rotor and the load mass. These are both included for generality.

The state variable block diagram of Fig. 3(b) follows from (11) and (12).

Fig. 4 shows the main hardware elements of the controlled plant together with the inverse and forward Clark-Park transformations, which are the same as for other vector control schemes. The subscript 'dem' indicates a 'demanded value' and the subscript, 'm', indicates a measured value. Details of the pulse width modulation (PWM) scheme are not given here as it is a standard one.

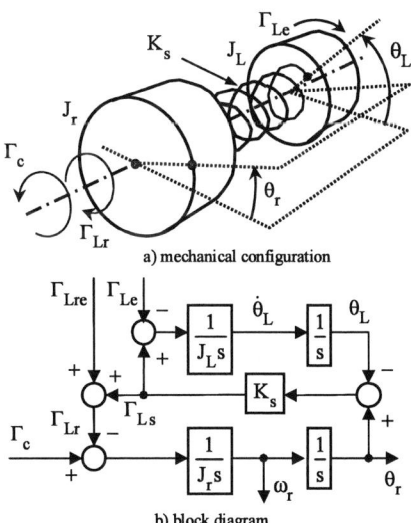

a) mechanical configuration

b) block diagram

Fig 3. State variable block diagram model of the driven mechanism.

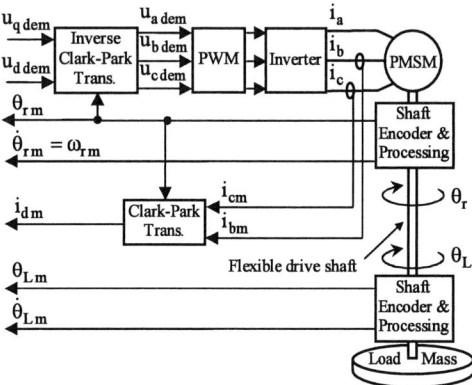

Fig. 4. Overall block diagram of the controlled plant.

It is sufficient to state that the stator voltages, u_a, u_b and u_c, are switched between the power supply voltage levels such that their short-term mean values are equal, respectively, to the demanded values, $u_{a\,dem}$, $u_{b\,dem}$ and $u_{c\,dem}$, which are generated from the demanded direct axis and quadrature axis stator voltage vector components, $u_{d\,dem}$ and $u_{q\,dem}$ via the inverse Clark-Park transformation as shown in Fig. 4. It is assumed that the shaft encoder processing provides speed measurements as in speed controlled drives as well as the angle measurements, as shown. The two control variables are $u_{d\,dem}$ and $u_{q\,dem}$, one controlled variable is $i_{d\,m}$ and the other controlled variable is either $\theta_{L\,m}$ or $\theta_{r\,m}$.

III. SLIDING MODE CONTROL SYSTEM DESIGN

A. The Settling Time Formula

The sliding mode controllers for the three controlled outputs will be designed to yield linear closed loop dynamics with non-overshooting step responses to 95% of the steady state outputs using the settling time formula [11]. The sliding surfaces will be hyper-planes in the output derivative state sub-space yielding linear closed loop systems with coincident poles at $s_{1,2,...,n} = -1/T_c$ and the corresponding settling time formula is:

$$T_s = 1.5(1+n)T_c \qquad (13)$$

B. General Form of Controller

Separate sliding mode controllers are designed, one dedicated to keeping the direct axis stator current value as close to $i_d = 0$ as possible, for the vector control, and two more, one for controlling the rotor angle, θ_r, if required and the other for controlling the load mass angle, θ_L, if required.

Fig. 5 shows the general structure of each of the three sliding mode control loops. The basic sliding mode control system of Fig. 1 would suffer from control chatter, i.e., high

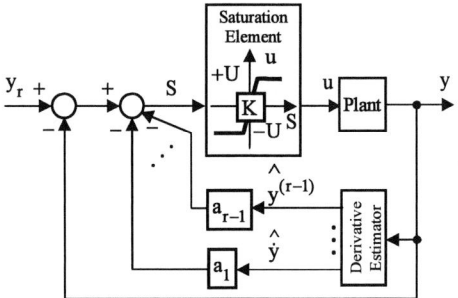

Fig 5. Structure of each sliding mode control loop.

frequency switching of the control variable, u, between its limits, $\pm U$, and this might interact adversely with the switching of the inverter, so a boundary layer is introduced [1] in which the switching element is replaced by a saturation element with the same control saturation limits having a finite but relatively high gain, K. The nominal closed loop transfer function of Fig. 5(b) is given by:

$$S \to 0 \Rightarrow y_r - \left(y + a_1\dot{y} + \ldots a_{r-1}y^{(r-1)}\right) = 0$$

$$\frac{y(s)}{y_r(s)} = \frac{1}{1 + a_1 s + \ldots + a_{r-1}s^{r-1}} \qquad (14)$$

Since the number of adjustable output derivative feedback gains is equal to the order, $r-1$, of this transfer function, the SMC loop may be designed by pole assignment using settling time formula (2) with $n = r-1$, so normalising with respect to the constant term, the desired closed loop transfer function is:

$$\frac{y(s)}{y_r(s)} = \left(\frac{1}{1 + T_c s}\right)^{r-1} = \left(\frac{1}{1 + \dfrac{T_s}{1.5r}s}\right)^{r-1} \qquad (15)$$

The design is then completed by equating corresponding coefficients in (14) and (15). This will be done for the three drive control loops below, once the relative degrees of the plant have been determined.

C. Plant Relative Degrees

The relative degree of a plant with respect to a given controlled output is the lowest derivative with respect to time that has direct algebraic dependence on any control variable. It is found by repeatedly differentiating the measurement equation and substituting for the derivatives of any state variables that appear on the right hand side of the resulting equations, using the state differential equations.

In the following, the relative degrees with respect to i_d, θ_r and θ_L are derived as these are the same as the required corresponding relative degrees with respect to $i_{d\,dem}$, $\theta_{r\,dem}$ and $\theta_{L\,dem}$.

1) Relative degree with respect to i_d : By inspection of (5), di_d/dt has direct algebraic dependence on u_d and therefore the relative degree is

$$r_i = 1 \qquad (16)$$

2) Relative degree with respect to θ_r : By inspection of (11), $\ddot{\theta}_r$ has no direct algebraic dependence on u_d or u_q, so making $\ddot{\theta}_r$ the subject of the equation and substituting for Γ_c using (9) yields:

$$\begin{aligned}\ddot{\theta}_r &= \left[J_r\left(H + Ki_d\right)i_q - \Gamma_{Lre} + K_s\left(\theta_L - \theta_r\right)\right]/J_r \\ &= \left(H + Ki_d\right)i_q + M\left[K_s\left(\theta_L - \theta_r\right) - \Gamma_{Lre}\right]\end{aligned} \qquad (17)$$

recalling that $M = 1/J_r$. Since only state variables and external disturbances appear on the RHS of (17), this equation is differentiated and substitutions made for di_d/dt and di_q/dt using (5) and (6) yielding:

$$\dddot{\theta}_r = \left(H + Ki_d\right)\frac{di_q}{dt} + Ki_q\frac{di_d}{dt} + M\left[K_s\left(\dot{\theta}_L - \dot{\theta}_r\right) - \dot{\Gamma}_{Lre}\right]$$

$$\begin{aligned}\Rightarrow \dddot{\theta}_r &= \left(H + Ki_d\right)\left[-\left(Ci_d + E\right)\omega_r - Di_q + Gu_q\right] \\ &\quad + Ki_q\left[-Ai_d + B\omega_r i_q + Fu_d\right] \\ &\quad + M\left[K_s\left(\dot{\theta}_L - \dot{\theta}_r\right) - \dot{\Gamma}_{Lre}\right]\end{aligned} \qquad (18)$$

Since, $\dddot{\theta}_r$ has direct algebraic dependence on both u_d and u_q, the rank is

$$r_r = 3 \qquad (19)$$

3) Relative degree with respect to θ_L : By inspection of (12), $\ddot{\theta}_r$ has no direct algebraic dependence on u_d or u_q, so making $\ddot{\theta}_r$ the subject of the equation and differentiating yields:

$$\ddot{\theta}_L = \left[K_s\left(\dot{\theta}_r - \dot{\theta}_L\right) - \dot{\Gamma}_{Le}\right]/J_L \qquad (20)$$

$\dot{\theta}_r$ and $\dot{\theta}_L$ are state variables and therefore $\ddot{\theta}_L$ has no direct algebraic dependence on u_d or u_q. Differentiating (20) and substituting for $\ddot{\theta}_r$ and $\ddot{\theta}_L$ using (11) and (12) yields:

$$\dddot{\theta}_L = \left\{K_s\left[\ddot{\theta}_r - \ddot{\theta}_L\right] - \ddot{\Gamma}_{Le}\right\}/J_L \qquad (21)$$

According to (17) and (12), there is no direct algebraic dependence of $\ddot{\theta}_r$ or $\ddot{\theta}_L$, and hence $\dddot{\theta}_L$ given by (21), on u_d or u_q. So (21) is differentiated and substitutions made for $\dddot{\theta}_r$ and $\dddot{\theta}_L$ using (18) and (20):

$$\ddddot{\theta}_L = \left\{K_s\left[\dddot{\theta}_r - \dddot{\theta}_L\right] - \dddot{\Gamma}_{Le}\right\}/J_L$$

$$-\left\{K_s\left[\begin{bmatrix}\left(H + Ki_d\right)\begin{bmatrix}-\left(Ci_d + E\right)\omega_r \\ -Di_q + Gu_q\end{bmatrix} \\ +Ki_q\left[-Ai_d + B\omega_r i_q + Fu_d\right] \\ +M\left[K_s\left(\dot{\theta}_L - \dot{\theta}_r\right) - \dot{\Gamma}_{Lre}\right] \\ -\left[K_s\left(\dot{\theta}_r - \dot{\theta}_L\right) - \dot{\Gamma}_{Le}\right]/J_L\end{bmatrix} - \ddot{\Gamma}_{Le}\right\}\middle/J_L\right. \qquad (22)$$

The direct algebraic dependence of $\dddot{\theta}_L$ on u_d and u_q is now apparent and therefore the rank is

$$r_L = 5 \qquad (23)$$

D. The Individual Sliding Mode Control Loops

Let T_{sr} and T_{sL} be, respectively, the required settling times of the rotor angle and load mass angle control loops. Then equating the denominators of (14) and (15) for each case yields the required output derivative feedback gains:

1) Direct axis stator current control loop: Here, $r_i = 1 \Rightarrow$

$$\frac{i_{dm}(s)}{i_{ddem}(s)} = 1 \qquad (24)$$

So in this trivial case, the sliding mode has zero order and therefore, in theory, has perfect following of the reference current with no dynamic lag. The sliding mode controller then reduces to a proportional controller with infinite gain, K. In practice K would be set to as high a value as possible, being limited by the finite sampling frequency of the digital implementation and any unmodelled plant dynamics.

1) Rotor angle control loop: Since $r_r = 3$, equating the characteristic polynomials of (14) and (15) yields:

$$1 + a_{1r}s + a_{2r}s^2 = \left(1 + \frac{2}{9}T_{sr}s\right)^2 = 1 + \frac{4}{9}T_{sr}s + \frac{4}{81}T_{sr}^2 \Rightarrow$$

$$\boxed{a_{1r} = \frac{4T_{sr}}{9}, \quad a_{2r} = \frac{4T_{sr}^2}{81}} \qquad (25)$$

1) Load mass angle control loop: In this case, $r_L = 5$, and equating the characteristic polynomials of (14) and (15) yields:

$$1 + a_1 s + a_2 s^2 + a_3 s^3 + a_4 s^4 = \left(1 + \frac{2T_{sL}}{15}s\right)^4 \Rightarrow$$

$$= 1 + 4.\frac{2T_{sL}}{15}s + 6.\frac{4T_{sL}^2}{225}s^2 + 4.\frac{8T_{sL}^3}{3375}s^3 + \frac{16T_{sL}^4}{50625}s^4$$

$$\boxed{\begin{aligned} a_{1L} &= \frac{8T_{sL}}{15}, \quad a_{2L} = \frac{24T_{sL}^2}{225}, \\ a_{3L} &= \frac{32T_{sL}^3}{3375}, \quad a_{4L} = \frac{16T_{sL}^4}{50625} \end{aligned}} \qquad (26)$$

Fig. 6 shows the overall control system block diagram with the three sliding mode controllers. The i_d control loop is operated permanently and either the θ_r or the θ_L control loop is operated according to the switch position. It should be noted that the term, $B\omega_r i_q$, in (5) may be regarded as a disturbance input to the i_d SMC loop and is counteracted by this loop while i_d is driven approximately

to zero. This minimises the interaction in the plant, as can be seen by inspection of (6) and (7) after setting $i_d = 0$. Hence the plant, although it is a multivariable one, i.e., more than one control input and more than one controlled output, can be controlled effectively by two separate single input, single output sliding mode controllers.

E. Output Derivative Estimation

Various software differentiation algorithms can be used for estimating the derivatives of the measured outputs to be controlled that avoid amplifying high frequency components of measurement noise. Conventional state observers can achieve this without introducing dynamic lag due to the filtering but are unsuitable here because an accurate plant model is needed, which would defeat the original objective of being able to work without it. A special 'high gain multiple integrator' observer, however, is a possible approach [12]. This incorporates a real plant model not necessarily resembling the plant but whose actuation error, $y - \hat{y}$, (where $\hat{y}$ is the estimate of y) is driven to small proportions by setting a relatively short observer settling time, requiring large eigenvalues yielding high correction loop gains. This observer still has measurement noise filtering properties. The simplest possible approach, however, is taken here, in the interests of minimising the computational demand on the digital processor of the drive controller. This ensures low-pass filtering of measurement noise regardless of the order of the derivative: Each pure differentiation is combined with a first order low pass filter, as shown in Fig 7. This does introduce dynamic lag in the estimates, which limits the value of K and therefore limits the robustness of the control system, but keeping the filtering time constant relatively small is predicted to yield acceptable results, according to the simulations presented.

IV. SIMULATIONS

The parameters of the non-salient PMSM are: winding resistance: $R_s = 33.3\,\Omega$; winding inductances: $L_d = L_q = 53.8\,\text{mH}$; permanent magnet flux: $\Psi_{PM} = 0.262\,\text{Wb}$; number of pole pairs: $p = 3$; rotor moment of inertia: $J_r = 0.0003\,\text{Kg}\,\text{m}^2$.

The parameters of the driven mechanical load are: Load mass moment of inertia: $J_L = 0.0003\,\text{Kg}\,\text{m}^2$; flexible drive shaft torsion spring constant: $K_s = 9\,\text{Nm}\,\text{rad}^{-1}$.

The controller parameters are set as follows: SMC loop settling times: $T_{si} = T_{sr} = T_{sL} = 0.2\,\text{s}$; Gain of saturation element: $K = 200$; Control saturation limit: $U = 360\,\text{V}$; $T_f = 10^{-4}\,\text{s}$.

In all the simulation results presented, $\Gamma_{Lr} = 0$ and the external load torque, Γ_{Le}, ramps from zero value at a constant rate of 20 Nm/s to a constant value of 2 Nm.

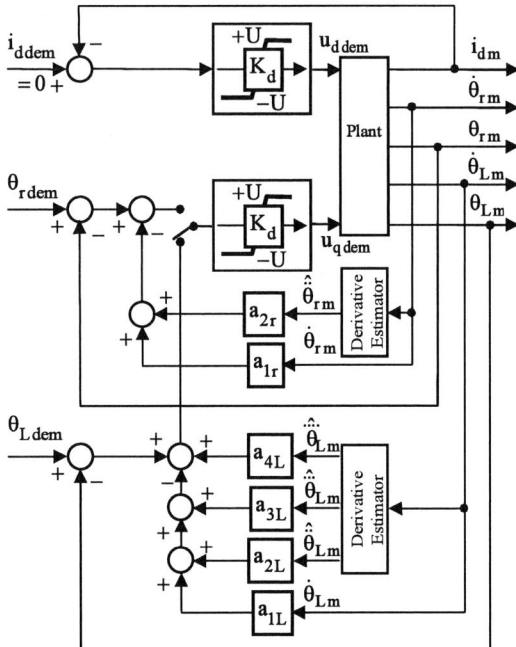

Fig 6. Overall control system block diagram.

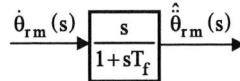

a) for rotor angle.

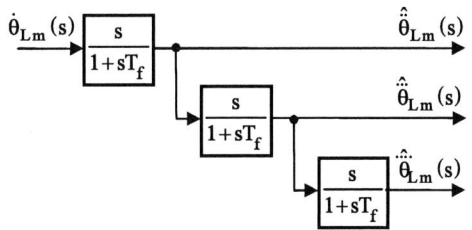

b) for load mass angle.

Fig.7. Model independent output derivative estimators incorporating measurement noise filtering

A. Rotor Position Control

A step rotor angle demand of $\theta_{r\,dem} = 10\,\text{rad}$ is applied. Fig. 9 shows the rotor angle response, $\theta_r(t)$, compared with the step response, $\theta_{rid}(t)$, of the ideal closed loop system with transfer function:

$$\frac{\theta_{rid}(s)}{\theta_{r\,dem}(s)} = \left(\frac{1}{1+s\,T_{sr}/6}\right)^3 \qquad (27)$$

does not have a very significant transient effect and this is attributed to the 'cushioning' effect of the torsion spring The simulated response follows the ideal one relatively closely on the scale of 10 rad., but the peak transient error

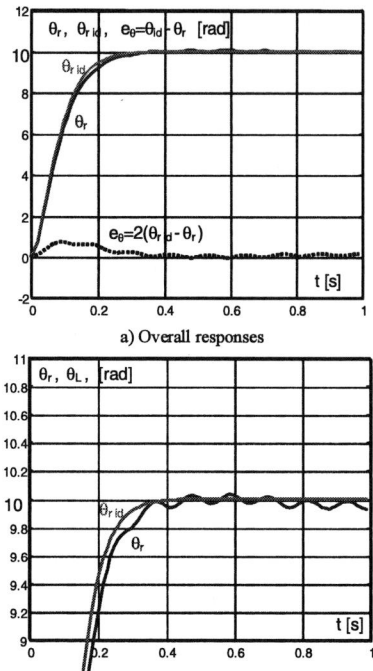

a) Overall responses

b) Magnified responses approaching the steady state

Fig. 9. Rotor control: ideal and simulated real rotor angles.

is significant as it is about 0.4 rad. The external load torque applied to the load mass at $t = 0.6\,\text{s}$, however, characteristic of the drive shaft, but a small steady state error is visible after the application of this external load torque. The oscillations of the rotor angle are due to the oscillating torsion spring torque, $\Gamma_{Ls}(t)$, indicated in Fig. 3, acting on the rotor due to the oscillating, uncontrolled load mass. This occurs due to the need for the gain, K, to be finite.

Fig. 10 shows the load mass oscillating without any significant damping while the rotor angle is controlled satisfactorily. The torsion spring torque, $\Gamma_{Ls}(t)$, acting on the PMSM rotor is treated as an external disturbance by the SMC loop and therefore a nearly equal and opposite component of the control torque, $\Gamma_c(t)$, counteracts $\Gamma_{Ls}(t)$ and allows the load mass to oscillate. Indeed, (12) may be rewritten as

$$\ddot{\theta}_L = -\frac{K_s}{J_L}\theta_L + \frac{1}{J_L}\left(K_s\theta_r - \Gamma_{Le}\right) \qquad (28)$$

and after settling the rotor control loop steady state condition of $\theta_r = \text{const.}$ (assuming perfect SMC operation) and $\Gamma_{Le} = \text{const.}$, (28) is the differential equation of simple harmonic motion at a frequency of $\sqrt{K_s/J_L}$ rad/s. In simple terms, the rotor angle SMC loop tightly controls the PMSM rotor, holding it fixed in the steady state while allowing the load mass to oscillate.

This is an example of the *zero dynamics* that occurs when applying sliding mode control to plants that are not of full rank, this dynamics being described by (28).

After application of the external load torque to the load mass, a small difference between the mean value of the oscillating $\theta_L(t)$ and the steady state value of $\theta_r(t)$ is visible and this is due to the twist angle of the drive shaft due to the constant steady value of the torsion torque.

Fig. 11 shows the oscillating spring torque referred to above together with the external load torque applied to the load mass.

Fig. 12 shows the corresponding direct and quadrature axis current components. The effectiveness of the sliding mode control system in keeping i_d nearly zero is apparent. As expected, the oscillations in $i_q(t)$ correspond to the oscillating control torque required to counteract the oscillating spring torque acting on the PMSM rotor.

B. Load Mass Position Control

A step load mass angle demand of $\theta_{L\,dem}=10\,\text{rad}$ is applied. Fig. 13 shows the rotor angle response, $\theta_r(t)$, compared with the step response, $\theta_{rid}(t)$, of the ideal closed loop system with transfer function:

$$\frac{\theta_{L\,id}(s)}{\theta_{L\,dem}(s)}=\left(\frac{1}{1+s\left(2T_{sL}/15\right)}\right)^4 \tag{27}$$

The initial transient error between these responses is more pronounced than for the rotor angle control and this is attributed to the increased order of the control loop and the gain of the saturation element remaining the same, but the control is still acceptable in being of a similar non-overshooting form.

Fig. 13 (b) clearly shows a small steady-state error due to the external load torque applied to the load mass and the finite value of the saturation element gain, K. Fig. 14 shows the corresponding rotor angle response. The steady state difference between the rotor and load mass angles after application of the external load torque to the load mass at $t=0.5\,\text{s}$ is due to the steady state constant torsion torque.

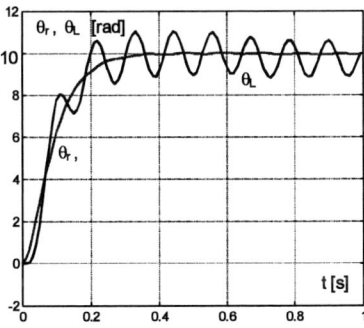

Fig. 10. Rotor control: rotor and load mass angles.

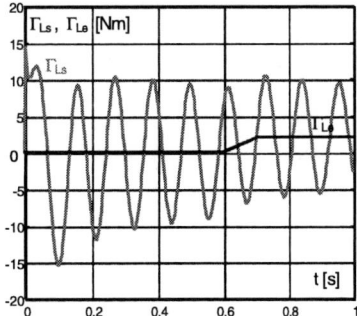

Fig. 11. Torsion spring torque and external load torque on load mass.

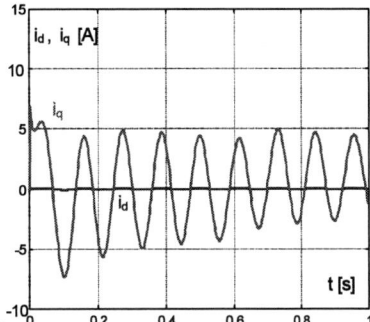

Fig. 12. Rotor control: direct and quadrature axis current components.

In order to achieve the specified non-overshooting step response, the shaft must transmit smooth accelerating and decelerating torques to the load mass, and counteract the external load torque. This can be seen in the spring torque variation of Fig. 15.

As for the rotor angle control, the sliding mode control system is effective in keeping the direct axis current to small proportions. The quadrature axis current behaves as would be expected in producing the accelerating and decelerating torques.

An important observation to be made is that the mechanical oscillations have been completely damped by the sliding mode control system. This is because the plant is of full rank with respect to the load mass angle measurement and therefore full state control is achieved without zero dynamics.

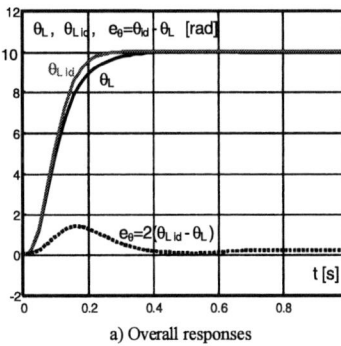

a) Overall responses

Fig. 13. Load mass control: ideal and simulated real load mass angles.

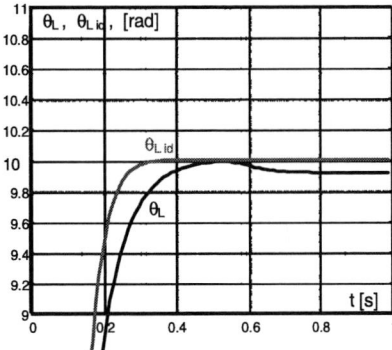

b) Magnified responses approaching the steady state

Fig. 13. Load mass control: ideal and simulated real load mass angles.

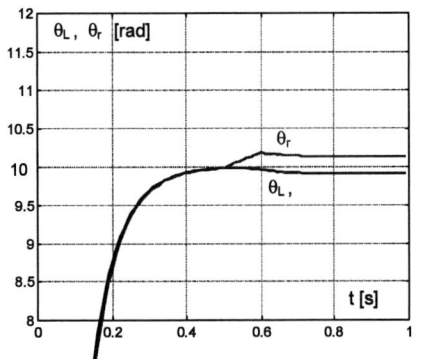

Fig. 14. Load mass angle control: rotor and load mass angles.

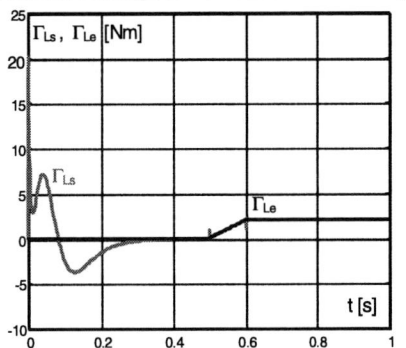

Fig. 15. Torsion spring torque and external load torque on load mass.

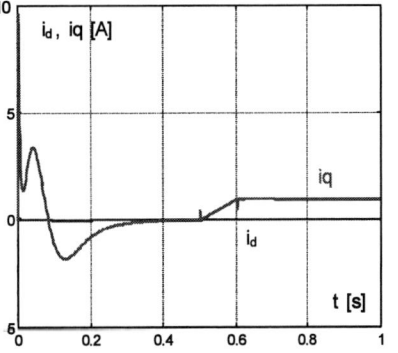

Fig. 16. Load mass control: direct and quadrature axis currents components

V. OVERALL CONCLUSIONS AND RECOMMENDATIONS

The simulations predict that the proposed sliding mode control system can be made to follow the ideal closed loop dynamics with moderate accuracy, for both rotor angle control and load mass angle control. As a preliminary result, this is acceptable considering that no motor or driven load parameters were needed. The vector control condition of keeping the direct axis current to small proportions is very effectively maintained.

It is recommended that the potential accuracy of the method is ascertained by exploring the design limits regarding sampling frequency, saturation element gain, and the derivative estimation filtering time constant, in the presence of measurement noise. Other derivative estimation methods should also be investigated, such as the high gain multiple integrator observer.

An interesting new line of investigation would be the application of the new SMC system to mechanical loads containing more than one significant vibration mode.

The results obtained here are sufficiently promising to warrant experimental trials, which will attract potential industrial users.

ACKNOWLEDGMENT

The authors wish to thank the Slovak Grant Agency VEGA for funding the project No.4087/07 'Servosystems with Rotational and Linear Motors without Position Sensor'.

REFERENCES

[1] V. I. Utkin, *Sliding Modes in Control Optimization*, Springer-Verlag, 1992, ISBN 9780387535166.

[2] J. Vittek and S. J. Dodds, *Forced Dynamics Control of Electric Drives*, EDIS, Žilina University Publishing Centre, 2003, ISBN 80-8070-087-7.

[3] O. Aguilar,. A. G. Loukianov and J. M.. Canedo,: "Observer-based Sliding Mode Control of Synchronous Motor", *Proceedings of IFAC Congress on Automatic Control*, Guadalajara, Mexico 2002, CD-Rom.

[4] E. Urlep and K. Jezernik, "Speed Sensorless Variable Structure Control of PMSM", *Proceedings of IEEE IECON 2006 Conference*, Paris, France, 2006, pp. 1194-1199.

[5] J. Arellano-Padilla, G. M. Asher and M. Summer, "Robust Fuzzy-Sliding Mode Position Controller for Motor Drives Operating Variable Loads", *Proceedings of EPE-PEMC 2002 Conference*, Cavtat, Croatia, CD-Rom.

[6] S. Brock, J. Deskur and K. Zawirski, "Robust Speed and Position Control of PMSM using Modified Sliding Mode Method", *Proceedings of EPE-PEMC 2000 Conference*, Kosice.

[7] J. H. Lee, J. S. Ko, S. K. Chung, D. S. Lee, J. J. Lee and M. J Young, "Continuous Variable Structure Controller for BLDSM Position Control with Prescribed Tracking Performance", *IEEE Trans. on Industrial Electronics*, vol. 41, No. 5, Oct. 1994.

[8] S. J. Dodds and J. Vittek, "Sliding mode control of PMSM Electric Drives with Flexible Coupling", *Proceedings of EDPE 2007*, Portoroz, Slovenia, CD-Rom.

[9] P. Korondi, H. Hashimoto and V. I. Utkin, "Direct Torsion Control of Flexible Shaft in an Observer-based Discrete Sliding Mode", *IEEE Transactions on Industrial Electronics*, y. 1998, vol. 45, No. 2, pp.291-296.

[10] S. J. Dodds, "Hyper Sliding Mode Control: A Novel Approach Achieving Robustness with Model Order Uncertainty", *Proceedings of AC&T 2006*, SCOT, University of East London, UK, ISBN 0-9550008-1-5.

[11] S. J. Dodds, "Settling time formulae for the design of control systems with linear closed loop dynamics", *Proceedings of AC&T 2008*, SCOT, University of East London, UK, ISBN 0-9550008-5-8.

[12] S. J. Dodds, "Observer Based Robust Control", *Proceedings of AC&T 2007*, SCOT, University of East London, UK, ISBN 0-9550008-3-1.

Sliding Mode Vector Control of PMSM Drives with Minimum Energy Position Following

Stephen J. Dodds

School of Computing and Technology, University of East London, UK, e-mail: *stephen.dodds@spacecon.co.uk*

Abstract—The original contribution of this paper is the direct use of a six switch inverter as the switching element of a multivariable sliding mode controller to achieve vector control for permanent magnet synchronous motor drives precisely realising a prescribed dynamic response to the rotor speed reference inputs. The extreme robustness against changes in the mechanical load parameters and external disturbance torques enables the user to set up the drive without any controller tuning, the only information required being the prescribed position step response settling time. The only information needed to design the controller is the relative degree (i.e., rank) of the plant with respect to the controlled outputs, i.e., the rotor speed and the direct axis current vector component. An outer position control loop is closed and a zero dynamic lag pre-compensator applied to achieve precise following of a pre-planned rest-to-rest manoeuvre that minimises the frictional energy loss for a given position change and manoeuvre time. Simulations predict that despite no knowledge of the load moment of inertia or the viscous friction coefficient, a) the precisely defined closed loop dynamics of the position step response is attained, b) precise following of pre-planned rest to rest manoeuvres is attained and c) step load torques cause negligible transient position errors.

Keywords—control of drive, highly dynamic drive, motion control, non-linear control, robustness, sliding mode control, synchronous motor, vector control.

I. INTRODUCTION

Standard vector controllers for permanent magnet synchronous motor (PMSM) drives employ proportional integral (PI) controllers for a) the direct axis stator current vector component, with zero reference input, and b) the rotor speed. For motion control applications, a third PI controller is employed to close the position control loop with the speed reference input as the control variable. These three PI controllers need tuning upon commissioning to attain satisfactory performance and re-adjustment when the dynamics of the mechanical load is changed. In general, this is time consuming and requires considerable experience and knowledge of control theory on the part of the user. Even once the control system has been satisfactorily tuned, a) a step external load torque usually causes a considerable transient position error and b) the dynamics of the closed loop system is not known in the form of a differential equation or transfer function. If following of a pre-planned position reference function is needed, then it would be possible to keep the dynamic following errors (i.e., the dynamic lag) to reasonable proportions by employing a derivative feed-forward pre-compensator in the path of the

reference input, but the derivative weightings would have to be determined by trial and error, again a time consuming task. All the aforementioned disadvantages are removed by means of the relatively simple vector sliding mode controller introduced in this paper.

The basic single input, single output sliding mode controller [1] is a bang-bang state feedback controller where the state variables are phase variables consisting of the controlled output and its derivatives, the order of the highest derivative being equal to the plant rank. The trajectories in the phase space are directed towards the switching boundary from both sides in the operational region of the plant. Under these circumstances, after the phase trajectory first reaches the switching boundary, the control variable switches continually, in theory at infinite frequency with a continually varying mark-space ratio such that the trajectory is constrained to *slide* in the boundary. The equation of the boundary is then the differential equation describing the behaviour of the closed loop system and the fact that this is independent of the plant and its disturbances explains the extreme robustness of this method. In practice, the control variable cannot switch at infinite frequency due to the finite sampling frequency of the digital processor and the presence of unmodelled plant dynamics, resulting in oscillation of the phase trajectory about the switching boundary with a zigzag motion, but if the sliding mode controller is correctly designed, the trajectory remains in a sufficiently close neighbourhood of the boundary for the closed loop dynamics to closely approximate the ideal one.

It is commonly stated that the aforementioned rapid switching of the control variable, referred to as control chatter, is undesirable due to the resulting mechanical vibrations and possible heating effects in the actuators, and various measures such as a boundary layer [1, 2] have been introduced for its elimination. In electric drive applications, however, the stator voltages, which are the control variables, have to operate in a switched mode to minimise power dissipation in the power electronic switches. For this reason, it is proposed to retain the basic form of sliding mode controller with switched control variables.

In the application of this paper, there are two output variables to be controlled by the sliding mode controller: the rotor speed, ω_r, and the direct axis stator current vector component, i_d. The three stator voltages, u_a, u_b and u_c, are the physical control variables but since a PMSM may be represented by one of its two-phase equivalent models, there are only two degrees of freedom of control represented by the two stator voltage vector components, so in effect there are only two control variables. Since there are more than one output variable to be controlled simultaneously, the term, *multivariable* sliding mode control is used. For the application in hand, the plant is of the general form:

$$\begin{cases} \dot{\mathbf{x}} = \mathbf{f}(\mathbf{x}, u_1, u_2) \\ y_1 = h_1(\mathbf{x}) \\ y_2 = h_2(\mathbf{x}) \end{cases} \tag{1}$$

where $\mathbf{x} \in \mathfrak{R}^n$ is the state vector, u_1, u_2 are the control variables, and y_1, y_2 are the controlled output variables. The functions on the right hand sides are smooth. Two switching functions, $S_1\left(y_{1\,dem}, y_1, \dot{y}_1, \ldots, y_1^{(r_1-1)}\right)$ and $S_2\left(y_{2\,dem}, y_2, \dot{y}_2, \ldots, y_2^{(r_2-1)}\right)$ are formed, where $y_{1\,dem}$ and $y_{2\,dem}$ are the demanded outputs that may be time varying and r_i is the relative degree (i.e., the rank) of the plant with respect to the output, y_i, $i = 1, 2$ such that over the range of normal plant operating states, the bang-bang control laws

$$\begin{cases} u_1 = u_{1\,max} \, \text{sgn}(S_1) \\ u_2 = u_{2\,max} \, \text{sgn}(S_2) \end{cases} \tag{2}$$

drive the sub-states, $\mathbf{y}_1 = \left(y_1, \dot{y}_1, \ldots, y_1^{(r_1-1)}\right)$ and $\mathbf{y}_2 = \left(y_2, \dot{y}_2, \ldots, y_2^{(r_2-1)}\right)$ towards the two switching boundaries:

$$\begin{cases} S_1\left(y_{1\,dem}, y_1, \dot{y}_1, \ldots, y_1^{(r_1-1)}\right) = 0 \\ S_2\left(y_{2\,dem}, y_2, \dot{y}_2, \ldots, y_2^{(r_2-1)}\right) = 0 \end{cases} \tag{3}$$

from both sides, where

$$\text{sgn}(S_i) \overset{\Delta}{=} \{-1 \text{ for } S_i < 0, \ 0 \text{ for } S_i = 0, \ +1 \text{ for } S_i > 0\}, i = 1, 2$$

. $u_{1\,max}$ and $u_{2\,max}$ are the control saturation magnitude limits. In the sliding mode, (3) is satisfied and these two differential equations describe the behaviour of the desired closed loop system.

Fig. 1 shows the general structure of the speed control loop incorporating the multivariable sliding mode controller.

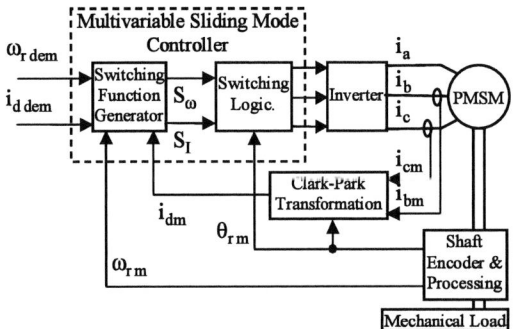

Fig. 1. General structure of sliding mode vector speed control system.

Here, i_a, i_b and i_c are the physical stator currents while i_{bm} and i_{cm} are stator current measurements, θ_{rm} is the rotor angle measurement, ω_{rm} and $\omega_{r\,dem}$ are, respectively, the rotor speed measurement and demand, i_{dm} and $i_{d\,dem}$ are, respectively measured and demanded direct axis stator current vector component, and finally S_ω and S_i are, respectively, the rotor speed and direct axis current switching function evaluations defined in the following section.

The system of Fig. 1 is designed to yield a linear closed loop dynamics with a known closed loop transfer function. Then a linear outer loop position controller is applied with $\omega_{r\,dem}$ as the control variable, also yielding a linear closed loop dynamics with a known transfer function of the general form

$$\frac{\theta_{rm}(s)}{\theta'_{r\,dem}(s)} = \left[\frac{1}{1 + s\dfrac{T_s}{1.5(1+n)}} \right]^n \tag{4}$$

where θ_r is the rotor angle, $\theta'_{r\,dem}$ is the demanded rotor angle (i.e., the reference input) and T_s is the settling time of the step response (5% criterion) [3]. In the example presented here, the order is $n = 3$. If $\theta_{r\,dem}(t)$ is a prescribed position function that has to be followed then this can be done with nominally zero dynamic lag by introducing a derivative feed forward pre-compensator in the reference input channel having the transfer function

$$\frac{\theta'_{r\,dem}(s)}{\theta_{r\,dem}(s)} = \left[1 + s\frac{T_s}{1.5(1+n)} \right]^n \tag{5}$$

this could be implemented either by pre-calculation of the reference input derivatives or by software differentiation.

As a contribution towards improving the environment, a closed loop control system has been presented in [4] aimed at minimising the energy loss in position control applications with zero initial and final velocities and a specified manoeuvre time, but this, although exhibiting robustness properties, does require the motor parameters for its design. The control system presented in this paper achieves the same without the need for motor parameters by using reference position functions of the same form as the position responses yielded by the control system of [4]. Maximising the manoeuvre time in general minimises the energy losses because the frictional power loss in driven mechanisms increases with the square of the velocity. An example of an application is a set of positioning mechanisms on a production line in the manufacturing industry that often needs coordination. If a given mechanism reaches the desired position before it is needed, then it will consume more energy than if it is controlled to reach the desired position just in time. Fig 2 shows the form of the position, $\theta_r(t)$, the velocity, $\omega_r(t)$ and the acceleration, $\alpha_r(t)$, of this minimum energy profile. It should be noted that it was proven in [4] that maximising the acceleration and

2560

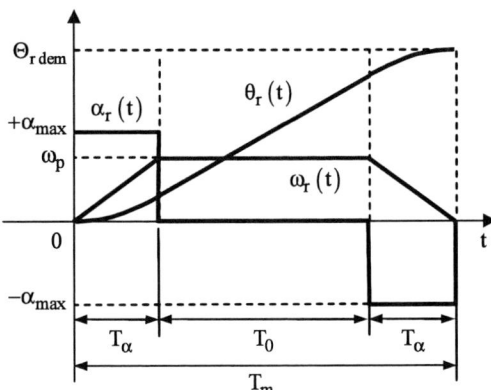

Fig. 2. Minimum energy rest-to-rest manoeuvre profile.

deceleration magnitudes minimises the frictional energy loss by minimising the magnitude of the peak velocity, ω_p, for a given position demand, $\Theta_{r\,dem}$, and manoeuvre time, T_m, but that this has not been rigorously proven to *absolutely* minimise the energy expenditure but has been chosen because a) it yields a substantially lower energy expenditure for a specified manoeuvre time than possible with conventional linear control techniques and b) it has a precisely defined and truly finite settling time that can be easily adjusted for coordination with other controlled degrees of freedom in a motion control system.

II. SLIDING MODE VECTOR CONTROL SYSTEM DESIGN

A. Model of PMSM and Mechanical Load

Although the parameters of the motor and its mechanical load are not needed for the control system design, the complete model is given here as it is used for the simulations and the relative degree determinations. The state differential equations

$$di_d/dt = -Ai_d + B\omega_r i_q + Fu_d \tag{6}$$

$$di_q/dt = -\left(Ci_d + E\right)\omega_r - Di_q + Gu_q \tag{7}$$

$$d\omega_r/dt = \left(H + Ki_d\right)i_q - M\Gamma_{Le} - F_v\omega_r \tag{8}$$

$$d\theta_r/dt = \omega_r \tag{9}$$

model the PMSM in the synchronously rotating d-q coordinate system, together with its mechanical load, where i_d, i_q and u_d, u_q are the stator current and voltage vector components, which are the control variables, ω_r and θ_r are the rotor angular velocity and angle, and Γ_{Le} is the external load torque. The constant coefficients are:

$$\begin{cases} A = R_s/L_d; \quad B = pL_q/L_d; \quad C = pL_d/L_q; \\ D = R_s/L_q; \quad E = p\Psi_{PM}/L_q; \quad F = 1/L_d; \quad G = 1/L_q; \\ H = 3p\Psi_{PM}/(2J); \quad K = 3p\left(L_d - L_q\right)/(2J); \quad M = 1/J \end{cases}$$

where Ψ_{PM} is the permanent magnet flux, R_s is the stator resistance, L_d and L_q are the direct and quadrature axis inductances, p is the number of pole pairs, $J = J_r + J_L$, where J_r is the rotor moment of inertia and J_L is the load moment of inertia referred to the rotor, and F_v is the viscous friction coefficient. It is evident from (8) that the control torque delivered by the motor is:

$$\Gamma_c = J\left(H + Ki_d\right)i_q \tag{10}$$

The controlled outputs are

$$y_1 = i_d \tag{11}$$

$$y_2 = \omega_r \tag{12}$$

B. Plant Relative Degrees

The relative degree of the plant with respect to the output, y_i is defined as the lowest order time derivative of y_i with direct algebraic dependence on any control variable.

Differentiating (11) w.r.t. time and substituting for di_d/dt using (6) yields:

$$\dot{y}_1 = -Ai_d + B\omega_r i_q + Fu_d \tag{13}$$

The RHS is algebraically dependent on u_d and therefore the rank with respect to y_1 is only

$$r_1 = 1 \tag{14}$$

Differentiating (7) w.r.t. time and substituting for $d\omega_r/dt$ using (8) yields

$$\dot{y}_2 = \left(H + Ki_d\right)i_q - M\Gamma_{Le} - F_v\omega_r \tag{15}$$

Neither of the two control variables appears on the r.h.s. and so (15) is differentiated and substitutions made for di_d/dt, di_q/dt and $d\omega_r/dt$ using (6), (7) and (8):

$$\begin{aligned} \ddot{y}_2 &= \left(H + Ki_d\right)di_q/dt + Ki_q\,di_d/dt - Md\Gamma_{Le}/dt - F_v\,d\omega_r/dt \\ &= \left(H + Ki_d\right)\left(-Ai_d + B\omega_r i_q + Fu_d\right) \\ &\quad + Ki_q\left(-\left(Ci_d + E\right)\omega_r - Di_q + Gu_q\right) - Md\Gamma_{Le}/dt \\ &\quad - F_v\left(\left(H + Ki_d\right)i_q - M\Gamma_{Le} - F_v\omega_r\right) \end{aligned}$$

In this case there are direct algebraic dependencies on both u_d and u_q. Therefore the rank with respect to y_2 is

$$r_2 = 2 \tag{16}$$

C. Switching Function Generator

The switching functions for the direct axis stator current vector component and the rotor speed will both be chosen as linear, which is usual practice [1].

1) Direct axis stator current: Since the rank is only $r_1 = 1$, no output derivatives are necessary and the switching function is trivial:

$$S_I = i_{d\,dem} - i_{dm} \tag{17}$$

2561

yielding a closed loop system without dynamics (i.e., of zero order), so $S_I \to 0 \Rightarrow i_{dm} = i_{ddem}$.

2) Rotor speed: Since the rank in this case is $r_2 = 2$, the first output derivative is needed and the switching function is as follows:

$$S_\omega = \omega_{rdem} - \left(\omega_{rm} + \frac{T_{s\omega}}{3}\dot\omega_{rm} \right) \tag{18}$$

where $T_{s\omega}$ is the step response settling time of the first order linear closed loop system in the sliding mode which has the transfer function

$$S_\omega \to 0 \Rightarrow \frac{\omega_r(s)}{\omega_{rdem}(s)} = \frac{1}{1 + s\,T_{s\omega}/3} \tag{19}$$

and To avoid amplification of high frequency components of measurement noise, an approximate derivative is formed by software differentiation combined with a first order low pass filter, so $\dot\omega_{rm}$ is replaced by an estimate given by:

$$\dot\omega_{re}(s) = \frac{s}{1 + sT_f}\omega_{rm}(s) \tag{20}$$

where T_f is the filtering time constant chosen sufficiently small for (19) to be a close approximation to the real closed loop performance, ensured by $T_f \ll T_{s\omega}/3$.

D. Switching Logic

u_d (controlling i_d) and u_q (controlling ω_r), switch between time varying values, $U_{di}(t)$ and $U_{qi}(t)$, dependent on the Clark-Park transformation matrix:

$$\begin{bmatrix} U_{di}(t) \\ U_{qi}(t) \end{bmatrix} = \begin{bmatrix} \cos(\omega_r t) & \frac{1}{\sqrt 3}\sin(\omega_r t) & -\frac{1}{\sqrt 3}\sin(\omega_r t) \\ -\sin(\omega_r t) & \frac{1}{\sqrt 3}\cos(\omega_r t) & -\frac{1}{\sqrt 3}\cos(\omega_r t) \end{bmatrix} \begin{bmatrix} U_{ai} \\ U_{bi} \\ U_{ci} \end{bmatrix} \tag{21}$$

where the subscript, i, denotes the ith active switching state of the inverter and:

$$\begin{bmatrix} U_{ai} \\ U_{bi} \\ U_{ci} \end{bmatrix} = \begin{bmatrix} 0 \\ +U_s \\ -U_s \end{bmatrix}_{i=1} = \begin{bmatrix} +U_s \\ 0 \\ -U_s \end{bmatrix}_{i=2} = \begin{bmatrix} +U_s \\ -U_s \\ 0 \end{bmatrix}_{i=3}$$
$$= \begin{bmatrix} 0 \\ -U_s \\ +U_s \end{bmatrix}_{i=4} = \begin{bmatrix} -U_s \\ 0 \\ +U_s \end{bmatrix}_{i=5} = \begin{bmatrix} -U_s \\ +U_s \\ 0 \end{bmatrix}_{i=6} \tag{22}$$

Fig. 3 shows the plot of $U_{di}(t)$ and $U_{qi}(t)$ for each of the six switching states. The electrical angle is divided into twelve segments over which $U_{di}(t)$ and $U_{qi}(t)$, $i = 1, 2, 3, 4, 5, 6$ keep the same sign and Table I shows the

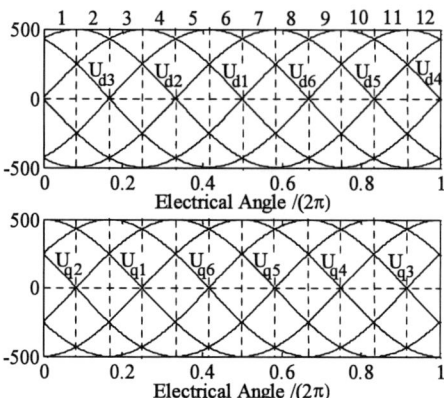

Fig. 3. Continuously varying switching voltage levels for d and q axes.

sign combinations in each of these segments for each of the six switching states. It is evident that for each segment, all four combinations of sign pairs of $U_{di}(t)$ and , i.e., $++, +-, -+, --$, are possible by appropriate choice of the inverter switching state and therefore simultaneous sliding mode control of i_d and ω_r is possible. The choice of switching state, however, is not unique and in this initial investigation, the choice yielding the largest control voltage magnitude is made, since this maximises the region of the switching manifold over which the condition for sliding motion holds, i.e., the phase trajectories are driven towards the manifold in a close neighbourhood of the manifold. This, of course, maximises the robustness of the control system but in a future investigation, it would be of interest to consider the lower voltage levels to obtain smoother stator current waveforms.

TABLE I.
CONTROL VARIABLE SIGNS FOR EACH SWITCHING STATE

	Inverter Switching State											
	Sw₁		Sw₂		Sw₃		Sw₄		Sw₅		Sw₆	
	Sign Pair		Sign Pair		Sign Pair		Sign Pair		Sign Pair		Sign Pair	
Segment	U_d	U_q	U_d	U_q	U_d	U_q	U_d	U_q	U_d	U_q	U_d	U_q
1	+	+	+	+	+	−	−	−	−	−	−	+
2	+	+	+	−	+	−	−	−	−	+	−	+
3	+	+	+	−	−	−	−	−	−	+	+	+
4	+	−	+	−	−	−	−	+	−	+	+	+
5	+	−	−	−	−	−	−	+	+	+	+	+
6	+	−	−	−	−	+	−	+	+	+	+	−
7				+	+	+	+	+	+	−		
8	−	−	−	+	−	+	+	+	+	−	+	−
9	−	−	−	+	+	+	+	+	+	−	−	−
10	−	+	−	+	+	+	+	−	+	−	−	−
11	−	+	+	+	+	+	+	−	−	−	−	−
12	−	+	+	+	+	−	+	−	−	−	−	+

2562

The switching state, i, is chosen to maximise

$$M = S_\omega U_{qi}(t) + S_i U_{di}(t) \qquad (23)$$

This ensures that the chosen $U_{qi}(t)$ has the same sign as S_ω given by (18) and that $U_{di}(t)$ has the same sign as S_I given by (17), as required to direct the phase trajectories towards the switching manifold. It follows that the maximum d-q axis control voltage magnitudes are automatically selected.

III. OUTER POSITION LOOP CONTROL DESIGN

The outer position loop controller is chosen as an IP controller with position feedback gain, K_θ, and integral gain, K_I, as shown in Fig. 4.

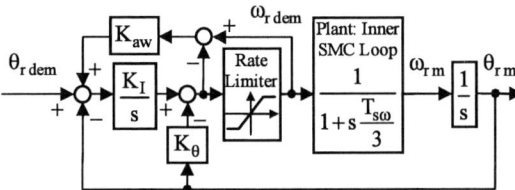

Fig. 4. Outer position control loop.

The plant for the position controller has the rotor speed demand, $\omega_{r\,dem}$, as the control variable and the measured rotor angle, θ_{rm}, as the controlled output. The integral term is included to ensure that no steady state errors can occur due to imperfect operation of the sliding mode speed controller. The anti-windup loop is included to implement the rate limit for energy saving purposes, when needed, if the system is operated by simply applying step reference position demands, The nominal plant transfer function consists of (19) followed by a kinematic integrator, as shown. Applying Mason's formula to this figure assuming that the rate limiter and therefore the anti-windup loop are inactive yields the following closed loop transfer function:

$$\frac{\theta_{rm}(s)}{\theta_{r\,dem}(s)} = \frac{3K_I/T_{s\omega}}{s^3 + (3/T_{s\omega})s^2 + (3K_\theta/T_{s\omega})s + 3K_I/T_{s\omega}} \qquad (22)$$

It is evident that this controller can be designed by pole assignment with $T_{s\omega}$, K_θ and K_I as adjustable parameters to yield the ideal closed loop transfer function of (4) with $n = 3$:

$$\frac{\theta_{rm}(s)}{\theta_{r\,dem}(s)} = \frac{(6/T_{s\theta})^3}{s^3 + 3(6/T_{s\theta})s^2 + 3(6/T_{s\theta})^2 s + (6/T_{s\theta})^3} \qquad (24)$$

Equations for the controller parameters in terms of the required settling time are then obtained by equating the corresponding transfer function coefficients of (22) and (23):

$$\boxed{T_{s\omega} = T_{s\theta}/6} \quad \boxed{K_\theta = 6/T_{s\theta}} \quad \boxed{K_I = 12/T_{s\theta}^2} \qquad (25)$$

IV. ZERO DYNAMIC LAG PRE-COMPENSATOR

To enable time varying position demands to be followed with negligible dynamic lag, a derivative feed-forward pre-compensator based on (5) is used with $n = 3$:

$$\frac{\theta'_{r\,dem}(s)}{\theta_{r\,dem}(s)} = \left(1 + s\frac{T_s}{6}\right)^3 = 1 + 3.\frac{T_s}{6}s + 3.\left(\frac{T_s}{6}\right)^2 s^2 + \left(\frac{T_s}{6}\right)^3 s^3 \qquad (26)$$

To ensure smooth operation of the software differentiation, several different algorithms are possible. The one chosen here approximates individual derivatives by introducing a high pass filter as in (20) so that (26) is replaced by:

$$\frac{\theta'_{r\,dem}(s)}{\theta_{r\,dem}(s)} = 1 + Q(s)\big[3 + Q(s)[3 + Q(s)]\big] \qquad (27)$$

where $Q(s) = \dfrac{T_s}{6}.\dfrac{s}{1 + sT_f}$.

V. SIMULATIONS

The parameters of the non-salient PMSM are as follows: Rated voltage: $U_0 = 430\,\mathrm{V}$; DC power supply voltage level: $U_s = U_0$; Power Rating: $P_r = 12000\,\mathrm{W}$; Base speed: $\omega_b = 150\,\mathrm{rad/s}$; Max. torque: $\Gamma_{max} = 40\,\mathrm{Nm}$; $\Psi_{PM} = 0.38\,\mathrm{Wb}$; $L_d = L_q = 5.4e-3\,\mathrm{H}$; $R_s = 0.1\,\Omega$; $p = 5$; $J_r = 0.03\,\mathrm{Kg\,m^2}$.

The controller parameters are set to the following values: Sampling frequency: $f_s = 50\,\mathrm{kHz}$. This sets the maximum switching frequency of the inverter switches. During the sliding motion, the actual period between successive switches will be integral multiples of $1/f_s = 20\,\mu s$; $T_{s\theta} = 0.5\,s$. The Matlab-Simulink simulations are carried out with Euler integration at a fixed step of $20\,\mu s$ so that the control and pre-compensation algorithms are as they would be in a hardware implementation. The filtering time constant used for the software differentiation is $T_f = 100\,\mu s$.

The coefficient of viscous friction yields a frictional power loss equal to 80% of the rated load power at the base speed: $F_e = 0.8P_r/\omega_b^2$.

The response, $\theta_r(t)$, to a step rotor angle demand of $\theta_{r\,dem} = 10\,\mathrm{rad}$ with a load moment of inertia of $J_L = 0.5\,\mathrm{kg\,m^2}$ and a step external load torque of $\Gamma_{Le} = 100\,\mathrm{Nm}$ applied at $t = 0.5\,s$ is shown in Figure 5. This is superimposed upon a plot of the step response of the ideal system with transfer function (23) and no difference between the two responses is visible on this scale, so the error between these responses is displayed in Fig. 6. This indicates a control accuracy better than 0.01 rad. and a remarkably high rejection of the large step load torque. Fig. 6 displays the electromagnetic torque developed by the motor, which clearly shows the acceleration and deceleration torque followed by a step in the torque that counteracts the external load torque. The oscillations are due to the power electronic switching.

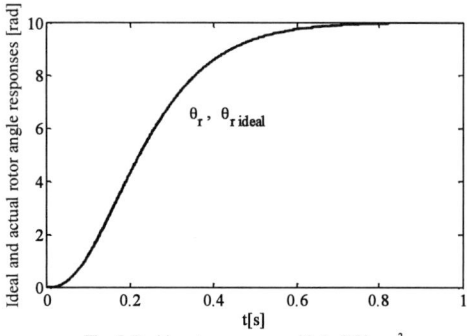

Fig. 5. Position step response with J_L=0.5 kg m^2.

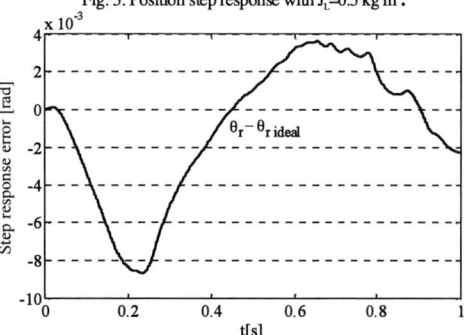

Fig. 6. Position error in step response (J_t=0.5 kg m^2).

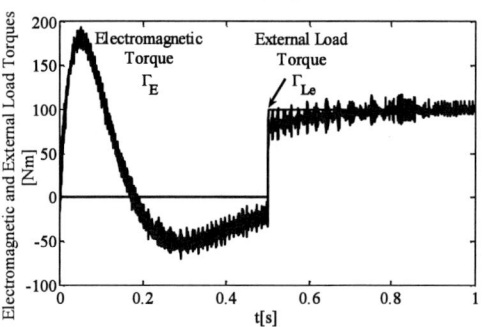

Fig. 7. External load torque and electromagnetic torque (J_t=0.5 kg m^2).

The operation of the new multivariable sliding mode controller is further checked by the switching function evaluations plotted in Fig. 8. Their behaviour resembles random walks and this is due to the discrete nature of the switching logic described in subsection II, C. The magnitude of variation of $S_\omega(t)$ is very small compared with the peak rotor speed of about 3 rad/s (i.e., the

maximum slope of $\theta_r(t)$ in Fig. 5. Also the magnitude of variation of $S_I(t)$ is very small compared with the stator current magnitudes which are of the order of tens of Amperes. This indicates satisfactory operation.

As would be expected with correct operation of the sliding mode controller, the control variables, in this case the stator voltages, would be expected to switch so frequently that on the 1 second time scale the graph would be completely filled between the control saturation limits. This proved to be the case and these graphs are not reproduced here since they convey no useful information beyond the confirmation of the rapid switching. So the aforementioned low pass filter with time constant, $T_f = 100\,\mu s$, was applied to the stator voltages switching between $\pm U_s$ to yield $u_{qf}(s) = \left[1/(1+sT_f)\right]u_q(s)$, $q = a,b,c$. The result is shown in Fig. 9 and is a good approximation to the equivalent control [1], which is the fictitious continuous control that if applied would hold the plant state precisely on the switching manifold. These filtered voltages show as three-phase oscillating variables, which occur due to the oscillating elements of the Clark-Park transformation matrix. The acceleration and deceleration is reflected by the initial increase in the frequency of the oscillations and the final decrease in the frequency. The corresponding unfiltered stator currents are shown in Fig. 10 and the expected changes in frequency are again visible. The sudden increase in amplitude of the three phase alternating currents at $t = 0.5\,s$ of course, produces the electromagnetic torque to counteract the step load torque.

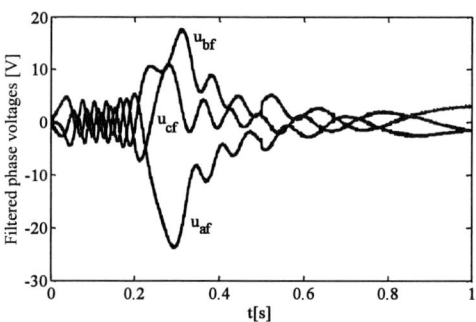

Fig. 9. Switched Phase voltages filtered with 0.1 s time constant (J_t=0.5 kg m^2).

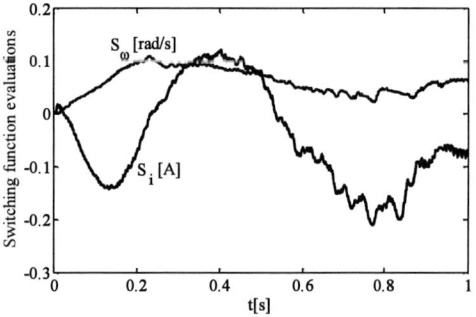

Fig. 8. Sliding mode control switching function values (J_t=0.5 kg m^2).

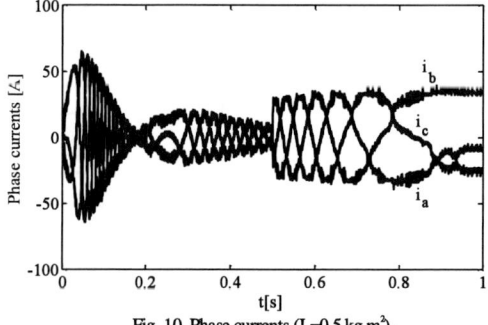

Fig. 10. Phase currents (J_t=0.5 kg m^2).

The corresponding stator current vector components of Fig. 11 show the effectiveness of the sliding mode controller in maintaining the direct current vector component, $i_d(t)$ near zero value, as required for the vector control. The quadrature axis current vector component, $i_q(t)$, reflects the electromagnetic torque variation of Fig. 7, as expected. The high frequency oscillations in the currents are due to the power electronic switching.

The robustness of the control system with respect to uncertainties in the plant parameters is already evident by virtue of the fact that the plant parameters are not needed for the control system design, but this fact is further substantiated by doubling the load moment of inertia to $J_L = 1\,\text{kg}\,\text{m}^2$ while the external load torque remains the same, i.e., a step of $\Gamma_{Le} = 100\,\text{Nm}$ applied at $t = 0.5\,\text{s}$ The result is indistinguishable from that of Fig. 5 and is therefore not presented but Fig. 12 shows the error between the actual and ideal step responses which, as for Fig. 6, indicates an accuracy better than 0.01 rad. and a very effective counteraction of the large step load torque.

Fig. 13 shows the external load torque and the electromagnetic torque and when this is compared with Fig. 7, the automatic increase in the magnitudes of the acceleration an deceleration torques to cater for the increased moment of inertia can be seen.

Next attention is focussed on control to minimise frictional energy loss, which is so important in the quest to improve the environment. Fig. 14 shows a comparison between two position step responses, one with a rate magnitude limit of $\omega_{r\,max} = 200\,\text{rad}/\text{s}$ which is not reached with the settling time of $T_{s\theta} = 1\,\text{s}$, and the other with a rate limit of $\omega_{r\,max} = 50\,\text{rad}/\text{s}$ which is clearly exercised. Fig. 15 shows the corresponding speed-time profiles and when the rate limit is being exercised, the form of the response resembles the minimum energy profile of Fig. 2. The frictional energy loss graphs of Fig. 16 indicate a very large energy saving of about 60%. When the rate limiting is brought into play, however, the settling time will be unknown and the pre-compensator of section IV would not be able to yield negligible dynamic lag. This rate limiting feature should therefore be regarded as a safety feature not normally exercised. The rate limiting to minimise frictional energy loss can be combined with high precision control with known manoeuvre times by employing a reference position function generator together with the zero dynamic lag pre-compensator to yield the motion profile of Fig. 2. Fig. 17 shows the result with the maximum acceleration magnitude set to $\alpha_{max} = 40\,\text{rad}/\text{s}/\text{s}$ (relatively small to clearly show the constant acceleration and deceleration segments. The manoeuvre time is set to $T_m = 1\,\text{s}$. Fig. 18 shows the error between the demanded position function and the achieved position and this indicates a similar accuracy to that obtained previously in the step responses, i.e., better than 0.01 rad.

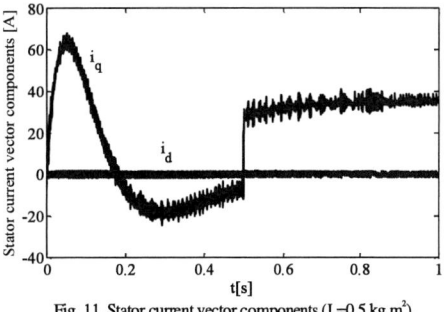

Fig. 11. Stator current vector components (J_t=0.5 kg m^2).

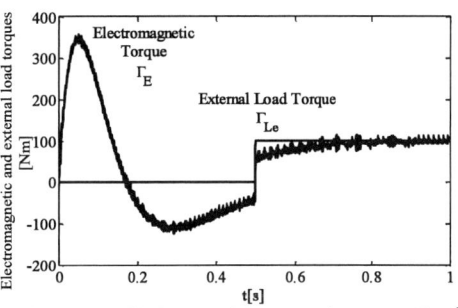

Fig. 12. Position error in step response (J_t=1 kg m^2).

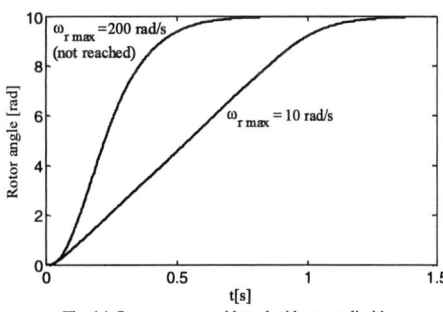

Fig. 13. External load torque and electromagnetic torque (J_t=1 kg m^2).

Fig. 14. Step responses with and without rate limiting.

Finally, a set of minimum energy rest-to-rest position manoeuvres is presented together with the frictional energy losses for a set of different manoeuvre times, the position demand being 10 rad in each case. The maximum acceleration and deceleration magnitude is set to $\alpha_{max} = 250\,\text{rad}/\text{s}/\text{s}$ which yields a maximum acceleration and deceleration torque of nearly $\Gamma_{max} = 250\,\text{Nm}$. Fig. 19 shows the position responses.

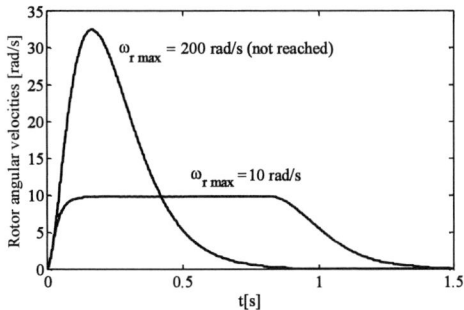

Fig. 15. Rotor speeds for step responses with and without rate limiting.

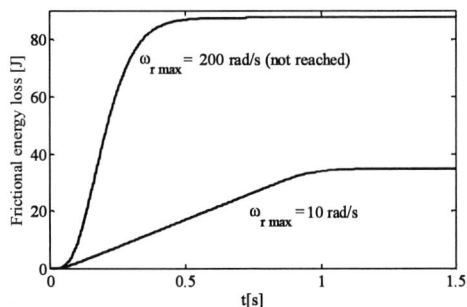

Fig. 16. Frictional energy losses with and without rate limiting.

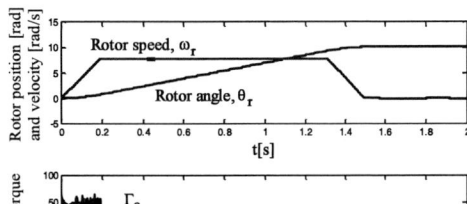

Fig. 17. Position and velocity response with electrical torque.

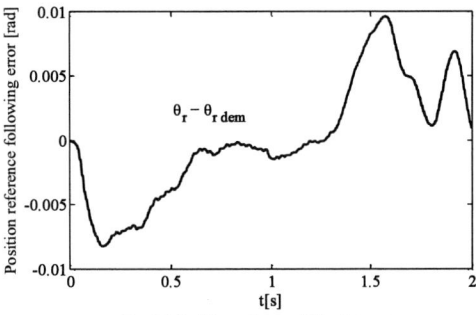

Fig. 18. Position reference following error.

The manoeuvre time of $T_m = 0.4\,s$ yields $T_0 = 0$ in Fig. 2 and the result is very close to the classical time optimal control for a double integrator plant [5]. The large penalty that must be paid in frictional energy loss for this shortest manoeuvre time is evident in Fig. 20. Accepting a longer settling time saves much energy and this is evident in Fig. 20.

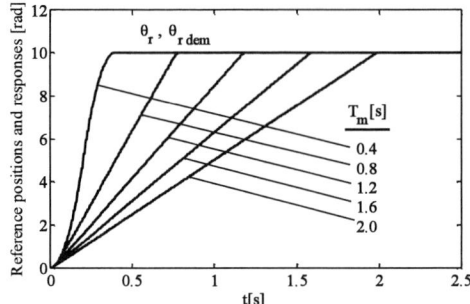

Fig. 19. Position references and responses for different manoeuvre times.

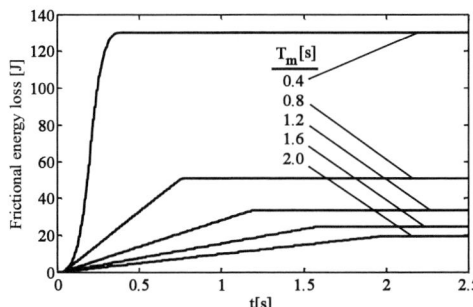

Fig. 20. Frictional energy losses for manoeuvres of Fig. 15.

VI. OVERALL CONCLUSIONS AND RECOMMENDATIONS

The simulations of the new sliding mode vector controller presented in this paper indicate very high robustness against external load torques and changes in the load moment of inertia. The achievement of the prescribed closed loop dynamic performance enables the drive to be used with a derivative feed-forward pre-compensator to follow arbitrarily time varying demanded position profiles, including the minimum energy one simulated. The drive would also be extremely simple to commission since there are no controller parameters to adjust. In view of the above, it is strongly recommended to carry out experiments with the new drive to identify any problems not revealed by the simulations. It would also be of interest to compare the qualities of the stator currents with standard PMSM vector controllers and consider variations on the switching strategy presented in section II C that yield smoother current waveforms, possibly by admitting the seventh switching state giving zero stator voltages.

REFERENCES

[1] V. I. Utkin, *Sliding Modes in Control Optimization*, Springer-Verlag, 1992, ISBN 9780387535166.

[2] J. Vittek and S. J. Dodds, *Forced Dynamics Control of Electric Drives*, EDIS, Žilina University Publishing Centre, 2003, ISBN 80-8070-087-7.

[3] S. J. Dodds, "Settling time formulae for the design of control systems with linear closed loop dynamics", *Proceedings of AC&T 2008*, SCOT, University of East London, UK, ISBN 0-9550008-5-8.

[4] S. J. Dodds, G. Sooriyakumar and R. Perryman, "Minimum energy forced dynamic position control of PMSM drives", *Proceedings of WSEAS Conference on Energy and Environment*, Cambridge, UK, 2008,

[5] E. P. Ryan, *Optimal Relay and Saturating Control System Synthesis*, Peter Peregrinus, ISBN: 0-90604-856-7 & 978-0-90604-856-6.

Author Index

A

Abbatelli, L. 61
Abbey, Chad 2178
Abdelhamid, Tamer H. 606
Abdellatif, Meriem 938
Abe, Seiya 393
Abourida, Simon 1077
Abroshan, Mohammad 1117
Abroushan, Mohammad 365
Abuishmais, Ibrahim 867
Abu-Rub, Haithem 1084, 1382
Adabi, Jafar 718, 903
Adamidis, Georgios 1840
Adamowicz, Marek 1729
Adzic, Evgenije 1957
Ahmadi, Muhammad 1847
Ahmed, M.M.R. 1866, 2472
Ahn, Jonng-Bo 2524
Ahn, ong-Bo 2492
Ait-Ahmed, Mourad 1740
Akhondi, Hamidreza 2071
Alarcón, E. 2108
Albert, Laurent 2037
Al-Diab, Ahmad 1710
Alexandrov, Alexandar 787
Al-Khayat, Nazar 2150
Allard, Bruno 2457
Al-Othman, A. K. 606
Amelon, Nicolas 1740
Anaya-Lara, O. 1784, 1941
Andersen, Michael A.E. 127
Ando, Kenji 614
Andrzejewski, Andrzej 1090
Areerak, K-N 2049
Arellano-Padilla, J. 1173
Arellano-Padilla, Jesus 769
Armstrong, S. 1688
Aroudi, A. El 2108
Aroudi, Abdelali El 2115, 2120
Arshad, Waqas M. 867
Asher, G. M. 2261
Asher, G.M. 2049
Asher, Greg 2300
Asiltürk, Ilhan 967
Aurel, Campeanu 893
Averberg, Andreas 213

B

Baalbergen, Freek J.F. 2170
Baghaee, H.R. 313, 629, 750
Bahri, I. 1365
Bailey, Chris 76
Bakas, Panagiotis 1840
Balazovic, Peter 1402

Balouktsis, Anastasios 1840
Baluta, Gh. 2043
Ban, Drago 818
Barai, Mukti 674
Barakat, Georges 1834, 810
Baranowski, Jerzy 1432, 1446
Barbosa, Fabián H. 637
Barlik, Roman 84
Barrero, R. 1512
Bartelt, R. 521
Baskys, Algirdas 1140
Bastiani, A. 1293
Baszynski, Marcin 1779
Bauer, Pavel 422
Bauer, Pavol 2170, 2354, 2368, 2371
Beck, Hans-Peter 1243
Bekbudov, Radiy 337
Bekishev, Anatoly 663
Bélanger, Jean 1077, 1475
Belfkira, Rachid 1834
Belkhodja, I. Slama 1149
Bellini, Armando 490
Bellmunt, Oriol 731
Belter, D. 1044
Benadero, Luis 2115
Bendkowski, Lukas 250
Benecke, Marcel 1280
Benkhoris, Mohamed-Fouad 1740
Bennani, A.Ben Abdelghani 1149
Beran, Leos 782
Bergas-Jane, Joan 731
Bergogne, Dominique 2457
Berthon, A. 1542
Bertoluzzo, Manuele 1491
Bertram, Torsten 1215
Betz, R.E. 1293
Bevilacqua, Pascal 2457
Bifaretti, Stefano 1771, 490, 561
Binder, Andreas 1625, 2385
Binkowski, T. 714
Birolleau, Damien 2037
Biswas, Jayanta 674
Bizon, Nicu 621
Blahník, Vojtech 1535
Blanco, M. 2481
Blazic, B. 2510
Böcker, J. 1598
Böcker, Joachim 159
Bodora, A. 326
Bogalecka, Elzbieta 1975, 804
Bojda, Petr 422
Bolgov, Viktor 154
Bolognani, Silverio 1097
Boora, Arash A 468, 723
Bossche, A. Van den. 1326
Botan, Corneliu 1111

Author Index

Botsali, FatihM. 949
Bouafia, Abdelouahab 703
Boucherit, M.S. 1987
Bouhalli, Nadia 281
Bozhko, S.V. 2049
Brand1tetter, Pavel 1375
Braslavsky, I.Ya. 1050
Breban, Stefan 1896
Brown, Neil L. 2150
Bruno, Francois 2205
Bucher, Alexander 244, 250
Buja, Giuseppe 1491
Bukatov, Alexander 1872
Bulic, Neven 556
Buonomo, S. 61
Buss, Martin 2312

C

C., Ilioudis Vasilios 1105
Caballero, M. 1555
Cabrita, Carlos M. P. 1646
Calado, Maria R. A. 1646
Camara, M.B. 1542
Cambál, Marek 982
Candusso, Denis 734
Cartes, D.A. 793
Case, Michael James 1798
Castaing, Ambroise 2464
Catalão, J. P. S. 1682
Cédl, Marek 1593, 372
Ceglia, Gerardo 268
Cepisca, Costin 1963, 908
Cernat, Mihai 1748
Cernohorský, Josef 1009
Cerovský, Zdenk 982
Cha, Gil-Ro 383
Champenois, Gérard 2015
Chan, Paul K.W. 1688
Chang, Hao-Chi 1652
Chang, Lon-Kou 320
Chang, Yuan-Chih 456
Chante, Jean-Pierre 2457
Charaabi, L. 1365
Chekhet, Eduard 307
Chen, Anyuan 799
Chen, Junling 2000
Chen, Yonggang 1981, 2000, 405, 515
Chen, Zhe 2325
Chen, Zong-Jie 1704
Cheng, K.W.E. 576
Cherif, M. Ghodbane 1149
Cheung, N. C. 1221
Chien, Sywe-Bin 1652
Chillet, Christian 2037
Chimento, F. 61

Chlodnicki, Zdzislaw 2150
Choi, Heung-Kwan 2524
Choi, Jaeho 2498
Choi, Uk-Don 1421
Chou, Ming-Chang 1652
Chrenko, Daniela 2156
Chrzan, Piotr J. 144
Chudzik, Piotr 1568
Chun, Tae-Won 1421
Clare, J. 1326
Clare, Jon C 207
Clare, Jon C. 229, 561
Clare, Jon 1771, 307
Comnac, Vasile 1748
Cook, B.J. 1293
Cook, D. 1326
Coquery, Gérard 2192
Correa, Pablo 451, 699
Courtecuisse, Vincent 1896, 2184
Cousineau, Marc 281
Cuk, Vladimir 1426
Cychowski, Marcin 2241
Czapp, Stanislaw 2059

D

Dabroom, A.M. 1337
Dakyo, B. 1911
Dannehl, J. 444
Darie, Eleonora 1963, 908
Darie, Emanuel 1963, 908
Davey, J. 1918
De Bernardinis, Alexandre 2192
De Castro, M.R. 2126
De Gersem, Herbert 2385
de Kock, H.W. 859
De Souza, Kleber C.A. 1951
Deaconu, Sorin 1409
Debowski, Andrzej 1568, 2289
Degeratu, Sonia 893
Delaney, Kieran 2241
Demenko, Andrzej 2412
Denny, Ernest Edward 1798
Depernet, Daniel 734
Derbel, Nabil 2120
Deskur, Jan 1204, 2227
Deuse, Jacques 2184
Dheilly, Nicolas 2457
Di, Lu 2205
Dianov, Anton 1002
Díaz, Nelson L. 637
Diblík, Martin 1676
Diguet, Marc 1382
Dilevs, Guntis 1811
Dimitrakakis, Georgios S. 1301
Dinkhauser, Vincenz 1819

Author Index

Dobrucky, Branislav 1402
Dockhorn, Matthias 1734
Dodds, Stephen J. 2551, 2559
Dodds, Stephen James 2543
Doebbelin, Reinhard 1280
Doi, Nobuaki 744
Dong, P. 576
Dong, Wei 1716
Dontchev, Dimitar 787
Drábek, Pavel 1593, 372
Draganov, Denis 1610
Dubowski, Marian Roch 1090
Dudak, Juraj 2368
Dudrik, Jaroslav 295
Duerbaum, Thomas 244, 250
Dufour, Christian 1077, 1475
Duke, Richard 528
Dumur, Guillaume 1475
Durovsky, Frantisek 961
Dybkowski, Mateusz 2211, 2306
Dzieniakowski, Maciej A. 2082

E

Eberhard, Andreas 1371
Eckel, Hans-Guenter 48
Edrington, C.S. 793
Egan, Michael G. 1249
Egorov, Mikhail 1257
Ehsan, Mehdi 1847
Eilenberger, Andreas 945
Elmoctar, Mohamed Y. Ould...........................810
Empringham, Lee 207, 229, 388
Endo, Tsunehiro 924
Eno, Otu A. 114
Erceg, Gorislav 556
Etxeberria-Otadui, I. 1555

F

Fabianowski, Jan...2082
Fabijanski, Pawel....................1040, 2055, 2087
Fahrni, C. 256
Fakham, Hicham 2142
Fan, Yue..1771
Farhangi, Sh. 173
Farshad, Siamak 1575
Fedák, Viliam 2354
Fedak, Viliam 961
Fedyczak, Zbigniew.................................165, 236
Feki, Moez 2120
Fernández, Herman 1947
Fernandez-Mola, Josep-Maria 731
Ferreira, Jan Abraham 187
Ferreira, Luís António Fialho Marcelino 2076
Fetyko, Jan..961
Filchev, T. 1326

Filho, Braz Jesus Cardoso...............................1345
Filka, Roman 1402
Fisher, R. 1293
Fleisch, Karl 48
Fodor, D. 2096
Foft, Jiři 1593
Foo, Gilbert...2269
Forster, Stefan...2420
Fotouhi, Reza 1575
Francois, Bruno 2142, 2184
Franke, W. Toke 69
Franko, Marek 2538
Friedli, T. 27
Fröhleke, Norbert..159
Fuchs, F.W. 444
Fuchs, Friedrich W. 1390, 1819, 69
Fujita, Y. 275
Fukushima, Kentaro..148
Funabashi, Toshihisa 2478, 2487
Funato, Hirohito 479
Funayama, Koichi..1020
Futami, Motoo 2337

G

Gabriela, Petropol Serb....................................893
Gan, W. C. 1221
Gao, Fanqiang..515
Gao, Q. 2261
Gao, Qiang..1058
García-Tabarés, L. 2481
Gardecki, Arkadiusz 1193
Gasiewski, Marcin 1562
Gaubert, Jean-Paul 703
Gavranic, Ivica 818
Gaztañaga, H. 1555
Gelezevicius, Vilius Antanas...........................1144
Gennadevich, Kiselev Michail..........................428
Gennadevich, Lepanov Michail.........................428
Gerada, C. 1173
Gerada, Chris 1058, 388, 769, 887
Ghaedi, Azam 1054
Gharehpetian, G.B. 313, 629, 750
Ghosh, Arindam 468, 723, 903
Gímenez, María Isabel.....................................1947
Giménez, María 268
Giral, Roberto 2115
Gizinski, Zygmunt 1562
Glasberger, Tomál 1268
Glavin, M.E. 1688
Glushkin, Evgeny 1872
Gnacinski, Piotr 826
Gobis, Vitoldas 1140
Goeldel, C. 2126
Gomis-Bellmunt, Oriol 1670
González-Hernández, S. 1784

Author Index

Gorbounov, Yassen............787
Goto, Hiroki............1163, 1168
Grabic, Stevan............1957
Grad, M.............714
Grecki, Filip............1440
Grigaitis, Arunas............1144
Grigans, Linards 2066
Grossi, Federica............874
Grzesiak, Lech M.............1071
Grzesik, B.............956
Gualous, H.............1542
Guo, Hai-Jiao............1163, 1168
Gustin, F.............1542
Gustin, Frederic 734
Guy, Owen J.............2464
Guzinski, Jaroslaw............1382, 994
Guzmán, Víctor.............1947, 268
Gwózdz, Michal............728

H

Haan, Sjoerd de............187
Habetler, Thomas G.............21
Hadas, Zdenek 1665
Hadjov, Kliment 787
Hájek, Vítezslav............2371
Halasz, S.............682
Halgos, Jan............2368
Hamada, Tomoyuki 1884
Hamar, J.............1755
Hameyer, K.............2393
Hameyer, Kay............2412
Harada, Yosuke............148
Hartansky, Rene............2368
Hartnett, Kevin J.............1249
Hasegawa, Masaru............614
Hashimoto, Seiji............932
Hayashi, Kenta............589
Hayashi, Yusuke............2445
Hayes, John G.............1249
Heising, C.............521
Helmut, Weiss 1722, 1934
Henrotte, F.............2393
Henze, Olaf............2385
Hercog, Darko............2349
Hicham, Fakham............2205
Himmelstoss, Felix. A.............331
Hiraki, Eiji............119, 1877
Hirokawa, Masahiko............393
Hissel, Daniel............2156
HISSEL, Daniel............734
Hmasic, N.............2134
Ho, S.L.............576
Hoffmann, Frank............1215
Hõimoja, H.............2005
Hõimoja, Hardi 1581

Hojo, Masahide............2487
Holtz, Joachim 1084
Holub, Marcin............195
Horen, Yoram 776
Horga, V.............2043
Horga, Vasile............1111
Hrasko, Martin............2538
Hu, Weihao............2325
Hubik, Vladimir............1620
Huiqing, Wen 1518, 417
Hurley, W.G 1688
Huttin, N.............1523

I

I., Margaris Nikolaos 1105
Iannuzzi, Diego............1469
Ibach, Robert 2082
Ibáñez, Fernando 268
Ichinokura, Osamu............1163, 1168, 758
Ichinose, Masaya 2337
Ide, Kazumasa 2337
Igic, P. M.............2464
Iida, Takahiko 595
Ikeda, Yoshiko............498
Ikhouane, Faycal............1670
Iman-Eini, H.............173
Inoue, Yukinori............1859
Ion, Petropol Serb............893
Iov, Florin............1771, 561
Ishikawa, Kazumi............1020
Iskhakov, Albert 663
Ito, Fumio 1309
Itoh, Jun-ichi............581
Itoi, M.............275
Ivanovic, Zoran............1957
Iwaji, Yoshitaka............924
Iwanski, Grzegorz............1440, 2164
Iwase, Yuta............2487
Izadbakhsh, Alireza............2102

J

Jalakas, T.............1263
Jalakas, Tanel 1257
Jan, Mucko............1316
Ján, Vittek............2219
Jang, Gil-Soo............2498
Jang, Su-Jin............1924
Jansen, Uwe............88
Janson, Kuno 154
Járdán, Rafael K.............916
Jardan, Rafael Kalman............2360
Jasim, O.............1173
Jasim, Omar 887
Jasinski, Marek............1904
Javurek, Jiri............1465

Author Index

Jedryczka, Cezary 2406
Jennings, Michael R. 2464
Jeon, Jin-Hong 2492, 2524
Jezernik, Karel 2283, 2349, 432
Ji, Young-Hyok 1929
Jian, Xiao 1722
Jin, Zhao 1128
Johnson, C Mark 76
Joós, Géza 2178
Joost, M. 1064
Judek, Slawomir 1497
Jufer, Marcel 1
Jun, Liu 1518, 417
Jung, Doo-Yong 1929
Jung, Yong-Chae 181, 1924, 1929, 383

K

Kalatchikov, P. 837
Kalisiak, Stanislaw 195
Kallaste, Ants 154
Kallenbach, E. 1598
Kalyoncu, Mete 1132, 949, 974
Kamata, Yuki 498
Kamiski, Bartlomiej 2378
Kamper, M.J. 859
Kampisios, Konstantinos 887
Kaneko, Daigo 924
Kanerva, Sami 867
Kaplon, Andrzej 377
Karaffy, Z. 2096
Karsli, Vedat M. 850
Karwowski, Krzysztof 1497
Kasa, Nobuyuki 595
Kasinski, A. 1044
Kasprowicz, Andrzej 1332
Katic, Vladimir 1957
Kato, Koji 581
Katsura, Seiichiro 1187, 1604, 1614
Kawamura, Atsuo 7, 924
Kayhan, Ince 1934
Kazimierz, Jaracz 912
Kazmierkowski, Marian P. 1548, 1904
Kelemen, Franjo 855
Kennel, R.M. 859
Kennel, Ralph 1239
Khaldi, B.S 1987
Kim, Eel-Hwan 2498
Kim, Heung-Gun 1421
Kim, Jae-Hong 2498
Kim, Jae-Hyung 1924, 1929
Kim, Jong-Yul 2492
Kim, Se-Ho 2498
Kim, Seul-Ki 2492, 2524
Kimura, Kensuke 1168
Kimura, Noriyuki 1884

Kinoshita, Hirotaka 2337
Kireev, V. 1598
Klimczak, Pawel 108
Klug, O. 2096
Klyachko, Leonid 663
Klytta, Marius 165
Knop, André 69
Kobayashi, Yukinori 479
Kobougias, Ioannis C. 1274
Koczara, Wlodzimierz 1440, 2150, 2164, 2254
Koda, Noriaki 1877
Kolar, J. W. 27
Kolesnikov, Artem 1872
Kolomeitsev, L. 1598
Kompa, K. 695
Komura, Akiyoshi 2337
Kondo, Masaki 1614
Koneke, Thies 1458
Kong, S.T. 43
Konstantinovich, Rozanov Yurie 428
Korondi, Peter 2360
Korotyeyev, Igor 236
Koskin, Y. 837
Kosmecki, Michal 1975
Kostylev, A.V. 1050
Kotodziejek, Piotr 804
Kouzou, A. 1987
Kowalski, Czeslaw T. 1359
Kraeftner, Wilhelm 331
Kraynov, D. 1598
Krettek, Johannes 1215
Krim, Fateh 703
Krismer, F. 27
Krykowski, K. 326
Krystkowiak, Michal 728
Krzeminski, Zbigniew 1382, 2294
Kubiak, Andrzej 2452
Kubin, Jiri 1815
Kubota, Sachio 1309
Kuchta, Jozef 2538
Kudarauskas, Sigitas 2200
Kuebrich, Daniel 244
Kuhn, Harald 1458
Kuisma, M. 1233
Kulka, Arkadiusz 657
Kumar, Dinesh 207
Kuperman, Alon 776
Kurokawa, Fujio 2434, 2504
Kürschner, Daniel 1696, 1734
Kuß, H. 695
Kusserow, Wolf 1239
Kütt, Lauri 154
kuwata, M. 275
Kyritsis, A.C. 1287

Author Index

L

Laczynski, Tomasz 569, 649
Lafoz, M. 2481
Lagoda, Ryszard 1040, 2055, 2087
Laloya, Eduardo............................845
Lange, E. 2393
Lapointe, Vincent 1077
Lastowiecki, Jozef 1440
Latka, M.......................................714
Latkovskis, Leonards 2066
Laugis, J. 1263
Laugis, Juhan 1017
Laur, R. 1064
Lazar, C. 2043
Lazar, Mihai.................................2457
Ledwich, Gerard 468, 723, 903
Lee, Joo-Hyuk...............................1924
Lee, Tzung-Lin.............................1704
Lehtla, Madis 1581
Lehtla, T......................................2011
Leidhold, Roberto 1353
Leszek, Szychta 2091
Leuchter, Jan 422
Levins, Nikolajs 1811
Lewandowski, Daniel 2289, 669
Lewicki, Arka diusz.......................1382
Lewis, A.W...................................1790
Leyva, R.......................................2108
Li, Kaihang 97
Li, Rongyuan 159
Li, Yaohua 1981, 2000, 405, 515
Li, Zixin............... 1981, 2000, 405, 515
Liaw, Chang-Ming 1652, 456
Lie, Xu 229
Liffran, Florent 409
Lillo, Liliana de 388
Lindemann, Andreas...........1280, 2420
Lingemann, M. 2134
Lis, Jacek D. 1359
Lisik, Zbigniew.............................2452
Lisowski, Grzegorz 669
Liu, Congwei 405
Llu, Li 793
Lladó, Juan...................................845
Lodzinski, Michal 2464
Lopez-de-Heredia, A. 1555
Lorenz, Robert D. 903
LU, Di..2142
Lu, Hua 76
Lu, Y...1221
Luft, Miroslaw..............................463
Luiz, Alex-Sander Amavel 1345
Luniewski, Piotr.............................88
Lyons, Brendan J. 1249
Lyskawinski, Wieslaw....................2406

M

Macek-Kaminska, Krystyna 1193
Madawala, U. K.............................139
Madawala, Udaya K. 1918
Maga, Dusan.................................2368
Mahmoudi, M.O. 1987
Mahyob, Amin..............................810
Mailat, Adrian..............................1748
Majidi, Behrooz 763
Maksimovic, Dragan 498
MAKYS, Pavol..............................2538
Malekian, Kaveh.......... 1117, 1123, 2071, 365, 763
Malska, W. 714
Man, T.K. 400, 475
Mandache, Lucian 1585
Mandra, Slawomir 1071
Mandrek, Slawomir 144
Marek, Stulrajter..........................2219
Margaliot, M. 260
Mariano, Sílvio José Pinto Simões 2076
Marouchos, Christos.......................1967
Martín-del-Brío, Bonifacio 845
Martínez, Abelardo 1947, 845
Martinez, Itziar 437
Martins, Denizar C. 1951
Masada, E. 1755
Mascibrodzki, Ireneusz...................1562
Mathis, W. 132
Matsui, Keiju 614
Matsui, Nobumasa 2504
Mawby, P.A. 2472
Mawby, Philip A. 2464
McEachern, Alex 1371
Mecke, Rudolf 1734
Melício, R. 1682
Mendes, V. M. F. 1682
Mertens, A. 132
Mertens, Axel 1458, 213, 569, 649
Meuret, R. 1523
Meynard, Thierry...........................281
Michalík, Jan 1535, 550
Michalke, N. 695
Mierlo, J. Van 1512
Milanovic, Miro............................301
Milimonfared, Jafar 1117, 2071, 365, 763
Mimura, Yasuhiro..........................2434
Mirsalim, M. 313, 629, 750
Mirsalim, Mojtaba 1123
Mirzaeva, G. 1155
Mishima, Tomokazu 119
Mitani, Tetsuya 2428
Mladenovic, I. 2022
Mohd, A. 2134
Mokrovica, Josipa..........................855
Mõlder, Heigo...............................154

Author Index

Molinas, Marta 2318
Möller, T. 2005
Mollov, Stefan V. 350
Molnár, Jan 1535, 550
Mondzik, Andrzej 345
Monmasson, E. 1365
Montesinos-Miracle, Daniel 1670, 731
Morel, Herve 2457
Moreno-Font, Vanessa 2115
Moreno-Goytia, E. 1784, 1941
Morimoto, Shigeo 1859
Morino, Kimio 2478
Morizane, Toshimitsu 1884
Morton, D. 2134
Mouni, Emile 2015
Mukhopadhyay, Siddhartha 485
Munk-Nielsen, Stig 108
Murata, Toshiaki 2337
Musallam, Mahera 76
Mustonen, P. 1233
Musumeci, S. 61
Muszynski, Roman 2227
Mutschler, Peter 1353
Müür, M. 2005
Mysinski, Wojciech 1321

N

Nagy, I. 1755
Nagy, Istvan 2360
Nagy, István 916
Naka, Toshiyuki 498
Nakagawa, Akio 498
Nakamura, Kazutoshi 498
Nakamura, Kenji 758
Nakaoka, M. 275
Nakaoka, Mutsuo 119
Nakayama, Hiroaki 1877
Nanakos, Anastasios Ch. 1827
Naouar, M-W. 1365
Narayanan, E.M. Sankara 43
Narjiss, Abdellah 734
Nasser, Mehdi 1896
Navarro, Daniel 437
Nawaz, Muhammed 2472
Nekoui, Mohammad Ali 1054
Ngwendson, L. 43
Ni, Bingchang 2331
Nichita, C. 1911
Nichita, Cristian 1834
Nicolae, Ileana-Diana 1585
Nicolae, Petre Marian 1585
Nicolae, Petre-Marian 1181
Niechaj, Marek 1890
Niemelä, Markku 1763
Nikolic, Aleksandar 1426

Nilssen, Robert 799
Ninomiya, Tamotsu 148, 393
Nishida, Yasuyuki 2530
Nishikata, Shoji 2343
Nishimiya, Ayumu 1163
Nishioka, Kunihiro 1309
Nitta, Mayumi 932
Noda, Shuji 1877
Norigoe, Isami 148
Novák, Jaroslav 982
Novák, Martin 982
Nowak, Lech 2400
Nowak, Mietek 84
Numata, Shigeo 2478
Nuutinen, Pasi 1763
Nyczkowski, Lukasz 740
Nymand, Morten 127
Nysveen, Arne 799

O

O'Sullivan, D.L. 1790
Ogiwara, H. 275
Ohashi, Hiromichi 2428, 2445, 54
Ohishi, Kiyoshi 1187, 1604, 1614
Ohsaki, H. 1755
Ohyama, Kazuhiro 2300
Okamatsu, Masashi 2434
Oleschuk, Valentin 1548
Omari, O. 2134
OMORI, Hideki 2530
Ondrusek, Cestmir 1665
ONEN, Umit 949
OPROESCU, Mihai 621
Orlik, B. 1064, 830
Orlowska-Kowalska, Teresa 2211, 2306
Ortjohann, E. 2134
Oyarbide, Estanislao 845

P

Pacas, Mario 2248
Pajchrowski, Tomasz 1198, 1204
Pakhomin, S. 1598
Palis, Frank 1610
Palis, Stefan 1660
Panoiu, Caius 1409
Panoiu, Manuela 1409
Papanikolaou, N.P. 1287
Papic, I. 2510
Paquin, Jean-Nicolas 1475
Park, JuneHo 2492
Park, Sang-Hoon 181, 383
Park, So-Ri 181
Parkatti, P. 201
Parker-Allotey, N-A. 2472
Patel, N. D. 139

Author Index

Patra, Pradipta..485
Patra", Amit 485
Pavelka, Jiri 221
Pavelka, Jirí 988
Pavlitov, Constantin...787
Pavlovsky, Martin..7
Pavol, Makys 2219
Pavoni, Alessandro 1491
Peftitsis, Dimosthenis 1840
Peltoniemi, Pasi 1763
Peplinski, Marcin..826
Pera, Marie-Cecile 2156
Perez, Francisco...845
Perez-Tomas, Amador 2464
Peric, Nedjeljko 2235
Peroutka, Zdenek 1268, 1529, 1535, 550
Peter, Bris 2219
Peter, Zaucher 1722
Petit, Marc 2184
Petrella, Roberto 1097
Petrisor, Anca 893
Piatek, Pawel 1446
Pietrzak-David, Maria..938
Piróg, Stanislaw 1779
Pittermann, Martin 1593, 372
Planson, Dominique..2457
Poljugan, Alen 1058
Pollán, Tomás 845
Popa, Anca Sorana 1225
Popa, Mircea 1225
Porada, Ryszard 740
Pospelov, Vladimir 663
Pronin, M. 837
Pugachevs, Vladislavs 1811
Pyrhönen, Juha...1763

Q

Quiroga, J...793

R

Rabkowski, Jacek 84
Raciti, A...61
Radomski, Grzegorz 504
Raducu, Marian...621
Radulescu, Mircea M..1896
Rafecas-Sabate, Josep 731
Rafiei, S.M.R..2102
Rahman, M.F...2269
Rahnamaee, Arash 1117, 365
Rao, Sachit..2312
Rathge, Christian 1696
Ratoi, Marcel 1111
Rawicki, Stanislaw 1481
Raynaud, Christophe...2457
Rednov, F...1598

Reghem, Pascal.. 1834, 810
Rerucha, Vladimir 422
Rezaei, Mohammad Mehdi...................................1123
Reznikov, B...260
Ribickis, Leonids 1811
Richter, F..1398
Riipinen, T...1233
Risteiu, Mircea 1243
Riz, A...2096
Roasto, I...2011
Robert, B.G.M...2126
Robert, Bruno Gerard Michel 2120
Robinson, Jonathan..2178
Robyns, B...1523
Robyns, Benoît 1896
Robyns, Benoit 2184
Rodic, Miran 2283
Rodriguez, E..2108
Rodriguez, Jose 451, 699
Rojas, A..1155
Rojko, Andreja 2349
Rolek, Jaroslaw...377
Rompelman, Otto..2354
Ronkowski, Mieczyslaw.......................................880
Rosin, A..2005
Rothenhagen, Kai 1390, 1904
Round, S. D..27
Ru1scin, Vladimír 295
Ruderman, A...260
Ruderman, Michael 1215
Rufer, A..256
Ruger, N. E...132
Rusinov, Radoslav 787
Rylko, Marek S...1249
Ryvkin, Sergey 1505
Rzasa, Janina 357

S

Saadi, S..1987
Sabirin, Chip Rinaldi 1625
Saito, Makoto 2439
Saito, Tsuyoshi 744
Sajkowski, M...956
Sakamoto, Kiyoshi...924
Sakamoto, Yosei 288
Salo, M...201
Salonen, Pasi...1763
Samanta, Susovon...485
Samuelsen, Dag 1416
Sanada, Masayuki 1859
Sánchez, Beatriz 845
Sánchez, Carlos 268
Sang-Joon, Lee 1002
Sang-Taek, Lee 1002
Sanjari, M. J.. 313, 629, 750

Author Index

San-Sebastian, J. 1555
Santo, António Espírito............... 1646
Sarraute, Emmanuel............... 281
Sasaki, Masahiro............... 2434
Sato, Muneo............... 1309
Saudemont, C............... 1523
Sayed, Mahmoud A. 542
Sayeef, S. 2269
Schallschmidt, Thomas 1610, 1660
Schanen, JL............... 173
Schmelter, A............... 2134
Schmid, Markus............... 244, 250
Schmidt, Istvan 1803
Schmidt-Obermoeller, Richard............... 1505
Schmitt, Günter............... 1239
Schneider, T............... 1598
Schnick, O. 132
Schrödl, Manfred............... 2275
Schroedl, Manfred............... 945
Schuffenhauer, U............... 695
Scollo, R............... 61
Sengupta, Sabyasachi 674
Seppä, L............... 1233
Shao, S............... 1293
Shapoval, Ivan 307
Sharma, R............... 1918
She, X............... 710
She, Yun 710
Shieh, Fa-Hwa 1652
Shimaoka, Yoshihiro 1309
Shimizu, Takaaki 498
Shimizu, Toshihisa 2428, 2445, 288, 600
Shimoda, Eisuke 2478
Shiraishi, Keiichi............... 2504
Shonin, O............... 837
Shoyama, Masahito............... 148, 393
Shyu, Juei Lung 643
Siatkowski, M............... 830
Silea, Ioan 1225
Silventoinen, P............... 1233
Simetzberger, Christian 2275
Simon, Miklós G............... 916
Singule, Vladislav 1620, 1665
Sinsukthavorn, W............... 2134
Siostrzonek, Tomasz............... 1779
Sîrbu, Ioana-Gabriela 1181, 1585
Siroky, Peter 2368
Sitar, Jan 2368
Sivkov, Oleg 221
Skovpen, Sergey 663
Skuta, Ondřej 1375
Slama-Belkhodja, I............... 1365
Slama-Belkhodja, Ilhem............... 938
Smet, Bart............... 102
Sobczuk, Dariusz............... 2378
Sobczynski, D............... 714

Sochacki, Mariusz 2452
Soltani, Hamid 718
Song, Sang-Hoon 383
Song, Seung-Ho 2498
Soroudi, Alireza 1847
Sosa-Ruiz, J............... 1941
Souad, Rafa............... 1209
Sourkounis, C............... 1398
Sourkounis, Constantinos 1633, 1710, 2331
Sozanski, Krzysztof Piotr 1995
Stadler, Paul Andreas 2543
Stala, Robert 1852, 345
Stamann, Mario 1660
Stanescu, Dan-Gabriel 1181
Staudt, V............... 521
Staudt, Volker............... 2371
Stefanutti, Fabio 1097
Steimel, A............... 521
Steimel, Andreas............... 1505, 2371
Stenzel, T............... 956
Stepanyuk, D.P............... 1050
Stepien, P............... 1293
Stocco, Piero 1097
Strac, Leonardo............... 855
Strzelecki, Ryszard Michal 1332
Strzelecki, Ryszard 1729
Stumpf, P............... 1755
Stumpf, Péter 916
Sugai, T............... 275
Sugimasa, Junji Tamura Masatoshi 2337
Suissa, Uri............... 776
Sulkowski, Waldemar 1416
Sumida, Yuichi 2434
Sumina, Damir............... 556
Sumiyoshi, Shinichiro 2530
Summers, T.J............... 1293
Sumner, M. 1173, 2261
Sumner, Mark 1058, 2300, 769
Sun, Z. G............... 1221
Susluoglu, Berrin............... 850
Suul, Jon Are 2318
Sveda, Martin 1620
Sweet, M............... 43
Sykulski, Jan K............... 2383
Szabat, Krzysztof............... 2211, 2241
Szamel, Laszlo 1033
Szczeniak, Pawel 165
Szczepankowski, Pawel............... 1332
Szczesniak, Pawel............... 236
Szelag, Wojciech 2406
Sziebig, Gabor 2360
Szmidt, Jan 2452
Szubert, Krzysztof 536
Szweda, Mariusz 826
Szychta, Elzbieta............... 463
Szychta, Leszek 463

Author Index

Szymanski, B. J. 695

T

Tackoen, X. 1512
Tae-Ho, Yoon 1002
Taguchi, Toyoki 498
Takahashi, Nobuo 1877
Takahashi, Rion 2337
Takao, Kazuto 2445, 54
Takeshita, Takaharu 542
Takeuchi, Nobuhito 614
Takeuchi, Toshihiro 924
Tan, Longcheng 1981, 2000, 405, 515
Tanabe, Takayuki 2478
Tanaka, Toshihiko 1877
Taniguchi, Katsunori 1884
Taniguchi, Satoshi 600
Tankari, A.M. 1911
Tao, Zhou 2205
Tapuchi, Saad 776
Tarczewski, Tomasz 1071
Tatakis, E.C. 1287
Tatakis, Emmanuel C. 1274, 1301, 1827
Tatsuta, Fujio 2343
Theodoridis, Michael P. 350
Thomas, D.W.P. 2049
Thomas, David W.P. 1716
Thompson, David S. 114
Tinkir, Mustafa 1132, 949, 974
Tnani, Slim 2015
Tournier, Dominique 2457
Tran, Quang-Vinh 1421
Trentin, Andrew 887
Trujillo, Cesar L. 637
Tsai, Jih-Run 1652
Tseng, K.J. 2516
Tsukakoshi, Kenta 148
Tsuruta, Yukinori 7
Tulbure, Adrian 1243
Turner, Robert W. 528
Tutaj, Andrzej 1432
Tuusa, H. 201

U

Ueda, Yoshinobu 2478, 2487
Ummaneni, Ravindra. B. 799
Undeland, Tore 2318, 657
Ünüvar, Ali 967
Urabe, R. 275
Urbanski, Konrad 1454
Utkin, Vadim 2312, 512

V

Väisänen, V. 1233

Valchev, V. 1326
van Duivenbode, Jeroen 102
Vasak, Mario 2235
Vedrana, Jerkovic 690
Vekic, Marko 1957
Vergnol, Arnaud 1896
Veszpremi, Karoly 1803
Vicuña, Javier 845
Villanueva, Elena 451
Villwock, Sebastian 2248
Vinnikov, D. 1263, 2011
Vinnikov, Dmitri 1257
Viscarret, U. 1555
Vittek, Jan 2551
Vladimír, Vavrus 2219
Vodovozo, Valery 1017
Vorontsov, A. 837
Vrana, Petr 1465

W

Wada, Keiji 2428, 288, 600
Walas, K. 1044
Walter, Julio 268
Walton, Simon 528
Wang, Ping 1981, 2000, 405, 515
Wang, Yi 187
Wang, Yue 2325
Wang, Zhaoan 2325
Weidinger, Thomas 2028
Weiland, Thomas 2385
Weindl, Ch. 2022
Werner, Timur 649
Wheeler, P. 1326
Wheeler, Patrick W. 207
Wheeler, Patrick W. 229
Wheeler, Patrick 388
Wiktor, Hudy 912
Willis, K. 1293
Winternheimer, Stefan 1872
Wisniewski, Janusz 2254
Wlas, Miroslaw 1084
Won, Chung-Yuen 181, 1924, 1929, 383
Wong, L.K. 400, 475
Wu, Dongming 97

X

XiaoyanHuang, ... 388
Xu, Wei 2000, 405
Xuhui, Wen 1518, 417
Xuhui, Zhang 1518, 417

Y

Yaguchi, Hiroyuki 1020
Yamanouchi, Wataru 1187

Author Index

Yang, Lingling 97
Yang, Liu 1128
Yang, Ru-Shiuan 320
Yin, Chunyan 76
Yokokura, Yuki 1187, 1604
Yokoyama, Tomoki 589, 744
Young-Kwan, Kim 1002
Yousefi, Ashkan 1847

Z

Zakrzewski, Zbigniew 1332
Zamma, Toshihiro 1020
Zanasi, Roberto 874
Zanchetta, Pericle 1716, 1771, 561, 887
Zare, Firuz 468, 718, 723, 903
Zaring, Carina 2472
Zarko, Damir 855
Zarko, Damirarko 818
Zaskalicka, Maria 899
Zaskalicky, Pavel 899
Zatocil, Heiko 1024
Zawirski, Krzysztof 1198, 1204, 1454
Zdenek, Jiri 1638
Zdravko, Valter 690
Zeljko, Spoljaric 690
Zeman, Karel 1529
Zeroug, Houcine 1209
Zhang, H. 1523
Zhang, S. 2516
Zhao, S. W. 1221
Zhou, Tao 2142
Zhu, Haibin 515
Zielinski, K. 1064
Zigic, Aleksandar 1426
Zinoviev, Genady Stepanovic 1332
Zlosnikas, Valerijus 1140
Zouhar, Jan 1665
Zulawnik, Marcin 1562
Zych, Michal 1562
Zymmer, Krzysztof 1332, 1562

9781424417414